Index of Applications

(continued on page A18)

Alice Kaseberg helps your students discover algebra.

INTERMEDIATE
ALGEBRA

A Just-in-Time Approach

EDITION
3

ALICE **KASEBERG**

Take a look through this special *Preview* section and see how Alice Kaseberg helps your students answer the age-old question:

"What is algebra good for?"

THOMSON

™

BROOKS/COLE

Alice Kaseberg's mathematical curiosity

Alice Kaseberg's 30 years of secondary school and community college teaching experience combines with her avid mathematical curiosity to produce a drive to answer the commonly asked question, "What is mathematics good for?" Her BA in Business Administration, MA in Mathematics, and MS in Engineering Science reflect her keen interest in both academics and mathematics applications. Her goal in writing textbooks is to help students succeed in mathematics, learn how to learn, and become actively aware of the mathematics in their daily lives.

She began her career as a teacher in Australia. Then, as a junior high school teacher, she class-tested materials from the *Mathematics Resource Project, Lane County Math Project, School Mathematics Project* (of Great Britain), and Bob Wirtz's *Drill and Practice at the Problem Solving Level*. Her mentors were early advocates of the discovery approach, laboratory activities, and membership/involvement in professional organizations.

Kaseberg has always reached out to mathematics students in a variety of situations. Her experience of teaching the full range of high school courses—remedial mathematics to Advanced Placement calculus—included working with both hearing impaired and English as a Second Language students. Always eager to take on new projects, she organized weekly problem-solving activities for the entire department. As a community college teacher (15 years), she has taught a range of courses from Beginning Algebra through Calculus, as well as Strength of Materials and Statics in Engineering. While on sabbatical from teaching, she worked 15 months for the EQUALS Project, Berkeley, California, writing *Odds-on-You* and coauthoring the *EQUALS Handbook*.

Discover Algebra

Kaseberg's approach gives students a powerful grasp of algebra.

Kaseberg presents an effective approach to the algebra curriculum. The first and second editions gained a strong following among instructors who found that Kaseberg's use of guided discovery and problem solving facilitated the learning of new concepts and strengthened skill retention. To borrow from S.I. Hayakawa in **The Aims and Tasks of General Semantics**, *"The key to survival . . . is flexibility; quickness in taking in relevant information and resourcefulness in devising ways of solving new problems as they arise."* Kaseberg's informal, interactive style makes algebra more accessible to students while maintaining a high level of mathematical accuracy.

To reduce preparation time for course leaders and facilitate use by adjuncts, the *Instructor's Resource Manual* is a valuable tool. The manual provides structured lesson and group-activity suggestions for each section in the textbook, incorporates materials from the textbook with supplemental projects and activities, suggests core homework assignments, and furnishes guided-discussion questions. This resource serves to bridge the gap between traditional pedagogy and the standards-based approach.

Foundations of the Kaseberg Approach

➤ **SKILLS FOR THE FUTURE:**
As teamwork and worker initiative become the norm in the workplace, the individual and group activities in the explorations and the projects provide students with experience in building good work habits and interpersonal skills. The projects also provide activities for student portfolios.

➤ **GUIDED DISCOVERY:**
Examples exploring new topics provide guided discovery of new concepts. Kaseberg places the solutions to the explorations at the end of the section to encourage instructors to use the textbook as part of the lecture materials in class or for students' independent study. Alternatively, the instructor can have students use the workbook for in-class exploration and discovery.

➤ **NUMBER SENSE:**
The emphasis on tables, number patterns (sequences), numerical investigations, and explorations preceding property statements establishes a foundation for number sense. These skills are extended to numerical approaches to finding equations.

➤ **SKILL BUILDING:**
Ample skill-building exercises give students valuable practice with fundamental algebraic concepts.

➤ **ARITHMETIC SKILLS:**
Practice with fractions, decimals, and percents is presented often and in subtle ways, in settings that connect with the algebra (tables, equations, number patterns) without being "beneath" the student.

1.1 Exercises

1. **Time Management** Return to Example 1. As directed, make a chart for your next seven days. Over the coming week, follow the four problem-solving steps. If you have family or other obligations, write in realistic times in the morning and evening.

2. **Time Management** To make sure that your course and work loads are sensible, check your plan for this term. Write down your number of credit hours and your number of credit hours by 3 and then add your number of hours of employment. For most people, 40 to 45 hours is a reasonable commitment to school and employment. If you are above 50 hours, do you have contingency plans for happenings outside your control?

In Exercises 3 to 8, match each definition or related statement with a vocabulary word. Record any unfamiliar words and their definitions in your study notes.

3. Choose from multiplicative inverses, additive inverses, integers, real numbers, irrational numbers

   a. All the numbers that can be located with a point on the number line

   b. The whole numbers and their opposites, $\{\ldots, -3, -2, -1, 0, 1, 2, 3, \ldots\}$

   c. Real numbers that cannot be written as rational numbers

   b. Two numbers that add to zero

   c. Two numbers that multiply to 1

   d. The set of natural numbers and zero, $\{0, 1, 2, 3, \ldots\}$

   e. The set of numbers that can be written $\frac{a}{b}$, a and b integers, b not equal to zero

5. Choose from associative property for addition, commutative property for multiplication, distributive property of multiplication over addition, factors, factoring

   a. $c(d + e) = cd + ce$

   b. Two numbers being multiplied

   c. $b \cdot c = c \cdot b$

   d. Changing the addition of ab and ac into the multiplication of a and $(b + c)$

   e. $a + (b + c) = (a + b) + c$

6. Choose from commutative property for addition, factored form of the distributive property, set, $\sqrt{-1}$, simplify

   a. $b \cdot c + b \cdot e = b(c + e)$

   b. To perform the indicated operation

   c. $a + b = b + a$

---

100 CHAPTER 2 | INEQUALITIES, FUNCTIONS AND LINEAR FUNCTIONS

Slope

EXAMPLE 5 Investigating slope
a. Write two new equations for the photocopy card—one if the cost changed to $0.15 per copy and another if the cost changed to $0.05 per copy.
b. Graph each equation on the original axes in Figure 11.
c. How does the graph change as the cost changes?

SOLUTION **a.** $y = 10 - 0.15x$
$y = 10 - 0.05x$

b. The lines are graphed in Figure 13.

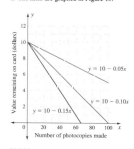

EXAMPLE 1 Solving problems by guess and check: time allocation   Jian's academic time includes the time spent attending class, plus 2 hours of study time for each hour spent in class. Each week, he spends twice as many hours on academics as he does at his job. He spends 28 more hours each week sleeping than at his job. He spends the remaining 44 hours of the week on other tasks. How many hours does Jian devote to sleep, work, and academics?
a. Make a table to find the hours spent at each activity.
b. Write equations that describe the problem.

SOLUTION **a.** *Understand and plan:* The three activities—sleep, work, and academics—make natural column headings for a guess-and-check table. The hours spent on other tasks and total hours form the two other columns. The total hours column should equal 168:

$$\frac{24 \text{ hours}}{1 \text{ day}} \cdot \frac{7 \text{ days}}{1 \text{ week}} = \frac{168 \text{ hours}}{1 \text{ week}}$$

Carry out the plan: A reasonable starting guess is 8 hours of sleep per night, which, times 7 nights per week, gives 56 hours of sleep per week. This guess results in 184 total hours for the week and so is too high. For our next guess, we drop the hours of sleep to 50. This guess results in 160 total hours and so is too low. Our third guess for sleep is 52 hours, a number between 50 and 56.

TABLE 7

Sleep	Work	Academics	Other	Total Hours
56	28	56	44	184 (too high)
50	22	44	44	160 (too low)
52	24	48	44	168 (correct)

As Table 7 shows, Jian sleeps 52 hours a week, works 24 hours, and spends 48 hours attending class and studying.

Check: The results satisfy the conditions in the problem. ✓

b. We will let $s =$ hours of sleep, $w =$ hours of work, and $a =$ hours on academics. Using these variables, we can describe the table setup with equations.

3

GEOMETRY CONNECTIONS:

Numerous applications connect algebra skills to geometry, including factoring to area, proportions to similar triangles, square roots to Pythagorean theorem, systems of equations to properties of geometric figures, and rational functions to area.

⇒ **UNIT ANALYSIS:**

The book presents a distinction between measure-to-measure conversions and rate-to-rate conversions. Units reinforce the properties of operations with rational expressions.

⇒ **POLYNOMIALS:**

A table method for multiplying and factoring polynomials provides a visualization of factoring by grouping and completing the square.

⇒ **FOURFOLD PRESENTATION:**

Kaseberg integrates tabular and graphical presentations with symbolic algebra and verbal descriptions to provide visual and numeric models of algebra and address the needs of a variety of learning styles.

⇒ **TABLES:**

Tables encourage students to guess-and-check, organize information, and observe patterns. Selecting key pieces of data and working backwards on the tables permit modeling linear and quadratic equations from the input-output table. The regular use of tables makes connections among related topics visually apparent.

⇒ **GRAPHING:**

Text boxes incorporated throughout the text highlight procedures for the TI-83 Plus graphing calculator.

⇒ **LINKED EXAMPLES:**

Because fewer is better, linked examples allow students to use the same setting in a variety of contexts, thus permitting them to apply new skills with minimal distraction from the details of the application.

⇒ **APPLICATIONS:**

Familiar real-world settings engage students and illustrate the relevance of algebra in their daily lives. Selected applications illustrate fundamental concepts in mathematics-based careers.

⇒ **EXERCISES:**

A variety of moderate-level problems provide long-term retention.

⇒ **REGULAR REVIEW AND FEEDBACK:**

Mid-chapter tests and cumulative reviews build confidence and give students the opportunity to check their progress.

---

4   Mid-Chapter Test

1. Give the word, expression, or equation that fits each:

   a. The highest or lowest point on a parabola

   b. The line of symmetry for $y = x^2$

   c. A polynomial with two terms

   d. The standard form for a quadratic equation

   e. The constant term in a quadratic equation is the place where the graph crosses the _____ axis.

2. Make a table to solve the following equations. Circle the places where you find the answers.

   a. $x^2 - 4x + 3 = 8$

   b. $x^2 - 4x + 3 = 0$

   c. $x^2 - 4x + 3 = -1$

   In Exercises 3 and 4, change the equation to standard quadratic form and find the coefficients a, b, and c.

   3. $y = \frac{1}{2} n(n + 1) + 1$

   4. $y = 1000(1 + x)^2$

   5. a. Make a table and graph for $A = 6x^2$, the surface area of a cube of side x. Let x be 0 to 8.

   b. If we double the side of the cube, what happens to the surface area?

Take a look at what's new to the Third Edition of Kaseberg's Intermediate Algebra.

➡ Every new student copy of the book is automatically packaged with FREE access to a powerful student mastery system. This suite includes *BCA Tutorial*, featuring interactive section-by-section tutorials with feedback, integrated video instruction, and access to the complete textbook, live tutoring through *vMentor*, and access to *InfoTrac® College Edition*. See pages 6-7 for details on these and other Technology Resources.

➡ Think of it as portable office hours! Every new student copy of the book is automatically packaged with *The Interactive Video Skillbuilder CD-ROM*. This valuable resource contains more than eight hours of video instruction, featuring a 10-question Web quiz per section (the results of which can be emailed to the instructor), a test for each chapter, with answers, and *MathCue* tutorial software. See page 7 for details.

➡ The new *Student Workbook*, with textbook examples for in-class work, permits students to work through explorations in class without the distraction of the textbook solutions. See page 9 for details.

Student Workbook

for Intermediate Algebra, A Just-in-Time Approach
Third Edition

➡ By user request, the number of **ERROR ANALYSIS PROBLEMS**, which encourage critical thinking, has been increased.

➡ A variety of **FUNCTION NOTATION** [$h(x)$, $g(x)$ as well as $f(x)$] appears throughout the text.

➡ **APPLICATIONS** have been updated throughout the text. New applications draw on a wide variety of fields including art, nutrition, music, ecology, biology, oceanography, and astronomy.

➡ All **GRAPHING CALCULATOR** material is standardized to the TI-83 Plus; introductions of graphing calculator skills include views of the appropriate screens.

➡ Available for packaging with the text, a new Chapter 11, *Counting and Probability*, focuses on these non-algebraic topics, which are being included in a growing number of exit or transfer exams, as well as in intermediate algebra curricula.

What's New?

Terrific technology resources support you and your students.

TECHNOLOGY RESOURCES FOR INSTRUCTORS

BCA Instructor Version

ISBN: 0-534-38727-6

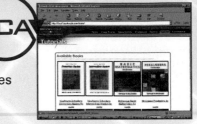

With a balance of efficiency and high performance, simplicity and versatility, *BCA* gives you the power to transform the learning and teaching experience. *BCA Instructor Version* comprises two components, *BCA Testing* and *BCA Tutorial*. *BCA Testing* is a revolutionary, Internet-ready, text-specific testing suite that allows instructors to customize exams and track student progress in an accessible, browser-based format. *BCA* offers full algorithmic generation of problems and free response mathematics. *BCA Tutorial* is a text-specific, interactive tutorial software program that is delivered via the Web (at **http://bca.brookscole.com**) and is offered in both student and instructor versions. The tracking program built into the instructor version of the software enables instructors to carefully monitor student progress. The complete integration of the testing, tutorial, and course management components simplifies your routine tasks. Results flow automatically to your gradebook and you can easily communicate to individuals, sections, or entire courses.

Free!

BCA has been updated with a diagnostic tool to assess and place students in any course and provide a personalized study plan. Other new updates:

■ Manipulate and format your tests with the new "rich text format" (RTF) conversion tool, which allows you to open tests in word processors like Microsoft® Word or StarOffice for formatting.

■ BCA now offers several new problem types to provide greater variety and more diverse challenges in your tests.

■ *Redesigned Tutorial* interface helps students learn math concepts faster.

Text-Specific Videotape Series

ISBN: 0-534-38721-7

This set of videotapes is available free upon adoption of the text. Each tape offers one chapter of the text and is broken down into 10- to 20-minute problem-solving lessons that cover each section of the chapter.

TECHNOLOGY RESOURCES FOR STUDENTS AND INSTRUCTORS

The Brooks/Cole Mathematics Resource Center

http://mathematics.brookscole.com

Accessible via the *Brooks/Cole Mathematics Resource Center*, the Book Companion Web Site accompanies the text, featuring *InfoTrac College Edition* activities, Internet projects, quizzes, and flash cards.

WebTutor™ ToolBox on WebCT and Blackboard

On WebCT packaged with the text: 0-534-12323-6
On Blackboard packaged with the text: 0-534-12314-7

Preloaded with content and available free via PIN code when packaged with Kaseberg's text, *WebTutor ToolBox* pairs all the content of the text's rich Book Companion Web Site with all the sophisticated course management functionality of a WebCT or Blackboard product. You can assign materials (including online quizzes) and have the results flow automatically to your gradebook. *ToolBox* is ready to use as soon as you log on—or, you can customize its preloaded content by uploading images and other resources, adding Web links, or creating your own practice materials. Students only have access to student resources on the Web site.

TECHNOLOGY RESOURCES FOR STUDENTS

New update 3.0!
BCA Tutorial Student Version

FREE access to this text-specific, interactive, Web-based tutorial system is included with this text. *BCA Tutorial Student Version* is browser-based, making it an intuitive mathematical guide. So sophisticated, it's simple, *BCA Tutorial* allows students to work with real math notation in real time, providing instant analysis and feedback. The tracking program built into the instructor version of the software enables instructors to carefully monitor student progress. Results flow automatically to your grade-book. Instructors can use *BCA* to easily communicate to individuals, sections, or entire courses. And, when students get stuck on a particular problem or concept, they need only log on to *vMentor*, accessed through *BCA Tutorial*, where they can talk (using their own computer microphones) to *vMentor* tutors who will skillfully guide them through the problem using the interactive whiteboard for illustration.

New!

A personalized study plan can be generated by a course-specific diagnostic built into *BCA Tutorial*. With a personalized study plan, students can focus their time where they need it the most.

Think of it as portable office hours!
Interactive Video Skillbuilder CD-ROM

ISBN: 0-534-38723-3

The *Interactive Video Skillbuilder CD-ROM* contains more than eight hours of video instruction. To help students evaluate their progress, each section contains a 10-question Web quiz (the results of which can be emailed to the instructor) and each chapter contains a chapter test, with answers to each problem on each test. Also includes *MathCue* tutorial and testing software. Keyed to the text, *MathCue* includes these components:

- *MathCue Skill Builder*—Presents problems to solve, evaluates answers and tutors students by displaying complete solutions with step-by-step explanations.
- *MathCue Quiz*—Allows students to generate large numbers of quiz problems keyed to problem types from each section of the book.
- *MathCue Chapter Test*—Also provides large numbers of problems keyed to problem types from each chapter.
- *MathCue Solution Finder*—This unique tool allows students to enter their own basic problems and receive step-by-step help as if they were working with a tutor.
- *Create customized MathCue sessions*—Students work default Skill Builders, Quizzes, and Tests—or customize sessions to include desired problems from one or more sections or chapters.
  - **Special note for instructors:** Instructors can create targeted, customized *MathCue* session files to send to students as e-mail attachments for use as make-up assignments, extra credit assignments, specialized review, quizzes, or other purposes.
- Print or e-mail score reports—Score reports for any *MathCue* session can be printed or sent to instructors via *MathCue's* secure e-mail score system.

Free!
InfoTrac® College Edition

In addition to robust tutorial services, your students also receive four months of anytime, anywhere access to *InfoTrac® College Edition*. This online library offers the full text of articles from almost 4,000 scholarly and popular publications, updated daily and going back as much as 22 years. Both adopters and their students receive unlimited access for four months.

Intermediate Algebra, Third Edition is also supported by a selection of outstanding print resources for you and your students.

FOR INSTRUCTORS

Instructor's Resource Manual— The essential Kaseberg resource!

ISBN: 0-534-38724-1

This resource serves to bridge the gap between traditional pedagogy and a standards-based approach. To reduce preparation time for course leaders and facilitate use by adjuncts, the *Instructor's Resource Manual* is a valuable tool. The manual provides structured lesson and group-activity suggestions for each section in the textbook, incorporates materials from the textbook with supplemental projects and activities, suggests core homework assignments, and furnishes guided-discussion questions. Key worksheets and overhead transparency masters follow the extensive discussion.

Annotated Instructor's Edition

ISBN: 0-534-39665-8

This special version of the complete student text contains a *Resource Integration Guide*—which begins on page 11 of this *Preview*—as well as answers printed next to the respective exercises. Graphs, tables, and other answers too long to appear next to their exercises are in a special answer section in the back of the text.

Complete Solutions Manual

ISBN: 0-534-39105-2

The *Complete Solutions Manual* provides worked-out solutions to all of the problems in the text.

Test Bank

ISBN: 0-534-38726-8

The test bank includes 8 tests per chapter as well as 3 final exams. The tests are made up of a combination of multiple-choice, free-response, true/false, and fill-in-the-blank questions.

FOR STUDENTS

NEW! Kaseberg provides even more support for student success!

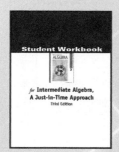

Student Workbook

ISBN: 0-534-39663-1

The new *Student Workbook*, with textbook examples for in-class work, permits students to work through explorations in class without the distraction of the textbook solutions. This invaluable resource also provides additional examples for trouble spots, calculator use, and, as appropriate, extensions. The workbook places pages directly into students' hands and eliminates the time and cost of photocopying.

Student Solutions Manual

ISBN: 0-534-38725-X

The *Student Solutions Manual* provides worked-out solutions to the odd-numbered problems in the text.

New!

Explorations in Beginning and Intermediate Algebra Using the TI-82/83/83-Plus/85/86 Graphing Calculator, Third Edition

Deborah J. Cochener and Bonnie M. Hodge, both of Austin Peay State University

ISBN: 0-534-40644-0

This user-friendly workbook improves both student understanding and retention of algebra concepts through a series of activities and guided explorations using the graphing calculator. An ideal supplement for any beginning or intermediate algebra course, *Explorations in Beginning and Intermediate Algebra, Third Edition*, is an ideal tool for integrating technology without sacrificing content. By clearly and succinctly teaching keystrokes, it allows class time to be devoted to investigations instead of how to use a graphing calculator.

Conquering Math Anxiety: A Self-Help Workbook (with CD-ROM), Second Edition

Cynthia A. Arem, Pima Community College

ISBN: 0-534-38634-2

This comprehensive workbook provides a variety of exercises and worksheets along with detailed explanations of methods to help "math-anxious" students deal with and overcome math fears. The author offers tips on specific strategies, as well as relaxation exercises. The book's major focus is to encourage students to take action. Expertly constructed hands-on activities help readers explore both the underlying causes of their problem and viable solutions. Many activities are followed by illustrated examples completed by other students. This edition now comes with a free relaxation CD-ROM and a detailed list of Internet Resources.

Mastering Mathematics: How to Be a Great Math Student, Third Edition

Richard Manning, Bryant College

ISBN: 0-534-34947-1

Providing solid tips for every stage of study, *Mastering Mathematics* stresses the importance of a positive attitude and gives students the tools to succeed in their math course.

Traditional Resources

Also available from Alice Kaseberg and Thomson • Brooks/Cole

New for 2004!

Introductory Algebra: A Just-in-Time Approach, Third Edition

ISBN: 0-534-38631-8
Annotated Instructor's Edition ISBN: 0-534-39666-6
©2004 / 688 pages / 8 1/2 x 11 / 4-color / Hardcover

Instructor's Resources

Instructor's Resource Manual
ISBN: 0-534-38731-4

Complete Solutions Manual
ISBN: 0-534-39100-1

BCA Instructor Version
ISBN: 0-534-38734-9

Test Bank
ISBN: 0-534-38733-0

Text-Specific Videos
ISBN: 0-534-38728-4

Student Resources

Interactive Video Skillbuilder CD-ROM
ISBN: 0-534-38730-6

BCA Tutorial Student Version
ISBN: 0-534-38729-2

Student Workbook
ISBN: 0-534-39662-3

Student Solutions Manual
ISBN: 0-534-38732-2

Contents

More Kaseberg

Chapter 1 ~ Problem Solving, Expressions, and Equations

Ideas for Instruction	Print Resources	Media Resources for Instructors	Media Resources for Students
Explorations in Beginning and Intermediate Algebra, Third Edition Activities 1–3, 7	**Instructors** **Test Bank** Forms A–H	**BCA Instructor Version** Contains BCA Testing and BCA Tutorial **http://bca.brookscole.com**	**BCA** **BCA Tutorial** Unlimited practice in questions and problems for students
Activities for Beginning and Intermediate Algebra Activities 1–3, 22–24	**Complete Solutions Manual** Chapter 1	**Web Site** Historical Notes, Math News, Careers, Web Quizzes, Internet Activities, *InfoTrac College Edition* exercises **http://mathematics.brookscole.com**	**vMentor** Customized interactive homework help and tutorial services online **http://bca.brookscole.com**
Math Facts, Second Edition Pages 15–16, 25–26, 103–110, 159–160	**Instructor's Resource Manual** Chapter 1	**Text-Specific Videos** Chapter 1	**Web Site** Historical Notes, Math News, Careers, Web Quizzes, Internet Activities, *InfoTrac College Edition* exercises **http://mathematics.brookscole.com**
Algebra Facts Pages 5–10, 17–18, 25–26, 65–67, 93–94, 111–112, 137–140, 147–150, 155–156	**Students** **Student Solutions Manual** Chapter 1	**InfoTrac® College Edition** **http://infotrac.thomsonlearning.com** **Graphing Calculator Videos** TI-83, TI-83 Plus, TI-89 Segments 1, 2 TI-86 Segments 1, 3	**Text-Specific Videos** Chapter 1 **Graphing Calculator Videos** TI-83, TI-83 Plus, TI-89 Segments 1, 2 TI-86 Segments 1, 3
Mastering Mathematics: How to Be a Great Math Student, Third Edition	**Student Workbook** Chapter 1	**Interactive Video Skillbuilder CD-ROM** Hours of video instruction, section quizzes, chapter tests, and *MathCue Tutorial*	**Interactive Video Skillbuilder CD-ROM** Hours of video instruction, section quizzes, chapter tests, and *MathCue Tutorial*
Conquering Math Anxiety, Second Edition		**WebTutor™ ToolBox** Chapter 1—Lecture notes, discussion threads, and quizzes on *WebCT* and *Blackboard*	**WebTutor™ ToolBox** Chapter 1—Lecture notes, discussion threads, and quizzes on *WebCT* and *Blackboard*

Chapter 2 ~ Inequalities, Functions, and Linear Functions

Ideas for Instruction	Print Resources	Media Resources for Instructors	Media Resources for Students
Explorations in Beginning and Intermediate Algebra, Third Edition Activities 7, 12, 16–17, 19	**Instructors** **Test Bank** Forms A–H	**BCA Instructor Version** Contains BCA Testing and BCA Tutorial **http://bca.brookscole.com**	**BCA** **BCA Tutorial** Unlimited practice in questions and problems for students
Activities for Beginning and Intermediate Algebra Activities 7–13	**Complete Solutions Manual** Chapter 2	**Web Site** Historical Notes, Math News, Careers, Web Quizzes, Internet Activities, *InfoTrac College Edition* exercises **http://mathematics.brookscole.com**	**vMentor** Customized interactive homework help and tutorial services online **http://bca.brookscole.com**
Algebra Facts Pages 63–67	**Instructor's Resource Manual** Chapter 2	**Text-Specific Videos** Chapter 2	**Web Site** Historical Notes, Math News, Careers, Web Quizzes, Internet Activities, *InfoTrac College Edition* exercises **http://mathematics.brookscole.com**
Mastering Mathematics: How to Be a Great Math Student, Third Edition	**Students** **Student Solutions Manual** Chapter 2	**InfoTrac® College Edition** **http://infotrac.thomsonlearning.com** **Graphing Calculator Videos** TI-83, TI-83 Plus, TI-89 Segment 2 TI-86 Segment 3	**Text-Specific Videos** Chapter 2 **Graphing Calculator Videos** TI-83, TI-83 Plus, TI-89 Segment 2 TI-86 Segment 3
Conquering Math Anxiety, Second Edition	**Student Workbook** Chapter 2	**Interactive Video Skillbuilder CD-ROM** Hours of video instruction, section quizzes, chapter tests, and *MathCue Tutorial*	**Interactive Video Skillbuilder CD-ROM** Hours of video instruction, section quizzes, chapter tests, and *MathCue Tutorial*
		WebTutor™ ToolBox Chapter 2—Lecture notes, discussion threads, and quizzes on *WebCT* and *Blackboard*	**WebTutor™ ToolBox** Chapter 2—Lecture notes, discussion threads, and quizzes on *WebCT* and *Blackboard*

Chapter 3 ~ Systems of Equations and Inequalities

Ideas for Instruction	Print Resources	Media Resources for Instructors	Media Resources for Students
Explorations in Beginning and Intermediate Algebra, Third Edition Activities 12–13 **Activities for Beginning and Intermediate Algebra** Activities 14–15 **Algebra Facts** Pages 131–134 **Mastering Mathematics: How to Be a Great Math Student, Third Edition** **Conquering Math Anxiety, Second Edition**	**Instructors** **Test Bank** Forms A–H **Complete Solutions Manual** Chapter 3 **Instructor's Resource Manual** Chapter 3 **Students** **Student Solutions Manual** Chapter 3 **Student Workbook** Chapter 3	**BCA Instructor Version** Contains BCA Testing and BCA Tutorial **http://bca.brookscole.com** **Web Site** Historical Notes, Math News, Careers, Web Quizzes, Internet Activities, *InfoTrac College Edition* exercises **http://mathematics.brookscole.com** **Text-Specific Videos** Chapter 3 **InfoTrac® College Edition** **http://infotrac.thomsonlearning.com** **Graphing Calculator Videos** TI-89 Segment 4 **Interactive Video Skillbuilder CD-ROM** Hours of video instruction, section quizzes, chapter tests, and *MathCue Tutorial* **WebTutor™ ToolBox** Chapter 3—Lecture notes, discussion threads, and quizzes on *WebCT* and *Blackboard*	**BCA** **BCA Tutorial** Unlimited practice in questions and problems for students **vMentor** Customized interactive homework help and tutorial services online **http://bca.brookscole.com** **Web Site** Historical Notes, Math News, Careers, Web Quizzes, Internet Activities, *InfoTrac College Edition* Exercises **http://mathematics.brookscole.com** **Text-Specific Videos** Chapter 3 **Graphing Calculator Videos** TI-89 Segment 4 **Interactive Video Skillbuilder CD-ROM** Hours of video instruction, section quizzes, chapter tests, and *MathCue Tutorial* **WebTutor™ ToolBox** Chapter 3—Lecture notes, discussion threads, and quizzes on *WebCT* and *Blackboard*

Chapter 4 ~ Quadratic and Polynomial Functions

Ideas for Instruction	Print Resources	Media Resources for Instructors	Media Resources for Students
Explorations in Beginning and Intermediate Algebra, Third Edition Activities 9–10, 20 **Activities for Beginning and Intermediate Algebra** Activities 4–6, 17 **Math Facts, Second Edition** Pages 57–60 **Algebra Facts** Pages 3–4, 11, 13–16, 43–52, 69–70, 75, 81–86, 95–96, 101–104, 143–144 **Mastering Mathematics: How to Be a Great Math Student, Third Edition** **Conquering Math Anxiety, Second Edition**	**Instructors** **Test Bank** Forms A–H **Complete Solutions Manual** Chapter 4 **Instructor's Resource Manual** Chapter 4 **Students** **Student Solutions Manual** Chapter 4 **Student Workbook** Chapter 4	**BCA Instructor Version** Contains BCA Testing and BCA Tutorial **http://bca.brookscole.com** **Web Site** Historical Notes, Math News, Careers, Web Quizzes, Internet Activities, *InfoTrac College Edition* exercises **http://mathematics.brookscole.com** **Text-Specific Videos** Chapter 4 **InfoTrac® College Edition** **http://infotrac.thomsonlearning.com** **Graphing Calculator Videos** TI-89 Segment 4 **Interactive Video Skillbuilder CD-ROM** Hours of video instruction, section quizzes, chapter tests, and *MathCue Tutorial* **WebTutor™ ToolBox** Chapter 4—Lecture notes, discussion threads, and quizzes on *WebCT* and *Blackboard*	**BCA** **BCA Tutorial** Unlimited practice in questions and problems for students **vMentor** Customized interactive homework help and tutorial services online **http://bca.brookscole.com** **Web Site** Historical Notes, Math News, Careers, Web Quizzes, Internet Activities, *InfoTrac College Edition* exercises **http://mathematics.brookscole.com** **Text-Specific Videos** Chapter 4 **Graphing Calculator Videos** TI-89 Segment 4 **Interactive Video Skillbuilder CD-ROM** Hours of video instruction, section quizzes, chapter tests, and *MathCue Tutorial* **WebTutor™ ToolBox** Chapter 4—Lecture notes, discussion threads, and quizzes on *WebCT* and *Blackboard*

Chapter 5 ~ Square Root Functions; Quadratic Equations and Inequalities

Ideas for Instruction	Print Resources	Media Resources for Instructors	Media Resources for Students
Explorations in Beginning and Intermediate Algebra, Third Edition Activities 5, 11 **Activities for Beginning and Intermediate Algebra** Activity 20 **Algebra Facts** Pages 19–20, 23–24, 99–108, 115–116 **Mastering Mathematics: How to Be a Great Math Student, Third Edition** **Conquering Math Anxiety, Second Edition**	**Instructors** **Test Bank** Forms A–H **Complete Solutions Manual** Chapter 5 **Instructor's Resource Manual** Chapter 5 **Students** **Student Solutions Manual** Chapter 5 **Student Workbook** Chapter 5	**BCA Instructor Version** Contains BCA Testing and BCA Tutorial **http://bca.brookscole.com** **Web Site** Historical Notes, Math News, Careers, Web Quizzes, Internet Activities, *InfoTrac College Edition* exercises **http://mathematics.brookscole.com** **Text-Specific Videos** Chapter 5 **InfoTrac® College Edition** **http://infotrac.thomsonlearning.com** **Graphing Calculator Videos** TI-89 Segment 4 **Interactive Video Skillbuilder CD-ROM** Hours of video instruction, section quizzes, chapter tests, and *MathCue Tutorial* **WebTutor™ ToolBox** Chapter 5—Lecture notes, discussion threads, and quizzes on *WebCT* and *Blackboard*	**BCA** **BCA Tutorial** Unlimited practice in questions and problems for students **vMentor** Customized interactive homework help and tutorial services online **http://bca.brookscole.com** **Web Site** Historical Notes, Math News, Careers, Web Quizzes, Internet Activities, *InfoTrac College Edition* exercises **http://mathematics.brookscole.com** **Text-Specific Videos** Chapter 5 **Graphing Calculator Videos** TI-89 Segment 4 **Interactive Video Skillbuilder CD-ROM** Hours of video instruction, section quizzes, chapter tests, and *MathCue Tutorial* **WebTutor™ ToolBox** Chapter 5—Lecture notes, discussion threads, and quizzes on *WebCT* and *Blackboard*

Chapter 6 ~ Quadratic Functions: Special Topics

Ideas for Instruction	Print Resources	Media Resources for Instructors	Media Resources for Students
Explorations in Beginning and Intermediate Algebra, Third Edition Activities 6, 20, 23 **Mastering Mathematics: How to Be a Great Math Student, Third Edition** **Conquering Math Anxiety, Second Edition**	**Instructors** **Test Bank** Forms A – H **Complete Solutions Manual** Chapter 6 **Instructor's Resource Manual** Chapter 6 **Students** **Student Solutions Manual** Chapter 6 **Student Workbook** Chapter 6	**BCA Instructor Version** Contains BCA Testing and BCA Tutorial **http://bca.brookscole.com** **Web Site** Historical Notes, Math News, Careers, Web Quizzes, Internet Activities, *InfoTrac College Edition* exercises **http://mathematics.brookscole.com** **Text-Specific Videos** Chapter 6 **InfoTrac® College Edition** **http://infotrac.thomsonlearning.com** **Graphing Calculator Videos** TI-83, TI-83 Plus, TI-89 Segment 2 TI-86 Segment 3 **Interactive Video Skillbuilder CD-ROM** Hours of video instruction, section quizzes, chapter tests, and *MathCue Tutorial* **WebTutor™ ToolBox** Chapter 6—Lecture notes, discussion threads, and quizzes on *WebCT* and *Blackboard*	**BCA** **BCA Tutorial** Unlimited practice in questions and problems for students **vMentor** Customized interactive homework help and tutorial services online **http://bca.brookscole.com** **Web Site** Historical Notes, Math News, Careers, Web Quizzes, Internet Activities, *InfoTrac College Edition* exercises **http://mathematics.brookscole.com** **Text-Specific Videos** Chapter 6 **Graphing Calculator Videos** TI-83, TI-83 Plus, TI-89 Segment 2 TI-86 Segment 3 **Interactive Video Skillbuilder CD-ROM** Hours of video instruction, section quizzes, chapter tests, and *MathCue Tutorial* **WebTutor™ ToolBox** Chapter 6—Lecture notes, discussion threads, and quizzes on *WebCT* and *Blackboard*

Chapter 7 ~ Rational Functions and Variation

Ideas for Instruction	Print Resources	Media Resources for Instructors	Media Resources for Students
Activities for Beginning and Intermediate Algebra Activities 18, 19 **Math Facts, Second Edition** Pages 129–130, 135–136 **Mastering Mathematics: How to Be a Great Math Student, Third Edition** **Conquering Math Anxiety, Second Edition**	**Instructors** **Test Bank** Forms A–H **Complete Solutions Manual** Chapter 7 **Instructor's Resource Manual** Chapter 7 **Students** **Student Solutions Manual** Chapter 7 **Student Workbook** Chapter 7	**BCA Instructor Version** Contains BCA Testing and BCA Tutorial **http://bca.brookscole.com** **Web Site** Historical Notes, Math News, Careers, Web Quizzes, Internet Activities, *InfoTrac College Edition* exercises **http://mathematics.brookscole.com** **Text-Specific Videos** Chapter 7 **InfoTrac® College Edition** **http://infotrac.thomsonlearning.com** **Interactive Video Skillbuilder CD-ROM** Hours of video instruction, section quizzes, chapter tests, and *MathCue Tutorial* **WebTutor™ ToolBox** Chapter 7—Lecture notes, discussion threads, and quizzes on *WebCT* and *Blackboard*	**BCA** **BCA Tutorial** Unlimited practice in questions and problems for students **vMentor** Customized interactive homework help and tutorial services online **http://bca.brookscole.com** **Web Site** Historical Notes, Math News, Careers, Web Quizzes, Internet Activities, *InfoTrac College Edition* exercises **http://mathematics.brookscole.com** **Text-Specific Videos** Chapter 7 **Interactive Video Skillbuilder CD-ROM** Hours of video instruction, section quizzes, chapter tests, and *MathCue Tutorial* **WebTutor™ ToolBox** Chapter 7—Lecture notes, discussion threads, and quizzes on *WebCT* and *Blackboard*

Chapter 8 ~ Exponents and Radicals

Ideas for Instruction	Print Resources	Media Resources for Instructors	Media Resources for Students
Explorations in Beginning and Intermediate Algebra, Third Edition Activities 4–5 **Math Facts, Second Edition** Pages 133–134, 143–144, 151–156, 167–168 **Algebra Facts** Pages 35–40, 55–56, 105–108, 115–116, 151–152 **Mastering Mathematics: How to Be a Great Math Student, Third Edition** **Conquering Math Anxiety, Second Edition**	**Instructors** **Test Bank** Forms A–H **Complete Solutions Manual** Chapter 8 **Instructor's Resource Manual** Chapter 8 **Students** **Student Solutions Manual** Chapter 8 **Student Workbook** Chapter 8	**BCA Instructor Version** Contains BCA Testing and BCA Tutorial **http://bca.brookscole.com** **Web Site** Historical Notes, Math News, Careers, Web Quizzes, Internet Activities, *InfoTrac College Edition* exercises **http://mathematics.brookscole.com** **Text-Specific Videos** Chapter 8 **InfoTrac® College Edition** **http://infotrac.thomsonlearning.com** **Interactive Video Skillbuilder CD-ROM** Hours of video instruction, section quizzes, chapter tests, and *MathCue Tutorial* **WebTutor™ ToolBox** Chapter 8—Lecture notes, discussion threads, and quizzes on *WebCT* and *Blackboard*	**BCA** **BCA Tutorial** Unlimited practice in questions and problems for students **vMentor** Customized interactive homework help and tutorial services online **http://bca.brookscole.com** **Web Site** Historical Notes, Math News, Careers, Web Quizzes, Internet Activities, *InfoTrac College Edition* exercises **http://mathematics.brookscole.com** **Text-Specific Videos** Chapter 8 **Interactive Video Skillbuilder CD-ROM** Hours of video instruction, section quizzes, chapter tests, and *MathCue Tutorial* **WebTutor™ ToolBox** Chapter 8—Lecture notes, discussion threads, and quizzes on *WebCT* and *Blackboard*

Ideas for Instruction	Print Resources	Media Resources for Instructors	Media Resources for Students
Explorations in Beginning and Intermediate Algebra, Third Edition Activity 22 **Mastering Mathematics: How to Be a Great Math Student, Third Edition** **Conquering Math Anxiety, Second Edition**	**Instructors** **Test Bank** Forms A–H **Complete Solutions Manual** Chapter 9 **Instructor's Resource Manual** Chapter 9 **Students** **Student Solutions Manual** Chapter 9 **Student Workbook** Chapter 9	**BCA Instructor Version** Contains BCA Testing and BCA Tutorial **http://bca.brookscole.com** **Web Site** Historical Notes, Math News, Careers, Web Quizzes, Internet Activities, *InfoTrac College Edition* Exercises **http://mathematics.brookscole.com** **Text-Specific Videos** Chapter 9 **InfoTrac® College Edition** **http://infotrac.thomsonlearning.com** **Interactive Video Skillbuilder CD-ROM** Hours of video instruction, section quizzes, chapter tests, and *MathCue Tutorial* **WebTutor™ ToolBox** Chapter 9—Lecture notes, discussion threads, and quizzes on *WebCT* and *Blackboard*	**BCA** **BCA Tutorial** Unlimited practice in questions and problems for students **vMentor** Customized interactive homework help and tutorial services online **http://bca.brookscole.com** **Web Site** Historical Notes, Math News, Careers, Web Quizzes, Internet Activities, *InfoTrac College Edition* exercises **http://mathematics.brookscole.com** **Text-Specific Videos** Chapter 9 **Interactive Video Skillbuilder CD-ROM** Hours of video instruction, section quizzes, chapter tests, and *MathCue Tutorial* **WebTutor™ ToolBox** Chapter 9—Lecture notes, discussion threads, and quizzes on *WebCT* and *Blackboard*

Ideas for Instruction	Print Resources	Media Resources for Instructors	Media Resources for Students
Mastering Mathematics: How to Be a Great Math Student, Third Edition **Conquering Math Anxiety, Second Edition**	**Instructors** **Test Bank** Forms A–H **Complete Solutions Manual** Chapter 10 **Instructor's Resource Manual** Chapter 10 **Students** **Student Solutions Manual** Chapter 10 **Student Workbook** Chapter 10	**BCA Instructor Version** Contains BCA Testing and BCA Tutorial **http://bca.brookscole.com** **Web Site (web Icon)** Historical Notes, Math News, Careers, Web Quizzes, Internet Activities, *InfoTrac College Edition* Exercises **http://mathematics.brookscole.com** **Text-Specific Videos** Chapter 10 **InfoTrac® College Edition** **http://infotrac.thomsonlearning.com** **Graphing Calculator Videos** TI-83, TI-83 Plus, TI-89 Segment 4 TI-86 Segment 5 **Interactive Video Skillbuilder CD-ROM** Hours of video instruction, section quizzes, chapter tests, and *MathCue Tutorial* **WebTutor™ ToolBox** Chapter 10—Lecture notes, discussion threads, and quizzes on *WebCT* and *Blackboard*	**BCA** **BCA Tutorial** Unlimited practice in questions and problems for students **vMentor** Customized interactive homework help and tutorial services online **http://bca.brookscole.com** **Web Site** Historical Notes, Math News, Careers, Web Quizzes, Internet Activities, *InfoTrac College Edition* exercises **http://mathematics.brookscole.com** **Text-Specific Videos** Chapter 10 **Graphing Calculator Videos** TI-83, TI-83 Plus, TI-89 Segment 4 TI-86 Segment 5 **Interactive Video Skillbuilder CD-ROM** Hours of video instruction, section quizzes, chapter tests, and *MathCue Tutorial* **WebTutor™ ToolBox** Chapter 10—Lecture notes, discussion threads, and quizzes on *WebCT* and *Blackboard*

Chapter 11 ~ Counting and Probability

Ideas for Instruction	Print Resources	Media Resources for Instructors	Media Resources for Students
Mastering Mathematics: How to Be a Great Math Student, Third Edition **Conquering Math Anxiety, Second Edition**	**Instructors** **Test Bank** Forms A–H **Complete Solutions Manual** Chapter 11 **Instructor's Resource Manual** Chapter 11 **Students** **Student Solutions Manual** Chapter 11 **Student Workbook** Chapter 11	**BCA Instructor Version** Contains BCA Testing and BCA Tutorial **http://bca.brookscole.com** **Web Site** Historical Notes, Math News, Careers, Web Quizzes, Internet Activities, *InfoTrac College Edition* exercises **http://mathematics.brookscole.com** **Text-Specific Videos** Chapter 11 **InfoTrac® College Edition** **http://infotrac.thomsonlearning.com** **Interactive Video Skillbuilder CD-ROM** Hours of video instruction, section quizzes, chapter tests, and *MathCue Tutorial* **WebTutor™ ToolBox** Chapter 11—Lecture notes, discussion threads, and quizzes on *WebCT* and *Blackboard*	**BCA** **BCA Tutorial** Unlimited practice in questions and problems for students **vMentor** Customized interactive homework help and tutorial services online **http://bca.brookscole.com** **Web Site** Historical Notes, Math News, Careers, Web Quizzes, Internet Activities, *InfoTrac College Edition* exercises **http://mathematics.brookscole.com** **Text-Specific Videos** Chapter 11 **Interactive Video Skillbuilder CD-ROM** Hours of video instruction, section quizzes, chapter tests, and *MathCue Tutorial* **WebTutor™ ToolBox** Chapter 11—Lecture notes, discussion threads, and quizzes on *WebCT* and *Blackboard*

www.brookscole.com

www.brookscole.com is the World Wide Web site for Brooks/Cole and is your direct source to dozens of online resources.

At *www.brookscole.com* you can find out about supplements, demonstration software, and student resources. You can also send email to many of our authors and preview new publications and exciting new technologies.

www.brookscole.com
Changing the way the world learns®

Annotated Instructor's Edition

INTERMEDIATE ALGEBRA
A Just-in-Time Approach
Third Edition

ALICE KASEBERG
formerly of Lane Community College

THOMSON
TM
BROOKS/COLE

Australia • Canada • Mexico • Singapore • Spain
United Kingdom • United States

THOMSON

BROOKS/COLE

Precalculus Editor ∎ *Jennifer Huber*
Publisher ∎ *Robert W. Pirtle*
Assistant Editor ∎ *Rebecca Subity*
Editorial Assistant ∎ *Carrie Dodson*
Technology Project Manager ∎ *Christopher Delgado*
Marketing Manager ∎ *Leah Thomson*
Marketing Assistant ∎ *Jessica Perry*
Advertising Project Manager ∎ *Bryan Vann*
Project Manager, Editorial Production ∎ *Harold Humphrey*
Print/Media Buyer ∎ *Jessica Reed*
Permissions Editor ∎ *Elizabeth Zuber*

Production Service ∎ *Lifland et al., Bookmakers*
Text Designer ∎ *Stephanie Kuhns, TECH·arts of Colorado, Inc.*
Art Editor ∎ *Sally Lifland*
Photo Researcher ∎ *Gail Magin*
Copy Editors ∎ *Sally Lifland, Gail Magin*
Interior Illustrators ∎ *Scientific Illustrators, Cyndie C. H. Wooley*
Cover Designer ∎ *Lisa Henry*
Cover Image ∎ *Paul Eekhoff/Masterfile*
Compositor ∎ *Better Graphics, Inc.*
Text and Cover Printer ∎ *Transcontinental Printing/Interglobe*

For more information about our products, contact us at:
Thomson Learning Academic Resource Center
1-800-423-0563

For permission to use material from this text, contact us by:
Phone: 1-800-730-2214   **Fax:** 1-800-730-2215
Web: http://www.thomsonrights.com

Library of Congress Control Number: 2003109637

Student Edition with InfoTrac College Edition: ISBN 0-534-38632-6

Annotated Instructor's Edition: ISBN 0-534-39665-8

Brooks/Cole–Thomson Learning
10 Davis Drive
Belmont, CA 94002
USA

Asia
Thomson Learning
5 Shenton Way #01-01
UIC Building
Singapore 068808

Australia/New Zealand
Thomson Learning
102 Dodds Street
Southbank, Victoria 3006
Australia

Canada
Nelson
1120 Birchmount Road
Toronto, Ontario M1K 5G4
Canada

Europe/Middle East/Africa
Thomson Learning
High Holborn House
50/51 Bedford Row
London WC1R 4LR
United Kingdom

Latin America
Thomson Learning
Seneca, 53
Colonia Polanco
11560 Mexico D.F.
Mexico

Spain/Portugal
Paraninfo
Calle/Magallanes, 25
28015 Madrid, Spain

Photo credits
p. 350: Fig. 1, courtesy of Marshall Space Flight Center/NASA; Fig. 2, courtesy of Goddard Space Flight Center/NASA.

To my grandparents, for their belief in education

My paternal grandparents chose tuition over farm payments, sending their son to a university in the depth of the 1930s depression. My maternal grandfather, a widower and civil servant, sent all four daughters through that same university, where the youngest one met and married the farmers' son.

Contents

3   Systems of Equations and Inequalities   149

The chapter presents three different ways to solve systems of two linear equations, as well as three different solution outcomes. (Chapter 10 presents a fourth way—matrices.) Exposure to equation solving continues with two methods of building equations: guess and check and quantity-value tables. Work is extended to solving systems of three equations and linear and absolute value inequalities.

4   Quadratic and Polynomial Functions   200

Work with the quadratic equation and quadratic modeling leads to operations with the polynomial function and three ways to solve quadratic equations: with tables, graphs, and factoring.

8 Exponents and Radicals   435

The presentation of operations with rational exponents and radicals builds on the discussion of exponents, scientific notation, and square roots in Chapter 5. Graphs and inverses once again play an important role in predicting solutions to equations.

9 Exponential and Logarithmic Functions   509

The type of change modeled by exponential growth is explored. Sequences are use to identify exponential functions and to write exponential equations. As the inverse to the exponential function, the logarithmic function is introduced and then used to solve equations. Detail is provided on exponential functions and logarithmic functions, their properties, and special applications.

10 Systems of Nonlinear Equations and Matrices   580

Techniques for solving systems of nonlinear equations and inequalities are applied to a final set of equations—the equations formed by the conic sections. Matrices permit a calculator approach to solving the systems of linear equations from Chapter 3.

Preface

Intermediate Algebra: A Just-in-Time Approach is based on

- the premise that concept development and understanding of mathematical thinking are facilitated by number patterns, problem solving, and discovery,
- agreement with the standards advocated by organizations such as the American Mathematical Association of Two Year Colleges (AMATYC) and the National Council of Teachers of Mathematics (NCTM),
- the availability of technology and its considered use, and
- the mastery of certain basic skills.

Intermediate Algebra: A Just-in-Time Approach is intended for students who have recently passed an Introductory Algebra course or who need to relearn or review forgotten algebra skills. The text is sufficiently complete in its presentation and practice of basic skills to accommodate the student with only an Introductory Algebra background. And the text is sufficiently different in its approach to provide new learning and thinking experiences for students with more algebra background. These students will be surprised to find an approach that is not a rehash of their high school algebra. Students and instructors alike will encounter new connections among topics, new methods, and new ways of learning.

Because students' levels of optimism and energy are highest at the beginning of the term, Chapter 1 introduces problem solving and connections among numeric, visual, verbal, and symbolic information (see the figure). From the beginning of this text, students are asked to use properties of real numbers to do arithmetic mentally, to integrate ideas from algebra and geometry (and, through projects, ideas from trigonometry, probability, and statistics), and to apply mathematics to settings from everyday life. Where possible, material is presented through a discovery approach: exploration, question, summary, example. This approach takes more words than the traditional mode of statement and example, but it greatly enhances learning.

Reviewers have noted that students have been quite satisfied with the motivation provided by the applications. The question "What is mathematics good for?" has simply not been an issue with users of the first and second editions of the text.

By reviewer and user request, the sections in this third edition have been made shorter and more uniform in length. Calculator screens now accompany most calculator references. In addition, systems of linear equations have been moved to Chapter 3 from Chapter 8 to permit more work with linear functions before quadratic functions are introduced.

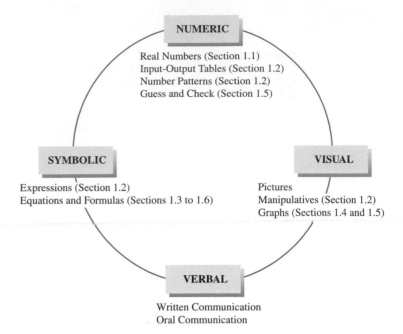

◼ Pedagogy

Objectives

The learning outcomes for each section are listed at the beginning of the section. They serve as a summary for both students and instructors and coordinate with the titles on the examples.

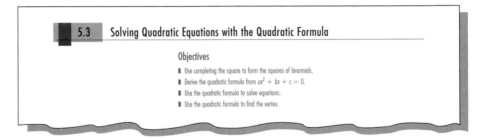

Warm-ups

The Warm-up at the beginning of each section is designed to serve as a class opener, reviewing important concepts, beginning exploration, and linking prior and upcoming topics. Warm-ups tend to be skill-oriented; they generally connect to the algebra needed to solve text examples. The answers to the Warm-up appear in the Answer Box at the end of the section.

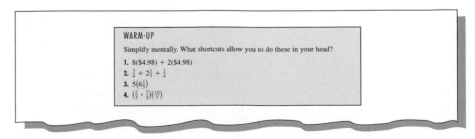

Small-Group Work

Some sections contain introductory questions or activities. These are intended to be done in class in small groups. In Section 1.1, for example, the Warm-up leads into Examples 6 and 7, where properties of numbers permit calculation of what at first appears to be difficult mental arithmetic. The examples demonstrate how each student may contribute to the class and, in turn, learn from others. It is important to emphasize that students improve their own understanding by helping others.

> **EXAMPLE 6** Using the properties of real numbers   Do these problems mentally by rearranging and regrouping the numbers.
>
> **a.** $\$4.25 + \$3.98 + \$3.75$ **b.** $-4.5 + 2.8 - 5.5 + 3.2$
>
> **c.** $1\frac{2}{3} - 2\frac{1}{2} + \frac{1}{3}$ **d.** $(-2)(7)(-5)$

Problem Solving

George Polya's four-step approach to problem solving—understanding the problem, making a plan, carrying out the plan, and checking the solution—is introduced in Section 1.1 and revisited where appropriate. The text then focuses on planning strategies. Section 1.2 introduces the strategies of *looking for a number pattern, making a table of inputs and outputs,* and *using manipulatives. Finding number patterns* continues in Sections 1.3, 2.4, 3.2, 4.1, and 9.1. *Breaking mind set* appears in Section 1.3. *Making a graph* first appears in Section 1.4. *Working backwards* is the fundamental idea used in solving equations and formulas in Section 1.5. *Choosing a test number or ordered pair and checking it* is used in Section 2.1 in drawing a line graph, in Section 3.5 in identifying half-planes for two-variable inequalities, and in Section 10.3 in solving systems of inequalities. *Guessing and checking,* which is a natural extension of choosing a test number for inequalities, is essential in writing equations in Section 1.5 and in building and solving systems of equations in Section 3.4. *Making a systematic list* is an essential component of factoring in Sections 4.3 and 4.4.

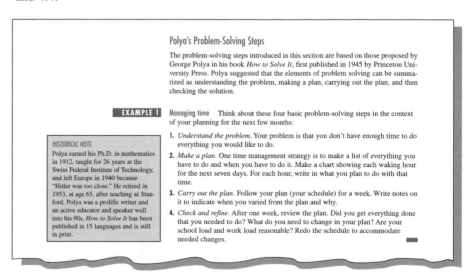

Polya's Problem-Solving Steps

The problem-solving steps introduced in this section are based on those proposed by George Polya in his book *How to Solve It,* first published in 1945 by Princeton University Press. Polya suggested that the elements of problem solving can be summarized as understanding the problem, making a plan, carrying out the plan, and then checking the solution.

EXAMPLE 1 Managing time   Think about these four basic problem-solving steps in the context of your planning for the next few months:

1. *Understand the problem.* Your problem is that you don't have enough time to do everything you would like to do.

2. *Make a plan.* One time management strategy is to make a list of everything you have to do and when you have to do it. Make a chart showing each waking hour for the next seven days. For each hour, write in what you plan to do with that time.

3. *Carry out the plan.* Follow your plan (your schedule) for a week. Write notes on it to indicate when you varied from the plan and why.

4. *Check and refine.* After one week, review the plan. Did you get everything done that you needed to do? What do you need to change in your plan? Are your school load and work load reasonable? Redo the schedule to accommodate needed changes.

HISTORICAL NOTE

Polya earned his Ph.D. in mathematics in 1912, taught for 26 years at the Swiss Federal Institute of Technology, and left Europe in 1940 because "Hitler was too close." He retired in 1953, at age 65, after teaching at Stanford. Polya was a prolific writer and an active educator and speaker well into his 90s. *How to Solve It* has been published in 15 languages and is still in print.

Explorations

Some examples are intended to be used in class for individual or group exploration. The solutions to these exploratory examples are included in the Answer Box at the end of the section.

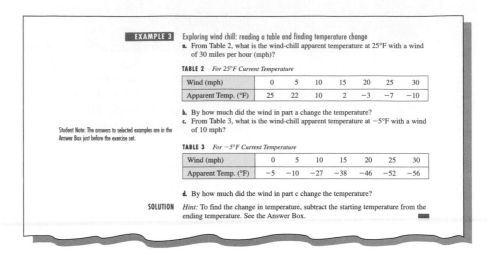

Hands-On Materials

The text supports use of an assortment of hands-on materials. Section 1.2 suggests algebra tiles for adding like terms; Section 9.1 uses paper folding; and Section 9.2 includes an exploration with coin tossing. Many of the projects in the Exercises involve hands-on materials.

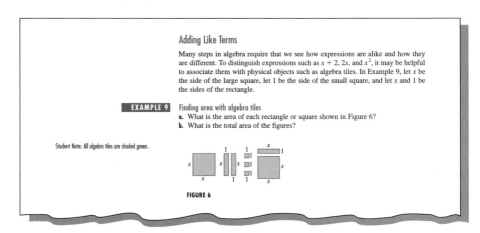

Examples

Each example begins with a title, which states the purpose of the example. Usually these titles relate back to the objectives for the section.

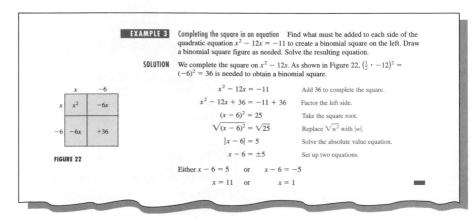

Think about it

"Think about it" questions are included within the reading material to encourage students to relate examples to prior material, to extend examples, and to practice verbalization skills. Answers to the questions are provided in the Answer Box.

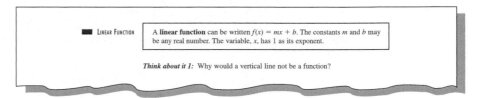

Sequences

Number patterns are used to identify and distinguish linear, quadratic, and exponential functions.

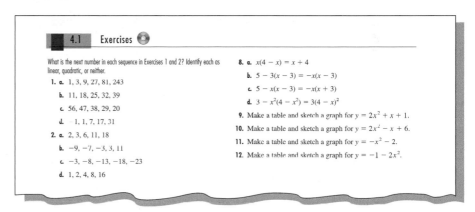

Reading Aids

An orange rectangle signals a transition for the reader. Such signals have been proven to aid readability. Here the orange rectangle leads the reader into an explanatory introduction to the next example.

> ■ At times, we want to describe regions in the coordinate plane with quadratic inequalities. After graphing the equation corresponding to the inequality, we can use test points to locate regions to be shaded.

Applications

To encourage creative thinking and depth in understanding, the text often poses a variety of questions about a single application setting. In addition, several applications, such as the cost of transcripts (introduced in Section 1.4), are repeated throughout the text so that students may observe the continuity and connections among topics.

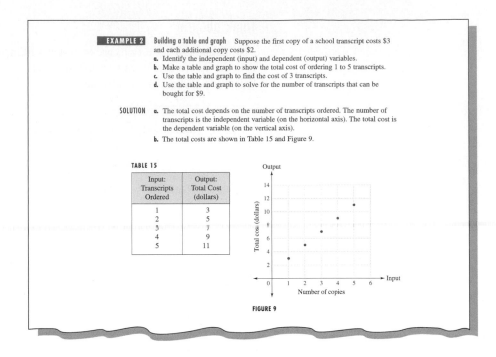

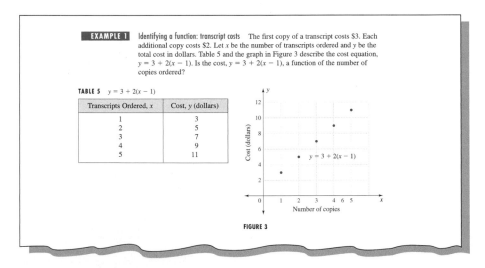

Input-Output Relationships and Functions

The text uses number patterns to introduce the concept of input-output relationships in Section 1.2 and then uses input-output relationships to lead to functions in Section 2.2.

Tables and Graphs

Extensive use is made of data in tabular form. Tables encourage organization of information and promote observation of patterns. They also prepare students for spreadsheet technology. Where appropriate, a graph is related to the table, to underscore the connections among algebra, geometry, statistics, and the real world. Numbers and their corresponding equations and graphs are employed to emphasize the fact that algebra is the transition language between arithmetic and analysis.

Aids are provided to help students read tables and graphs. For example, where students might think a parabolic graph showed a path, the horizontal axis is numbered with digital clocks.

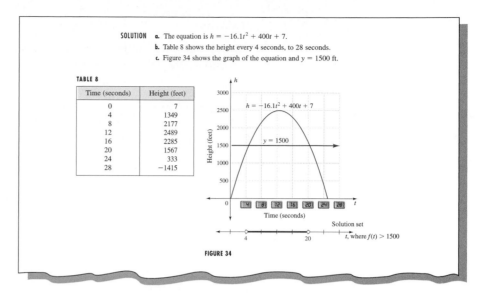

SOLUTION   **a.** The equation is $h = -16.1t^2 + 400t + 7$.

b. Table 8 shows the height every 4 seconds, to 28 seconds.

c. Figure 34 shows the graph of the equation and $y = 1500$ ft.

TABLE 8

Time (seconds)	Height (feet)
0	7
4	1349
8	2177
12	2489
16	2285
20	1567
24	333
28	−1415

FIGURE 34

Calculator Techniques

The graphing calculator is essential. Graphs and tables are provided in many examples so that students who are just learning the technology are not handicapped by a lack of understanding of the calculator.

Suggestions for use of graphing calculators in general are included throughout the text in Graphing Calculator technique boxes or tips identified by the graphing calculator icon. For a quick reference to techniques, look under Graphing Calculator in the Glossary/Index.

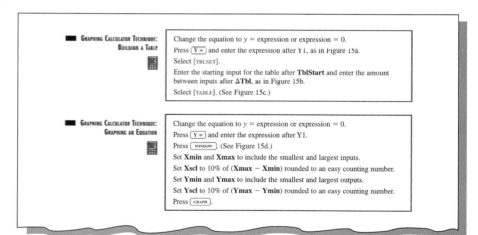

GRAPHING CALCULATOR TECHNIQUE: BUILDING A TABLE

Change the equation to $y = $ expression or expression $ = 0$.

Press $\boxed{Y=}$ and enter the expression after Y1, as in Figure 15a.

Select [TBLSET].

Enter the starting input for the table after **TblStart** and enter the amount between inputs after **ΔTbl**, as in Figure 15b.

Select [TABLE]. (See Figure 15c.)

GRAPHING CALCULATOR TECHNIQUE: GRAPHING AN EQUATION

Change the equation to $y = $ expression or expression $ = 0$.

Press $\boxed{Y=}$ and enter the expression after Y1.

Press $\boxed{\text{WINDOW}}$. (See Figure 15d.)

Set **Xmin** and **Xmax** to include the smallest and largest inputs.

Set **Xscl** to 10% of (**Xmax** − **Xmin**) rounded to an easy counting number.

Set **Ymin** and **Ymax** to include the smallest and largest outputs.

Set **Yscl** to 10% of (**Ymax** − **Ymin**) rounded to an easy counting number.

Press $\boxed{\text{GRAPH}}$.

Answer Boxes

Answers to the Warm-up and Explorations, as well as the "Think about it" questions, are placed in the Answer Box at the end of the section (just before the exercises). By providing answers as feedback, the Answer Box permits the text to be used in class or as a laboratory manual for group work or independent study.

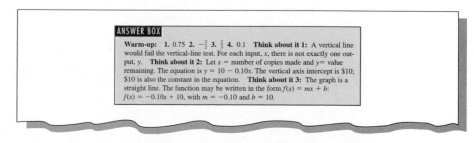

ANSWER BOX

Warm-up: **1.** 0.75 **2.** $-\frac{3}{5}$ **3.** $\frac{2}{3}$ **4.** 0.1 **Think about it 1:** A vertical line would fail the vertical-line test. For each input, x, there is not exactly one output, y. **Think about it 2:** Let $x = $ number of copies made and $y = $ value remaining. The equation is $y = 10 - 0.10x$. The vertical axis intercept is $10; $10 is also the constant in the equation. **Think about it 3:** The graph is a straight line. The function may be written in the form $f(x) = mx + b$: $f(x) = -0.10x + 10$, with $m = -0.10$ and $b = 10$.

Exercises

Tables and graphs in the exercises give students practice in skills such as solving equations. They also help students to learn graphing technology.

21. Refer to the graph of $f(x) = x^2 - 2x - 8$ in the figure.

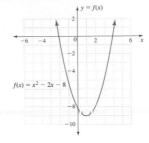

$f(x) = x^2 - 2x - 8$

a. Find the x-intercept points.

b. Find the y-intercept point.

c. Find the equation for the axis of symmetry.

e. Solve $-2x^2 + x + 1 = -2$.

f. Solve $-2x^2 + x + 1 = -5$.

g. Solve $-2x^2 + x + 1 = 0$; compare your answers with those in part a.

h. What is the range for $g(x)$?

23. Make a table and graph $f(x) = -x^2 - 2x + 8$.

a. Find the x-intercept points.

b. Find the y-intercept point.

c. Find the equation for the axis of symmetry.

d. Find the vertex.

e. Solve $-x^2 - 2x + 8 = 0$; compare your answers with those in part a.

f. Solve $-x^2 - 2x + 8 = 8$.

g. Solve $-x^2 - 2x + 8 = 10$.

h. Solve $-x^2 - 2x + 8 = 5$.

Projects

Projects are intended for group work or for individual effort. They may be more complicated problems related to the topic at hand, activity-based problems using manipulatives, or real-world applications that require research outside class. Selected projects might be due a week or so after homework is completed for any given section. Projects suited to in-class group work are marked with an asterisk in the Index of Projects, included in the Annotated Instructor's Edition and repeated in the *Instructor's Resource Manual*.

28. Draw sketches of quadratic functions that explain why quadratic equations can have only 0, 1, or 2 possible solutions.

29. Explain the relationship between the vertex of a parabolic graph and the set of outputs, or range, for the function.

30. Describe the relationship between the value of a in $y = ax^2 + bx + c$ and whether the parabola opens up or opens down.

31. If a graph of a quadratic function has the y-axis as an axis of symmetry and has $x = 4$ as one x-intercept, what will be another x-intercept?

32. If a graph of $f(x)$ has the y-axis as a line of symmetry, what may be said about $f(a)$ and $f(-a)$?

■ **Projects**

33. Positions on a Parabola   Match each phrase with one of these three positions on a parabola: x-intercept, y-intercept, or vertex. Make a sketch to clarify your answer. Answers may vary depending on your sketch.

a. The highest profit in a parabolic business profit curve

b. The place where a ball on a parabolic path falls to the ground when thrown

c. The start-up costs in a business, where the total cost graph is a parabola

d. The time it takes for a porpoise to re-enter the water after a parabolic leap

e. The down payment in a purchase plan where the total cost graph is a parabola

f. The highest point on a jet of water

g. The number of sales required to change from loss to profit on a parabolic business income curve

h. The maximum height in the parabolic path of a ball

i. The initial height of a golf tee, where the path of the ball is a parabola

j. The lowest point in a parabolic cable on a suspension bridge

k. The initial cost in a parabolic cost curve

l. The point where a diver enters the water, when the curve shows his parabolic path

34. Estimating Reaction Time   Turn your hand so that you can catch a ruler between your thumb and fingers. Have someone hold a ruler just above finger level and drop it between your outstretched thumb and fingers (see the figure). Catch the ruler. Use the point on the ruler at which your fingers catch it as an estimate of distance d in inches. Use $d = \frac{1}{2} gt^2$, with $g \approx 32.2$ ft/sec^2, to estimate your reaction time. (*Hint:* How many inches in a foot?)

Mid-Chapter Test

To keep students engaged and build their confidence, a Mid-Chapter Test is included. This test gives students the opportunity to check their progress. All answers appear in the back of the book, in the answer section.

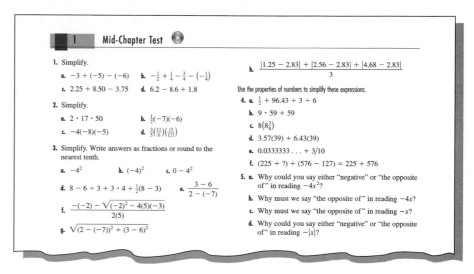

Chapter Summary, Review Exercises, and Chapter Test

Every chapter ends with a Chapter Summary, Review Exercises, and a Chapter Test. The student is provided with answers to the odd-numbered Review Exercises and answers to all of the Chapter Test questions.

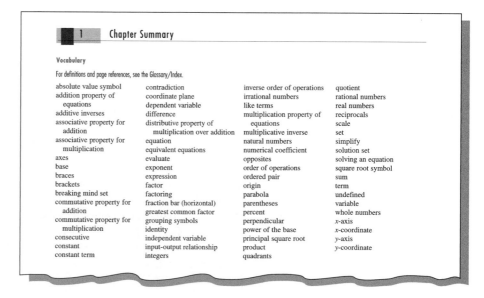

Cumulative Review

To help students maintain skills, a set of Cumulative Review Exercises is placed at the end of selected chapters.

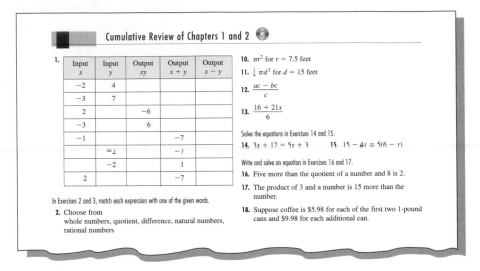

Final Exam Review

A Final Exam Review, containing exercises requiring both short and long answers, follows the last chapter. The review is divided by chapter and may be used as a source of additional exercises or cumulative review material.

Glossary/Index

For the convenience of students and instructors, essential vocabulary is defined and referenced by page in the combined Glossary/Index. Vocabulary is considered essential if it is listed in the Chapter Summary. Nonessential terms are referenced only by page in the Glossary/Index.

Acknowledgments

Reviewers and class-testers are the appreciated but unrecognized co-authors of a single-author textbook. Their comments, questions, and suggestions have contributed significantly to this and earlier editions. First let me thank the reviewers of this edition:

Anthony Aidoo, *Eastern Connecticut State University*
Kenneth R. Anderson, *Chemeketa Community College*
Mayme Kay Banasiak, *University of Tennessee–Chattanooga*
Timothy J. Biehler, *Finger Lakes Community College*
Jennifer Dollar, *Grand Rapids Community College*
Elizabeth Hodes, *Santa Barbara City College*
Susan Poston, *Chemeketa Community College*

Then let me thank those who provided valuable input through the focus group and Web survey:

Barbara S. Allen, *Delta College*
Chris Allgyer, *Mountain Empire Community College*
Wayne Barber, *Chemeketa Community College*
Judy Chilcott, *Truckee Meadows Community College*

Terry Darling, *Lewis and Clark Community College*
Hortencia I. Garcia, *Eastern Connecticut State University*
Tom Grogan, *Cincinnati State University*
Judith Hector, *Walters State Community College*
Stephen Paul Hess, *Grand Rapids Community College*
Tracey Hoy, *College of Lake County*
Doug Mace, *Kirtland Community College*
Annette Magyar, *Southwestern Michigan University*
Tammi Marshall, *Cuyamaca College*
William S. Newhall, *Truckee Meadows Community College*
Michael Norris, *Los Medanos College*
William A. Prescott, Jr., *University of Maine at Machias*
Carol Rardin, *Central Wyoming College*
Marcell Romancky, *Kirtland Community College*
Michael Satnik, *Central Washington University*
Barbara D. Sehr, *Indiana University–Kokomo*
Sister Monica Zore, *Marian College*

Finally, I would like to thank the following reviewers and class-testers for their helpful comments and significant contributions to the first and second editions:

Rick Armstrong, *Florissant Valley Community College*
Joann Bossenbroek, *Columbus State Community College*
Bradd Clark, *University of Southwestern Louisiana*
Robert Davidson, *East Tennessee State University*
Marsha Davis, *Eastern Connecticut State University*
John Dersch, *Grand Rapids Community College*
Margaret Donaldson, *Eastern Tennessee State University*
Megan Florence, *Albuquerque Technical-Vocational Institute*
Judy Kasabian, *El Camino College*
Linda Knauer, *Pellissippi State Technical Community College*
Jaclyn LeFebure, *Illinois Central College*
Lalo Mata, *Hartnell College*
Marveen McCready, *Chemeketa Community College*
Greg Perkins, *Hartnell College*
Pat Stone, *Tomball College*
Katherine Struve, *Columbus State Community College*
Deborah White, *College of the Redwoods*
Robert Wynegar, *University of Tennessee, Chattanooga*

This project started ten years ago, and it still has the full support of my husband, Rob Bowie. To say that I could not have done it without him is a total understatement. A special thank you to Jennifer Huber, Leah Thomson, and everyone at Brooks/Cole. Sally Lifland and her crew at Lifland et al., Bookmakers continue to make my best efforts so much better.

To the Student

Getting an education is one of the most valuable activities you will carry out in your life. Although poor economic circumstances may require you to relocate or retrain, you will not lose your education in a stock market crash. Your education will give you the skills to learn, again and again.

Whether your current educational goal is obtaining a degree or advancing in your chosen career, you should know exactly why you are taking this course and why it is important for you to learn and succeed. If your goal and purpose are not clear, you may find it difficult to make the daily commitment needed to study, learn, and succeed.

What to Expect from This Text

Intermediate Algebra: A Just-in-Time Approach presents the skills, concepts, problem-solving opportunities, and applications now considered fundamental to an Intermediate Algebra course. "Just in time," an industrial engineering term that is applied to delivery of inventory or manufacturing materials as needed, describes this text in two ways. First, it refers to the book's approach to the study of algebra, in which you work on real-world problems, learning algebraic principles and procedures just in time as you need them. Second, it describes the book's curriculum, which allows you to concentrate on what you need to know to use modern technology, just in time for this twenty-first century.

Alternative Approaches

Consider how you best learn directions to a friend's house—in words over the phone (verbally), from a map (visually), or from having been there (kinesthetically). As a student of algebra, you may prefer words (a verbal approach); drawings, pictures, and graphs (a visual approach); working with objects in your hands (a kinesthetic approach); or a combination of these approaches. The best way to achieve success in mathematics is to learn new concepts using the approach that you find easiest and then reinforce your learning with other approaches.

To help you achieve success with algebra, *Intermediate Algebra: A Just-in-Time Approach* presents concepts in as many ways as space permits. Because of the variety of approaches, those of you who have had algebra before will find that some concepts are presented quite differently than the way you learned them the first time. Acknowledging alternative approaches is a way of validating your own discoveries.

Independent Thinking

Although you will find the examples helpful, this text is designed to encourage your independent thinking. Look for patterns and relationships; discover concepts for yourself; seek out applications that are meaningful to you. Try to work through examples and explorations on your own first, before you look at the solution. The more involvement you have with the material, the more useful it will be and the longer you will remember it.

Groups

You are encouraged to work with others throughout this course. One of the most important benefits of working on mathematics with other people in and out of class is that you clarify your own understanding when you explain an idea to someone else. This is especially true for the kinesthetic learner.

▰ Beginning the New Term: First Day

Problem Solving

Section 1.1 introduces problem-solving steps in a mathematical context. Right now, think about these four basic problem-solving steps in the context of your planning for the next few months:

1. *Understand the problem.* Your problem is that you don't have enough time to do everything you would like to do.

2. *Make a plan.* One time management strategy is to make a list of everything you have to do and when you have to do it. Make a chart showing each waking hour for the next seven days. For each hour, write in what you plan to do with that time. (There are many problem-solving strategies for planning. For a more complete listing, see Problem Solving in the Preface.)

3. *Carry out the plan.* Follow your plan (your schedule) for a week. Write notes on it to indicate when you varied from the plan and why.

4. *Check and refine.* After one week, review the plan. Did you get everything done that you needed to do? What do you need to change in your plan? Are your school load and work load reasonable? Redo the schedule to accommodate needed changes.

Time Management

Make sure your course load is sensible. Exceptionally few people can productively manage 60 hours per week of classes, study, and work. Check how sensible your plan for this term is:

> Multiply your number of credits hours by 3, and then add your number of hours of employment.

For most people, 40 to 45 hours is a reasonable commitment to school and employment.

Make or buy a calendar for the term, with spaces large enough for noting assignments, tests, appointments, and errands. Keep the calendar with you at all times.

Are You in the Right Course?

Each mathematics course has one or more prerequisite courses. Having passed a placement test does not ensure that you are prepared to succeed. If you have studied the background material recently, then usually with time, effort, confidence, and patience you will be able to learn the new material. If you have had a semester or a quarter or a summer break since your last math course, it is necessary to review. If the review provided in the text is not sufficient for you to recall, say, operations with fractions, you should immediately seek advice from your instructor or outside help.

Make sure your course load is sensible.

Have you studied the prerequisite material recently?

Use your prior book as a reference. If you took the prerequisite course more than a year ago, you may want to retake it before going on.

After the First Class

As you review your course syllabus, write test dates and other deadlines on your calendar. If you are working or taking classes at two schools, make sure there are no schedule conflicts with final exams. Talk with your instructor this week to resolve any scheduling problems.

Beginning the New Term: First Week

Here are a number of ways to be successful as a student.

Get a Good Start

Have homework ready to turn in as you walk into class.

To succeed, you need to attend class, read the book before class, and do the homework in a timely manner. Plan your study time. Some students set up their schedules to have the hour after math class free, to review notes and start the assignment.

Success also depends on being prepared with the proper equipment: an appropriate calculator, a six-inch ruler also marked in centimeters, and graph paper. Do all your graphs on graph paper.

Keep in mind that your first homework paper is a "grade application," just as a cover letter and resume are part of a job application. First impressions count. Neatness and completeness make a lasting impression on the instructor. So does having homework ready to turn in as you walk into class.

Stick to Your Plan

Successful students plan their time carefully. Because studying is most productively done in the daytime and in hour-long segments, plan to study between classes. Unless you have a pressing need to leave campus, stay an extra hour and study again after your last class. Not only will a schedule help you be more efficient; it will also remind you of your priorities and prevent you from avoiding tasks that need to be done.

Prepare for the Next Day's Class

Each of the following steps will get you progressively more prepared for your next class.

- Skim the textbook section first. Read objectives. List unfamiliar words and identify new skills or concepts.
- Scan the section and look for definitions of vocabulary.
- Write vocabulary words and definitions on note cards.
- Outline the section, including summaries of skills and applications. (For your convenience in outlining, the objectives, headings, and example titles are in color in the text.)
- Read through the steps in several examples.
- Try the homework ahead of time.

Take Notes

Observe the five R's of note-taking (the Cornell system):

Don't recopy notes; summarize them.

1. *Record*. Write down the ideas and concepts in a lecture. (Reading the text ahead of time will help identify these items.) Don't recopy notes.

2. *Reduce.* Summarize notes immediately after class (or as soon as possible); highlight important items.

3. *Recite.* Say out loud in your own words the main ideas and concepts.

4. *Reflect.* Think about how the material fits in with what you already know.

5. *Review.* Once a week, go over the ideas from each class so far in the term.

■ Beginning the New Term: First Month

Homework

Do the homework as one of the steps in your learning—not just as a requirement of the course. Work on assignments as soon as possible, right after class or early in the day or weekend. This gives you the option of going back later and spending more time on a difficult exercise. During long study periods, build in breaks to keep yourself fresh: Work for an hour, do another subject for an hour, and then come back.

Make the homework meaningful. Write notes to yourself on homework papers. Highlight exercises that were difficult and that you want to review again later. Highlight formulas or key steps. Re-read the objectives. Summarize the definitions and solution methods in your own words to be sure you understand. Describe how the current section fits in with prior sections.

Highlight exercises that were difficult and that you want to review again later.

Using the Answer Box and Answer Section Effectively

Practice working quickly. Do not work with the answers in front of you. Wait to check your answers until you have finished several exercises or half or more of the homework assignment. Let your own reasoning tell you whether something is correct.

Stuck on the Homework?

Suppose you took notes in class, read the section, and still are stumped by an exercise. If you understand the directions but can't get the problem to work, try it again on a clean sheet of paper. If you don't know how to do an exercise, summarize the relevant information and drawings and go on to another exercise. Be sure to read the exercise aloud before you give up. Sometimes we hear things that we miss when reading.

Sometimes we get too close to a problem and overlook the obvious. A fresh point of view may help. Come back later. If necessary, call another student. If you are off-campus, call your instructor during office hours to get a hint or suggested strategy. Many instructors also welcome e-mail questions.

Obtain help from your teacher or from the resource center as you need it. Don't wait until just before a test.

Falling Behind?

Do current assignments first.

If you find yourself falling behind, let your instructor know that you are trying to catch up. Set up a plan that allows two to three days for each missed assignment. Most important, do current assignments first, even if you have to skip a few problems because you missed material. Work immediately after the class session. By doing the current assignment first, you will stay with the class. If you gradually complete missed work, within a reasonable amount of time you will be completely caught up. Do not skip class because you are behind or confused.

Forgetting Material?

Many students select one or two exercises from each section and write them on 3″ × 5″ cards, with complete solutions on the back. These "flash cards" may then be shuffled and practiced at any time for review. Cards provide an excellent way to study for tests and the final exam. Include vocabulary words in your card set.

■ Strategies for Taking and Learning from Tests

Prepare Yourself Academically

1. Attend class, and do the homework completely and regularly. If there are exercises on the homework that you do not know how to do, get help from a classmate, the teacher, or another appropriate source.

Work under time pressure on a regular basis.

2. Work under time pressure on a regular basis. Set yourself a limited amount of time to do portions of the homework. Use a time limit when doing review exercises or the practice tests at the middle and end of each chapter. Working in one- or two-hour blocks of time is usually more productive than spending all afternoon and evening on math one day a week.

Prepare Yourself Physically

3. Get a good night's sleep. Being rested helps you think clearly, even if you know less material.

Prepare Yourself Mentally

Psych yourself up! This is especially important if you have your test later in the day.

Get everything ready for the next day.

4. If you have a test at 8:00 a.m., use the last few minutes before bed to get everything ready for the next day. Make your lunch, set your books or pack on a chair by the door, set out your umbrella or appropriate weather gear, and make sure you have change for the bus or train or that your car's tires, battery, and gasoline level are okay.

5. Plan 10 or 15 minutes of quiet time before the test. Try to arrive early, if possible.

6. Mentally picture yourself taking the test.

   a. Imagine writing your name on the test.

   b. Imagine reading through the test completely to see where the instructor put various types of questions.

   c. Imagine writing notes, formulas, or reminders to yourself on the test.

   d. Imagine working your favorite type of problem first.

Take the Test Right

7. Arrive early. Be ready—pencil sharpened and homework papers ready to turn in.

8. Concentrate on doing the steps that you imagined in item 6 above.

Work problems you know how to solve first.

9. Work quickly and carefully through those problems you know how to solve. Don't spend over two minutes on one problem until you have tried every problem.

10. Be confident that, having prepared for the test, you can succeed.

Learn from the Test

11. After you turned in the test, did you remember information that would have helped you on the test? Would reading through the test more thoroughly at the start have given you time to recall information you needed?

12. Before you forget, look up and write down anything that you needed to know for the test but did not know.

13. When you get the test back, look at each item you missed. Which ones did you know how to do, and which ones did you not know how to do? Figure out what caused you to miss the ones you knew how to do. Carefully re-work on paper the ones you did not know how to do, getting help as needed.

14. Write down what you will do differently in preparing for the next test.

Problem Solving, Expressions, and Equations

How do you feel on a cold day when the wind is blowing? On a cold day when the wind is calm? Table 1 shows the temperature you feel (apparent temperature) as a result of the cooling provided by the wind. On a hot day, this cooling effect can be created with a fan. Which causes a greater drop in apparent temperature, a 30-mile-per-hour wind at 25°F or a 10-mile-per-hour wind at −5°F?

This chapter introduces algebra in four ways: symbolically (with the traditional algebraic notation), numerically (with number patterns, input-output tables, and other numerical methods), visually (with graphs and models), and verbally (in words with phrases and sentences). We review operations with and properties of real numbers. We evaluate and simplify expressions and solve formulas, as well as equations in one variable.

TABLE 1   *Wind-Chill Apparent Temperature Chart*

Wind Speed (miles per hour)

T \ S	0	5	10	15	20	25	30
35	35	32	22	16	11	8	5
30	30	27	16	9	4	0	−2
25	25	22	10	2	−3	−7	−10
20	20	16	4	−5	−10	−14	−18
15	15	11	−3	−11	−17	−22	−25
10	10	6	−9	−18	−25	−29	−33
5	5	1	−15	−25	−32	−37	−41
0	0	−5	−21	−31	−39	−44	−48
−5	−5	−10	−27	−38	−46	−52	−56

Current Temperature (°Fahrenheit)

1

1.1 Problem Solving and Number Sense

Student Note: Use the objectives to help you consider these questions:
What is it that I need to learn from this section?
Do I know it?
How does this section fit in with prior concepts?

Objectives

■ Make a time management plan, following Polya's problem-solving steps.

■ Identify sets of numbers.

■ Add, subtract, multiply, and divide integers and real numbers.

■ Identify properties of real numbers and use them to simplify expressions.

■ Apply the order of operations to expressions including square root, exponents, and absolute value.

Student Note: Each section begins with a set of exercises called a Warm-up. The Warm-up reviews skills or concepts needed within the section. Answers to the Warm-up are provided in the Answer Box just before the Exercises at the end of the section.

WARM-UP   Answers may vary. Use pairs or groups to encourage other descriptions besides those given here and in Examples 6 and 7.

Simplify mentally. What shortcuts allow you to do these in your head?

1. $8(\$4.98) + 2(\$4.98)$ $\$49.80$; factor: $\$4.98(8 + 2)$

2. $\frac{3}{4} + 2\frac{1}{3} + \frac{1}{4}$ $3\frac{1}{3}$; commutative property for addition: $\frac{3}{4} + \frac{1}{4} + 2\frac{1}{3}$

3. $5\left(6\frac{1}{5}\right)$ 31; distributive property of multiplication over addition: $5\left(6 + \frac{1}{5}\right) = 30 + 1$

4. $\left(\frac{2}{3} \cdot \frac{7}{5}\right)\left(\frac{10}{7}\right)$ $\frac{4}{3}$; associative property for multiplication: $\frac{2}{3}\left(\frac{7}{5} \cdot \frac{10}{7}\right) = \frac{2}{3} \cdot \frac{2}{1}$

THIS SECTION REVIEWS the names of the various sets of real numbers, operations with positive and negative real numbers, properties of real numbers, and order of operations. The descriptions that follow may seem boring, but the vocabulary of number sets and their properties make it possible to give definitions, describe patterns, develop number sense, write equations, and find solutions—in other words, to do algebra!

HISTORICAL NOTE
Polya earned his Ph.D. in mathematics in 1912, taught for 26 years at the Swiss Federal Institute of Technology, and left Europe in 1940 because "Hitler was too close." He retired in 1953, at age 65, after teaching at Stanford. Polya was a prolific writer and an active educator and speaker well into his 90s. *How to Solve It* has been published in 15 languages and is still in print.

Polya's Problem-Solving Steps

The problem-solving steps introduced in this section are based on those proposed by George Polya in his book *How to Solve It*, first published in 1945 by Princeton University Press. Polya suggested that the elements of problem solving can be summarized as understanding the problem, making a plan, carrying out the plan, and then checking the solution.

EXAMPLE 1   Managing time   Think about these four basic problem-solving steps in the context of your planning for the next few months:

1. *Understand the problem.* Your problem is that you don't have enough time to do everything you would like to do.

2. *Make a plan.* One time management strategy is to make a list of everything you have to do and when you have to do it. Make a chart showing each waking hour for the next seven days. For each hour, write in what you plan to do with that time.

3. *Carry out the plan.* Follow your plan (your schedule) for a week. Write notes on it to indicate when you varied from the plan and why.

4. *Check and refine.* After one week, review the plan. Did you get everything done that you needed to do? What do you need to change in your plan? Are your school load and work load reasonable? Redo the schedule to accommodate needed changes.

Sets of Numbers

A **set** is *a collection of objects or numbers*. The **real numbers** are *the set of numbers that can be located with a point on the number line*. The numbers shown in Figure 1, $-3\frac{1}{2}$, $-\sqrt{4}$, -0.5, $\sqrt{2}$, and π, are all real numbers.

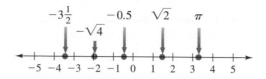

FIGURE 1 Real number line

Many of the sets of numbers contained in the real numbers can be described by a listing of the numbers placed in braces, { }. The **natural numbers** are *the set containing the numbers for counting*, $\{1, 2, 3, \ldots\}$. When we *add zero to the set of natural numbers*, we have the set of **whole numbers**, $\{0, 1, 2, 3, \ldots\}$. The positive and negative numbers in Table 1 are integers. The **integers** are *the whole numbers and their opposites*, $\{\ldots, -3, -2, -1, 0, 1, 2, 3, \ldots\}$.

By placing each integer over 1, $\{\ldots, -\frac{3}{1}, -\frac{2}{1}, -\frac{1}{1}, \frac{0}{1}, \frac{1}{1}, \frac{2}{1}, \frac{3}{1}, \ldots\}$, we show that integers can be written as rational numbers. **Rational numbers** *can be written* $\frac{a}{b}$, *where a and b are integers and b is not zero.*

The **principal square root** of x, $\sqrt{x}$, is *the positive real number that, when multiplied by itself, gives the radicand, x*. For example, $\sqrt{3} \cdot \sqrt{3} = 3$. The $\sqrt{}$ symbol is called the square root sign or radical. Some square roots, such as $\sqrt{4} = 2$ and $\sqrt{9} = 3$, are rational; others are irrational. The **irrational numbers** are *the set of numbers that cannot be written as a rational number*. Irrational numbers include pi (π), $\sqrt{2}$, $\sqrt{3}$, and $\sqrt{5}$.

Here is a summary of the real numbers.

▮ SET OF REAL NUMBERS

> Sets of rational numbers (which can be written $\frac{a}{b}$ with a and b integers, b not zero) include
>
> - natural numbers $\{1, 2, 3, \ldots\}$
> - whole numbers $\{0, 1, 2, 3, \ldots\}$
> - integers $\{\ldots, -3, -2, -1, 0, 1, 2, 3, \ldots\}$
> - square roots with exact decimal values $\{\sqrt{1}, \sqrt{4}, \sqrt{9}, \sqrt{16}, \ldots\}$
>
> Sets of irrational numbers (which cannot be written $\frac{a}{b}$ with a and b integers, b not zero) include
>
> - square roots without exact decimal values $\{\sqrt{2}, \sqrt{3}, \sqrt{5}, \ldots\}$
> - pi (π), equal to $3.14159265358\ldots$

Student Note: Try the problem posed in the example before reading the solution. You might want to cover up the solution with paper or a bookmark.

The square root of a negative number, such as $\sqrt{-4}$, is not a real number and is **undefined** (*has no value or meaning*) when we work within the set of real numbers. Note that $-\sqrt{4}$ is defined; $-\sqrt{4} = -2$.

EXAMPLE 2

Identifying sets of numbers   Write the names of all the sets of numbers to which each number belongs.

a. 5   **b.** $\frac{1}{4}$ **c.** $\sqrt{49}$ **d.** $\sqrt{11}$ **e.** 1.5

SOLUTION
a. 5 belongs to the sets of real, rational, integer, whole, and natural numbers.
b. $\frac{1}{4}$ belongs to the sets of real and rational numbers.
c. $\sqrt{49}$ equals 7 and belongs to the sets of real, rational, integer, whole, and natural numbers.
d. $\sqrt{11}$ does not have an exact square root and is irrational and real.
e. 1.5 equals $\frac{3}{2}$ and belongs to the sets of real and rational numbers.

Operations with Positive and Negative Real Numbers

As mentioned earlier, the positive and negative numbers in the wind-chill chart in Table 1 are integers. The exploration in Example 3 is designed to start you thinking about operations with integers.

EXAMPLE 3

Exploring wind chill: reading a table and finding temperature change
a. From Table 2, what is the wind-chill apparent temperature at 25°F with a wind of 30 miles per hour (mph)?

TABLE 2 *For 25°F Current Temperature*

Wind (mph)	0	5	10	15	20	25	30
Apparent Temp. (°F)	25	22	10	2	−3	−7	−10

b. By how much did the wind in part a change the temperature?
c. From Table 3, what is the wind-chill apparent temperature at −5°F with a wind of 10 mph?

Student Note: The answers to selected examples are in the Answer Box just before the exercise set.

TABLE 3 *For −5°F Current Temperature*

Wind (mph)	0	5	10	15	20	25	30
Apparent Temp. (°F)	−5	−10	−27	−38	−46	−52	−56

d. By how much did the wind in part c change the temperature?

SOLUTION
Hint: To find the change in temperature, subtract the starting temperature from the ending temperature. See the Answer Box.

Parts b and d in Example 3 required subtraction of integers. We can model this subtraction of integers by finding the starting temperature on a number line (thermometer) and counting the number of degrees to the ending temperature. (See Figure 2.) If the temperature falls, the change is negative; if the temperature rises, the change is positive.

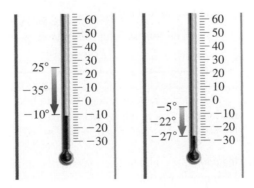

FIGURE 2

Here is a summary of the operations with positive and negative real numbers.

■ SUMMARY OF OPERATIONS WITH POSITIVE OR NEGATIVE REAL NUMBERS

Addition of Positive and Negative Real Numbers
If the signs are alike, add the numbers and place the common sign on the answer:

a. $-6 + (-8) = -14$ **b.** $21 + 17 = 38$

If the signs are different, subtract the number portion and place the sign of the number farthest from zero on the answer:

a. $15 + (-10) = +(15 - 10) = 5$ **b.** $4 + (-5) = -(5 - 4) = -1$

Subtraction
Two numbers are **opposites**, or **additive inverses**, *if they add to zero*. Opposites are the same distance from zero on the number line and have opposite signs. Subtraction is the inverse operation to addition. To subtract two numbers, change subtraction of a number to addition of the opposite number:

a. $6 - 5 = 6 + (-5) = 1$
b. $6 - (-4) = 6 + (+4) = 10$
c. $-11 - 25 = -11 + (-25) = -36$
d. $-27 - (-5) = -27 + (+5) = -27 + 5 = -22$

Multiplication of Positive and Negative Real Numbers
If two numbers have like signs, their product is positive:

a. $(-2)(-3) = +6$ **b.** $3(5) = 15$

If two numbers have unlike signs, their product is negative:

a. $(-4)(+3) = -12$ **b.** $(+6)(-3) = -18$

The product of zero and any number is zero:

a. $0(6) - 6(0) - 0$ **b.** $0(-3) = -3(0) = 0$

Division
Two numbers are **reciprocals** *if their product is* 1. A reciprocal is also known as a **multiplicative inverse**. Division is the inverse operation to multiplication. To divide two numbers, change division by a number to multiplication by the reciprocal:

a. $-5 \div \frac{1}{3} = -5 \cdot \frac{3}{1} = -15$ **b.** $6 \div \frac{2}{3} = 6 \cdot \frac{3}{2} = \frac{18}{2} = 9$

The inverse of an operation performed after a given operation will nullify the original operation:

a. $6 + 4 - 4 = 6$ **b.** $5 \cdot 8 \div 8 = 5$

Mathematicians use **simplify** as *a shorthand way to write directions*. Simplify can mean "to perform the indicated operations," as in Example 4.

EXAMPLE 4 Practicing operations with real numbers   Simplify by performing these operations.

a. $4 + (-3)$ **b.** $4(-3)$ **c.** $4 \div \dfrac{-4}{5}$ **d.** $2 \cdot (-4)$

e. $(-6) + (-9)$ **f.** $(-6) \div \dfrac{-3}{4}$ **g.** $-6 - (-3)$ **h.** $-6 - 3$

i. $-3 + 2 - 2$ **j.** $-6 \div 2 \cdot 2$

SOLUTION   **a.** $4 + (-3) = 1$ **b.** $4(-3) = -12$

c. $4 \div \dfrac{-4}{5} = 4 \cdot \dfrac{-5}{4} = \dfrac{4 \cdot (-5)}{4} = -5$

d. $2 \cdot (-4) = -8$

e. $(-6) + (-9) = -15$

f. $-6 \div \dfrac{-3}{4} = -6 \cdot \dfrac{-4}{3} = \dfrac{-6 \cdot (-4)}{3} = \dfrac{24}{3} = 8$

g. $-6 - (-3) = -6 + (+3) = -3$

h. $-6 - 3 = -6 + (-3) = -9$

i. $-3 + 2 - 2 = -3$

j. $-6 \div 2 \cdot 2 = -6$

MAKING A DISTINCTION: SUBTRACTION, NEGATIVE, AND OPPOSITE

> The symbol "−" is used in three different ways:
>
> **1.** To indicate *subtraction*, as in 6 − 7, read "6 subtract 7" or "6 minus 7."
>
> **2.** To indicate that a number is *negative*, as in −6, read "negative 6."
>
> **3.** To indicate the *opposite*, as in −(6), read "the opposite of 6."
>
> For −6, it is correct to say either "negative 6" or "the opposite of 6." However, if the "−" sign is in front of a set of parentheses (or other grouping symbol) or a letter, as in −a, we must say "the opposite," rather than "negative," unless we can prove that the expression is always negative.

EXAMPLE 5   Distinguishing subtraction, negative, and opposite   Write each in words.

a. -7 **b.** $0 - 3$ **c.** $-(-7)$ **d.** $6 - (-2)$

SOLUTION   **a.** negative seven (or opposite of 7)

b. zero subtract 3 (or zero minus 3)

c. the opposite of negative 7

d. 6 subtract negative 2 (or 6 minus negative 2)

 Calculator note: The (−) key is used both for a negative sign and for the opposite, but never for subtraction. Explore the results from entering each part of Example 5 with the subtraction key and the (−) key. Make notes on your observations.

Properties of Real Numbers

ASSOCIATIVE AND COMMUTATIVE PROPERTIES   Real numbers have properties for addition and multiplication that permit us to rearrange terms or to change the order in which we add or multiply. We used many of the following properties in the Warm-up. These properties are true for all real numbers, a, b, and c.

ASSOCIATIVE PROPERTY FOR ADDITION

> $a + (b + c) = (a + b) + c$
> $2 + (3 + 4) = (2 + 3) + 4$

ASSOCIATIVE PROPERTY FOR MULTIPLICATION

> $a \cdot (b \cdot c) = (a \cdot b) \cdot c$
> $3 \cdot (4 \cdot 5) = (3 \cdot 4) \cdot 5$

The associative property for addition permits us to regroup or add numbers in a convenient order. The associative property for multiplication permits regrouping to multiply numbers in a convenient order.

■ COMMUTATIVE PROPERTY FOR ADDITION

$$a + b = b + a$$
$$6 + 8 = 8 + 6$$

■ COMMUTATIVE PROPERTY FOR MULTIPLICATION

$$a \cdot b = b \cdot a$$
$$6 \cdot 8 = 8 \cdot 6$$

The commutative property for addition and the commutative property for multiplication allow us to rearrange the terms in an expression.

EXAMPLE 6

Use pairs or groups to discuss Examples 6 and 7.

Using the properties of real numbers   Do these problems mentally by rearranging and regrouping the numbers.

a. $4.25 + $3.98 + $3.75   **b.** $-4.5 + 2.8 - 5.5 + 3.2$

c. $1\frac{2}{3} - 2\frac{1}{2} + \frac{1}{3}$ **d.** $(-2)(7)(-5)$

SOLUTION   **a.** $4.25 + $3.98 + $3.75 = $4.25 + $3.75 + $3.98   Rearrange.

$\qquad\qquad\qquad\qquad\qquad = $8.00 + 3.98 Regroup.

$\qquad\qquad\qquad\qquad\qquad = 11.98

b. $-4.5 + 2.8 - 5.5 + 3.2 = -4.5 - 5.5 + 2.8 + 3.2$ Rearrange.

$\qquad\qquad\qquad\qquad\qquad = -10 + 6$ Regroup.

$\qquad\qquad\qquad\qquad\qquad = -4$

c. $1\frac{2}{3} - 2\frac{1}{2} + \frac{1}{3} = 1\frac{2}{3} + \frac{1}{3} - 2\frac{1}{2}$ Rearrange.

$\qquad\qquad\qquad\quad = 2 - 2\frac{1}{2}$ Regroup.

$\qquad\qquad\qquad\quad = -\frac{1}{2}$

d. $(-2)(7)(-5) = (-2)(-5)(7)$ Rearrange.

$\qquad\qquad\qquad = 10 \cdot 7$ Regroup.

$\qquad\qquad\qquad = 70$ ■

DISTRIBUTIVE PROPERTY   Another property of all real numbers, the distributive property, justifies multiplying out parentheses and factoring.

■ DISTRIBUTIVE PROPERTY OF MULTIPLICATION OVER ADDITION

$$a(b + c) = ab + ac$$
$$3(4 + 5) = 3 \cdot 4 + 3 \cdot 5$$

Because a is multiplying both b and c, we say that this property distributes a over the addition of b and c. Because all subtraction can be changed to addition of the opposite, the distributive property applies to subtraction.

The addition (or subtraction) sign in each part of Example 7 is hidden.

EXAMPLE 7

Applying the distributive property   Do these multiplication problems mentally.

a. $6\left(5\frac{1}{2}\right)$ **b.** $8\left(7\frac{1}{4}\right)$ **c.** $7(406)$ **d.** $5(499)$

SOLUTION   **a.** $6\left(5\frac{1}{2}\right) = 6\left(5 + \frac{1}{2}\right)$ Change to addition.

$\qquad\qquad\quad = 30 + 3$ Distribute the 6.

$\qquad\qquad\quad = 33$

b. $8\left(7\frac{1}{4}\right) = 8\left(7 + \frac{1}{4}\right)$ Change to addition.

$\qquad\qquad\quad = 56 + 2$ Distribute the 8.

$\qquad\qquad\quad = 58$

c. $7(406) = 7(400 + 6)$ Change to addition.

        $= 2800 + 42$ Distribute the 7.

        $= 2842$

d. $5(499) = 5(500 - 1)$ Change to subtraction.

        $= 2500 - 5$ Distribute the 5.

        $= 2495$

FACTORING   **Factors** are *the numbers being multiplied.* In $a(b + c)$, there are two factors: a and $b + c$. In $3 \cdot 5 \cdot 7$, there are three factors: 3, 5, and 7.

     When we reverse the distributive property, $ab + ac = a(b + c)$, we are factoring. In this case, **factoring** is *changing the addition of ab and ac into the multiplication of a and (b + c).*

DISTRIBUTIVE PROPERTY IN FACTORED FORM	$ab + ac = a(b + c)$ $15 + 24 = 3(5 + 8)$

EXAMPLE 8   Factoring   Apply factoring to do these mentally.

   a. $9 \cdot 105 - 9 \cdot 5$ **b.** $8 \cdot \$2.98 + 2 \cdot \$2.98$

SOLUTION   **a.** $9 \cdot 105 - 9 \cdot 5 = 9(105 - 5)$ Factor the 9.

                 $= 9(100)$ Multiply.

                 $= 900$

   b. $8 \cdot \$2.98 + 2 \cdot \$2.98 = \$2.98(8 + 2)$ Factor the $2.98.

                       $= \$2.98(10)$ Multiply.

                       $= \$29.80$

Think about it 1: Were there larger numbers than 9 and $2.98 that might have been factored from each part of Example 8? Why were 9 and $2.98 chosen?

Student Note: On arithmetic calculators and in several computer programming languages, the order of operations is not that used in algebra. Hence, the order of operations is an agreement, not a property of numbers. If you try $5 - 4 \cdot 3$ on an arithmetic calculator, you will get 3. If you try $5 - 4 \cdot 3$ on a scientific or graphing calculator, you may get -7.

Order of Operations

The **order of operations** is *an agreement about the order in which we work with numbers in algebra.*

ORDER OF OPERATIONS	**1.** Calculate expressions within parentheses and other grouping symbols. **2.** Calculate exponents and square roots. **3.** Do the remaining multiplications and divisions in order of appearance, left to right. **4.** Do the remaining additions and subtractions in order of appearance, left to right.

Note: $(-7)^2 = (-7)(-7)$ and $72^2 = 72 \cdot 72$.

EXAMPLE 9   Applying the order of operations   Simplify the following expressions.

   a. Perimeter of a photograph, in inches: $P = 2(5 + 7)$

   b. Outside surface area of a shoe box without lid, in square inches: $S = 2 \cdot 4 \cdot 6 + 2 \cdot 4 \cdot 14 + 1 \cdot 6 \cdot 14$

c. Volume of a CD box, in cubic centimeters: $V = 1.0(14.2)(12.4)$

d. From solving an equation: $d = (-7)^2 - 4(4)(-2)$

e. Distance around a running track, in meters: $D = 2(100) + 2\pi(31.831)$

f. A puzzle: $A = 3 - \frac{8}{2} + 6 \cdot 3 + 3$

SOLUTION   a. $P = 2(5 + 7)$ Add first.

$P = 2(12)$ Multiply second.

$P = 24$ in.

Student Note: The comments to the right tell you what to do to the given step to get to the next step.

b. $S = 2 \cdot 4 \cdot 6 + 2 \cdot 4 \cdot 14 + 1 \cdot 6 \cdot 14$ Multiply first.

$S = 48 + 112 + 84$ Add second.

$S = 244$ in^2

c. $V = 1.0(14.2)(12.4) = 176.08$ cm^3

d. $d = (-7)^2 - 4(4)(-2)$ Do the exponent first.

$d = 49 - 4(4)(-2)$ Do multiplication from left to right.

$d = 49 - (-32)$ Do subtraction.

$d = 81$

e. $D = 2(100) + 2\pi(31.831)$ Multiply first.

$D \approx 200 + 200$ Add second.

$D \approx 400$ m

f. $A = 3 - \frac{8}{2} + 6 \cdot 3 + 3$ Do division, then multiplication.

$A = 3 - 4 + 18 + 3$ Do subtraction.

$A = -1 + 18 + 3$ Do addition.

$A = 20$

In parts b and f of Example 9, we used a dot for multiplication of positive integers. In parts c, d, and e, we used parentheses for multiplication of numbers containing negative signs or decimals. There is no rule for this, but the choice makes the multiplications easier to read.

GROUPING SYMBOLS   **Grouping symbols**, such as parentheses, *indicate when an operation is to be done out of its normal order.* Grouping symbols are first in the order of operations. When several groupings are needed, we use brackets or braces. The preferred way to write multiple symbols is {[( )]}, with **parentheses** (*rounded*) on the inside, **brackets** (*square-shaped*) used next, and **braces** (*like little wires*) used outermost. If the context is not confusing, double parentheses, (( )), are acceptable.

Think about it 2: What is the value of $5 - \{2 - [3 - (4 + 5)]\}$?

If an absolute value, square root, or fraction contains operations, it is a grouping symbol. Note that we *place parentheses around numbers and operations within grouping symbols when using a calculator.*

The **absolute value symbol** is used in $|5 - 8|$. Find $5 - 8$ inside the absolute value, and then find the absolute value: $|-3| = 3$. On a calculator, write abs $(5 - 8)$. Absolute value is under [2nd] [CATALOG].

The **square root symbol**, or radical sign, is used in $\sqrt{1 + 3}$. Add 1 and 3, and then take the square root: $\sqrt{4} = 2$. On a calculator, write $\sqrt{(1 + 3)}$.

The **horizontal fraction bar** is a grouping symbol in $\dfrac{8 - \sqrt{36}}{3 - 5}$. Find the values of the numerator and denominator separately, and then divide the two parts:

$$\frac{8 - \sqrt{36}}{3 - 5} = \frac{8 - 6}{-2} = -1$$

On a calculator, write $(8 - \sqrt{36})/(3 - 5)$.

Think about it 3: If the calculator places an opening parenthesis after the square root, you need to include a closing parenthesis: $(8 - \sqrt{(36)})/(3 - 5)$. Describe how 11 is obtained as a wrong answer from this entry: $(8 - \sqrt{(36)}/(3 - 5)$.

EXAMPLE 10   Using grouping symbols   Simplify the following expressions.

a. Slope of a line: $m = \dfrac{-2 - 8}{1 - (-4)}$

b. Body mass index: $i = \dfrac{160(704.5)}{72^2}$

c. Distance formula: $d = \sqrt{(1 - (-4))^2 + (-2 - 8)^2}$

d. Mean absolute deviation: $d = \dfrac{|9.2 - 9.5| + |9.3 - 9.5| + |9.9 - 9.5|}{3}$

e. From solving an equation: $x = \dfrac{-(-7) + \sqrt{(-7)^2 - 4(4)(-2)}}{2(4)}$

f. A puzzle: $A = 5\{6 - 4[3 - 8(5 - 7) + 1] - 2\}$

SOLUTION   **a.** $m = \dfrac{-2 - 8}{1 - (-4)}$ Find the numerator and denominator.

$\quad\quad m = \dfrac{-10}{5}$ Divide.

$\quad\quad m = -2$

b. $i = \dfrac{160(704.5)}{72^2}$ Find the numerator and denominator.

$\quad\quad i = \dfrac{112{,}720}{5184}$ Divide.

$\quad\quad i \approx 21.7$

c. $d = \sqrt{(1 - (-4))^2 + (-2 - 8)^2}$ Simplify inside the parentheses.

$\quad\quad d = \sqrt{5^2 + (-10)^2}$ Apply exponents.

$\quad\quad d = \sqrt{25 + 100}$ Simplify inside the radical.

$\quad\quad d = \sqrt{125} \approx 11.2$ Take the square root.

d. $d = \dfrac{|9.2 - 9.5| + |9.3 - 9.5| + |9.9 - 9.5|}{3}$ Do the absolute values.

$\quad\quad d = \dfrac{0.3 + 0.2 + 0.4}{3}$ Add the numerator.

$\quad\quad d = \dfrac{0.9}{3} = 0.3$ Do the division.

e. $x = \dfrac{-(-7) + \sqrt{(-7)^2 - 4(4)(-2)}}{2(4)}$ Simplify under the radical.

$\quad\quad x = \dfrac{-(-7) + \sqrt{49 - (-32)}}{2(4)}$ Simplify the numerator and denominator.

$\quad\quad x = \dfrac{7 + \sqrt{81}}{8}$ Take the square root.

$\quad\quad x = \dfrac{7 + 9}{8} = \dfrac{16}{8} = 2$ Divide and simplify.

f. $A = 5\{6 - 4[3 - 8(5 - 7) + 1] - 2\}$ Simplify inside the parentheses.

$\quad\quad A = 5\{6 - 4[3 - 8(-2) + 1] - 2\}$ Multiply by 8; simplify inside the brackets.

$\quad\quad A = 5\{6 - 4[20] - 2\}$ Multiply by 4; simplify inside the braces.

$\quad\quad A = 5\{-76\}$ Multiply by 5.

$\quad\quad A = -380$

ANSWER BOX

Warm-up: **1.** $49.80; factor: $4.98(8 + 2) **2.** $3\frac{1}{3}$; commutative property for addition: $\frac{3}{4} + \frac{1}{4} + 2\frac{1}{3}$ **3.** 31; distributive property of multiplication over addition: $5(6 + \frac{1}{5}) = 30 + 1$ **4.** $\frac{4}{3}$; associative property for multiplication: $\frac{2}{3}(\frac{7}{5} \cdot \frac{10}{7}) = \frac{2}{3} \cdot \frac{2}{1}$ **Example 3: a.** $-10°F$ **b.** Subtract the starting temperature from the ending temperature: $-10° - 25° = -35°$. **c.** $-27°F$ **d.** Subtract the starting temperature from the ending temperature: $-27° - (-5°) = -27° + (+5°) = -27° + 5° = -22°$ **Think about it 1:** 105 and 5 have 5 as a common factor; 8 and 2 have 2 as a common factor. The numbers placed into parentheses were 100 and 10, both easy to multiply by. **Think about it 2:** $5 - \{2 - [3 - (4 + 5)]\} = 5 - \{2 - [3 - 9]\} = 5 - \{2 - [-6]\} = -3$. **Think about it 3:** $\sqrt{36}$ is divided by -2. The result, -3, is subtracted from 8.

1.1   Exercises 🔘

1. **Time Management** Return to Example 1. As directed, make a chart for your next seven days. Over the coming week, follow the four problem-solving steps. If you have family or other obligations, write in realistic times in the morning and evening.

2. **Time Management** To make sure that your course and work loads are sensible, check your plan for this term. Write down your number of credit hours and your number of hours of employment. Multiply your number of credit hours by 3 and then add your number of hours of employment. For most people, 40 to 45 hours is a reasonable commitment to school and employment. If you are above 50 hours, do you have contingency plans for happenings outside your control?

In Exercises 3 to 8, match each definition or related statement with a vocabulary word. Record any unfamiliar words and their definitions in your study notes.

3. Choose from
   multiplicative inverses, additive inverses, integers, real numbers, irrational numbers

   a. All the numbers that can be located with a point on the number line   real numbers

   b. The whole numbers and their opposites, $\{\ldots, -3, -2, -1, 0, 1, 2, 3, \ldots\}$ integers

   c. Real numbers that cannot be written as rational numbers   irrational numbers

   d. Two numbers that multiply to 1   multiplicative inverses

   e. Two numbers that add to zero   additive inverses

4. Choose from
   whole numbers, natural numbers, rational numbers, opposites, reciprocals

   a. The set of numbers for counting, $\{1, 2, 3, \ldots\}$
   natural numbers

   b. Two numbers that add to zero   opposites

   c. Two numbers that multiply to 1   reciprocals

   d. The set of natural numbers and zero, $\{0, 1, 2, 3, \ldots\}$
   whole numbers

   e. The set of numbers that can be written $\frac{a}{b}$, a and b integers, b not equal to zero   rational numbers

5. Choose from
   associative property for addition, commutative property for multiplication, distributive property of multiplication over addition, factors, factoring

   a. $c(d + e) = cd + ce$
   distributive property of multiplication over addition
   b. Two numbers being multiplied   factors

   c. $b \cdot c = c \cdot b$ commutative property for multiplication

   d. Changing the addition of ab and ac into the multiplication of a and $(b + c)$ factoring

   e. $a + (b + c) = (a + b) + c$
   associative property for addition

6. Choose from
   commutative property for addition, factored form of the distributive property, set, $\sqrt{-1}$, simplify

   a. $b \cdot c + b \cdot e = b(c + e)$
   factored form of the distributive property
   b. To perform the indicated operation   simplify

   c. $a + b = b + a$ commutative property for addition

   d. A collection of objects or numbers   set

   e. Is undefined in the real numbers $\sqrt{-1}$

7. Choose from
   positive, negative, zero, one

   a. The answer when a negative and a positive number are multiplied   negative

b. The answer when zero and one are added   one

c. The answer when two different negative numbers are divided   positive

d. The answer when a number and its reciprocal are multiplied   one

e. The answer when a number and its opposite are added   zero

8. Choose from positive, negative, zero

a. The answer when two negative numbers are multiplied   positive

b. The answer when two negative numbers are added   negative

c. The answer when two positive numbers are added   positive

d. The answer when zero and any number are multiplied   zero

e. The answer when two positive numbers are multiplied   positive

In Exercises 9 to 14, which name does not describe the given number?

9. $\frac{1}{2}$: rational, integer, real   integer

10. 1.5: natural, rational, real   natural

11. $\sqrt{9}$: whole, real, irrational   irrational

12. $\sqrt{5}$: irrational, real, integer   integer

13. 0: integer, whole, natural   natural

14. -1: natural, integer, rational   natural

15. Which set of numbers is the same as the positive integers?   natural numbers

16. Which set of numbers is the same as the set containing zero and the positive integers?   whole numbers

In Exercises 17 and 18, assume that the order for subtraction and division of m and n is $m - n$ and $m \div n$.

17. Add, subtract, multiply, and divide these pairs of numbers.

a. 4 and -12
$-8, 16, -48, -\frac{1}{3}$

b. $\frac{1}{2}$ and $\frac{1}{3}$ $\frac{5}{6}, \frac{1}{6}, \frac{1}{6}, \frac{3}{2}$

c. $-\frac{2}{3}$ and $\frac{1}{6}$
$-\frac{1}{2}, -\frac{5}{6}, -\frac{1}{9}, -4$

d. 0.75 and -0.2
$0.55, 0.95, -0.15, 3.75$

e. -0.45 and -1.5
$-1.95, 1.05, 0.675, 0.3$

f. $-\frac{5}{8}$ and $-\frac{10}{9}$
$-1\frac{53}{72}, \frac{35}{72}, \frac{25}{36}, \frac{9}{16}$

18. Add, subtract, multiply, and divide these pairs of numbers.

a. -15 and -3
$-18, -12, 45, 5$

b. -6 and 2 $-4, -8, -12, -3$

c. $\frac{1}{4}$ and $-\frac{1}{2}$
$-\frac{1}{4}, \frac{3}{4}, -\frac{1}{8}, -\frac{1}{2}$

d. $-\frac{5}{12}$ and $-\frac{1}{3}$ $-\frac{3}{4}, -\frac{1}{12}, \frac{5}{36}, 1\frac{1}{4}$

e. 2.5 and -0.25
$2.25, 2.75, -0.625, -10$

f. $-\frac{3}{4}$ and $\frac{5}{6}$
$\frac{1}{12}, -1\frac{7}{12}, -\frac{5}{8}, -\frac{9}{10}$

19. Simplify.

a. $8 - (-6) + (-2)$ 12

b. $15 + (-9) - (-8)$ 14

c. $-3.6 - (-4.8) + (4.3)$ 5.5

d. $-8.3 + (-3.8) - (-2.1)$ -10

e. $4\frac{1}{4} - 5\frac{1}{2} + 2\frac{1}{4}$ 1

f. $6\frac{1}{4} + 4\frac{1}{2} - 3\frac{1}{2}$ $7\frac{1}{4}$

20. Simplify.

a. $-4 - (-3) + (-8)$ -9

b. $12 - (-4) + (-2)$ 14

c. $4.5 + (-3.2) - (-2.8)$ 4.1

d. $6.2 - (-1.4) + (-3.5)$ 4.1

e. $2\frac{1}{4} - 3\frac{1}{2} + 4\frac{1}{4}$ 3

f. $-3\frac{1}{2} + 2\frac{1}{4} - 5\frac{1}{4}$ $-6\frac{1}{2}$

21. Simplify.

a. $\sqrt{-9}$ undefined

b. $\sqrt{9}$ 3

c. $-\sqrt{9}$ -3

22. Simplify.

a. $-\sqrt{16}$ -4

b. $\sqrt{-16}$ undefined

c. $\sqrt{16}$ 4

In Exercises 23 and 24, write in words.

23. **a.** $-(-3)$ opposite of negative three

b. -2 negative two

c. $5 - (-3)$ five subtract negative three

24. **a.** $3 - (-4)$ three subtract negative four

b. -4 negative four

c. $-(-5)$ opposite of negative five

In Exercises 25 and 26, do the operations, naming each property (associative property for addition, associative property for multiplication, commutative property for addition, or commutative property for multiplication) as it is used.

25. **a.** $4 + (-2) + (-8) + 6$ 0

b. $-5(7)(-2)(3)$ 210

c. $5 + (-2) + 15 + 12$ 30

d. $\frac{1}{2}(19)(20)$ 190

26. **a.** $-6 + 7 + 23 + 16$ 40

b. $6(5)(-2)(-7)$ 420

c. $9(25)(4)\left(\frac{1}{2}\right)$ 450

d. $8 + (-7) + (-3) + 22$ 20

In Exercises 27 and 28, simplify mentally using number properties as shortcuts.

27. **a.** $19 \cdot 92 + 19 \cdot 8$ 1900

b. $20\left(\frac{1}{5} + 0.5\right)$ 14

c. $299 \cdot 5$ 1495

d. $107 \cdot 29 - 7 \cdot 29$ 2900

e. $8\left(4\frac{1}{4}\right)$ 34

28. **a.** $28 \cdot 96 + 28 \cdot 4$ 2800

b. $28\left(0.25 + \frac{1}{7}\right)$ 11

c. $399 \cdot 3$ 1197

d. $205 \cdot 17 - 5 \cdot 17$ 3400

e. $12\left(3\frac{1}{3}\right)$ 40

29. Simplify. Round to the nearest tenth.

   a. At a range of 90 meters, the area of a circular archery target, in square centimeters: $A = \pi(122 \div 2)^2$
     $A \approx 11{,}689.9$ cm²

   b. Surface area of a can of pumpkin, in square inches: $S = 2\pi \cdot 2^2 + 2\pi(2)(4.5)$ $S \approx 81.7$ in²

   c. From solving an equation: $d = (-5)^2 - 4(6)(-7)$
     $d = 193$

   d. A puzzle: $x = 9 - 4(9 - 2) + 8 \div 4 - 2 \cdot 5$
     $x = -27$

30. Simplify. Round to the nearest tenth.

   a. Area of Olympic basketball's 3-second restricted zone (a trapezoid), in square meters:
   $A = \frac{1}{2}(5.8)(3.6 + 6.0)$ $A \approx 27.8$ m²

   b. Volume of a handball, in cubic inches:
   $V = \frac{4}{3}\pi(1.875 \div 2)^3$ $V \approx 3.5$ in³

   c. From solving an equation: $d = (-6)^2 - 4(8)(-3)$
     $d = 132$

   d. A puzzle: $x = 12 - 8 \div 4 + 5 \cdot 6 - 3^3$
     $x = 13$

31. Simplify.

   a. $m = \dfrac{4 - 8}{6 - (-2)}$ $-\frac{1}{2}$ **b.** $c = \sqrt{3^2 + 4^2}$ 5

   c. $F = \dfrac{9}{5}(100) + 32$ 212 **d.** $C = \dfrac{5}{9}(212 - 32)$ 100

   e. $h = \dfrac{4}{125}(-20)^2$ 12.8

32. Simplify.

   a. $m = \dfrac{7 - 3}{2 - (-4)}$ $\frac{2}{3}$ **b.** $c = \sqrt{6^2 + 8^2}$ 10

   c. $F = \dfrac{9}{5}(37) + 32$ 98.6 **d.** $C = \dfrac{5}{9}(98.6 - 32)$ 37

   e. $h = \dfrac{4}{125}(-15)^2$ 7.2

33. Simplify.

   a. $x = \dfrac{3 - 12}{3^2 - 6}$ $x = -3$

   b. $d = \sqrt{(6 - (-2))^2 + (3 - (-3))^2}$ $d = 10$

   c. $m = \dfrac{3 - (-3)}{6 - (-2)}$ $m = \frac{3}{4}$

   d. $x = \dfrac{-(-2) + \sqrt{(-2)^2 - 4(3)(-5)}}{2(3)}$ $x = 1\frac{2}{3}$

   e. $d = \{|4.05 - 7.12| + |7.51 - 7.12| + |9.80 - 7.12|\} \div 3$ $d = 2.04\overline{6}$

34. Simplify.

   a. $x = \dfrac{4 - 12}{2^3 - 6}$ $x = -4$

   b. $d = \sqrt{(14 - 2)^2 + (2 - (-3))^2}$ $d = 13$

   c. $m = \dfrac{2 - (-3)}{14 - 2}$ $m = \frac{5}{12}$

   d. $x = \dfrac{-(-2) - \sqrt{(-2)^2 - 4(3)(-5)}}{2(3)}$ $x = -1$

   e. $d = \{|6.76 - 7.91| + |7.14 - 7.91| + |9.83 - 7.91|\} \div 3$ $d = 1.28$

35. The statement $4(x - 3) = 4 \cdot x - 4 \cdot 3 = 4x - 12$ contains a subtraction. Explain why we can apply the distributive property of multiplication over addition to an expression containing a subtraction.   Possible answer: Subtraction can be restated as addition of the opposite.

36. The statement $\dfrac{x + 5}{2} = \dfrac{x}{2} + \dfrac{5}{2}$ contains a division. Explain why we can apply the distributive property of multiplication over addition to an expression containing a division.   Possible answer: Division can be restated as multiplication of the reciprocal.

▮ **Applications**

37. From past experience with mathematics, name the sources of the equations in parts a, b, c, and d of Exercise 31; parts b, c, and d of Exercise 33; and, using part d of Exercise 33 as a hint, part c of Exercise 29.   slope, Pythagorean theorem, Celsius to Fahrenheit, Fahrenheit to

38. Table 2 is repeated here.   Celsius, distance formula between two ordered pairs, slope, quadratic formula, discriminant for quadratic formula

TABLE 2   *For 25°F Current Temperature*

Wind (mph)	0	5	10	15	20	25	30
Apparent Temp. (°F)	25	22	10	2	−3	−7	−10

   a. Suppose you step from a protected outdoor area into a 30-mile-per-hour wind. What is the temperature change? $-10 - 25 = -35°$

   b. Suppose the wind is blowing 10 miles per hour and you are riding a bicycle 15 miles per hour into the wind. What is the change in temperature due to the wind? What is the change in temperature due to the wind and the bicycle speed together? $10 - 25 = -15°$; $-7 - 25 = -32°$

1.2   Input-Output Tables and Algebraic Techniques

Objectives

- Identify variables, constants, and numerical coefficients in algebraic expressions.
- Build input-output tables.
- Identify bases and apply exponents.
- Apply the distributive property in expressions.
- Add like terms.
- Use the simplification property of fractions to find equivalent fractions.
- Evaluate expressions and formulas.

WARM-UP

How does each number shown below the design describe the design? Draw the next design in the pattern. What is the next number in each pattern?

1. Pattern is 1, 3, 5, 7, 9, _11_.   number of dots

   1   3   5   7   9

2. Pattern is 4, 7, 10, 13, _16_.   number of line segments

   4   7   10   13

3. Pattern is 1, 3, 6, 10, _15_.   number of dots in a triangular array

   1   3   6   10

4. Pattern is 2, 6, 12, 20, _30_.   number of dots in a rectangular array

   2   6   12   20

5. Pattern is 1, 4, 9, 16, _25_.   number of dots in a square array

   1   4   9   16

THE REVIEW IN SECTION 1.1 focused on numbers. This section reviews input-ouput tables, algebraic expressions, and algebraic techniques. Some of the examples and illustrations may suggest new ways of thinking about algebraic notation.

Algebraic Notation

The numbers and letters in algebraic notation have special names.

■ SELECTED WORDS IN ALGEBRAIC NOTATION

Group activity: List ways variables are used. Partial answer: In formulas, $v = lwh$; in an arithmetic process, $a + b + c$; in functions; in identities, $a + b = b + a$; as an unknown, $y + 4 = 2$; as parameters, $y = mx + b$.

> A **variable** is a letter or symbol that can represent any number from some set.
>
> A **constant** is a number, letter, or symbol whose value is fixed.
>
> A **numerical coefficient** is the sign and number multiplying the variable(s).

An **expression** is *any combination of signs, numbers, constants, and variables with operations such as addition, subtraction, multiplication, or division.*

EXAMPLE 1 Identifying constants, variables, and numerical coefficients   For each expression listed, name the constants, the variables, and the numerical coefficients.

   a. $2x + 6$ **b.** $1.49x - 0.30$ **c.** $7 - x$

SOLUTION **a.** 2 is a constant and the numerical coefficient of the variable x; 6 is a constant.

   b. 1.49 is a constant and the numerical coefficient of the variable x; -0.30 is a constant. The subtraction sign goes with the 0.30.

   c. 7 is a constant. The numerical coefficient of the variable x is -1. There is an assumed 1 in front of the x, and the subtraction sign goes with the variable x.

We can use algebraic notation to state the rules for the patterns in the Warm-up. In this and the next section, you will encounter the algebraic notation describing all these patterns. A helpful way to relate number patterns and their rules is to organize them in input-output tables.

Input-Output Tables

An **input-output relationship** is a *rule, given in words, numbers, symbols, or algebraic equations, that associates sets of numbers.* The rule describes how to get from one set of numbers (the inputs) to the other set of numbers (the outputs).

In the Warm-up Exercises, the inputs are the position, or order, in which the designs appear (1, 2, 3, 4, . . .). The outputs are the numbers in the patterns. To build an input-output table, we list the inputs and outputs in columns.

EXAMPLE 2 Building input-output tables   Write the number patterns from Warm-up Exercises 1 and 2 in input-output tables. Suggest ways of relating the inputs and outputs.

1   3   5   7   9   11

FIGURE 3

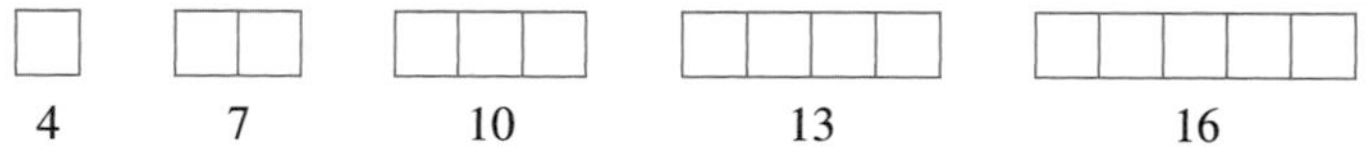

4   7   10   13   16

FIGURE 4

SOLUTION Table 4 is the input-output table for Figure 3; Table 5 is the input-output table for Figure 4.

TABLE 4

Input: Position	Output: Number of Dots
1	1
2	3
3	5
4	7
5	9
6	11

TABLE 5

Input: Position	Output: Number of Line Segments
1	4
2	7
3	10
4	13
5	16

In Table 4, the outputs are about twice the inputs. We might predict that the rule for the table would include the input multiplied by 2.

In Table 5, the outputs are about three times the inputs. We might predict that the rule for the table would include the input multiplied by 3.

See the Answer Box for the rules.

EXAMPLE 3   Building an input-output table

a. Make an input-output table for the pattern in Warm-up Exercise 3 (repeated in Figure 5).

b. Find the change in inputs and change in outputs.

c. Suggest a rule.

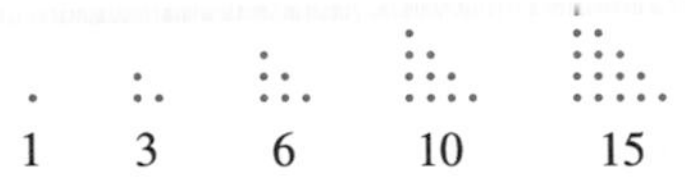

```
 .     . .    . .     . . .     . . . .
      . .    . . .   . . . .   . . . . .
             . . .   . . . .   . . . . .
 1     3      6       10        15
```

FIGURE 5

SOLUTION   a. To build the input-output table in Table 6, we use Figure 5 to match the number of dots (outputs) with a position (inputs).

TABLE 6

Change in Input	Input: Position	Output: Number of Dots	Change in Output
	1	1	
1 ⟨	2	3	⟩ 2
1 ⟨	3	6	⟩ 3
1 ⟨	4	10	⟩ 4
1 ⟨	5	15	⟩ 5

Student Note: Pause, and find the changes in inputs and outputs for the tables in Example 2. Compare these changes with the ones in Example 3. More on the changes later.

b. We find the change in the outputs. In the table, we write the change in the input numbers in a column to the left of the inputs and the change in the output numbers in a column to the right of the outputs. The outputs are not a consistent amount apart.

c. The output numbers get large quickly, so we might assume that the input is being multiplied, but not by the same number each time. Because multiplication is associated with the area of rectangles (see Example 9), the triangular arrangement of the dots suggests multiplication followed by division by 2. See the Answer Box for the rule.

The rule in Example 3 requires an exponent.

Bases and Exponents

The **exponent** is *the small raised number or expression to the right* of a number or expression called the base. The **base** is *the number to which the exponent is applied*. In 4^2, the 4 is the base and the 2 is the exponent. In this case, the exponent means to multiply the base by itself: $4 \cdot 4 = 16$.

Together the base and exponent are called the **power of the base**. The expression 4^2 is called the *second power of four*. The expression 4^3 is called the *third power of* *four*. In 4^3, the 4 is the base and the 3 is the exponent. The exponent means to write the base three times and then multiply. In general,

▮ POSITIVE INTEGER EXPONENTS

> When the exponent is a positive integer, n, the base, b, is written n times:
>
> $$b^n = \underbrace{b \cdot b \cdot b \cdot \cdots \cdot b}_{n \text{ factors}}$$

In letters, $x \cdot x = x^2$ and $x \cdot x^2 = x^3$. In words, x^2 is x-squared, or the square of x; x^3 is x-cubed, or the cube of x.

Think about it 1: Make a list of the squares of x for $x =$ the first five natural numbers.

EXAMPLE 4 **Identifying bases and applying exponents** Name the base in each expression and then write the expression without exponents.

a. $(-3)^2$ **b.** -3^2 **c.** $0 - 3^2$ **d.** $2x^2$

SOLUTION **a.** The base is -3:

$$(-3)^2 = (-3)(-3) = +9$$

b. The base is 3:

$$-3^2 = -(3 \cdot 3) = -9$$

c. The base is 3:

$$0 - 3^2 = 0 - 9 = -9$$

d. The base is x:

$$2x^2 = 2xx$$

▬

Think about it 2: How is $(2x)^2$ the same as or different from $2x^2$?

Distributive Property and Factoring

DISTRIBUTIVE PROPERTY The distributive property, $a(b + c) = ab + ac$, justifies multiplying out parentheses and factoring algebraic expressions.

EXAMPLE 5 **Applying the distributive property** Multiply in these problems.

a. Find the area of the rectangle, where length and width are in feet:

Student Note: All two-dimensional geometric figures (except algebra tiles) are shaded blue.

$x + 4$

2 ▭

b. Using the table form for multiplication is like finding a rectangular area. Multiply the numbers on the left and top to obtain the inner "area" (in the white area).

Multiply	x^2	$+3x$	$+2$
x			

SOLUTION **a.**

	$x + 4$
2	$2x + 8$

Thus, 2 ft $\cdot$ (x ft + 4 ft) = $2x$ ft^2 + 8 ft^2.

b.

Multiply	x^2	$+3x$	$+2$
x	x^3	$+3x^2$	$+2x$

Thus, $x(x^2 + 3x + 2) = x^3 + 3x^2 + 2x$.

EXAMPLE 6 Building an input-output table   Build an input-output table for $x(x + 1)$ with the first four natural numbers as inputs. Repeat for $x^2 + x$. Which pattern in the Warm-up is this?

SOLUTION   The inputs and outputs appear in Tables 7 and 8. The pattern 2, 6, 12, 20 is in Warm-up Exercise 4, the rectangular array.

TABLE 7

Input x	Output $x(x + 1)$
1	$1(2) = 2$
2	$2(3) = 6$
3	$3(4) = 12$
4	$4(5) = 20$

TABLE 8

Input x	Output $x^2 + x$
1	$1 + 1 = 2$
2	$4 + 2 = 6$
3	$9 + 3 = 12$
4	$16 + 4 = 20$

FACTORING   **Factors** are *the numbers being multiplied.* In $x(x + 1)$, there are two factors: x and $x + 1$. In abc, there are factors: a, b, and c.

EXAMPLE 7 Identifying factors   How many factors are in each of these expressions?
a. $xy(z + 1)$ **b.** $a(b + c + d)$ **c.** $abcd$

SOLUTION   **a.** Three factors: x, y, and $z + 1$

b. Two factors: a and $b + c + d$

c. Four factors: a, b, c, and d

Recall that when we reverse the distributive property, $ab + ac = a(b + c)$, we are factoring. In this case, **factoring** is *changing the addition of ab and ac into the multiplication of a and* ($b + c$). The letter a is called the **greatest common factor**, *the largest number that divides both terms.*

Because factoring suggests a process, the word *factor*, like *simplify*, is a common direction in examples and exercises. Its meaning will vary somewhat with the topic being studied.

EXAMPLE 8 Factoring   Name the greatest common factor and then use it to factor.
a. Change to a product: $9a - 6$.
b. Find the dimensions (length and width); the area is given.

	$6x^2 + 12x$

c.

Factor			
	$2x^2$	$+6x$	$+2$

SOLUTION   **a.** 3 is the greatest common factor:

$$9a - 6 = 3(3a - 2)$$

b. $6x$ is the greatest common factor:

$$x + 2$$

$6x$	$6x^2 + 12x$

Thus, $6x^2 + 12x = 6x(x + 2)$.

c. 2 is the greatest common factor:

Factor	x^2	$+3x$	$+1$
2	$2x^2$	$+6x$	$+2$

Thus, $2x^2 + 6x + 2 = 2(x^2 + 3x + 1)$.   ▬

Think about it 3: In part b of Example 8, what are other possible dimensions? Why might the one shown in the solution be preferred?

Adding Like Terms

Many steps in algebra require that we see how expressions are alike and how they are different. To distinguish expressions such as $x + 2$, $2x$, and x^2, it may be helpful to associate them with physical objects such as algebra tiles. In Example 9, let x be the side of the large square, let 1 be the side of the small square, and let x and 1 be the sides of the rectangle.

EXAMPLE 9   Finding area with algebra tiles
a. What is the area of each rectangle or square shown in Figure 6?
b. What is the total area of the figures?

FIGURE 6

SOLUTION   **a.** The large squares have area x^2 square units. The rectangles have area $x \cdot 1 = x$ square units. The small squares have area $1 \cdot 1 = 1$ square unit.

b. There are two x^2 tiles, three x tiles, and three 1 tiles, so the total area is $2x^2 + 3x + 3$.   ▬

In writing the total area of the tiles in Example 9, it is sensible to count the x^2 tiles together because they describe the same shape. The rectangles are a different shape, so they cannot be counted with either the x^2 tiles or the small square tiles.

LIKE TERMS   A **term** is *the product of a number (with or without a sign) and one or more variables (with or without exponents).* **Like terms** *have identical variable factors.* The large square tiles model one set of like terms. The rectangles model another set of like terms. The small square tiles model still a third set of like terms.

When we count the like terms (like tiles) and write a single term to describe the total count, we are *adding like terms*.

EXAMPLE 10   Adding like terms   Find the like terms and add them.

a. $a + 2ab + 3ab - 4ba + 5b^2$ b. $x^2 + 2x - 3x^2 + 4x - 5$

c. $2\pi r^2 + 2\pi rh$ d. $x^3 - 2x^2 + x - x^2 + 2x - 1$

SOLUTION   a. $a + ab + 5b^2$. The terms containing ab and ba have the same variables as factors and are like terms.

b. $-2x^2 + 6x - 5$

c. Not like terms: 2 and π are constants, the variables are r^2 and rh and are not identical factors.

d. $x^3 - 3x^2 + 3x - 1$

The -5 in part b and the -1 in part d *have no variables* and thus are called **constant terms**.

Simplifying Expressions

Recall that **simplify** is a short way of writing directions. Simplify has several meanings. In the last section, simplify meant to *apply the properties of real numbers or the order of operations to an expression.* Simplify also means *to add like terms* or *to change fractions to lowest terms* using the simplification property of fractions.

When convenient, we write the reciprocal of a as $1/a$, $a \neq 0$. Because the product of a number and its reciprocal is 1, the expression

$$a \cdot \frac{1}{a} = \frac{a}{a} = 1$$

leads to the simplification property of fractions.

SIMPLIFICATION PROPERTY OF FRACTIONS

For all real numbers, $a \neq 0$, $c \neq 0$,

$$\frac{ab}{ac} = \boxed{\frac{a}{a}} \cdot \frac{b}{c} = 1 \cdot \frac{b}{c} = \frac{b}{c}$$

The simplification property says that if the numerator and denominator of any fraction contain common factors, these factors may be eliminated, giving a simpler fraction of the same value.

To simplify a fraction, we change the numerator and denominator to factors and eliminate common factors of the form a/a. This is called *changing to lowest terms.*

EXAMPLE 11   Changing to lowest terms   Identify the common factors in the numerator and denominator of each fraction, and simplify to lowest terms.

a. $\dfrac{450}{60}$ b. $\dfrac{25}{36}$ c. $\dfrac{abc}{cde}$ d. $\dfrac{xy + y^2}{y}$ e. $\dfrac{x + 3}{3}$

SOLUTION   a. $\dfrac{450}{60} = \dfrac{2 \cdot 3 \cdot 3 \cdot 5 \cdot 5}{2 \cdot 2 \cdot 3 \cdot 5} = \dfrac{15}{2}$

b. $\dfrac{25}{36} = \dfrac{5 \cdot 5}{6 \cdot 6} = \dfrac{5 \cdot 5}{2 \cdot 2 \cdot 3 \cdot 3}$

There are no common factors in 25 and 36, so $\frac{25}{36}$ cannot be simplified.

c. $\dfrac{a \cdot b \cdot c}{c \cdot d \cdot e} = \dfrac{ab}{de}$

d. $\dfrac{\cancel{y}(x + y)}{\cancel{y}} = \dfrac{x + y}{1} = x + y$

e. $\dfrac{x + 3}{3}$; the numerator and denominator have no common factors and so the

fraction cannot be simplified. ▬

Parts b and e of Example 11 illustrate that *if there are no common factors in the numerator and denominator, the fraction cannot be simplified.*

Evaluating Expressions and Formulas

Like *simplify*, *evaluate* is another short way of writing directions. To **evaluate**, *substitute numbers in place of the variables in expressions or formulas.* Many applications requiring an input-output table have a rule relating the input and output.

EXAMPLE 12   **Evaluating an expression**   Set up an input-output table for $x = 1$ to $x = 4$. Evaluate the expression $(x^2 + x)/2$ and match with a pattern in the Warm-up.

SOLUTION   The outputs in Table 9 match the pattern in Warm-up Exercise 3.

TABLE 9

Input x	Output $(x^2 + x)/2$
1	$(1 + 1)/2 = 1$
2	$(4 + 2)/2 = 3$
3	$(9 + 3)/2 = 6$
4	$(16 + 4)/2 = 10$

▬

FORMULAS   The following steps summarize evaluating a formula.

▬ EVALUATING FORMULAS

> To evaluate a formula:
>
> **1.** State the formula and the information given.
>
> **2.** Substitute the given information into the formula.
>
> **3.** Estimate the result.
>
> **4.** Find the result, using a calculator as needed. Note any special keystrokes.

EXAMPLE 13   **Evaluating formulas**   Look in Appendix 1, as needed, for the formulas. Use a calculator, indicating any special keystrokes.

a. Find the area of a trapezoid whose parallel sides are 5 feet and 8 feet and height is 4 feet.

b. Find the volume of a golf ball (sphere) with a diameter of 1.68 inches.

c. Find S where $S = \dfrac{n}{2}[2a + (n - 1)d]$, $n = 50$, $a = 1$, and $d = 3$.

SOLUTION   **a.** If a and b are the parallel sides and h is the height,

$$A = \tfrac{1}{2}h(a + b), \ a = 5 \text{ ft}, \ b = 8 \text{ ft}, \ h = 4 \text{ ft}$$

$$A = \tfrac{1}{2}(4 \text{ ft})(5 \text{ ft} + 8 \text{ ft}) \qquad\qquad \text{Substitute.}$$

We can do this problem mentally, omitting the estimate and calculator steps.

$$A = \tfrac{1}{2}(4)(5 + 8) \text{ ft}^2$$

$$A = 26 \text{ ft}^2$$

b. $V = \dfrac{4\pi}{3}\left(\dfrac{d}{2}\right)^3$, $d = 1.68$ in.

$$V = \dfrac{4\pi}{3}\left(\dfrac{1.68 \text{ in.}}{2}\right)^3 \qquad \text{Substitute.}$$

For $\pi \approx 3$ and $(1.68/2) \approx 1$,

$$\dfrac{4(3)}{3}(1)^3 \approx 4 \qquad\qquad \begin{array}{l}\text{Estimate by rounding to}\\ \text{the nearest whole number.}\end{array}$$

$$4 \; \boxed{\text{2nd}} \; \pi/3 * (1.68/2) \; \boxed{\wedge} \; 3 \; \boxed{\text{ENTER}} \qquad \text{Enter into calculator.}$$

$$V - 2.48 \text{ in}^3$$

Note the multiplication sign between the division by 3 and the parentheses around 1.68/2. Without the multiplication sign, some calculators recognize an "implied" multiplication first. These calculators multiply 3 times (1.68/2)^3 first and then do the division between 4π and 3.

c. $S = \dfrac{n}{2}[2a + (n-1)d]$, $n = 50$, $a = 1$, $d = 3$

$$S = \tfrac{50}{2}[2(1) + (50 - 1)\cdot 3] \qquad \text{Substitute.}$$

$$\tfrac{1}{2}(50)(50)(3) = \tfrac{1}{2}(7500) = 3750 \qquad \text{Estimate.}$$

$$S = 50/2 * (2 + (49) * 3) \; \boxed{\text{ENTER}} \qquad \text{Enter into calculator.}$$

$$S = 3725$$

ANSWER BOX

Warm-up: 1. number of dots; 11 **2.** number of line segments; 16 **3.** number of dots in triangular array; 15 **4.** number of dots in rectangular array; 30 **5.** number of dots in square array; 25 **Example 2:** The rule in Table 4 is "one less than twice the input." The rule in Table 5 is "three times the input plus one." **Example 3:** The rule is "multiply the input by one more than the input, and then divide by two." **Think about it 1:** 1, 4, 9, 16, 25, as in Warm-up Exercise 5. **Think about it 2:** $(2x)^2$ has a base of $2x$. $(2x)^2 = 2x \cdot 2x = 4x^2$ and does not equal $2x^2$. **Think about it 3:** $6(x^2 + 2x)$ and $x(6x + 12)$ are two possible pairs of dimensions. The dimensions $6x(x + 2)$ have the greatest common factor, $6x$.

1.2 Exercises

In Exercises 1 and 2, identify the variables, numerical coefficients, and constant term in each expression.

1. a. $2x + 3$ $\;x, 2, 3$ **b.** πr^2 $\;r, \pi, \text{none}$

c. $4 - x$ $\;x, -1, 4$ **d.** $x^2 + x - 1$ $\;x, 1 \text{ and } 1, -1$

2. a. $3x - 2$ $\;x, 3, -2$ **b.** $2\pi r$ $\;r, 2\pi, \text{none}$

c. $9 - x$ $\;x, -1, 9$ **d.** $x^2 - x + 1$ $\;x, 1 \text{ and } -1, 1$

3. Explain how these dots show the pattern formed by $2x - 1$. Visualize x columns of 2 dots with one dot missing.

4. Explain how these dots show the pattern formed by $3x + 1$. Visualize x columns of 3 dots with one extra dot.

In Exercises 5 to 8,
(a) Make an input-output table for the pattern.
(b) Find the change in the outputs for each table.
(c) Describe each table with a rule.

5. The input is the set of natural numbers. The output is the number of dots. The output is 30 for the input 10.

See Answer Section.

Input: 1   2   3   4

6. The input is the set of natural numbers. The output is the number of line segments. The output is 21 for the input 10.   See Additional Answers.

Input: 1   2   3   4

7. The input is the set of natural numbers. The output is the number of line segments. The output is 29 for the input 10.   See Answer Section.

Input: 1   2   3   4   5

8. The input is the set of natural numbers. The output is the number of line segments. The output is 39 for the input 10.   See Additional Answers.

Input: 1   2   3   4

Write parts a, b, and c of Exercises 9 and 10 in words; then answer the questions in parts d and e.

9. a. $4 - (-c)$ four subtract the opposite of c

 b. $-z^3$ the opposite of the cube of z (or z-cubed)

 c. $-z^2$ negative z-squared or the opposite of z-squared

 d. Make an input-output table for part b. Let $z = -2$, $-1, 0, 1, 2$. Is the output always negative?
 $-z^3 = +8, +1, 0, -1, -8$. Output may be positive or negative.
 e. Why is there a choice of "negative" or "the opposite of" in part c? $-z^2$ will be negative because z^2 is always positive for any real number z.

10. a. $-y^3$ the opposite of y-cubed

 b. $3 - (-x)$ three subtract the opposite of x

 c. $-x^2$ the opposite of the square of x

 d. Is the answer in part a always negative? Explain.   No; if y is a negative number, y^3 is negative, and the opposite of a negative number is positive.
 e. Make an input-output table for part c. Let $x = -2$, $-1, 0, 1, 2$. Is the output always negative?
 $-x^2 = -4, -1, 0, -1, -4$. Except for $x^2 = 0$, outputs are negative.

11. If possible, simplify or write without parentheses.

 a. $(-3)^2$ 9   **b.** $(-2)^3$ -8 **c.** -3^2 -9

 d. $(3x)^2$ $9x^2$ **e.** $3x^2$ $3x^2$ **f.** $0 - 3x^2$ $-3x^2$

12. If possible, simplify or write without parentheses.

 a. -4^2 -16 **b.** $(-4)^2$ 16   **c.** $(-4x)^2$ $16x^2$

 d. $-4x^2$ $-4x^2$ **e.** $(-2)^4$ 16   **f.** $0 - 4x^2$ $-4x^2$

13. Identify the base in each part of Exercise 11.
 a. -3 b. -2 c. 3 d. $3x$ e. x f. x
14. Identify the base in each part of Exercise 12.
 a. 4 b. -4 c. $-4x$ d. x e. -2 f. x

In Exercises 15 and 16, simplify by removing parentheses.

15. a. $(-2x)^3$ $-8x^3$ **b.** $-(2x)^3$ $-8x^3$ **c.** $(-3y)^4$ $81y^4$

 d. $-(xy)^4$ $-x^4y^4$ **e.** $(-xy)^4$ x^4y^4 **f.** $0 - (-2y)^3$

16. a. $(3y)^3$ $27y^3$ **b.** $(-3y)^3$ $-27y^3$ **c.** $-(3y)^3$ $-27y^3$ $8y^3$

 d. $(2xy)^3$ $8x^3y^3$ **e.** $(-xy)^2$ x^2y^2 **f.** $0 - (-2x)^2$ $-4x^2$

Exercises 17 to 22 contain visual models of the distributive property. Multiply or factor as indicated. Use the greatest common factor when factoring.

17. a.

Multiply	x	-2
$3x$	$3x^2$	$-6x$

 b.

Factor	a	$-b$
$2a$	$2a^2$	$-2ab$

18. a.

Multiply	y	$+3$
$-2y$	$-2y^2$	$-6y$

 b.

Factor	4	$-x$
$2x$	$8x$	$-2x^2$

19. a. Find the area.

$2x + 4$

| 3 | $6x + 12$ |

 b. Factor to find the dimensions.

$3y + 2$

| 5 | $15y + 10$ |

20. a. Find the area.

$a + 3$

| 4 | $4a + 12$ |

 b. Factor to find the dimensions.

$1 + rt$

| p | $p + prt$ |

21. a. Find the area.

$x + 5$

| x | $x^2 + 5x$ |

 b. Factor to find the dimensions.

$a + c$

| b | $ab + bc$ |

22. a. Find the area.

$y + 10$

| y | $y^2 + 10y$ |

b. Factor to find the dimensions.

In Exercises 23 and 24, multiply to remove the parentheses.

23. a. $-2(3 - 4x)$ $-6 + 8x$ **b.** $-a(b - c)$ $-ab + ac$

    c. $x(x^2 - 2x - 3)$ $x^3 - 2x^2 - 3x$

24. a. $-6(x - 4)$ $-6x + 24$ **b.** $-x(x - 3)$ $-x^2 + 3x$

    c. $a(a^2 - 3a + 2)$ $a^3 - 3a^2 + 2a$

In Exercises 25 to 30, name the greatest common factor and then factor.

25. a. $6x - 54$ **b.** $15x - 225$ **c.** $6a^2 - 8a$
    $6, 6(x - 9)$ $15, 15(x - 15)$ $2a, 2a(3a - 4)$

26. a. $5x - 75$ **b.** $11x - 121$ **c.** $10b^2 - 15b$
    $5, 5(x - 15)$ $11, 11(x - 11)$ $5b, 5b(2b - 3)$

27. a. $ab - b^2$ **b.** $xy - x^2y$ **c.** $(xy)^2 - xy^2$
    $b(a - b)$ $xy(1 - x)$ $xy^2(x - 1)$

28. a. $ac - a$ **b.** $x^2y - xy^2$ **c.** $x^2y - (xy)^2$
    $a(c - 1)$ $xy(x - y)$ $x^2y(1 - y)$

29. a. $6ab - 12a + 3$ **b.** $4ac + 12a - 16$
    $3(2ab - 4a + 1)$ $4(ac + 3a - 4)$

30. a. $8ac + 4c - 12$ **b.** $9ab - 15b + 6$
    $4(2ac + c - 3)$ $3(3ab - 5b + 2)$

31. Add like terms in the figure below.
$3x^2 + 5x + 6$

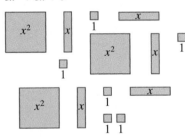

32. Add like terms in the figure below.
$3x^2 + 2x + 2$

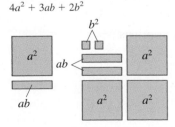

33. Add like terms in the figure below.
$4a^2 + 3ab + 2b^2$

34. Add like terms in the figure below.
$2a^2 + 5ab + 6b^2$

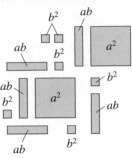

In Exercises 35 to 38, add like terms.

35. a. $3a - 2b + 4b - 2a$ $a + 2b$

    b. $6x^2 + 2x - 4x^2 - 3$ $2x^2 + 2x - 3$

36. a. $4ab - 2ac + 3ba - 4ac$ $7ab - 6ac$

    b. $6x^2 + 4x - 4x - x^2$ $5x^2$

37. a. $6ac - 3ab + 2ca + 7bc$ $8ac - 3ab + 7bc$

    b. $x(4 + 2x) - x(x - 4)$ $x^2 + 8x$

38. a. $7b - 4c + 3a - 6b$ $3a + b - 4c$

    b. $x(3x - 1) - x(4 - 2x)$ $5x^2 - 5x$

In Exercises 39 to 44, name the greatest common factor and simplify the fraction by changing to lowest terms.

39. a. $\dfrac{15}{45}$ $15, \frac{1}{3}$ **b.** $\dfrac{48}{160}$ $16, \frac{3}{10}$ **c.** $\dfrac{5280}{3600}$ $240, \frac{22}{15}$

40. a. $\dfrac{84}{35}$ $7, \frac{12}{5}$ **b.** $\dfrac{52}{39}$ $13, \frac{4}{3}$ **c.** $\dfrac{1024}{256}$ $256, 4$

41. a. $\dfrac{abe}{ben}$ $be, \frac{a}{n}$ **b.** $\dfrac{ab + bc}{b}$ $b, a + c$ **c.** $\dfrac{x - 2}{2}$ cannot be simplified

42. a. $\dfrac{cot}{ox}$ $o, \frac{ct}{x}$ **b.** $\dfrac{ab + ac}{a}$ $a, b + c$ **c.** $\dfrac{x - 2}{x}$ cannot be simplified

43. a. $\dfrac{xy + xz}{xy}$ $x, \frac{y + z}{y}$ **b.** $\dfrac{x^2 + 2x}{2x}$ $x, \frac{x + 2}{2}$ **c.** $\dfrac{xy^2 - 2x^2y}{2xy}$ $xy, \frac{y - 2x}{2}$

44. a. $\dfrac{ab + bc}{ab}$ $b, \frac{a + c}{a}$ **b.** $\dfrac{y^2 + 3y}{3y}$ $y, \frac{y + 3}{3}$ **c.** $\dfrac{2ab^2 - 2a^2b}{2ab}$ $2ab, b - a$

Evaluate the expressions in Exercises 45 to 50 for the given numbers.

45. a. $x^2 + 2x + 1$ for $x = 7$ **b.** $(x + 1)^2$ for $x = 7$
    64 64

46. a. $x^2 + 9$ for $x = 5$ **b.** $(x + 3)^2$ for $x = 5$
    34 64

47. a. $a^2 - b^2$ for $a = 9$, $b = 7$
    32

    b. $(a - b)^2$ for $a = 9$, $b = 7$
    4

48. a. $a^2 + 2ab + b^2$ for $a = 12, b = 3$ 225

    b. $(a + b)^2$ for $a = 12, b = 3$ 225

49. a. $9 - 3(x - 2)$ for $x = 8$ -9 **b.** $6(x - 2)$ for $x = 8$ 36

50. a. $x^2 - x(4 - x)$ for $x = 5$ 30

    b. $x(4 - x)$ for $x = 5$ -5

Evaluate the formulas in Exercises 51 to 54. Round to the nearest tenth.

51. Celsius temperature: $C = \frac{5}{9}(F - 32)$

    a. $F = 32$ **b.** $F = -40$ **c.** $F = 98.6$
      $C = 0$ $C = -40$ $C = 37$

52. Fahrenheit temperature: $F = \frac{9}{5}C + 32$

    a. $C = 100$ **b.** $C = 37$ **c.** $C = 20$
      $F = 212$ $F = 98.6$ $F = 68$

53. Volume of a sphere: $V = \frac{4}{3}\pi r^3$

    a. $r = 1.5$ cm   **b.** $r = 3$ cm   **c.** $r = 6$ cm
      $V = 14.1$ cm^3 $V = 113.1$ cm^3 $V = 904.8$ cm^3

54. Surface area of a sphere: $S = 4\pi r^2$

    a. $r = 1.5$ cm   **b.** $r = 3$ cm   **c.** $r = 6$ cm
      $S = 28.3$ cm^2 $S = 113.1$ cm^2 $S = 452.4$ cm^2

Evaluate the formulas in Exercises 55 to 58 for the given numbers. Round to the nearest tenth.

55. Sum, arithmetic sequence: $S = \frac{n}{2}[2a + (n - 1)d]$

    a. $n = 100, a = 1, d = 2$ $S = 10{,}000$

    b. $n = 100, a = 2, d = 2$ $S = 10{,}100$

    c. $n = 100, a = 5, d = 5$ $S = 25{,}250$

56. Last term, arithmetic sequence: $L = a + (n - 1)d$

    a. $n = 100, a = 1, d = 2$ $L = 199$

    b. $n = 100, a = 2, d = 2$ $L = 200$

    c. $n = 100, a = 5, d = 5$ $L = 500$

57. Distance between points (a, b) and (c, d):
$D = \sqrt{(a - c)^2 + (b - d)^2}$

    a. $(6, 3)$ and $(-2, -3)$ 10

    b. $(14, 2)$ and $(2, -3)$ 13

    c. $(-4, 3)$ and $(11, -5)$ 17

58. Area of a triangle with sides a, b, and c:
$S = \frac{1}{2}(a + b + c)$ and
$A = \sqrt{S(S - a)(S - b)(S - c)}$

    a. $a = 3, b = 4, c = 5$; find S first. $S = 6, A = 6$

    b. $a = 6, b = 8, c = 10$; find S first. $S = 12, A = 24$

    c. $a = 5, b = 12, c = 13$; find S first. $S = 15, A = 30$

59. Answer the following question carefully. Which is larger: a or the opposite of a? for $a > 0, a > -a$; for $a < 0$, $-a > a$

60. Calculator Exploration Antarctic explorers James Siple and Charles Passel are credited with developing the wind-chill apparent temperature chart (Tables 1, 2, and 3 in Section 1.1) in the 1940s. The National Weather Service updated the formula effective 11/01/01. We will use the new formula, containing rational exponents, in Section 8.3. The Siple and Passel formula is

$$W = 91.4 + (T - 91.4) \cdot (0.478 + 0.301\sqrt{S} - 0.02S)$$

W is the apparent (or wind-chill) temperature in degrees Fahrenheit, T is the temperature at zero wind in degrees Fahrenheit, and S is the wind speed in miles per hour.

    a. Use the formula and $T = 25°F$ with $S = 10$ mph to check the numbers in Table 2, which is repeated here.

    There are no answers to this exploration. Students use wind chill to explore use of their calculator.

TABLE 2   *For 25°F Current Temperature*

Wind Speed (mph) S	0	5	10	15	20	25	30
Apparent Temp. (°F) W	25	22	10	2	-3	-7	-10

Here is the formula for $S = 10$ mph:

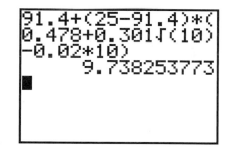

    b. Repeat part a for $S = 15$ mph to find $W \approx 2°F$.

    c. Did you retype the entire formula in part b?

The calculator is an excellent tool for repetitive substitutions. Press [2nd] [ENTER]. The formula for $T = 25$ and $S = 15$ should appear. Replace $S = 15$ with $S = 20$ and press [ENTER].

d. Check the entries in the $T = 20°F$ row of the wind-chill chart (Table 1) on page 1. Let the wind speeds be $S = 10, 15$, and 20. [INS] or the [DEL] key may be needed to enter temperatures or wind speeds having more or less than two digits.

e. To obtain a complete row, select a current temperature—say, $T = 25$. Go to [Y=]. After Y_1, enter the temperature formula, with X replacing S and your

choice of T. Press [2nd] [TBLSET], and let **TblStart** = 5 and Δ**Tbl** = 5. Press [2nd] [TABLE] to see the wind speed under X and the apparent temperatures under Y_1.

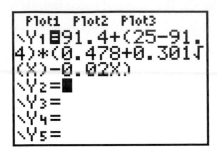

1.3   Problem Solving and Writing Algebraic Notation

Objectives

▪ Apply problem solving to finding rules for tables.

▪ Identify the names of results of operations.

▪ Identify special language used to describe operations in word problems.

▪ Translate word sentences into and from algebraic notation.

▪ Write equations from word problems, including percent problems.

WARM-UP

In Exercises 1 to 3, using scratch paper, place the patterns in an input-output table. State your assumptions. What are likely to be the next numbers in the tables?

1. $-5, -4, -3, -2 \ldots$ For an input of natural number 5, the output is -1.

2. $8, 16, 24, 32, \ldots$ For an input of natural number 5, the output is 40.

3. $5, 3, 1, -1, \ldots$ For an input of natural number 5, the output is -3.

For Exercises 4 and 5, guess the next input and output entries.

4.

Input	Output
0	-4
2	-3
4	-2
6	-1

8, 0

5. *Celsius and Fahrenheit Temperatures*

Input °C	Output °F
0	32
20	68
40	104
60	140

80, 176

6. How are Exercises 4 and 5 different from Exercises 1–3? Answers vary; inputs are not consecutive natural numbers; changes in inputs are equal but not 1.

IN THIS SECTION, we use problem solving to find rules for selected patterns from tables. We review specialized vocabulary for describing operations and their results. We practice writing equations in both words and algebraic notation.

Problem Solving

Section 1.1 listed Polya's problem-solving steps in the context of time management. We use the steps in the context of finding rules from the Warm-up number patterns.

EXAMPLE 1   Applying problem-solving steps
a. Use the problem-solving steps to find the rule for the numbers in Warm-up Exercise 1.
b. Use the rule to predict the 100th number.

SOLUTION   a. To *understand the problem*, we must consider the assumptions. We assume that the natural numbers are the inputs and the given numbers are the outputs, as shown in Table 10.

TABLE 10

Input	Output
1	-5
2	-4
3	-3
4	-2

Our *plan* is to find a pattern, either from past experience or by looking for changes in the numbers. We will calculate the changes in both the input and output numbers to look for a pattern. In order to keep track of the sets of numbers in Table 11, we will call the inputs x and the outputs y and label the changes in terms of x or y.

TABLE 11

Change in x		Input x	Output y		Change in y
		1	-5		
1	⟨	2	-4	⟩	1
1	⟨	3	-3	⟩	1
1	⟨	4	-2	⟩	1

As we *carry out the plan*, we look to see that we are satisfying the assumptions. Table 11 shows that as the input rises by 1, the output rises by 1. The input and output are staying a constant difference apart. We look at that difference:

$$1 - (-5) = 6$$

$$2 - (-4) = 6$$

$$3 - (-3) = 6$$

$$4 - (-2) = 6$$

All pairs of inputs and outputs are 6 apart. In terms of the letters, our rule might be $x - y = 6$ or $y = x - 6$.

Finally, we *check the solution* by returning to the original problem. Are there other processes that confirm the result? We found the next pair of numbers to be $x = 5$ and $y = -1$. We check these in the rule:

$$x - y = 6$$

$$5 - (-1) = 6 \qquad \text{Check.}$$

Another check: Sometimes it is helpful to find the output when the input is zero. Working backwards in the table gives $y = -6$ when $x = 0$. Let $x = 0$ in the rule:

$$x - y = 6$$

$$0 - y = 6$$

$$y = -6 \qquad \text{Check.}$$

b. The 100th number would have 100 listed under input x.

$$x - y = 6 \qquad \text{Substitute 100 for } x.$$

$$100 - y = 6 \qquad \text{Solve mentally.}$$

$$y = 94$$

▬

EXAMPLE 2 Applying problem-solving steps
a. Use the problem-solving steps to find the rule for the numbers in Warm-up Exercise 2.
b. Use the rule to predict the 100th number.

SOLUTION **a.** *Understand:* Assume the inputs are natural numbers.

Plan: Make a table showing changes in x and in y.

Carry out the plan: Enter the inputs (1, 2, 3, 4) and the outputs (8, 16, 24, 32). Write the changes in Table 12.

TABLE 12

Change in x		Input x		Output y		Change in y
		1		8		
1	⟨	2		16	⟩	8
1	⟨	3		24	⟩	8
1	⟨	4		32	⟩	8

We observe that for every change of 1 in the input, the output changes by 8. In fact, the outputs are all multiples of 8. The rule may be

$$y = 8x$$

Check: We found the next pair of numbers to be $x = 5$ and $y = 40$. We check these in the rule:

$$y = 8x$$

$$40 = 8(5) \qquad \text{Check.}$$

Working backwards in the table gives $x = 0$ and $y = 0$.

$0 = 8(0)$ Check.

b. The 100th number would have 100 listed under input x.

$y = 8x$ Substitute 100 for x.

$y = 8(100)$

$y = 800$

EXAMPLE 3 Applying problem-solving steps
a. Use the problem-solving steps to find the rule for the numbers in Warm-up Exercise 3.
b. Use the rule to predict the 100th number.

SOLUTION **a.** *Understand:*   Assume the inputs are natural numbers.

Plan: Make a table showing changes in x and in y.

Carry out the plan: Enter the inputs (1, 2, 3, 4) and the outputs (5, 3, 1, -1). Write the changes in Table 13.

TABLE 13

Change in x		Input x	Output y		Change in y
1	⟨	1	5	⟩	2
1	⟨	2	3	⟩	-2
1	⟨	3	1	⟩	-2
		4	-1		

We observe that for every change of 1 in the input, the output decreases by 2. This suggests $y = -2x$, but something still needs to be added because $y = -2x$ is consistently smaller than the outputs. (Check this yourself.) If we work backwards up the table, we find that $x = 0$ matches with $y = 7$. The rule may be

$y = -2x + 7$

Check: We found the next pair of numbers to be $x = 5$ and $y = -3$. We check these in the rule:

$y = -2x + 7$

$-1 = -2(5) + 7$

$-3 = -10 + 7$ Check.

b. The 100th number would have 100 listed under input x.

$y = -2x + 7$ Substitute 100 for x.

$y = -2(100) + 7$

$y = -193$

BREAKING MIND SET   One of the most important things about problem solving is to focus on understanding the task rather than memorizing processes. Understand that we are looking for how the input and output are related. **Breaking mind set** is

Looking ahead to functions: Working backwards up the table builds intuition about $f(0)$ being the output y when $x = 0$.

thinking outside the expected. After three examples, we expect tidy tables and subtracting to find changes. No one said we had to list the tables vertically or use **consecutive** (*following after each other without interruption*) natural numbers as inputs. Look at the "sideways" table in Example 4 (Table 14) and answer the questions mentally.

EXAMPLE 4   Breaking mind set
a. What number goes below the 4?
b. What expression goes below *n*?
c. Describe the rule for the table, using the words *input* and *output*.

TABLE 14

Input	1	4	5	8	n
Output	9	?	13	16	?

SOLUTION   Hints: This pattern is not from the Warm-up Exercises. Both Example 1 and Example 2 are somewhat related. See the Answer Box for the solution.   ▬

Equations

When *x* represents an input and *y* represents an output, we may write the rules in algebraic sentences—that is, equations. An **equation** is *a statement of equality between two quantities.* Equations are **equivalent** *when they describe the same pattern or when they have exactly the same sets of numbers as solutions.* The *sets of numbers* are called **solution sets**.

EXAMPLE 5   Identifying equivalent equations   Which equations are not equivalent to the rule for Example 1, $x - y = 6$?
a. $y = x - 6$
b. $y + x = 6$
c. $y + x = -6$
d. $y + 6 = x$
e. $y - x = -6$

SOLUTION   See the Answer Box.   ▬

The following box describes building equations for selected number patterns such as those in Examples 1 to 3.

▬ EQUATIONS FROM PATTERNS:
A SPECIAL CASE

> Place the pattern into an input-output table with consecutive natural numbers as inputs.
>
> The constant change in output is the numerical coefficient of *x*.
>
> The output for $x = 0$ is a constant in the equation.

The box describes a special case because the inputs are consecutive natural numbers and change by 1 and the outputs change by a constant number.

Think about it 1: **a.** How might the directions in the box change for cases where the inputs are not consecutive natural numbers, as in Warm-up Exercises 4 and 5? Break mind set. Spend some time on this on your own.
b. Will the directions work for the pattern 1, 4, 9, 16, 25?

With number patterns, we frequently think of the rules in words before changing them to equations. With applications, mathematical relationships are almost always written in words.

Vocabulary for Writing and Describing Equations

We now review the words and symbols needed to describe operations and rules in algebraic notation.

■ RESULTS OF OPERATIONS

> A **sum** is the answer to an addition problem.
>
> A **difference** is the answer to a subtraction problem.
>
> A **product** is the answer to a multiplication problem.
>
> A **quotient** is the answer to a divison problem.

Unless otherwise indicated, assume that the order of the numbers for differences and quotients is that listed in the problem. The difference between a and b is $a - b$; the quotient of a and b is $a \div b$.

EXAMPLE 6 **Writing equations from words** Match each sentence with a table; then write the sentences as equations. Let x be the input variable. Let y be the output variable. Use the equal sign, =, to replace "is."

a. The output is the sum of the input and six.
b. The output is the quotient of the input and eight.
c. The difference between the input and the output is four.
d. The product of the input and the output is twelve.

1.

x	y
-2	-6
0	-4
3	-1

2.

x	y
-2	-6
3	4
-4	-3

3.

x	y
-2	4
0	6
2	8

4.

x	y
-8	-1
0	0
16	2

SOLUTION **a.** $3, y = x + 6$ **b.** $4, y = \dfrac{x}{8}$

c. $1, x - y = 4$ **d.** $2, x \cdot y = 12$ ■

Sum, difference, product, and *quotient* name the results of operations. The operations themselves may be described by a variety of terms.

ADDITION AND SUBTRACTION *Plus, added to, greater than, more than,* and *increased by* are words and phrases that describe addition. Subtraction words include *less, less than, fewer, minus,* and *decreased by*. If the word *than* appears in a subtraction setting, the order of the numbers is the reverse of the word order; for example, six less than the input, n, is $n - 6$.

EXAMPLE 7 **Writing equations from words** Match each sentence with a table; then write the sentences as equations. Let x be the input variable and y be the output variable. Use the equal sign, =, to replace "is."

a. The input increased by the output is five.
b. The input is decreased by three to give the output.
c. The sum of the input and six is the output decreased by three.
d. The input is three less than the output.

1.		
	x	y
	-2	-5
	1	-2
	4	1

2.		
	x	y
	-2	7
	0	9
	5	14

3.		
	x	y
	-2	1
	0	3
	5	8

4.		
	x	y
	-2	7
	2	3
	6	-1

SOLUTION **a.** $4, x \mid y - 5$ **b.** $1, x - 3 = y$

c. $2, x + 6 = y - 3$ **d.** $3, x = y - 3$ ▬

MULTIPLICATION AND DIVISION *Double a number* and *twice a number* mean multiplication by 2, while *triple a number* means multiplication by 3. The words *at . . . each* suggest multiplication; for example, "5 candied apples at \$3.59 each" means to multiply 5 times \$3.59. The word *of* means multiplication in fraction, decimal, and percent notation; for example, $\frac{1}{4}$ *of* 8 means $\frac{1}{4}$ times 8. *Half of twelve* means $\frac{1}{2}$ times 12. *Divide 8 in half* means $8 \div 2 = 4$, whereas *divide 8 by a half* means $8 \div \frac{1}{2} = 8 \cdot 2 = 16$.

Think about it 2: How is it possible for the second and fourth tables in Example 7 to begin with the same pair of numbers and yet describe different rules?

EXAMPLE 8 **Writing equations from words** Match each sentence with a table; then write the sentence as an equation.
a. The sum of twice the input and six is the output.
b. The output is the difference between half of the input and four.
c. Double the input decreased by four is half the output.
d. The difference between seven and triple the input is the output.

1.		
	x	y
	0	7
	2	1
	3	-2

2.		
	x	y
	-4	-2
	-1	4
	1	8

3.		
	x	y
	0	-8
	2	0
	5	12

4.		
	x	y
	-2	-5
	0	-4
	2	-3

SOLUTION **a.** $2, 2x + 6 = y$ **b.** $4, y = \frac{1}{2}x - 4$

c. $3, 2x - 4 = \frac{1}{2}y$ **d.** $1, 7 - 3x = y$ ▬

When an equation includes two operations or a set of parentheses, it may be necessary to use the word *sum*, *difference*, *product*, or *quotient* or to reorder the terms to avoid confusion.

EXAMPLE 9 **Writing sentences from equations**   Write the equations in words, using the suggested words when given.

a. $y = x - 2$, less than

b. $y = 5/x + 2$, sum, quotient

c. $y = 5/(x + 2)$, quotient, sum

d. $y = 3 - x/3$

e. $y = \frac{1}{2}x(x + 1)$

SOLUTION Answers may vary.

a. The output is two less than the input.

b. The output is the sum of two and the quotient of five and the input.

c. The output is the quotient of five and the sum of the input and two.

d. The output is the difference between three and the quotient of the input and three.

e. The output is the product of half the input with the sum of the input and one.

Applications

INDEPENDENT AND DEPENDENT VARIABLES   To find input and output in application settings, we look for how one piece of information depends on another. For example, the total cost of a purchase depends on the number of items purchased. The number purchased is the **independent variable**, or *input variable*. The total cost is the **dependent variable**, or *output variable*.

EXAMPLE 10 **Identifying independent and dependent variables**   Choose an independent variable (input) and dependent variable (output) in each setting.

Student Note: Match variables with phrases having number value. Let $x =$ Tim's age or $y =$ Jane's height. Don't use $x =$ Tim or $y =$ Jane.

a. A student's grade, G, in algebra depends on number, N, of classes attended.

b. The cost, c, of college tuition depends on the number, n, of credit hours taken.

c. Area of a circle of radius r: $A = \pi r^2$

d. Bowling handicap for an average below 200: $H = 0.80(200 - A)$

SOLUTION a. The independent variable is number of classes attended, N; the dependent variable is grades, G.

b. The independent variable is number of credit hours, n; the dependent variable is cost, c.

c. The independent variable is radius, r; the dependent variable is area, A.

d. The independent variable is the average, A; the dependent variable is the handicap, H.

PERCENTS   Percent problems are one of the most common applications of equations. All percents in an equation must be written as decimals or fractions. Remember that **percent** means *division by 100*; that is, $15\% = 15/100$, $0.5\% = 0.5/100$, and $200\% = 200/100$.

- To change a percent to a decimal, remove the percent sign and place the number over 100. Divide, as necessary.

- To change a percent to a fraction, remove the percent sign, place the number over 100, and simplify to lowest terms.

EXAMPLE 11 **Writing equations for percent problems**   Write an equation for each setting. Let x represent the independent variable and y the dependent variable.

a. About 20% of a salesperson's working hours are productive hours.

    b. The property tax on a home is 1.15567% of the value of the home.

    c. A corporate executive is paid 16,000% of the annual income of an hourly worker.

    d. The total cost of a meal is a tip of 15% of the cost and a sales tax of 8% of the cost added to the cost.

SOLUTION   **a.** $y = 0.20x$; $y =$ number of productive hours, $x =$ number of working hours

    b. $y = 0.0115567x$; $y =$ amount of property tax, $x =$ value of home

    c. $y = 160x$; $y =$ executive's income, $x =$ worker's annual income

    d. $y = 0.15x + 0.08x + x$; $y =$ total cost, $x =$ cost of meal

---

ANSWER BOX

Warm-up: 1. For an input of natural number 5, the output is -1. **2.** input 5, output 40 **3.** input 5, output -3 **4.** input 8, output 0 **5.** input 80, output 176 **6.** Answers vary; the inputs are not consecutive natural numbers. The changes in inputs are equal but not 1. **Example 4:** Example 1 involves subtraction between the input and output numbers, and Example 2 has an 8 in it. **a.** The output for 4 is 12. **b.** The output for n is $n + 8$. **c.** The rule is to add eight to the input to obtain the output. **Example 5:** We might test two or three pair of inputs and outputs from the table. We might solve each equation for y and compare them (more on equation solving in Section 1.6). Equations b and c are not equivalent to the others. **Think about it 1: a.** Sorry, no answer, but a hint: change in inputs. **b.** No, the pattern 1, 4, 9, 16, 25, . . . does not have a constant change in output. Its rule is $y = x^2$. **Think about it 2:** For $x = -2$, both $x + 6 = y - 3$ and $x + y = 5$ give $y = 7$. Another equation that fits $x = -2$ and $y = 7$ is $x \cdot y = -14$. Can you find another?

1.3   Exercises

In Exercises 1 to 8, write each pattern in a table, find the changes in inputs and outputs, and write an equation for the table.

1. 2, 5, 8, 11, . . . $y = 3x - 1$

2. 5, 8, 11, 14, . . .
$y = 3x + 2$

3. 3, 5, 7, 9, . . . $y = 2x + 1$

4. 1, 3, 5, 7, . . .
$y = 2x - 1$

5. 2, 4, 6, 8, . . . $y = 2x$

6. 3, 6, 9, 12, . . . $y = 3x$

7. 2, 5, 10, 17, . . . $y = x^2 + 1$

8. 0, 3, 8, 15, . . .
$y = x^2 - 1$

◼ Breaking Mind Set

In Exercises 9 to 12, mentally find the indicated outputs. A good strategy is to try for a few minutes and then come back later.

9. $2, \frac{1}{2}n$

Input	0	2	4	5	n
Output	0	1	?	2.5	?

10. $4, n - 1$

Input	-1	0	5	-3	n
Output	-2	-1	?	-4	?

11. $8, n^3$

Input	0	1	2	3	n
Output	0	1	?	27	?

12. $1, n^2$

Input	-1	0	1	$\frac{1}{2}$	n
Output	1	0	?	$\frac{1}{4}$	?

In Exercises 13 to 16, which equation is not equivalent to the others?

13. $y - 2 = x + 5$, $x + y = 7$, $x - y = -7$ $x + y = 7$

14. $x - 3 = y$, $x - y = 3$, $y + x = -3$ $y + x = -3$

15. $x + y = 8$, $y - x = -8$, $y = -x + 8$ $y - x = -8$

16. $y + 3 - x = 0$, $y + 4 = x + 1$, $y = 3 - x$ $y = 3 - x$

In Exercises 17 to 28, write the expressions in words, using x, *sum*, *difference*, *product*, and/or *quotient*.

17. $x - 7$ the difference between x and seven

18. $5 + x$ the sum of five and x

19. $x/7$ the quotient of x and seven

20. $5x$ the product of five and x

21. $7 - x$ the difference between seven and x

22. $x/5$ the quotient of x and five

23. $7/x$ the quotient of seven and x

24. $5 - x$ the difference between five and x

25. $7/(x + 2)$ the quotient of seven and the sum of x and two

26. $5/(2 - x)$ the quotient of five and the difference between two and x

27. $(x + 2)/(x - 2)$ the quotient of the sum of x and two and the difference between x and two

28. $(x - 5)/(x + 5)$ the quotient of the difference between x and five and the sum of x and five

In Exercises 29 to 36, write equations from the sentences.

29. Ten decreased by twice the input gives the output.
$10 - 2x = y$

30. Twice the input increased by five gives the output.
$2x + 5 = y$

31. The output is ten more than half the input. $y = \frac{1}{2}x + 10$

32. The output is half the input decreased by six. $y = \frac{1}{2}x - 6$

33. Two more than the quotient of the input and fifteen gives the output. $\frac{x}{15} + 2 = y$

34. The output is sixteen less than the product of three and the input. $y = 3x - 16$

35. The output is fourteen fewer than the product of seven and the input. $y = 7x - 14$

36. Twelve added to the quotient of four and the input is the output. $\frac{4}{x} + 12 = y$

In Exercises 37 to 44, write the algebraic statements in words. Let x be the input and y be the output. Answers vary

37. $y = 2x - 5$

38. $y = \frac{1}{2}x + 7$

39. $x + y = 11$

40. $x - y = 40$

41. $2x - y = 5$

42. $3x + y = 10$

43. $\frac{x}{8} = y$

44. $y = \frac{x}{5} + 2$

Name the independent and dependent variables in each formula or sentence in Exercises 45 to 49.

45. Area of a square of side s: $A = s^2$ $s; A$

46. Volume of a sphere of radius r: $V = \frac{4}{3}\pi r^3$ $r; V$

47. Area of an equilateral triangle of side s: $A = \dfrac{s^2\sqrt{3}}{4}$ $s; A$

48. Energy released from material of mass m (c is a constant): $E = mc^2$ $m; E$

49. The exercise heart rate, E, depends on the person's age, A. $A; E$

In Exercises 50 to 54, say which number should be the independent variable and which number should be the dependent variable.

50. The cost of riding the local transit system depends on the number of trips made.
number of trips; cost of riding local transit

51. The cost of shipping a package depends on the weight of the package. weight of package; cost of shipping

52. The income tax rate depends on the adjusted gross income. adjusted gross income; income tax rate

53. The number of kittens sitting on an older computer monitor; the temperature of the room A cooler day might bring more kittens: temperature; number of kittens

54. The number of puppies in a store window; the length of time a shopper looks in the window Shoppers might look longer if there were more puppies: number of puppies; length of time

In Exercises 55 to 60, write the application in an equation.

55. The output is the total cost of x cans of soup at $1.29, less a $0.45 manufacturer's coupon. $y = 1.29x - 0.45$

56. The output is the total cost of renting the Rug Doctor cleaning system for x hours at $8.95 per hour, plus $12 in cleaning solution. $y = 8.95x + 12$

57. You put $1.25 into a parking meter. The output is the value of your time remaining on the meter after x minutes at $0.05 per minute. $y = 1.25 - 0.05x$

58. The output is the amount remaining on a $20 phone card after you talk for x minutes at $0.35 per minute.
$y = 20 - 0.35x$

59. The output is the total cost of taking x credit hours at $125 per hour tuition and paying $85 in fees.
$y = 125x + 85$

60. The cell phone charge is $9.50 per month and $0.05 per minute. What is the total cost for x minutes of service each month? $y = 0.05x + 9.50$

In Exercises 61 and 62, change to percents.

61. a. $1/6$ 16.7% **b.** $3/5$ 60% **c.** 20 2000% **d.** 0.05 5%

62. a. $5/6$ 83.3% **b.** $2/5$ 40% **c.** 15 1500% **d.** 0.025 2.5%

In Exercises 63 and 64, change to decimals.

63. a. 20% 0.2 **b.** 12.5% 0.125 **c.** 0.005% 0.00005 **d.** 500% 5

64. a. 15% 0.15 **b.** 0.01% 0.0001 **c.** 6.25% 0.0625 **d.** 1000% 10

In Exercises 65 to 72, define variables and write an equation.

65. Daily ski lift tickets cost about 5% of the cost of a season ticket. $y = 0.05x$; $x =$ season ticket cost, $y =$ daily cost

66. Youth (7–12) season ski lift tickets cost about 25% of the cost of an adult season ticket.
$y = 0.25x$; $x =$ adult season ticket cost, $y =$ youth season ticket cost

67. FICA (Social Security) withholding is 6.2% of your wages. $y = 0.062x$; $x =$ amount of wages, $y =$ amount of tax

68. FICA Medicare takes 1.45% of your wages.
$y = 0.0145x$; $x =$ amount of wages, $y =$ amount of tax

69. One of the borrowing costs of a real estate loan is 1% of the amount of money borrowed.
$y = 0.01x$; $x =$ amount of loan, $y =$ one borrowing cost

70. Your net return after a check-cashing store takes 2.5% of the amount of the check. $y = x - 0.025x$ (or $y = 0.075x$); x = amount of check, y = amount received

71. Your net pay is your wages less the sum of 7.65% of your wages for Social Security, 20% for federal taxes, and n% for state and local taxes. $y = x - [0.0765x + 0.20x + (n/100)x]$; x = amount of wages, y = net pay

72. a. The total cost of a purchase includes a $5 discount and a sales tax of 8.75% of the original cost of the items. $y = x - 5 + 0.0875x$; x = cost of items, y = amount paid

    b. If the discount is taken before the tax is calculated, how will the equation change? $y = x - 5 + 0.0875(x - 5)$

∎ **More Breaking Mind Set Problems**

73. A group of two mothers and two daughters go shopping. How is it possible that the group has only three people? a woman with her mother and her daughter

74. Without lifting your pencil, connect all nine dots with exactly four straight lines. See Additional Answers.

∎ **Project**

75. Words for Powers Here are the words for some common powers:

x^2	x-squared, the square of x
x^3	x-cubed, the cube of x
3^2	the second power of 3
2^4	the fourth power of 2

Write these expressions or equations in algebraic notation.

a. Five divided by the square of the sum of two and x
$5/(2 + x)^2$

b. The sum of x and y cubed $x + y^3$

c. The cube of the sum of x and y $(x + y)^3$

d. The difference between the square of b and $4ac$
$b^2 - 4ac$

e. The sum of the square of a and the square of b
$a^2 + b^2$

f. The sixth power of two equals the square of the third power of two $2^6 = (2^3)^2$

Write these expressions in words.

g. $x^3 + 2x^2 + 1$ the sum of the cube of x, the product of two and the square of x, and one

h. $2 - (x + 1)^2$ the difference between two and the square of the sum of x and one

i. $(2 + x)^2$ the square of the sum of two and x

j. $2 + x^2$ the sum of two and the square of x

k. $3/(2 + x)^3$ the quotient of three and the cube of the sum of two and x

l. $3/2 + x^3$ the sum of the cube of x and the quotient of three and two

---

1   Mid-Chapter Test

1. Simplify.

  a. $-3 + (-5) - (-6)$ -2 **b.** $-\frac{1}{2} + \frac{1}{4} - \frac{3}{4} - \left(-\frac{1}{4}\right)$ $-\frac{3}{4}$

  c. $2.25 + 8.50 - 3.75$ 7   **d.** $6.2 - 8.6 + 1.8$ -0.6

2. Simplify.

  a. $2 \cdot 17 \cdot 50$ 1700   **b.** $\frac{1}{2}(-7)(-6)$ 21

  c. $-4(-8)(-5)$ -160 **d.** $\frac{3}{5}\left(\frac{11}{6}\right)\left(\frac{15}{121}\right)$ $\frac{3}{22}$

3. Simplify. Write answers as fractions or round to the nearest tenth.

  a. -4^2 -16 **b.** $(-4)^2$ 16   **c.** $0 - 4^2$ -16

  d. $8 - 6 \div 3 + 3 \cdot 4 + \frac{1}{2}(8 - 3)$ $20\frac{1}{2}$ **e.** $\dfrac{3 - 6}{2 - (-7)}$ $-\frac{1}{3}$

  f. $\dfrac{-(-2) - \sqrt{(-2)^2 - 4(5)(-3)}}{2(5)}$ $-\frac{3}{5}$

  g. $\sqrt{(2 - (-7))^2 + (3 - 6)^2}$ $3\sqrt{10} \approx 9.5$

  h. $\dfrac{|1.25 - 2.83| + |2.56 - 2.83| + |4.68 - 2.83|}{3}$ $\frac{37}{30} \approx 1.2$

Use the properties of numbers to simplify these expressions.

4. a. $\frac{1}{2} + 96.43 + 3 \div 6$ 97.43

  b. $9 \cdot 59 + 59$ 590

  c. $8\left(8\frac{3}{8}\right)$ 67

  d. $3.57(39) + 6.43(39)$ 390

  e. $0.0333333\ldots + 3/10$ $\frac{1}{3}$ (hint: $\frac{3}{10} = 0.3$)

  f. $(225 + ?) + (576 - 127) = 225 + 576$ 127

5. a. Why could you say either "negative" or "the opposite of" in reading $-4x^2$? $-4x^2$ will be negative because $4x^2$ is always positive.

  b. Why must we say "the opposite of" in reading $-4x$? x could be positive or negative.

  c. Why must we say "the opposite of" in reading $-x$? x could be positive or negative.

  d. Why could you say either "negative" or "the opposite of" in reading $-|x|$? $-|x|$ will be negative because $|x|$ is always positive.

6. Multiply.

   a. $3(x + 2)$ $3x + 6$ **b.** $-3(x - 2)$ $-3x + 6$

   c. $7 - 4(x + 5)$ $-4x - 13$ **d.** $8 - 2(4 - x)$ $2x$

7. Name the greatest common factor and then factor.

   a. $4x - 18$ $2, 2(2x - 9)$ **b.** $x^2 + 5x$ $x, x(x + 5)$

   c. $mn^2 - np^2$ $n, n(mn - p^2)$ **d.** $63x^2y - 49xy^2$

                                         $7xy, 7xy(9x - 7y)$

8. Name the variables, constant term, and numerical coefficients.

   a. $-4 - x$ $x; -4; -1$ **b.** $2x - x^2$ $x;$ none; 2 and -1

9. Remove parentheses and add like terms.

   a. $3(x^2 + 3x - 4) - x(x^2 + 3x - 4)$ $-x^3 + 13x - 12$

   b. $a(a^2 - 2ab + b^2) - b(a^2 + 2ab - b^2)$

       $a^3 - 3a^2b - ab^2 + b^3$

10. Simplify.

   a. $\dfrac{x^2yz}{xy^2z}$ $\dfrac{x}{y}$ **b.** $\dfrac{x^2yz}{(xy)^2z}$ $\dfrac{1}{y}$ **c.** $\dfrac{xy - yz}{xy}$ $\dfrac{x - z}{x}$

11. For each formula, name the independent and dependent variables. Then evaluate the formula for the given values of the variables.

   a. Height of an equilateral triangle of side s: $h = \dfrac{s\sqrt{3}}{2}$,

      $s = 4$ yd $s; h; h = 2\sqrt{3}$ yd

   b. Area of a circle of side d: $A = \pi\left(\dfrac{d}{2}\right)^2$, $d = 5$ cm

      $d; A; A = \frac{25}{4}\pi$ cm^2 or $A = 6.25\pi$ cm^2

12. Name the independent and dependent variables in each setting.

   a. The number of quarters in a parking meter and the paid parking time   number of quarters; parking time

   b. The distance traveled while jogging n miles per hour

     speed; distance traveled

   c. The size of a piece of luggage and the cost to take it on a certain airline   size of luggage; cost

13. Write an equation for each sentence.

   a. The total cost with sales tax is 8.75% of the price plus the price. $C = x + 0.0875x$ or $C = 1.0875x$

   b. The amount owed on a credit card with 0% APR for 12 months on transferred money is the amount transferred, x dollars, plus a fee of 3% of the amount transferred. $A = x + 0.03x$ or $A = 1.03x$

   c. The output is the total cost of x cans of frozen juice at $1.49 each, less $0.30 for a manufacturer's coupon.

     $C = 1.49x - 0.30$

14. With the animal pen design shown, two square pens are added at once on the same side.

Number of Pairs of Pens	Total Number of Panels
1	7
2	12
3	17
4	22
10	52
50	252
100	502

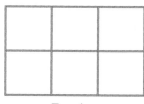

Top view

   a. Complete the table.

   b. Let x be the number of pairs of pens. Let y be the total number of panels needed to build the pens. Find a rule for the pattern in the table. $y = 5x + 2$

In Exercises 15 to 22, match each of the word phrases with one of the algebraic expressions.

15. Four increased by double a number   j

16. Five times a number plus three   a

17. Five less than three times a number   g

18. Three more than half a number   h

19. Twice a number less four   i

20. Two less than four times a number   b

21. Five decreased by three times a number   e

22. One-half less four times a number   c

   a. $5x + 3$

   b. $4x - 2$

   c. $\frac{1}{2} - 4x$

   d. $3 + 2(x - 2)$

   e. $5 - 3x$

   f. $4 + 0.5(x - 2)$

   g. $3x - 5$

   h. $0.5x + 3$

   i. $2x - 4$

   j. $4 + 2x$

1.4 Coordinate Graphs

Objectives

■ Identify horizontal and vertical axes, quadrants, the origin, and ordered pairs.
■ Graph data and describe patterns in graphs.
■ Build input-output tables and graphs from equations.
■ Graph ordered pairs from input-output tables.

WARM-UP

1. Suppose the first copy of a school transcript costs $3 and each additional copy costs $2. What is the cost of two transcripts? five transcripts? $5; $11

2. The height above the road of the cable for a suspension bridge in The Netherlands is described by the equation $y = \frac{4}{125}x^2$, where x is the distance from the center of the bridge. What is the height of the cable at a distance of 10 feet from the center? 20 feet? 3.2 ft; 12.8 ft

3. List the powers of 2 from 2^1 to 2^8. 2, 4, 8, 16, 32, 64, 128, 256

THIS SECTION STARTS WITH a summary of the vocabulary and concepts needed to graph data on rectangular coordinate axes. The section then introduces building graphs from tables and equations.

Rectangular Coordinate Graphs

Have you ever used a map to find your way around an unfamiliar city or region? Does your car or van have a global positioning system (GPS)? Rectangular coordinate grids are the basis for maps. The GPS uses satellite technology to locate points on a worldwide grid. When we locate points, lines, or curves on a grid in mathematics, we are graphing.

VOCABULARY   The following paragraphs summarize the basic vocabulary and concepts needed for coordinate graphing.

In rectangular coordinate graphing, the **axes** are *two number lines placed at right angles so that they cross at zero*. The axes are **perpendicular** because *they cross at a 90° angle*. The axes lie on *a flat surface* called the **coordinate plane**.

Quadrants are *the four regions separated by the axes*. The quadrants, shown in Figure 7, are numbered counterclockwise because of applications in engineering and physics.

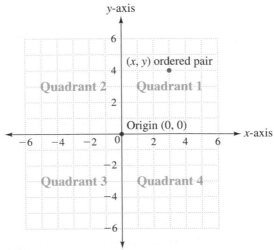

FIGURE 7

An **ordered pair** is *a pair of numbers (x, y) that uniquely describes each point in the coordinate plane.* The numbers in the ordered pair may be any real numbers (positive, negative, or zero). The **origin**, (0, 0), is at *the intersection of the two axes.*

LOCATING POINTS ON A GRAPH   In the ordered pair (*x*, *y*), the first number *x*, or the **x-coordinate**, *gives a position in the horizontal direction.* The second number *y*, or the **y-coordinate**, *gives a position in the vertical direction.*

The point labeled *A* in quadrant 1 of Figure 8 has the ordered pair (3, 2). Horizontally, it is 3 units to the right from the origin. Vertically, it is up 2 units from the origin.

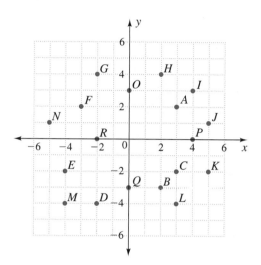

FIGURE 8

The point with the ordered pair (2, −3) is in quadrant 4. Horizontally, it is 2 units to the right from the origin. Vertically, it is down 3 units from the origin. The point (2, −3) is labeled *B* in Figure 8.

EXAMPLE 1   Finding and writing ordered pairs

a. Using Figure 8, locate the points named by these ordered pairs:

(2, 4), (4, 3), (−2, 4), (−2, −4), (3, −2), (−3, 2), (−4, −2)

b. Name the ordered pairs labeled *J* to *R* in Figure 8.

SOLUTION   **a.** (2, 4) is *H*; (4, 3) is *I*; (−2, 4) is *G*; (−2, −4) is *D*; (3, −2) is *C*; (−3, 2) is *F*; (−4, −2) is *E*.

b. *J* is (5, 1); *K* is (5, −2); *L* is (3, −4); *M* is (−4, −4); *N* is (−5, 1); *O* is (0, 3); *P* is (4, 0); *Q* is (0, −3); *R* is (−2, 0).   ▬

The ordered pair (*x*, *y*) is also called the Cartesian coordinates, after René Descartes (1596–1650). Descartes, a French philosopher and soldier, is credited with devising analytic geometry, a blending of algebra and geometry with graphing on the coordinate axes. The graphing calculator has given his contribution to mathematics even more significance to our study.

PLOTTING POINTS IN APPLICATION SETTINGS   The inputs and outputs for a particular application form ordered pairs. The independent variable (input) is always placed on the horizontal axis. The dependent variable (output) is placed on the vertical axis. In Example 2, we return to the Warm-up.

EXAMPLE 2   Building a table and graph   Suppose the first copy of a school transcript costs $3 and each additional copy costs $2.

a. Identify the independent (input) and dependent (output) variables.

b. Make a table and graph to show the total cost of ordering 1 to 5 transcripts.

c. Use the table and graph to find the cost of 3 transcripts.

d. Use the table and graph to solve for the number of transcripts that can be bought for $9.

SOLUTION **a.** The total cost depends on the number of transcripts ordered. The number of transcripts is the independent variable (on the horizontal axis). The total cost is the dependent variable (on the vertical axis).

b. The total costs are shown in Table 15 and Figure 9.

TABLE 15

Input: Transcripts Ordered	Output: Total Cost (dollars)
1	3
2	5
3	7
4	9
5	11

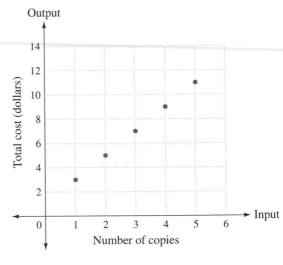

FIGURE 9

c. In Table 15, three transcripts matches with $7. In Figure 9, find 3 on the horizontal axis, and then look for the point on the graph where $x = 3$. Again, for 3 transcripts, the total cost is $7.

d. In Table 15, the $9 cost matches with 4 transcripts. In Figure 9, find $9 on the vertical axis, and then look for the point on the graph where $y = 9$. ▬

Think about it 1: Is the concept of buying $2\frac{1}{2}$ copies of a transcript meaningful? Would it be appropriate to draw a line through the points in Example 2?

Graphing from an Equation

The rules and patterns for input-output tables (see Section 1.2, page 15) are most commonly written as equations. In the next few examples, we are given the equation.

An input-output table is a listing of some of the ordered pairs for an equation. A graph shows all points (and hence ordered pairs) for an equation in the region pictured by the graph.

EXAMPLE 3 **Graphing from equations** For the two equations given, make a table with integer inputs, x, from -2 to 2, and graph the resulting input-output pairs. If the points form a recognizable pattern, connect them and identify the resulting shape.

a. $y = 2x + 1$ **b.** $y = x^2$

SOLUTION   **a.** The table and graph are shown in Table 16 and Figure 10.

Suggestions for calculator technique are on page 44.

TABLE 16

Input x	Output $y = 2x + 1$
-2	$2(-2) + 1 = -3$
-1	$2(-1) + 1 = -1$
0	$2(0) + 1 = 1$
1	$2(1) + 1 = 3$
2	$2(2) + 1 = 5$

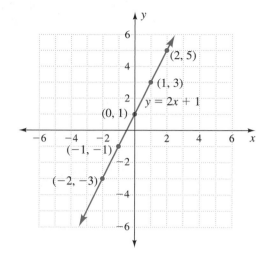

FIGURE 10

b. The table and graph are shown in Table 17 and Figure 11.

TABLE 17

Input x	Output $y = x^2$
-2	$(-2)^2 = 4$
-1	$(-1)^2 = 1$
0	$(0)^2 = 0$
1	$(1)^2 = 1$
2	$(2)^2 = 4$

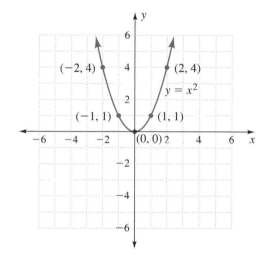

FIGURE 11

The points in the figure for part a appear to lie on a straight line. We connect the points left to right and extend the resulting line with arrows to indicate its infinite length. The points in the figure for part b appear to lie on a smooth curve called a **parabola**. We again connect the points left to right and extend the curve with arrows to indicate that it continues infinitely.   ▬

Think about it 2: Look at the graph for $y = 2x + 1$. Find an ordered pair for a new point on the graph and, with substitution, show that the ordered pair makes the equation true.

LABELING THE AXES   The axes on coordinate graphs may be labeled with any names or letters. If no names or letters are given in an application, use x and y. Because x and y are commonly placed on the axes, the *horizontal axis* is often called the **x-axis** and the *vertical axis* is often called the **y-axis**.

If available, always include units (such as $, pounds, months, or years). Write them in parentheses beside or below the name of the axis.

When writing numbers on the axes, choose an appropriate **scale**, or *spacing between the numbers*. One estimate for scale is

$$\frac{\text{Highest number} - \text{lowest number}}{10}$$

Round to the nearest easy counting number.

In a horizontal input-output table, the inputs are always on the top row and the outputs on the bottom row.

EXAMPLE 4 Setting up axes

a. Make a table for $y = 2^x$ for x between -3 and 5. Check the first four outputs on the table with your calculator. Use 2 ⌃ (−) 3, etc.

b. Find an appropriate scale for the vertical axis.

c. Graph $y = 2^x$.

SOLUTION **a.** The inputs and outputs are in Table 18.

TABLE 18

x	-3	-2	-1	0	1	2	3	4	5
$y = 2^x$	0.125	0.25	0.5	1	2	4	8	16	32

b. We can number the horizontal axis with the integers. We need to use a different scale for the vertical axis. The highest output number is 32. The lowest is about 0.

$$\frac{\text{Highest number} - \text{lowest number}}{10} = \frac{32 - 0}{10} \approx 3$$

An estimate for the vertical scale is 3. However, because all the outputs are even, we will use a vertical scale of 4 so that we can place the points on the grid lines.

c. The graph is in Figure 12.

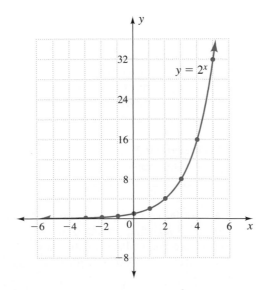

Student Note: The grid has a horizontal scale of 1 and a vertical scale of 4, but some of the grid numbers are not shown, to prevent the graph from becoming cluttered.

FIGURE 12

In Example 5, vertical cables connect a bridge to a suspended cable. (See Figure 13.) The vertical cables transform the suspended cable into a parabola. When the origin, (0, 0), is placed at the center of the bridge, the equation for the cable's height above the bridge is $y = \frac{4}{125}x^2$. (You will be able to find this equation yourself after you study Chapter 4.) The negative x's represent distances to the left of the center of the bridge.

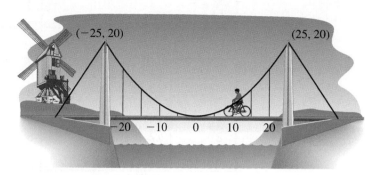

FIGURE 13

EXAMPLE 5 Building tables and graphs from equations: suspension bridge cables A 50-foot suspension bridge across a canal in The Netherlands has a cable anchored 20 feet above the road surface at each end of the bridge. We assume that the center of the bridge is at the origin of a pair of axes.

The equation of the cable's height above the road is $y = \frac{4}{125}x^2$ for x between -25 and 25. Use the equation to make a table and graph for inputs from -20 to 20 in steps of 10.

SOLUTION The table and graph are shown in Table 19 and Figure 14. The width of the bridge (50 feet) suggest a horizontal scale of 50/10 = 5. The height of the bridge (20 feet) and output numbers in the table suggest a vertical scale of 4.

TABLE 19 $y = \frac{4}{125}x^2$

Distance from Bridge Center, x (feet)	Height, y (feet)
-20	12.8
-10	3.2
0	0.0
10	3.2
20	12.8

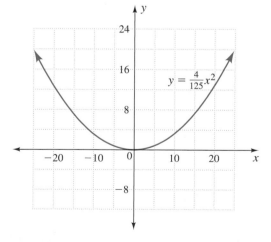

FIGURE 14

Repeat Example 5 on your calculator and compare your work with the screens in Figure 15.

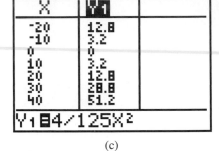

(a)   (b)

(c)   (d)

FIGURE 15

 GRAPHING CALCULATOR TECHNIQUE:
BUILDING A TABLE

Change the equation to y = expression or expression = 0.

Press ⬚ Y = ⬚ and enter the expression after Y$_1$, as in Figure 15a.

Select [TBLSET].

Enter the starting input for the table after **TblStart** and enter the amount between inputs after Δ**Tbl**, as in Figure 15b.

Select [TABLE]. (See Figure 15c.)

GRAPHING CALCULATOR TECHNIQUE:
GRAPHING AN EQUATION

Change the equation to y = expression or expression = 0.

Press ⬚ Y = ⬚ and enter the expression after Y$_1$.

Press ⬚ WINDOW ⬚. (See Figure 15d.)

Set **Xmin** and **Xmax** to include the smallest and largest inputs.

Set **Xscl** to 10% of (**Xmax** − **Xmin**) rounded to an easy counting number.

Set **Ymin** and **Ymax** to include the smallest and largest outputs.

Set **Yscl** to 10% of (**Ymax** − **Ymin**) rounded to an easy counting number.

Press ⬚ GRAPH ⬚.

We can summarize our graphing steps thus far in terms of problem solving:

STEPS FOR GRAPHING FROM
AN EQUATION OR DATA

Understand the problem. Choose the independent and dependent variables.

Plan. Choose inputs for a table and scales for axes.

Carry out the plan. Make a table, if needed. Graph ordered pairs. Look for patterns. Consider whether dots should be connected.

Check. Substitute to see that ordered pairs fit the problem situation or satisfy the equation. Consider whether the graph has a reasonable shape. Consider how new information might change the graph. Make observations about the results.

Graphing Data without an Equation

The next example returns to the wind-chill apparent temperature chart, introduced on the chapter opener page. Tables 20 and 21 show wind speed with apparent temperature; they were used to compare temperature changes in Example 3 of Section 1.1. Return to the original chart on the chapter opener page and identify the rows given here.

TABLE 20   *For 25°F Current Temperature*

Wind (mph)	0	5	10	15	20	25	30
Apparent Temp. (°F)	25	22	10	2	−3	−7	−10

TABLE 21   *For −5°F Current Temperature*

Wind (mph)	0	5	10	15	20	25	30
Apparent Temp. (°F)	−5	−10	−27	−38	−46	−52	−56

The rows in Table 20 form the top row and the row beginning with 25 in Table 1 (page 1). The rows in Table 21 form the top row and the row beginning with −5 in Table 1.

EXAMPLE 6   **Graphing data**   Set up one set of axes and then graph Table 20 and Table 21 on these axes.

SOLUTION   *Understand:* Because the apparent temperature depends on the wind, let wind be the independent variable (input) and apparent temperature be the dependent variable (output).

Plan: The horizontal (input) axis could match the evenly spaced wind speeds (0, 5, 10, . . .). Depending on the size of the grid on the graph paper, the scale could be either 5 or 10. The scale on the vertical (output) axis should include the numbers from both tables, ranging from 25°F to −56°F.

$$\frac{\text{Highest number} - \text{lowest number}}{10} = \frac{25 - (-56)}{10} \approx 8$$

There is no reason to use a scale of 8, so round to 10 to make the scale similar to the scale on the horizontal axis.

Carry out the plan: The graphs are shown in Figure 16.

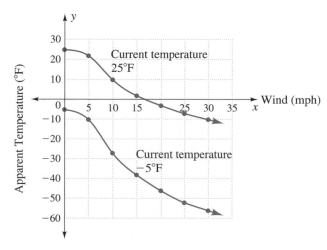

FIGURE 16

Check: The two graphs follow a similar pattern, but not exactly the same pattern. No points are out of the basic alignment. Graphing additional current temperature rows might reveal or confirm other observations. See the Project in Exercise 42.

▬

Repeat Example 6 on a calculator. For the extra set of data, place *x* in L1, *y* for 25°F in L2, and *y* for −5°F in L3. Use two stat plots: Plot1 with L1 and L2 and Plot2 with L1 and L3.

▬ GRAPHING CALCULATOR
TECHNIQUE: PLOTTING ORDERED PAIRS
WITH THE STATISTICS FUNCTION

Clear or shut off equations in (Y =) and clear prior lists.

Enter data: Place *x* in one list and *y* in a second list, as in Figure 17a.

Set the window, as shown in Figure 17b:

   Set **Xmin** and **Xmax** to include the smallest and largest *x*.

   Set **Xscl** to 10% of (**Xmax** − **Xmin**) rounded to an easy counting number.

   Repeat for **Ymin**, **Ymax**, and **Yscl**.

Choose the statistical plot: (2nd) [STAT PLOT]. Press (ENTER) twice to turn it on. (See Figure 17c.)

   Choose the type of graph.

   Choose the source of list for *x* and for *y*.

   Choose the type of mark for the graph.

Press (GRAPH) to plot the data. (See Figure 17d.)

Note: When you have finished, go back to [STAT PLOT] and turn it off. Leaving a statistical plot on may create an "Invalid Dimension" in the future and keep you from doing other graphing.

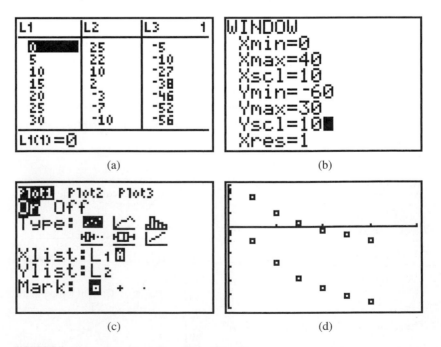

(a)   (b)

(c)   (d)

FIGURE 17 In (d), 0 to 40 on X, −60 to 30 on Y

ANSWER BOX

Warm-up: 1. $5, $11 **2.** 3.2 feet, 12.8 feet **3.** 2, 4, 8, 16, 32, 64, 128, 256
Think about it 1: No; only whole numbers of transcripts are reasonable. A line through the points on the graph would associate an output with $2\frac{1}{2}$ copies, 3.14 copies, and so forth. **Think about it 2:** One point on the line is $\left(-\frac{1}{2}, 0\right)$. Substitute $\left(-\frac{1}{2}, 0\right)$ into $y = 2x + 1$: $0 = 2\left(-\frac{1}{2}\right) + 1$, a true statement.

1.4 Exercises

1. Draw a set of axes and label these: See Answer Section.

  a. vertical axis   **b.** quadrant 4

  c. origin   **d.** $(-1, 3)$

  e. $(-4, 0)$

2. Draw a set of axes and label these: See Additional Answers.

  a. quadrant 2   **b.** horizontal axis

  c. $(4, -1)$ **d.** $(0, -2)$

  e. origin

In Exercises 3 and 4, match each definition or related statement with a vocabulary word.

3. Choose from
   evaluate, scale, independent variable, dependent variable, solution set

  a. All the numbers that make an equation true
    solution set
  b. Substitute for variables in an expression   evaluate

  c. The distance between numbers labeled on the axes
    scale
  d. What is y if y depends on x?   dependent variable

  e. What is x if y depends on x?   independent variable

4. Choose from
   origin, (x, y), x-axis, perpendicular, y-axis

  a. The vertical axis representing output y-axis

  b. The name of the point $(0, 0)$ where the axes cross
    origin
  c. An ordered pair (x, y)

  d. The horizontal axis representing input x-axis

  e. Two lines forming a right angle   perpendicular

For Exercises 5 to 8, refer to the Celsius and Fahrenheit temperatures shown in the graph below.

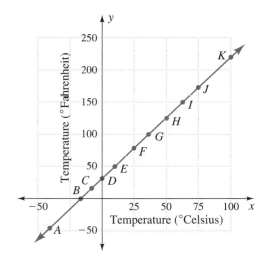

In Exercises 5 and 6, locate the points by naming the closest letter.

5. a. $(75, 167)$ J **b.** $(10, 50)$ E

  c. Normal body temperature: $(37, 98.6)$ G

  d. Temperature at which water boils: $(100, 212)$ K

6. a. $(50, 122)$ H **b.** $(-10, 14)$ C

  c. Temperature at which water freezes: $(0, 32)$ D

In Exercises 7 and 8, estimate the coordinates of the points from the graph. Your estimate should be within five units of the answer.

7. a. Point B $(-18, 0)$

  b. The point at which the Celsius temperature equals the Fahrenheit temperature $A(-40, -40)$

8. a. Point F $(25, 77)$ **b.** Point I $(66, 150)$

For Exercises 9 to 18, make an input-output table for each equation. Use as inputs the integers from −3 to 3.

9. $y = 3x - 2$
See Answer Section.

10. $y = 2x + 3$
See Additional Answers.

11. $y = 4 - 2x$
See Answer Section.

12. $y = 3 - x$
See Additional Answers.

13. $y = -2 - x$
See Answer Section.

14. $y = 3 - 2x$
See Additional Answers.

15. $y = 2x - x^2$
See Answer Section.

16. $y = x^2 - x$
See Additional Answers.

17. $y = x^2 + 2$
See Answer Section.

18. $y = 2 - x^2$
See Additional Answers.

For Exercises 19 to 28, use graph paper and label the x-axis from −4 to 4 and the y-axis from −12 to 12.

19. Graph Exercise 9.
See Answer Section.

20. Graph Exercise 10.
See Additional Answers.

21. Graph Exercise 11.
See Answer Section.

22. Graph Exercise 12.
See Additional Answers.

23. Graph Exercise 13.
See Answer Section.

24. Graph Exercise 14.
See Additional Answers.

25. Graph Exercise 15.
See Answer Section.

26. Graph Exercise 16.
See Additional Answers.

27. Graph Exercise 17.
See Answer Section.

28. Graph Exercise 18.
See Additional Answers.

29. a. Which of the graphs in Exercises 19, 21, 23, 25, and 27 are straight lines?   19, 21, 23

b. Which of the graphs in Exercises 19, 21, 23, 25, and 27 are parabolas?   25, 27

c. Which equation, $y = ax + b$ or $y = ax^2 + bx + c$, makes a straight line? $y = ax + b$

30. a. Which of the graphs in Exercises 20, 22, 24, 26, and 28 are straight lines?   20, 22, 24

b. Which of the graphs in Exercises 20, 22, 24, 26, and 28 are parabolas?   26, 28

c. Which equation, $y = ax + b$ or $y = ax^2 + bx + c$, makes a parabola? $y = ax^2 + bx + c$

In Exercises 31 to 36, make a table and graph from each equation. Calculating and plotting these by hand is important in developing number sense and graphing skills.

31. $y = 3^x$ See Answer Section.   **32.** $y = x^3$ See Additional Answers.

33. $x^2 + y^2 = 25$ (Find 12 ordered pairs. Try $x = 0$, $y = 3$, and $x = 3$.)   See Answer Section.

34. $x^2 + y^2 = 169$ (Find 12 ordered pairs. Try $y = 0$, $x = 5$, and $y = 5$.)   See Additional Answers.

35. $x = y^2$ (Find two y values for $x = 4$, 9, and 16.)
See Answer Section.

36. $y = |x|$ See Additional Answers.

37. Graph, comment on your results, and then practice with a calculator table and graph.   See Answer Section.

Arrangements of Roses

Number of Roses	Cost
3	$13
6	$23
12	$38

38. Graph, comment on your results, and then practice with a calculator table and graph. See Additional Answers.

Economy Three-Ring Notebooks

Binder Size (inches)	Cost
$\frac{1}{2}$	$1.98
1	$2.98
$1\frac{1}{2}$	$3.49
2	$3.98
$2\frac{1}{2}$	$5.49

39. Graph, comment on your results, and then practice with a calculator table and graph.   See Answer Section.

Selected Western Ski Resorts	Acreage Served by Lifts	Adult Season Pass 2002–03
Aspen/Snowmass, CO*	3,683	$1,399
Mt. Bachelor, OR	3,683	$ 860
Squaw Valley, CA*	4,200	$1,339
Steamboat Springs, CO*	2,939	$ 885
Sun Valley, ID*	2,054	$1,750

Register Guard, Eugene, OR, 09/01/02.
*Rates quoted are for preseason purchase.

40. The following are the dimensions of souvenir bags from national two-year college mathematics conferences: 1997 Conference: $5'' \times 12'' \times 12''$; 2000 Conference: $4'' \times 16'' \times 12''$; 2001 Conference: $5\frac{1}{2}'' \times 12'' \times 14''$; 2002 Conference: $5\frac{1}{2}'' \times 12'' \times 15\frac{1}{4}''$. Make a table and graph for years after 1996 and the volumes of the bags. Comment on your results and then practice with a calculator graph. See Additional Answers.

■ **Project**

41. "Four" Equations Place the input-output pairs into a table. Use the operations $+$, $-$, $\times$, and $\div$ with x, y, and 4 to write the equation for each table. Graph your equation and check that the points lie on the graph.

a. $(1, 4)$, $(2, 2)$, $\left(3, \frac{4}{3}\right)$ $xy = 4$

b. $\left(1, \frac{1}{4}\right)$, $\left(2, \frac{1}{2}\right)$, $\left(3, \frac{3}{4}\right)$ $x/4 = y$

c. $(1, 4)$, $(2, 8)$, $(3, 12)$ $4x = y$

d. $(1, 5)$, $(2, 6)$, $(3, 7)$ $x + 4 = y$

e. $(1, 3)$, $(2, 2)$, $(3, 1)$ $x + y = 4$

f. $(1, -3)$, $(2, -2)$, $(3, -1)$ $x - 4 = y$

42. Wind-Chill Temperatures with Current Temperature Held Constant Return to Example 6, page 45.

a. Make an enlarged copy of the graph in Figure 16. Include the original graphs for 25°F and −5°F but increase the vertical axis to 35. Now make additional graphs on the same axes.

b. Return to Table 1 on page 1. Start with a current temperature of 35°F. Plot the apparent temperature for different wind speeds—that is, (0, 35), (5, 32), (10, 22), (15, 16), (20, 11), (25, 8), and (30, 5). Label this graph with 35°F.

c. On the same axes, plot two other graphs for current temperatures of 15°F and 5°F. Label each graph with its current temperature.   See Additional Answers.

d. List your observations about the graphs.
Shapes are similar; drop in apparent temperature becomes less rapid for higher wind speeds, more rapid for lower current temperatures.

1.5   Solving Equations with a Table and Graph

Objectives

■ Solve equations from a table.

■ Write solutions in a solution set.

■ Solve equations from a graph.

■ Use guess and check to solve a word problem and build an equation.

WARM-UP

1. List the squares of the numbers from 1 to 15.   1, 4, 9, 16, 25, 36, 49, 64, 81, 100, 121, 144, 169, 196, 225

2. List the powers of 2 from 2^1 to 2^8.   2, 4, 8, 16, 32, 64, 128, 256

3. What is the value of each of the following?
   a. $(-2)^2$ 4   **b.** -2^2 −4   **c.** $0 - 2^2$ −4   **d.** $0 - (-2)^2$ −4

SOLVING EQUATIONS IS a fundamental skill in traditional algebra. This section summarizes tabular and graphic techniques for finding solutions.

Solving Equations from a Table

Tables and graphs are powerful means to visually check the reasonableness of results, to make observations, and to enhance traditional algebra skills. In fact, while some simple-looking equations of the form $2^x = x^2$ cannot be solved by traditional algebra, a table gives good approximations and a graph gives very accurate results.

EXAMPLE 1   **Solving from a table**   Using a table for $y = 6 - 2x$, find x when $y = -2$. Check the solution.

SOLUTION   Table 22 contains inputs and outputs for $y = 6 - 2x$. Look for the input, x, that matches the output $y = -2$.

TABLE 22

Input x	Output y
1	4
2	2
3	0
4	−2

The input is $x = 4$.

Check: $6 - 2(4) \stackrel{?}{=} -2$

$6 - 8 = -2$ ✓

The symbol $\stackrel{?}{=}$ is used with each check.

EXAMPLE 2   Solving from a table   Using a table, solve $4x - 8 = 4$.

SOLUTION   Table 23 contains inputs and outputs for $y = 4x - 8$. Look for the input, x, that matches the output $y = 4$.

TABLE 23

Input x	Output y
1	-4
2	0
3	4
4	8

The input is $x = 3$.

Check: $4(3) - 8 \stackrel{?}{=} 4$

$12 - 8 = 4$ ✓

EXAMPLE 3   Solving from a table   Using a table for $y = 2^x$, solve the first equation and make estimates of the solutions to the next two equations. Refine your estimates to the nearest tenth, using guess and check on a calculator.

a. $2^x = 8$ **b.** $2^x = 6$ **c.** $2^x = 12$

SOLUTION   Table 24 contains inputs and outputs for $y = 2^x$.

TABLE 24

Input x	Output y
1	2
2	4
3	8
4	16

a. $x = 3$
   Check: $2^3 = 8$

b. $x \approx 2.6$
   Check: $2^{2.6} \approx 6$

c. $x \approx 3.6$
   Check: $2^{3.6} \approx 12$

Solving Equations from a Table and a Graph

EXAMPLE 4   Solving from a table and graph   Solve $5 - 1.5x = 8$ from a table and a graph.

SOLUTION   Table 25 contains inputs and outputs for $y = 5 - 1.5x$. The output $y = 8$ does not appear in the table, but the table suggests that there could be such an output for an input smaller than 1. We draw a graph from the table. On the graph, an output of $y = 8$ matches an input of $x = -2$.

TABLE 25

Input x	Output $y = 5 - 1.5x$
1	3.5
2	2
3	0.5
4	-1

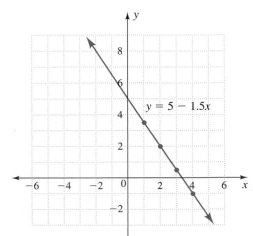

$y = 5 - 1.5x$

Check: $5 - 1.5(-2) \stackrel{?}{=} 8$

$5 - (-3) = 8$ ✓   **FIGURE 18**

In Section 1.3, we worked backwards on an input-output table to find the output when $x = 0$. We can find other values as well.

EXAMPLE 5   Working backwards on a table   Use changes on the input-output table in Example 4 to find the output for $x = -2$.

SOLUTION   In Table 26, we add changes columns to the table from Example 4. Next, we work backwards to find outputs for $x = 0$, $x = -1$, and $x = -2$. For $x = -2$, $y = 8$, confirming our graphical result in Example 4.

TABLE 26

Change in x	Input x	Output $y = 5 - 1.5x$	Change in y
1 ⟨	−2	8	⟩ −1.5
1 ⟨	−1	6.5	⟩ −1.5
1 ⟨	0	4	⟩ −1.5
1 ⟨	1	3.5	⟩ −1.5
1 ⟨	2	2	⟩ −1.5
1 ⟨	3	0.5	⟩ −1.5
	4	−1	

Solution Set

In Examples 1 to 4, there was one solution. For many equations, there is more than one solution. For other equations, there is no solution. Solutions may be listed in a solution set.

DEFINITION OF SOLUTION SET

> The **solution set** to an equation is the set of all numbers that, when substituted for the variable(s), make a true statement.

In Example 6, the equations have different numbers of solutions.

EXAMPLE 6   Solving equations with tables and graphs: suspension bridge cables, continued   Solve the following three equations to find the position, x, on the road surface for the given cable height.

a. The cable height is 8 feet: $\frac{4}{125}x^2 = 8$

b. The cable touches the surface: $\frac{4}{125}x^2 = 0$

c. The cable height is −4 feet (that is, 4 feet below the surface): $\frac{4}{125}x^2 = -4$

SOLUTION   The table and graph from Example 5 of Section 1.4 are reproduced here as Table 27 and Figure 19.

TABLE 27 $y = \frac{4}{125}x^2$

Distance from Bridge Center, x (feet)	Height, y (feet)
−20	12.8
−10	3.2
0	0.0
10	3.2
20	12.8

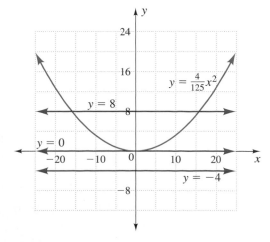

FIGURE 19

a. To solve $\frac{4}{125}x^2 = 8$, we look in the y column in Table 27 for the first pair of numbers between which 8 falls: 12.8 and 3.2. The input must be between $x = -20$ and $x = -10$. We might estimate $x = -15$ ft:

$$\frac{4}{125}(-15)^2 = 7.2 \qquad \text{Too low}$$

We try $x = -16$:

$$\frac{4}{125}(-16)^2 = 8.192$$

The output is close to 8 at $x = -16$. We write $x \approx -16$.

In the graph in Figure 19, the line $y = 8$ intersects the curve $y = \frac{4}{125}x^2$ just to the left of $x = -15$. From the graph, $x \approx -16$.

There is a second position in the table where the output is 8 and a second point on the graph where $y = 8$. By reasoning similar to that above, a second solution is at $x \approx 16$. The set of approximate solutions is $\{-16, 16\}$.

b. From Table 27, the solution to $\frac{4}{125}x^2 = 0$ is $x = 0$. To solve $\frac{4}{125}x^2 = 0$ from the graph in Figure 19, we find $y = 0$ (the x-axis) and look for where $y = 0$ intersects the graph. The solution is at the origin, $(x, y) = (0, 0)$. The solution set is $\{0\}$.

c. The solution to $\frac{4}{125}x^2 = -4$ cannot be found from Table 27. There are no negative outputs from the equation $y = \frac{4}{125}x^2$. The line $y = -4$ does not intersect the graph of $y = \frac{4}{125}x^2$. There are no real-number solutions to $\frac{4}{125}x^2 = -4$. *The solution set is empty.* We write either an empty set $\{\ \}$ or the symbol $\varnothing$.

___

Think about it 1: Do $\{0\}$ and $\{\ \}$ have the same meaning?

EXAMPLE 7   Using a calculator   Solve part a of Example 6 on a graphing calculator:
a. by building a table
b. by graphing

SOLUTION   **a.** To solve by creating a table, place the equation $y = \frac{4}{125}x^2$ in $\boxed{\text{Y} =}$. Go to [TBLSET]. Use -20 for **TblStart**, and let 10 be the change (**ΔTbl**). Display the calculator table, and compare it with Table 27. Now go back to [TBLSET] and change to a **TblStart** of -17 and a change of 0.5. Display the calculator table, and compare it with Figure 20. The output is close to 8 at $x = -16$.

X	Y1
-17	9.248
-16.5	8.712
-16	8.192
-15.5	7.688
-15	7.2
-14.5	6.728
-14	6.272

Y1◻4/125X²

FIGURE 20

To obtain a closer solution, reset the change to a smaller value. Follow similar steps to find the second solution, $x \approx 16$.

b. To solve by graphing, place the equation $y = \frac{4}{125}x^2$ in $\boxed{\text{Y} =}$. Enter $y = 8$ as a second equation. Go to $\boxed{\text{WINDOW}}$. Set the minimum for x at -25 and the maximum at 25, with a scale of 5. Set the minimum for y at -8 and the maximum at

24, with a scale of 4. Graph, and compare the result with Figure 21. Trace to each intersection of $y = \frac{4}{125}x^2$ and $y = 8$.

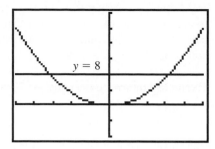

FIGURE 21 −25 to 25 on X, −8 to 24 on Y

The intersections both suggest that the approximate solution set is $\{-16, 16\}$.

Example 8 shows how graphing prompts us to look for multiple solutions.

EXAMPLE 8 Solving from a table and a graph

a. Solve $x^3 - x - 1 = -1$.

b. How might we be confident that we have found all the solutions?

SOLUTION **a.** Table 28 has inputs starting at 1. The table shows a solution at $x = 1$. However, in Figure 22, the graph of $y = x^3 - x - 1$ bends. When we also graph $y = -1$, its line passes three times through the graph for $y = x^3 - x - 1$.

TABLE 28

x	y
1	−1
2	5
3	23
4	59

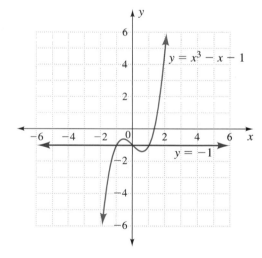

FIGURE 22

We expand Table 28, this time in a calculator table (Figure 23), starting with $x = -3$ and using 1 for the change inputs.

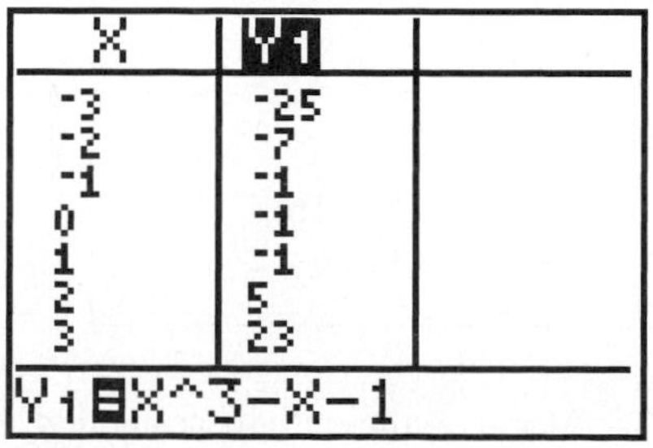

X	Y1	
-3	-25	
-2	-7	
-1	-1	
0	-1	
1	-1	
2	5	
3	23	

Y1∎X^3−X−1

FIGURE 23

The calculator table in Figure 23 shows three solutions, $x = -1$, $x = 0$, and $x = 1$, in agreement with the graph. The solution set is $\{-1, 0, 1\}$.

b. For values of x greater than 3, the x^3 term gets large quickly, causing the graph to continue upward. For x smaller than -3, the negative values in the x^3 term cause the graph to continue downward. It is probable that there are no other solutions to $x^3 - x - 1 = -1$. ▬

Example 9 returns to the equation $2^x = x^2$, mentioned at the beginning of the section.

EXAMPLE 9 **Solving an equation with tables and graphs**   Solve the equation $2^x = x^2$.

a. Complete Tables 29 and 30. For what values of x are the outputs the same? The first four rows of the first table are completed in case you have forgotten the meaning of negative and zero exponents.

TABLE 29

x	$y = 2^x$
-3	$2^{-3} = \frac{1}{8} = 0.125$
-2	$2^{-2} = \frac{1}{4} = 0.25$
-1	$2^{-1} = \frac{1}{2} = 0.50$
0	$2^0 = 1$
1	
2	
3	
4	
5	

TABLE 30

x	$y = x^2$
-3	
-2	
-1	
0	
1	
2	
3	
4	
5	

b. On the same axes, graph $y = 2^x$ and $y = x^2$ for integer inputs -3 to 5. What are the points of intersection?

c. Write the solution set to $2^x = x^2$.

SOLUTION   **a.** The completed versions are shown in Tables 31 and 32.

TABLE 31

x	$y = 2^x$
-3	$2^{-3} = \frac{1}{8} = 0.125$
-2	$2^{-2} = \frac{1}{4} = 0.25$
-1	$2^{-1} = \frac{1}{2} = 0.50$
0	$2^0 = 1$
1	$2^1 = 2$
2	$2^2 = 4$
3	$2^3 = 8$
4	$2^4 = 16$
5	$2^5 = 32$

TABLE 32

x	$y = x^2$
-3	9
-2	4
-1	1
0	0
1	1
2	4
3	9
4	16
5	25

At $x = 2$, the outputs are both 4. At $x = 4$, the outputs are both 16. The tables show two solutions: $x = 2$ and $x = 4$. Is this all of the solutions?

b. Figure 24 shows the two points of intersection: $(2, 4)$ and $(4, 16)$. Like the tables, the graph gives the solutions $x = 2$ and $x = 4$. However, there is a third point of intersection with x between -1 and 0. This intersection is slightly closer to $x = -1$ than to $x = 0$. We might estimate the intersection as closer to $-\frac{3}{4}$ than $-\frac{1}{2}$ and write $x \approx -0.75$.

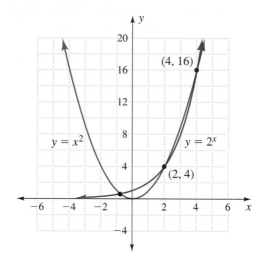

FIGURE 24

Check: $2^{(-0.75)} \stackrel{?}{=} (-0.75)^2$

$0.595 \approx 0.563$ ✓

c. The set of approximate solutions is $\{-0.75, 2, 4\}$.

Think about it 2: Graph the equations $y = x^2$ and $y = 2^x$ on a graphing calculator. Trace and zoom to find the solution between -1 and 0. How might we have noticed this solution from the tables?

Building and Solving Equations with Guess and Check

Our last example shows use of a table and guess and check to find an answer to a word problem and to build an equation for the word problem. When guess and check fails to give a solution, the work provides an estimate of the solution.

POLYA'S PROBLEM-SOLVING STEPS   A good guess-and-check process follows Polya's four problem-solving steps: understand, plan, carry out the plan, and check.

- We move toward *understanding* by reading carefully.
- We *plan* by considering what might be reasonable inputs and preparing a table in which to record our guesses.
- We *carry out the plan* by working through the problem with the chosen inputs.
- We *check* by comparing the result with the conditions or requirements of the original problem.

In Example 10, we use a table to organize our steps and record our guesses so that we learn from each guess. *Each row in the table represents a guess.*

EXAMPLE 10   Building equations with guess and check: appliance repair cost   J's Appliance Repair charges a \$22 service call fee and \$28 per hour for the repair. A repair required \$85 in parts. How long was the repair job if the total bill was \$205? Set up a guess and check. Solve the problem. Write the equation.

SOLUTION   The phrases "how long" and "\$28 per hour" tell us that the independent variable should be the number of hours. In Table 33, we set up hours as the variable to guess.

To write the equation, we let x be the number of hours and work through the table with x. We add the three costs—the charge at \$28 per hour, the service call fee, and the parts—to obtain the total cost:

$$28x + 22 + 85 = 205$$

TABLE 33

Guess (hours)	Charge at $28 per Hour (dollars)	Service Call Fee (dollars)	Parts (dollars)	Total Cost (dollars)
2	$2(28) = 56$	22	85	$56 + 22 + 85 = 163$ too low
4	$4(28) = 112$	22	85	$112 + 22 + 85 = 219$ too high
3	$3(28) = 84$	22	85	$84 + 22 + 85 = 191$ too low
3.5	$3.5(28) = 98$	22	85	$98 + 22 + 85 = 205$ correct
x	$x(28)$	22	85	205

Think about it 3: In Example 10, how do we know that a guess of 10 hours is too large?

GUESSING STRATEGIES   Example 10 illustrates two important points about guessing stategies.

First, *any number can be used to get started*. Although making an estimate of the answer is a good idea, the first guess does not need to be close to the correct answer. If you have trouble estimating in an unfamiliar situation, any number—such as the day of the month on which your birthday occurs—will get you started. The first guess will show whether the column headings are reasonable and whether the solution to the problem is larger or smaller than the guess.

Second, *doubling or halving may be useful in guessing possible solutions*. In Example 10, we doubled 2 hours (too low) to get 4 hours (too high). To find a middle number, we halved the distance between 2 and 4 and selected 3 as the next guess.

ANSWER BOX

Warm-up: 1. 1, 4, 9, 16, 25, 36, 49, 64, 81, 100, 121, 144, 169, 196, 225 **2.** 2, 4, 8, 16, 32, 64, 128, 256 **3. a.** 4 **b.** -4 **c.** -4 **d.** -4 **Think about it 1:** No; $\{0\}$ is the set containing the number 0, whereas $\{\ \}$ is a set with nothing in it. **Think about it 2:** $x \approx -0.767$. At $x = -1$, the output for x^2 is larger than that for 2^x. At $x = 0$, the output for x^2 is smaller than that for 2^x. Thus, a solution must lie between $x = -1$ and $x = 0$. **Think about it 3:** 10 hours at $28 per hour is $280, larger than the total bill of $208.

1.5   Exercises

In each of Exercises 1 to 4, make a table for the first equation. Use the table to solve for x in the other two equations.

1. $y = 3x - 4$; $3x - 4 = -1$, $3x - 4 = 5$ $x = 1, x = 3$

2. $y = 5x - 7$; $5x - 7 = 3$, $5x - 7 = 13$ $x = 2, x = 4$

3. $y = 4 - 3x$; $4 - 3x = -14$, $4 - 3x = 16$ $x = 6, x = -4$

4. $y = 2 - 3x$; $2 - 3x = 8$, $2 - 3x = 14$ $x = -2, x = -4$

In Exercises 5 and 6, solve with a table.

5. a. $1 - 0.25x = 2$ **b.** $1 - 0.25x = 0.75$
         $x = -4$ $x = 1$

6. a. $3 - \frac{1}{2}x = 5$ **b.** $3 - \frac{1}{2}x = 1$
         $x = -4$ $x = 4$

In Exercises 7 to 10, solve with a graph.

7. a. $3 - x = 1.5$ **b.** $3 - x = -0.5$
         $x = 1.5$ $x = 3.5$

8. a. $4 - 2x = 3$
  $x = 0.5$
9. a. $\frac{1}{2}x + 2 = 1$
  $x = -2$
10. a. $\frac{1}{2}x - 1 = -2$
  $x = -2$

b. $4 - 2x = 7$
  $x = -1.5$
b. $\frac{1}{2}x + 2 = 0.5$
  $x = -3$
b. $\frac{1}{2}x - 1 = -0.5$
  $x = 1$

In Exercises 11 to 16, solve with a table. Check the number of solutions with a calculator graph.

11. a. $x^2 + x = 0$
  $x = 0$ or $x = -1$
c. $x^2 + x = -1$
  no real-number solution
12. a. $3x - x^2 = 0$
  $x = 0$ or $x = 3$
c. $3x - x^2 = 3$
  no real-number solution
13. a. $4 - x^2 = 0$
  $x = 2$ or $x = -2$
c. $4 - x^2 = -5$
  $x = -3$ or $x = 3$
14. a. $x^2 + 4 = 0$
  no real-number solution
c. $x^2 + 4 = 4$
  $x = 0$
15. a. $2 - x - x^2 = 2$
  $x = -1$ or $x = 0$
c. $2 - x - x^2 = 4$
  no real-number solution
16. a. $9 - x^2 = 5$
  $x = -2$ or $x = 2$
c. $9 - x^2 = 0$
  $x = -3$ or $x = 3$

b. $x^2 + x = 6$
  $x = -3$ or $x = 2$
d. $x^2 + x = -0.25$
  $x = -0.5$
b. $3x - x^2 = -4$
  $x = -1$ or $x = 4$
d. $3x - x^2 = 2.25$
  $x = 1.5$
b. $4 - x^2 = 5$
  no real-number solution
d. $4 - x^2 = 4$
  $x = 0$
b. $x^2 + 4 = 13$
  $x = -3$ or $x = 3$
d. $x^2 + 4 = 8$
  $x = 2$ or $x = -2$
b. $2 - x - x^2 = 0$
  $x = -2$ or $x = 1$
d. $2 - x - x^2 = -4$
  $x = -3$ or $x = 2$
b. $9 - x^2 = 8$
  $x = -1$ or $x = 1$
d. $9 - x^2 = 10$
  no real-number solution

In Exercises 17 to 20, solve with a calculator table. Check with a calculator graph as needed. Write answers in a solution set.

17. a. $2 + x - 2x^2 = 2$
  $\{0, \frac{1}{2}\}$
c. $2 + x - 2x^2 = -4$
  $\{-1.5, 2\}$
18. a. $2 - x - 3x^2 = 2$
  $\{-\frac{1}{3}, 0\}$
c. $2 - x - 3x^2 = 3$
  $\{\ \}$
19. a. $2 - 2x - 3x^2 = 1$
  $\{-1, \frac{1}{3}\}$
c. $2 - 2x - 3x^2 = 2\frac{1}{3}$
  $\{-\frac{1}{3}\}$
20. a. $2 - 3x - 2x^2 = 4$
  $\{\ \}$
c. $2 - 3x - 2x^2 = 2$
  $\{-1\frac{1}{2}, 0\}$

b. $2 + x - 2x^2 = 0$
  $\approx \{1.3, -0.8\}$
d. $2 + x - 2x^2 = 3$
  $\{\ \}$
b. $2 - x - 3x^2 = 0$
  $\{-1, \frac{2}{3}\}$
d. $2 - x - 3x^2 = -2$
  $\{-1\frac{1}{3}, 1\}$
b. $2 - 2x - 3x^2 = -3$
  $\{-1\frac{2}{3}, 1\}$
d. $2 - 2x - 3x^2 = 2$
  $\{-\frac{2}{3}, 0\}$
b. $2 - 3x - 2x^2 = 3$
  $\{-1, -\frac{1}{2}\}$
d. $2 - 3x - 2x^2 = 0$
  $\{-2, \frac{1}{2}\}$

The body mass index compares weight and height. An index in the range of 19 to 24 is recommended for good health. In Exercises 21 to 24, use the formula

$$I = \frac{W(704.5)}{H^2}$$ to complete the tables for the given heights. The index, I, is in

terms of weight in pounds, W, and height in inches, H. (Hint: Use the calculator table with the height substituted for H.) Round to the nearest tenth.

21. For height 62 inches:

Weight (lb)	Index
100	18.3
110	20.2
120	22.0
130	23.8
140	25.7

22. For height 66 inches:

Weight (lb)	Index
110	17.8
120	19.4
130	21.0
140	22.6
150	24.3

23. For height 70 inches:

Weight (lb)	Index
130	18.7
140	20.1
150	21.6
160	23.0
170	24.4

24. For height 74 inches:

Weight (lb)	Index
150	19.3
160	20.6
170	21.9
180	23.2
190	24.4

Use the tables in Exercises 21 to 24 to solve the equations in Exercises 25 to 28.

25. a. $23.8 = \dfrac{W(704.5)}{62^2}$
  $W = 130$ lb
b. $20.2 = \dfrac{W(704.5)}{62^2}$
  $W = 110$ lb

26. a. $21.0 = \dfrac{W(704.5)}{66^2}$
  $W = 130$ lb
b. $24.3 = \dfrac{W(704.5)}{66^2}$
  $W = 150$ lb

27. a. $23.0 = \dfrac{W(704.5)}{70^2}$
  $W = 160$ lb
b. $24.4 = \dfrac{W(704.5)}{70^2}$
  $W = 170$ lb

28. a. $19.3 = \dfrac{W(704.5)}{74^2}$
  $W = 150$ lb
b. $23.2 = \dfrac{W(704.5)}{74^2}$
  $W = 180$ lb

29. Explain how to use a table to evaluate the equation $y = 3x + 4$ at $x = 2$. Find output for input 2.

30. Explain how to use a graph to evaluate the equation $y = 3x + 4$ at $x = 2$. Find y-coordinate for point of intersection of $y = 3x + 4$ and $x = 2$.

31. Explain how to use a graph to solve the equation $3x + 4 = 7$. Find x-coordinate for point of intersection of $y = 3x + 4$ and $y = 7$.

32. Explain how to use a table to solve the equation $3x + 4 = 7$. Find input for output 7.

Complete each sentence in Exercises 33 to 38 with "finding the dependent variable" or "finding the independent variable."

33. Solving the equation $3x + 4 = -2$ is ___.
  finding the independent variable
34. Locating the output in a table when we are given the input is ___. finding the dependent variable

35. Locating y, given the graph and x, is ___.
  finding the dependent variable
36. Evaluating the equation $y = 3x + 4$ for $x = -2$ is ___.
  finding the dependent variable
37. Locating x, given the graph and y, is ___.
  finding the independent variable
38. Locating the input in a table when we are given the output is ___. finding the independent variable

In Exercises 39 to 44, set up a guess-and-check table to solve each puzzle problem without an equation. In the last row, let your guess be x.

39. Because of a delay in service, the store manager agrees to sell a new washing machine for $350. This includes the delivery charge ($24.95) and 7% sales tax; no tax is paid on the delivery charge. How much should he record as the purchase price? $303.79

40. Janelle makes $\frac{1}{2}$-ounce chocolate pieces for a business friend. She charges \$0.45 each to make the chocolates and to wrap them in foil. Special labels for the chocolates cost \$8.50 per 1000 labels. She can buy the labels only in sets of 1000 (and discards labels that are not used). How many chocolates can she make for the friend's budget of \$500?   1073 pieces

41. A truck is rented for three days. The charge per day is \$37. The cost per mile is \$0.25. The rental agency gives a 5% discount because of a misaligned headlamp. If the payment after the discount is \$215.65, how many miles were driven?   464 miles

42. Three sides of a triangle add to 91. One side is three more than six times the smallest. The third side is three less than seven times the smallest. What is the measure of each side?   6.5, 42, 42.5

43. A \$1 million estate is divided three ways. The first part is \$50,000 more than the second part. The third part is \$30,000 less than the first part. How many dollars are in each part?   \$360,000; \$310,000; \$330,000

44. The total cost of a meal includes tip, sales tax, and the cost of the meal itself. DeMya spent \$33.88 altogether. Her tip was 15% of the price of the meal. The tip was 2.5 times the sales tax. What was the price of the meal?   \$28.00

Find the solution set for Exercises 45 to 48 with a calculator table. Use guess and check to obtain solutions to the nearest tenth. Use a calculator graph to check the number of solutions.

45. a. $x^3 + 3x^2 - x = 3$
$\{-3, -1, 1\}$
b. $x^3 + 3x^2 - x = 6$
$\{-2.3, -2, 1.3\}$
c. $x^3 + 3x^2 - x = 0$
$\{-3.3, 0, 0.3\}$
d. $x^3 + 3x^2 - x = 7$
$\{1.4\}$

46. a. $x^3 - 3x^2 + x = -1$
$\{-0.4, 1, 2.4\}$
b. $x^3 - 3x^2 + x = -2$
$\{-0.6, 1.6, 2\}$
c. $x^3 - 3x^2 + x = 0$
$\{0, 0.4, 2.6\}$
d. $x^3 - 3x^2 + x = 3$
$\{3\}$

47. a. $x^3 + 3x^2 - 1 = -1$
$\{-3, 0\}$
b. $x^3 + 3x^2 - 1 = 3$
$\{-2, 1\}$
c. $x^3 + 3x^2 - 1 = 1$
$\{-2.7, -1, 0.7\}$
d. $x^3 + 3x^2 - 1 = 0$
$\{-2.9, \ 0.7, 0.5\}$

48. a. $x^3 + 3x^2 + 1 = 3$
$\{-2.7, -1, 0.7\}$
b. $x^3 + 3x^2 + 1 = 0$
$\{-3.1\}$
c. $x^3 + 3x^2 + 1 = 2$
$\{-2.9, -0.7, 0.5\}$
d. $x^3 + 3x^2 + 1 = 5$
$\{-2, 1\}$

In Exercises 49 to 52, find the intersections by graphing the left and right sides of each equation. The inputs for each point of intersection form the solution set. Round to the nearest hundredth.

49. $2^{(x-1)} = x^2$ $\{-0.58, 1, 6.32\}$ **50.** $3^x - 1 = x^3$ $\{-0.85, 0, 2, 3.22\}$;
see Answer Section.   see Additional Answers.

51. $3^{(x+1)} = 3 - x^2$ **52.** $2^{(x+1)} = 4 - x^2$
$\{-1.57, 0\}$; see Answer Section. $\{-1.86, 0.77\}$; see Additional Answers.

Parentheses have been added to exponent expressions to assist in graphing on the calculator.

1.6   Solving Equations and Formulas

Objectives

▮ Simplify expressions containing inverse operations.
▮ Apply the addition property of equations.
▮ Apply the multiplication property of equations.
▮ Solve equations using inverses.
▮ Solve formulas using inverses.

WARM-UP

From Example 5, Section 1.3: Which equations in Exercises 1 to 5 are not equivalent to the rule $x - y = 6$?

1. $y = x - 6$ **2.** $y + x = 6$ **3.** $y + x = -6$
                         not equivalent   not equivalent
4. $y + 6 = x$ **5.** $y - x = -6$

Simplify or remove parentheses in Exercises 6 to 10.

6. $9 - 4(x - 6)$ $33 - 4x$ **7.** $5 - 2(7 - x)$ $-9 + 2x$

8. $\frac{5}{9}(212 - 32)$ 100   **9.** $\frac{9}{5}(100) + 32$ 212

10. $\frac{n}{2}(2a + (n - 1)d)$, where $a = 5$, $d = 5$, and $n = 10$ 275

THIS SECTION REVIEWS inverse operations. It shows how the order of operations and inverse operations may help us plan the steps to follow in solving an equation or formula. We apply Polya's problem-solving steps in solving and checking equations and formulas.

Inverse Operations and Inverses

Inverse operations permit us to transform the equations in Warm-up Exercises 1–5 and compare them with $x - y = 6$.

▪ **INVERSE OPERATIONS**

An operation is undone by its inverse operation.

Subtraction is the inverse operation to addition:

$$a + b - b = a$$

Division is the inverse operation to multiplication:

$$a \cdot b \div b = a$$

Inverse operations permit us to change subtraction to addition of the opposite and to change division to multiplication by the reciprocal. Inverse operations permit us to transform the equations in Warm-up Exercises 1–5 and compare them with $x - y = 6$.

EXAMPLE 1   **Applying inverse operations**   Simplify by first changing to the inverse operation.

  a. $-8 - 4$

  b. $-8 \div (-4)$

  c. $-8 \div \frac{1}{4}$

  d. $8 - \left(-\frac{1}{4}\right)$

  e. $8 \div -\frac{1}{4}$

SOLUTION   **a.** $-8 - 4 = -8 + (-4) = -12$

  b. $-8 \div (-4) = -8 \cdot \left(-\frac{1}{4}\right) = +\frac{8}{4} = 2$

  c. $-8 \div \frac{1}{4} = -8 \cdot 4 = -32$

  d. $8 - \left(-\frac{1}{4}\right) = 8 + \frac{1}{4} = 8\frac{1}{4}$

  e. $8 \div -\frac{1}{4} = 8 \cdot (-4) = -32$ ▬

Because subtraction is the inverse operation to addition, *opposites* (numbers with opposite signs) are also called *additive inverses*.

▪ **ADDITIVE INVERSES**

Additive inverses add to zero:

$$a + (-a) = 0$$

A *reciprocal* is the number that multiplies x to give 1. The reciprocal of x is $\frac{1}{x}$, $x \neq 0$. The reciprocal of $\frac{a}{b}$ is $\frac{b}{a}$, $a \neq 0$, $b \neq 0$. Because division is the inverse operation to multiplication, reciprocals are also called *multiplicative inverses*.

■ MULTIPLICATIVE INVERSES

Multiplicative inverses multiply to 1:

$$x \cdot \frac{1}{x} = 1 \quad \text{and} \quad \frac{a}{b} \cdot \frac{b}{a} = 1, \quad x \neq 0, a \neq 0, b \neq 0$$

EXAMPLE 2 Using inverses Complete these statements.

a. $8 + \Box = 0$

b. $4 \cdot \Box = 1$

c. $\frac{5}{8} \cdot \Box = 1$

d. $\frac{2}{3} + \Box = 0$

e. $\frac{1}{3} \cdot \Box = 1$

f. $1\frac{1}{2} \cdot \Box = 1$

SOLUTION See the Answer Box. ■

Solving Equations with the Inverse Order of Operations

Computers and calculators can now solve complicated equations using algebraic notation. You might ask, *What is left to learn?* The answer is that you need to

- know what it means to solve an equation.
- understand basic properties of equation solving.
- estimate solutions.
- plan and think step by step through a sequence of operations.
- solve common formulas.

   In Section 1.5, we solved equations from tables and graphs. In Section 1.5, **solving an equation** meant *finding an input when given an output.* In this section, **solving an equation** means *using algebraic notation to isolate a variable on one side of the equation.* This process is based on the properties of equations.

PROPERTIES OF EQUATIONS   Recall that **equivalent equations** *have the same set of solutions.* We obtain equivalent equations using the properties of equations.

■ ADDITION PROPERTY OF EQUATIONS

Adding the same number to both sides of an equation produces an equivalent equation:

$$\text{If} \quad a = b, \quad \text{then} \quad a + c = b + c.$$

Because subtraction may be written as adding the opposite, this property also applies to subtraction:

$$\text{If} \quad a = b, \quad \text{then} \quad a - c = b - \text{c}.$$

■ MULTIPLICATION PROPERTY OF EQUATIONS

Multiplying both sides of an equation by the same nonzero number produces an equivalent equation:

$$\text{If} \quad a = b, \quad \text{then} \quad ac = bc \quad \text{for } c \neq 0.$$

Because division may be written as multiplying by the reciprocal, this property also applies to division:

$$\text{If} \quad a = b, \quad \text{then} \quad \frac{a}{c} = \frac{b}{c} \quad \text{for } c \neq 0.$$

EXAMPLE 3 **Solving equations with the properties of equations** Solve these equations using the addition and multiplication properties of equations.

a. $x - 3 = -4$

b. $\dfrac{x}{2} = 8$

SOLUTION   **a.** $x - 3 = -4$ Add 3 to both sides.

$x - 3 + 3 = -4 + 3$ Note: $-3 + 3 = 0$

$x = -1$

Check: $-1 - 3 \overset{?}{=} -4$

$-4 = -4 \checkmark$

b. Multiply both sides by the denominator (or least common denominator) to eliminate the denominator.

$\dfrac{x}{2} = 8$ Multiply both sides by 2.

$2\left(\dfrac{x}{2}\right) = 2(8)$ Note: $2 \cdot \frac{1}{2} = 1$

$x = 16$

Check: $\dfrac{16}{2} \overset{?}{=} 8$

$8 = 8 \checkmark$

Think about it: Describe the role of inverses in solving the equations in Example 3.

Checking a solution should give an identity, $a = a$. An **identity** is *an algebraic sentence that is true for all inputs to the variables*. A **contradiction** is *an algebraic sentence that is false for all values of the variables*. A wrong answer substituted into an equation gives a contradiction. If you obtain $x = 2$ for $2x = 2(x + 2)$, substituting 2 for x gives $4 = 8$, a contradiction.

CHECKING A SOLUTION

> If a number or expression, x, is substituted into an equation or formula and the resulting equation simplifies to an identity, then x is a solution to the equation.

MAKING A PLAN   In Example 3, only one step—one inverse operation—was required to solve each equation for x. When there is more than one operation in an equation, we need to know *which step to do first to solve the equation*.

Example 4 illustrates how inverse operations can apply to a sequence of several operations.

EXAMPLE 4 **Applying inverse operations to daily life** Here are three steps we might take in leaving home on a cold morning:

Put on coat.   Fasten buttons.   Tie scarf.

What is the result when we reverse the order of the steps and do the inverse, or opposite, operation at each step?

SOLUTION The reverse order with inverse operations is

Untie scarf.   Unbutton coat.   Take off coat.

These are the steps we take when we arrive at our destination.   ▬

▬ By looking at the steps in the original activity (putting on, fastening, and tying), we can plan what to do to reverse the activity (untying, unfastening, taking off). This reversal of steps with the opposite operation is the basis for our plan in solving equations.

We will use the phrase **inverse order of operations** to describe *the reverse order of operations with a sequence of inverse operations.* In solving the equations in Examples 5 and 6, we list the order of operations on x and then write a plan using the inverse order of operations.

EXAMPLE 5 **Making a plan** Solve $5x + 6 = 9$ for x with the inverse order of operations.
a. *Understand:* Make an estimate.
b. *Plan:* Identify the order of operations on x in the equation $5x + 6 = 9$. List the inverse order of operations.
c. *Carry out the plan.*
d. *Check.*

SOLUTION **a.** *Understand:* Because $5(1) + 6 = 11$, $x \approx 1$ in $5x + 6 = 9$.

b. *Plan:* The x is multiplied by 5, and then 6 is added. The inverse order of operations is to subtract 6 and then divide by 5.

c. *Carry out the plan:*

$$5x + 6 = 9 \qquad \text{Subtract 6 from both sides.}$$
$$5x + 6 - 6 = 9 - 6$$
$$5x = 3 \qquad \text{Divide by 5 on both sides.}$$
$$\frac{5x}{5} = \frac{3}{5}$$
$$x = 0.6$$

d. *Check:* $5(0.6) + 6 \overset{?}{=} 9$
$$3 + 6 \overset{?}{=} 9$$
$$9 = 9 \checkmark$$ ▬

▬ In Example 5, we simplified $5x/5$ to x. We will use this type of simplification frequently in the following examples.

EXAMPLE 6 **Solving with a plan** Solve $9 - 4(x - 6) = 21$ for x with the inverse order of operations.
a. *Understand:* Make an estimate.
b. *Plan:* List the order of operations on x. List the inverse order of operations.
c. *Carry out the plan.*
d. *Check.*

SOLUTION **a.** *Understand:* Because subtraction is the same as adding the opposite, the equation $9 - 4(x - 6) = 21$ can be written $9 + (-4)(x - 6) = 21$. The value of

$(x - 6)$ needs to be negative in order to add an expression to 9 to get 21. Thus, x is smaller than 6.

b. *Plan:* The order of operations on x is

Subtract 6 from x.   Multiply the result by -4.   Add 9.

The inverse order of operations is

Subtract 9.   Divide the result by -4.   Add 6.

c. *Carry out the plan:*

$$9 + (-4)(x - 6) = 21 \qquad \text{Subtract 9 from both sides.}$$
$$9 + (-4)(x - 6) - 9 = 21 - 9$$
$$-4(x - 6) = 12 \qquad \text{Divide by } -4 \text{ on both sides.}$$
$$\frac{-4(x - 6)}{-4} = \frac{12}{-4}$$
$$x - 6 = -3 \qquad \text{Add 6 on both sides.}$$
$$x - 6 + 6 = -3 + 6$$
$$x = 3$$

d. *Check:* $9 - 4(3 - 6) \stackrel{?}{=} 21$
$$21 = 21 \quad \checkmark$$

Solving Equations

Inverse operations tell us that division by a/b is the same as multiplying by the reciprocal, b/a. We apply this fact in planning Example 7.

EXAMPLE 7 Solving equations   Solve $\frac{5}{9}(x - 32) = 100$ for x.

SOLUTION
$$\frac{5}{9}(x - 32) = 100 \qquad \text{Multiply by } \frac{9}{5} \text{ on both sides.}$$
$$\frac{9}{5} \cdot \frac{5}{9}(x - 32) = 100 \cdot \frac{9}{5}$$
$$x - 32 = 180 \qquad \text{Add 32 on both sides.}$$
$$x - 32 + 32 = 180 + 32$$
$$x = 212$$

Check: $\left(\frac{5}{9}\right)(212 - 32) \stackrel{?}{=} 100$
$$100 = 100 \quad \checkmark$$

EXAMPLE 8 Solving equations   Solve $3x + 8 = 2 - 3x$ for x.

SOLUTION   We choose steps that keep a positive coefficient on the variable term.

$$3x + 8 = 2 - 3x \qquad \text{Add } 3x \text{ to both sides.}$$
$$6x + 8 = 2 \qquad \text{Subtract 8 from both sides.}$$
$$6x = -6 \qquad \text{Divide both sides by 6.}$$
$$x = -1$$

Check: $3(-1) + 8 \stackrel{?}{=} 2 - 3(-1)$
$$-3 + 8 \stackrel{?}{=} 2 + 3$$
$$5 = 5 \quad \checkmark$$

It may be convenient or necessary to simplify one side of the equation first.

EXAMPLE 9 **Solving equations** Solve $2x = 2(x + 2)$ for x.

SOLUTION

$$2x = 2(x + 2) \quad \text{Simplify the right side.}$$

$$2x = 2x + 4 \quad \text{Subtract } 2x \text{ from each side.}$$

$$0 = 4$$

There is no solution. The algebraic sentence $0 = 4$ is always false.

The statement $0 = 4$ is a contradiction. Finding a contradiction when we solve an equation indicates that there is no solution.

Solving Formulas

The inverse order of operations works especially well in solving formulas. Solving formulas may be troublesome because the solution is an expression, often no simpler than the original formula.

In the following examples, we continue to perform operations on both sides.

EXAMPLE 10 **Solving formulas with a plan** Foresters measure the circumference (C) of a tree to find the radius (r) for estimating tree volume. Solve $C = 2\pi r$ for r, using the inverse order of operations.

SOLUTION ***Plan:*** The r is multiplied by 2π. To solve, we divide by 2π.

$$C = 2\pi r \quad \text{Divide by } 2\pi.$$

$$\frac{C}{2\pi} = \frac{2\pi r}{2\pi}$$

$$\frac{C}{2\pi} = r$$

Check: $C \stackrel{?}{=} 2\pi\left(\dfrac{C}{2\pi}\right)$

$$C = C \quad \checkmark$$

EXAMPLE 11 **Solving with a plan** Airlines and post offices have restrictions on the sizes of suitcases and boxes. Most involve finding the perimeter (P) of a rectangle with length (l) and width (w). Solve $P = 2l + 2w$ for w.

SOLUTION ***Plan:*** The w is multiplied by 2, and then $2l$ is added. To solve, we subtract $2l$ and divide by 2.

$$P = 2l + 2w \quad \text{Subtract } 2l.$$

$$P - 2l = 2l + 2w - 2l$$

$$P - 2l = 2w \quad \text{Divide by 2.}$$

$$\frac{P - 2l}{2} = \frac{2w}{2}$$

$$\frac{P - 2l}{2} = w$$

Check: $P \overset{?}{=} 2l + 2\left(\dfrac{P-2l}{2}\right)$

$\qquad\quad P \overset{?}{=} 2l + P - 2l$

$\qquad\quad P = P \quad \checkmark$

To solve the formula in Example 12, we have a choice. We may use the process for fractions shown in Example 7, or we may consider the factor $n/2$ as a multiplication by n and then a division by 2. Example 12 uses the latter procedure. Try the fraction form as an exercise.

EXAMPLE 12 **Solving with a plan**   Solve $S = \dfrac{n}{2}(2a + (n-1)d)$ for d.

SOLUTION   *Plan:* The order of operations on d is

Multiply by $(n-1)$.   Add $2a$.   Multiply the result by n.   Divide by 2.

Suggest that students practice this example on a separate piece of paper.

The inverse order of operations is

Multiply by 2.   Divide by n.   Subtract $2a$.   Divide by $(n-1)$.

$$S = \frac{n}{2}[2a + (n-1)d] \qquad \text{Multiply by 2.}$$

$$2S = n[2a + (n-1)d] \qquad \text{Divide by } n.$$

$$\frac{2S}{n} = 2a + (n-1)d \qquad \text{Subtract } 2a.$$

$$\frac{2S}{n} - 2a = (n-1)d \qquad \text{Divide by } n-1.$$

$$\frac{\dfrac{2S}{n} - 2a}{n-1} = d$$

The check is left as an exercise.

Although the formula for d can be simplified, simplifying is not recommended at this time. Ways to simplify such expressions will be introduced in Chapter 6.

▮ One important concept bears repeating: *To solve a formula means to isolate a particular variable on one side. The variable should not appear anywhere on the other side.* The thermodynamics formula $E = (T_h - T_c)/T_h$ contains the variable T_h twice. It would not be correct to solve for T_h and obtain

$$T_h = \frac{T_h - T_c}{E}$$

The variable T_h appears on both sides. Example 13 shows a correct solution of the formula for T_h. The solution is considerably more difficult than those in the other examples and is shown to illustrate the careful work needed in order to isolate one variable when it appears twice in a formula.

EXAMPLE 13 **Solving when the variable appears twice**   Solve $E = \dfrac{T_h - T_c}{T_h}$ for T_h.

SOLUTION *Plan:* Because T_h appears twice, we must rearrange the equation until T_h appears only once. We then apply the inverse order of operations.

Carry out the plan: We rearrange the equation until T_h appears only once.

$$E = \frac{T_h - T_c}{T_h}$$ Multiply both sides by T_h.

$$ET_h = T_h - T_c$$ Subtract T_h from both sides.

$$ET_h - T_h = -T_c$$ Factor T_h on the left side.

$$T_h(E - 1) = -T_c$$ Apply the inverse order of operations: divide by $(E - 1)$.

$$\frac{T_h(E - 1)}{E - 1} = \frac{-T_c}{E - 1}$$ Simplify.

$$T_h = \frac{-T_c}{E - 1}$$ Clear the negative sign: multiply by $\frac{-1}{-1}$.

$$T_h = \frac{-T_c(-1)}{(E - 1)(-1)}$$

$$T_h = \frac{T_c}{1 - E}$$

The check is left as an exercise.

On a symbolic calculator, replace T_c with C and T_h with H.

Example 13 is a good problem to use to test the symbolic manipulation feature on advanced calculators.

ANSWER BOX

Warm-up: Equations 2 and 3 are not equivalent. **6.** $33 - 4x$ **7.** $-9 + 2x$
8. 100 **9.** 212 **10.** 275 **Example 2: a.** -8 **b.** $\frac{1}{4}$ **c.** $\frac{8}{5}$ **d.** $-\frac{2}{3}$ **e.** 3 **f.** $\frac{2}{3}$
Think about it: In each equation, we performed the opposite operation with the inverse number to solve for x.

1.6   Exercises

1. Simplify by first changing to the inverse operation.

a. $-4 - 6$ $\ \ -10$ **b.** $-5 - (-3)$ -2 **c.** $-\frac{1}{2} - \left(-\frac{3}{2}\right)$ $\ 1$

d. $100 \cdot \frac{1}{4}$ $\ \ 25$ **e.** $\frac{5}{8} \div \frac{5}{8}$ $\ 1$ **f.** $-6 \div \frac{1}{3}$ $\ -18$

2. Simplify by first changing to the inverse operation.

a. $5 - (-6)$ $\ \ 11$ **b.** $-3 - (-3)$ $\ \ 0$ **c.** $\frac{1}{4} - \frac{3}{4}$ $\ \ -\frac{1}{2}$

d. $4 \div \frac{1}{4}$ $\ \ 16$ **e.** $-9 \div \frac{2}{3}$ $\ -13\frac{1}{2}$ **f.** $\frac{3}{4} \div \frac{3}{4}$ $\ 1$

3. Complete these statements.

a. $-3 + \square = 0$ **b.** $5 \cdot \square = 1$ $\frac{1}{5}$ **c.** $-2 + \square = 1$
$\quad 3$

d. $8 \cdot \square = 0$ **e.** $\frac{7}{8} \cdot \square = 1$ $\frac{8}{7}$ **f.** $\frac{3}{8} + \square = 0$
$\quad 0$ $\qquad\qquad\qquad\qquad\qquad\qquad -\frac{3}{8}$

4. Complete these statements.

a. $7 \cdot \square = 1$ $\frac{1}{7}$ **b.** $7 + \square = 0$ $\ -7$ **c.** $7 + \square = 1$ $\ -6$

d. $-\frac{2}{3} \cdot \square = 1$ **e.** $-12 + \square = 0$ **f.** $\frac{5}{8} + \square = 0$
$\quad -\frac{3}{2}$ $\qquad\qquad\quad 12$ $\qquad\qquad\qquad -\frac{5}{8}$

5. Solve these equations.

a. $x - 8 = -4$ **b.** $x + 3 = -6$ **c.** $x - 6 = -8$
$\quad x = 4$ $\qquad\quad x = -9$ $\qquad\quad x = -2$

d. $\frac{x}{2} = -6$ **e.** $\frac{3x}{4} = 21$ **f.** $\frac{3x}{8} = 12$
$\quad x = -12$ $\qquad\quad x = 28$ $\qquad\quad x = 32$

g. $-\frac{2}{3}x = 24$ **h.** $-\frac{1}{4}x = 8$ **i.** $\frac{7}{8}x = 35$
$\quad x = -36$ $\qquad\quad x = -32$ $\qquad\quad x = 40$

6. Solve these equations.

a. $x + 5 = -2$ **b.** $x - 12 = -3$ **c.** $x - 4 = -9$
$\quad x = -7$ $\qquad\quad x = 9$ $\qquad\quad x = -5$

d. $\frac{x}{4} = 12$ **e.** $\frac{3x}{4} = 27$ **f.** $-\frac{x}{8} = 40$
$\quad x = 48$ $\qquad\quad x = 36$ $\qquad\quad x = -320$

g. $\frac{2}{3}x = 18$ **h.** $-\frac{3}{8}x = 24$ **i.** $\frac{5}{8}x = 40$
$\quad x = 27$ $\qquad\quad x = -64$ $\qquad\quad x = 64$

In Exercises 7 to 12, which activities have a meaningful inverse order of operations? Describe the original activity and the inverse, if it exists. In which is the order not important?

7. Put on shirt, put on vest, put on jacket. Take off jacket, take off vest, take off shirt; dressing and undressing

8. Shut off car, remove key, open door. Close door, insert key, start car; arriving and departing

9. Turn on camera, aim, take picture. Inverse not meaningful; taking pictures

10. Fold clothes, put dishes away. Order not important

11. Dig hole, pile firewood, turn on sprinkler. Order not important

12. Put on sock, put on shoe, tie shoelace. Untie lace, take off shoe, take off sock; going barefoot

Write a plan and solve the equations in Exercises 13 to 16.

13. a. $2x - 4 = 8$ $x = 6$ **b.** $2x - 4 = 6$ $x = 5$

   c. $2x - 4 = -6$ $x = -1$ **d.** $2x - 4 = 0$ $x = 2$

14. a. $\frac{1}{2}x + 1 = 5$ $x = 8$ **b.** $\frac{1}{2}x + 1 = 3$ $x = 4$

   c. $\frac{1}{2}x + 1 = 0$ $x = -2$ **d.** $\frac{1}{2}x + 1 = -4$ $x = -10$

15. a. $\frac{5}{9}(x - 32) = -25$ $x = -13$ **b.** $\frac{5}{9}(x - 32) = 10$ $x = 50$

   c. $\frac{5}{9}(x - 32) = 5$ $x = 41$ **d.** $\frac{5}{9}(x - 32) = -30$ $x = -22$

16. a. $0.80(200 - A) = 88$ $A = 90$ **b.** $0.80(200 - A) = 56$ $A = 130$

   c. $0.80(200 - A) = 36$ $A - 155$ **d.** $0.80(200 - A) = 20$ $A = 175$

In Exercises 17 to 24, solve the equation for x.

17. $3x + 3 = 0$ $x = -1$ **18.** $3x + 3 = -9$ $x = -4$

19. $\frac{x}{3} - 1 = 4$ $x = 15$ **20.** $\frac{2x}{3} - 1 = 11$ $x = 18$

21. $\frac{3x}{4} + 5 = 23$ $x = 24$ **22.** $\frac{3x}{4} - 8 = 19$ $x = 36$

23. $\frac{3}{8}x - 4 = 8$ $x = 32$ **24.** $\frac{3}{8}x + 3 = 24$ $x = 56$

In Exercises 25 to 30, which algebraic sentences are identities (I) and which are contradictions (C)?

25. a. $4 = 4$ I   **b.** $8 = 3$ C

26. a. $6 = 0$ C   **b.** $R = R$ I

27. a. $x - 2 = 2x - x$ C   **b.** $x^2 = x \cdot x$ I

28. a. $2x = x + x$ I   **b.** $x + 4 = x - 4$ C

29. a. $x(x^2) = x^3$ I   **b.** $\frac{x}{4} = x\left(\frac{1}{2} - 0.25\right)$ I

30. a. $x - 1 = x$ C   **b.** $\frac{1}{2}x = x\left(0.75 - \frac{1}{4}\right)$ I

In Exercises 31 to 50, simplify one side of the equation before solving.

31. $x - 2(4 - x) = -17$ $x = -3$ **32.** $x - 2(4 - x) = 13$ $x = 7$

33. $x - 3(4 + x) = 6$ $x = -9$ **34.** $x - 3(4 + x) = -8$ $x = -2$

35. $x + 3(5 - x) = -9$ $x = 12$ **36.** $x + 3(5 - x) = 21$ $x = -3$

37. $8 - 3(9 - x) = -14.5$ $x = 1.5$ **38.** $7 - 2(8 - x) = -3.4$ $x = 2.8$

39. $14 - 9(x + 2) = 45.5$ $x = -5.5$ **40.** $17 - 7(x + 5) = 5.8$ $x = -3.4$

41. $37.45 = x + 0.07x$ $x = 35$ **42.** $50.40 = x + 0.05x$ $x = 48$

43. $46.17 = x + 0.15x + 0.065x$ $x = 38$

44. $63.70 = x + 0.15x + 0.075x$ $x = 52$

45. $2(x + 3) = 1 + 4x$ $x = 2.5$

46. $3(x - 3) = 3 + 11x$ $x = -1.5$

47. $4 - (x - 2) = 2(x + 6)$ $x = -2$

48. $3 - 2(x + 1) = -4x - 5$ $x = -3$

49. $2 - 3(x + 1) = x - 3$ $x = \frac{1}{2}$

50. $3 - (x - 1) = 6 + 5x$ $x = -\frac{1}{3}$

In Exercises 51 to 70, solve the formula for the indicated variable.

51. Distance: $D = rt$ for t $t = \frac{D}{r}$

52. Circumference: $C = \pi d$ for d $d = \frac{C}{\pi}$

53. Linear equation: $y = mx + b$ for b $b = y - mx$

54. Electronics: $I = \frac{E}{R}$ for E $E = IR$

55. Volume of sphere: $V = \frac{4\pi r^3}{3}$ for r^3 $r^3 = \frac{3V}{4\pi}$

56. Surface area of sphere: $A = 4\pi r^2$ for r^2 $r^2 = \frac{A}{4\pi}$

57. Thread on a screw: $pN = 1$ for p $p = \frac{1}{N}$

58. Area of triangle: $A - \frac{1}{2}bh$ for b $b = \frac{2A}{h}$

59. Area of trapezoid: $A = \frac{1}{2}h(a + b)$ for b $b = \frac{2A}{h} - a$

60. Geometric sequence: $a_n = a_1 r^{n-1}$ for a_1 $a_1 = \frac{a_n}{r^{n-1}}$

61. Arithmetic sequence: $a_n = a_1 + (n - 1)d$ for a_1 $a_1 = a_n - (n - 1)d$

62. Sum of geometric sequence: $S = \frac{a_1}{r - 1}$ for a_1 $a_1 = S(r - 1)$

63. Average: $A = \frac{a + b + c}{3}$ for a $a = 3A - b - c$

64. Arithmetic sequence: $a_n = a_1 + (n - 1)d$ for d $d = \frac{a_n - a_1}{n - 1}$

65. Sum of arithmetic sequence: $S = \frac{1}{2}n(a_1 + a_n)$ for a_1 $a_1 = \frac{2S}{n} - a_n$

66. Cost of skating party for 10 or more people: $C = 75 + 3.50(x - 10)$ for x $x = \frac{C - 75}{3.5} + 10$

67. Thermodynamics: $E = \frac{T_h - T_c}{T_h}$ for T_c $T_c = -ET_h + T_h$

68. Sound absorption: $T = \frac{0.16V}{S_e}$ for V $V = \frac{TS_e}{0.16}$

69. Sound velocity: $V = 344 + 0.6(T - 20)$ for T $T = \frac{V - 344}{0.6} + 20$

70. Musical acoustics: $f = (2n - 1)\left(\frac{V}{4L}\right)$ for V $V = \frac{4fL}{2n - 1}$

In Exercises 71 to 76, solve the formula for the indicated variable. Hint: Multiply both sides by the denominator.

71. $I = \dfrac{E}{R}$ for R $R = \dfrac{E}{I}$

72. $r = \dfrac{D}{t}$ for t $t = \dfrac{D}{r}$

73. $T = \dfrac{0.16V}{S_e}$ for S_e $S_e = \dfrac{0.16V}{T}$

74. $f = (2n - 1)\left(\dfrac{V}{4L}\right)$ for L $L = (2n - 1) \cdot \dfrac{V}{4f}$

75. $S = \dfrac{a_1}{r - 1}$ for r $r = \dfrac{a_1 + S}{S}$

76. $c = \dfrac{a}{a + b}$ for a $a = \dfrac{-bc}{c - 1}$ or $a = \dfrac{bc}{1 - c}$

For Exercises 77 to 81, check the formula solution by substituting the second formula into the first formula.

77. $D = rt$ for r; $r = \dfrac{D}{t}$

78. $I = prt$ for r; $r = \dfrac{I}{pt}$

79. $A = \dfrac{1}{2}h(a + b)$ for h; $h = \dfrac{2A}{a + b}$

80. $S = \dfrac{n}{2}[2a + (n - 1)d]$ for d; $d = \dfrac{\dfrac{2S}{n} - 2a}{n - 1}$

81. $E = \dfrac{T_h - T_c}{T_h}$ for T_h; $T_h = \dfrac{T_c}{1 - E}$

Hint: Division by a/b is equivalent to multiplication by b/a.

82. Repeat Example 12, solving $S = \dfrac{n}{2}[2a + (n - 1)d]$ for d, but this time undo the multiplication by $n/2$ by multiplying both sides by the reciprocal, $2/n$. $d = \dfrac{\dfrac{2S}{n} - 2a}{n - 1}$

■ Project

83. Inverse Operations

a. List the reverse order with inverse operations for these steps. What do the original steps describe? What does the inverse describe?

   Turn on tape player.   Open tape box.

   Take out tape.   Close tape box.

   Insert tape into tape player.   Press play.

   playing a cassette tape; removing a tape and putting it back in the box

b. Find and summarize a job-related order of operations and a meaningful inverse order of operations (example: delivery of petroleum-based fuel to a storage facility, including safety and environmental steps). Indicate any steps that must be taken that are not in the inverse.

■ 1   Chapter Summary

Vocabulary

For definitions and page references, see the Glossary/Index.

absolute value symbol	contradiction	inverse order of operations	quotient
addition property of equations	coordinate plane	irrational numbers	rational numbers
additive inverses	dependent variable	like terms	real numbers
associative property for addition	difference	multiplication property of equations	reciprocals
associative property for multiplication	distributive property of multiplication over addition	multiplicative inverse	scale
axes	equation	natural numbers	set
base	equivalent equations	numerical coefficient	simplify
braces	evaluate	opposites	solution set
brackets	exponent	order of operations	solving an equation
breaking mind set	expression	ordered pair	square root symbol
commutative property for addition	factor	origin	sum
commutative property for multiplication	factoring	parabola	term
consecutive	fraction bar (horizontal)	parentheses	undefined
constant	greatest common factor	percent	variable
constant term	grouping symbols	perpendicular	whole numbers
	identity	power of the base	x-axis
	independent variable	principal square root	x-coordinate
	input-output relationship	product	y-axis
	integers	quadrants	y-coordinate

Concepts

See the vocabulary list for important definitions. In this chapter, algebra is presented visually, numerically, and symbolically (with algebraic notation).

1.1 ■■ Problem Solving and Number Sense

Polya's problem-solving steps are to (1) understand the problem, (2) make a plan, (3) carry out the plan, and (4) check the solution and extend it to other situations.

Identify definitions and sets of numbers, properties of real numbers, and the order of operations.

Simplify expressions by applying the properties of real numbers or doing operations.

The square root of a negative number is undefined in the set of real numbers.

1.2 ■■ Input-Output Tables and Algebraic Techniques

The input-output relationship or rule for a table tells how to get from the input number to the output number across the table.

Simplify expressions by adding like terms, applying exponents, or changing fractions to lowest terms.

Evaluate formulas by stating the formula and the given information, substituting the information into the formula, estimating, and finding the result.

1.3 ■■ Problem Solving and Writing Algebraic Notation

Look for patterns, look at prior problems, and break mind set to find rules for input-output tables.

Use correct vocabulary in writing equations or expressions involving operations.

Identify independent and dependent variables in an application setting.

Write equations containing percents.

1.4 ■■ Coordinate Graphs

When graphing, place inputs (independent variables) on the horizontal axis and outputs (dependent variables) on the vertical axis.

To estimate the appropriate scale for a graph, find 10% of the difference between the highest number and lowest number to be graphed.

1.5 ■■ Solving Equations with a Table and Graph

Tables and graphs provide estimates of solutions, if not exact solutions. Graphs suggest the number of possible solutions.

Some equations cannot be solved with algebraic symbols yet can be solved with a graph.

Guess and check in a table is an effective way to understand a problem, solve a problem, or write equations to assist in solving a problem.

1.6 ■■ Solving Equations and Formulas

An operation is undone by its inverse operation

Subtraction and addition are inverse operations; multiplication and division are inverse operations.

A number or expression is the solution to an equation (or formula) if its substitution into the equation (or formula) results in $a = a$, an identity. A number or expression is not a solution to an equation (or formula) if its substitution into the equation (or formula) results in a contradiction.

---

 1 Review Exercises

For Exercises 1 to 12, use the vocabulary words from page 68.

1. List five properties of real numbers.   associative properties for addition and multiplication, commutative properties for addition and multiplication, distributive property for multiplication over addition

2. List five sets of numbers included in the real numbers.   integers, irrational numbers, natural numbers, rational numbers, whole numbers

3. List six grouping symbols.   absolute value, braces, brackets, fraction bar, parentheses, square root

4. List terms related to rectangular coordinate graphing. Draw a picture and use the terms to label the picture.   axes, coordinate plane, ordered pair, origin, quadrants, scale, *x*- and *y*-axes

5. List two terms that describe input and output variables.   independent variable, dependent variable

6. List seven terms describing numbers or letters or groupings of numbers and letters in algebraic notation.   base, constant, exponent, expression, numerical coefficient, term, variable, power of the base

7. List four terms describing results from operations.   sum, difference, product, quotient

8. List three terms that have to do with solving equations.   equivalent equations, reverse order of operations, solution set

9. List three terms that describe inputs and outputs.   input-output relationships, independent variable, dependent variable

10. List two terms that describe 2 and -2 as a pair of numbers.   additive inverses, opposite numbers

11. List two terms that describe $\frac{1}{2}$ and 2 as a pair of numbers.   multiplicative inverses, reciprocals

12. List three terms that describe answers or results from solving equations.   solution set, identity, contradiction

For Exercises 13 to 16, complete the tables.

	Input x	Input y	Output $x + y$	Output $x - y$
13. a.	-7	3	-4	-10
b.	3	-7	-4	10
c.	7	3	10	4
14. a.	-3	7	4	-10
b.	3	7	10	-4
c.	-3	-7	-10	4

	Input x	Input y	Output $x \cdot y$	Output $x + y$
15. a.	3	4	12	7
b.	-2	3	-6	1
c.	-7	-3	21	-10
d.	-3	5	-15	2
16. a.	2	-3	-6	-1
b.	4	-3	-12	1
c.	-1	-2	2	-3
d.	3	-1	-3	2

Show how the real-number properties let us do the computations in Exercises 17 to 20 mentally. Tell what property you use.   Answers may vary.

17. $\left(\frac{2}{3} + 1\frac{1}{2}\right) + \frac{1}{2}$ $2\frac{2}{3}$; associative property for addition

18. $\frac{3}{8} + \frac{1}{3} + \frac{5}{8}$ $1\frac{1}{3}$; commutative property for addition

19. $25 \cdot 13 \cdot 4$ 1300; commutative property for multiplication

20. $5(6 \cdot 7) = (5 \cdot 6) \cdot 7$
210; associative property for multiplication

Simplify the expressions in Exercises 21 to 26.

21. Puzzle problem: $-2^2 + 5 - 3(4 - 5)$ 4

22. Puzzle problem: $3\{6 - 2[8 - 3(9 - 12) + 3] - 1\}$
-105
23. Surface area of a metal can:
$2\pi(1.5 \text{ in.})^2 + 2\pi(1.5 \text{ in.})(4 \text{ in.})$ $16.5\pi \text{ in}^2$

24. Area of a trapezoid: $\frac{1}{2}(5 \text{ in.})(4.5 \text{ in.} + 7.6 \text{ in.})$ 30.25 in^2

25. Eighth term of an arithmetic sequence:
$\$20.00 + (8 - 1)(-\$1.75)$ $\$7.75$

26. Sum of eight terms of an arithmetic sequence:
$\frac{1}{2}(8)(\$20 + \$7.75)$ $\$111.00$

In Exercises 27 and 28, write the algebraic notation in words.

27. a. $3 - 2x$ The difference between three and two times x **b.** $3(x - 5)$ Three times the difference between x and five

28. a. $4x + 5$
The sum of four times x and five   **b.** $5(x + 3)$
Five times the sum of x and three

In Exercises 29 and 30, write the word phrases in algebraic notation.

29. Eight more than the quotient of a number and 15 $\frac{x}{15} + 8$

30. The product of six and the difference between 5 and a number $6(5 - x)$

In Exercises 31 and 32, eliminate parentheses as needed and add like terms.

31. $9 - 3(2x - 5y) - 4(5x - 3y)$ $-26x + 27y + 9$

32. $a(b^2 + 3b - 1) - 3b(a^2 - 2a + 1) - ab(a - b)$
$2ab^2 + 9ab - 4a^2b - a - 3b$

Evaluate the expressions in Exercises 33 and 34.

33. $a^2 - b^2$ for $a = -2$ and $b = -5$ -21

34. $a^2 + 2ab + b^2$ for $a = -3$ and $b = -4$ 49

Evaluate the formulas in Exercises 35 and 36.

35. $A = \pi r^2$ for $r = 3.5$ in. $A = 12.25\pi \text{ in}^2$

36. $A = \frac{1}{2}bh$ for $b = 4.5$ cm and $h = 6.5$ cm $A = 14.625 \text{ cm}^2$

Build an input-output table and a graph for the equations in Exercises 37 to 40.

37. $y = -x$ See Answer Section.

38. $y = 4 - 2x$ See Additional Answers.

39. $y = 30 - 3.5x$ See Answer Section.

40. $y = 150x + 300$ See Additional Answers.

In Exercises 41 to 46, solve the equations as directed.

41. $2x + 2(1.5) = 11$ for x $x = 4$

42. $25 = \frac{1}{2}(2.5)(x)$ for x $x = 20$

43. $40 = \frac{5}{9}(F - 32)$ for F $F = 104$

44. $-40 = \frac{5}{9}(F - 32)$ for F $F = -40$

45. $20 + (n - 1)(-1.75) = 0.75$ for n $n = 12$

46. $60 + 3.75(n - 5) = 150$ for n $n = 29$

In Exercises 47 to 50, solve the formulas for the indicated letter.

47. $I = Prt$ for r $r = \frac{1}{Pt}$

48. $A = \frac{1}{2}bh$ for b $b = \frac{2A}{h}$

49. $C = a + bY$ for b $b = \frac{C - a}{Y}$

50. $y = mx + b$ for b $b = y - mx$

In Exercises 51 and 52, find the greatest common factor for the terms shown and then write the expression as a product.

51. $6x^2 + 15x$
$3x, 3x(2x + 5)$

52. $12a^2b - 16ab^2$
$4ab, 4ab(3a - 4b)$

Solve the equations in Exercises 53 to 55 from the graphs in the figure and with algebraic notation.

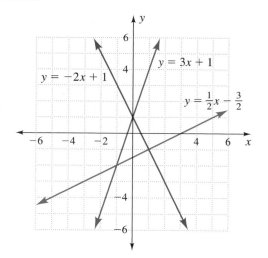

53. $3x + 1 = -2x + 1$ $x = 0$

54. $\frac{1}{2}x - \frac{3}{2} = -2x + 1$ $x = 1$

55. $3x + 1 = \frac{1}{2}x - \frac{3}{2}$ $x = -1$

56. Draw a picture or describe how these sets of numbers are related to each other: natural numbers, rational numbers, real numbers, integers, whole numbers.

57. Make a table and find a rule for each pattern.

   a. $7, 4, 1, -2, -5, \ldots$ $y = -3x + 10$

   b. $-20, -13, -6, 1, 8, \ldots$ $y = 7x - 27$

58. Name the letter in the figure that labels each of the following ordered pairs and estimate y for $x = 37$.

   a. $(-40, -40)$ A

   b. $(0, 32)$ D

   c. $(100, 212)$ K

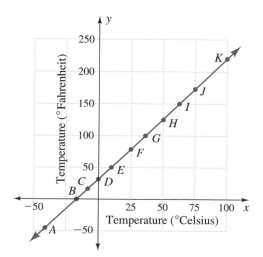

For $x = 37$, y is about 100 (or 98.6 exactly).

59. Graph $y = x^2$ and $y = -2x + 3$ on your calculator. Solve $x^2 = -2x + 3$. $x = -3$ and $x = 1$

60. The total commission on a small e-Bay sale is $0.30 plus 1.5% of the final sale price. Write an equation to find the total commission. What is the final sale price for a total commission of $0.75? $C = 0.30 + 0.015x$; $x = \$30$

61. The minimum balance in an escrow account (to pay taxes and insurance) for a mortgage loan must be at least 2/12 of the total annual payments from the account. Write an equation for the minimum balance for the account. Round decimals to the nearest thousandth. Find total annual payments if the escrow balance is $367.40.
E = escrow balance, x = total payments from account; $E = 0.167x$, $x = \$2200$

62. Error Analysis Simplify by hand and then on your calculator.

   a. $(6 - \sqrt{(9)}/(8 - 5)$ 5   **b.** $(6 - \sqrt{(9)})/(8 - 5)$ 1

   c. Find one error in either part a or part b. Explain how it changes the results. What assumptions will the calculator make?   In a, one set of parentheses is incomplete, so calculator assumes a parenthesis at the end, making numerator $\sqrt{9}$, not $6 - \sqrt{9}$.

Use guess and check in a table to solve the problems in Exercises 63 and 64.

63. Body mass index is calculated by multiplying weight in pounds by 704.5 and then dividing by the square of height in inches. A healthy range for body mass index is 19 to 24. To the nearest pound, for what weight will a 5-foot 5-inch person have an index of 19? An index of 24? $\approx$114 lb, $\approx$144 lb

64. The figure shows an angle of x degrees in a circle with radius r inches. The length of the arc opposite angle x is S, in inches. To find S, multiply π, r, and x and then divide by 180. In a circle with radius 10 inches, for what angle, x, will S be approximataely 15.7 inches? In the same circle, for what x will S exactly equal the radius, 10 inches? $\approx$90°, $\approx$57.3°

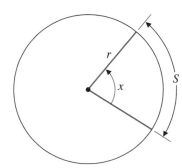

1 Chapter Test

1. Another name for additive inverse is _____.   opposite

2. A(n) _____ is used to locate a point on the coordinate graph.   ordered pair

3. A(n) _____ variable is the input, x, or set of numbers on the horizontal axis.   independent

4. Draw a pair of rectangular coordinate axes. Show an example of and/or label the following: origin, quadrant 4, y-axis.   See Answer Section.

5. Add, subtract, multiply, and divide (in the order given).

 a. -27 and -3 **b.** -1.5 and 0.5
   $-30, -24, 81, 9$ $-1, -2, -0.75, -3$
 c. $1\frac{1}{2}$ and $-\frac{1}{4}$
   $1\frac{1}{4}, 1\frac{3}{4}, -\frac{3}{8}, -6$

6. Describe how a property of the real numbers lets us do these computations mentally.

 a. $(348 + 295) + 105$ 748; associative property for addition

 b. $230 + 689 + 70$ 989; commutative property for addition

7. Simplify these expressions.

 a. $4\{2 - 3[4 + 5(2 - 5)] + 5\}$ 160

 b. -3^2 -9

 c. $(-3)^2$ 9

 d. $\frac{1}{2} \cdot 25(17 + 23)$ 500

 e. $3x - (8 - 3x)$ $6x - 8$

8. Write in words: $4(x - 3)$.   four times the difference between x and 3

9. Solve $A = \dfrac{a + b + c}{3}$ for b. $b = 3A - a - c$

10. Solve $A = \dfrac{h}{2}(a + b)$ for b. $b = \dfrac{2A}{h} - a$

11. Find the greatest common factor in $15x^2y - 39xy^2$. $3xy$

12. Use the graph in the figure to solve the following equations.

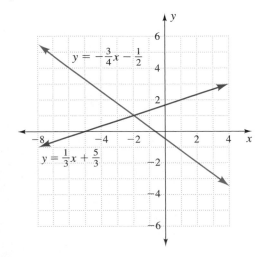

a. $\frac{1}{3}x + \frac{5}{3} = 0$ $x = -5$ **b.** $-\frac{3}{4}x - \frac{1}{2} = -2$ $x = 2$

c. $\frac{1}{3}x + \frac{5}{3} = -\frac{3}{4}x - \frac{1}{2}$ $x = -2$

13. Solve parts a, b, and c in Exercise 12 using algebraic notation.

14. Give an example of a whole number that is not a positive integer or a natural number.   zero

15. Evaluate the formula in Exercise 10 for $h = 1.1$ m, $a = 1.4$ m, and $b = 2.6$ m.   2.2 m²

16. Place into an input-output table: (3, 1), (6, 2), and (9, 3). For what x will $y = 2\frac{1}{3}$? $x = 7$

17. Find the rule for $-1, -5, -9, -13, -17, \ldots$. $y = -4x + 3$

18. Problem solving: Suggest how one could change decimal parts of a year into months—for example, 0.669 of a year.   1 year = 12 months, so multiply the decimal part by 12 to obtain 8 months.

19. A one-bedroom assisted living apartment is $2875 per month. An additional $800 is charged each month for personal care. Write an equation for the total cost of x months. If there is no change in cost, in how many years will life savings of $250,000 be used up? $C = 3675x$; 68 months or 5 years 8 months

20. Simplify by hand and then on your calculator:

 a. $5 - \sqrt{}(4)/2 - 3$ 1   **b.** $5 - \sqrt{}(4)/(2 - 3$ 7

 c. Find one error in either part a or part b. Explain how it changes the results. What assumptions will the calculator make?   In b, the denominator set of parentheses is incomplete, so the calculator assumes a parenthesis at the end of the expression, dividing $\sqrt{4}$ by $2 - 3 = -1$ instead of by 2.

21. The basic charge for renting a car for three days is $45.59 per day. The company adds a fuel charge of $0.149 per mile, and then a 7.25% sales tax. On top of this, a $0.97 per day vehicle license fee is added. The total paid is $159.03. How many miles was the car driven? Solve with a guess-and-check table.   ≈59 miles

Inequalities, Functions, and Linear Functions

More than 1250 people died in the United States during the heat wave of 1980. East of the Rocky Mountains, summer combines high temperatures with high humidities, causing even greater heat. The heat index chart in Table 1 shows the apparent temperature for various air temperatures (in °F) and relative humidities (%). The heat index (or apparent-temperature) values were found for shady, light-wind conditions. Exposure to full sunshine can increase the heat index values by up to 15°. Apparent temperatures above 105° are considered dangerous for lengthy exposure or physical activity.

This chapter covers special input-output relationships called functions and ways to write these relationships in function notation. The focus is on functions related to straight lines: linear functions. We look at number patterns, special straight lines (horizontal lines and the line $y = x$), and absolute value in terms of the function concept. The chapter opens with inequalities, line graphs, and intervals.

TABLE 1 *Apparent Temperature*

Relative Humidity (%)

Air Temperature (°F)	0	10	20	30	40	50	60	70	80	90	100
125	111	123	141								
120	107	116	130	148							
115	103	111	120	135	151						
110	99	105	112	123	137	150					
105	95	100	105	113	123	135	149				
100	91	95	99	104	110	120	132	144			
95	87	90	93	96	101	107	114	124	136		
90	83	85	87	90	93	96	100	106	113	122	
85	78	80	82	84	86	88	90	93	97	102	108
80	73	75	77	78	79	81	82	85	86	88	91
75	69	70	72	73	74	75	76	77	78	79	80

Source: *Heat Wave*, U.S. Department of Commerce, National Oceanic and Atmospheric Administration, NOAA/PA 85001.

2.1 Inequalities, Line Graphs, and Intervals

Objectives

■ Identify inequality signs for given statements.

■ Given an inequality or a line graph, find the other.

■ Given two inequality statements, write a compound inequality.

■ Given an inequality, a line graph, or an interval, find the other two.

■ Write equations with inequalities as conditions.

Inequalities are used in a variety of ways in this chapter: writing domain and range, evaluating functions, describing conditional settings.

WARM-UP

1. A birthday party costs $50 for up to 10 children and $4.50 for each additional child. What is the cost for 15 children? $72.50

2. An air compressor rents for $50 for up to 6 hours and $25 for each additional hour or part of an hour. What is the cost for 12 hours? $200

3. A telephone call costs $1.50 for the first minute and $0.10 for each additional minute. What is the cost of a 30-minute call? $4.40

IN THIS SECTION, we review inequalities, line graphs, and intervals. Inequalities will be applied to logic, compound statements, and conditional inputs. Intervals will be used to describe calculator inputs and outputs.

Inequalities

Have you ever rented a DVD and had to pay a late fee when you returned it? If so, you have experienced a conditional setting.

A **condition** is *a limitation on the inputs, derived from the problem setting, that determines what equation is used to solve the problem.* We write conditions with inequalities. Suppose rental of a DVD costs $3.00 for two days and a $4 late fee for each additional day. If x is the input, the DVD rental has the conditions that inputs are $0 \leq x \leq 2$ days or $x > 2$ days. (More on conditional settings in Examples 8, 9, and 10.)

An **inequality** is *a statement that one quantity is greater than or less than another quantity.* The chart below gives a summary of the inequality symbols.

Inequality Symbol	Meaning
$<$	is less than (is to the left of on the number line)
$>$	is greater than (is to the right of on the number line)
$\leq$	is less than or equal to
$\geq$	is greater than or equal to

When the number line is vertical, $<$ implies *below* and $>$ implies *above*.

EXAMPLE 1   Using inequality symbols   Describe how the numbers are related. Tell what inequality symbols make each of the statements true.

a. $7 \ \square \ 6$ b. $-\frac{1}{2} \ \square \ -\frac{1}{4}$ c. $0.1 \ \square \ 0.05$ d. $-3 \ \square \ -3$

SOLUTION   a. 7 is greater than 6; $>$ or $\geq$ b. $-\frac{1}{2}$ is less than $-\frac{1}{4}$; $<$ or $\leq$

c. 0.1 is greater than 0.05; $>$ or $\geq$ d. -3 is equal to -3; $\geq$ or $\leq$

Line Graphs

We must be able to read and describe sets of numbers on a number line in order to effectively use a graphing calculator. Inequalities provide a convenient way to describe these sets of numbers. The equality sign in $\leq$ and $\geq$ means that we include the number in our set, which we indicate on a graph by using a solid dot. For the other inequality signs, $<$ and $>$, we use a small circle to indicate exclusion of the number.*

EXAMPLE 2 Recognizing line graphs Match the inequality and the graph of the set of numbers it describes.

a. $x > 2$ **b.** $x < 0$ **c.** $y > -2$ **d.** $y < 0$

e. $x \geq -2$ **f.** $y \leq 2$ **g.** $y > 2$ **h.** $x \leq 2$

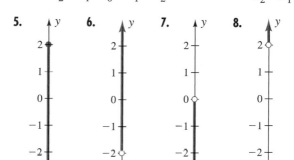

SOLUTION   **a.** 3   **b.** 2   **c.** 6   **d.** 7   **e.** 1   **f.** 5   **g.** 8   **h.** 4

LOGIC AND INEQUALITIES The definitions of the inequalities $\leq$ and $\geq$ use the word *or*. Many definitions and other mathematical statements include the logic phrase *a or b* or *a and b*.

LOGIC: *a* OR *b* VERSUS *a* AND *b*

> The phrase *a or b* is true if either the condition *a* is true or the condition *b* is true or if both conditions are true. The inequality $a \leq b$ is true if either *a* is less than *b* or *a* is equal to *b*, because the definition of the symbol $\leq$ contains the word *or*.
>
> The phrase *a and b* is true if both *a* and *b* are true. Because a number cannot be both greater than and equal to another number, the word *and* cannot be used in describing the inequality $\geq$.

The line graphs for $x \geq 0$ and for $x < 3$ are shown in the first two number lines in Figure 1. Together, $x \geq 0$ *and* $x < 3$ require that we include only numbers in both sets (the numbers larger than or equal to zero and at the same time smaller than 3), as shown in the third number line of Figure 1. This inequality may be written as one statement, $0 \leq x < 3$, which is read *x is between 0 and 3, including 0*.

---

*A bracket,] or [, may be used instead of a dot, and a parenthesis,) or (, may be used instead of a small circle.

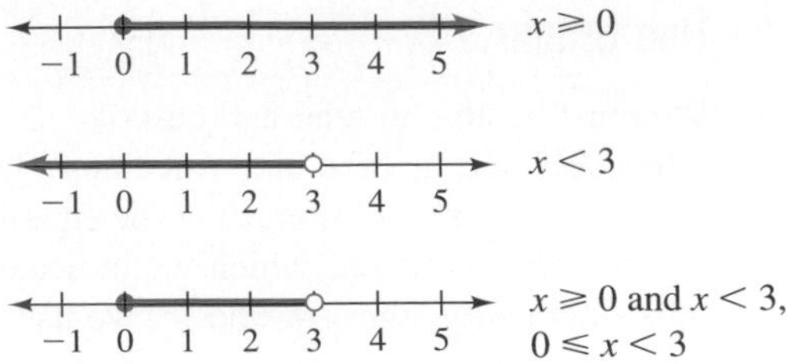

$x \geq 0$

$x < 3$

$x \geq 0$ and $x < 3$,
$0 \leq x < 3$

FIGURE 1

Example 3 illustrates the use of *and* and *or* with inequalities.

EXAMPLE 3   **Recognizing line graphs**   Match the inequality statement with its graph. Then describe the inequalities in parts a to h in words.

a. $x > 3$ or $x < 1$ **b.** $-1 < x < 3$

c. $x < 3$ or $x > 5$ **d.** $1 \leq x \leq 5$

e. $x < 5$ or $x > 3$ **f.** $x \leq 3$ or $x \geq 5$

g. $x \leq 1$ or $x \geq 3$ **h.** $x < 5$ and $x > -1$

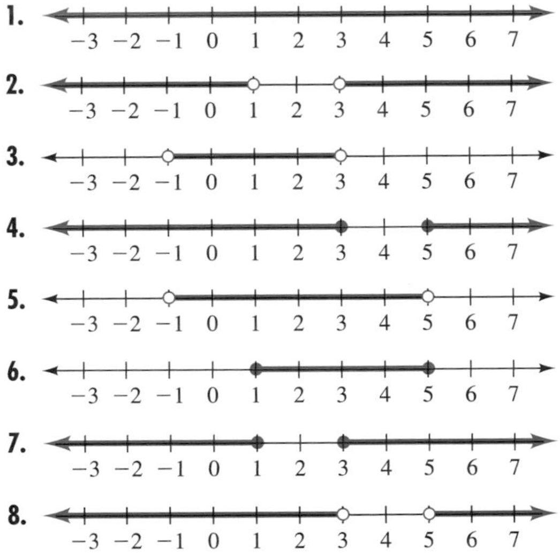

1.

2.

3.

4.

5.

6.

7.

8.

SOLUTION   **a.** 2; the set of numbers less than 1 or greater than 3

b. 3; the set of numbers between -1 and 3

c. 8; the set of numbers less than 3 or greater than 5

d. 6; the set of numbers between 1 and 5, including 1 and 5

e. 1; the set of numbers less than 5 or greater than 3. The line graph in 1 includes all numbers because the set of numbers to the right of 3 is joined with the set of numbers to the left of 5, to cover the entire number line.

f. 4; the set of numbers less than or equal to 3 or greater than or equal to 5

g. 7; the set of numbers less than or equal to 1 or greater than or equal to 3

h. 5; the set of numbers between -1 and 5

COMPOUND INEQUALITIES   The statement $0 \leq x < 3$ in Figure 1 is a compound inequality. A **compound inequality** *combines two inequality statements into one statement with two inequality signs in the same direction.* The combined inequality must be true even if the middle number or expression is left out. When we leave x out, $0 < 3$ is still true. The inequality signs in a compound inequality must always have the same direction. The statement $0 < x > -3$ is not acceptable.

EXAMPLE 4 Writing compound inequalities   Write each compound inequality as two inequality statements. Write each pair of inequalities as a compound inequality.

    a. $-1 < x < 3$ **b.** $1 \leq x \leq 5$ **c.** $x < 5$ and $x > -1$

    d. The set of numbers greater than or equal to negative one and less than or equal to five

    e. The set of numbers between negative one and four

    f. The set of numbers between one and five, including one and five

SOLUTION **a.** $-1 < x$ and $x < 3$. Note that $-1 < x$ is also $x > -1$.

    b. $1 \leq x$ and $x \leq 5$. Note that $1 \leq x$ is also $x \geq 1$.

    c. $-1 < x < 5$ **d.** $-1 \leq x \leq 5$

    e. $-1 < x$ and $x < 4$ **f.** $1 \leq x$ and $x \leq 5$

Interval Notation

We now consider interval notation, which you may find easier to use than inequality notation.

    An **interval** is *a set containing all the numbers between its endpoints as well as one endpoint, both endpoints, or neither endpoint*. Interval notation indicates the inclusion or exclusion of endpoints by the use of brackets or parentheses. Brackets, [ ], indicate that the endpoints are included in the set. Parentheses, ( ), are used when the endpoints are excluded from the set. We may mix brackets and parentheses in one interval, as shown in Table 2.

TABLE 2

Inequality	Interval	Words	Line Graph
$2 \leq x \leq 5$	[2, 5]	Set of numbers between 2 and 5, including 2 and 5	
$2 \leq x < 5$	[2, 5)	Set of numbers between 2 and 5, including 2	
$2 < x \leq 5$	(2, 5]	Set of numbers between 2 and 5, including 5	
$2 < x < 5$	(2, 5)	Set of numbers between 2 and 5	

EXAMPLE 5 Writing intervals   Write each as an interval.

    a. $-2 < x \leq 4$

    b. $-3 \leq x < 8$

    c. The set of numbers between -3 and 6, including -3

    d. The set of numbers between 7 and 11, including 7 and 11

    e.

    f.

SOLUTION **a.** $(-2, 4]$ **b.** $[-3, 8)$ **c.** $[-3, 6)$

    d. $[7, 11]$ **e.** $(8, 11]$ **f.** $(-4, 0)$

Think about it 1: In part f of Example 5, the answer was $(-4, 0)$. What else does the notation $(-4, 0)$ describe?

Inequalities and intervals give shorter ways to write calculator window information.

EXAMPLE 6 **Writing inequalities and intervals** Write the following calculator window settings as inequalities and intervals.

  a. $\mathbf{Xmin} = -10, \mathbf{Xmax} = 10$

  b. $\mathbf{Xmin} = 0, \mathbf{Xmax} = 20$

  c. $\mathbf{Ymin} = -4, \mathbf{Ymax} = 24$

SOLUTION **a.** $-10 \le x \le 10$, x on the interval $[-10, 10]$

  b. $0 \le x \le 20$, x on the interval $[0, 20]$

  c. $-4 \le y \le 24$, y on the interval $[-4, 24]$ ▬

▮ To write inequalities such as $x > 5$ or $x \le 2$ with interval notation, we need an **infinity sign**, ∞. **Infinite** means *without bound*. We draw arrows on the ends of a line or axis to indicate that the line goes on without bound. A positive sign before the infinity sign means that a horizontal line extends infinitely to the right; a negative sign indicates that it extends infinitely to the left (see Figure 2).

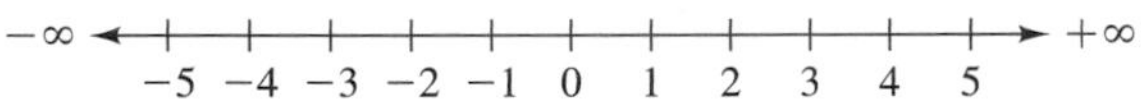

$$-\infty \; \longleftarrow\!\!\!\begin{array}{c}|\quad|\quad|\quad|\quad|\quad|\quad|\quad|\quad|\quad|\quad|\\ -5\;\; -4\;\; -3\;\; -2\;\; -1\;\; 0\;\; 1\;\; 2\;\; 3\;\; 4\;\; 5\end{array}\!\!\!\longrightarrow \; +\infty$$

FIGURE 2

Because infinity is not a number, it cannot be included in a set of numbers. As shown in the intervals in Table 3, we always use a parenthesis next to an infinity sign.

TABLE 3

Inequality	Interval	Words	Line Graph
$x \ge 5$	$[5, +\infty)$	Set of numbers greater than 5, including 5	
$x < 2$	$(-\infty, 2)$	Set of numbers less than 2	
$-\infty < x < +\infty$	$(-\infty, +\infty)$	Set of all real numbers, $\mathbb{R}$	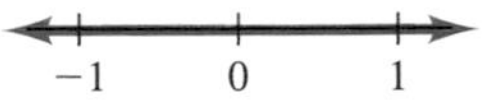

EXAMPLE 7 **Writing intervals** Write each as an interval.

  a. $x \le 6$

  b. $x < -3$

  c. The set of numbers greater than negative two

  d. The set of numbers less than or equal to four

  e.

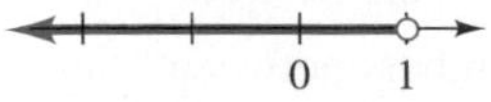

  f.

SOLUTION **a.** $(-\infty, 6]$ **b.** $(-\infty, -3)$ **c.** $(-2, +\infty)$

  d. $(-\infty, 4]$ **e.** $(-\infty, 1)$ **f.** $[-3, +\infty)$ ▬

Applications with Conditions

CONDITIONAL SETTINGS When application settings indicate rules that change for different inputs, we use inequalities to describe the conditions.

EXAMPLE 8 Finding conditions in problem settings   How are these examples the same? Use inequalities to describe the conditions on the inputs.
a. Birthday party: $50 for up to 10 children, $4.50 for each additional child
b. Air compressor rental: $50 for up to 6 hours, $25 for each additional hour
c. Telephone call: $1.50 for the first minute, $0.10 for each additional minute

SOLUTION   In each setting, there is a flat fee followed by a charge for additional children or units of time. If x is the input,

a. $0 < x \le 10$ children; $x > 10$ children

b. $0 < x \le 6$ hours; $x > 6$ hours

c. $0 < x \le 1$ minute; $x > 1$ minute ▬

EQUATIONS WITH CONDITIONS   Making a few entries in an input-output table can help us write equations.

EXAMPLE 9 Writing equations for conditional settings   Table 4 shows the DVD rental cost for 1 to 5 days where renting a DVD costs $3.00 for two days and a $4 late fee for each additional day. Write two equations that describe the total cost of the rental.

TABLE 4

Number of Days	Total Cost (dollars)
1	3
2	3
3	$3 + 4(1) = 7$
4	$3 + 4(2) = 11$
5	$3 + 4(3) = 15$

SOLUTION   The total cost is a flat fee for the first two days and then rises by $4 for each additional day. Let x be the total number of rental days and y be the total cost.

$$y = 3, \qquad\qquad 0 < x \le 2$$
$$y = 3 + 4(x - 2), \quad x > 2$$

The inequality following the equation shows the days to which the equation applies. ▬

Think about it 2: Why is this equation wrong: $y = 3 + 4x$ for $x > 2$?

Student Note: We graph conditional equations in Section 2.6.

EXAMPLE 10 Writing equations for conditional settings   Write a set of equations for each of these settings.
a. Birthday party: $50 for up to 10 children, $4.50 for each additional child
b. Air compressor rental: $50 for up to 6 hours, $25 for each additional hour or part of an hour
c. Telephone call: $1.50 for the first minute, $0.10 for each additional minute or part of a minute

SOLUTION   **a.** Let x = total number of children; let y = total cost in dollars.

$$y = 50, \qquad\qquad 0 < x \le 10$$
$$y = 50 + 4.50(x - 10), \quad x > 10$$

b. Let x = total hours of rental. The phrase *part of an hour* means that any fraction of an hour is rounded up to the next highest hour. Let y = total cost in dollars.

$$y = 50, \qquad\qquad 0 < x \le 6$$

$$y = 50 + 25(x - 6), \quad x > 6$$

c. Let x = total minutes talked. The phrase *part of a minute* means that any fraction of a minute is rounded up to the next highest minute. Let y = total cost in dollars.

$$y = 1.50, \qquad\qquad 0 < x \le 1$$

$$y = 1.50 + 0.10(x - 1), \quad x > 1$$

ANSWER BOX

Warm-up: **1.** \$72.50 **2.** \$200 **3.** \$4.40 **Think about it 1:** The ordered pair (x, y), where $x = -4$, $y = 0$. To avoid confusion, we often say or write *interval* or *ordered pair* before the notation. **Think about it 2:** For 3 days ($x = 3$), the rental charge would be $3 + 3(4) = \$15$. The \$4 is paid only for the number of days after the second day, $x - 2$, not all the days, x.

2.1 Exercises

Write the value of each side of the statements in Exercises 1 and 2. Then write $>$, $=$, or $<$ between the values.

1. a. $\frac{1}{2} + \frac{1}{2} \boxed{>} \frac{1}{2} \cdot \frac{1}{2}$ **b.** $\frac{1}{2} \div \frac{1}{2} \boxed{=} \frac{1}{2} + \frac{1}{2}$

c. $\frac{1}{4} + \frac{1}{4} \boxed{>} \frac{1}{2} \cdot \frac{1}{2}$ **d.** $\frac{1}{2} - \frac{1}{4} \boxed{<} \frac{7}{8} - \frac{3}{8}$

e. $0.3 \cdot 0.5 \boxed{<} 0.4 \cdot 0.4$

f. $1.5 - 3.5 \boxed{>} -2.5 - 0.5$

2. a. $\frac{3}{4} \div \frac{1}{4} \boxed{>} \frac{2}{3} \div \frac{1}{3}$ **b.** $\frac{5}{8} - \frac{1}{4} \boxed{>} \frac{1}{4} \cdot \frac{1}{2}$

c. $\frac{3}{4} + \frac{1}{4} \boxed{=} \frac{2}{3} + \frac{1}{3}$ **d.** $\frac{3}{8} + \frac{1}{8} \boxed{>} \frac{2}{3} - \frac{1}{3}$

e. $0.8 + 0.7 \boxed{<} 2.5 - 0.8$

f. $0.3 - 0.7 \boxed{>} 0.5 - 1.1$

In Exercises 3 to 16, complete the table for each inequality. Assume the variable is x unless otherwise noted. Answers may vary.

	Inequality	Line Graph	Inequality in Words
3.	$x \le 2$		x is less than or equal to 2
4.	$y > 2$		y is greater than 2
5.	$-1 < x < 5$		x is between -1 and 5
6.	$y \le 1$ or $y \ge 3$		y is less than or equal to 1 or y is greater than or equal to 3
7.	$x \ge -1$		x is greater than or equal to -1
8.	$y < -2$		y is less than -2
9.	$-2 < x < 4$		x is between -2 and 4

	Inequality	Line Graph	Inequality in Words
10.	$x \le 1$ or $x \ge 3$	−4 −3 −2 −1 0 1 2 3 4 5 6	x is less than or equal to 1 or x is greater than or equal to 3
11.	$x < 2$ or $x > 4$	−4 −3 −2 −1 0 1 2 3 4 5 6	x is less than 2 or x is greater than 4
12.	$-1 < x < 4$	−4 −3 −2 −1 0 1 2 3 4 5 6	x is between −1 and 4
13.	$x < -3$ or $x > 2$	−6 −5 −4 −3 −2 −1 0 1 2 3 4	x is less than −3 or x is greater than 2
14.	$-\infty < x < +\infty$	−6 −5 −4 −3 −2 −1 0 1 2 3 4	x is any real number
15.	$x \le -4$ or $x \ge 1$	−6 −5 −4 −3 −2 −1 0 1 2 3 4	x is less than or equal to −4 or x is greater than or equal to 1
16.	$-2 \le x \le 5$	−5 −4 −3 −2 −1 0 1 2 3 4 5	x is between −2 and 5, including −2 and 5

In Exercises 17 to 26, write each compound inequality as two inequality statements. If appropriate, write each pair of inequalities as a compound inequality.

17. $-3 < x < 4$
$x > -3$ and $x < 4$
18. $x < 4$ and $x \ge 1$
$1 \le x < 4$
19. $x > 4$ or $x < 1$
not appropriate
20. $x > 1$ or $x > 3$
not appropriate
21. $x > -1$ and $x \le 5$
$-1 < x \le 5$
22. $2 \le x \le 5$
$x \ge 2$ and $x \le 5$
23. $-1 \le x < 1$
$x \ge -1$ and $x < 1$
24. $x > -3$ and $x < 1$
$-3 < x < 1$
25. $3 < x$ and $4 > x$
$3 < x < 4$
26. $x > -2$ and $x \le 5$
$-2 < x \le 5$

In Exercises 27 and 28, write the calculator window settings as inequalities and intervals.

27. a. **Xmin** $= -25$, **Xmax** $= 15$
$-25 \le x \le 15$, x on the interval $[-25, 15]$
b. **Ymin** $= -10$, **Ymax** $= 20$
$-10 \le y \le 20$, y on the interval $[-10, 20]$
28. a. **Xmin** $= -20$, **Xmax** $= 20$
$-20 \le x \le 20$, x on the interval $[-20, 20]$
b. **Ymin** $= 0$, **Ymax** $= 30$
$0 \le y \le 30$, y on the interval $[0, 30]$

Complete this chart for Exercises 29 to 40. Assume the variable is x.

	Inequality	Interval	Words	Line Graph
29.	$-3 < x < 5$	$(-3, 5)$	Set of numbers greater than −3 and less than 5	−3 0 5
30.	$-1 < x \le 3$	$(-1, 3]$	Set of numbers greater than −1 and less than 3, including 3	−1 0 3
31.	$-4 < x \le 2$	$(-4, 2]$	Set of numbers greater than −4 and less than 2, including 2	−4 0 2
32.	$-2 \le x \le 4$	$[-2, 4]$	Set of numbers greater than −2 and less than 4, including −2 and 4	−2 0 4
33.	$x > 5$	$(5, \infty)$	Set of numbers greater than 5	0 5
34.	$x \le 5$	$(-\infty, 5]$	Set of numbers less than or equal to 5	0 5
35.	$x < -2$	$(-\infty, -2)$	Set of numbers less than −2	−2 0
36.	$x \ge -2$	$[-2, \infty)$	Set of numbers greater than or equal to −2	−2 0
37.	$x \le -3$	$(-\infty, -3]$	Set of numbers less than or equal to −3	−3 0

	Inequality	Interval	Words	Line Graph
38.	$x < 3$	$(-\infty, 3)$	Set of numbers less than 3	
39.	$x \geq 4$	$[4, +\infty)$	Set of numbers greater than or equal to 4	
40.	$x > -1$	$(-1, +\infty)$	Set of numbers greater than -1	

In Exercises 41 to 50, write equations with conditions on the inputs.

41. A fax costs \$4 for up to two pages and \$0.50 for each page thereafter. What is the total cost, y, of x pages?
$y = 4 + 0.5(x - 2)$ for $x > 2$; $y = 4$ for $0 < x \leq 2$

42. A college transcript costs \$5 for the first copy and \$1 for each additional copy. If x is the total number of copies, what is the total cost, y? $y = 5 + 1(x - 1)$ for $x > 1$; $y = 5$ for $x = 1$

43. Rental of a weed cutter costs \$20 for 3 hours plus \$5 for each additional hour or part of an hour. What is the total cost, y, of x hours of rental? $y = 20 + 5(x - 3)$ for $x > 3$, x rounded up to next integer; $y = 20$ for $0 < x \leq 3$

44. Rental of a pressure washer costs \$30 for the first 4 hours and \$10 per additional hour or part of an hour. If x is the total hours rented, what is the total cost, y? $y = 30 + 10(x - 4)$ for $x > 4$, x rounded up to next integer; $y = 30$ for $0 < x \leq 4$

45. A rental car costs \$65 for the first 100 miles and \$0.15 per mile thereafter. If x is the total miles driven, what is the total cost, y? $y = 65 + 0.15(x - 100)$ for $x > 100$; $y = 65$ for $0 < x \leq 100$

46. A taxi cab costs \$5 to hire and ride the first $\frac{1}{2}$ mile. Each additional $\frac{1}{10}$ mile costs \$0.75. If x is the total miles traveled, what is the total cost, y? $y = 5 + 7.5(x - 0.5)$ for $x > 0.5$; $y = 5$ for $0 < x \leq 0.5$

47. A party at an ice rink costs \$85 for up to 10 skaters and \$4.75 for each additional skater. What is the total cost, y, for x skaters? $y = 85 + 4.75(x - 10)$ for $x > 10$; $y = 85$ for $0 < x \leq 10$

48. A party at a roller skating rink costs \$40 for 10 skaters and \$4.50 for each additional skater. What is the total cost, y, for x skaters? $y = 40 + 4.5(x - 10)$ for $x > 10$; $y = 40$ for $0 < x \leq 10$

49. A cell phone costs \$19.95 each month for the first 100 minutes and \$0.25 for each additional minute or portion of a minute. What is the total cost, y, for x minutes a month? $y = 19.95 + 0.25(x - 100)$ for $x > 100$, x rounded up to next integer; $y = 19.95$ for $0 < x \leq 100$

50. An e-mail account costs \$9.95 each month for the first 10 hours and \$2 for each additional hour or part of an hour. What is the total cost, y, for x hours a month?
$y = 9.95 + 2(x - 10)$ for $x > 10$, x rounded up to next integer; $y = 9.95$ for $0 \leq x \leq 10$

In Exercises 51 and 52, match each definition or related statement with a vocabulary word. Record any unfamiliar words and their definitions in your study notes.

51. Choose from
inequality, logic phrase "a or b," $\geq$, $<$

   a. is less than $<$

   b. A statement that one quantity is greater than or less than another quantity   inequality

   c. is greater than or equal to $\geq$

   d. Either a is true or b is true or both are true.
   logic phrase "a or b"

52. Choose from
unacceptable statement, dot on a number line, compound inequality, logic phrase "a and b," small circle on a number line

   a. Both a and b are true.   logic phrase "a and b"

   b. The number at this position on the number line is excluded from the set.   small circle on a number line

   c. The number at this position on the number line is included in the set.   dot on a number line

   d. It combines two inequality statements into one statement with two inequality signs in the same direction.
   compound inequality

   e. $6 > 4 < 5$
   unacceptable statement

2.2	Functions

Objectives

- Identify a function.
- Apply the vertical-line test.
- Write a function in function notation.
- Evaluate functions written in function notation.
- Find the domain and range of a function.

WARM-UP

1. Complete the table for the equation $x = y^2$.

Parabola Opening to the Right, $x = y^2$

x	y
0	0
1	1
1	-1
4	2
4	-2

In Exercises 2 and 3, make a table for integer inputs on the interval $[-2, 1]$.

2. $y = 5x$ Outputs are $-10, -5, 0, 5.$

3. $y = x - 4$ Outputs are $-6, -5, \quad 4, \quad 3.$

THIS SECTION INTRODUCES functions and function notation. We examine tables and graphs of functions as well as methods of identifying functions. We find the domain and range of functions.

Functions

For years, researchers have sought a cure for cancer, ways to encourage people to recycle, and methods to direct government spending to increase employment. A key element in this research is finding an association between action and results. In mathematics, we have a similar goal: identifying associations where, if we input x, only one outcome, y, can result.

■ FUNCTION

A **function** is a relationship or association where for each value of input x there is exactly one value of output y.

Chapter 1 indicated that in some cases the output depends on the input, or y *depends on x*. Now we can use the term *function* to say that the output is a function of the input, or *y is a function of x.*

IDENTIFYING FUNCTIONS   In Examples 1 and 2, we look at tables and graphs to find out which relationship is a function and can be written *y is a function of x.* Both a table and a graph are helpful in finding whether *there is exactly one y for each x.*

EXAMPLE 1 Identifying a function: transcript costs   The first copy of a transcript costs $3. Each additional copy costs $2. Let x be the number of transcripts ordered and y be the total cost in dollars. Table 5 and the graph in Figure 3 describe the cost equation, $y = 3 + 2(x - 1)$. Is the cost, $y = 3 + 2(x - 1)$, a function of the number of copies ordered?

TABLE 5 $y = 3 + 2(x - 1)$

Transcripts Ordered, x	Cost, y (dollars)
1	3
2	5
3	7
4	9
5	11

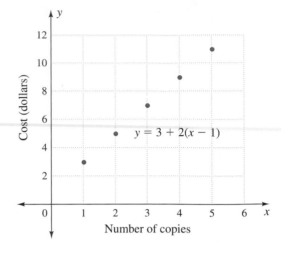

FIGURE 3

SOLUTION   For each number of transcripts ordered there is exactly one cost. For each x there is only one y. This satisfies the requirement that for each input there is exactly one output. Thus, total cost is a function of the number of transcripts ordered.   ■

EXAMPLE 2 Identifying a function   One parabola opening to the right is described by Table 6 and the graph in Figure 4. Is the parabola in this position a function?

TABLE 6 $x = y^2$

Input, x	Output, y
0	0
1	1
1	−1
4	2
4	−2

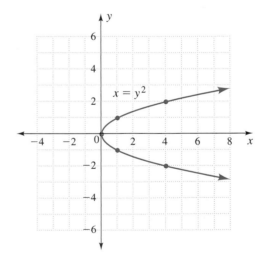

FIGURE 4

SOLUTION   At $x = 1$ in Table 6, y may be either 1 or −1. At $x = 4$, y may be either 2 or −2. Only at $x = 0$ is there one output, $y = 0$, for one input. All other positive inputs of x have two outputs of y.

For $x = 1$ in Figure 4, there are two points that make the equation true: (1, 1) and (1, −1). Similarly, at $x = 4$, there are two points that make the equation true: (4, 2) and (4, −2). Because a function requires exactly one y for each x, this parabola is not the graph of a function.   ■

VERTICAL-LINE TEST   The function definition indicates that *a function must have exactly one output for each input.* If a table shows two outputs for any input, the rule for the table is not a function.

■ THE TWO-OUTPUT TEST

> A table represents a function if every input has no more than one output.

In Table 6 in Example 2, there are two cases in which identical inputs have two different outputs.

If a graph shows two points with the same input, the rule for the graph is not a function. We can identify a function from its graph with the vertical-line test.

■ VERTICAL-LINE TEST

> A graph shows a function if every vertical line intersects the graph no more than once.

A parabola opening to the right, as shown in Figure 5, does not represent a function. At least one vertical line cuts the parabola in two places.

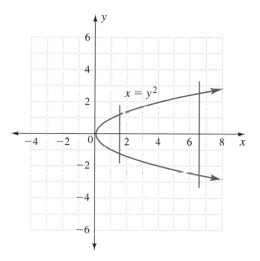

FIGURE 5 $x = y^2$ fails vertical-line test

EXAMPLE 3   Identifying functions from graphs   Use the vertical-line test to identify which of the following graphs are functions.

a.

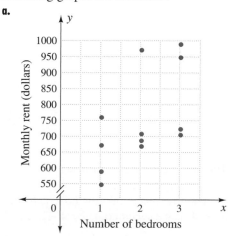

Apartment rent

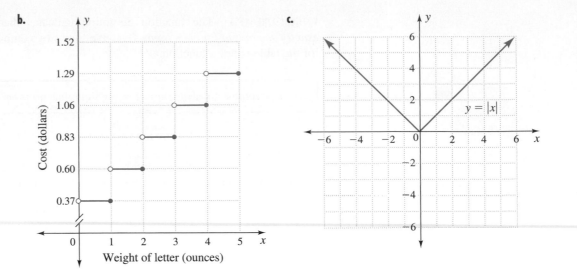

First-class postage, 2003

SOLUTION   **a.** The graph for apartment rent shows more than one output (y) for at least one input (x). At least one vertical line passes through more than one point. It is not a function.

b. The first-class postage cost is a function of weight in ounces. A vertical line appears to pass through two points, but one point is a dot and the other an open circle. Dots are included in the graph; open circles are excluded from the graph. The graph shows a function.

c. The graph of $y = |x|$, the absolute value of x, is intersected exactly once by any vertical line. The absolute value of x is a function. ▬

Think about it 1: How would the graph of more current data compare with the graph in part a?

Visualizing Functions

There are two common ways to visualize a function. One is with a function machine, and the other is through mapping.

FUNCTION MACHINE   A function machine is a mysterious box, containing a rule that takes an input and, after applying the rule, produces an output. Figure 6 illustrates a function machine for the rule $y = 2x$.

FIGURE 6 Function as a machine in a box

The first object into the box, 7, is the first out, $2 \cdot 7$. The second object into the box, 4, is the second out, $2 \cdot 4$. The third object, b, comes out as $2 \cdot b$.

EXAMPLE 4 Visualizing functions What is the rule for each box?

a.

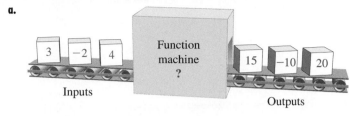

b.

SOLUTION *Hints:* Part a is a multiplication; part b is a subtraction. See the Answer Box. ▬

Student Note: Learn whichever visualization—function machine or mapping—makes more sense to you.

MAPPING Mapping is a way of showing a rule connecting inputs and outputs. For a function, each input is connected to a single output. Figure 7 illustrates a mapping for the function $y = 2x$ and for the rule $x = y^2$ (not a function).

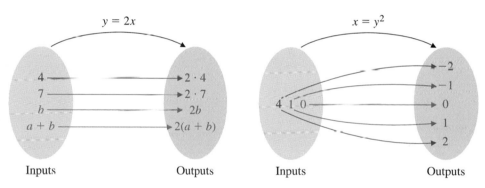

FIGURE 7 Rules as a mapping from input set to output set

EXAMPLE 5 Visualizing functions Use the rule to match each input with its output.

a. $y = x^2$ b. $y = 2x - 1$

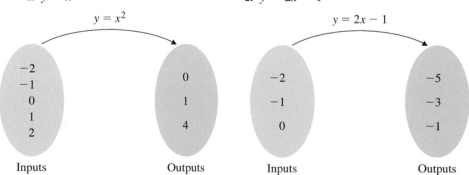

SOLUTION The matches are listed as ordered pairs.

a. $(-2, 4)$, $(-1, 1)$, $(0, 0)$, $(1, 1)$, $(2, 4)$. Some of the outputs are used twice, but the rule, $y = x^2$, is a function because each input matches with exactly one output. The mapping is shown in Figure 8.

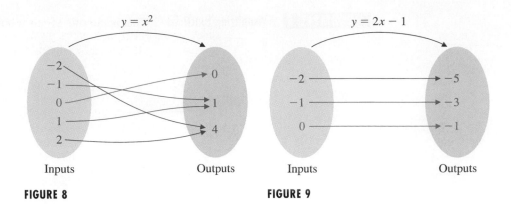

FIGURE 8   **FIGURE 9**

b. $(-2, -5), (-1, -3), (0, -1)$. Each input matches with exactly one output. The mapping is shown in Figure 9.

Function Notation

We write functions in algebraic notation by indicating the name of the function (a letter, usually f) and the independent variable (the input, usually x).

 FUNCTION NOTATION

> The notation $f(x)$ means *a function of x* and is read *f of x*.
>
> The symbols $f(x)$ do not imply multiplication, and *of* in *f of x* does not mean multiplication.

The parentheses in $f(x)$ contain the variable or variables. Although $f(x)$ is commonly used to describe the output to a function, other letters may also be used: $A(x)$, $g(x)$, $G(x)$, etc.

EXAMPLE 6 **Writing function notation** Define an appropriate variable, identify the function, and then write each sentence in function notation.
a. The area (in square units) of a triangle with base 10 units is 5 times the length of the altitude.
b. The cost of a transcript is $3 for the first copy and $2 for each additional copy.
c. The sales tax on a purchase is 8.5% of the price in dollars.
d. The equation $y = x^2$ describes a function of x.
e. The equation $y = 2x - 1$ describes a function of x.

SOLUTION **a.** Let x be the length of the altitude, A be the area function; $A(x) = 5x$ square units.

b. Let n be the number of transcripts ordered, c be the total cost function; $c(n) = 3 + 2(n - 1)$ dollars.

c. Let c be the cost of the item, t be the tax function; $t(c) = 0.085c$ dollars.

d. Let x be the input, f be the squaring function; $f(x) = x^2$.

e. Let x be the input, f be the function doubling a number and subtracting one; $f(x) = 2x - 1$.

Some functions are usually written with capital letters, such as the A used for the area function in Example 6. Others use lowercase letters, such as the c and t in Example 6. In chemistry, medicine, and studies of the environment, pH is used to describe acidity or alkalinity. Although the parentheses are left out of pH, it is still a function. In mathematics, we would write pH as $p(H)$. (More on the pH function beginning with Chapter 9 opener.) A function on a calculator appears either as a symbol, $\boxed{x^2}$, or with multiple letters preceding the opening parenthesis, **abs(**. Be patient. In time, you will be as comfortable using this variety of notations as writing $\frac{1}{2}$ as $\frac{2}{4}$, $\frac{5}{10}$, 0.5, and 50%.

CALCULATOR FUNCTIONS   Calculator functions are used in two different ways. To use some functions, we first enter either a number or a variable and then choose the function. With these functions, the parentheses do not show on the view screen.

EXAMPLE 7   Using calculator functions   Follow the steps to evaluate each function on your calculator.

a. Reciprocal function: 4 $\boxed{x^{-1}}$ $\boxed{\text{ENTER}}$

b. Change-to-a-fraction function: .25 $\boxed{\text{MATH}}$ **1: ▶ Frac** $\boxed{\text{ENTER}}$

c. Squaring function: 4 $\boxed{x^2}$ $\boxed{\text{ENTER}}$

SOLUTION   a. .25 (that is, $\frac{1}{4}$ as a decimal). To save space on a limited screen, the calculator leaves off the zero before a decimal point. However, it is *always* preferable to write decimals with a zero: 0.25, not .25.

b. $\frac{1}{4}$

c. 16

To use other calculator functions we first choose the function and then enter either a number or a variable in the parentheses that follow. These functions always give the first parenthesis in $f(x)$. Closing with an ending parenthesis $\boxed{)}$ is a good habit and essential for complicated expressions.

EXAMPLE 8   Using calculator functions   Follow the steps to evaluate each function on your calculator.

a. Square root function: $\boxed{\text{2nd}}$ $[\sqrt{}]$ 1/4 $\boxed{)}$ $\boxed{\text{ENTER}}$

b. 10^x function: $\boxed{\text{2nd}}$ $[10^x]$ 2 $\boxed{)}$ $\boxed{\text{ENTER}}$

c. Absolute value function: $\boxed{\text{2nd}}$ [CATALOG] $\boxed{\text{ENTER}}$ $\boxed{(-)}$ 2 $\boxed{)}$ $\boxed{\text{ENTER}}$

SOLUTION   a. .5 $\left(\text{that is, } 0.5 \text{ or } \frac{1}{2}\right)$

b. 100

c. 2

Think about it 2: Investigate using variables in calculator functions. Enter 3 $\boxed{\text{STO>}}$ $\boxed{\text{X,T,θ,n}}$ and then redo each of the functions in Examples 7 and 8 with X instead of a number.

Evaluating Functions

One of the most important reasons for studying functions is that the notation, $f(x)$, is a convenient way to indicate the evaluation of a function for a certain number or expression. *Function notation has the advantage of allowing us to simultaneously name the rule and the input.*

EVALUATING FUNCTIONS GIVEN AS EXPRESSIONS   In Example 9, we return to the suspension bridge from Chapter 1 (see Figure 10) and find the height in feet of cable at several distances from the center of the bridge. When we write $f(x) = \frac{4}{125} x^2$ and ask for $f(-25)$, we want the x in $\frac{4}{125} x^2$ replaced by -25. Thus,

$$f(-25) = \frac{4}{125}(-25)^2 = \frac{4}{125}(625) = 20 \text{ ft}$$

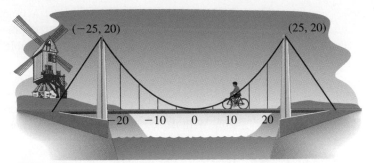

FIGURE 10

EXAMPLE 9 Evaluating functions: tabulating suspension bridge cable heights Find $f(-15), f(-5)$, $f(0), f(5), f(15)$, and $f(25)$, where the height in feet of the cable is given by the function $f(x) = \frac{4}{125}x^2$. Summarize these results in a table. Comment about any patterns in the table.

SOLUTION
$f(-15) = \frac{4}{125}(-15)^2 = \frac{4}{125}(225) = 7.2$ ft

$f(-5) = \frac{4}{125}(-5)^2 = \frac{4}{125}(25) = 0.8$ ft

$f(0) = \frac{4}{125}(0)^2 = 0$ ft

$f(5) = \frac{4}{125}(5)^2 = \frac{4}{125}(25) = 0.8$ ft

$f(15) = \frac{4}{125}(15)^2 = \frac{4}{125}(225) = 7.2$ ft

$f(25) = \frac{4}{125}(25)^2 = \frac{4}{125}(625) = 20$ ft

TABLE 7 $f(x) = \frac{4}{125}x^2$

Input, x	Output, $f(x)$
-25	20
-15	7.2
-5	0.8
0	0
5	0.8
15	7.2
25	20

The height of the cable is the same for equal distances from the center of the bridge. This is apparent from the illustration in Figure 10 as well as from Table 7. The outputs match because the inputs are squared. Thus, the output for -15 is the same as the output for 15.

Because the output is 7.2 for inputs $x = -15$ and $x = 15$, we can state this equality as $f(-15) = f(15)$. Thus, *function notation provides a convenient way to describe the equality of outputs.*

EVALUATING FUNCTIONS WITH VARIABLES AS INPUTS The inputs to a function are not limited to numbers. In Example 10, the inputs are variables.

EXAMPLE 10 Evaluating functions where inputs are variables If $g(x) = x^2 - 2x - 3$, find the following. The answers to these evaluations will be expressions.

a. $g(a)$ **b.** $g(\pi)$ **c.** $g(\square)$

d. $g(-x)$ **e.** $g(a + b)$ **f.** $g(x + h)$

SOLUTION
a. $g(a) = a^2 - 2a - 3$

b. $g(\pi) = \pi^2 - 2\pi - 3$

c. $g(\square) = \square^2 - 2\square - 3$

d. $g(-x) = (-x)^2 - 2(-x) - 3 = x^2 + 2x - 3$

e. $g(a + b) = (a + b)^2 - 2(a + b) - 3$
$$= a^2 + 2ab + b^2 - 2a - 2b - 3$$

f. $g(x + h) = (x + h)^2 - 2(x + h) - 3$
$$= x^2 + 2xh + h^2 - 2x - 2h - 3$$ ▬

Caution: Three cautions are needed about Example 10.

Student Note: For more on squaring binomials, see page 283.

- The notation $g(a + b)$ is function notation, with $a + b$ replacing the x in $g(x)$. The symbol $g(a + b)$ looks like part of the distributive property, $c(a + b) = ca + cb$, but it is not. One way to determine what a symbol means is by looking at the context. Compare the expression to others in the same problem.

- The solutions to parts e and f contain the squares of binomials, $(a + b)^2$ and $(x + h)^2$. Carefully review multiplying these:

$$(a + b)^2 = (a + b)(a + b) = a^2 + ab + ab + b^2 = a^2 + 2ab + b^2$$

Similarly,

$$(x + h)^2 = (x + h)(x + h) = x^2 + 2hx + h^2$$

- The distributive property of multiplication over addition is used in the solutions to parts e and f. Look for $-2(a + b)$ in the solution to part e and for $-2(x + h)$ in the solution to part f.

Domain and Range

For many functions, there are limitations on the inputs and/or outputs. In order to describe the restrictions on the sets of inputs or outputs of a function, we give the sets names: domain and range.

▬ DOMAIN AND RANGE

> The **domain** is the set of all inputs to a function. The **range** is the set of all outputs.

EXAMPLE 11

Finding domain and range What are the domain and range for each of these problem settings? Use inequalities for part b.
a. The transcript costs in Example 1, page 84
b. The suspension bridge cable heights in Example 9, page 90

SOLUTION

a. The domain, or set of inputs, for the transcript costs is the set of positive integers. The range, or set of outputs, is the set of odd positive integers greater than 1.

b. The height function for the bridge cable has a real number domain between -25 and $+25$, $-25 \leq x \leq 25$. The range is the set of real numbers between 0 and 20, $0 \leq f(x) \leq 20$. ▬

The *set of numbers including zero and the positive real numbers* is a common domain or range and is given a special name: **non-negative real numbers**. When the domain or range includes the *set of all real numbers*, we use the symbol $\mathbb{R}$. For example,

The domain of $f(x) = 5x$ is $\mathbb{R}$.

The range of $f(x) = 5x$ is $\mathbb{R}$.

The words *domain* and *range* are not so strange if we consider what is being described. *Domain* represents the set of starting numbers, as in starting out from home. The word *domain* has the same root as domicile (legal word for home), domestic (pertaining to one's home or homeland), and dominion (territory under one ruler). *Range* represents the set of outcomes. The word *range* implies outside the home, as in "out on the range" or allowing an animal "to range."

Application

You probably have noticed that a hot day feels even hotter when the humidity is high. What may surprise you is that very low humidity will make a hot day seem cooler than the actual air temperature.

   The increase (or decrease) in temperature is measurable and is the basis for the heat index chart in Table 1 on the chapter opening page, where apparent temperature is related to relative humidity and air temperature. Example 12 shows data from the heat index chart for an air temperature of 85°F.

EXAMPLE 12   Describing a relationship: heat index   Table 8 shows the apparent temperature for several humidity levels at an air temperature of 85°F.

TABLE 8   *Apparent Temperatures for 85°F Air Temperature*

Relative Humidity, x (%)	0	10	20	30	40	50	60	70	80	90	100
Apparent Temperature (°F)	78	80	82	84	86	88	90	93	97	102	108

   a. Estimate the relative humidity when the apparent temperature is the same as the air temperature, 85°F.
   b. For what levels of humidity is the apparent temperature cooler than the air temperature?
   c. For what levels of humidity is the apparent temperature warmer than 105°F? (This temperature greatly increases risk of sunstroke or heatstroke.)
   d. Does the chart describe a function?
   e. What is the domain of the function?
   f. State the relationship in the table using the word *function*.
   g. What other factors might change the apparent temperature?

SOLUTION   **a.** approximately 35% (between 30% and 40% humidity)

   b. approximately $x < 35\%$

   c. approximately $x > 95\%$

   d. Yes; for each input x, there is exactly one output y.

   e. $0\% \le x \le 100\%$

   f. The apparent temperature for an 85°F air temperature is a function of the relative humidity.

   g. Wind, standing in the shade or sun, color and type of clothing   ▬

2.2 Exercises

In Exercises 1 to 6:
(a) Evaluate the equation for the given numbers. Record your answers in an input-output table. Assume x = input, y = output.
(b) Is the equation a function? If so, write the function in function notation.

1. The total cost of x compact disks at \$15 each with one \$4 discount coupon is $y = 15x - 4$, where y is in dollars. Evaluate for $x = 1, 2, 3, 4$.   11, 26, 41, 56; $f(x) = 15x - 4$

2. The equation for the outer edge of the base of a cylinder of radius 3 feet is $x^2 + y^2 = 9$. Evaluate for $y = 3$, $-3, 0$.   0, 0, 3 or -3; not a function

3. $x = y^2 + 2$; evaluate for $y = -4, -2, 0, 2, 4$.
   18, 6, 2, 6, 18; not a function

4. $3x^2 = y$; evaluate for $y = 27, 12, 3, 0$.
   3 or -3, 2 or -2, 1 or -1, 0; $f(x) = 3x^2$

5. $y = 25 - x^2$ for x in the set $\{-5, -2, 0, 2, 5\}$
   0, 21, 25, 21, 0; $f(x) = 25 - x^2$

6. $x = 25 - y^2$ for x in the set $\{25, 16, 9, 0\}$
   0, 3 or -3, 4 or -4, 5 or -5; not a function

Which of Exercises 7 to 12 represent functions, $y = f(x)$?

7. x = weight of a first-class letter
y = postage cost of the letter in 2003, when the rates were 37 cents for the first ounce and 23 cents for each additional ounce up to 11 ounces   function

8. x = number of gallons of gasoline   function
y = cost of the gasoline at \$1.69 per gallon

9. x = height of a child in third grade
y = weight of the child   not a function

10. x = shoe size
y = cost of the shoe   not a function

11. x = number of homework assignments completed by a student   not a function
y = the student's score on a test

12. x = state   function
y = state capital of x

In Exercises 13 to 18, explain why the vertical-line test tells us whether or not each graph is a function. State specific numbers in your explanation.

13.

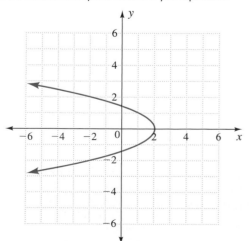

not a function

14.

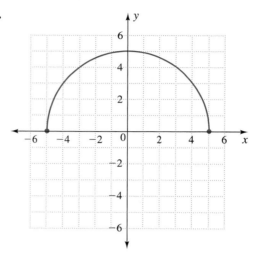

function

15.

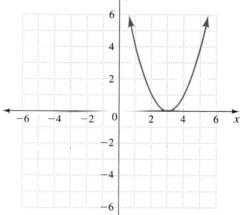

function

16.

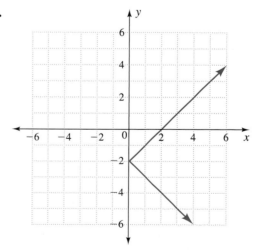

not a function

17.

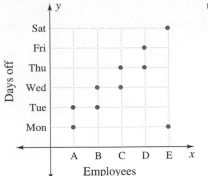

not a function

18.

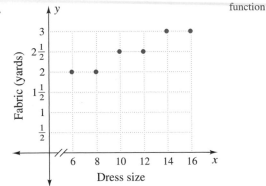

function

In Exercises 19 to 22, guess a rule that associates the input set with the output set.

19. $\{3, 4, 5, 6\}$ with $\{5, 6, 7, 8\}$ $y = x + 2$

20. $\{3, 4, 5, 6\}$ with $\{1.5, 2, 2.5, 3\}$ $y = \frac{1}{2}x$

21. $\{3, 4, 5, 6\}$ with $\{1, \frac{4}{3}, \frac{5}{3}, 2\}$ $y = \frac{1}{3}x$

22. $\{3, 4, 5, 6\}$ with $\{0, 1, 2, 3\}$ $y = x - 3$

In Exercises 23 to 27, copy the expression twice. Show how parentheses would have to be placed to force the calculator to obtain each answer. Based on the order of operations, which is the correct answer?

23. $\sqrt{100} - 36$ $\sqrt{(100 - 36)}$ or $\sqrt{(100 - 36}$; $\sqrt{(100)} - 36$; b

   a. 8   **b.** -26

24. $4/5^{-1}$ $(4/5)^{-1}$; $4/(5^{-1})$; b

   a. 1.25   **b.** 20

25. $2 + 3^2$ $(2 + 3)^2$; $2 + (3^2)$ or $2 + 3^2$; b

   a. 25   **b.** 11

26. 10^2 + 3 (Use the 10^x key.)

          10^2 + 3 or (10^2) + 3; 10^(2 + 3) or 10^(2 + 3; a
   a. 103   **b.** 100000

27. $|3 - 4|$ abs(3) $-$ 4; abs(3 $-$ 4) or abs (3 $-$ 4; b

   a. -1 **b.** 1

28. If you omit one of a pair of parentheses, where does the calculator place it? at the end of the expression

In Exercises 29 to 34, evaluate the function for integers on the interval $[-2, 4]$. Record the results in a table.

29. $g(x) = 5 + 2(x - 3)$ **30.** $f(x) = 25 + 4(x - 10)$
      See Answer Section.   See Additional Answers.
31. $f(x) = x^2 - 2x - 3$ **32.** $g(x) = x^2 + 4x - 5$
      See Answer Section.   See Additional Answers.
33. $g(x) = 8 - x^2$ **34.** $f(x) = 6 - x - x^2$
      See Answer Section.   See Additional Answers.

Evaluate the transcript cost function $f(x) = 3 + 2(x - 1)$ for the inputs in Exercises 35 and 36.

35. a. $f(1)$ 3   **b.** $f(5)$ 11

   c. $f(n)$ $2n + 1$ **d.** $f(n + m)$ $2n + 2m + 1$

36. a. $f(3)$ 7   **b.** $f(4)$ 9

   c. $f(m)$ $2m + 1$ **d.** $f(x - h)$ $2x - 2h + 1$

Evaluate the parabolic function $g(x) = x^2 + x - 2$ for the inputs in Exercises 37 and 38.

37. a. $g(3)$ 10   **b.** $g(1)$ 0

   c. $g(\square)$ $\square^2 + \square - 2$ **d.** $g(n)$ $n^2 + n - 2$

   e. $g(n - m)$ $n^2 - 2nm + m^2 + n - m - 2$

38. a. $g(-2)$ 0   **b.** $g(2)$ 4

   c. $g(\pi)$ ≈ 11.011 **d.** $g(-n)$ $n^2 - n - 2$

   e. $g(x + h)$ $x^2 + 2xh + h^2 + x + h - 2$

In Exercises 39 to 42, match the sets of inputs and outputs given with one of the graphs in Exercises 13 to 16.

39. $-5 \le x \le 5$ and $0 \le y \le 5$ Exercise 14

40. $x \ge 0$ and y is any real number $\mathbb{R}$ Exercise 16

41. x is any real number $\mathbb{R}$ and $y \ge 0$ Exercise 15

42. $x \le 2$ and y is any real number $\mathbb{R}$ Exercise 13

In Exercises 43 to 48, identify the domain and range. A table or graph may be helpful.

43. $f(x) = x^2$ x is any real number; $f(x) \ge 0$

44. $g(x) = 1 - x^2$ x is any real number; $g(x) \le 1$

45. $h(x) = x^2 - 2$ x is any real number; $h(x) \ge -2$

46. $f(x) = x^2 + 2$ x is any real number; $f(x) \ge 2$

47. $g(x) = 6 - x^2$ x is any real number; $g(x) \le 6$

48. $h(x) = 4 - x^2$ x is any real number; $h(x) \le 4$

For each inequality in Exercises 49 to 56, assume x refers to inputs and y refers to outputs.
(a) Is the set a domain or range?
(b) Say whether each inequality includes positive numbers, negative numbers, negative numbers plus zero, or non-negative numbers.
(c) Write each inequality as an interval.

49. $x \le 0$ domain; negative numbers plus zero; $(-\infty, 0]$

50. $x < 0$ domain; negative numbers; $(-\infty, 0)$

51. $y > 0$ range; positive numbers; $(0, \infty)$

52. $y \geq 0$ range; non-negative numbers; $[0, +\infty)$

53. $x > 0$ domain; positive numbers; $(0, \infty)$

54. $y < 0$ range; negative numbers; $(-\infty, 0)$

55. $y \leq 0$ range; negative numbers plus zero; $(-\infty, 0]$

56. $x \geq 0$ domain; non-negative numbers; $[0, +\infty)$

In Exercises 57 to 60, draw pairs of mapping ovals (see Example 5). Place the inputs $-1, 0, 1$ in the first oval. Place the outputs $-1, 0, 1$ in the second oval. Draw lines showing these relationships. Explain why each is or is not a function.

57. If $x = -1, y = 1$.   If $x = 0, y = 0$.   If $x = 1, y = 1$.
See Answer Section; function

58. If $x = -1, y = -1$ or 1.   If $x = 0, y = 0$.   For $x = 1$, no y is defined.   See Additional Answers; not a function

59. If $x = -1, y = 0$.   If $x = 0, y = -1$ or 1.   If $x = 1, y = 0$.   See Answer Section; not a function

60. If $x = -1, y = -1$.   If $x = 0, y = 0$.   If $x = 1, y = 1$.
See Additional Answers; function

The **relevant domain** is the allowable input set for a given problem setting. Find the relevant domain for each of Exercises 61 to 66.

61. The area of a circle as a function of the length of its radius, $A(r) = \pi r^2$ $r > 0$

62. The area of a square as a function of the length of its side, $A(x) = x^2$ $x > 0$

63. The perimeter of a square as a function of the length of its side, $P(x) = 4x$ $x > 0$

64. The circumference of a circle as a function of the length of its radius, $C(r) = 2\pi r$ $r > 0$

65. The amount earned in a day at $6.50 per hour
See below.

66. The distance traveled in a car in three hours as a function of rate (speed)   rates between 0 and the legal speed limit (55 or 65 or 70); higher speeds limited by production capability of cars

A computer must be programmed to expect certain inputs—that is, relevant domains. For Exercises 67 to 70, give the number of digits expected for the relevant domains in programs seeking these data.

67. A telephone number plus area code   10

68. A telephone area code   3

69. A social security number   9

70. A postal zip code (There may be two answers.)   5 or 9

71. Discuss the difference between having integers and real numbers as the domain. Use Exercises 1, 14, 15, and 18 to illustrate your points.   When inputs are CDs or dress sizes, fractional and decimal values are meaningless; in other cases, inputs may be

72. Explain what $f(2)$ represents.   fractional or decimal values. the output of the function f when the input is 2

73. Use function notation to write "the output of the function g when the input is $a + b$." $g(a + b)$

65. variable—8 to 10 hours if working full-time, more if working overtime, less if working part-time; no less than 0, no more than 24 hours

Solve each of these equations with algebraic notation, and check with the tables in your answers to Exercises 29 and 30. Extend your tables as needed.

74. a. $5 + 2(x - 3) = -3$ -1

b. $5 + 2(x - 3) = 5$ 3

75. a. $25 + 4(x - 10) = -15$ 0

b. $25 + 4(x - 10) = -23$ -2

Solve Exercises 76 to 79 with the tables in your answers to Exercises 31 to 34. It may be necessary to look for patterns and extend the table.

76. a. $x^2 + 4x - 5 = -8$ $x = -3$ or $x = -1$

b. $x^2 + 4x - 5 = -5$ $x = 0$ or $x = -4$

77. a. $x^2 - 2x - 3 = 0$ $x = 3$ or $x = -1$

b. $x^2 - 2x - 3 = 21$ $x = 6$ or $x = -4$

78. a. $6 - x - x^2 = 6$ $x = 0$ or $x = -1$

b. $6 - x - x^2 = 0$ $x = -3$ or $x = 2$

79. a. $8 - x^2 = 4$ $x = -2$ or $x = 2$

b. $8 - x^2 = 7$ $x = -1$ or $x = 1$

■■■ **Projects**

80. Heat Index

a. Graph the table in Example 12.

b. In Table 1 on the chapter opener page, find the row with the 85°F air temperature data.

c. Select another row (for example, air temperature 110°F). Using relative humidity as the input and apparent temperature as the output, graph these data on the same set of axes. You may need to extend the y-axis.

d. Repeat for another row (for example, air temperature 75°F).

e. When the apparent temperature is greater than 124°F, there is extreme danger of heatstroke or sunstroke. Draw this line on your graph and label appropriately.
See Additional Answers.

81. Bus Stop   What are the pros and cons of placing a bus stop before an intersection? After an intersection? Placement of a bus stop seems to be a function of what factors?

82. Doors   The direction in which a door opens is a function of its location. Describe the direction in which these doors open: exterior house door, movie theater door, bedroom door, bathroom door, closet door, cupboard door, door at the top of a staircase, door at the bottom of a staircase. State any reasons why the direction may be customary or mandatory.

2.3   Linear Functions

Objectives

■ Identify a linear function.

■ Find x- and y-intercepts of the graph of a linear function.

■ Find the slope of a linear function from data points, a table, and a graph.

■ Find the meaning of slope in an application setting.

■ Identify increasing and decreasing linear functions.

■ Graph linear functions, given intercepts and/or slope.

WARM-UP

Simplify these expressions.

1. $\dfrac{-7.5 - (-11.25)}{10 - 5}$ 0.75

2. $\dfrac{3 - 6}{6 - 4}$ $-\frac{3}{2}$

3. $\dfrac{3 - 1}{6 - 3}$ $\frac{2}{3}$

4. $\dfrac{70 - 50}{400 - 200}$ 0.1

THIS SECTION INTRODUCES linear functions, x-intercepts and y-intercepts, and slope. We use *increasing* and *decreasing* to describe the direction of change shown by a line.

Linear Equations

In Chapter 1 and thus far in Chapter 2, we have had a variety of applications of mathematics. In some applications, we worked without equations (wind chill and heat index apparent temperatures). In others, we wrote equations from problem settings (transcript cost and birthday party costs). In still others, we were given equations (suspension bridge cable height). Now, starting with linear equations, we begin to organize equations into categories and explore ways of finding equations from data.

■ LINEAR EQUATION IN ONE VARIABLE

> A **linear equation in one variable** can be written
>
> $$ax + b = 0$$
>
> where a and b are any real number and a is not zero.

We work with linear equations in one variable when we solve equations.

■ LINEAR EQUATION IN TWO VARIABLES

> A **linear equation in two variables** can be written
>
> $$ax + by = c$$
>
> where a, b, and c are any real number and a and b are not both zero.

We work with linear equations in two variables when we graph. The equations given in the definitions are said to be in **standard form**.

EXAMPLE 1 Identifying linear equations   Is the equation linear? If so, is it linear in one variable or two variables? Name the variables. Place each linear equation in standard form. Give the values of a, b, and c.

a. $x = 4$ **b.** $2x - 3 = 0$ **c.** $C = 2\pi r$

d. $x^2 = 4$ **e.** $3n - 2 = 11n + 14$ **f.** $2\pi r = 8$

g. $y = 2x + 3$ **h.** $C = x + 0.15x$ **i.** $4x^2 - 2x(2x + 1) = 10$

SOLUTION The equations in parts e, f, and i first need to be simplified and the variables placed in alphabetical order.

a. one variable, x; $x - 4 = 0$; $a = 1$, $b = -4$

b. one variable, x; $2x - 3 = 0$; $a = 2$, $b = -3$

c. two variables, C and r; $C - 2\pi r = 0$; $a = 1$, $b = -2\pi$, $c = 0$

d. not linear

e. one variable, n; $n + 2 = 0$; $a = 1$, $b = 2$

f. one variable, r; $\pi r - 4 = 0$; $a = \pi$, $b = -4$

g. two variables, x and y; $2x - y = -3$; $a = 2$, $b = -1$, $c = -3$

h. two variables, C and x; $C - 1.15x = 0$; $a = 1$, $b = -1.15$, $c = 0$

i. one variable, x; $x + 5 = 0$; $a = 1$, $b = 5$

Linear Functions

A linear function is a function whose graph is a straight line. With the exception of the equation of a vertical line, $x = c$, all linear equations $ax + by = c$ may also be written as linear functions. We usually write $f(x) = mx + b$ for linear functions because this form excludes vertical lines. (More on vertical lines in Section 2.5.)

LINEAR FUNCTION

> A **linear function** can be written $f(x) = mx + b$. The constants m and b may be any real number. The variable, x, has 1 as its exponent.

Think about it 1: Why would a vertical line not be a function?

To identify a linear function, use algebraic notation to obtain $f(x) = mx + b$.

EXAMPLE 2 Identifying linear functions   Which of the following are linear functions?

a. $f(x) = 10 - 0.10x$ **b.** $f(x) = 3 + 2(x - 1)$ **c.** $f(x) = 4 - x^2$

SOLUTION **a.** In $f(x) = 10 - 0.10x$, $m = -0.10$ and $b = 10$. The function is linear.

b. $f(x) = 3 + 2(x - 1)$ Multiply with the distributive property.

    $f(x) = 3 + 2x - 2$ Add like terms.

    $f(x) = 2x + 1$ The function is linear, with $m = 2$ and $b = 1$.

c. $f(x) = 4 - x^2$ is not linear because x^2 does not have 1 as an exponent.

Intercepts

APPLICATION: PHOTOCOPY CARD   A prepaid photocopy card costs $10. Each copy costs $0.10. The cost per copy is subtracted from the value of the card as copies are made. The graph for the value on the photocopy card after x copies have been made is shown in Figure 11. The graph crosses both horizontal and vertical axes. We now look at these crossing points more formally.

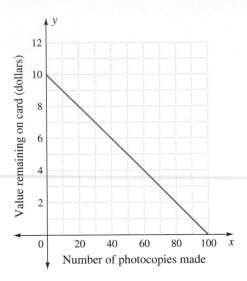

FIGURE 11

THE HORIZONTAL INTERCEPT   A graph crosses the horizontal axis (x-axis) at the horizontal intercept (x-intercept). In Figure 12, the point $(a, 0)$ is a horizontal intercept point.

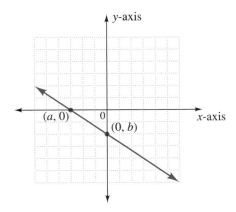

FIGURE 12

HORIZONTAL INTERCEPT   The horizontal intercept point, $(a, 0)$, is the intersection of a graph with the horizontal axis, or x-axis. The letter a denotes the horizontal intercept number, or **horizontal intercept (x-intercept)**.

THE VERTICAL INTERCEPT   A graph crosses the vertical axis (y-axis) at the vertical intercept (y-intercept). In Figure 12, the point $(0, b)$ is a vertical intercept point.

VERTICAL INTERCEPT   The vertical intercept point, $(0, b)$, is the intersection of a graph with the vertical axis. The letter b denotes the vertical intercept number, or **vertical intercept (y-intercept)**.

EXAMPLE 3 Finding the intercepts

a. What is the horizontal intercept point in the photocopy card graph in Figure 11? What does the intercept mean in the problem situation?

b. What is the vertical intercept point in the photocopy card graph? What does the intercept mean in the problem situation?

SOLUTION a. The graph crosses the horizontal axis at (100, 0). The horizontal intercept point is (100, 0). The intercept, 100, is the number of photocopies that can be made from a prepaid $10 card at $0.10 per copy.

b. The graph crosses the vertical axis at (0, 10). The vertical intercept point is (0, 10). The intercept, 10, is the value in dollars of the photocopy card when it is purchased. ▬

Think about it 2: From the word description, write an equation for the photocopy card where the value remaining on the card depends on the number of copies that have been made. Does any of the intercept information appear in the equation?

INTERCEPTS AND FUNCTION NOTATION Function notation gives a convenient way to describe finding the intercepts. In function notation, an ordered pair is $(x, f(x))$.

> The horizontal axis intercept point has $f(x) = 0$. To find the horizontal or x-intercept, set the function equal to zero and solve for x.
>
> The vertical axis intercept point, $(0, b)$, has $f(0) = b$. To find the vertical or y-intercept, evaluate $f(0)$.

EXAMPLE 4 Finding intercepts The value on the photocopy card after x copies have been made is $f(x) = 10 - 0.10x$.

a. Find the horizontal intercept with $f(x) = 0$.

b. Find the vertical intercept with $f(0)$.

c. Do the results agree with the graph?

SOLUTION a.
$$f(x) = 10 - 0.10x \qquad \text{Replace } f(x) \text{ with } 0.$$
$$0 = 10 - 0.10x \qquad \text{Subtract 10 from both sides.}$$
$$-10 = -0.10x \qquad \text{Divide both sides by } -0.10.$$
$$100 = x$$

The horizontal intercept point is (100, 0).

b.
$$f(x) = 10 - 0.10x \qquad \text{Find } f(0) \text{ by evaluating for } x = 0.$$
$$f(0) = 10 - 0.10(0) \qquad \text{Simplify.}$$
$$f(0) = 10$$

The vertical intercept point is (0, 10).

c. The horizontal intercept, $x = 100$, is the number of copies that can be made from a $10 prepaid card. The vertical intercept, $y = 10$, is the cost of the card. The results agree with the graph (see Figure 11). ▬

Think about it 3: Give two ways we know that $f(x) = 10 - 0.10x$ is a linear function.

Think about it 4: **a.** Is it possible for the graph of a function to intersect the vertical axis in two places?

b. Is it possible for the graph of a function to intersect the horizontal axis in two places?

Slope

EXAMPLE 5 Investigating slope

a. Write two new equations for the photocopy card—one if the cost changed to $0.15 per copy and another if the cost changed to $0.05 per copy.

b. Graph each equation on the original axes in Figure 11.

c. How does the graph change as the cost changes?

SOLUTION   **a.** $y = 10 - 0.15x$

$y = 10 - 0.05x$

b. The lines are graphed in Figure 13.

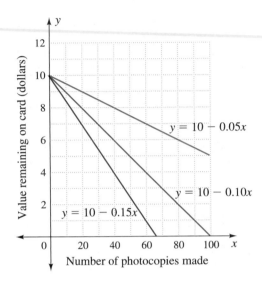

FIGURE 13

c. If the cost increases from $0.10 to $0.15, the graph gets steeper.
If the cost decreases from $0.10 to $0.05, the graph gets flatter.

The questions investigated in Example 5 focus on the change in the graph. We summarize these changes with the concept of slope. The slope of a linear function refers to the steepness of its graph.

▋ SLOPE | **Slope** is a rate of change. |

To obtain the rate of change, or slope, we divide the change in output by the corresponding change in input. In terms of variables, slope is the ratio of the change in y to the change in x:

$$\text{Slope} = \frac{\text{change in output}}{\text{change in input}} = \frac{\Delta y}{\Delta x} = \frac{y_2 - y_1}{x_2 - x_1}$$

The small triangle Δ is the Greek letter *delta* and means *the change in* the variable that follows it. The formula for slope reflects the fact that subtraction is the operation we associate with change.

The small numbers to the right of the variables are subscripts. **Subscripts** are *numbers of reduced size placed below and to the right of variables to distinguish a particular item from a group of similar items.* We generally assume that the point described by (x_1, y_1) is to the left of the one described by (x_2, y_2).

In Example 6, we return to the photocopy card setting, naming the function $V(x)$ instead of $f(x)$.

EXAMPLE 6 Finding slope from a table and graph: photocopy card The prepaid photocopy card costs $10. Each copy costs $0.10. The cost per copy is deducted from the value of the card as copies are made. Table 9 and the graph in Figure 14 show the value remaining on the card after several copies have been made. Calculate the slope, first from the table and then from the graph.

TABLE 9 *Photocopy Card*

Copies, x	Remaining Value, $y = V(x)$ (dollars)
0	10.00
10	9.00
20	8.00
30	7.00

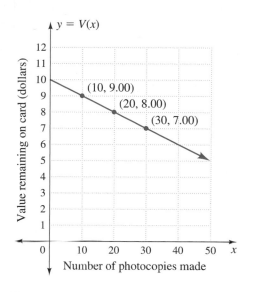

FIGURE 14

SOLUTION In Table 10, the change in y divided by the change in x is

$$m = \frac{\Delta y}{\Delta x} = \frac{-1.00}{10} = -0.10$$

Thus, the slope is $-\$0.10$ per copy. The slope is negative because it is the cost of each copy subtracted from the card's value.

TABLE 10 *Photocopy Card*

Δx	Copies, x	Remaining Value, $y = V(x)$ (dollars)	Δy	$\dfrac{\Delta y}{\Delta x}$
10 ⟨	0	10.00		
10 ⟨	10	9.00	⟩ −1.00	−0.10
10 ⟨	20	8.00	⟩ −1.00	−0.10
	30	7.00	⟩ −1.00	−0.10

In Figure 14, the line drops $1.00 for each 10 copies made. Thus, the change in y over the change in x is $-1.00/10 = -0.10$, the same as the value from the table.

In both the table and the graph in Example 6, the slope was the same, no matter what two points were chosen. This unchanging slope is an important property of straight lines.

SLOPE OF LINEAR FUNCTIONS

If the slope between any two points of a function is constant, the function is linear.

You may find it helpful to think of slope from a graph as rise over run:

$$\frac{\text{Rise}}{\text{Run}} = \text{slope}$$

The change in y (rise) divided by the change in x (run) will give the same slope as before. Rise over run may also help identify the units on slope.

■ **UNITS ON SLOPE**

> To find the units on slope, divide the units on the output axis by the units on the input axis.

In Example 6, the units on slope are cents per copy.
In Example 7, we name the function $P(x)$ instead of $f(x)$.

EXAMPLE 7 Finding slope: chocolate bar profit   Suppose deluxe chocolate bars for a fundraiser are sold at a profit of $0.75 each. Before these deluxe bars can be sold, an initial investment of $15 must be made for a display case. Calculate the slope of the line representing the total profit if x bars are sold. Use both Table 11 and the graph in Figure 15.

TABLE 11 *Profit on Sales*

Bars Sold, x	Total Profit, $y = P(x)$ (dollars)
0	−15.00
5	−11.25
10	−7.50
15	−3.75
20	0

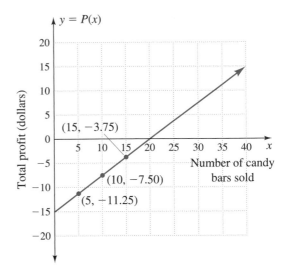

FIGURE 15

SOLUTION   **TABLE 12** *Profit on Sales*

Δx	x	$y = P(x)$	Δy	$\Delta y/\Delta x$
5	0	−15.00	3.75	0.75
5	5	−11.25	3.75	0.75
5	10	−7.50	3.75	0.75
5	15	−3.75	3.75	0.75
	20	0		

In Table 12, the change in y divided by the change in x is

$$\frac{3.75}{5} = 0.75$$

The slope is the $0.75 profit per chocolate bar.

If we use the points $(5, -11.25)$ and $(10, -7.50)$ to find the slope from the graph, we have

$$\frac{y_2 - y_1}{x_2 - x_1} = \frac{-7.50 - (-11.25)}{10 - 5} = \frac{3.75}{5} = 0.75$$

As found from the table, the slope is the $0.75 profit per bar sold. ▬

Slope has many applications. In construction, slope describes the steepness of a roof. In highway design, the slope of a road is measured by *grade*. An 8% grade, or 8/100 slope, is considered quite steep. In geology, the steepness of a layer of rock is called *dip*. Dip is based on the angle down from horizontal. In economics, the steepness of a cost curve is called *marginal cost*. In investments, the change in stock prices from minute to minute is "watched" by computer programs, and changes of specified size can lead to an automatic buy or sell order.

Increasing and Decreasing Functions

When *a graph rises from left to right*, it represents an **increasing function**. When *a graph falls from left to right*, it represents a **decreasing function**. In the chocolate bar example, the graph increased from left to right and the slope was positive. In the photocopy card example, the graph was a decreasing function from left to right and the slope was negative.

▬ SLOPE AND LINEAR FUNCTIONS

> An increasing linear function has a positive slope, and a decreasing linear function has a negative slope.

Graphing with Intercepts and Slope

We use information about intercepts and slope to graph more quickly and accurately. Note how the sign of the slope (and its connection with increasing or decreasing functions) tells us in what direction to draw the line.

EXAMPLE 8 Graphing with intercepts and slope   Graph the linear functions satisfying the facts given:

a. y-intercept $= 3$, slope $= -\frac{2}{3}$ **b.** slope $= -\frac{1}{2}$, x-intercept $= -1$

SOLUTION **a.** We plot $(0, 3)$ for the y-intercept (Figure 16). The slope is negative, so the function decreases as we go from left to right. We count down 2 units from $(0, 3)$ and right 3 units, ending at $(3, 1)$. We draw a line passing through $(0, 3)$ and $(3, 1)$.

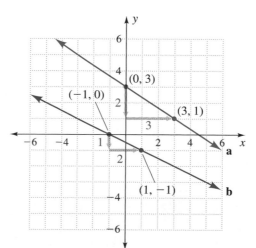

FIGURE 16

b. We plot $(-1, 0)$ for the x-intercept (Figure 16). The slope is negative, so the function decreases as we go from left to right. We count down 1 unit from $(-1, 0)$ and right 2 units, ending at $(1, -1)$.

EXAMPLE 9 **Graphing with intercepts and slope**   Graph the linear functions satisfying the facts given:

a. x-intercept $= -2$, slope $= \frac{4}{5}$ **b.** y-intercept $= -2$, x-intercept $= 2$

SOLUTION   **a.** We plot $(-2, 0)$ for the x-intercept (Figure 17). The slope is positive, so the function increases as we go from left to right. We count up 4 units and right 5 units, ending at $(3, 4)$. We draw a line passing through $(-2, 0)$ and $(3, 4)$.

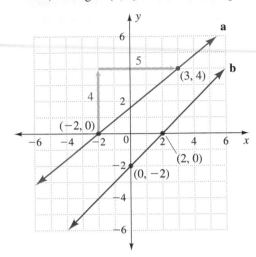

FIGURE 17

b. We plot $(0, -2)$ for the y-intercept (Figure 17). We plot $(2, 0)$ for the x-intercept and draw a line through the points.

EXAMPLE 10 **Graphing with intercepts and slope**   Graph the linear functions satisfying the facts given:

a. slope $= -2$, y-intercept $= -1$ **b.** slope $= 1$, x-intercept $= 0$

SOLUTION   **a.** We plot $(0, -1)$ for the y-intercept (Figure 18). The slope -2 is the same as $-2/1$. The slope is negative, so the function decreases as we go from left to right. We go down 2 units and right 1 unit, ending at $(1, -3)$. We draw a line passing through $(0, -1)$ and $(1, -3)$.

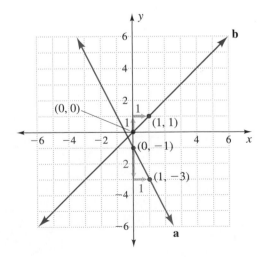

FIGURE 18

b. We plot (0, 0), the origin, for the *x*-intercept (Figure 18). The slope 1 is the same as 1/1. The slope is positive, so the function increases as we go from left to right. We count up 1 unit and right 1 unit ending at (1, 1). We draw a line through (0, 0) and (1, 1). ▬

> ## ANSWER BOX
>
> **Warm-up:** **1.** 0.75 **2.** $-\frac{3}{2}$ **3.** $\frac{2}{3}$ **4.** 0.1 **Think about it 1:** A vertical line would fail the vertical-line test. For each input, *x*, there is not exactly one output, *y*. **Think about it 2:** Let *x* = number of copies made and *y* = value remaining. The equation is $y = 10 - 0.10x$. The vertical axis intercept is \$10; \$10 is also the constant in the equation. **Think about it 3:** The graph is a straight line. The function may be written in the form $f(x) = mx + b$: $f(x) = -0.10x + 10$, with $m = -0.10$ and $b = 10$. **Think about it 4:**
> **a.** No; if a graph has two intersections with the vertical axis, it has two outputs for one input—that is, two *y*-values for $x = 0$. In that case, it cannot be a function. **b.** A function can have two (or more) *x*-intercepts. For example, the graph of the function $f(x) = x^2 - 4$ has two horizontal axis intercepts, which can be found by solving $x^2 - 4 = 0$ (obtaining $x = 2$ and $x = -2$).

2.3 Exercises

Identify those equations in Exercises 1 to 10 that are linear in either one or two variables and place them in standard form ($ax + b = 0$ or $ax + by = c$).

1. $y = 3x - 8$ $3x - y = 8$

2. $x = 5 + y$ $x - y = 5$

3. $C = x + 0.15x + 0.06x$ $C - 1.21x = 0$

4. $3x = 4$ $3x - 4 = 0$

5. $x^2 = 0$ not linear

6. $4n - 5 = 10n + 7$ $n + 2 = 0$

7. $2\pi r = 11$ $2\pi r - 11 = 0$

8. $6x^2 - 2x(3x - 4) = 7$ $8x - 7 = 0$

9. $t - 6n = 4n - 2$ $10n - t = 2$

10. $P = 4L$ $4L - P = 0$

Which of the functions in Exercises 11 to 20 are linear functions?

11. $f(x) = \frac{5}{9}(x - 32)$ linear

12. $f(x) = \frac{9}{5}x + 32$ linear

13. $C(x) = 2\pi x$ linear

14. $A(x) = \pi x^2$ not linear

15. $f(x) = x^2 + 2x$ not linear

16. $f(x) = 3x - x^2$ not linear

17. $v(t) = gt + v_0$, where g and v_0 are constants   linear

18. $h(t) = \frac{1}{2}gt^2 + v_0 t + h_0$, where g, v_0, and h_0 are constants
not linear

19. $h(x) = ax^2 + bx + c$, where a, b, and c are constants
not linear

20. $s(x) = 2ax + b$, where a and b are constants
linear

In Exercises 21 to 24, name the *x*- and *y*-intercepts.

21. 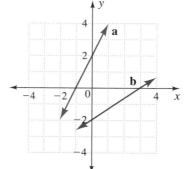   a: $-1, 2$; b: $3, -2$

22. 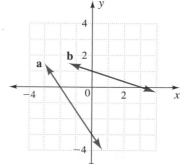   a: $-2, -3$; b: $3, 1$

23.

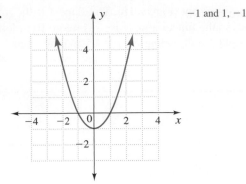

-1 and $1, -1$

24.

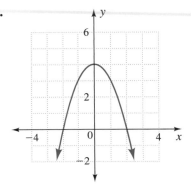

-2 and $2, 4$

Match each description in Exercises 25 to 28 with one of the lines graphed in Exercises 21 and 22.

25. This function decreases less rapidly than the other one.
22b

26. This function increases less rapidly than the other one.
21b

27. This function increases more rapidly than the other one.
21a

28. This function decreases more rapidly than the other one.
22a

Match the descriptions in Exercises 29 and 30 with the graphs in Exercises 23 and 24.

29. This function increases for $x < 0$ and decreases for $x > 0$. 24

30. This function decreases for $x < 0$ and increases for $x > 0$. 23

31. What is the slope of each line in Exercise 21? a: 2; b: $\frac{2}{3}$

32. What is the slope of each line in Exercise 22?
a: $-\frac{3}{2}$; b: $-\frac{1}{3}$

In Exercises 33 to 46, find the slope of a line that passes through the given points.

33. $(4, 5)$ and $(-2, 7)$ $-\frac{1}{3}$ **34.** $(3, -5)$ and $(6, 2)$ $\frac{7}{3}$

35. $(-3, -5)$ and $(5, -1)$ $\frac{1}{2}$ **36.** $(-4, -2)$ and $(6, -4)$
$-\frac{1}{5}$

37. $(5, 0)$ and $(0, -3)$ $\frac{3}{5}$ **38.** $(0, 4)$ and $(-5, 0)$ $\frac{4}{5}$

39. $(-2, 0)$ and $(0, -3)$ $-\frac{3}{2}$ **40.** $(4, 0)$ and $(0, 3)$ $-\frac{3}{4}$

41. $(1.5, 5.3)$ and $(2.6, -4.2)$ $-\frac{95}{11}$

42. $(-1.4, 2.3)$ and $(-0.1, 5.1)$ $\frac{28}{13}$

43. $\left(-1\frac{3}{4}, 4\frac{1}{2}\right)$ and $\left(2\frac{1}{2}, -1\frac{1}{4}\right)$ $-\frac{23}{17}$

44. $\left(2\frac{2}{3}, -1\frac{1}{3}\right)$ and $\left(3\frac{1}{3}, 1\frac{2}{3}\right)$ $\frac{9}{2}$

45. $\left(-1\frac{1}{3}, 3\frac{2}{3}\right)$ and $\left(1\frac{2}{3}, 2\frac{1}{3}\right)$ $-\frac{4}{9}$

46. $\left(2\frac{1}{2}, -1\frac{3}{4}\right)$ and $\left(3\frac{3}{4}, -2\frac{1}{2}\right)$ $-\frac{3}{5}$

In Exercises 47 to 52, decide if the table defines a linear function. If the function is linear,
(a) Find the slope and the units on the slope.
(b) Name the x-intercept point and its meaning in the problem setting.
(c) Name the y-intercept point and its meaning.

47. *Sales Tax Table*

Sales, x (dollars)	Sales Tax, y (dollars)
10	0.60
20	1.20
30	1.80
40	2.40

$0.06 tax/$ sales; $(0, 0)$, for \$0 of sales, tax is \$0; $(0, 0)$, there is 0 sales tax if there are 0 sales.

48. *Economy Theater Tickets*

Tickets, x	Cost, y (dollars)
1	4.00
2	8.00
3	12.00
4	16.00

\$4/ticket; $(0, 0)$, if no tickets are purchased, cost is \$0; $(0, 0)$, cost of \$0 occurs when no tickets are purchased.

49. *Mass Transit Ticket Value*

Trips, x	Value, y (dollars)
0	20.00
1	19.25
2	18.50
3	17.75

$-$\$0.75/trip; $\left(26\frac{2}{3}, 0\right)$, maximum number of trips is 26; $(0, 20)$, original value of ticket is \$20.00.

50. *Rock Falling from 100-meter Cliff*

Time, t (seconds)	Height, y (meters)
0	100
1	95.1
2	80.4
3	55.9
4	21.5

not linear

51. *Throwing a Ball*

Time, t (seconds)	Height, y (feet)
0	6.0
0.5	46.0
1.0	77.9
1.5	101.8
2.0	117.6

not linear

52. *Ice Cream Parlour Gift Certificate*

Cones, x	Value, y (dollars)
0	15.00
1	13.50
2	12.00
3	10.50

−$1.50/cone; (10, 0), maximum number of cones that can be purchased with gift certificate is 10; (0, 15), original value of gift certificate is $15.00.

In Exercises 53 to 56, sketch a graph satisfying the facts given. Place each pair of lines on one set of axes.

53. a. x-intercept 2, slope $-\frac{3}{2}$ See Answer Section.

b. y-intercept 3, slope 1

54. a. y-intercept -2, slope 3   See Additional Answers.

b. x-intercept $-$ 2, slope 2

55. a. x-intercept -2, slope $\frac{2}{3}$ See Answer Section.

b. y-intercept -3, x-intercept 2

56. a. y-intercept $\frac{3}{2}$, x-intercept $\frac{5}{2}$ See Additional Answers.

b. x-intercept -1, slope $-\frac{3}{4}$

57. Suppose we know the x-intercept, a, and y-intercept, b, of a line. Describe how we might use the intercepts a and b to find the slope of the line.   Slope is $-\frac{b}{a}$.

58. Describe how to draw the graph of a line, given one intercept and the slope.   Plot the intercept. Find a second point by counting up with the numerator value (down if it's negative) and counting right with the denominator value.

Think about the definition of slope. In Exercises 59 to 62, name the next point to the right of the given point having integer coefficients. Do not use a graph.

59. A line passing through (4, 5) with slope $-\frac{2}{3}$ (7, 3)

60. A line passing through $(-2, 4)$ with slope -3 $(-1, 1)$

61. A line passing through $(-4, 1)$ with slope $\frac{3}{5}$ (1, 4)

62. A line passing through $(3, -2)$ with slope $\frac{4}{3}$ (6, 2)

63. Make a table with three ordered pairs and find the slopes for the alternative photocopy card equations.

a. $y = 10 - 0.05x$
slope $= -0.05$

b. $y = 10 - 0.15x$
slope $= -0.15$

64. What is the slope of the line relating Celsius and Fahrenheit temperatures? State your choices for x and y. (Water freezes at 0°C and 32°F. Water boils at 100°C and 212°F.)   for $x = °C, y = °F, \frac{9}{5}$

■ **Projects**

65. Golden Gate Bridge Construction Data*

	Concrete Quantities	
	Cubic Yards	**Cubic Meters**
San Francisco pier and fender	130,000	99,400
Marin pier	23,500	17,970
Anchorages, pylons, and cable housings	182,000	139,160
Approaches	28,500	21,800
Paving	25,000	19,115

a. Let $x =$ cubic yards and $y =$ cubic meters. Make a table, listing the ordered pairs from smallest to largest, and find Δx, Δy, and slope.   slope ≈ 0.765

b. What can you conclude about cubic yards and cubic meters?   They are linearly related.

c. What is the meaning of the slope?   the conversion rate: 0.765 cubic meter in a cubic yard

	Structural Steel Quantities	
	Tons	**Kilograms**
Main towers	44,400	40,280,000
Suspended structure	24,000	21,771,000
Anchorages	4,400	3,991,000
Approaches	10,200	9,250,000

d. Let $x =$ tons and $y =$ kilograms (kg). Make a table, listing the ordered pairs from smallest to largest, and find Δx, Δy, and slope.   slope < 907

e. What can you conclude about tons and kilograms?   They are linearly related.

f. What is the meaning of the slope?   the conversion rate: 907 kilograms in a ton

66. Trigonometry and Slope   In construction, staircase and ramp slopes are often described in terms of degrees. For example, the preferred steepness of a staircase in a home is from 30° to 35° up from horizontal, whereas for an access ramp it is 7°. We use the [TAN] key on a calculator to change an angle in degrees into a slope. Set the calculator to degree measure under [MODE].

If your graphing calculator is correctly set, [TAN] 45 = 1. If you get [TAN] 45 = 1.619775 or [TAN] 45 = 0.85408, the calculator is not set to degree measure.

Use a calculator to find the slopes of these settings:

a. A 35° staircase ≈ 0.70 **b.** A 30° staircase ≈ 0.58

c. A 7° ramp ≈ 0.12

*Thanks to Diane Kinney, Brookings OR for the idea for this problem. Data are from www.goldengatebridge.org/research/factsGGBDesign.html.

To change slope or percent grade to degrees, use ⌈2nd⌉ ⌈TAN⌉. Make sure the calculator is set to degree measure. Verify that ⌈2nd⌉⌈TAN⌉ 1 = 45. If you get 0.785398 or 50, the calculator is not set to degree measure.

Find the angle measures for these:

d. A very steep road grade: 8% or 0.08 ≈4.6°

e. The freeway grade descending northbound from Siskiyou Summit on Interstate 5 in southern Oregon: 6%   ≈3.4°

f. As of 1973, the steepest standard gauge railroad gradient by adhesion, achieved by the Guatemalan State Electric Railway between the Samala River Bridge and Zunil: 1:11 ≈5.2°

Research question: Find the steepness of Lombard Street in San Francisco, Lookout Mountain in Chattanooga, or a well-known grade in your city or a nearby locality. Write it as both a percent grade and an angle.

---

2   Mid-Chapter Test

1. Show each of these inequalities on a number line. Write each as an interval.   See Answer Section.

a. $-4 < x$ $(-4, +\infty)$

b. $6 > x$ $(-\infty, 6)$

c. $-3 \le x \le 2$ $[-3, 2]$

d. $3 < x \le 6$ $(3, 6]$

e. $x > 3$ or $x < -2$
$(-\infty, -2), (3, +\infty)$

f. $x > -2$ or $x \le 3$
$\mathbb{R}$ or $(-\infty, +\infty)$

2. Write each as an inequality and an interval.

a. Inputs between -1 and 1, including the endpoints
$-1 \le x \le 1; [-1, 1]$

b. Inputs greater than -3, including -3
$x \ge -3; [-3, +\infty)$

c. Outputs greater than or equal to -2 $y \ge -2; [-2, +\infty)$

d. Outputs less than 4 or greater than -2
all real numbers, $\mathbb{R}$; $(-\infty, +\infty)$

e. Outputs less than or equal to 4 and greater than -2
$-2 < y \le 4; (-2, 4]$

3. Write each in words.

a. $[-1, 3)$ the set of numbers between -1 and 3, including -1

b. $-4 < x < -1$ the set of inputs between -4 and -1

c. $(-\infty, -2]$ the set of numbers less than or equal to -2

d. $y \le -1$ the set of outputs less than or equal to -1

4. Calls on a cell phone cost $16.45 for the first 30 minutes and $0.29 for each additional minute. Write equations for the total cost, with conditions on the input.
$y = 16.45 + 0.29(x - 30)$ for $x > 30$; $y = 16.45$ for $0 < x \le 30$

Complete the statements in Exercises 5 to 7.

5. The set of numbers $x \ge 0$ is called __non-negative__.

6. The set of inputs to a function is called the __domain__.

7. The ordered pair describing the intersection of a graph and the vertical axis is written __$(0, y)$__.

8. Evaluate the function $f(x) = 3x - 5$ for the indicated inputs.

a. $f(1)$ -2

b. $f(3)$ 4

c. $f(-5)$ -20

d. $f(a)$ $3a - 5$

e. $f(a + b)$ $3(a + b) - 5$ or $3a + 3b - 5$

9. Evaluate the function $f(x) = x^2 - x$ for the inputs listed in Exercise 8.
0; 6; 30; $a^2 - a$; $(a + b)^2 - (a + b)$ or $a^2 + 2ab + b^2 - a - b$

For each graph in Exercises 10 to 12, answer the following questions:
(a) What is the domain? Write as both an inequality and an interval.
(b) What is the range? Write as both an inequality and an interval.
(c) Does the graph describe a function?

10.

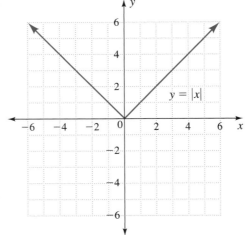

all real numbers, $-\infty < x < +\infty$, $(-\infty, +\infty)$; non-negative, $y \ge 0$, $[0, \infty)$; yes

11.

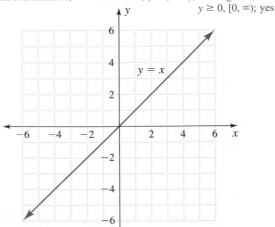

all real numbers, $-\infty < x < +\infty$, $(-\infty, +\infty)$; all real numbers, $-\infty < y < +\infty$, $(-\infty, +\infty)$; yes

12. circle $-5 \le x \le 1$, $[-5, 1]$; $-3 \le y \le 3$, $[-3, 3]$; no

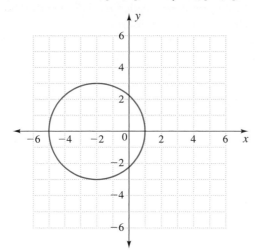

13. Write each linear equation in standard form, $ax + by = c$.

a. $y = 5$ $0x + 1y = 5$

b. $x = 4$ $1x + 0y = 4$

c. $2\pi r = 7$ $2\pi r + 0y = 7$

d. $y = x^2 + 3x$ not linear

14. Find the horizontal and vertical intercepts for each equation.

a. $3x + 4y = 12$ $(4, 0); (0, 3)$

b. $F = \frac{9}{5}C + 32$ (Assume C is the independent variable and F the dependent variable.) $(-17.78, 0); (0, 32)$

15. Find the slope and vertical axis intercept from each table.

a.

Input	Output
1	2
2	-1
3	-4
4	-7

slope $-\frac{3}{1}$, intercept $(0, 5)$

b. *Moorage Fees (Boston #1)*

Boat Length (feet)	Cost (dollars)
25	92.50
36	131
42	152

slope \$3.50 per foot, intercept $(0, 5)$

16. On one set of axes, sketch the linear functions satisfying the facts given.

a. y-intercept 4, slope $-\frac{2}{3}$

b. x-intercept -2, slope -3

See Answer Section.

2.4 Modeling with a Linear Function

Objectives

▌ Find a linear equation using the slope-intercept equation.

▌ Find a linear equation using the point-slope equation.

▌ Find a linear equation using a table.

▌ Find a linear equation using calculator regression.

WARM-UP

1. Use $y = mx + b$, with $m = -\frac{3}{2}$, $x = -4$, and $y = 1$, to solve for b. $b = -5$

2. Find the slope for lines passing through these points: $(11, 2.16)$ and $(17, 3.3)$. $\frac{0.19}{1}$

3. Change to $y = mx + b$ form: $y - 6 = -\frac{3}{2}(x - 4)$ $y = -\frac{3}{2}x + 12$

In Exercises 4 to 6, what is the next number in the number pattern? What is the rule for each pattern?

4. 8, 16, 24, 32, ___40___ $y = 8x$

5. 8, 9, 10, 11, ___12___ $y = x + 7$

6. 8, 9.5, 11, 12.5, ___14___ $y = 1.5x + 6.5$

IF WE HAVE REASON to believe that the data in a problem are linearly related, we can *find the equation describing the function*. This process is called **modeling with a linear function**. To *model* means to give shape or form to information.

This section introduces four ways to obtain the equation describing a linear function. We use information about slopes and intercepts or slopes and ordered pairs to find linear equations from word problems or graphs. We fit a line to data with a table and with a calculator. Finally, we resolve the question of why constant differences lead to the equation of a number pattern.

Slope-Intercept Equation

All linear functions may be simplified to a common form, the slope-intercept equation.

SLOPE-INTERCEPT EQUATION

> The **slope-intercept equation** of a straight line is
>
> $$y = mx + b$$
>
> where m is the slope and b is the y-intercept, $f(0)$.

We have written many equations in this form already. In Example 1, we return to two linear functions introduced in Section 2.3 and preview one used in Section 2.5.

EXAMPLE 1 Finding slope and y-intercept from an equation   Name the slope and y-intercept in each of these equations, and tell the meaning of each in the problem setting.
a. Photocopy card: $y = 10 - 0.10x$, where x = number of copies and y = value in dollars
b. Chocolate bar profit: $y = 0.75x - 15$, where x = number of bars and y = profit in dollars
c. Car rental cost: $y = 0.10x + 30$, where x = number of miles and y = cost in dollars

SOLUTION As needed, we rewrite the equation to fit $y = mx + b$.
a. We rewrite $y = 10 - 0.10x$ as $y = -0.10x + 10$. The slope is $-\$0.10$ per photocopy; the y-intercept is $\$10$, the initial cost of the card.

b. The slope is $\$0.75$ profit per bar; the y-intercept is $\$15$, the cost of the display.

c. The slope is $\$0.10$ per mile; the y-intercept is $\$30$, the basic cost.   ▬

The process in Example 2 is the opposite of the process in Example 1. The information is given in numbers so that you can build the equations.

EXAMPLE 2 Finding the equation given the slope and y-intercept   Use the slope and y-intercept to write the linear equation for each of these cases.
a. $m = 5, b = \frac{1}{2}$
b. $m = \frac{1}{2}, b = 5$
c. $m = 5, b = 0$

SOLUTION We substitute the numbers for m and b into $y = mx + b$.

a. $y = 5x + \frac{1}{2}$

b. $y = \frac{1}{2}x + 5$

c. $y = 5x + 0$, or $y = 5x$ ▬

Think about it 1: Solve $y = mx + b$ for b. What does this equation give?

▨ To find the equation of a line using the slope-intercept equation, we use the slope formula and then $b = y - mx$, obtained from $y = mx + b$.

▨ USING THE SLOPE-INTERCEPT EQUATION

Using a slope formula and then a formula for b seems more natural to most students. It parallels linear regression with a graphing calculator. The calculator finds a and b for $y = ax + b$.

---

1. Find the slope of the line through (x_1, y_1) and (x_2, y_2):

$$m = \frac{y_2 - y_1}{x_2 - x_1}$$

2. Find the y-intercept of the line:

$$b = y_1 - mx_1$$

3. Place m and b into $y = mx + b$.

EXAMPLE 3 **Using the slope-intercept equation** Find the equation of a line through $(-4, 1)$ and $(0, -5)$.

SOLUTION We use the steps outlined above.

Step 1: The slope is

$$m = \frac{y_2 - y_1}{x_2 - x_1} = \frac{-5 - 1}{0 - (-4)} = -\frac{6}{4} = -\frac{3}{2}$$

Step 2: The y-intercept is $b = y - mx$. We let $(x, y) = (-4, 1)$ and $m = -\frac{3}{2}$ from part a.

$$b = 1 - \left(-\tfrac{3}{2}\right)(-4) = 1 - 6 = -5$$

Step 3: We place m and b into the slope-intercept equation.

$$y = mx + b$$
$$y = -\tfrac{3}{2} x + (-5)$$
$$y = -\tfrac{3}{2} x - 5$$

▬

Think about it 2: Why was it not necessary to calculate the y-intercept in step 2?

Point-Slope Equation

Although the slope-intercept equation can be used for any nonvertical linear data, finding the y-intercept is sometimes difficult. In such cases, using the point-slope equation may be easier.

▨ POINT-SLOPE EQUATION

The point-slope equation is optional because we can enter the two points into a graphing calculator, perform a linear regression, and obtain the linear equation in slope-intercept form.

The **point-slope equation** is

$$y - y_1 = m(x - x_1)$$

where m is the slope and (x_1, y_1) is a point on the line.

Example 4 suggests how the point-slope equation might have been discovered.

EXAMPLE 4 **Using the point-slope equation: long-distance telephone costs** It costs \$2.16 to make an 11-minute long-distance call and \$3.30 to make a 17-minute long-distance call.
  a. Explain how to assign the variables x and y to the problem setting. Write the information as two ordered pairs.
  b. Draw a graph, and connect the data points with a straight line.

c. Use the two points to find the slope.

d. Find another expression for the slope: Use one point, such as (17, 3.30), and (x, y) in the slope formula. Set this expression equal to the slope in part c. Solve for the $y = mx + b$ form.

e. Does the y-intercept from the graph agree with the one from the equation in part d?

SOLUTION   a. The cost of a call is a function of the length of the call. Because y is a function of x, we let cost in dollars be y and time in minutes be x. The ordered pairs are (11, 2.16) and (17, 3.30).

b. The graph is shown in Figure 19.

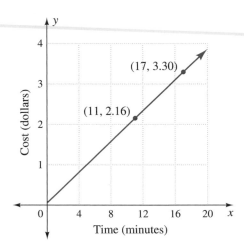

FIGURE 19

c. We find the slope as follows:

$$m = \frac{y_2 - y_1}{x_2 - x_1} = \frac{3.30 - 2.16}{17 - 11} = \frac{1.14}{6} = \frac{0.19}{1}, \text{ or } \$0.19 \text{ per min}$$

d. Using the point (17, 3.30) and (x, y) in the slope formula, we have

$$m = \frac{y_2 - y_1}{x_2 - x_1} = \frac{y - 3.30}{x - 17}$$

The two slope formulas are equal, so we set them equal and solve for y:

Student Note: The second step shows the point-slope form of the linear equation.

$$\frac{y - 3.30}{x - 17} = \frac{0.19}{1} \qquad \text{Multiply both sides by } (x - 17).$$

$$y - 3.30 = 0.19(x - 17) \qquad \text{Distribute 0.19 over } (x - 17).$$

$$y - 3.30 = 0.19x - 3.23 \qquad \text{Add 3.30 to both sides.}$$

$$y = 0.19x - 3.23 + 3.30 \qquad \text{Simplify.}$$

$$y = 0.19x + 0.07$$

e. The y-intercept is $0.07. The graph agrees, showing a small positive intercept.

Think about it 3: Use the slope, $m = 0.19$, and the data point (11, 2.16) to find the y-intercept by substituting into $b = y_1 - mx_1$. Does your result agree with the y-intercept found in Example 4?

■ USING THE POINT-SLOPE EQUATION

> **1.** Find the slope:
>
> $$m = \frac{y_2 - y_1}{x_2 - x_1}$$
>
> **2.** Substitute the slope and one point (x_1, y_1) into $y - y_1 = m(x - x_1)$.
> **3.** Solve for y.

EXAMPLE 5 Using the point-slope equation   Find the equation of the line passing through $(4, 6)$ and $(6, 3)$.

SOLUTION   *Step 1:* The slope is

$$\frac{y_2 - y_1}{x_2 - x_1} = \frac{3 - 6}{6 - 4} = -\frac{3}{2}$$

Step 2: We substitute the slope, $-\frac{3}{2}$, and $(4, 6)$ into $y - y_1 = m(x - x_1)$:

$$y - 6 = -\frac{3}{2}(x - 4)$$

Step 3: We solve for y:

$$y - 6 = -\frac{3}{2}x + 6$$
$$y = -\frac{3}{2}x + 12$$

Fitting an Equation with a Table

Most sets of data have far more than the two ordered pairs we saw in Examples 3 to 5. Nevertheless, when a graph suggests that data might be linearly related, we can find an equation that approximately "fits" the data. We will start with a table to estimate an equation, $y = mx + b$.

In Example 6, look for how we estimate an average to find a slope and look for the effect of having $\Delta x = 10$ between the first data point $(x = 10, y = 3.29)$ and the y-intercept, $f(0)$, which is at $x = 0$.

EXAMPLE 6 Finding a linear equation with a table   A local discount store has these prices on medium-duty extension cords (16/3 SJTW): 10 feet, \$3.29; 22 feet, \$3.99; 50 feet, \$5.99; 100 feet, \$8.99.

For students who are not verbally adept, the table model provides a visual method of finding linear equations. It reinforces function notation, especially the vertical intercept, $f(0)$.

a. Graph the data.
b. Suggest whether a linear function might describe the data.
c. Use a table to estimate the slope and y-intercept.
d. What does the equation mean in the problem setting?

SOLUTION   **a.** The graph is in Figure 20.

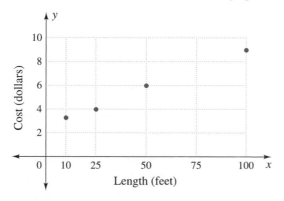

FIGURE 20

TABLE 13   *Medium-Duty Extension Cords (16/3 SJTW)*

Δx	Length (feet)	Cost (dollars)	Δy	Slope, $\Delta y/\Delta x$
10 ⟨	0	2.69	⟩ 10(0.06) = 0.60	0.06
15 ⟨	10	3.29	⟩ 0.70	0.0467
25 ⟨	25	3.99	⟩ 2.00	0.08
50 ⟨	50	5.99	⟩ 3.00	0.06
	100	8.99		

b. Relatively speaking, the 25-foot cord price is lower than the other three, but one can visualize a line passing near all four points.

c. Table 13 shows Δy, Δx, and slope.

The slopes are 0.0467, 0.08, and 0.06, so we estimate an average slope of 0.06. Working backwards in the table gives $\Delta y = 10(0.06) = 0.60$ between $x = 10$ and $x = 0$. **Because slope is the unit change, we must have slope times the number of units x of change to get Δy.** Subtracting 0.60 from 3.29 gives the y-intercept, 2.69. A linear function that approximates the data is

$$y = 0.06x + 2.69$$

d. The equation suggests that all the extension cords have a common cost (parts, assembly packaging, and shipping) of $2.69 plus a cost of $0.06 per foot (materials).

In Example 7, again watch for how we estimate the slope and adjust for the 12-unit spacing between $x = 12$ and $x = 0$ in finding the corresponding Δy.

EXAMPLE 7 **Finding a linear equation with a table** A pet store has these prices on TetraMin tropical flakes: 12 grams, $1.99; 28 grams, $2.99; 62 grams, $5.99.
a. Graph the data.
b. Suggest whether a linear function might describe the data.
c. Use an input-output table to estimate the slope and y-intercept.
d. What does the equation mean in the problem setting?

SOLUTION **a.** The graph is in Figure 21.

b. We can visualize a line passing near the three points.

c. Table 14 shows Δy, Δx, and slope.

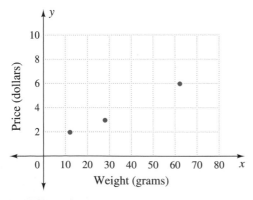

FIGURE 21

TABLE 14 *TetraMin Tropical Flakes*

Δx	Size (grams)	Cost (dollars)	Δy	Slope, $\Delta y/\Delta x$
12 ⟨	0	1.09	⟩ 12(0.075) = 0.90	0.075
16 ⟨	12	1.99	⟩ 1.00	0.0625
34 ⟨	28	2.99	⟩ 3.00	0.088
	62	5.99		

The slopes are 0.0625 and 0.088, so we estimate an average slope of 0.075. Working backwards in the table gives $\Delta y = 12(0.075) = 0.90$ between $x = 12$ and $x = 0$. The y-intercept is $1.99 - 0.90 = 1.09$. A linear function that approximates the data is

$$y = 0.075x + 1.09$$

d. The equation suggests that the basic cost of the fish food (packaging, shipping) is $1.09 and the contents add $0.075 per gram.

■ SUMMARY: FINDING A LINEAR
EQUATION FROM A TABLE

1. Graph the data to see whether they have a linear shape.
2. Use the change in x and change in y to find the slope between ordered pairs. Estimate an average slope, m, for the table.
3. Using the slope, work backwards in the table to find the output, b, for $x = 0$. In working backwards, if the first input is not $x = 1$, multiply the slope by Δx to get Δy.
4. State the equation in the form $y = mx + b$.
5. Graph the equation to see that it is reasonably close to the data.

We will graph the equations from Examples 6 and 7 in Examples 10 and 11.

A SPECIAL CASE: FINDING EQUATIONS FOR NUMBER PATTERNS WITH CONSTANT DIFFERENCES   In Section 1.3, we explored rules for number patterns. We placed the number pattern in the output column of an input-output table, and the natural numbers (1, 2, 3, 4, . . .) in the input column. With the table process from Examples 6 and 7, we find exact rules for number patterns with constant differences.

In Example 8, look for how $\Delta x = 1$ makes the constant difference, Δy, equal to the slope of a linear equation—and shortens the work in the table.

EXAMPLE 8   **Finding an equation with a table**   Place the pattern starting 8, 9.5, 11, 12.5, . . . into an input-output table. Find Δx and Δy. Verify that the slope, $\Delta y/\Delta x$, equals $\Delta y/1$, the change in y. Work backwards in the table to find the number where $x = 0$ (the y-intercept). Write the rule (equation) for the pattern.

SOLUTION   **TABLE 15**

Δx	x	y	Δy
1 ⟨	0	6.5	⟩ 1.5
1 ⟨	1	8	⟩ 1.5
1 ⟨	2	9.5	⟩ 1.5
1 ⟨	3	11	⟩ 1.5
	4	12.5	

The slope in Table 15 is constant:

$$m = \frac{\Delta y}{\Delta x} = \frac{1.5}{1} = 1.5$$

The data are linear. Working backwards in the chart to $x = 0$ gives the y-intercept, b. At $x = 0$, $y = 6.5$. The equation as $y = mx + b$ is

$$y = 1.5x + 6.5$$

Check: In Warm-up Exercise 6, we found 14 to be the output for $x = 5$:

$$14 \stackrel{?}{=} 1.5(5) + 6.5$$

$$14 = 14 \ \checkmark$$

EXAMPLE 9   **Finding an equation**   Place the pattern starting 8, 16, 24, 32, . . . into an input-output table. Find Δx and Δy. If appropriate, find the slope, $\Delta y/\Delta x$, and the number where $x = 0$ (the y-intercept). Write the rule (equation) for the pattern.

SOLUTION   **TABLE 16**

Δx	x	y	Δy
1	0	0	
1	1	8	8
1	2	16	8
1	3	24	8
	4	32	8

The slope in Table 16 is constant:

$$m = \frac{\Delta y}{\Delta x} = \frac{8}{1} = 8$$

The data are linear. Working backwards in this chart to $x = 0$ gives the y-intercept, b. At $x = 0$, $y = 0$. The equation as $y = mx + b$ is

$$y = 8x + 0$$

Check: In Warm-up Exercise 4, we found 40 to be the output for $x = 5$:

$$40 \overset{?}{=} 8(5) + 0$$

$$40 = 40 \ \checkmark$$

An arithmetic sequence is another name for a number pattern that can be modeled with a linear equation. An **arithmetic sequence** is *a function with the natural numbers (or positive integers) as its set of inputs, and numbers with a constant difference between the numbers as its set of outputs.* The arithmetic sequences always have $\Delta x = 1$ and $\Delta y =$ a constant.

Think about it 4: Using a table, find the equation for the arithmetic sequence in Warm-up Exercise 5: 8, 9, 10, 11, ...

Fitting an Equation with Linear Regression

The graphing calculator statistical functions include a feature called **linear regression** or LinReg. Although you may be surprised at the accuracy of the table solutions found thus far, there is no question that for large amounts of data or "messy" numbers, a calculator approach is welcome.

■■ GRAPHING CALCULATOR TECHNIQUE: LINEAR REGRESSION

1. Begin by pressing (STAT)(ENTER).
2. Clear prior lists L1 and L2. To clear, move the cursor up to L1, press (CLEAR), and then move the cursor down; the list will be emptied. Repeat as needed for other lists.
3. Enter the new data, usually in L1 and L2. Set up (WINDOW) and graph using (2nd) [STAT PLOT].
4. Select linear regression: (STAT) [CALC] 4. Specify the location of the lists, usually L1 and L2, and press (ENTER). Record information about a, b, and r.

Because LinReg uses $y = ax + b$ instead of $y = mx + b$, use a in place of m. (More about r after we use LinReg in Examples 10 and 11 to redo Examples 6 and 7.)

EXAMPLE 10   Finding an equation with linear regression   Using linear regression, fit a linear equation to the extension cord cost data: (10 ft, $3.29), (25 ft, $3.99), (50 ft, $5.99), (100 ft, $8.99).

a. Enter the data into L1 and L2. Graph the data on the calculator to verify that a linear function is appropriate. Perform the linear regression. State the equation with a and b rounded to three decimal places and r rounded to four decimal places.

b. Graph the regression equation with the data and graph $y = 0.06x + 2.69$ from Example 6.

c. Compare the regression equation with the equation from Example 6.

SOLUTION   **a.** The regression results are in Figure 22: $y = 0.065x + 2.577$, $r = 0.9979$.

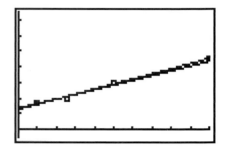

FIGURE 22   **FIGURE 23**   0 to 100 on X, -1 to 15 on Y

b. The data and graphs of the equations are in Figure 23. Both graphs are close to all data points.

c. From Example 6, $y = 0.06x + 2.69$. The slopes are within a half cent, and the y-intercepts differ by 12 cents.

EXAMPLE 11   Finding an equation with linear regression   Using linear regression, fit a linear equation to the fish food cost data: (12 grams, $1.99), (28 grams, $2.99), (62 grams, $5.99).

a. Enter the data into L1 and L2. Graph the data on the calculator to verify that a linear function is appropriate. Perform the linear regression. State the equation with a and b rounded to three decimal places and r rounded to four decimal places.

b. Graph the regression equation with the data and graph $y = 0.075x + 1.09$ from Example 7.

c. Compare the regression equation with the equation from Example 7.

SOLUTION   **a.** The regression results are in Figure 24: $y = 0.081x + 0.893$, $r = 0.9971$.

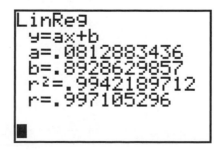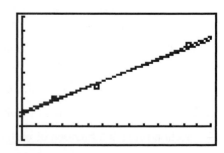

FIGURE 24   **FIGURE 25** -1 to 70 on X, -1 to 8 on Y

b. The data and graphs of the equations are in Figure 25. Both graphs are close to all data points.

c. From Example 7, $y = 0.075x + 1.09$. The slopes are within 1 cent, and the y-intercepts are within 20 cents.

COEFFICIENT OF CORRELATION   The calculator calculates the **coefficient of correlation**, r, to *determine how well the data fit a straight line* If $r = 1$ or $r = -1$, the data are perfectly linear. If $r = 0$, the data are not linear. The inequality $-1 \le r \le 1$ describes the range for r. Figure 26 shows the range with sample sets of data. When the coefficient of correlation is close to -1 or $+1$, we may have a good fit to a line.

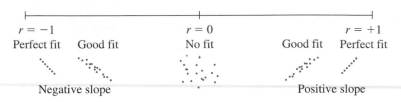

| $r = -1$ | | $r = 0$ | | $r = +1$ |
| Perfect fit | Good fit | No fit | Good fit | Perfect fit |

Negative slope   Positive slope

FIGURE 26 Coefficient of correlation

Figure 26 shows that a negative coefficient of correlation matches with a negatively sloped line and a positive coefficient of correlation indicates a positively sloped line. The project in Exercise 63 explores sets of data with a coefficient of correlation near zero.

ANSWER BOX

Warm-up: **1.** $b = -5$ **2.** $\frac{0.19}{1}$ **3.** $y = -\frac{3}{2}x + 12$ **4.** 40 **5.** 12 **6.** 14
Think about it 1: Subtract mx from each side of $y = mx + b$ to obtain $b = y - mx$. This equation gives the y-intercept of the graph of $y = mx + b$.
Think about it 2: The ordered pair $(0, -5)$ is a y-intercept point. From that we know that $b = -5$.   **Think about it 3:** $b = y_1 - mx_1$, $b = 2.16 - (0.19)(11)$, $b = 0.07$, as before. It does not matter which ordered pair we use in finding the equation.   **Think about it 4:** $y = 1x + 7$

2.4   Exercises

Name the slope and vertical axis intercept, or y-intercept, for the equations in Exercises 1 to 10.

1. Amount of sales tax: $y = 0.055x$ $0.055 per dollar; $0

2. Total cost for a pizza party: $y = 3.50x$ $3.50 per person; $0

3. Total cost of a party at a wave pool:
$y = 3.00x + 10$ $3.00 per person; $10

4. Total cost of a party at a bowling alley:
$y = 5.50x + 15$ $5.50 per person; $15

5. Circumference is a function of radius: $C = 2\pi r$ $2\pi; 0$

6. Circumference is a function of diameter: $C = \pi d$ $\pi; 0$

7. Friction force is a function of normal force N: $F = \mu N$
$\mu; 0$

8. Distance traveled is a function of time: $d = 55t$ $55; 0$

9. Consumption (spending) is a function of income Y:
$C = a + bY$ $b; a$

10. Total cost is a function of the number of units produced:
$y = Vx + F$ $V; F$

Build a linear equation from the information given in Exercises 11 to 20.

11. Slope is 8; y-intercept is -4. $y = 8x - 4$

12. Slope is -4; y-intercept is $\frac{1}{2}$. $y = -4x + \frac{1}{2}$

13. Slope is $\frac{1}{2}$; y-intercept is -8. $y = \frac{1}{2}x - 8$

14. Slope is 0; y-intercept is -4. $y = -4$

15. Slope is -2; y-intercept is 0. $y = -2x$

16. Slope is $\frac{1}{4}$; y-intercept is -2. $y = \frac{1}{4}x - 2$

17. $(3, 6)$ and $(0, -2)$ are on the line. $y = \frac{8}{3}x - 2$

18. $(-3, 2)$ and $(0, 4)$ are on the line. $y = \frac{2}{3}x + 4$

19. $(-2, 4)$ and $(5, -3)$ are on the line. $y = -x + 2$

20. $(4, -3)$ and $(-2, 1)$ are on the line. $y = -\frac{2}{3}x - \frac{1}{3}$

21. According to a newspaper medical columnist, aerobic exercise is safe and effective when your pulse rate is between 50 and 70% of your maximum pulse rate. Your maximum pulse rate is found by subtracting your age from 220.*

a. Which is correct: Pulse rate is a function of age, or age is a function of pulse rate?
Pulse rate is a function of age.

b. In what range is your safe exercise pulse rate?

c. Write a formula that describes the lower (50%) safe exercise pulse rate. $P = 0.5(220 - x)$

d. Write a formula that describes the upper (70%) safe exercise pulse rate. $P = 0.7(220 - x)$

e. What range of pulse rates is safe for a 50-year-old?
85 to 119

f. For what age is the pulse rate between 95 and 133?
30

22. A budget checking account costs $5.00 per month for the first 15 checks and $0.50 per check for each additional check. Zero is the minimum balance. The total cost per month is a function of the number x of checks written.

a. What is the cost for 10 checks? $5.00

b. What is the cost for 25 checks? $10.00

c. Write the cost function for 0 to 15 checks.
$C = 5.00 for $0 < x \le 15$

d. What is the rate of change, or slope, of the function for 0 to 15 checks? 0

e. What is the y-intercept? $5.00

f. What is the meaning, if any, of the y-intercept?
cost per month

g. Write the cost function for more than 15 checks.
$C = 0.50(x - 15)$ for $x > 15$, C in $

h. What is the rate of change, or slope, of the function for more than 15 checks? $0.50 per check

Exercises 23 and 24 refer to accounting, where total cost, C, is related to variable cost per unit, V, and fixed cost, F, by the equation $C = Vx + F$.

23. The bank charges 2.5% of the amount of the loan (called *points*) plus $300 in fees. What is the fixed cost? What is the variable cost per dollar borrowed? What is the cost function to borrow x dollars? (This cost does not include the interest paid on the loan!)
$300; $0.025; $C = 0.025x + 300$, C in $

24. Suppose blue jeans cost $17.80 per pair plus a $250 order handling fee. What is the fixed cost? What is the variable cost per pair? What is the cost function to buy x pairs of jeans? $250; $17.80; $C = 17.80x + 250$, C in $

In Exercises 25 to 30, find the equations of lines through the given points.

25. (2, 5) and (5, 4)
$y = -\frac{1}{3}x + \frac{17}{3}$

26. (1, 2) and (4, 1)
$y = -\frac{1}{3}x + \frac{7}{3}$

27. (5, 4) and (4, 1)
$y = 3x - 11$

28. (2, 5) and (1, 2)
$y = 3x - 1$

29. (4, 1) and (2, 5)
$y = -2x + 9$

30. (4, 1) and (−2, −2)
$y = \frac{1}{2}x - 1$

In Exercises 31 and 32, total cost, C, is related to variable cost per unit, V, and fixed cost, F, by the equation $C = Vx + F$.

31. Maria, owner of the Runner's Outlet, orders 250 pairs of shoes for $11,000. Three months later, under the same purchase plan, she orders 300 pairs of shoes for $13,185. What is the cost function? What is the fixed cost? What is the variable cost per pair?
$C = 43.70x + 75$, C in $; $75; $43.70

32. Yuan predicts that manufacturing 10,000 calculators will cost $125,900, while manufacturing 15,000 calculators will cost $177,350. What is the manufacturing cost function? What is the fixed cost? What is the variable cost per unit? $C = 10.29x + 23,000$, C in $; 23,000; $10.29

33. Let $(C, F) = (x, y)$. Use the coordinates (100, 212) and (0, 32) to derive the temperature conversion formula for Fahrenheit in terms of Celsius. $F = \frac{9}{5}C + 32$

34. Let $(F, C) = (x, y)$. Use the coordinates (212, 100) and (32, 0) to derive the temperature conversion formula for Celsius in terms of Fahrenheit.
$C = \frac{5}{9}F - \frac{160}{9}$ or $C = \frac{5}{9}(F - 32)$

In Exercises 35 to 38, use the point-slope equation.

35. Fall sale prices of garden plants are $0.99 for 4-inch pots and $1.99 for 6-inch pots. Fit a linear equation.
$y = 0.5x - 1.01$

36. Tall kitchen garbage bags are $1.99 for 15 bags and $4.79 for 60 bags. Fit a linear equation.
$y = 0.062x + 1.057$

37. Atta cat food is $4.99 for 7 lb and $7.99 for 16 lb. Fit a linear equation. $y = x/3 + 2.66$

38. Whiskas ground food with Bits of Beef is $0.69 for a 350-gram can and $1.00 for a 468-gram can. Fit a linear equation. $y = 0.0026x - 0.229$

In Exercises 39 to 44, use a table to fit a linear equation to the data and then use linear regression to fit an equation.

39. Pyrex glass measuring cups (capacity, price): 1 cup, $2.79; 2 cups, $3.29; 4 cups, $4.79; 8 cups, $6.79
average slope 0.60, $y = 0.60x + 2.19$; $y = 0.578x + 2.247$, $r = 0.9962$

40. Mirro sauce pans (capacity, price): 1 quart, $7.97; 2 quarts, $9.97; 3 quarts, $10.97
average slope 1.5, $y = 1.50x + 6.47$; $y = 1.5x + 6.637$, $r = 0.9820$

41. Fujifilm CD-Rs (number, price): 5, $3.99; 30, $14.99; 50, $19.99 average slope 0.35, intercept $3.99 - 5(0.35) = 2.24$, $y = 0.35x + 2.24$; $y = 0.359x + 2.818$, $r = 0.9889$

42. WearEver saute pans (diameter, price): 6 inches, $11.99; 10 inches, $15.99; 12 inches, $19.99 average slope 1.5, intercept $11.99 - 6(1.5) = 2.99$, $y = 1.5x + 2.99$; $y = 1.286x + 3.99$, $r = 0.9820$

43. Atta Boy dog food (size, cost): 4 kg, $4.99; 8 kg, $6.99; 19.28 kg, $9.99 average slope 0.38, intercept $4.99 - 0.38(4) = 3.47$, $y = 0.38x + 3.47$; $y = 0.314x + 4.051$, $r = 0.9882$

44. Duraco plastic planters (size, cost): 4.5 inches, $1.62; 6.5 inches, $2.83; 8.5 inches, $3.98 average slope 0.6, $y = 0.6x - 1.08$; $y = 0.59x - 1.025$, $r = 0.9999$

In Exercises 45 to 50, use a table to fit an equation to the sequence of numbers.

45. −8, −3, 2, 7, 12, . . . $y = 5x - 13$

46. −7, −3, 1, 5, 9, . . . $y = 4x - 11$

47. 9, 11, 13, 15, 17, . . . $y = 2x + 7$

*From *To Your Good Health*, Dr. Paul Donohue, © 1992, North American Syndicate.

48. 7, 10, 13, 16, 19, . . . $y = 3x + 4$

49. 2, 8, 14, 20, 26, . . . $y = 6x - 4$

50. 8, 1, −6, −13, −20, . . . $y = 15 - 7x$

51. Danielson snap-on fishing floats (diameter, price): 1.25 inches, $0.29; 1.5 inches, $0.39; 1.75 inches, $0.49; 2 inches, $0.59. Check suitability for a linear regression and fit a linear equation. $y = 0.4x - 0.21, r \approx 1$

52. The Clean Card provides 76 minutes of long-distance telephone calls for $5, 153 minutes for $10, and 306 minutes for $20. Check suitability for a linear regression and fit a linear equation. $y = 0.065x + 0.033, r \approx 1$

53. Suggest why $r \approx 1$ in Exercise 51.   See below.

54. Suggest why $r \approx 1$ in Exercise 52.   See below.

55. For the NCAA basketball tournament, the #1 team is matched with the #16 team in a region, the #2 team is matched with the #15 team, the #3 team with the #14 team, and so forth. Let x be the ranking of the first team in the match-up and y be the ranking of the opponent. Is the match-up a linear function? If so, give its equation.
linear; $y = -x + 17$

56. Large eggs weigh a minimum of 50 grams each and cost $0.99 per dozen. Extra-large eggs weigh a minimum of 56 grams each and cost $1.09 per dozen. Jumbo eggs weigh a minimum of 63 grams each and cost $1.19 per dozen. Find total weight for each dozen. Fit a linear equation for the cost of the dozen eggs as a function of weight. $y \approx 0.0013x + 0.225, x$ in g, y in $

57. Hand-packed ice cream costs $2.75 for 12 ounces, $4.75 for 24 ounces, and $7.95 for 50 ounces. Draw a graph for cost as a function of weight. Fit a linear equation.
See Answer Section; $y \approx 0.135x + 1.29, x$ in oz, y in $

58. Income eligibility guidelines for free weatherization services by a local utility are based on the number of people in the household and annual income: (1, $16,680), (2, $19,080), (3, $21,420), (4, $23,820), (5, $25,740). Fit a linear equation.
$y \approx 2286x + 14490, x$ in no. of people, y in $

59. Sale prices in 1984 for round diamond pendants set in 14K gold are listed below. Let x = carat value and y = price. Fit a linear equation.
$y \approx 1051.3x - 32.9$

Size in Carats	0.05	0.10	0.15	0.20	0.25	0.30	0.50
Price	$39	$79	$99	$159	$229	$299	$495

53. $0.10 is added to the price for each quarter-inch increase in diameter. Each ordered pair exactly fits the price increase rule. Note that 0.10 to $\frac{1}{4}$ = 0.40 to 1, which is the slope.

54. The cost per minute is 0.0654 or 0.0658 for the three ordered pairs. There is only a 3-cent y-intercept.

60. Describe how you can use the slope a/b of a line and an ordered pair (c, d) for a point on the line to find an ordered pair for another point on the line. Your answer can include sentences, a picture, and/or an expression.

61. How do the input-output table and changes in x and y provide the information needed to find the equation for any linear function?
The slope is $m = \Delta y/\Delta x$, and the output for $x = 0$ is b in $y = mx + b$.

Projects

62. Arithmetic Sequences   In Examples 8 and 9, we used a table to find the linear function, $y = mx + b$, modeling an arithmetic sequence. Another way to build the equation for an arithmetic sequence uses just the first term, a, and the difference $(d = \Delta y)$: $y = a + (x - 1)d$. Use this new formula to find the equation for each sequence below.

a. 8, 9.5, 11, 12.5, . . . $y = 1.5x + 6.5$

b. 8, 16, 24, 32, . . . $y = 8x$

c. 8, 9, 10, 11, . . . $y = x + 7$

d. 3, 7, 11, 15, 19, . . . $y = 4x - 1$

e. 12, 5, −2, −9, −16, . . . $y = 19 - 7x$

f. 8, 3, −2, −7, −12, . . . $y = 13 - 5x$

Why won't the rule for sequences work for all linear functions?   The inputs must be natural numbers having $\Delta x = 1$, and the outputs must be evenly spaced (a constant difference).

63. Coefficient of Correlation   Sketch a graph of each of the following sets of data.

a. $(1, 2), (1, 0), (-1, 2),$ and $(-1, 0)$ $y = 0x + 1, r = 0;$ data are in a rectangle, not clustered around a straight line.

b. $(1, 1.5), (1, 0.5), (-1, 2),$ and $(-1, 0)$ See below.

c. $(1, 1.5), (1.1, 0.5), (-1, 2),$ and $(-1.1, 0)$
$y = 0.0113x + 1, r = 0.015;$ now we have a slight slope.

d. $(1, 1.5), (1.5, 0.5), (-1, 2),$ and $(-1.5, 0)$ See below.

e. $(1, 1.5), (3, 0.5), (-1, 2),$ and $(-3, 0)$
$y = 0.05x + 1, r = 0.1414$
Perform a linear regression on each set of data. Record the equation as well as r, the coefficient of correlation. Explain why the equation is reasonable. Suggest why r is reasonable.

f. Guess and check to change the x coordinate in the second and fourth data points to obtain $r = 0.2$.
$(1, 1.5), (9.9, 0.5), (-1, 2),$ and $(-9.9, 0)$ gives $y \approx 0.0225x + 1$ and $r = 0.200002525.$

64. Line of Best Fit   An alternative method of finding an equation from data starts with a graph of the data. Select two data points and draw a line that "fits" the set of points. Use the point-slope equation to find the equation of the line drawn. Select two exercises for which you drew a graph and use this method to fit an equation to the data. Compare your results.

63b. $y = 0x + 1, r = 0;$ data are in a trapezoid shape—still two outputs for $x = 1$ and two for $x = -1$. Next set modifies this.

63d. $y = 0.038x + 1, r = 0.062;$ now the points stretch farther apart, giving more of a clustering along the line. Still, the data are hardly linear.

2.5 Special Lines

Objectives

▪ Identify horizontal and vertical lines by their slopes.

▪ Find the equations of horizontal and vertical lines.

▪ Identify parallel lines by slope.

▪ Identify perpendicular lines by slope.

▪ Find equations of lines parallel or perpendicular to another line.

WARM-UP

Find the slopes of lines passing through the following ordered pairs.

1. $(-6, 4), (-6, -6)$ undefined   **2.** $(0, 4), (-6, 4)$ 0

3. $(200, 70), (400, 90)$ $\frac{1}{10}$ **4.** $(1, 4), (4, 6)$ $\frac{2}{3}$

5. $(1, 4), (3, 1)$ $-\frac{3}{2}$

IN THIS SECTION, we look at horizontal, vertical, parallel, and perpendicular lines. Slope plays an important role in distinguishing each type of line.

Horizontal and Vertical Lines

In Example 1, we find the slope of a horizontal line and a vertical line.

EXAMPLE 1   **Exploring slopes of horizontal and vertical lines**   Figure 27 shows a horizontal line, $y = 4$, and a vertical line, $x = -6$. Use the indicated ordered pairs to calculate the slope of each line.

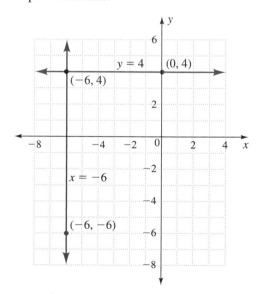

FIGURE 27

SOLUTION   The slope of the horizontal line, $y = 4$, through $(-6, 4)$ and $(0, 4)$ is

$$\frac{y_2 - y_1}{x_2 - x_1} = \frac{4 - 4}{0 - (-6)} = \frac{0}{6} = 0$$

The change in y is zero, so the slope of $y = 4$ is zero.

The slope of the vertical line, $x = -6$, through $(-6, -6)$ and $(-6, 4)$ is

$$\frac{y_2 - y_1}{x_2 - x_1} = \frac{4 - (-6)}{-6 - (-6)} = \frac{4 + 6}{-6 + 6} = \frac{10}{0}$$

Because the fraction has a zero denominator, the slope of the vertical line, $x = -6$, is undefined.

■ EQUATIONS OF HORIZONTAL AND VERTICAL LINES

A horizontal line, with zero slope, has equation $y = 0x + b$ or $y = b$.

A vertical line, with undefined slope, has equation $x = a$.

EXAMPLE 2 Graphing and finding the equations of horizontal and vertical lines A line passes through the given point with the indicated slope. Graph the line and give its equation.

a. Point $(3, 2)$, slope zero
b. Point $(3, 2)$, undefined slope
c. Point $(0, -2)$, slope zero
d. Point $(0, 2)$, undefined slope
e. Point $(-3, 0)$, undefined slope

SOLUTION Graphs of parts a and b are in Figure 28. Graphs of parts c, d, and e are in Figure 29.

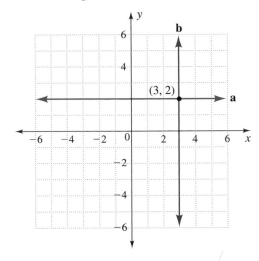

FIGURE 28

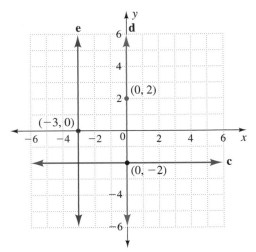

FIGURE 29

a. $y = 0x + 2$, $y = 2$

b. $x = 3$

c. $y = 0x - 2$, $y = -2$

d. $x = 0$, the vertical axis

e. $x = -3$

Parallel Lines

In Examples 3 and 4, we review the role of the slope and y-intercept in the orientation and position of a graph.

EXAMPLE 3 Comparing graphs: catering a wedding reception Suppose a caterer has three menus, costing $18 per person, $21 per person, and $25 per person. Write and graph the total cost equations for a $450 wedding cake and each menu choice. What effect does the menu choice have on the graph?

SOLUTION   The cost equations are

$$y = 18x + 450, \qquad y = 21x + 450, \qquad \text{and} \qquad y = 25x + 450$$

All three graphs pass through the same y-intercept, \$450. As shown in Figure 30, the higher the cost per person (the slope, m), the steeper the graph.

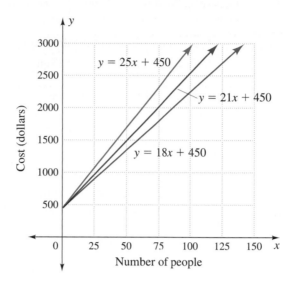

FIGURE 30

EXAMPLE 4   **Comparing graphs: more catering**   Suppose a caterer has three different cakes but only one menu price, \$18 per person. The three cakes cost \$450, \$550, and \$800. Write and graph the total cost equations. What effects do the cost of the cake and the menu price have on the graphs?

SOLUTION   The cost equations are

$$y = 18x + 450, \qquad y = 18x + 550, \qquad \text{and} \qquad y = 18x + 800$$

The graphs pass through different y-intercepts. As shown in Figure 31, the higher the cost of the cake, the higher the y-intercept, b. The menu cost (slope) stays constant, so the graphs are parallel.

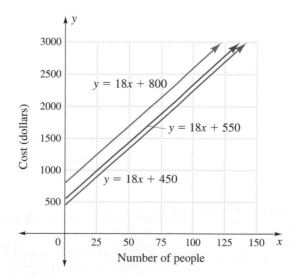

FIGURE 31

■ PARALLEL LINES

> Two lines are parallel if and only if their slopes are equal.

Note: The logic for "*a* and *b*" and "*a* or *b*" is presented in Section 2.1. Another logical phrase is "*a* if and only if *b*." This statement indicates that if "if *a* then *b*" is true, then so is "if *b* then *a*." The statement "two lines are parallel if and only if their slopes are equal" means two things:

> If two lines are parallel, then their slopes are equal.

> If the slopes of two lines are equal, then the lines are parallel.

▬ Parallel lines have special meaning in applications.

Here is a sample program for the slope, *M*, through (*A, B*) and (*C, D*):

PROGRAM: SLOPE
:<Prompt> A,B,C,D
:(D − B)/(C − A) [STO▶] M
:<Disp>"M", M

Modify it to include the *y*-intercept, *I*, with

:B − M*A [STO▶] I
:<Disp>"I", I

< > means that the command is in the [PRGM] menu.

EXAMPLE 5 | Writing cost functions and finding slopes: car rental costs   The total cost of renting a car is a function of the basic charge, insurance, and number of miles driven. Suppose a compact car costs $0.10 per mile plus $30 in basic charges and insurance. A mid-size car costs $50 in basic charges and insurance plus $0.10 per mile.

a. Write a cost function for the compact car rental, and use it to find the cost for a 200-mile trip and a 400-mile trip.
b. Write a cost function for the mid-size car rental, and use it to find the cost for a 200-mile trip and a 400-mile trip.
c. Graph the cost functions.
d. What is the slope of each line? What is the meaning of the slope?
e. Why are the lines parallel?

SOLUTION | **a.** The compact car costs $30 + $0.10 per mile for *x* miles, or $C = 30 + 0.10x$ dollars. The 200-mile trip costs $30 + 0.10(200) = 50$, or $50. The 400-mile trip costs $30 + 0.10(400) = 70$, or $70.

b. The mid-size car costs $50 + $0.10 per mile for *x* miles, or $C = 50 + 0.10x$ dollars. The 200-mile trip costs $50 + 0.10(200) = 70$, or $70. The 400-mile trip costs $50 + 0.10(400) = 90$, or $90.

c. In Figure 32, the bottom line is for the compact, and the top line is for the mid-size car.

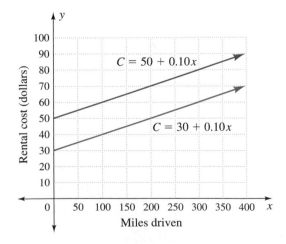

FIGURE 32

d. The slope of the graph for the compact car is given by

$$\text{Slope} = \frac{\text{change in output}}{\text{change in input}} = \frac{\$70 - \$50}{(400 - 200)\ \text{mi}}$$

$$= \frac{\$20}{200\ \text{mi}} = \frac{\$0.10}{\text{mi}}$$

The slope of the graph for the mid-size car is given by

$$\text{Slope} = \frac{\text{change in output}}{\text{change in input}} = \frac{\$90 - \$70}{(400 - 200)\ \text{mi}}$$

$$= \frac{\$20}{200\ \text{mi}} = \frac{\$0.10}{\text{mi}}$$

The slopes are both $0.10 per mile—the same for each car.

e. The lines are parallel because their slopes—the cost per mile—are the same.

Think about it 1: What do the vertical axis intercepts in Example 5 mean?

Perpendicular Lines

Perpendicular lines intersect at a right angle. There are two ways lines can be perpendicular.

First, the lines can be parallel to the horizontal and vertical axes. (See Example 1.)

■ SLOPES OF PERPENDICULAR LINES 1

> Two lines are perpendicular if one has a zero slope and the other has an undefined slope.

Second, the lines can be perpendicular if their slopes are *negative reciprocals*.

■ SLOPES OF PERPENDICULAR LINES 2A

> If $\dfrac{a}{b}$ is the slope of one line, $\dfrac{-b}{a}$ is the slope of a perpendicular line.

We note that negative reciprocals multiply to -1:

$$\frac{a}{b} \cdot \frac{-b}{a} = \frac{-ab}{ab} = -1$$

Furthermore, because their slope is undefined, we must exclude vertical lines from the second way to find perpendicular lines. **Nonvertical** means *not vertical.*

■ SLOPES OF PERPENDICULAR LINES 2B

> Nonvertical lines are perpendicular if and only if their slopes multiply to -1. The slopes are then negative reciprocals.

EXAMPLE 6 Identifying parallel and perpendicular lines by their slope   Use Figure 33 to find the slopes of these line segments:

a. *AB*

b. *BC*

c. *CD*

d. *AD*

State which pairs are parallel and which are perpendicular.

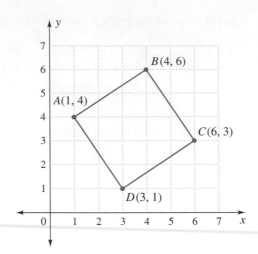

FIGURE 33

SOLUTION   **a.** The slope of AB is

Student Note: For convenience in finding the slope, we take (x_2, y_2) to be to the left of (x_1, y_1) in these solutions.

$$\frac{y_2 - y_1}{x_2 - x_1} = \frac{6 - 4}{4 - 1} = \frac{2}{3}$$

b. The slope of BC is

$$\frac{y_2 - y_1}{x_2 - x_1} = \frac{3 - 6}{6 - 4} = \frac{-3}{2}$$

c. The slope of CD is

$$\frac{y_2 - y_1}{x_2 - x_1} = \frac{3 - 1}{6 - 3} = \frac{2}{3}$$

d. The slope of AD is

$$\frac{y_2 - y_1}{x_2 - x_1} = \frac{1 - 4}{3 - 1} = \frac{-3}{2}$$

There are two sets of parallel line segments: AB and CD, AD and BC. There are four ways to pair the perpendicular line segments: AB and BC, AD and CD, AB and AD, BC and CD.

Think about it 2: What is the name of the shape $ABCD$ in Figure 33? Explain your reasoning.

Finding Equations of Lines Parallel or Perpendicular to a Given Line

The point-slope equation is useful in finding equations of lines parallel to or perpendicular to another line. For Examples 7 and 8, we return to the setting in Example 6.

EXAMPLE 7   Finding the equation of a line through a given point and parallel to another line
Find the equation of a line passing through $(3, 1)$ parallel to the graph of $y = -\frac{3}{2}x + 12$.

SOLUTION   ***Step 1:*** We find the slope from the equation $y = -\frac{3}{2}x + 12$: $m = -\frac{3}{2}$.

Step 2: We substitute the slope, $-\frac{3}{2}$, and the point (3, 1) into the point-slope equation:

$$y - y_1 = m(x - x_1)$$
$$y - 1 = -\frac{3}{2}(x - 3)$$

Step 3: We solve for *y*:

$$y - 1 = -\frac{3}{2}x + \frac{9}{2}$$
$$y = -\frac{3}{2}x + \frac{11}{2}$$

EXAMPLE 8 Finding the equation of a line through a given point and perpendicular to another line
Find the equation of a line passing through (3, 1) perpendicular to the graph of $y = -\frac{3}{2}x + 12$.

SOLUTION We follow the steps for using the point-slope equation.

Step 1: We find the slope from the equation $y = -\frac{3}{2}x + 12$; $m = -\frac{3}{2}$. The slope of a perpendicular line is the negative reciprocal of $-\frac{3}{2}$ and is $m = \frac{2}{3}$.

Step 2: We substitute the slope, $\frac{2}{3}$, and the point (3, 1) into the point-slope equation:

$$y - y_1 = m(x - x_1)$$
$$y - 1 = \frac{2}{3}(x - 3)$$

Step 3: We solve for *y*:

$$y - 1 - \frac{2}{3}x - 2$$
$$y - \frac{2}{3}x - 1$$

Think about it 3: The data in Examples 7 and 8 are from Figure 33 in Example 6. Match the equations $y = -3x/2 + 12$, $y = -3x/2 + 11/2$, and $y = 2x/3 - 1$ with the segments they contain (*AB*, *BC*, *CD*, or *DA*). What is the equation of the line passing through the remaining segment in the square?

ANSWER BOX

Warm-up: **1.** undefined **2.** zero **3.** $\frac{1}{10}$ **4.** $\frac{2}{3}$ **5.** $-\frac{3}{2}$ **Think about it 1:** The intercepts are the basic charges (without the per-mile cost) for the rental cars. **Think about it 2:** Square; the sides are perpendicular and, by the Pythagorean theorem, of the same length. **Think about it 3:** $y = -3x/2 + 12$ contains *BC*; $y = -3x/2 + 11/2$ contains *AD*; and $y = 2x/3 - 1$ contains *CD*. The line through *AB* is $y = 2x/3 + 10/3$.

| 2.5 | Exercises | |

In Exercises 1 to 4, find the slope of a line that passes through the given points. Find the equation of the line.

1. (4, −2) and (4, 0)
undefined, $x = 4$

3. (−2, −3) and (4, −3)
0, $y = -3$

2. (−2, 2) and (2, 2)
0, $y = 2$

4. (−2, 5) and (−2, 2)
undefined, $x = -2$

In Exercises 5 to 10, find the equation of the line passing through the given point with the indicated slope.

5. Point (4, −3), zero slope $y = -3$

6. Point (−2, 3), undefined slope $x = -2$

7. Point $(-1, -2)$, undefined slope $x = -1$

8. Point $(-1, -4)$, zero slope $y = -4$

9. Origin, zero slope $y = 0$

10. Origin, undefined slope $x = 0$

In Exercises 11 and 12, show that the product of each number and its negative reciprocal is -1.

11. a. $\frac{3}{4}$　　　　　**b.** $-3\frac{1}{2}$　　**c.** 4.5

　　　d. 0.10　　　　**e.** $-p/q$

12. a. $-\frac{4}{5}$　　　　**b.** $2\frac{1}{4}$　　**c.** -0.25

　　　d. -0.3　　　　**e.** n/m

In Exercises 13 to 20, do the pairs of equations have graphs that are parallel, perpendicular, or neither?

13. a. $2x = y - 2(3 - x)$ perpendicular

　　b. $y - 4 = 2x + y$

14. a. $2x = y - 4x - 2$ parallel

　　b. $-2(3x) = 5 - y$

15. a. $x = -6y$ perpendicular

　　b. $3x = y - 3x - 4$

16. a. $x + y = 4$ perpendicular

　　b. $x - y = 2$

17. a. $x = 4$ parallel

　　b. $y = x + y - 5$

18. a. $2x + y = 2(x - 1)$ parallel

　　b. $x + y = 6 + x$

19. a. $y + 2x = 2$ neither

　　b. $y = 2(x - 1)$

20. a. $y = 3 - x$ perpendicular

　　b. $y - x = 3$

In Exercises 21 to 26, write equations and tell whether the graphs would be parallel lines.

21. Tickets to a concert cost $78, $98, or $108 each. Equations are for the total cost, C, in dollars, for the number, x, of people sitting in the same-priced section.
$C = 78x, C = 98x, C = 108x$; not parallel

22. Different-sized rental trucks cost $19.99, $29.99, or $39.99 plus $0.10 per mile. Equations are for the total rental cost, C, in dollars, for x miles driven.
$C = 0.10x + 19.99, C = 0.10x + 29.99, C = 0.10x + 39.99$; parallel

23. A no-fee telephone card is available for $5.00, $10.00, or $20.00. Each calls costs $0.15 per minute. Equations are for the value remaining, V, on each card after x minutes. $V = 5.00 - 0.15x, V = 10.00 - 0.15x, V = 20.00 - 0.15x$; parallel

24. Suppose first-class postage stamps are purchased individually. Equations are for the change returned, R, upon buying x stamps with $5.00, $10.00, and $20.00.   If the postage cost is $0.37, R = 5.00 - 0.37x, R = 10.00 - 0.37x, R = 20.00 - 0.37x$; parallel

25. First-class postage stamps are available in rolls of 100, sheets of 50, or packets of 20. Equations are for the total cost, C, of purchasing a number, x, of each type of packaging.   See below.

26. Older riding horses cost $750, $1000, or $1250 to buy and $125 per month to board. Equations are for the total cost, C, for owning a horse for x months.
$C = 750 + 125x, C = 1000 + 125x, C = 1250 + 125x$; parallel

In Exercises 27 to 36, find the equation of a line fitting the information given.

27. Parallel to $2x + 3y = 6$ and passing through the origin
$y = -\frac{2}{3}x$

28. Parallel to $3x - 2y = 6$ and passing through the origin
$y = \frac{3}{2}x$

29. Perpendicular to $y = -\frac{1}{2}x + 3$ and passing through the origin $y = 2x$

30. Perpendicular to $y = \frac{3}{4}x - 2$ and passing through the origin $y = -\frac{4}{3}x$

31. Perpendicular to $y = \frac{5}{8}x - 3$ and passing through $(2, 3)$ $y = -\frac{8}{5}x + \frac{31}{5}$

32. Perpendicular to $y = -\frac{1}{4}x + 2$ and passing through $(3, 1)$ $y = 4x - 11$

33. Parallel to $4x - 3y = 12$ and passing through $(-2, 1)$
$y = \frac{4}{3}x + \frac{11}{3}$

34. Parallel to $3x + 4y = 12$ and passing through $(-1, 3)$
$y = -\frac{3}{4}x + \frac{9}{4}$

35. Perpendicular to $5x - 2y = 8$ and passing through $(2, -1)$ $y = -\frac{2}{5}x - \frac{1}{5}$

36. Perpendicular to $2x + 5y = 15$ and passing through $(1, 4)$ $y = \frac{5}{2}x + \frac{3}{2}$

37. Find the slopes of the line segments connecting the points in the figure on the next page. What can you conclude about the line segments? $\frac{1}{2}, -2, \frac{1}{2}, -2$; opposite lines are parallel; adjacent lines are perpendicular.

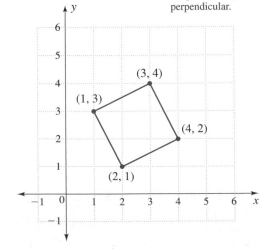

25. Although the postage cost appears to be the slope, buying in groups of stamps changes the slope. If the postage cost is $0.37, C = 100(0.37)x, C = 50(0.37)x, C = 20(0.37)x$; not parallel

38. Find the slopes of the line segments connecting the points in the figure. What can you conclude about the line segments? $-\frac{1}{3}, 3, -\frac{1}{3}, 3$; opposite lines are parallel; adjacent lines are perpendicular.

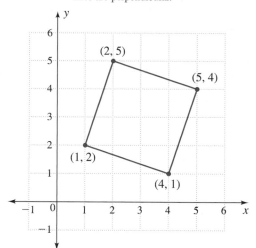

39. Draw the diagonals in the figure in Exercise 37. Use the points below to calculate the slopes for the diagonals. What do you observe? See Answer Section; diagonals are perpendicular.

a. (2, 1) and (3, 4) **b.** (1, 3) and (4, 2)
$\quad$ 3 $\quad\quad\quad\quad\quad\quad\quad\quad -\frac{1}{3}$

40. Draw the diagonals in the figure in Exercise 38. Use the points below to calculate the slopes for the diagonals. What do you observe? See Additional Answers; diagonals are perpendicular.

a. (2, 5) and (4, 1) **b.** (1, 2) and (5, 4)
$\quad -2 \quad\quad\quad\quad\quad\quad\quad\quad \frac{1}{2}$

41. How might we use the slope to tell if a rectangle is formed by a four-sided figure on a rectangular coordinate graph? Try your description first on (−2, 0), (0, 5), (2, 0), and (0, −3), and then on (−1, −1), (1, −2), (3, 2), and (1, 3). Opposite sides should have same slopes; adjacent sides, negative reciprocal slopes; is not rectangle; is rectangle.

42. The equation of a horizontal line, $y = b$, names the [vertical or horizontal] axis intercept. The equation of a vertical line, $x = a$, names the [vertical or horizontal] axis intercept.

43. Translate into two statements: Nonvertical lines are perpendicular if and only if their slopes multiply to −1. Nonvertical lines are perpendicular if their slopes multiply to −1. If nonvertical lines are perpendicular, then their slopes multiply to −1.

■ **Projects**

44. Perpendicular Lines and Scale on Axes The perpendicular lines in this section have been drawn on axes with equal scale. Investigate different scales for these perpendicular lines.

a. Graph the equations $y = 5x$ and $y = -\frac{1}{5}x$ for x on the interval [−1, 5] and y on [−5, 25]. See Additional Answers.

b. Graph the equations in part a for x on the interval [−5, 25] and y on [−5, 25]. See Additonal Answers.

In parts c and d, repeat the equations in part a but graph with a calculator.

c. Instead of setting 〔WINDOW〕, choose 〔ZOOM〕 and then **6:ZStandard**. What are the contents of 〔WINDOW〕? Press 〔GRAPH〕. Do the graphs appear to meet at right angles? Although the numbers on the axes are the same, are the scales the same? **Xmin** = −10, **Xmax** = 10, **Xscl** = 1, **Ymin** = −10, **Ymax** = 10, **Yscl** = 1; no; no

d. Instead of setting 〔WINDOW〕, choose 〔ZOOM〕 and then **5:ZSquare**. What are the contents of 〔WINDOW〕? Press 〔GRAPH〕. Do the graphs appear to meet at right angles? If prior window was that set in part c, **Xmin** ≈ −15.16, **Xmax** ≈ 15.16, **Xscl** = 1, **Ymin** = −10, **Ymax** = 10, **Yscl** = 1; yes.

e. Explain what **5:ZSquare** does. See below.

45. Circle A circle is placed on axes, with the center at the origin.

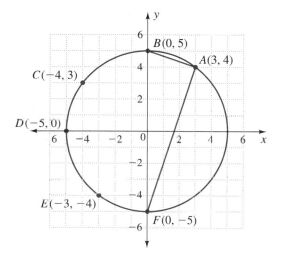

In parts a to c, find the slopes of the lines between the ordered pairs and state a relationship.

a. (0, 5) and (3, 4) with (0, −5) and (3, 4)
$-\frac{1}{3}$, 3, negative reciprocals

b. (0, 5) and (−4, 3) with (0, −5) and (−4, 3)
$\frac{1}{2}$, −2, negative reciprocals

c. (0, 5) and (−5, 0) with (0, −5) and (−5, 0)
1, −1, negative reciprocals

d. What can be said about the angles *BAF*, *BCF*, and *BDF*? right angles

e. Reinforce your answer to part d by finding these slopes: (0, 5) and (−3, −4) compared with (0, −5) and (−3, −4). 3, $-\frac{1}{3}$, negative reciprocals

f. *BF* is the diagonal of the circle with endpoints *B* and *F*. Points *A*, *C*, *D*, and *E* are on the circumference. Write a concluding statement describing all angles such as those in part d. A right angle is formed by connecting a point on the circumference with the endpoints of the diameter.

44e. It sets the scale on the *x*-axis to eliminate the distortion of the rectangular viewing window. The re-set is relative to the previous **Ymin** and **Ymax** settings.

2.6 Special Functions

Objectives

■ Identify and graph constant and identity functions.

■ Identify and graph absolute value functions.

■ Solve absolute value equations.

■ Graph equations resulting in dot graphs.

■ Graph equations resulting in step graphs.

WARM-UP

1. Simplify $2(x + 1) - x - 2$. $\quad x$

2. Simplify $2(x^2 + 2x - 1) - x(2x + 3) + 2$. $\quad x$

3. Solve $x + 3 = 2$. $\quad x = -1$

4. Solve $x + 3 = -2$. $\quad x = -5$

THIS SECTION INTRODUCES three special functions: the constant function, identity function, and absolute value function. Solutions to absolute value equations are included. The discussion of linear functions is extended to those with conditions that yield dot graphs or step graphs.

Constant Function

We now consider a special linear function: the constant function.

EXAMPLE 1 Exploring the constant function: university tuition   Suppose you attend a university with a fixed tuition of $8000 per semester. What is the cost for taking 10, 20, and 30 credit hours during the semester? Draw a graph of tuition as a function of number of credit hours taken. What do you observe about the graph?

SOLUTION The tuition cost remains $8000, regardless of the number of credit hours taken. Thus, the graph in Figure 34 passes through the points (10, 8000), (20, 8000), and (30, 8000). The graph is a horizontal line.

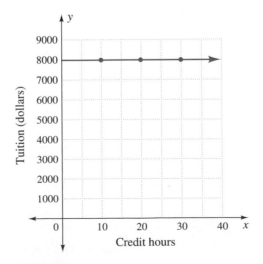

FIGURE 34

The fact that the university's tuition is the same no matter how many credit hours are taken makes the graph a constant function.

■ CONSTANT FUNCTION

> A **constant function** has a fixed output for all inputs.

In Example 2, we calculate the slope and find the y-intercept, domain, and range for a constant function.

EXAMPLE 2　Finding features of a constant function

a. List the ordered pairs from Table 17, and find the corresponding points in Figure 35.
b. Calculate the slope of this constant function.
c. What is the y-intercept for $f(x) = 3$?
d. What are the domain and range for the function?

TABLE 17

Input x	Output $f(x) = 3$
-4	3
-2	3
0	3
5	3

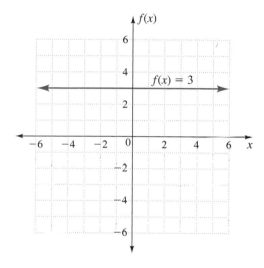

FIGURE 35　Constant function

SOLUTION　**a.** The listed coordinates are $(-4, 3)$, $(-2, 3)$, $(0, 3)$, and $(5, 3)$. The points all lie on the horizontal line.

b. The slope between $(-2, 3)$ and $(5, 3)$ is given by

$$\frac{y_2 - y_1}{x_2 - x_1} = \frac{3 - 3}{5 - (-2)} = \frac{0}{7} = 0$$

The slope is zero.

c. The y-intercept for the graph of $f(x) = 3$ is $y = 3$.

d. The set of inputs, or domain, for the function is all real numbers, $\mathbb{R}$. The set of outputs, or range, is a constant, 3. ■

■ PROPERTIES OF A CONSTANT FUNCTION

> Every horizontal line is a constant function.
>
> The slope of the constant function is zero.
>
> The constant function $f(x) = c$ intersects the vertical axis at $(0, c)$.
>
> The domain (unless limited by an application) of the constant function is the set of all real numbers, $\mathbb{R}$.
>
> The range of the constant function $f(x) = c$ is the y-intercept value: the single number c.

Think about it 1: What are the domain and range for university tuition in Example 1?

Identity Function

Have you heard the expressions "what you see is what you get" and "garbage in, garbage out"? In mathematics, the identity function behaves according to the principle "what you put in is what you get out." Each output of the identity function is identical to the corresponding input.

■■■ IDENTITY FUNCTION

> The identity function is $f(x) = x$. The **identity function** has each output equal to its corresponding input.

The graph of $f(x) = x$ is the line $y = x$; see Table 18 and Figure 36.

TABLE 18

Input x	Output $f(x) = x$
−2	−2
−1	−1
0	0
1	1

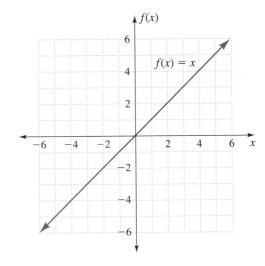

FIGURE 36   Identity function

EXAMPLE 3   **Identifying constant and identity functions**   Which situations describe identity functions? Which describe constant functions?

a. Input: Number of local telephone calls made for a $21.95 per month fee
   Output: Total cost of local telephone service

b. Input: Charge balance on a credit card
   Output: The payment due in order to avoid paying interest or finance charges

c. Input: The amount of money put into a "no charge" change machine
   Output: The amount of money received in change from the machine

d. Input: Number of cups of coffee consumed at a restaurant with a $3 "bottomless cup"
   Output: Total cost of the coffee

SOLUTION

a. Constant function; the total cost is fixed for any number of local calls.

b. Identity function; the payment due equals the charge balance.

c. Identity function; the money put in equals the change received.

d. Constant function; the $3 cost is fixed for all the coffee you can drink.   ■■■

■■■   The graph of the identity function (see Figure 36) extends infinitely to the left and to the right. The set of inputs, or domain, is the set of all real numbers, $\mathbb{R}$. Because the outputs to the identity function exactly match the inputs, the range is also the set of all real numbers, $\mathbb{R}$. The domain and/or range may be limited for certain applications.

■ PROPERTIES OF THE IDENTITY FUNCTION

The graph of the identity function is the line $y = x$.

The identity function passes through the origin, $(0, 0)$.

Both the domain and the range (unless limited by an application) of the identity function are the set of all real numbers, $\mathbb{R}$.

■ MAKING A DISTINCTION: IDENTITY AND IDENTITY FUNCTION

An identity is an algebraic sentence, such as $a = a$, that is true for all inputs to the variables.

The identity function is the function that associates each input with itself (as the output), as in $f(x) = x$.

EXAMPLE 4 Distinguishing identities from the identity function   Is the example given an identity or an identity function?

a. $8 = 8$ is found when checking an equation.

b. $y = 2(x^2 + 2x - 1) - x(2x + 3) + 2$

c. $4 = 4$ is found when solving an equation.

d. $g(x) = 2(x + 1) - x - 2$

SOLUTION   **a.** Identity

b. After simplifying, $y = x$, the identity function.

c. Identity

d. After simplifying, $g(x) = x$, the identity function.   ▬

Absolute Value Function

In measuring the distance between exits on a freeway or in calculating daily heating or air conditioning requirements, we must have a positive outcome. The absolute value function is a tool for obtaining positive outcomes.

■ ABSOLUTE VALUE FUNCTION

The **absolute value function** gives the distance a number is from zero on the number line. The function is written $f(x) = |x|$.

EXAMPLE 5 Finding the domain and range   What are the domain and range for the absolute value function, $f(x) = |x|$?

SOLUTION   The absolute value function is shown in Table 19 and Figure 37.

TABLE 19

Input x	Output $f(x) = |x|$
-5	5
-3	3
-1	1
0	0
1	1
3	3
5	5

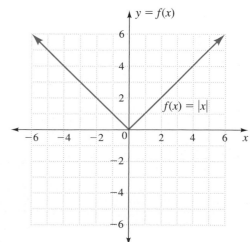

FIGURE 37

Any real number is a defined input, so the domain is the set of all real numbers, $\mathbb{R}$.

There are no negative outputs, so the range is positive or zero: $f(x) \geq 0$. ▬

The absolute value graph is made up of two parts. Example 6 explores the equation of each part.

EXAMPLE 6 **Exploring the absolute value graph** Use the absolute value graph in Figure 37 to write the equation and input condition for

a. the left side of the graph   **b.** the right side of the graph

SOLUTION   **a.** The left part of the graph of $y = |x|$, in Figure 38(a), has a slope of -1 and a y-intercept at zero. This part is linear for all negative inputs, $x < 0$. The equation is $y = -1x + 0$, or $y = -x$ for $x < 0$.

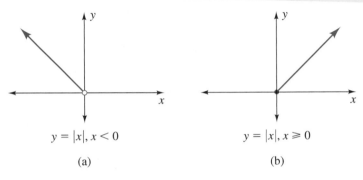

$$y = |x|, x < 0$$

(a)

$$y = |x|, x \geq 0$$

(b)

FIGURE 38

b. The right part of the graph of $y = |x|$, in Figure 38(b), has a slope of 1 and a y-intercept at zero. This part is linear for all non-negative inputs, $x \geq 0$. The equation is $y = 1x + 0$, or $y = x$ for $x \geq 0$. ▬

The equations in Example 6 lead to the formal definition of absolute value.

▬ DEFINITION OF ABSOLUTE VALUE

The **absolute value** of x is x whenever the input is zero or positive and the opposite of x whenever the input is negative.

Symbolically, $y = |x|$ is defined as

$$y = \begin{cases} x & \text{if } x \geq 0 \\ -x & \text{if } x < 0 \end{cases}$$

As a function, $f(x) = |x|$ is defined as

$$f(x) = \begin{cases} x & \text{if } x \geq 0 \\ -x & \text{if } x < 0 \end{cases}$$

It is incorrect to think of x as being positive and $-x$ as being negative.

EXAMPLE 7 **Finding the opposite of x** Evaluate $-x$ for

a. $x = 4$

b. $x = -2$

SOLUTION   **a.** For $x = 4$, $-x = -(4) = -4$.

b. For $x = -2$, $-x = -(-2) = 2$.

Note that $-x$ is positive when x is negative. ▬

EXAMPLE 8   Solving absolute value equations   Solve $|x + 3| = 2$. Confirm the results with a graph.

SOLUTION   Because the expression inside the absolute value sign may be either positive or negative, we have two outcomes.

If $x + 3$ is positive,

$$x + 3 = 2, \qquad x = -1$$

If $x + 3$ is negative,

$$x + 3 = -2, \qquad x = -5$$

Check: $|-1 + 3| = |2| = 2$ ✓
$|-5 + 3| = |-2| = 2$ ✓

The graphs of $y = |x + 3|$ and $y = 2$ are shown in Figure 39. The line $y = 2$ intersects $y = |x + 3|$ at $x = -1$ and $x = -5$, confirming the results found symbolically.

Student Note: Recall that absolute value is abbreviated abs on most calculators and is listed under (**2nd**) [CATALOG].

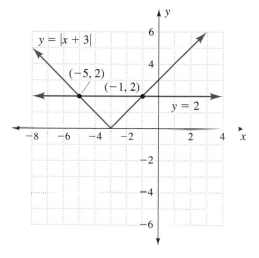

FIGURE 39

Graphs of Linear Equations in Conditional Settings

Student Note: Conditional settings were introduced in Section 2.1.

The absolute value definition is an equation based on whether the a inside $|a|$ is positive, zero, or negative—that is, the **conditions** that $a > 0$, $a = 0$, or $a < 0$. The graph of absolute value contains two linear pieces.

In many settings, the equations may look linear but conditions either cause the points on the graph not to be connected (resulting in a dot graph) or create a graph formed by short horizontal line segments (resulting in a step graph).

DOT GRAPHS   Because they relate to number of days and number of children, the settings in Examples 9 and 10 are meaningful only for positive integer inputs, and their graphs are dots.

EXAMPLE 9   Graphing equations for conditional settings   A popular arcade game costs $3.00 for two plays and $4.00 for each additional consecutive play. The equations are

Student Note: To get the introductory rate of two plays for $3, the player must go to the end of the waiting line.

$$y = 3, \qquad\qquad 0 < x \le 2$$
$$y = 3 + 4(x - 2), \quad x > 2$$

Graph the equations.

SOLUTION   The graph is shown in Figure 40.

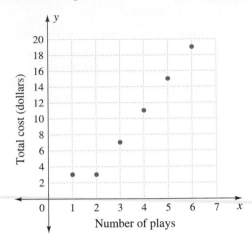

FIGURE 40

EXAMPLE 10   **Graphing equations for conditional settings**   A birthday party costs $50 for up to 10 children and $4.50 for each additional child. Graph the party cost equations:

$$y = 50, \qquad\qquad 0 < x \le 10$$
$$y = 50 + 4.50(x - 10), \quad x > 10$$

SOLUTION   The graph is shown in Figure 41.

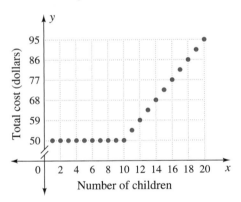

FIGURE 41

Think about it 2: What is the meaning of the double slash, $/\!/$, on the vertical axis in the party cost graph in Figure 41? What is the spacing between the remaining numbers on the vertical axis, and why might that spacing have been chosen?

◼ DOT GRAPHS | A **dot graph** is formed when inputs to an equation are limited to integers.

STEP GRAPHS   The inputs to the settings in Examples 11 and 12 are not limited to positive integers. However, because fractional inputs are rounded, the graphs are step graphs—graphs formed by short, horizontal line segments.

EXAMPLE 11   **Graphing equations for conditional settings**   An air compressor rents for $50 for up to 6 hours and $25 for each additional hour or part of an hour. The rental cost equations are

$$y = 50, \qquad\qquad 0 < x \le 6$$
$$y = 50 + 25(x - 6), \quad x > 6$$

Student Note: Remember that we round x before substituting, here and in the telephone call equations below.

Graph the equations.

SOLUTION   The graph is shown in Figure 42.

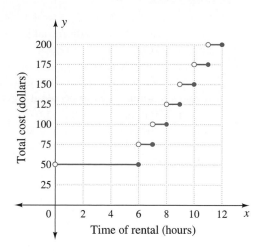

FIGURE 42

EXAMPLE 12   **Graphing equations for conditional settings**   A telephone call costs $1.50 for the first minute and $0.10 for each additional minute or part of a minute. The telephone call equations are

$$y = 1.50, \qquad\qquad x \le 1$$
$$y = 1.50 + 0.10(x - 1), \quad x > 1$$

Graph the equations.

SOLUTION   The graph is shown in Figure 43.

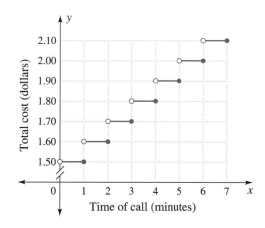

FIGURE 43

Think about it 3: Describe situations in which the small circle above zero in the graphs in Figures 42 and 43 might be a dot—that is, situations in which there would be an output value when the input was zero.

◼ STEP GRAPHS

> A **step graph** is formed when fractional or decimal inputs are rounded to integers before they are substituted into the equation.

Think about it 4: In what way are all the graphs in Figures 40 through 43 the same?

ANSWER BOX

Warm-up: 1. x **2.** x **3.** $x = -1$ **4.** $x = -5$ **Think about it 1:**
Because tuition is paid only for taking a positive number of credits, the
domain must be restricted to positive numbers. Additional information about
inputs (whether there is an upper limit on the number of credits, whether the
university offers half-credit courses, etc.) may be needed to identify the highest domain number and to decide if fractions or decimals are permitted in the
domain. The range is constant: $8000. **Think about it 2:** The double slash
means that the spacing between 0 and 50 is not the same as the spacing
between other numbers on the vertical axis. The other numbers are spaced
9 units apart because $2(4.50) = 9$. **Think about it 3:** There could be a
charge for placing the rental order, whether or not you actually took the
machine. There could be a connection charge for placing the phone call, even
if no one answered the telephone. **Think about it 4:** Each starts with a horizontal section. The rest of the graph rises to the right, either in a line or as a
series of steps.

2.6 Exercises

In Exercises 1 to 4, graph the functions. State the domain and range of each.

1. $f(x) = 2$
See Answer Section; $\mathbb{R}$; 2

2. $f(x) = -3$
See Additional Answers; $\mathbb{R}$; -3

3. $f(x) = -2$
See Answer Section; $\mathbb{R}$; -2

4. $f(x) = \pi$
See Additional Answers; $\mathbb{R}$; π

In Exercises 5 to 12, which situations describe identity functions?
Which describe constant functions? Explain your reasoning.

5. Input: Number of visits to a county fair on a $15 season
pass
Output: Total cost for visits to the fair constant function

6. Input: Amount of money loaned without interest
Output: Amount of money paid back, if loan is repaid in
full identity function

7. Input: Number of horseback rides per month
Output: Monthly payment of $100 for unlimited rides
constant function

8. Input: Number of helpings eaten
Output: Total cost for one person of eating at a restaurant that advertises "All You Can Eat for $10.95"
constant function

9. You follow the rule "Eat an apple a day to keep the doctor away."
Input: Days since you started to follow the rule
Output: Number of apples eaten each day
constant function

10. You follow the rule "Eat an apple a day to keep the doctor away."
Input: Days since you started to follow the rule
Output: Total number of apples eaten
identity function

11. Input: Money put into a postage machine at the post
office
Output: Total value of the stamps and change received
identity function

12. Input: The key you want to press on the computer keyboard
Output: The character appears that you wanted to enter
identity function

In Exercises 13 to 22, is the example an identity, an identity function, or
neither?

13. $3 = 3$ is found when checking an equation.
identity

14. $a(b + c) = ab + ac$ identity

15. $-a(b - c) = -ab - bc$ neither

16. $f(x) = x \div a \cdot x$ neither

17. $h(n) = n$ identity function

18. $g(t) = t^2 - 2\left(\frac{1}{2} t^2 - 1\right) - 2$ neither

19. $a + b + c = a + c + b$ identity

20. $12 = 12$ is found when solving an equation.
identity

21. $y = x + a - a$
identity function

22. $a = a$ is found when checking a formula.
identity

The *line of symmetry* is a line cutting a figure into two parts that would fold exactly
onto each other.

23. a. For $f(x) = |x - 2|$, find $f(-3), f(-1), f(0), f(1)$,
$f(3)$, and $f(5)$. Graph $f(x)$. 5, 3, 2, 1, 1, 3; see Answer Section.

b. Use function notation to say that 0 and 4 give the
same output to $f(x)$. $f(0) = f(4) = 2$

c. Name the line of symmetry of $f(x) = |x - 2|$. $x = 2$

24. a. For $f(x) = |x + 2|$, find $f(-3), f(-1), f(0), f(1)$,
$f(3)$, and $f(5)$. Graph $f(x)$.
1, 1, 2, 3, 5, 7; see Additional Answers.

b. Use function notation to say that -3 and -1 give the same output to $f(x)$. $\;f(-3) = f(-1) = 1$

c. Name the line of symmetry of $f(x) = |x + 2|$.
$x = -2$

Graph the equations in Exercises 25 to 32. Give the domain and range of each.

25. $y = |x + 3|$
See Answer Section.

26. $y = |x - 3|$
See Additional Answers.

27. $y = |x| + 3$
See Answer Section.

28. $y = |x| - 3$
See Additional Answers.

29. $y = -|x - 3|$
See Answer Section.

30. $y = -|x + 3|$
See Additional Answers.

31. $y = -|x| - 3$
See Answer Section.

32. $y = -|x| + 3$
See Additional Answers.

For Exercises 33 to 35, use the graph in the figure below.

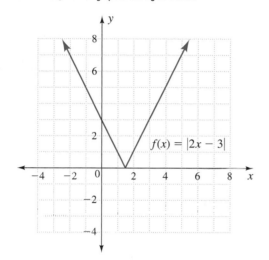

33. Solve these equations from the graph.

a. $|2x - 3| = 5$
$x = -1$, or $x = 4$

b. $|2x - 3| = 3$ $\;x = 0$ or $x = 3$

c. $|2x - 3| = 0$ $\;x = \frac{3}{2}$

d. $|2x - 3| = -2$ no solution

34. Solve these equations from the graph.

a. $|2x - 3| = 6$
$x = -1.5$ or $x = 4.5$

b. $|2x - 3| = 7$
$x = -2$ or $x = 5$

c. $|2x - 3| = 8$
$x = -2.5$ or $x = 5.5$

d. $|2x - 3| = -4$
no solution

35. a. What is the slope of the left side of the graph of $f(x) = |2x - 3|$? -2

b. What is the slope of the right side? 2

c. What is the y-intercept of $f(x) = |2x - 3|$? What input gives the y-intercept? $3, 0$

d. Write a linear equation for the piece of the graph containing $(0, 3)$. State the equation as subject to $x \leq 1.5$. $\;y = -2x + 3, x \leq 1.5$

e. Find $f(1.5)$ with your equation in part d. $\;0$

f. Write a linear equation for the piece of the graph containing $(3, 3)$. State the equation as subject to $x \geq 1.5$. $\;y = 2x - 3, x \geq 1.5$

g. Find $f(1.5)$ with your equation in part f. $\;0$

For Exercises 36 to 38, use the graph in the figure below. Extend the graph as needed.

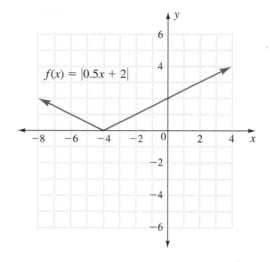

36. Solve these equations from the graph.

a. $|0.5x + 2| = 2$
$x = 0$ or $x = -8$

b. $|0.5x + 2| = 0$
$x = -4$

c. $|0.5x + 2| = -2$
no solution

d. $|0.5x + 2| = 5$
$x = 6$ or $x = -14$

37. Solve these equations from the graph.

a. $|0.5x + 2| = 3$
$x = -10$ or $x = 2$

b. $|0.5x + 2| = -1$
no solution

c. $|0.5x + 2| = 1$
$x = -6$ or $x = -2$

d. $|0.5x + 2| = 4$
$x = 4$ or $x = -12$

38. a. What is the slope of the left side of the graph of $f(x) = |0.5x + 2|$? -0.5

b. What is the slope of the right side? 0.5

c. What is the y-intercept of the function? 2

d. Write a linear equation for the piece of the graph containing $(0, 2)$. State the equation as subject to $x \geq -4$. $\;y = 0.5x + 2, x \geq -4$

e. Find $f(-4)$ with your equation in part d. $\;0$

f. Write a linear equation for the piece of the graph containing $(-6, 1)$. State the equation as subject to $x \leq -4$. $\;y = -0.5x - 2, x \leq -4$

g. Find $f(-4)$ with your equation in part f. $\;0$

Solve the equations in Exercises 39 to 46.

39. $|x| = 4$ $\;\{\pm 4\}$

40. $|x| = 8$ $\;\{\pm 8\}$

41. $|x + 2| = 3$ $\;\{-5, 1\}$

42. $|x - 3| = 4$ $\;\{-1, 7\}$

43. $|x - 5| = 2$ $\;\{3, 7\}$

44. $|x + 4| = 3$ $\;\{-7, -1\}$

45. $|x - 4| = 2$ $\;\{2, 6\}$

46. $|x + 1| = 4$ $\;\{-5, 3\}$

47. Match the absolute value expressions (a, b, c, and d) with the expressions needed on a graphing calculator (1, 2, 3, and 4).

a. $|x + 2|$ 2

b. $|x| + 2$ 3

c. $\dfrac{1}{|x + 2|}$ 4

d. $\dfrac{1}{|x| + 2}$ 1

1. $1 \div (\text{abs}(x) + 2)$

2. $\text{abs}(x + 2)$

3. $\text{abs}(x) + 2$

4. $1 \div \text{abs}(x + 2)$

48. In calculating degree-days for estimating heating and air conditioning requirements, a meteorologist uses $f(T) = |72 - T|$. Find $f(68)$, $f(76)$, $f(-20)$, and $f(103)$.
4, 4, 92, 31

49. Following is a list of possible exit numbers for cities along Route 35.

Laredo	0
San Antonio	153
Austin	230
Dallas	423
Gainesville	482

Exit numbers are determined by the number of miles from the exit to the southern border of Texas. Find the driving distance.

a. From San Antonio to Dallas 270 mi

b. From Austin to Gainesville 252 mi

If x_1 and x_2 represent exit numbers, write a formula for finding the distance, D, between exits.
$D = |x_1 - x_2|$

50. Postal Rates Research

a. Complete this sentence, using the current postal rates for first-class postage: Mailing a letter costs _____ cents for the first ounce and _____ cents for each additional ounce or portion thereof up to _____ ounces.

b. Draw a graph of the total postage cost for 0 to 8 ounces.

In Exercises 51 to 58,
(a) Explain whether a dot graph or a step graph is appropriate.
(b) Draw a graph for the equations. (Hint: In Exercises 53 to 56, use // and do not start numbering of the axes at the origin.)

51. A fax costs $4 for up to two pages and $0.50 for each page thereafter. The total cost, y, is for x pages.

$$y = 4 + 0.5(x - 2), \quad x > 2$$
$$y = 4, \qquad\qquad 0 < x \le 2$$
dot; see Answer Section.

52. A college transcript costs $5 for the first copy and $1 for each additional copy. The total cost, y, is for x copies.

$$y = 5 + 1(x - 1), \quad x > 1$$
$$y = 5, \qquad\qquad x = 1$$
dot; see Additional Answers.

53. Rental of a weed cutter costs $20 for 3 hours plus $5 for each additional hour or part of an hour. The total cost, y, is for an x-hour rental.

$$y = 20 + 5(x - 3), \quad x > 3$$
$$y = 20, \qquad\qquad 0 < x \le 3$$
step; see Answer Section.

54. Rental of a pressure washer costs $30 for the first 4 hours and $10 per additional hour or part of an hour. The total cost, y, is for an x-hour rental.

$$y = 30 + 10(x - 4), \quad x > 4$$
$$y = 30, \qquad\qquad 0 < x \le 4$$
step; see Additional Answers.

55. A party at an ice rink costs $85 for up to 10 skaters and $4.75 for each additional skater. The total cost, y, is for x skaters.

$$y = 85 + 4.75(x - 10), \quad x > 10$$
$$y = 85, \qquad\qquad 0 < x \le 10$$
dot; see Answer Section.

56. A party at a roller skating rink costs $40 for 10 skaters and $4.50 for each additional skater. The total cost, y, is for x skaters.

$$y = 40 + 4.5(x - 10), \quad x > 10$$
$$y = 40, \qquad\qquad 0 < x \le 10$$
dot; see Additional Answers.

57. A cell phone costs $19.95 each month for the first 100 minutes and $0.25 for each additional minute or portion of a minute. The total cost, y, is for x minutes a month.

$$y = 19.95 + 0.25(x - 100), \quad x > 100$$
$$y = 19.95, \qquad\qquad 0 \le x \le 100$$
step; see Answer Section.

58. An e-mail account costs $9.95 each month for the first 10 hours and $2.00 for each additional hour or part of an hour. The total cost, y, is for x hours a month.

$$y = 9.95 + 2(x - 10), \quad x > 10$$
$$y = 9.95, \qquad\qquad 0 \le x \le 10$$
step; see Additional Answers.

59. What phrasing helps identify a step graph?
part of an hour, portion of a minute

60. What characterizes settings that produce dot graphs?
Inputs are whole objects: pages, copies, children.

■ Project

61. Cutting a Square in Half Suppose a square is drawn with corners at (0, 0), (4, 0), (4, 4), and (0, 4). Write equations for as many lines as possible that divide the area of the square in half. Note any constant functions or identity functions or equations that are not functions.

Equations should include $y = x$ (identity), $y = 2$ (constant), $x = 2$ (not a function), $y = -x + 4$, and several others such as $y = \frac{1}{2}x + 1$, $y = -\frac{1}{2}x + 3$, and $y = -2x + 6$.

2 Chapter Summary

Vocabulary

For definitions and page references, see the Glossary/Index.

absolute value	horizontal intercept	linear function	slope-intercept equation
absolute value function	identity function	linear regression	standard form
arithmetic sequence	increasing function	modeling with a linear	step graph
coefficient of correlation	inequality	function	subscripts
compound inequality	infinite	non-negative real numbers	two-output test
condition	infinity sign	nonvertical	vertical intercept
constant function	interval	parallel lines	vertical-line test
decreasing function	line of best fit	point-slope equation	*x*-intercept
domain	linear equation in one variable	range	*y*-intercept
dot graph	linear equation in two	relevant domain	
function	variables	slope	

Concepts

See the list of vocabulary for important definitions.

2.1 ■ Inequalities, Line Graphs, and Intervals

The phrase *a and b* is true if both *a* and *b* are true.

The phrase *a or b* is true if either the condition *a* is true or the condition *b* is true or if both conditions are true.

The compound inequality $a < x < c$ may be written as two inequalities, $a < x$ and $x < c$.

Interval notation may look like the coordinates of a point. Read carefully when you see (a, b) to see whether the reference is to an ordered pair (a, b) or an interval (a, b) describing the set $a < x < b$.

See Table 20 for symbols used in inequalities and intervals.

2.2 ■ Functions

We may describe a function with a graph, a written rule, a listing of ordered pairs, or an equation.

When graphing a function, place the inputs on the horizontal axis and the outputs on the vertical axis.

A relationship is identified as a function from its ordered pairs, its table (one output for each input), or its graph (passing the vertical-line test).

Functions are not limited to mathematical relationships.

Functions may be visualized as putting numbers or objects through a function machine or as mapping one set of numbers or objects onto another set.

We evaluate functions by substituting the input for x into the function described by $f(x)$.

We can describe relationships between outputs to functions with notation such as $f(a)$ and $f(b)$: in $f(x) = x^2$, $f(-2) = f(2)$.

The domain and range of a mathematical function may be limited in application settings to a relevant domain and relevant range.

TABLE 20

Inequality Symbol	Interval Notation	Word Meaning	Line Graph Notation
<	(,)	is less than	small circle or (,)
>	(,)	is greater than	small circle or (,)
=		is equal to	dot
≤	[,]	is less than or equal to	dot or [,]
≥	[,]	is greater than or equal to	dot or [,]
$+\infty$	, $+\infty$)	positive infinity	→
$-\infty$	$(-\infty,$	negative infinity	←

2.3 ∎ Linear Functions

Graphs cross the axes at intercept points $(a, 0)$ and $(0, b)$. These intercepts, a and b, have special significance in applications. In function notation, a is found by solving $f(x) = 0$ and b is found by $f(0)$.

A function is a linear function if and only if its slope is constant. To find the units on slope, divide the units on the output axis by the units on the input axis.

We describe the change in a function's graph as *increasing* or *decreasing*.

If the slope of a line is positive, the function is an increasing function; if the slope of a line is negative, the function is a decreasing function.

2.4 ∎ Modeling with a Linear Function

To build an equation with the slope-intercept equation $y = mx + b$, find the slope, m, and the y-intercept, b.

To build an equation with the point-slope equation, substitute one point, (x_1, y_1), and the slope, m, into $y - y_1 = m(x - x_1)$.

To build an equation from a table, estimate an average slope from $\Delta y / \Delta x$. If $x = 0$ does not appear in the table, estimate the y-intercept with $b = y - mx$; that is, multiply the estimated slope times the first input, x, and subtract from the first output, y. Write the equation as $y = mx + b$.

To find an equation with calculator regression, enter the data, set a window appropriate to the data, plot the data to see which regression is most appropriate, and choose the regression from the statistical options. Check by graphing the regression equation with the data.

2.5 ∎ Special Lines

Vertical lines, with equation $x = c$, are the only linear equations that are not also linear functions, $f(x) = mx + b$.

A horizontal line has a zero slope.

A vertical line has an undefined slope.

Parallel lines have the same slope.

Lines with slopes that are negative reciprocals—that multiply to -1—are perpendicular.

The logical phrase "if and only if" means that both "if a then b" and "if b then a" are true.

2.6 ∎ Special Functions

A constant function has a horizontal graph and an equation $y = c$, where c is a real number.

The graph of the identity function is the line $y = x$.

The absolute value function describes the distance a number is from zero on the number line. Expressions in an absolute value function must be placed in parentheses when they are entered into a calculator.

To solve $|a| = n$, work two cases:

1. For positive a, solve $a = n$.

2. For negative a, solve $a = -n$.

Restrictions on inputs or conditions on equations may result in graphs that change direction, dot graphs, or step graphs.

---

 2 Review Exercises

Find words in the Vocabulary list that match each of the descriptions in Exercises 1 to 16. If more than one answer is needed, the number in parentheses tells how many are needed.

1. Used to find out if a graph is a function
vertical-line test

2. Set of all inputs; set of all outputs (2)
domain; range

3. A graph for which the inputs are only integers
dot graph

4. A graph for which the inputs are rounded off before the outputs are found
step graph

5. Limits on inputs due to an application setting
relevant domain

6. Limits on outputs due to an application setting
relevant range

7. Set of data with constant slope
linear function

8. Used in finding a linear equation on a graphing calculator (2)   coefficient of correlation, linear regression

9. A function where the output exactly matches the input
identity function

10. A function with only one output (regardless of input)
constant function

11. A function with a positive output for any real-number input   absolute value function (squaring function is not in list)

12. Numbers that are zero or positive
non-negative numbers

13. Numbers that are small and placed below and to the right of letters or variables
subscripts

14. Number or expression that describes the rate of change of a graph   slope

15. Ways to find a linear equation (4)
regression, point-slope, slope-intercept, sequence

16. Functions whose graphs are rising, falling, or horizontal when viewed left to right (3)
increasing function, decreasing function, constant function

17. Write each word phrase as an inequality and an interval; then draw a line graph.

    a. The set of numbers between -8 and -4, including -4 $-8 < x \leq -4$; $(-8, -4]$; see Answer Section.

    b. The set of numbers less than 7 or greater than -2
    $-\infty < x < +\infty$; $(-\infty, +\infty)$; see Answer Section.

    c. The set of numbers less than 7 and greater than -2 $-2 < x < 7$; $(-2, 7)$; see Answer Section.

    d. The set of numbers greater than -3
    $x > -3$; $(-3, +\infty)$; see Answer Section.

    e. The positive real numbers
    $x > 0$; $(0, +\infty)$; see Answer Section.

    f. The non-negative real numbers
    $x \geq 0$; $[0, +\infty)$; see Answer Section.

18. Write a word phrase, an inequality, and an interval for each line graph.

    a.
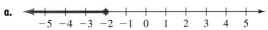
Set of numbers less than or equal to -2; $x \leq -2$; $(-\infty, -2]$

    b.
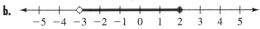
Set of numbers between -3 and 2, including 2; $-3 < x \leq 2$; $(-3, 2]$

    c.
Set of numbers greater than -1; $x > -1$; $(-1, +\infty)$

    d.
Set of numbers less than or equal to zero or the numbers greater than 2; $x \leq 0$ or $x > 2$; $(-\infty, 0]$ or $(2, +\infty)$

In Exercises 19 to 22, answer the following questions:
(a) What is the domain?
(b) What is the range?
(c) Does the graph describe a function?

19. ellipse $-6.2 \leq x \leq 2.2$; $-3 \leq y \leq 3$; no

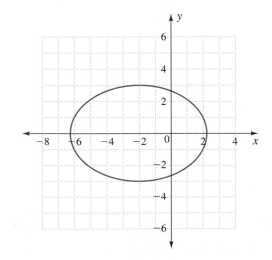

20.
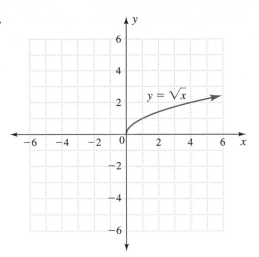
$x \geq 0$; $y \geq 0$; yes

21.
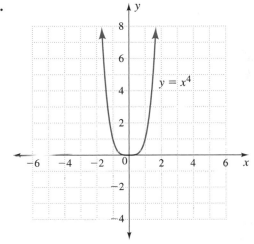
all real numbers; $y \geq 0$; yes

22.
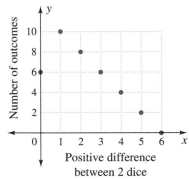
Positive difference between 2 dice

$\{0, 1, 2, 3, 4, 5, 6\}$; $\{0, 2, 4, 6, 8, 10\}$; yes

23. Use the graph of f in the figure to evaluate the expressions and answer the questions. The graph does not intersect the x-axis.

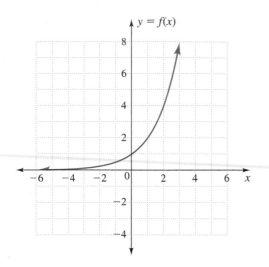

$y = f(x)$

a. $f(0)$ 1   **b.** $f(2)$ 4   **c.** $f(-1)$ $\frac{1}{2}$

d. $f(3)$ 8   **e.** $f(1)$ 2

f. For what inputs is $f(x) \geq 2$? $x \geq 1$

g. For what inputs is $f(x) < 0$? none

h. For what inputs is $f(x) \leq 4$? $x \leq 2$

i. What is the y-intercept? 1

j. What is the set of possible inputs (domain) for $f(x)$?
all real numbers

k. What is the set of possible outputs (range) for $f(x)$?
$y > 0$

24. In the figure in Exercise 23, what inputs x give these function values?

a. $f(x) = 1$ {0}   **b.** $f(x) = 3$ {≈1.5}

c. $f(x) = \frac{1}{2}$ {−1}   **d.** $f(x) = 4$ {2}

e. $f(x) = -2$ { }

In Exercises 25 to 32, evaluate the function $f(x) = 2x^2 - 3x + 1$ for the indicated inputs.

25. $f(1)$ 0   **26.** $f(2)$ 3

27. $f(0.5)$ 0   **28.** $f(1.5)$ 1

29. $f(-2)$ 15   **30.** $f(-0.5)$ 3

31. $f(\square)$ $2\square^2 - 3\square + 1$ **32.** $f(-a)$ $2a^2 + 3a + 1$

33. Find the equation of a line perpendicular to $y = 3x + 4$ through

a. the origin $y = -\frac{1}{3}x$ **b.** (3, 4) $y = -\frac{1}{3}x + 5$

34. Find the equation of a line parallel to $y = \frac{2}{3}x - 4$ through

a. the origin $y = \frac{2}{3}x$ **b.** (4, 3) $y = \frac{2}{3}x + \frac{1}{3}$

Make a table in which the columns are headed Ordered Pairs, Slope, Equation, Horizontal or Vertical, x-intercept, y-intercept. Complete the table for each set of ordered pairs in Exercises 35 to 42.

35. (1, 3) and (2, 2)
-1; $y = -x + 4$; neither; 4; 4

36. (2, 3) and (2, 2)
undefined; $x = 2$; vertical; 2; none

37. (−3, 3) and (0, 0)
-1; $y = -x$; neither; 0; 0

38. (−3, 3) and (3, 3)
0; $y = 3$; horizontal; none; 3

39. (−1, −1) and (1, 3)
2; $y = 2x + 1$; neither; $-\frac{1}{2}$; 1

40. (2, 3) and (−3, 4)
$-\frac{1}{5}$; $y = -\frac{1}{5}x + 3\frac{2}{5}$; neither; 17; $3\frac{2}{5}$

41. (−3, 3) and (2, 3)
0; $y = 3$; horizontal; none; 3

42. (0, 4) and (−1, −1)
5; $y = 5x + 4$; neither; $-\frac{4}{5}$; 4

43. In Exercises 35, 37, 39, and 41, are any of the lines formed by the ordered pairs parallel? Perpendicular?
Exercises 35 & 37, parallel; none perpendicular;

44. In Exercises 36, 38, 40, and 42, are any of the lines formed by the ordered pairs parallel? Perpendicular?
none parallel; Exercises 36 & 38, 40 & 42, perpendicular

In Exercises 45 to 48:

(a) Write an equation.

(b) Indicate the slope and y-intercept, and explain what, if anything, each means.

45. Marcia pays a sales tax of 6.5% on a price x. What is the tax paid? $y = 0.065x$; $0.065 tax/$1 purchased; $0, no tax if nothing is purchased

46. Don's party at a restaurant costs $10.95 per person. What is the total cost?
$y = 10.95x$; $10.95/person; $0, no cost if no one attends

47. An annual inspection for a Beechcraft Skipper airplane costs $500 plus $45 per hour for any repairs needed. Parts are extra. What is the total cost without parts?
$y = 500 + 45x$; $45/hr of repair; $500, basic inspection cost

48. David's minimum monthly cell phone bill is $19 plus $0.20 per minute in send mode. Long-distance charges are extra. What is his monthly bill without long-distance charges? $y = 19 + 0.20x$; $0.20/min; $19, basic monthly charge

In Exercises 49 and 50, use calculator regression to find a linear equation.

49. Time and total score for a computer solitaire game are given by the following data points: (125, 6286), (122, 6391), (118, 6573), and (117, 6618). Let $x =$ time (in seconds) and $y =$ total score. $y \approx 11{,}528 - 42x$

50. The bonus points awarded for the times in the solitaire game in Exercise 49 are as follows: (125, 5600), (122, 5705), (118, 5915), and (117, 5950). Let $x =$ time to complete a game (in seconds) and $y =$ bonus points.
$y \approx 11{,}244 - 45x$

51. From the table, find an equation for mooring a boat.
slope $1.50 per foot, intercept (0, 8), $y = 1.50x + 8$

Moorage Fees (Boston #2)

Boat Length (feet)	Cost (dollars)
25	45.50
36	62
42	71

In Exercises 52 to 54, what is the rule for each pattern?

52. 18, 16, 14, 12, 10, . . . $y = -2x + 20$

53. 31, 27, 23, 19, 15, . . . $y = -4x + 35$

54. $-5, -2, 1, 4, 7, \ldots$ $y = 3x - 8$

In Exercises 55 to 60, write an equation. Identify each as an increasing, decreasing, or constant function.

55. A ski resort advertises a monthly pass for $350. What is the total cost to Jason if x is the number of ski trips he makes each month? $C = \$350$; constant

56. The annual fee for an athletic club is $350. No charge is made for each use of the facility. What is the total annual cost if Chae Hyok uses the club x times? $C = \$350$; constant

57. A restaurant advertises "Anything You Want to Eat for $5.95 per Pound." What is the total cost to Kellie if x is the number of pounds she eats? $C = 5.95x$, C in $; increasing

58. The annual fee for an athletic club is $350. Each use costs an additional $5. What is the total yearly cost Chamique pays if she uses the club x times? $C = 5x + 350$, C in $; increasing

59. Paul pays $350 in advance on his account at the athletic club. Each time he uses the club, $5 is deducted from the account. What value remains in his account after x visits to the club? $V = 350 - 5x$, V in $; decreasing

60. Kristen buys a $35 mass transportation ticket. Each ride costs $1.50, deducted automatically from the ticket when she leaves the station. What value remains on the ticket after x rides? $V = 35 - 1.50x$, V in $; decreasing

For each situation in Exercises 61 to 64, define variables, write an equation, and say whether the graph will be a dot graph or a step graph. Test your equation with a number to see that it makes sense in the problem setting.

61. The total cost of a party at an ice rink that charges $85 for up to ten people and $4.75 for each additional person
x = no. of people, y = total cost; $y = 85 + 4.75(x - 10)$, $x > 10$; $y = 85$, $0 < x \leq 10$; dot graph

62. The total cost of a party at a roller skating ring that charges $40 for up to 10 children and $4.50 for each additional child x = no. of children, y = total cost; $y = 40 + 4.5(x - 10)$, $x > 10$; $y = 40$, $0 < x \leq 10$; dot graph

63. The total cost of a long-distance telephone call at $0.26 for the first minute or part thereof and $0.19 per minute for each additional minute or part thereof x = no. of min, y = cost; $y = 0.26$, $0 < x \leq 1$; $y = 0.26 + 0.19(x - 1)$, $x > 1$; step graph

64. The total cost of a long-distance telephone call at $1.08 for the first 2 minutes or part thereof and $0.63 for each additional minute or part thereof x = no. of min, y = cost; $y = 1.08$, $0 < x \leq 2$; $y = 1.08 + 0.63(x - 2)$, $x > 2$; step graph

65. Graph $y = |x - 1|$. From your graph and with algebraic notation, solve these equations. See Answer Section.

a. $|x - 1| = 4$ $\{-3, 5\}$ **b.** $|x - 1| = 2$ $\{-1, 3\}$

c. $|x - 1| = 0$ $\{1\}$ **d.** $|x - 1| = -2$ $\{\ \}$

66. Graph the conditional equations
$$y = 1 - x \text{ for } x < 1$$
and
$$y = x - 1 \text{ for } x \geq 1$$
See Additional Answers; the graph is for $y = |x - 1|$.

In Exercises 67 to 74, give the domain and range.

67. $y = 365$ all real numbers; 365

68. $y = 365x$ all real numbers; all real numbers

69. $y = |x - 1|$ all real numbers; $y > 0$

70. $y = |x| - 1$ all real numbers; $y \geq -1$

71. $y = x$ all real numbers; all real numbers

72. $y = |x + 1|$ all real numbers; $y \geq 0$

73. $y = |x| + 1$ all real numbers; $y \geq 1$

74. $y = 1$ all real numbers; 1

75. Solve $|0.5x - 3| = 4$. $x = -2, x = 14$

76. Solve $|5 - 0.2x| = 4$. $x = 5, x = 45$

---

2   Chapter Test

1. Write an inequality and an interval and draw a line graph for each description:

a. The set of numbers less than or equal to 5
$x \leq 5$; $(-\infty, 5]$; see Answer Section.

b. The set of numbers greater than -2 and less than 5 $-2 < x < 5$; $(-2, 5)$; see Answer Section.

c. The set of numbers greater than -2 or less than 5
all real numbers; $(-\infty, +\infty)$; see Answer Section.

2. Which are functions?

a.

Input: Name	Output: Sport
Ali	Basketball
Ali	Golf
Tarom	Volleyball
Dean	Badminton

not a function

b.

Input: Name	Output: Registration Day
Terrie	Sept. 1
Jessica	Sept. 3
Dianne	Sept. 1
Hee-Jin	Sept. 2

function

c.

Input	Output
2	3
3	3
4	3
5	3

function

d.

Input	Output
2	4
2	6
3	7
4	3

not a function

3. If $f(x) = 3x^2 - 2x - 4$, find

a. $f(-2)$ 12

b. $f(0)$ -4

c. $f(2)$ 4

4. a. Find the slope of the line passing through $(-2, 4)$ and $(5, 2)$. $-\frac{2}{7}$

b. Find the equation of the line in part a.
$y = -\frac{2}{7}x + \frac{24}{7}$

c. Find the slope of a line parallel to the line in part a.
$-\frac{2}{7}$

d. Find the equation of a line through $(2, -1)$ and perpendicular to the line in part a.
$y = \frac{7}{2}x - 8$

5. Fill in the missing word or choose the correct word from those given in brackets.

a. The slope of a horizontal line is ___zero___.

b. A line that falls from left to right has a [__negative__, zero, positive] slope and is said to be a(an) [increasing, __decreasing__] function.

c. If the slope of a graph between all pairs of points is constant, the graph is a __linear__ function.

d. A horizontal linear graph is also called a __constant__ function.

e. Linear equations have a __constant__ slope.

f. The set of inputs to a sequence or number pattern is the __positive__ integers or natural numbers.

6. Alicia's taxi ride cost $2.50 plus $7.00 per mile.

a. Write a linear equation to describe the total cost of an x-mile ride. $y = 7x + 2.5$, y in $

b. What is the slope? $7 per mile

7. Heidi's long-distance telephone call cost $0.13 for 1 minute. A 19-minute call to the same location on another day cost $2.11.

a. Identify reasonable inputs and outputs, and define the variables. time in minutes $= x$; cost in dollars $= y$

b. Write the data as two ordered pairs.
(1, 0.13), (19, 2.11)

c. Fit a linear equation to the data.
$y = 0.11x + 0.02$

8. Sale prices in 1984 for solitaire diamond engagement rings set in 14K gold are listed below. Let $x =$ points and $y =$ price. One carat $= 100$ points. Fit a linear equation with calculator regression.
$y \approx 10.1x - 13.8$

Size in Points	10	20	25	33	50	100
Price	$99	$179	$249	$299	$499	$999

9. Give the next number in each sequence. Find the rule for each.

a. 10, 18, 26, 34, 42, ___50; $y = 8x + 2$___

b. $-16, -9, -2, 5, 12,$ ___19; $y = 7x - 23$___

In Exercises 10 to 15, sketch a group to fit the equation or description.

10. $y = |x| - 3$ See Answer Section.

11. $y = x$ See Answer Section.

12. College tuition is $300 per credit up to 12 credits, with no additional charge for up to a total maximum load of 24 credits. Fractions of credits are possible.
See Answer Section.

13. College tuition is $250 per credit. Maximum load is 24 credits. Fractions of credits are possible.
See Answer Section.

14. Filling an 8-inch-tall cylindrical bucket with a garden hose in 2 minutes, with the depth of the water in the bucket as a function of time. Explain your graph.
See Answer Section.

15. The total monthly cost for Ki to ride the bus if a monthly bus pass costs $35 and he rides the bus x times.
See Answer Section.

16. Make a table for the total cost of 1 to 6 copies of a transcript if the cost is $5 (total) for up to two copies and $2 for each additional copy. Graph your function. Explain why the points should or should not be connected.
See Answer Section.

17. Solve for x: $|2x + 2.5| = 5$.
$x = -3.75$, $x = 1.25$

Cumulative Review of Chapters 1 and 2

1.

Input x	Input y	Output xy	Output $x + y$	Output $x - y$
−2	4	−8	2	−6
−3	7	−21	4	−10
2	−3	−6	−1	5
−3	−2	6	−5	−1
−1	−6	6	−7	5
−5	−2	10	−7	−3
3	−2	−6	1	5
2	−9	−18	−7	11

In Exercises 2 and 3, match each expression with one of the given words.

2. Choose from

whole numbers, quotient, difference, natural numbers, rational numbers

a. The answer to a subtraction problem   difference

b. The set of numbers $\{1, 2, 3, 4, \ldots\}$ natural numbers

c. The set of numbers that may be written in the form a/b, where a and b are integers and $b \neq 0$
rational numbers
d. The answer to a division problem
quotient
e. The set of numbers $\{0, 1, 2, 3, 4, \ldots\}$
whole numbers

3. Choose from

factors, sets, opposites, reciprocals, factoring

a. Two numbers, n and $-n$, that add to zero
opposites
b. Two numbers or expressions, a and b, that are multiplied to obtain the product ab factors

c. Two numbers, n and $1/n$, that multiply to 1   reciprocals

d. Removing a common factor from two or more terms
factoring
e. Collections of objects or numbers   sets

Complete the sentences in Exercises 4 to 7 to make true statements. Choose from subtraction to addition of the opposite number, a product to a sum, division to multiplication by the reciprocal, a sum to a product.

4. The multiplication $a(b + c)$ changes <u>a product to a sum</u>.

5. Factoring $ab + ac$ changes <u>a sum to a product</u>.

6. To subtract real numbers, we may change ——.
subtraction to addition of the opposite number
7. To divide real numbers, we may change ——
division to multiplication by the reciprocal

Simplify the expressions in Exercises 8 to 13.

8. $3[4 - 2(5 - 8) - 6]$ 12

9. $a(b + c) - b(a + c) + c(a - b)$ $2ac - 2bc$

10. πr^2 for $r = 7.5$ feet 56.25π ft^2

11. $\frac{1}{4}\pi d^2$ for $d = 15$ feet 56.25π ft^2

12. $\dfrac{ac - bc}{c}$ $a - b$

13. $\dfrac{16 + 21x}{6}$ does not simplify

Solve the equations in Exercises 14 and 15.

14. $3x + 12 = 5x + 3$ **15.** $15 - 4x = 5(6 - x)$
$x = 4.5$ $x = 15$

Write and solve an equation in Exercises 16 and 17.

16. Five more than the quotient of a number and 8 is 2.
$\frac{x}{8} + 5 = 2; x = -24$
17. The product of 3 and a number is 15 more than the number. $3x = x + 15; x = 7.5$

18. Suppose coffee is $5.98 for each of the first two 1-pound cans and $9.98 for each additional can.

a. Make a table and graph for the total cost of 1 to 5 cans.   See Additional Answers.

b. Discuss whether or not the points should be connected.   not connected, for you cannot purchase just part of a can

c. Does the total cost equation describe a function? If so, what is the domain of the function?
function; whole numbers

In Exercises 19 to 22, find $f(-1)$, $f(0)$, $f(1)$, and $f(2)$.

19. $f(x) = x + 2$ 1, 2, 3, 4   **20.** $f(x) = 2x$ −2, 0, 2, 4

21. $f(x) = x^2$ 1, 0, 1, 4   **22.** $f(x) = 3 - x$ 4, 3, 2, 1

What are the slope and the equation for each pair of points in Exercises 23 and 24?

23. $(4, -3)$ and $(-6, 1)$ **24.** $(-2, -7)$ and $(2, -1)$
$-\frac{2}{5}; y = -\frac{2}{5}x - \frac{7}{5}$ $\frac{3}{2}; y = \frac{3}{2}x - 4$
25. What is the equation of a line passing through the origin that is perpendicular to $y = 3x - 5$?
$y = -\frac{1}{3}x$

In Exercises 26 to 29, give the next set of ordered pairs for each list, and describe the function with a rule in either words or symbols.

26. $(0, 0), (1, 1), (2, 4), (3, 9)$ $(4, 16); f(x) = x^2$

27. $(0, 0), (1, 1), (2, 2), (3, 3)$ $(4, 4); f(x) = x$

28. $(0, 0), (1, 2), (2, 4), (3, 6)$ $(4, 8); f(x) = 2x$

29. $(0, 4), (1, 4), (2, 4), (3, 4)$ $(4, 4); f(x) = 4$

In Exercises 30 and 31, write an equation.

30. Elise's car rental costs $60 plus $0.07 for each mile driven. What is the total cost of the rental?
$y = 0.07x + 60$, x in mi, y in $
31. Marianne's conference exhibit costs $500 plus $65 per worker at the exhibit. What is the total cost of the exhibit? $y = 65x + 500$, x in no. of workers, y in $

32. Use the graph of f in the figure to evaluate the expressions and answer the questions.

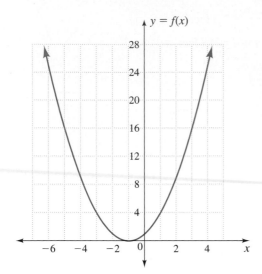

$y = f(x)$

a. $f(0)$ 1   **b.** $f(1)$ 4   **c.** $f(-2)$ 1

d. $f(3)$ 16   **e.** $f(-1)$ 0

f. For what inputs is $f(x) \geq 16$? $x \leq -5$ or $x \geq 3$

g. For what inputs is $f(x) < 0$? none

h. For what inputs is $f(x) < 4$? $-3 < x < 1$

i. What is the y-intercept? 1

j. What is the set of possible inputs (domain) for $f(x)$?
all real numbers

k. What is the set of possible outputs (range) for $f(x)$?
$y \geq 0$

33. In the figure in Exercise 32, what inputs x give these function values?

a. $f(x) = 0$ $\{-1\}$ **b.** $f(x) = 9$ $\{-4, 2\}$

c. $f(x) = 16$ $\{-5, 3\}$ **d.** $f(x) = 4$ $\{-3, 1\}$

e. $f(x) = -2$ $\{\ \}$

34. Use a calculator to graph the data in the table and fit a linear equation.*
$y \approx 17.2 - 5.2x$

Altitude, x (kilometers)	Summer Temperature, $f(x)$ (°C)
0	15.7
0.5	14.5
1.0	12.0
1.5	10.0
2.0	7.5
3.0	2.4
4.0	-3.0
5.0	-8.9
6.0	-15.1

*Data in Exercise 34 are from *Handbook of Chemistry and Physics*, 36th ed. (Chemical Rubber Publishing Company, Cleveland, Ohio, © 1954), p. 3093.

Systems of Equations and Inequalities

The three chocolate rabbits shown in Figure 1 are selections at Janelle's candy store. Although pricing usually depends on weight, some smaller styles of rabbits require more labor and are priced accordingly. The number made of each style depends on historic demand and anticipated sales. We will return to the chocolate rabbits in Section 3.4.

The chapter begins with the substitution and elimination methods of solving systems of equations. We review graphical solutions in Section 3.2 and obtain equations from quantity-value tables in Section 3.3. We solve systems of three or more linear equations in Section 3.4. The chapter closes with systems of inequalities (Section 3.5).

FIGURE 1

3.1 Solving Systems of Two Linear Equations by Substitution or Elimination

Objectives

▪ Solve an equation in two variables for one variable.

▪ Find the solution to a system of equations in two variables by substitution.

▪ Find the solution to a system of equations in two variables by elimination.

▪ Solve geometry problems by writing and solving systems of equations.

WARM-UP

1. Solve for x in terms of y: $x = -9 - 2y$

$$2y + x = -9$$

2. Solve for y in terms of x: $y = \dfrac{-9 - x}{2}$

$$2y + x = -9$$

3. Change to standard form (in one step): $-2x + y = 4$

$$y = 2x + 4$$

4. Change to standard form (in one step): $x - y = 10$

$$x = 10 + y$$

IN THIS SECTION, we solve systems of equations by substitution and elimination. We solve applications that can be translated into algebraic notation directly from the word problem or that rely on a fact from geometry.

Systems of Equations

A bottle and cork together cost $1.10.

The bottle costs $1.00 more than the cork.

What is the cost of each?

The bottle and cork problem is a famous puzzle, well known for tricking people into answering without doing the mathematics. No, the answer is not $1.00 and $0.10. Try the math now or in Exercise 43.

The bottle and cork problem can be translated into a system of equations. A **system of equations** is *a set of two or more equations to be solved for values of the variables that satisfy all of the equations, if such values exist. The set of numbers that make all equations true* is the **solution to the system**.

Systems of equations are commonly used in geometry, science, business, and economics to describe relationships and conditions. The systems we study in this chapter will suggest some of these applications.

Computer and calculator technology now allows us to solve systems of linear equations with ease. Our purpose here is to examine basic equation solving so as to better understand the work done by the technology.

■ SOLUTION METHODS FOR
SYSTEMS OF EQUATIONS

We can solve systems of equations in many ways:

1. Guess and check (Section 1.5)

2. Substitution (Section 3.1)

3. Elimination (Section 3.1)

4. Graphing and calculator graphing (Section 3.2)

5. Calculator matrices (Sections 10.4 and 10.5)

Substitution Method

The substitution method for solving equations is particularly valuable because it applies to both linear and nonlinear equations. (We will work with nonlinear equations in Chapter 10.) The **substitution method** of solving equations obtains its name from the replacement process: *we replace variables with equivalent expressions or numbers in order to eliminate one variable.*

SOLVING FOR ONE VARIABLE   The first step in solving a system of equations by substitution is to solve an equation for one of its variables.

EXAMPLE 1   **Solving for one variable**   Solve each equation for the indicated variable.

a. $x - 2y = 6$ for y

b. $2y - 3x = -6 - y$ for y

SOLUTION   **a.** We solve $x - 2y = 6$ for y as follows:

$$x - 2y = 6 \qquad \text{Subtract } x \text{ from both sides.}$$

$$-2y = 6 - x \qquad \text{Divide both sides by } -2.$$

$$y = \frac{6 - x}{-2} \qquad \text{Multiply both numerator and denominator by } -1.$$

$$y = \frac{x - 6}{2}$$

b. We solve $2y - 3x = -6 - y$ for y as follows:

$$2y - 3x = -6 - y \qquad \text{Add } y \text{ to both sides.}$$

$$3y - 3x = -6 \qquad \text{Add } 3x \text{ to both sides.}$$

$$3y = 3x - 6 \qquad \text{Divide both sides by 3.}$$

$$y = x - 2$$

■

Caution: In part b of Example 1, it would not have been correct to leave y terms on both sides.

■ DEFINITION OF "IN TERMS OF"

To solve for the variable y **in terms of** x means to isolate y on one side of the equation and obtain an expression containing only x and constants on the other side of the equal sign.

MAKING THE SUBSTITUTION   The second step in the substitution method for a system of equations is to substitute the expression from the first step into the second equation, as shown in Example 2.

EXAMPLE 2   Solving by substitution   Solve this system of linear equations by substitution:

$$2y + x = -9 \qquad (1)$$

$$4y - 3x = 12 \qquad (2)$$

SOLUTION   We solve the first equation for x:

$$2y + x = -9 \qquad \text{Subtract } 2y \text{ from both sides.}$$

$$x = -9 - 2y$$

Then we substitute $-9 - 2y$ for x in the second equation:

$$4y - 3x = 12$$

$$4y - 3(-9 - 2y) = 12 \qquad \text{Apply the distributive property.}$$

$$4y + 27 + 6y = 12 \qquad \text{Subtract 27 from both sides and add } y \text{ terms.}$$

$$10y = -15 \qquad \text{Divide by 10.}$$

$$y = -1.5$$

Then we find the value of x:

$$2(-1.5) + x = -9 \qquad \text{Substitute } -1.5 \text{ for } y \text{ in the first equation.}$$

$$-3 + x = -9 \qquad \text{Add 3 to both sides.}$$

$$x = -6$$

Student Note: Any form of the first equation can be used. Using the original form ensures that we do not repeat an error made in solving for x.

The solution to the system is $x = -6$ and $y = -1.5$.

Check (in both equations): $2(-1.5) + (-6) \stackrel{?}{=} -9$ ✓
$4(-1.5) - 3(-6) \stackrel{?}{=} 12$ ✓

Think about it 1: In Example 2, why do we choose to solve the first equation for x?

■ SOLVING TWO EQUATIONS BY SUBSTITUTION

> **1.** Solve one equation for a first variable.
>
> **2.** Substitute the resulting expression into the second equation. The second equation now has one variable. Solve.
>
> **3.** Substitute the resulting number into the first equation and evaluate. State the values of both variables, including units.
>
> **4.** Check the solution.

Elimination Method

In solving equations by the elimination method, we generally use the standard form of a linear equation.

■ STANDARD FORM OF A LINEAR EQUATION

> All linear equations in two variables may be written as
>
> $$ax + by = c$$
>
> where a, b, and c are any real numbers.

The **elimination method** of solving a system of equations is *a process in which one variable is removed from the system of equations by adding (or subtracting) two* *equations*. Elimination depends on the equality of the left and right sides of an equation, as shown in Example 3.

EXAMPLE 3 Exploring the addition property of equations

a. Add 3 to both sides of the equation $x - 3 = 8$. Do you obtain an equation with the same solution as the original equation?

b. Add x to both sides of the equation $2x = 9 - x$. Do you obtain an equation with the same solution as the original equation?

c. Suppose $x = 3$. Add x to the left side of $y - x = 4$ and add 3 to the right side. Have you added equal values to both sides of the equation?

d. Suppose $x + y = -2$. Add $x + y$ to the left side of $x - y = 10$ and add -2 to the right side. Have you added equal values to both sides of the equation?

SOLUTION **a.** $x - 3 + 3 = 8 + 3$

$x = 11$

Adding 3 to both sides creates an equation with the same solution as the original equation.

b. $2x + x = 9 - x + x$

$3x = 9$

$x = 3$

Adding x to both sides creates an equation with the same solution as the original equation.

c. $y - x + x = 4 + 3$

$y = 7$

Because $x = 3$, we added equal values to both sides.

d. $x - y + (x + y) = 10 + (-2)$

$2x = 8$

Because $x + y = -2$, we added equal values to both sides. ▬

The addition property of equations states that if $a = b$, then $a + c = b + c$. As we are reminded in Example 3, the letter c can be any number or expression. The addition property of equations is the basis for solving by elimination.

EXAMPLE 4 Solving by elimination   Solve this system of linear equations by elimination:

$$y = 2x + 4 \quad (1)$$

$$x = 10 + y \quad (2)$$

SOLUTION We arrange each equation in standard form.

Equation 1:

$$y = 2x + 4 \qquad \text{Subtract } 2x \text{ from both sides.}$$

$$-2x + y = 4$$

Equation 2:

$$x = 10 + y \qquad \text{Subtract } y \text{ from both sides.}$$

$$x - y = 10$$

Then we write the equations together, one above the other:

$$-2x + y = 4 \qquad (1)$$

$$\underline{x - y = 10} \qquad (2) \qquad \text{Add the equations.}$$

$$-x = 14$$

$$x = -14$$

Substituting $x = -14$ into the first equation, we can solve for y:

$$y = 2(-14) + 4$$

$$y = -24$$

The solution to the system is $x = -14$ and $y = -24$.

Check: $(-24) \overset{?}{=} 2(-14) + 4$ ✓
$(-14) \overset{?}{=} 10 + (-24)$ ✓   ▬

Adding the equations together is productive only if the terms for one variable have opposite coefficients. Then adding the terms yields zero, eliminating that variable.

▬ ELIMINATING LIKE TERMS | If two equations in standard form contain like terms with opposite coefficients, then adding the equations will eliminate those terms.

▬ In Example 5, we multiply both sides of the second equation by -3, the opposite of the numerical coefficient on x in the first equation. This creates equations with opposite coefficients on the x-terms.

EXAMPLE 5   Solving by elimination   Solve these equations:

$$3x + 2y = 8 \qquad (1)$$

$$x + 5y = -6 \qquad (2)$$

SOLUTION   We obtain -3 as the coefficient on x in the second equation.

$$-3(x + 5y) = -3(-6) \qquad \text{-3 times the second equation}$$

$$-3x - 15y = 18$$

$$\underline{3x + 2y = 8} \qquad \text{Add the first equation.}$$

$$0 - 13y = 26 \qquad \text{Divide by -13.}$$

$$y = -2$$

$$3x + 2(-2) = 8 \qquad \text{Substitute $y = -2$ in the first equation.}$$

$$x = 4$$

The solution to the system is $x = 4$ and $y = -2$.

Check: $3(4) + 2(-2) \overset{?}{=} 8$ ✓
$4 + 5(-2) \overset{?}{=} -6$ ✓   ▬

▬ In Example 6, both equations must be multiplied to produce like terms with opposite coefficients.

EXAMPLE 6   Solving by elimination   Solve these equations:

$$4x + 2y = -35 \qquad (1)$$

$$-3x + 5y = 81.5 \qquad (2)$$

SOLUTION   We obtain 12 and -12 as the respective coefficients on x:

$$3(4x + 2y) = 3(-35) \quad \text{3 times (1)}$$

$$4(-3x + 5y) = 4(81.5) \quad \text{4 times (2)}$$

$$\begin{array}{rl} 12x + 6y = & -105 \\ -12x + 20y = & 326 \end{array} \quad \text{Add the equations.}$$

$$26y = 221 \quad \text{Divide by 26.}$$

$$y = 8.5$$

$$4x + 2(8.5) = -35 \quad \text{Substitute } y = 8.5 \text{ in the first equation.}$$

$$x = -13$$

The solution to the system is $x = -13$ and $y = 8.5$.

Check: $4(-13) + 2(8.5) \stackrel{?}{=} -35$ ✓
$-3(-13) + 5(8.5) \stackrel{?}{=} 81.5$ ✓

Think about it 2: Which variable would have been eliminated by adding the equations in Example 6 if we had multiplied the first equation by -5 and the second by 2?

■ Solving Two Equations by Elimination

> **1.** Arrange the equations in standard form.
>
> **2.** Multiply one or both equations by numbers that create opposite values (additive inverses) of the coefficients on one pair of like terms.
>
> **3.** Add the equations to eliminate one variable, and solve for the variable that remains.
>
> **4.** Use substitution to find the other variable. State the values of both variables, including units.
>
> **5.** Check the solution.

Applications

Applications such as the bottle and cork problem should be "translated" directly from sentences into a system of equations. In solving all such problems, we use three steps.

Plan: Identify the unknowns and their units, choose variables, note facts not given in the problem, and state a possible solution method.

Carry out the plan: Write the equations and solve them. State the answer in terms of the unknowns and their units.

Check: Substitute to show that the solutions work in both equations.

EXAMPLE 7   **Comparing compensation packages**   Salespeople are often paid a small salary plus a percent commission on sales. Find the level of sales for which the two compensation packages are equal: Package A is $250 plus 10% of sales; Package B is $550 plus 6% of sales. Find the income earned at that level of sales.

SOLUTION   ***Plan:*** Write equations with I as income in dollars and x as sales in dollars. Change percents to decimals. Solve for sales, making the two incomes equal. Then substitute to find income.

Carry out the plan:

Package A: $I = 250 + 0.10x$

Package B: $I = 550 + 0.06x$

We substitute the equation for Package B into the equation for Package A, replacing the variable I:

$550 + 0.06x = 250 + 0.10x$ Subtract 250 from both sides.

$300 + 0.06x = 0.10x$ Subtract $0.06x$ from both sides.

$300 = 0.04x$ Divide both sides by 0.04.

$7500 = x$

Then we substitute sales, x, into the equation for Package A:

$I = 250 + 0.10(7500) = 1000$

For sales of \$7500, each package gives \$1000 in income.

Check:

$I = 550 + 0.06(7500) = 1000$ Package B gives the same income. ✓ ▬

GEOMETRY PROBLEMS Geometry problems may be troublesome because they rely on facts that are not stated in the problem. Always list the facts as part of the plan.

Useful Facts:

The perimeter of a rectangle is $2l + 2w$, where l = length, w = width.

The sum of the measures of the angles of a triangle is 180°.

A right angle measures 90°.

Supplementary angles *have measures that add to 180°.* In Figure 2, *a* and *b* are supplementary angles.

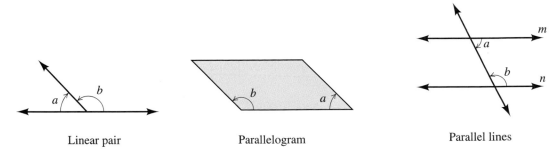

Linear pair Parallelogram Parallel lines

FIGURE 2

Complementary angles *have measures that add to 90°.* In Figure 3, *c* and *d* are complementary angles.

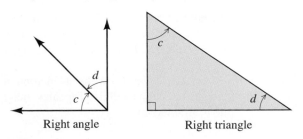

Right angle Right triangle

FIGURE 3

EXAMPLE 8 **Finding angle measures** One angle measures 15° more than another. Find their measures if the angles are complementary.

SOLUTION *Plan:* Let the angle measures be x and y in degrees. Let x be the measure of the larger angle. The angle measures add to 90°. Solve for one angle measure; substitute to find the other.

Carry out the plan:

$$x + y = 90 \qquad (1)$$

$$x - 15 = y \qquad (2)$$

Before we solve the equations, we change to standard form.

$$
\begin{array}{rl}
x + y = & 90 \\
\underline{x - y = \quad 15} & \text{Add the equations.} \\
2x \quad = & 105 \qquad \text{Divide by 2.} \\
x = & 52.5° \qquad \text{Substitute 52.5 for } x \text{ in the original second equation.} \\
52.5 - 15 = y & \\
37.5° = y &
\end{array}
$$

Check: $52.5 + 37.5 = 90$ Substitute for x and y in the first equation. ✓ ▬

> **ANSWER BOX**
>
> **Warm-up: 1.** $x = -9 - 2y$ **2.** $y = \dfrac{-9 - x}{2}$ **3.** $-2x + y = 4$
>
> **4.** $x - y = 10$ **Think about it 1:** The coefficient on x is 1, so solving for x saves doing a division and avoids fractions.   **Think about it 2:** The y would be eliminated.

3.1 Exercises

In Exercises 1 to 8, solve the equations for the indicated variable.

1. $3x + y = 4$ for y $y = -3x + 4$

2. $5x - y = 7$ for y $y = 5x - 7$

3. $x - 3y = 7$ for x $x = 3y + 7$

4. $2y - x = 6$ for x $x = 2y - 6$

5. $4y - 2x = 5$ for x $x = 2y - \frac{5}{2}$

6. $3y - 2x = 11$ for x $x = \frac{3}{2}y - \frac{11}{2}$

7. $4x - 2y = -3$ for y $y = 2x + \frac{3}{2}$

8. $2x - 3y = -5$ for y $y = \frac{2}{3}x + \frac{5}{3}$

Solve the systems of equations in Exercises 9 to 16 by substitution.

9. $y = 3x - 4$
$3y - 2x = 9$
$x = 3, y = 5$

10. $2y = 3x - 10$
$2.5x + y = 3$
$x = 2, y = -2$

11. $0.6x + y = 3$
$1.4x + y + 3 = 0$
$x = -7.5, y = 7.5$

12. $5y + 2x - 9 = 0$
$y + 1.2x = 3$
$x = 1.5, y = 1.2$

13. $x + 2y = 9.4$
$3x - 5y = 4$
$x = 5, y = 2.2$

14. $x - 3y = -1.4$
$2x - 7y + 1.6 = 0$
$x = -5, y = -1.2$

15. $y = 1.8x - 6.6$
$y + 1.4x = 3$
$x = 3, y = -1.2$

16. $x + 3y = 14.5$
$2x - y = 1$
$x = 2.5, y = 4$

Solve the systems of equations in Exercises 17 to 24 by elimination.

17. $3y + 2x = -2$
$3y - 5x = 26$
$x = -4, y = 2$

18. $2x + 3y = 10.9$
$2x + 16.1 = 3y$
$x = -1.3, y = 4.5$

19. $5x = 14.7 + 6y$
$2x + 6y = -4.2$
$x = 1.5, y = -1.2$

20. $3x + 4y = 4$
$2x + y = 5$
$x = 3.2, y = -1.4$

21. $2x + 4y = 0$
$3x - 2y = 24$
$x = 6, y = -3$

22. $3x + 2y = 2$
$-4x + 3y = 37$
$x = -4, y = 7$

23. $5x - 2y = -9$
$6x + 7y = 8$
$x = -1, y = 2$

24. $4x - 5y = 7$
$3x - 2y = 0$
$x = -2, y = -3$

Solve the systems of equations in Exercises 25 to 32. Choose either substitution or elimination, as appropriate.

25. $A + B = 18$
$A - B = -58$
$A = -20, B = 38$

26. $B - C = 17$
$B + 2C = -19$
$B = 5, C = -12$

27. $C + D = 2$
$2C + D = -11$
$C = -13, D = 15$

28. $D + 5E = 54$
$7D = E$ $D = 1.5, E = 10.5$

29. $4E = F$
$3E + 4F = 34.2$
$E = 1.8, F = 7.2$

30. $\frac{1}{2}F = 2G$
$F + G = 1.5$
$F = 1.2, G = 0.3$

31. $3G + 2H = 0$
$H + G = 0.8$
$G = -1.6, H = 2.4$

32. $2H + 3J = 5.4$
$5H = 2J - 11.2$
$H = -1.2, J = 2.6$

In Exercises 33 to 38, what question might be asked about the sentence? What facts must be added by the reader because they are not stated in the sentence?

33. Samantha has 15 dimes and 12 quarters.
What is total value of money? Dimes are $0.10; quarters, $0.25.

34. Jessika has 25 pennies and 12 nickels.
What is total value of money? Pennies are $0.01; nickels, $0.05.

35. An isosceles triangle has one angle 10° larger than the two equal angles.   What is measure of each angle? Sum of three interior angles in triangle is 180°.

36. A parallelogram has one interior angle three times another interior angle.   What are measures of angles? Sum of interior angles is 360°; adjacent angles add to 180°.

37. A rectangle has adjacent sides measuring 5 inches and 10 inches.   What is the perimeter? What are the measures of the other two sides? Opposite sides of a rectangle are equal.

38. A right triangle has its smaller angles equal.
What is the measure of each angle? The right angle is 90°. The sum of the three angle measures is 180°. The two smaller angles are complementary.

In Exercises 39 to 42, find the level of sales for which the two compensation packages are equal. Find the income at that level of sales.

Package 1 is $250 plus 10% of sales.
Package 2 is $350 plus 8% of sales.
Package 3 is $550 plus 4% of sales.
Package 4 is $750 plus 4% of sales.
Package 5 is $550 plus 6% of sales.

39. Packages 3 and 1
$5000, $750

40. Packages 4 and 1
≈$8333, $1083

41. Packages 4 and 2
$10,000, $1150

42. Packages 5 and 2
$10,000, $1150

In Exercises 43 to 54, set up a plan to solve the exercise, carry out your plan, and check.

43. A bottle and cork together cost $1.10. The bottle costs $1.00 more than the cork. What is the cost of each?
$1.05, $0.05

44. A deluxe box of chocolates cost $45. The fancy box costs $40 less than the chocolates it contains. What is the cost of each? $42.50, $2.50

45. The length and width of a photo add to 20 inches. The length is four more than the width. Find the length and width.   12 in., 8 in.

46. The length and width of a pitcher's plate in baseball add to 30 inches. The length is 18 inches more than the width. Find the length and width.   24 in., 6 in.

47. The perimeter of a rectangular soccer field is 380 yards. The length is 30 yards more than the width. Find the length and width.   110 yd, 80 yd

48. The perimeter of a rectangular tennis court is 228 feet. The width is 3 feet less than half the length. Find the length and width.   78 ft, 36 ft

49. Two angles are complementary. The difference between their measures is 25°. Find the angle measures.
57.5°, 32.5°

50. Two angles are supplementary. One is 35° more than the other. Find the angle measures.   107.5°, 72.5°

51. Two angles are supplementary. A third of the larger angle measure added to the smaller angle measure gives 90°. Find the angle measures.   135°, 45°

52. Two angles are complementary. When the larger is added to four times the smaller, the result is 180°. Find the angle measures.   60°, 30°

53. Two angles are complementary. When the larger is doubled, the angles become supplementary. Find the angle measures.   90°, 0°; impossible with nonzero angles

54. Two angles are complementary. When the larger is doubled and the smaller is tripled, the resulting angles are supplementary. Find the angle measures.
90°, 0°; impossible with nonzero angles

55. Solve for y only:
$acx + bcy = ce$
$acx + ady = af$
$y = \dfrac{ce - af}{bc - ad}$ or $y = \dfrac{af - ce}{ad - bc}$

56. Solve for x only:
$adx + bdy = de$
$bcx + bdy = bf$
$x = \dfrac{de - bf}{ad - bc}$ or $x = \dfrac{bf - de}{bc - ad}$

57. Explain when using substitution may be more convenient than using elimination.
One variable has coefficient 1 and is easy to solve for.

58. Explain when using elimination may be more convenient than using substitution.
Coefficients on same variable in two equations are opposites or can be made so.

3.2 Solving Systems of Two Linear Equations by Graphing

Objectives

▮ Solve systems of linear equations by graphing.

▮ Identify systems of two linear equations having a single solution, an infinite number of solutions, or no solution.

▮ Explain the algebraic results from solving systems containing equations whose graphs are parallel or coincident.

WARM-UP

These exercises review material from Section 2.5.

1. Find the equation of a line parallel to $y = -\frac{3}{2}x + 2$ that passes through the point (2, 5). $\ y = -\frac{3}{2}x + 8$

2. Find the equation of a line perpendicular to $y = -\frac{3}{2}x + 2$ that passes through the point (3, 0). $\ y = \frac{2}{3}x - 2$

IN THIS SECTION, we use graphs of linear equations to understand why some systems of linear equations have a unique solution while others have no solution.

Graphing

Graphing gives a method of finding the numerical solution to a system of linear equations and lends a visual meaning to the solution. *Because every point on the graph of an equation makes the equation true, the point of intersection of two graphs is the ordered pair that makes both equations true.*

An advantage of the graphical solution is that we are reminded by the ordered pair itself that a system of two linear equations has two variables (unknowns) and must be solved for both variables. A common error is to solve a system for only one of the two variables.

In a graph drawn by hand, the point of intersection can be located with only limited accuracy. Intersection features of graphing calculators and built-in solvers in computer programs permit us to locate coordinates of the point of intersection that are correct to many decimal places.

EXAMPLE 1   **Solving with a graph**   Solve the following system of linear equations from a calculator graph:

$$3y + 6 = 2x \qquad (1)$$
$$2y + 3x = 16 \qquad (2)$$

SOLUTION   First, we solve the equations for y.

Equation 1:

$$3y + 6 = 2x \qquad \text{Subtract 6.}$$
$$3y = 2x - 6 \qquad \text{Divide by 3.}$$
$$y = \tfrac{2}{3}x - 2$$

Equation 2:

$$2y + 3x = 16 \qquad \text{Subtract } 3x.$$

$$2y = -3x + 16 \qquad \text{Divide by 2.}$$

$$y = -\frac{3}{2}x + 8$$

The graphs are shown in Figure 4a. We enter the equations in the calculator under $\boxed{Y=}$, graph, and trace to estimate the intersection on the graph. We can then confirm the intersection with the $\boxed{\text{2nd}}$ [CALC] **5 : intersect** option. To the nearest thousandth, the intersection is (4.615, 1.077); see Figure 4b.

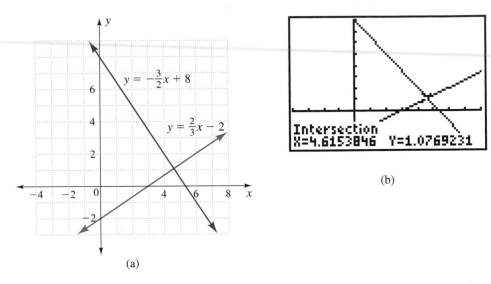

(a)

(b)

FIGURE 4 In (b), −4 to 8 on X, −3 to 8 on Y

 GRAPHING CALCULATOR TECHNIQUE: SOLVING TWO EQUATIONS GRAPHICALLY

> Solve each equation for y as a function of x. If the equations are not in x and y, solve each for one variable in terms of the second variable. Enter the two functions of x as Y_1 and Y_2 under $\boxed{Y=}$. Set the viewing window. Graph, and verify that the point of intersection, if any, shows in the viewing window.
>
> Use the **5 : intersect** option under $\boxed{\text{2nd}}$ [CALC] to find the point of intersection of the two functions.

We now examine two special geometric and algebraic results for a system of two linear equations.

Coincident Lines

Two lines that are described by equivalent equations are **coincident lines**. Because coincident lines have all points in common, we say that *a system of equations whose graphs form coincident lines has an infinite number of solutions*. Nonvertical coincident lines have the same slope and the same y-intercept. They are easy to identify because we obtain the same equation when we transform the equations to $y = mx + b$ form.

EXAMPLE 2 **Identifying coincident lines** Solve each equation in this system for y. What can you conclude about the graphs of the equations?

$$2x - 3y = 6 \qquad (1)$$

$$y + 2 = \frac{2}{3}x \qquad (2)$$

SOLUTION Equation 1:

$$2x - 3y = 6 \qquad \text{Subtract 6 and add } 3y \text{ on both sides.}$$

$$2x - 6 = 3y \qquad \text{Divide by 3.}$$

$$\tfrac{2}{3}x - 2 = y$$

Equation 2:

$$y + 2 = \tfrac{2}{3}x \qquad \text{Subtract 2.}$$

$$y = \tfrac{2}{3}x - 2$$

The equations are both $y = \tfrac{2}{3}x - 2$, so they describe the same line (see Figure 5).

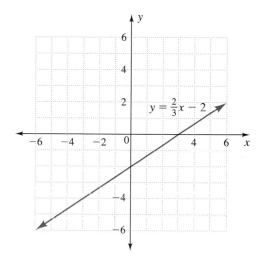

FIGURE 5

▮ In Example 3, we examine the algebraic results from solving a system containing equations with coincident graphs.

EXAMPLE 3 **Finding the algebraic results for coincident lines** Solve the system of equations in Example 2 by substitution.

$$2x - 3y = 6 \qquad (1)$$

$$y + 2 = \tfrac{2}{3}x \qquad (2)$$

SOLUTION Equation 2:

$$y + 2 = \tfrac{2}{3}x \qquad \text{Solve for } y.$$

$$y = \tfrac{2}{3}x - 2 \qquad \text{Substitute into Equation 1.}$$

Equation 1:

$$2x - 3y = 6 \qquad \text{Let } y = \tfrac{2}{3}x - 2.$$

$$2x - 3(\tfrac{2}{3}x - 2) = 6 \qquad \text{Simplify.}$$

$$2x - 2x + 6 = 6 \qquad \text{Add like terms.}$$

$$6 = 6 \qquad \text{Identity}$$

The variable drops out, and the remaining statement, $6 = 6$, is an identity and is always true. This result indicates that all coordinate pairs (x, y) that make the first equation true also make the other equation true.

Think about it 1: Solve Example 3 by elimination. Are the results the same?

■■ COINCIDENT LINES

A true statement such as $0 = 0$ in the solution to a system of equations implies that there are an infinite number of solutions. In the case of two equations in two variables, the graphs are coincident lines.

Parallel Lines

Parallel lines provide a second special case in working with a system of two linear equations. *Two lines in the coordinate plane that have no intersection* are **parallel lines**. Because parallel lines have no point of intersection, we say that a *system of equations whose graphs are parallel lines has no real-number solution.* Nonvertical parallel lines have the same slope and different y-intercepts.

EXAMPLE 4   Identifying parallel lines   Solve each equation in this system for y. What can you conclude about the graphs of the equations?

$$3x + 2y = 4 \qquad (1)$$

$$8 - \tfrac{3}{2}x = y \qquad (2)$$

SOLUTION   Equation 1:

$$3x + 2y = 4 \qquad \text{Subtract } 3x.$$

$$2y = -3x + 4 \qquad \text{Divide by 2.}$$

$$y = -\tfrac{3}{2}x + 2$$

Equation 2:

$$y = -\tfrac{3}{2}x + 8$$

The equations have the same slope, $-\tfrac{3}{2}$, but different y-intercepts. The lines are parallel (see Figure 6).

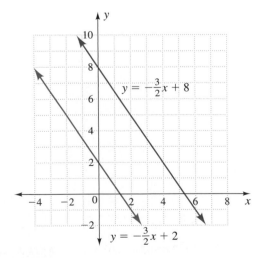

FIGURE 6

■ In Example 5, we examine the algebraic results from solving a system of linear equations whose graphs are parallel lines.

EXAMPLE 5   Finding the algebraic results for parallel lines   Solve the system of equations in Example 4 by elimination.

$$3x + 2y = 4 \qquad (1)$$

$$8 - \tfrac{3}{2}x = y \qquad (2)$$

SOLUTION   Equation 1:

$$3x + 2y = 4$$

Equation 2:

$8 - \tfrac{3}{2}x = y$	Multiply by 2 to clear the denominator.
$16 - 3x = 2y$	Change to standard form.
$-3x - 2y = -16$	(Note that the signs were left negative.)
$\underline{3x + 2y = 4}$	Add the first equation.
$0 = -12$	Contradiction

The variables drop out and the remaining statement is a contradiction, always false, implying that there are no solutions to the system of equations. Because the equations describe parallel lines (Example 4), it is reasonable that there is no solution.   ■

Think about it 2: Solve Example 5 by substitution. Are the results the same?

■ **PARALLEL LINES**

> A false statement such as $0 = 1$ in the solution to a system of equations implies that there are no real-number solutions. In the case of two equations in two variables, the graphs are parallel.

■ **GRAPHING A SYSTEM OF TWO LINEAR EQUATIONS**

> In graphing a system of two linear equations, we have three possible outcomes (see the figures below):
>
> **1.** The equations describe lines intersecting in exactly one point (point *A*). The system of equations has exactly one solution—the ordered pair at the point of intersection.
>
> **2.** The equations describe the same line. The graphs are coincident (line *BC*). The system has an infinite number of solutions because ordered pairs for all points on the graph of the first equation satisfy the second equation.
>
> **3.** The equations describe parallel lines (lines *DE* and *FG*). The graphs have no point of intersection; hence, the system has no real-number solution.
>
>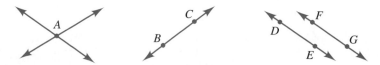

Table 1 summarizes the geometry and algebra associated with solving a system of two linear equations.

TABLE 1

Geometry	Intersecting lines	Coincident lines	Parallel lines
Algebra	The equations can be solved for x and y.	The variables drop out, and the remaining statement is true.	The variables drop out, and the remaining statement is false.
Numerical Result	The unique solution is given by the ordered pair (x, y).	An identity, $a = a$, results.	A contradiction (such as $0 = 1$) results.
Solution	One ordered pair is the solution.	An infinite number of ordered pairs satisfy the system.	There is no real-number solution to the system.

ANSWER BOX

Warm-up: 1. $y = -\frac{3}{2}x + 8$ **2.** $y = \frac{2}{3}x - 2$ **Think about it 1:** Multiplying Equation 2 by the denominator 3 gives an equation identical to the first. Subtraction yields $0 = 0$. This always-true statement agrees with the results in Example 3. The lines are coincident.   **Think about it 2:** Substitute Equation 2 into Equation 1. Simplify. When we add like terms, the variable x drops out, leaving $16 = 4$. This false statement agrees with the results in Example 5. The lines are parallel.

3.2   Exercises

Solve the systems of equations in Exercises 1 to 16 by graphing. For each system that describes coincident or parallel lines, solve again using substitution or elimination. Describe the numerical result: an identity or a contradiction.

1. $x + y = 5$

$x - y = -3$ See Answer Section; $(1, 4)$

2. $x - y = 6$ See Additional Answers; parallel;

$y - 3 = x$ variables drop out, contradiction

3. $2x - 3 = 2y$ See Answer Section; coincident;

$x - y = 1.5$ variables drop out, identity

4. $y - x = 4$

$y + x = 6$ See Additional Answers; $(1, 5)$

5. $2y = 1.5x + 1$ See Answer Section; parallel;

$4y - 3x + 8 = 0$ variables drop out, contradiction

6. $y - 2x = 3$ See Additional Answers; $\left(-\frac{5}{3}, -\frac{1}{3}\right)$

$y + x = -2$

7. $x + 2y = 4$ See Answer Section; $\left(\frac{10}{3}, \frac{1}{3}\right)$

$x - y = 3$

8. $6y - 4x = 3$ See Additional Answers; coincident;

$y = \frac{2x}{3} + \frac{1}{2}$ variables drop out, identity

9. $2x - y = 4$ See Answer Section; $\left(\frac{1}{2}, -3\right)$

$4x + y = -1$

10. $y = \frac{2x}{3} - 2$ See Additional Answers; parallel; variables drop out, contradiction

$3y - 2x = 4$

11. $y = \frac{x}{3} + 2$ See Answer Section; coincident; variables drop out, identity

$3y - 6 = x$

12. $3x + 2y = -2$ See Additional Answers; $\left(-1, \frac{1}{2}\right)$

$3y + x = \frac{1}{2}$

13. $x + y = 4$ See Answer Section; parallel; variables drop out, contradiction

$x + 2 = -y$

14. $4x - 2y = 8$ See Additional Answers; coincident; variables drop out, identity

$2x = y + 4$

15. $x - 2y = 5$ See Answer Section; coincident; variables drop out, identity

$2x - 10 = 4y$

16. $x - y = 6$ See Additional Answers; parallel; variables drop out, contradiction

$x - 10 = y$

In Exercises 17 to 20, find the level of sales for which the two compensation packages are equal. Find the total income at that sales level.

Package 1 is $250 plus 10% of sales.
Package 2 is $350 plus 8% of sales.
Package 3 is $550 plus 4% of sales.
Package 4 is $750 plus 4% of sales.
Package 5 is $550 plus 6% of sales.

17. Packages 1 and 2 $5000; $750

18. Packages 2 and 3 $5000; $750

19. Packages 3 and 4 none; 4 is always better.

20. Packages 3 and 5 zero sales; $550

21. If we increase the percent of sales in a compensation package, how does the graph change? becomes steeper

22. If we increase the fixed amount in a compensation package, how does the graph change?
shifts up to new vertical intercept

23. Which compensation packages have the same vertical axis intercept when graphed? packages 3 and 5

24. Which compensation packages give parallel lines when graphed? packages 3 and 4

25. Describe how you might write a pair of equations with no solution. Write equations for pair of parallel lines.

26. Describe how you might write a pair of equations with an infinite number of solutions. See below.

27. Can all equations whose graphs are straight lines be written in the standard form, $ax + by = c$? yes

28. Write the equation of the vertical line $x = 2$ in standard form, and identify a, b, and c. $x + 0y = 2; a = 1, b = 0, c = 2$

26. Write one equation, multiply (or divide) both sides by same number to create equivalent equation; write pair of equations whose graphs have same slope and y-intercept.

29. Write the equation of the horizontal line $y = 3$ in standard form, and identify a, b, and c.
$0x + y = 3; a = 0, b = 1, c = 3$

30. Can all equations whose graphs are straight lines be written in the slope-intercept form of a linear equation, $y = mx + b$? no; $x = c$ for any c.

31. In parts a to d, think about real numbers.

a. What number n is the *additive identity*—that is, $a + n = a$? $n = 0$

b. What number n is the *multiplicative identity*—that is, $a \cdot n = a$? $n = 1$

c. What is the product of a number and its multiplicative inverse? the identity, 1

d. What linear function is the *identity function*? (*Hint:* See Section 2.6.) $y = x$ or $f(x) = x$

▮ **Project**

32. Hiker's Dilemma One beautiful morning, Kristin starts at the rim of the Grand Canyon and hikes down to Phantom Ranch on the Colorado River, arriving in the afternoon. The next morning, she leaves the ranch and returns to the rim, on the same trail. See Additional Answers.

a. Explain why there is a point on the trail at which she arrives at the same time on both days. Although it does not matter, you may assume a uniform rate of travel walking down the trail and another uniform rate walking up the trail.

b. Describe times of departure and arrival that would make it impossible for her to arrive at one point at the same time on both days.

Hint: Instead of thinking about one hiker, consider two hikers, one starting at the rim and one at the ranch. Would they meet somewhere along the trail?

▮ **3.3   Quantity-Value Tables**

Mathematics is too often perceived as a jumble of numbers and useless information. This section gives students a valuable tool for organizing their thinking. Spend less time on skills, if you must, but don't skip this section.

Objectives

▮ Identify quantity and value information.

▮ Solve quantity and value problems by writing and solving systems of equations.

WARM-UP

1. What is the value of 4 dimes, 5 nickels, and 3 pennies? $0.68

2. Write these percents as decimals:

a. 16%   **b.** 5%   **c.** 7%   **d.** 7.5%   **e.** 8.9%
0.16   0.05   0.07   0.075   0.089

3. A mixture problem from dosage computation in a nursing program: How much hydrogen peroxide in a 3% solution (household strength) must be blended with water to obtain 1.5 liters of a 0.4% solution? How much water? See Example 5.

IN SECTION 3.1, sentences in the word problems could be directly translated into equations. When word problems cannot be directly translated, quantity-value tables can be used to organize information for building equations.

Quantity and Value

Word problem facts are frequently related to quantity and value or quantity and rate. **Quantity** *answers the question How many? or How much?* **Value** may be *monetary worth.* Value is most commonly given as a rate. The word **per** identifies *a value or rate*, such as percent, cents per quarter, kilometers per hour, or cost per item.

EXAMPLE 1 Identifying quantity and value   Which of these describe quantities? Which describe values or rates?
a. 4 nickels
b. 5 cents per nickel
c. 10 field goals in football
d. 3 points per field goal in football
e. 5 points for first place, 3 points for second place
f. $9.99 per pound
g. 2 meters
h. 21% interest on a credit card
i. 1.25% interest on a savings account

SOLUTION   Parts a, c, and g describe quantities. The rest are values.   ▬

Quantity-Value Tables

The author uses quantity-value tables in teaching centroid and center of mass applications in both engineering and calculus.

The **quantity-value table** is *characterized by multiplying the quantity Q of each item by the value V (or rate) of each item to obtain the item's total value, $Q \cdot V$.*

We use quantity-value tables to organize data on grade point average in Examples 2 and 3. The value or rate may be the number assigned to a specific outcome or activity. For example, the grade point for a specific letter grade is a rate based on an outcome: 4 points per credit hour for an A, 3 points per credit hour for a B, 2 points per credit hour for a C, and so forth.

EXAMPLE 2 Setting up a quantity-value table: GPA calculations   What is a student's grade point average (GPA) if she received an A in a 3-credit-hour psychology course, a B in a 4-credit-hour trigonometry course, and a C in a 4-credit-hour computer science course? Show the information in a quantity-value table.

SOLUTION   In Table 2, we set up a table with the quantity, the value, and their product in separate columns.

TABLE 2

Subject and Grade	Quantity (credit hours)	Value (points per credit hour)	$Q \cdot V$ (grade points)
Psychology, A	3	4	12
Trigonometry, B	4	3	12
Computer science, C	4	2	8
Total	11		32

GPA ≈ 2.91

To calculate the grade point average, we add the items in the $Q \cdot V$ column in Table 2 and divide by the total credit hours:

$$\text{GPA} = \frac{\text{total } Q \cdot V}{\text{total } Q} = \frac{32 \text{ points}}{11 \text{ credit hours}} \approx 2.91 \text{ points per credit hour}$$

If we were to place the GPA into the table, it would go into the last row of the value column, making the product across the row equal to the sum of the $Q \cdot V$ column.

▬ In Example 3, look carefully at how the two expressions for the lower right corner are used to build the equation.

EXAMPLE 3 **Finding a missing quantity, given an average value: GPAs, continued** A student with a 2.25 GPA has 44 credit hours. Using the grade rates from Example 2, find how many credit hours of Bs he needs to raise his GPA to a 2.5. (*Hint:* Let x be the number of credit hours of Bs.)

SOLUTION **Plan:** The desired GPA, 2.50, is placed in the last row of the value column. The sum of the last column (the total points from current and future grades) gives one expression for the lower right corner. The product of the total quantity of credit hours and the desired GPA gives a second expression for the lower right corner.

TABLE 3 *Grade Point Average*

Item	Quantity (credit hours)	Value (points per credit hour)	$Q \cdot V$ (grade points)
Current GPA status	44	2.25	44(2.25)
Future B grades	x	3.00	$3x$
Total	44 + x	2.50 average	(Two expressions: the sum down and the product across)

Carry out the plan: The two expressions for the lower right corner of Table 3 are *the total down*, $44(2.25) + 3x$, and *the product across*, $(44 + x)(2.50)$. We set them equal and solve for x:

$$44(2.25) + 3x = (44 + x)(2.50)$$
$$99 + 3x = 110 + 2.50x$$
$$0.5x = 11$$
$$x = 22 \text{ credit hours}$$

Check: $44(2.25) + 3(22) \overset{?}{=} (44 + 22)(2.50)$ ✓ ▬

Percents as Rates

Examples 4, 5, and 6 include percents as rates. We change the percents to decimals.

EXAMPLE 4 **Setting up the quantity-value table: interest on debt** Suppose Heidi has a $12,500 student loan at 7.5%, $1200 in credit card debt at 21%, and an $8000 car loan at 8.9%. How much in total has she borrowed? What is the total annual interest paid? What is the average annual rate of interest for the loans? Organize the information in a quantity-value table.

SOLUTION   The information is organized in Table 4.

TABLE 4

Type of Loan	Quantity: Amount of Loan (in dollars)	Value: Rate of Interest	$Q \cdot V$: Interest Paid (in dollars)
Student loan	12,500	0.075	937.50
Credit card balance	1200	0.21	252.00
Car loan	8000	0.089	712.00
Total	21,700		1901.50

The average annual rate, ≈0.088, goes here.

Heidi has borrowed $21,700. The total annual interest paid is $1901.50. The average annual interest rate is calculated by dividing the total interest paid by the total amount of the loans:

$$\text{Average rate} = \frac{\text{interest paid}}{\text{total loans}} = \frac{1901.50}{21,700.00} \approx 0.088$$

The average annual interest rate paid on the loans is about 8.8%. The 8.8% could be placed in the last row of the value column in Table 4. The product across the last row equals the sum of the $Q \cdot V$ column.   ▬

In Example 5, we return to the mixture problem in the Warm-up.

EXAMPLE 5   Setting up equations with a quantity-value table: mixtures   How much hydrogen peroxide in a 3% solution (household strength) must be blended with water to obtain 1.5 liters of a 0.4% solution? How much water?

SOLUTION   The quantities of hydrogen peroxide and water are unknown. The total quantity in Table 5 is 1.5 liters (L). The value of the hydrogen peroxide is 3%. The value of the water is a usual trouble spot. It is 0%.

TABLE 5

Item	Quantity	Value	$Q \cdot V$
Hydrogen peroxide	x	0.03	$0.03x$
Water	y	0	$0y$
Total	1.5	Blend: 0.004	1.5(0.004)

Equation 1 is from the quantity column:

$$x + y = 1.5$$

Equation 2 is from the $Q \cdot V$ column (after we multiply across the total row):

$0.03x + 0y = 1.5(0.004)$	Simplify.
$0.03x = 0.006$	Divide by 0.03.
$x = 0.2 \text{ L}$	Hydrogen peroxide

We replace x in Equation 1 with 0.2:

$$0.2 + y = 1.5$$

$$y = 1.3 \text{ L} \qquad \text{Water}$$

Check: Place $x = 0.2$ and $y = 1.3$ into the table, and find the corresponding values for $Q \cdot V$. Add down the $Q \cdot V$ column and multiply across the total row to see that both approaches lead to the same number in the lower right corner. ▬

EXAMPLE 6 Setting up equations with a quantity-value table: dog food blend   How many pounds of dog food A, containing 16% protein, need to be blended with dog food B, containing 5% protein, to obtain 500 pounds of a blend with 7% protein? Set up equations and solve with substitution. Then check with a graph.

SOLUTION We will let x be the number of pounds of dog food A and y be the number of pounds of dog food B.

TABLE 6

Dog Food	Quantity (pounds)	Value (percent protein, converted to decimal)	$Q \cdot V$ (pounds of protein)
A	x	0.16	$0.16x$
B	y	0.05	$0.05y$
Total	500	0.07	$(500)(0.07) = 35$

The average, or the blended value, is 7%. We multiply 500 pounds by 7% to obtain the sum of the $Q \cdot V$ column in Table 6: 35 pounds.

There are two equations within the table. From the quantity column, we have

$$x + y = 500 \qquad (1)$$

From the $Q \cdot V$ column, we have

$$0.16x + 0.05y = 35 \qquad (2)$$

We obtain $y = 500 - x$ from the first equation, and we substitute into the second equation:

$$0.16x + 0.05y = 35 \qquad \text{Substitute } y = 500 - x.$$

$$0.16x + 0.05(500 - x) = 35 \qquad \text{Apply the distributive property.}$$

$$0.16x + 25 - 0.05x = 35 \qquad \text{Add like terms, and subtract 25 from both sides.}$$

$$0.11x = 10 \qquad \text{Divide by 0.11.}$$

$$x \approx 90.9 \text{ lb of dog food A}$$

$$y = 500 - x \qquad \text{Equation 1}$$

$$y \approx 409.1 \text{ lb of dog food B}$$

Because the blend has a slightly higher protein level than dog food B, a relatively smaller amount of dog food A is needed.

The graphs of the two equations are shown in Figure 7. The graphs intersect near (100, 400), in agreement with the symbolic work.

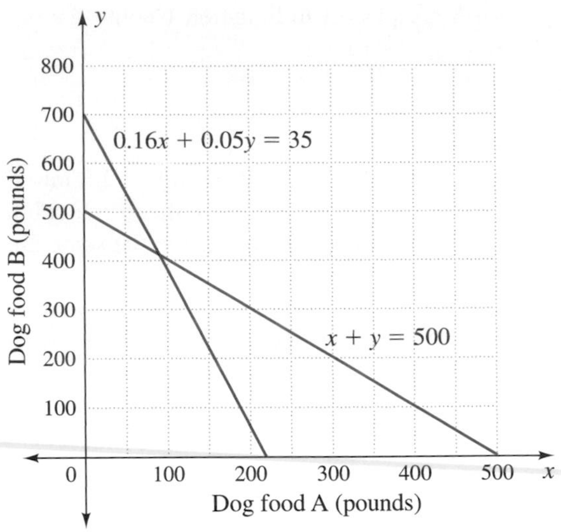

FIGURE 7

■ SUMMARY: QUANTITY-VALUE TABLES

Student Note: Find your own shortcuts. Whether or not you actually create the tables, organize information into quantity and value to clarify your thinking in setting up equations.

- Adding the items in the quantity column gives the total quantity.

- Adding the items in the $Q \cdot V$ column gives the same number as multiplying the total quantity by the average value (across the last row).

- The total of the $Q \cdot V$ column divided by the total of the quantity column gives an average value. The last number in the value column is an average value and is not found by adding the items in the value column.

ANSWER BOX

Warm-up: 1. $0.68   **2. a.** 0.16 **b.** 0.05 **c.** 0.07 **d.** 0.075 **e.** 0.089
3. See Example 5.

3.3   Exercises

Identify the quantity and value words and numbers in Exercises 1 to 6.

1. Invest $1600 at 5% annual interest and $12,000 at 8% annual interest. $1600, $12,000; 5%, 8% annual interest

2. Blend 200 kilograms of Salvadoran coffee at $15.40 per kilogram with 300 kilograms of Guatemalan coffee at $18.70 per kilogram.
200 kg, 300 kg; $15.40/kg, $18.70/kg

3. Earn $6.50 per hour for 20 hours at the first job and $7.25 per hour for 15 hours at the second job.
20 hr, 15 hr; $6.50/hr, $7.25/hr

4. Sell 200 five-pound bags of onions for $2.98 each and 100 ten-pound bags for $4.49 each.
200 bags, 100 bags; $2.98/bag, $4.49/bag

5. Blend 100 milliliters of 8% solution with 1000 milliliters of water at 0% solution.
100 mL, 1000 mL; 8%, 0% solution

6. Earn 10 points for first place, 5 points for second place, and 1 point for third place. Team A had 3 first-place finishes and 2 third-place finishes. Team B had 1 first-place finish and 5 second-place finishes.
3 first and 2 third, 1 first and 5 second; 10, 5, 1 points

In Exercises 7 to 28, set up equations. Quantity-value tables may be helpful. Solve by any method: substitution, elimination, or graphing.

7. What is the GPA of a student who earns an A in computer science (4 credit hours), a B in psychology (3 credit hours), a C in English composition (3 credit hours), and a B in physical education (1 credit hour)?
3.09 points/credit hour

8. What is the GPA of a student with 22 credit hours at a grade of A, 37 credit hours at a grade of B, and 16 credit hours at a grade of C? 3.08 points/credit hour

9. Suppose that a student has a 3.08 GPA with 75 credit hours. He wants to raise his GPA to a 3.25. How many credit hours at a grade of A does he need? 17 credit hours

10. A student has 3 credit hours at a grade of A, 30 credit hours at a grade of B, and 5 credit hours at a grade of C. How many more credit hours at a grade of B does she need to raise her GPA to a 3.00? no real-number solution

11. How many kilograms of Honduran coffee at $18.70 per kilogram need to be blended with Indonesian coffee at $23.65 per kilogram to produce 200 kilograms of a blend selling for $19.80 per kilogram? ≈156 kg, ≈44 kg

12. How many kilograms of Honduran coffee at $18.70 per kilogram need to be blended with Indonesian coffee at $23.65 per kilogram to produce 200 kilograms of a blend selling for $22.00 per kilogram? ≈67 kg, ≈133 kg

13. Suppose 9 dimes and 13 quarters weigh 3.5 ounces. On the same scale, 12 dimes and 9 quarters weigh 3 ounces. Estimate the weight of each coin. ≈0.1 oz, ≈0.2 oz

14. Suppose 15 nickels and 7 copper-clad zinc pennies weigh 3.5 ounces. On the same scale, 5 nickels and 16 pennies weigh 2.5 ounces. Estimate the weight of each coin. ≈0.19 oz, ≈0.10 oz

15. An inheritance of $75,000 is to be split and invested at 6% and 9% interest to produce an average 8% interest rate. How much money is to be invested at each rate? $25,000, $50,000

16. A personal bank loan costs 12% interest, and a credit card loan costs 20% interest. How should $10,000 be borrowed to keep the interest rate at 15%? $6250, $3750

17. A small business averages 13% interest on loans totaling $10,000. Credit card debt is 21%, and bank interest is 11%. How much has been borrowed from each source? $2000, $8000

18. A money market account pays 5% interest, and a savings account pays 3% interest. How much is saved in each account if the average interest on the $2400 total savings is $4.5%? $1800, $600

19. Farzaneh earns $1782 interest on a total of $23,600 placed in two investments. The investments earn 4.5% and 8.5%. How much money is in each investment? $5600, $18,000

20. Polly earns $2343 interest on a total of $29,800 placed in two investments. The investments earn 3.5% and 8.5%. How much money is in each investment? $3800, $26,000

21. A doctor prescribes 0.5 liter of a 5% glucose solution. How much distilled water (0% glucose) and 50% glucose solution must be blended to obtain the proper solution? 0.45 L, 0.05 L

22. A nurse's aide must prepare 4000 milliliters of a 0.5% potassium permanganate solution for an astringent. How much distilled water and 4% potassium permanganate solution must be blended? 3500 mL, 500 mL

23. How much hydrogen peroxide in a 3% solution must be blended with hydrogen peroxide in a 20% solution to obtain 1000 liters of a 6.4% solution? 800 L, 200 L

24. How much 24% carbamide peroxide solution must be blended with 8% carbamide peroxide solution to obtain 500 liters of a 10% carbamide peroxide solution? 62.5 L, 437.5 L

25. How much water and 20% hydrogen peroxide solution should be blended by a manufacturer to obtain 1000 gallons of a 3% hydrogen peroxide solution? 850 gal, 150 gal

26. How much anhydrous glycerol containing no carbamide peroxide must be blended with a 24% carbamide peroxide solution to produce 200 gallons of a 10% carbamide peroxide solution suitable for treating mouth sores? ≈116.7 gal, ≈83.3 gal

27. How many pounds of dog food A, containing 15% protein, need to be blended with dog food B, containing 10% protein, to obtain 1000 pounds of a blend with 12% protein? 400 lb, 600 lb

28. How many pounds of cat food A, containing 8% protein, need to be blended with cat food B, containing 13% protein, to obtain 500 pounds of a blend with 9% protein? 400 lb, 100 lb

---

3 | Mid-Chapter Test

In Exercises 1 and 2, solve the equation for y.

1. $5x - 2y = 4$
$y = \frac{5}{2}x - 2$

2. $2x - \frac{1}{2}y = 5$
$y = 4x - 10$

In Exercises 3 to 6, solve the system of equations by substitution. Show your work. If there is no unique intersection, describe the graph.

3. $2y - 3x = 8$
$y = 1.5x - 3$
contradiction; parallel lines

4. $3y = 2x - 6$ $(2, -\frac{2}{3})$
$y = -\frac{5}{6}x + 1$

5. $y = -0.8x - 1.8$
$y = -1.2x + 3$
$(12, -11.4)$

6. $y = 2.5x + 4$
$2y - 5x = 8$
identity; coincident lines

7. Solve by elimination.
$x + y = \frac{1}{2}$
$x - y = \frac{3}{4}$
$x = \frac{5}{8}, y = -\frac{1}{8}$

8. Solve by elimination.
$2x - 12 = -3y$
$6 - 4y = 3x$
$x = -30, y = 24$

9. What system of equations is solved by the graph of the two lines in the figure? What is the solution?

$y = -x + 2, y = \frac{2}{3}x + 2; x = 0, y = 2$

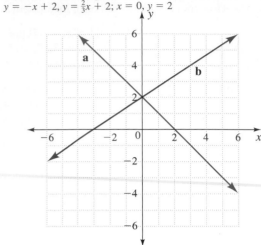

10. Two angles are supplementary. The difference between the angle measures is 90°. What are the angle measures?
135°; 45°

11. Wilt Chamberlain and Elgin Baylor scored a total of 54,568 points in their careers in the National Basketball Association. Chamberlain scored 8270 more points than Baylor. How many did each score?
31,419; 23,149

In Exercises 12 and 13, set up a quantity-value table, write equations, and solve the problem.

12. Helen works at one job 20 hours a week at $6 per hour. If she wants to raise her average hourly rate to $8 per hour, how many hours must she work at another job paying $12 per hour?   10 hr

13. A nurse needs 1 pint of 2% hydrogen peroxide solution for a topical antiseptic. The over-the-counter bottle contains a 3% solution. How much water and 3% hydrogen peroxide solution need to be blended?
$\frac{1}{3}$ pt, $\frac{2}{3}$ pt

3.4   Solving Systems of Three Linear Equations

Objectives

- Solve problems by guess and check.
- Solve systems of three equations by substitution.
- Solve systems of three equations by elimination.
- Identify systems with an infinite number of solutions and with no solutions.
- Solve application problems.

WARM-UP

Translate each sentence into an equation.

1. There are five times as many medium rabbits, m, as large rabbits, l. $m = 5l$

2. There are three times as many small rabbits, s, as medium rabbits, m.
$s = 3m$

3. There are a third as many quarters, q, as dimes, d. $3q = d$ or $q = \frac{1}{3}d$

THIS SECTION INTRODUCES systems of linear equations containing three or more variables. We summarize several strategies for building and solving these systems of equations. An additional method, with matrices on the calculator, is in Section 10.5. We look at possible graphs of these systems and examine the formal names for systems of equations having no solution, one solution, and an infinite number of solutions.

Linear Equations

The linear equation $y = mx + b$ or $ax + by = c$ is said to be of **degree 1** because *the highest exponent on the variables in any term is 1*. Although the graphs of these equations are straight lines, the word *linear* refers not to the straight line but to the exponent 1 on the variables.

If a, b, c, d, and e are constants, $ax + by + cz = d$ is a linear equation in three variables (x, y, and z) and $aw + bx + cy + dz = e$ is a linear equation in four variables (w, x, y, and z). A **linear equation** is *any equation that is of the first degree in its variable(s)*.

Guess and Check

Organizing information about a problem and guessing the solution may improve our understanding. If we don't solve the problem by guessing, we are frequently able to set up equations that we can use to solve the problem.

In Example 1, we will use a table to organize our steps and record our guesses so that we learn from each guess. *Each row in the table represents a guess.*

EXAMPLE 1

Solving problems by guess and check: time allocation   Jian's academic time includes the time spent attending class, plus 2 hours of study time for each hour spent in class. Each week, he spends twice as many hours on academics as he does at work. He spends 28 more hours each week sleeping than at work. He spends the remaining 44 hours of the week on other tasks. How many hours does Jian devote to sleep, work, and academics?

a. Make a table to find the hours spent at each activity.

b. Write equations that describe the problem.

SOLUTION

a. *Understand and plan:* The three activities—sleep, work, and academics—make natural column headings for a guess-and-check table. The hours spent on other tasks and total hours form the two other columns. The total hours column should equal 168:

$$\frac{24 \text{ hours}}{1 \text{ day}} \cdot \frac{7 \text{ days}}{1 \text{ week}} = \frac{168 \text{ hours}}{1 \text{ week}}$$

Carry out the plan: A reasonable starting guess is 8 hours of sleep per night, which, times 7 nights per week, gives 56 hours of sleep per week. This guess results in 184 total hours for the week and so is too high. For our next guess, we drop the hours of sleep to 50. This guess results in 160 total hours and so is too low. Our third guess for sleep is 52 hours, a number between 50 and 56.

TABLE 7

Sleep	Work	Academics	Other	Total Hours
56	28	56	44	184 (too high)
50	22	44	44	160 (too low)
52	24	48	44	168 (correct)

As Table 7 shows, Jian sleeps 52 hours a week, works 24 hours, and spends 48 hours attending class and studying.

Check: The results satisfy the conditions in the problem. ✓

b. We will let s = hours of sleep, w = hours of work, and a = hours on academics. Using these variables, we can describe the table setup with equations.

$s - 28 = w$	From the first two columns
$a = 2w$	From the second and third columns
$s + w + a + 44 = 168$	From each row

Had we not solved the problem with the table, we could have solved these equations. ▬

Think about it 1: If x equals the number of hours spent sleeping, what one equation, in terms of x, describes Example 1?

The problem in Example 2 does not have whole-number answers, so continued guessing is not productive. We use the patterns formed within the table to find equations.

EXAMPLE 2 Setting up equations by guess and check: another time allocation   Tana's priority is to get through school. She plans to spend twice as many hours in class as she does at work. She studies 3 hours for every hour in class. To maintain her health and sanity, she wants to limit the total time allocated to work and school to 60 hours per week. How many hours of classes, study, and work should she plan?
a. Make a guess-and-check table with four entries.
b. Write equations that describe the problem.

SOLUTION **a.** ***Understand and plan:*** The hours allocated to work, class, and study make natural column headings in a table. The total time should be 60 hours.

Carry out the plan: We start Table 8 with a guess of 10 hours of work per week. This guess leads us to 90 total hours, which is too high. We cut the number of work hours in half and obtain a total of 45 hours, which is too low. Subsequent guesses of 8 and 7 work hours get us closer to the 60 hours.

TABLE 8

Work	Class	Study	Total
10	20	60	90 (too high)
5	10	30	45 (too low)
8	16	48	72 (too high)
7	14	42	63 (too high)

b. Letting w = hours of work, c = hours in class, and s = hours of study, we can describe the table with equations.

$$c = 2w$$ From the first and second columns

$$s = 3c$$ From the second and third columns

$$w + c + s = 60$$ From each row

We will solve this system of equations in Example 3. ▬

Think about it 2: If x = the number of hours spent working, what one equation, in terms of x, describes Example 2?

Substitution

In Example 3, we solve the system of three equations from Example 2. Substitution permits us to replace two of the three variables in one equation.

EXAMPLE 3 Solving by substitution Solve the system of equations in Example 2 by substitution:

$$c = 2w \quad (1)$$
$$s = 3c \quad (2)$$
$$c + s + w = 60 \quad (3)$$

SOLUTION Each equation contains c.

$$c = 2w \qquad \text{Solve (1) for } w \text{ in terms of } c.$$

$$\frac{c}{2} = w$$

$$c + s + w = 60 \qquad \text{Substitute } \frac{c}{2} = w \text{ and } s = 3c \text{ into (3).}$$

$$c + 3c + \frac{c}{2} = 60$$

$$4.5c = 60$$

$$c = 13\frac{1}{3} \text{ hr}$$

We now find w and s, given $c = 13\frac{1}{3}$ hr.

$$w = \frac{13\frac{1}{3}}{2} = 6\frac{2}{3} \text{ hr} \qquad (1)$$

$$s = 3\left(13\frac{1}{3}\right) = 40 \text{ hr} \qquad (2)$$

Substitution may be used to change a system of three equations to a system of two equations, as shown in Example 4.

EXAMPLE 4 Solving by substitution Solve the following system of equations by substitution:

$$a + b + c = 3 \qquad (1)$$
$$2a + 3b + c = 13 \qquad (2)$$
$$2a - b = 0 \qquad (3)$$

SOLUTION Because the third equation is missing a variable, we can solve that equation for one variable and substitute into the other two equations, creating two equations in two unknowns.

$$2a - b = 0 \qquad \text{Solve (3) for } b.$$
$$2a = b$$

We replace b with $2a$ in the first and second equations.

$$a + 2a + c = 3 \quad \text{gives} \quad 3a + c = 3 \qquad (4)$$
$$2a + 3(2a) + c = 13 \quad \text{gives} \quad 8a + c = 13 \qquad (5)$$

We solve the fourth equation for c, getting $c = 3 - 3a$, and then substitute for c in the fifth equation.

$$8a + (3 - 3a) = 13 \qquad \text{Simplify.}$$
$$5a + 3 = 13 \qquad \text{Solve for } a.$$
$$a = 2$$

We substitute $a = 2$ into the third equation, $2a - b = 0$, to find $b = 4$. We substitute $a = 2$ and $b = 4$ into the second equation, $a + b + c = 3$, to find $c = -3$. The solution to the system is $a = 2$, $b = 4$, and $c = -3$.

Check: $2 + 4 + (-3) \stackrel{?}{=} 3$ ✓

$\qquad\quad 2(2) + 3(4) + (-3) \stackrel{?}{=} 13$ ✓

$\qquad\quad 2(2) - (4) \stackrel{?}{=} 0$ ✓

Examples 3 and 4 suggest the following strategies for solving three equations by substitution:

■ SOLVING THREE EQUATIONS
BY SUBSTITUTION

> • Write two equations in terms of the same variable and substitute them into the third equation. Solve the remaining one equation in one unknown.
> • Solve one equation in terms of a single variable and substitute into each of the other two equations. Solve the remaining two equations in two unknowns.

Elimination

■ SOLVING SYSTEMS OF THREE LINEAR
EQUATIONS BY ELIMINATION

> **1.** Change the equations to standard form.
> **2.** From the three equations, form two pairs from which to eliminate the same variable term.
> **3.** For the equations in the first pair, find multipliers that change the coefficients on the term to be eliminated to opposite numbers. Then add the equations to get an equation in two variables.
> **4.** Do the same for the equations in the second pair.
> **5.** Apply the elimination process to the two equations from steps 3 and 4, solving as in Section 3.1.

EXAMPLE 5 **Solving by elimination** Solve the system of equations by elimination.

$$2a + 3b - c = 1 \qquad (1)$$
$$3a - b + 2c = 14 \qquad (2)$$
$$a - 2b + 3c = 11 \qquad (3)$$

SOLUTION The equations are in standard form. (If they were not, we would restate them here in standard form.) Because each variable appears with a coefficient of 1, any variable could be eliminated. We start by pairing the equations to eliminate b.

In pair 1, we eliminate b terms by adding $-3b$ to $3b$.

$$\begin{array}{ll} 2a + 3b - \;\; c = \;\; 1 & (1) \\ \underline{9a - 3b + 6c = 42} & \text{Add 3 times (2).} \\ 11a \qquad\;\; + 5c = 43 & (4) \end{array}$$

In pair 2, we eliminate b terms by adding $-2b$ to $2b$.

$$\begin{array}{ll} -6a + 2b - 4c = -28 & \text{-2 times (2)} \\ \underline{\;\;\; a - 2b + 3c = \;\;\; 11} & \text{Add (3).} \\ -5a \qquad\quad - c = -17 & (5) \end{array}$$

In pair 3, we eliminate c terms in Equations 4 and 5 by adding $-5c$ to $5c$.

$$11a + 5c = 43 \qquad (4)$$
$$\underline{-25a - 5c = -85} \qquad \text{Add 5 times (5).}$$
$$-14a = -42 \qquad \text{Solve for } a.$$
$$a = 3$$

$$-5(3) - c = -17 \qquad \text{Substitute } a = 3 \text{ in (5) and solve for } c.$$
$$c = 2$$

$$2(3) + 3b - (2) = 1 \qquad \text{Substitute } a = 3 \text{ and } c = 2 \text{ in (1) and solve for } b.$$
$$b = -1$$

The solution to the system is $a = 3$, $b = -1$, $c = 2$.

Check: $\qquad\qquad 2(3) + 3(-1) - (2) \overset{?}{=} 1 \qquad$ Equation 1 ✓

$\qquad\qquad\qquad\quad 3(3) - (-1) + 2(2) \overset{?}{=} 14 \qquad$ Equation 2 ✓

$\qquad\qquad\qquad\quad (3) - 2(-1) + 3(2) \overset{?}{=} 11 \qquad$ Equation 3 ✓ ▬

EXAMPLE 6 Solving by elimination   Solve the system of equations by elimination.

$$3x + y + 2z = -7 \qquad (1)$$
$$4x + 3y - z = 3 \qquad (2)$$
$$5x + 5y - 3z = 11 \qquad (3)$$

SOLUTION Observe how we state the purpose of the pairing to keep us focused.
In pair 1, we eliminate z terms by adding $-2z$ to $2z$.

$$3x + y + 2z = -7 \qquad (1)$$
$$\underline{8x + 6y - 2z = 6} \qquad \text{Add 2 times (2).}$$
$$11x + 7y = -1 \qquad (4)$$

In pair 2, we eliminate z terms by adding $-3z$ to $3z$.

$$-12x - 9y + 3z = -9 \qquad -3 \text{ times (2)}$$
$$\underline{5x + 5y - 3z = 11} \qquad \text{Add (3).}$$
$$-7x - 4y = 2 \qquad (5)$$

In pair 3, we eliminate x terms in Equations 4 and 5 by adding $-77x$ to $77x$.

$$77x + 49y = -7 \qquad 7 \text{ times (4)}$$
$$\underline{-77x - 44y = 22} \qquad \text{Add 11 times (5).}$$
$$5y = 15 \qquad \text{Solve for } y.$$
$$y = 3$$

$$-7x - 4(3) = 2 \qquad \text{Substitute } y = 3 \text{ into (5) and solve for } x.$$
$$x = -2$$

$$3(-2) + (3) + 2z = -7 \qquad \text{Substitute } x = -2 \text{ and } y = 3 \text{ into (1) and solve for } z.$$
$$z = -2$$

The solution to the system is $x = -2$, $y = 3$, $z = -2$.

Check:
$$3(-2) + (3) + 2(-2) \stackrel{?}{=} -7 \quad (1) \quad \checkmark$$
$$4(-2) + 3(3) - (-2) \stackrel{?}{=} 3 \quad (2) \quad \checkmark$$
$$5(-2) + 5(3) - 3(-2) \stackrel{?}{=} 11 \quad (3) \quad \checkmark$$

Graphs and Types of Systems

The graphs of linear equations in three variables are **planes**, or *infinite flat surfaces*. The *points on the planes are described by three numbers*, or **ordered triples**. When we graph three planes on a three-dimensional graph, there are many possible arrangements (see Figure 8). As a result, we need to use more general vocabulary than the words *intersecting*, *parallel*, and *coincident*, which summarized the graphs of two-variable linear equations.

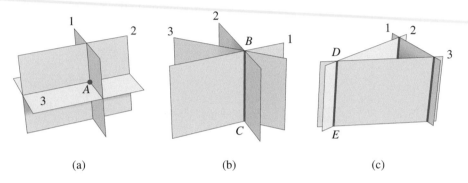

(a)   (b)   (c)

FIGURE 8

We describe linear equations in terms of the number of solutions that the system contains.

1. If *a system of linear equations has exactly one solution*, the equations are **consistent**.

If the system of linear equations doesn't have exactly one solution, there are two possibilities:

2. If *a system of linear equations has no solution*, the equations are **inconsistent**.

3. If *a system of equations has an infinite number of solutions*, the equations are **dependent**.

EXAMPLE 7   Identifying systems of linear equations   Identify as consistent, inconsistent, or dependent the two-variable linear equations whose graphs are
a. parallel lines
b. coincident lines
c. intersecting lines

SOLUTION   **a.** Parallel lines have no intersection, and hence their equations have no solution. The equations describing parallel lines are inconsistent.

b. Coincident lines have all points in common, an infinite number of solutions, so their equations are dependent.

c. Intersecting lines have exactly one point of intersection, so their equations are consistent.

EXAMPLE 8   Identifying types of equations   Based on the intersections (or lack of intersections) of the three planes shown, identify the graphs in Figure 8 as representing consistent, inconsistent, or dependent linear equations.

SOLUTION   In Figure 8(a), there is one point of intersection, point A. The equations describing these graphs are consistent.

In Figure 8(b), there is a line of intersection, BC, for all three planes. The line contains an infinite number of points and, hence, an infinite number of solutions. The equations describing these graphs are dependent.

In Figure 8(c), although there is an intersection line, DE, the line is not part of the plane labeled with a 1. There is no common intersection of all three planes and, hence, no solution to the system. The equations describing the graphs are inconsistent.   ▬

Examples 9 and 10 explore an inconsistent system and a dependent system.

EXAMPLE 9   Exploring an inconsistent system   Solve the system of equations

$$x + y - 2z = 2 \qquad (1)$$
$$x - z = 4 \qquad (2)$$
$$y - z = 1 \qquad (3)$$

SOLUTION   The variable z appears in all three equations. We solve the second and third equations in terms of z and substitute for x and y in the first equation.

$$x = 4 + z \qquad\qquad\qquad (2)$$
$$y = 1 + z \qquad\qquad\qquad (3)$$
$$x + y - 2z = 2 \qquad (1)$$
$$(4 + z) + (1 + z) - 2z = 2$$
$$4 + z + 1 + z - 2z = 2 \qquad \text{The variable } z \text{ drops out.}$$
$$5 = 2 \qquad \text{Contradiction}$$

We obtain a contradiction. There is no real-number solution to the system, so the equations are inconsistent.   ▬

When the variable drops out, leaving a contradiction, the system has no solutions and the equations are inconsistent. The graphs of the equations may be parallel planes, or they may be planes that intersect in such a way that they have no common point in all three planes, as in Figure 8(c).

EXAMPLE 10   Exploring a dependent system   Solve the system of equations

$$x - y - 2z = 3 \qquad (1)$$
$$x - z = 5 \qquad (2)$$
$$y + z = 2 \qquad (3)$$

SOLUTION   The variable z appears in all three equations. We solve the second and third equations in terms of z and substitute for x and y in the first equation.

$$x = 5 + z \qquad\qquad\qquad (2)$$
$$y = 2 - z \qquad\qquad\qquad (3)$$
$$x - y - 2z = 3 \qquad (1)$$
$$(5 + z) - (2 - z) - 2z = 3$$
$$5 + z - 2 + z - 2z = 3 \qquad \text{The variable } z \text{ drops out.}$$
$$3 = 3 \qquad \text{Identity}$$

We obtain an identity. There are an infinite number of solutions to the system; hence, the equations are dependent.

When the variable drops out, leaving an identity, the system has an infinite number of solutions and the equations are dependent. In the three-variable case, the three equations may describe three planes intersecting in a line, as in Figure 8(b), or two or more equations may describe the same plane.

Table 9 summarizes the geometry and algebra associated with solving a system of three linear equations.

TABLE 9

	Consistent Equations	Inconsistent Equations	Dependent Equations
Geometry	Planes intersect in one common point.	There is no common intersection.	There is linear or planar intersection.
Algebra	Equations can be solved to yield unique x, y, and z values.	The variables drop out, and the resulting statement is false.	The variables drop out, and the resulting statement is true.
Numerical Result	The unique solution is given by the ordered triple (x, y, z).	A contradiction results.	An identity, $a = a$, results.
Solution	There is one solution, an ordered triple.	There is no solution.	An infinite number of ordered triples satisfy the system.

Applications

The applications in Examples 11 and 12 review writing equations when sentences include the phrase "times as many."

COIN PUZZLE   The coin puzzle in Example 11 illustrates that some information may not be given in a problem.

EXAMPLE 11

Setting up equations: counting coins   Alison has 61 coins altogether for a total of $4.69. She has one-third as many quarters as dimes. She also has some pennies. Set up a system of equations that could be used to find the number of each kind of coin.

SOLUTION   We let p = number of pennies, d = number of dimes, and q = number of quarters. There are three variables, so we need to set up three equations. The system of equations is

Student Note: In many problems, remember that there may be both quantity and value.

$$p + d + q = 61$$ This equation gives the quantity of coins.

$$0.01p + 0.10d + 0.25q = 4.69$$ This equation gives the monetary value.

$$3q = d$$ If she has one-third as many quarters as dimes, then the number of quarters can be multiplied by 3 to get the number of dimes.

The solution to Example 11 is left as an exercise.

Think about it 3: We used the equation $3q = d$ to describe dimes in terms of quarters. What equation describes quarters in terms of dimes?

CANDY PRODUCTION   In Example 12, we return to the chocolate rabbits of the chapter opening. Fenton is Janelle's son.

EXAMPLE 12   Setting up equations: chocolate rabbits   Fenton makes three sizes of chocolate rabbits (see Figure 1 on page 149). From prior sales records, he estimates that the store can sell five times as many small rabbits as medium rabbits and five times as many medium rabbits as large rabbits. If he wants to make 620 rabbits altogether, how many of each should he make? Identify variables and set up a system of equations.

SOLUTION   We will let s = number of small rabbits, m = number of medium rabbits, and l = number of large rabbits. The system of equations is $s + m + l = 620$, $s = 5m$, and $m = 5l$.

The solution is left as an exercise.

ANSWER BOX

Warm-up: 1. $m = 5l$ **2.** $s = 3m$ **3.** $3q = d$ or $q = \frac{1}{3}d$ **Think about it 1:** $x + (x - 28) + 2(x - 28) + 44 = 168$ **Think about it 2:** $x + 2x + 3(2x) = 60$ **Think about it 3:** $q = \frac{1}{3}d$

3.4   Exercises

Set up guess-and-check tables for Exercises 1 to 6. Make three to four guesses, and then write equations. Solve your equations using any method.

1. Celine works twice as many hours as she spends in class. She studies 3 hours for every hour in class. To maintain her health and sanity, she limits her total time commitment to work and school to 60 hours per week. How many hours does she plan to spend on each activity: attending class, studying, and working?
$w = 2c, s = 3c, c + s + w = 60; c = 10$ hr, $s = 30$ hr, $w = 20$ hr

2. Vaughn works 35 hours per week. He studies 3 hours for every hour in class. To maintain his health and sanity, he limits his total time commitment to work and school to 60 hours per week. How many hours does he spend attending class and studying?
$35 + c + s = 60, s = 3c; c = 6\frac{1}{4}$ hr, $s = 18\frac{3}{4}$ hr

3. The two equal angles in an isosceles triangle are each 21° smaller than the third angle. What is the measure of each angle?
$x = y, z - 21 = x, x + y + z = 180; 53°, 53°, 74°$

4. The largest angle in an isosceles triangle is 38° greater than each of the two equal angles. What is the measure of each angle?
$x = y, z - 38 = x, x + y + z = 180; 85\frac{1}{3}°, 47\frac{1}{3}°, 47\frac{1}{3}°$

5. An ounce of macadamia nuts (salted and roasted in oil) contains 222 calories. There are twice as many grams of carbohydrates as protein. The total weight of the protein, fat, and carbohydrates is 28 grams. There are 4 calories per gram of protein and carbohydrates and 9 calories per gram of fat. How many grams are there each of carbohydrates, fat, and protein?
$c = 2p, c + f + p = 28, 4c + 9f + 4p = 222; c = 4$ g, $f = 22$ g, $p = 2$ g

6. A loosely packed cup of raisins contains 489 calories. It contains 4 more grams of protein than of fat. There are 121 grams of carbohydrates, fat, and protein altogether. There are 4 calories per gram of protein and carbohydrates and 9 calories per gram of fat. How many grams are there each of carbohydrates, fat, and protein?
$p = f + 4, c + f + p = 121, 4c + 9f + 4p = 489; c = 115$ g, $f = 1$ g, $p = 5$ g

Solve the systems of equations in Exercises 7 to 10 by substitution.

7. $2a + 3b + 2c = 1$
$3a + 2c = 10$
$a + b = 1$
$a = 3, b = -2, c = \frac{1}{2}$

8. $3a + 4b + c = -2$
$2a - c = -10$
$4b - 4a = 15$
$a = -3, b = \frac{3}{4}, c = 4$

9. $2a + b - 4c = 17$
$5a - 2b = 27$
$b + 3c + 7 = 0$
$a = 5, b = -1, c = -2$

10. $4a + b - c = -12$
$b + c = 2$
$2a + c = 5$
$a = -\frac{1}{2}, b = -4, c = 6$

Solve the systems of equations in Exercises 11 to 14 by substituting before using elimination on the two remaining equations.

11. $x + y + z = 2$
$x + 3z = 0$
$2z = 7 + x + y$
$x = -9, y = 8, z = 3$

12. $2x + y + z = 9$
$x = 3 + y + z$
$3x = 2z$
$x = 4, y = -5, z = 6$

13. $2x + 3y - z = 11$
$3x + y + 2z = 13$
$2y = z + 7$
$x = -1, y = 6, z = 5$

14. $3x = 2y + z$
$4x + z + 17 = 3y$
$2z + 5y + 4 = 0$
$x = -1, y = 2, z = -7$

Solve the systems of equations in Exercises 15 to 22 by elimination.

15. $x + y - z = 2$
$x - y + z = 3$
$-x + y + z = 4$
$x = 2.5, y = 3, z = 3.5$

16. $a + b + c = 4$
$a - b - c = 0$
$-a + b - c = 2$
$a = 2, b = 3, c = -1$

17. $2a + 3b + c = -4$

$3a + 2b + 4c = 19$

$a - b - 3c = -19$

$a = -1, b = -3, c = 7$

18. $3x - y + 2z = 39$

$2x - 2y - 3z = 18$

$4x + 3y + z = 15$

$x = 8, y = -7, z = 4$

19. $a + b + c = 10$

$-a + b = 4$

$2a - b + c = 0$

$a = 2, b = 6, c = 2$

20. $2x - y = 3$

$x + 2z = 4$

$-2x + y + z = 1$

$x = -4, y = -11, z = 4$

21. $x + y - z = 2$

$x - z = 4$

$x + y = 3$

$x = 5, y = -2, z = 1$

22. $x - z = 3$

$x + 3y + 2z = 6$

$-2x + y = 3$

$x = \frac{1}{3}, y = \frac{11}{3}, z = -\frac{8}{3}$

In Exercises 23 to 30, solve the systems of three equations. Distinguish between inconsistent and dependent equations.

23. $a + 2b + 3c = 2$

$4a + 5b + 6c = 3$

$a + b + c = 1$

contradiction, inconsistent

24. $a + 2b + 3c = 2$

$4a + 5b + 6c = 5$

$a + b + c = 1$

identity, dependent

25. $2a + b + c = 3$

$a + 2b - 2c = 2$

$a - b + 3c = 1$

identity, dependent

26. $3y + 5z = 0$

$3x + 7z = 0$

$x + y + 4z = 3$

contradiction, inconsistent

27. $2x - y + z = 1$

$x + z = -2$

$x - y = 3$

identity, dependent

28. $a - b + c = 4$

$a - b - c = 0$

$-a + b - c = 2$

contradiction, inconsistent

29. $2x + y - z = 3$

$5x - 4z = 0$

$5y + 3z = 0$

contradiction, inconsistent

30. $-2x + 3z = 0$

$-x + y + 3z = -1$

$x + y = -1$

identity, dependent

31. Tell whether the three planes in each figure have a single point of intersection, no common point of intersection, or a line or plane of intersection. Identify each figure as being the graph of consistent, inconsistent, or dependent equations. Assume that planes 1, 2, and 3 in part c and planes 1 and 2 in part d are parallel.

a.

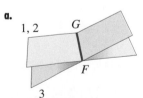

1, 2   G

F

3

plane; dependent

b.

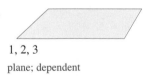

1, 2, 3

plane; dependent

c.

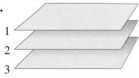

1

2

3

no common point of intersection; inconsistent

d.

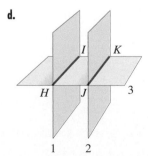

I   K

H   J   3

1   2

no common point of intersection; inconsistent

32. Identify each of the following as describing consistent, inconsistent, and/or dependent equations.

a. The three equations have an infinite number of solutions. dependent

b. The three equations have no solution. inconsistent

c. The three equations have exactly one solution. consistent

d. An algebraic solution yields a false statement. inconsistent

e. An algebraic solution yields a true statement. dependent

f. The planes do not intersect. inconsistent

g. The planes intersect in a common point. consistent

h. The planes intersect in a common line. dependent

i. The planes intersect in a common plane. dependent

33. Using any method, solve the coin puzzle system of equations: 29 pennies, 24 dimes, 8 quarters

$$p + d + q = 61$$
$$0.01p + 0.10d + 0.25q = 4.69$$
$$3q = d$$

34. Using any method, solve the candy production system of equations: 500 small, 100 medium, 20 large

$$s + m + l = 620$$
$$s - 5m = 0$$
$$m = 5l$$

Set up systems of equations in Exercises 35 to 42, and solve.

35. In 1987, Ayako Okamoto and Curtis Strange were the leading money winners in their respective professional golf associations. Together, they won $412,545 more than Corey Pavin won in 1991. Strange won $459,907 more than Okamoto. Okamoto and Pavin together won $1,445,464. How much did each win? $O + S = P + 412,545$; $S = O + 459,907$; $O + P = 1,445,464$; $466,034$; $925,941$; $979,430$

36. In 1978, Nancy Lopez and Tom Watson were the leading money winners in their respective professional golf associations. Together, they won $43,750 less than Betsy King won in 1993. King won $406,179 more than Lopez. Watson won $233,563 less than King. How much did each win? $L + W = K - 43,750$; $K = L + 406,179$; $W + 233,563 = K$; $189,813$; $362,429$; $595,992$

37. A banana contains 121 calories. It contains equal amounts of protein and fat. There are 29 grams of carbohydrates, fat, and protein altogether. There are 4 calories per gram of protein and carbohydrates and 9 calories per gram of fat. How many grams are there each of carbohydrates, fat, and protein? $f = p$, $c + f + p = 29$, $4c + 9f + 4p = 121$; $c = 27$ g, $f = 1$ g, $p = 1$ g

38. An ounce of sunflower seeds contains 170 calories. There is 1 gram more of protein than of carbohydrates. The total weight of the protein, fat, and carbohydrates is

25 grams. There are 4 calories per gram of protein and carbohydrates and 9 calories per gram of fat. How many grams are there each of carbohydrates, fat, and protein?
$p = c + 1, c + f + p = 25, 4c + 9f + 4p = 170; c = 5$ g, $f = 14$ g, $p = 6$ g

39. An 8-ounce container of lowfat yogurt contains 148 calories. The cup contains a total of 32 grams of protein, fat, and carbohydrates. There are 12 more grams of carbohydrates than of fat. Protein and carbohydrates have 4 calories per gram; fat has 9 calories per gram. How many grams each of protein, fat, and carbohydrates are there?
$p + f + c = 32, c = f + 12, 4p + 9f + 4c = 148; p = 12$ g, $f = 4$ g, $c = 16$ g

40. An ounce of dry-roasted cashew nuts (salted) contains 169 calories. There are 5 more grams of carbohydrates than of protein. The total weight of protein, fat, and carbohydrates is 26 grams. Protein and carbohydrates have 4 calories per gram; fat has 9 calories per gram. How many grams each of protein, fat, and carbohydrates are there?
$p + f + c = 26, c = p + 5, 4p + 9f + 4c = 169; p = 4$ g, $f = 13$ g, $c = 9$ g

41. Shelby has \$5.80 in nickels, dimes, and quarters in her piggy bank. There are 44 coins altogether, with 3 more quarters than dimes. How many of each type of coin does she have?
$n + d + q = 44, q = d + 3, 0.05n + 0.10d + 0.25q = 5.80; 17, 12, 15$

42. Zachary has nickels, dimes, and quarters in his piggy bank. He has 58 coins with a total value of \$6.50. There are twice as many dimes as nickels. How many of each coin docs he have?
$n + d + q = 58, d = 2n, 0.05n + 0.10d + 0.25q = 6.50; 16, 32, 10$

43. Explain why a system of equations cannot be both inconsistent and dependent.
Cannot have both no solution and an infinite number of solutions

■ Project

44. Systems of Four or Five Equations Write and solve a system of equations for each of these two settings.

a. Vienna blend coffee is a mixture of Ethiopian (\$10.95 per pound), French roast (\$8.95), Guatemalan (\$8.50), and light Colombian (\$8.25). The blend calls for twice as much light Colombian as Guatemalan and four times as much French roast as Ethiopian. The cost of the blended coffee is \$8.59 per pound. The person blending the coffee wants to use 50-pound bags, so round answers to the nearest 50 pounds. How many pounds of each are used to make 1000 pounds of Vienna blend?
50 lb; 200 lb; 250 lb; 500 lb

b. "Four ingredients," a stir-fry favorite, contains shrimp, barbequed pork, beef, and chicken sauteed with vegetables. The weight of the uncooked ingredients is 20 ounces. The weight of the pork and beef together is the same as the weight of the vegetables. The chicken weighs twice as much as the beef. The shrimp, beef, and vegetables together weigh 12 ounces. The shrimp and chicken together weigh 10 ounces. How many ounces are there of each ingredient? 4 oz; 2 oz; 3 oz; 6 oz; 5 oz

---

3.5   Solving Linear and Absolute Value Inequalities

Objectives

■ Solve inequalities in one variable with graphs and with algebraic notation.

■ Find a number-line solution to one-variable inequalities.

■ Solve absolute value inequalities with graphs and with algebraic notation.

■ Find a number line solution to absolute value inequalities.

WARM-UP

Perform the indicated operation on each true statement. Consider what inequality sign must go between the numbers in your answers.

1. $-5 < 8$ Multiply both sides by 2. $-10 < 16$

2. $-4 < 3$ Add -5 to both sides. $-9 < -2$

3. $-2 < 4$ Subtract 10 from both sides. $-12 < -6$

4. $-3 < 2$ Multiply both sides by -4. $12 > -8$

5. $-6 < -2$ Divide both sides by -2. $3 > 1$

IN THIS SECTION, we solve inequalities in one variable and inequalities containing absolute values with graphs and algebraic notation.

Solving Inequalities in One Variable by Graphing

Has a grocery checker ever asked you if you wanted cash back when you used your credit card to pay for groceries? This is a common "courtesy" in many stores. Are they really doing you a favor? Consider how much the cash you get back is costing you! The following fee schedules appear on credit card disclosure statements:

First U.S.A. Bank Liquid Card. ATM cash advance: 3% of the amount of the advance, but not less than $10. All other cash advances: 3% of the advance, but not less than $15.

First U.S.A. Bank Platinum Card. All cash advances: 3% of the amount of the advance, but not less than $5.00.

Sears Gold MasterCard. All cash access: 3% of the transaction amount but not less than $5.

PADI Platinum Plus Card. Bank and ATM cash advance: 3% (minimum $5). Credit card cash advance checks and balance transfers: 3% of each cash advance (minimum $5, maximum $40).

BMW Platinum Visa. Cash advance: the greater of 2.5% or $5.00.

Chase Manhattan Bank USA. Cash advance: 3% of each cash advance ($5 minimum).

(By the way, if you don't pay off the cash advance with your next billing statement, the interest rates are usually 19.99% APR but can be as high as 23.99% APR.)

EXAMPLE 1 Solving inequalities with a graph: credit card cash advance checks
a. Write three equations describing the fees charged for credit card cash advance checks on the PADI (Professional Association of Diving Instructors) Platinum Plus card.
b. Write inequalities to describe the advances for which the minimum and maximum fees are paid.
c. With a calculator, graph the equations in part a.
d. On the calculator graph, solve for the amounts at which the minimum charge ($5) ends and the maximum charge ($40) begins.
e. Summarize the fee schedule.

SOLUTION
a. Let x = amount of cash advance in dollars and y = fee in dollars.

$$y = 0.03x \qquad \text{3\% changed to a decimal}$$

$$y = 5 \qquad \text{The constant function for the \$5 minimum fee}$$

$$y = 40 \qquad \text{The constant function for the \$40 maximum fee}$$

b. $5 < 0.03x \qquad$ Minimum fee

$\quad 0.03x > 40 \qquad$ Maximum fee

c. The three equations are graphed in Figure 9. The calculator screen appears in Figure 10.

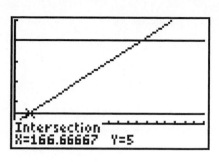

FIGURE 10 -10 to 2000 on X,
-10 to 50 on Y

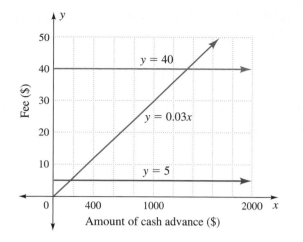

FIGURE 9

d. $y = 5$ and $y = 0.03x$ intersect at $x = \$166.67$. (See Figure 10.) $y = 40$ and $y = 0.03x$ intersect at $x = \$1333.33$.

e. For a cash advance of $0 to $166.67, $5 is more than $0.03x$, so you pay the minimum, $5. For a cash advance between $166.67 to $1333.33, $5 < 0.03x < \$40$, so you pay $0.03x$. For a cash advance above $1333.33, $40 is less than $0.03x$, so you pay the maximum, $40.   ▬

The solution to an inequality is another inequality. Thus, we frequently draw the solution set on a number line and then write it as an inequality.

▬ SOLVING AN INEQUALITY

1. Graph each side of the inequality separately.

2. To find the starting point for the solution set, locate the x-coordinate of the point of intersection of the graphs. Place a small circle (for $<$ or $>$) or a dot (for $\leq$ or $\geq$) on the number line.

3. To find the direction in which to draw the solution set, look at the inequality sign.

a. For an inequality containing $<$ or $\leq$, read the inequality as "is below." Thus, solutions for $a < b$ lie in the region where the graph of $y = a$ is *below* the graph of $y = b$.

b. For an inequality containing $>$ or $\geq$, read the inequality as "is above." Thus, solutions for $a > b$ lie in the region where the graph of $y = a$ is *above* the graph of $y = b$.

On the number line, draw an arrow satisfying the condition in a or b above.

4. Write the solution set as an inequality.

5. Check with a test point. Select an input that makes the solution set true. Substitute that input into the original inequality.

EXAMPLE 2   Solving inequalities with a graph   Solve these inequalities.
a. $3 - x > -2$ **b.** $2x - 3 < 3 - x$

SOLUTION   **a.** The graph of $y = 3 - x$ intersects $y = -2$ at $(5, -2)$. The graph of $y = 3 - x$ is above $y = -2$ to the left of $x = 5$. Thus, $x < 5$ solves $3 - x > -2$. We check with a test point: $x = 4$ satisfies $x < 5$, and $3 - 4 > -2$ is true. Thus, the solution set is as shown in Figure 11.

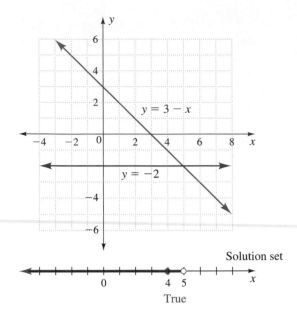

FIGURE 11

b. The graph of $y = 2x - 3$ intersects $y = 3 - x$ at $(2, 1)$. The graph of $y = 2x - 3$ is below the graph of $y = 3 - x$ to the left of $x = 2$. Thus, $x < 2$ solves $2x - 3 < 3 - x$. We check with a test point: $x = 1$ satisfies $x < 2$, and $2(1) - 3 < 3 - 1$ is true. Thus, the solution set is as shown in Figure 12.

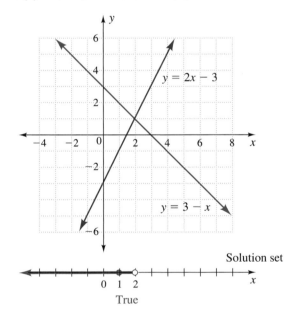

FIGURE 12

APPLICATION: WEDDING RECEPTION   The caterer for a wedding reception has three menu options: \$18 per person, \$21 per person, and \$25.00 per person. The total amount paid to the caterer will be the cost of a wedding cake (\$450) plus the cost of the menu choice. The total budget for food is \$3000. The number of people who could attend under each option is found by solving three inequalities:

$$18x + 450 \leq 3000$$

$$21x + 450 \leq 3000$$

$$25x + 450 \leq 3000$$

EXAMPLE 3 Solving inequalities with a graph   Use Figure 13 to solve the inequalities graphically.
a. $18x + 450 \leq 3000$ **b.** $21x + 450 \leq 3000$ **c.** $25x + 450 \leq 3000$

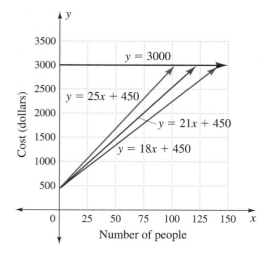

FIGURE 13

SOLUTION   **a.** We trace along the line $y = 18x + 450$ to the intersection with $y = 3000$. The x-coordinate of the point of intersection is about 140. The graph of $y = 18x + 450$ is below $y = 3000$ to the left of the point of intersection, so the inequality $18x + 450 \leq 3000$ is true for $x \leq 140$.

b. We trace along the line $y = 21x + 450$ to the intersection with $y = 3000$. The x-coordinate of the point of intersection is about 120. The graph of $y - 21x + 450$ is below $y = 3000$ to the left of the point of intersection, so the inequality $21x + 450 \leq 3000$ is true for $x \leq 120$.

c. Similarly, with the point of intersection about 100, the inequality $25x + 450 \leq 3000$ is true for $x \leq 100$.

GRAPHING CALCULATOR SOLUTION   Repeating the process on a graphing calculator, we can find more accurate solutions: $x \leq 141$, $x \leq 121$, and $x \leq 102$, respectively.   ▬

Solving Linear Inequalities in One Variable with Algebraic Notation

For the most part, solving inequalities involves the same properties as solving equations. Steps in solving inequalities make equivalent inequalities. **Equivalent inequalities** *have the same solution set.*

▬ **ADDITION PROPERTY OF INEQUALITIES**

> When a number is added to (or subtracted from) each side of an inequality, the result is an equivalent inequality. The direction of the inequality is not changed.
>   If $a < b$ and c is any real number,
>
> $a + c < b + c$

▬ **MULTIPLICATION PROPERTY OF INEQUALITIES 1**

> When an inequality is multiplied or divided by a positive number, the result is an equivalent inequality. The direction of the inequality sign is not changed.
>   If $a < b$ and $c > 0$,
>
> $ac < bc$

These properties hold for all other inequalities: $\leq$, $>$, and $\geq$.
    We now return to our caterer situation.

EXAMPLE 4 Solving an inequality with algebraic notation Solve $21x + 450 \leq 3000$.

SOLUTION

$$21x + 450 \leq 3000 \qquad \text{Subtract 450 from both sides.}$$

$$21x \leq 2550 \qquad \text{Divide both sides by 21.}$$

$$x \leq 121.4$$

The number of people who could attend at $21 per person is $x \leq 121$. ▬

The exceptions to solving inequalities are what cause difficulty. We explored the exceptions in the Warm-up.

▮ MULTIPLICATION PROPERTY
OF INEQUALITIES 2

When each side of an inequality is multiplied or divided by a negative number, reverse the direction of the inequality sign.
   If $a < b$ and $c < 0$, then

$$ac > bc$$

A change in the direction of an inequality sign is needed only for multiplication and division by a negative. For all other additions, subtractions, multiplications, and divisions, the inequality sign is unchanged.

APPLICATION: PHOTOCOPY CARD   A prepaid photocopy card costs $10. Each copy costs $0.10. The cost per copy is subtracted from the value of the card as copies are made. The value on the photocopy card after x copies have been made is $y = 10 - 0.10x$.

EXAMPLE 5 Solving an inequality: photocopy card
a. Write an inequality that can be solved to find the number of copies that have been made if the value on the card is under $7.
b. Solve the inequality with a graph.
c. Solve the inequality with algebraic notation.

SOLUTION **a.** $10 - 0.10x < 7$
b. Figure 14 shows the graph of $y = 10 - 0.10x$ intersecting with $y = 7$ at $(30, 7)$. The value on the card, the graph of $y = 10 - 0.10x$, is below $y = 7$ to the right of the intersection—that is, for $x > 30$. Thus, for more than 30 copies, the card's value will be under $7.

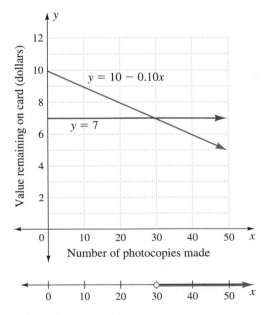

FIGURE 14

c. The value on the card is to be less than \$7, so we write $10 - 0.10x < 7$.

$$10 - 0.10x < 7 \qquad \text{Subtract 10 from both sides.}$$

$$10 - 0.10x - 10 < 7 - 10 \qquad \text{Simplify.}$$

$$-0.10x < -3 \qquad \text{Divide both sides by } -0.10, \text{ and reverse the inequality sign because of the division by a negative.}$$

$$\frac{-0.10x}{-0.10} > \frac{-3}{-0.10} \qquad \text{Simplify.}$$

$$x > 30$$

❚ **A STRATEGY FOR INEQUALITIES**

> Keeping a positive coefficient on the variable term avoids changing the inequality sign.

In Example 6, we add the term with a negative numerical coefficient to both sides to avoid division by a negative number.

EXAMPLE 6   **Keeping a positive coefficient on the variable term**   Solve $10 - 0.10x < 7$ by first adding $0.10x$ to both sides.

SOLUTION

$$10 - 0.10x < 7 \qquad \text{Add } 0.10x \text{ to both sides.}$$

$$10 < 7 + 0.10x \qquad \text{Subtract 7 from both sides.}$$

$$3 < 0.10x \qquad \text{Divide both sides by } 0.10.$$

$$\frac{3}{0.10} < \frac{0.10x}{0.10} \qquad \text{Simplify.}$$

$$30 < x$$

Student Note: A common application of algebra is in rearranging equations to match an answer. Always first consider your answer to be correct, and then determine what might have been done by someone else to obtain the answer given.

The result, $x > 30$, agrees with that in Example 5.

Think about it 1: With algebraic notation, solve the equivalent inequality for the photocopy machine, $7 > 10 - 0.10x$.

COMPOUND INEQUALITIES   We return to compound inequalities in preparation for solving absolute value inequalities. Many answers to absolute value inequalities may be written as compound inequalities. When solving compound inequalities, apply each equation-solving operation to all three parts of the inequality statement.

EXAMPLE 7   **Solving compound inequalities**   Solve each compound inequality.

a. $11 > 5 - 3x > -7$

b. $14 > 6 + 2(1 - x) > 2$

SOLUTION   **a.** $11 > 5 - 3x > -7$ Subtract 5 from each part.

$$6 > -3x > -12 \qquad \text{Divide by } -3 \text{ in each part and reverse the inequality signs.}$$

$$-2 < x < 4$$

Check: $x = 0$ satisfies the condition, and $11 > 5 - 3(0) > -7$ is true. ✓

b. $14 > 6 + 2(1 - x) > 2$ Subtract 6 from each part.

$$8 > 2(1 - x) > -4 \qquad \text{Divide by 2 in each part.}$$

$$4 > 1 - x > -2 \qquad \text{Subtract 1 from each part.}$$

$$3 > -x > -3 \qquad \text{Divide by } -1 \text{ in each part and reverse the inequality signs.}$$

$$-3 < x < 3$$

Check: $x = 0$ satisfies the condition, and $14 > 6 + 2(1 - 0) > 2$ is true. ✓ ❚

Think about it 2: Return to Example 1. Write the limitations on the use of $y = 0.03x$ as a compound inequality and solve.

Solving Inequalities Containing Absolute Value

To solve inequalities containing absolute value symbols, we add one step to the work done thus far: We change the absolute value expression into two inequalities. The two inequalities come from the definition of absolute value:

$$|a| = a \quad \text{if } a > 0$$

$$|a| = -a \quad \text{if } a < 0$$

EXAMPLE 8 Solving absolute value inequalities   Solve $|3 - 4x| < 13$.

SOLUTION   The two inequalities are

$$3 - 4x < 13 \qquad \text{If } 3 - 4x \text{ is positive}$$

$$-(3 - 4x) < 13 \qquad \text{If } 3 - 4x \text{ is negative}$$

Inequality 1:

$$3 - 4x < 13 \qquad \text{Subtract 3 from both sides.}$$

$$-4x < 10 \qquad \text{Divide by } -4 \text{ on both sides and reverse the inequality sign.}$$

$$x > \frac{10}{-4} \qquad \text{Simplify.}$$

$$x > -2.5$$

Inequality 2:

$$-(3 - 4x) < 13 \qquad \text{Multiply out the parentheses.}$$

$$-3 + 4x < 13 \qquad \text{Add 3 on both sides.}$$

$$4x < 16 \qquad \text{Divide by 4 on both sides.}$$

$$x < 4$$

Summarizing the result as a compound inequality, we say that the solution to $|3 - 4x| < 13$ is

$$-2.5 < x < 4$$

EXAMPLE 9 Solving absolute value inequalities with a graph
  a. Solve $|3 - 4x| < 13$ from a graph.
  b. Draw a number line showing the solution.
  c. Solve $|3 - 4x| < 13$ with a calculator graph.
  d. View the calculator table for $y = |3 - 4x|$ and compare the solution found in part a with the table.

SOLUTION   **a.** Figure 15 shows a horizontal line, $y = 13$, intersecting the graph at $x = -2.5$ and $x = 4$. The graph is below $y = 13$ between these intersections. The solution is

$$-2.5 < x < 4$$

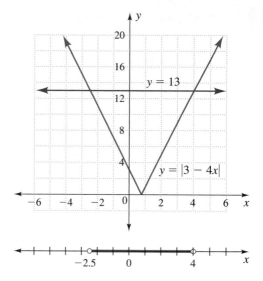

FIGURE 15

b. The compound-inequality solution is shown on the number line below the graph in Figure 15.

c. The first of the two intersections is shown in Figure 16.

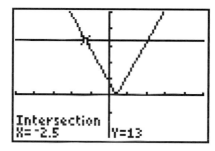

FIGURE 16 -10 to 10 on X, -10 to 250 on Y

d. With $Y_1 = $ abs($3 - 4X$), the calculator table in Figure 17 shows that outputs of less than 13 under Y_1 correspond to X values between -2.5 and 4.

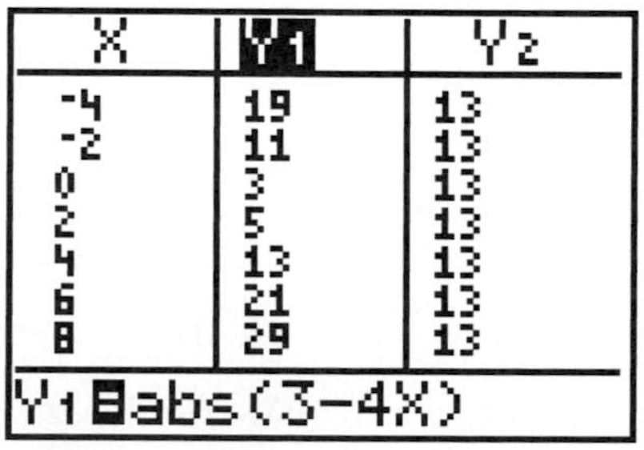

X	**Y1**	Y2
-4	19	13
-2	11	13
0	3	13
2	5	13
4	13	13
6	21	13
8	29	13

Y1⊟abs(3-4X)

FIGURE 17

ANSWER BOX

Warm-up: 1. $-10 < 16$ **2.** $-9 < -2$ **3.** $-12 < -6$ **4.** $12 > -8$ (must reverse inequality sign to obtain a true statement)   **5.** $3 > 1$ (must reverse inequality sign to obtain a true statement)   **Think about it 1:** $7 > 10 - 0.10x, 0.10x + 7 > 10, 0.10x > 3, 0.10x/0.10 > 3/0.10, x > 30$ **Think about it 2:** $5 < 0.03x < 40, 5/0.03 < x < 40/0.03, 166.67 < x < 1333.33$

3.5   Exercises

In Exercises 1 to 14, use the graphs to solve the inequalities and then solve each inequality with algebraic notation.

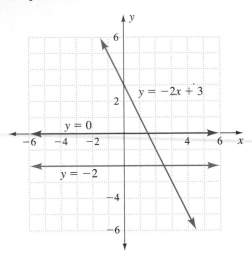

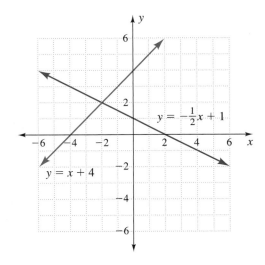

1. $-2 < -2x + 3$ $y = -2$ is below $y = -2x + 3$ for $x < 2.5$

2. $-2 > -2x + 3$ $y = -2$ is above $y = -2x + 3$ for $x > 2.5$

3. $-2x + 3 > 0$ $y = -2x + 3$ is above $y = 0$ for $x < 1.5$

4. $-2x + 3 < 0$ $y = -2x + 3$ is below $y = 0$ for $x > 1.5$

5. $x + 4 > -\frac{1}{2}x + 1$ $y = x + 4$ is above $y = -\frac{1}{2}x + 1$ for $x > -2$

6. $-\frac{1}{2}x + 1 > x + 4$ $y = -\frac{1}{2}x + 1$ is above $y = x + 4$ for $x < -2$

7. $0 < x + 4$ $y = 0$ is below $y = x + 4$ for $x > -4$

8. $x + 4 > 0$ $y = x + 4$ is above $y = 0$ for $x > -4$

9. $-\frac{1}{2}x + 1 \leq 0$ $y = -\frac{1}{2}x + 1$ is below or same as $y = 0$ for $x \geq 2$

10. $-\frac{1}{2}x + 1 > 0$ $y = -\frac{1}{2}x + 1$ is above $y = 0$ for $x < 2$

11. $-2x + 3 \geq 3$ $y = -2x + 3$ is above or same as $y = 3$ for $x \leq 0$

12. $-2x + 3 < -1$ $y = -2x + 3$ is below $y = -1$ for $x > 2$

13. $x + 4 \leq 4$ $y = x + 4$ is below or same as $y = 4$ for $x \leq 0$

14. $x + 4 \geq -2$ $y = x + 4$ is above or same as $y = -2$ for $x \geq -6$

In Exercises 15 to 24, graph the left and right sides of the inequalities and solve each inequality from the graph. Record the solution on a number line.

15. $1 - 2x < 3$ $x > -1$; see Answer Section.

16. $4 < x - 2$ $x > 6$; see Additional Answers.

17. $-2 < 1 - 2x$ $x < \frac{3}{2}$; see Answer Section.

18. $x - 2 > -2$ $x > 0$; see Additional Answers.

19. $x < 1 - 2x$ $x < \frac{1}{3}$; see Answer Section.

20. $x - 2 < -x$ $x < 1$; see Additional Answers.

21. $x \geq 1 - 2x$ $x \geq \frac{1}{3}$; see Answer Section.

22. $x - 2 \geq -x$ $x \geq 1$; see Additional Answers.

23. $x + 2 \leq 1 - 2x$ $x \leq -\frac{1}{3}$; see Answer Section.

24. $2 - x \leq x - 2$ $x \geq 2$; see Additional Answers.

25. Graph each side of $x - 2 < x$. What do you observe about the solutions to this inequality?
See Answer Section; $y = x - 2$ is below $y = x$ for all real numbers.

26. Graph each side of $1 - 2x \leq 4 - 2x$. What do you observe about the solutions to this inequality?
See Additional Answers; $y = 1 - 2x$ is below $y = 4 - 2x$ for all real numbers.

27. Use your graph in Exercise 25 to write an inequality that is never true.
For example, $x - 2 > x$

28. Use your graph in Exercise 26 to write an inequality that is never true.
For example, $1 - 2x \geq 4 - 2x$

Solve the inequalities in Exercises 29 to 48 using algebraic notations.

29. $1 - 2x < x$ $x > \frac{1}{3}$

30. $-2x + 3 > x$ $x < 1$

31. $-\frac{1}{2}x + 1 < 4 - x$ $x < 6$

32. $4 - x < -2x + 3$ $x < -1$

33. $0.4x - 6 \leq 4$ $x \leq 25$

34. $0.6x + 3 \geq 6$ $x \geq 5$

35. $8 - 0.75x > 11$ $x < -4$

36. $4 - 0.25x < 5.75$ $x > -7$

37. $3(1 - x) - 1 < 2.5x - 9$ $x > 2$

38. $2(x - 3) > 3(4 - x)$ $x > 3.6$

39. $5 - 3.5(x + 1) \geq 4 - 1.5x$ $x \leq -1.25$

40. $5.2 - 1.2(x - 3) \leq 9.4 + 2.8x$ $x \geq -0.15$

41. $10 < 5x < 20$ $2 < x < 4$

42. $3 < x - 2 < 7$ $5 < x < 9$

43. $10 < -5x < 20$ $-2 > x > -4$

44. $-4 < x + 6 < -1$ $-10 < x < -7$

45. $-3 < 2 - x < 4$ $5 > x > -2$

46. $-5 < 3 - 2x < -1$ $4 > x > 2$

47. $-16 \leq 4 - 5x \leq 29$ $-5 \leq x \leq 4$

48. $26 \geq 6 - 2.5x \geq 11$ $-8 \leq x \leq -2$

49. Solve these absolute value equations and inequalities from the graph shown in the figure.

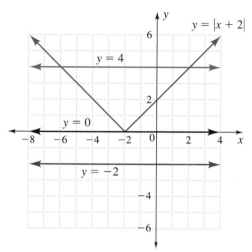

a. $|x + 2| = 0$ $x = -2$ **b.** $|x + 2| > 0$ all real numbers

c. $|x + 2| < 4$
$-6 < x < 2$

d. $|x + 2| > -2$
all real numbers

e. $|x + 2| < -2$
no solution

f. $|x + 2| = 3$
$x = 1$ or $x = -5$

g. $|x + 2| > 3$
$x < -5$ or $x > 1$

50. Solve these absolute value equations and inequalities from the graph shown in the figure.

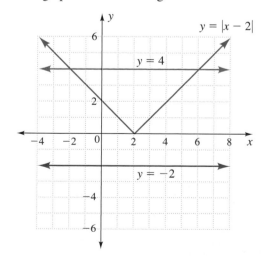

a. $|x - 2| = 0$ $x = 2$ **b.** $|x - 2| \geq 0$ all real numbers

c. $|x - 2| > 4$
$x < -2$ or $x > 6$

d. $|x - 2| < -2$ no solution

e. $|x - 2| > -2$
all real numbers

f. $|x - 2| = 3$
$x = -1$ or $x = 5$

g. $|x - 2| < 3$ $-1 < x < 5$

In Exercises 51 to 58, solve with algebraic notation. If possible, rewrite the answer as a compound inequality.

51. $|2x - 3| > 7$ $x < -2$ or $x > 5$

52. $|3x - 2| < 7$ $x < 3$ and $x > -1\frac{2}{3}$; $-1\frac{2}{3} < x < 3$

53. $|5 - 4x| < 11$ $x > -1.5$ and $x < 4$; $-1.5 < x < 4$

54. $|2 - 5x| \geq 10$ $x \leq -1.6$ or $x \geq 2.4$

55. $|3 - \frac{1}{2}x| > 4.5$ $x < -3$ or $x > 15$

56. $|4 - \frac{1}{3}x| < 6$ $x > -6$ and $x < 30$; $-6 < x < 30$

57. $|0.5 - 0.25x| \leq 2$ $x \geq -6$ and $x \leq 10$; $-6 \leq x \leq 10$

58. $|0.4 - 0.2x| > 1.2$ $x < -4$ or $x > 8$

59. The BMW Platinum card plan charges 2.5% or $5.00, whichever is greater, for a cash advance. Write an inequality to find the cash advances for which the $5 fee will be charged. Solve the inequality.
$0.025x \leq 5, x \leq 200$

60. The First U.S.A. Bank Liquid card plan charges an ATM cash advance fee of 3% of the amount of the advance, but not less than $10. Write an inequality to find the cash advances for which the $10 fee will be charged. Solve the inequality. $0.03x \leq 10, x \leq 333.33$

61. The First U.S.A. Bank Platinum card plan charges 2% of the amount of each purchase of foreign currency. The minimum fee is $5 and maximum fee is $25. Write a compound inequality to find the amounts of foreign currency for which the 2% fee applies. Solve the inequality. $5 \leq 0.02x \leq 25, 250 \leq x \leq 1250$

62. The First U.S.A. Bank Liquid card plan charges 3% of the amount of each purchase of foreign currency. The minimum fee is $5 and maximum fee is $50. Write a compound inequality to find the amounts of foreign currency for which the 3% fee applies. Solve the inequality. $5 \leq 0.03x \leq 50, 166.67 \leq x \leq 1666.67$

63. Evergreen Film charges $3.75 plus $0.50 per photo for processing color film. Quick Stop charges $9.00 plus $0.25 per photo. Write an inequality to find the number of photos for which Evergreen's processing will be cheaper. Solve the inequality.
$3.75 + 0.50x < 9 + 0.25x; x < 21$ photos

64. Mall Photo offers small photo sheets at $10 each plus a $12 sitting fee. Store Photo charges $12 per small photo sheet plus a $9.99 sitting fee. Write an inequality to find the number of photos sheets for which Mall Photo will be cheaper. Solve the inequality. $10x + 12 < 12x + 9.99$; $x > 1$ sheet (only at $x = 1$ will Store Photo be cheaper)

65. A birthday party at Downtown Pizza costs $30 plus $6 per guest. A birthday party at City Center Pizza costs $50 plus $5 per guest. Write an inequality to find the number of guests for which Downtown Pizza will be cheaper. Solve the inequality. $30 + 6x < 50 + 5x; x < 20$ guests

66. A swim party at Green Acres pool costs $35 plus $3 per guest. A party at Pine Bowl costs $15 plus $5.50 per guest. Write an inequality to find the number of guests for which the Pine Bowl party will be cheaper. Solve the inequality. $15 + 5.50x < 35 + 3x; x < 8$ guests

67. How might Exercise 65 be solved by reasoning?
See below.

68. How might Exercise 66 be solved by reasoning?
See below.

69. Explain how you would check your work in solving a compound inequality.
Choose a number satisfying answer and see if it works.

70. Explain how the graph of the inside expression, such as $y = x + 5$ in $4 < x + 5 < 6$, could be used to solve the inequality.
Look for x where $y = x + 5$ is above $y = 4$ and below $y = 6$.

71. Explain why we need to reverse the inequality sign when we multiply an inequality on both sides by a negative. Illustrate your explanation with a numerical example as needed. Negative changes relative value of sides: for example, $20 > 10 \Rightarrow (-1)(20)$ ⟦?⟧ $(-1)(10) \Rightarrow -20 < -10$

72. Explain how we can avoid multiplication or division by a negative when solving an inequality such as $3 - 2x > 5$. Add $2x$ to both sides first.

73. Explain why some endpoints on line graphs are dots and others are small circles. Dot indicates that value is a solution; small circle indicates that value is excluded.

67. There is a $20 difference in basic costs. The per-guest charge differs by $1. It will take $20/1 = 20$ guests for the party costs to be equal.

Project

74. Checks and Debits Bank USA charges $9.00 per month for an unlimited number of checks each month and no minimum balance.
Safe Bank charges $1.25 per month for a debit card and $0.50 per card use with no minimum balance.
Student's Credit Union charges $3.00 per month for up to 15 checks and $0.75 for each additional check, with no minimum balance.

a. Write an equation for cost per month for a Bank USA account. What kind of function is it?
$y = 9$, a constant function

b. Write an equation for the monthly cost of a Safe Bank card. $y = 1.25 + 0.50x$

c. Write the credit union charges as two equations with inequality conditions. Compare your equations with costs for 14, 15, and 16 checks to make sure that the equations are correct.
$y = 3$ for $x \le 15$, $y = 3 + 0.75(x - 15)$ for $x > 15$

d. For what number of uses is the Safe Bank card cheaper than checks with Bank USA?
$x \le 15$

e. For what number of checks is Bank USA cheaper than the Student's Credit Union?
$x > 23$

f. For what number of checks is the Student's Credit Union cheaper than the Safe Bank card?
$x < 38$

68. There is a $20 difference in basic costs. The per-guest charge differs by $2.50. It will take $20/2.50 = 8$ guests for the costs to be equal.

---

3 Chapter Summary

Vocabulary

For definitions and page references, see the Glossary/Index.

addition property of inequalities	dependent equations	ordered triples	standard form of a linear equation
coincident lines	elimination method	parallel lines	substitution method
complementary angles	equivalent inequalities	per	supplementary angles
compound inequalities	inconsistent equations	planes	system of equations
consistent equations	in terms of	quantity	value
degree 1	linear equation	quantity-value tables	
	multiplication properties of inequalities	solution to the system	

Concepts

3.1 ▬ Solving Systems of Two Linear Equations by Substitution or Elimination

To solve equations by substitution:

1. Solve one equation for a first variable.

2. Substitute the resulting expression into the second equation. The second equation now has one variable. Solve for the second variable.

3. Substitute the value of the second variable into the first equation, and find the value of the first variable.

4. Check the solutions.

To solve equations by elimination:

1. Arrange the equations in standard form.

2. Multiply one or both equations by numbers that create opposite values (additive inverses) of the coefficients on one pair of like terms.

3. Add the equations to eliminate one variable, and solve for the variable that remains.

4. Use substitution to find the other variable.

5. Check the solutions.

3.2 ■ Solving Systems of Two Linear Equations by Graphing

A true statement, or identity, such as $0 = 0$ in the solution to a system of equations implies that there are an infinite number of solutions. In the case of two linear equations in two variables, the lines are coincident.

A false statement, or contradiction, such as $0 = 1$ in the solution to a system of equations implies that there are no real-number solutions. In the case of two linear equations in two variables, the lines are parallel.

3.3 ■ Quantity-Value Tables

In quantity-value tables:

- Adding the items in the quantity column gives the total quantity.

- Adding the items in the $Q \cdot V$ column gives the same number as multiplying the total quantity by the average value (across the last row).

- The total of the $Q \cdot V$ column divided by the total of the quantity column gives an average value. The last number in the value column is not found by adding the items in the value column.

3.4 ■ Solving Systems of Three Linear Equations

To solve by elimination:

1. Change the equations to standard form.

2. From the three equations, form two pairs from which to eliminate the same variable term.

3. For the equations in the first pair, find multipliers that change the coefficients on the term to be eliminated to

opposite numbers. Then add the equations to get an equation in two variables.

4. Do the same for the equations in the second pair.

5. Apply the elimination process to the two equations from steps 3 and 4, solving as in Section 3.1.

A system of equations is consistent if it has exactly one solution.

A system of equations is inconsistent if it has no solution.

A system of equations is dependent if it has an infinite number of solutions.

3.5 ■ Solving Linear and Absolute Value Inequalities

The solution set to inequalities in one variable may be shown on a number line.

A number may be added to or subtracted from both sides of an inequality without changing the direction of the inequality.

A positive number may be multiplied or divided on both sides of an inequality without changing the direction of the inequality. A negative number multiplied or divided on both sides of an inequality will change the direction of the inequality.

The solution set to an absolute value inequality may be shown on a number line.

To solve absolute value inequalities with algebraic notation, write two expressions from $|a| < k$: $-a < k$ and $a < k$. Solve each for the unknown variable. (For $|a| > k$, write $-a > k$ or $a > k$ and solve.) Some solutions may be written as compound inequalities.

---

3 Review Exercises

In Exercises 1 to 14, solve the system of equations by substitution or elimination.

1. $x + 2y = -4$
$3x - 4y = 4$
$x = -0.8, y = -1.6$

2. $2x - 3y = 13$
$4x - y = 11$
$x = 2, y = -3$

3. $y = 2x - 1$
$\frac{1}{2}(y + 1) = x$
identity (infinite number of solutions), coincident lines

4. $2x + 3y = 6$
$3y = 2(3 - x)$
identity (infinite number of solutions), coincident lines

5. $y = 3x - 1$
$y - 2 = 3x$
contradiction ({ } or ∅), parallel lines

6. $3x + 5y = 5$
$2x - 4y = 18$
$x = 5, y = -2$

7. $5x - 102 = 9y$
$3.5x + 1.5y = 9$
$x = 6, y = -8$

8. $1.5x = 3.5y + 19$
$5.5x + 1.5y = 41$
$x = 8, y = -2$

9. $y = -x/3 - 1/3$
$2.5x - 4y = 20.5$
$x = 5, y = -2$

10. $-3x + 4y = 5$
$y = \frac{3}{4}x + 2$
contradiction ({ } or ∅), parallel lines

11. $0.3x - 0.4y = 1.4$
$0.2x - 0.5y = 0.7$
$x = 6, y = 1$

12. $1.3x - 0.7y = 5$
$2x - 3y = 2.5$
$x = 5.3, y = 2.7$

13. $\frac{1}{2}x + \frac{3}{4}y = 2$
$\frac{1}{3}x - \frac{1}{2}y = -1$
$x = \frac{1}{2}, y = 2\frac{1}{3}$

14. $\frac{1}{4}x - \frac{1}{2}y = -1$
$\frac{1}{2}x + \frac{3}{5}y = 2$
$x = 1, y = 2.5$

In Exercises 15 to 20, solve the systems of three equations using either substitution or elimination.

15. $a + b + c = 0$
$a - b + c = 6$
$a - 2b - c = 7$
$a = 2, b = -3, c = 1$

16. $x + 2y - z = -1$
$-3x - y + z = -4$
$x + 3z = -3$
$x = 1.5, y = -2, z = -1.5$

17. $3x - y + z = 4$
$-x + z = 2$
$-y + 4z = 10$
identity, infinite number of solutions (dependent)

18. $2x - 4z = 20$
$5y + 3z = -19$
$-2x + 3y = -14$
$x = 4, y = -2, z = -3$

19. $2x + y - z = 3$
$x + y + 2z = 4$
$x - y - 3z = 6$
$x = 6.2, y = -7, z = 2.4$

20. $x + y + z = 180$
$x = 2y$
$x = z$
$x = 72, y = 36, z = 72$

In Exercises 21 to 24, give the equations of the two lines and the solution of the system of equations formed by the lines.

21.

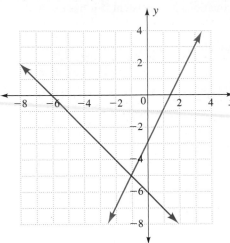

$y = 2x - 3, y = -x - 6; x = -1, y = -5$

22.

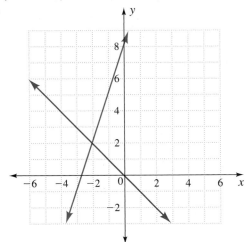

$y = -x, y = 3x + 8; x = -2, y = 2$

23.

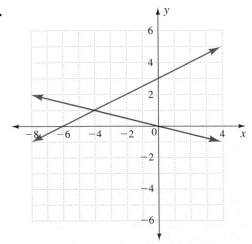

$y = \frac{1}{2}x + 3, y = -\frac{1}{4}x; x = -4, y = 1$

24.

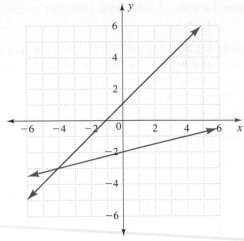

$y = x + 1, y = \frac{1}{4}x - 2; x = -4, y = -3$

Solve Exercises 25 to 28 with a quantity-value table.

25. A student has a 2.80 GPA with 20 credit hours. How many credit hours at a grade of B+ (3.25 points per credit hour) are needed to reach a 3.00 GPA?
16 credit hours

26. A student has a 2.80 GPA with 20 credit hours. How many credit hours at a grade of A− (3.75 points per credit hour) are needed to reach a 3.00 GPA?
$5\frac{1}{3}$ credit hours (or 6 if fractional credit hours are not available)

27. How many pounds of dog food C containing 13% protein and dog food D containing 8% protein need to be blended to obtain 1000 pounds of dog food containing 9% protein? 200 pounds, 800 pounds

28. An order for disinfectant requires 2000 milliliters of a 3% Lysol solution from a 25% stock solution. How much water and stock solution need to be blended?
1760 mL, 240 mL

In Exercises 29 to 32, using guess and check and set up equations as necessary. Find the length, width, and height of the boxes.

29. A box has rectangular sides. The sum of the length, width, and height is 12 inches. The length is 2 inches more than the height. The width is half the sum of the length and the height. 5 in., 4 in., 3 in.

30. A box has rectangular sides. The sum of the length, width, and height is 19 inches. The sum of the width and the height is 3 more than the length. The width is 2 less than the length. 8 in., 6 in., 5 in.

31. A box with rectangular sides has a length equal to the sum of the width and the height. The width is 2 inches more than the height. The sum of the length, width, and height is 20 inches. 10 in., 6 in., 4 in.

32. A box has rectangular sides. The sum of the length, width, and height is 21 inches. The length is six times the sum of the width and the height. The width is $\frac{1}{3}$ inch less than the height. 18 in., $\frac{4}{3}$ in., $\frac{5}{3}$ in.

In Exercises 33 to 43, solve with systems of equations.

33. Jeanne Maiden-Naccarato and Tish Johnson lead the Women's International Bowling Congress with the most 300 games (a perfect score in bowling). Jeanne has 2 more perfect games than Tish. Together they have 40 perfect games. How many does each have? 21, 19

34. Two angles are supplementary. The measure of the first angle less the complement of the second angle is 90°. What is the measure of each angle? infinite number of solutions (any pair of angle measures adding to 180°)

35. Two angles are supplementary. The measure of one angle is five times the measure of the other. What is the measure of each angle? 150°, 30°

36. Two angles are complementary. Four times the measure of the larger angle less half the measure of the smaller angle is 180°. What is the measure of each angle? 50°, 40°

37. Two angles are complementary. The sum of twice the measure of each angle is 180°. What is the measure of each angle? infinite number of solutions (any pair of angle measures adding to 90°)

38. A cup of raisins has 215 total milligrams in calcium, phosphorus, and iron. The amount of phosphorus is 1 milligram less than twice the amount of calcium. There is 47 times as much phosphorus as iron. How many milligrams are there of each substance? 71 mg, 141 mg, 3 mg

39. A cup of skim milk contains 201 calories. The cup contains a total of 49 grams of protein, fat, and carbohydrates. There are 10 more grams of carbohydrates than of protein. Protein and carbohydrates have 4 calories per gram; fat has 9 calories per gram. How many grams each of protein, fat, and carbohydrates are there? 19 g, 1 g, 29 g

40. A 1-ounce snack bag of walnuts contains 184 calories. Protein and carbohydrates have 4 calories per gram; fat has 9 calories per gram. The total weight of protein, fat, and carbohydrates is 26 grams. There are 9 more grams of fat than of protein. How many grams each of protein, fat, and carbohydrates are there? 7 g, 16 g, 3 g

41. Audrey has $12.40 in her piggy bank. She has 74 coins altogether, in nickels, dimes, and quarters. The number of quarters is 5 less than double the number of nickels. How many of each coin does she have? 23, 10, 41

42. Sandra has $86 in her billfold. The bills consist of $10, $5, and $1 denominations. The number of $5 bills is 2 less than the total number of $10 and $1 bills. If she has 18 bills altogether, how many of each does she have? 4, 8, 6

43. Anders purchased textbooks, paperback books, and notebooks at the bookstore. Individual prices of the items are $26.00, $3.50, and $1.50, respectively. Individual weights of the items are 2 pounds, 1 pound, and 0.5 pound, respectively. The number of notebooks is the same as the total number of textbooks and paperback books. The total cost of the purchase was $152.50. The total weight was 17 pounds. How many of each item were purchased? 5, 3, 8

44. You solve a system of three equations and in the process find the statement $1 = 0$. With regard to the system, indicate true or false:

a. The algebra yielded a contradiction. true

b. The graphs intersect in a plane or in a line. false

c. The graphs intersect in a unique point. false

d. The system is dependent. false

e. The system is consistent. false

f. The system is inconsistent. true

g. The algebra yielded an identity. false

45. Use the graph in the figure to solve the following inequalities. Show the solutions to the inequalities on a number line. Then solve using algebraic notation.

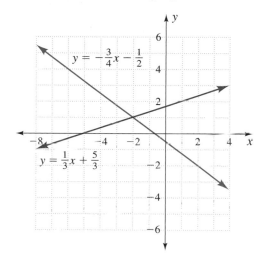

a. $-\frac{3}{4}x - \frac{1}{2} < -2$ $x > 2$; see Answer Section.

b. $\frac{1}{3}x + \frac{5}{3} < -\frac{3}{4}x - \frac{1}{2}$ $x < -2$; see Answer Section.

c. $\frac{1}{3}x + \frac{5}{3} < 2$ $x < 1$; see Answer Section.

Solve the inequalities in Exercises 46 to 52 from the graphs in the figure. Then solve using algebraic notation.

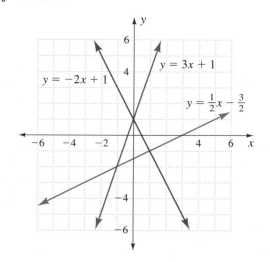

46. $3x + 1 > 4$ $x > 1$

47. $-2x + 1 \le 3$ $x \ge -1$

48. $\frac{1}{2}x - \frac{3}{2} \ge 0$ $x \ge 3$

49. $\frac{1}{2}x - \frac{3}{2} < -1$ $x < 1$

50. $3x + 1 > -2x + 1$
$\quad x > 0$

51. $\frac{1}{2}x - \frac{3}{2} \le -2x + 1$
$\quad x \le 1$

52. $\frac{1}{2}x - \frac{3}{2} \ge 3x + 1$
$\qquad\quad x \le -1$

53. Body Mass Index

    a. Solve the inequality $147.7 < x < 186.5$

$$19 < \frac{x(704.5)}{74^2} < 24$$

    to find the healthy weight range in pounds for a 6-foot 2-inch person. Round to tenths of a pound.

    b. Solve the inequality

$$19 < \frac{x(704.5)}{62^2} < 24$$

    to find the healthy weight range in pounds for a 5-foot 2-inch person. Round to tenths of a pound.
    $103.7 < x < 131.0$

54. Solve each compound inequality for x.

    a. $1 < 5 + \frac{2}{3}x < 13$ $-6 < x < 12$

    b. $-6 < 4 + 2x < 18$ $-5 < x < 7$

    c. $-9 < 3 - 4x < 11$ $-2 < x < 3$

    d. $0 < 2 - \frac{1}{2}x < 6$ $-8 < x < 4$

55. Write and solve a compound inequality: For what amounts of sales, x, will a 5% sales tax be between \$8 and \$9? $8 < 0.05x < 9;\ 160 < x < 180$

56. Write and solve a compound inequality: For what amounts of meal costs, x, will a 15% tip be between \$4.50 and \$12? $4.5 < 0.15x < 12;\ 30 < x < 80$

In Exercises 57 to 60, solve for x.

57. $|4x - 3| < 6$
$-\frac{3}{4} < x < \frac{9}{4}$

58. $|2x - 1| > 5$
$\qquad\quad x < -2 \text{ or } x > 3$

59. $\left|\frac{3}{4}x - 5\right| > 1$
$\ x < \frac{16}{3} \text{ or } x > 8$

60. $\left|\frac{1}{2}x + 2\right| < 4$
$\quad -12 < x < 4$

---

3   Chapter Test

1. Solve by substitution. $x = 108,\ y = 78$

$$3x - 4y = 12$$
$$x - \frac{3}{2}y = -9$$

2. Solve by elimination.   contradiction ($\{\ \}$ or $\varnothing$), parallel lines

$$5x - 2y = 5$$
$$5x = 2(2 + y)$$

In Exercises 3 to 6, solve the systems by any method. Show sketches if you do the problem graphically, or show the algebra if you do the problem symbolically.

3. $2x - 3y = 6$

$\ 4x + 5y = -32$
$\ x = -3,\ y = -4$

4. $x + 2y = -8$

$\ 0.5x + y = -4$
 See below.

5. $2x + y + 2z = 5$

$\ x + 3z = 4$

$\ 4x + 3y = 0$
 contradiction ($\{\ \}$), inconsistent

6. $3x + y + z = 8$

$\ 2x - y - 2z = 17$

$\ -x - y + 3z = -9$
$\ x = 4.5,\ y = -3,\ z = -2.5$

7. Solve by guess and check or reasoning: The length of a box equals the sum of the width and the height. The width is 8 inches more than the height. The sum of the length, width, and height is 24 inches. Find the length, width, and height.   12 in., 10 in., 2 in.

8. Carol Heiss won the World Championship in figure skating for 5 straight years. Peggy Fleming won the championship for 3 straight years. Peggy's first championship came 10 years after Carol's. The sum of their first winning years is 3922. In what years did they first win?
1956, 1966

9. An eye lubricant is to contain 1.4% polyvinyl alcohol. If 400 gallons are to be made, how much distilled water and 16% polyvinyl solution need to be blended by the manufacturer?   365 gal, 35 gal

10. Riley has a total of 59 pennies, nickels, and quarters. He has 10 more quarters than nickels and a total of \$6.35. How many of each coin does he have?   25, 12, 22

11. Marcilio buys cans of juice at \$0.89 each, uniform apples at \$0.59 each, and packages of Oreos® at \$3.98 each. Individual weights are 17 ounces each for the juice, 8 ounces each for the apples, and 16 ounces per package of Oreos. The total weight of his purchase is 158 ounces. The total cost of the purchase is \$15.07. He buys 11 items altogether. How many of each item does he purchase?   6, 3, 2

12. You solve a system of three equations and in the process find the statement $0 = 0$. With regard to the system, indicate true or false:

    a. The graphs intersect in a plane or in a line.   true

    b. The system is dependent.   true

    c. The system contains parallel planes.   false

    d. The algebra yielded a contradiction.   false

4. identity (infinite number of solutions), coincident lines

13. Give the equations of the two lines and the solution of the system of equations formed by the lines.

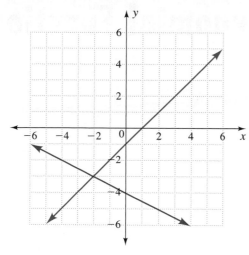

$y = x - 1, y = -\frac{1}{2}x - 4; (-2, -3)$

14. Solve for x: $\quad -8 < x < 4$

$$-\frac{1}{2} < \frac{1}{2} - \frac{1}{4}x < 2\frac{1}{2}$$

15. Solve for x: $\quad -4 < x < 12$

$$\left|2 - \frac{1}{2}x\right| < 4$$

16. Body Mass Index Solve the inequality

$$19 \le \frac{x(704.5)}{68^2} \le 24$$

to find the healthy weight range in pounds for a 5-foot 8-inch person. Round to tenths of a pound.

124.7 lb $\le x \le$ 157.5 lb

CHAPTER 4

Quadratic and Polynomial Functions

Have you ever tried to use a drinking fountain when the water flow failed to reach an appropriate height? Suppose, for sanitary and comfortable drinking, the top of the parabola formed by the water flowing from a drinking fountain needs to be 5 inches from the nozzle horizontally and 6 inches above the nozzle vertically (see Figure 1). In this chapter, we explore how knowing the width and height of the stream of water is sufficient to fit a quadratic equation to the parabola it forms.

This chapter introduces quadratic and polynomial functions. We return to tables to distinguish linear, quadratic, and other number patterns (sequences) and to find and graph equations. We review polynomial operations, including factoring. We solve quadratic equations by factoring.

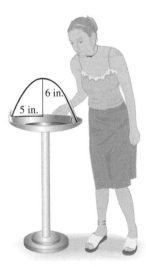

FIGURE 1

4.1 Quadratic Functions and Solving Quadratic Equations with Tables and Graphs

Objectives

- ▮ Identify a quadratic function by number pattern.
- ▮ Identify a quadratic function from the equation form.
- ▮ Identify the input variable as well as coefficients a, b, and c in quadratic functions.
- ▮ Graph quadratic functions and find the domain and range.
- ▮ Find the intercepts, vertex, and axis of symmetry for selected functions.
- ▮ Solve a quadratic equation with a table or graph.

WARM-UP

In Exercises 1 to 3, make an input-output table for each function using inputs -3 to 3.

1. $f(x) = x^2 + x$ $f(x) = 6, 2, 0, 0, 2, 6, 12$
2. $f(x) = 4 - x^2$ $f(x) = -5, 0, 3, 4, 3, 0, -5$
3. $f(x) = 3x^2 + 2x + 5$ $f(x) = 26, 13, 6, 5, 10, 21, 38$

IN THIS SECTION, we find number patterns for quadratic functions, which we use to define the quadratic function. In addition, we examine tables and graphs of quadratic functions. The characteristic parabolic shape of the quadratic function, the intercepts, and symmetry permit us to draw selected graphs with limited information. Tables and graphs allow us to solve quadratic equations using the same skills we use in solving linear and other types of equations.

Patterns

For convenience, we call *the set of differences between consecutive terms* in a sequence simply **first differences**. Then, *the differences between the differences of consecutive terms* we call the **second differences**.

EXAMPLE 1   **Exploring differences**   Make an input-output table with inputs -3 to 3. Find the first and second differences between consecutive terms in the outputs of these functions. Are any of the functions linear?

a. $f(x) = x^2 + x$ **b.** $f(x) = 4 - x^2$
c. $f(x) = 3x^2 + 2x + 5$

SOLUTION   **a.**   Table 1 contains values for $f(x) = x^2 + x$.

TABLE 1

Input	Output	First Difference	Second Difference
-3	6		
		$\rangle$ -4	
-2	2		$\rangle$ $+2$
		$\rangle$ -2	
-1	0		$\rangle$ $+2$
		$\rangle$ 0	
0	0		$\rangle$ $+2$
		$\rangle$ 2	
1	2		$\rangle$ $+2$
		$\rangle$ 4	
2	6		$\rangle$ $+2$
		$\rangle$ 6	
3	12		

b. Table 2 contains values for $f(x) = 4 - x^2$.

c. Table 3 contains values for $f(x) = 3x^2 + 2x + 5$.

TABLE 2

Input	Output	First Difference	Second Difference
-3	-5		
		$+5$	
-2	0		-2
		$+3$	
-1	3		-2
		$+1$	
0	4		-2
		-1	
1	3		-2
		-3	
2	0		-2
		-5	
3	-5		

TABLE 3

Input	Output	First Difference	Second Difference
-3	26		
		-13	
-2	13		$+6$
		-7	
-1	6		$+6$
		-1	
0	5		$+6$
		$+5$	
1	10		$+6$
		$+11$	
2	21		$+6$
		$+17$	
3	38		

None of the functions have a constant first difference. None are linear. All have a constant second difference. ▬

The second differences reveal a numerical distinction between linear and quadratic functions. For inputs 1, 2, 3, . . . , **quadratic functions** and **quadratic sequences** have *constant second differences*. The name *quadratic* comes from the Latin word *quad*, meaning square, which refers to the exponent in x^2.

Quadratic Functions

The rules for the functions in Example 1, $f(x) = x^2 + x$, $f(x) = 4 - x^2$, and $f(x) = 3x^2 + 2x + 5$, all contain terms with the exponent 2 as the highest power. They are *functions of degree 2*, or **quadratic functions**.

▬ DEFINITION OF QUADRATIC FUNCTION

> **Quadratic functions** are functions of degree 2 that may be written in the form
>
> $$f(x) = ax^2 + bx + c$$
>
> where a, b, and c are real numbers, $a \neq 0$. The input variable is x.

EXAMPLE 2

Finding the coefficients of terms in quadratic functions   Identify the input variable as well as a, b, and c in these functions.

a. $f(x) = x^2$

b. $f(x) = x(x + 1)$

c. Area of circle: $A = \pi r^2$

d. Approximate length of skid marks (in feet) at s miles per hour on a dry concrete road surface: $L = s^2/24$

e. Height of an object in vertical motion, where g is the acceleration due to gravity, v_0 is the initial velocity, and h_0 is the initial height: $h = -\frac{1}{2}gt^2 + v_0 t + h_0$

SOLUTION

a. In $f(x) = x^2$, the input variable is x. We assume there to be a 1 before the x^2, so $a = 1$. There is nothing to match with the other two terms $bx + c$, so $b = 0$ and $c = 0$.

b. We change $f(x) = x(x + 1)$ to the form $f(x) = x^2 + x$. The input variable is x. Thus, $a = 1$, $b = 1$, and $c = 0$.

c. The area of a circle is a function of radius r, so r is the input variable, $a = \pi$, $b = 0$, and $c = 0$.

d. The length of the skid mark is a function of speed s, so the input variable is s. Thus, $a = \frac{1}{24}$, $b = 0$, $c = 0$.

e. The height is a function of time t. Thus, the input variable is t, $a = -\frac{1}{2}g$, $b = v_0$, and $c = h_0$.

Quadratic equations in one variable are written in standard form as $ax^2 + bx + c = 0$, with $a > 0$.

EXAMPLE 3   **Identifying quadratic equations in one variable**   Which of the following equations are quadratic? Simplify and identify the coefficients a, b, and c.

a. $x^2 = 2$ **b.** $x^2 - 1 = x^2 + x$

c. $5x^2 - 2(x - 3) = 7$ **d.** $4x^2 + 3x(x + 1) = 10$

SOLUTION   **a.** $x^2 = 2$ Subtract 2 from both sides.

$x^2 - 2 = 0$ $a = 1, b = 0, c = -2$

Quadratic

b. $x^2 - 1 = x^2 + x$ Subtract $x^2 - 1$ from both sides.

$0 = x + 1$

Linear

c. $5x^2 - 2(x - 3) = 7$ Multiply out the parentheses.

$5x^2 - 2x + 6 = 7$ Subtract 7 from both sides.

$5x^2 - 2x - 1 = 0$ $a = 5, b = -2, c = -1$

Quadratic

d. $4x^2 + 3x(x + 1) = 10$ Multiply out the parentheses.

$4x^2 + 3x^2 + 3x = 10$ Add like terms and subtract 10 from both sides.

$7x^2 + 3x - 10 = 0$ $a = 7, b = 3, c = -10$

Quadratic

Graphing Quadratic Functions

POINT PLOTTING   A linear function can be graphed with two points. But it takes several points to make a good graph of a quadratic function, because the graph is a type of *curve* called a **parabola**. The cable in the bridge in Figure 2 is a portion of a parabola. In Example 4, the plot of seven points barely gives a suggestion of the parabolic graph.

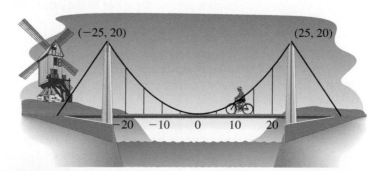

FIGURE 2

EXAMPLE 4   Graphing a quadratic function by plotting points
  a. Use Table 3 in Example 1 to plots points for the function $f(x) = 3x^2 + 2x + 5$.
  b. Graph the function with a calculator.
  c. Compare Table 3 with Tables 1 and 2. How are the tables different?

SOLUTION   **a.** The points are plotted in Figure 3. It is, at best, difficult to see how a smooth parabolic curve is to be drawn through the points. For example, is the lowest point between $x = -1$ and $x = 0$, at $x = 0$, or between $x = 0$ and $x = 1$?

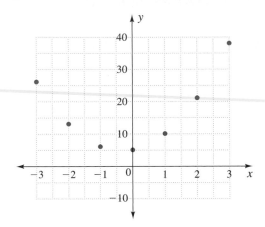

FIGURE 3

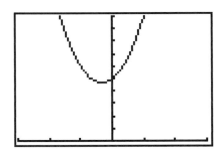

FIGURE 4 -3 to 3 on X, 0 to 10 on Y

b. The calculator graph in Figure 4, with x on the interval $[-3, 3]$ and y on $[0, 10]$, shows a smooth curve with the lowest point between $x = -1$ and $x = 0$. Additional ordered pairs with fractional inputs would be helpful in drawing this graph by hand.

c. Most of the outputs in Tables 2 and 3, for the other two quadratic functions, appear in pairs.   ▬

Think about it 1: Find matching output pairs for $f(x) = 3x^2 + 2x + 5$ by using a fractional value for **ΔTbl** in the calculator table.

The following shortcuts for sketching graphs are based on tables in which outputs appear in pairs or on vertical or horizontal intercepts. These shortcuts will not help in graphing functions like the one in Example 4, but for many purposes drawing a quick sketch by hand is adequate.

INTERCEPTS   Some of the first points we find in graphing a quadratic function are its intercepts with the axes. Recall that the **x-intercept points**, if they exist, are *the points where a graph crosses or touches the x-axis*. Because $y = 0$ everywhere on the x-axis, the x-intercepts are described as the inputs for which the output is zero, $f(x) = 0$. The **y-intercept point** is *the point where a graph crosses or touches the y-axis*. Because $x = 0$ everywhere on the y-axis, the y-intercept is $f(0)$, the function evaluated at zero.

EXAMPLE 5 Graphing quadratic functions and finding intercepts
a. Use Table 4 to graph the quadratic function $f(x) = x^2 + x$.
b. Identify the x- and y-intercept points on the table and graph.

SOLUTION **a.** The graph in Figure 5 was drawn with the data from Table 4.

TABLE 4

Input x	Output $f(x) = x^2 + x$
-3	6
-2	2
-1	0
0	0
1	2
2	6
3	12

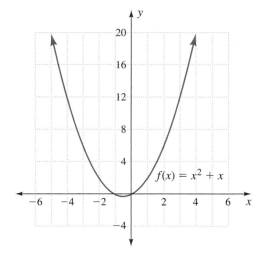

FIGURE 5

b. The table and the graph show two x-intercept points, $(-1, 0)$ and $(0, 0)$. These ordered pairs satisfy $f(x) = 0$.

$$(-1)^2 + (-1) \overset{?}{=} 0 \quad \checkmark$$

$$(0)^2 + (0) \overset{?}{=} 0 \quad \checkmark$$

The origin, $(0, 0)$, is also the y-intercept point $f(0)$. ▬

AXIS OF SYMMETRY   The parabolic graph of a quadratic function has an **axis**, or **line**, **of symmetry**—*a line across which the graph can be folded so that points on one side of the graph match with points on the other side of the graph*. In Example 5, the symmetry is evident in Table 4, where the output numbers appear in pairs as we move up and down on the table from $f(x) = 0$, the lowest output shown. The symmetry is also evident in the graph in Figure 5 in that folding the graph by matching the x-intercepts ($x = -1$ and $x = 0$) would create an axis of symmetry.

EXAMPLE 6 Using the axis of symmetry to find other ordered pairs   Use the axis of symmetry in Table 4 to predict the output for $x = -4$.

SOLUTION Symmetry indicates that the 0, 2, 6, 12 pattern for the output numbers in the table will be repeated in both directions. Thus, at $x = -4$, $y = 12$. The ordered pair $(-4, 12)$ lies on the graph. ▬

Think about it 2: Name another function that has a vertical axis of symmetry. Could a function have a horizontal axis of symmetry?

VERTEX   The *highest or lowest point on a parabola* is called the **vertex**. In Figure 5, the vertex is the lowest point, located on the graph between $x = -1$ and $x = 0$. The concept of symmetry allows us to find the coordinates of the vertex for the function $f(x) = x^2 + x$ in Example 7.

EXAMPLE 7 Finding the vertex
a. Find the equation of the axis of symmetry for $f(x) = x^2 + x$. (See Figure 6.)
b. Find the ordered pair for the vertex.

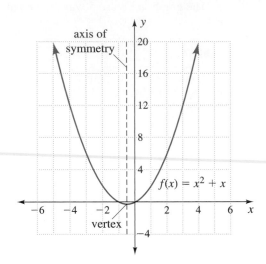

FIGURE 6

SOLUTION **a.** Table 4 in Example 5 shows the x-intercepts $(-1, 0)$ and $(0, 0)$ to have the lowest outputs. The vertex must lie halfway between these two lowest outputs, on the axis of symmetry. The line halfway between $x = -1$ and $x = 0$ is $x = -\frac{1}{2}$, or $x = -0.5$. Thus, the axis of symmetry has the equation $x = -0.5$.

b. To find the vertex, we substitute $x = -0.5$ into the function:

$$f(x) = x^2 + x$$

$$f(-0.5) = (-0.5)^2 + (-0.5)$$

$$= 0.25 - 0.50$$

$$= -0.25$$

The coordinates of the vertex are $(-0.5, -0.25)$, or $\left(-\frac{1}{2}, -\frac{1}{4}\right)$ as fractions. ▬

The pairing of the outputs in the table for a quadratic function and the symmetry of the quadratic function graph lead us to conclude that *the axis of symmetry passes through the vertex of a parabola.*

EXAMPLE 8 Graphing a quadratic function   Make a table and graph for the quadratic function $f(x) = 4 - x^2$ using the following steps.
a. Start a table with 0, 1, 2, and 3 as inputs. Plot these points and note the y-intercept.
b. Look for patterns that show symmetry, and make other entries in the table without evaluating the function.
c. If possible, identify where $f(x) = 0$ and plot the x-intercept points.
d. Use the x-intercepts or any other two equal outputs to locate the line of symmetry and the vertex.
e. Draw the parabola.

SOLUTION **a.** Table 5 shows outputs for inputs 0 to 3. These ordered pairs are plotted in Figure 7.

TABLE 5

Input x	Output $f(x) = 4 - x^2$
0	4
1	3
2	0
3	-5

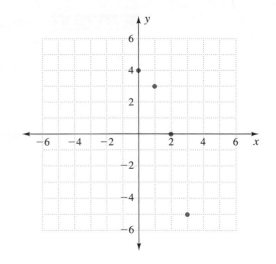

FIGURE 7

b. The output numbers for $x = 0$ and $x = 1$ are the closest together, so we extend the table to negative x values to look for symmetry.

$$f(-1) = 4 - (-1)^2 = 3$$

We observe that $f(-1) = f(1) = 3$, so we can use the symmetry pattern to extend the table for $f(-2)$ and $f(-3)$. The extended table is shown in Table 6.

c. The x-intercepts, where $f(x) = 0$, are at $x = -2$ and $x = 2$.

TABLE 6

Input x	Output $f(x) = 4 - x^2$
-3	-5
-2	0
-1	3
0	4
1	3
2	0
3	-5

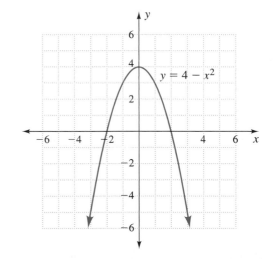

FIGURE 8

d. Note that only $x = 0$ has 4 as an output. The ordered pair $(0, 4)$ is the vertex. That $x = 0$ is halfway between each pair of inputs with equal outputs confirms that $(0, 4)$ is the vertex. The line of symmetry, $x = 0$, passes through $(0, 4)$, the vertex.

e. The graph of $f(x) = 4 - x^2$ is shown in Figure 8. ▬

Think about it 3: How is the graph of $f(x) = 4 - x^2$ the same as and different from that of $f(x) = x^2 + x$?

The vertex, if available, helps identify the range of a quadratic function.

EXAMPLE 9   Finding domain and range   What are the domain and range for these functions?

a. $f(x) = x^2 + x$ **b.** $f(x) = 4 - x^2$

SOLUTION   Both functions have any real number as input, so their domains are all real numbers. Both functions have ranges—sets of outputs—that start at their vertex.

a. The graph of $f(x) = x^2 + x$ turns up, with its vertex at the bottom. The vertex for $f(x) = x^2 + x$ is $(-0.5, -0.25)$, so the range is all real numbers -0.25 or larger, or $y \geq -0.25$.

b. The graph of $f(x) = 4 - x^2$ turns down, with its vertex at the top. The vertex for $f(x) = 4 - x^2$ is $(0, 4)$, so the range is all real numbers 4 or smaller, or $y \leq 4$.

Table 7 lists important features of the graph of a quadratic function.

TABLE 7

Name	Definition	Table	Graph
x-intercept points (where $y = 0$ or $f(x) = 0$, if they exist)	places where graph crosses x-axis	x · y / · 0	$(\_, 0)$
y-intercept point (where $x = 0$ or at $f(0)$)	places where graph crosses y-axis	x · y / 0 ·	$(0, \_)$
vertex (lies on the axis of symmetry)	highest or lowest point	The x-coordinate is halfway between ordered pairs with equal outputs.	

Solving Quadratic Equations with Tables and Graphs

Tables and graphs of quadratic functions may be used to solve the quadratic equations.

 SOLVING QUADRATIC EQUATIONS $f(x) = n$

> Solve for x by locating the output n in the table for $f(x)$.
> Solve for x by finding the point of intersection of $y = f(x)$ and $y = n$.

EXAMPLE 10   Solving quadratic equations with a table and graph   Solve these equations with Table 8 and Figure 9.

a. $x^2 + x = 0$
b. $x^2 + x = 6$
c. $x^2 + x = -4$

TABLE 8

Input x	Output $f(x) = x^2 + x$
-3	6
-2	2
-1	0
0	0
1	2
2	6
3	12

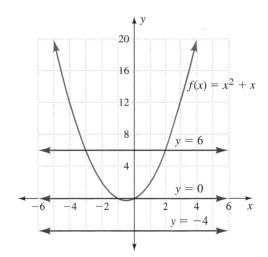

FIGURE 9

SOLUTION   **a.** $x^2 + x = 0$ in two locations in Table 8: at $x = -1$ and $x = 0$. The graph of $y = 0$, the x-axis, intersects the graph of $f(x) = x^2 + x$ twice: at $x = -1$ and $x = 0$.

     b. $x^2 + x = 6$ in two locations in Table 8: at $x = -3$ and $x = 2$. The graph of $y = 6$ intersects the graph of $f(x) = x^2 + x$ twice: at $x = -3$ and $x = 2$.

     c. The graph of $f(x) = x^2 + x$ does not intersect the line $y = -4$. There are no real-number solutions to the equation $x^2 + x = -4$. ■

■ SUMMARY: SOLVING
QUADRATIC EQUATIONS

A quadratic equation $ax^2 + bx + c = 0$ or $ax^2 + bx + c = d$ may have zero, one, or two solutions.

- Solve $ax^2 + bx + c = 0$ with a table by finding the input x where $f(x) = 0$ or $y = 0$.
- Solve $ax^2 + bx + c = d$ with a table by finding the input x where $f(x) = d$ or $y = d$.
- Solve $ax^2 + bx + c = 0$ with a graph by finding the x-intercepts.
- Solve $ax^2 + bx + c = d$ with a graph by finding intersections of $y = d$ and the graph of the corresponding quadratic function.

ANSWER BOX

Warm-up: Values of $f(x)$ for x on the interval $[-3, 3]$: **1.** 6, 2, 0, 0, 2, 6, 12   **2.** $-5, 0, 3, 4, 3, 0, -5$ **3.** 26, 13, 6, 5, 10, 21, 38   **Think about it 1:** **TblStart** $= -3$ and **ΔTbl** $= 1/3$ (1 divided by 3).   **Think about it 2:** There are various possible answers, but $y = |x|$ is one example. A function cannot have a horizontal line of symmetry because the vertical-line test would fail if there were two output values for some inputs.   **Think about it 3:** The two graphs are parabolas, each with a line of symmetry. The graph of $f(x) = 4 - x^2$ opens down, whereas the graph of $f(x) = x^2 + x$ opens up.

4.1 Exercises

What is the next number in each sequence in Exercises 1 and 2? Identify each as linear, quadratic, or neither.

1. a. 1, 3, 9, 27, 81, 243   729; neither

  b. 11, 18, 25, 32, 39   46; linear

  c. 56, 47, 38, 29, 20   11; linear

  d. −1, 1, 7, 17, 31   49; quadratic

2. a. 2, 3, 6, 11, 18   27; quadratic

  b. −9, −7, −3, 3, 11   21; quadratic

  c. −3, −8, −13, −18, −23   −28; linear

  d. 1, 2, 4, 8, 16   32; neither

In Exercises 3 and 4, identify the equations as linear, quadratic, or neither.

3. a. $y = 1 - x^3$ neither

  b. $y = x^5 + 1$ neither

  c. $y = x^2 + x + 3$ quadratic

  d. $y = x - 1$ linear

4. a. $y = 1 - x^2 - x^4$ neither

  b. $y = x + 1$ linear

  c. $y = 5 - x^2$ quadratic

  d. $y = x^3 + x + 1$ neither

In Exercises 5 and 6, name the input variable and the numbers representing a, b, and c when the function is written in **standard form**, $f(x) = ax^2 + bx + c$.

5. a. Surface area of a cylinder of height 3 meters:
  $f(r) = \pi r^2 + 2\pi r(3 \text{ meters})$ $r; \pi, 2\pi(3), 0$

  b. Length of a pendulum: $f(T) = \dfrac{g}{4\pi^2}T^2$ $T; g/4\pi^2, 0, 0$

  c. Triangular numbers: $y = \frac{1}{2}x(x + 1)$ $x; \frac{1}{2}, \frac{1}{2}, 0$

6. a. Surface area of a sphere: $f(r) = 4\pi r^2$ $r; 4\pi, 0, 0$

  b. Triangular numbers: $y = \dfrac{n^2 + 3n + 2}{2}$ $n; \frac{1}{2}, \frac{3}{2}, 1$

  c. Height of a bridge cable: $h(x) = \frac{4}{125}x^2$ $x; \frac{4}{125}, 0, 0$

In Exercises 7 and 8, which equations are quadratic? Simplify and identify the coefficients a, b, and c.

7. a. $x^3 = 8$ not quadratic

  b. $x - 4 = x(x + 4)$ quadratic; 1, 3, 4

  c. $x - 2(x - 4) = x(x - 4)$ quadratic; 1, −3, −8

  d. $1 + x = x - 2(x + 1)$ not quadratic

8. a. $x(4 - x) = x + 4$ quadratic; 1, −3, 4

  b. $5 - 3(x - 3) = -x(x - 3)$ quadratic; 1, −6, 14

  c. $5 - x(x - 3) = -x(x + 3)$ not quadratic

  d. $3 - x^2(4 - x^2) = 3(4 - x)^2$ not quadratic

9. Make a table and sketch a graph for $y = 2x^2 + x + 1$.
  For x on $[-3, 3]$, y is 16, 7, 2, 1, 4, 11, 22; see Answer Section.

10. Make a table and sketch a graph for $y = 2x^2 - x + 6$.
  For x on $[-3, 3]$, y is 27, 16, 9, 6, 7, 12, 21; see Additional Answers.

11. Make a table and sketch a graph for $y = -x^2 - 2$.
  For x on $[-3, 3]$, y is −11, −6, −3, −2, −3, −6, −11; see Answer Section.

12. Make a table and sketch a graph for $y = -1 - 2x^2$.
  For x on $[-3, 3]$, y is −19, −9, −3, −1, −3, −9, −19; see Additional Answers.

In Exercises 13 and 14, identify the coordinates of the x- and y-intercepts and the vertex from the table.

13.

x	$f(x) = x^2 - 4x + 3$	
−2	15	
−1	8	
0	3	y-intercept
1	0	x-intercept
2	−1	vertex
3	0	x-intercept
4	3	
5	8	

14.

x	$f(x) = x^2 + x - 6$	
−3	0	x-intercept
−2	−4	
−1	−6	vertex at $\left(-\frac{1}{2}, -6\frac{1}{4}\right)$
0	−6	y-intercept
1	−4	
2	0	x-intercept
3	6	
4	14	

15. Graph the function in Exercise 13.   See Answer Section.

16. Graph the function in Exercise 14.   See Additional Answers.

17. From the table in Exercise 13, what is $f(6)$?   15

18. From the table in Exercise 14, what is $f(-5)$?   14

In Exercises 19 and 20, solve the equations from the tables in Exercises 13 and 14. Write the answers in set notation.

19. a. $x^2 - 4x + 3 = 8$
  $\{-1, 5\}$

  b. $x^2 - 4x + 3 = 0$
  $\{1, 3\}$

  c. $x^2 - 4x + 3 = 15$
  $\{-2, 6\}$

  d. $x^2 - 4x + 3 = -5$
  $\{\ \}$ or $\varnothing$

20. a. $x^2 + x - 6 = 14$ $\{-5, 4\}$

   b. $x^2 + x - 6 = 0$ $\{-3, 2\}$

   c. $x^2 + x - 6 = -4$ $\{-2, 1\}$

   d. $x^2 + x - 6 = -10$ $\{\ \}$ or $\varnothing$

21. Refer to the graph of $f(x) = x^2 - 2x - 8$ in the figure.

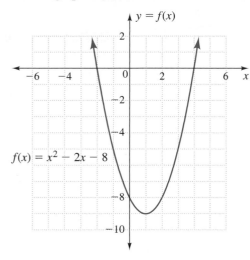

a. Find the x-intercept points. $(-2, 0), (4, 0)$

b. Find the y-intercept point. $(0, -8)$

c. Find the equation for the axis of symmetry. $x = 1$

d. Find the vertex. $(1, -9)$

e. Solve $x^2 - 2x - 8 = -9$. $\{1\}$

f. Solve $x^2 - 2x - 8 = -5$. $\{-1, 3\}$

g. Solve $x^2 - 2x - 8 = 0$; compare your answers with those in part a. $\{-2, 4\}$

h. What is the range for $f(x)$? $y \geq -9$

22. Refer to the graph of $g(x) = -2x^2 + x + 1$ in the figure.

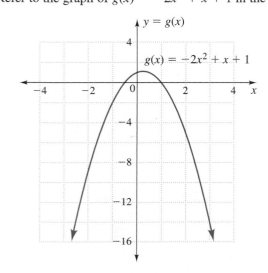

a. Find the x-intercept points. $(-\frac{1}{2}, 0), (1, 0)$

b. Find the y-intercept point. $(0, 1)$

c. Find the equation for the axis of symmetry. $x = \frac{1}{4}$

d. Find the vertex. $(\frac{1}{4}, 1\frac{1}{8})$

e. Solve $-2x^2 + x + 1 = -2$. $\{-1, 1\frac{1}{2}\}$

f. Solve $-2x^2 + x + 1 = -5$. $\{-1\frac{1}{2}, 2\}$

g. Solve $-2x^2 + x + 1 = 0$; compare your answers with those in part a. $\{-\frac{1}{2}, 1\}$

h. What is the range for $g(x)$? $y \leq 1\frac{1}{8}$

23. Make a table and graph $f(x) = -x^2 - 2x + 8$.
For x in $[-4, 2]$, y is 0, 5, 8, 9, 8, 5, 0; see Answer Section.

a. Find the x-intercept points.
$(-4, 0), (2, 0)$

b. Find the y-intercept point.
$(0, 8)$

c. Find the equation for the axis of symmetry.
$x = -1$

d. Find the vertex.
$(-1, 9)$

e. Solve $-x^2 - 2x + 8 = 0$; compare your answers with those in part a. $\{-4, 2\}$

f. Solve $-x^2 - 2x + 8 = 8$. $\{-2, 0\}$

g. Solve $-x^2 - 2x + 8 - 10$. $\{\ \}$ or $\varnothing$

h. Solve $-x^2 - 2x + 8 = 5$. $\{-3, 1\}$

i. What is the range for $f(x)$? $y \leq 9$

24. Make a table and graph $f(x) = x^2 + 5x + 6$.
For x in $[-5, 1]$, y is 6, 2, 0, 0, 2, 6, 12; see Additional Answers.

a. Find the x-intercept points.
$(-3, 0), (-2, 0)$

b. Find the y-intercept point.
$(0, 6)$

c. Find the equation for the axis of symmetry.
$x = -2.5$

d. Find the vertex.
$(-2.5, -0.25)$

e. Solve $x^2 + 5x + 6 = 6$.
$\{-5, 0\}$

f. Solve $x^2 + 5x + 6 = 2$.
$\{-4, -1\}$

g. Solve $x^2 + 5x + 6 = 0$; compare your answers with those in part a. $\{-3, -2\}$

h. Solve $x^2 + 5x + 6 = -2$. $\{\ \}$ or $\varnothing$

i. What is the range for $f(x)$? $y \geq -0.25$

In Exercises 25 and 26, make a table and graph and use the graph to answer the questions.

25. The formula for the surface area of a cylinder is $A = 2\pi r^2 + 2\pi rh$. This cylinder might be a metal can (see the figure). Let $h = 6$ in. and $r = 0$ to 6 in.
For r in $[0, 6]$, A is 0, 44, 101, 170, 251, 346, 452; see Answer Section.

a. For what radius will the surface area be 200 square inches? Using calculator π, ≈ 3.4 in.

b. For what radius will the surface area be 400 square inches? ≈ 5.5 in.

c. Why is the radius in part b not double the radius in part a? top and bottom surface area varies with square of radius

26. The formula for the amount of money in an account earning interest compounded yearly for 2 years is $A = P(1 + r)^2$. Let $P = \$1000$ and r be 0 to 20%. See Additional Answers.

a. For what interest rate, r, will the account contain 1100 at the end of the 2 years? $\approx 5\%$

b. For what interest rate, r, will the account contain 1200 at the end of the 2 years? $\approx 9.5\%$

c. For what interest rate, r, will the account contain 1400 at the end of 2 years? $\approx 18\%$

27. If $a = 0$ in $y = ax^2 + bx + c$, what is the resulting equation? a linear equation

28. Draw sketches of quadratic functions that explain why quadratic equations can have only 0, 1, or 2 possible solutions.

29. Explain the relationship between the vertex of a parabolic graph and the set of outputs, or range, for the function. Range is bounded by vertex's y-coordinate.

30. Describe the relationship between the value of a in $y = ax^2 + bx + c$ and whether the parabola opens up or opens down. Graph opens up if a is positive, down if a is negative.

31. If a graph of a quadratic function has the y-axis as an axis of symmetry and has $x = 4$ as one x-intercept, what will be another x-intercept? $x = -4$

32. If a graph of $f(x)$ has the y-axis as a line of symmetry, what may be said about $f(a)$ and $f(-a)$? They are equal.

▪ Projects

33. Positions on a Parabola Match each phrase with one of these three positions on a parabola: x-intercept, y-intercept, or vertex. Make a sketch to clarify your answer. Answers may vary depending on your sketch.

a. The highest profit in a parabolic business profit curve vertex

b. The place where a ball on a parabolic path falls to the ground when thrown x-intercept

c. The start-up costs in a business, where the total cost graph is a parabola y-intercept

d. The time it takes for a porpoise to re-enter the water after a parabolic leap x-intercept

e. The down payment in a purchase plan where the total cost graph is a parabola y-intercept

f. The highest point on a parabolic jet of water vertex

g. The number of sales required to change from loss to profit on a parabolic business income curve x-intercept

h. The maximum height in the parabolic path of a ball vertex

i. The initial height of a golf tee, where the path of the ball is a parabola y-intercept

j. The lowest point in a parabolic cable on a suspension bridge vertex

k. The initial cost in a parabolic cost curve y-intercept

l. The point where a diver enters the water, when the curve shows his parabolic path x-intercept

34. Estimating Reaction Time Turn your hand so that you can catch a ruler between your thumb and fingers. Have someone hold a ruler just above finger level and drop it between your outstretched thumb and fingers (see the figure). Catch the ruler. Use the point on the ruler at which your fingers catch it as an estimate of distance d in inches. Use $d = \frac{1}{2} gt^2$, with $g \approx 32.2$ ft/sec^2, to estimate your reaction time. (*Hint:* How many inches in a foot?)

4.2   Modeling Quadratic Functions

Objectives

■ Use a table and differences to find a quadratic function.

■ Use calculator regression to find a quadratic function.

WARM-UP

1. Place the number sequence 1, 4, 7, 10, 13, . . . in an input-output table. Is a linear function appropriate?   See Example 1.

2. Place the number sequence 1, 4, 9, 16, 25, . . . in an input-output table.

   a. Find the differences.   See Example 2.

   b. Is the sequence linear or quadratic?

   c. Work backwards in the table to find $f(0)$.

   d. From observation, find the function that describes the sequence.

Answer the following questions for the quadratic function $f(x) = ax^2 + bx + c$.

3. What is $f(0)$? c

4. What is $f(0)$ on the graph of $f(x)$? y-intercept

5. What is $f(1)$? $a + b + c$

To acquaint students with using a system of equations to find a quadratic function, have them do Exercise 31 and then practice with Exercises 25 to 28.

IN THIS SECTION, we build quadratic equations of the form $y = ax^2 + bx + c$ with tables and with calculator regression. The table is limited in application, but it shows that regression equations do relate to number values and are not just "calculator magic."

Student Note: The word *function* describes the type of relationship as well as the rule for the relationship in algebraic notation. However, because the result calculators provide is called a regression equation rather than a regression function, also look for the word *equation* used to describe the rule.

Building a Quadratic Function from a Table

To build a function, we need to find its parameters. A **parameter** is *a letter representing a number that changes the orientation and position of the graph.* The slope m and y-intercept b in the linear equation $y = mx + b$ are parameters. The letters a, b, and c in $y = ax^2 + bx + c$ are all parameters.

In Example 1, we return to Warm-up Exercise 1 and review modeling a linear function (from Section 2.4).

EXAMPLE 1   Finding a linear function from a table

   a. Place the number sequence 1, 4, 7, 10, 13, . . . in an input-output table.

   b. Find the differences.

   c. Is the pattern linear or quadratic?

   d. Work backwards in the table to find $f(0)$.

   e. What is the slope?

   f. What is the equation or rule describing the sequence?

SOLUTION   **a.** Note that the inputs in Table 9 are the natural numbers, 1, 2, 3, ... We leave the first row blank so that we can work backwards to find $f(0)$ in later steps.

TABLE 9

Input	Output	First Difference
0	−2	
1	1	3
2	4	3
3	7	3
4	10	3
5	13	3

b. The first differences, 3, are constant.

c. The sequence can be described by a linear function.

d. We work backwards to find $f(0)$, the vertical axis intercept. We subtract 3 from 1 and obtain −2 as the output matching with $x = 0$.

e. The slope is $3/1$ because natural number inputs have $\Delta x = 1$.

f. $y = mx + b$ Substitute $b = -2$ from part d and $m = 3$ from part e.

$y = 3x - 2$

Check: Let $Y_1 = 3x - 2$ in ⟨Y=⟩. In [TBLSET], let **TblStart** $= 1$ and **ΔTbl** $= 1$. View [TABLE] to see the number pattern under Y_1.   ▬

For the linear function in Example 1, we needed two parameters, m and b (slope and vertical axis intercept), from the table. For the quadratic function $f(x) = ax^2 + bx + c$, we need three parameters: a, b, and c. In Example 2, we return to the setting in Warm-up Exercise 2.

EXAMPLE 2   Exploring a, the coefficient of x^2, and c, the constant term
a. Place the number sequence 1, 4, 9, 16, 25, . . . in an input-output table.
b. Find the first and second differences.
c. From the table, decide whether the sequence is linear or quadratic.
d. From observation, find the rule for the sequence.
e. What is parameter a? Compare a to the second differences.
f. Go to Example 1 of Section 4.1 on page 201. For each function, compare a to the second differences. How is a related to the second differences?
g. Work backwards in the table to find $c = f(0)$.

SOLUTION   **a.** We leave the first row of Table 10 blank so that we can work backwards to find $f(0)$ in later steps.

TABLE 10

Input	Output	First Difference	Second Difference
0	0 = c		
1	1	1	+2
2	4	3	+2 = 2a
3	9	5	+2
4	16	7	+2
5	25	9	

b. The first differences change. The second differences, 2, are constant.

c. The sequence is quadratic.

d. The outputs are the squares of the inputs. The rule is $y = x^2$.

e. In $y = x^2$, $a = 1$. The second difference is 2, or twice 1. *It appears that the constant second difference is 2a.*

f. From Example 1 of Section 4.1,

for $f(x) = x^2 + x$, $a = 1$ and the second difference is $+2$.

for $f(x) = 4 - x^2$, $a = -1$ and the second difference is -2.

for $f(x) = 3x^2 + 2x + 5$, $a = 3$ and the second difference is 6.

Each function has a constant second difference equal to 2a.

g. Working backwards in the table, we subtract the second difference, 2, from 3 to get 1 as the first difference and then subtract 1 from the output 1 to obtain 0 for $f(0)$. Thus, $c = 0$. ∎

Example 2 shows how a table can be used to find parameters a and c in $f(x) = ax^2 + bx + c$. The steps in Example 2 gave us $y = 1x^2 + bx + 0$. Only parameter b remains to be found. One additional fact will permit us to find b:

Each ordered pair in the table gives a fact about the function y = f(x).

EXAMPLE 3 Finding b, the coefficient of x

a. In $y = 1x^2 + bx + 0$, substitute for $x = 1$ and $y = f(1)$ and solve for b.

b. Restate the equation $y = ax^2 + bx + c$, substituting for a, b, and c.

SOLUTION **a.** From Table 10, we know that at $x = 1$, $y = f(1) = 1$.

$$y = 1x^2 + bx + 0 \qquad \text{Let } x = 1, y = 1.$$

$$1 = 1(1^2) + b(1) + 0 \qquad \text{Simplify.}$$

$$1 = 1 + b \qquad \text{Solve for } b.$$

$$0 = b$$

b. $y = 1x^2 + 0x + 0$ or $y = x^2$, as expected.

Check: Let $Y_1 = x^2$ in $\boxed{Y=}$. In [TBLSET], let **TblStart** $= 1$ and **ΔTbl** $= 1$. View [TABLE] to see the number pattern under Y_1. ∎

FINDING A QUADRATIC FUNCTION FROM A TABLE

> To find a, b, and c in $f(x) = ax^2 + bx + c$ when given a quadratic sequence:
>
> **1.** List the sequence in a table and calculate the first and second differences.
>
> **2.** Find a, given that the constant second difference is $2a$.
>
> **3.** Find c by working backwards from the second difference row to the $f(0)$ term matching $x = 0$. This is also the vertical axis or y-intercept.
>
> **4.** Write $y = ax^2 + bx + c$. Find $f(1)$ from the table. Substitute for a, c, $x = 1$, and $y = f(1)$. Solve for b.
>
> **5.** Write the function $f(x) = ax^2 + bx + c$, substituting for a, b, and c.
>
> To check, place the function in Y_1 under $\boxed{Y=}$ on the calculator. Under [TBLSET], let **TblStart** $= 1$ and **ΔTbl** $= 1$. View [TABLE] to see the sequence under Y_1.

EXAMPLE 4 Finding quadratic functions from a table   Use an input-output table and differences to find the quadratic function for the sequence 6, 21, 40, 63, 90, ...

SOLUTION *Step 1:*   Table 11 shows that 4 is the constant second difference. The sequence can be described with a quadratic function.

TABLE 11

Input	Output	First Difference	Second Difference
0	$-5 = c$		
		11	
1	6		+4
		15	
2	21		$+4 = 2a$
		19	
3	40		+4
		23	
4	63		+4
		27	
5	90		

Step 2: Because $2a = 4$, the common second difference, we know that $a = 2$.

Step 3: Working backwards from the second differences to $f(0)$ gives $c = -5$.

Step 4: From the table, we know that at $x = 1, f(1) = 6$. Thus, for $x = 1, y = 6$.

$$y = ax^2 + bx + c \qquad \text{Let } a = 2, c = -5, x = 1, y = 6.$$

$$6 = 2(1^2) + b(1) - 5 \qquad \text{Solve for } b.$$

$$b = 9$$

Step 5: The quadratic function is $f(x) = 2x^2 + 9x - 5$.

Check: Let $Y_1 = 2x^2 + 9x - 5$. In [TBLSET], let **TblStart** = 1 and **ΔTbl** = 1. View [TABLE] to see the sequence under Y_1. ▬

EXAMPLE 5   **Finding quadratic functions from a table**   Use an input-output table and differences to find the quadratic function for the sequence 4, 2, −2, −8, −16, ...

SOLUTION   *Step 1:* Table 12 shows that −2 is the constant second difference. The sequence can be described with a quadratic function.

TABLE 11

Input	Output	First Difference	Second Difference
0	$4 = c$		
		0	
1	4		−2
		−2	
2	2		$-2 = 2a$
		−4	
3	−2		−2
		−6	
4	−8		−2
		−8	
5	−16		

Step 2: Because $2a = -2$, the common second difference, we know that $a = -1$.

Step 3: Working backwards from the second differences to $f(0)$ gives $c = 4$.

Step 4: From the table, we know that at $x = 1, f(1) = 4$. Thus, for $x = 1, y = 4$.

$$y = ax^2 + bx + c \qquad \text{Let } a = -1, c = 4, x = 1, y = 4.$$

$$4 = -1(1^2) + b(1) + 4 \qquad \text{Solve for } b.$$

$$b = 1$$

Step 5: The quadratic function is $f(x) = -1x^2 + 1x + 4$.

Check: Let $Y_1 = -1x^2 + 1x + 4$. In [TBLSET], let **TblStart** = 1 and **ΔTbl** = 1. View [TABLE] to see the sequence under Y_1. ▬

Think about it 1: Instead of using $f(1) = 4$ in Example 5, use $f(2) = 2$ to solve for b. Next use $f(3) = -2$ to solve for b. Do you still get $b = 1$? Why is $f(1)$ easiest to use?

Finding a Quadratic Function with a Calculator

We now consider a calculator method for finding a quadratic function. The calculator method may be used on approximate or experimental data. We first enter the data into the statistical lists on a graphing calculator. Then we choose the calculator option for **quadratic regression**, or *fitting a quadratic function*.

GRAPHING CALCULATOR TECHNIQUE:
FITTING A QUADRATIC FUNCTION

> Enter the data into lists L1 and L2 under (STAT) EDIT.
>
> Find the calculation of quadratic regression, under (STAT) [CALC] **5 : QuadReg.**
>
> Press 5 (ENTER) to run the regression.
>
> Record the values of a, b, c, and R^2 for $y = ax^2 + bx + c$.

There is a unique (single) straight line that passes through every two points. There is a unique parabola that passes through any set of three points.[1] Calculator quadratic regression fits a quadratic function *exactly* to three points, whether or not the result is meaningful or appropriate in the problem setting. The $R^2 = 1$ listed after parameters a, b, and c in QuadReg indicates the exact fit.

EXAMPLE 6

Fitting a function with a calculator: pricing products Suppose you are hired as sales manager for a discount store. The previous manager left no information, and you want to re-create pricing rules. The store sells 3.5 ounces of Pond's cold cream for $3.99, 6.1 ounces for $5.49, and 9.5 ounces for $6.39.
a. Place the size and cost data into L1 and L2.
b. Graph the data. Suggest why a linear or quadratic function might be more suitable.
c. Use linear regression to fit a linear function to the data.
d. Use quadratic regression to fit a quadratic function to the data.
e. Graph each regression equation with the data.
f. Use each regression equation to find a cost for a 12-ounce size. Discuss your results.

Student Note: The regression equation is a function, but we will use the word *equation* to conform to calculator terminology.

SOLUTION **a.** The data appear in Figure 10a.

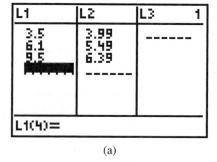
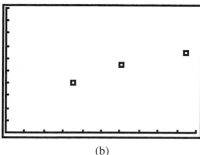

(a) (b)

FIGURE 10 In (b), 0 to 10 on X, 0 to 10 on Y

---

[1]If two points have the same input—for example, $(0, 3)$ and $(0, 5)$—the parabola will be "sideways" and not a function. The equation cannot be found with methods of this section.

b. The graph of the data is in Figure 10b. A linear function seems more suitable, because as the size gets larger, the price should increase. The data appear to have a slightly curved shape.

c. $a = 0.393$, $b = 2.786$, $r = 0.9759$; $y = 0.393x + 2.786$

d. $a = -0.052$, $b = 1.076$, $c = 0.860$, $R^2 = 1$; $y = -0.052x^2 + 1.076x + 0.860$

e. The regression equations appear in Figure 11a. Their graphs are in Figure 11b.

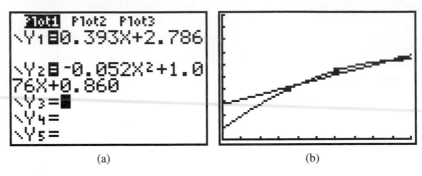

(a)   (b)

FIGURE 11   In (b), 0 to 10 on X, 0 to 10 on Y

f. The cost for the 12-ounce size would be $7.50 from the linear regression equation and $6.28 from the quadratic equation. Because the parabola's vertex is near 10.3 ounces, any larger size would be priced lower than $6.39. This is not reasonable.

Think about it 2: What is the relevant domain for the quadratic regression equation used to estimate cost in Example 6? For the linear equation?

In Example 7, we have more than three data points, so the quadratic regression is not necessarily an exact fit.

EXAMPLE 7   Fitting a function with a calculator: pricing products   A discount store sells a 6-inch-deep Duraco planter fir $2.69, an 8-inch planter for $3.97, a 10-inch planter for $5.99, and a 12-inch planter for $8.97.

a. Place the size and cost data into L1 and L2.
b. Graph the data. How does the graph suggest whether a linear or quadratic function might be more suitable?
c. Use linear regression to fit a linear function to the data.
d. Use quadratic regression to fit a quadratic function to the data.
e. Graph each regression equation with the data.
f. Use each regression equation to find costs for 14- and 16-inch pots. Discuss your results. Assume the diameter equals the depth.

SOLUTION   **a.** The data appear in Figure 12a.

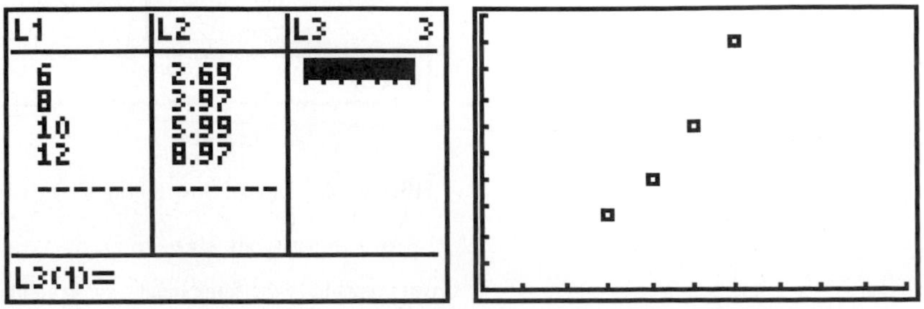

(a)   (b)

FIGURE 12   In (b), 0 to 20 on X, 0 to 10 on Y

b. The graph of the data is in Figure 12b. A quadratic function seems more suitable, because the data form a slight curve.

c. $a = 1.043$, $b = -3.982$, $r = 0.9837$; $y = 1.043x - 3.982$

d. $a = 0.1063$, $b = -0.8695$, $c = 4.093$, $R^2 = 0.9999$;
$y = 0.1063x^2 - 0.8695x + 4.093$

e. In Figures 13a and 13b, which show the regression equations and their calculator graphs, the linear regression equation is in Y_1 and the quadratic regression equation in Y_2.

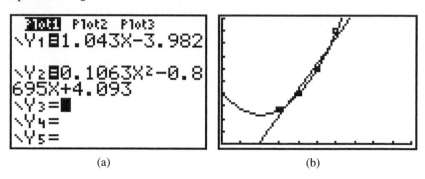

(a)   (b)

FIGURE 13   In (b), 0 to 20 on X, 0 to 10 on Y

f. The linear regression equation sets the 14-inch planter at $10.62 and the 16-inch planter at $12.71. The quadratic regression equation sets the 14-inch planter at $12.75 and the 16-inch planter at $17.38. The amount of plastic in the planter increases much faster than the radius. If the cost is based on the amount of plastic used, the quadratic equation is more reasonable. There is a close quadratic fit, with R^2 almost equal to 1.   ▬

■ RETRIEVING AND EVALUATING
REGRESSION EQUATIONS

To retrieve a regression equation in [Y =], place the cursor beside Y_1, press [VARS] for variables, choose **5** for Statistics, press the right arrow twice to highlight **EQ** for equation, and press [ENTER] to choose the first item, **RegEQ** (regression equation).

To graph a second regression equation from [Y =], place the cursor beside Y_2 and repeat the above steps.

To evaluate the regression equations for additional inputs, press [TBLSET] and move the cursor to highlight **Ask** beside **Indpnt** (the independent variable). Press [ENTER]. Select [TABLE] and place the new inputs under X. New outputs will appear under Y_1 and Y_2.

Application: Reverse Engineering

Parabolas—which are the curves formed when quadratic functions are graphed—describe both natural and structural shapes, from the flow of water from a drinking fountain or fire hose to bridge arches and suspension bridge cables. In Example 8, we apply **reverse engineering**, *taking measurements of structures and fitting equations*. The example and related exercises provide practice in writing ordered pairs for key points on a figure, as well as in using the calculator to find quadratic equations.

EXAMPLE 8

Student Note: We fit a regression equation here to find quadratic functions.

Finding ordered pairs and equations   A parabola is formed by the water flow from a drinking fountain, as shown in Figure 14.

a. Place the origin at point A. Find the ordered pairs for the lettered points B and C. Fit an equation through the points A, B, and C.

b. Place the origin at point B. Find the ordered pairs for the lettered points A and C. Fit an equation through the points A, B, and C.

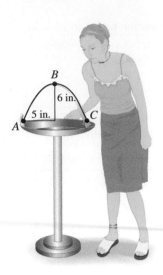

FIGURE 14

c. How are the equations in parts a and b the same? Different?

d. Repeat the process for an origin at point C.

SOLUTION **a.** From $A(0, 0)$, we move right 5 units and up 6 units to B. B is $(5, 6)$, as shown in Figure 15a. From A, we move right 10 units to C. C is $(10, 0)$. We place the three ordered pairs under $\boxed{\text{STAT}}$ EDIT so that the x-value is in L1 and the y-value is in L2. When we calculate a quadratic regression on L1 and L2, we get $y = -0.24x^2 + 2.4x$.

b. From $B(0, 0)$, we move left 5 units and down 6 units to A. A is $(-5, -6)$, as shown in Figure 15b. From B, we move right 5 units and down 6 units. C is $(5, -6)$. Repeating the steps in part a, we have $y = -0.24x^2$.

c. Both equations have $a = -0.24$, or $a = -6/25$, and no constant term. The first equation has a coefficient of 2.4 on x.

d. If we start from $C(0, 0)$, A is $(-10, 0)$ and B is $(-5, 6)$, as shown in Figure 15c. The equation is $y = -0.24x^2 - 2.4x$. The equation is the same as in part a except for the opposite sign on the coefficient of x.

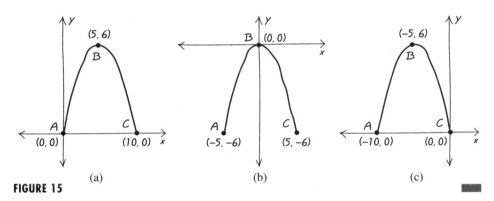

(a)   (b)   (c)

FIGURE 15

ANSWER BOX

Warm-up: **1.** See Example 1. **2.** See Example 2. **3.** c **4.** y-intercept **5.** $a + b + c$ **Think about it 1:** From the table, at $x = 2$, $f(2) = 2$. In $y = ax^2 + bx + c$, let $a = -1$, $c = 4$, $x = 2$, $y = 2$. Check that $2 = -1(2^2) + b(2) + 4$ solves to $b = 1$. From the table, at $x = 3$, $f(3) = -2$. In $y = ax^2 + bx + c$, let $a = -1$, $c = 4$, $x = 3$, $y = -2$. Check that $-2 = -1(3^2) + b(3) + 4$ solves to $b = 1$. Multiplying by 1 or 1^2 is easiest. **Think about it 2:** The values for which the quadratic function is sensible might be as low as 1 or 2 ounces and as high as the vertex, or about 10.3 ounces. For the line, much higher inputs would be sensible; the upper limit on input depends on how much

---

4.2   **Exercises**

In Exercises 1 to 14, use first and second differences to find out whether each sequence may be described with a linear function, a quadratic function, or neither. Use the table method to fit a linear or quadratic equation. Check with calculator regression.

1. 7, 16, 27, 40, 55, . . .
    quadratic; $y = x^2 + 6x$
3. 4, 8, 12, 16, 20, . . .
    linear; $y = 4x$
5. 8, 27, 50, 77, 108, . . .
    quadratic; $y = 2x^2 + 13x - 7$

2. 1, 8, 27, 64, 125, . . .
    neither
4. 4, 12, 24, 40, 60, . . .
    quadratic; $y = 2x^2 + 2x$
6. 12, 23, 34, 45, 56, . . .
    linear; $y = 11x + 1$

7. 5, 12, 21, 32, 45, . . .
    quadratic; $y = x^2 + 4x$
9. 13, 25, 37, 49, 61, . . .
    linear; $y = 12x + 1$
11. 1, 1, 2, 3, 5, 8, . . . neither

12. 26, 35, 44, 53, 62, . . . linear; $y = 9x + 17$

13. 9, 16, 21, 24, 25, . . . quadratic; $y = -x^2 + 10x$

14. 3, 4, 7, 11, 18, . . . neither

8. 7, 14, 21, 28, 35, . . .
    linear; $y = 7x$
10. 36, 44, 50, 54, 56, . . .
    quadratic; $y = -x^2 + 11x + 26$

The number patterns in Exercises 15 to 18 appear on adjacent circles on the *Science News* (April 21, 2001) cover design shown in the figure. Use the table method to fit a quadratic function to the pattern, and give the next three numbers.

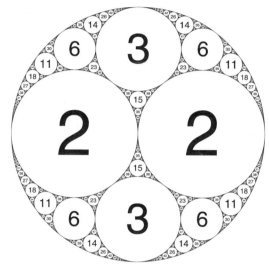

Courtesy of Allan R. Wilks, AT&T Labs–Research

15. 2, 3, 6, 11, 18, 27, . . . $y = x^2 - 2x + 3$; 38, 51, 66

16. 2, 6, 14, 26, 42, 62, . . . $y = 2x^2 - 2x + 2$; 86, 114, 146

17. 3, 15, 35, 63, 99, . . . $y = 4x^2 - 1$; 143, 195, 255

18. 6, 23, 50, 87, . . . $y = 5x^2 + 2x - 1$; 134, 191, 258

In Exercises 19 to 22, place the size and cost data in L1 and L2 under ⬚STAT⬚ EDIT and graph the data.
(a) How does the graph suggest a linear or a quadratic function?
(b) Use linear regression to fit an equation to the data. State the equation.
(c) Use quadratic regression to fit an equation to the data. State the equation.
(d) Use each regression equation to find an output for the indicated input. Discuss your results.
(Hint: Graph each regression equation with the data. Note the position of the points relative to each graph.)

19. Noxzema cream: 2.5 ounces for $1.57, 10.75 ounces for $3.29, and 14 ounces for $4.49. Estimate the cost of 16 ounces. $4.81, $5.38; see Answer Section.

20. Pyrex glass storage bowls, with volume in cups: $1\frac{2}{3}$-cup size for $3.99, 5-cup size for $5.99, 12-cup size for $7.99. Estimate the cost of an 8-cup bowl.
$6.65, $7.21; see Additional Answers.

21. AKRO MILS plastic indoor planting pots: 6-inch depth for $0.99, 8-inch depth for $1.69, 10-inch depth for $2.69, and 12-inch depth for $3.99. Estimate the cost for a 16-inch depth. $5.84; $7.49; see Answer Section.

22. Durable 3-ring notebooks: 1-inch width for $3.29, 1.5-inch width for $3.99, 2-inch width for $4.99, and 3-inch width for $6.99. Estimate the price for a $\frac{1}{2}$-inch width and a 4-inch width.
$2.23, $2.57; $8.81, $9.44; see Additional Answers.

23. When a small rock is dropped from the top of an abandoned mine shaft, time after release and distance from bottom are as follows: 1 sec, 560 ft; 2 sec, 512 ft; 3 sec, 432 ft; 4 sec, 320 ft.

a. Graph on a calculator. Fit both a linear and a quadratic regression equation.
$y = -80x + 656, y = -16x^2 + 576$
b. What are the meanings of the slope and intercept in the linear equation?
drop of 80 ft per sec from starting height of 656 ft
c. What is the speed of the rock when first dropped? What does this say about the distance traveled in the first second compared with later seconds? 0 ft per sec. It cannot travel as far in the first second as in later seconds; graph is not linear.
d. From the quadratic equation, what is the meaning of the constant term? starting height, 576 ft
e. Based on the quadratic equation, in how many seconds will the rock hit the bottom? 6 sec

24. Graph the data from the table below on a calculator. Explain why quadratic regression may be appropriate. Fit an equation. What common formula generates the data? Points seem to follow parabolic shape;
$y \approx 3.14x^2 - 0.007x + 0.012$ circle area $= \pi r^2$

Input	Output
2	12.57
3	28.27
4	50.27
5	78.54

Exercises 25 to 28 apply reverse engineering. Use calculator quadratic regression to find all equations. Pay close attention to how equations are alike or different.

25. See the concrete arch bridge in the figure.

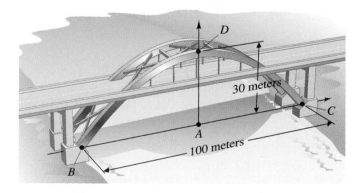

a. Place the origin at *A*. Find the ordered pairs for the lettered points on the arch (*B*, *C*, and *D*). Fit an equation through the arch (*B*, *C*, and *D*).
$B(-50, 0), C(50, 0), D(0, 30); y = -0.012x^2 + 30$
b. Place the origin at *B*, *C*, and *D*, respectively. For each origin, find the ordered pairs for the remaining lettered points on the arch. Fit an equation through the arch. See Answer Section.

26. See the suspension bridge in the figure.

a. Place the origin at *A*. Find the ordered pairs for the lettered points on the cable (*B*, *C*, and *D*). Fit an equation to the cable (*B*, *C*, and *D*).
B(0, 20), *C*(25, 0), *D*(50, 20); $y = 0.032x^2 - 1.6x + 20$

b. Place the origin at *B*, *C*, *D*, and *E*, respectively. For each origin, find the ordered pairs for the remaining lettered points on the cable. Fit an equation to the cable. See Additional Answers.

27. See the parabolic water spray from the fireboat in the figure. Place the origin at point *A*. Find the ordered pairs for the remaining lettered points on the water spray. Fit an equation to the water spray. Repeat for each other lettered point in the figure. See Answer Section.

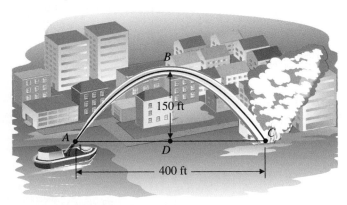

28. See the parabolic archway in the figure. Place the origin at point *A*. Find the ordered pairs for the remaining lettered points on the archway. Fit an equation to the archway. Repeat for each other lettered point in the figure.
See Additional Answers.

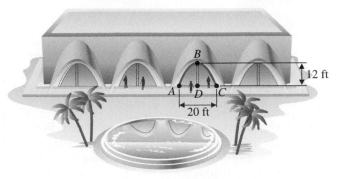

29. Algebraic Differences

a. First consider a quadratic function: Copy the table. Add two more columns for first and second differences. Calculate the first and second differences and write them in the two columns.

x	$f(x) = ax^2 + bx + c$	1st	2nd
1	$a + b + c$		
2	$4a + 2b + c$	$⟩ 3a + b$	$⟩ 2a$
3	$9a + 3b + c$	$⟩ 5a + b$	$⟩ 2a$
4	$16a + 4b + c$	$⟩ 7a + b$	$⟩ 2a$
5	$25a + 5b + c$	$⟩ 9a + b$	$⟩ 2a$

b. Are the second differences constant? What fact have we proved? Yes; the second difference, $2a$, is constant.

c. Is the lead term of the row of first differences $3a + b$? How might this help you write the function?
Yes; could use this to find b.

d. Next consider a linear function. Make a table with inputs 1, 2, 3, 4, and 5 and outputs

$$f(x) = ax + b$$

e. Add another column to the table for first differences. Calculate the first differences and write them in this column. Explain how the results may be used to write a linear function. See Answer Section.

■ Projects

30. Multiplication Table   This project may be started now and finished after the review of factoring in Sections 4.3 and 4.4. Make two photocopies of the following multiplication table. On the first photocopy, do parts a to e.

1	2	3	4	5	6	7	8	9	10	11	12
2	4	6	8	10	12	14	16	18	20	22	24
3	6	9	12	15	18	21	24	27	30	33	36
4	8	12	16	20	24	28	32	36	40	44	48
5	10	15	20	25	30	35	40	45	50	55	60
6	12	18	24	30	36	42	48	54	60	66	72
7	14	21	28	35	42	49	56	63	70	77	84
8	16	24	32	40	48	56	64	72	80	88	96
9	18	27	36	45	54	63	72	81	90	99	108
10	20	30	40	50	60	70	80	90	100	110	120
11	22	33	44	55	66	77	88	99	110	121	132
12	24	36	48	60	72	84	96	108	120	132	144

a. Using two different colors, shade the squares containing the numbers in the sequences for x^2 and $x(x + 1)$.
center diagonal, upper left to lower right; diagonal just above center

b. Use regression to find the rules for the sequences starting with the second and third numbers in the x^2 diagonal (4, 9, 16, 25, 36, . . . and 9, 16, 25, 36, 49, . . .). Predict the rule for 49, 64, 81, 100, 121, ... $(x + 1)^2, (x + 2)^2, (x + 6)^2$

c. Shade the squares and use regression to find the rules for 3, 8, 15, 24, 35, . . . and 4, 10, 18, 28, 40, ... Factor the rules. Predict the rules for sequences starting with the second number in each set.
$(x + 2)x, (x + 3)x; (x + 3)(x + 1), (x + 4)(x + 1)$

d. Organize the rules in parts a, b, and c so that you can predict the rule for a sequence on any diagonal starting in the first, second, or third row.
See Additional Answers.

e. Consider diagonals going right to left. Shade and find rules in factored form for 12, 22, 30, 36, 40, . . . and 11, 20, 27, 32, 35, ... Predict and find the rule for a sequence on an adjacent diagonal. How do the rules differ from those in parts a to d? $x(13 - x), x(12 - x)$; contain $-x^2$ or one negative coefficient on x in factored form

On the second photocopy, do parts f to h.

f. In chess, a knight moves two squares forward and one square to the side. Shade 1, 6, 15, 28, 45, . . . and 2, 8, 18, 32, 50, ... The numbers appear in a knight's sequence from left to right. What indicates that the sequences are quadratic? Find the rules and write in factored form.
constant second differences; $x(2x - 1), 2x^2$

g. These knight's sequences are adjacent to those in part f: 2, 9, 20, 35, 54, . . . ; 4, 12, 24, 40, 60, . . . ; and 3,

12, 25, 42, 63, ... Predict the rules before using regression. How do the rules in parts f and g differ from those in parts a to d?
$(2x - 1)(x + 1), 2x(x + 1), (2x - 1)(x + 2)$; 2 appears on x in one factor.

h. Explore the rules for three sequences of another type. Your exploration could extend the table to negative products.

31. Systems of Equations As mentioned in the text, three ordered pairs with different inputs can be made into exactly one quadratic function. Choose two sets of three ordered pairs and find the quadratic function that fits each set. Use the process described in steps a to d.

a. Choose three arbitrary ordered pairs with different inputs—say, (0, 1), (1, 5), and (4, 3).

b. Translate the ordered pairs into function notation: $f(0) = 1, f(1) = 5$, and $f(4) = 3$.

c. Substitute each ordered pair into $f(x) = ax^2 + bx + c$.

$1 = a(0^2) + b(0) + c$	Simplifies to $1 = c$.
$5 = a(1^2) + b(1) + c$	Simplifies to $5 = a + b + c$.
$3 = a(4^2) + b(4) + c$	Simplifies to $3 = 16a + 4b + c$.

d. Solve the system of equations for parameters a, b, and c. In this case, $a = -1\frac{1}{6}, b = 5\frac{1}{6}$, and $c = 1$ or $y = -\frac{7}{6}x^2 + \frac{31}{6}x + 1$. Check by entering the function in Y_1 and observing the ordered pairs in [TABLE].
Students can check their solution with quadratic regression on a calculator.

4.3	**Polynomial Functions and Operations**

Objectives

▮ Identify a polynomial and types of polynomials.

▮ Add and subtract polynomials.

▮ Multiply monomials, binomials, and trinomials.

▮ Factor trinomials.

▮ Find common monomial factors.

WARM-UP

List all the pairs of positive integer factors of these numbers.
Example: $28 = 1 \cdot 28 = 2 \cdot 14 = 4 \cdot 7$

1. 60 $1 \cdot 60, 2 \cdot 30, 3 \cdot 20, 4 \cdot 15, 5 \cdot 12, 6 \cdot 10$

2. 12 $1 \cdot 12, 2 \cdot 6, 3 \cdot 4$

3. 24 $1 \cdot 24, 2 \cdot 12, 3 \cdot 8, 4 \cdot 6$

4. 80 $1 \cdot 80, 2 \cdot 40, 4 \cdot 20, 5 \cdot 16, 8 \cdot 10$

5. 100 $1 \cdot 100, 2 \cdot 50, 4 \cdot 25, 5 \cdot 20, 10 \cdot 10$

6. 180 $1 \cdot 180, 2 \cdot 90, 3 \cdot 60, 4 \cdot 45, 5 \cdot 36, 6 \cdot 30, 9 \cdot 20, 10 \cdot 18, 12 \cdot 15$

IN THIS SECTION, we review polynomials and their addition, subtraction, multiplication, and factorization. The multiplying and factoring are presented in table form. The table method is likely to be new to you.

Polynomial Functions

The factored polynomial equations will be studied in Section 4.5.

Lindsey has designed a parabolic arched entry for the front of a museum (Figure 16). She needs to calculate its equation so that she can write specifications for the height at uniform intervals across the entry. To create the equation, she uses a basic equation form for a parabola, $y = a(x - x_1)(x - x_2)$. The variables x_1 and x_2 are x-intercepts.

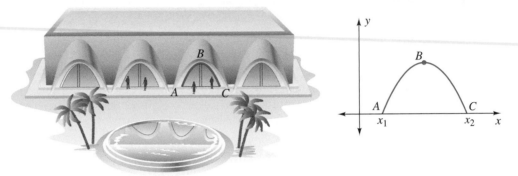

FIGURE 16

The equation $y = a(x - x_1)(x - x_2)$ is an example of a polynomial equation in factored form. The right side of the equation contains three factors a, $(x - x_1)$, and $(x - x_2)$, hence the name *factored form*. After numbers are placed in the equation for x_1 and x_2, the factors in a polynomial are usually multiplied out, leaving a sum of terms. A polynomial is defined by the nature of these terms.

DEFINITION OF POLYNOMIAL FUNCTION

> A **polynomial** is an expression written as a sum of terms. The terms in a polynomial are limited to a number or a product of a number and one or more variables having non-negative integers as exponents. The general **polynomial function** is of the form
>
> $$f(x) = a_n x^n + a_{n-1} x^{n-1} + a_{n-2} x^{n-2} + \cdots + a_1 x^1 + a_0$$
>
> The degree is n, $n \geq 0$.

Recall that non-negative means zero or positive. Polynomials include expressions such as $x^2 - y^2$, πr^2, 5, and $x^3 + 8$.

All polynomials have a finite number of terms. Common polynomials are named for the number of terms they contain. *One-, two-,* and *three-term polynomial expressions* are called **monomials**, **binomials**, and **trinomials**, respectively.

EXAMPLE 1   **Naming polynomials**   Identify the expression on the right side of each function as a monomial, a binomial, a trinomial, or not a polynomial and give its degree.
a. Bridge-cable equation: $y = \frac{4}{125} x^2$
b. Square root equation: $y = \sqrt{x}$
c. Perimeter of a rectangle formula: $P = 2l + 2w$
d. Vertical motion equation: $h = -\frac{1}{2} gt^2 + v_0 t + h_0$, where g, v_0, and h_0 are constants
e. Absolute value function: $y = |x|$
f. $y = x^{-2} + 4x^{-1} + 4$

SOLUTION

a. $\frac{4}{125}x^2$ is a monomial; the x is raised to the exponent 2, a positive integer. Its degree is 2.

b. $\sqrt{x}$ is not a polynomial; neither zero nor a positive integer exponent will give the square root of x.

c. $2l + 2w$ is a binomial in two variables; the l and the w are both raised to the exponent 1. Its degree is 1.

d. $-\frac{1}{2}gt^2 + v_0 t + h_0$ is a trinomial in the variable t. Its degree is 2.

e. The absolute value of x, $|x|$, is not a polynomial; it cannot be created with a zero or positive exponent.

f. $x^{-2} + 4x^{-1} + 4$ has negative exponents and thus is not a polynomial. ▬

Addition and Subtraction of Polynomials

We add and subtract polynomials by combining like terms. As mentioned in Chapter 1, *like terms* are expressions with identical variable factors:

$$3x^2, \tfrac{1}{2}x^2, \text{ and } 0.25x^2 \text{ are like terms.}$$

$$-2x^2y, -\tfrac{1}{4}x^2y, \text{ and } 6.3x^2y \text{ are like terms.}$$

EXAMPLE 2 Adding like terms   Add like terms in these polynomial expressions.

a. $4x + 3y - 6x + 2y$ **b.** $3x^2 - \frac{1}{2}x^2 + 0.25x^2$

c. $4xy - xy + 2.5xy$ **d.** $-2x^2y - \frac{1}{4}x^2y + 6.3x^2y$

SOLUTION

a. $4x + 3y - 6x + 2y = -2x + 5y$

b. $3x^2 - \frac{1}{2}x^2 + 0.25x^2 = 3x^2 - 0.5x^2 + 0.25x^2 = 2.75x^2$

c. Because $-xy$ means $-1xy$, we may write

$$4xy - 1xy + 2.5xy = 5.5xy$$

d. $-2x^2y - \frac{1}{4}x^2y + 6.3x^2y = -2x^2y - 0.25x^2y + 6.3x^2y = 4.05x^2y$ ▬

▬ In Example 3, we apply the distributive property, $a(b + c) = ab + ac$, to multiply out expressions before adding like terms.

EXAMPLE 3 Multiplying and adding like terms   Add or subtract these expressions, as indicated.

a. $x(x^2 + 6x + 9) + 3(x^2 + 6x + 9)$

b. $3x(2x + 3) - 2(2x + 3)$

c. $x(x^2 - 2xy + y^2) - y(x^2 - 2xy + y^2)$

SOLUTION   We multiply with the distributive property and then rearrange the terms to place like terms together.

a. $x(x^2 + 6x + 9) + 3(x^2 + 6x + 9) = x^3 + 6x^2 + 9x + 3x^2 + 18x + 27$
$$= x^3 + 6x^2 + 3x^2 + 9x + 18x + 27$$
$$= x^3 + 9x^2 + 27x + 27$$

b. $3x(2x + 3) - 2(2x + 3) = 6x^2 + 9x - 4x - 6$
$$= 6x^2 + 5x - 6$$

c. $x(x^2 - 2xy + y^2) - y(x^2 - 2xy + y^2)$
$$= x^3 - 2x^2y + xy^2 - x^2y + 2xy^2 - y^3$$
$$= x^3 - 2x^2y - x^2y + xy^2 + 2xy^2 - y^3$$
$$= x^3 - 3x^2y + 3xy^2 - y^3$$ ▬

Note that Example 3 uses these exponent facts:

$$x \cdot x = x^2 \qquad \text{and} \qquad x^2 \cdot x = x^3$$

Also note that the answers are written in *descending order of exponents*. When there are two variables, the terms are arranged in descending order of exponents on the variable occurring first in the alphabet.

Multiplication of Polynomials

Student Note: Know how to multiply? Be patient. Look for patterns.

The algebraic capabilities of newer calculators allow us to leave much of the tedious work to the calculator, once we understand the structure of multiplying (and factoring). Because of this technological development, the approach to multiplication (and factorization) presented here is based on tables.

To multiply a binomial times another binomial or a trinomial, use the distributive property twice.

◼ MULTIPLICATION OF TWO BINOMIALS

> To multiply the binomials $(ax + b)$ $(cx + d)$, add the product $ax(cx + d)$ to the product $b(cx + d)$.

The table for multiplication leads to a visual version of factoring by grouping.

A table organizes the products so that all the terms and patterns are easy to see.

Multiply	cx	$+d$	
ax	acx^2	$+adx$	This row shows $ax(cx + d)$.
$+b$	$+bcx$	$+bd$	This row shows $b(cx + d)$.

The product is $(ax + b)(cx + d) = acx^2 + adx + bcx + bd$.

◼ In Figure 16, suppose the x-intercepts of the parabolic museum arch are $x_1 = 10$ and $x_2 = 30$. The equation of the parabola becomes

$$y = a(x - x_1)(x - x_2) = a(x - 10)(x - 30)$$

Example 4 shows that the equation of the parabola may also be written in the form $y = a(x^2 - 40x + 300)$.

EXAMPLE 4   **Multiplying binomials**   Multiply $(x - 10)(x - 30)$ with a table.

Multiply	x	-10
x		
-30		

SOLUTION   The table shows that the distributive property is used twice, once for $x(x - 10)$ and once for $-30(x - 10)$. The entries within the white region are the products.

Multiply	x	-10
x	x^2	$-10x$
-30	$-30x$	$+300$

The product is
$$(x - 10)(x - 30)$$
$$= x^2 + ((-10x) + (-30x)) + 300$$
$$= x^2 - 40x + 300 \qquad ◼$$

EXAMPLE 5 Multiplying binomials   Multiply $(2x + 3)(3x - 2)$ in a table.

SOLUTION   We place the binomials on the left and across the top of the table:

Multiply	$3x$	-2
$2x$	$6x^2$	$-4x$
$+3$	$+9x$	-6

The product is
$$(2x + 3)(3x - 2)$$
$$= 6x^2 + ((-4x) + 9x) - 6$$
$$= 6x^2 + 5x - 6$$

∎ A common mental shortcut for multiplying binomials $(ax + b)(cx + d)$ is the FOIL method: take the product of the "first terms," $ax \cdot cx$; add the product of the "outside terms," $ax \cdot d$; add the product of the "inside terms," $b \cdot cx$; and then add the product of the "last terms," $b \cdot d$. The disadvantage of the FOIL method is that it is limited to the product of binomials. The table method may be used for any product of polynomials; see Example 6 for a product of a binomial and a trinomial.

EXAMPLE 6 Multiplying a binomial and a trinomial   Multiply $(x + 3)(x^2 + 6x + 9)$ in a table.

SOLUTION   Although the positions do not matter, in this text we will place the binomial on the left and the trinomial across the top.

Multiply	x^2	$+6x$	$+9$
x	x^3	$+6x^2$	$+9x$
$+3$	$+3x^2$	$+18x$	$+27$

The product is
$$(x + 3)(x^2 + 6x + 9)$$
$$= x^3 + (6x^2 + 3x^2)$$
$$+ (9x + 18x) + 27$$
$$= x^3 + 9x^2 + 27x + 27$$

The middle step in the solution on the right is not needed. The table groups like terms. We usually write the answer, $x^3 + 9x^2 + 27x + 27$, directly from the table.

EXAMPLE 7 Multiplying two trinomials   Multiply $(x^2 - 2x - 1)(x^2 + 2x - 1)$ in a table.

Multiply	x^2	$+2x$	-1
x^2			
$-2x$			
-1			

SOLUTION   The table, which has nine entries, shows a clear organization of like terms, four of which add to zero.

Multiply	x^2	$+2x$	-1
x^2	x^4	$+2x^3$	$-x^2$
$-2x$	$-2x^3$	$-4x^2$	$+2x$
-1	$-x^2$	$-2x$	$+1$

The product is
$$(x^2 - 2x - 1)(x^2 + 2x - 1)$$
$$= x^4 - 6x^2 + 1$$

EXAMPLE 8 Exploration   As mentioned, the reason for using table multiplication is to identify patterns. Stop now and examine the tables in Examples 4 and 5. Consider these questions:

a. Where in the table is the product of the first terms?

b. Where are like terms located?

c. Where is the product of the last terms?

d. What do the products of the terms in the diagonals have in common?

Do the patterns extend to the tables in Examples 6 and 7?

SOLUTION   When the terms being multiplied are arranged in descending order of the exponent on x, the following patterns appear:

a. The first entry (upper left corner) in the white part of each table is the product of the first terms.

b. Like terms are on a diagonal.

c. The last entry (lower right corner) in the white part is the product of the last terms.

d. The product of the two terms on one diagonal is equal to the product of the two terms on the other diagonal. In Example 4,

$$300(x^2) = 300x^2 \quad \text{and} \quad (-10x)(-30x) = 300x^2$$

In Example 5,

$$6x^2(-6) = -36x^2 \quad \text{and} \quad (-4x)(+9x) = -36x^2$$

In Examples 6 and 7, like terms in adjacent columns are on a diagonal. In Example 6, the diagonal products for adjacent columns are equal. (Can you find a way to make the products equal in Example 7?)   ▬

The pattern in part d above—equal diagonal products—is a pattern that does not show up in other ways of multiplying polynomials. The equal diagonal products will be central to the table-based factoring process.

You will be asked to prove equal diagonal products for binomials in the exercises.

Factoring Polynomials with Tables

Student Note: If you can factor $2x^2 - 11x - 6$, continue to rely on your own method. Learn enough about the table method to try a few exercises, help other students, and complete the square in Section 5.3.

Factoring is the reverse operation to multiplication of polynomials. The basis for this reversible relationship is the *distributive property*, $a(b + c) = ab + ac$. Through factoring, the sum $ab + ac$ is changed back to the product $a(b + c)$. Trinomial factoring changes the sum of three terms back to a product $(ax + b)(cx + d)$. The table method provides a systematic way to factor trinomials.

Example 9 refers to the **greatest common factor**, *the largest number that divides evenly into two other numbers*. The greatest common factor of 16 and 24 is 4. The greatest common factor is abbreviated gcf.

EXAMPLE 9   Factoring a trinomial   Use the patterns from multiplication to enter information from $2x^2 - 11x - 6$ into a table. Use other patterns to build the white part of the table, and then factor to find the binomials on the left and the top of the table.

SOLUTION   We enter the first and last terms directly into the table.

Factor		
	$2x^2$	
		-6

sum $-11x$

product $-12x^2$

The sum of the other diagonal is $-11x$, and the product of the diagonal terms is $-12x^2$. Because the two products of the diagonals are equal, we need to look for terms that multiply to $-12x^2$ and add to $-11x$. Without signs, the possible terms are

$$12x \text{ and } 1x \qquad 6x \text{ and } 2x \qquad 4x \text{ and } 3x$$

Only $-12x$ and $1x$ add to $-11x$, so we can place these two terms *in any order* in the other diagonal of the table.

Factor		
	$2x^2$	$-12x$
	$+1x$	-6

Next, we complete the top row by removing the *greatest common factor* (gcf) from the first row of the white part of the table. We then use the top row to find the missing factor in the left column. We use the number in the bottom right corner as a check.

Factor	x	-6
2x	$2x^2$	$-12x$
+1	$+1x$	-6

Write the gcf here ⟶

The result is $2x^2 - 11x - 6 = (2x + 1)(x - 6)$. ▬

EXAMPLE 10 **Factoring a trinomial**   Factor $12x^2 - 8x - 15$.

SOLUTION We first enter $12x^2$ and -15 into the table.

Factor		
	$12x^2$	
		-15

sum $-8x$

product $-180x^2$

The product along the diagonal is $-180x^2$, which must be factored into a pair of terms adding to $-8x$. A list of all factors may not be needed but is included here for completeness:

$1x$ and $180x$ $2x$ and $90x$ $3x$ and $60x$ $4x$ and $45x$ $5x$ and $36x$

$6x$ and $30x$ $10x$ and $18x$ $9x$ and $20x$ $12x$ and $15x$

The factors $-18x$ and $10x$ give the required sum and are placed in the table.

Factor		
	$12x^2$	$-18x$
	$+10x$	-15

We identify the gcf of the first row of the white part of the table and use it to complete the top row. Next, we find the lower left factor. We then check with the lower right number.

Factor	$2x$	-3
6x	$12x^2$	$-18x$
+5	$+10x$	-15

Write the gcf here. ⟶

The result is $12x^2 - 8x - 15 = (6x + 5)(2x - 3)$. ▬

▬ Regardless of the factoring method selected, we must look for a **common monomial factor**, *an expression that divides each term evenly*, before factoring. The monomial $5x$ is the common monomial factor in $20x^2 - 15x$. *If there is a common monomial factor, it should be removed from an expression before any other factoring is done.*

Example 11 illustrates a factorization where the common monomial factor contains the variable x. You may recall a shortcut that could be used to factor the expression in Example 11—more on shortcuts in Section 4.4.

EXAMPLE 11 **Factoring** Factor $12x^3 - 75x$.

SOLUTION First we factor $3x$, the greatest common monomial factor, from each term:

$$12x^3 - 75x = 3x(4x^2 - 25)$$

The second expression, $4x^2 - 25$, may be factored again. We enter $4x^2$ and -25 into the table. This time, the sum of the diagonal terms is zero because the expression $4x^2 - 25$ is missing its middle term. The product of the diagonal terms is $-100x^2$, and its factors, $-10x$ and $10x$, add to zero. We enter them into the table and factor.

Factor	$2x$	$+5$
$2x$	$4x^2$	$+10x$
-5	$-10x$	-25

Thus,

$$12x^3 - 75x = 3x(4x^2 - 25) = 3x(2x - 5)(2x + 5)$$

Think about it: What is the same and what is different about a greatest common factor and a common monomial factor?

The factors in our answers to Examples 9 to 11 provide a complete factorization of the original expression. A polynomial expression is *completely factored* when all of the factors are prime. A **prime number** and a **prime factor** are *divisible only by 1 and the number or factor itself.*

SUMMARY: FACTORING BY TABLE

To factor $adx^2 + dbx + dc$:

1. First remove common monomial factor d so that $ax^2 + bx + c$ remains:

$$d(ax^2 + bx + c)$$

2. Enter ax^2 and c into one diagonal.

3. Multiply ax^2 and c to obtain the diagonal product.

4. Find factors of the diagonal product that add to bx.

5. Enter the factors into the other diagonal.

6. Factor out the greatest common factor, find the other factors, and write the binomial pair.

7. Multiply factors to check that the white "area" is $ax^2 + bx + c$.

Use the table method in the exercises until you understand the process. Afterward, use guess and check or whatever method seems appropriate.

ANSWER BOX

Warm-up: **1.** $1 \cdot 60, 2 \cdot 30, 3 \cdot 20, 4 \cdot 15, 5 \cdot 12, 6 \cdot 10$ **2.** $1 \cdot 12,$ $2 \cdot 6, 3 \cdot 4$ **3.** $1 \cdot 24, 2 \cdot 12, 3 \cdot 8, 4 \cdot 6$ **4.** $1 \cdot 80, 2 \cdot 40, 4 \cdot 20, 5 \cdot 16,$ $8 \cdot 10$ **5.** $1 \cdot 100, 2 \cdot 50, 4 \cdot 25, 5 \cdot 20, 10 \cdot 10$ **6.** $1 \cdot 180, 2 \cdot 90,$ $3 \cdot 60, 4 \cdot 45, 5 \cdot 36, 6 \cdot 30, 9 \cdot 20, 10 \cdot 18, 12 \cdot 15$ **Think about it:** The expressions have almost the same meaning. *Greatest common factor* is more general. A gcf can be numbers or variables, but it can also include square roots or negative powers of variables. For example, $\sqrt{y} + x\sqrt{y}$ has $\sqrt{y}$ as a common factor; $y^{-1} + xy^{-1}$ has y^{-1} as a common factor. Neither $\sqrt{y}$ nor y^{-1} is a monomial. A common monomial factor is a gcf that is also a polynomial expression. The distinction is made in this section to make the directions easier to read.

4.3 Exercises

List all the pairs of positive integer factors of the numbers in Exercises 1 and 2.

1. a. 48 $1 \cdot 48, 2 \cdot 24, 3 \cdot 16, 4 \cdot 12, 6 \cdot 8$

b. 36 $1 \cdot 36, 2 \cdot 18, 3 \cdot 12, 4 \cdot 9, 6 \cdot 6$

c. 72 $1 \cdot 72, 2 \cdot 36, 3 \cdot 24, 4 \cdot 18, 6 \cdot 12, 8 \cdot 9$

2. a. 27 $1 \cdot 27, 3 \cdot 9$

b. 90 $1 \cdot 90, 2 \cdot 45, 3 \cdot 30, 5 \cdot 18, 6 \cdot 15, 9 \cdot 10$

c. 20 $1 \cdot 20, 2 \cdot 10, 4 \cdot 5$

Identify each expression in Exercises 3 to 6 as monomial, binomial, trinomial, or not polynomial.

3. a. $\sqrt{y} + x\sqrt{y}$ not polynomial **b.** $x^2 + y^2$ binomial

4. a. $y^{-1} + xy^{-1}$ not polynomial **b.** $3a + 3b$ binomial

5. a. $2\pi r$ monomial **b.** $-\frac{1}{2}gt^2 + vt + h$ trinomial

6. a. $x^2y + xy^2 + y^3$ trinomial **b.** $|x^2 + y^2|$ not polynomial

Completing the tables in Exercises 7 and 8 will improve your ability to guess and check some sets of factors.

7.

m	n	$m + n$	$m \cdot n$
-3	-5	-8	15
3	4	7	12
2	6	8	12
3	5	8	15
-4	-6	-10	24
-2	-12	-14	24
-3	-8	-11	24
2	-6	-4	-12

8.

m	n	$m + n$	$m \cdot n$
1	15	16	15
4	6	10	24
3	8	11	24
2	12	14	24
-4	6	2	-24
2	-12	-10	-24
-3	8	5	-24
2	-10	-8	-20

In Exercises 9 to 18, add or subtract the polynomials. Multiply as needed.

9. a. $(x^3 + 2x^2 + 4x) - (2x^2 + 4x + 8)$ $x^3 - 8$

b. $(x^3 - 6x^2 + 9x) - (3x^2 - 18x + 27)$
$x^3 - 9x^2 + 27x - 27$

10. a. $(x + 6) + (3x - 2)$ $4x + 4$

b. $(x - 4) + (2x + 3)$ $3x - 1$

11. a. $(x - 6) + (2x + 5)$ $3x - 1$

b. $(3x - 1) + (4x - 5)$ $7x - 6$

12. a. $4x^2 + 2x - 3 - 3(x^2 - 2x - 1)$ $x^2 + 8x$

b. $3x^2 - 2x + 1 - 2(x^2 - x - 1)$ $x^2 + 3$

13. a. $x(x - 2) + 2(x - 2)$ $x^2 - 4$

b. $4(3 - x) + x(3 - x)$ $12 - x - x^2$

14. a. $(x^3 - 3x^2 + 9x) + (3x^2 - 9x + 27)$ $x^3 + 27$

b. $(x^3 + 4x^2 + 4x) - (2x^2 + 8x + 8)$ $x^3 + 2x^2 - 4x - 8$

15. a. $(x^3 + 2x^2 + x) + (x^2 + 2x + 1)$ $x^3 + 3x^2 + 3x + 1$

b. $(a^3 - a^2b + ab^2) + (a^2b - ab^2 + b^3)$
$a^3 + b^3$

16. a. $(x^3 + x^2 + x) - (x^2 + x + 1)$ $x^3 - 1$

   b. $(a^3 + a^2b + ab^2) - (a^2b + ab^2 + b^3)$ $a^3 - b^3$

17. $x(x^2 - 2xy + y^2) - y(x^2 - 2xy + y^2)$ $x^3 - 3x^2y + 3xy^2 - y^3$

18. $x(x^2 - xy + y^2) + y(x^2 - xy + y^2)$ $x^3 + y^3$

Complete the tables in Exercises 19 to 24.

19.

Multiply	$2x$	$+3$
$3x$	$6x^2$	$9x$
-1	$-2x$	-3

$6x^2 + 7x - 3$

20.

Multiply	$2x$	$+1$
$4x$	$8x^2$	$4x$
-9	$-18x$	-9

$8x^2 - 14x - 9$

21.

Multiply	$3x$	$+1$
$3x$	$9x^2$	$3x$
$+1$	$3x$	1

$9x^2 + 6x + 1$

22.

Multiply	$4x$	$+3$
$4x$	$16x^2$	$12x$
$+3$	$12x$	9

$16x^2 + 24x + 9$

23.

Multiply	x^2	$-2x$	$+4$
x	x^3	$-2x^2$	$4x$
$+2$	$2x^2$	$-4x$	8

$x^3 + 8$

24.

Multiply	x^2	$-6x$	$+9$
x	x^3	$-6x^2$	$9x$
-3	$-3x^2$	$18x$	-27

$x^3 - 9x^2 + 27x - 27$

Multiply the expressions in Exercises 25 to 48. Use a table as needed.

25. $(x + 6)(x - 3)$
   $x^2 + 3x - 18$
26. $(x - 3)(x + 4)$
   $x^2 + x - 12$
27. $(x - 6)(x + 3)$
   $x^2 - 3x - 18$
28. $(x - 6)(x + 2)$
   $x^2 - 4x - 12$
29. $(x - 9)(x - 2)$
   $x^2 - 11x + 18$
30. $(x + 9)(x + 2)$
   $x^2 + 11x + 18$
31. $(x + 4)(x - 4)$
   $x^2 - 16$
32. $(x - 4)(x - 4)$
   $x^2 - 8x + 16$
33. $(x - 5)(x - 5)$
   $x^2 - 10x + 25$
34. $(x + 5)(x - 5)$
   $x^2 - 25$
35. $(2x - 3)(x + 4)$
   $2x^2 + 5x - 12$
36. $(2x + 1)(x - 12)$
   $2x^2 - 23x - 12$

37. $(2x + 3)(x - 4)$
   $2x^2 - 5x - 12$
38. $(2x - 3)(x - 4)$
   $2x^2 - 11x + 12$
39. $(3x - 1)(3x + 2)$
   $9x^2 + 3x - 2$
40. $(3x + 4)(2x - 3)$
   $6x^2 - x - 12$
41. $(x + 2)^2$
   $x^2 + 4x + 4$
42. $(x - 1)^2$
   $x^2 - 2x + 1$
43. $(2x - 1)(2x + 1)$
   $4x^2 - 1$
44. $(3x + 2)(3x - 2)$
   $9x^2 - 4$
45. $(x - 2)(x^2 + 2x + 4)$
   $x^3 - 8$
46. $(x + 3)(x^2 - 3x + 9)$
   $x^3 + 27$
47. $(x^2 - 3x - 2)(x^2 + 3x - 2)$ $x^4 - 13x^2 + 4$
48. $(x^2 + 4x + 2)(x^2 - 4x + 2)$ $x^4 - 12x^2 + 4$

Complete the tables in Exercises 49 to 52.

49.

Factor	$6x$	1
$2x$	$12x^2$	$+2x$
-3	$-18x$	-3

$(2x - 3)(6x + 1)$

50.

Factor	$6x$	-5
$3x$	$18x^2$	$-15x$
1	$+6x$	-5

$(3x + 1)(6x - 5)$

51.

Factor	x	-3
$3x$	$3x^2$	$-9x$
-4	$-4x$	$+12$

$(3x - 4)(x - 3)$

52.

Factor	x	-3
$2x$	$2x^2$	$-6x$
5	$+5x$	-15

$(2x + 5)(x - 3)$

Factor the expressions in Exercises 53 to 84. Use a table, guess and check, or another method.

53. $x^2 + 7x + 12$
   $(x + 4)(x + 3)$
54. $x^2 - 9x + 18$
   $(x - 3)(x - 6)$
55. $x^2 + x - 12$
   $(x - 3)(x + 4)$
56. $x^2 - 8x + 12$
   $(x - 6)(x - 2)$
57. $x^2 - 7x + 12$
   $(x - 3)(x - 4)$
58. $x^2 + 4x - 12$
   $(x + 6)(x - 2)$
59. $x^2 - 11x - 12$
   $(x - 12)(x + 1)$
60. $x^2 + 13x + 12$
   $(x + 1)(x + 12)$
61. $x^2 + 3x - 28$
   $(x + 7)(x - 4)$
62. $x^2 - 2x - 15$
   $(x - 5)(x + 3)$
63. $6x^2 + 19x + 10$
   $(2x + 5)(3x + 2)$
64. $6x^2 + 17x + 10$
   $(6x + 5)(x + 2)$
65. $6x^2 + 11x - 10$
   $(2x + 5)(3x - 2)$
66. $6x^2 + 7x - 10$
   $(6x - 5)(x + 2)$
67. $6x^2 - 17x + 10$
   $(6x - 5)(x - 2)$
68. $6x^2 - 11x - 10$
   $(2x - 5)(3x + 2)$
69. $15x^2 - x - 6$
   $(3x - 2)(5x + 3)$
70. $10x^2 + x - 3$
   $(2x - 1)(5x + 3)$
71. $6x^2 + 13x + 5$
   $(2x + 1)(3x + 5)$
72. $10x^2 + 13x - 3$
   $(5x - 1)(2x + 3)$
73. $10x^2 + 61x + 6$
   $(10x + 1)(x + 6)$
74. $10x^2 + 4x - 6$
   $2(5x - 3)(x + 1)$

75. $6x^2 - 32x + 10$
$2(3x - 1)(x - 5)$

76. $15x^2 - 9x - 6$
$3(5x + 2)(x - 1)$

77. $x^3 + 2x^2 + 4x$
$x(x^2 + 2x + 4)$

78. $2x^2 + 4x + 8$
$2(x^2 + 2x + 4)$

79. $x^3 - 6x^2 + 9x$
$x(x - 3)^2$

80. $3x^2 - 18x + 27$
$3(x - 3)^2$

81. $a^3 - a^2b + ab^2$
$a(a^2 - ab + b^2)$

82. $a^2b - ab^2 + b^3$
$b(a^2 - ab + b^2)$

83. $a^3 + a^2b + ab^2$
$a(a^2 + ab + b^2)$

84. $a^2b + ab^2 + b^3$
$b(a^2 + ab + b^2)$

85. Describe how Exercises 21, 22, 41, and 42 are alike. Name other exercises that are like these.
all in form $(a \pm b)^2$; Exercises 32 and 33

86. Describe how Exercises 31, 43, and 44 are alike. Name other exercises that are like these.
all in form $(a + b)(a - b)$; Exercise 34

Factor the expressions in Exercises 87 to 94. Look for patterns.

87. $3x^2 - 48$
$3(x - 4)(x + 4)$

88. $15x^2 - 60$
$15(x - 2)(x + 2)$

89. $12x^2 - 27$
$3(2x - 3)(2x + 3)$

90. $24 - 6x^2$
$6(2 - x)(2 + x)$

91. $20x^2 - 45$
$5(2x - 3)(2x + 3)$

92. $2x^2 - 18$
$2(x - 3)(x + 3)$

93. $28 - 63x^2$
$7(2 - 3x)(2 + 3x)$

94. $18x^2 - 8$
$2(3x - 2)(3x + 2)$

95. Match each definition with the appropriate word. Give an example for each. Choose from
square, binomial, monomial, factors, perfect cube, trinomial, exponent

a. Two or more numbers or expressions being multiplied
factors

b. A one-term polynomial
monomial

c. A two-term polynomial
binomial

d. A three-term polynomial
trinomial

e. The result of multiplying a number times itself
square

f. The result of multiplying three identical factors
perfect cube

96. Proof Show that the diagonals of the table for $(ax + b)(cx + d)$ always multiply to the same product. Suppose that a, b, c, and d are any real numbers. Start with a table for the product $(ax + b)(cx + d)$.
See Additional Answers.

97. Multiply these trinomials.

a. $(x^2 + x + 1)(x^2 - x - 1)$ $x^4 - x^2 - 2x - 1$

b. $(x^2 - x + 1)(x^2 + x - 1)$ $x^4 - x^2 + 2x - 1$

c. $(x^2 + x + 1)(x^2 + x - 1)$ $x^4 + 2x^3 + x^2 - 1$

d. $(x^2 - x + 1)(x^2 - x - 1)$ $x^4 - 2x^3 + x^2 - 1$

e. $(x^2 + x + 1)(x^2 + x + 1)$ $x^4 + 2x^3 + 3x^2 + 2x + 1$

■ **Projects**

98. Consecutive Numbers If you multiply three consecutive whole numbers, what will always be true about the product? Think about the product algebraically:

$$x(x + 1)(x + 2)$$

Extend to the product of four consecutive whole numbers. See Additional Answers.

99. Factoring Patterns Give several numbers that may be used in place of the b in the expression in order to make a trinomial that factors.

a. $x^2 + bx + 12$
$\pm 13, \pm 8, \pm 7$

b. $x^2 + bx - 15$
$\pm 14, \pm 2$

c. $x^2 + bx - 20$
$\pm 19, \pm 8, \pm 1$

d. $x^2 + bx + 18$
$\pm 19, \pm 11, \pm 9$

100. Mouse From a 1942 letter to my grandfather from another engineer: The circumference of a circle with radius r is $C = 2\pi r$. Suppose the circumference of the earth at the equator is $C_1 = 2\pi r_1$. Suppose 1 foot is added to a string that fits tightly about the equator and that string is held evenly above the surface of the earth. The new circumference is $C_2 = C_1 + 1$ foot. Without finding a number for either r_1 or the new radius r_2, find an expression that tells whether a mouse could run under the lengthened string. Check the result: Fit a string around a basketball or large medicine ball, add a foot, and then hold the string evenly around the ball. Recalculate with numbers if you still don't believe the results. See Additional Answers.

---

4 **Mid-Chapter Test**

1. Give the word, expression, or equation that fits each:

a. The highest or lowest point on a parabola vertex

b. The line of symmetry for $y = x^2$ y-axis or $x = 0$

c. A polynomial with two terms binomial

d. The standard form for a quadratic function
$y = ax^2 + bx + c$

e. The constant term in a quadratic function is the place where the graph crosses the ___y___ axis.

2. Make a table to solve the following equations. Circle the places where you find the answers.

a. $x^2 - 4x + 3 = 8$ $x = -1$ or $x = 5$

b. $x^2 - 4x + 3 = 0$ $x = 1$ or $x = 3$

c. $x^2 - 4x + 3 = -1$ $x = 2$

In Exercises 3 and 4, change the function to standard quadratic form and find the coefficients a, b, and c.

3. $y = \frac{1}{2}n(n + 1) + 1$ $a = \frac{1}{2}, b = \frac{1}{2}, c = 1$

4. $y = 1000(1 + x)^2$ $a = 1000, b = 2000, c = 1000$

5. a. Make a table and graph for $A = 6x^2$, the surface area of a cube of side x. Let x be 0 to 8. See Answer Section.

b. If we double the side of the cube, what happens to the surface area? It is 4 times the original.

6. Identify each sequence as linear, or quadratic, or neither, and then find the function for each.

   a. $15, 8, 3, 0, -1, \ldots$ quadratic; $y = x^2 - 10x + 24$

   b. $4, 9, 14, 19, 24, \ldots$ linear; $y = 5x - 1$

   c. $-3, 4, 11, 18, 25, \ldots$ linear; $y = 7x - 10$

   d. $1, -1, 1, -1, 1, \ldots$
   neither; $(-1)^n$ or $y = 1$ for odd x, $y = -1$ for even x

7. Use calculator regression to fit a function to this table. Explain your choice of functions.

x	$f(x)$
2	19.620
3	44.145
4	78.480
5	122.625

$f(x) \approx 4.9x^2$; quadratic regression because second differences are a constant 9.81

8. Simplify the expressions.

   a. $x^3 - 6x^2 + 9x - x^2 + 18x - 27$ $x^3 - 7x^2 + 27x - 27$

   b. $x^3 + 3x^2 + 9x + (-3x^2 - 9x - 27)$ $x^3 - 27$

   c. $16a + 4b + c - (9a + 3b + c)$ $7a + b$

   d. $9a + 3b + c - (4a + 2b + c)$ $5a + b$

9. Multiply the expressions.

   a. $(x - 5)(x + 5)$ $x^2 - 25$

   b. $(1 - x)(1 - x)$ $x^2 - 2x + 1$

   c. $(1 - x)(x + 3)$ $-x^2 - 2x + 3$

   d. $(2x + 3)(3 - 2x)$ $-4x^2 + 9$

10. Factor the expressions.

   a. $3x^2 + 5x - 12$ $(3x - 4)(x + 3)$

   b. $x^2 - x$ $x(x - 1)$

   c. $6x^2 + 5x - 4$ $(2x - 1)(3x + 4)$

   d. $3x^3 + 6x^2 + 3x$ $3x(x + 1)^2$

11. Suppose we stand on the ground, $h_0 = 0$, and throw a ball straight up with an initial velocity of 60 ft/sec.

   a. Write an equation describing the height in terms of time. Use the vertical motion equation, $h = -\frac{1}{2} gt^2 + v_0 t + h_0$. Let the acceleration due to gravity be $g = 32$ ft/sec^2. $h = -16t^2 + 60t$

   b. Make a table and graph for $t = 0, 1, 2, 3, 4$, using a calculator as needed.   See Answer Section.

   c. What are the x-intercepts? Explain why there are two answers. $(0, 0)$, $(3.75, 0)$; first is at instant ball is thrown.

   d. Solve the equation for $h = 56$ ft. Use a graph and/or table as needed. Explain your answer(s). $\{1.75, 2\}$; on the way up and down

   e. What is the highest point the ball reaches?   56.25 ft

4.4   Special Products and Factors and Their Graphs

Objectives

■ Factor perfect square trinomials.

■ Factor differences of squares.

■ Identify graphs related to squares of binomials and differences of squares.

■ Factor sums and differences of cubes.

WARM-UP

Multiply these expressions.

  1. $(x - 3)(x - 3)$ $x^2 - 6x + 9$

  2. $(x - 3)(x + 3)$ $x^2 - 9$

  3. $(x + 3)(x + 3)$ $x^2 + 6x + 9$

  4. $(x - 3)(x^2 + 3x + 9)$ $x^3 - 27$

  5. $(x - 3)(x^2 - 6x + 9)$ $x^3 - 9x^2 + 27x - 27$

  6. $(a + b)(a - b)$ $a^2 - b^2$

  7. $(a - b)(a - b)$ $a^2 - 2ab + b^2$

  8. $(a + b)(a + b)$ $a^2 + 2ab + b^2$

  9. $(a + b)(a^2 - ab + b^2)$ $a^3 + b^3$

  10. $(a + b)(a^2 + 2ab + b^2)$ $a^3 + 3a^2b + 3ab^2 + b^3$

IN THIS SECTION, we review two special products from multiplication of binomials: squares of binomials and differences of squares. We examine the graphs of the expressions.

Special Products: Binomials

SQUARES OF BINOMIALS   The products in the Warm-up contain several patterns. One of the patterns is the perfect square trinomial, obtained by squaring a binomial.
The **square of a binomial** (also known as a **binomial square**) is

$$(a + b)(a + b) \quad \text{or} \quad (a + b)^2$$

We may interpret the square of a binomial as the area of a square with sides $a + b$, as shown in Figure 17. The product $(a + b)^2$ expands to $a^2 + 2ab + b^2$, a **perfect square trinomial**. The middle term of $a^2 + 2ab + b^2$, $2ab$, is the sum of the diagonal terms in the square, $ab + ab$.

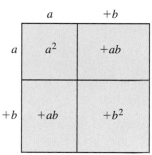

binomial square

FIGURE 17

EXAMPLE 1   Factoring   Factor each expression.
a. $x^2 + 10x + 25$
b. $1 - 8x + 16x^2$
c. $x^2 + 2xy + y^2$
d. $25a^2 + 10ab + b^2$

SOLUTION   **a.** $x^2 + 10x + 25 = (x + 5)(x + 5)$

b. $1 - 8x + 16x^2 = (1 - 4x)(1 - 4x)$

c. $x^2 + 2xy + y^2 = (x + y)(x + y)$

d. $25a^2 + 10ab + b^2 = (5a + b)(5a + b)$

EXAMPLE 2   Exploring graphs and solutions
a. Factor $x^2 + 6x + 9$.
b. Graph $y = x^2$ and $y = (x + 3)^2$. Compare their positions.
c. Where is the vertex for $y = (x + 3)^2$?
d. Solve $(x + 3)^2 = 0$ from the graph.

SOLUTION   **a.** $x^2 + 6x + 9 = (x + 3)^2$

b. The graphs are in Figure 18. The graph of $y = (x + 3)^2$ is shifted 3 units to the left of that of $y = x^2$.

c. The vertex is on the x-axis at $(-3, 0)$.

d. $y = (x + 3)^2$ touches the x-axis at $x = -3$. Thus, $(x + 3)^2 = 0$ for $x = -3$. The x-coordinate of the vertex solves the equation.

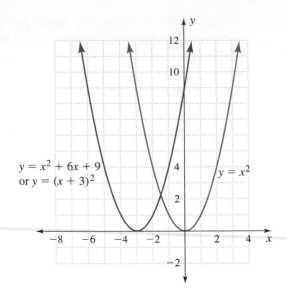

FIGURE 18

EXAMPLE 3 Exploring solutions and graphs

a. Factor $x^2 - 6x + 9$.

b. Graph $y = x^2$ and $y = (x - 3)^2$. Compare their positions.

c. Where is the vertex for $y = (x - 3)^2$?

d. Solve $(x - 3)^2 = 0$ from the graph.

SOLUTION **a.** $x^2 - 6x + 9 = (x - 3)^2$

b. The graphs are in Figure 19. The graph of $y = (x - 3)^2$ is shifted 3 units to the right of that of $y = x^2$.

c. The vertex is on the x-axis at $(3, 0)$.

d. $y = (x - 3)^2$ touches the x-axis at $x = 3$. Thus, $(x - 3)^2 = 0$ for $x = 3$. The x-coordinate of the vertex solves the equation.

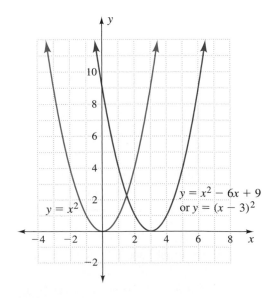

FIGURE 19

The equations in Examples 2 and 3 contain a binomial square. Although there is only one number solving each original equation, we say the solution is a double root. In $(x - r)(x - r) = 0$, *the $(x - r)$ factor appears twice and the solution, $x = r$, is a* **double root.** Each graph is a **horizontal shift,** *a movement left or right,* from the graph of $y = x^2$.

▰ GRAPHS AND SOLUTIONS OF EQUATIONS WITH BINOMIAL SQUARES

> The graph of $y = (x - r)^2$ has its vertex on the x-axis at $(r, 0)$.
>
> The graph of $y = (x - r)^2$ is shifted horizontally r units from that of $y = x^2$.
>
> The solution to $(x - r)^2 = 0$ is $x = r$, a double root.

DIFFERENCES OF SQUARES   The *product* $(a + b)(a - b) = a^2 - b^2$ is called a **difference of squares.**

Multiply	a	$+b$
a	a^2	$+ab$
$-b$	$-ab$	$+b^2$

Diagonal sum = 0

Note that the products on the diagonal add to zero, leaving the difference of two squared terms as the answer.

EXAMPLE 4   Factoring   Factor each expression.

a. $x^2 - 25$

b. $1 - 16x^2$

c. $x^2 - y^2$

d. $25a^2 - b^2$

SOLUTION   **a.** $x^2 - 25 = (x + 5)(x - 5)$

b. $1 - 16x^2 = (1 + 4x)(1 - 4x)$

c. $x^2 - y^2 = (x + y)(x - y)$

d. $25a^2 - b^2 = (5a + b)(5a - b)$ ▰

EXAMPLE 5   Exploring graphs and solutions

a. Factor $x^2 - 9$.

b. Graph $y = x^2$ and $y = x^2 - 9$. Compare their positions.

c. Where is the vertex?

d. Solve $x^2 - 9 = 0$ from the graph.

SOLUTION   **a.** $x^2 - 9 = (x + 3)(x - 3)$

b. The graphs are in Figure 20. The graph of $y = x^2 - 9$ is shifted 9 units below that of $y = x^2$.

c. The vertex is on the y-axis, $(0, -9)$.

d. $y = x^2 - 9$ crosses the x-axis at $x = -3$ and $x = 3$. Thus, $x^2 - 9 = 0$ for $x = -3$ and $x = 3$.

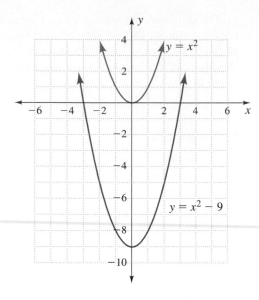

FIGURE 20

EXAMPLE 6   Exploring graphs and solutions
a. Factor $x^2 - 1$.

b. Graph $y = x^2$ and $y = x^2 - 1$. Compare their positions.

c. Where is the vertex?

d. Solve $x^2 - 1 = 0$ from the graph.

SOLUTION   a. $x^2 - 1 = (x + 1)(x - 1)$

b. The graphs are in Figure 21. The graph of $y = x^2 - 1$ is shifted 1 unit below that of $y = x^2$.

c. The vertex is on the y-axis, $(0, -1)$.

d. $y = x^2 - 1$ crosses the x-axis at $x = -1$ and $x = 1$. Thus, $x^2 - 1 = 0$ for $x = -1$ and $x = 1$.

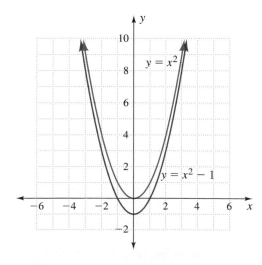

FIGURE 21

The graphs in Examples 5 and 6 are **vertical shifts**, *movements up or down* from the graph of $y = x^2$.

■ GRAPHS AND SOLUTIONS OF EQUATIONS
WITH DIFFERENCES OF SQUARES

> For c, a positive number,
>
> The graph of $y = x^2 - c^2$ has its vertex on the y-axis at $(0, -c^2)$.
>
> The graph of $y = x^2 - c^2$ is shifted vertically c^2 units from that of $y = x^2$.
>
> The two solutions to $x^2 - c^2 = 0$ are $x = c$ and $x = -c$.

Special Products: Cubic Expressions

When *the degree in a polynomial is 3*, we say the polynomial is a **cubic expression**.

EXAMPLE 7 Multiplying polynomials Multiply these expressions.

    a. $(x - 2)(x^2 + 2x + 4)$ **b.** $(x + 1)(x^2 - x + 1)$

SOLUTION   **a.**

Multiply	x^2	$+2x$	$+4$
x	x^3	$+2x^2$	$+4x$
-2	$-2x^2$	$-4x$	-8

Diagonal sums $= 0$

b.

Multiply	x^2	$-x$	$+1$
x	x^3	$-x^2$	$+x$
$+1$	$+x^2$	$-x$	$+1$

Diagonal sums $= 0$

The products are

$$(x - 2)(x^2 + 2x + 4) = x^3 - 8 \quad \text{and} \quad (x + 1)(x^2 - x + 1) = x^3 + 1$$

SUMS AND DIFFERENCES OF CUBES   The expression $a^3 + b^3$ is called a **sum of cubes**, and the expression $a^3 - b^3$ is called a **difference of cubes**. The factors creating the sums and differences of cubes may be summarized with these products.

■ SUM AND DIFFERENCE OF CUBES

> The product $(x + y)(x^2 - xy + y^2)$ gives the sum of cubes, $x^3 + y^3$.
>
> The product $(x - y)(x^2 + xy + y^2)$ gives the difference of cubes, $x^3 - y^3$.

If you have difficulty remembering the second factor for a sum or difference of cubes, you can use a table to find it. Observe that the tables in Example 8, like the products in Example 7, contain two rows and three columns.

EXAMPLE 8 Factoring cubic expressions Use a table to find the second factor in each of these expressions.

    a. $x^3 - 27 = (x - 3)(\underline{\hspace{1.5cm}})$

    b. $a^3 + b^3 = (a + b)(\underline{\hspace{1.5cm}})$

SOLUTION   **a.** We enter the factor $x - 3$ on the left side of the table, and we enter x^3 and -27 as the first and last entries inside the table.

Factor			
x	x^3		
-3			-27

Then we find the remaining factors and table entries. We start with the factor above x^3 and then the table entry below x^3. Note that the diagonal terms must add to zero because the expression $x^3 - 27$ contains no middle terms.

Factor	x^2	$+3x$	$+9$
x	x^3	$+3x^2$	$+9x$
-3	$-3x^2$	$-9x$	-27

The factorization is $x^3 - 27 = (x - 3)(x^2 + 3x + 9)$.

b. We enter the factor $a + b$ on the left side of the table, and we enter a^3 and b^3 as the first and last entries inside the table.

Factor			
a	a^3		
$+b$			$+b^3$

Then we find the remaining factors and table entries. We start with the factor above a^3 and then the table entry below a^3. Note that the diagonal terms must add to zero because the expression $a^3 + b^3$ contains no middle terms.

Factor	a^2	$-ab$	$+b^2$
a	a^3	$-a^2b$	$+ab^2$
$+b$	$+a^2b$	$-ab^2$	$+b^3$

The factorization is $a^3 + b^3 = (a + b)(a^2 - ab + b^2)$.   ■

EXAMPLE 9   **Graphing cubic polynomials**   Graph each product from Example 7 together with $y = x^3$. Compare the graphs.

a. $y = x^3 - 8$ **b.** $y = x^3 + 1$

SOLUTION   The graphs of $y = x^3 - 8$ and $y = x^3 + 1$ are shown in Figures 22 and 23, respectively. The graph of $y = x^3 - 8$ is shifted 8 units below that of $y = x^3$. The graph of $y = x^3 + 1$ is shifted 1 unit above that of $y = x^3$.

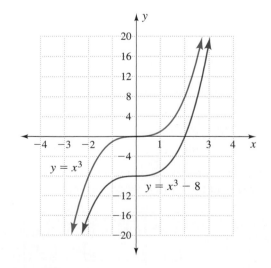

FIGURE 22

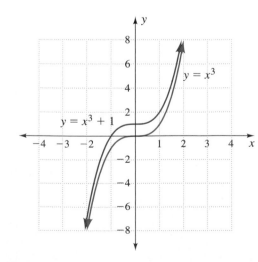

FIGURE 23   ■

The graphs of the functions $f(x) = x^3 \pm c$ are vertical shifts of the graph of $f(x) = x^3$. Verify the vertical shifts in Example 9 on a graphing calculator by using the up and down cursor keys to compare the outputs.

Application: Mental Mathematics

Circus managers of the past and television hosts of the present have occasionally brought on "mathematical geniuses" who amaze people with feats of mental mathematics based on simple shortcuts easily explained by the special products in this section.

EXAMPLE 10   **Exploring mental multiplication**   Find the products in parts a to d in your head.

a. $5^2, 4 \times 6$

b. $8^2, 7 \times 9$

c. $9^2, 8 \times 10$

d. $10^2, 9 \times 11$

e. Write an expression in algebraic notation describing the products above, where x is the number being squared.

f. What special product from this section describes how the numbers are related?

g. Mentally multiply 99×101.

SOLUTION   a. $25, 24$

b. $64, 63$

c. $81, 80$

d. $100, 99$

e. $x^2, (x - 1)(x + 1)$

f. The factorization of the difference of squares: $x^2 - 1 = (x - 1)(x + 1)$

g. Think $100^2 - 1 = 9999$.

ANSWER BOX

Warm-up: **1.** $x^2 - 6x + 9$ **2.** $x^2 - 9$ **3.** $x^2 + 6x + 9$ **4.** $x^3 - 27$
5. $x^3 - 9x^2 + 27x - 27$ **6.** $a^2 - b^2$ **7.** $a^2 - 2ab + b^2$
8. $a^2 + 2ab + b^2$ **9.** $a^3 + b^3$ **10.** $a^3 + 3a^2b + 3ab^2 + b^3$

4.4   Exercises

Multiply the expressions in Exercises 1 to 10. Identify answers that are perfect square trinomials (pst) or differences of squares (ds).

1. $(x - 3)(x + 4)$
$x^2 + x - 12$

3. $(2x + 3)(x - 4)$
$2x^2 - 5x - 12$

5. $(x + 3)(x + 3)$
$x^2 + 6x + 9$; pst

7. $(x + 6)(x - 6)$
$x^2 - 36$; ds

9. $(2x - 5)^2$
$4x^2 - 20x + 25$; pst

2. $(2x + 1)(x - 12)$
$2x^2 - 23x - 12$

4. $(x - 3)(x - 3)$
$x^2 - 6x + 9$; pst

6. $(x - 8)(x + 8)$
$x^2 - 64$; ds

8. $(x - 4)(x - 4)$
$x^2 - 8x + 16$; pst

10. $(x + 5)^2$
$x^2 + 10x + 25$; pst

In Exercises 11 to 20, fill in numbers that make the expressions equal to binomial squares.

11. $x^2 + \underline{8} x + 16 = (x + 4)^2$

12. $x^2 + \underline{6} x + 9 = (x + 3)^2$

13. $x^2 - \underline{14} x + 49 = (x - \underline{7})^2$

14. $x^2 - \underline{1} x + \underline{\frac{1}{4}} = (x - \frac{1}{2})^2$

15. $x^2 - 5x + \underline{6.25} = (x - 2.5)^2$

16. $x^2 - 3x + \underline{\ 2\frac{1}{4}\ } = \left(x - 1\frac{1}{2}\right)^2$

17. $x^2 + 7x + \underline{\ 12.25\ } = (x + \underline{\ 3.5\ })^2$

18. $x^2 + 9x + \underline{\ 20\frac{1}{4}\ } = \left(x + \underline{\ 4\frac{1}{2}\ }\right)^2$

19. $x^2 - 24x + \underline{\ 144\ } = (x - \underline{\ 12\ })^2$

20. $x^2 + bx + \underline{\ \frac{b^2}{4}\ } = \left(x + \underline{\ \frac{b}{2}\ }\right)^2$

Factor the expressions in Exercises 21 to 30.

21. $x^2 + 4x + 4$
$(x + 2)^2$

22. $y^2 - 8y + 16$
$(y - 4)^2$

23. $n^2 + 2n + 1$
$(n + 1)^2$

24. $1 + 6n + 9n^2$
$(1 + 3n)^2$

25. $49x^2 + 14x + 1$
$(7x + 1)^2$

26. $s^2 - 10st + 25t^2$
$(s - 5t)^2$

27. $9x^2 - 6xy + y^2$
$(3x - y)^2$

28. $36a^2 + 12ab + b^2$
$(6a + b)^2$

29. $x^2 + 5x + 6\frac{1}{4}$
$\left(x + 2\frac{1}{2}\right)^2$

30. $x^2 + 3x + 2.25$
$(x + 1.5)^2$

Factor the expressions in Exercises 31 to 44.

31. $x^2 - 4$
$(x + 2)(x - 2)$

32. $y^2 - 16$
$(y - 4)(y + 4)$

33. $n^2 + 16$
cannot be factored

34. $1 - 9n^2$
$(1 + 3n)(1 - 3n)$

35. $49x^2 - 1$
$(7x + 1)(7x - 1)$

36. $s^2 - 25t^2$
$(s - 5t)(s + 5t)$

37. $y^2 - 36z^2$
$(y - 6z)(y + 6z)$

38. $x^2 + 64$
cannot be factored

39. $9x^2 - y^2$
$(3x - y)(3x + y)$

40. $36a^2 - b^2$
$(6a + b)(6a - b)$

41. $x^2 - 144$
$(x + 12)(x - 12)$

42. $25x^2 - 64$
$(5x + 8)(5x - 8)$

43. $0.25x^2 - 0.01$
$(0.5x + 0.1)(0.5x - 0.1)$

44. $0.01x^2 - 0.36$
$(0.1x + 0.6)(0.1x - 0.6)$

45. Use the table to help explain that $n^2 + 16$ cannot be factored.

Multiply	n	
n	n^2	
		$+16$

The diagonal must both add to zero and multiply to $+16n^2$; not possible.

46. Use the table to help explain that $x^2 + 64$ cannot be factored.

Multiply	x	
x	x^2	
		$+64$

The diagonal must both add to zero and multiply to $+64x^2$; not possible.

47. Using a graphing calculator, graph $y = (x + 5)^2$. Compare the position of the graph with that of $y = x^2$. What is the x-intercept point?
shift left 5 units; $(-5, 0)$

48. Describe how you would find the graph of $y = (x - 4)^2$ given the graph of $y = x^2$.
shift right 4 units

49. Describe how to find the graph of $y = (2x - 5)^2$ from the graph of $y = (2x)^2$. Check with a graphing calculator. shift $y = (2x)^2$ right 2.5 units

In Exercises 50 to 53, consider these results: If $y = (x - 3)^2$, the vertex of the parabola will be on the x-axis at $x = 3$. If $y = (x + 3)^2 = [x - (-3)]^2$, the vertex will be on the x-axis at $x = -3$.

50. a. Describe how to find the vertex for $y = (x - r)^2$.
on x-axis at r
b. Apply your rule to $y = (x + 1)^2$.
$(-1, 0)$
c. Apply your rule to $y = (x - 4)^2$.
$(4, 0)$

51. a. Compare the vertex of the graph of $y = (3x)^2$ with the vertex of the graph of $y = (3x + 6)^2$. Describe how to find the vertex in terms of s and t for $y = (sx - t)^2$.
$(0, 0), (-2, 0)$; let $sx - t = 0$, solve for x; on x-axis at t/s
b. Apply your rule to $y = (2x + 1)^2$. $\left(-\frac{1}{2}, 0\right)$

c. Apply your rule to $y = (3x - 4)^2$. $\left(\frac{4}{3}, 0\right)$

52. a. Describe the shift of $y = x^2$ as left or right in terms of the value of r in $y = (x - r)^2$.
right $|r|$ units when $r > 0$, left $|r|$ units when $r < 0$
b. Apply your rule to $y = (x + 2)^2$.
shift left 2 units, as $x + 2 = x - (-2)$
c. Apply your rule to $y = (x - 1)^2$. shift right 1 unit

53. a. Describe the shift of $y = x^2$ as left or right in terms of the values of s and t in $y = (sx - t)^2$, $s > 0$.
right $|t/s|$ units when $t > 0$, left $|t/s|$ units when $t < 0$
b. Apply your rule to $y = (2x - 1)^2$. shift right $\frac{1}{2}$ unit

c. Apply your rule to $y = (2x + 3)^2$. shift left $\frac{3}{2}$ units

54. Using a graphing calculator, graph $y = x^2 - 16$. Compare the position of the graph with that of $y = x^2$. Describe the x-intercept points.
shift down 16 units; $\{-4, 4\}$

55. Describe how you would find the graph of $y = (x - 8)(x + 8)$ given the graph of $y = x^2$.
shift down 64 units

56. Does the coefficient on x in $y = (ax - 2)(ax + 2)$ matter in finding the y-intercept?
No; y-intercept is $(0, -4)$.

Multiply the expressions in Exercises 57 to 62.

57. $(x + 1)(x^2 - x + 1)$ $x^3 + 1$

58. $(x - 1)(x^2 - 2x + 1)$ $x^3 - 3x^2 + 3x - 1$

59. $(x - 3)(x^2 - 6x + 9)$ $x^3 - 9x^2 + 27x - 27$

60. $(x + 2)(x^2 + 4x + 4)$ $x^3 + 6x^2 + 12x + 8$

61. $(x + 1)(x^2 - 2x + 1)$ $x^3 - x^2 - x + 1$

62. $(x - 1)(x^2 + x + 1)$ $x^3 - 1$

Use the tables in Exercises 63 and 64 to answer the questions.

63. $x^3 + 64$ equals $(x + 4)$ multiplied by what trinomial?
$x^2 - 4x + 16$

Factor	x^2	$-4x$	$+16$
x	x^3	$-4x^2$	$+16x$
$+4$	$+4x^2$	$-16x$	$+64$

64. $x^3 - 125$ equals $(x - 5)$ multiplied by what trinomial?
$x^2 + 5x + 25$

Factor	x^2	$+5x$	$+25$
x	x^3	$+5x^2$	$+25x$
-5	$-5x^2$	$-25x$	-125

Factor the expressions in Exercises 65 and 68.

65. a. $x^3 + 27$
$(x + 3)(x^2 - 3x + 9)$
b. $x^3 + 8$
$(x + 2)(x^2 - 2x + 4)$
c. $x^3 - 64$
$(x - 4)(x^2 + 4x + 16)$

66. a. $x^3 + 1$
$(x + 1)(x^2 - x + 1)$
b. $x^3 + 125$
$(x + 5)(x^2 - 5x + 25)$
c. $x^3 - 1000$
$(x - 10)(x^2 + 10x + 100)$

67. a. $x^3 - 0.001$ $(x - 0.1)(x^2 + 0.1x + 0.01)$

b. $x^3 - \frac{1}{27}$ $\left(x - \frac{1}{3}\right)\left(x^2 + \frac{1}{3}x + \frac{1}{9}\right)$

68. a. $x^3 + \frac{1}{8}$ $\left(x + \frac{1}{2}\right)\left(x^2 - \frac{1}{2}x + \frac{1}{4}\right)$

b. $x^3 - 0.008$ $(x - 0.2)(x^2 + 0.2x + 0.04)$

69. Compare $y = (x + 2)^3$ with $y = x^3 + 8$ on a graphing calculator. Are they the same? Compare each to $y = x^3$.
no; shifted left 2 units; moved up 8 units

70. Compare $y = (x + 1)^3$ with $y = x^3 + 1$ on a graphing calculator. Are they the same? Compare each to $y = x^3$.
no, shifted left 1 unit; shifted up 1 unit

■ **Applications**

71. Traffic Circles The area of the concrete around a traffic circle is $A = \pi R^2 - \pi r^2$, where R is the outer radius for the concrete and r is the radius of the inner planted area.

a. Factor the expression on the right. $\pi(R - r)(R + r)$

b. Explain which factor is the width of the concrete.
$R - r$, the distance between the outer and inner radii

72. Picture Frames A square frame of outer dimensions N by N evenly surrounds a square mirror of dimensions n by n.

a. Write an expression for the area of the frame. $N^2 - n^2$

b. Factor the expression in part a. $(N - n)(N + n)$

c. What is the width of the frame? $\frac{1}{2}(N - n)$

Exercises 73 to 78 return to the mental arithmetic application in Example 10.

73. Give three examples of multiplications fitting the rule $x^2 - 1 = (x - 1)(x + 1)$.

74. With a calculator, verify that $85^2 - 1 = 84 \times 86$.
$7225 - 1 = 7224$

75. Multiply mentally.

a. 19×21 399

b. 39×41 1599

76. Multiply mentally.

a. 24×26 624
b. 29×31
899

The idea behind the pattern $x^2 - 1 = (x - 1)(x + 1)$ works for other differences of squares besides $x^2 - 1$. What pattern is used for Exercises 77 and 78?

77. a. $10^2, 8 \times 12$
b. $20^2, 18 \times 22$
c. $100^2, 98 \times 102$ $x^2 - 4 = (x - 2)(x + 2)$

78. a. $10^2, 7 \times 13$
b. $20^2, 17 \times 23$
c. $100^2, 97 \times 103$ $x^2 - 9 = (x - 3)(x + 3)$

■ **Projects**

79. Powers of Binomials Multiply out the powers of binomials in parts a to c. See Additional Answers.

a. $(x + 1)^2$ to $(x + 1)^4$

b. $(x - 1)^2$ to $(x - 1)^4$

c. $(x + y)^2$ to $(x + y)^4$

d. Record patterns you find.

e. Research Pascal's triangle. Relate it to the powers of binomials.

80. Cubic Function Differences

a. Use a graphing calculator table or spreadsheet to evaluate these cubic functions for $x = 1, 2, 3, 4, 5$. List outputs in a sequence. Take the difference between terms, repeat as necessary, and discuss patterns in the differences.

$$y = x^3$$
$$y = 2x^3 + x$$
$$y = 0.5x^3 + x^2$$

b. How are the patterns in part a like and unlike linear and quadratic sequences? Predict a relationship between the nth row of differences and a polynomial equation of degree n.

c. Use your observations to find which two of these three sequences are from cubic polynomials.

$$3, 10, 29, 66, 127, \ldots$$
$$3, 9, 27, 81, 243, \ldots$$
$$-4, 3, 22, 59, 120, \ldots$$

d. Finally, consider finding a cubic function. Make a table with inputs 1, 2, 3, 4, and 5 and outputs

$$f(x) = ax^3 + bx^2 + cx + d$$

Set up columns to calculate the first, second, and third differences. Explain how the results may be used to write a cubic equation for any cubic sequence. See Additional Answers.

| 4.5 | **Solving and Modeling Quadratic Equations with Factors** |

Objectives

■ Solve a quadratic equation by factoring.

■ Find x-intercepts by factoring.

■ Use x-intercepts and one other point to find a quadratic function.

WARM-UP

Multiply these expressions.

1. $(3x + 5)(2x - 3)$ $6x^2 + x - 15$

2. $(3x - 2)(4x + 5)$ $12x^2 + 7x - 10$

3. $(2x - 3)(6x + 5)$ $12x^2 - 8x - 15$

4. $(2x + 1)(4x + 9)$ $8x^2 + 22x + 9$

5. $(x + 3)^2$ $x^2 + 6x + 9$

6. $\frac{100}{12}(0.12x^2 - 2.4x + 9)$ $x^2 - 20x + 75$

Factor these expressions.

7. $x^2 - x - 6$ $(x - 3)(x + 2)$

8. $x^2 + 6x + 5$ $(x + 5)(x + 1)$

9. $x^2 - 20x + 75$ $(x - 5)(x - 15)$

10. $x^2 - 25y^2$ $(x - 5y)(x + 5y)$

11. Use calculator quadratic regression to fit an equation to $(0, 0)$, $(10, 12)$, and $(20, 0)$. $y = -0.12x^2 + 2.4x$

IN THIS SECTION, we continue to look at ways of solving and modeling quadratic equations.

Solving by Factoring

Factoring provides an algebraic method of solving quadratic equations. The solution-by-factoring method requires an understanding of the zero product rule.

■ ZERO PRODUCT RULE

If the product of two expressions is zero, then either one or the other expression is zero. Symbolically,

$$\text{if} \quad A \cdot B = 0, \quad \text{then either} \quad A = 0 \quad \text{or} \quad B = 0$$

EXAMPLE 1   Exploring the zero product rule   Describe the role of zero in these equations and expressions.

a. $A \cdot (-5) = 0$

b. $5 \cdot B = 0$

c. $(x + 2)(x - 3)$ if $x = 3$

d. $(x + 2)(x - 3)$ if $x = -2$

SOLUTION   **a.** $A = 0$ because the other factor, -5, is not zero.

b. $B = 0$ because the other factor, 5, is not zero.

c. $(3 + 2)(3 - 3) = 5 \cdot 0 = 0$

The product $(x + 2)(x - 3)$ is zero if $x = 3$.

d. $(-2 + 2)(-2 - 3) = 0(-5) = 0$
The product $(x + 2)(x - 3)$ is zero if $x = -2$.

Parts a and b illustrate the fact that, according to the zero product rule, if one factor of $ab = 0$ is not zero, the other factor is zero. Parts c and d illustrate the fact that if one factor is zero, the product is zero. This fact is useful in checking solutions.

■ SOLVING BY FACTORING

> **1.** Write the equation in $= 0$ form.
>
> **2.** Factor.
>
> **3.** Using the zero product rule, set each factor equal to zero, and solve for the variable.
>
> **4.** Check the solutions.

The zero product rule indicates that if we factor an expression, we may solve for inputs that cause the expression to have a zero output.

EXAMPLE 2   **Solving by factoring**   Solve $x^2 - x - 6 = 0$ by factoring. Compare the solutions with the x-intercepts in Figure 24.

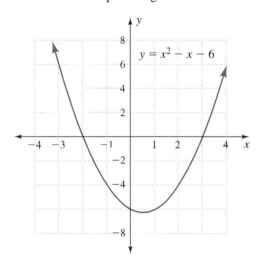

FIGURE 24

SOLUTION

$$x^2 - x - 6 = 0 \qquad \text{Factor.}$$

$$(x + 2)(x - 3) = 0 \qquad \text{Apply the zero product rule.}$$

$$\text{Either } (x + 2) = 0 \quad \text{or} \quad (x - 3) = 0 \qquad \text{Solve the factor equations.}$$

$$x = -2 \quad \text{or} \qquad x = 3$$

Check: See parts c and d of Example 1.

In Figure 24, the graph of $y = x^2 - x - 6$ crosses the x-axis at $x = -2$ and $x = 3$. The x-intercepts are solutions to the equation when $y = 0$.

■ WRITING TWO SOLUTIONS TO AN EQUATION

> In solving by factoring, each factor is the source of a solution. Because x can represent only one input at a time, use the phrasing *either . . . or* in the solution.

The *either . . . or* phrasing is important. Use it in all solutions by factoring.

EXAMPLE 3 Solving by factoring   Estimate the solutions to $6x^2 + x - 2 = 0$ from Table 13 and the graph in Figure 25. Solve by factoring.

TABLE 13

To see the symmetry in this table, set the calculator table to **TblStart** $= -1.5$ and **ΔTbl** $= 1 \div 6$.

x	$f(x) = 6x^2 + x - 2$
-1.5	10
-1	3
-0.5	-1
0	-2
0.5	0
1	5
1.5	13

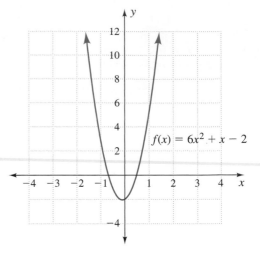

FIGURE 25

SOLUTION   In the table, $f(x) = 0$ between $x = -1$ and $x = -0.5$ and at $x = 0.5$. The graph agrees with these solutions in that it intersects the x-axis between -1 and -0.5 and again at $x = 0.5$. The solutions to the equation should agree.

$$6x^2 + x - 2 = 0 \qquad \text{Factor.}$$

$$(3x + 2)(2x - 1) = 0 \qquad \text{Apply the zero product rule.}$$

$$\text{Either } 3x + 2 = 0 \qquad \text{or} \quad 2x - 1 = 0 \qquad \text{Solve the factor equations.}$$

$$3x = -2 \quad \text{or} \qquad 2x = 1$$

$$x = -\tfrac{2}{3} \quad \text{or} \qquad x = \tfrac{1}{2}$$

The solution $x = -\tfrac{2}{3}$ is between $x = -1$ and $x = -0.5$, as predicted. ■

■   To use the zero product rule, we must have $f(x) = 0$. Note that $a \cdot b = c$, $c \neq 0$, does *not* imply that $a = c$ or $b = c$. Thus, in the first step of the solution to Example 4, we change the equation to $= 0$ form.

EXAMPLE 4 Solving by factoring   Solve $3x(3x - 4) = -4$ by factoring.

SOLUTION

$$3x(3x - 4) = -4 \qquad \text{Multiply out parentheses and add 4 to both sides.}$$

$$9x^2 - 12x + 4 = 0 \qquad \text{Factor.}$$

$$(3x - 2)(3x - 2) = 0 \qquad \text{Apply the zero product rule.}$$

$$\text{Either } 3x - 2 = 0 \quad \text{or} \quad 3x - 2 = 0 \qquad \text{Solve the factor equations.}$$

$$x = \tfrac{2}{3} \quad \text{or} \qquad x = \tfrac{2}{3}$$

The equation is a perfect trinomial square. The solution, $x = \tfrac{2}{3}$, is a double root.

Check: For $x = \tfrac{2}{3}$: $3\left(\tfrac{2}{3}\right)\left(3\left(\tfrac{2}{3}\right) - 4\right) \overset{?}{=} -4$ ✓   ■

EXAMPLE 5 Solving by factoring Solve $3x(x + 1) - 5 = 31$ by factoring.

SOLUTION

$$3x(x + 1) - 5 = 31$$ Multiply out parentheses and subtract 31 on both sides.

$$3x^2 + 3x - 5 - 31 = 0$$ Combine like terms.

$$3x^2 + 3x - 36 = 0$$ Factor the greatest common factor.

$$3(x^2 + x - 12) = 0$$ Factor.

$$3(x + 4)(x - 3) = 0$$ Apply the zero product rule.

Either $x + 4 = 0$ or $x - 3 = 0$ Solve the factor equations.

$$x = -4 \quad \text{or} \quad x = 3$$

The equation has two solutions.

Check: For $x = -4$: $3(-4)(-4 + 1) - 5 \stackrel{?}{=} 31$
$$-12(-3) - 5 = 31 \checkmark$$
For $x = 3$: $3(3)(3 + 1) - 5 \stackrel{?}{=} 31$
$$9(4) - 5 = 31 \checkmark$$

Instead of factoring out the greatest common factor, 3, from $3x^2 + 3x - 36 = 0$ in Example 5, we might have divided both sides by 3. Because $0/3 = 0$, the right side remains zero and the zero product rule still applies.

■ Example 6 contains a suggestion for factoring with larger numbers. The example is about banker's boxes—boxes specially designed to store either letter- or legal-size file folders.

EXAMPLE 6 Solving by factoring: storage boxes The width of banker's boxes can be found from the equation $10.5x^2 + 31.5x = 1890$. The length is 3 inches more than the width. Find the length and width of these storage boxes.

SOLUTION First we change the equation to $f(x) = 0$ form.

$$10.5x^2 + 31.5x = 1890$$ Subtract 1890 from both sides.

$$10.5x^2 + 31.5x - 1890 = 0$$ Divide both sides by 10.5.

$$x^2 + 3x - 180 = 0$$

Then, rather than list the factors in order to find the correct ones, we look at the middle term. The $3x$ indicates two factors close together. Our first guess is $10 \times 18 = 180$, but 10 and 18 differ by 8. Our second guess is $12 \times 15 = 180$; 12 and 15 differ by 3.

$$(x + 15)(x - 12) = 0$$ Apply the zero product rule.

Either $x = -15$ or $x = 12$ Neither the length nor the width can be negative.

If the width is $x = 12$ inches, the length is $x + 3 = 15$ inches.

Check: For $x = 12$: $10.5(12^2) + 31.5(12) \stackrel{?}{=} 1890$
$$1512 + 378 = 1890 \checkmark$$

Caution: The solutions to the equation in Example 6 included $x = -15$. The length of the box was 15. Consider this a coincidence. Never drop the negative sign from $x = -15$ to find the other unknown. Instead, look for another fact.

We return to building a quadratic function, our topic in Section 4.2, and look at another method.

Building a Quadratic Function from x-intercepts

The relationship among the solutions to an equation, its factors, and the graph is given by the factor theorem, illustrated in Figure 26.

■ FACTOR THEOREM

> Suppose that f is a polynomial function and x_1 is a constant. If $x - x_1$ is a factor of the polynomial, then $x = x_1$ is a solution to the equation $f(x) = 0$. Also, if $x = x_1$ is a solution to the equation $f(x) = 0$, then $x - x_1$ is a factor of $f(x)$.

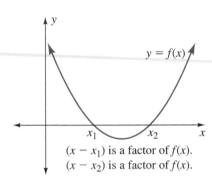

$(x - x_1)$ is a factor of $f(x)$.
$(x - x_2)$ is a factor of $f(x)$.

FIGURE 26

Student Note: As with regression equations, we will see the word *equation* rather than *function* in these examples.

If we know the x-intercepts (say, x_1 and x_2), we can write the quadratic equation as $y = a(x - x_1)(x - x_2)$. Knowing one additional ordered pair, (x, y), permits us to solve for a.

EXAMPLE 7

Finding a quadratic equation Use the equation $y = a(x - x_1)(x - x_2)$ to find the equation of a graph passing through $(-2, 0)$, $(5, 0)$, and $(4, 3)$.

SOLUTION

The ordered pairs $(-2, 0)$ and $(5, 0)$ are x-intercepts.

$$y = a(x - x_1)(x - x_2) \qquad \text{Substitute } x_1 = -2 \text{ and } x_2 = 5.$$
$$y = a(x + 2)(x - 5) \qquad \text{To find } a, \text{ substitute } (4, 3).$$
$$3 = a(4 + 2)(4 - 5) \qquad \text{Solve for } a.$$
$$3 = a(6)(-1)$$
$$-\tfrac{1}{2} = a \qquad \text{Substitute } a \text{ into the equation resulting from step 1.}$$
$$y = -\tfrac{1}{2}(x + 2)(x - 5)$$

■

■ BUILDING AN EQUATION FROM x-INTERCEPTS
AND ONE ORDERED PAIR, (x, y)

> **1.** Place x-intercepts x_1 and x_2 into $y = a(x - x_1)(x - x_2)$.
> **2.** Substitute the ordered pair, (x, y), into the equation resulting from step 1. Solve for a.
> **3.** Substitute a into the equation resulting from step 1.

Substituting x-intercepts is most useful when we are able to choose a convenient origin in a figure, as we can in Example 8.

In Example 8, we use intercepts to derive the equation for the parabolic arches in a museum entryway.

EXAMPLE 8

Finding a quadratic equation: museum arches The parabolic arches for an art museum, shown in Figure 27, must be built with 20 feet between A and C and a height of 12 feet. For an origin at A, find the quadratic equation satisfying the given dimensions.

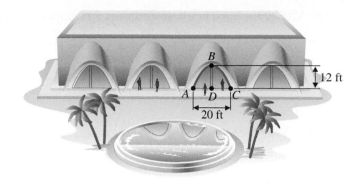

FIGURE 27

SOLUTION A is (0, 0), B is (10, 12), and C is (20, 0).

$$y = a(x - x_1)(x - x_2) \qquad \text{Substitute } x_1 = 0 \text{ and } x_2 = 20.$$

$$y = a(x - 0)(x - 20) \qquad \text{Multiply factors.}$$

$$y = a(x^2 - 20x)$$

To find a, we substitute (10, 12) for x and y in the equation:

$$12 = a(10^2 - 20(10)) \qquad \text{Simplify.}$$

$$12 = -100a \qquad \text{Solve for } a.$$

$$a = -\frac{12}{100} = -0.12$$

The equation is $y = -0.12(x^2 - 20x)$, or $y = \ 0.12x^2 + 2.4x$

In Example 8, the placement of the origin at point A made it possible for points A and C to be x-intercept points. Well-reasoned placement of axes in a problem situation can simplify both calculating information and finding equations.

EXAMPLE 9 **Solving by factoring: museum accessibility**   A collector in Greece has offered to loan the museum a large sculpture. The shipping box is rectangular in cross section and mounted on wheels. From Example 8, the equation for the height of the parabolic entry (labeled with letters A, B, and C) is

$$h(x) = -0.12x^2 + 2.4x$$

Both x and $h(x)$ are in feet. The variable x is the distance along the floor from the lower left corner of the entry—point A in both Figure 27 and Figure 28. What width box can pass through the door if the box (including wheels) is 9 feet high?

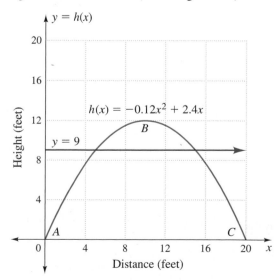

FIGURE 28

SOLUTION   If the height of the box is 9 feet, we need to have $h(x) = 9$. Setting up the equation to factor, we have

$$h(x) = 9$$

$$-0.12x^2 + 2.4x = 9 \qquad \text{Multiply both sides by } -1 \text{ and add 9 on each side.}$$

$$0.12x^2 - 2.4x + 9 = 0 \qquad \text{Multiply by 100 to remove decimals.}$$

$$12x^2 - 240x + 900 = 0 \qquad \text{Divide by 12, a common factor.}$$

$$x^2 - 20x + 75 = 0 \qquad \text{Factor.}$$

$$(x - 15)(x - 5) = 0 \qquad \text{Apply the zero product rule.}$$

$$\text{Either } x - 15 = 0 \quad \text{or} \quad x - 5 = 0 \qquad \text{Solve the factor equations.}$$

$$x = 15 \quad \text{or} \qquad x = 5$$

The coordinates of the arch at the 9-foot height are (5, 9) and (15, 9). There is 10 feet between the coordinates at that height, so a box 9 feet tall and 10 feet wide will fit through the door.   ▬

The limitation of factoring as a solution tool is that the quadratic equations in many applications are much too complicated. In Example 9, we transformed the equation into a factorable form. If we had changed the height of the shipping box to 8 or 10 feet, the equation could not have been factored.

ANSWER BOX

Warm-up: **1.** $6x^2 + x - 15$ **2.** $12x^2 + 7x - 10$ **3.** $12x^2 - 8x - 15$
4. $8x^2 + 22x + 9$ **5.** $x^2 + 6x + 9$ **6.** $x^2 - 20x + 75$ **7.** $(x - 3)(x + 2)$
8. $(x + 5)(x + 1)$ **9.** $(x - 5)(x - 15)$ **10.** $(x - 5y)(x + 5y)$
11. $y = -0.12x^2 + 2.4x$

4.5   Exercises 💿

In Exercises 1 to 24, solve by factoring. Indicate any differences of squares (ds) or perfect square trinomials (pst) and double roots.

1. $x^2 - x - 6 = 0$ $\{-2, 3\}$

2. $x^2 + x - 12 = 0$ $\{-4, 3\}$

3. $2x^2 + x - 1 = 0$ $\{-1, \frac{1}{2}\}$

4. $2x^2 + x - 3 = 0$ $\{-1\frac{1}{2}, 1\}$

5. $x^2 - 121 = 0$ $\{\pm 11\}$; ds

6. $x^2 - 144 = 0$ $\{\pm 12\}$; ds

7. $3x^2 - 48 = 0$ $\{\pm 4\}$; ds

8. $9x^2 - 81 = 0$ $\{\pm 3\}$; ds

9. $4x^2 - 2x = 0$ $\{0, \frac{1}{2}\}$

10. $9x^2 - 3x = 0$ $\{0, \frac{1}{3}\}$

11. $2x^2 = 6x$ $\{0, 3\}$

12. $3x^2 = 6x$ $\{0, 2\}$

13. $2x^2 - 6x + 4 = 0$ $\{1, 2\}$

14. $3x^2 - 15x + 18 = 0$ $\{2, 3\}$

15. $2x^2 - 11x + 12 = x - 6$ $\{3\}$; pst, double root

16. $3x^2 + 10x + 13 = 1 - 2x$ $\{-2\}$; pst, double root

17. $\dfrac{x^2}{2} - \dfrac{3x}{2} = 5$ $\{-2, 5\}$

18. $\dfrac{x^2}{3} - \dfrac{x}{3} = 4$ $\{-3, 4\}$

19. $0.25x^2 + 0.50x - 6 = 0$ $\{-6, 4\}$

20. $0.5x^2 - 1.5x - 14 = 0$ $\{-4, 7\}$

21. $0.4x^2 + 0.2x = 0.2$ $\{-1, \frac{1}{2}\}$

22. $0.5x^2 + 0.25x = 0.75$ $\{-\frac{3}{2}, 1\}$

23. $x^2 - \dfrac{5x}{3} = \dfrac{2}{3}$ $\{-\frac{1}{3}, 2\}$

24. $x^2 - \dfrac{7x}{3} = 2$ $\{-\frac{2}{3}, 3\}$

In Exercises 25 to 30, name the x-intercepts for each equation without graphing the equation.

25. $y = (x - 4)(x + 3)$
4, −3

26. $y = (x - 2)(x + 5)$
2, −5

27. $y = (2x - 5)(x + 3)$
$\frac{5}{2}$, −3

28. $y = (3x + 1)(x - 4)$
$-\frac{1}{3}$, 4

29. $y = (3x - 4)(2x + 1)$
$\frac{4}{3}$, $-\frac{1}{2}$

30. $y = (2x - 3)(4x + 1)$
$\frac{3}{2}$, $-\frac{1}{4}$

In Exercises 31 to 34, find a if the graph of the equation is to pass through the given point.

31. The graph of $y = a(x + 1)(x + 5)$ passes through $(-4, 3)$.
$a = -1$

32. The graph of $y = a(x + 1)(x + 5)$ passes through $(-2, 6)$.
$a = -2$

33. The graph of $y = a(x + 1)(x + 5)$ passes through $(0, 7.5)$.
$a = 1.5$

34. The graph of $y = a(x + 1)(x + 5)$ passes through $(-2, -3)$. $a = 1$

In Exercises 35 to 44, use the factored equation method to write a quadratic equation passing through the points. Check with a quadratic regression.

35. $(3, 0), (-2, 0), (-3, 12)$ $y = 2x^2 - 2x - 12$

36. $(3, 0), (-2, 0), (0, -6)$ $y = x^2 - x - 6$

37. $(0, 12), (3, 0), (-2, 0)$ $y = -2x^2 + 2x + 12$

38. $(-2, 0), (1, 18), (3, 0)$ $y = -3x^2 + 3x + 18$

39. $(-3, 0), (5, 0), (3, 6)$ $y = -0.5x^2 + x + 7.5$

40. $(-2, 0), (-1, 10), (4, 0)$ $y = -2x^2 + 4x + 16$

41. $(-2, 0), (2, -24), (5, 0)$ $y = 2x^2 - 6x - 20$

42. $(-2, 12), (-4, 0), (2, 0)$ $y = -1.5x^2 - 3x + 12$

43. $(10, 0), (30, 0), (20, 12)$ $y = -0.12x^2 + 4.8x - 36$

44. $(0, 0), (10, 15), (20, 0)$ $y = -0.15x^2 + 3x$

45. A parabolic entry design has the equation $y = -\frac{3}{40}(x^2 - 50x + 400)$. Return the equation to the form $y = a(x - x_1)(x - x_2)$, and then explain how to find the vertex on the parabola. $y = -\frac{3}{40}(x - 40)(x - 10)$; average x-intercept values, substitute to find y.

46. A parabolic entry design has the equation $y = -0.025x^2 + 2x - 30$. Change the equation to the form $y = a(x - x_1)(x - x_2)$, and then explain how to find the vertex on the parabola. $y = -0.025(x - 60) \times (x - 20)$; average x-intercept values, substitute to find y.

47. Explain how to find the x-intercepts from an equation that has been factored, such as $(ax + b)(cx + d) = 0$.
Use zero product rule and solve factor equations.

48. Error Analysis   A student divides both sides of $2x^2 = 8x$ by $2x$, obtaining $x = 4$. What is wrong? What is the correct method?
Lost one solution, $x = 0$. Should factor $2x^2 - 8x = 0$ and solve.

In Exercises 49 to 52, choose any two values for a and find different quadratic equations with the given solution.

49. $\{-2, -1\}$ $0 = a(x + 2)(x + 1)$, a any real number

50. $\{-3, -2\}$ $0 = a(x + 3)(x + 2)$, a any real number

51. $\{-\frac{3}{2}, \frac{1}{2}\}$ $0 = a(2x + 3)(2x - 1)$, a any real number

52. $\{-\frac{1}{3}, \frac{2}{3}\}$ $0 = a(3x + 1)(3x - 2)$, a any real number

▪ **Applications**

53. While working on a cooling tower, a construction worker tosses a candy bar to a friend at the base (height = 0 feet) of the tower. The formula describing height in feet as a function of time in seconds is $h(t) = -\frac{1}{2}(32)t^2 + v_0t + h_0$. The height of the worker is $h_0 = 320$ feet. The speed of the candy bar after t seconds is $v = -32t + v_0$. Find the time needed for the candy bar to hit the ground and the speed of the candy bar just before it is caught, and note any safety precautions needed. Comment on the surprising results.

a. Candy bar is first tossed up; $v_0 = 16$ ft/sec (10.9 mph). $t = 5$ sec, $v = -144$ ft/sec; baseball catcher's mitt or bucket of foam, protective face mask, and helmet are suggested.

b. Candy bar is thrown down; $v_0 = -16$ ft/sec.
$t = 4$ sec, $v = -144$ ft/sec; same precautions. Speed is same in both settings.

Use the following figure for Exercises 54 and 55. The figure has squares in the corners.

(a) Write an expression for the area of the inner part of the figure.

(b) Set your expression equal to zero and solve.

(c) Explain why only one of the two algebraic answers makes sense.

(d) Answer the question.

54. Suppose the figure represents a sheet of plywood 4 feet by 8 feet. For what value of x will the inner area be 45 square feet?   none; see Additional Answers.

55. Suppose the figure represents a sheet of notebook paper $8\frac{1}{2}$ inches by 11 inches. For what value of x will the inner area be 51 square inches?
1.25 in.; see Answer Section.

Exercises 56 and 57 are a factoring challenge. These equations for museum door height can be solved by eliminating the decimals, removing common factors, and using a systematic process such as a factoring table.

56. $-0.12x^2 + 2.4x = 5.25$ $x = 2.5$ and $x = 17.5$

57. $-0.12x^2 + 2.4x = 11.25$ $x = 7.5$ and $x = 12.5$

4 Chapter Summary

Vocabulary

axis (line) of symmetry
binomial
binomial square
common monomial factor
cubic expression
difference of cubes
difference of squares
double root
factor theorem

first differences
greatest common
   factor
horizontal shift
monomial
parabola
parameter
perfect square trinomial
polynomial

polynomial function
prime factor
prime number
quadratic function
quadratic regression
quadratic sequence
reverse engineering
second differences
square of a binomial

standard form of a quadratic
   equation
sum of cubes
trinomial
vertex
vertical shifts
x-intercept point
y-intercept point
zero product rule

Concepts

4.1 ▬ Quadratic Functions and Solving Quadratic Equations with Tables and Graphs

Quadratic functions are functions of degree 2 written in the form $y = ax^2 + bx + c$, where a, b, and c are real numbers, $a \neq 0$.

A sequence of numbers is the output of a quadratic function if the second differences are constant. A sequence of numbers is the output to a linear equation if the first differences are constant.

See Table 14 for important points on the graph of a quadratic function:

A quadratic equation $ax^2 + bx + c = 0$ or $ax^2 + bx + c = d$ may have zero, one, or two real-number solutions.

When solving $ax^2 + bx + c = d$ with a graph, find the intersections of $y = d$ and the graph of the corresponding quadratic function. When solving $ax^2 + bx + c = d$ with a table, find the input x where the output $f(x) = 0$ or $y = d$.

4.2 ▬ Modeling Quadratic Functions

Use the difference method to find a, b, and c in $f(x) = ax^2 + bx + c$ from a quadratic sequence:

1. List the sequence in a table and calculate the first and second differences.

2. Find a, given that the constant second difference is $2a$.

3. Work backwards from the second difference row to find c, the $f(0)$ term before the first term, $f(1)$.

4. Use $f(1)$, $x = 1$, a, and c to find b in $f(x) = ax^2 + bx + c$.

5. State the function $f(x) = ax^2 + bx + c$, substituting for a, b, and c.

Fit a quadratic function to data if the data form a sequence with constant second differences or if the graph of the data has a parabolic or approximately parabolic shape. Use quadratic regression to find quadratic functions.

TABLE 14

Name	Definition	Table	Graph
x-intercept points (where $y = 0$ or $f(x) = 0$, if they exist)	places where graph crosses x-axis	(table: x, y; 0)	(_, 0)
y-intercept point (where $x = 0$ or at $f(0)$)	places where graph crosses y-axis	(table: x, y; 0)	(0, _)
vertex (lies on the axis of symmetry)	highest or lowest point	The x-coordinate is halfway between ordered pairs with equal outputs.	

4.3 ▦ Polynomial Functions and Operations

Name a polynomial according to the number of terms it has.

Arrange the terms in a polynomial in descending order of exponents on the alphabetically first variable.

To factor $adx^2 + dbx + dc$ by table:

1. First remove common monomial factor d so that $ax^2 + bx + c$ remains:

$$d(ax^2 + bx + c)$$

2. Enter ax^2 and c into one diagonal.

3. Multiply ax^2 and c to obtain the diagonal product.

4. Find factors of the diagonal product that add to bx.

5. Enter these factors into the other diagonal.

6. Factor out the greatest common factor, find the other factors, and write the binomial pair.

7. Multiply factors to check that the inner "area" is $ax^2 + bx + c$.

Factor		
gcf	first term ax^2	
		last term c

diagonal sum bx

diagonal product acx^2

4.4 ▦ Special Products and Factors and Their Graphs

The square of a binomial, or binomial square, multiplies to a perfect square trinomial:

$$(a + b)^2 = a^2 + 2ab + b^2$$

For equations containing binomial squares,

1. The graph of $y = (x - r)^2$ has its vertex on the x-axis at $(r, 0)$.

2. The graph of $y = (x - r)^2$ is shifted horizontally r units from that of $y = x^2$.

3. The solution to $(x - r)^2 = 0$ is $x = r$, a double root.

The expression $a^2 - b^2$ is a difference of squares:

$$(a - b)(a + b) = a^2 - b^2$$

For equations containing differences of squares, where c is a positive number,

1. The graph of $y = x^2 - c^2$ has its vertex on the y-axis at $(0, -c^2)$.

2. The graph of $y = x^2 - c^2$ is shifted vertically c^2 units from that of $y = x^2$.

3. The two solutions to $x^2 - c^2 = 0$ are $x = c$ and $x = -c$.

Sum of cubes: $x^3 + y^3 = (x + y)(x^2 - xy + y^2)$

Difference of cubes: $x^3 - y^3 = (x - y)(x^2 + xy + y^2)$

The graph of $y = x^3 \pm c$ is shifted vertically from that of $y = x^3$.

4.5 ▦ Solving and Modeling Quadratic Equations with Factors

The following arc four techniques for solving quadratic equations $ax^2 + bx + c = 0$ when $f(x) = ax^2 + bx + c$.

1. Guess numbers for x, and check until the output $f(x)$ is zero.

2. Draw a graph and find x-intercepts (if they exist) where $f(x) = 0$ (Section 4.1).

3. Make a table and find x, where $f(x) = 0$ or $y = 0$ if it exists (Section 4.1).

4. Factor and use the zero product rule (if the product of two expressions is zero, then either one or the other expression is zero) (Section 4.5).

To find a quadratic equation with x-intercepts and one other point, use $y = a(x - x_1)(x - x_2)$.

---

4 Review Exercises

Place each equation in standard form and find the parameters *a*, *b*, and *c*.

1. $y = 3(x - 3)^2 - 4$
   $y = 3x^2 - 18x + 23$; $a = 3, b = -18, c = 23$

2. $y = -4(x + 2)^2 + 5$
   $y = -4x^2 - 16x - 11$; $a = -4, b = -16, c = -11$

3. $x(x - 3) = x + 7$
   $x^2 - 4x - 7 = 0$; $a = 1, b = -4, c = -7$

4. $x - 4x(2 - x) = x - x^2$
   $5x^2 - 8x = 0$; $a = 5, b = -8, c = 0$

In Exercises 5 to 7 tell whether the statement is true or false.

5. The vertex of the graph of a quadratic function is the location of the maximum or minimum y-value. true

6. The x-coordinate of the vertex of the graph of a quadratic function is halfway between the x-intercepts of the graph. true

7. The vertex of the parabola lies on the axis of symmetry of the graph. true

8. Fill in the blank: The solutions to the quadratic equation, $f(x) = 0$, give the _x-intercepts_ on the graph.

9. a. Make a table of values for $f(x) = x^2 - x - 6$. Use inputs -2 to $+2$. See Answer Section.

   b. Identify the x-intercepts. $(-2, 0), (3, 0)$

   c. Find the equation for the axis of symmetry.
$x = 0.5$

   d. What is the vertex? $(0.5, -6.25)$

   e. Graph the data. See Answer Section.

   f. Solve $x^2 - x - 6 = 6$. $\{-3, 4\}$

   g. Solve $x^2 - x - 6 = 0$. $\{-2, 3\}$

   h. Solve $x^2 - x - 6 = -7$. $\{\ \}$ or $\varnothing$

   i. State the range of $f(x)$. $y \geq -6.25$

10. a. Make a table of values for $f(x) = x^2 + x - 12$. Use inputs -2 to $+2$. See Additional Answers.

   b. Identify the x-intercepts. $(-4, 0), (3, 0)$

   c. Find the equation for the axis of symmetry. $x = -0.5$

   d. What is the vertex? $(-0.5, -12.25)$

   e. Graph the data. See Additional Answers.

   f. Solve $x^2 + x - 12 = -6$. $\{-3, 2\}$

   g. Solve $x^2 + x - 12 = 0$. $\{-4, 3\}$

   h. Solve $x^2 + x - 12 = -10$. $\{-2, 1\}$

   i. State the range of $f(x)$. $y \geq -12.25$

In Exercises 11 and 12, make a table. Round to whole numbers.

11. Surface area of spheres with radius 0 to 10 inches:
$A = 4\pi r^2$ See Answer Section.

12. Decline in purchasing power due to inflation rate r per year over 2 years: $A = P(1 - r)^2$. Let r be 0 to 8%, with income of $P = \$20{,}000$. See Additional Answers.

Use the graphs in Exercises 13 and 14 to solve the equations. Estimate x, as needed, to the nearest 0.5.

13. a. $2x^2 - 8x + 4 = 0$ $\approx\{0.5, 3.5\}$

   b. $2x^2 - 8x + 4 = 4$ $\{0, 4\}$

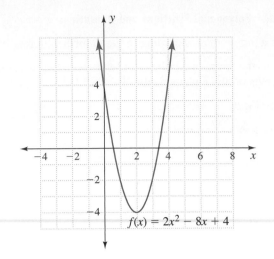

$f(x) = 2x^2 - 8x + 4$

14. a. $-2x^2 - 4x - 1 = 0$ $\approx\{-0.5, -1.5\}$

   b. $-2x^2 - 4x - 1 = -7$ $\{-3, 1\}$

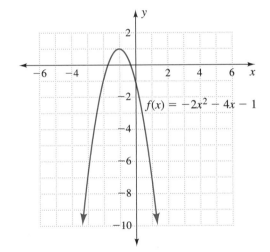

$f(x) = -2x^2 - 4x - 1$

15. Use the graph in Exercise 13 to write an equation that has

   a. One solution $2x^2 - 8x + 4 = -4$

   b. No solutions any $2x^2 - 8x + 4 < -4$

   c. Three solutions not possible

16. Use the graph in Exercise 14 to write an equation that has

   a. One solution $-2x^2 - 4x - 1 = 1$

   b. No solutions any $-2x^2 - 4x - 1 > 1$

   c. Three solutions not possible

Find out whether the sequences in Exercises 17 to 24 are linear, quadratic, or neither. Fit an equation to the linear or quadratic sequences.

17. 6, 15, 28, 45, 66, . . . $y = 2x^2 + 3x + 1$

18. $-1, 3, 13, 29, 51, \ldots$
$y = 3x^2 - 5x + 1$
19. $-1, 2, 5, 8, 11, \ldots$
$y = 3x - 4$
20. $8, 2, -4, -10, -16, \ldots$
$y = 14 - 6x$
21. $15, 12, 7, 0, -9, \ldots$
$y = 16 - x^2$
22. $16, 15, 8, -11, -48, \ldots$
neither
23. $7, 9, 5, -11, -45, \ldots$
neither
24. $24, 21, 16, 9, 0, \ldots$
$y = 25 - x^2$

25. The table below gives the boiling point of water at various elevations above sea level. The differences in boiling points are due to atmospheric pressure.

Boiling Point of Water

Elevation above Sea Level (feet), x	Boiling Temperature (°C), y
0	100
5,280 (Denver)	95
14,500 (Mt. Whitney)	85
29,500 (Mt. Everest)	71

Data from G. Tyler Miller and David G. Lygre, *Chemistry*, Wadsworth Publishing Co., 1991, p. 128.

a. Fit a linear function to the data.
$y \approx 99.93 - 0.000989x$
b. Fit a quadratic function to the data.
$y \approx 0.00000000276x^2 - 0.00107x + 100.2$
c. Graph the functions with the data, and discuss which you think is a better description.
See Answer Section; for the interval, both are good.
d. Use both functions to predict the boiling point of water near the Dead Sea, 1312 feet below sea level.
$y \approx 101.2°C; y \approx 101.6°C$
e. Suppose the boiling temperature were near 0°C at 30 miles above sea level. (There are 5280 feet per mile.) How would this change your response to part c? *Hint:* Extend the x-axis to 500,000 feet.
See below.

26. The table below gives the surface area of a sphere for various radii.

Radius	Surface Area of Sphere
1	12.57
2	50.27
3	113.10
4	201.06

a. Fit a linear function to the data. $y \approx 62.83x - 62.825$
b. Fit a quadratic function to the data.
$y \approx 12.565x^2 + 0.005x$
c. Discuss which you think is a better description. (*Hint:* Use differences.) quadratic; second differences are constant.

25. e. The straight line does not fit this new fact. The parabola fits the new fact but then turns upward after about 190,000 feet, which is also not realistic.

d. Use both functions to predict the surface of a sphere for a radius of 10. Which is correct?
quadratic; $y \approx 1256.55$; linear; $y \approx 565.5$; quadratic
e. If we double the radius from 2 units to 4 units, what happens to the surface area of the sphere?
The surface area is about 4 times larger.

27. A parabolic jet of water from a fire hose is to reach a maximum height of 100 feet when it is a horizontal distance of 30 feet from its origin. Make a sketch and label coordinates. Find the equation of the jet of water.
$y = -0.111x^2 + 6.667x$
28. A parabolic jet of water from a fire hose is to reach a maximum height of 75 feet when it is a horizontal distance of 25 feet from its origin. Make a sketch and label coordinates. Find the equation of the jet of water.
$y = -0.12x^2 + 6x$
29. Identify each of the following expressions as monomial, binomial, trinomial, or not a polynomial.

a. $x^3 - y^3$ binomial
b. $\sqrt{x} - \sqrt{y}$ not a polynomial
c. $x^2 + x^{-1}$ not a polynomial
d. $3x + 3$ binomial
e. $3x^3 - 3x^2 + 3x$ trinomial
f. $|x + 1|$ not a polynomial

30. Simplify.

a. $(x^2 + 3x) + (x^3 - 4x^2 - 5x)$ $x^3 - 3x^2 - 2x$
b. $(x^2 + 3x) - (x^3 - 4x^2 - 5x)$ $-x^3 + 5x^2 + 8x$
c. $x^2 - x + 1 - x^2 + x + 1$ 2
d. $x^2 - x + 1 - (x^2 + x + 1)$ $-2x$
e. $14 - 6(x + 3)$ $-6x \quad 4$
f. $12 - 7(x - 2)$ $-7x + 26$

31. Multiply these expressions.

a. $(x - 3)(x + 3)$ $x^2 - 9$
b. $(2x - 5)(2x - 5)$ $4x^2 - 20x + 25$
c. $(x - 1)(x^2 + x + 1)$ $x^3 - 1$
d. $(n + 4)(n + 4)$ $n^2 + 8n + 16$
e. $(2x - 3)(3x + 4)$ $6x^2 - x - 12$
f. $(x^2 + x - 1)(x^2 - x - 1)$ $x^4 - 3x^2 + 1$

32. Multiply these expressions.

a. $(x - 5)(x - 5)$ $x^2 - 10x + 25$
b. $(x + 2)(x^2 - 2x + 4)$ $x^3 + 8$
c. $(n - 2)(n + 2)$ $n^2 - 4$
d. $(3x + 4)(3x + 4)$ $9x^2 + 24x + 16$
e. $(3x - 2)(4x + 3)$ $12x^2 + x - 6$
f. $(x^2 - 3x + 3)(x^2 + 3x - 3)$ $x^4 - 9x^2 + 18x - 9$

33. Factor these expressions.

a. $x^2 + 3x - 4$
$(x + 4)(x - 1)$

b. $2x^2 - 3x$
$x(2x - 3)$

c. $2x^2 + x - 3$
$(2x + 3)(x - 1)$

d. $9x^2 + 12x + 4$
$(3x + 2)^2$

e. $x^2 - x$
$x(x - 1)$

f. $3x^2 + 6x + 3$
$3(x + 1)^2$

34. Factor these expressions.

a. $x^2 + x$
$x(x + 1)$

b. $3x^2 - x - 4$
$(x + 1)(3x - 4)$

c. $2x^2 - 9x + 10$
$(2x - 5)(x - 2)$

d. $3x^2 + 10x + 8$
$(3x + 4)(x + 2)$

e. $x^2 + x - 6$
$(x + 3)(x - 2)$

f. $4x^2 - 16x + 16$
$4(x - 2)^2$

35. Multiply, and give the name of any special expressions formed.

a. $(2x + 3)(2x - 3)$ $4x^2 - 9$; difference of squares

b. $(2x - 3)(2x - 3)$ $4x^2 - 12x + 9$; perfect square trinomial

c. $(2x + 3)(2x + 3)$ $4x^2 + 12x + 9$; perfect square trinomial

d. $(3x - 2)(2x + 3)$ $6x^2 + 5x - 6$

In Exercises 36 to 40, fill in numbers that make these expressions binomial squares.

36. $x^2 - \underline{30}\, x + 225 = (x - \underline{15}\,)^2$

37. $x^2 + 10x + \underline{25} = (x + \underline{5}\,)^2$

38. $x^2 + 20x + \underline{100} = (x + \underline{10}\,)^2$

39. $x^2 + \underline{26}\, x + 169 = (x + \underline{13}\,)^2$

40. $x^2 + \underline{16}\, x + 64 = (x + \underline{8}\,)^2$

Factor the expressions in Exercises 41 to 48.

41. $x^2 - 49$ $(x + 7)(x - 7)$

42. $4x^2 - 9$ $(2x - 3)(2x + 3)$

43. $4x^2 - 4x + 1$
$(2x - 1)^2$

44. $x^2 + 24x + 144$
$(x + 12)^2$

45. $x^2 + 16$
cannot be factored

46. $x^2 - 225$
$(x - 15)(x + 15)$

47. $4x^2 + 12x + 9$
$(2x + 3)^2$

48. $x^2 + 36$
cannot be factored

Multiply the expressions in Exercises 49 to 52.

49. $(a - b)(a^2 - 2ab + b^2)$ $a^3 - 3a^2b + 3ab^2 - b^3$

50. $(a - b)(a^2 + ab + b^2)$ $a^3 - b^3$

51. $(a + b)(a^2 - ab + b^2)$ $a^3 + b^3$

52. $(a + b)(a^2 + 2ab + b^2)$ $a^3 + 3a^2b + 3ab^2 + b^3$

Factor the expressions in Exercises 53 to 58.

53. $a^3 - 8$ $(a - 2)(a^2 + 2a + 4)$

54. $x^3 - 125$
$(x - 5)(x^2 + 5x + 25)$

55. $x^3 + 27$
$(x + 3)(x^2 - 3x + 9)$

56. $a^3 + 1$
$(a + 1)(a^2 - a + 1)$

57. $x^3 - 1000$
$(x - 10)(x^2 + 10x + 100)$

58. $a^3 + 64$
$(a + 4)(a^2 - 4a + 16)$

59. Suppose the triangle shown in the figure is an equilateral triangle containing another equilateral triangle. The area for such a triangle is given by $A = \sqrt{3}$ times $s^2/4$, where

s is the length of the side of the triangle. Write an expression for the shaded area. Factor the expression.

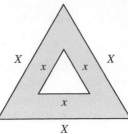

$A = \dfrac{\sqrt{3}\, X^2}{4} - \dfrac{\sqrt{3}x^2}{4} = \dfrac{\sqrt{3}}{4}(X - x)(X + x)$

60. Suppose the triangle shown in the figure contains another triangle of the same shape. The area of such a triangle is half the base times the height. Write an expression for the shaded area. Factor the expression.

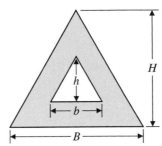

$A = \frac{1}{2}BH - \frac{1}{2}bh = \frac{1}{2}(BH - bh)$; the algebraic expression does not factor.

In Exercises 61 to 74, solve the equations by factoring.

61. $x^2 + 3x - 4 = 0$
$\{-4, 1\}$

62. $x^2 - 5x + 6 = 0$
$\{2, 3\}$

63. $x^2 + 16 = 8x$
$\{4\}$

64. $x^2 = 6x - 9$
$\{3\}$

65. $9x^2 + 12x + 4 = 0$ $\{-\frac{2}{3}\}$

66. $3x^2 + x - 4 = 0$ $\{-1\frac{1}{3}, 1\}$

67. $2x^2 + x = 3$ $\{-1\frac{1}{2}, 1\}$

68. $9x^2 = 4$ $\{\pm\frac{2}{3}\}$

69. $6x^2 + 5x + 2 = 8$ $\{-\frac{3}{2}, \frac{2}{3}\}$

70. $4x(x - 5) = 15 - 3x$
$\{-\frac{3}{4}, 5\}$

71. $24(x + 6) = -x^2$ $\{-12\}$

72. $9x^2 + 1 = 6x$ $\{\frac{1}{3}\}$

73. $4x(x - 1) = 9 - 4x$ $\{\pm\frac{3}{2}\}$

74. $5x(5x + 2) = 2(8 + 5x)$
$\{\pm\frac{4}{5}\}$

In Exercises 75 and 76, find at least one quadratic equation whose graph passes through the x-axis at the given points.

75. $x = 4$ and $x = -2$ $y = a(x - 4)(x + 2)$ for any a

76. $x = -3$ and $x = 1$ $y = a(x + 3)(x - 1)$ for any a

In Exercises 77 and 78, find a quadratic equation whose graph passes through the given points.

77. $(1, 3)$ and the intercepts in Exercise 75 $y = -\frac{1}{3}x^2 + \frac{2}{3}x + \frac{8}{3}$

78. $(-1, 4)$ and the intercepts in Exercise 76
$y = -x^2 - 2x + 3$

In Exercises 79 to 82, choose any two values for a and find different quadratic equations with the given solutions.

79. $\{-2, 4\}$ $0 = a(x + 2)(x - 4)$, a any real number

80. $\{-3, 5\}$ $0 = a(x + 3)(x - 5)$, a any real number

81. $\{-\frac{2}{5}, \frac{1}{5}\}$ $0 = a(5x + 2)(5x - 1)$, a any real number

82. $\{-\frac{3}{4}, \frac{1}{4}\}$ $0 = a(4x + 3)(4x - 1)$, a any real number

83. How does the graph of $y = x^2 - 9$ show that $x^2 - 9 = 0$ has two real-number solutions while that of $y = x^2 + 9$ shows that $x^2 + 9 = 0$ has no real-number solutions?
  $y = x^2 - 9$ crosses x-axis twice; $y = x^2 + 9$ does not cross x-axis.

84. How does the graph of $y = (x + 4)^2$ indicate that $(x + 4)^2 = 0$ has one real-number solution?
  Vertex of $y = (x + 4)^2$ is on x-axis.

85. Explain why the left side of $(x + 4)^2 = 0$ has two factors and yet the equation has only one real-number solution. What do we call this solution?
  Factors are the same; double root

86. Explain how the graph of $f(x) = (x - 2)(x^2 + 2x + 4)$ is the same as and different from the graph of $f(x) = x^3$.
  same shape, shifted down 8 units

87. Compare the graph of $y = (x + 2)^2$ with that of $y = x^2$. Discuss vertices and intercepts.
  2 units to left; $(-2, 0)$ vs. $(0, 0)$; $(0, 4)$ vs. $(0, 0)$

---

4 Chapter Test

1. Below is a table for $y = x^2 + 5x - 6$. Use the table and reasoning to find *all* the solutions to the following equations. Show clearly how you use the table and any numbers you add to it.

x	y
-8	18
-6	0
-4	-10
-2	-12
0	-6
2	8
4	30

a. $x^2 + 5x - 6 = -10$
  $\{-4, -1\}$
b. $x^2 + 5x - 6 = 8$
  $\{-7, 2\}$
c. $x^2 + 5x - 6 = 0$
  $\{-6, 1\}$
d. $x^2 + 5x - 6 = -20$
  $\{\ \}$
e. Find the equation for the axis of symmetry.
  $x = -2.5$
f. What are the coordinates of the vertex?
  $(-2.5, -12.25)$
g. What are the domain and range for $f(x)$?
  $\mathbb{R}; y \geq -12.25$

2. Give the next number in each sequence. Identify which is linear (an arithmetic sequence) and which is quadratic; find the rule for each.

a. $-6, 2, 12, 24, 38, \ldots$
  54; quadratic; $y = x^2 + 5x - 12$
b. $23, 15, 7, -1, -9, \ldots$
  -17; linear; $y = -8x + 31$

Multiply the expressions in Exercises 3 to 9. Identify the product as a perfect square trinomial, difference of squares, or other.

3. $(x - 6)(x - 2)$
  $x^2 - 8x + 12$; other
4. $(x - 12)(x + 1)$
  $x^2 - 11x - 12$; other
5. $(x + 12)(x + 1)$
  $x^2 + 13x + 12$; other
6. $(2x - 3)(x + 4)$
  $2x^2 + 5x - 12$; other
7. $(x - 2)^2$
  $x^2 - 4x + 4$; perfect square trinomial
8. $(2x + 1)(2x - 1)$
  $4x^2 - 1$; difference of squares
9. $(x - 2)(x^2 + 2x + 4)$ $x^3 - 8$; other

Factor the expressions in Exercises 10 to 15.

10. $2x^2 - 7x + 6$
  $(2x - 3)(x - 2)$
11. $x^2 - 6x + 9$
  $(x - 3)^2$
12. $x^2 - 49$
  $(x - 7)(x + 7)$
13. $12x^2 + 30x - 18$
  $6(2x - 1)(x + 3)$
14. $x^3 - 27$
  $(x - 3)(x^2 + 3x + 9)$
15. $x^3 + 125$
  $(x + 5)(x^2 - 5x + 25)$

In Exercises 16 to 18, solve by factoring.

16. $2x^2 - 7x + 6 = 0$ $\{\frac{3}{2}, 2\}$
17. $x^2 = 4(3x + 7)$ $\{-2, 14\}$
18. $4x^2 = 2x + 30$ $\{-\frac{5}{2}, 3\}$

19. A quadratic graph has x-intercepts at -3 and 4 as well as a y-intercept at 2. What is its equation?
  $y = -\frac{1}{6}x^2 + \frac{1}{6}x + 2$

20. Enter data as years since 1964—that is, $(0, 61.9)$, $(4, 60.9)$, etc.

Voter Turnout Since 1964

Year	1964	1968	1972	1976	1980
Percent	61.9	60.9	55.2	53.5	52.8
Year	1984	1988	1992	1996	2000
Percent	53.3	50.3	55.2	49.0	50.3

Source: Sidlow and Henschen, *America at Odds*, Wadsworth, 2002, p. 213.

a. Fit a linear regression equation and explain what it says about an election in 2008 (44 years after 1964). Round answers to three nonzero digits.
  $y = -0.298x + 59.6$; voter turnout will decline to 46.5%.
b. Fit a quadratic regression equation and explain what it says about an election in 2008. Round to three nonzero digits.
  $y = 0.0111x^2 - 0.697x + 61.7$; voter turnout will have bottomed out and will rise to 52.5%.

21. Quadratic functions can model the path of water in a drinking fountain, the path of a ball thrown across a field, or a cable in a suspension bridge. Describe how the x-intercepts, the y-intercept, and the vertex of a graph relate to one of these applications.

Cumulative Review of Chapters 1 to 4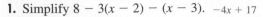

1. Simplify $8 - 3(x - 2) - (x - 3)$. $-4x + 17$

2. Solve $3x - 3(2x - 1) = 27$. $x = -8$

3. Find a linear function through the origin that is parallel to $2y = 3x + 4$. $y = \frac{3}{2}x$

4. Suppose the amount earned is a function of hours worked and wage per hour. Write an equation and solve it for each situation.

  a. How many hours will Lynn have to work at $5.50 per hour to match what Tamika earns working 40 hours at $8.00 per hour? $5.5x = 40(8)$; 58.2 hr

  b. How many hours will Shea have to work at $6.00 per hour to pay Kyra for a 3-hour automobile repair at $35 per hour? $6x = 3(35)$; 17.5 hr

5. Solve for n:

  a. $c = nt^2$ $n = \frac{c}{t^2}$

  b. $s = 2a + d(n - 1)$ $n = \frac{s - 2a + d}{d}$

6. Solve for x: $\dfrac{x + 7}{2} > x + 6$. $x < -5$

7. Evaluate $f(x) = 3x^2 - 5x + 2$ for each of the following.

  a. $f(-1)$ 10

  b. $f(0)$ 4

  c. $f(2)$ 2

  d. $f(x + 1)$ $3(x + 1)^2 - 5(x + 1) + 2$ or $3x^2 + x$

8. Give the next number in each pattern. Identify whether the sequence is linear or quadratic and fit an equation to each.

  a. 5, 12, 21, 32, 45, . . . 60; quadratic; $y = x^2 + 4x$

  b. 8, 15, 22, 29, 36, . . . 43; linear; $y = 7x + 1$

  c. 6, 12, 18, 24, 30, . . . 36; linear; $y = 6x$

  d. 4, 15, 30, 49, 72, . . . 99; quadratic; $y = 2x^2 + 5x - 3$

9. Find the equation of a parabola passing through the set of points (3, 0), (−4, 0), and (2, −4). $y = \frac{2}{3}x^2 + \frac{2}{3}x - 8$

10. Explain why these can be done mentally:

  a. $6\left(5\frac{2}{3}\right)$ distributive property of multiplication over addition: $6(5) + 6\left(\frac{2}{3}\right) = 30 + 4 = 34$

  b. $\frac{2}{7} + \frac{3}{5} + \frac{5}{7}$ commutative property of addition: $\frac{2}{7} + \frac{5}{7} + \frac{3}{5} = 1\frac{3}{5}$

11. What is the rule for this table? Explain which clues were most helpful in guessing.

x	4	2	7	? 1	10	? 8
$f(x)$	? 19	7	52	4	103	67

$f(x) = x^2 + 3$

12. Solve for x: $5 > 4 - x > 2$. $-1 < x < 2$

13. Solve $|2x - 1| > 6$. $x < -2.5$ or $x > 3.5$

14. Solve the system of equations by elimination:

$$2x - 3y = 16$$
$$3x = 10 + y$$

$x = 2, y = -4$

15. The measures of the three angles of a triangle add to 180°. The measure of the smallest angle added to three times the measure of the middle angle equals the measure of the largest angle. The measure of the middle angle is four times the measure of the smallest angle. What are the measures of the three angles? 10°, 40°, 130°

16. Sketch a graph of these ordered pairs and connect them in order:

$$(-3, 2), (-2, -3), (3, -2), (2, 3), (-3, 2)$$

Name the shape and tell why you know it is that shape. square; sides have the same length and alternate sides are perpendicular.

17. Use the graph below of $f(x) = x^2 - 2x - 3$ to solve the following.

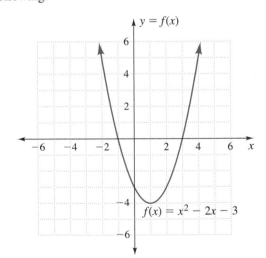

  a. $x^2 - 2x - 3 = 0$ {−1, 3}

  b. $x^2 - 2x - 3 = -3$ {0, 2}

18. Solve these equations:

 a. $x^2 + 6x - 16 = 0$ $\quad \{2, -8\}$

 b. $5x^2 + x - 4 = 0$ $\{-1, \frac{3}{7}\}$

 c. $7x^2 + 4x = 3$ $\quad \{-1, \frac{4}{5}\}$

 d. $6x^2 + 3 = 11x$ $\quad \{\frac{1}{3}, 1\frac{1}{2}\}$

19. Multiply.

 a. $(5x - 3)(3x - 5)$ $\quad 15x^2 - 34x + 15$

 b. $(3 - x)(3 + x)$ $\quad 9 - x^2$

20. Factor.

 a. $16 - x^2$ $\quad (4 - x)(4 + x)$

 b. $x^3 - 64$ $\quad (x - 4)(x^2 + 4x + 16)$

CHAPTER 5

Square Root Functions; Quadratic Equations and Inequalities

The clavadistas (cliff divers) from La Quebrada at Acapulco, Mexico, dive from a height of 35 meters above the water and need to move horizontally away from the cliff, as shown on the left in Figure 1. The path of their dives, a parabola, may be described with equations in this chapter.

Olympic platform divers attempt to keep their line of motion as vertical as possible from the end of the board to the water's surface, as shown on the right in Figure 1. The path of their dives is vertical, not parabolic. However, the position of the diver with respect to time may be modeled by the parabolas in this chapter.

This chapter continues the discussion of quadratic functions; we graph, apply, and solve quadratic equations. Within the material, we review square roots and the Pythagorean theorem.

FIGURE 1

5.1   The Square Root Function and the Pythagorean Theorem

Objectives

▮ Explain the difference between square roots and principal square roots.

▮ Identify a square root function and its domain and range.

▮ Simplify expressions containing square roots.

▮ Apply the Pythagorean theorem when appropriate.

Students who know the answers to Warm-up Exercises 1 to 4 will understand the concepts in this section more quickly.

WARM-UP

1. List the squares of the numbers from 1 to 16. 1, 4, 9, 16, 25, 36, 49, 64, 81, 100, 121, 144, 169, 196, 225, 256

2. List the squares of the numbers 20, 25, 30, 40, and 50. 400, 625, 900, 1600, 2500

3. List the squares of the numbers $\frac{1}{2}$, $\frac{1}{3}$, and $\frac{1}{4}$. $\frac{1}{4}$, $\frac{1}{9}$, $\frac{1}{16}$

4. List the squares of the numbers 0.1, 0.2, and 0.3. 0.01, 0.04, 0.09

5. List the squares of the numbers 1.5, 2.5, 3.5, and 6.5. 2.25, 6.25, 12.25, 42.25

IN THIS SECTION, we find square roots and apply square root properties. We then practice square root skills with the Pythagorean theorem and related applications.

Square Roots

Have you ever looked at the f-stop numbers on a camera lens (Figure 2) and wondered how they were related? The material in this section will help you understand the sequence of f-stops: 1.4, 2, 2.8, 4, 5.6, 8, 11, 16. We begin by examining the meaning of *square root*. We will return to f-stops in the exercises.

FIGURE 2

The **square root** of x is *the real number that, multiplied by itself, produces* x. Each positive number has two real square roots, one positive and one negative. Because $3 \cdot 3 = 9$ and $(-3)(-3) = 9$, 3 and -3 are both square roots of 9. Thus, if

$x^2 = 9$, then $x = \pm 3$. The **plus or minus sign**, $\pm$, indicates two numbers, *the positive and the negative of the number that follows the sign.*

EXAMPLE 1 Finding square roots   For what real numbers are these equations true?

a. $x^2 = 16$ **b.** $x^2 = 81$ **c.** $x^2 = -9$ **d.** $x^2 = \frac{1}{4}$

SOLUTION   The numbers that make the equation true are the square roots of the number on the right side of the equation.

a. $x = \pm 4$

b. $x = \pm 9$

c. No real number, when squared, gives -9.

d. $x = \pm \frac{1}{2}$

 To solve $x^2 = k$ with a calculator graph, find the intersections of the graphs of $y = x^2$ and $y = k$.

SQUARE ROOT SYMBOL   Did you notice that no square root symbol was used in the last several paragraphs? The square root symbol, $\sqrt{}$, has a special meaning, which we now define.

■ **DEFINITION OF PRINCIPAL SQUARE ROOT**

> The **principal square root** of x is the *positive* real number that, when multiplied by itself, produces x.

The symbol $\sqrt{x}$ refers only to the positive square root and is read "the principal square root of x." Although it is correct to say that 3 and -3 are square roots of 9, it is not correct to write $\sqrt{9} = \pm 3$, because the symbol $\sqrt{}$ stands for the positive root.

■ **MAKING A DISTINCTION**

> When evaluating $\sqrt{x}$, we have one outcome.
>
> When solving equations of the form $x^2 = k$, $k \geq 0$, we have two solutions:
> $$x = +\sqrt{k} \qquad \text{and} \qquad x = -\sqrt{k}$$

Student Note: Unless directed otherwise, round decimals to three decimal places.

In Chapter 1, we defined an irrational number as a number that cannot be written as a rational number a/b, where a and b are integers and $b \neq 0$. Irrational numbers are nonrepeating, nonterminating decimals. Irrational numbers commonly arise when we take square roots.

EXAMPLE 2 Finding square roots   Find these square roots, using a calculator as needed. Round the decimals to three decimal places. Identify each root as rational or irrational.

a. $\sqrt{2}$ **b.** $\sqrt{2.25}$ **c.** $\sqrt{3}$ **d.** $\sqrt{4}$ **e.** $\sqrt{4.41}$
f. $\sqrt{5}$ **g.** $\sqrt{6.25}$ **h.** $\sqrt{8}$ **i.** $\sqrt{-9}$ **j.** $\sqrt{128}$

SOLUTION   **a.** $\sqrt{2} \approx 1.414$, irrational   **b.** $\sqrt{2.25} = 1.5$, rational

c. $\sqrt{3} \approx 1.732$, irrational   **d.** $\sqrt{4} = 2$, rational

e. $\sqrt{4.41} = 2.1$, rational   **f.** $\sqrt{5} \approx 2.236$, irrational

g. $\sqrt{6.25} = 2.5$, rational   **h.** $\sqrt{8} \approx 2.828$, irrational

i. Not defined in the set of real numbers   **j.** $\sqrt{128} \approx 11.314$, irrational

Think about it 1: Do any of the answers in Example 2 resemble camera f-stops?

CALCULATIONS WITH SQUARE ROOTS   When using a calculator to evaluate expressions, follow the order of operations and remember that grouping symbols include fraction bars. Wait until the last operation is finished before rounding answers.

EXAMPLE 3   **Calculating square root expressions**   Estimate the value of each expression, and then evaluate the expression with a calculator.

a. $\dfrac{-2 + \sqrt{5}}{2}$
b. $\dfrac{3 - \sqrt{2}}{2}$

SOLUTION   The estimations are important because it is easy to incorrectly enter the expressions on a calculator. Use parentheses so that the entire numerator is evaluated before division by the denominator, as shown in Figure 3.

a. $\dfrac{-2 + \sqrt{5}}{2}$ is between $\dfrac{-2 + 2}{2} = 0$ and $\dfrac{-2 + 3}{2} = \dfrac{1}{2}$.

$$\dfrac{-2 + \sqrt{5}}{2} = (-2 + \sqrt{5}) \div 2 \approx 0.118$$

b. $\dfrac{3 - \sqrt{2}}{2}$ is between $\dfrac{3 - 1}{2} = 1$ and $\dfrac{3 - 2}{2} = \dfrac{1}{2}$.

$$\dfrac{3 - \sqrt{2}}{2} = (3 - \sqrt{2}) \div 2 \approx 0.793$$

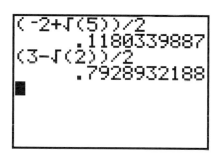

FIGURE 3   ▬

EXAMPLE 4   **Exploring square root expressions**   Estimate the value of each expression, and then evaluate the expression with a calculator. When you have finished, look for expressions that are equal and try to recall any properties from prior courses that explain why these expressions are equal.

a. $\sqrt{32}$
b. $3\sqrt{2}$
c. $\sqrt{18}$
d. $4\sqrt{2}$

SOLUTION   a. $\sqrt{32}$ is between $\sqrt{25}$ and $\sqrt{36}$, so $\sqrt{32}$ is between 5 and 6.

$$\sqrt{32} \approx 5.657$$

b. $\sqrt{2}$ is between 1 and 2, so $3\sqrt{2}$ is between 3 and 6.

$$3\sqrt{2} \approx 4.243$$

c. $\sqrt{18}$ is between $\sqrt{16}$ and $\sqrt{25}$, so $\sqrt{18}$ is between 4 and 5.

$$\sqrt{18} \approx 4.243$$

d. $\sqrt{2}$ is between 1 and 2, so $4\sqrt{2}$ is between 4 and 8.

$$4\sqrt{2} \approx 5.657$$

We have two pairs of equal answers, implying that $\sqrt{18} = 3\sqrt{2}$ and $\sqrt{32} = 4\sqrt{2}$. The expressions are equal because of the product property of square roots.

PROPERTIES OF SQUARE ROOTS   Both numerical and variable square root expressions may be simplified by using the properties of square roots.

PRODUCT PROPERTY OF SQUARE ROOTS

> For a and b positive real numbers,
>
> $$\sqrt{a \cdot b} = \sqrt{a} \cdot \sqrt{b}$$

QUOTIENT PROPERTY OF SQUARE ROOTS

> For a and b positive real numbers and $b \neq 0$,
>
> $$\sqrt{\frac{a}{b}} = \frac{\sqrt{a}}{\sqrt{b}}$$

The first two parts of Example 5 show why the expressions in parts a and c of Example 4 are equal.

EXAMPLE 5   **Applying the product property**   Use the product property of square roots to simplify these expressions without a calculator.

a. $\sqrt{32}$ **b.** $\sqrt{18}$ **c.** $\sqrt{12} \cdot \sqrt{3}$ **d.** $\dfrac{3 - \sqrt{18}}{3}$ **e.** $\dfrac{4 + \sqrt{12}}{8}$

SOLUTION   In parts a and b, we factor the radicand into a perfect square and another number. Then we take the square root of the square factor.

a. $\sqrt{32} = \sqrt{16 \cdot 2} = \sqrt{16} \cdot \sqrt{2}$ Product property

$\qquad = 4\sqrt{2}$

b. $\sqrt{18} = \sqrt{9 \cdot 2} = \sqrt{9} \cdot \sqrt{2}$ Product property

$\qquad = 3\sqrt{2}$

In part c, we use the product property in reverse.

c. $\sqrt{12} \cdot \sqrt{3} = \sqrt{12 \cdot 3}$ Product property

$\qquad\qquad = \sqrt{36} = 6$

In parts d and e, the product property sets up an expression ready to be factored.

d. $\dfrac{3 - \sqrt{18}}{3} = \dfrac{3 - \sqrt{9 \cdot 2}}{3} = \dfrac{3 - 3\sqrt{2}}{3} = \dfrac{3(1 - \sqrt{2})}{3}$

$\qquad\qquad\qquad = 1 - \sqrt{2}$

e. $\dfrac{4 + \sqrt{12}}{8} = \dfrac{4 + \sqrt{4 \cdot 3}}{8} = \dfrac{4 + 2\sqrt{3}}{8} = \dfrac{2(2 + \sqrt{3})}{2 \cdot 4}$

$\qquad\qquad\qquad\qquad = \dfrac{2 + \sqrt{3}}{4}$

Think about it 2:   The 2 and the 4 in the final step in part e of Example 5 do not simplify. Explain why.

EXAMPLE 6   **Applying the quotient property**   Use the quotient property of square roots to simplify these expressions.

a. $\sqrt{\dfrac{25}{9}}$ **b.** $\dfrac{\sqrt{18}}{\sqrt{2}}$

SOLUTION   **a.** $\sqrt{\dfrac{25}{9}} = \dfrac{\sqrt{25}}{\sqrt{9}} = \dfrac{5}{3}$

        b. $\dfrac{\sqrt{18}}{\sqrt{2}} = \sqrt{\dfrac{18}{2}} = \sqrt{9} = 3$

■ The *squares of the integers*, known as the **perfect squares** (1, 4, 9, 16, 25, . . .), and the even powers of ten may be combined with the product and quotient properties to find exact square roots.

EXAMPLE 7 **Finding exact square roots** Find the square root and show the power of ten as needed.

    a. $\sqrt{9}$ **b.** $\sqrt{900}$ **c.** $\sqrt{90{,}000}$ **d.** $\sqrt{0.0009}$ **e.** $\sqrt{\dfrac{9}{10{,}000}}$

SOLUTION   **a.** $\sqrt{9} = 3$

        b. $\sqrt{900} = \sqrt{9} \cdot \sqrt{100} = 3 \cdot 10 = 30$

        c. $\sqrt{90{,}000} = \sqrt{9} \cdot \sqrt{10{,}000} = 3 \cdot 100 = 300$

        d. $\sqrt{0.0009} = \sqrt{9} \cdot \sqrt{0.0001} = 3 \cdot 0.01 = 0.03$

        e. $\sqrt{\dfrac{9}{10{,}000}} = \dfrac{\sqrt{9}}{\sqrt{10{,}000}} = \dfrac{3}{100} = 0.03$ ■

EXAMPLE 8 **Finding exact square roots** Find these square roots without a calculator.

    a. $\sqrt{144}$ **b.** $\sqrt{1.44}$ **c.** $\sqrt{14{,}400}$ **d.** $\sqrt{\dfrac{1}{121}}$

    e. $\sqrt{2500}$ **f.** $\sqrt{0.25}$ **g.** $\sqrt{\dfrac{1}{25}}$ **h.** $\sqrt{1.69}$

SOLUTION   **a.** 12   **b.** 1.2   **c.** 120   **d.** $\frac{1}{11}$

        e. 50   **f.** 0.5   **g.** $\frac{1}{5}$ **h.** 1.3 ■

The Square Root Function

The *square root symbol* is also known as the **radical sign** (or radical). The *real number under the radical* is called the **radicand**.

principal square root of 25

$$\sqrt{25} = 5$$

radical   radicand

■ **DEFINITION OF THE SQUARE ROOT FUNCTION**

> The **square root function** is the function that gives the principal square root of a real number. The square root function is written $f(x) = \sqrt{x}$, $x \geq 0$.

EXAMPLE 9 **Finding domain and range**
    a. Graph $f(x) = \sqrt{x}$ for integer inputs between -2 and 5.
    b. Find the domain and range.

SOLUTION   **a.** Table 1 and Figure 4 show ordered pairs and a graph for $f(x) = \sqrt{x}$. Use a calculator to check the table and graph.

TABLE 1

Input x	Ouput $f(x) = \sqrt{x}$
-2	undefined in $\mathbb{R}$
-1	undefined in $\mathbb{R}$
0	0
1	1
2	1.414
3	1.732
4	2
5	2.236

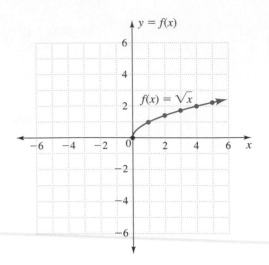

FIGURE 4

b. The domain for the square root function is $x \geq 0$. The output column in Table 1 indicates that the square root function has no output for negative inputs. The graph in Figure 4 shows no points to the left of the vertical axis. Both the table and the graph indicate that the function is undefined when $x < 0$.

The range for the square root function is $f(x) \geq 0$. Because $\sqrt{x}$ is the principal or positive square root, the output for $f(x) = \sqrt{x}$ is always non-negative. This is in agreement with both the positive outputs in Table 1 and the first-quadrant graph in Figure 4. ▬

Think about it 3: How does the graph in Figure 4 show that $\sqrt{4} = 2$, not ± 2?

The Pythagorean Theorem

Have you ever wondered how a carpenter finds out whether the corners of the forms for a rectangular foundation are right angles? As you will see in Example 13, the carpenter applies a method based on the Pythagorean theorem.

The Pythagorean theorem relates the length of the *perpendicular sides* (**legs**) of a right triangle to the length of the *longest side* (**hypotenuse**).* The legs and hypotenuse are marked on the right triangle in Figure 5. It is customary to label the sides with lowercase letters and the angles with uppercase letters. Leg a is opposite angle A, leg b is opposite angle B, and the hypotenuse c is opposite right angle C.

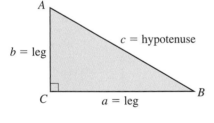

FIGURE 5

EXAMPLE 10 Exploring the Pythagorean theorem   Complete Table 2.

TABLE 2

Triangle	Leg a	Leg b	Hypotenuse c	a^2	b^2	$a^2 + b^2$	c^2
Figure 6							
Figure 7							
Figure 8							

a. In the first three columns of the table, enter the lengths of the sides of the right triangles shown in Figures 6, 7, and 8.

*The Greek mathematician and philosopher Pythagoras (ca. 582–500 B.C.) is credited with being the first to record a proof of this relationship for right triangles, but historical documents from China indicate that the relationship was known long before Pythagoras's time.

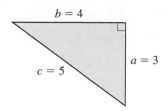

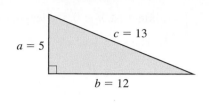

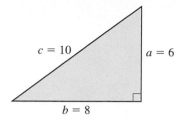

FIGURE 6   **FIGURE 7**   **FIGURE 8**

b. Fill in the remaining four columns.

c. Compare the last two columns.

SOLUTION   Table 3 contains the measures of the sides of three right triangles and the squares of the sides.

TABLE 3

Triangle	Leg a	Leg b	Hypotenuse c	a^2	b^2	$a^2 + b^2$	c^2
Figure 6	3	4	5	9	16	25	25
Figure 7	5	12	13	25	144	169	169
Figure 8	6	8	10	36	64	100	100

The Pythagorean theorem formula appears in the last two columns, where

$$a^2 + b^2 = c^2$$

PYTHAGOREAN THEOREM

If a triangle is a right triangle, then the sum of the squares of the lengths of the two shorter sides (legs) is equal to the square of the length of the longest side (hypotenuse).

Note that the theorem refers to *the sum of the squares of the lengths of the two shorter sides (legs)*. Unless the corners of the triangles are labeled with letters, a and b are arbitrarily assigned to the legs—the two perpendicular sides.

EXAMPLE 11   Finding missing sides   Identify the perpendicular sides in the right triangles below. What is the length of the missing side n in each drawing?

a.

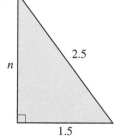

b.

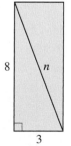

c.

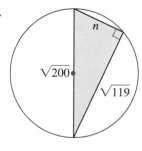

SOLUTION   Because n describes a physical object, $n \geq 0$.

a. The sides labeled n and 1.5 are perpendicular. The hypotenuse is 2.5.

$$a^2 + b^2 = c^2 \qquad \text{Substitute } a = n, b = 1.5, c = 2.5.$$
$$n^2 + 1.5^2 = 2.5^2$$
$$n^2 + 2.25 = 6.25$$
$$n^2 = 4 \qquad \text{Find the square root.}$$
$$n = \pm 2$$

Because $n \geq 0$, the length of the leg is 2 units.

b. The sides labeled 3 and 8 are perpendicular. The hypotenuse is n.

$$a^2 + b^2 = c^2 \quad \text{Substitute } a = 3, b = 8, c = n.$$
$$3^2 + 8^2 = n^2$$
$$9 + 64 = n^2$$
$$73 = n^2 \quad \text{Find the square root.}$$
$$\pm\sqrt{73} = n$$

Because $n \geq 0$, the length of the hypotenuse is $\sqrt{73}$ units. If the square root is irrational, it is permissible to retain the radical sign or to use a calculator to obtain a decimal approximation. In this case, $n = \sqrt{73} \approx 8.544$, rounded to the nearest thousandth.

c. The sides labeled n and $\sqrt{119}$ are perpendicular, and the hypotenuse, $\sqrt{200}$, is the diameter of the circle.

$$a^2 + b^2 = c^2 \quad \text{Substitute } a = n, b = \sqrt{119}, c = \sqrt{200}.$$
$$n^2 + (\sqrt{119})^2 = (\sqrt{200})^2$$
$$n^2 + 119 = 200$$
$$n^2 = 81 \quad \text{Find the square root.}$$
$$n = \pm 9$$

Because $n \geq 0$, the length of the leg is 9 units.

Changing the positions of the *if* and *then* statements in the Pythagorean theorem gives the converse of the Pythagorean theorem. The converse is also true.

CONVERSE OF THE PYTHAGOREAN THEOREM

> If the sum of the squares of the lengths of the two shorter sides (legs) is equal to the square of the length of the longest side (hypotenuse), then the triangle is a right triangle.

The converse is used to find out whether a given triangle is a right triangle.

EXAMPLE 12 **Identifying right triangles** Each triangle below is drawn to look like a right triangle. Which *are* right triangles?

a.

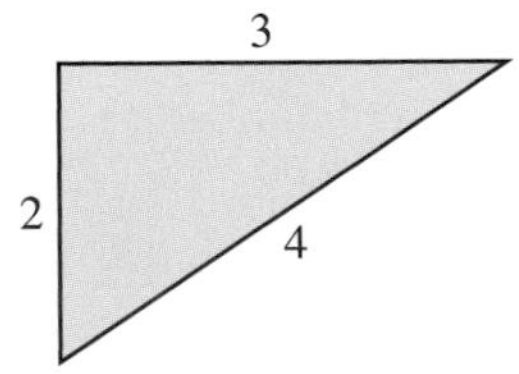

b.

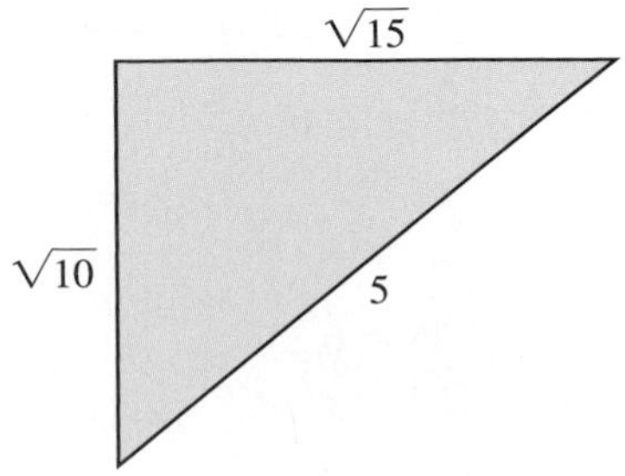

SOLUTION   The $\overset{?}{=}$ sign means that we are checking for equality.

a. The shorter sides are 2 and 3. The third side is $c = 4$.

$$a^2 + b^2 = c^2 \qquad \text{Substitute } a = 2, b = 3, c = 4.$$

$$2^2 + 3^2 \overset{?}{=} 4^2$$

$$4 + 9 \overset{?}{=} 16$$

$$13 \neq 16$$

This triangle is *not* a right triangle.

b. The shorter sides are $\sqrt{10}$ and $\sqrt{15}$. The third side is $c = 5$.

$$a^2 + b^2 = c^2 \qquad \text{Substitute } a = \sqrt{10}, b = \sqrt{15}, c = 5.$$

$$(\sqrt{10})^2 + (\sqrt{15})^2 \overset{?}{=} 5^2$$

$$10 + 15 \overset{?}{=} 25$$

$$25 = 25 \qquad \checkmark$$

This triangle is a right triangle.   ▬

Applications

The Pythagorean theorem has many practical applications. In work with physical objects or geometric shapes, when only a positive answer is sensible, the negative square root is omitted.

EXAMPLE 13   **Applying the Pythagorean theorem: foundations**   Carpenters know that the diagonals of a square or rectangle are equal. They use this fact to check that foundation forms have been built correctly. If the dimensions for a rectangular addition to a house are 16 feet by 20 feet, how long should the diagonal be? (See Figure 9.)

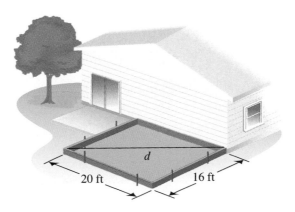

FIGURE 9

SOLUTION   Because the 16 feet and the 20 feet represent the sides of the rectangle and hence are perpendicular, the diagonal will form the hypotenuse c, $c \geq 0$.

$$a^2 + b^2 = c^2 \qquad \text{Substitute } a = 16, b = 20.$$

$$16^2 + 20^2 = c^2$$

$$656 = c^2 \qquad \text{Find the square root.}$$

$$25.6 \approx c$$

The diagonal is approximately 25.6 feet. The carpenter knows that the other diagonal is the hypotenuse of an identical triangle, and so it too must measure 25.6 feet. ▬

▬ In Figure 10, the transmission tower is shown first in side view, with three supporting cables. The top view is from the sky, looking down the tower, and shows the positions of the cables attached to the tower. In Example 14, we find the lengths of the cables.

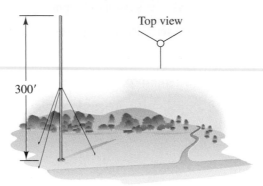

Top view

300′

FIGURE 10

EXAMPLE 14 Applying the Pythagorean theorem: tower cables   The 300-foot tower in Figure 10 is to be braced by three cables attached halfway up the tower. The cables are to be attached to the ground at a distance from the tower equal to one-fourth the height of the tower. What is the total length of cable needed to brace the tower?

SOLUTION We assume that the tower is held at a right angle to the ground. The 300-foot tower will be braced at a height of

$$\tfrac{1}{2}(300) = 150 \text{ ft}$$

The cable will be attached to the ground at

$$\tfrac{1}{4}(300) = 75 \text{ ft} \quad \text{from the base of the tower}$$

We label the perpendicular sides of the right triangle 150 ft and 75 ft (see Figure 11). The hypotenuse is labeled c.

$$a^2 + b^2 = c^2 \qquad \text{Substitute } a = 150, b = 75.$$
$$150^2 + 75^2 = c^2$$
$$28,125 = c^2 \qquad \text{Find the square root.}$$
$$168 \text{ ft} \approx c$$

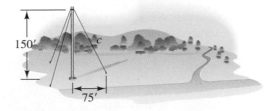

150′

c

75′

FIGURE 11

There are three cables, so

$$3(168 \text{ ft}) = 504 \text{ ft}$$

is needed. The person ordering the cable will need to round up to allow some extra for fastening the cable to the ground supports and the tower. ▬

ANSWER BOX

Warm-up: **1.** 1, 4, 9, 16, 25, 36, 49, 64, 81, 100, 121, 144, 169, 196, 225, 256 **2.** 400, 625, 900, 1600, 2500 **3.** $\frac{1}{4}, \frac{1}{9}, \frac{1}{16}$ **4.** 0.01, 0.04, 0.09 **5.** 2.25, 6.25, 12.25, 42.25 **Think about it 1:** $\sqrt{2} \approx 1.414$ is close to the f-stop 1.4; $\sqrt{4} = 2$ is equal to the next f-stop; $\sqrt{8} \approx 2.828$ is close to the f-stop 2.8; $\sqrt{128} \approx 11.314$ is close to the f-stop 11. **Think about it 2:** The 2 is not a factor of the numerator. Only common factors can be simplified in changing a fraction to lowest terms. **Think about it 3:** The graph is in the first quadrant only. If, for the input 4, the output were ±2, the graph would be in the first and fourth quadrants.

5.1 Exercises

In Exercises 1 to 8, for what real numbers are the equations true?

1. $x^2 = 225$ $x = \pm 15$

2. $x^2 = 169$ $x = \pm 13$

3. $x^2 = 121$ $x = \pm 11$

4. $x^2 = 64$ $x = \pm 8$

5. $x^2 = 10{,}000$ $x = \pm 100$

6. $x^2 = 900$ $x = \pm 30$

7. $x^2 = -36$ no real number

8. $x^2 = -4$ no real number

In Exercises 9 to 18, simplify without a calculator.

9. 15^2 225

10. 25^2 625

11. $\sqrt{49}$ 7

12. $\sqrt{144}$ 12

13. 11^2 121

14. 16^2 256

15. $\sqrt{0.01}$ 0.1

16. $\sqrt{0.0001}$ 0.01

17. $\sqrt{1.96}$ 1.4

18. $\sqrt{0.16}$ 0.4

In Exercises 19 to 30, use the product property of square roots to simplify.

19. $\sqrt{75}$ $5\sqrt{3}$

20. $\sqrt{18}$ $3\sqrt{2}$

21. $\sqrt{40}$ $2\sqrt{10}$

22. $\sqrt{20}$ $2\sqrt{5}$

23. $\sqrt{200}$ $10\sqrt{2}$

24. $\sqrt{90}$ $3\sqrt{10}$

25. $\sqrt{150}$ $5\sqrt{6}$

26. $\sqrt{45}$ $3\sqrt{5}$

27. $\sqrt{80}$ $4\sqrt{5}$

28. $\sqrt{98}$ $7\sqrt{2}$

29. $\sqrt{32}$ $4\sqrt{2}$

30. $\sqrt{300}$ $10\sqrt{3}$

In Exercises 31 to 34, estimate and then evaluate the expressions using a calculator. Round to the nearest thousandth.

31. $1 + \sqrt{2}$ 2.414

32. $1 - 2\sqrt{2}$ −1.828

33. $\dfrac{1 - \sqrt{3}}{2}$ −0.366

34. $1 - \sqrt{2}$ −0.414

In Exercises 35 to 38, estimate and then evaluate the expressions. Round to the nearest thousandth. Use the product property of square roots to show why the answers are equal to those in Exercises 31 to 34, respectively.

35. $\dfrac{2 + \sqrt{8}}{2}$ 2.414

36. $\dfrac{2 - \sqrt{32}}{2}$ −1.828

37. $\dfrac{3 - \sqrt{27}}{6}$ −0.366

38. $\dfrac{5 - \sqrt{50}}{5}$ −0.414

In Exercises 39 to 44, simplify the radical expressions without a calculator.

39. $\dfrac{4 + \sqrt{8}}{4}$ $\dfrac{2 + \sqrt{2}}{2}$

40. $\dfrac{2 - \sqrt{2}}{2}$ simplified

41. $\dfrac{2 + \sqrt{2}}{2}$ simplified

42. $\dfrac{4 - \sqrt{8}}{4}$ $\dfrac{2 - \sqrt{2}}{2}$

43. $\dfrac{4\sqrt{8}}{4}$ $2\sqrt{2}$

44. $\dfrac{3\sqrt{6}}{3}$ $\sqrt{6}$

Find the square roots in Exercises 45 and 46 without a calculator.

45. a. $\sqrt{0.81}$ 0.9

b. $\sqrt{0.0025}$ 0.05

c. $\sqrt{\dfrac{1}{36}}$ $\dfrac{1}{6}$

d. $\sqrt{\dfrac{121}{49}}$ $\dfrac{11}{7}$

e. $\sqrt{2.25}$ 1.5

f. $\sqrt{640{,}000}$ 800

g. $\dfrac{\sqrt{48}}{\sqrt{3}}$ 4

h. $\dfrac{\sqrt{162}}{\sqrt{2}}$ 9

46. a. $\sqrt{0.64}$ 0.8

b. $\sqrt{0.0009}$ 0.03

c. $\sqrt{\dfrac{1}{81}}$ $\dfrac{1}{9}$

d. $\sqrt{\dfrac{169}{25}}$ $\dfrac{13}{5}$

e. $\sqrt{6.25}$ 2.5

f. $\sqrt{360{,}000}$ 600

g. $\dfrac{\sqrt{75}}{\sqrt{3}}$ 5

h. $\dfrac{\sqrt{128}}{\sqrt{2}}$ 8

47. These questions refer to the graphs of $x = y^2$ on page 84 and $y = \sqrt{x}$ on page 266.

   a. How are the graphs alike? How are they different?
     identical for $y \geq 0$; $y = \sqrt{x}$ is a function, $x = y^2$ is not.

   b. List three number facts that show why the graphs have points in common.
     $4 = 2^2$, $2 = \sqrt{4}$; $9 = 3^2$, $3 = \sqrt{9}$; $16 = 4^2$, $4 = \sqrt{16}$

   c. What restriction on $y = \sqrt{x}$ indicates why its graph is different from that of $x = y^2$?
     $y = \sqrt{x}$ has restriction $y \geq 0$; $x = y^2$ does not.

   d. Why are there no points on either graph in quadrants 2 and 3?
     x cannot be a negative number if y is a real number.

48. True or false:

   a. The graph of $y = \sqrt{x}$ passes the vertical line test.   true

   b. The graph of $y = \sqrt{x}$ is half a parabola with vertex at $(0, 0)$.   true

   c. The ordered pair $(4, -2)$ lies on the graph of $y = \sqrt{x}$.   false

   d. The ordered pair $\left(\frac{1}{2}, \frac{1}{4}\right)$ lies on the graph of $y = \sqrt{x}$.   false

49. Showing your steps carefully, use the product property of square roots and factoring to simplify this expression:

$$\dfrac{\sqrt{125} + \sqrt{45} - (\sqrt{75} + \sqrt{27})}{\sqrt{5} - \sqrt{3}}$$ 8

50. Complete the table, and describe the mathematical pattern behind the f-stops on a camera.

f-stop	Radical form
1.4	$\sqrt{2}$
2	$\sqrt{4} = 2$
2.8	$\sqrt{8} = 2\sqrt{2}$
4	$\sqrt{16} = 4$
5.6	$\sqrt{32} = \underline{4}\ \sqrt{2}$
8	$\sqrt{64} = 8$
11	$\sqrt{128} = \underline{8}\ \sqrt{2}$
16	$\sqrt{256} = 16$

▬ **Error Analysis**

51. What is wrong with this simplification?

$$\dfrac{6 - 2\sqrt{3}}{2} = 3 - 2\sqrt{3}$$ Factor 2 from numerator before changing to lowest terms.

52. What is wrong with this simplification?

$$\dfrac{\sqrt{35}}{7} = \sqrt{5}$$ Division is incorrect because 7 is not under the radical sign.

In Exercises 53 to 58, solve for the missing side x.

53. a.

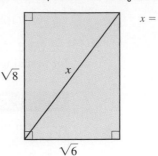

$x = \sqrt{14}$

   b.

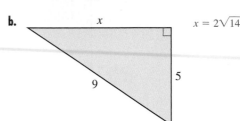

$x = 2\sqrt{14}$

54. a.

$x = 5\sqrt{2}$

   b.

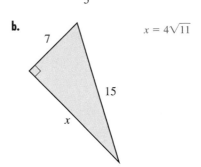

$x = 4\sqrt{11}$

55. a.   **b.**

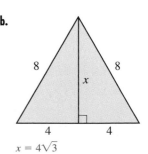

$x = 3\sqrt{3}$

$x = 4\sqrt{3}$

56. a.   **b.**

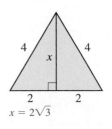

$x = 5\sqrt{3}$

$x = 2\sqrt{3}$

57. a.

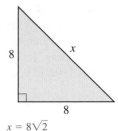

$x = 8\sqrt{2}$

b.

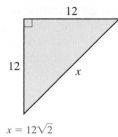

$x = 12\sqrt{2}$

c.

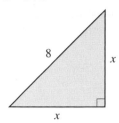

$x = 4\sqrt{2}$

58. a.

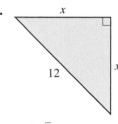

$x = 6\sqrt{2}$

b.

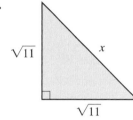

$x = \sqrt{22}$

c.

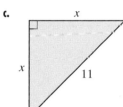

$x = \dfrac{11\sqrt{2}}{2}$

59. If the legs are equal in length, multiply by $\dfrac{\sqrt{2}}{\_\_\_}$ to get the length of the hypotenuse.

60. If the shorter leg is half the hypotenuse, multiply by $\dfrac{\sqrt{3}}{\_\_\_}$ to get the length of the other leg.

Show which of the sets of numbers in Exercises 61 to 70 could be lengths of sides of right triangles.

61. 12, 16, 20
right triangle

62. 15, 20, 25
right triangle

63. 4.5, 6, 7.5
right triangle

64. 10.5, 14, 17.5
right triangle

65. 8, 9, 12
not a right triangle

66. 11, 12, 16
not a right triangle

67. 16, 30, 34
right triangle

68. 11, 60, 61
right triangle

69. $\sqrt{5}, \sqrt{11}, 4$
right triangle

70. $\sqrt{3}, \sqrt{6}, 3$
right triangle

Standard measuring tapes show inches, not decimal portions of feet. In Exercises 71 and 72, change decimal portions of feet to inches. Multiply the decimal by 12 inches/1 foot.

71. a. 0.97 ft 11.6 in.   **b.** 0.05 ft 0.6 in.

72. a. 0.08 ft 1 in.   **b.** 0.6 ft 7.2 in.

In Exercises 73 to 76, find the diagonal (in feet and inches) of a room with the given dimensions.

73. 12 feet by 12 feet ≈16 ft 11.6 in.

74. 11 feet by 28 feet ≈30 ft 1 in.

75. 13 feet by 19 feet ≈23 ft 0.3 in.

76. 15 feet by 20 feet 25 ft

77. A support wire for a 50-foot-tall radio antenna is to be fastened halfway up the antenna. The other end of the wire is to be attached to the ground, 16 feet from the base of the tower. Rounding up to the nearest tenth of a foot, how long a wire is needed? 29.7 ft

78. A support wire for a 30-foot electric pole is to be fastened two-thirds of the distance up the pole. The other end of the wire is to be attached to the ground, 14 feet from the base of the pole. Rounding up to the nearest tenth of a foot, how long a wire is needed? 24.4 ft

A ladder is in a safe position if the height it reaches on the wall is four times the distance of the base from the wall. To the nearest tenth of a foot, find the length of the ladder in Exercises 79 and 80.

79. The ladder must reach a 16-foot height. 16.5 ft

80. The ladder must reach a 20-foot height. 20.6 ft

81. Both the Pythagorean theorem and its converse are true, so the two statements can be written as one using the phase *if and only if*. Write such a statement. A triangle is a right triangle if and only if the sum of the squares of the lengths of the two shorter sides is equal to the square of the length of the longest side.

■■■■ **Projects**

82. Binomial Squares Use the figure to help prove the Pythagorean theorem by showing that $a^2 + b^2 = c^2$ is true for any a, b, and c that satisfy the conditions of the theorem. The conditions are that a and b are legs of a right triangle and c is the hypotenuse. We are given that $\triangle PQR$ is a right triangle.

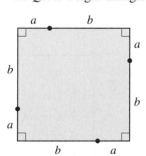
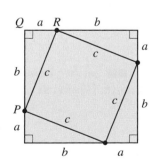

a. What is the area of the first square, which has $a + b$ on each side? $(a + b)^2$

b. Write the area of each of the five parts in the second figure, and then add the areas. $4\left(\frac{1}{2} ab\right) + c^2$

c. Because the two figures are both squares with sides $a + b$, set the expressions for their areas equal and show that $a^2 + b^2 = c^2$.

$(a + b)^2 = 4\left(\frac{1}{2} ab\right) + c^2,\ a^2 + 2ab + b^2 = 2ab + c^2,\ a^2 + b^2 = c^2$

83. Pythagorean Theorem and Solving Quadratic Equations Prove that the numbers 3, 4, and 5 are the *only* *consecutive positive integers* that satisfy the Pythagorean theorem.

a. Let x, $x + 1$, and $x + 2$ be any three consecutive numbers. Write the Pythagorean theorem with this consecutive number notation, and simplify.
$x^2 + (x + 1)^2 = (x + 2)^2; x^2 - 2x - 3 = 0$

b. Find the solutions to the resulting equation by graphing and factoring. $x = -1, x = 3$

c. Explain why the set $\{3, 4, 5\}$ is the only possible solution.
$\{-1, 0, 1\}$ satisfies the conditions but does not form a right triangle.

84. Triangle Investigation We have seen that if $a^2 + b^2 = c^2$, the triangle is a right triangle. What can be said about a triangle if $a^2 + b^2 > c^2$? What can be said about a triangle if $a^2 + b^2 < c^2$? Draw several triangles, with sides in centimeters, to illustrate your conclusions.
Angle between a and b is $<90°$; angle between a and b is $>90°$.

85. F-stops One set of numbers on a camera lens is the f-stops: 1.4, 2, 2.8, 4, 5.6, 8, 11, 16. Exercise 50 lists the mathematical sequence generating the numbers. Research the photographic meaning. Why do these numbers appear on the camera?

5.2 Solving Quadratic Equations with Square Roots

Objectives

▮ Solve a quadratic equation by taking the square root.

▮ Find x-intercepts for $y = ax^2 + c$ with square roots.

▮ Apply the Pythagorean theorem in finding facts about special triangles.

▮ Solve a formula with squares and square roots.

WARM-UP

1. Multiply $(a + b)(a + b)$. $a^2 + 2ab + b^2$

2. Factor $x^2 + 6x + 9$. $(x + 3)^2$

3. Factor $4x^2 - 12x + 9$. $(2x - 3)^2$

4. Solve $|x + 3| = 2$ $x = -1$ or $x = -5$

5. Solve $|2x - 3| = 2$. $x = \frac{5}{2}$ or $x = \frac{1}{2}$

IN THIS SECTION, we continue looking at ways of solving quadratic equations. A list of solution methods for quadratic equations is given here for your reference.

▮ **SUMMARY: TECHNIQUES FOR SOLVING QUADRATIC EQUATIONS** $ax^2 + bx + c = 0$

When $f(x) = ax^2 + bx + c$,

1. Guess numbers for x, and check until the output $f(x)$ is zero.

2. Draw a graph and find x-intercepts, where $f(x) = 0$ (Sections 4.1 and 4.4).

3. Make a table and find x, where $f(x) = 0$ or $y = 0$ (Section 4.1).

4. Factor and use the zero product rule (Section 4.5).

5. If $b = 0$, solve for x^2 and take the square root of both sides (Section 5.2).

6. In $f(x) = d$, if $f(x)$ is a perfect square, take the square root of both sides (Section 5.2).

7. Complete the square (Section 5.3).

8. Use the quadratic formula (Section 5.3).

9. Use the solve feature on a programmable calculator or computer (Section 5.3).

Absolute Value and Square Roots

In Example 1, we review solving equations with tables and graphs while justifying the two solutions to $x^2 = 4$. When we solve $x^2 = 4$, we take the principal square root of both sides and obtain $\sqrt{x^2} = 2$.

EXAMPLE 1   **Exploring the graph of $y = \sqrt{x^2}$**   Solve $\sqrt{x^2} = 2$ by table and graph. What other equation could be used to represent the graph of $y = \sqrt{x^2}$?

SOLUTION   Table 4 shows $y = \sqrt{x^2}$ for several inputs. To solve $\sqrt{x^2} = 2$, we look for the value 2 in the second column of Table 4. The value 2 appears twice, so there are two solutions to $\sqrt{x^2} = 2$: $x = -2$ and $x = 2$, or $x = \pm 2$.

The graphs of $y = 2$ and $y = \sqrt{x^2}$ are shown in Figure 12. The graph of $y = \sqrt{x^2}$ appears to be the same as the graph of the absolute value of x, $y = |x|$. (You may want to check the graph on your own graphing calculator.) The graph shows that $\sqrt{x^2}$ is not x. The graph of $y = x$ is a straight line.

Figure 12 shows the solutions, located at the intersections of the graph of $y = 2$ with the graph of $y = \sqrt{x^2}$. We again conclude that the solutions to $\sqrt{x^2} = 2$ are $x = -2$ and $x = 2$, or $x = \pm 2$.

TABLE 4

x	$y = \sqrt{x^2}$
-3	3
-2	2
-1	1
0	0
1	1
2	2

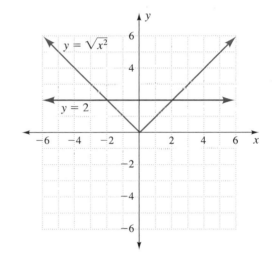

FIGURE 12

Because $y = \sqrt{x^2}$ and $y = |x|$ have the same graph, the source of the two solutions in $\sqrt{x^2} = 2$ is the absolute value, not the square root.

▬ $\sqrt{w^2}$ **and** $|w|$ | In solving equations, $\sqrt{w^2}$ can be replaced by $|w|$.

Solving by Taking Square Roots

When $ax^2 + bx + c$ is a perfect square trinomial, the quadratic equation $ax^2 + bx + c = d$ can be solved with the technique of *taking the square root of both sides of an equation*.

Consider the equation $x^2 + 6x + 9 = 4$. The expression $x^2 + 6x + 9$ is equivalent to $(x + 3)^2$, a *binomial square*. A binomial square permits us to apply square roots to solve the quadratic equation.

EXAMPLE 2   **Solving by taking the square root**   Solve $x^2 + 6x + 9 = 4$.

SOLUTION

$$x^2 + 6x + 9 = 4 \qquad \text{Factor the left side.}$$

$$(x + 3)^2 = 4 \qquad \text{Take the square root of both sides.}$$

$$\sqrt{(x + 3)^2} = \sqrt{4} \qquad \sqrt{w^2} = |w|$$

$$|x + 3| = 2 \qquad x + 3 \text{ may be positive or negative.}$$

$$x + 3 = \pm 2$$

$$\text{Either } x + 3 = 2 \quad \text{or} \quad x + 3 = -2$$

$$x = -1 \quad \text{or} \quad x = -5$$

Check: $(-1)^2 + 6(-1) + 9 \overset{?}{=} 4 \; \checkmark$

$\quad\quad\quad (-5)^2 + 6(-5) + 9 \overset{?}{=} 4 \; \checkmark$

Example 2 illustrates the process we will need in order to derive a general solution to quadratic equations in Section 5.3. As with factoring, if the equation solves quickly with the square root method, use square roots. If not, try another method.

EXAMPLE 3 **Solving by taking the square root** Solve $4x^2 - 12x + 9 = 4$.

$$4x^2 - 12x + 9 = 4 \qquad \text{Factor the left side.}$$

$$(2x - 3)^2 = 4 \qquad \text{Take the square root of both sides.}$$

$$\sqrt{(2x - 3)^2} = \sqrt{4} \qquad \text{Simplify with } \sqrt{w^2} = |w|.$$

$$|2x - 3| = 2 \qquad 2x - 3 \text{ may be positive or negative.}$$

$$2x - 3 = \pm 2$$

$$\text{Either } 2x - 3 = 2 \quad \text{or} \quad 2x - 3 = -2$$

$$x = \tfrac{5}{2} \quad \text{or} \quad x = \tfrac{1}{2}$$

Check: $4\left(\tfrac{5}{2}\right)^2 - 12\left(\tfrac{5}{2}\right) + 9 \overset{?}{=} 4 \; \checkmark$

$\quad\quad\quad 4\left(\tfrac{1}{2}\right)^2 - 12\left(\tfrac{1}{2}\right) + 9 \overset{?}{=} 4 \; \checkmark$

Solving with square roots is most easily done when $b = 0$ in $ax^2 + bx + c$.

EXAMPLE 4 **Solving by taking the square root** Solve $75x^2 - 12 = 0$.

SOLUTION

$$75x^2 - 12 = 0 \qquad \text{Add 12 to both sides.}$$

$$75x^2 = 12 \qquad \text{Divide both sides by 75.}$$

$$x^2 = \tfrac{12}{75} \qquad \text{Simplify the fraction to lowest terms.}$$

$$x^2 = \tfrac{4}{25} \qquad \text{Take the square root of both sides.}$$

$$\sqrt{x^2} = \frac{\sqrt{4}}{\sqrt{25}} \qquad \text{Apply the quotient property and } \sqrt{w^2} = |w|.$$

$$|x| = \tfrac{2}{5} \qquad \text{Inside the absolute value, } x \text{ is } \tfrac{2}{5} \text{ or } x \text{ is } -\tfrac{2}{5}.$$

$$x = \pm \tfrac{2}{5}$$

Check: $75\left(\tfrac{2}{5}\right)^2 - 12 \overset{?}{=} 0 \; \checkmark$

$\quad\quad\quad 75\left(-\tfrac{2}{5}\right)^2 - 12 \overset{?}{=} 0 \; \checkmark$

Special Triangles

In the next two examples, we apply the Pythagorean theorem to derive information about special triangles. Focus on how the square roots and the Pythagorean theorem

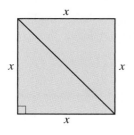

FIGURE 13

are used rather than memorizing the geometric relationships themselves. When the variable w in $\sqrt{w^2}$ is positive, we don't need the absolute value.

ISOSCELES RIGHT TRIANGLES   When a square is cut with a diagonal line (see Figure 13), two identical triangles are formed. These triangles are unique in at least three ways:

1. The angle measures of each triangle are 45°, 45°, and 90°.

2. Each *triangle has two equal sides* and thus is an **isosceles triangle**.

3. Each *triangle has a right angle along with the two equal angles* and thus is an **isosceles right triangle**.

The hypotenuse (here, the diagonal of the square) could be the distance from home plate to second base in baseball or the distance traveled when we cut across an empty block instead of taking the sidewalks around the block.

EXAMPLE 5 Applying the Pythagorean theorem: isosceles right triangles   Find the hypotenuse of each isosceles right triangle. Compare the length of the hypotenuse to the length of the leg.

a.   **b.**

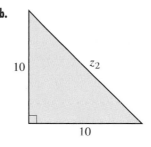

SOLUTION   The equal legs form the base and height of each triangle; $z_1 > 0$ and $z_2 > 0$.

a. For the smaller triangle,

$$z_1{}^2 = 5^2 + 5^2$$
$$z_1{}^2 = 25 + 25$$
$$z_1 = \sqrt{50} = 5\sqrt{2}$$

The length of the hypotenuse of the triangle is $\sqrt{2}$ times the length of the leg.

b. For the larger triangle,

$$z_2{}^2 = 10^2 + 10^2$$
$$z_2{}^2 = 100 + 100$$
$$z_2 = \sqrt{200} = 10\sqrt{2}$$

The length of the hypotenuse of the triangle is $\sqrt{2}$ times the length of the leg.

RIGHT TRIANGLES WITHIN EQUILATERAL TRIANGLES   Another important triangle is formed by drawing the height in an equilateral triangle. The **height of a triangle** (also called the **altitude**) is *the perpendicular distance from a vertex (corner) to the opposite side*. The **equilateral triangle** has several special properties:

1. The triangle has three equal sides.

2. The height of the equilateral triangle is a line of symmetry, cutting the triangle into two identical triangles.

3. The two triangles have angle measures 30°, 60°, and 90°, as shown in Figure 14, and are therefore right triangles.

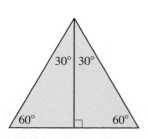

FIGURE 14

EXAMPLE 6 | **Applying the Pythagorean theorem: equilateral triangles**   If the side of the equilateral triangle in Figure 15 is 2 inches, what is the height of the triangle?

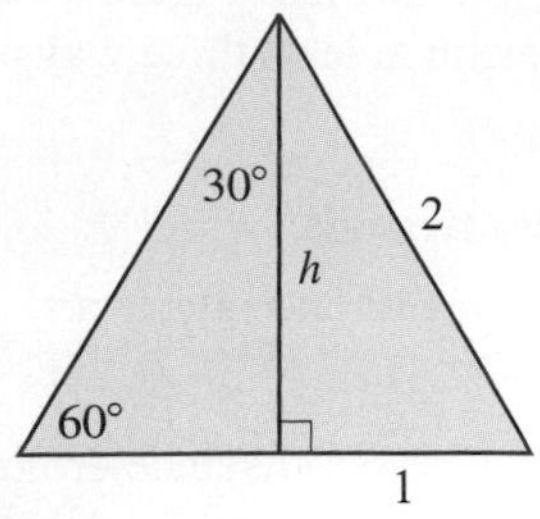

SOLUTION   The height cuts the triangle symmetrically into two identical triangles, so the base of the right triangle is half the side, or 1 inch. From the right triangle, we have

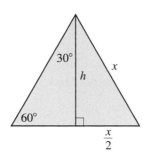

$$a^2 + b^2 = c^2 \qquad \text{Substitute } a = 1, b = h, c = 2; h > 0.$$

$$1^2 + h^2 = 2^2$$

$$1 + h^2 = 4$$

$$h^2 = 4 - 1$$

$$h^2 = 3 \qquad \text{Find the square root.}$$

$$h = \sqrt{3} \approx 1.732 \text{ in.}$$

FIGURE 15

One of the most powerful uses of algebra is in finding general results or formulas, as required in proving relationships. In Example 7, we use the Pythagorean theorem to solve for the height of any equilateral triangle.

EXAMPLE 7 | **Applying the Pythagorean theorem to geometry**   Find the height of any equilateral triangle in terms of its side (see Figure 16). Let x be the length of the side of the triangle.

FIGURE 16

SOLUTION   Remember that $h > 0$ and $x > 0$.

$$a^2 + b^2 = c^2 \qquad \text{Substitute } a = h, b = \frac{x}{2}, c = x.$$

$$h^2 + \left(\frac{x}{2}\right)^2 = x^2 \qquad \text{Remove parentheses.}$$

$$h^2 + \frac{x^2}{4} = x^2 \qquad \text{Subtract } \frac{x^2}{4} \text{ from both sides.}$$

Student Note: See Section 7.6 for more on this step.

$$h^2 = x^2 - \frac{x^2}{4} \qquad \text{Find a common denominator.}$$

$$h^2 = \frac{4x^2 - x^2}{4} \qquad \text{Simplify.}$$

$$h^2 = \frac{3x^2}{4} \qquad \text{Find the square root.}$$

$$h = \frac{x\sqrt{3}}{2}, x > 0, h > 0$$

It is customary to write $\dfrac{\sqrt{3}x}{2}$ rather than $\dfrac{x\sqrt{3}}{2}$. However, carelessly drawn radical signs can cause the x to incorrectly appear under the radical sign. It is recommended that you write variables before the radical unless you are told to do otherwise.

Solving Formulas Containing Squares and Square Roots

We can apply our work with squares and square roots to formulas. Remember that a solution plan is important in solving formulas.

EXAMPLE 8 **Solving formulas containing squared expressions**   Solve the formula for the area of a circle, $A = \pi r^2$, for r. Make an inverse order of operations plan.

SOLUTION   ***Plan:*** The order of operations on r is to square and then multiply by π. The inverse order of operations is to divide by π and then take the square root.

$$A = \pi r^2 \qquad \text{Divide by } \pi.$$

$$\frac{A}{\pi} = r^2 \qquad \text{Take the square root.}$$

$$\sqrt{\frac{A}{\pi}} = r$$

The absolute value symbol is not needed around r because r represents the radius of a circle and is already positive.

Check: $A \overset{?}{=} \pi\left(\sqrt{\dfrac{A}{\pi}}\right)^2$ ✓

When we solved $A = \pi r^2$ for r in Example 8, we obtained a radical expression, $r = \sqrt{A/\pi}$. With some restrictions, squares and square roots are inversely related, as are (1) multiplication and division and (2) addition and subtraction.

EXAMPLE 9 **Solving formulas containing square roots: distance to the horizon**   On the moon, the approximate distance in miles seen to the horizon from a height h, in feet, is $d = \sqrt{3h/8}$. Solve for h.

SOLUTION   ***Plan:*** The variable h is multiplied by 3 and divided by 8, and then the square root is applied. The inverse order of operations is to square, multiply by 8, and divide by 3.

$$d = \sqrt{\frac{3h}{8}} \qquad \text{Square both sides.}$$

$$d^2 = \frac{3h}{8} \qquad \text{Multiply by 8.}$$

$$8d^2 = 3h \qquad \text{Divide by 3.}$$

$$\frac{8d^2}{3} = h$$

Check: $d \overset{?}{=} \sqrt{\dfrac{3\left(\dfrac{8d^2}{3}\right)}{8}}$ ✓

Remember that the distance d is positive, so $\sqrt{d^2} = d$ without absolute value.

5.2 Exercises

1. In Example 2, we took the square root of $(x + 3)^2 = 4$ and found $|x + 3| = 2$.

   a. On the calculator, graph $y = (x + 3)^2$ and $y = 4$. Find the intersections. $x = -5, x = -1$

   b. On the calculator, graph $y = |x + 3|$ and $y = 2$. Find the intersections. $x = -5, x = -1$

   c. Graph $y = \sqrt{((x + 3)^2)}$. Explain the result.
     same as for $y = |x + 3|$; $\sqrt{(w^2)} = |w|$

2. In Example 3, we took the square root of $(2x - 3)^2 = 4$ and found $|2x - 3| = 2$.

   a. On the calculator, graph $y = (2x - 3)^2$ and $y = 4$. Find the intersections. $x = \frac{1}{2}, x = \frac{5}{2}$

   b. On the calculator, graph $y = |2x - 3|$ and $y = 2$. Find the intersections. $x = \frac{1}{2}, x = \frac{5}{2}$

   c. Graph $y = \sqrt{(2x + 3)^2}$. Explain the result.
     same as for $y = |2x - 3|$; $\sqrt{(w^2)} = |w|$

Graph the expressions in Exercises 3 to 6 with a graphing calculator, and record the graphs. Parentheses are needed with both types of expressions.

3. $y = \sqrt{(x - 4)^2}$ and $y = |x - 4|$ See Answer Section.

4. $y = \sqrt{(x + 1)^2}$ and $y = |x + 1|$ See Additional Answers.

5. $y = \sqrt{(x - 1)^2}$ and $y = |x - 1|$ See Answer Section.

6. $y = \sqrt{(x + 4)^2}$ and $y = |x + 4|$ See Additional Answers.

In Exercises 7 to 28, solve by taking square roots.

7. $x^2 + 4x + 4 = 9$
   $\{-5, 1\}$

8. $x^2 - 6x + 9 = 25$
   $\{-2, 8\}$

9. $x^2 - 10x + 25 = 36$
   $\{-1, 11\}$

10. $x^2 + 8x + 16 = 4$
   $\{-6, -2\}$

11. $x^2 + 8x + 16 = 9$
   $\{-7, -1\}$

12. $x^2 + 2x + 1 = 16$
   $\{-5, 3\}$

13. $x^2 - 4x + 4 = 16$
   $\{-2, 6\}$

14. $x^2 - 8x + 16 = 100$
   $\{-6, 14\}$

15. $4x^2 - 1 = 0$ $\{-\frac{1}{2}, \frac{1}{2}\}$

16. $9x^2 - 4 = 0$ $\{-\frac{2}{3}, \frac{2}{3}\}$

17. $4x^2 + 4x + 1 = 0$ $\{-\frac{1}{2}\}$

18. $25x^2 + 10x + 1 = 9$
   $\{-\frac{4}{5}, \frac{2}{5}\}$

19. $4x^2 - 16 = 0$ $\{-2, 2\}$

20. $12x^2 - 3 = 0$
   $\{-\frac{1}{2}, \frac{1}{2}\}$

21. $x^2 = 225$ $\{\pm 15\}$

22. $x^2 = 169$ $\{\pm 13\}$

23. $x^2 - 121 = 0$ $\{\pm 11\}$

24. $x^2 - 144 = 0$ $\{\pm 12\}$

25. $3x^2 - 48 = 0$ $\{\pm 4\}$

26. $9x^2 - 81 = 0$ $\{\pm 3\}$

27. $5x^2 - 45 = 0$ $\{\pm 3\}$

28. $12x^2 - 48 = 0$ $\{\pm 2\}$

In Exercises 29 to 32, solve. Evaluate roots and round to the nearest tenth.

29. $16x^2 - 48 = 0$ $\{\pm 1.7\}$

30. $8x^2 - 56 = 0$ $\{\pm 2.6\}$

31. $9x^2 - 45 = 0$ $\{\pm 2.2\}$

32. $12x^2 - 36 = 0$ $\{\pm 1.7\}$

In Exercises 33 to 38, find the length of the hypotenuse of an isosceles right triangle with legs of the length given. Sketch a triangle and show your work. Check with the results of Example 5.

33. 20 meters $20\sqrt{2}$ m

34. 12 meters $12\sqrt{2}$ m

35. $\sqrt{2}$ feet   2 ft

36. $\sqrt{8}$ inches   4 in.

37. $\sqrt{18}$ inches   6 in.

38. $\sqrt{32}$ feet   8 ft

In Exercises 39 to 42, find the height of the triangles described. Sketch a triangle and show your work.

39. Side of the equilateral triangle is 8 inches. $4\sqrt{3}$ in.

40. Side of the equilateral triangle is 12 inches. $6\sqrt{3}$ in.

41. Side of the equilateral triangle is 5 inches. $\frac{5\sqrt{3}}{2}$ in.

42. Side of the equilateral triangle is 10 inches. $5\sqrt{3}$ in.

43. Use the results of Example 7 to find the area of an equilateral triangle in terms of its side x by substituting into $A = \frac{1}{2}bh$. $A = \frac{x^2\sqrt{3}}{4}$

44. Suppose we double the lengths of the legs of a right triangle. Use the Pythagorean theorem to show how the length of the hypotenuse will change.   Let $C =$ new hypotenuse. $(2a)^2 + (2b)^2 = C^2$, $4(c^2) = C^2$, $2c = C$; hypotenuse will double.

45. Suppose $x =$ length of the legs of an isosceles right triangle. Use the Pythagorean theorem to find a formula for the length of the hypotenuse.
   $x^2 + x^2 = h^2$, $2x^2 = h^2$, $h = x\sqrt{2}$

46. Find the area of an isosceles right triangle with perpendicular sides of length 5 meters. Find the area of another with perpendicular sides of length 10 meters. Compare the two areas. *Hint:* See Example 5.
   12.5 m²; 50 m²; larger area is 4 times the smaller area.

47. In many applications, the area of a circle is given in terms of diameter instead of radius. This is because the diameter of a cylinder in a combustion engine, the diameter of a log, or the diameter of a highway culvert is more accurately measured than the radius.

   a. The area of a circle in terms of diameter is $A = \pi d^2/4$. Solve for diameter in terms of area.
   $d = 2\sqrt{A/\pi}$

   b. Evaluate $\pi/4$, and round to four decimal places. Observe the location of the four digits on the calculator keyboard. What is it about the digits in $\pi/4$ that makes $A = \pi d^2/4$ an easy formula to use?
   0.7854; keys form a square.

For Exercises 48 to 50, use the formulas in Exercise 47.

48. A round drainage pipe (culvert) under a highway needs to have a cross-sectional area of 255 square inches. What diameter culvert is needed? ≈ 18 in.

49. The cross-sectional area of a water pipe is 5026 square inches. To the nearest tenth, what is the diameter of the pipe in inches and in feet? (This pipe was estimated to supply 60 gallons of water per person per day to a city of 1 million inhabitants in 1909.)   80 in.; 6.7 ft

50. A household water supply pipe has a cross-sectional area of 0.7854 square inch. What is the diameter of the pipe? 1 in.

51. The surface area of a spherical balloon is $A = 4\pi r^2$. To the nearest tenth, what is the radius of the balloon when the surface area is 100 square inches? $r \approx 2.8$ in.

52. The surface area of a cube with side length x is $A = 6x^2$. To the nearest tenth, what is the edge of the cube if the area is 100 square inches? $x \approx 4.1$ in.

53. The **vertical motion equation** for the height of an object is $h = -\frac{1}{2}gt^2 + v_0 t + h_0$. If an object is dropped, v_0 is zero.

   a. Show that the length of time required for the object to hit the ground when dropped from a height h_0 is

$$t = \sqrt{\frac{2h_0}{g}}$$

   (*Hint:* When the object hits the ground, the height h will be zero.)

In parts b, c, and d, use the vertical motion formula. The acceleration due to gravity is $g \approx 9.81$ m/sec^2, or $g \approx 32.2$ ft/sec^2. Round to the nearest tenth.

   b. What is the time needed for an object dropped from 555 feet to reach the ground? (555 feet is the height of the Washington Monument in our nation's capital.) 5.9 sec

   c. A worker on a cooling tower building project drops a metal bolt from 49.05 meters. How long will it take to reach the ground? 3.2 sec

   d. What is the time needed for a metal bolt dropped from the top of the Warsaw (Poland) Radio mast to hit the ground? The mast is 646.4 meters high. 11.5 sec

54. In auto accident investigations, the speed r of a car in miles per hour is estimated by measuring the length of the skid marks. Round to the nearest whole number. For a wet concrete road and skid marks of length L, measured in feet, the formula for speed is $r = \sqrt{12L}$.

   a. Solve the formula for L. $L = r^2/12$

   b. If a car is traveling at 55 miles per hour, how long will the skid marks be? 252 ft

   c. If a car made a 100-foot skid mark, how fast was it traveling? 35 mph

For a dry concrete road and skid marks of length L, measured in feet, the formula for speed is $r = \sqrt{24L}$.

   d. Solve the formula for L. $L = r^2/24$

   e. If a car is traveling at 45 miles per hour, how long will the skid marks be? 84 ft

   f. If a car made a 100-foot skid mark, how fast was it traveling? 49 mph

   g. For a speed r, the length of a skid mark on wet pavement is __2__ times the length of a skid mark on dry pavement. For a skid mark of length L, the speed on dry pavement is __$\sqrt{2}$__ times the speed on wet pavement.

55. On Earth, the approximate distance in miles seen to the horizon from a height h, in feet, is $d = \sqrt{3h/2}$. Round answers to the nearest tenth.

   a. Solve for the height, h. $h = \dfrac{2d^2}{3}$

   b. Find the distance seen on Earth from a height of 500 feet. 27.4 mi

   c. Find the distance seen on the moon from a height of 500 feet (see Example 9). 13.7 mi

   d. Find the height on Earth needed to see a distance of 9 miles. 54 ft

   e. Find the height on the moon needed to see a distance of 9 miles. 216 ft

   f. The curvature of the moon is greater, so we see __half__ as far on the moon for a given height and must be __4__ times higher on the moon to see the same distance.

◼ **Error Analysis** Zero product rule applied incorrectly; must start with $ax^2 + bx + c = 0$ or take square root of both sides.

56. A student solves $x^2 + 6x + 9 = 4$ by factoring the left side and setting each factor equal to 4. What is wrong?

57. The calculator shows "abs(" when absolute value is chosen. Explain what happens when the closing parenthesis,), is omitted. It is placed at the end of the expression.

58. Do $f(x) = $ abs $x - 2$ and $f(x) = $ abs$(x - 2)$ give the same graphs? If not, how is the first entered on the calculator? no; abs$(x) - 2$

◼ **Project**

59. Simplifying Square Root Expressions Match the expressions in parts a to f with their simplified form. Choose from

$$a\sqrt{b},\ a^2\sqrt{b},\ |a|\sqrt{b},\ \sqrt{ab},\ ab,\ b\sqrt{a},\ |b|\sqrt{a},\ b^2\sqrt{a}$$

   a. $\sqrt{b^2 a}$, b is positive or zero $b\sqrt{a}$

   b. $\sqrt{a^2 b}$, a is any real number $|a|\sqrt{b}$

   c. $\sqrt{a^2 b}$, a is positive or zero $a\sqrt{b}$

   d. $\sqrt{ab^4}$ $b^2\sqrt{a}$

   e. $\sqrt{a^4 b}$ $a^2\sqrt{b}$

   f. $\sqrt{ab^2}$, b is any real number $|b|\sqrt{a}$

5 **Mid-Chapter Test**

1. Factor the radicands and change to $a\sqrt{b}$ form.

   a. $\sqrt{75}$ $5\sqrt{3}$ **b.** $\sqrt{18}$ $3\sqrt{2}$

   c. $\sqrt{32}$ $4\sqrt{2}$ **d.** $\sqrt{20}$ $2\sqrt{5}$

2. Match the square root or radical expression to one of the following:

$$\sqrt{27}, \quad \sqrt{48}, \quad \sqrt{50}, \quad \sqrt{12}, \quad \sqrt{8}, \quad \sqrt{28}, \quad \sqrt{72}$$

   a. $2\sqrt{2}$ $\sqrt{8}$ **b.** $5\sqrt{2}$ $\sqrt{50}$ **c.** $4\sqrt{3}$ $\sqrt{48}$

   d. $6\sqrt{2}$ $\sqrt{72}$ **e.** $3\sqrt{3}$ $\sqrt{27}$ **f.** $2\sqrt{3}$ $\sqrt{12}$

3. Simplify without a calculator.

   a. $\sqrt{2500}$ 50 **b.** $\sqrt{0.0025}$ 0.05

   c. $\sqrt{2.25}$ 1.5 **d.** $\sqrt{22,500}$ 150

4. Simplify without a calculator.

   a. $\sqrt{\dfrac{49}{64}}$ $\dfrac{7}{8}$ **b.** $\dfrac{\sqrt{24}}{\sqrt{6}}$ 2

   c. $\dfrac{\sqrt{48}}{\sqrt{27}}$ $\dfrac{4}{3}$ **d.** $\dfrac{1}{\sqrt{81}}$ $\dfrac{1}{9}$

5. Solve $x^2 = 16$ and $x = \sqrt{16}$. Explain why the solution sets are not the same.
$x^2 = 16$ has two solutions, $\{-4, 4\}$; $x = \sqrt{16}$ has one, $\{4\}$.

6. Evaluate with a calculator. Round to the nearest thousandth.

   a. $\dfrac{3 + \sqrt{6}}{3}$ 1.816 **b.** $\dfrac{\sqrt{30} - 5}{10}$ 0.048

7. Why is $(3 + \sqrt(6)/3) \approx 3.82$ incorrect for part a of Exercise 6?
Closing parenthesis needed after $\sqrt(6)$; last parenthesis is meaningless.

8. Which of these sets of numbers could be the lengths of the sides of a right triangle?

   a. $\{7.5, 10, 12.5\}$ yes   **b.** $\{8, 15, 17\}$ yes

   c. $\{6, 8, 12\}$ no

9. Solve for the missing side in each figure.

   a.

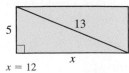

$x = 12$

   b.

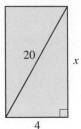

$x = 8\sqrt{6}$

   c.

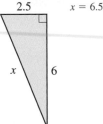

$x = 6.5$

10. Solve for x.

   a. $|x - 2| = 3$ $\{-1, 5\}$ **b.** $|x + 5| = 2$ $\{-7, -3\}$

11. Solve by taking the square root.

   a. $x^2 = 121$ $\{\pm 11\}$ **b.** $x^2 = 0.04$ $\{\pm 0.2\}$

   c. $(x + 1)^2 = 4$ $\{-3, 1\}$ **d.** $(x + 4)^2 = 25$ $\{-9, 1\}$

   e. $x^2 + 4x + 4 = 9$ $\{-5, 1\}$ **f.** $x^2 + 6x + 9 = 1$ $\{-4, -2\}$

12. An isosceles triangle has legs with measure $\sqrt{8}$. What is the length of the hypotenuse?   4

13. An equilateral triangle has sides of length 6. Sketch the triangle, and draw in the height.

   a. True or false: The height is an axis of symmetry. true

   b. True or false: The height cuts the base into two equal parts. true

   c. True or false: The height is longer than the sides of the original triangle. false

   d. Find the length of the height. $3\sqrt{3}$

14. Solve for t: $h = -\frac{1}{2} gt^2$, where g is a negative number.
$t = \sqrt{-2h/g}$

15. Solve for A: $d = \sqrt{4A/\pi}$, where A is the area of a circle and d is the diameter. $A = \pi d^2/4$

5.3 Solving Quadratic Equations with the Quadratic Formula

Objectives

▮ Use completing the square to form the squares of binomials.

▮ Derive the quadratic formula from $ax^2 + bx + c = 0$.

▮ Use the quadratic formula to solve equations.

▮ Use the quadratic formula to find the vertex.

WARM-UP

Multiply these binomials.

1. $(x - 2)(x - 2)$ $x^2 - 4x + 4$

2. $(x + 3)(x + 3)$ $x^2 + 6x + 9$

3. $\left(x - \frac{5}{2}\right)^2$ $x^2 - 5x + \frac{25}{4}$

4. $\left(x - \frac{7}{6}\right)^2$ $x^2 - \frac{7}{3}x + \frac{49}{36}$

5. $\left(x + \dfrac{b}{2a}\right)^2$ $x^2 + \frac{bx}{a} + \frac{b^2}{4a^2}$

Factor each expression into the square of a binomial.

6. $x^2 - 12x + 36$ $(x - 6)^2$

7. $a^2 - 5a + 6.25$ $(a - 2.5)^2$

8. $x^2 + 3x + 2.25$ $(x + 1.5)^2$

Fill in the missing parts.

9. $(x + 2)^2 = (x + 2)(x + 2) = x^2 + 4x + \underline{\quad 4 \quad}$

10. $(x - 5)^2 = (x - 5)(x - 5) = x^2 - \underline{\quad 10 \quad}x + \underline{\quad 25 \quad}$

11. $(x + \underline{\quad 6 \quad})^2 = x^2 + 12x + \underline{\quad 36 \quad}$

12. $(x + \underline{\quad 2.5 \quad})^2 = x^2 + 5x + \underline{\quad 6.25 \quad}$

TABLES, GRAPHS, FACTORING, AND SQUARE ROOTS have all been used to solve quadratic equations. None of the methods work as well for all quadratic equations as the quadratic formula. In this section, you will learn how to complete the square (an essential step in solving for the quadratic formula), how to solve for the quadratic formula, and how to use the quadratic formula.

Completing the Square

In this transition from the table (Section 4.3) to an area model for completing the square, the square is emphasized visually.

When we multiply the binomial $(x + y)$ by itself, the *square of the binomial gives a three-term expression* called a **perfect square trinomial**: $(x + y)^2 = x^2 + 2xy + y^2$. We may think of the square of a binomial in terms of the area of a square with sides $x + y$, as in Figure 17.

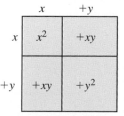

$$x^2 + 2xy + y^2 = (x + y)^2$$

binomial square

FIGURE 17

In Example 1, we find what must be added to an expression to build a binomial square.

EXAMPLE 1   Exploration: finding a binomial square   What must be added to the expressions in Figure 18 in order to obtain a binomial square? Draw the resulting binomial square, and describe it in symbols.

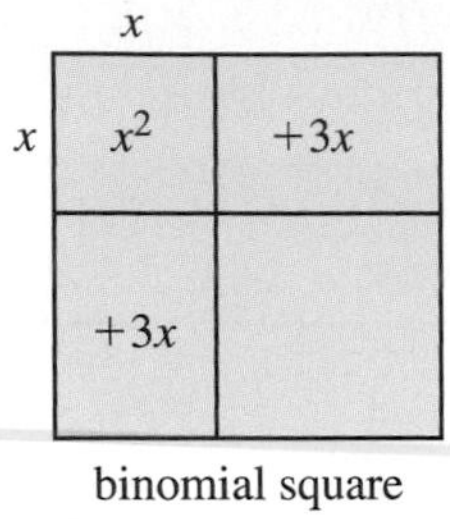

$$x^2 + 3x + 3x + \underline{\hspace{1cm}} = (x + \underline{\hspace{1cm}})^2$$

binomial square

FIGURE 18

SOLUTION   We start by finding the factors, or sides of the square. In Figure 19, the factors (sides) are both $x + 3$. These factors produce a 9 in the lower right corner, so a 9 would be added to make a square. The completed square is

$$x^2 + 6x + 9 = (x + 3)^2$$

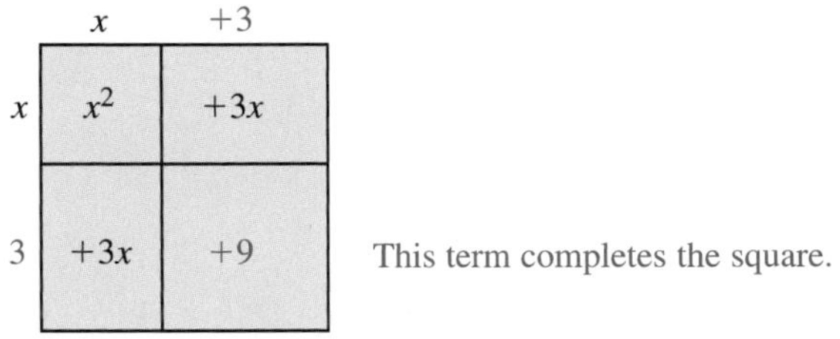

This term completes the square.

FIGURE 19

DEFINITION OF COMPLETING THE SQUARE

> **Completing the square** is the process of finding the term to add to $x^2 + bx$ to build a perfect square trinomial or binomial square.

In a binomial square, the diagonal terms (lower left to upper right) are equal and may be found by taking half of the second term.

EXAMPLE 2   Completing the square with a square   Complete the square on $x^2 - 5x$. Draw the resulting binomial square and describe it in symbols.

SOLUTION   Half of $-5x$ is $-\frac{5}{2}x$, so

$$x^2 - 5x = x^2 - \frac{5}{2}x - \frac{5}{2}x$$

To complete the square, we enter $x^2 - \frac{5}{2}x - \frac{5}{2}x$ into Figure 20. In Figure 21, we factor to find the sides of the square.

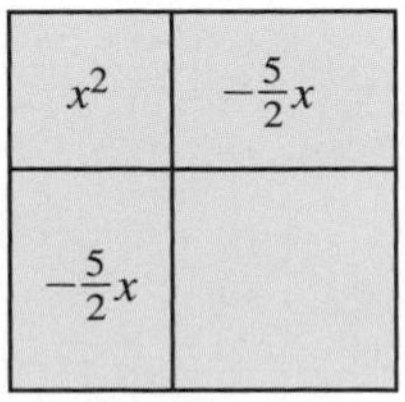

Half the coefficient of x

This term completes the square.

FIGURE 20 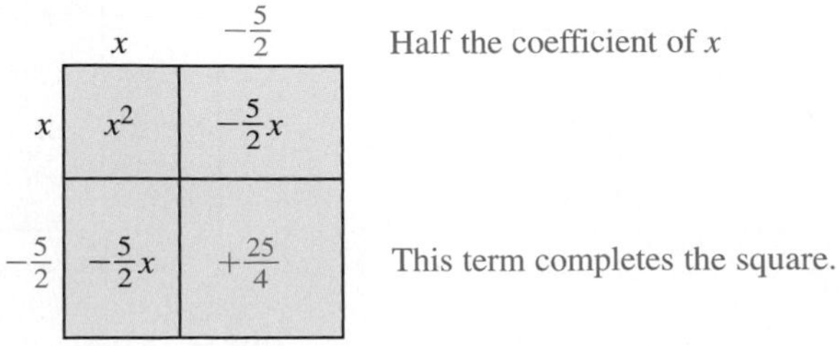   **FIGURE 21**

The square of $-\frac{5}{2}$ is $\frac{25}{4}$, which is added to make a square. The completed square is

$$x^2 - 5x + \tfrac{25}{4} = \left(x - \tfrac{5}{2}\right)^2$$

Completing the Square in an Equation

To complete the square in an equation, we build a binomial square on one side. We keep the equation balanced by adding the same term to both sides.

EXAMPLE 3   **Completing the square in an equation**   Find what must be added to each side of the quadratic equation $x^2 - 12x = -11$ to create a binomial square on the left. Draw a binomial square figure as needed. Solve the resulting equation.

SOLUTION   We complete the square on $x^2 - 12x$. As shown in Figure 22, $\left(\frac{1}{2} \cdot -12\right)^2 = (-6)^2 = 36$ is needed to obtain a binomial square.

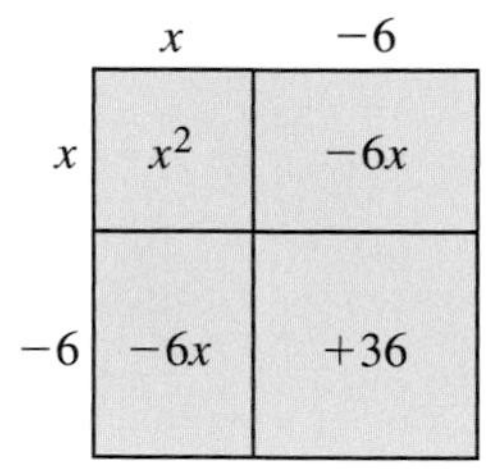

	x	-6
x	x^2	$-6x$
-6	$-6x$	$+36$

FIGURE 22

$$x^2 - 12x = -11 \qquad \text{Add 36 to complete the square.}$$
$$x^2 - 12x + 36 = -11 + 36 \qquad \text{Factor the left side.}$$
$$(x - 6)^2 = 25 \qquad \text{Take the square root.}$$
$$\sqrt{(x-6)^2} = \sqrt{25} \qquad \text{Replace } \sqrt{w^2} \text{ with } |w|.$$
$$|x - 6| = 5 \qquad \text{Solve the absolute value equation.}$$
$$x - 6 = \pm 5 \qquad \text{Set up two equations.}$$

Either $x - 6 = 5$ or $x - 6 = -5$

$x = 11$ or $x = 1$

The purpose of introducing the completing the square process is to derive a general rule, the quadratic formula, not to solve equations by completing the square. The following summary indicates how to complete the square in an equation when $a \neq 1$.

SUMMARY: COMPLETING THE SQUARE IN AN EQUATION

For the quadratic equation $ax^2 + bx + c = 0$,

1. Subtract c from both sides.

2. Divide all terms by a, the coefficient of x^2.

3. a. Find half the coefficient of the x term to get $\dfrac{b}{2a}$.

   b. Square the result to get $\left(\dfrac{b}{2a}\right)^2$.

   c. Add $\left(\dfrac{b}{2a}\right)^2$ to both sides of the equation.

4. Write the left side as the square of a binomial.

EXAMPLE 4   **Completing the square in an equation**   Find what must be added to both sides of the quadratic equation $3x^2 - 7x - 20 = 0$ to create a binomial square on the left. Draw a binomial square figure.

SOLUTION $3x^2 - 7x - 20 = 0$ We need $x^2 + \dfrac{b}{a}x$ to complete the square.

Step 1: $3x^2 - 7x = 20$ Add 20 to both sides.

Step 2: $x^2 - \frac{7}{3}x = \frac{20}{3}$ Divide both sides by 3.

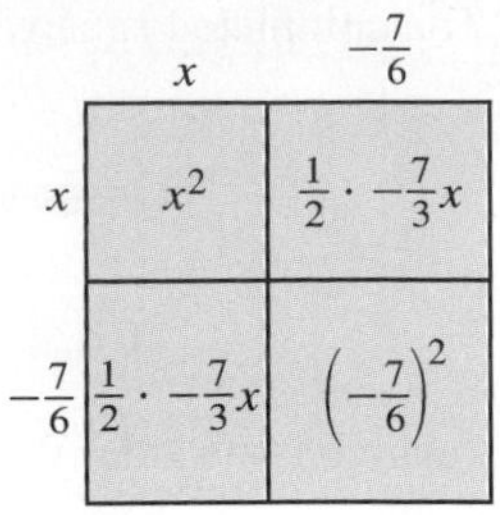

FIGURE 23

The completion of the square is shown in Figure 23.

Step 3: $\quad x^2 - \frac{7}{3}x + \left(-\frac{7}{6}\right)^2 = \frac{20}{3} + \left(-\frac{7}{6}\right)^2$ Complete the square. Add the square of half the coefficient of x, $\left(\frac{1}{2} \cdot -\frac{7}{3}\right)^2$ or $\left(-\frac{7}{6}\right)^2$, to both sides.

Step 4: $\quad\quad\quad\quad \left(x - \frac{7}{6}\right)^2 = \frac{20}{3} + \left(-\frac{7}{6}\right)^2$ Write the left side of step 3 as a binomial square.

The dimensions of the square in Figure 23 appear on the left side of step 4. See the Answer Box for the rest of the steps in solving the equation.   ▬

EXAMPLE 5   **Completing the square in an equation**   Find what must be added to each side of the quadratic equation $ax^2 + bx + c = 0$ to create a binomial square on the left. Draw a binomial square figure.

SOLUTION $\quad\quad\quad\quad ax^2 + bx + c = 0$ We need $x^2 + \dfrac{b}{a}x$ to complete the square.

Step 1: $\quad\quad\quad ax^2 + bx = -c$ Subtract c from both sides.

Step 2: $\quad\quad\quad x^2 + \dfrac{b}{a}x = -\dfrac{c}{a}$ Divide both sides by a.

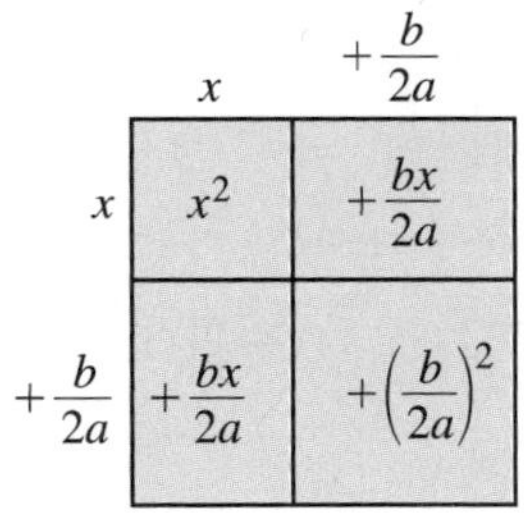

FIGURE 24

The completion of the square is shown in Figure 24.

Step 3: $\quad x^2 + \dfrac{bx}{a} + \left(\dfrac{b}{2a}\right)^2 = -\dfrac{c}{a} + \left(\dfrac{b}{2a}\right)^2$ Complete the square. Add the square of half the coefficient of x, $\left(\dfrac{1}{2} \cdot \dfrac{b}{a}\right)^2$ or $\left(\dfrac{b}{2a}\right)^2$, to both sides.

Step 4: $\quad\quad\quad \left(x + \dfrac{b}{2a}\right)^2 = -\dfrac{c}{a} + \left(\dfrac{b}{2a}\right)^2$ Write the left side of step 3 as a binomial square.

The dimensions of the square in Figure 24 appear on the left side of step 4.   ▬

The Quadratic Formula

▬ **QUADRATIC FORMULA**

> If an equation can be written $ax^2 + bx + c = 0$, $a \neq 0$, then its solutions, if they exist, are
>
> $$x = \frac{-b \pm \sqrt{b^2 - 4ac}}{2a}$$

The quadratic formula is found by applying equation-solving steps to the equation $ax^2 + bx + c = 0$. The solution requires *completing the square* as well as taking square roots and solving an absolute value equation. We will do the solution in four parts.

EXAMPLE 6   **Solving $ax^2 + bx + c = 0$, Part 1**   Explain what has been done in these two steps in solving the quadratic equation.

$$ax^2 + bx + c = 0$$

Step 1: $\quad\quad ax^2 + bx = -c$

Step 2: $\quad\quad x^2 + \dfrac{b}{a}x = -\dfrac{c}{a}$

SOLUTION   *Step 1:* Subtract c from both sides of $ax^2 + bx + c = 0$.

Step 2: Divide both sides of the equation from step 1 by a, $a \neq 0$.   ▬

Part 2 The next two steps in solving the quadratic equation require *completing the square* and writing the result as the *square of a binomial*. The mathematics behind these two steps appeared in steps 3 and 4 of Example 5.

Step 3: $x^2 + \dfrac{b}{a}x + \left(\dfrac{b}{2a}\right)^2 = \left(\dfrac{b}{2a}\right)^2 - \dfrac{c}{a}$ *Complete the square* on the left side of the equation from step 2, and add the result, $(b/2a)^2$, to both sides.

Step 4: $\left(x + \dfrac{b}{2a}\right)^2 = \dfrac{b^2}{4a^2} - \dfrac{c}{a}$ Write the left side of the equation from step 3 as the *square of a binomial*; expand the expression on the right side.

The next steps are done to the right side only and involve adding fractions.

EXAMPLE 7   Solving $ax^2 + bx + c = 0$, **Part 3**   Explain what has been done in these two steps in solving the quadratic equation.

Step 5: $\left(x + \dfrac{b}{2a}\right)^2 = \dfrac{b^2}{4a^2} - \dfrac{4ac}{4a^2}$

Step 6: $\left(x + \dfrac{b}{2a}\right)^2 = \dfrac{b^2 - 4ac}{4a^2}$

SOLUTION   *Step 5:* Set up a common denominator for the fractions on the right side of the equation from step 4.

Step 6: Add the fractions on the right side of the equation from step 5.   ■

The fourth part of solving the quadratic equation starts with steps used in solving equations by taking square roots (Section 5.2).

EXAMPLE 8   Solving $ax^2 + bx + c = 0$, **Part 4**   Explain what has been done in these steps in solving the quadratic equation.

Step 7: $\sqrt{\left(x + \dfrac{b}{2a}\right)^2} = \dfrac{\sqrt{b^2 - 4ac}}{\sqrt{4a^2}}$

Step 8: $\left|x + \dfrac{b}{2a}\right| = \dfrac{\sqrt{b^2 - 4ac}}{2a}$

Step 9: $x + \dfrac{b}{2a} = \pm\dfrac{\sqrt{b^2 - 4ac}}{2a}$

Step 10: $x = -\dfrac{b}{2a} \pm \dfrac{\sqrt{b^2 - 4ac}}{2a}$

Step 11: $x = \dfrac{-b \pm \sqrt{b^2 - 4ac}}{2a}$

SOLUTION   *Step 7:* Take the square root on both sides of the equation from step 6, applying the quotient rule on the right side.

Step 8: In the equation from step 7, replace $\sqrt{\left(x + \dfrac{b}{2a}\right)^2}$ with $\left|x + \dfrac{b}{2a}\right|$ and $\sqrt{4a^2}$ with $2a$.

Step 9: Solve the absolute value equation from step 8.

Step 10: Subtract $b/2a$ from both sides of the equation from step 9.

Step 11: Add the fractions on the right side of the equation from step 10.   ■

Because a, b, and c are not specified, the process in Examples 6 to 8 has solved *all* quadratic equations for x and derived the quadratic formula.

Applying the Quadratic Formula

EXAMPLE 9   **Applying the quadratic formula**   Use the quadratic formula to solve these equations.

a. $4x^2 = 5x + 1$

b. $2x^2 + 4x = -3$

SOLUTION   **a.** The equation $4x^2 = 5x + 1$ must be rearranged to equal zero. In $4x^2 - 5x - 1 = 0$, $a = 4$, $b = -5$, and $c = -1$.

$$x = \frac{-b \pm \sqrt{b^2 - 4ac}}{2a} = \frac{-(-5) \pm \sqrt{(-5)^2 - 4(4)(-1)}}{2(4)}$$

$$= \frac{-(-5) \pm \sqrt{25 + 16}}{8}$$

$$= \frac{5 \pm \sqrt{41}}{8}$$

The solutions are irrational numbers:

$$x = \frac{5 + \sqrt{41}}{8} \approx 1.425 \qquad \text{or} \qquad x = \frac{5 - \sqrt{41}}{8} \approx -0.175$$

b. The equation $2x^2 + 4x = -3$ must be rearranged to equal zero. In $2x^2 + 4x + 3 = 0$, $a = 2$, $b = 4$, and $c = 3$.

$$x = \frac{-b \pm \sqrt{b^2 - 4ac}}{2a} = \frac{-4 \pm \sqrt{4^2 - 4(2)(3)}}{2(2)}$$

$$= \frac{-4 \pm \sqrt{16 - 24}}{4}$$

$$= \frac{-4 \pm \sqrt{-8}}{4}$$

There are no real-number solutions because the square root of a negative number is not a real number.   ▬

In each part of Example 9, the answer had a different form. Part a had irrational solutions. Part b had no real-number solution. In Section 6.2, after complex numbers are introduced, we will solve equations like the one in part b. We now compare the graphs related to the equations in Example 9.

EXAMPLE 10   **Comparing quadratic graphs and numbers of solutions**   Graph these equations, and identify the solutions to $y = 0$ on the graphs.

a. $y = 4x^2 - 5x - 1$

b. $y = 2x^2 + 4x + 3$

SOLUTION   The solutions to $y = 0$ are the x-intercepts.

a. The graph in Figure 25 has two x-intercepts. This confirms the solutions in part a of Example 9: $4x^2 - 5x - 1 = 0$ near $x = 1\frac{1}{2}$ and between $x = -\frac{1}{2}$ and $x = 0$.

b. The graph in Figure 26 does not cross the x-axis, so it has no x-intercepts. This confirms that the equation $2x^2 + 4x + 3 = 0$ has no real-number solutions.

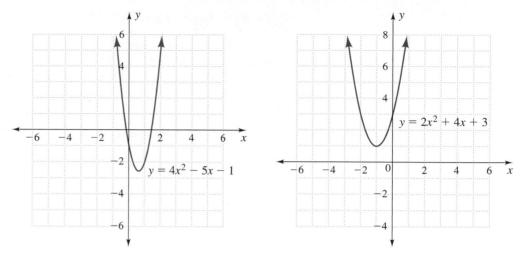

FIGURE 25   **FIGURE 26**

Finding the Vertex from the Quadratic Formula

The quadratic formula in the form

$$x = -\frac{b}{2a} \pm \frac{\sqrt{b^2 - 4ac}}{2a}$$

not only shows why the vertex is symmetrically placed between the x-intercepts but also gives a formula for the x-coordinate of the vertex. In Figure 27, the distance between the axis of symmetry and each x-intercept is $\sqrt{b^2 - 4ac}/2a$. The axis of symmetry, which passes through the vertex V, has equation $x = -b/2a$. This same expression is also the x-coordinate of the vertex.

Because the x-coordinate of the vertex is $x = -b/2a$, the y-coordinate of the vertex may be described with function notation $y = f(-b/2a)$.

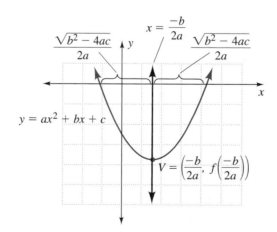

FIGURE 27

▮ FINDING THE VERTEX

For the function $f(x) = ax^2 + bx + c$,
• The **x-coordinate of the vertex** is a portion of the quadratic formula, $x = -b/2a$.
• The **y-coordinate of the vertex** is the output $f(-b/2a)$.

Application: Height of a Diver

As shown in the chapter opener, many dives require an Olympic diver to jump up as vertically as possible in order to just miss the platform on the way past. The diver's height above the water at any given time (see Figure 28) can be approximated by a quadratic equation called the vertical motion equation.

■ VERTICAL MOTION EQUATION

In terms of time t, the height h of an object in vertical motion is

$$h = -\tfrac{1}{2} gt^2 + v_0 t + h_0$$

where g is the acceleration due to gravity (9.81 m/sec² or 32.2 ft/sec²), v_0 is the initial upward velocity (in m/sec or ft/sec), and h_0 is the initial height (in meters or feet).

EXAMPLE 11   Using the vertical motion equation: Olympic diving   A competitor stands on a diving platform 32.8 feet (10 meters) above the water. She jumps from the platform with an initial upward velocity, v_0, of 6 ft/sec.

a. Write a vertical motion equation describing her height h above the water in terms of time t. Build a table for the equation, with inputs from $t = 0$ to 2 seconds, every two tenths of a second. Graph the equation. Check the table and graph with your calculator.

Use the calculator to answer the questions in parts b to d.

b. After how many seconds does she reach the water?
c. After how many seconds is her height above the water 23 feet?
d. What is the maximum height she attains?

FIGURE 28

SOLUTION   a. The diver's height at any time is based on $h = -\tfrac{1}{2}gt^2 + v_0 t + h_0$.

$$h = -\frac{1}{2}\left(\frac{32.2 \text{ ft}}{\text{sec}^2}\right)t^2 + \frac{6 \text{ ft}}{\text{sec}}t + 32.8 \text{ ft}$$

$$= -16.1t^2 + 6t + 32.8$$

Because t is in seconds, the seconds in the denominators will be eliminated when the time is substituted into the equation. The final answer will be in feet. Table 5 shows the diver's height at time t. The graph in Figure 29 also shows her height above the water at time t. Remember that the graph does *not* show a picture of the dive because the horizontal axis is time, not distance.

TABLE 5

Time (seconds)	Height (feet) $h = -16.1t^2 + 6t + 32.8$
0	32.8
0.2	33.4
0.4	32.6
0.6	30.6
0.8	27.3
1.0	22.7
1.2	16.8
1.4	9.6
1.6	1.2
1.8	−8.6
2.0	−19.6

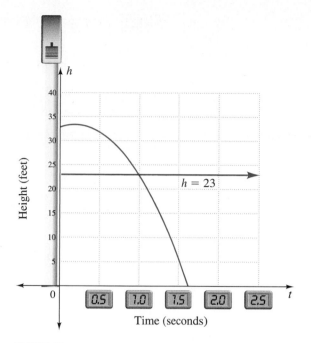

FIGURE 29

b. The diver enters the water at $h = 0$; so, using the equation from part a, we have

$$0 = -16.1t^2 + 6t + 32.8$$

A zero output in Table 5 would be located between $t = 1.6$ and $t = 1.8$ sec. The graph crosses the horizontal axis about halfway between 1.5 and 1.75 seconds. A time of 1.65 seconds is a reasonable solution.

Let $a = -16.1$, $b = 6$, and $c = 32.8$ in the quadratic formula. A convenient form for entering the formula into a calculator is

$$t = (-b \pm \sqrt{(b^2 - 4ac))}/(2a)$$
$$= [(-)]6 + \sqrt{(6^2 - 4 * [(-)]16.1 * 32.8))}/(2 * [(-)]16.1) \approx -1.25$$

We discard this negative time and press ⟨**2nd**⟩ ⟨**ENTER**⟩ to return to the expression. Changing to a subtraction after -6, we have

$$t = (-6 - \sqrt{(6^2 - 4 * -16.1 * 32.8))}/(2 * -16.1) \approx 1.63$$

It takes about 1.63 seconds to reach the water.

c. Substituting 23 feet for h gives

$$23 = -16.1t^2 + 6t + 32.8$$

Changing the equation to equal-zero form, we have

$$0 = -16.1t^2 + 6t + 9.8$$

where $a = -16.1$, $b = 6$, $c = 9.8$. Applying the quadratic formula yields

$$t = (-b \pm \sqrt{(b^2 - 4ac))}/(2a)$$
$$= ([(-)]6 + \sqrt{(6^2 - 4 * [(-)]16.1 * 9.8))}/(2 * [(-)]16.1) \approx 1.0$$

Pressing [2nd] [ENTER] to return to the expression, we change to a subtraction after -6:

$$t = (-6 - \sqrt{(6^2 - 4 * -16.1 * 9.8)})/(2 * -16.1) \approx -0.6$$

This negative time is discarded. The diver takes 1.0 second to reach $h = 23$ ft.

d. The maximum height is at the vertex:

$$x = -\frac{b}{2a} \approx 0.186 \text{ sec}$$

and

$$f\left(\frac{-b}{2a}\right) \approx 33.4 \text{ ft}$$

■

A calculator program for the quadratic formula and vertex appears below.

■ **GRAPHING CALCULATOR TECHNIQUE:**
QUADRATIC FORMULA

The following program gives the vertex and solutions for a quadratic equation. Use the negative key, not subtraction, for $-B$ and $-D$.

[PRGM] QUAD
⟨Prompt⟩ A, B, C
[(−)]B/(2A) [STO] V
⟨Disp⟩ "VERTEX X Y", V, AV² + BV + C
B² − 4AC [STO] D
⟨If⟩ D < 0
⟨Goto⟩ 1
([(−)]B + √(D))/(2A) [STO] R
([(−)]B − √(D))/(2A) [STO] S
⟨Disp⟩ "SOLUTIONS", R, S
⟨Goto⟩ 2
⟨Lbl⟩ 1
⟨Disp⟩ "COMPLEX SOLUTIONS"
⟨Disp⟩ [(−)]B/(2A) + i√([(−)]D)/(2A), [(−)]B/(2A) − i√([(−)]D)/(2A)
⟨Stop⟩
⟨Lbl⟩ 2
⟨Stop⟩

⟨ ⟩ means that the command is in the [PRGM] menu.

ANSWER BOX

Warm-up: **1.** $x^2 - 4x + 4$ **2.** $x^2 + 6x + 9$ **3.** $x^2 - 5x + \frac{25}{4}$
4. $x^2 - \frac{7}{3}x + \frac{49}{36}$ **5.** $x^2 + bx/a + b^2/4a^2$ **6.** $(x - 6)^2$ **7.** $(a - 2.5)^2$
8. $(x + 1.5)^2$ **9.** $(x + 2)(x + 2) = x^2 + 4x + 4$ **10.** $(x - 5)(x - 5) =$
$x^2 - 10x + 25$ **11.** $(x + 6)^2 = x^2 + 12x + 36$ **12.** $(x + 2.5)^2 = x^2 +$
$5x + 6.25$ **Example 4:** $|x - \frac{7}{6}| = \sqrt{\frac{20}{3} + \frac{49}{36}}$; $|x - \frac{7}{6}| = \frac{17}{6}$; $x = 4$ or $x = -1\frac{2}{3}$

5.3 Exercises

Multiply out the expressions in Exercises 1 to 8.

1. $(x + 3)^2$ $x^2 + 6x + 9$

2. $(x + 5)^2$ $x^2 + 10x + 25$

3. $(x - 6)^2$ $x^2 - 12x + 36$

4. $(x - 7)^2$ $x^2 - 14x + 49$

5. $\left(x - \frac{1}{2}\right)^2$ $x^2 - x + \frac{1}{4}$

6. $\left(x + \frac{1}{4}\right)^2$ $x^2 + \frac{1}{2}x + \frac{1}{16}$

7. $(2x - 3)^2$ $4x^2 - 12x + 9$

8. $(3x + 1)^2$ $9x^2 + 6x + 1$

Write each expression in Exercises 9 to 16 as a binomial square.

9. $x^2 - 4x + 4$ $(x - 2)^2$

10. $x^2 - 10x + 25$ $(x - 5)^2$

11. $x^2 + 8x + 16$ $(x + 4)^2$

12. $x^2 + 4x + 4$ $(x + 2)^2$

13. $x^2 + 18x + 81$ $(x + 9)^2$

14. $x^2 + 6x + 9$ $(x + 3)^2$

15. $x^2 - 16x + 64$ $(x - 8)^2$

16. $x^2 - 2x + 1$ $(x - 1)^2$

Complete the squares in Exercises 17 to 20. Write the square of a binomial and the perfect square trinomial described by each completed square.

17.

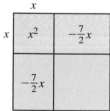

$(a + 7)^2 = a^2 + 14a + 49$
Missing entries: $+7, +7, 49$

18.

a · a^2 · $+\frac{9}{2}a$ / $+\frac{9}{2}a$

$\left(a + \frac{9}{2}\right)^2 = a^2 + 9a + \frac{81}{4}$
Missing entries: $+\frac{9}{2}, +\frac{9}{2}, \frac{81}{4}$

19.

x · x^2 · $-\frac{7}{2}x$ / $-\frac{7}{2}x$

$\left(x - \frac{7}{2}\right)^2 = x^2 - 7x + \frac{49}{4}$
Missing entries: $-\frac{7}{2}, -\frac{7}{2}, \frac{49}{4}$

20.

x · x^2 · $-\frac{3}{2}x$ / $-\frac{3}{2}x$

$\left(x - \frac{3}{2}\right)^2 = x^2 - 3x + \frac{9}{4}$
Missing entries: $-\frac{3}{2}, -\frac{3}{2}, \frac{9}{4}$

Complete the square on the equations in Exercises 21 to 28. Identify the square of the binomial that you obtain on the left side.

21. $x^2 + 22x = 23$ $x^2 + 22x + 121 = 144; (x + 11)^2 = 144$

22. $x^2 + 20x = -19$ $x^2 + 20x + 100 = 81; (x + 10)^2 = 81$

23. $x^2 + 12x = 13$ $x^2 + 12x + 36 = 49; (x + 6)^2 = 49$

24. $x^2 - 8x = -7$ $x^2 - 8x + 16 = 9; (x - 4)^2 = 9$

25. $x^2 - 11x = -18$ $x^2 - 11x + \frac{121}{4} = \frac{49}{4}; \left(x - \frac{11}{2}\right)^2 = \frac{49}{4}$

26. $x^2 + 13x = -12$ $x^2 + 13x + \frac{169}{4} = \frac{121}{4}; \left(x + \frac{13}{2}\right)^2 = \frac{121}{4}$

27. $x^2 + 15x = -14$ $x^2 + 15x + \frac{225}{4} = \frac{169}{4}; \left(x + \frac{15}{2}\right)^2 = \frac{169}{4}$

28. $x^2 + 25x = -100$ $x^2 + 25x + \frac{625}{4} = \frac{225}{4}; \left(x + \frac{25}{2}\right)^2 = \frac{225}{4}$

In Exercises 29 to 42, use factoring or the quadratic formula to solve the equations. If there are real-number solutions, indicate whether they are rational or irrational. Confirm your answers with a graphing calculator. (Hint: To learn the quadratic formula quickly, write it each time you use it.)

29. $x^2 + 6x - 8 = 0$ $\{-3 \pm \sqrt{17}\}$; irrational

30. $x^2 + 3x - 12 = 0$ $\left\{\frac{-3 \pm \sqrt{57}}{2}\right\}$; irrational

31. $x^2 - 6x + 8 = 0$ $\{4, 2\}$; rational

32. $x^2 - 11x - 12 = 0$ $\{12, -1\}$; rational

33. $4x^2 - 4x + 1 = 0$ $\left\{\frac{1}{2}\right\}$; rational

34. $9x^2 - 6x + 1 = 0$ $\left\{\frac{1}{3}\right\}$; rational

35. $2x^2 + 3x = -1$ $\left\{-1, -\frac{1}{2}\right\}$; rational

36. $5x^2 + 3x = -3$ $\left\{\frac{-3 \pm \sqrt{-51}}{10}\right\}$; not real

37. $2x^2 = -2 - 3x$ $\left\{\frac{-3 \pm \sqrt{-7}}{4}\right\}$; not real

38. $2x^2 = 2 - 3x$ $\left\{-2, \frac{1}{2}\right\}$; rational

39. $5x = 3 + 2x^2$ $\left\{1, \frac{3}{2}\right\}$; rational

40. $5x - 3 = -2x^2$ $\left\{-3, \frac{1}{2}\right\}$; rational

41. $x^2 - 20x + 60 = 10 - x^2$ $\{5\}$; rational

42. $2x^2 - 6x + 30 = 3 + 12x - x^2$ $\{3\}$; rational

In Exercises 43 to 48, complete the quadratic formula without a calculator. Identify a, b, and c, and write the original equation.

43. $x = \dfrac{-(-4) - \sqrt{(-4)^2 - 4(1)(-5)}}{2(1)}$ $x = -1; 1, -4, -5;$ $x^2 - 4x - 5 = 0$

44. $x = \dfrac{-(-4) + \sqrt{(-4)^2 - 4(1)(-12)}}{2(1)}$ $x = 6; 1, -4, -12;$ $x^2 - 4x - 12 = 0$

45. $x = \dfrac{-5 + \sqrt{5^2 - 4(2)(-12)}}{2(2)}$ $x = \frac{3}{2}; 2, 5, -12;$ $2x^2 + 5x - 12 = 0$

46. $x = \dfrac{-5 - \sqrt{5^2 - 4(7)(-2)}}{2(7)}$ $x = -1; 7, 5, -2;$ $7x^2 + 5x - 2 = 0$

47. $x = \dfrac{-1 + \sqrt{1 - 4(3)(-4)}}{2(3)}$ $x = 1; 3, 1, -4; 3x^2 + x - 4 = 0$

48. $x = \dfrac{-(-5) - \sqrt{(-5)^2 - 4(6)(1)}}{2(6)}$ $x = \frac{1}{3}; 6, -5, 1;$ $6x^2 - 5x + 1 = 0$

49. Explain how to find the x-coordinate of the vertex from the quadratic formula.

$x = -b/2a$, leading term of quadratic formula

50. Explain how to find the y-coordinate of the vertex given the equation of the axis of symmetry of the graph.

$x = -b/2a$ describes the x-coordinate; find $f(-b/2a)$.

In Exercises 51 to 54, find the vertex for the graph of each equation. Round to three decimal places.

51. $y = 2x^2 + 3x + 5$ $(-0.75, 3.875)$

52. $y = 5x^2 + 4x + 3$ $(-0.4, 2.2)$

53. $y = 5x^2 + 3x - 3$ $(-0.3, -3.45)$

54. $y = 3x^2 + 4x + 5$ $(-0.667, 3.667)$

Solve the equations in Exercises 55 to 58 using the quadratic formula. The equations are from the art museum parabolic arch formula. Round to the nearest thousandth.

55. $-0.12x^2 + 2.4x = 8$ $\{4.226, 15.774\}$

56. $-0.12x^2 + 2.4x = 9$ $\{5, 15\}$

57. $-0.12x^2 + 2.4x = 10$ $\{5.918, 14.082\}$

58. $-0.12x^2 + 2.4x = 11$ $\{7.113, 12.887\}$

Exercises 59 to 62 refer to the vertical motion in Example 11.

59. a. What is the meaning of the y-intercept in the graph?   initial height of 32.8 ft

b. Does the part of the parabola to the left of the y-axis have any meaning in the problem setting?   no

c. At what two times is the diver at 32.8 feet?
0 and ≈ 0.4 sec

60. a. Estimate the time taken to travel between the highest point in the dive and the 16-foot height. ≈ 1 sec

b. Estimate the time taken to travel between the 16-foot height and the water surface. ≈ 0.4 sec

c. The distances in parts a and b are approximately the same. Why is there such a large difference in time?   Velocity increases as diver nears water.

61. Suppose another Olympic diver has an initial jump velocity $v_0 = 4$ ft/sec. Using the new velocity in the vertical motion equation in Example 11, solve for $h = 0$ to find her total time in the air. Compare this time with that of the original diver in the example.
≈ 1.55 sec

62. Suppose a third Olympic diver has an initial jump velocity $v_0 = 8$ ft/sec. Using the new velocity in the vertical motion equation in Example 11, solve for $h = 0$ to find her total time in the air. Compare this time with that of the original diver in the example.
≈ 1.7 sec

The cliff diving from La Quebrada at Acapulco, Mexico, is a popular tourist attraction. The clavadistas (cliff divers) dive from an initial height h_0 of 35 meters above the water (see the figure). In Exercises 63 to 65, assume that $h = 0$ at the water

level. In the metric system, the acceleration due to gravity is $g \approx 9.81$ m/sec^2. Round to the nearest tenth of a second.

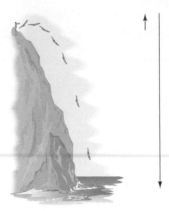

parabolic path   vertical motion

63. Suppose a diver runs toward the cliff's edge and dives without any upward motion. This means there is no initial upward velocity v_0. Write a vertical motion equation to estimate the number of seconds required to reach the water. Solve the equation. $0 = -\frac{1}{2}(9.81)t^2 + 35$, $t = 2.7$ sec

64. Suppose a diver dives with an initial upward velocity of 2 meters per second. Write a vertical motion equation to estimate the number of seconds required to reach the water. Solve the equation.
$0 = -\frac{1}{2}(9.81)t^2 + 2t + 35$, $t = 2.9$ sec

65. Suppose a diver dives with an initial upward velocity of 1.5 meters per second. Write a vertical motion equation to estimate the number of seconds required to reach the water. Solve the equation.
$0 = -\frac{1}{2}(9.81)t^2 + 1.5t + 35$, $t = 2.8$ sec

66. For a 10-meter-board diver, height above the surface of the water is given by $h = -\frac{1}{2}(9.81)t^2 + v_0 t + 10$. Round to three digits.

a. Find the vertex for $v_0 = 2$ m/sec. $(0.204$ sec, 10.2 m$)$

b. Find the vertex for $v_0 = 4$ m/sec. $(0.408$ sec, 10.8 m$)$

c. Find the vertex for $v_0 = 8$ m/sec. $(0.815$ sec, 13.3 m$)$

d. Compare the x-coordinates of the vertices as the initial velocity doubles.   Time also doubles.

e. How will the maximum height of the diver change as the initial velocity changes?   An equation fit to the vertices shows that height changes quadratically: $h \approx \frac{1}{2}(9.81)t^2 + 10$

▬ **Error Analysis**

67. One statement is true; the rest are false. Find the true statement and explain why the others are false.

a. $\sqrt{16x^2} = 4x$ $\sqrt{w^2} = |w|$, not w

b. $\sqrt{16 + x^2} = 4 + x$ There is no square root of sums property.

c. $\sqrt{4x^2 - 9y^2} = 2x - 3y$ There is no square root of differences property.

d. $\sqrt{4x^2 - 4x + 1} = |2x - 1|$ true

68. What is wrong with these keystrokes for evaluating the quadratic formula for $2x^2 - 7x + 3 = 0$?

a. Replacing $\pm$ with $+$: Expression should start with $-(-7)$ and needs $)$ before division sign.

$(-7 + \sqrt{((-7)^2 - 4(2)(3))/(2(2))}$

b. Replacing $\pm$ with $-$:
$(-7 - \sqrt{(-7^2 - 4 * 2 * 3))/(2 * 2)}$

Expression should start with $-(-7)$ and needs $(-7)^2$ under radical sign.

69. An alternative derivation of the quadratic formula is shown below. Describe what was done at each step.

$ax^2 + bx + c = 0$ Standard form

$4a^2x^2 + 4abx + 4ac = 0$ Multiply both sides by $4a$.

$4a^2x^2 + 4abx = -4ac$ Subtract $4ac$ from both sides.

$4a^2x^2 + 4abx + b^2 = b^2 - 4ac$ Add b^2 to both sides.

$(2ax + b)^2 = b^2 - 4ac$ Factor left side.

$\sqrt{(2ax + b)^2} = \pm\sqrt{b^2 - 4ac}$ Take square root of both sides.

$2ax + b = \sqrt{b^2 - 4ac}$ or $2ax + b = -\sqrt{b^2 - 4ac}$
Simplify left side, giving both solutions.

$2ax = -b + \sqrt{b^2 - 4ac}$ or $2ax = -b - \sqrt{b^2 - 4ac}$
Subtract b from both sides.

$x = \dfrac{-b + \sqrt{b^2 - 4ac}}{2a}$ or $x = \dfrac{-b - \sqrt{b^2 - 4ac}}{2a}$
Divide both sides by $2a$.

■ **Project**

70. Generalizing the Quadratic Formula Apply the quadratic formula; follow the suggested steps.

a. In $2\square^2 + 3\square - 5 = 0$, solve for $\square$. $\{-2.5, 1\}$

b. In $2\square^2 + 3\square - 2 = 0$, solve for $\square$. $\{-2, 0.5\}$

c. In $x^4 - 2x^2 + 1 = 0$, substitute A for x^2; solve for A and then for x. $A = 1, x = \pm 1$

d. In $x^4 - 1 = 0$, substitute A for x^2; solve for A and then for x. $A = \pm 1, x = \pm 1$

e. In $16x^4 - 8x^2 + 1 = 0$, solve for x^2 and then for x. $x^2 = 0.25, x = \pm 0.5$

f. In $x^4 + 5x^2 - 36 = 0$, solve for x^2 and then for x. $x^2 = -9, x^2 = 4, x = \pm 2$

g. What can you conclude about the quadratic formula and equations in parts d to f?

If the equations contain only x^4 and x^2 and constants, we can apply the quadratic formula.

5.4   Solving Quadratic Inequalities

Objectives

■ Solve inequalities with a table or a graph.

■ Solve application problems involving inequalities.

WARM-UP

Complete this table.

	Inequality	Interval	Words	Line Graph
1.	$-3 < x \le 4$	$(-3, 4]$	Set of numbers greater than -3 and less than 4, including 4	
2.	$x \ge -3$	$[-3, +\infty)$	Set of numbers greater than -3, including -3	
3.	$x \le -2$	$(-\infty, -2]$	Set of numbers less than -2, including -2	
4.	$x < 2$	$(-\infty, 2)$	Set of numbers less than 2	
5.	$x < 4$ or $x > 8$	$(-\infty, 4)$ or $(8, +\infty)$	Set of numbers less than 4 or greater than 8	

IN THIS SECTION, we use tables and graphs to aid us in solving quadratic inequalities.

Using a Table or Graph to Solve Inequalities

There are many times when we are interested in finding as a solution a set of numbers, instead of a single number. For example, to earn a passing grade we need any score over 70%. For paying our bills, any amount in our checking account over the amount owed (plus the minimum balance) is satisfactory. We describe these sets of numbers with inequalities or intervals.

EXAMPLE 1   **Solving inequalities with tables, graphs, and x-intercepts**   Build a table and graph for the function $f(x) = x^2 - 3x - 10$, with inputs on the interval $x = -4$ to $x = 7$. Using inequalities and graphs, describe where

a. $f(x) = 0$ **b.** $f(x) > 0$ **c.** $f(x) < 0$

SOLUTION   Table 6 contains data for $f(x) = x^2 - 3x - 10$, and Figure 30 contains the graph.

TABLE 6

Input x	Ouput $f(x) = x^2 - 3x - 10$
-4	18
-3	8
-2	0
-1	-6
0	-10
1	-12
2	-12
3	-10
4	-6
5	0
6	8
7	18

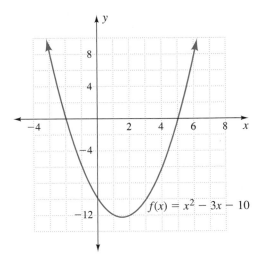

FIGURE 30

a. The x-intercepts, where $f(x) = 0$, are $x = -2$ and $x = 5$. In Figure 30, the top line graph shows the x-intercepts.

b. Table 6 indicates that the output is positive, $f(x) > 0$, for $x < -2$ or for $x > 5$. The graph is above the x-axis when $x < -2$ or $x > 5$. This indicates where $f(x) > 0$. Thus,

$$x^2 - 3x - 10 > 0 \qquad \text{for } x < -2 \quad \text{or} \quad x > 5$$

The middle line graph in Figure 30 shows the solutions: the intervals $(-\infty, -2)$ and $(5, +\infty)$.

c. For inputs between $x = -2$ and $x = 5$, the table output is negative, $f(x) < 0$. The graph is below the x-axis for x between -2 and 5. This indicates that $f(x) < 0$. Thus,

$$x^2 - 3x - 10 < 0 \qquad \text{for } -2 < x < 5$$

The bottom line graph in Figure 30 shows the solutions: the interval $(-2, 5)$. ▬

EXAMPLE 2 Solving inequalities with tables, graphs, and x-intercepts Build a table and a graph for $f(x) = 2x^2 - 3x - 5$ for the integers on the interval $[-3, 4]$. Find the intercepts, and determine for what inputs

a. $2x^2 - 3x - 5 < 0$ **b.** $2x^2 - 3x - 5 \leq 0$

SOLUTION Table 7 contains data for $f(x) = 2x^2 - 3x - 5$. The graph is in Figure 31.

Table 7 shows only one x-intercept, $x = -1$. The graph in Figure 31 shows two intercepts. To solve the inequality $2x^2 - 3x - 5 < 0$, we need both x-intercepts. We could use the quadratic formula, but the factors of $2x^2 - 3x - 5$ are $(x + 1)$ and $(2x - 5)$, so the other x-intercept is where $2x - 5 = 0$, $x = 2.5$.

TABLE 7

Input x	Ouput $f(x) = 2x^2 - 3x - 5$
-3	22
-2	9
-1	0
0	-5
1	-6
2	-3
3	4
4	15

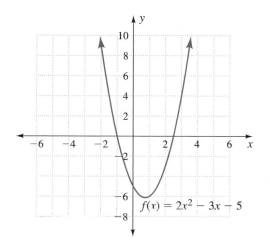

$f(x) < 0$

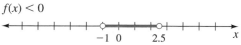

$f(x) \leq 0$

FIGURE 31

a. The inequality $2x^2 - 3x - 5 < 0$ is true for all numbers between $x = -1$ and $x = 2.5$, or $-1 < x < 2.5$. The solution is shown in the top line graph in Figure 31.

b. To solve $2x^2 - 3x - 5 \leq 0$, we include the intercepts on our interval because the intercepts themselves make the inequality true. We replace the $<$ signs with $\leq$ to include the intercepts: $-1 \leq x \leq 2.5$. The solution is shown in the bottom line graph in Figure 31. ▬

In Examples 1 and 2, the boundary points for the solutions were the x-intercepts. When an inequality has variables (or numbers) on both sides, we can either (1) graph both sides of the inequality and find intersections or (2) use algebra to obtain a zero on one side and find x-intercepts. In Examples 3 and 4, we use graphing to solve the problem and algebraic methods to check our solutions.

EXAMPLE 3 **Solving quadratic inequalities** Solve $2x^2 - 5x + 4 \le -x + 20$.

a. Using a calculator, graph each side of the inequality.

b. Find the intersections.

c. Draw a line graph showing the values of x that make the inequality true. Check the intersections by solving by factoring (if possible). Check the line graph by substituting in a test point.

SOLUTION

a. The graphs of $y = 2x^2 - 5x + 4$ and $y = -x + 20$ are shown in Figure 32.

b. Using [2nd] [CALC] **5 : intersect**, we locate where the graphs cross: $x = -2$ and $x = 4$.

c. The graph of the first equation is below the graph of the second equation between -2 and 4, so $-2 \le x \le 4$, as shown in the line graph in Figure 32.

Check intersections:

$$2x^2 - 5x + 4 \le -x + 20 \qquad \text{Add } x - 20 \text{ to both sides.}$$

$$2x^2 - 5x + 4 + x - 20 \le 0 \qquad \text{Combine like terms.}$$

$$2x^2 - 4x - 16 \le 0 \qquad \text{Factor.}$$

$$2(x - 4)(x + 2) \le 0$$

The expression is zero at $x = -2$ or $x = 4$.

Check line graph: Zero is in the interval $(-2, 4)$, so $x = 0$ should make the inequality true.

$$2(0)^2 - 5(0) + 4 \overset{?}{\le} -(0) + 20$$

$$4 \le 20 \;\checkmark$$

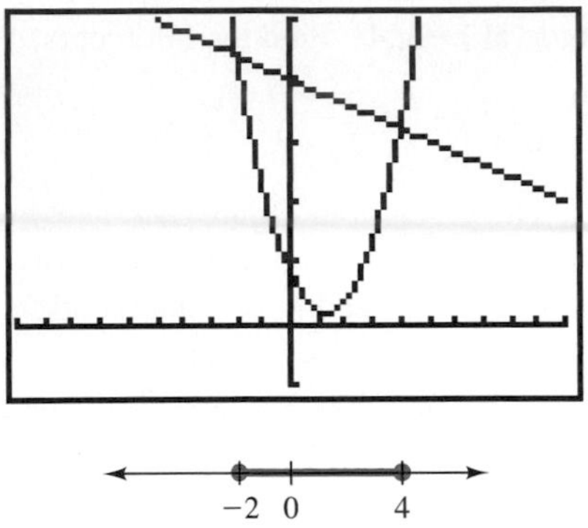

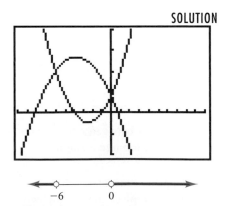

FIGURE 32 -10 to 10 on X, -5 to 25 on Y

EXAMPLE 4 **Solving quadratic inequalities** Solve $x^2 + 5x + 4 > -\frac{2}{3}x^2 - 5x + 4$.

a. Using a calculator, graph each side of the inequality.

b. Find the intersections.

c. Draw a line graph showing the values of x that make the inequality true. Check the intersections by solving by factoring (if possible). Check the line graph by substituting in a test point.

SOLUTION

a. The graphs of $y = x^2 + 5x + 4$ and $y = -\frac{2}{3}x^2 - 5x + 4$ are shown in Figure 33.

b. Using [2nd] [CALC] **5 : intersect**, we locate where the graphs cross: $x = -6$ and $x = 0$.

c. The graph of the first equation is above the graph of the second equation to the left of $x = -6$ or $x < -6$ and to the right of $x = 0$ or $x > 0$, as shown in the line graph in Figure 33.

Check intersections:

$$x^2 + 5x + 4 \overset{?}{>} -\frac{2}{3}x^2 - 5x + 4 \quad \text{Add } \frac{2}{3}x^2 + 5x - 4 \text{ to both sides.}$$

$$\frac{5}{3}x^2 + 10x > 0 \qquad \text{Multiply by 3 to clear the denominator.}$$

$$5x^2 + 30x > 0 \qquad \text{Factor.}$$

$$5x(x + 6) > 0$$

The expression is zero at $x = 0$ or $x = -6$.

FIGURE 33 -10 to 10 on X, -10 to 20 on Y

Check line graph: Three is in the interval $(0, +\infty)$, so $x = 3$ should make the inequality true.

$$(3)^2 + 5(3) + 4 \overset{?}{>} -\frac{2}{3}(3)^2 - 5(3) + 4$$

$$28 > -17 \;\checkmark$$

Applications

EXAMPLE 5 Solving inequalities with a table and graph: height of an emergency flare   An emergency rescue flare is fired straight up through a layer of fog 1500 feet thick. The flare's height in terms of time is found from the vertical motion equation $h = -\frac{1}{2}gt^2 + v_0t + h_0$. Suppose h_0 is the height of the hand holding the flare gun, 7 feet, and the acceleration due to gravity is $g = 32.2$ ft/sec^2. The flare is fired with an initial velocity v_0 of 400 ft/sec.

a. Set up an equation describing the height of the flare.

b. Make a table for the height every 4 seconds until the flare returns to the ground.

c. Graph the equation for the height of the flare and the equation for the fog layer, $y = 1500$ ft.

d. Write an inequality that tells when the height of the flare will be above 1500 feet.

e. From the table and graph estimate the time interval during which the flare will be visible to a plane flying above the fog.

f. Check the endpoints on the time interval by using the quadratic formula to solve the inequality in part d for intercepts.

SOLUTION **a.** The equation is $h = -16.1t^2 + 400t + 7$.

b. Table 8 shows the height every 4 seconds, to 28 seconds.

c. Figure 34 shows the graph of the equation and $y = 1500$ ft.

TABLE 8

Time (seconds)	Height (feet)
0	7
4	1349
8	2177
12	2489
16	2285
20	1567
24	333
28	−1415

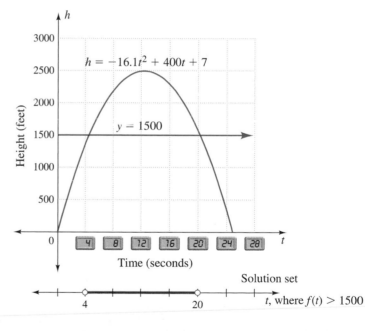

FIGURE 34

d. The flare will be above 1500 feet for $-16.1t^2 + 400t + 7 > 1500$.

e. The table indicates that the flare will be above 1500 feet between $t \approx 4$ and $t \approx 20$ sec, or $4 < t < 20$. From the graph, we observe the height of the flare and confirm our estimate, the interval (4, 20) seconds. The flare will be visible for about 16 seconds.

f. *Check:*

$$-16.1t^2 + 400t + 7 > 1500 \qquad \text{Subtract 1500 from both sides.}$$

$$-16.1t^2 + 400t - 1493 > 0$$

We let $a = -16.1$, $b = 400$, and $c = -1493$ and use the quadratic formula to find the intercepts:

$$x = \frac{-400 \pm \sqrt{400^2 - 4(-16.1)(-1493)}}{2(-16.1)}$$

$$= \frac{-400 \pm \sqrt{63850.8}}{-32.2}$$

$$x \approx 4.6 \text{ sec} \quad \text{or} \quad x \approx 20.3 \text{ sec}$$

Our interval is reasonable.

Think about it 1: Does the parabola in Figure 34 show the actual path of the flare?

COST AND REVENUE CURVES   The graphs in the next two examples are presented to show general ideas, not graphing of data. This style of graph is typical in economics and business textbooks.

Cost describes *money to be paid*; **revenue** describes *money to be received. Revenues greater than cost* create **profit**. *Revenues less than cost* create **loss**.

EXAMPLE 6   Solving inequality problems from a graph: cost and revenue curves   In Figure 35, the revenue curve gives sales volume in dollars for the business year starting July 1 and ending the following June 30. The cost curve shows the costs incurred during the year to make those sales. The intercepts for the cost curve are $x = 0$ and $x = 365$.

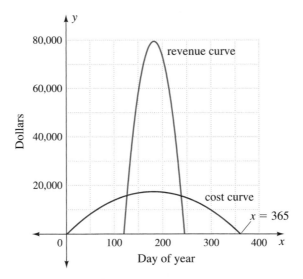

FIGURE 35

a. Estimate the time period (day _____ to day _____) during which revenue is received.

b. Estimate the time period (day _____ to day _____) during which revenue is greater than cost, thus creating a profit.

c. Assume each curve is a parabola, and estimate its vertex.

SOLUTION
a. The revenue curve intersects the x-axis from approximately day 120 to day 245. Thus, revenue is received for the period $120 < x < 245$.

b. The revenue curve is above the cost curve from approximately day 125 to day 240, or for the period $125 < x < 240$.

c. The parabolic curves appear to be symmetric to the beginning and ending of the business year, so the x-coordinate of the vertex for each curve is $\frac{365}{2}$, or about 182. For the revenue curve, the vertex is approximately $(182, 80,000)$. For the cost curve, the vertex is approximately $(182, 17,500)$. ▬

SUPPLY AND DEMAND CURVES   In agricultural planning, those months for which food demand curves are above food supply are months in which food must be imported or taken from storage. In Example 7, we look at supply (production) and demand (consumption) for an agricultural product.

EXAMPLE 7   Solving inequality problems from a graph: agricultural supply and demand curves   Figure 36 shows a tropical country's consumption of rice and its twice yearly production of rice. The production (harvest) takes a month to complete.

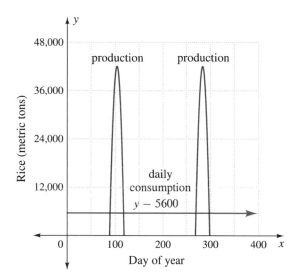

FIGURE 36

a. For what days in the year is production greater than consumption?
b. Assume the two curves are parabolas. Estimate the vertices.

SOLUTION
a. Production is greater than consumption from approximately day 90 to day 120 and day 270 to day 300.

b. The first vertex is at approximately $(105, 42,000)$. The second vertex is at approximately $(285, 42,000)$. ▬

Think about it 2: Rice stores easily between harvests. What would it mean to this tropical country if the food represented in Figure 36 could not be stored for future use?

ANSWER BOX **Warm-up:**

	Inequality	Interval	Words	Line Graph
1.	$-3 < x \leq 4$	$(-3, 4]$	Set of numbers greater than -3 and less than 4, including 4	◀─○┼┼┼┼┼┼●─▶ -3 0 4
2.	$x \geq -3$	$[-3, +\infty)$	Set of numbers greater than -3, including -3	◀─●┼┼┼┼┼┼┼─▶ -3 0
3.	$x \leq -2$	$(-\infty, -2]$	Set of numbers less than -2, including -2	◀━━━●┼┼┼┼─▶ -4 -2 0
4.	$x < 2$	$(-\infty, 2)$	Set of numbers less than 2	◀━┼┼┼┼┼○─▶ 0 2
5.	$x < 4$ or $x > 8$	$(-\infty, 4)$ or $(8, +\infty)$	Set of numbers less than 4 or greater than 8	◀━┼┼○━○━┼─▶ 0 4 8

Think about it 1: No; the horizontal axis would need to be distance, not time, to show the actual path. The graph shows height above ground at a given time. **Think about it 2:** If the food could not be stored, it would need to be sold or exported immediately, and the demand (consumption needs) during the rest of the year would have to be met by imports.

5.4 Exercises

For Exercises 1 to 8, complete the table.

	Inequality	Interval	Words	Line Graph
1.	$x < -3$ or $x > 2$	$(-\infty, -3)$ or $(2, +\infty)$	Set of numbers less than -3 or greater than 2	◀─○┼○─▶ -3 0 2
2.	$-5 \leq x < 3$	$[-5, 3)$	Set of numbers greater than or equal to -5 and less than 3	◀─●┼○─▶ -5 0 3
3.	$x \geq 4$	$[4, +\infty)$	Set of numbers greater than or equal to 4	◀━┼━●━┼━▶ 4 5
4.	$x < 4$	$(-\infty, 4)$	Set of numbers less than 4	◀━━━○━┼━▶ 0 4
5.	$x > -3$	$(-3, +\infty)$	Set of numbers greater than -3	◀─○┼─▶ -3 0
6.	$x \leq -1$	$(-\infty, -1]$	Set of numbers less than or equal to -1	◀━━●┼─▶ -1 0
7.	$x \leq 4$	$(-\infty, 4]$	Set of numbers less than or equal to 4	◀━━━●─▶ 0 4
8.	$x > -1$	$(-1, +\infty)$	Set of numbers greater than -1	◀─○┼─▶ -1 0

In Exercises 9 to 30, find for what inputs each inequality is true. Use factoring to find the x-intercepts.

9. $x^2 - 5x + 6 > 0$
   $x < 2$ or $x > 3$

10. $x^2 - 5x + 6 \leq 0$
   $2 \leq x \leq 3$

11. $x^2 - 5x - 14 \leq 0$
   $-2 \leq x \leq 7$

12. $x^2 - 5x + 4 < 0$
   $1 < x < 4$

13. $x^2 - 5x - 6 \geq 0$
   $x \leq -1$ or $x \geq 6$

14. $x^2 - 5x - 24 > 0$
   $x < -3$ or $x > 8$

15. $x^2 + 6x + 9 > 0$
   $\mathbb{R}, x \neq -3$

16. $x^2 - 6x + 9 < 0$
   { }

17. $4x^2 - 4x + 1 < 0$
   { }

18. $4x^2 + 4x + 1 \geq 0$
   $\mathbb{R}$

19. $x^2 + 2x - 15 < 0$
$-5 < x < 3$

20. $x^2 - 8x + 15 \geq 0$
$x \leq 3 \text{ or } x \geq 5$

21. $x^2 + 10x - 24 \geq 0$
$x \leq -12 \text{ or } x \geq 2$

22. $x^2 - 10x + 24 < 0$
$4 < x < 6$

23. $x^2 - 3x - 18 \leq 0$
$-3 \leq x \leq 6$

24. $x^2 + 7x - 18 > 0$
$x < -9 \text{ or } x > 2$

25. $x^2 + 17x - 18 > 0$
$x < -18 \text{ or } x > 1$

26. $x^2 + 9x + 18 \leq 0$
$-6 \leq x \leq -3$

27. $x^2 + 8x + 16 \geq 0$
$\mathbb{R}$

28. $x^2 - 10x + 25 < 0$
$\{ \} \text{ or } \varnothing$

29. $x^2 - 14x + 49 < 0$
$\{ \} \text{ or } \varnothing$

30. $x^2 + 12x + 36 > 0$
$\mathbb{R}, x \neq -6$

In Exercises 31 to 42, find for what inputs each inequality is true. Write the answers as intervals.

31. $x^2 - x > x + 3$ $(-\infty, -1) \text{ or } (3, +\infty)$

32. $3 - x > x^2 + x$ $(-3, 1)$

33. $x^2 + 2x - 9 \leq 1 - x$ $[-5, 2]$

34. $2x + 7 < x^2 - x - 3$ $(-\infty, -2) \text{ or } (5, +\infty)$

35. $x^2 + x < x - 1$ $\{ \} \text{ or } \varnothing$

36. $x - x^2 > 2x + 1$ $\{ \} \text{ or } \varnothing$

37. $x^2 - x \geq x - 1$ $(-\infty, +\infty)$

38. $x - x^2 \geq 2x$ $[-1, 0]$

39. $2x^2 - 2 < x^2 - x$ $(-2, 1)$

40. $2x^2 + 1 > x - x^2$ $(-\infty, +\infty)$

41. $3x^2 - 3 \geq 4x - x^2$ $\left(-\infty, -\frac{1}{2}\right] \text{ or } \left[\frac{3}{2}, +\infty\right)$

42. $x^2 - 9x - 6 \geq 12 - x^2$ $\left(-\infty, -\frac{3}{2}\right] \text{ or } [6, +\infty)$

43. The curved line in the figure shows the path of a kicked ball. The ball is kicked with an initial velocity of 60 ft/sec, at an angle of 45° relative to the ground. The straight line is the path the ball would follow if there were no gravity.

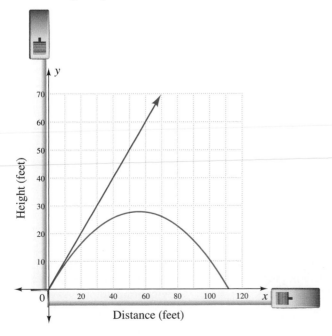

a. The ball is to pass over a horizontal bar 10 feet high. For what horizontal distances will the ball go over the bar? $\approx 12 < x < 100$

b. The ball may be knocked to the ground by the opposing team if it is below 10 feet. For what horizontal distance is the ball in danger of being knocked down?
$\approx 0 < x < 12 \text{ or } 100 < x < 112$

c. What is the highest point reached by the ball? ≈ 28 ft

d. Does the graph show the length of time the ball is in the air? Explain your reasoning.
No; it shows the ball's path; both axes are in feet.

e. If there were no gravity, what would be the height of the ball when it was horizontally 50 feet from the kicker? 50 ft

f. For what horizontal distances is the ball over 20 feet high? $\approx 25 < x < 85$

44. From their truck, an inspection team can inspect suspension bridge cable within 20 feet of the road surface. The parabolic cable has its vertex at the road surface in the center of the bridge. Its equation is $y = \frac{2}{125} x^2$.

a. For what horizontal interval on the 80-foot-long bridge can the inspection be accomplished from the truck? $-35.4 < x < 35.4$, where $x = 0$ is at center of bridge

b. The inspection team obtains a new truck with a hydraulic basket that can carry a worker high enough to inspect up to 25 feet above the roadway. For what horizontal interval can the inspection now be accomplished? $\approx -39.5 < x < 39.5$

45. Refer to Example 6.

a. What type of product might have revenue (sales) and costs as described?
seasonal products such as ski equipment, heating oil

b. Between what calendar dates do the sales occur?
$\approx$Nov. 1 to March 31

c. What date is the 182nd day from July 1?
Dec. 29

d. What date is the 120th day from July 1?
Oct. 28

The areas under the graphs give the total revenue and total cost for the year. To approximate the area for this portion of a parabola, use the formula $A = \frac{2}{3}bh$, where b is the estimated distance between the x-intercepts and h is the perpendicular distance from the vertex to the x-axis.

e. Use the area to estimate the total revenue for the year.
$6,600,000

g. Use the area to estimate the total costs for the year.
$4,250,000

h. Will there be an overall profit (total revenue − total cost > 0) for the year? yes

46. Refer to Example 7.

a. The average consumption of rice in this country is 5600 metric tons per day. What is the consumption for the year? What part of the graph represents the yearly rice consumption?
$\approx$2,044,000 metric tons; area under daily consumption

b. The area under the two parabolic graphs gives the total rice production for the year. Find the total production. (*Hint:* Approximate the areas with $A = \frac{2}{3}bh$, where b is the estimated distance between the x-intercepts and h is the perpendicular distance from the vertex to the x-axis.) $\approx$1,680,000 metric tons

c. If annual consumption is larger than annual production, the country must import rice. Estimate the amount of rice imported by comparing the answers to parts a and b. 364,000 metric tons

▉ **Project**

47. Comedy Wild West* A movie special-effects technician places a hole and a plug in a water tank so that when the sheriff fires at a cattle rustler and misses, it appears that a bullet pierces the tank and releases a spray of water (see the figure). The hole is 9 feet above the level of the head of the cattle rustler. The rustler is to be hit with a spray of water. The tank has an open top, and the water level is 6 feet above the bullet hole. How far from the tank can the rustler be and still get wet? Repeat for a water level 4 feet above the hole. Summarize with an inequality for the 4-foot to 6-foot water levels.

Useful facts
Horizontal motion: The equation $d = v_x t$ tells how far out the water travels.
Vertical motion: The equation $h = -\frac{1}{2}gt^2 + v_y t + h_0$, where $g = 32.2$ ft/sec^2, is a source of the time, t.
Horizontal water speed from a hole in the side of a container is given by $v_x = \sqrt{2gn}$, where n is the distance between the hole and the top of the water.

State any assumptions, including why $v_y = 0$, the meaning of $h - h_0$, and how rapidly or slowly the water level in the tank might drop.   See Additional Answers.

*Adapted from Serway and Faughn, "Shoot-Out at the Old Water Tank," in *College Physics*, 6th edition, Brooks/Cole, 2003, p. 278

▉ **5** **Chapter Summary**

Vocabulary

completing the square
converse of the
   Pythagorean theorem
cost
equilateral triangle
height of a triangle
   (altitude)
hypotenuse

isosceles right triangle
isosceles triangle
legs
loss
perfect square
perfect square trinomial
plus or minus sign
principal square root

product property of square
   roots
profit
Pythagorean theorem
quadratic formula
quotient property of square
   roots
radical sign

radicand
revenue
square root
square root function
vertical motion equation
x-coordinate of the vertex
y-coordinate of the vertex

Concepts

5.1 ▮ The Square Root Function and the Pythagorean Theorem

$x^2 = n$ has two solutions, $x = \pm \sqrt{n}$.

$x = \sqrt{n}$ has one solution, the principal square root of n.

The square root of a negative number is undefined in the set of real numbers.

5.2 ▮ Solving Quadratic Equations with Square Roots

Taking the square root of the square of an expression results in absolute value: $\sqrt{w^2} = |w|$.

Solving $|x| = d$ yields $x = \pm d$.

For $ax^2 + bx + c = 0$, if $b = 0$, solve for x^2 and take the square root of both sides. The equation $x^2 = k$, $k \geq 0$ has two solutions: $x = +\sqrt{k}$ and $x = -\sqrt{k}$.

For $a^2 + 2ab + b^2 = d$, replace the perfect square trinomial on the left with $(a + b)^2$ and take the square root of both sides.

Discard negative solutions to quadratic equations when working with geometric figures.

5.3 ▮ Solving Quadratic Equations with the Quadratic Formula

To complete the square in a quadratic equation $ax^2 + bx + c = 0$, $a \neq 0$, subtract c from both sides, divide all terms by a, and add $(b/2a)^2$ to both sides.

The quadratic formula gives two solutions to $ax^2 + bx + c = 0$, if they exist:

$$x = \frac{-b \pm \sqrt{b^2 - 4ac}}{2a}$$

Written as separate terms, the quadratic formula shows the equation of the axis of symmetry and thus the x-coordinate of the vertex:

$$x = -\frac{b}{2a} \pm \frac{\sqrt{b^2 - 4ac}}{2a}$$

The x-coordinate of the vertex is $x = -b/2a$. The y-coordinate of the vertex is $f(-b/2a)$.

5.4 ▮ Solving Quadratic Inequalities

1. Graph each side of the inequality.
2. Find the intersections.
3. Write inequalities or intervals or draw a line graph showing the values of x that make the inequality true. Check the intersections or intercepts by factoring or with the quadratic formula. Check the inequalities by substituting in a test point.

▮ 5 ▮ Review Exercises

Simplify the radical expressions in Exercises 1 to 4.

1. a. $\sqrt{75}$ $5\sqrt{3}$ **b.** $\sqrt{8}$ $2\sqrt{2}$

 c. $\sqrt{32}$ $4\sqrt{2}$

2. a. $\sqrt{50}$ $5\sqrt{2}$ **b.** $\sqrt{18}$ $3\sqrt{2}$

 c. $\sqrt{72}$ $6\sqrt{2}$

3. a. $\dfrac{3 - 3\sqrt{6}}{3}$ $1 - \sqrt{6}$ **b.** $\dfrac{3 + 3\sqrt{6}}{3}$ $1 + \sqrt{6}$

4. a. $\dfrac{3 + \sqrt{18}}{3}$ $1 + \sqrt{2}$ **b.** $\dfrac{3 + \sqrt{12}}{2}$ $\dfrac{3 + 2\sqrt{3}}{2}$

5. Simplify.

 a. $\sqrt{14,400}$ 120   **b.** $\sqrt{1.44}$ 1.2

 c. $\sqrt{0.0121}$ 0.11   **d.** $\sqrt{0.00000001}$ 0.0001

6. Simplify.

 a. $\sqrt{0.0064}$ 0.08   **b.** $\sqrt{62,500}$ 250

 c. $\sqrt{1,000,000}$ 1000   **d.** $\sqrt{0.0225}$ 0.15

7. Show that each of the following sets of numbers could be the lengths of the sides of a right triangle.

 a. $\{7.5, 10, 12.5\}$
  $12.5^2 = 7.5^2 + 10^2$
 b. $\{18, 24, 30\}$
  $30^2 = 18^2 + 24^2$
 c. $\{\sqrt{5}, \sqrt{8}, \sqrt{13}\}$
  $(\sqrt{13})^2 = (\sqrt{5})^2 + (\sqrt{8})^2$
 d. $\{4, \sqrt{20}, 6\}$
  $6^2 = 4^2 + (\sqrt{20})^2$

8. Solve for x.

 a.

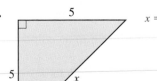
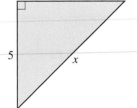

 $x = 5\sqrt{2}$

 b.

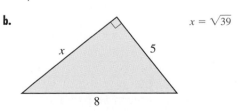

 $x = \sqrt{39}$

c.

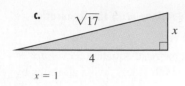

$x = 1$

d.

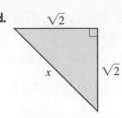

$x = 2$

e.

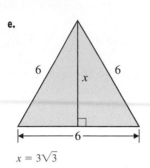

$x = 3\sqrt{3}$

f.

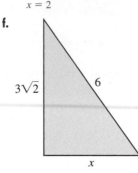

$x = 3\sqrt{2}$

9. Find the length of the hypotenuse z of an isosceles right triangle with legs x (see the figure). Write z in terms of x. Divide the expression for z by x to compare the lengths, hypotenuse to leg. $z = x\sqrt{2}; \frac{z}{x} = \sqrt{2}$

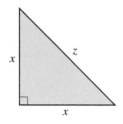

10. Solve for x.

   a. $x^2 = 225$ $\{\pm 15\}$ **b.** $x = \sqrt{225}$ $\{15\}$

   c. $x^2 = 0.81$ $\{\pm 0.9\}$ **d.** $x = \sqrt{0.81}$ $\{0.9\}$

   e. $|x + 3| = 2$ $\{-1, -5\}$ **f.** $|x - 5| = 3$ $\{2, 8\}$

11. True or false: $\sqrt{x}$ can be negative. Explain how your answer can be shown by the graph of $y = \sqrt{x}$.
false; there is no graph in the third or fourth quadrant.

12. The period of a pendulum is $T = 2\pi\sqrt{L/g}$. Solve the formula for L, the length of the pendulum.
$L = gT^2/4\pi^2$

13. An approximation to the perimeter of an ellipse is

$C = 2\pi\sqrt{\dfrac{a^2 + b^2}{2}}$. Solve the formula for a. $a = \sqrt{\dfrac{C^2 - 2\pi^2 b^2}{2\pi^2}}$

14. Complete the square on these equations.

   a. $x^2 - 8x + \underline{\ \ 16\ \ } = 3 + \underline{\ \ 16\ \ }$

   b. $x^2 - 10x + \underline{\ \ 25\ \ } = 2 + \underline{\ \ 25\ \ }$

   c. $x^2 - 3x + \underline{\ \ \frac{9}{4}\ \ } = 4 + \underline{\ \ \frac{9}{4}\ \ }$

   d. $x^2 + 7x + \underline{\ \ \frac{49}{4}\ \ } = 5 + \underline{\ \ \frac{49}{4}\ \ }$

15. Carry out these steps in solving $ax^2 + bx + c = 0$ for x to obtain the quadratic formula.

   a. Expand the expression $(x + b/2a)^2$. $x^2 + bx/a + b^2/4a^2$

   b. Combine the fractions: $b^2/4a^2 - c/a$. $(b^2 - 4ac)/4a^2$

   c. Simplify the expression $\sqrt{(x + b/2a)^2}$. $|x + b/2a|$

16. State the quadratic formula as it might appear on a calculator, adding any extra parentheses needed.
   For $ax^2 + bx + c = 0, x = (-b \pm \sqrt{(b^2 - 4ac)})/(2a)$

In Exercises 17 to 28, solve the equations by taking the square root or using the quadratic formula. Pretend you receive twice as much credit for taking the square root as for using the quadratic formula.

17. $3x^2 = 75$ $\{\pm 5\}$

18. $4x^2 + 8x + 4 = 0$ $\{-1\}$

19. $-\frac{1}{2}(9.81)x^2 + 6x + 10 = 11$ $\approx\{0.199, 1.024\}$

20. $-\frac{1}{2}(9.81)x^2 + 6x + 10 = 0$ $\approx\{-0.942, 2.165\}$

21. $(x - 5)^2 = 1$ $\{4, 6\}$

22. $9x^2 = 4$ $\{\pm\frac{2}{3}\}$

23. $9x^2 - 18x - 25 = 47$ $\{-2, 4\}$

24. $(x + 6)^2 = 9$ $\{-9, -3\}$

25. $4x^2 + 4x + 1 = 25$ $\{-3, 2\}$

26. $9x^2 + 6x + 1 = 9$ $\{-\frac{4}{3}, \frac{2}{3}\}$

27. $3x^2 - 6x + 3 = 0$ $\{1\}$

28. $25x^2 - 5x = 0$ $\{0, \frac{1}{5}\}$

29. How many real-number solutions does the equation $5x^2 + 7x = -4$ have? Explain in terms of a graph and again using the quadratic formula. No real-number solution; no x-intercepts; negative number under the radical sign

30. What was done in each step in solving $x^2 - x - 1 = 0$ for x by completing the square? (*Note:* The coefficient of x^2 is 1, so no division is necessary.)

$$x^2 - x - 1 = 0$$

$$x^2 - x = 1 \quad \text{Add 1 to both sides.}$$

$$x^2 - 1x + \tfrac{1}{4} = 1 + \tfrac{1}{4} \quad \text{Add } \tfrac{1}{4} \text{ to both sides.}$$

$$\left(x - \tfrac{1}{2}\right)^2 = 1 + \tfrac{1}{4} \quad \text{Factor left side.}$$

$$\left(x - \tfrac{1}{2}\right)^2 = \tfrac{5}{4} \quad \text{Add terms on right side.}$$

$$\sqrt{\left(x - \tfrac{1}{2}\right)^2} = \sqrt{\tfrac{5}{4}} \quad \text{Take square root.}$$

$$\left|x - \tfrac{1}{2}\right| = \dfrac{\sqrt{5}}{2} \quad \text{Use def. of principal square root.}$$

$$x - \tfrac{1}{2} = \pm\dfrac{\sqrt{5}}{2} \quad \text{Use def. of absolute value.}$$

$$x = \tfrac{1}{2} \pm \dfrac{\sqrt{5}}{2} \quad \begin{array}{l}\text{Add } \tfrac{1}{2} \text{ to both sides, evaluate,}\\ \text{and round.}\end{array}$$

Either $x \approx 1.618$ or $x \approx -0.618$.

31. Suppose we stand on the ground, $h_0 = 0$, and throw a ball straight up with an initial velocity of 72 ft/sec.

   a. Write an equation describing height in terms of time. Use the vertical motion equation, $h = -\frac{1}{2}gt^2 + v_0t + h_0$. Let the acceleration due to gravity be $g = 32$ ft/sec². $h = -16t^2 + 72t$

   b. What are the x-intercepts? Explain why there are two answers.   (0, 0), (4.5, 0); first is when ball is thrown, second is when it comes back down.

   c. Solve the equation for $h = 80$ ft. Explain your answer(s).   {2, 2.5}; on the way up and down

   d. What is the highest point the ball reaches?   81 ft

   e. Write an inequality for the length of time the ball will be above 80 feet. Solve the inequality. During this time, how far has the ball traveled? $-16t^2 + 72t > 80$; $2 < t < 2.5$; 2 ft (1 up and 1 down)

   f. Write an inequality for the length of time the ball will be below 32 feet. Solve the inequality. During this time, how far has the ball traveled? $-16t^2 + 72t < 32$; $0 < t < 0.5$ and $4 < t < 4.5$; 64 ft

   g. Compare your answers to parts e and f: The ball travels __2__ feet in 0.5 second at the top of the parabola and __32__ feet in 0.5 second near the ground.   Speed slows at the top.

32. Use the graph and equation in the figure to answer these questions.

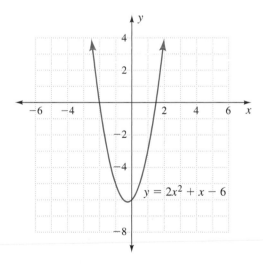

$y = 2x^2 + x - 6$

   a. What are the solutions to $y = 0$?   {−2, 1.5}

   b. What is the equation of the axis of symmetry of the graph? $x = -\frac{1}{4}$

   c. What is the vertex of the graph? $\left(-\frac{1}{4}, -6\frac{1}{8}\right)$

   d. For what inputs is $2x^2 + x - 6 > 0$? $x < -2$ or $x > 1.5$

   e. Give a value of y such that $y = 2x^2 + x - 6$ has no real-number solutions.   Possible answer: $y = -10$

   f. Between what two values of y will $y = 2x^2 + x - 6$ have two negative real-number solutions? $-6\frac{1}{8} < y < -6$

33. Find the vertex for each of these functions:

   a. $g(x) = x^2 + 3x + 4$ $\left(-\frac{3}{2}, \frac{7}{4}\right)$

   b. $h(x) = 0.5x^2 - 2x + 1$ $(2, -1)$

   c. $k(x) = 3x^2 - 4x$ $\left(\frac{2}{3}, -\frac{4}{3}\right)$

34. Does the constant term, c, influence the axis of symmetry of a quadratic graph?   No; axis of symmetry is $x = -b/2a$ and does not contain c.

In Exercises 35 to 46, find the set of inputs that satisfy each inequality. Round to the nearest thousandth.

35. $x^2 + 6x - 8 \leq 0$ $-7.123 \leq x \leq 1.123$

36. $x^2 + 6x - 7 \geq 0$ $x \leq -7$ or $x \geq 1$

37. $1 - 4x^2 \leq 0$ $x \leq -\frac{1}{2}$ or $x \geq \frac{1}{2}$

38. $2x^2 - 5x - 3 > 0$ $x < -\frac{1}{2}$ or $x > 3$

39. $3x^2 + 5x - 3 > 19$ $x < -3.667$ or $x > 2$

40. $-4x^2 + 5x + 2 < -7$ $x < -1$ or $x > 2\frac{1}{4}$

41. $x^2 + 5x + 2 > x - 3$ for all x, $\mathbb{R}$

42. $x^2 - 3 > 5 - x^2$ $x < -2$ or $x > 2$

43. $x^2 + 3 < 2 - x^2$ no real-number solution

44. $x^2 + 3x + 2 > 2 - x^2$ $x < -1.5$ or $x > 0$

45. $x^2 + 5x + 2 \leq 2 - x$ $-6 \leq x \leq 0$

46. $4x^2 + 6x - 3 \leq 1 + x - 2x^2$ $-\frac{4}{3} \leq x \leq \frac{1}{2}$

47. From their truck, an inspection team can inspect suspension bridge cable within 20 feet of the road surface. The equation of the parabolic suspension cable is $y = 0.0375(x - 20)^2 + 10$. For what horizontal interval on the 40-foot-long bridge can the inspection be accomplished from the truck? $\approx 3.7 < x < 36.3$

5 Chapter Test

1. State the quadratic formula for solving $ax^2 + bx + c = 0$. $x = \dfrac{-b + \sqrt{b^2 - 4ac}}{2a}$

2. Simplify.

   a. $\sqrt{45}$ $3\sqrt{5}$ **b.** $\sqrt{98}$ $7\sqrt{2}$

3. Simplify without a calculator: $(6 + \sqrt{18})/12$. $(2 + \sqrt{2})/4$

Solve the equations in Exercises 4 to 8 by the indicated method.

4. $x^2 = 196$ by taking the square root $\{\pm 14\}$

5. $(x - 4)^2 = 25$ by taking the square root $\{-1, 9\}$

6. $x^2 + 10x + \underline{}^{25} = \underline{}^{25} + 24$ by completing the square and taking the square root $\{-12, 2\}$

7. $0.04x^2 - 169 = 0$ by taking the square root $x = \pm 65$

8. $x^2 + 2x + 1 = 16$ by taking the square root $\{-5, 3\}$

9. Solve $|x - 2| = 1$. $\{1, 3\}$

10. Solve $2x^2 - 7x + 6 = 0$ with the quadratic formula. Show all your steps. $\{1.5, 2\}$

11. Solve for x:

    a.

    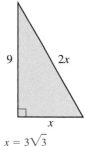

    $x = 3\sqrt{3}$

    b.

    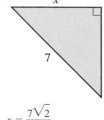

    $x = \dfrac{7\sqrt{2}}{2}$

12. Use any method to find the x-intercepts, if they exist, and then find the vertex for the graph of $f(x) = x^2 - x - 2$. $\{-1, 2\}; \left(\frac{1}{2}, -2\frac{1}{4}\right)$

Solve inequalities in Exercises 13 to 17.

13. $x^2 + 5x - 6 \leq 0$
    $-6 \leq x \leq 1$

14. $x^2 + 5x - 6 > 0$
    $x < -6$ or $x > 1$

15. $x^2 + x < x + 1$
    $-1 < x < 1$

16. $4 - x^2 < x^2 + 6$
    for all x, $\mathbb{R}$

17. $5x^2 + 15x - 16 \geq 4 - 4x - x^2$
    $x \leq -4$ or $x \geq \frac{5}{6}$

18. **a.** Make a table and a graph for the annual salary A if r is the percent raise per year for 2 years and P is the starting salary in the formula $A = P(1 + r)^2$. Suppose the starting salary is \$20,000. Let r be 0 to 10%. See Answer Section.

    b. For what percent raise will the salary be \$22,898 after 2 years? 7%

    c. Extend your graph; for what percent raise will the salary be \$30,000? $\approx 22.5\%$

19. The velocity (speed) of a falling object, without air resistance, is $v = \sqrt{2gs}$. Solve for s, the distance fallen. $s = v^2/2g$

20. A 10-year-old cuts across an empty square city block rather than taking the sidewalk around the block. If she takes 212 steps to cut diagonally across the block, approximately how many steps would she have to take if she stayed on the sidewalk? 300 steps

21. Compare the solutions to $\sqrt{x^2} = 2$ and $x = \sqrt{4}$. Include a graph with your explanation.
    $\sqrt{x^2} = 2, x = \pm 2; x = \sqrt{4}, x = 2$; see Answer Section.

Quadratic Functions: Special Topics

The water jets sent up by the fireboat in Figure 1 form several different parabolas. The parabolas may be described by quadratic functions of the form $f(x) = ax^2 + bx + c$. In Section 6.1, we investigate the role of a, b, and c in controlling the shape of a parabola such as the water jet. Solving equations with no real-number solutions is introduced in Section 6.2. We continue with graphs and translations of the vertex as we examine the vertex form of the quadratic equation in Section 6.3 and its relation to the maximum and minimum values in Section 6.4.

FIGURE 1

6.1 The Roles of a, b, and c in Graphing Quadratic Functions

Topics in Chapter 6 (delayed, in many curricula, until college algebra) may be omitted in deference to more timely concepts such as exponential functions. Section 6.3 stresses completing the square and, because of the introduction of vertex in Section 5.3, is not needed for maximum and minimum in Section 6.4.

Objectives

■ Find the effect of a on the graph of $f(x) = ax^2$.

■ Find the effect of b on the graph of $f(x) = x^2 + bx$.

■ Find the effect of c on the graph of $f(x) = x^2 + c$.

WARM-UP

1. Complete the table.

x	x^2	$-x^2$
-4	16	-16
-2	4	-4
1	1	-1
3	9	-9
5	25	-25

2. a. In what quadrants are the ordered pairs (x, x^2)? first and second

   b. In what quadrants are the ordered pairs $(x, -x^2)$? third and fourth

3. Suggest how the negative sign in $y = -x^2$ changes the graph of $y = x^2$.

4. Complete the table.   It changes the orientation of the parabola from turning up to turning down.

x	x^2	$2x^2$	$\frac{1}{2}x^2$
-2	4	8	2
-1	1	2	$\frac{1}{2}$
0	0	0	0
1	1	2	$\frac{1}{2}$
2	4	8	2

doubles x^2

5. Describe how the 2 makes the $2x^2$ column different from the x^2 column.

6. Describe how the $\frac{1}{2}$ makes the $\frac{1}{2}x^2$ column different from the x^2 column.

takes half of x^2

IN THIS SECTION, we examine the graphs of quadratic functions in more detail. We consider the roles played by the parameters a, b, and c in determining the orientation, shape, and position of the graph of a quadratic function.

Review of Graphing a Quadratic Function

Have you ever wondered why the game of golf requires so many clubs? The length of a golf club, its weight, and the angle of the striking face are designed to produce a particular path of a golf ball, as shown in Figure 2. Selection of the right club and consistency in its use are prerequisites to playing the game of golf well.

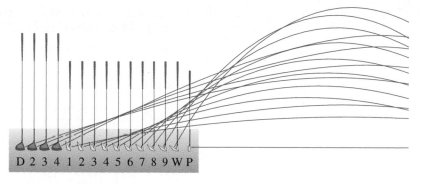

FIGURE 2 Golf clubs

In Example 1, the parabolic model used to approximate the path of a golf ball requires trigonometric coefficients, so we will not show how the function was obtained.

EXAMPLE 1 Reviewing features of the graph of a quadratic function: path of a golf ball Suppose we use the quadratic function $f(x) = -0.0005812x^2 + 0.1763x + 15$ to model the path of a golf ball hit upward at a 10° angle from a tee (starting point) 15 feet above the fairway. The equation includes a 169-ft/sec initial velocity.

a. Make a table and graph for $f(x)$ using 50-foot increments for x.
b. Find the x-intercepts, $f(x) = 0$, and their meaning.
c. Find the y-intercept and its meaning.
d. Find the vertex.

SOLUTION **a.** The table is shown in Table 1 and the graph in Figure 3. The path of the ball is really flatter than it appears, because the scales on the x- and y-axes in Figure 3 are not the same (see Exercise 54).

TABLE 1 $f(x) = -0.0005812x^2 + 0.1763x + 15$

x	$f(x)$
0	15
50	22.4
100	26.8
150	28.4
200	27.0
250	22.8
300	15.6
350	5.5
400	−7.5

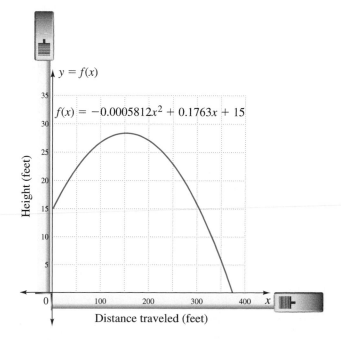

FIGURE 3

b. The quadratic formula with $a = -0.0005812$, $b = 0.1763$, and $c = 15$ gives $x \approx -69.3$ and $x \approx 372.6$ as the x-intercepts. Because the function represents the path of a ball, we disregard $x \approx -69.3$. The positive x-intercept, $x \approx 372.6$ ft, gives the position where the ball strikes the ground.

c. The y-intercept is $(0, 15)$, or $f(0) = 15$ ft. The y-intercept represents the initial height of the ball at the tee.

d. The x-coordinate of the vertex is

$$x = -\frac{b}{2a} = -\frac{0.1763}{2(0.0005812)} \approx 151.7 \text{ ft}$$

The y-coordinate of the vertex is $f(-b/2a)$.

$$f(x) = -0.0005812x^2 + 0.1763x + 15$$

$$f(151.7) = -0.0005812(151.7)^2 + 0.1763(151.7) + 15$$

$$\approx 28.4 \text{ ft}$$

Thus, the vertex is approximately $(151.7, 28.4)$.

The Shape and Position of Quadratic Function Graphs

How high the golf ball reaches (the vertex) and how far it flies before hitting the ground (the x-intercept) are determined by how hard and at what angle the ball is hit, as well as how high the tee is initially. The following examples suggest the connection of these factors to a, b, and c in the function $y = ax^2 + bx + c$.

ROLE OF PARAMETER a   Examples 2 and 3 investigate the role of the parameter a on the graph.

EXAMPLE 2   Exploring the sign on a

a. Explain how the negative sign in front of the x^2 in Table 2 changes the numbers from those generated by x^2.

b. Explain how the graph of $y = -x^2$ compares to that of $y = x^2$.

c. Test your conclusions with the equation and graph in Example 1 for the path of the golf ball.

SOLUTION   **a.** The outputs for the expression $-x^2$ have the opposite sign from the outputs for x^2.

Student Note: The *orientation* is the direction in which something faces.

b. The graph of $y = -x^2$ is upside down compared with the graph of $y = x^2$ (see Figure 4). The negative sign changes the orientation of the parabola, from opening up for $y = x^2$ to opening down for $y = -x^2$.

TABLE 2

x	x^2	$-x^2$
-4	16	-16
-2	4	-4
1	1	-1
3	9	-9
5	25	-25

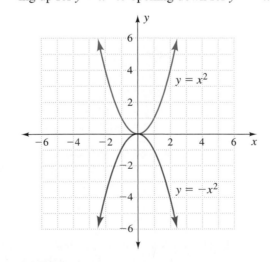

FIGURE 4

c. The function in Example 1 has a negative sign on the x^2 term, and the graph opens down. ◼

Example 2 suggests that a controls the orientation of the parabola.

◼ THE SIGN ON a AND THE DIRECTION OF THE PARABOLA

> A positive a on the x^2 term causes the parabola to open up, $\cup$. A negative a on the x^2 term causes the parabola to open down, $\cap$.

Example 3 examines the steepness of parabolic graphs.

◼ **EXAMPLE 3** Exploring the value of a

a. Graph $y = 2x^2$, $y = x^2$, and $y = \frac{1}{2}x^2$ (from Table 3), and compare the graphs.

b. Comment on the value of a in the golf ball equation.

SOLUTION **a.** In Figure 5, the graph of $y = 2x^2$ is the steepest. The graph of $y = \frac{1}{2}x^2$ is flatter than the other two. The numbers 2 and $\frac{1}{2}$ appear to change the shape of the parabola.

TABLE 3

x	x^2	$2x^2$	$\frac{1}{2}x^2$
-2	4	8	2
-1	1	2	$\frac{1}{2}$
0	0	0	0
1	1	2	$\frac{1}{2}$
2	4	8	2

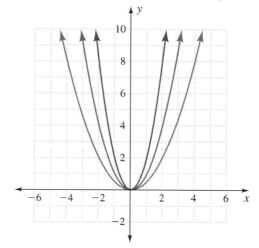

FIGURE 5

b. For the golf ball function, a is close to zero. If the graph were drawn with the same scale on the x- and y-axes, the path of the ball would be nearly flat. ◼

Example 3 suggests that a controls the shape of the parabola.

◼ THE SIZE OF a AND THE STEEPNESS OF THE PARABOLA

> The coefficient on the x^2 term, the parameter a, controls the shape of the graph. If a is larger than 1, the graph is steeper than the graph of $y = x^2$. If a is between zero and 1, the graph is flatter than the graph of $y = x^2$.

ROLE OF PARAMETER b Example 4 investigates one effect of the parameter b on the graph of $y = x^2 + bx$.

◼ **EXAMPLE 4** Exploring b Graph $y = x^2$, $y = x^2 + 1x$, $y = x^2 + 2x$, and $y = x^2 + 3x$. Explain how the b in $y = x^2 + bx$ changes the graph from that of $y = x^2$.

SOLUTION The graphs are shown in Figure 6. The graphs are the same shape as that of $y = x^2$, but their vertices have moved away from the origin.

$$— y = x^2 \qquad — y = x^2 + 1x$$
$$— y = x^2 + 2x \qquad — y = x^2 + 3x$$

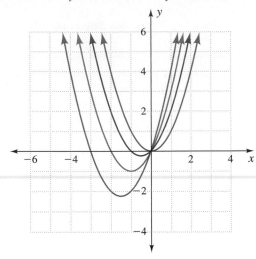

FIGURE 6

THE ROLE OF b

> The parameter b in $y = x^2 + bx$ contributes to a change in the position of the vertex of the parabola from that of $y = x^2$.

Although our conclusion that b changes the position of the vertex is correct, both a and c also affect the position of the vertex. No simple generalization about b is possible.

ROLE OF PARAMETER c Example 5 investigates the impact of the parameter c on the graph.

EXAMPLE 5   Exploring c Use the graph in Figure 7 to answer these questions.
a. Explain how adding a constant to x^2 changes the graph. Explain what the parameter c represents in the graph.
b. Test your conclusion with the golf ball function in Example 1.

$$— y = x^2 \qquad — y = x^2 + 2 \qquad — y = x^2 - 1$$

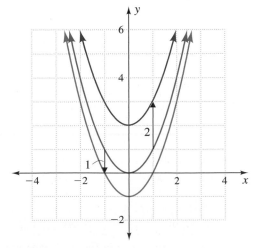

FIGURE 7

SOLUTION   **a.** The graph of $y = x^2 + 2$ is shifted up 2 units from that of $y = x^2$. The graph of $y = x^2 - 1$ is 1 unit lower than that of $y = x^2$. In $y = x^2 + 2$, the y-intercept is 2. In $y = x^2 - 1$, the y-intercept is -1. Adding a number to x^2 shifts the graph vertically. The number also becomes the y-intercept.

b. In Example 1, the function $f(x) = -0.0005812x^2 + 0.1763x + 15$ has 15 as its y-intercept.   ▬

Example 5 suggests that the parameter c contributes to the vertical position of the parabola.

▬ **THE ROLE OF *c***

> The parameter c in $y = ax^2 + bx + c$ is the output when $x = 0$. The ordered pair $(0, c)$ is the y-intercept point of the graph of a quadratic function.
>
> If the equation is of the form $y = ax^2 + c$, then $|c|$ is the distance (number of units) the parabola $y = ax^2$ is shifted vertically (up or down) and the vertex is $(0, c)$.

Application: Fireboat

Our conclusions can be applied to the path of the water jet coming from the cannons on the fireboats shown in Figure 1 (on page 309) and Figure 8.

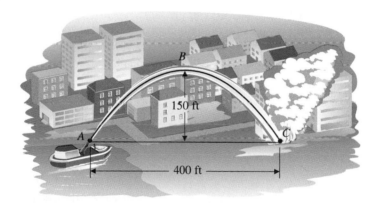

FIGURE 8

EXAMPLE 6   Predicting parameters: fireboat water jet
a. Letting the origin be at the water jet cannon, find the x-intercept points, y-intercept point, and vertex of the water spray between the boat and the fire in Figure 8.
b. Assume that the path of the water is parabolic. Predict the sign on a and the value of c in an equation that fits the path. Find the equation of the path of the water spray using $y = a(x - x_1)(x - x_2)$.

SOLUTION   **a.** If the origin $(0, 0)$ is placed at the water cannon in Figure 8, the x-intercept points of the water spray graph represent the starting point of the water as well as the position at which the water returns to the same level as the cannon, $(400, 0)$. The y-intercept point is also at the origin. The vertex of the water spray is at its highest point, $(200, 150)$.

b. The graph appears parabolic. The parabola turns down, so we expect a negative coefficient for x^2. The curve passes through the origin, so $c = 0$. We let (x, y) be $(200, 150)$ in $y = a(x - 0)(x - 400)$ and solve for a. We substitute a back into the equation and obtain $y = -0.00375x^2 + 1.5x + 0$.   ▬

Student Note: Check your results with quadratic regression.

6.1   Exercises

1. In Example 6, explain how the equation tells us that the path of the water is a parabola that turns down.
   a is negative.

2. In Example 6, explain how the graph indicates that the coefficient on the x^2 term is negative.   It turns down.

3. In Example 6, explain how *a* indicates that the graph of the water jet is flatter than that of $y = -x^2$.
   $|a| = 0.00375, |a| < 1$

4. In the vertical motion equation $h = -\frac{1}{2}gt^2 + v_0 t + h_0$, *g* is either 9.81 m/sec² or 32.2 ft/sec². Compare the shape of the graph with that of $y = -t^2$.
   a is negative, $|a| > 1$, so the curve will be steeper.

5. In Example 6, the graph passes through the origin. What is the *y*-intercept? What is *c*?   0; 0

6. In Example 1, if the tee (starting point) were at the same level as the fairway, the path would pass through the origin. What is the value of *c*?   0

7. If any quadratic graph passes through the origin, what is its *y*-intercept and, hence, the value of *c*?   0; 0

8. Explain how $x = -b/2a$ tells us that all functions of the form $y = ax^2 + c$ have vertices on the *y*-axis.
   $b = 0, x = -\frac{0}{2}a$, so $x = 0$.

In Exercises 9 and 10, match the situation described with one of the following equations:

$$y = -0.16x^2 + 0.09x, \qquad y = \tfrac{4}{125}x^2$$

Explain how you found the correct equation.

9. The equation of a parabolic cable supporting a bridge
   $y = \frac{4}{125}x^2$; positive coefficient of x^2

10. The equation of the path of a cliff diver
   $y = -0.16x^2 + 0.09x$; negative coefficient of x^2

11. Graph $y = \frac{1}{2}x^2$ and $y = \frac{1}{2}x^2 + 3$. Draw arrows to show the shift in the graph of $y = \frac{1}{2}x^2$ to the position of the second graph.   See Answer Section.

12. Graph $y = \frac{1}{2}x^2$ and $y = \frac{1}{2}x^2 - 2$. Draw arrows to show the shift in the graph of $y = \frac{1}{2}x^2$ to the position of the second graph.   See Additional Answers.

13. Graph $y = 2x^2$ and $y = 2x^2 - 3$. Draw arrows to show the shift in the graph of $y = 2x^2$ to the position of the second graph.   See Answer Section.

14. Graph $y = 2x^2$ and $y = 2x^2 + 1$. Draw arrows to show the shift in the graph of $y = 2x^2$ to the position of the second graph.   See Additional Answers.

In Exercises 15 to 22, identify the function whose graph will make a steeper parabola.

15. $f(x) = 2x^2$ or $g(x) = \pi x^2$ $g(x)$

16. $g(x) = \pi x^2$ or $h(x) = 3x^2$ $g(x)$

17. $j(x) = 4x^2$ or $g(x) = \pi x^2$ $j(x)$

18. $k(x) = 3.5x^2$ or $g(x) = \pi x^2$ $k(x)$

19. $f(x) = \frac{2}{3}x^2$ or $h(x) = \frac{3}{4}x^2$ $h(x)$

20. $g(x) = 2.299x^2$ or $h(x) = 2.32x^2$ $h(x)$

21. $p(x) = 2.41x^2$ or $q(x) = 2.288x^2$ $p(x)$

22. $r(x) = \frac{3}{8}x^2$ or $s(x) = \frac{1}{3}x^2$ $r(x)$

23. Choose the correct word, *steeper* or *flatter*, to complete this sentence: A coefficient on x^2 larger than 1 gives a graph ‾‾steeper‾‾ than the graph of $y = x^2$.

24. Choose the correct word, *larger* or *smaller*, to complete this sentence: A positive coefficient ‾‾smaller‾‾ than 1 gives a graph flatter than the graph of $y = x^2$.

Use the properties of parameters *a* and *c* to sketch the graph of each equation. Selected factoring to find *x*-intercepts may be helpful. No calculators!

25. $y = x^2 - 1x$
   See Answer Section.

26. $y = x^2 + 2$
   See Additional Answers.

27. $y = -x^2$
   See Answer Section.

28. $y = \frac{1}{2}x^2$
   See Additional Answers.

29. $y = 2x^2$
   See Answer Section.

30. $y = -x^2 + 1x$
   See Additional Answers.

31. $y = x^2 + 1$
   See Answer Section.

32. $y = x^2 - 2x$
   See Additional Answers.

33. $y = -x^2 + 2x$
   See Answer Section.

34. $y = x^2 - 1$
   See Additional Answers.

35. $y = x^2 - 2$
   See Answer Section.

36. $y = -2x^2$
   See Additional Answers.

37. $y = -\frac{1}{2}x^2$
   See Answer Section.

38. $y = -x^2 - 2$
   See Additional Answers.

39. $y = 1 - x^2$
   See Answer Section.

40. $y = -x^2 - 3x$
   See Additional Answers.

41. $y = 3x^2 + 1$
   See Answer Section.

42. $y = \frac{1}{3}x^2 - 1$
   See Additional Answers.

43. $y = x^2 - 3x$
   See Answer Section.

44. $y = -3x^2 + 1$
   See Additional Answers.

45. $y = \frac{1}{3}x^2 - 2$
   See Answer Section.

46. $y = 3x^2 + 3$
   See Additional Answers.

Eight quadratic functions, with different function names, are listed below. Find which functions satisfy the conditions in Exercises 47 to 52. It is not necessary to graph the functions.

$$f(x) = 2x^2 + 2x - 3 \qquad g(x) = 0.5x^2 + 3x - 2$$

$$h(x) = \frac{1}{2}x^2 + 2x \qquad j(x) = 2(x - 1)^2$$

$$k(x) = -2x^2 - 1 \qquad m(x) = \frac{x^2 + 2x}{2}$$

$$p(x) = -\frac{1}{2}(x + 1)^2 \qquad q(x) = \frac{(x + 1)(x + 2)}{2}$$

47. Name the functions that have the same coefficient on x^2 as $g(x)$. $h(x)$, $m(x)$, $q(x)$

48. Name the functions that have the same coefficient on x^2 as $f(x)$. $j(x)$

49. Name the functions that have a parabolic graph that turns up. $f(x)$, $g(x)$, $h(x)$, $j(x)$, $m(x)$, $q(x)$

50. Name the functions that have a parabolic graph that passes through the origin. $h(x)$, $m(x)$

51. Name the functions that have a parabolic graph with $(0, 2)$ as its y-intercept. $j(x)$

52. Name the functions that have a parabolic graph that turns down. $k(x)$, $p(x)$

53. In the vertical motion equation, if the motion starts at $h = 0$ and returns to $h = 0$, then the vertex, in terms of v_0 and g, is _____. $(v_0/g, v_0^2/2g)$

54. The equation in Example 1 is $f(x) = -0.0005812x^2 + 0.1763x + 15$. Enter $f(x)$ in $\boxed{Y=}$, with X on $[0, 400]$ and Y on $[0, 35]$. Press $\boxed{\text{ZOOM}}$ and choose **5 : ZSquare**. Explain what you see. Record the numbers now under $\boxed{\text{WINDOW}}$.   path of ball without distortion; **Xmin** = 0, **Xmax** = 400, **Ymin** = −114, **Ymax** = 149

55. A graph has the same shape as $y = x^2$. Its vertex is $(0, -3)$. What is its equation? (There are two possibilities.) $y = x^2 - 3$ or $y = -x^2 - 3$

56. A graph has the same shape as $y = x^2$. Its vertex is $(0, 4)$. What is its equation? (There are two possibilities.) $y = x^2 + 4$ or $y = -x^2 + 4$

57. Suppose the origin in the fireboat problem (Example 6) is placed 20 feet directly below the cannon at the water line.

a. What are the coordinates for the cannon, the highest point on the spray of water, and the fire?
$(0, 20)$, $(200, 170)$, $(400, 20)$

b. Predict the new equation of the water spray. Confirm your solution by fitting a quadratic equation through the three points. $y = -0.00375x^2 + 1.5x + 20$

58. Suppose the origin in the fireboat problem (Example 6) is placed at the fire and the remaining coordinates are adjusted to be in the second quadrant or on the negative x-axis.

a. What are the coordinates for the cannon, the highest point on the spray of water, and the fire?
$(-400, 0)$, $(-200, 150)$, $(0, 0)$

b. Predict the new equation of the water spray. Confirm your solution by fitting a quadratic equation through the three points. $y = -0.00375x^2 - 1.5x$

▮▮▮ Project

59. Rotating Liquids Research

a. Fill a jar slightly less than ▮▮▮ jar at the center of an old p▮▮▮ pottery wheel, or a revolvin▮▮▮ What shape does the water f▮▮▮ turned? How does the shape ▮▮▮ rotation changes?

b. In recent years, the rotating concept was adapted to vats of molten glass, which were cooled as they rotated in order to create large reflecting telescope lenses that required little grinding. The same idea is also behind using rotating vats of mercury to form liquid telescope reflectors. Research and write a report on one of these techniques.

6.2   Complex Numbers and Solving Polynomial Equations

Objectives

■ Change square roots of negative numbers into complex-number notation.

■ Identify complex conjugates.

■ Add, subtract, and multiply complex numbers.

■ Multiply and simplify binomials containing irrational and complex numbers.

■ Use the discriminant, $b^2 - 4ac$, to identify the number and type of roots of quadratic equations.

■ Find complex solutions to polynomial equations.

WARM-UP

1. Simplify these expressions.

    a. $(-1)^0$ 1   **b.** $(-1)^1$ −1   **c.** $(-1)^3$ −1

    d. $(-1)^2$ 1   **e.** $(-1)^8$ 1   **f.** $(-1)^4$ 1

2. Multiply and simplify these expressions.

    a. $\sqrt{2} \cdot \sqrt{32}$ 8   **b.** $\sqrt{8} \cdot \sqrt{2}$ 4

    c. $\sqrt{3} \cdot \sqrt{27}$ 9   **d.** $\sqrt{20} \cdot \sqrt{5}$ 10

3. Look for perfect square factors and simplify, using $\sqrt{a \cdot b} = \sqrt{a} \cdot \sqrt{b}$.

    Example: $\sqrt{52} = \sqrt{4 \cdot 13} = \sqrt{4} \cdot \sqrt{13} = 2\sqrt{13}$

    a. $\sqrt{12}$ $2\sqrt{3}$ **b.** $\sqrt{24}$ $2\sqrt{6}$

    c. $\sqrt{45}$ $3\sqrt{5}$ **d.** $\sqrt{8}$ $2\sqrt{2}$

THROUGHOUT OUR WORK with quadratic equations, the intersection of a parabolic graph with the x-axis has given the solutions to $ax^2 + bx + c = 0$. This section answers the question *What solutions are obtained when the graph does not intersect the x-axis?* The imaginary unit and the complex number system will be introduced, as well as operations with complex numbers.

The Imaginary Unit

Student Note: Imaginary numbers may be graphed on an Argand diagram, which has real numbers on the horizontal axis and imaginary numbers on the vertical axis.

In solving $x^2 + 1 = 0$, we obtain $x^2 = -1$, or $x = \pm \sqrt{-1}$. Although the square roots of negative numbers are not real numbers and cannot be graphed on the rectangular coordinate axes, mathematicians have nevertheless investigated the properties of these numbers. In order to do so, they defined a new symbol, the imaginary unit i.

■ **DEFINITION OF IMAGINARY UNIT**

$$i = \sqrt{-1} \qquad \text{or} \qquad i^2 = -1$$

where i is the **imaginary unit.**

The solutions to $x^2 + 1 = 0$ are $x = \pm \sqrt{-1}$, or $x = \pm i$.

    The imaginary unit is important in the study of alternating current in electricity. Because the letter i is already used for something else in electronics, the imaginary unit is j, or the j-operator, as in $z = R + j\omega L$ or $z = R - j/\omega C$. In mathematics, however, we traditionally use i for the imaginary unit.

■ Our product and quotient properties of square roots apply only to the real numbers. To work with square roots containing negative numbers, we first change them to products of a real number and the imaginary unit.

> If a is real and positive,
>
> $$\sqrt{-a} = i\sqrt{a}$$

EXAMPLE 1 **Changing square root notation to imaginary units** Change these numbers into the product of a real number and the imaginary unit.

a. $\sqrt{-4}$ b. $\sqrt{-8}$ c. $\sqrt{-32}$

SOLUTION a. $\sqrt{-4} = i\sqrt{4} = i \cdot 2 = 2i$

Student Note: Simplification using the product rule is easiest if you look for the largest possible perfect square. For example, use $\sqrt{32} = \sqrt{16} \cdot \sqrt{2}$, not $\sqrt{32} = \sqrt{4} \cdot \sqrt{8}$.

b. $\sqrt{-8} = i\sqrt{8} = i\sqrt{4 \cdot 2} = i\sqrt{4}\sqrt{2} = 2i\sqrt{2}$

c. $\sqrt{-32} = i\sqrt{32} = i\sqrt{16}\sqrt{2} = 4i\sqrt{2}$

Where convenient, the square roots of negatives may be written with decimals:

$$\sqrt{-8} \approx 2.828i \qquad \text{and} \qquad \sqrt{-32} \approx 5.657i$$

Caution: Although placing the i after the radical sign is acceptable, it may lead to errors if the i appears to be under the radical. When a number or letter is multiplying a radical, place it in front of the radical.

The Complex Number System

When we combine *the real numbers with the imaginary unit*, we create the **complex number system**. Complex numbers may be written $a + bi$, where a and b are any real numbers, including zero. In a **complex number** $a + bi$, a is the real part and bi is the imaginary part. All real numbers a are complex numbers with a zero imaginary part: $a + 0i$.

EXAMPLE 2 **Writing numbers in complex number notation** Write the following as complex numbers.

a. $5i$ b. $\sqrt{-27}$ c. 15

d. $\dfrac{6 + 8i}{2}$ e. $\dfrac{\pm\sqrt{-8}}{2}$ f. $\dfrac{-4 \pm \sqrt{-8}}{4}$

SOLUTION a. $5i = 0 + 5i$

b. $\sqrt{-27} = i\sqrt{27} = i\sqrt{9}\sqrt{3} = 3i\sqrt{3},$ or $0 + 3i\sqrt{3}$

c. $15 = 15 + 0i$

d. $\dfrac{6 + 8i}{2} = \dfrac{6}{2} + \dfrac{8i}{2} = 3 + 4i$

e. $\dfrac{\pm\sqrt{-8}}{2} = \dfrac{\pm i\sqrt{8}}{2} = \dfrac{\pm i\sqrt{4}\sqrt{2}}{2}$

$= \dfrac{\pm i \cdot \boxed{2} \cdot \sqrt{2}}{\boxed{2}} = \pm i\sqrt{2},$ or $0 \pm i\sqrt{2}$

f. $\dfrac{-4 \pm \sqrt{-8}}{4} = -\dfrac{4}{4} \pm \dfrac{i\sqrt{8}}{4} = -1 \pm \dfrac{i\sqrt{4}\sqrt{2}}{4}$

$= -1 \pm \dfrac{2i\sqrt{2}}{4} = -1 \pm \dfrac{i\sqrt{2}}{2}$

The two expressions may be written separately: $-1 + \dfrac{i\sqrt{2}}{2}$ and $-1 - \dfrac{i\sqrt{2}}{2}$.

The real part of each answer is -1. The decimal form may be simpler: $-1 \pm 0.707i$.

Complex Conjugates

Recall that for $x^2 = -1$, $x = \pm i$. Complex-number solutions always occur in pairs. Other pairs are $x = i\sqrt{2}$ and $x = -i\sqrt{2}$ or $x = -1 + \dfrac{i\sqrt{2}}{2}$ and $x = -1 - \dfrac{i\sqrt{2}}{2}$. These pairs are given a special name—complex conjugates.

■ DEFINITION OF COMPLEX CONJUGATES

> If a and b are real numbers, expressions of the form $a + bi$ and $a - bi$ are **complex conjugates**.

Complex conjugates are complex numbers with the same real-number part and opposite imaginary parts.

EXAMPLE 3 Writing complex conjugates Name the complex conjugate for each expression.

a. i **b.** $4 + i$ **c.** $5 - 2i$

SOLUTION *Hint:* To obtain a complex conjugate, replace i with $-i$ or $-i$ with i. See the Answer Box.   ■

Think about it: What part of the quadratic formula shows that the complex solutions to a quadratic equation always appear as complex conjugates?

We now turn to operations with complex numbers. Look for special results when we add, subtract, or multiply complex conjugates.

Operations with Complex Numbers

We define addition and subtraction of complex numbers as follows.

■ ADDITION OF COMPLEX NUMBERS

> $$(a + bi) + (c + di) = (a + c) + (b + d)i$$

■ SUBTRACTION OF COMPLEX NUMBERS

> $$(a + bi) - (c + di) = (a - c) + (b - d)i$$

Addition and subtraction with complex numbers are similar to the operations with polynomials. Both operations may be thought of as adding like terms. We add and subtract like numbers, real to real and imaginary to imaginary.

EXAMPLE 4 Adding and subtracting complex numbers Add or subtract these complex numbers. If any are complex conjugates, note anything special about the results.

a. $(2 + 3i) + (2 - 3i)$ **b.** $(3 - 5i) - (3 + 5i)$

c. $(4i + 6) - (7i + 2) - (3i - 5)$

SOLUTION **a.** $(2 + 3i) + (2 - 3i) = (2 + 2) + (3 - 3)i = 4 + 0i = 4$
These numbers are complex conjugates. Their sum is a real number.

b. $(3 - 5i) - (3 + 5i) = (3 - 3) + (-5 - 5)i = 0 - 10i = -10i$
The difference of these complex conjugates has only an imaginary part.

c. $(4i + 6) - (7i + 2) - (3i - 5) = 6 - 2 - (-5) + (4 - 7 - 3)i$

$$= 9 - 6i$$ ■

■ Before we define multiplication with complex numbers, we return to our definition of the imaginary unit and note the statement $i^2 = -1$.

▬ DEFINITION OF IMAGINARY UNIT

$$i = \sqrt{-1} \quad \text{or} \quad i^2 = -1$$

where i is the **imaginary unit**.

The $i^2 = -1$ is used to simplify after multiplying complex numbers.

▬ MULTIPLICATION OF COMPLEX NUMBERS

$$(a + bi)(c + di) = ac + adi + bci + bdi^2$$
$$= ac - bd + (ad + bc)i$$

The multiplication of complex numbers is similar to the multiplication of binomials $(a + b)(c + d)$. Try a favorite method, or use a table as shown in Example 5.

EXAMPLE 5 **Multiplying complex numbers**   Multiply these complex numbers. If any are complex conjugates, note anything special about the products.

a. $(3 + 4i)(3 + 4i)$

b. $(3 - 4i)^2$

c. $(\sqrt{3} + i)(\sqrt{3} - i)$

SOLUTION We multiply each in a table.

a.

Multiply	3	$+4i$
3	9	$+12i$
$+4i$	$+12i$	$+16i^2$

$$(3 + 4i)(3 + 4i) = 9 + 24i + 16i^2 = 9 + 24i - 16$$

because $i^2 = -1$. Thus, $(3 + 4i)(3 + 4i) = -7 + 24i$.

b.

Multiply	3	$-4i$
3	9	$-12i$
$-4i$	$-12i$	$+16i^2$

$$(3 - 4i)(3 - 4i) = 9 - 24i + 16i^2 = 9 - 24i - 16 = -7 - 24i$$

Student Note: In part c, the diagonal terms add to zero.

c.

Multiply	$\sqrt{3}$	$+i$
$\sqrt{3}$	3	$+i\sqrt{3}$
$-i$	$-i\sqrt{3}$	$-i^2$

$$(\sqrt{3} + i)(\sqrt{3} - i) = 3 - i^2 = 3 - (-1) = 4$$

These numbers are complex conjugates. Their product is a real number. ▬

Solving Polynomial Equations

We now return to solving equations when the solution set may contain complex numbers.

QUADRATIC EQUATIONS AND THE DISCRIMINANT   In Example 6, we examine three quadratic equations, one of which has no real-number solutions. The example shows how we can predict the type of solutions a quadratic equation may have.

EXAMPLE 6 **Finding numbers of solutions** Graph each function. Comment on the position of the graph relative to the x-axis. Use the quadratic formula to find the solution to each equation when $f(x) = 0$. Comment on the number and types of solutions.

a. $f(x) = x^2 + 2x + 1$

b. $f(x) = x^2 + 2$

c. $f(x) = x^2 - 2x$

SOLUTION The functions are graphed in Figure 9.

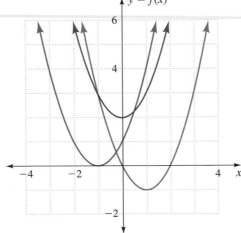

FIGURE 9

a. The graph of $f(x) = x^2 + 2x + 1 = (x + 1)^2$ has one x-intercept: $x = -1$, a double root. The solution to $x^2 + 2x + 1 = 0$ with the quadratic formula is

$$x = \frac{-b \pm \sqrt{b^2 - 4ac}}{2a} = \frac{-2 \pm \sqrt{2^2 - 4 \cdot (1) \cdot (1)}}{2(1)} = \frac{-2 \pm 0}{2} = -1$$

In this problem, $\sqrt{b^2 - 4ac} = 0$, so the $\pm$ has no impact on the answer. Thus, there is only one real-number solution: $x = -1$. The solution is a double root.

b. The graph of $f(x) = x^2 + 2$ has no x-intercepts. The solution to $x^2 + 2 = 0$ with the quadratic formula is

$$x = \frac{-b \pm \sqrt{b^2 - 4ac}}{2a} = \frac{-0 \pm \sqrt{0^2 - 4 \cdot (1) \cdot (2)}}{2(1)} = \frac{-0 \pm \sqrt{-8}}{2}$$

$$= \pm \frac{i\sqrt{8}}{2} = \pm \frac{2i\sqrt{2}}{2} = \pm i\sqrt{2}$$

In this problem, $\sqrt{b^2 - 4ac} = \sqrt{-8}$ is undefined in the real numbers, so there are no real-number solutions. The solutions are complex conjugates.

c. The graph of $f(x) = x^2 - 2x$ has two x-intercepts: $x = 0$ and $x = 2$. The solution to $x^2 - 2x = 0$ with the quadratic formula is

$$x = \frac{-b \pm \sqrt{b^2 - 4ac}}{2a} = \frac{-(-2) \pm \sqrt{(-2)^2 - 4 \cdot (1) \cdot (0)}}{2(1)} = \frac{2 \pm 2}{2}$$

In this problem, $\sqrt{b^2 - 4ac} = 2$, so the $\pm$ gives two real-number solutions: $x = 0$ and $x = 2$.

The square root portion of the quadratic formula, $\sqrt{b^2 - 4ac}$, controls the nature of the solutions. The expression $b^2 - 4ac$ is called the **discriminant**. The following statements summarize the role of $b^2 - 4ac$ in the number and types of solutions to $ax^2 + bx + c = 0$ and the graph of $f(x) = ax^2 + bx + c$.

■ NAMING THE SOLUTIONS WITH THE DISCRIMINANT, $b^2 - 4ac$

For quadratic equations in the form $ax^2 + bx + c = 0$,

- If $b^2 - 4ac$ is positive, there are two real-number solutions. The graph of $f(x)$ passes through the x-axis twice.
- If $b^2 - 4ac$ is zero, there is one real-number solution, a double root. The graph of $f(x)$ touches the x-axis once.
- If $b^2 - 4ac$ is negative, there are no real-number solutions. The solutions are complex numbers (complex conjugates). The graph of $f(x)$ does not touch the x-axis.

POLYNOMIAL EQUATIONS WITH DEGREE GREATER THAN 2   Recall that linear (degree 1) equations, $y = mx + b$, have one solution to $y = 0$, whereas quadratic (degree 2) equations, $y = ax^2 + bx + c$, have two solutions to $y = 0$ (two real-number solutions, a double root, or two complex-number solutions).

In Example 7, we solve a cubic equation—that is, a polynomial with degree 3.

EXAMPLE 7   Solving equations with complex solutions

a. Use a graph to discuss the number of real-number solutions and the number of complex-number solutions to $x^3 - 1 = 0$.

b. Solve $x^3 - 1 = 0$ algebraically.

SOLUTION   **a.** Only one x-intercept appears in the graph of $y = x^3 - 1$ (see Figure 10). We anticipate finding only one real-number solution.

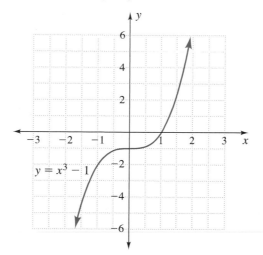

FIGURE 10

b.
$$x^3 - 1 = 0 \qquad \text{Factor with difference of cubes.}$$
$$(x - 1)(x^2 + x + 1) = 0 \qquad \text{Apply the zero product rule.}$$
$$\text{Either} \quad x - 1 = 0 \quad \text{or} \quad x^2 + x + 1 = 0$$

The solution to the first equation is $x = 1$. We need the quadratic formula to solve the second factor equation.

$$x = \frac{-b \pm \sqrt{b^2 - 4ac}}{2a} = \frac{-1 \pm \sqrt{1^2 - 4(1)(1)}}{2(1)} = \frac{-1 \pm \sqrt{-3}}{2} = \frac{-1}{2} \pm \frac{i\sqrt{3}}{2}$$

There are three solutions: one real number, $x = 1$, and two complex numbers,

$$x = -\frac{1}{2} + \frac{i\sqrt{3}}{2} \quad \text{and} \quad x = -\frac{1}{2} - \frac{i\sqrt{3}}{2}$$

We can write the answers in decimal form:

$$x = 1, \quad x \approx -0.5 + 0.866i, \quad \text{and} \quad x \approx -0.5 - 0.866i \qquad \blacksquare$$

The examples thus far suggest the following conclusion.

■ COUNTING SOLUTIONS TO $f(x) = 0$

> If we include complex-number solutions and repetitions from double roots, the degree of the polynomial function $f(x)$ indicates the total number of solutions to $f(x) = 0$.

Just because we can count the solutions does not mean we can find them. It is important to note that *only the real-number solutions appear on a rectangular coordinate graph*. Other solutions may be difficult or impossible to find if they cannot be found through factoring or computerized guess and check.

EXAMPLE 8

Solving equations with complex solutions
a. Use a graph to discuss the number of real-number solutions to $x^4 - 1 = 0$.
b. Solve $x^4 - 1 = 0$ algebraically, and indicate the number of real- and complex-number solutions.

SOLUTION

a. Two real-number solutions to $x^4 - 1 = 0$, $x = 1$ and $x = -1$, appear as x-intercepts in the graph of $y = x^4 - 1$ (see Figure 11).

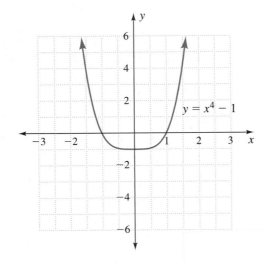

FIGURE 11

b.
$$x^4 -$$
$$(x^2 - 1)(x^2 + 1$$
$$(x - 1)(x + 1)(x^2 + 1$$
$$\text{Either} \quad x - 1 = 0, \quad x +$$
$$x = 1,$$

There are four solutions altogetl

Certain other polynomial equations are solvable. From graphs of polynomials of higher degrees, we may find x-intercepts and corresponding factors. Using polynomial division (Section 7.5), we can then divide to find the remaining factors and solutions.

ANSWER BOX

Warm-up: **1. a.** 1 **b.** −1 **c.** −1 **d.** 1 **e.** 1 **f.** 1 **2. a.** 8 **b.** 4
c. 9 **d.** 10 **3. a.** $2\sqrt{3}$ **b.** $2\sqrt{6}$ **c.** $3\sqrt{5}$ **d.** $2\sqrt{2}$ **Example 3: a.** $-i$
b. $4 - i$ **c.** $5 + 2i$ **Think about it:** The square root is preceded by $\pm$, so
each time the square root is nonzero, there will be two solutions. When complex, the solutions will be of the form $a \pm bi$, complex conjugates.

6.2 Exercises

In Exercises 1 to 4, change each to an expression containing a product of a real number and the imaginary unit. Leave answers in either decimal or simplified radical form.

1. a. $\sqrt{-20}$ $2i\sqrt{5}$ **b.** $\sqrt{-40}$ $2i\sqrt{10}$ **c.** $\sqrt{-72}$ $6i\sqrt{2}$

2. a. $\sqrt{-63}$ $3i\sqrt{7}$ **b.** $\sqrt{-75}$ $5i\sqrt{3}$ **c.** $\sqrt{-80}$ $4i\sqrt{5}$

3. a. $\sqrt{-3}$ $i\sqrt{3}$ **b.** $\sqrt{-54}$ $3i\sqrt{6}$ **c.** $\sqrt{-27}$ $3i\sqrt{3}$

4. a. $\sqrt{-12}$ $2i\sqrt{3}$ **b.** $\sqrt{-48}$ $4i\sqrt{3}$ **c.** $\sqrt{-24}$ $2i\sqrt{6}$

Change each expression in Exercises 5 to 8 to a complex number.

5. a. 16 $16 + 0i$

 b. $\sqrt{-16}$ $0 + 4i$

 c. $\sqrt{36} + \sqrt{-6}$ $6 + i\sqrt{6}$

6. a. $\sqrt{25}$ $5 + 0i$

 b. $\sqrt{-11}$ $0 + i\sqrt{11}$

 c. $8 - \sqrt{-49}$ $8 - 7i$

7. a. $\dfrac{6 + \sqrt{-12}}{2}$ $3 + i\sqrt{3}$

 b. $\dfrac{2 - \sqrt{-24}}{4}$ $\frac{1}{2} - \frac{1}{2}i\sqrt{6}$

8. a. $\dfrac{5 - \sqrt{-50}}{5}$ $1 - i\sqrt{2}$

 b. $\dfrac{3 + \sqrt{-18}}{6}$ $\frac{1}{2} + \frac{i\sqrt{2}}{2}$

In Exercises 9 to 16, give the conjugate of each complex number.

9. $2 + 3i$
 $2 - 3i$
10. $4i - 2$
 $-2 - 4i$
11. $i + 1$
 $1 - i$
12. $2i + 2$
 $2 - 2i$
13. $2i - 3$
 $-3 - 2i$
14. $3 - i$
 $3 + i$
15. $4 - 2i$
 $4 + 2i$
16. $3 + 2i$
 $3 - 2i$

Simplify the expressions in Exercises 17 to 30.

17. $2i + 3i - 4i - 5i + 6i + 7i - 8i - 9i$ $-8i$

18. $3i - (-5i)$ $8i$

19. $6 + 4i - 7i + 5 - 12i$ $11 - 15i$

20. $a + bi - a + bi$ $2bi$

21. $-a + bi - (a + bi)$
 $-2a$

22. $a - bi + (bi - a)$
 0

23. $3 - 4(5 - i)$ $-17 + 4i$

24. $(4 - 3i) + (2 - 6i)$
 $6 - 9i$

25. $(6 + 4i) - (3 - 2i)$
 $3 + 6i$

26. $(5 - i) - (4 + 3i)$
 $1 - 4i$

27. $(7 - 2i) + (6 - 4i)$
 $13 - 6i$

28. $3i - 4(5 - i)$
 $-20 + 7i$

29. $4i - 3(5 - i)$
 $-15 + 7i$

30. $4 - 3(5 - i)$
 $-11 + 3i$

Multiply the conjugate expressions in Exercises 31 to 34. Compare and explain the results.

31. a. $(4 - 3i)(4 + 3i)$ 25

 b. $(3 + 4i)(3 - 4i)$ 25

32. a. $(5 + 12i)(5 - 12i)$ 169

 b. $(12 - 5i)(12 + 5i)$ 169

33. a. $(\sqrt{2} + i)(\sqrt{2} - i)$ 3

 b. $(1 - i\sqrt{2})(1 + i\sqrt{2})$ 3

34. a. $(3 + i\sqrt{2})(3 - i\sqrt{2})$ 11

 b. $(\sqrt{2} + 3i)(\sqrt{2} - 3i)$ 11

Use the discriminant to predict the number and types of solutions to the equations in Exercises 35 to 44.

35. $4x^2 + 7x - 2 = 0$
 2 real solutions
36. $2x^2 + 3x - 5 = 0$
 2 real solutions
37. $5x^2 + 4x + 1 = 0$
 2 complex solutions
38. $x^2 + 2x + 5 = 0$
 2 complex solutions
39. $4x^2 + 4x + 1 = 0$
 1 real solution
40. $9x^2 - 6x + 1 = 0$
 1 real solution
41. $2x^2 + 6x + 5 = 0$
 2 complex solutions
42. $2x^2 + 2x + 5 = 0$
 2 complex solutions
43. $\frac{1}{2}x^2 - 3 = 0$
 2 real solutions
44. $\frac{1}{2}x^2 - 5 = 0$
 2 real solutions

In Exercises 45 to 58, solve for all solutions, real or complex.

45. $x^2 - 8 = 0$ $\{\pm 2\sqrt{2}\} \approx \{\pm 2.8\}$

46. $x^2 - 27 = 0$ $\{\pm 3\sqrt{3}\} \approx \{\pm 5.2\}$

47. $x^2 + 2x + 4 = 0$ $\{-1 \pm i\sqrt{3}\} \approx \{-1 \pm 1.7i\}$

48. $x^2 + x + 1 = 0$ $\left\{-\frac{1}{2} \pm i\frac{\sqrt{3}}{2}\right\} \approx \{-0.5 \pm 0.87i\}$

49. $x^2 - 4x + 8 = 0$ $\{2 \pm 2i\}$

50. $x^2 - 3x + 6 = 0$ $\left\{\frac{3}{2} \pm i\frac{\sqrt{15}}{2}\right\} \approx \{1.5 \pm 1.9i\}$

51. $x^2 + 3x = 4$ $\{-4, 1\}$

52. $x^2 - 3x = 10$ $\{-2, 5\}$

53. $x^2 - 4x + 4 = 0$ $\{2\}$

54. $x^2 + 6x + 9 = 0$ $\{-3\}$

55. $(x + 1)^2 + 1 = 0$ $\{-1 \pm i\}$

56 $(x - 2)^2 + 1 = 0$ $\{2 \pm i\}$

57. $2x^2 + 10x + 13 = 0$ $\{-2.5 \pm 0.5i\}$

58. $13x^2 + 2x + 2 = 0$ $\left\{-\frac{1}{13} \pm \frac{5}{13}i\right\} \approx \{-0.08 \pm 0.38i\}$

The tables in Exercises 59 and 60 are a helpful reminder in factoring sums and differences of cubes if you forget the formulas.

59. Factor $x^3 - 8$. $(x - 2)(x^2 + 2x + 4)$

Factor	x^2	$+2x$	$+4$
x	x^3	$+2x^2$	$+4x$
-2	$-2x^2$	$-4x$	-8

60. Factor $x^3 + 64$. $(x + 4)(x^2 - 4x + 16)$

Factor	x^2	$-4x$	$+16$
x	x^3	$-4x^2$	$+16x$
$+4$	$+4x^2$	$-16x$	$+64$

Solve for all solutions, real or complex, in Exercises 61 to 72.

61. $x^3 + 8 = 0$ $\{-2, 1 \pm i\sqrt{3}\} \approx \{-2, 1 \pm 1.7i\}$

62. $x^4 - 16 = 0$ $\{\pm 2, \pm 2i\}$

63. $x^4 - 81 = 0$ $\{\pm 3, \pm 3i\}$

64. $x^3 + 1 = 0$ $\left\{-1, \frac{1}{2} \pm i\frac{\sqrt{3}}{2}\right\} \approx \{-1, 0.5 \pm 0.87i\}$

65. $x^3 + 3x^2 + 2x = 0$ $\{-2, -1, 0\}$

66. $x^3 + 3x^2 + x = 0$ $\left\{-\frac{3}{2} \pm \frac{\sqrt{5}}{2}, 0\right\} \approx \{-2.6, -0.38, 0\}$

67. $x^3 - 27 = 0$ $\{3, (-3 \pm 3i\sqrt{3})/2\}$ or $\{3, -1.5 \pm 2.6i\}$

68. $x^3 - 64 = 0$ $\{4, -2 \pm 2i\sqrt{3}\}$ or $\{4, -2 \pm 3.5i\}$

69. $x^3 - 4x^2 = 0$ $\{0 \text{ (double root)}, 4\}$

70. $x^3 + 4x = 0$ $\{0, \pm 2i\}$

71. $x^2 - 8x^5 = 0$
$\{0 \text{ (double root)}, \frac{1}{2}, (-1 \pm i\sqrt{3})/4\}$ or $\{0, \frac{1}{2}, -0.25 \pm 0.43i\}$

72. $4x^5 - 5x^3 + x = 0$
$\{0, \pm 1, \pm\frac{1}{2}\}$

Projects

73. Complex Cubes Multiply the expressions. What do you observe about the results?

   a. $(1 + i\sqrt{3})^3$ -8 **b.** $(-1 - i\sqrt{3})^3$ 8

   c. $\left(-\frac{3}{2} + \frac{3i\sqrt{3}}{2}\right)^3$ 27 **d.** $\left(-\frac{1}{2} + \frac{i\sqrt{3}}{2}\right)^3$ 1

The cubes give integer answers, so the number being cubed is a complex cube root.

74. Complex Conjugates What conclusions can be drawn about these operations with complex conjugates?

   a. $(a + bi) + (a - bi)$ gives a real number

   b. $(a + bi) - (a - bi)$ gives an imaginary number

   c. $(a + bi)(a - bi)$ gives a real number

The complex conjugate is used to eliminate a complex number from a denominator. Multiply the top and bottom of each of the following fractions by the complex conjugate of the denominator.

   d. $\dfrac{1}{3i}$ $-\frac{1}{3}i$ **e.** $\dfrac{2}{i}$ $-2i$ **f.** $\dfrac{3}{-2i}$ $\frac{3}{2}i$

   g. $\dfrac{1}{2 + 3i}$ $\frac{2}{13} - \frac{3}{13}i$ **h.** $\dfrac{3}{2 + i}$ $\frac{6}{5} - \frac{3}{5}i$ **i.** $\dfrac{4}{3 - 2i}$ $\frac{12}{13} + \frac{8}{13}i$

75. Powers of i Use $i^1 = i$ and $i^2 = -1$ to investigate the powers of i. Look for a pattern. Use your pattern to predict $i^{16}, i^{24}, i^{35}, i^{46},$ and i^{97}. (*Hint:* $i^{11} = -i$ and $i^{32} = 1$.)

All powers equal one of $i^1 = i, i^2 = -1, i^3 = -i, i^4 = +1$. Divide the exponent by 4; the remainder indicates the power. 1, 1, $-i$, -1, i

---

6   Mid-Chapter Test

1. Describe how the -2 in $y = -2x^2$ changes the graph of $y = x^2$.   opens down and is steeper

2. Explain how the -4 in $y = x^2 - 4$ changes the position, vertex, and intercepts from those of $y = x^2$.
4 units below; $(0, -4)$ vs. $(0, 0)$; $(-2, 0), (2, 0)$ vs. $(0, 0)$

Sketch graphs for these equations.

3. $y = 3 - x^2$
See Answer Section.

4. $y = x^2 - x$
See Answer Section.

5. $y = x^2 + 3$
See Answer Section.

6. $y = -\frac{1}{2}x^2$
See Answer Section.

7. Write as complex numbers.

a. $\sqrt{-16}$ $0 + 4i$ **b.** 4 $4 + 0i$

c. $3 + \sqrt{-4}$ $3 + 2i$ **d.** $\sqrt{-56}$ $0 + 2i\sqrt{14}$

8. Simplify.

a. $\dfrac{4 - \sqrt{-48}}{2}$ $2 - 2i\sqrt{3}$ **b.** $\dfrac{9 + \sqrt{-18}}{6}$ $\frac{3 + i\sqrt{2}}{2}$

9. Simplify.

a. $(3 + 2i) - (6 + 2i)$ -3 **b.** $(4 - 2i) - 2(3 - 4i)$ $-2 + 6i$

10. Multiply, and simplify using properties of complex numbers.

a. $(3 - 2i)(3 + 2i)$ 13 **b.** $(2 - i)(2 + i)$ 5

c. $(2 - 3i)(2 + 3i)$ 13

11. a. Compare your answers to parts a and c in Problem 10. What is unusual about the answers? both 13; reversing a and b in $(a + bi)(a - bi)$ gives same real-number product.

b. Find the products both 5

$$(1 + 2i)(1 - 2i) \quad \text{and} \quad (2 + i)(2 - i)$$

12. Evaluate the discriminant to find how many real-number solutions exist to $f(x) = 0$, where $f(x) = -x^2 + 2x + 3$. Find the solutions. $16; 2; \{-1, 3\}$

13. Find all real- or complex-number solutions to $f(x) = 0$, where $f(x) = (x - 2)(x^2 + 2x + 4)$.
$\{2, -1 \pm i\sqrt{3}\}$ or $\{2, -1 \pm 1.7i\}$

14. Solve $x^2 + x + 1 = 0$. $\{(-1 \pm i\sqrt{3})/2\}$ or $\{-0.5 \pm 0.87i\}$

15. Solve $x^2 - 9x^4 = 0$. $\{0 \text{ (double root)}, \pm\frac{1}{3}\}$

6.3   Shifts, Vertex Form, and Applications of Quadratic Functions

Objectives

- Identify horizontal or vertical shifts of $y = x^2$.
- Use the vertex form to predict the graph of a quadratic equation.
- Find the vertex from a quadratic equation in vertex form.
- Find the equation of a parabola from its vertex and one other point.
- Use completing the square to change a quadratic equation to vertex form.

WARM-UP

1. Complete the table.

x	x^2	$(x - 2)^2$	$(x + 1)^2$
-2	4	16	1
-1	1	9	0
0	0	4	1
1	1	1	4
2	4	0	9

2. Expand these binomial squares.

a. $(x - 2)^2$ $x^2 - 4x + 4$ **b.** $(x + 1)^2$ $x^2 + 2x + 1$

c. $(x - 1.5)^2$ $x^2 - 3x + 2.25$ **d.** $(2x + 1)^2$ $4x^2 + 4x + 1$

3. What must be added to make each of these expressions the square of a binomial?

a. $x^2 + 6x$ 9   **b.** $x^2 - 5x$ 6.25

THE VERTEX OF A STREAM OF WATER from a drinking fountain (see Figure 12) is likely to be the point of interest for obtaining a drink of water. In this section, we obtain the quadratic equation $y = ax^2 + bx + c$ of a parabolic graph, using only the vertex and one other point. A review of horizontal and vertical shifts introduces the vertex form of a quadratic equation. We complete the square to change a quadratic equation into vertex form.

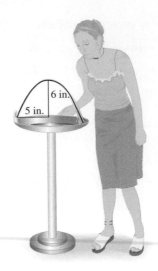

FIGURE 12

Horizontal Shifts

In Section 4.4, we graphed equations containing the squares of binomials, such as $y = (x + 3)^2$ and $y = (x - 3)^2$, and observed the following.

■ HORIZONTAL SHIFT

> If the quadratic equation can be written as $y = (x - g)^2$, the vertex is at $(g, 0)$. The graph of $y = (x - g)^2$ is *shifted horizontally g units* from the graph of $y = x^2$.

The formal name given to the horizontal shift of $y = x^2$ to $y = (x - g)^2$ is **translation**, *shifting a graph parallel to one of the axes.* Because $a = 1$ (in $y = ax^2 + bx + c$), the shapes of the graphs will not change.

EXAMPLE 1 **Predicting horizontal shifts** Predict the graphs for the following equations, and then check by graphing.

　　a. $y = (x + 1)^2$ 　　　　**b.** $y = (x - 2)^2$

SOLUTION **a.** When we let $x + 1 = 0$, we obtain $x = -1$, the x-intercept. The point $(-1, 0)$ is 1 unit to the left of the origin. Thus, the graph of $y = (x + 1)^2 = [x - (-1)]^2$ is shifted 1 unit to the left of that of $y = x^2$, as shown in Figure 13.

b. When we let $x - 2 = 0$, we obtain $x = 2$, the x-intercept. The point $(2, 0)$ is 2 units to the right of the origin. Thus, the graph of $y = (x - 2)^2$ is shifted 2 units to the right of that of $y = x^2$, as shown in Figure 14.

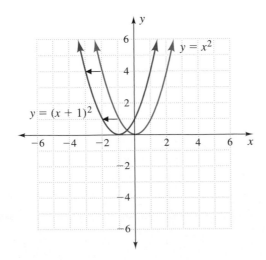

FIGURE 13

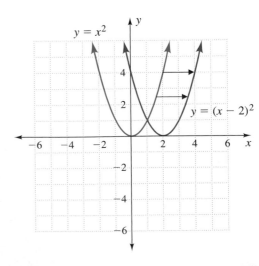

FIGURE 14

Think about it 1: For $h(x) = (x - g)^2$, tell how you might know whether the graph shifts to the right or to the left of $f(x) = x^2$.

Vertical Shifts

In Section 4.4, we graphed equations containing differences of squares, such as $y = x^2 - 1$ and $y = x^2 - 4$. In Section 6.1, we extended the graphing to equations containing either the addition or the subtraction of a constant, $y = x^2 \pm j$, and observed the following.

■ VERTICAL SHIFT	If the quadratic equation can be written as $y = x^2 + j$, the vertex is at $(0, j)$. The parabola for $y = x^2 + j$ is *shifted vertically $\lvert j \rvert$ units* from the graph of $y = x^2$.

EXAMPLE 2   Predicting vertical shifts   Predict the graphs for the following equations, and then check by graphing.

    a. $y = x^2 + 1$ **b.** $y = x^2 - 2$

SOLUTION   **a.** The equation shows 1 added to x^2. The vertex $(0,1)$ is 1 unit above the origin. Thus, the graph of $y = x^2 + 1$ is shifted 1 unit above that of $y = x^2$, as shown in Figure 15.

    b. The equation shows 2 subtracted from x^2. The vertex $(0, -2)$ is 2 units below the origin. Thus, the graph of $y = x^2 - 2$ is shifted 2 units below that of $y = x^2$, as shown in Figure 16.

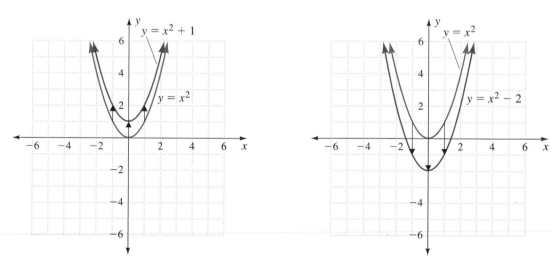

FIGURE 15   **FIGURE 16**

Think about it 2: For $g(x) = x^2 + j$, tell how you might know whether the graph shifts up or down from $f(x) = x^2$.

Quadratic Equations in Vertex Form

In Example 1, we obtained a horizontal shift from equations containing a binomial square. In Example 2, we obtained a vertical shift from equations having a constant added or subtracted. In Example 3, we explore the graph that results from both shifts.

EXAMPLE 3   **Combining horizontal and vertical shifts**   Graph $y = (x - 2)^2 + 3$, and describe how the graph might be obtained from $y = x^2$.

SOLUTION   The graph of $y = (x - 2)^2 + 3$ has the vertex at $x = 2$ and $y = 3$ (see Figure 17). The vertex shifted to the right 2 units because of the $(x - 2)^2$ and then shifted up 3 units because 3 was added to $(x - 2)^2$.

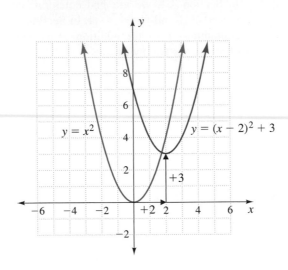

FIGURE 17

Quadratic equations written as in Example 3 are said to be in vertex form.

DEFINITION OF VERTEX FORM

> The **vertex form of a quadratic equation** is $y = a(x - h)^2 + k$, where the vertex coordinates are (h, k).

The letter a is the same as in $y = ax^2 + bx + c$. The formula shows a horizontal and vertical translation, or shift, of the vertex point $(0, 0)$ on $y = x^2$ to a new vertex (h, k), as shown in Figure 18.

Student Note: The signs on h and k are not known, so Figure 18 represents any shift from $(0, 0)$ to (h, k).

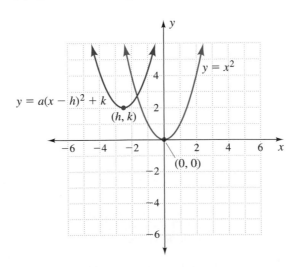

FIGURE 18

EXAMPLE 4   **Finding the vertex from vertex form and graphing**
a. Find the vertex from the vertex form of $y = (x + 1)^2 - 3$.
b. Use shifts to graph $y = (x + 1)^2 - 3$.

SOLUTION **a.** We change the equation to $y = (x - h)^2 + k$ form:

$$y = (x + 1)^2 - 3$$

$$y = [x - (-1)]^2 + (-3)$$

Since $h = -1$ and $k = -3$, the vertex is at $(-1, -3)$.

b. The graph of $y = (x + 1)^2 - 3$ has the same shape as that of $y = x^2$. Starting at the origin, the vertex shifted to the left 1 unit because of the $(x + 1)^2$ and then shifted down 3 units because 3 was subtracted from $(x + 1)^2$. The resulting graph has vertex $(-1, -3)$, as shown in Figure 19.

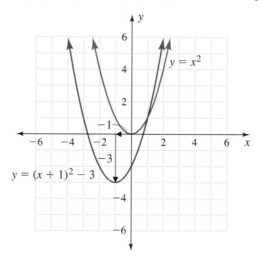

FIGURE 19

Applications

In Examples 5 and 6, we use the vertex form to find equations for application settings. In these applications, we are given the vertex and one other point and need to find the equation for the parabola.

BUILDING A QUADRATIC EQUATION FROM ITS VERTEX AND ONE OTHER POINT

1. Start with the vertex form, $y = a(x - h)^2 + k$.

2. Using the vertex (h, k), substitute for h and k in $y = a(x - h)^2 + k$.

3. Using the ordered pair for another point on the graph, substitute for x and y.

4. Solve for a.

5. Substitute a and the vertex into the vertex form.

EXAMPLE 5   **Building an equation from a vertex and one other point: a water fountain's stream**   Suppose a parabolic stream of water reaches 10 inches across a drinking fountain (see Figure 20). The top of the water stream is 6 inches higher than the nozzle where the water emerges. Fit an equation to the stream of water.

SOLUTION Let point A be the origin. The vertex is $(5, 6)$, point B. Use the vertex form.

$$y = a(x - h)^2 + k \qquad \text{Substitute the vertex, } (h, k) = (5, 6).$$

$$y = a(x - 5)^2 + 6 \qquad \text{Because the origin, } (0, 0), \text{ lies on the curve and makes the}$$
$$\qquad\qquad\qquad\qquad \text{equation true, substitute } x = 0, y = 0 \text{ into the equation.}$$
$$0 = a(0 - 5)^2 + 6 \qquad \text{Solve for } a.$$

$$-6 = 25a$$

$$-\frac{6}{25} = a$$

$$a = -0.24$$

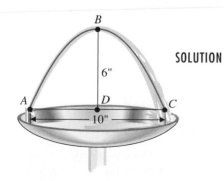

FIGURE 20

Now we substitute $a = -0.24$ and the vertex into the vertex form.

$$y = -0.24(x - 5)^2 + 6 \quad \text{or} \quad y = -0.24x^2 + 2.4x$$

The coefficient a is negative, in agreement with the path of the water.

EXAMPLE 6   **Building an equation from a vertex and one other point: suspension bridge cable**   Suppose the lowest point on a parabolic cable for a suspension bridge is to be 10 feet above the roadbed. The supports for the cable are 40 feet apart (horizontally) and rise 25 feet above the roadbed, as shown in Figure 21. Using any convenient origin, assign coordinates to points in the picture. Fit an equation to the cable, using these coordinates.

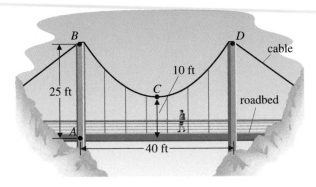

FIGURE 21

SOLUTION   Let the origin be at point A. Other locations for the origin are possible. Then the vertex of the cable is at $(20,10)$, point C. One other point on the cable is at the support $(0, 25)$, point B.

$$y = a(x - h)^2 + k \qquad \text{Substitute the vertex, } (h, k) = (20, 10), \text{ into the vertex form.}$$

$$y = a(x - 20)^2 + 10 \qquad \text{Substitute point } B, (0, 25).$$

$$25 = a(0 - 20)^2 + 10 \qquad \text{Solve for } a.$$

$$15 = a(400)$$

$$\tfrac{15}{400} = a$$

$$a = 0.0375$$

Now we substitute a and the vertex into the vertex form.

$$y = 0.0375(x - 20)^2 + 10 \quad \text{or} \quad y = 0.0375x^2 - 1.5x + 25$$

The coefficient a is positive, in agreement with the upward direction of the parabolic cable.

This material on completing the square to change to vertex form is optional. The formula $x = -b/2a$, introduced in Section 5.3, permits finding the vertex without completing the square. Students find the formula approach more sensible and useful than completing the square.

Completing the Square to Change to Vertex Form

We can change a quadratic equation into vertex form either by completing the square on the equation in standard form, $y = ax^2 + bx + c$, or by finding $h = -b/2a$ and $k = f(-b/2a)$ and placing them into the vertex form.

If $a = 1$, first we apply the process used in Section 5.3 to complete the square on a quadratic equation. Then we change to the less general vertex form, $y = (x - g)^2 + j$.

CHANGING A QUADRATIC EQUATION TO VERTEX FORM, $a = 1$

1. Write the equation as $y = x^2 + bx + c$.
2. Subtract the constant term, c, from both sides.
3. Complete the square on the right, adding the required number to both sides.
4. Change to squared form on the right.
5. Change to vertex form, $y = (x - g)^2 + j$.

The vertex (g, j) can now be read from the equation.

EXAMPLE 7   **Changing a quadratic equation to vertex form**   Complete the square to find the vertex form of the equation, and identify the vertex. Check by graphing the original equation.

a. $y = x^2 + 6x + 5$ **b.** $y = x^2 - 5x - 1$

SOLUTION   **a.**

$$y = x^2 + 6x + 5 \qquad \text{Subtract the constant term, 5, from both sides.}$$
$$y - 5 = x^2 + 6x \qquad \text{Complete the square on the right.}$$
$$y - 5 + 9 = x^2 + 6x + 9 \qquad \text{Change to squared form on the right.}$$
$$y + 4 = (x + 3)^2 \qquad \text{Change to vertex form.}$$
$$y = (x + 3)^2 - 4$$

The vertex is at $(-3, -4)$. The graph is shown in Figure 22.

b.

$$y = x^2 - 5x - 1 \qquad \text{Add 1 to both sides.}$$
$$y + 1 = x^2 - 5x \qquad \text{Complete the square on the right.}$$
$$y + 1 + (2.5)^2 = x^2 - 5x + (2.5)^2 \qquad \text{Simplify.}$$
$$y + 7.25 = x^2 - 5x + 6.25 \qquad \text{Change to squared form on the right.}$$
$$y + 7.25 = (x - 2.5)^2 \qquad \text{Change to vertex form.}$$
$$y = (x - 2.5)^2 - 7.25$$

The vertex is at $(2.5, -7.25)$. The graph is shown in Figure 22.

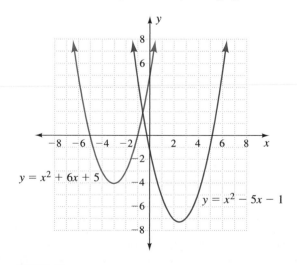

$y = x^2 + 6x + 5$

$y = x^2 - 5x - 1$

FIGURE 22

ANSWER BOX

Warm-up: **1.**

x	x^2	$(x-2)^2$	$(x+1)^2$
-2	4	16	1
-1	1	9	0
0	0	4	1
1	1	1	4
2	4	0	9

2. a. $x^2 - 4x + 4$ **b.** $x^2 + 2x + 1$ **c.** $x^2 - 3x + 2.25$ **d.** $4x^2 + 4x + 1$
3. a. 9   **b.** 6.25   **Think about it 1:** For $(x - g)^2$, the vertex is at
$x - g = 0$, or $x = g$. The graph shifts to the vertex position. Thus, for $g > 0$,
the shift is to the right; for $g < 0$, the shift is to the left.   **Think about
it 2:** For $x^2 + j$, the vertex is at $(0, j)$, so the shift is up. For $x^2 - j$, which is
equal to $x^2 + (-j)$, the vertex is at $(0, -j)$, so the shift is down.

6.3   Exercises

Each of the graphs in Exercises 1 to 8 has the same steepness as the graphs of
$y = x^2$ and $y = -x^2$. Identify the equation of each graph. Check with a graphing
calculator.   a: $y = (x + 2)^2$,   b: $y = (x - 3)^2$

a: $y = x^2 + 2$,   b: $y = x^2 - 1$

1.

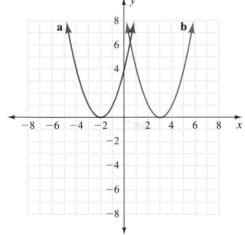

3.

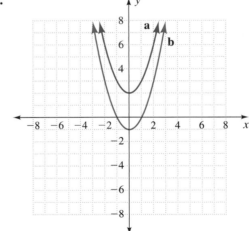

2.

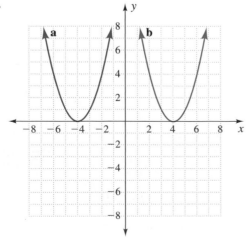

a: $y = (x + 4)^2$,   b: $y = (x - 4)^2$

4.

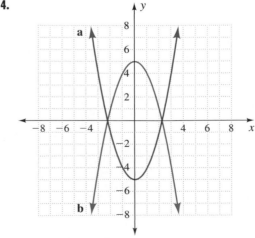

a: $y = x^2 - 5$,   b: $y = -x^2 + 5$

a: $y = x^2 - 1$,
5. b: $y = -x^2 - 3$

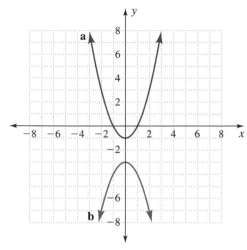

a: $y = x^2 + 3$,
6. b: $y = x^2 + 1$

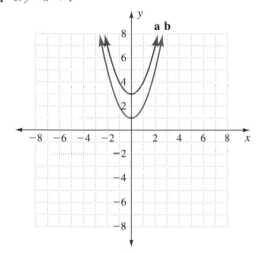

a: $y = (x + 1)^2 - 2$,
7. b: $y = (x - 2)^2 + 1$

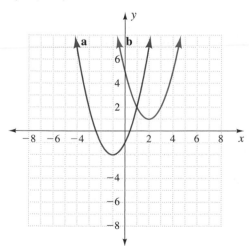

8.

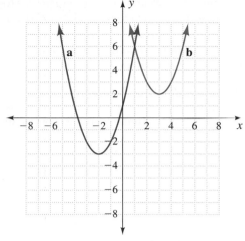

a: $y = (x + 2)^2 - 3$,　　b: $y = (x - 3)^2 + 2$

In Exercises 9 to 16, describe how the graph of each equation might be obtained from $y = x^2$. Use complete sentences, and describe all shifts or translations. Sketch your graph, and check with a graphing calculator.

9. $y = (x + 2)^2$　　Shift left 2 units.

10. $y = (x - 3)^2$　　Shift right 3 units.

11. $y = (x - 4)^2$　　Shift right 4 units.

12. $y = (x + 4)^2$　　Shift left 4 units.

13. $y = (x - 3)^2 + 4$　　Shift right 3 units, up 4 units.

14. $y + 3 = (x + 2)^2$　　Shift left 2 units, down 3 units.

15. $y - 3 = (x + 4)^2$　　Shift left 4 units, up 3 units.

16. $y = (x - 4)^2 - 3$　　Shift right 4 units, down 3 units.

Identify the vertex of each equation in Exercises 17 to 22.

17. $y = (x - 1)^2$　　$(1, 0)$

18. $y = (x + 2)^2$　　$(-2, 0)$

19. $y = (x + 3)^2 + 4$　　$(-3, 4)$

20. $y = (x - 4)^2 + 3$　　$(4, 3)$

21. $y = (x - 2)^2 - 3$　　$(2, -3)$

22. $y = (x + 1)^2 - 2$　　$(-1, -2)$

In Exercises 23 to 28, find the equation of the parabola, given the vertex V and another point P.

23. $V(3, 1)$, $P(-1, 3)$　　$y = \frac{1}{8}(x - 3)^2 + 1$

24. $V(2, -1)$, $P(-2, 3)$

25. $V(-1, -1)$, $P(1, 7)$

26. $V(-2, 1)$, $P(-3, -$

27. $V(3, -2)$, $P(5, -1($

28. $V(-3, 2)$, $P(-2, 4)$

29. A Water Fountain's Stream Use the symmetry property of parabolas to find the ordered pair for C on the stream of water in Example 5, and use quadratic regression on the three points to find the equation of the stream of water. *Possible answer: $(10, 0)$; $y = -0.24x^2 + 2.4x$*

30. Suspension Bridge Cable Use the symmetry property of parabolas to find the ordered pair for D on the cable in Example 6, and use quadratic regression on the three points to find the equation of the cable. *Possible answer: $(40, 25)$; $y = 0.0375x^2 - 1.5x + 25$*

31. Suspension Bridge Cable Suppose we change the location of the origin in Example 6 to the top of the support tower, currently labeled B.

   a. Write the resulting coordinates for A and C. *$(0, -25)$, $(20, -15)$*

   b. Use the vertex form to find the resulting equation. *$y = 0.0375x^2 - 1.5x$ or $y = 0.0375(x - 20)^2 - 15$*

   c. Compare your equation with that found in Example 6, and explain the difference in the equations. *vertical shift of 25*

32. Suspension Bridge Cable, Again Suppose we change the location of the origin in Example 6 to the vertex of the cable, currently labeled C.

   a. Write the resulting coordinates for A and B. *$(-20, -10)$, $(-20, 15)$*

   b. Use the vertex form to find the resulting equation. *$y = 0.0375x^2$*

   c. Compare your equation with that found in Example 6, and explain the difference in the equations. *horizontal shift of 20, vertical shift of 10*

33. Water Jet Equation A jet of water from a fireboat is to reach a height of 100 feet when it is a horizontal distance of 30 feet from its origin. Find the equation of the jet of water. *$y = -\frac{1}{9}(x - 30)^2 + 100$ or $y = -\frac{1}{9}x^2 + \frac{20}{3}x$*

34. Another Water Jet Equation A jet of water from a fire-boat is to reach a height of 200 feet when it is a horizontal distance of 40 feet from its origin. Find the equation of the jet of water. *$y = -\frac{1}{8}(x - 40)^2 + 200$ or $y = -\frac{1}{8}x^2 + 10x$*

In Exercises 35 to 42, change the equations to vertex form by completing the square. Check by graphing the original and vertex forms with a graphing calculator.

35. $y = x^2 + 10x + 25$ *$y = (x + 5)^2$*

36. $y = x^2 + 4x + 4$ *$y = (x + 2)^2$*

37. $y = x^2 - 6x + 9$ *$y = (x - 3)^2$*

38. $y = x^2 - 10x + 25$ *$y = (x - 5)^2$*

39. $y = x^2 + 10x + 30$ *$y = (x + 5)^2 + 5$*

40. $y = x^2 + 4x - 5$ *$y = (x + 2)^2 - 9$*

41. $y = x^2 - 6x + 8$ *$y = (x - 3)^2 - 1$*

42. $y = x^2 - 10x + 20$ *$y = (x - 5)^2 - 5$*

43. Use $h = -b/2a$ and $k = f(-b/2a)$ to find the vertex. Write the equations in vertex form.

   a. $y = -0.025x^2 + 10x + 25$ *$y = -0.025(x - 200)^2 + 1025$*

   b. $y = -0.125x^2 - 5x + 10$ *$y = -0.125(x + 20)^2 + 60$*

   c. $y = 0.15x^2 - 6x + 15$ *$y = 0.15(x - 20)^2 - 45$*

   d. $y = 0.75x^2 + 4.5x - 5$ *$y = 0.75(x + 3)^2 - 11.75$*

   e. $y = 0.9x^2 - 2.7x - 15$ *$y = 0.9(x - 1.5)^2 - 17.025$*

■ Projects

44. Absolute Value and Function Shifts Graph the following equations on a graphing calculator. Compare the positions of the pairs of graphs. How do the changes in position compare with the movements of parabolas in this section? Write a quadratic equation that shows the same change in position with $y = x^2$.

   a. $y = |x|$ and $y = |x + 2|$ *shift left 2; same; $y = (x + 2)^2$*

   b. $y = |x|$ and $y = |x| + 2$ *shift up 2; same; $y = x^2 + 2$*

   c. $y = |x|$ and $y = |x| - 2$ *shift down 2; same; $y = x^2 - 2$*

   d. $y = |x|$ and $y = |x - 2|$ *shift right 2; same; $y = (x - 2)^2$*

   How will the graphs of other functions change?

   e. $y = \sqrt{x}$ and $y = \sqrt{(x + 2)}$ *$y = \sqrt{x}$ shifts left 2 units*

   f. $y = \sqrt{x}$ and $y = \sqrt{(x)} + 2$ *$y = \sqrt{x}$ shifts up 2 units*

   g. $f(x)$ and $f(x - 2)$ *$y = f(x)$ shifts right 2 units*

   h. $f(x)$ and $f(x) - 2$ *$y = f(x)$ shifts down 2 units*

45. Measuring a Fountain's Stream Using the source of water as the origin, measure the stream of water coming from a drinking fountain turned on full force. Find its equation.

46. A Summer or Warm Day Investigation In writing, describe an experiment that will solve the following problem. List the equipment needed. Check your plan with your instructor before doing the project.

   Find the equations for water coming from a garden hose held at various angles. If you can measure the height, what angle gives the highest stream of water? What angle gives the greatest distance reached by the stream of water?

   Include a record of 15 to 20 angles, the heights, and the distance reached by the water at each angle. Plot the angles on the x-axis and the distance on the y-axis.

6.4 Solving Minimum and Maximum Problems

Objectives

▮ Find the vertex from a graph, table, and formula.

▮ Solve minimum and maximum application problems.

WARM-UP

Use the quadratic formula to find the x-intercepts of each function. Simplify and reduce any fractions, but leave your answers in the form

$$x = -\frac{b}{2a} \pm \frac{\sqrt{b^2 - 4ac}}{2a}$$

1. $f(x) = x^2 + 6x - 16$ -3 ± 5 **2.** $f(x) = x^2 - 2x$ 1 ± 1

3. $f(x) = x^2 - 2x - 8$ 1 ± 3 **4.** $f(x) = x^2 + 4x - 21$ -2 ± 5

5. $f(x) = 2x^2 + 3x - 4$ $-\frac{3}{4} \pm \dfrac{\sqrt{41}}{4}$

THIS SECTION FOCUSES on calculating the vertex and solving minimum and maximum problems.

Finding the Vertex

Have you ever looked up facts in the *Guinness Book of Records*? This book is filled with facts about greatest and smallest values. In mathematics applications, the vertex is the position on parabolic graphs where the **maximum** (*greatest*) and **minimum** (*smallest*) values are obtained.

▬ Thus far, we have the following results about the vertex of a parabola.

▬ SUMMARY: VERTEX OF THE PARABOLIC GRAPH OF $y = ax^2 + bx + c$

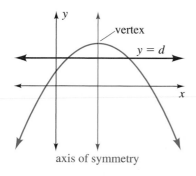

FIGURE 23

- Once symmetry is observed in an input-output table, the vertex can be located at the highest or lowest output.
- The vertex is the highest or lowest point on a parabolic graph (Figure 23).
- The vertex lies on the axis of symmetry of a parabolic graph (Figure 23).
- The x-coordinate of the vertex is the midpoint between the x-intercepts [solutions to $f(x) = 0$].
- The x-coordinate of the vertex is the midpoint of any horizontal segment, $y = d$, that intersects the parabola in two points (Figure 23).
- The vertex has the ordered pair (x, y), where $x = -b/2a$ and $y = f(-b/2a)$.
- The vertex is (h, k) when we have completed the square on a quadratic equation and changed it into vertex form,

  $$y = a(x - h)^2 + k$$

Applications

BRIDGE CLEARANCE   Not only does a parabolic arch provide a pleasing and naturally strong bridge support; it is also a convenient shape for passage by ships.

Given the basic equation, we can use it to find the vertex. Having the vertex then permits us to find other important information, such as the clearance between the bridge and the water in Example 1.

EXAMPLE 1 Finding the vertex: bridge arch   The arch supporting the bridge in Figure 24 has the equation $y = -0.0022x^2 + 1.578x$ relative to the origin shown, at the left bank of the river. Assume the variables x and y are in feet.

(0, 0)

FIGURE 24

a. Find the vertex.
b. Explain how to estimate the maximum clearance between the bridge and the river.
c. Explain how we know the base of the arch passes through the origin.

SOLUTION   a. The equation $y = -0.0022x^2 + 1.578x$ indicates that $a = -0.0022$ and $b = 1.578$. The x-coordinate of the vertex is

$$x = -\frac{b}{2a} = -\frac{1.578}{2(-0.0022)} \approx 359 \text{ ft}$$

The y-coordinate of the vertex is

$$y = f\left(-\frac{b}{2a}\right) = -0.0022(359)^2 + 1.578(359) \approx 283 \text{ ft}$$

The vertex is at approximately (359, 283) relative to the origin, as shown.

b. Given that the origin is at the river level, the y-coordinate of the vertex, 283 feet, gives an estimate of the vertical distance between the highest point on the arch and the river. The river may rise or fall, changing the clearance.

c. There is no constant term, c, in the equation.   ▬

PATH OF A BALL   In Example 2, we work with a parabola that represents the path of a ball. In this example, the horizontal and vertical axes are both labeled in feet.

EXAMPLE 2 Finding the vertex: path of a soccer ball   Orlando kicks a soccer ball at a 45° angle relative to the ground. If we ignore wind and air resistance, the path the ball follows is given by the quadratic equation

$$y = -\frac{g}{v^2}x^2 + x$$

The input x is the horizontal distance traveled by the ball, and the output y is the vertical position. The constant g is the average acceleration due to gravity, 32.2 ft/sec², and v is the initial velocity from the kick. Suppose the ball is kicked with a 80-ft/sec velocity. What is the maximum height reached by the soccer ball?

SOLUTION   We substitute $v = 80$ ft/sec and $g = 32.2$ ft/sec² into the given quadratic equation and simplify to identify a, b, and c.

$$y = -\frac{g}{v^2}x^2 + x = -\frac{(32.2)}{80^2}x^2 + x \approx -0.005x^2 + x$$

The coefficient of x^2, $a = -0.005$, is rounded to the nearest thousandth. The equation is graphed in Figure 25.

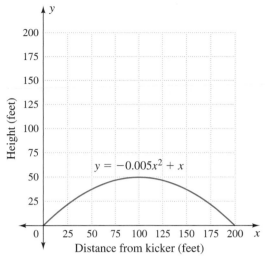

FIGURE 25

Because $a \approx -0.005$, $b = 1$, and $c = 0$, the x-coordinate of the vertex, in feet, is

$$x = -\frac{b}{2a} = -\frac{1}{2(-0.005)} = 100 \text{ ft}$$

The y-coordinate of the vertex is the maximum height reached by the ball. This height is $f(-b/2a)$, or $f(100)$:

$$f(100) = -0.005(100)^2 + 1(100) = 50 \text{ ft}$$

These results look much more accurate than they are. Remember that the coefficient a was rounded. ▬

TRANSITION CURVES   Have you ever driven over a hill so fast that the motion seemed to lift you out of your car seat? Civil engineers try to avoid such a possibility when they calculate the **transition curve**, *the roadbed design over a hill or between two hills*. Transition curves based on quadratic equations permit comfortable and safe travel.

The vertex of a transition curve between two hills is the minimum point, and it indicates the site for the storm water drain in Example 3.

In Section 10.5, students are given the equation for the slope of $y = ax^2 + bx + c$:

$$m = 2ax + b$$

They use slopes and systems of equations to find the transition curve.

EXAMPLE 3   Finding the vertex: roadbed transition curve   A roadbed passes through point G on one hill and point H on the next hill. A side view of the required roadbed is shown in Figure 26. Suppose point G has coordinates $(0, 1000)$, with units in feet. Point H has an x-coordinate of 1500 feet because it is 1500 feet away from the y-axis.

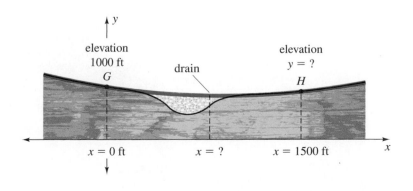

FIGURE 26

The roadbed transition curve to meet these requirements is given by

$$y = \frac{1}{60000} x^2 - \frac{3}{100} x + 1000$$

a. Find the y-coordinate for point H in Figure 26.

b. Find the vertex (storm drain location) of the parabolic roadbed.

c. Use a calculator to make a table with inputs from 0 to 1500 feet in steps of 300. Confirm that your answers to parts a and b are correct.

d. Find how long a water puddle would form if the storm drain were placed 1.5 feet in elevation above the vertex. Assume the water has nowhere else to run.

SOLUTION   **a.** To find the elevation at point H, we find $f(1500)$ in the transition curve equation.

$$y = \frac{1}{60000} x^2 - \frac{3}{100} x + 1000 = \frac{1}{60000} (1500)^2 - \frac{3}{100} (1500) + 1000$$

$$= 992.5 \text{ ft}$$

(It may be necessary to place parentheses around the fractions in a calculator.)

b. The x-coordinate of the vertex is $x = -b/2a$.

$$x = -\frac{b}{2a} = -\frac{-\frac{3}{100}}{2(\frac{1}{60000})} = \frac{3}{100} \cdot \frac{60000}{2} = 900$$

The y-coordinate of the vertex is $f(900)$.

$$y = \frac{1}{60000} x^2 - \frac{3}{100} x + 1000 = \frac{1}{60000} (900)^2 - \frac{3}{100} (900) + 1000$$

$$= 986.5 \text{ ft}$$

X	Y1	
0	1000	
300	992.5	
600	988	
900	986.5	
1200	988	
1500	992.5	
1800	1000	

X=0

FIGURE 27

c. We place the transition curve in the calculator as Y1 and evaluate for inputs in the table. The symmetry in outputs in Figure 27 is at the input 900. Thus, the vertex of the quadratic equation, the location of the storm water drain, is at (900, 986.5). The output at $x = 1500$ agrees with our results in part a.

d. If the water had nowhere else to run, it would back up to the 600- and 1200-foot marks relative to the x-axis (see Figure 27). The puddle would be 600 feet long.   ▬

Applications That Need Equations: Maximum Area

When no equations are given, we have to read the problem carefully and list relevant equations. We may need to solve one equation for a variable and substitute the result into another equation. Example 4 shows how the maximum or minimum question tells which equation should receive the substitution.

EXAMPLE 4   Finding the vertex: garden area   Mei has 80 feet of wire bird-netting to use around her vegetable garden to keep out slugs and cats. The garden is to be rectangular (Figure 28). What length and width give the maximum area for the available netting?

FIGURE 28

SOLUTION **Plan:** The perimeter of the rectangular garden is given by $P = 2l + 2w$. The area of the garden is given by $A = l \cdot w$. To find maximum area, we substitute into the area equation.

Carry out the plan: The perimeter of the garden is given by $P = 2l + 2w$. The perimeter is 80 feet. Substituting $P = 80$ ft and solving for l yields

$$80 = 2l + 2w$$

$$l = 40 - w$$

The area of the garden is given by $A = lw$. Substituting $l = 40 - w$ for l in the area formula yields

$$A = lw$$

$$A = (40 - w)w$$

$$A = 40w - w^2$$

This is a quadratic equation, so its vertex will give the maximum area as a function of width. The vertex is located where

$$w = -\frac{b}{2a} = -\frac{40}{2(-1)} = 20 \text{ ft}$$

Thus,

$$A = 40w - w^2 = 40(20) - 20^2 = 400 \text{ ft}^2$$

The width is 20 feet, and from $l = 40 - w$ we find that the length is also 20 feet. Thus, the maximum area rectangular garden with a fixed perimeter of 80 feet is a square.

Check: Substituting in the answers $l = 20$ and $w = 20$ does not really check for the maximum area. Checking nearby values may be better. Because $l = 40 - w$, let $w = 19$ and $l = 21$. Then $A = 19 \cdot 21 = 399$, a smaller area than 400. We might try a calculator table: Let $Y1 = 40 - x$ and $Y2 = x \cdot Y1$ and view width (X), length (Y1), and area (Y2) in the table. ▬

In Example 4, you may be uncomfortable with the idea that the answer is a square when the problem asked for a rectangle. Keep in mind that a square is a rectangle with four equal sides. *Rectangle* is the more generic name.

ANSWER BOX

Warm-up: **1.** -3 ± 5 **2.** 1 ± 1 **3.** 1 ± 3 **4.** -2 ± 5

5. $-\frac{3}{4} \pm \frac{\sqrt{41}}{4}$

6.4 Exercises

In Exercises 1 to 4, use factoring to find the *x*-intercepts, and then use the intercepts to find the vertex for each equation.

1. $y = x^2 - 2x$
 $(0, 0), (2, 0); (1, -1)$

2. $y = x^2 - 2x - 8$
 $(4, 0), (-2, 0); (1, -9)$

3. $y = -x^2 - 4x + 21$
 $(-7, 0), (3, 0); (-2, 25)$

4. $y = x^2 + 4x - 21$
 $(-7, 0), (3, 0); (-2, -25)$

Use

$$x = -\frac{b}{2a} \quad \text{and} \quad y = f\left(-\frac{b}{2a}\right)$$

to find the vertex for each quadratic function in Exercises 5 to 14.

5. $f(x) = x^2 + 8x + 15$ $(-4, -1)$

6. $f(x) = x^2 - 6x + 8$ $(3, -1)$

7. $f(x) = x^2 - 4x + 5$ $(2, 1)$

8. $f(x) = x^2 + 4x + 7$ $(-2, 3)$

9. $f(x) = 2x^2 - x - 3$ $\left(\frac{1}{4}, -\frac{25}{8}\right)$

10. $f(x) = 2x^2 + 5x - 3$ $\left(-\frac{5}{4}, -\frac{49}{8}\right)$

11. $f(x) = 3x^2 + 6x - 2$ $(-1, -5)$

12. $f(x) = 3x^2 + 12x + 5$ $(-2, -7)$

13. $f(x) = -x^2 + 3x - 2$ $(1.5, 0.25)$

14. $f(x) = -x^2 + 2x - 3$ $(1, -2)$

In Exercises 15 to 18, find the vertex for each equation and then find the indicated distances.

15. A bridge support arch has the equation $y = -\frac{2}{225} x^2 + \frac{4}{3} x + 10$, with x in feet. Find the clearance between the arch vertex and the water if $(75, 60)$

   a. the water is at the level of the x-axis. 60 ft

   b. the water is at the level of the y-intercept. 50 ft

16. A bridge support arch has the equation $y = -\frac{2}{1125} x^2 + \frac{8}{15} x + 25$, with x in feet. Find the clearance between the arch vertex and the water if $(150, 65)$

   a. the water is at the level of the x-axis. 65 ft

   b. the water is at the level of the y-intercept. 40 ft

17. A suspension bridge cable has the equation $y = 0.001x^2 - 0.4x + 50$, with x in feet. Find the distance between the roadbed or power line and the vertex if $(200, 10)$

   a. the roadbed passes through the origin. 10 ft

   b. the power line passes midway between the origin and the y-intercept. 15 ft

18. A suspension bridge cable has the equation $y = \frac{9}{8000} x^2 - \frac{9}{20} x + 60$, with x in feet. Find the distance between the roadbed or power line and the vertex if $(200, 15)$

   a. the roadbed passes through the origin. 15 ft

   b. the power line passes 20 feet above the origin. 5 ft

Exercises 19 to 26 refer to the formula in Example 2. Round answers to nearest tenth.

19. The path of a soccer ball is given by $y \approx -0.005x^2 + x$, with x in feet. Suggest three ways to find the horizontal distance it travels before returning to the ground, and use one of these ways to find the distance. factor to find
x-intercepts, use quadratic formula, double horizontal distance to vertex; 200 ft

20. The soccer ball passed through the origin. Explain how we know this and what it means in the problem setting.
There is a zero y-intercept; ball started on the ground.

21. Suppose Rosa kicks a soccer ball at a 45° angle with an initial velocity of 90 ft/sec. Predict the maximum height of the ball and the horizontal distance traveled by the ball before it returns to the ground. 62.9 ft; 251.6 ft

22. Suppose Arne kicks a soccer ball at a 45° angle with an initial velocity of 60 ft/sec. Predict the maximum height of the ball and the horizontal distance traveled by the ball before it returns to the ground. 28.0 ft; 111.8 ft

23. A signal flare with a parabolic flight path needs to be seen over a 500-foot-high obstacle. Assume the parabolic path is at a 45° angle. Will it reach sufficient height if it is fired with an initial velocity of 200 ft/sec? What is the highest point it reaches? no; 310.6 ft

24. Use guess and check, a table, or a graph to determine the initial velocity required to reach 500 feet in Exercise 23. 253.8 ft/sec

25. Sketch a set of three graphs that show the path of a ball (or other object) hit from (0, 0) at a 45° angle with initial velocities of 50, 100, and 200 ft/sec. Use the graphs to finish these statements: See Answer Section.

   a. When the initial velocity is doubled, the height of the vertex above the ground is <u>quadrupled</u>.

   b. When the initial velocity is doubled, the distance the ball travels is <u>quadrupled.</u>

26. Set up three tables showing the coordinates of the vertex and x-intercepts for the path of a ball (or other object) hit from (0, 0) at a 45° angle with initial velocities of 40, 80, and 160 ft/sec. Use the tables to finish the statements in parts a and b in Exercise 25.
See Additional Answers; quadrupled; quadrupled

In Exercises 27 and 32, an object is thrown or launched straight up. Its height above ground level relative to the time in the air behaves according to the equation $h = -0.5gt^2 + v_0 t + h_0$, where g is acceleration due to gravity (32.2 ft/sec²), v_0 is the initial velocity, and h_0 is the initial height. Round answers to the nearest tenth.

27. What is the maximum height for a ball thrown with an initial velocity of 40 ft/sec? Assume the ball leaves the hand at 5.5 feet. 30.3 ft

28. What is the maximum height for a ball thrown with an initial velocity of 30 ft/sec? Assume the ball leaves the hand at 4.5 feet. 18.5 ft

29. A ball of fireworks is shot vertically from the ground with an initial velocity of 115 ft/sec. The fuse is set to go off at the maximum height. Find the time until the fireworks burst and the height at which they burst.
at the vertex: (3.6 sec, 205.4 ft)

30. A ball of fireworks is shot vertically from a point on the Brooklyn Bridge 266 feet above the water. The initial velocity is 120 ft/sec. The fuse is set to go off at the maximum height. Find the time until the fireworks burst and the height at which they burst.
at the vertex: (3.7 sec, 489.6 ft)

31. Explain why the initial height, h_0, does not affect the time taken to reach the maximum height.
At the vertex, $t = -b/2a$ and does not contain $h_0 = c$.

32. Explain how we know that the vertical motion equation does not give the path of the object. The input is time, not feet; a path requires feet for both input and output.

In Exercises 33 to 38, the road grades connecting two hills are designed with a parabolic curve. Engineers need to find the minimum point for a storm drain.

33. In the figure, the horizontal distance between points K and L is 2000 feet. The elevation of point K is 700 feet. The equation for the road grade is

$$y = \frac{7}{400,000} x^2 - \frac{1}{25} x + 700$$

Find the elevation of point L and the location for the storm drain. 690 ft; (1143, 677)

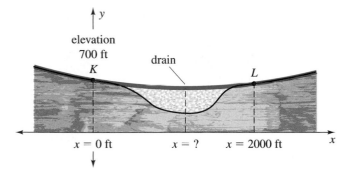

34. In the figure in Exercise 33, a decreased slope at point K and an increased slope at point L result in a new equation for the road grade:

$$y = \frac{1}{50,000} x^2 - \frac{3}{100} x + 700$$

Find the elevation of point L and the location for the storm drain. 720 ft; (750, 689)

35. In the figure, the horizontal distance between point M and point N is 2400 feet. The elevation of point N is 1200 feet. The equation for the road grade is

$$y = \frac{11}{480,000} x^2 - \frac{1}{20} x + 1188$$

Find the elevation of point M and the location for the storm drain. 1188 ft; (1091, 1161)

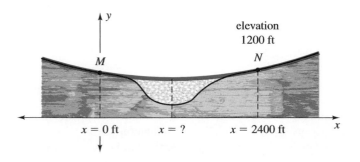

36. In the figure in Exercise 35, a decreased slope at point M and a decreased slope at point N result in a new equation for the road grade:

$$y = \frac{1}{80,000} x^2 - \frac{1}{25} x + 1224$$

Find the elevation of point M and the location for the storm drain. 1224 ft; (1600, 1192)

37. For the highway roadbed in Example 3, the transition curve between the hills was given by

$$y = \frac{1}{60000} x^2 - \frac{3}{100} x + 1000$$

with a storm drain at the vertex, at (900, 986.5) feet. Suppose the storm drain plugs and water backs up onto the roadway to a depth of 0.1 foot, or an elevation of 986.6 feet (see the figure). Solve the inequality

$$\frac{1}{60000} x^2 - \frac{3}{100} x + 1000 < 986.6$$

to see how long a puddle is formed by the water. ≈155 ft

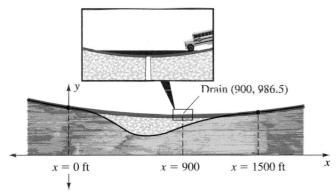

38. Repeat Exercise 37 but with water accumulating to a depth of 0.25 foot (3 inches) on the roadway. ≈245 ft

39. Miguel has 60 feet of deer fencing for a garden next to his house. The garden is to be rectangular, with the house forming one side as shown in the figure. What is the largest possible rectangular area? What length and width give the largest possible area?
$A = 450$ ft^2; 30 ft, 15 ft

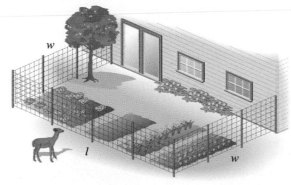

40. Rita has 90 feet of fencing for a corral next to her barn. The corral is to be rectangular, with the barn forming one side, similar to the garden in Exercise 39. What is the largest possible rectangular area? What length and width give the largest possible area?
$A = 1012.5$ ft^2; 45 ft, 22.5 ft

41. Juan has 120 feet of fencing with which to make a movable pen as shown in the figure. Find the largest possible fenced area. Find the length and width that give the largest area. $A = 600$ ft^2; 20 ft, 30 ft

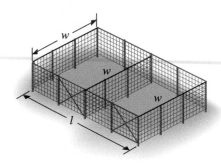

42. Rachel is building pens along an existing fence as shown in the figure. She has 150 feet of fencing. Find the largest possible fenced area. Find the length and width that give the largest area. $A = 1125$ ft^2; 15 ft, 75 ft

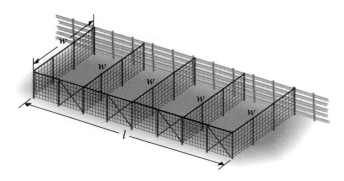

43. Describe how to determine the second x-intercept if we know one x-intercept $(x, 0)$ and the vertex (h, k).
distance $|h - x|$ from vertex on opposite side

44. Substitute $x = -b/2a$ into $y = ax^2 + bx + c$ and find a formula for the y-coordinate of the vertex. Simplify your formula with a common denominator. Where have you seen parts of this formula before?

■ Project $y = \dfrac{4ac - b^2}{4a}$; somewhat like quadratic formula

45. Shipping Box The museum director has asked you to find the dimensions (height and width, to the nearest tenth of a foot) of the box with the largest rectangular cross section that could pass through the parabolic doorway in the figure. The wheels supporting the box are recessed and will not interfere with the cross section's area.

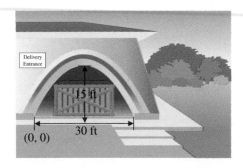

At the floor, the width of the parabola is 30 feet. The height of the parabola is 15 feet. Assume the origin is on the floor at the lower left. Use these suggestions to get you started:

a. Find the equation for the parabola. $y = -x^2/15 + 2x$

b. Use the quadratic formula (or calculator **intersect**) to find the intersections of selected heights with the graph of the parabola. For example, solve for x when $y = 14.4$ ft. Use the x's to find the width of the opening at height $y = 14.4$ ft. Multiply width by height to find the area. $x_1 = 12$ ft, $x_2 = 18$ ft, $w = 6$ ft, $A = 86.4$ ft^2

c. Repeat step b for $y = 12.6$ ft.
$x_1 = 9$ ft, $x_2 = 21$ ft, $w = 12$ ft, $A = 151.2$ ft^2

d. Now you are on your own to find the maximum area.
$h = 10.1$ ft, $w = 17.3$ ft, $A = 174.8$ ft^2

■ 6 **Chapter Summary**

Vocabulary

complex conjugate	discriminant	minimum	vertex form of a quadratic
complex number	horizontal shift	transition curve	equation
complex number	imaginary unit	translation	vertical shift
system	maximum		

Concepts

6.1 ▬ The Roles of *a*, *b*, and *c* in Graphing Quadratic Functions

The parameters a, b, and c in $f(x) = ax^2 + bx + c$ interact to control the shape, orientation, and position of the graph of $f(x)$.

A positive a on the x^2 term causes the parabola to open up, $\cup$. A negative a on the x^2 term causes the parabola to open down, $\cap$.

The coefficient on the x^2 term, the parameter a, controls the steepness of the graph. If a is larger than 1, the graph is steeper than the graph of $y = x^2$. If a is between zero and 1, the graph is flatter than the graph of $y = x^2$.

The parameter b in $y = x^2 + bx$ contributes to a change in the position of the vertex of the parabola from that of $y = x^2$.

The parameter c in $y = ax^2 + bx + c$ is the output when $x = 0$. The coordinate $(0, c)$ is the y-intercept point of the graph of a quadratic function. If the equation is of the form $y = ax^2 + c$, then c is the distance (number of units) the parabola $y = ax^2$ is shifted vertically (up or down).

6.2 ▬ Complex Numbers and Solving Polynomial Equations

If n is a positive number, then $\sqrt{-n} = i\sqrt{n}$, where i is the imaginary unit.

Addition of complex numbers:

$$(a + bi) + (c + di) = (a + c) + (b + d)i$$

Subtraction of complex numbers:

$$(a + bi) - (c + di) = (a - c) + (b - d)i$$

Multiplication of complex numbers:

$$(a + bi)(c + di) = ac + adi + bci + bdi^2$$
$$= ac - bd + (ad + bc)i$$

If the discriminant $b^2 - 4ac$ is positive, there are two real-number solutions to the quadratic equation $ax^2 + bx + c = 0$. The graph of $f(x) = ax^2 + bx + c$ passes through the x-axis twice.

If $b^2 - 4ac$ is zero, there is one real-number solution, a double root. The graph of $f(x)$ touches the x-axis once.

If $b^2 - 4ac$ is negative, there are two complex-number solutions. The graph of $f(x)$ does not touch the x-axis.

Only the real numbers appear on a rectangular coordinate graph.

The degree of a polynomial equation indicates the maximum number of real-number solutions or the total number of real-number and complex-number solutions.

6.3 ▬ Shifts, Vertex Form, and Applications of Quadratic Functions

If the quadratic equation can be written as $y = (x - g)^2$, the vertex is at $(g, 0)$. The graph of $y = (x - g)^2$ is shifted horizontally g units from the graph of $y = x^2$.

If the quadratic equation can be written as $y = x^2 + j$, the vertex is at $(0, j)$. The parabola for $y = x^2 + j$ is shifted vertically $|j|$ units from the graph of $y = x^2$.

The vertex is (h, k) when we have completed the square on a quadratic equation and changed it into vertex form,

$$y = a(x - h)^2 + k$$

To obtain vertex form (when $a = 1$) by completing the square, write the equation as $y = x^2 + bx + c$. Subtract the constant term, c, from both sides. Complete the square on the right, adding the required number to both sides. Change to squared form on the right. Change to vertex form, $y = (x - g)^2 + j$. The vertex is (g, j).

6.4 ▬ Solving Minimum and Maximum Problems

The vertex is the position on parabolic graphs where the maximum (greatest) and minimum (smallest) values are obtained.

For the function $f(x) = ax^2 + bx + c$, the x-coordinate of the vertex is a portion of the quadratic formula, $x = -b/2a$. The y-coordinate of the vertex is the output, $f(-b/2a)$.

---

6   Review Exercises

1. How does the graph of $y = 3x^2$ differ from that of $y = x^2$? steeper

2. How does the graph of $y = -x^2$ differ from that of $y = x^2$? upside down

3. How does the graph of $y = x^2 - 9$ differ from that of $y = x^2$? shifted down 9 units

4. How does the graph of $y = x^2 + 4$ differ from that of $y = x^2$? shifted up 4 units

Sketch graphs for the equations in Exercises 5 to 10.

5. $y = 3 - x^2$ See Answer Section.

6. $y = x^2 - x$ See Additional Answers.

7. $y = x^2 + 3$ See Answer Section.

8. $y = -\frac{1}{2}x^2$ See Additional Answers.

9. $y = \frac{1}{2}x^2 - 4$ See Answer Section.

10. $y = 2x^2 + 3$ See Additional Answers.

11. Match each equation with a reasonable setting. Choose from

golf ball path, vertical motion of fireworks, Olympic diver vertical motion, bridge suspension cable

Tell what parameter or parameters help you choose.

a. $h = -4.905t^2 + 5t + 10$ Olympic diver; a implies a very steep curve, $c \neq 0$ is diver's initial height in vertical motion.

b. $h = -4.905t^2 + 115t$ fireworks; a implies a very steep curve, $c = 0$ starts at ground, t in vertical motion, b is large initial velocity.

c. $y = 0.03125x^2 - 2.5x + 60$
bridge cable; parabola turns up for $a > 0$.

d. $y = -0.0005812x^2 + 0.1763x + 10$
golf ball; curve turns down for $a < 0$, a implies a very flat curve.

12. Which function gives the steeper curve?

a. $y = \pi r^2$ or $y = 4r^2$ $y = 4r^2$

b. $y = 0.031x^2$ or $y = x^2/30$ $y = x^2/30$

c. $y = -2x^2$ or $y = -1.5x^2$ $y = -2x^2$

Change each radical expression in Exercises 13 and 14 to a product of a real number and the imaginary unit.

13. a. $\sqrt{-16}$ $4i$ **b.** $\sqrt{-50}$ $5i\sqrt{2} \approx 7.07i$

14. a. $\sqrt{-64}$ $8i$ **b.** $\sqrt{-18}$ $3i\sqrt{2} \approx 4.2i$

In Exercises 15 and 16, simplify the complex numbers.

15. a. $\dfrac{8 + 6i}{2}$ $4 + 3i$ **b.** $\dfrac{6 + 3i}{3}$ $2 + i$

16. a. $\dfrac{4 - \sqrt{-48}}{2}$ $2 - 2i\sqrt{3}$ **b.** $\dfrac{9 + \sqrt{-18}}{6}$ $\dfrac{3 + i\sqrt{2}}{2}$

17. Simplify.

a. $(3 + 2i) - (6 + 2i)$ -3 **b.** $(4 - 2i) - 2(3 - 4i)$ $-2 + 6i$

18. Name the complex conjugates.

a. $4 - 3i$ **b.** $2i - 6$ **c.** $-5i$
$4 + 3i$ $-6 - 2i$ $5i$

In Exercises 19 and 20, multiply and simplify the expressions.

19. a. $(4 + 3i)(4 + 3i)$ **b.** $(4 - 3i)(4 + 3i)$ 25
$7 + 24i$
c. $(3 - 4i)(3 + 4i)$ 25

20. a. $(1 - 3i)(1 + 3i)$ 10 **b.** $(3 - i)(3 + i)$ 10

c. $(1 + 3i)(1 + 3i)$ $-8 + 6i$

For Exercises 21 to 23, choose the phrase that best completes each sentence:
(a) then the quadratic equation has one real root (a double root).
(b) then the quadratic equation has two real-number roots.
(c) then the quadratic equation has two complex-number roots.

21. If $b^2 - 4ac$ is negative, ___c___ .

22. If $b^2 - 4ac$ is zero, ___a___ .

23. If $b^2 - 4ac$ is positive, ___b___ .

Which description matches each statement in Exercises 24 to 26?
(a) The graph of the quadratic equation just touches the x-axis.
(b) The graph of the quadratic equation intersects the x-axis twice.
(c) The graph of the quadratic equation does not intersect the x-axis.

24. $b^2 - 4ac < 0$ c   **25.** $b^2 - 4ac > 0$ b

26. $b^2 - 4ac = 0$ a

Evalute $b^2 - 4ac$ for the quadratic equations in Exercises 27 to 31. Indicate the number and types of solutions to $f(x) = 0$ and the number of x-intercepts in the graph of $f(x)$.

27. $f(x) = -2x^2 + x + 3$ 25; 2 real; 2

28. $f(x) = -1x^2 + 2x - 1$ 0; 1 double root; 1

29. $f(x) = 2x^2 + 3x + 4$ -23; 2 complex; 0

30. $f(x) = 3x^2 + 2x + 1$ -8; 2 complex; 0

31. $f(x) = x^2 + 6x + 9$ 0; 1 double root; 1

In Exercises 32 to 36, solve $f(x) = 0$ for the function in the given exercises.

32. Exercise 27 $\{-1, 1.5\}$

33. Exercise 28 $\{1\}$ (double root)

34. Exercise 29 $\{\frac{3}{4} \pm i\sqrt{23}/4\}$

35. Exercise 30 $\{-\frac{1}{3} \pm i\sqrt{2}/3\}$

36. Exercise 31 $\{-3\}$ (double root)

In Exercises 37 and 38, write an output $y = n$ such that $f(x) = n$ has the number of real-number solutions indicated, if possible. Use the graph and draw the horizontal line representing each output.
(a) no solutions   (b) 1 solution
(c) 2 solutions   (d) 3 solutions
(e) 4 solutions

37. $f(x) = x^3 + 2x^2 - 5x - 6$

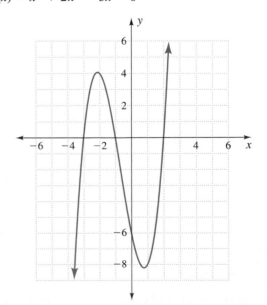

not possible; possible answers: $y = 5$; $y = 4$; $y = 2$; not possible

38. $f(x) = 2x^4 + x^3 - 9x^2 - 4x + 4$

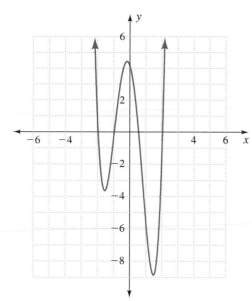

possible answers: $y = -9$; $y \approx -9$; $y = -6$; $y \approx -3.5$; $y = 1$

Use the graphs in Exercises 37 and 38 to solve the equations in Exercises 39 and 40. Estimate x, as needed, to the nearest 0.5.

39. a. $x^3 + 2x^2 - 5x - 6 = 0$ $\{-3, -1, 2\}$

   b. $x^3 + 2x^2 - 5x - 6 = -6$ $\{-3.5, 1.5, 0\}$

40. a. $2x^4 + x^3 - 9x^2 - 4x + 4 = 0$ $\{-2, -1, 0.5, 2\}$

   b. $2x^4 + x^3 - 9x^2 - 4x + 4 = -10$ not possible

In Exercises 41 to 48, find all solutions to $f(x) = 0$, real or complex.

41. $f(x) = x^3 - 27$ $\left\{3, -\frac{3}{2} \pm 3i\sqrt{3}/2\right\}$

42. $f(x) = x^3 + 8$ $\{-2, 1 \pm i\sqrt{3}\}$

43. $f(x) = x^3 + x^2 + x$ $\left\{0, -\frac{1}{2} \pm i\sqrt{3}/2\right\}$

44. $f(x) = x^3 + 3x^2 + 4x$ $\left\{0, -\frac{3}{2} \pm i\sqrt{7}/2\right\}$

45. $f(x) = 2x^3 - 2x^2$ $\{0 \text{ (double root)}, 1\}$

46. $f(x) = x^3 + 9x$ $\{0, \pm3i\}$

47. $f(x) = x^4 - 16$ $\{\pm2, \pm2i\}$

48. $f(x) = x^2 - 9x^4$ $\left\{0 \text{ (double root)}, \pm\frac{1}{3}\right\}$

Use the graph of $y = x^2$ and shifts to graph the equations in Exercises 49 to 52.

49. $y = (x - 3)^2 + 2$
See Answer Section.

50. $y = (x + 3)^2 + 1$
See Additional Answers.

51. $y = (x + 2)^2 - 1$
See Answer Section.

52. $y = (x - 4)^2 - 2$
See Additional Answers.

In Exercises 53 to 58, find the equation of the graph passing through the given points.

53. Vertex (8, 2) and the origin $y = -\frac{1}{32}(x - 8)^2 + 2$

54. Vertex (4, −3) and the origin $y = \frac{3}{16}(x - 4)^2 - 3$

55. Vertex at the origin and (8, 2) $y = x^2/32$

56. Vertex at the origin and (4, −3) $y = -3x^2/16$

57. Vertex (−2, 3) and (2, −3) $y = -\frac{3}{8}(x + 2)^2 + 3$

58. Vertex (2, −3) and (−2, 3) $y = \frac{3}{8}(x - 2)^2 - 3$

59. A stream of water from a fireboat cannon must reach a height of 250 feet at a horizontal distance of 100 feet from the cannon. Find the equation of the stream of water if the origin is at the cannon.
$y = -0.025(x - 100)^2 + 250$

60. A suspension bridge cable spans 500 feet with a height of 100 feet. Find the equation of the cable if the origin is at the top of the left support. $y = 0.0016(x - 250)^2 - 100$

In Exercises 61 and 62, complete the square to find the vertex form of the equation.

61. $y = x^2 + 4x + 9$ $y = (x + 2)^2 + 5$

62. $y = x^2 - 6x + 6$ $y = (x - 3)^2 - 3$

In Exercises 63 to 66, use any method to find the x-intercepts, if they exist, and then find the vertex for the graph of each function.

63. $f(x) = x^2 - x - 2$ $\{-1, 2\}$; $(0.5, -2.25)$

64. $f(x) = x^2 - x - 6$ $\{2, 3\}$; $(0.5, -6.25)$

65. $y = 2x^2 - 4x + 5$ no x-intercepts; $(1, 3)$

66. $y = 3x^2 - 6x + 5$ no x-intercepts; $(1, 2)$

67. Suppose an emergency rescue flare is fired into the air at a 45° angle. The path of the flare is

$$y = -\frac{g}{v^2}x^2 + x$$

The input x is the horizontal distance traveled by the flare, and the output y is the vertical position. The constant g is the average acceleration due to gravity, 32.2 ft/sec², and v is the initial velocity of the flare in feet per second. Find the vertex of the path of the flare for each of these initial velocities of the flare.

   a. 132 ft/sec (the speed of a very fast baseball pitch)
$\approx(271, 135)$

   b. 220 ft/sec (the speed of a golf club in the hands of a professional golfer) $\approx(752, 376)$

   c. 400 ft/sec $\approx(2484, 1242)$

   d. 500 ft/sec $\approx(3882, 1941)$

68. A jet of water has the equation $y = -0.12x^2 + 6x$, with x in feet.

   a. Find the maximum height reached. 75 ft

   b. Find the horizontal distance for which the jet of water will be over 60 feet. $\approx13.8 \text{ ft} < x < 36.2 \text{ ft}$

69. A jet of water has the equation $y = -\frac{1}{9}x^2 + \frac{20}{3}x$, with x in feet.

   a. Find the maximum height reached.
100 ft

   b. Find the horizontal distance for which the jet of water will be below 50 feet. $\approx x < 8.8 \text{ ft or } x > 51.2 \text{ ft}$

70. The arch supporting a bridge is marked with the origin at the lower left point. The height of the arch is 40 feet, and the span across the stream is 150 feet.

a. Find the equation of the arch. $y = -0.0071x^2 + 1.067x$

b. Find the vertex of the arch. $(75, 40)$

c. If the arch is supported by concrete piers rising 12 feet above the water level, find the maximum clearance for sailboat masts. $12 + 40 = 52$ ft

71. A roadbed transition curve is $y = \frac{1}{48000} x^2 - \frac{1}{50} x + 700$.

a. Locate the minimum point for the drain.
$(480, 695.2)$

b. If the drain plugs and water rises on the road, how long will a 0.3-foot-deep puddle reach?
between $x = 360$ ft and $x = 600$ ft, or 240 ft

72. A wrestling mat has a perimeter of 157 feet and has the maximum possible area for a rectangular shape. Write the area as a function of only the width, and show that the vertex of the graph of the area function gives the maximum area. $A = 78.5w - w^2$; vertex $(39.25, 1541)$; maximum area is square with $w = l = 39.25$; $39.25^2 \approx 1541$

73. A boxing ring has a perimeter of 80 feet and has the maximum possible area for a rectangular shape. Write the area as a function of only the width, and show that the vertex of the graph of the area function gives the maximum area.
$A = 40w - w^2$; vertex $(20, 400)$; maximum area is square with $w = l = 20$; $20^2 = 400$

---

6   Chapter Test

1. Compare the graph of $y = x^2 + 2$ with the graph of $y = x^2$. Discuss intercepts and vertices.
shifted up 2 units; no x-intercepts; vertex at $(0, 2)$

2. Explain how the negative sign in $y = -x^2$ changes its graph from that of $y = x^2$. upside down

3. Which of the two equations

$$y = x - \frac{32.2}{2500} x^2 \quad \text{or} \quad y = \frac{4}{125} x^2$$

could describe the path of a ball through the air? Explain.
the first, because its graph opens down

4. Change the radical expression to a product of a real number and the imaginary unit.

a. $\sqrt{-36}$ $6i$ **b.** $\sqrt{-75}$ $5i\sqrt{3} \approx 8.7i$

5. Simplify these complex numbers.

a. $\dfrac{5 + 10i}{5}$ $1 + 2i$ **b.** $\dfrac{8 + 4i}{2}$ $4 + 2i$

6. Multiply and simplify the expressions.

a. $(1 + 5i)(1 - 5i)$ 26   **b.** $(5 - i)(5 + i)$ 26

c. $(1 - 5i)(1 - 5i)$ $-24 - 10i$

7. Simplify $5(2 - 3i) - 3(6 - 7i)$. $-8 + 6i$

8. a. Evaluate the discriminant for $f(x) = -1x^2 + 2x + 3$.
16

b. Indicate the number and type of solutions to $f(x) = 0$ and the number of x-intercepts in the graph of $f(x)$.
2 real; 2

c. Solve $f(x) = 0$.
$\{-1, 3\}$

9. Find all solutions to $f(x) = 0$, real or complex, for $f(x) = (x - 1)(x^2 + x + 1)$.
$\left\{1, -\frac{1}{2} \pm i\sqrt{3}/2\right\}$

10. Describe the graph of $y = (x - 3)^2 - 1$ in terms of shifts of $y = x^2$.
shifted right 3 units, down 1 unit

11. A parabola has its vertex at $(-1, -8)$ and contains the point $(2, 19)$. Find the equation. $y = 3x^2 + 6x - 5$

12. Complete the square to change $y = x^2 + 4x - 1$ into vertex form. $y = (x + 2)^2 - 5$

13. The vertical motion of an Olympic diver from the 3-meter board is described by $h = -\frac{1}{2}(9.81)t^2 + 6t + 3$, with h in meters and t in seconds.

a. Find her maximum height above the water and how long it takes her to reach that height.
$(0.61 \text{ sec}, 4.83 \text{ m})$

b. Find the length of time before she hits the water.
1.6 sec

14. You have 15 yards of fencing and wish to make a rectangular kennel for your dog. You build the kennel next to your garage so that you need to enclose only three sides, as shown in the figure.

a. Write an equation that describes the area y in terms of the width w and the length l. $y = lw$

b. Eliminate l from your area equation with a substitution. $y = 15w - 2w^2$

c. Find the vertex of the graph of the area function, and explain how it shows the maximum area inside the kennel. $\left(\frac{15}{4}, 28\frac{1}{8}\right)$; area, y, is largest at peak of graph.

Cumulative Review of Chapters 1 to 6

1. Multiply mentally: $8\left(3\frac{1}{4}\right)$. $24 + 2 = 26$

2. Multiply mentally: $(25\%)(6)(8)(4)$.
25% of $4 = 1$; $1 \cdot 6 \cdot 8 = 48$

3. Solve $5 - 3(x - 2) = 3(x - 1)$. $x = 2\frac{1}{3}$

4. Find a linear function through $(0, 2)$ that is parallel to $3x - 2y = 6$. $y = 3x/2 + 2$

5. Simplify.

   a. $\sqrt{(-2)^2}$ 2

   b. $\sqrt{-2^2}$ $2i$

   c. $\sqrt{-2}$ $i\sqrt{2}$

   d. $(\sqrt{-2})^2$ -2

6. Factor the radicand and simplify to $a\sqrt{b}$.

   a. $\sqrt{108}$ $6\sqrt{3}$

   b. $\sqrt{24}$ $2\sqrt{6}$

7. Multiply $(3 - i)(3 + i)$. 10

8. Evaluate $f(x) = |x - 4|$ for each of the following.

   a. $f(-1)$ 5

   b. $f(a + b)$ $|a + b - 4|$

9. Evaluate $f(x) = x^2$ for each of the following.

   a. $f(x + 1)$ $x^2 + 2x + 1$

   b. $f(a + b)$ $a^2 + 2ab + b^2$

10. Show whether sides of these lengths form a right triangle: 4.5, 20, 20.5.   yes; $4.5^2 + 20^2 = 20.5^2$

11. Give the next number in each sequence. Tell whether the sequence is linear, quadratic, or other. Fit an equation to linear and quadratic patterns.

   a. 3, 0, 1, 6, 15   28; quadratic; $y = 2x^2 - 9x + 10$

   b. 8, 11, 14, 17, 20   23; linear; $y = 3x + 5$

   c. 3, 4, 7, 11, 18   29; other

Solve the inequalities in Exercises 12 to 15.

12. $\dfrac{x - 6}{2} < x + 1$ $x > -8$

13. $x^2 - 2x - 3 \geq -3$ $x \leq 0$ or $x \geq 2$

14. $x^2 - 1 > 1 - x^2$ $x < -1$ or $x > 1$

15. $x^2 - 4x + 4 \geq 8 - x$ $x \leq -1$ or $x \geq 4$

16. Solve $|x - 5| = 8$. $x = -3, x = 13$

17. Solve $|x + 6| < 5$. $-11 < x < -1$

18. Solve by substitution. $x = -4, y = 5$

   $x + 2y = 6$

   $3x - 4y = -32$

19. Solve by elimination. $a = \frac{1}{2}, b = 3$

   $2a - 3b = -8$

   $4a + 5b = 17$

20. Two angles are supplementary. The difference in their measures is 25°. Find the angle measures. $102.5°, 77.5°$

21. Andrew has $450 on a credit card at 22% interest. At what interest rate must he borrow $1200 in order to have an average interest rate of 10%?   5.5%

22. Find the equation of the parabola with vertex $(2, 4)$ and containing $(-3, 2)$. $y = -\frac{2}{25}(x - 2)^2 + 4$ or $y = -\frac{2}{25}x^2 + \frac{8}{25}x + \frac{92}{25}$

23. Complete $x^2 + 3x + \underline{\frac{9}{4}} = (x + \underline{\frac{3}{2}})^2$ to make a true statement about a perfect square trinomial.

24. Match the equations $y = \sqrt{x}$, $y = x^2$, and $y = x$ with the graphs in the figure. Only the first-quadrant portions of the graphs are shown. Give one fact about each graph that explains your choice.

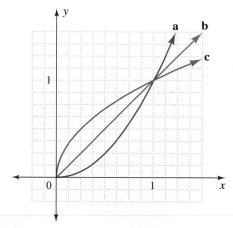

a. $y = x^2$ because $(\frac{1}{2})^2 = \frac{1}{4}$. b. $y = x$ because $(\frac{1}{2}, \frac{1}{2})$ lies on the linear graph. c. $y = \sqrt{x}$ because $\sqrt{\frac{1}{4}} = \frac{1}{2}$.

CHAPTER 7

Rational Functions and Variation

Both the galaxy in Figure 1 and the hurricane in Figure 2 are in rotational motion, and both have a rate of turning that is dependent on how compact their structure is relative to the center of rotation. For the galaxy, the more densely packed the stars are toward the center of the galaxy, the faster the galaxy rotates. For the hurricane, the smaller the "eye" is, the closer the cloud mass and moisture are to the center of rotation and the faster the accompanying winds. The relationship between position relative to the center of rotation and rate of turning is described as an inverse variation. Inverse variation is discussed in Section 7.3.

The chapter opens with a review of ratio and proportion and introduces direct variation in Section 7.2. Sections 7.4, 7.5, and 7.6 examine algebraic skills with rational expressions. The chapter closes with a discussion of solving rational equations in Section 7.7.

FIGURE 1 Galaxy M81 in Ursa Major   **FIGURE 2** A hurricane photographed over three days

7.1 Ratios, Units of Measure, and Proportions

Objectives

■ Find when ratios are equivalent.

■ Simplify ratios.

■ Simplify expressions containing units, and set up a unit analysis.

■ Solve word problems and similar triangle problems with proportions.

WARM-UP

Factor and simplify to lowest terms.

1. $\frac{12}{15}$ $\frac{4}{5}$ **2.** $\frac{35}{21}$ $\frac{5}{3}$ **3.** $\frac{63}{36}$ $\frac{7}{4}$ **4.** $\frac{48}{30}$ $\frac{8}{5}$ **5.** $\frac{150}{200}$ $\frac{3}{4}$ **6.** $\frac{12}{56}$ $\frac{3}{14}$

IN THIS SECTION, we review ratios and their simplification when units are included. We work with the unit analysis method as well as with proportions and applications to similar triangles.

Ratios

A **ratio** is *the quotient of two quantities*. Ratios compare like or unlike quantities. Recall that the slope of a linear equation is the ratio of rise to run, or change in *y* to change in *x*.

Think about it 1: The fraction $\frac{5}{0}$ is undefined, as is the slope of a vertical line. Give examples of situations where a ratio of 5 to 0 is meaningful.

If two ratios divide to the same decimal or simplify to the same fraction, they are equivalent.

DEFINITION OF EQUIVALENT RATIOS

> Two ratios are **equivalent ratios** if they simplify to the same number.

EXAMPLE 1   Exploring pi   The number pi is the ratio of the circumference to the diameter of a circle. Many arithmetic textbooks use the fraction $\frac{22}{7}$ as an approximation to pi. Find out whether $\frac{22}{7}$ and $\frac{2218}{706}$ are equivalent ratios. Compare them to the value of pi programmed into your calculator.

SOLUTION   See the Answer Box.

   Simplifying ratios that contain only numbers or units is exactly like simplifying fractions: We look for common factors in the numerator and denominator and eliminate those factors based on the fact that $a/a = 1$.

SIMPLIFYING EXPRESSIONS

- Multiplication by 1 does not change number values:

$$a \cdot 1 = a$$

- The number 1 appears in many forms—numbers, variables, and units:

$$\frac{25}{25}, \quad \frac{a}{a}, \quad \frac{m}{m}, \quad \frac{feet}{foot}, \quad \frac{pounds}{pound}$$

- We can simplify within a product:

$$\frac{3}{4} \cdot \frac{5}{3} = \frac{5}{4}, \quad \frac{\$7.50}{1 \text{ hour}} \cdot \frac{8 \text{ hours}}{1 \text{ day}} = \frac{\$60}{1 \text{ day}}$$

EXAMPLE 2 Simplifying expressions containing units   Simplify.

a. $\dfrac{50 \text{ m}^3}{25 \text{ m}}$ **b.** $\dfrac{300 \text{ foot} \cdot \text{pounds}}{10 \text{ pounds}}$ **c.** $\dfrac{30 \text{ degree} \cdot \text{days}}{6 \text{ days}}$

SOLUTION   **a.** $\dfrac{50 \text{ m}^3}{25 \text{ m}} = \dfrac{25 \cdot 2 \, m \cdot \text{m} \cdot \text{m}}{25 \, m} = \dfrac{2 \text{ m}^2}{1}$

b. $\dfrac{300 \text{ foot} \cdot \text{pounds}}{10 \text{ pounds}} = \dfrac{30 \cdot 10 \text{ foot} \cdot \text{pounds}}{10 \text{ pounds}} = 30 \text{ feet}$

Student Note: Engineers and mechanics use the foot-pound.

The foot-pound is a unit of measure resulting from the product of feet and pounds. It is preferable to use a dot to show the multiplication rather than the customary dash (foot-pound).

c. $\dfrac{30 \text{ degree} \cdot \text{days}}{6 \text{ days}} = \dfrac{5 \cdot 6 \text{ degree} \cdot \text{days}}{6 \text{ days}} = 5 \text{ degrees}$

The degree·day is used by public utilities to describe changes from day to day in power consumption.   ▬

Unit Analysis

Unit analysis is *a method for changing from one unit of measure to another or for changing from one rate to another.* A key idea in unit analysis is that facts such as 12 inches = 1 foot are formed into a fraction worth 1. The numerator and denominator represent different ways of expressing the same measure. That is,

$$\dfrac{12 \text{ inches}}{1 \text{ foot}} = 1, \qquad \dfrac{5280 \text{ feet}}{1 \text{ mile}} = 1, \qquad \text{and} \qquad \dfrac{60 \text{ seconds}}{1 \text{ minute}} = 1$$

UNIT TO UNIT   To set up a unit analysis, we arrange the facts into fractions so that each unit of measure appears once in a numerator and once in a denominator. Where possible, we use the four problem-solving steps: understand, plan, carry out the plan, and check.

EXAMPLE 3 Changing from one unit to another   Convert 10 miles into inches.

SOLUTION   *Understand*: The first step is to read the problem and identify the question. Here we start with miles and want to end with inches. We list the units-of-measure facts that relate the starting and ending units:

12 inches = 1 foot

5280 feet = 1 mile

Plan: We will start with 10 miles (marked in yellow) and arrange the facts as fractions so that other units are eliminated and only inches remain.

Carry out the plan:

$$\dfrac{10 \text{ mi}}{1} \cdot \dfrac{5280 \text{ ft}}{1 \text{ mi}} \cdot \dfrac{12 \text{ in.}}{1 \text{ ft}} = 633,600 \text{ in.}$$

Check: We must make sure that all facts are correctly written, the unwanted units are eliminated, and the answer is reasonable. As we might expect, we obtain a large number of inches in a mile.   ▬

▬ *If a ratio contains units of measure of the same type* (length, mass, capacity), *the units should be made the same before the ratio is simplified or is compared with another ratio.* The facts in Table 1 may be useful in simplifying ratios of units. More facts are listed on the inside cover.

TABLE 1   *Measurement Facts*

1000 milliliters = 1 liter	16 ounces = 1 pound
1000 grams = 1 kilogram	1 yard = 36 inches
100 centimeters = 1 meter	1 yard = 3 feet
1000 meters = 1 kilometer	1 mile = 5280 feet
	16 tablespoons = 1 cup
	4 cups = 1 quart
	4 quarts = 1 gallon

EXAMPLE 4   Simplifying ratios   Use unit analysis to eliminate units, and simplify.

a. 150 centimeters to 2 meters

b. 2500 grams to 1 kilogram

c. 2 tablespoons to 1 gallon

d. n inches to m feet

e. x cups to y quarts

SOLUTION

a. $\dfrac{150 \text{ cm}}{2 \text{ m}} \cdot \dfrac{1 \text{ m}}{100 \text{ cm}} = \dfrac{150}{200} = \dfrac{3}{4}$

b. $\dfrac{2500 \text{ g}}{1 \text{ kg}} \cdot \dfrac{1 \text{ kg}}{1000 \text{ g}} = \dfrac{2500}{1000} = \dfrac{5}{2}$

c. $\dfrac{2 \text{ tbsp}}{1 \text{ gal}} \cdot \dfrac{1 \text{ cup}}{16 \text{ tbsp}} \cdot \dfrac{1 \text{ qt}}{4 \text{ cups}} \cdot \dfrac{1 \text{ gal}}{4 \text{ qt}} = \dfrac{2}{256} = \dfrac{1}{128}$

d. $\dfrac{n \text{ in.}}{m \text{ ft}} \cdot \dfrac{1 \text{ ft}}{12 \text{ in.}} = \dfrac{n}{12m}$

e. $\dfrac{x \text{ cups}}{y \text{ qt}} \cdot \dfrac{1 \text{ qt}}{4 \text{ cups}} = \dfrac{x}{4y}$

UNIT ANALYSIS

1. Identify the units of measure to be changed, and identify the units needed in the answer.

2. List facts that contain the starting units of measure and the ending units. List facts needed to relate the starting and ending units.

3. Write the starting units. Using your list of facts, set up a product of fractions so that each unit of measure appears once in the numerator and once in the denominator.

4. Use $a/a = 1$ to eliminate the unwanted units of measure, and then calculate with the numbers.

Unit analysis is especially helpful for changing units in area and volume problems.

EXAMPLE 5   Changing from one unit to another   A rectangular floor is 13 feet by 18 feet. Find its area in square yards.

SOLUTION   *Understand*: The area is the product of width and length. Our unit fact is

3 feet = 1 yard

Plan: We multiply to find the area in square feet and use the fact that 3 feet = 1 yard twice to obtain square yards.

Carry out the plan:

$$\text{Area} = \frac{13 \,\cancel{ft}}{1} \cdot \frac{18 \,\cancel{ft}}{1} \cdot \frac{1 \text{ yd}}{3 \,\cancel{ft}} \cdot \frac{1 \text{ yd}}{3 \,\cancel{ft}} = \frac{13 \cdot \overset{2}{\cancel{18}} \text{ yd}^2}{\underset{1}{\cancel{9}}} = 26 \text{ yd}^2$$

Check: We can change the room size to yards first and then find the area.

$$13 \text{ ft} \cdot \frac{1 \text{ yd}}{3 \text{ ft}} = \frac{13}{3} \text{ yd}$$

$$18 \text{ ft} \cdot \frac{1 \text{ yd}}{3 \text{ ft}} = 6 \text{ yd}$$

$$\frac{13}{\underset{1}{\cancel{3}}} \text{ yd} \cdot \frac{\overset{2}{\cancel{6}}}{1} \text{ yd} = 26 \text{ yd}^2$$

∎

If you change units in your head, the method suggested here can help you put your mental process into written form.

RATE TO RATE   A **rate** is *a comparison of a quantity of one unit to a quantity of another unit*. In Example 6, we change from one rate to another. Changing rates usually means changing two units at a time.

EXAMPLE 6   **Changing rates: snail's pace**   Convert a snail's pace, 6 inches per minute, to miles per hour.

SOLUTION   *Understand*: Here we start with inches per minute and want to end with miles per hour. We need to change inches to miles and minutes to hours. The facts are

$$12 \text{ inches} = 1 \text{ foot}$$

$$5280 \text{ feet} = 1 \text{ mile}$$

$$60 \text{ minutes} = 1 \text{ hour}$$

Plan: We start with a fraction containing inches over minutes and arrange the facts as fractions so that other units are eliminated and only miles and hours remain.

Carry out the plan:

$$\frac{6 \,\cancel{in.}}{1 \,\cancel{min}} \cdot \frac{1 \,\cancel{ft}}{12 \,\cancel{in.}} \cdot \frac{1 \text{ mi}}{5280 \,\cancel{ft}} \cdot \frac{60 \,\cancel{min}}{1 \text{ hr}} = \frac{6 \cdot 1 \cdot 1 \cdot 60}{1 \cdot 12 \cdot 5280 \cdot 1} \, \frac{\text{mi}}{\text{hr}}$$

$$\approx 0.00568 \text{ mile per hour}$$

Check: We must make sure that all facts are correctly written, the unwanted units are eliminated, and the answer is reasonable. As we might expect, the snail's pace is slower than 1 mile per hour, so the answer is reasonable.   ∎

When changing both units in a rate, it is convenient to put all fractions that change the denominator units to the left of the given rate and all fractions that change the numerator units to the right. This placement of the fractions puts like units close together. In Example 6, we might have written

$$\frac{60 \text{ min}}{1 \text{ hr}} \cdot \frac{6 \text{ in.}}{1 \text{ min}} \cdot \frac{1 \text{ ft}}{12 \text{ in.}} \cdot \frac{1 \text{ mi}}{5280 \text{ ft}}$$

Think about it 2: Why can we arrange the unit fractions in any order?

PROBLEM SOLVING   We apply the unit analysis approach to problems with units by setting up rates or facts as fractions, as shown in Example 7.

EXAMPLE 7   Solving problems with units: volume of shower water   An energy-efficient showerhead permits a flow of 2 gallons per minute. How many gallons of water are used in a 30-day month by a person taking a 5-minute shower every other day?

SOLUTION   *Understand*: We start by listing key phrases describing the facts:

2 gallons per minute

30 days per month

5 minutes per shower

1 shower every 2 days

Plan: We are looking for gallons per month, so we start with a fraction containing gallons on the top and look for facts that eliminate all other units except months.

Carry out the plan:

$$\frac{2 \text{ gallons}}{1 \text{ minute}} \cdot \frac{5 \text{ minutes}}{\text{shower}} \cdot \frac{1 \text{ shower}}{2 \text{ days}} \cdot \frac{30 \text{ days}}{1 \text{ month}} = \frac{2(5)(30)}{2} \frac{\text{gallons}}{\text{month}}$$

$$= 150 \text{ gallons per month}$$

Check: We must confirm that all facts are correctly written and that unwanted units are eliminated.   ▬

Proportions

The main floor of the house in Figure 3 is 70 inches above sidewalk level. A slope ratio of 1 foot to 8 feet makes a steep but adequate wheelchair ramp (see Figure 4). We need proportions to find the number of horizontal feet needed for the ramp.

FIGURE 3

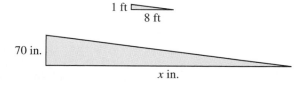

FIGURE 4

Two equal ratios form a **proportion**

$$\frac{a}{b} = \frac{c}{d}, \quad b \neq 0, d \neq 0$$

To solve for a variable in a proportion, we can multiply by the product of the denominators, *bd*.

In Example 8, look for phrases that give the relevant facts and then rewrite the phrases into a proportion.

EXAMPLE 8   Using a proportion: wheelchair ramp   How many horizontal feet of ramp are needed to rise 70 inches above sidewalk level? Use a slope ratio of 1 foot to 8 feet.

SOLUTION   *Phrases*: The slope ratio of rise to run is 1 to 8.
        The ramp ratio is 70 inches to *x* inches.

Proportion: We set up a proportion based on the slope, rise over run:

$$\frac{1}{8} = \frac{70 \text{ in.}}{x} \qquad \text{Multiply both sides by } 8x.$$

$$\frac{8x}{8} = \frac{70 \text{ in.} \cdot 8x}{x} \qquad \text{Simplify.}$$

$$1x = 560 \text{ in.}$$

We next change from inches to feet:

$$x = 560 \text{ in.} \cdot \frac{1 \text{ ft}}{12 \text{ in.}} = 46\tfrac{2}{3} \text{ ft}$$

The ramp must run nearly 47 feet around the house in order to rise from the sidewalk to the main floor.   ▬

The triangles in Figure 4 representing ramp slope are similar triangles.

SIMILAR TRIANGLES   **Similar triangles** *have the same shape but different sizes.* Figure 5 shows two similar triangles.

The similar triangles in Figure 5 are arranged to show how the sides of the smaller triangle and the sides of the larger triangle compare. When triangles are arranged in this way, we say that the corresponding sides are lined up. **Corresponding sides** are *the sides that are in the same position compared to the other parts of the triangles.*

Corresponding angles are *the angles that are in the same position compared to the other parts of the triangles.* As we change the size of similar triangles, the angles stay the same size.

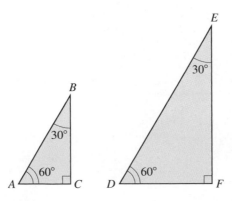

FIGURE 5

Similar triangles have corresponding angles that are equal.

EXAMPLE 9   Finding corresponding angles and sides   Identify the corresponding angles and sides in Figure 5.

SOLUTION   The equal angles in the two triangles are in the same position, so the corresponding angles are A and D, at 60°; B and E, at 30°; and C and F, at 90°. Side AB corresponds to side DE, side AC to DF, and side BC to EF.   ◼

   ▨ As we enlarge a triangle by, say, doubling one side, the other sides grow proportionately and also double.

Corresponding sides of similar triangles are proportional.

EXAMPLE 10   Writing proportions for sides of similar triangles   Write three different proportions showing that the ratios of the sides of the similar triangles in Figure 6 are the same.

SOLUTION

$$\frac{4}{8} = \frac{10}{20} \qquad \text{Both ratios simplify to } \tfrac{1}{2}.$$

$$\frac{4}{8} = \frac{12}{24} \qquad \text{Both ratios simplify to } \tfrac{1}{2}.$$

$$\frac{20}{10} = \frac{24}{12} \qquad \text{Both ratios simplify to } \tfrac{2}{1}.$$

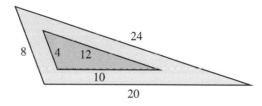

FIGURE 6

Think about it 3: What other proportions can be written from the triangles in Figure 6?

EXAMPLE 11   Applying proportions to similar triangles   Use proportions to find the lengths of the indicated sides in Figure 7.

SOLUTION   To find side z in Figure 7, we use the proportion

$$\frac{5\sqrt{2}}{5} = \frac{z}{3}$$

We solve the proportion by multiplying:

$$\frac{5\sqrt{2}}{5} = \frac{z}{3} \qquad \text{Multiply by } 5 \cdot 3.$$

$$5 \cdot 3 \cdot \frac{5\sqrt{2}}{5} = \frac{z}{3} \cdot 5 \cdot 3 \qquad \text{Simplify.}$$

$$3 \cdot 5\sqrt{2} = 5 \cdot z \qquad \text{Divide by 5.}$$

$$3\sqrt{2} = z$$

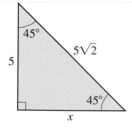

FIGURE 7

Both side x and side y are missing. We may use the Pythagorean theorem to find x and then a proportion to find y. To find x, we use

$a^2 + b^2 = c^2$ Substitute the lengths of the sides.

$x^2 + 5^2 = (5\sqrt{2})^2$ Simplify square root expressions.

$x^2 + 25 = 50$ Subtract 25 from both sides.

$x^2 = 25$ Take the square root.

$x = 5$

To find y, we use the proportion

$$\frac{5}{5} = \frac{3}{y}$$

Thus, $y = 3$.

Think about it 4: From Example 11, what geometry fact can we guess about the lengths of sides opposite equal angles in a triangle?

OVERLAPPING SIMILAR TRIANGLES   In many situations, similar triangles overlap. In Figure 8, the light from a streetlamp casts a shadow, b_2, beyond a person of height h_2. The triangle with the shadow and person as sides is similar to the triangle with the ground, b_1, and the streetlamp, h_1, as sides.

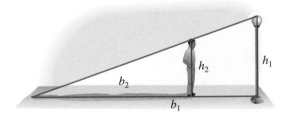

FIGURE 8

EXAMPLE 12 Identifying proportions in overlapping triangles: shadows   Describe the relationship among sides b_1, b_2, h_1, and h_2 in Figure 8. Write a sentence and a proportion.

SOLUTION   The bases b_1 and b_2 and heights h_1 and h_2 are corresponding sides of similar triangles and are proportional. Thus,

$$\frac{h_1}{b_1} = \frac{h_2}{b_2}$$

EXAMPLE 13 Identifying proportions in overlapping triangles: more shadows   Use a proportion to find the height h of the streetlamp shown in Figure 9.

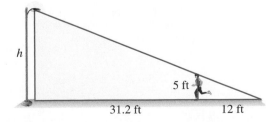

FIGURE 9

SOLUTION In this figure, the similar triangles overlap. The base of the large triangle is 43.2 ft. We build a proportion with the ratios of height to base:

$$\frac{h}{43.2 \text{ ft}} = \frac{5 \text{ ft}}{12 \text{ ft}}$$ Multiply by (12 ft · 43.2 ft).

$$12 \text{ ft} \cdot 43.2 \text{ ft} \cdot \frac{h}{43.2 \text{ ft}} = \frac{5}{12 \text{ ft}} \cdot 43.2 \text{ ft} \cdot 12 \text{ ft}$$ Simplify.

$$12h \text{ ft} = 5(43.2) \text{ ft}^2$$ Divide by 12 ft.

$$h = 18 \text{ ft}$$

The streetlamp is 18 feet in height.

ANSWER BOX

Warm-up: **1.** $\frac{4}{5}$ **2.** $\frac{5}{3}$ **3.** $\frac{7}{4}$ **4.** $\frac{8}{5}$ **5.** $\frac{3}{4}$ **6.** $\frac{3}{14}$ **Think about it 1:** A sports team may have a 5 to 0 ratio of wins to losses. A score in soccer might be 5 to 0. **Example 1:** $\frac{22}{7} \approx 3.142\,857$ and $\frac{2218}{706} \approx 3.141\,643$. The ratios do not have the same decimal value, so they are not equal. Both ratios may be used to approximate the constant pi (π), but neither is exactly pi. Rounded to fourteen decimal places, pi is 3.141 592 653 589 79. **Think about it 2:** The commutative property of multiplication allows us to multiply in any order. **Think about it 3:** Some other proportions are $\frac{8}{4} = \frac{20}{10}$, $\frac{24}{12} = \frac{8}{4}$, and $\frac{10}{20} = \frac{12}{24}$. **Think about it 4:** The lengths of sides that are opposite equal angles in a triangle are equal.

7.1 Exercises

Simplify the expressions in Exercises 1 to 8.

1. $\dfrac{186 \text{ cm}^2}{6 \text{ cm}}$ 31 cm

2. $\dfrac{125 \text{ ft}^3}{25 \text{ ft}}$ 5 ft²

3. $\dfrac{36 \text{ in.}}{1728 \text{ in}^3}$ $\frac{1}{48 \text{ in}^2}$

4. $\dfrac{54 \text{ yd}^3}{9 \text{ yd}}$ 6 yd²

5. $\dfrac{1500 \text{ foot} \cdot \text{pounds}}{25 \text{ feet}}$ 60 lb

6. $\dfrac{1024 \text{ degree} \cdot \text{gallons}}{128 \text{ degrees}}$ 8 gal

7. $\dfrac{1200 \text{ kilowatt hours}}{24 \text{ hours}}$ 50 kw

8. $\dfrac{4800 \text{ kilowatt hours}}{60 \text{ kilowatts}}$ 80 hr

In Exercises 9 to 18, arrange the listed facts into a unit analysis and solve the problem.

9. 300 milliliters is how many liters?
1 liter = 1000 milliliters $300 \text{ mL} \cdot \dfrac{1 \text{ L}}{1000 \text{ mL}} = 0.3 \text{ L}$

10. 160 pounds is how many kilograms?
2.2 pounds is 1 kilogram. $160 \text{ lb} \cdot \dfrac{1 \text{ kg}}{2.2 \text{ lb}} \approx 72.7 \text{ kg}$

11. 25 feet is how many meters?
1 foot = 12 inches
1 meter is 39.37 inches. $25 \text{ ft} \cdot \dfrac{12 \text{ in.}}{1 \text{ ft}} \cdot \dfrac{1 \text{ m}}{39.37 \text{ in.}} \approx 7.62 \text{ m}$

12. 55 gallons is how many pints?
1 gallon = 4 quarts
2 pints = 1 quart $55 \text{ gal} \cdot \dfrac{4 \text{ qt}}{1 \text{ gal}} \cdot \dfrac{2 \text{ pt}}{1 \text{ qt}} = 440 \text{ pt}$

13. 150 cubic feet is how many cubic yards?
1 yard = 3 feet $150 \text{ ft}^3 \cdot \left(\dfrac{1 \text{ yd}}{3 \text{ ft}}\right)^3 \approx 5.6 \text{ yd}^3$

14. 48 square inches is how many square feet?
1 foot = 12 inches $48 \text{ in}^2 \cdot \left(\dfrac{1 \text{ ft}}{12 \text{ in.}}\right)^2 \approx \dfrac{1}{3} \text{ ft}^2$

15. 100 cubic inches is how many cubic feet?
12 inches = 1 foot $100 \text{ in}^3 \cdot \left(\dfrac{1 \text{ ft}}{12 \text{ in.}}\right)^3 \approx 0.058 \text{ ft}^3$

16. 1600 square centimeters is how many square meters?
1 meter = 100 centimeters $1600 \text{ cm}^2 \cdot \left(\dfrac{1 \text{ m}}{100 \text{ cm}}\right)^2 = 0.16 \text{ m}^2$

17. 200 milliliters of water is how many grams?
1 kilogram is 1000 milliliters of water.
1000 grams = 1 kilogram $200 \text{ mL} \cdot \dfrac{1 \text{ kg}}{1000 \text{ mL}} \cdot \dfrac{1000 \text{ g}}{1 \text{ kg}} = 200 \text{ g}$

18. 300 grams of water is how many milliliters? See the facts in Exercise 17. $300 \text{ g} \cdot \dfrac{1 \text{ kg}}{1000 \text{ g}} \cdot \dfrac{1000 \text{ mL}}{1 \text{ kg}} = 300 \text{ mL}$

Simplify the expressions in Exercises 19 to 22.

19. a. $\dfrac{\frac{1}{2} \text{ foot}}{2 \text{ inches}}$ $\frac{3}{1}$ **b.** $\dfrac{3000 \text{ grams}}{6 \text{ kilograms}}$ $\frac{1}{2}$ **c.** $\dfrac{32 \text{ ounces}}{6 \text{ pounds}}$ $\frac{1}{3}$

20. a. $\dfrac{1 \text{ foot}}{4 \text{ inches}}$ $\frac{3}{1}$ **b.** $\dfrac{2 \text{ meters}}{150 \text{ centimeters}}$ $\frac{4}{3}$ **c.** $\dfrac{1500 \text{ meters}}{1 \text{ kilometer}}$ $\frac{3}{2}$

21. a. $\dfrac{300 \text{ milliliters}}{30 \text{ liters}}$ $\frac{1}{100}$ **b.** $\dfrac{2 \text{ years}}{180 \text{ months}}$ $\frac{2}{15}$ **c.** $\dfrac{40 \text{ minutes}}{\frac{1}{4} \text{ hour}}$ $\frac{8}{3}$

22. a. $\dfrac{2 \text{ liters}}{300 \text{ milliliters}}$ $\frac{20}{3}$ **b.** $\dfrac{4 \text{ years}}{150 \text{ months}}$ $\frac{8}{25}$ **c.** $\dfrac{12 \text{ minutes}}{\frac{1}{2} \text{ hour}}$ $\frac{2}{5}$

In Exercises 23 to 28, arrange the facts into a unit analysis and solve the problem.

23. 55 feet per second is how many miles per hour?

1 mile = 5280 feet
1 minute = 60 seconds
1 hour = 60 minutes

$\dfrac{55 \text{ ft}}{1 \text{ sec}} \cdot \dfrac{1 \text{ mi}}{5280 \text{ ft}} \cdot \dfrac{60 \text{ sec}}{1 \text{ min}} \cdot \dfrac{60 \text{ min}}{1 \text{ hr}} =$

37.5 mi/hr or 37.5 mph

24. 40 miles per hour is how many feet per second?

1 mile = 5280 feet
1 minute = 60 seconds
1 hour = 60 minutes

$\dfrac{40 \text{ mi}}{1 \text{ hr}} \cdot \dfrac{5280 \text{ ft}}{1 \text{ mi}} \cdot \dfrac{1 \text{ hr}}{60 \text{ min}} \cdot \dfrac{1 \text{ min}}{60 \text{ sec}} \approx$

59 ft/sec

25. One gallon for five miles is how many dollars per day of driving?

1 hour to travel 55 miles
1 gallon is $1.35.
1 driving day is 10 hours.

$\dfrac{1 \text{ gal}}{5 \text{ mi}} \cdot \dfrac{55 \text{ mi}}{1 \text{ hr}} \cdot \dfrac{\$1.35}{1 \text{ gal}} \cdot \dfrac{10 \text{ hr}}{1 \text{ day}} =$

$148.50/day

26. 240 milliliters in 12 hours is how many microdrops per minute?

60 microdrops = 1 milliliter
1 hour = 60 minutes

$\dfrac{240 \text{ mL}}{12 \text{ hr}} \cdot \dfrac{60 \text{ microdrops}}{1 \text{ mL}} \cdot \dfrac{1 \text{ hr}}{60 \text{ min}} =$

20 microdrops/min

27. Prescription dosage for young children may be based on their age relative to 150 months, as 150 months is considered "adult" for many prescriptions. If the adult dosage is 500 milligrams, how many milligrams should a 1-year-old infant receive?

1 year = 12 months $1 \text{ yr} \cdot \dfrac{12 \text{ mo}}{1 \text{ yr}} \cdot \dfrac{\text{adult}}{150 \text{ mo}} \cdot \dfrac{500 \text{ mg}}{\text{adult}} = 40 \text{ mg}$

28. Suppose that when a bale of peat moss is opened and loosened, it expands to three times its original volume. The bale originally contains 4 cubic feet of peat moss. How many bales of peat moss will be needed to cover to a depth of 2 inches a garden plot that is 15 feet by 40 feet?

12 inches = 1 foot $\dfrac{15 \text{ ft}}{1} \cdot \dfrac{40 \text{ ft}}{1} \cdot \dfrac{2 \text{ in.}}{1} \cdot \dfrac{1 \text{ ft}}{12 \text{ in.}} \cdot \dfrac{1 \text{ bale}}{4 \text{ ft}^3 \cdot 3} =$

$8\frac{1}{3}$ bales

Solve the proportions in Exercises 29 and 30. Round to the nearest tenth.

29. a. $\dfrac{3}{x} = \dfrac{5}{14}$ **b.** $\dfrac{4}{25} = \dfrac{x}{15}$ **c.** $\dfrac{7}{24} = \dfrac{16}{x}$

    $x = 8.4$ $x = 2.4$ $x = 54.9$

30. a. $\dfrac{4}{13} = \dfrac{x}{7}$ **b.** $\dfrac{x}{15} = \dfrac{8}{35}$ **c.** $\dfrac{24}{x} = \dfrac{32}{9}$

    $x = 2.2$ $x = 3.4$ $x = 6.75$

31. An access ramp is to have a rise-to-run ratio of 1 to 12. How long a horizontal distance (in feet) is needed for the ramp to rise 15 inches?   15 ft

32. An access ramp covers a horizontal distance of 60 feet. Its slope is 1 to 12. What is its vertical rise?   5 ft

In Exercises 33 and 34, find the missing sides for the similar triangles shown.

33.

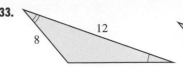
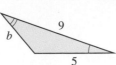

$a = 6\frac{2}{3}, b = 6$

34.

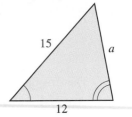

$a = 10, b = 6.4$

In Exercises 35 to 40, set up proportions to find the lengths of the sides marked with a letter in the figure.

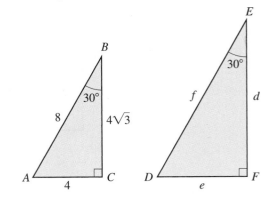

35. Let $e = 6$.
$d = 6\sqrt{3}, f = 12$

36. Let $f = 10$.
$d = 5\sqrt{3}, e = 5$

37. Let $d = 10$.
$e = 10\sqrt{3}/3, f = 20\sqrt{3}/3$

38. Let $d = 8$.
$e = 8\sqrt{3}/3, f = 16\sqrt{3}/3$

39. Let $e = \sqrt{5}$.
$d = \sqrt{15}, f = 2\sqrt{5}$

40. Let $f = 2\sqrt{3}$.
$d = 3, e = \sqrt{3}$

In Exercises 41 and 42, set up proportions and/or apply the Pythagorean theorem to find the lengths of the sides marked with a letter.

41.

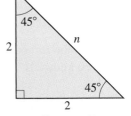
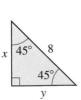

$n = 2\sqrt{2}, x = 4\sqrt{2}, y = 4\sqrt{2}$

42.

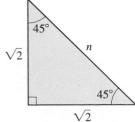

$n = 2, x = \dfrac{5\sqrt{2}}{2}, y = \dfrac{5\sqrt{2}}{2}$

The ratios of base to height for similar triangles are equal. With the help of the figure below, set up proportions to describe the situations in Exercises 43 to 46 and answer the questions. Round your answers to the nearest tenth.

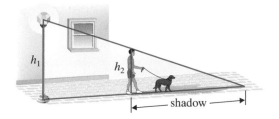

43. A 25-foot streetlight casts light past a 5-foot-tall person, causing a shadow. The person is standing 18 feet from the base of the streetlight. How long is the shadow?
4.5 ft

44. A small tree has been planted 10 feet from a streetlight. The streetlight is 20 feet tall. The light creates a tree shadow 16 feet long. What is the height of the tree?
12.3 ft

45. A gorilla is walking away from a streetlight. At the instant the gorilla is 12 feet from the base of the streetlight, the gorilla's shadow is 5 feet long. The streetlight is 20 feet tall. What is the height of the gorilla?
5.9 ft

46. A 5-foot-tall parking meter is located 16 feet from the base of a 20-foot streetlight. How long is the parking meter's shadow?   5.3 ft

■ Error Analysis

In Exercises 47 to 50, explain what has been done wrong in each solution.

47. Simplify the ratio 6 ft/2 yd:

$$\frac{6\text{ ft}}{2\text{ yd}} = \frac{3}{1}$$

unit conversion needed

48. Simplify the ratio 3 qt/2 gal:

$$\frac{3\text{ qt}}{2\text{ gal}} = \frac{3\text{ qt}}{2\text{ gal}} \cdot \frac{4\text{ qt}}{1\text{ gal}} = \frac{12}{2} = \frac{6}{1}$$

unit conversion upside down

49. Solve the proportion $\frac{4}{15} = \frac{6}{x}$:

$$\frac{4}{15} = \frac{6}{x}, \quad 4x = 6$$

did not multiply each side by 15

50. Solve the proportion $\frac{4}{15} = \frac{6}{x}$:

$$\frac{4}{15} = \frac{6}{x}, \quad \frac{90}{4x}$$ wrote multiplication results as
fraction instead of equation

■ Projects

51. Oil Spill The 1989 Exxon Valdez oil spill in Alaska released 11 million gallons of oil into Prince William Sound. Over how many square miles would this much oil spread if it were uniformly the thickness of a sheet of 20-pound photocopy paper? (Useful facts: 1 gallon is 231 cubic inches, 500 sheets of 20-pound photocopy paper is 2 inches thick, 1 mile is 5280 feet.)
≈158 square miles

52. Astronomy and the Small Angle Equation The small angle equation is

$$\frac{\alpha}{206{,}265\text{ sec}} = \frac{d}{D}$$

where α is an arc angle measured in seconds (see the figure). One degree of arc angle is 60 minutes of arc, or 3600 seconds of arc. The distance from the viewer to the object is D. The diameter of the object is d.

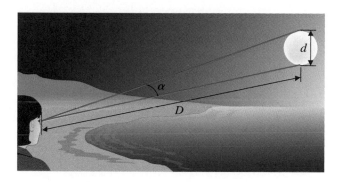

a. The moon has a diameter of approximately 3480 kilometers and is at a distance of about 384,000 kilometers from Earth. Find α, the viewing angle at which we see the moon.
1869.277 sec ~ 31.155 min ~ 0.519°

b. Design a tool for sighting objects and estimating the arc angle.

c. Use your tool to estimate the arc angle for five distant objects of known size (such as a parked compact car 16 feet long). Use the formula to predict how far away the car is.

d. The small angle equation comes from a proportion based on the ratio between the 360 degrees in a circle and the circumference of a circle, $2\pi r$. In this situation, the radius of a circle is D, so the circumference is $2\pi D$ (see the figure).

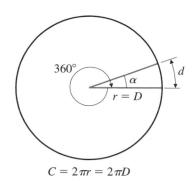

$$C = 2\pi r = 2\pi D$$

The small angle equation may be found by starting with

$$\frac{360°}{\text{circumference}} = \frac{\text{angle in seconds}}{\text{length of distant object}}$$

$$\frac{360°}{2\pi D} = \frac{\alpha \text{ seconds}}{d}$$

Show the steps needed to rearrange the above equation into the same form as the small angle equation. It is necessary to apply unit analysis to change degrees into seconds and to obtain the constant 206,265.

7.2   Proportions and Direct Variation

Objectives

■ Find when linear data represent a direct variation.

■ Translate word expressions into linear, quadratic, or joint variation.

WARM-UP

Assume these sets of ordered pairs make linear functions. Find the slope of each function. Label the slope with appropriate units.

1. (2 cans, $0.98) and (15 cans, $7.35) $0.49 per can

2. (5 minutes, $2.10) and (14 minutes, $4.44) $0.26 per minute

3. (12 feet, 12 hours) and (32 feet, 17 hours) $\frac{1}{4}$ hour per foot

4. (15 minutes, $12.45) and (8 minutes, $6.64) $0.83 per minute

---

IN THIS SECTION, we use proportions and linear equations to examine the concept of direct variation. We discuss why proportions cannot be applied to all linear situations, and we consider quadratic and joint variation. The emphasis is on vocabulary and new ways of describing familiar relationships.

Proportions and Linear Functions

You should not assume that all linear functions represent proportional data. For linear data to be proportional, the ratio of the output to the input, y/x, for each data pair must be the same.

EXAMPLE 1   **Identifying proportional data**   Which of these linear data sets represent proportional relationships?

a. Two cans of refried beans cost 98 cents. Fifteen cans of refried beans cost $7.35.

b. One Friday, a long-distance call costs $2.10 for 5 minutes. Another Friday at the same time, a call to the same place costs $4.44 for 14 minutes.

SOLUTION   **a.** $\dfrac{\$0.98}{2 \text{ cans}} = \$0.49 \text{ per can}$; $\dfrac{\$7.35}{15 \text{ cans}} = \$0.49 \text{ per can}$

The data form equal ratios. A proportion is appropriate.

b. $\dfrac{\$2.10}{5 \text{ min}} = \$0.42 \text{ per minute}$; $\dfrac{\$4.44}{14 \text{ min}} \approx \$0.32 \text{ per minute}$

The data do not form equal ratios.

We may identify proportionality by graphing linear data.

EXAMPLE 2 Identifying proportional data with a graph   The data from parts a and b of Example 1 are graphed in Figures 10 and 11. Use the answers from Example 1 to tell which graph shows proportional linear data, and then identify the difference between the two graphs.

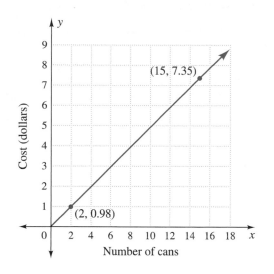

FIGURE 10

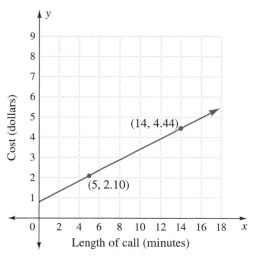

FIGURE 11

SOLUTION   Part a of Example 1 contains sets of proportional data. The graph in Figure 10 passes through the origin.

Part b does not deal with proportional data. The graph in Figure 11 does not pass through the origin.

EXAMPLE 3 Identifying proportional data with an equation   Find the linear equations for parts a and b of Example 1.

SOLUTION   **a.** For (2 cans, $0.98) and (15 cans, $7.35), the slope is

$$\frac{y_2 - y_1}{x_2 - x_1} = \frac{7.35 - 0.98}{15 - 2} = \frac{6.37}{13} = \$0.49 \text{ per can}$$

The slope of the line is the same as the equal ratios of the data. The vertical axis intercept is

$$b = y_1 - mx_1 = 0.98 - 0.49(2) = \$0$$

The linear equation is $y = 0.49x + 0$. The equation has a zero constant term. Its graph passes through the origin. The data are proportional.

b. For (5 minutes, $2.10) and (14 minutes, $4.44), the slope is

$$\frac{y_2 - y_1}{x_2 - x_1} = \frac{4.44 - 2.10}{14 - 5} = \frac{2.34}{9} = \$0.26 \text{ per minute}$$

The slope of the line is different from either of the unequal ratios of the data. The vertical axis intercept is

$$b = y_1 - mx_1 = 2.10 - 0.26(5) = \$0.80$$

The linear equation is $y = 0.26x + 0.80$. The equation has an $0.80 initial charge. Its graph does not pass through the origin. The data are not proportional.

Direct Variation and Linear Variation

In Examples 1 to 3, we saw that the cost of the refried beans is proportional to the number of cans purchased. Buying zero cans cost nothing. The equation $y = 0.49x$ has no constant term, and its graph passes through the origin. To summarize, we say, "The cost *varies directly* with the number purchased."

Direct variation occurs *when the ratio of outputs to inputs is constant*:

$$\frac{y_1}{x_1} = \frac{y_2}{x_2}$$

Direct variation applies to more than just linear equations (see the discussions of quadratic variation and joint variation later in this section). To restrict the variation to linear functions, we say, "The cost *varies linearly* with the number purchased."

■■■ LINEAR VARIATION

> If there is a constant k such that $y = kx$, we say y *varies linearly as x*. The ratio of outputs to inputs is k for all ordered pairs, and the line connecting the data points passes through the origin.

For linear data, linear variation occurs when the data are proportional.

EXAMPLE 4 Identifying linear variation   Using ratios, graphs, and equations, show which of these linear data sets represent linear variation (that is, are proportional).

a. Repairing a 12-foot length of sidewalk takes 12 hours. The same contractor takes 17 hours to repair a 32-foot length of sidewalk.

b. One 15-minute phone call costs $12.45. Another call to the same place at the same time of day costs $6.64 for 8 minutes.

SOLUTION   **a.** $\dfrac{12 \text{ hr}}{12 \text{ ft}} = 1$ hour per foot; $\dfrac{17 \text{ hr}}{32 \text{ ft}} \approx 0.53$ hour per foot

The data do not form equal ratios.

   The graph in Figure 12 does not pass through the origin. The slope is

$$\frac{y_2 - y_1}{x_2 - x_1} = \frac{17 - 12}{32 - 12} = \frac{5}{20} = \frac{1}{4} \text{ hour per foot}$$

The vertical axis intercept is

$$b = y_1 - mx_1 = 12 - \tfrac{1}{4}(12) = 9 \text{ hr}$$

   The linear equation is $y = \tfrac{1}{4}x + 9$. The equation has a 9-hour fixed time regardless of the length of the repair. The data are not proportional. The data do not vary linearly.

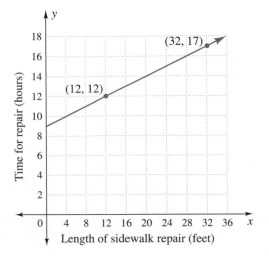

FIGURE 12

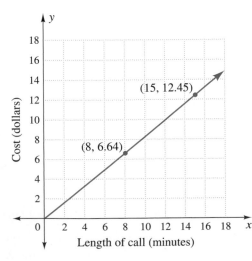

FIGURE 13

b. $\dfrac{\$12.45}{15 \text{ min}} = \$0.83$ per minute; $\dfrac{\$6.64}{8 \text{ min}} = \$0.83$ per minute

The data form equal ratios.

The graph in Figure 13 passes through the origin. The slope is

$$\frac{y_2 - y_1}{x_2 - x_1} = \frac{12.45 - 6.64}{15 - 8} = \frac{5.81}{7} = 0.83 \text{ dollar per minute}$$

The slope is the same as the equal ratios of the data. The vertical axis intercept is

$$b = y_1 - mx_1 = 12.45 - 0.83(15) = \$0$$

The linear equation is $y = 0.83x$. The equation has a zero constant term. There is no initial charge for placing the call. The data are proportional. The data vary linearly. ▬

Think about it: How is the ratio of outputs to inputs, y/x, different from the slope ratio? When a linear graph passes through the origin, why is the ratio of outputs to inputs, y/x, the same as the slope?

▬ There are two common ways of writing linear variation. The *ratio form* is $y/x = k$. The *function form* is $f(x) = kx$ or $y = kx$. The *constant ratio*, y/x, is known as the **constant of variation** (or **constant of proportionality**).

EXAMPLE 5 **Describing variation**   Describe the situation as a linear variation, and identify the constant of variation.
a. The circumference of a circle is pi times the diameter.
b. A can of corn costs $0.69. The output is the total cost of x cans.

SOLUTION **a.** $C = \pi d$; C varies linearly with d. The constant of variation is π.

b. Cost varies linearly with the number of cans purchased. The constant of variation is $0.69 per can. ▬

In many applications, the constant of variation has a special name. In physics, the amount of stretch in a spring is described by the spring constant. In engineering work with forces, k is the coefficient of friction. In temperature properties of materials science, k is the coefficient of thermal expansion.

The constant of variation in $y = kx$ is identified by the letter k, not m, for two reasons:

1. Only linear functions based on proportional data vary directly.

2. The constant of variation also appears in direct variation for nonlinear functions. Nonlinear functions exhibiting direct variation include $f(x) = kx^2$ and $f(x) = k\sqrt{x}$.

Quadratic Variation

The function $f(x) = kx^2$ describes quadratic variation. In **quadratic variation**, the output varies with the square of the input, so *the ratio of the output and the square of the input is constant*, $y/x^2 = k$.

▬ **QUADRATIC VARIATION**

> If there is a constant k such that $y = kx^2$, we say y *varies with the square of* x. The constant k is the constant of variation.

The area of a circle provides one example of quadratic variation. The area varies with the square of the diameter, $A = (\pi/4)d^2$.

EXAMPLE 6 **Applying quadratic variation: area of a circle**
a. The diameter of one automobile engine cylinder bore is 83 millimeters. Find the area of the circle associated with this diameter.

b. The diameter of the head of a pin is $\frac{1}{8}$ inch. Find the area of the circle associated with this diameter.

c. What is the constant of variation?

SOLUTION   **a.** $A = (\pi/4)d^2 \approx 0.7854(83 \text{ mm})^2 \approx 5411 \text{ mm}^2$

b. $A = (\pi/4)d^2 \approx 0.7854(0.125 \text{ in.})^2 \approx 0.0123 \text{ in}^2$

c. $\dfrac{\pi}{4}$ is the constant of variation.   ▬

The motion of a falling object also behaves with quadratic variation.

EXAMPLE 7   Applying quadratic variation: distance an object falls   The distance an object falls in t seconds is $d = \frac{1}{2} gt^2$. The constant g is the acceleration due to gravity, $g \approx 32.2 \text{ ft/sec}^2$.

a. Write a sentence describing the variation. Name the constant of variation.

b. Make a table for d, the distance fallen, with t as input. Let t be on the interval 0 to 8 seconds.

c. Graph the data from the table.

SOLUTION   **a.** The distance fallen varies with the square of the time in motion. The constant of variation is $\frac{1}{2} g$.

b. *Distance Fallen*

Time, t (seconds)	Distance, d (feet)
0	0
1	16.1
2	64.4
3	144.9
4	257.6
5	402.5
6	579.6
7	788.9
8	1030.4

c.

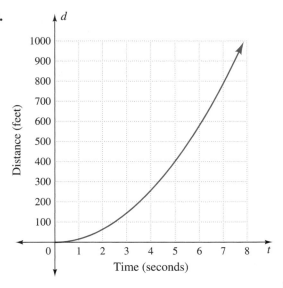

Quadratic variation is limited to quadratic equations without linear and constant terms.

EXAMPLE 8   Identifying quadratic variation   Which of these equations describe quadratic variation?

a. Vertical motion equation: $h = -\frac{1}{2} gt^2 + v_0 t + h_0$

b. Standard form of a quadratic equation: $y = ax^2 + bx + c$

c. Area of a square of side x: $A = x^2$

d. $y = (x + 3)^2$

SOLUTION   **a.** Because the equation contains the terms $v_0 t$ and h_0, it is not of the form $y = kx^2$.

b. Because the equation contains the terms bx and c, it is not of the form $y = kx^2$.

c. Quadratic variation with $k = 1$

d. Because the equation expands to $y = x^2 + 6x + 9$, it is not of the form $y = kx^2$.   ▬

VARIATION AND SYSTEMS OF EQUATIONS   In some settings, we need to know only the type of variation in order to compare outcomes. Examples 9 and 10 show that *we can eliminate k using systems of equations without knowing or finding the constant of variation.*

EXAMPLE 9   **Comparing data using systems of equations**   The area of a circle varies with the square of the radius. Compare the area of an 8-inch-radius circle with the area of a 4-inch-radius circle.

SOLUTION   ***Plan***: We need to compare the area, y_1, for the 8-inch radius with the area, y_2, for the 4-inch radius.

Carry out the plan:

$y_1 = kx_1^2$ and $y_2 = kx_2^2$	Write equations for two data points.
$y_1 = k(8)^2$ and $y_2 = k(4)^2$	Substitute known data.
$y_1 = k \cdot 64$ and $\dfrac{y_2}{16} = k$	Solve the second equation for k and substitute for k in the first equation.
$y_1 = \dfrac{y_2}{16} \cdot 64$	Simplify.
$y_1 = 4y_2$	

The area of the 8-inch-radius circle is four times the area of the 4-inch-radius circle. ▬

EXAMPLE 10   **Comparing data using systems of equations**   The distance an object falls varies with the square of the time in motion. Compare the distance fallen in 30 seconds with the distance fallen in 20 seconds.

SOLUTION   ***Plan***: We need to compare the distance, y_1, for 30-second motion with the distance, y_2, for 20-second motion.

Carry out the plan:

$y_1 = kx_1^2$ and $y_2 = kx_2^2$	Write equations for two data points.
$y_1 = k(30)^2$ and $y_2 = k(20)^2$	Substitute known data.
$y_1 = k \cdot 900$ and $\dfrac{y_2}{400} = k$	Solve the second equation for k and substitute for k in the first equation.
$y_1 = \dfrac{y_2}{400} \cdot 900$	Simplify.
$y_1 = \frac{9}{4} y_2$ or $y_1 = 2.25y_2$	

The distance traveled in 30 seconds is 2.25 times the distance traveled in 20 seconds. ▬

Joint Variation

Joint variation describes the product of two or more input variables.

▬ **JOINT VARIATION**

> If there is a constant k such that $y = kwx$, we say *y varies jointly with w and x*. The constant k is the constant of variation.

In joint variation, there may be multiple inputs contributing to an output.

EXAMPLE 11 **Applying joint variation** Translate each of these variations into an equation. Identify the constant of variation in each setting.

    a. The volume of a cylinder varies jointly with pi times the square of the radius of the base and the height.

    b. The sales tax on a purchase of apples varies jointly with the price, c, per pound and number, w, of pounds purchased. The sales tax rate is 7%.

SOLUTION **a.** $V = \pi r^2 h$; the constant of variation is π.

Student Note: The variables in part a are in the customary order.

    b. $T = 0.07cw$; the constant of variation is the tax rate, 0.07. ▬

In Example 12, we find the constant of variation for the volume of a pyramid. The **volume of a pyramid** *varies jointly with the area of its base and its height*:

$$V = kBh$$

EXAMPLE 12 **Applying joint variation: volume of a gardening pyramid** A clear plastic pyramid is used to cover tomatoes in a garden and speed their growth (Figure 14). The plastic cover has a square base of 2.5 feet on each side and a height of 3 feet. The volume of the pyramid is 6.25 cubic feet. Find the constant of variation for the volume of a pyramid.

SOLUTION The volume is $V = kBh$, where B is the area of the base and h is the height. For the plastic cover,

$$6.25 = k(2.5^2)3$$
$$6.25 = k(6.25)3$$
$$k = \tfrac{1}{3}$$

FIGURE 14 The volume formula for a pyramid is $V = \tfrac{1}{3}Bh$. ▬

ANSWER BOX

Warm-up: **1.** $0.49 per can **2.** $0.26 per minute **3.** $\tfrac{1}{4}$ hour per foot **4.** $0.83 per minute **Think about it:** The slope is the change in y over the change in x, so two ordered pairs are required to calculate it. The constant of variation is the ratio y to x of a single ordered pair. In proportional data, the ratio y/x is the same as the slope because the two ordered pairs on the line are $(0, 0)$, the origin, and (x, y):

$$\text{Slope} = m = \frac{y_2 - y_1}{x_2 - x_1} = \frac{y - 0}{x - 0} = \frac{y}{x}$$

7.2 Exercises

In Exercises 1 to 6,

(a) Identify variables for input and output. For each situation, write a linear equation, $y = mx$ or $y = mx + b$.

(b) State whether the relationship is a direct variation ($y = kx$).

(c) Explain the meaning of the nonzero y-intercept for those situations in which the data are not proportional.

1. A 10-pound weight stretches a spring 4 inches. A 15-pound weight stretches the same spring 6 inches.
    x = weight in lb, y = distance stretched in in.; $y = \tfrac{2}{5}x$; direct

2. A first-class letter costs $0.60 for 2 ounces and $1.06 for 4 ounces.
    x = weight in oz, y = cost in $; $y = 0.23x + 0.14$; not prop.; no meaning

3. Tuition and fees are $185 for 2 credit hours and $345 for 4 credit hours.
    x = no. of credit hr, y = cost in $; $y = 80x + 25$; not prop.; student fees

4. A bicyclist travels 15 miles in 1 hour and 45 miles in 3 hours. x = time in hr, y = distance in mi; $y = 15x$; direct

5. Tax on a $16 purchase is $1.20. Tax on a $30 purchase is $2.25. x = purchase price, y = tax; $y = 0.075x$; direct

6. Making 3 dozen cookies takes 2 hours. Making 6 dozen cookies takes 3 hours. x = no. of dozens of cookies, y = time in hr; $y = \frac{1}{3}x + 1$; not prop; time to mix dough

In Exercises 7 to 12, find the constant of variation k, and indicate the units of measure associated with k. Then answer the question.

7. The distance a car travels varies directly with the time traveled. Suppose (time, distance) = (3, 165). How far will the car travel in 8 hours? Assume distance is in miles and time is in hours. k = 55 mi/hr; 440 mi

8. The amount of money earned varies directly with the time worked. Suppose (time, earnings) = (12, 67.80). How much will be earned in 40 hours? Assume time is in hours and earnings is in dollars. k = $5.65/hr; $226.00

9. The total cost of compact discs varies directly with the number purchased. Suppose (number, total cost) = (8, 115.92). What is the cost of 10 discs? k = $14.49/CD; $144.90

10. The cost of attending a concert varies directly with the number of tickets purchased. Suppose (number, total cost) = (9, 351). What is the cost of 10 tickets? k = $39/ticket; $390

11. The weight of a pancake is proportional to the square of its radius. Suppose a 7-inch-diameter pancake weighs 4.5 ounces. What will a 3-inch-diameter pancake of the same thickness weigh? $k \approx 0.367$ oz/in^2; ≈ 0.827 oz

12. The surface area of a sphere is proportional to the square of the radius. Suppose (radius, area) = (5, 314.16). What is the surface area of a sphere with an 8-inch radius? $k \approx 12.57$ in^2/in.; 804 in^2

Complete each sentence in Exercises 13 to 16 by choosing the correct phrase or equation from within the brackets.

13. Linear equations that can be written as $y = mx$ [are or are not] linear variations.

14. Linear equations that can be written as $y = mx + b, b \neq 0$ [are or are not] linear variations.

15. If a set of data is proportional, then its equation is of the form [$y = mx$ or $y = mx + b, b \neq 0$].

16. If a linear set of data is not proportional, then its equation is of the form [$y = mx$ or $y = mx + b, b \neq 0$].

In Exercises 17 and 18, identify the constant of variation.

17. **a.** $A = \frac{1}{2}bh$ $k = \frac{1}{2}$

 b. $A = lw$ $k = 1$

 c. $V = \frac{1}{3}\pi r^2 h$ $k = \frac{1}{3}\pi$

18. **a.** $I = \$1000rt$ $k = \$1000$

 b. $D = rt$ $k = 1$

 c. $V = \pi r^2 h$ $k = \pi$

In Exercises 19 to 26, translate the sentence into a variation equation with constant of variation k. Research, as needed, to find the exact constant of variation.

19. The circumference of a circle varies directly with the diameter. $C = kd; k = \pi$

20. The circumference of a circle varies directly with the radius. $C = kr; k = 2\pi$

21. The area of a circle varies directly with the square of the radius. $A = kr^2; k = \pi$

22. The area of a square varies with the square of its side. $A = ks^2, k = 1$

23. The area of a rectangle varies jointly with length and width. $A = klw; k = 1$

24. Distance traveled varies jointly with rate and time. $D = krt; k = 1$

25. The volume of a cylinder varies jointly with the height and the square of the radius. $V = khr^2; k = \pi$

26. The surface area of a cube varies with the square of the side of the cube. $A = ks^2; k = 6$

Exercises 27 to 32 show that it is not always necessary to find the constant of variation to compare data.

27. The circumference of a circle varies directly with the diameter. Compare the circumference of an 8-foot-diameter circle with that of a 3-foot-diameter circle. $\frac{8}{3}$ of the length

28. The interest earned on a simple-interest savings account varies directly with the amount of money in the account. Compare the interest earned on a $500 account with the interest earned on a $200 account. 2.5 times the amount

29   tter.
                                                           om,
                                                           rgy
                                                           t.
                                                           of a

30   the
                                                           for a
                                                           r a

3   e road
                                                           , s, in
                                                           at

32. The area of a square varies directly with the square of the length of a side. Compare the area of a square with 7-inch sides with the area of a square with 2-inch sides. 12.25 times the area

33. Use the results from Example 12 to estimate the original volume of the Great Pyramid of Cheops (see the figure). The square base of the pyramid had an area of approximately 53,095 square meters, and the original height was approximately 146.6 meters.
$\approx 2{,}594{,}576$ m^3

7.3 Inverse Variation and Related Graphs

Objectives

- Explore inverse variation.
- Identify variation as direct or inverse.
- Find the constant of variation.
- Use the product form of inverse variation in applications.
- Graph inverse variation equations and identify features.
- Apply inverse square variation.

WARM-UP

Make a list of pairs of whole-number factors of these numbers. Be systematic; start the first one with $1 \cdot 120$, $2 \cdot 60$, and so on.

1. 120 $1 \cdot 120, 2 \cdot 60, 3 \cdot 40,$ **2.** 64 $1 \cdot 64, 2 \cdot 32,$ **3.** 400$1 \cdot 400, 2 \cdot 200, 4 \cdot 100,$
$4 \cdot 30, 5 \cdot 24, 6 \cdot 20, 8 \cdot 15, 10 \cdot 12$ $4 \cdot 16, 8 \cdot 8$ $5 \cdot 80, 8 \cdot 50, 10 \cdot 40, 16 \cdot 25, 20 \cdot 20$

IN THIS SECTION, we examine inverse variation, inverse proportions, the behavior of the graphs of inverse equations, and inverse square variation.

Inverse Variation

The photographs on the chapter opening page show a hurricane and a galaxy. The pictures were chosen because of the similarity in form of these two different objects. As vastly different as a hurricane and a galaxy of stars are, both have a rate of turning that is dependent on how compact their structure is relative to the center of rotation. We describe the relationship between the position of hurricane clouds and galaxy stars relative to their centers and their rate of turning as an inverse variation.

In Example 1, we use numbers to investigate an inverse variation setting.

EXAMPLE 1   **Exploring driving speed**   Suppose Indy car driver Sara Fisher has 120 miles to drive—how long will it take her? Record six rates and times in a table. Graph your data with rate on the horizontal axis and time on the vertical axis. What equation relates the distance, rate, and time?

SOLUTION The time it takes for Fisher to drive 120 miles depends on how fast she drives. In an Indy car street race, the 120 miles might take only an hour (see Table 2). At congested freeway speeds, 2 hours might be reasonable. At slower city traffic speeds, the trip may take 4 hours.

TABLE 2

Rate (miles per hour)	Time (hours)	Distance (miles)
120	1	120
60	2	120
40	3	120
30	4	120
20	6	120
10	12	120

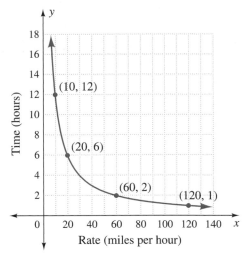

FIGURE 15

Because the distance is constant, both the rate and the time change. As the rate in miles per hour increases, the time in hours to travel 120 miles decreases. This relationship is shown most clearly by the graph in Figure 15.

We write this relationship as

$$r \cdot t = 120 \qquad \text{or} \qquad t = \frac{120}{r}$$

In more formal language, we say that the time needed to drive 120 miles varies inversely with the rate. ▄

INVERSE VARIATION

> If there is a constant k such that $xy = k$, or $y = k/x$, we say y *varies inversely with x.* The constant k is the constant of variation.

In Example 1, the distance, 120 miles, is the constant of variation. Look for the constant of variation in Example 2.

EXAMPLE 2 Identifying inverse variation: lottery jackpot Suppose the current lottery jackpot is $24 million. Susumu has a winning ticket but hears that there may be up to 24 winners.

a. Make a table and a graph of the number of winning tickets and the prize received for each winning ticket.

b. Write an equation describing the fact that the amount won varies inversely with the number of winning tickets.

c. Identify the constant of variation.

SOLUTION **a.** The number of winning tickets, x, and the prize received, y, are given in Table 3 and are graphed in Figure 16. A negative number of winners is meaningless, so we graph only first-quadrant numbers.

TABLE 3 *Lottery Jackpot with Different Numbers of Winning Tickets*

Number of Tickets	Prize (dollars)	Lottery Jackpot (dollars)
1	24 million	24 million
2	12 million	24 million
4	6 million	24 million
6	4 million	24 million
12	2 million	24 million
24	1 million	24 million

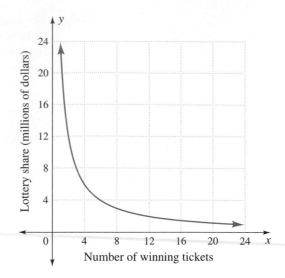

FIGURE 16

b. If x is the number of winning tickets and y is the price received, then the equation is $xy = 24$ million, or $y = 24$ million$/x$.

c. The constant of variation is the $24 million jackpot. The number remains constant regardless of the number of winners. ▬

Think about it 1: How many winners would reduce the lottery prize to $1? Where would such a point be located on a graph such as Figure 16, where 2 inches represents 24 winners?

Direct versus Inverse Variation

To distinguish direct variation from inverse variation, look for a constant ratio or constant product.

▬ **DISTINGUISHING DIRECT AND INVERSE VARIATION**

For data (x_1, y_1) and (x_2, y_2):

Direct variation has a constant ratio $y_1/x_1 = y_2/x_2$, or $y/x = k$, the variation constant.

Inverse variation has a constant product $x_1y_1 = x_2y_2$, or $x \cdot y = k$, the variation constant.

EXAMPLE 3 **Identifying direct and inverse variation** Look for a constant ratio or constant product in each table, and identify the table as showing direct or inverse variation. What is the meaning, if any, of the constant of variation?

a. *Servings in a Gallon of Milk*

Size of Glass (ounces)	Number of Glasses
6	$21\frac{1}{3}$
8	16
12	$10\frac{2}{3}$

b. *Distance Traveled in a Car*

Time (hours)	Distance (miles)
2	110
3	165
4	220

c. *Vacation on a Budget*

Vacation Days	Budget ($ per day)
5	200
10	100
15	$66\frac{2}{3}$

SOLUTION **a.** The product is

$$x \cdot y = 6 \cdot 21\tfrac{1}{3} = 8 \cdot 16 = 12 \cdot 10\tfrac{2}{3} = 128$$
$$k = 128$$

There is a constant product. A total of 128 ounces of milk is available in a gallon of milk. The table shows inverse variation.

b. The ratio is

$$\frac{y}{x} = \frac{110}{2} = \frac{165}{3} = \frac{220}{4} = \frac{55}{1}$$
$$k = \frac{55}{1}$$

The constant ratio in the table is $\frac{55}{1}$. For $k = \frac{55}{1}$, the distance increases by 55 miles with each hour traveled. The table shows direct variation.

c. The product is

$$x \cdot y = 5 \cdot 200 = 10 \cdot 100 = 15 \cdot 66\tfrac{2}{3} = 1000$$
$$k = 1000$$

There is a constant product. A total of $1000 is available for the vacation. The table shows inverse variation. ■

Inverse Variation: Proportion Form

In Examples 1 and 2, the constants of variation, 120 miles and $24 million, were given. The meaning of the constants was quite clear. In the settings below, the value and meaning of the constant of variation may not be as obvious.

EXAMPLE 4 Exploring inverse variation: balance   The equipment needed for this exploration is a 12-inch ruler, three coins of the same value, and a pencil. Place the pencil under the 6-inch mark on the 12-inch ruler (see Figure 17). The ruler should balance, with neither end touching the table top.

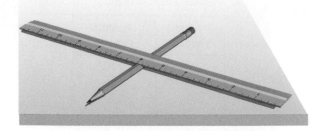

FIGURE 17

a. Center a stack of two coins at the 4-inch mark. Where will the third coin need to be placed to bring the ruler into balance?

b. Repeat part a, placing the two coins at the $3\frac{1}{2}$-inch mark.

c. Explain your results.

SOLUTION   **a.** When the third coin is placed 4 inches from the center (at the 10-inch mark), it will balance two coins placed 2 inches from the center (at the 4-inch mark). See Figure 18.

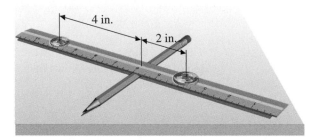

FIGURE 18

b. When the third coin is placed 5 inches from the center (at the 11-inch mark), it will balance two coins placed $2\frac{1}{2}$ inches from the center.

c. The single coin is placed twice as far from the center of the ruler as the two coins because it weighs half as much. *The distance the coins are placed from the center varies inversely with the number of coins.*   ▬

Example 5 shows how we can use the data from Example 4 to explain our exploration results and to find the constant of variation.

EXAMPLE 5   **Finding the constant of variation**   Write an equation relating the numbers and positions of the coins in Example 4. Identify the constant of variation.

SOLUTION   **a.**   2 coins(2 inches from center) = 1 coin(x inches from center)

$$4 \text{ coin} \cdot \text{inches} = x \text{ coin} \cdot \text{inches}$$

$$x = 4$$

Check: 2 coins(2 in.) $\stackrel{?}{=}$ 1 coin(4 in.) ✓

The constant of variation k is 4 coin·inches.

b.   2 coins(2.5 inches from center) = 1 coin(x inches from center)

$$5 \text{ coin} \cdot \text{inches} = x \text{ coin} \cdot \text{inches}$$

$$x = 5$$

Check: 2 coins(2.5 in.) $\stackrel{?}{=}$ 1 coin(5 in.) ✓

The constant of variation k is 5 coin·inches.

▬ For balance on the ruler, *the product of the number, x_1, of coins on one side and their distance, y_1, from the center is equal to the product of the number, x_2, of coins on the other side and their distance, y_2, from the center.* That is,

$$x_1 \cdot y_1 = x_2 \cdot y_2$$

The constant product $x \cdot y = k$ is the constant of variation.

When the focus of the problem is on the inverse relationship in the data and not on the constant of variation, we describe the relationship with the phrase *inversely proportional.*

▬ **INVERSE PROPORTIONS**

> Quantities are *inversely proportional* if their product is constant: $x \cdot y = k$.
> For paired data, such as (x_1, y_1) and (x_2, y_2), being inversely proportional
> means that $x_1 \cdot y_1 = x_2 \cdot y_2$. The constant product k is called the *constant of
> variation* (or *constant of proportionality*).

The constant product $x_1 \cdot y_1 = x_2 \cdot y_2$ doesn't look like a proportion, as it is not in the form

$$\frac{a}{b} = \frac{c}{d}$$

In the exercises, we will change the constant product form into the alternative proportion form

$$\frac{y_1}{x_2} = \frac{y_2}{x_1} \qquad \text{Inverse variation}$$

For now, observe that the denominators are the opposite of what they are in the constant ratios for direct variation,

$$\frac{y_1}{x_1} = \frac{y_2}{x_2} \qquad \text{Direct variation}$$

In Example 6, the setting is children on a seesaw rather than coins on a ruler. *The product of one child's weight and distance from the center is equal to the product of the other child's weight and distance from the center.*

EXAMPLE 6 Finding the constant of variation: balancing a seesaw   Use an inverse proportion to describe the relationship between the weight and the position of the children in balance on the seesaw in Figure 19. State the constant of variation.

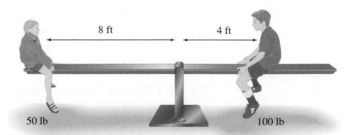

50 lb   8 ft   4 ft   100 lb

FIGURE 19

SOLUTION   100 pounds · 4 feet = 50 pounds · 8 feet

400 pound·feet = 400 pound·feet

The position on the seesaw is inversely proportional to the weight of the child. The constant of variation for these positions is $k = 400$ pound·feet. ▬

Because either the phrasing *varies inversely* or the phrasing *inversely proportional* may be used in application settings, you should be familiar with both sets of vocabulary.

Inverse Variation Graphs

We return to graphing to further examine inverse variation. The graphs for Fisher's driving rate over 120 miles (Figure 15) and for the prize from a $24 million lottery (Figure 16) are portions of a graph called a *hyperbola*.

UNDEFINED VALUES   In Example 7, we examine the behavior of a hyperbola on both sides of an input that makes the equation undefined.

EXAMPLE 7   Investigating outputs near a zero in the denominator

a. Make tables and a graph of $y = 2/x$ for both negative and positive inputs.

b. Describe the behavior of $y = 2/x$ near $x = 0$ and why it occurs.

SOLUTION   **a.** Tables 4 and 5 show outputs for $y = 2/x$ as x gets close to zero. When graphed, the data in each table form one side of the hyperbola (Figure 20).

TABLE 4 $y = 2/x$ as x approaches 0 *from the left*

Input, x	Output, $y = 2/x$
−4	$2/{-4} = -0.5$
−2	$2/{-2} = -1.0$
−1	$2/{-1} = -2.0$
−0.5	$2/{-0.5} = -4.0$
−0.25	$2/{-0.25} = -8.0$
−0.1	$2/{-0.1} = -20.0$
−0.001	$2/{-0.001} = -2000$

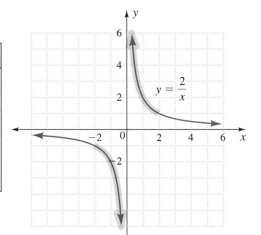

FIGURE 20

TABLE 5 $y = 2/x$ as x approaches 0 *from the right*

Input, x	Output, $y = 2/x$
4	$2/4 = 0.5$
2	$2/2 = 1.0$
1	$2/1 = 2.0$
0.5	$2/0.5 = 4.0$
0.25	$2/0.25 = 8.0$
0.1	$2/0.1 = 20.0$
0.001	$2/0.001 = 2000$

b. The hyperbola is quite steep between $x = -1$ and $x = 0$ and between $x = 0$ and $x = 1$. If we trace the third-quadrant curve from left to right, the inputs x approach zero and the curve turns downward. The curve never touches the y-axis.

If we trace the first-quadrant graph from right to left, the inputs x approach zero and the curve rises. The curve gets closer to but never touches the y-axis. At $x = 0$, the equation becomes $y = \frac{2}{0}$ and is undefined. There is no point on the graph for $x = 0$. ▬

The hyperbolic graph of an inverse variation equation, $y = k/x$, is characterized by a nearly vertical graph whenever the denominator approaches zero.

▬ GRAPHICAL BEHAVIOR
NEAR A ZERO DENOMINATOR

The graph of $y = k/x$ becomes nearly vertical whenever the denominator approaches zero.

Graphing calculator note: Calculators are designed to reject an attempt to divide by zero. If you set **Xmin** $= -9.4$ and **Xmax** $= 9.4$ and trace along the graph of $y = 2/x$ (Figure 21), the output will be blank when the input creates a zero denominator. An error will appear in the calculator table at $x = 0$ (Figure 22).

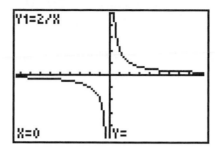

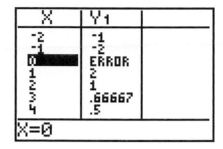

FIGURE 21 -9.4 to 9.4 or X,
$\quad\quad\quad\quad\quad -5$ to 5.4 or Y

FIGURE 22

INPUTS THAT APPROACH $-\infty$ OR $+\infty$ As Fisher increases the speed of her car, the time it takes to drive 120 miles drops toward zero. In the lottery example, as the number of winning tickets increases, the prize approaches zero. In Example 8, we look at similar behavior in the graph of $y = 2/x$. We look at the outputs first as the input x approaches negative infinity and then as the input x approaches positive infinity.

EXAMPLE 8   Investigating outputs for inputs approaching infinity   Describe the behavior of $y = 2/x$ for x smaller than -4 and for x larger than 4.

SOLUTION   As Tables 6 and 7 indicate, the expression $2/x$ becomes close to zero as we input numbers larger than 4 and smaller than -4. For these same inputs, the graph in Figure 23 approaches $y = 0$, the x-axis.

TABLE 6 $y = 2/x$ as x approaches $-\infty$

Input, x	Output, $y = 2/x$
-4	$2/-4 = -0.5$
-5	$2/-5 = -0.4$
-10	$2/-10 = -0.2$
-100	$2/-100 = -0.02$
-1000	$2/-1000 = -0.002$
$-10,000$	$2/-10,000 = -0.0002$

TABLE 7 $y = 2/x$ as x approaches ∞

Input, x	Output, $y = 2/x$
4	$2/4 = 0.5$
5	$2/5 = 0.4$
10	$2/10 = 0.2$
100	$2/100 = 0.02$
1000	$2/1000 = 0.002$
10,000	$2/10,000 = 0.0002$

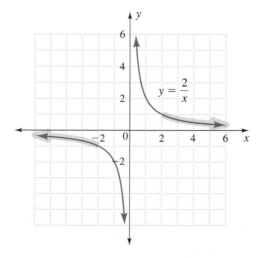

FIGURE 23

GRAPHICAL BEHAVIOR AS x GOES TO $+\infty$ OR $-\infty$

For large x (toward $+\infty$) or small x (toward $-\infty$), the graph of $y = k/x$ approaches the x-axis.

SYMMETRY   The hyperbola in Figure 24 has two axes of symmetry, the dashed lines $y = x$ and $y = -x$.

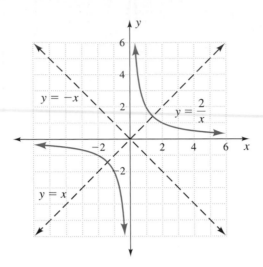

$y = -x$

$y = \dfrac{2}{x}$

$y = x$

FIGURE 24

  Graphing calculator note: The symmetry in the hyperbola may be difficult to observe because of distortion in the viewing window. You can adjust the display with the **5 : ZSquare** option under ⌊ **ZOOM** ⌋.

Inverse Square Variation

The amount of light energy reaching a surface per unit time is called the *intensity* of the light. Figure 25 shows light striking a surface at three different distances from the source.

Student Note: Explore this relationship with a flashlight on a wall or do the project in Exercise 42.

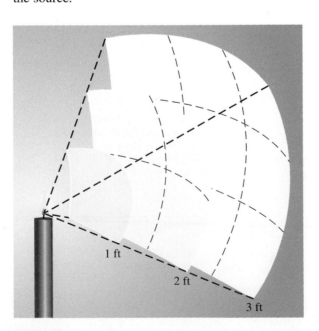

1 ft

2 ft

3 ft

FIGURE 25

The first square in Figure 25 shows the light striking an area of 1 square foot at a distance of 1 foot from a candle. This same amount of light must spread over 4 square feet at a distance of 2 feet from the candle. Thus, at 2 feet, the intensity of light is $\frac{1}{4}$ of that reaching the original square at 1 foot. The light must spread over 9 square feet at a distance of 3 feet from the candle. Thus, at 3 feet, the intensity of light is $\frac{1}{9}$ of that reaching the original square at 1 foot. *The intensity of light is inversely proportional to the square of the distance from the light source:* $y = k/d^2$.

EXAMPLE 9   **Comparing data using systems of equations**   Nellie places her reading chair 10 feet from a light source. Compare the intensity of light on her reading materials, which are at a distance of 10 feet, with the intensity of light 1 foot from the light source.

SOLUTION   *Plan*: We need to compare the light intensity, y_1, at 10 feet with the light intensity, y_2, at 1 foot.

Carry out the plan:

$$y_1 = \frac{k}{x_1^2} \quad \text{and} \quad y_2 = \frac{k}{x_2^2} \qquad \text{Write equations for two data points.}$$

$$y_1 = \frac{k}{10^2} \quad \text{and} \quad y_2 = \frac{k}{1^2} \qquad \text{Substitute known data.}$$

$$y_1 = \frac{k}{100} \quad \text{and} \quad y_2 = \frac{k}{1} \qquad \begin{array}{l}\text{Solve the second equation for } k \text{ and substitute} \\ \text{for } k \text{ in the first equation.}\end{array}$$

$$y_1 = \frac{y_2}{100}$$

The light intensity at 10 feet is $\frac{1}{100}$ the light intensity at 1 foot.   ▬

Think about it 2: Predict an equation for the inverse square proportion. Check it by substituting information from Example 9.

ANSWER BOX

Warm-up: 1. $1 \cdot 120, 2 \cdot 60, 3 \cdot 40, 4 \cdot 30, 5 \cdot 24, 6 \cdot 20, 8 \cdot 15,$ $10 \cdot 12$ **2.** $1 \cdot 64, 2 \cdot 32, 4 \cdot 16, 8 \cdot 8$ **3.** $1 \cdot 400, 2 \cdot 200, 4 \cdot 100,$ $5 \cdot 80, 8 \cdot 50, 10 \cdot 40, 16 \cdot 25, 20 \cdot 20$ **Think about it 1:** The lottery prize would be \$1 if there were 24 million winners. The point (24 million, \$1) would be located 2,000,000 inches to the right, just above the horizontal axis. Use unit analysis to change 2,000,000 inches into an appropriate unit of measure. For a further challenge, calculate an estimate of the distance the graph would be above the x-axis.   **Think about it 2:** $x_1^2 \cdot y_1 = x_2^2 \cdot y_2;$ $10^2 y_1 = 1^2 y_2.$

7.3   Exercises

Describe each of the situations in Exercises 1 to 6 with an inverse variation equation. Note the units on your variables, and identify the constant of variation.

1. The number of points given to each problem on a 100-point test varies inversely with the number of problems on the test.

    $y = 100/x$; $y =$ points/problem, $x =$ no. of problems; $k = 100$ points

2. The time required to drill a 1200-foot well varies inversely with the speed of the drill in feet per minute.

    $y = 1200/x$; $y =$ min, $x =$ ft/min; $k = 1200$ ft

3. The number of years a landfill can be used varies inversely with the number of cubic yards of trash deposited each year. The landfill has 1 million cubic yards of space.

    $y = 1,000,000/x$; $y =$ yr, $x =$ yd^3/yr; $k = 1,000,000$ yd^3

4. The number of years the world resources of silver will last varies inversely with the number of metric tons used per year. Estimated world supply in 2001 was 430,000 metric tons.

    $y = 430,000/x$; $y =$ yr, $x =$ metric tons/yr; $k = 430,000$ metric tons

5. The number of years the world resources of copper will last varies inversely with the number of metric tons used per year. Estimated world supply in 2001 was 650 million metric tons.

$y = 650,000,000/x$; y = yr, x = metric tons/yr; $k = 650,000,000$ metric tons

6. Assuming no interest or other funds are added, the number of years a $100,000 savings account will last after retirement varies inversely with the number of dollars spent per year. $y = 100,000/x$, y = yr, x = $/yr; $k = $100,000$

In Exercises 7 to 16, identify the data as reflecting direct variation or inverse variation or neither, and identify the constant of variation.

7. *Limited Resource Usage*

Production	Life Span
160 tons/day	35 years
200 tons/day	28 years

inverse variation;
$k = 5600 \frac{\text{tons}}{\text{day}} \cdot \text{years}$

8. *Real Cost of Purchase*

Times to Wear Blouse	Cost per Wearing
2	$26.00
10	$ 5.20

inverse variation;
$k = 52

9. *Bulk Purchase Cost*

Number of Items	Total Cost
160	$35.00
200	$43.75

direct variation;
$k = $0.21875/\text{item}$

10. *Clothing Purchase*

Number of Jeans	Total Cost
2	$ 56
10	$280

direct variation;
$k = $28/\text{pair of jeans}$

11. *Bicycle Transportation*

Rate	Time
15 mi/hr	6 hr
25 mi/hr	3.6 hr

inverse variation; $k = 90$ mi

12. *Bicycle Transportation*

Rate	Distance
15 mi/hr	75 mi
25 mi/hr	125 mi

direct variation; $k = 5$ hr

13. *Work Output*

Number of Workers	Time to Re-roof House
2	6 days
6	2 days

inverse variation;
$k = 12$ worker·days

14. *Moped Transportation*

Time of Travel	Distance Traveled
3 hours	90 miles
6 hours	150 miles

neither

15. *Student Loans*

Number of Semesters	Loan Amount
2	$2200
3	$3300

direct variation;
$k = $1100/\text{semester}$

16. *Vacation Costs*

Length of Vacation	Total Cost
3 days	$ 540
8 days	$1440

direct variation; $k = $180/\text{day}$

In Exercises 17 to 24, set up and solve an inverse proportion. Identify the constant of variation for each setting.

17. Jehan weighs 35 pounds and sits 6 feet from the balance point of a seesaw. She just balances Dalia, sitting 4 feet from the balance point. How heavy is Dalia?
$x = 52.5$ lb; $k = 210$ lb·ft

18. Shu-Ju bicycles 15 miles per hour for 3 hours. Su Lin bicycles the same route in 3 hours and 15 minutes. What is Su Lin's rate? $R \approx 13.8$ mi/hr; $k = 45$ mi

19. A batch of stew serves 80 people at $\frac{3}{4}$ cup per person. How many people may be served at 1 cup per person?
$x = 60$ people; $k = 60$ cups

20. A landfill will be in service 10 years if each household throws away a full 90-gallon container per week. If the same households recycle and reduce the throw-away portion to 30 gallons per week, how long will the landfill be in service? $x = 30$ yr; $k = 900$ (gal/week)·year

21. The world reserves of lead will last 71 years if 1,820,000 metric tons are used each year. How many years will the reserves last if 1,300,000 metric tons are used each year? $x = 99.4$ yr; $k = 129,220,000$ metric tons

22. If the entire world production of rice in 2002 had been evenly divided among the world's population, each person would have received about 0.26 kilogram (half a pound) of rice per day. The world population in 2002 was 6.234 billion. How much rice will there be per person per day in 2025 with the same annual production if the population increases to 7.84 billion?
$x = 0.207$ kg/day; $k = 1.62$ billion kg/day

23. If the yellow stripe on a road is 4 inches wide, a gallon of paint will make a stripe 900 feet long. If the stripe is reduced to 3 inches wide, how long a stripe will the gallon paint? $x = 1200$ ft/gal; $k = 300$ ft^2/gal

24. At \$40 per ticket, an athletic office can sell out its 75,000-seat stadium. In order to maintain the same total revenue, how many tickets would have to be sold at \$64 per ticket? $t = 46,875$; $k = \$3$ million

25. Graph the inverse square function, $f(x) = 1/x^2$.
See Answer Section.
  a. What is $f(0)$? undefined

  b. What is the equation of the axis of symmetry? $x = 0$

  c. Describe the behavior of the graph for $x < -3$ and $x > 3$. approaches x-axis

26. Graph the reciprocal function, $f(x) = 1/x$.
See Additional Answers.
  a. What is $f(0)$? undefined

  b. What are the equations of the axes of symmetry?
  $y = x, y = -x$
  c. Describe the behavior of the graph for $x < -4$.
  approaches x-axis from below
  d. Describe the behavior of the graph for $x > 4$.
  approaches x-axis from above

27. Graph $y = 1/x$ and $y = -1/x$ on the same axes. Compare the positions of the graphs.
  reflected over x-axis; see Answer Section.

28. a. Graph $y = 2/x$, $y = 8/x$, and $y = 16/x$ on the same axes. See Additional Answers.

  b. What is the same about the graphs?
  in 1st and 3rd quadrants
  c. What is different? $2/x$, $8/x$, $16/x$ progressively farther from the axes; $16/x$ above $8/x$ and $2/x$ in 1st quadrant, below in 3rd
  d. What is the constant of variation for each equation?
  the numerator
  e. How does increasing the constant of variation change the graph? increases distance from x-axis

  f. Predict the position of $y = 1/x$ relative to the other graphs. closer to x-axis

  g. Predict the position of $y = 32/x$ relative to the other graphs. above in 1st quadrant, below in 3rd

  h. Describe the behavior of the graphs as x gets larger than 50. get close to x-axis

29. Write one inverse variation equation for (x_1, y_1) and another one for (x_2, y_2). Use substitution to eliminate k and find the constant product form $x_1 \cdot y_1 = x_2 \cdot y_2$.
$y_1 = \dfrac{k}{x_1}$, $y_2 = \dfrac{k}{x_2}$; $k = x_2 y_2$; multiply $y_1 = \dfrac{x_2 y_2}{x_1}$ by x_1.

30. What algebraic step is needed to change the constant product $x_1 \cdot y_1 = x_2 \cdot y_2$ into the inverse proportion $y_1/x_2 = y_2/x_1$? Divide by $x_1 \cdot x_2$.

31. Eliminate the denominators from the proportion form of a direct variation $y_1/x_1 = y_2/x_2$. How is the result different from the constant product of inverse variation?
$x_2 y_1 = x_1 y_2$; subscripts are not alike on each side.

32. Show whether $x_1/x_2 = y_1/y_2$ is direct or inverse variation.
direct; multiply by y_2 and divide by x_1.

33. Are there algebraic steps to change $x_1 y_1 = x_2 y_2$ into $y_1/x_1 = y_2/x_2$? Explain.
No; first is inverse variation and second is direct.

■ Inverse Square Variation

34. The force due to gravitational attraction between planets varies inversely with the square of the distance between the planets. Which equation would describe gravitational attraction? c

  a. $F = km_1 m_2 d^2$ **b.** $F = \dfrac{km_1 m_2}{d}$

  c. $F = \dfrac{km_1 m_2}{d^2}$

35. Under certain circumstances, the intensity of a magnetic field varies inversely with the cube of the distance from the magnet. Which equation would describe the intensity? b

  a. $H = \dfrac{M}{r^2}$ **b.** $H = \dfrac{M}{r^3}$

  c. $H = Mr^3$ **d.** $H = Mr^2$

36. If the intensity of light relative to the distance from its source is described by $y = k/d^2$, how many times greater is the intensity of light reaching 3 feet from a light source than that of light reaching 6 feet from the same source? 4

37. If the intensity of light relative to the distance from its source is described by $y = k/d^2$, how many times greater is the intensity of light reaching 2 feet from a light source than that of light reaching 6 feet from the same source? 9

■ Problems for Extra Practice

38. a. The legal freeway speed is 65 mph, and you have 20 miles left to travel. How long in minutes will it take you to drive 20 miles at the posted speed?
  ≈ 18.5 min
  b. Make a table for the number of minutes it will take you to drive the 20 miles at 60, 65, 70, 75, and 80 mph. 20, 18.5, 17.1, 16, 15 min

  c. How fast would you need to drive to cut 5 minutes off the driving time at 65 mph? 89 mph

39. The force you need to loosen a jar lid is inversely proportional to the distance your grip is from the center of the lid. It takes 10 pounds of force to open a lid with a gripper handle that places the force 3 inches from the center (see the figure). How many pounds of force are needed to open the lid without the gripper if the lid is held 1 inch from the center?

a. Identify (x_1, y_1) and (x_2, y_2). (10 lb, 3 in.), (*x*, 1 in.)

b. Construct and solve an inverse proportion.
(10 lb)(3 in.) = *x*(1 in.), 30 lb
c. Identify the constant of variation. 30 lb·in.

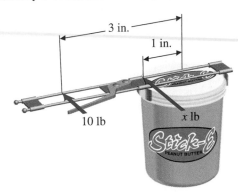

3 in.

1 in.

10 lb   *x* lb

40. a. The base unit of a portable telephone is plugged into an electrical outlet with a small box. The box is a transformer that changes 120-volt electricity (V_1) into 12-volt electricity (V_2). The output current (I_2) from the transformer is 100 mA. What is the input current (I_1) when the voltage and current are related by the formula $V_1/V_2 = I_2/I_1$?   10 mA

b. Are voltage (V) and current (I) directly or inversely proportional in this problem setting? Show why.
inversely
c. What is the constant of variation?   1.2 volt·amp
(because $m = 10^{-3}$)

▇ Projects

41. Chemistry Formulas Chemistry and physics have several laws relating pressure, volume, and temperature of gases, as well as concentration and volume. In parts a to d, use multiplication to transform the proportions into formulas without denominators. Identify as direct or inverse variations.

a. $\dfrac{V_1}{T_1} = \dfrac{V_2}{T_2}$ **b.** $\dfrac{V_1}{C_2} = \dfrac{V_2}{C_1}$ $C_1V_1 = C_2V_2$; inverse

$T_2V_1 = T_1V_2$; direct

c. $\dfrac{P_1}{V_2} = \dfrac{P_2}{V_1}$ $P_1V_1 = V_2P_2$; inverse   **d.** $\dfrac{P_1V_1}{T_1} = \dfrac{P_2V_2}{T_2}$

$P_1T_2V_1 = P_2T_1V_2$; P and V inverse, P and T direct

In parts e to g, change the formulas back into proportion form.

e. $C_1V_1 = C_2V_2$; make a proportion with the Cs on top.
$\dfrac{C_2}{V_1} = \dfrac{C_1}{V_2}$

f. $V_1T_2 = V_2T_1$; make a proportion with the Ts on top.
$\dfrac{T_1}{V_1} = \dfrac{T_2}{V_2}$

g. $P_1V_1 = P_2V_2$; make a proportion with the Vs on top.
$\dfrac{V_2}{P_1} = \dfrac{V_1}{P_2}$

42. Measuring Light Intensity Light intensity is measured by the formula $I = k/d^2$, where k is a constant and $d \neq 0$. Obtain a light meter or other device for measuring light intensity. In a darkened room, measure light intensity at several distances from a single light source (such as a bare light bulb).

a. Construct a three-column table. Record your distance, d, and intensity, I, data in the first two columns.

b. In the third column, calculate $k = I \cdot d^2$.

c. Plot your data on a graph, with distance on the horizontal axis and intensity on the vertical axis.

d. Comment on how accurately your experiment reflects the formula. In a perfect experiment, how should the k numbers in the table compare? k is a constant.

e. What factors might change the accuracy of your results?

43. Skater's Spin How is an ice skater's spin similar to the motion of a hurricane or galaxy? Use sketches to illustrate your answer. To see a single, double, and triple axel in skating, visit http://btc.montana.edu/olympics/physbio/biomechanics/cam03.html.

Angular velocity is inversely proportional to the distance of the mass from the center of rotation.

7.4 Simplification, Multiplication, and Division of Rational Expressions

Factoring and operations are introduced and practiced at a low level of complexity. The steps in the operations on rational expressions are reinforced with substantial work involving units of measure.

Objectives

▪ Find when a rational expression is not defined.

▪ Simplify rational expressions.

▪ Multiply and divide rational expressions.

▪ Simplify complex rational expressions using reciprocals.

▪ Multiply and divide expressions containing units of measure.

WARM-UP

Factor.

1. $x^2 - 16$ $(x - 4)(x + 4)$ **2.** $x^2 - 4x - 5$ $(x - 5)(x + 1)$

3. $2x^2 + 9x + 10$ $(2x + 5)(x + 2)$ **4.** $x^2 - 3x$ $x(x - 3)$

5. $x^2 + x - 6$ $(x + 3)(x - 2)$ **6.** $3x^2 + 6x$ $3x(x + 2)$

7. $x^2 + 4x + 3$ $(x + 1)(x + 3)$ **8.** $x^2 + 10x + 25$ $(x + 5)^2$

9. $6x^2 + 4x$ $2x(3x + 2)$ **10.** $4x^2 - 25$ $(2x - 5)(2x + 5)$

11. $5 + 4x - x^2$ $(5 - x)(1 + x)$ **12.** $4 + 3x - x^2$ $(4 - x)(1 + x)$

THIS AND THE NEXT TWO SECTIONS focus on operations with rational expressions. Work with units of measure is included.

EXAMPLE 1   **Working with rational expressions: comparative ages**   Table 8 lists comparative ages for a mother and child if the mother is 20 years old when her child is born.

TABLE 8

Students may prefer to use personal data for the exploration of comparative ages. The exploration is continued in the exercises with graphing activities. Unlike the graphs in Section 7.3, which approach $y = 0$, the x-axis, as x gets large, the graph of y_3 approaches the line $y = 1$ as x gets large. We can see this by dividing both terms of the expression $(x + 20)$ by x, to yield $1 + 20/x$. As x gets large, $20/x$ goes to zero and the expression approaches 1. Students can understand these ideas without formal limit theory.

Child's Age, x	Mother's Age, y_1	Differences in Ages, y_2	Quotient of Ages (mother to child), y_3
0	20	20	
1	21	20	
2	22		
3			
4			

a. Find the missing values in Table 8.

b. What kind of function is the difference in ages?

c. What result do we get as the first entry in the column for the quotient of the mother's age to the child's age? Why?

d. As the child's age increases, what happens to the quotient of ages? What is the quotient of ages when the child is 50?

e. If the child's age is x, write equations to describe the other columns, y_1, y_2, and y_3.

SOLUTION   **a.** The mother's age increases by one each year: 20, 21, 22, 23, 24. The difference in ages is always 20. The quotients of ages are $\frac{20}{0}$ (which is undefined), $\frac{21}{1} = 21$, $\frac{22}{2} = 11$, $\frac{23}{3} = 7\frac{2}{3}$, and $\frac{24}{4} = 6$.

b. Since the child's age and the mother's age are always 20 years apart, the difference in ages is a constant function, $y = 20$.

c. The first entry in the column for the quotient of ages is undefined because we are dividing by a child's age of zero.

d. The quotient of the ages gets smaller but stays above 1. When the child's age is 50, the quotient is

$$\frac{50 + 20}{50} = \frac{70}{50} = 1.4$$

e. If the child's age is x, $y_1 = 20 + x$, $y_2 = 20$, and $y_3 = (x + 20)/x$.　　■

Think about it 1: When is the quotient of ages $\frac{2}{1}$? Will the quotient of ages ever reach $\frac{1}{1}$?

Rational Functions

Recall that **rational numbers** are *the set of numbers that may be written as the ratio of two integers a/b, with b $\neq$ 0.* Rational functions and expressions are closely related to fractions.

■ RATIONAL FUNCTION

> The rational function $f(x)$ is a function that may be written as the ratio of two polynomials $p(x)/q(x)$, where the denominator is not zero.

Application of function concepts to rational functions is treated lightly. See page 400 for asymptotes.

As suggested by the definition, the *domain* of the rational function excludes inputs that make the denominator zero; see Example 2. The *range* of the rational function will become more apparent when we divide $p(x)/q(x)$ in Section 7.5.

A **rational expression** is formed by *the quotient of two polynomials.* The quotients $1/x$, $2/x$, and k/x^2 are rational expressions found in the inverse variation equations in Section 7.3.

■ EXAMPLE 2

Finding when a rational expression is undefined　For what values of the variables is each expression undefined?

a. $\dfrac{a + 3}{3 - a}$　　　　　　　　**b.** $\dfrac{a + b}{a}$

c. $\dfrac{a}{a - b}$　　　　　　　　**d.** $\dfrac{x}{x^2 - 3x - 4}$

SOLUTION　**a.** $\dfrac{a + 3}{3 - a}$ has a zero denominator if $a = 3$.

b. $\dfrac{a + b}{a}$ has a zero denominator if $a = 0$.

c. $\dfrac{a}{a - b}$ has a zero denominator if $a = b$.

d. $\dfrac{x}{x^2 - 3x - 4} = \dfrac{x}{(x + 1)(x - 4)}$ has a zero denominator if $x = -1$ or $x = 4$.　　■

Simplification of Rational Expressions

SIMPLIFYING TO 1　When we simplify a rational expression, we factor the numerator and denominator and remove the common factors, such as $a/a = 1$. If there are no common factors, the expression cannot be simplified.

EXAMPLE 3   Simplifying rational expressions   Simplify the following expressions. Factor as needed. Any variable making a zero denominator must be excluded from possible inputs.

a. $\dfrac{b^2}{2b}$, $b \neq 0$

b. $\dfrac{6xy}{3x^2}$, $x \neq 0$

c. $\dfrac{x + 2}{(x + 2)(x - 3)}$, $x \neq -2$, $x \neq 3$

d. $\dfrac{a + a^2}{1 + a}$, $a \neq -1$

e. $\dfrac{x^2 - 4x - 5}{x^2 + 2x + 1}$, $x \neq -1$

SOLUTION   a. $\dfrac{b^2}{2b} = \dfrac{b \cdot \boxed{b}}{2 \cdot \boxed{b}} = \dfrac{b}{2}$

b. $\dfrac{6xy}{3x^2} = \dfrac{2 \cdot \cancel{3} \cdot \cancel{x} \cdot y}{\cancel{3} \cdot \cancel{x} \cdot x} = \dfrac{2y}{x}$

c. $\dfrac{x + 2}{(x + 2)(x - 3)} = \dfrac{\cancel{(x + 2)}}{\cancel{(x + 2)}(x - 3)} = \dfrac{1}{x - 3}$

d. $\dfrac{a + a^2}{1 + a} = \dfrac{a\cancel{(1 + a)}}{\cancel{(1 + a)}} = a$

e. $\dfrac{x^2 - 4x - 5}{x^2 + 2x + 1} = \dfrac{(x - 5)\cancel{(x + 1)}}{(x + 1)\cancel{(x + 1)}} = \dfrac{x - 5}{x + 1}$

A COMMON ERROR   Example 4 illustrates a common student error: eliminating terms.

EXAMPLE 4   Finding errors in simplifying   What is wrong with these simplifications? Explain the correct process.

a. $\dfrac{3 + 4}{4} = \dfrac{3 + \cancel{4}}{\cancel{4}}$

b. $\dfrac{x^2 + x - 2}{x + 2} = \dfrac{x^2 + \cancel{x} - \cancel{2}}{\cancel{x} + \cancel{2}}$

SOLUTION   The expressions are not factored, and several *terms* have been eliminated. The correct process is to factor and, if possible, eliminate common *factors* from the numerator and denominator.

a. $\dfrac{3 + 4}{4} = \dfrac{7}{4}$

b. $\dfrac{x^2 + x - 2}{x + 2} = \dfrac{(x - 1)\cancel{(x + 2)}}{\cancel{(x + 2)}} = x - 1$

Placing parentheses around binomials, as in the denominator of part b, may help prevent the error of eliminating terms.

SIMPLIFYING TO −1   Example 5 illustrates various algebraic ways of dividing opposite numbers in problems such as

$$\dfrac{5 - 8}{8 - 5} = \dfrac{-3}{3} = -1$$

EXAMPLE 5   Simplifying rational expressions   Simplify $\dfrac{x - y}{y - x}$.

SOLUTION   There are at least three ways to simplify the expression.

Method 1: We multiply the numerator and denominator of the fraction by −1:

$$\dfrac{x - y}{y - x} = \dfrac{(-1)(x - y)}{(-1)(y - x)} = \dfrac{(-1)\cancel{(x - y)}}{\cancel{(x - y)}} = -1$$

We obtain common factors by distributing -1 in only the denominator.

Method 2: We multiply either the numerator or the denominator by $(-1)(-1)$. Because $(-1)(-1) = 1$, this multiplication will not change the fraction.

$$\frac{x - y}{y - x} = \frac{(-1)(-1)(x - y)}{(y - x)} = \frac{(-1)(y - x)}{(y - x)} = -1$$

We obtain common factors by distributing one of the -1 factors in the numerator.

Method 3: We factor -1 from either the numerator or the denominator:

$$\frac{x - y}{y - x} = \frac{(x - y)}{(-1)(-y + x)} = \frac{(x - y)}{(-1)(x - y)} = \frac{1}{-1} = -1$$

Think about it 2: Simplify the expression in Example 5 two more ways. First, multiply the denominator by $(-1)(-1)$ and simplify as in method 2. Second, factor -1 from the numerator and simplify as in method 3.

It does not matter which method you use to simplify expressions in Example 5. Choose a method that makes sense and consistently gives you the correct result. Following are three important observations about Example 5.

The factors $(x - y)$ and $(y - x)$ are additive inverses, or opposites. **Additive inverses** *add to zero.*

A rational expression $\dfrac{x - y}{y - x}$, containing additive inverses in the numerator and denominator, simplifies to -1.

A negative sign on a fraction may be placed in any of three positions—in the numerator, in the denominator, or before the fraction. Thus, these fractions all have the same value:

$$\frac{-a}{b} = \frac{a}{-b} = -\frac{a}{b}, \quad b \neq 0$$

Caution: Expressions such as $x - y$ and $x + y$ are not additive inverses. The sum of $x - y$ and $x + y$ is $2x$, not zero.

EXAMPLE 6 Simplifying with additive inverses   Which expressions simplify to -1?

a. $\dfrac{5 - x}{x - 5}$ **b.** $\dfrac{2 - x}{x + 2}$ **c.** $\dfrac{x + 3}{x - 3}$

SOLUTION We add to show whether the expressions contain additive inverses (opposites).

a. $5 - x + x - 5 = 0$
The numerator and denominator are additive inverses. The expression simplifies to -1.

b. $2 - x + x + 2 = 4$
The numerator and denominator are not additive inverses. No other simplification is possible.

c. $x + 3 + x - 3 = 2x$
The numerator and denominator are not additive inverses. No other simplification is possible.

We can summarize our simplification as follows.

> When the numerator and denominator of a rational expression are the same, $\dfrac{a}{a}$, the expression simplifies to 1.
>
> When the numerator and denominator of a rational expression are additive inverses, $\dfrac{a - b}{b - a}$, the expression simplifies to -1.

Multiplication of Rational Expressions

Recall that to multiply fractions we factor, simplify, and then multiply across the numerators and the denominators:

$$\frac{12}{9} \cdot \frac{5}{4} = \frac{3 \cdot 4 \cdot 5}{3 \cdot 3 \cdot 4} = \frac{5}{3}$$

As in multiplying fractions, we factor rational expressions and simplify before multiplying. In some cases, no multiplication will need to be done.

EXAMPLE 7   Multiplying rational expressions   Multiply $\dfrac{5x}{y} \cdot \dfrac{y^2}{10x^2}, x \neq 0, y \neq 0.$

SOLUTION $\dfrac{5x}{y} \cdot \dfrac{y^2}{10x^2} = \dfrac{5xy^2}{10x^2y} = \dfrac{5 \cdot x \cdot y \cdot y}{2 \cdot 5 \cdot x \cdot x \cdot y} = \dfrac{y}{2x}$

In this solution, we write the numerators and denominators as products, factor them completely, and eliminate common factors.

MULTIPLYING RATIONAL EXPRESSIONS

> **1.** Write the product with a single numerator and denominator:
>
> $$\frac{a}{b} \text{ times } \frac{c}{d} = \frac{a \cdot c}{b \cdot d}, \quad b \neq 0, d \neq 0$$
>
> **2.** Factor as needed.
>
> **3.** Eliminate common factors from the numerator and denominator. *Do not eliminate common terms.*
>
> **4.** Multiply remaining factors in the numerator and in the denominator.

Your instructor may direct you to leave your answers in factored form to show that no further common factors need to be eliminated.

In Example 8, the expressions contain addition or subtraction operations. To eliminate common factors, we must factor the expressions into the product of a monomial and a binomial, such as $a(b + c)$, or the product of two binomials, $(a + b)(c + d)$.

EXAMPLE 8   Multiplying rational expressions   Multiply $\dfrac{x^2 + 2x}{x + 6} \cdot \dfrac{4x^2 - 25}{2x^2 + 9x + 10}, x \neq -6,$ $x \neq -\frac{5}{2}, x \neq -2.$

SOLUTION $\dfrac{x^2 + 2x}{x + 6} \cdot \dfrac{4x^2 - 25}{2x^2 + 9x + 10} = \dfrac{x(x + 2)(2x - 5)(2x + 5)}{(x + 6)(x + 2)(2x + 5)} = \dfrac{x(2x - 5)}{x + 6}$

We write the two fractions as one fraction in factored form and then eliminate common factors.

Caution: If we multiply the numerators (or denominators) in Example 8, the resulting expression is not easily simplified:

$$\frac{x^2 + 2x}{x + 6} \cdot \frac{4x^2 - 25}{2x^2 + 9x + 10} = \frac{4x^4 + 8x^3 - 25x^2 - 50x}{2x^3 + 21x^2 + 64x + 60}$$

If your homework solutions contain expressions like the one on the right above, go back to the original exercise statement, factor it, and eliminate common factors from the numerator and denominator before multiplying.

Division of Rational Expressions

The fact that dividing a dollar into 4 equal parts is the same as finding $\frac{1}{4}$ of a dollar reminds us that division by a number a is the same as multiplication by the reciprocal $1/a$:

$$\$1.00 \div 4 = \$1.00 \cdot \tfrac{1}{4} = \$0.25$$

◼ DIVISION OF FRACTIONS

> To divide, multiply the first fraction by the reciprocal of the second:
>
> $$\frac{a}{b} \text{ divided by } \frac{c}{d} = \frac{a}{b} \text{ times } \frac{d}{c} = \frac{ad}{bc}, \quad b \neq 0, c \neq 0, d \neq 0$$

Rational expressions have the same relationship between multiplication and division as fractions do. Examples 9 and 10 illustrate division of rational expressions and remind us to simplify the expressions by factoring.

EXAMPLE 9 Dividing rational expressions   Divide $\dfrac{a^2 + ab}{a^2 - b^2} \div \dfrac{a}{b}$. Assume no zero denominators.

SOLUTION

$$\frac{a^2 + ab}{a^2 - b^2} \div \frac{a}{b} \qquad \text{Change division to multiplication.}$$

$$= \frac{a^2 + ab}{a^2 - b^2} \cdot \frac{b}{a} \qquad \text{Factor.}$$

$$= \frac{a(a + b) \cdot b}{(a - b)(a + b) \cdot a} \qquad \text{Eliminate common factors.}$$

$$= \frac{b}{a - b}$$

EXAMPLE 10 Dividing rational expressions   Divide $\dfrac{x^2 - 4x - 5}{x - 3} \div \dfrac{5 + 4x - x^2}{x^2 - 9}$. Assume no zero denominators.

SOLUTION

$$\frac{x^2 - 4x - 5}{x - 3} \div \frac{5 + 4x - x^2}{x^2 - 9} \qquad \text{Change to multiplication.}$$

$$= \frac{x^2 - 4x - 5}{x - 3} \cdot \frac{x^2 - 9}{5 + 4x - x^2} \qquad \text{Factor.}$$

$$= \frac{(x - 5)(x + 1)(x - 3)(x + 3)}{(x - 3)(5 - x)(1 + x)} \qquad \text{Eliminate common factors.}$$

$$= \frac{(x - 5)(x + 3)}{(5 - x)} \qquad \text{Look for additive inverses.}$$

$$= (-1)(x + 3)$$

Complex Rational Expressions: Method 1, Reciprocals

A second method, using an LCD, is presented in Section 7.6.

The technique of changing division to multiplication by a reciprocal may be applied to simplifying complex rational expressions and expressions containing units of measure.

A fraction that contains a rational expression in either the numerator or the denominator is called a **complex rational expression**. When the numerator, the denominator, or both are simple fractions, we may replace the fraction bar with a division sign.

EXAMPLE 11 Simplifying complex rational expressions   Simplify $\dfrac{\dfrac{a}{b}}{\dfrac{c}{d}}$. Assume no zero denominators.

SOLUTION

$$\frac{\dfrac{a}{b}}{\dfrac{c}{d}} = \frac{a}{b} \div \frac{c}{d} = \frac{a}{b} \cdot \frac{d}{c} = \frac{ad}{bc}$$

We simplify the complex rational expression by writing it as a division problem with the longer fraction bar replaced by a division sign. The division is then changed to multiplication by the reciprocal.

EXAMPLE 12 Simplifying complex rational expressions   Simplify $\dfrac{\dfrac{x+2}{x^2}}{\dfrac{x^2-4}{x}}$.

SOLUTION

$$\frac{\dfrac{x+2}{x^2}}{\dfrac{x^2-4}{x}}$$ Replace the middle fraction bar with division.

$$= \frac{x+2}{x^2} \div \frac{x^2-4}{x}$$ Change to multiplication by the reciprocal.

$$= \frac{x+2}{x^2} \cdot \frac{x}{x^2-4}$$ Factor.

$$= \frac{(x+2) \cdot x}{x \cdot x \cdot (x+2)(x-2)}$$ Eliminate common factors.

$$= \frac{1}{x(x-2)}$$

Simplifying units of measure is one of the most useful applications of changing division to multiplication.

EXAMPLE 13 Simplifying with units   Predict the units obtained from each expression. Simplify by first writing each rate as a fraction.

a. $\dfrac{\$2.70 \text{ per dozen}}{12 \text{ cookies per dozen}}$

b. $\dfrac{24 \text{ capsules per box}}{4 \text{ capsules per day}}$

SOLUTION   **a.** We anticipate that the dozens will be eliminated, leaving the units as dollars per cookie:

$$\frac{\$2.70 \text{ per dozen}}{12 \text{ cookies per dozen}} = \frac{\dfrac{\$2.70}{1 \text{ dozen}}}{\dfrac{12 \text{ cookies}}{1 \text{ dozen}}} = \frac{\$2.70}{1 \text{ dozen}} \cdot \frac{1 \text{ dozen}}{12 \text{ cookies}}$$

$$= \frac{\$2.70}{12 \text{ cookies}} = \$0.225 \text{ per cookie}$$

Our prediction is correct.

b. Following the pattern in part a, we might predict that the capsules will be eliminated, leaving the units as boxes per day:

$$\frac{24 \text{ capsules per box}}{4 \text{ capsules per day}} = \frac{\dfrac{24 \text{ capsules}}{1 \text{ box}}}{\dfrac{4 \text{ capsules}}{1 \text{ day}}} = \frac{24 \text{ capsules}}{1 \text{ box}} \cdot \frac{1 \text{ day}}{4 \text{ capsules}}$$

$$= \frac{24 \text{ days}}{4 \text{ boxes}} = 6 \text{ days per box}$$

Our prediction is not correct. The units are the reciprocal of our prediction.

ANSWER BOX

Warm-up: **1.** $(x - 4)(x + 4)$ **2.** $(x - 5)(x + 1)$ **3.** $(2x + 5)(x + 2)$
4. $x(x - 3)$ **5.** $(x + 3)(x - 2)$ **6.** $3x(x + 2)$ **7.** $(x + 1)(x + 3)$
8. $(x + 5)^2$ **9.** $2x(3x + 2)$ **10.** $(2x - 5)(2x + 5)$ **11.** $(5 - x)(1 + x)$
12. $(4 - x)(1 + x)$ **Think about it 1:** The quotient of ages is $\frac{2}{1}$ when the
mother reaches twice what her age was when the child was born. The
quotient will never reach $\frac{1}{1}$ because the mother will always be 20 years older.
Think about it 2:

$$\frac{x - y}{y - x} = \frac{(x - y)}{(-1)(-1)(y - x)} = \frac{(x - y)}{(-1)(x - y)} = \frac{1}{-1} = -1$$

$$\frac{x - y}{y - x} = \frac{(-1)(-x + y)}{y - x} = \frac{(-1)(y - x)}{(y - x)} = -1$$

7.4   Exercises

For what numbers will the expressions in Exercises 1 to 6 be undefined?

1. $\dfrac{4 - x}{x + 3}$ $x = -3$

2. $\dfrac{x + 3}{2 - x}$ $x = 2$

3. $\dfrac{a + 3}{4 - a}$ $a = 4$

4. $\dfrac{3 - b}{b^2 + 3b + 2}$ $x = -2, x = -1$

5. $\dfrac{x(x - 3)}{x^2 - 3x + 28}$ defined for all $\mathbb{R}$

6. $\dfrac{x(x + 1)}{x^2 - 2x - 8}$ $x = -2, x = 4$

Give the additive inverse of each expression in Exercises 7 and 8.

7. a. $x + y$ $-x - y$ **b.** $-x + y$ $x - y$ **c.** $y - x$ $x - y$

8. a. $-x + y$ $x - y$ **b.** $-x - y$ $x + y$ **c.** $x - y$ $y - x$

What expression should be placed in the denominator or numerator in Exercises 9 and 10 to make a true statement? State any restrictions on the variables, as needed, to prevent undefined expressions.

9. a. $\dfrac{x - 3}{x - 3} = 1$ $x \neq 3$ **b.** $\dfrac{x + 2}{-2 - x} = -1$ $x \neq -2$

c. $\dfrac{b - a}{b - a} = 1$ $b \neq a$ **d.** $\dfrac{3 - x}{x - 3} = -1$ $x \neq 3$

10. a. $\dfrac{x - 3}{3 - x} = -1$ $x \neq 3$ **b.** $\dfrac{x + 2}{x + 2} = 1$ $x \neq -2$

c. $\dfrac{b - a}{a - b} = -1$ $a \neq b$ **d.** $\dfrac{3 - x}{3 - x} = 1$ $x \neq 3$

In Exercises 11 to 20, factor and simplify. State any restrictions on the variables.

11. a. $\dfrac{16a^2}{10a}$ $\dfrac{8a}{5}, a \neq 0$ **b.** $\dfrac{2x^2 y}{10xy^2}$ $\dfrac{x}{5y}, x \neq 0, y \neq 0$

c. $\dfrac{-49x^3 y^5}{63x^2 y^2}$ $\dfrac{-7xy^3}{9}, x \neq 0, y \neq 0$ **d.** $\dfrac{64a^3 b^2 c}{-48a^5 b^3}$ $\dfrac{-4c}{3a^2 b}, a \neq 0, b \neq 0$

12. a. $\dfrac{12k^2}{4k^3}$ $\dfrac{3}{k}, k \neq 0$ **b.** $\dfrac{8xy^2}{2xy}$ $4y, x \neq 0, y \neq 0$

c. $\dfrac{-72x^3 y^5}{42x^4 y^2}$ $\dfrac{-12y^3}{7x}, x \neq 0, y \neq 0$ **d.** $\dfrac{54a^4 b^3 c^3}{-36ab^4 c}$ $\dfrac{-3a^3 c^2}{2b}, a \neq 0, b \neq 0, c \neq 0$

13. a. $\dfrac{xy}{2x + y}$ simplified; $y \neq -2x$ **b.** $\dfrac{xy}{2x + xy}$ $\dfrac{y}{2 + y}, x \neq 0, y \neq -2$

14. a. $\dfrac{y^2}{6x^2 + y}$ simplified; $y \neq -6x^2$

b. $\dfrac{xy}{6x^2 + 2xy}$ $\dfrac{y}{6x + 2y}, x \neq 0, y \neq -3x$

15. a. $\dfrac{3 - x}{(x + 3)(x - 3)}$ $-\dfrac{1}{x + 3}, x \neq \pm 3$

b. $\dfrac{x^2 - 9}{x^2 + 5x + 6}$ $\dfrac{x - 3}{x + 2}, x \neq -3, -2$

16. a. $\dfrac{x - 1}{(1 - x)(x + 5)}$ $\dfrac{-1}{x + 5}, x \neq -5, 1$

b. $\dfrac{x^2 - 16}{x^2 + 3x - 4}$ $\dfrac{x - 4}{x - 1}, x \neq -4, 1$

17. a. $\dfrac{2ac + 4bc}{4ad + 8bd}$ $\dfrac{c}{2d}, a \neq -2b, d \neq 0$

b. $\dfrac{6x^2 + 3x}{12x^2 - 6x}$ $\dfrac{2x + 1}{2(2x - 1)}, x \neq 0, \frac{1}{2}$

18. a. $\dfrac{3ab + 3ac}{5b^2 + 5bc}$ $\dfrac{3a}{5b}, b \neq -c, 0$

b. $\dfrac{5x^2 - 10x}{2x^2 - 4x}$ $\dfrac{5}{2}, x \neq 0, 2$

19. a. $\dfrac{x^2 + x - 6}{2 - x}$ $-(x + 3), x \neq 2$

b. $\dfrac{x - 3}{6 - 2x}$ $-\dfrac{1}{2}, x \neq 3$

20. a. $\dfrac{x^2 + x - 2}{2x + 4}$ $\dfrac{x - 1}{2}, x \neq -2$

b. $\dfrac{x - 4}{12 - 3x}$ $-\dfrac{1}{3}, x \neq 4$

In Exercises 21 to 34, multiply or divide as indicated. Assume no zero denominators.

21. a. $\dfrac{1}{x} \cdot \dfrac{x^2}{1}$ x

b. $\dfrac{1}{a} \div \dfrac{a^2b^2}{1}$ $\dfrac{1}{a^3b^2}$

c. $\dfrac{a}{b} \div \dfrac{a^2b}{b^2}$ $\dfrac{1}{a}$

d. $\dfrac{b}{a} \div \dfrac{a^2}{b^2}$ $\dfrac{b^3}{a^3}$

e. $\dfrac{x}{y} \div \dfrac{x^3}{y^2}$ $\dfrac{y}{x^2}$

f. $\dfrac{18}{cd} \div \dfrac{4d^2}{9c}$ $\dfrac{81}{2d^3}$

22. a. $\dfrac{1}{x} \cdot \dfrac{x^3}{1}$ x^2

b. $\dfrac{1}{b} \div \dfrac{a^2b^2}{1}$ $\dfrac{1}{a^2b^3}$

c. $\dfrac{x^2}{y} \div \dfrac{xy^2}{x^3}$ $\dfrac{x^4}{y^3}$

d. $\dfrac{a}{b} \cdot \dfrac{b^2}{a^2}$ $\dfrac{b}{a}$

e. $\dfrac{x^2}{y^3} \div \dfrac{x}{y}$ $\dfrac{x}{y^2}$

f. $\dfrac{28}{xy} \div \dfrac{63x^3}{2y^2}$ $\dfrac{8y}{9x^4}$

23. a. $\dfrac{a^2 + 7a + 12}{a^2 - 4} \div \dfrac{a^2 + 4a}{a - 2}$ $\dfrac{a + 3}{a(a + 2)}$

b. $\dfrac{x^2 - 2x}{x} \cdot \dfrac{x^2 - 1}{x^2 - 3x + 2}$ $x + 1$

24. a. $\dfrac{b - 3}{b^2 - 4b + 3} \div \dfrac{b^2 - b}{b - 1}$ $\dfrac{1}{b(b - 1)}$

b. $\dfrac{x^2 - 6x + 9}{x^2 + 3x} \div \dfrac{x - 3}{x + 3}$ $\dfrac{x - 3}{x}$

25. a. $\dfrac{x^2 + 3x}{x} \cdot \dfrac{x^2 - x - 6}{x^2 - 9}$ $x + 2$

b. $\dfrac{4a^2 + 4a + 1}{4 - 9a^2} \div \dfrac{4a^2 + 2a}{3a - 2}$ $\dfrac{-2a - 1}{2a(2 + 3a)}$

26. a. $\dfrac{b^2 + 2b + 1}{b + 1} \cdot \dfrac{b^2}{b^2 + b}$ b

b. $\dfrac{x^2 - 4}{x - 2} \div \dfrac{1}{x^2 - x}$ $(x + 2)(x - 1)x$

27. a. $\dfrac{x + 2}{x^2 - 4x + 4} \cdot \dfrac{x^2 - 2x}{x + 2}$ $\dfrac{x}{x - 2}$

b. $\dfrac{a^2 - 5a}{a^2 + 5a} \div \dfrac{a^2 - 10a + 25}{a}$ $\dfrac{a}{(a + 5)(a - 5)}$

28. a. $\dfrac{x^2 - 6x + 9}{x^2 + 3x} \div \dfrac{x^2 - 9}{x}$ $\dfrac{x - 3}{(x + 3)^2}$

b. $\dfrac{x^2 - 2x}{3x^2 - 5x - 2} \cdot \dfrac{9x^2 - 4}{9x^2 - 12x + 4}$ $\dfrac{x(3x + 2)}{(3x + 1)(3x - 2)}$

29. a. $\dfrac{a^2 - b^2}{a^2 + 2ab + b^2} \cdot \dfrac{a + b}{a^3 - b^3}$ $\dfrac{1}{a^2 + ab + b^2}$

b. $\dfrac{x^2 - 2xy + y^2}{x^2 - y^2} \cdot \dfrac{x - y}{y - x}$ $\dfrac{-(x - y)}{x + y}$ or $\dfrac{y - x}{x + y}$

30. a. $\dfrac{a^2 - 2ab + b^2}{a^2 - b^2} \cdot \dfrac{a^3 + b^3}{a - b}$ $a^2 - ab + b^2$

b. $\dfrac{y^2 - x^2}{x^2 + 2xy + y^2} \cdot \dfrac{x + y}{x - y}$ -1

31. a. $\dfrac{a^3 - b^3}{a + b} \cdot \dfrac{a^2 + 2ab + b^2}{a^2 - b^2}$ $a^2 + ab + b^2$

b. $\dfrac{x^2 + 2xy + y^2}{x^3 + y^3} \cdot \dfrac{x^2 - y^2}{x + y}$ $\dfrac{(x + y)(x - y)}{x^2 - xy + y^2}$ or $\dfrac{x^2 - y^2}{x^2 - xy + y^2}$

32. a. $\dfrac{a^3 + b^3}{a^2 - b^2} \cdot \dfrac{b - a}{a + b}$ $\dfrac{-(a^2 - ab + b^2)}{a + b}$

b. $\dfrac{y - x}{x^2 + 2xy + y^2} \cdot \dfrac{x + y}{x^3 - y^3}$ $\dfrac{-1}{(x + y)(x^2 + xy + y^2)}$

33. a. $\dfrac{6 + x - 2x^2}{9 - 4x^2} \cdot \dfrac{15 - 7x - 2x^2}{x^2 - 2x}$ $\dfrac{-(x + 5)}{x}$

b. $\dfrac{25 - 9x^2}{6x^2 + 7x - 5} \div \dfrac{5 + 7x - 6x^2}{2 + 2x - 4x^2}$ $\dfrac{-2(x - 1)}{2x - 1}$

34. a. $\dfrac{3x^2 - 5x + 2}{6x^2 + x - 5} \cdot \dfrac{18x^2 - 3x - 10}{8 - 18x^2}$ $\dfrac{-(x-1)}{2(x+1)}$

b. $\dfrac{6x^2 - 13x + 6}{14x^2 - 25x + 6} \div \dfrac{14 - 21x}{49x^2 + 7x - 6}$ $\dfrac{-(7x+3)}{7}$

In Exercises 35 and 36, write each phrase as a complex rational expression and then divide. Assume that neither a nor b is zero.

35. a. The quotient of $\dfrac{1}{a}$ and $\dfrac{1}{b}$ $\dfrac{\frac{1}{a}}{\frac{1}{b}} = \dfrac{b}{a}$

b. The quotient of $\dfrac{1}{b}$ and a $\dfrac{\frac{1}{b}}{a} = \dfrac{1}{ab}$

c. The quotient of $\dfrac{1}{b}$ and b $\dfrac{\frac{1}{b}}{b} = \dfrac{1}{b^2}$

d. The quotient of $\dfrac{a}{b}$ and $\dfrac{1}{b}$ $\dfrac{\frac{a}{b}}{\frac{1}{b}} = a$

36. a. The quotient of $\dfrac{1}{b}$ and $\dfrac{1}{a}$ $\dfrac{\frac{1}{b}}{\frac{1}{a}} = \dfrac{a}{b}$

b. The quotient of $\dfrac{a}{b}$ and $\dfrac{1}{a}$ $\dfrac{\frac{a}{b}}{\frac{1}{a}} = \dfrac{a^2}{b}$

c. The quotient of a and $\dfrac{1}{a}$ $\dfrac{a}{\frac{1}{a}} = a^2$

d. The quotient of $\dfrac{1}{b}$ and $\dfrac{a}{b}$ $\dfrac{\frac{1}{b}}{\frac{a}{b}} = \dfrac{1}{a}$

Simplify the expressions in Exercises 37 to 40.

37. $\dfrac{\frac{x^2 - 3x + 2}{x}}{\frac{x^2 - 1}{x^2}}$ $\dfrac{x(x-2)}{x+1}$ or $\dfrac{x^2 - 2x}{x+1}$

38. $\dfrac{\frac{x^2 - 4}{x^2}}{\frac{x^2 + 5x + 6}{x}}$ $\dfrac{x-2}{x(x+3)}$ or $\dfrac{x-2}{x^2 + 3x}$

39. $\dfrac{\frac{x - 2}{x^2 + 3x - 4}}{\frac{2 - x}{x - 1}}$ $\dfrac{-1}{x+4}$

40. $\dfrac{\frac{x - 1}{x^2 - 9}}{\frac{x^2 + 2x - 3}{3 - x}}$ $\dfrac{-1}{(x+3)^2}$

Simplify the expressions in Exercises 41 to 50. The word "per" means division and may be replaced by a fraction bar.

41. $\dfrac{5280 \text{ feet}}{88 \text{ feet per second}}$ 60 sec

42. $\dfrac{93{,}000{,}000 \text{ miles}}{186{,}000 \text{ miles per second}}$ 500 sec

43. $\dfrac{65 \text{ miles per hour}}{15 \text{ miles per gallon}}$ $4\frac{1}{3}$ gal/hr

44. $\dfrac{500 \text{ miles per hour}}{200 \text{ gallons per hour}}$ 2.5 mi/gal

45. $\dfrac{24 \text{ cans per case}}{\$3.98 \text{ per case}}$ ≈ 6 cans/\$

46. $\dfrac{250 \text{ vitamins per bottle}}{7 \text{ vitamins per week}}$ ≈ 35 weeks/bottle (medications not rounded up to next week)

47. $\dfrac{8 \text{ stitches per inch}}{\frac{1 \text{ foot}}{12 \text{ inches}}}$ 96 stitches/ft

48. $\dfrac{140 \text{ heartbeats per minute}}{0.25 \text{ mile per minute}}$ 560 heartbeats/min

49. $\dfrac{95 \text{ words per minute}}{300 \text{ words per page}}$ ≈ 0.3 page/min

50. $\dfrac{440 \text{ cycles per second}}{344 \text{ meters per second}}$ ≈ 1.3 cycles/m

In Exercises 51 to 54, choose an appropriate operation (multiplication or division) to eliminate the common unit so as to create a sensible result.

51. Diesel fuel for a truck costs \$1.40 per gallon, and the truck gets 7 miles per gallon. division; \$0.20/mi or 5 mi/\$

52. A student takes 15 credit hours per term and pays \$80 per credit hour. multiplication; \$1200/term

53. Take classes 4 days a week and spend 2 hours per day between classes having coffee. multiplication; spend 8 hr/wk having coffee

54. Frozen Soy Dreams bars are 8 per package and cost \$3.60 per package. division; \$0.45/bar or 2.2 bars/\$

▮ **Error Analysis**

Describe errors in the simplification of the first expression in Exercises 55 to 58.

55. $\dfrac{x^2 + x + 2}{x} = x^2 + 2$ Terms, not factors, were simplified.

56. $\dfrac{(x-1)(x-2)}{2-x} = x - 1$ $\dfrac{x-2}{2-x} = -1$, not 1

57. $\dfrac{x^2 - 2x + 1}{x - 2} = \dfrac{x(x-2) + 1}{(x-2)}$ Numerator factors to $(x-1)(x-1)$; 1 is also divided by $x - 2$.

58. $\dfrac{x \cdot x + 2 \cdot 2}{x - 2} = x - 2$ 2 and x are not common factors; $x^2 + 4$ does not factor.

59. When we multiply $a \cdot b \cdot c$, why do we not multiply $a \cdot b$ and $a \cdot c$? The result has an extra factor of a; multiplication is not distributive over multiplication.

In Exercises 60 and 61, indicate true or false and explain.

60. $\dfrac{4}{5} \div \dfrac{2}{3} = \dfrac{4 \div 2}{5 \div 3} = \dfrac{2}{\frac{5}{3}} = \dfrac{6}{5}$ true; always works; may give fraction in numerator or denominator

61. $\dfrac{5}{6} \div \dfrac{1}{2} = \dfrac{5 \div 1}{6 \div 2} = \dfrac{5}{3}$ true; always works; may give fraction in numerator or denominator

▮ **Projects**

62. Graphs of Rational Functions Graph the expressions in Exercises 15a, 15b, and 19a both before and after simplifying. Use parentheses to separate the numerator from the denominator. Then answer the following questions for each expression.

a. Is there any major difference between the graphs before and after simplifying?

b. Does the expression give a straight line? If so, what is the slope of the line?

c. Are there any nearly vertical parts? Where? Why?

d. Are there any holes in the graphs? To make the holes show up, set a viewing window that gives x in tenths. Where are the holes? Why?
See Additional Answers.

63. Comparative Ages Return to the equations in part e of Example 1, page 384. Enter the equations into a graphing calculator, with $\boxed{Y =}$.

a. Look at the columns for Y1, Y2, and Y3 in the graphing calculator table. Does the table agree with that in the example? yes

b. What causes the error in the entry for $x = 0$ under Y3? 0 in denominator

c. Find the child's age when the quotient of ages is $\frac{2}{1}$.
20

d. Move the cursor down the column for Y3. What number does this column seem to be approaching?
1

e. Look at the expression that describes the quotient of ages, Y3. Change it into two fractions, and simplify if possible. How does the result explain your answer to part d? Y3 = $1 + 20/x$; as x gets large, $20/x$ gets small.

f. Set an appropriate viewing window, and graph the three functions. Use a window setting that gives integers for x as you trace. List the dimensions of your window.

g. Trace the graphs, and confirm that the values agree with those in Table 8. Name the function formed by each graph. Y1 is linear; Y2 is constant; Y3 is rational.

h. Trace Y3 for values of x between 0 and 1. Reset the window as needed. What happens to Y3 as x gets close to zero? increases rapidly

i. Trace Y3 to find the coordinates for $x = 0$. What does the calculator show for coordinates at $x = 0$? Why does this happen?
no value given; function is undefined at $x = 0$.

j. Trace Y3 into negative values of x. Do these have any meaning in the problem situation? no

---

 7 Mid-Chapter Test

Simplify the expressions in Exercises 1 to 13.

1. $\dfrac{32}{24}$ $\frac{4}{3}$

2. $\dfrac{3}{10} \cdot \dfrac{5}{18}$ $\frac{1}{12}$

3. $\dfrac{abc}{aces}$ $\frac{b}{es}$

4. $\dfrac{3}{10} \div \dfrac{5}{18}$ $1\frac{2}{25}$

5. $\dfrac{12cd^2}{8c^2 d}$ $\frac{3d}{2c}$

6. $\dfrac{(x-2)(x+3)}{(x+3)}$ $x - 2$

7. $\dfrac{a-6}{a^2 - 3a - 18}$ $\frac{1}{a+3}$

8. $\dfrac{b + bc}{b}$ $1 + c$

9. $\dfrac{12x^2 - 43x - 20}{x^2 - 16}$ $\frac{12x + 5}{x + 4}$

10. $\dfrac{ab^2}{c^2} \div \dfrac{ac}{b}$ $\frac{b^3}{c^3}$

11. $\dfrac{x^2 - 16}{x^2 + 6x + 8} \cdot \dfrac{x+2}{x-2}$ $\frac{x-4}{x-2}$

12. $\dfrac{\dfrac{x+1}{x}}{\dfrac{x^2}{x^2 - 1}}$ $x(x - 1)$ or $x^2 - x$

13. $\dfrac{x^3 - 64}{x^2 + 6x + 8} \div \dfrac{4 - x}{x + 2}$ $\frac{-(x^2 + 4x + 16)}{x + 4}$

14. For what values of x is $f(x) = \dfrac{x}{(x + 2)(2x^2 - 7x + 3)}$ undefined? $x = -2, x = \frac{1}{2}, x = 3$

Simplify the expressions in Exercises 15 and 16.

15. $\dfrac{2\frac{2}{3} \text{ yards}}{5 \text{ feet}}$ $\frac{8}{5}$

16. $\dfrac{150 \text{ months}}{15 \text{ years}}$ $\frac{5}{6}$

17. Change 1,000,000 tablespoons into gallons, using the following facts.

$$1 \text{ quart} = \tfrac{1}{4} \text{ gallon}$$
$$2 \text{ cups} = 1 \text{ pint}$$
$$16 \text{ tablespoons} = 1 \text{ cup}$$
$$1 \text{ quart} = 2 \text{ pints} \qquad 3906.25 \text{ gal}$$

18. An airplane flies 450 miles per hour and costs $1500 per hour to operate. By eliminating hours, we obtain what two facts? 0.3 mi/$ and $3.33/mi

In Exercises 19 and 20, solve the proportions. Round to the nearest hundredth.

19. $\dfrac{a}{15} = \dfrac{12}{35}$ $a = 5.14$

20. $\dfrac{x+1}{8} = \dfrac{2x - 1}{14}$ $x = 11$

21. Solve for the missing sides in these similar triangles.
$n = 8, p = 13.2$

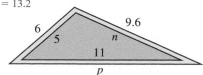

In Exercises 22 and 23, find the equation of the line through the ordered pairs. State whether the inputs and outputs are proportional.

22. $(-4, -6), (10, 15)$ $y = \frac{3}{2}x$; prop.

23. (1 minute, $0.60), (3 minutes, $1.50)
$y = 0.45x + 0.15$; not prop.

24. a. What is the constant of variation in the figure below?

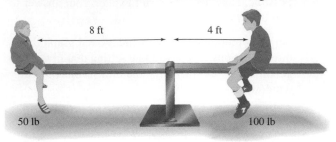

8 ft 4 ft

50 lb 100 lb

400 lb·ft

b. If the 100-pound child is replaced by an 80-pound child, where should the 80-pound child sit?
5 ft from pivot point

25. Suppose that the number of cookies per child varies inversely with the number of children present. There are 100 cookies altogether. Describe the situation with an inverse variation equation. Identify the constant of variation.
$y = 100/x$; y = no. of cookies/child, x = no. of children; $k = 100$ cookies

26. The price of a round pizza varies with the square of its diameter, $p = kd^2$. If a 13-inch pizza costs $12, what will a 16-inch pizza cost? $18.18

7.5 Division of Polynomials and Related Graphs

Objectives

▮ Graph selected rational expressions.

▮ Divide polynomial expressions using long division.

▮ Find whether a divisor is a factor of the dividend.

▮ Use long division to factor cubic expressions.

> **WARM-UP**
>
> Divide these numbers with long division.
>
> **1.** $11\overline{)1331}$ 121
> **2.** $35\overline{)7213}$ $206\frac{3}{35}$
>
> Subtract these expressions.
>
> **3.** $\begin{array}{r} 2x^2 + 3x \\ -(2x^2 + 2x) \\ \hline x \end{array}$
> **4.** $\begin{array}{r} x^3 + 3x^2 \\ -(x^3 + x^2) \\ \hline 2x^2 \end{array}$
>
> **5.** $\begin{array}{r} x^3 + 3x^2 \\ -(x^3 - x^2) \\ \hline 4x^2 \end{array}$
> **6.** $\begin{array}{r} x^2 - 3x \\ -(x^2 - 4x) \\ \hline x \end{array}$

IN THIS SECTION, look for how division of polynomials with algebra is similar to division with numbers, how division relates to finding factors, and how division verifies factoring formulas. Also look for some unusual features in the graphs.

Division Vocabulary and Review

In our work in this section, we will use the following words to describe the parts of a division problem:

$$\text{divisor}\overline{)\text{dividend}} \quad \text{quotient + remainder}$$

When a division problem is written as a fraction, the dividend becomes the numerator and the divisor becomes the denominator:

$$\frac{\text{numerator}}{\text{denominator}} = \frac{\text{dividend}}{\text{divisor}}$$

EXAMPLE 1 Reviewing long division   Divide 1331 by 11 using long division.

SOLUTION   We use a five-step process repeated three times.

$$
\begin{array}{r}
121 \\
11\overline{)1331} \\
\underline{11} \\
023 \\
\underline{22} \\
011 \\
\underline{11} \\
0
\end{array}
$$

Enter estimates in the quotient at left.
1. Estimate 13/11. 2. Enter 1 in quotient.
3. Multiply 11 by estimate. 4. Subtract.
5. Bring down next number. 1. Estimate 23/11. 2. Enter 2 in quotient.
3. Multiply 11 by estimate. 4. Subtract.
5. Bring down next number. 1. Estimate 11/11. 2. Enter 1 in quotient.
3. Multiply 11 by estimate. 4. Subtract.
5. Bring down next number. Stop.

The division has a zero remainder, showing that $1331/11 = 121$.

Check: $11 \cdot 121 = 1331$ ✓   ■

Graphing to Predict the Quotient

EXAMPLE 2 Exploring the quotient of a rational expression

a. Make a table and a graph for $y = (x^3 + 3x^2 + 3x + 1) \div (x + 1)$.
b. Find the equation of the graph. Check with calculator regression.
c. What is the degree of the numerator? the denominator? the equation of the graph? How do the degrees predict the quotient?
d. Why does the graph have a gap?

SOLUTION   **a.** Table 9 contains values for $(x^3 + 3x^2 + 3x + 1) \div (x + 1)$. The table is symmetric about $x = -1$. Figure 26 contains the graph of the quotient from $y = (x^3 + 3x^2 + 3x + 1) \div (x + 1)$.

b. The graph appears to be a parabola with the same shape as $y = x^2$. The vertex at $(-1, 0)$ means a horizontal shift to the left by 1 unit and suggests the equation $y = (x + 1)^2$. Applying quadratic regression on a calculator, we obtain $y = x^2 + 2x + 1$.

c. The degree of the numerator is 3; the degree of the denominator is 1; the degree of the equation is 2. We subtract the degrees to find the degree of the quotient.

 d. When we trace to exactly $x = -1$ on a graphing calculator, there is a blank for y and a *hole in the graph*. This is because the original expression is undefined at $x = -1$. If we reset the window to **Xmin** $= -4.7$ and **Xmax** $= 4.7$, then a trace will show no output for $x = -1$.

TABLE 9

x	$\dfrac{x^3 + 3x^2 + 3x + 1}{x + 1}$
-4	9
-3	4
-2	1
-1	undefined
0	1
1	4
2	9

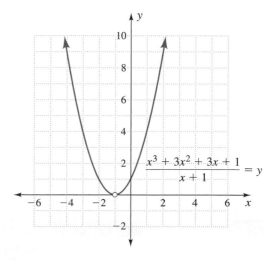

FIGURE 26   ■

Division with a Zero Remainder

In Example 3, we apply long division to the rational expression in Example 2. Note the similarity between the division process here and the division with numbers in Example 1.

EXAMPLE 3 **Finding a quotient with long division** Divide $x^3 + 3x^2 + 3x + 1$ by $x + 1$. Compare the quotient with the graphical result in Example 2.

SOLUTION

$$
\begin{array}{r}
x^2 + 2x + 1 \\
x + 1 \overline{\smash)\ x^3 + 3x^2 + 3x + 1} \\
\underline{-(x^3 + \ x^2)} \\
2x^2 + 3x \\
\underline{-(2x^2 + 2x)} \\
1x + 1 \\
\underline{-(1x + 1)} \\
0
\end{array}
$$

Enter estimates in the quotient at left.

1. Estimate x^3/x. 2. Enter x^2 in quotient.

3. Multiply $x + 1$ by estimate. 4. Subtract.

5. Bring down next term. 1. Estimate $2x^2/x$. 2. Enter $+2x$ in quotient.

3. Multiply $x + 1$ by estimate. 4. Subtract.

5. Bring down next term. 1. Estimate $1x/x$. 2. Enter $+1$ in quotient.

3. Multiply $x + 1$ by estimate. 4. Subtract.

5. Bring down next term. Stop.

The division indicates that the original expression and the quotient (with the restriction on x) are equivalent:

$$\frac{x^3 + 3x^2 + 3x + 1}{x + 1} = x^2 + 2x + 1, \quad x \neq -1$$

The quotient agrees with the parabolic graph in Example 2.

Check: $(x + 1)(x^2 + 2x + 1) = x^3 + 3x^2 + 3x + 1$ ✓ ▬

In Example 3, the division came out even—with a zero remainder—leading us to conclude that $x^3 + 3x^2 + 3x + 1$ factors into $(x + 1)(x^2 + 2x + 1)$.

Think about it 1: Write $x^3 + 3x^2 + 3x + 1$ as the product of three factors.

▬ SUMMARY: DIVIDING POLYNOMIAL EXPRESSIONS WITH LONG DIVISION

- Arrange the divisor and the dividend in descending order of exponents.
- Replace missing terms in the dividend with terms having a zero numerical coefficient.
- Divide, using the five steps:
  1. Estimate how many times the first term of the divisor divides into the first term of the dividend.
  2. Enter the estimate in the quotient.
  3. Multiply the divisor by the estimate, and place the product below the like terms in the dividend.
  4. Subtract.
  5. Bring down the next term.
- Repeat steps 1 to 5, as needed.

In Example 4, we predict the quotient with a table and graph. In Example 5, we use the polynomial division process for the same rational expression.

EXAMPLE 4 Exploring the graph of a rational expression

a. Make a table and a graph for $y = (x^3 - 1) \div (x - 1)$. What happens at $x = 1$?

b. Predict the degree of the resulting equation from the graph, and fit an equation with calculator regression.

SOLUTION **a.** Table 10 shows a symmetry of outputs for pairs of inputs. The graph in Figure 27 appears to be a parabola. If you plot the graph on a calculator, using **Xmin** $= -4.7$ and **Xmax** $= 4.7$, you can trace integer inputs to find the hole that corresponds to the undefined value at $x = 1$.

Because the remainder is zero, the range for this function is limited.

TABLE 10

x	$\dfrac{x^3 - 1}{x - 1}$
-3	7
-2	3
-1	1
0	1
1	undefined
2	7
3	13

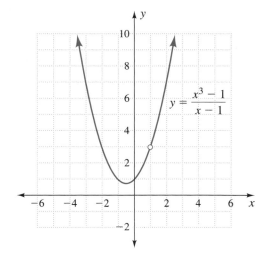

FIGURE 27

b. The parabolic shape of the graph in Figure 27 suggests a quotient of degree 2, or a quadratic expression. Applying quadratic regression to the table data, we obtain a quotient equation $y = x^2 + x + 1$. ▬

Think about it 2: What is the value of $\dfrac{x^3 - 1}{x - 1}$ as x takes on decimal values close to 1?

In Example 5, we do the division on the expression graphed in Example 4. *Note how we adjust the dividend expression in order to carry out the division steps.*

EXAMPLE 5 Finding a quotient with long division   Divide $x^3 - 1$ by $x - 1$.

SOLUTION
$$
\begin{array}{r}
x^2 \\
x - 1 \overline{)\ \ x^3 - 1} \\
\underline{-(x^3 - x^2)} \\
\end{array}
$$

Enter estimates in quotient at left.
1. Estimate x^3/x. 2. Enter x^2 in quotient.
3. Multiply $x - 1$ by estimate.
4. Subtraction is not possible because 1 and x^2 are not like terms.

To provide like terms for the subtraction, we must add terms with a coefficient of zero, $0x^2$ and $0x$, between the x^3 and the 1 in the dividend.

$$
\begin{array}{r}
x^2 + \ x + 1 \\
x - 1 \overline{)\ \ x^3 + 0x^2 + 0x - 1} \\
\underline{-(x^3 - x^2)} \\
x^2 + 0x \\
\underline{-(x^2 - x)} \\
+x - 1 \\
\underline{-(x - 1)} \\
0
\end{array}
$$

Enter estimates in quotient at left.
1. Estimate x^3/x. 2. Enter x^2 in quotient.
3. Multiply $x - 1$ by estimate. 4. Subtract.
5. Bring down next term. 1. Estimate x^2/x. 2. Enter $+x$ in quotient.
3. Multiply $x - 1$ by estimate. 4. Subtract.
5. Bring down next term. 1. Estimate x/x. 2. Enter $+1$ in quotient.
3. Multiply $x - 1$ by estimate. 4. Subtract.
5. Bring down next term. Stop.

The division indicates that the original expression and the quotient with the restriction on x are equivalent:

$$\frac{x^3 - 1}{x - 1} = x^2 + x + 1, \quad x \neq 1$$

Check: $(x - 1)(x^2 + x + 1) = x^3 - 1$ ✓ ▬

The division in Example 5 shows that $x^3 - 1$ factors to $(x - 1)(x^2 + x + 1)$. In the examples thus far, we have obtained a zero remainder for each division. *When there is no remainder, we say the divisor is a factor of the dividend.* This idea can be represented visually as follows:

$$\text{factor} \overline{)\text{dividend}}^{\,\text{quotient} + \text{zero}}$$

▮ ZERO REMAINDERS AND FACTORS

> A zero remainder from division indicates that the divisor is a factor of the dividend.
>
> When the division is written as a fraction, a zero remainder indicates that the numerator and the denominator have a common factor.
>
> A zero remainder indicates that the graph will contain a hole with an x-coordinate that makes a zero denominator in the original fraction.

Division with a Nonzero Remainder

In Examples 6 and 7, we repeat the table, graph, and polynomial division process for an expression that does not come out even after division; it has a nonzero remainder. We start with the division.

EXAMPLE 6   Finding a quotient and remainder with long division   Predict the degree of the quotient, and then divide $x^3 - 2x^2 - 5x + 6$ by $x + 1$.

SOLUTION   The quotient will be of degree 2, a quadratic expression.

$$
\begin{array}{r}
x^2 - 3x - 2 \\
x + 1 \overline{)\ x^3 - 2x^2 - 5x + 6} \\
\underline{-(x^3 +\ x^2)} \\
-3x^2 - 5x \\
\underline{-(-3x^2 - 3x)} \\
-2x + 6 \\
\underline{-(-2x - 2)} \\
8
\end{array}
$$

This time, the remainder is not zero. We have

$$\frac{x^3 - 2x^2 - 5x + 6}{x + 1} = x^2 - 3x - 2 + \frac{8}{x + 1}, \quad x \neq -1$$ ▬

In Example 6, we wrote the remainder, 8, over the divisor, $x + 1$. This is a standard form for nonzero remainders:

$$\text{divisor} \overline{)\text{dividend}}^{\,\text{quotient} + \frac{\text{remainder}}{\text{divisor}}}$$

EXAMPLE 7   **Graphing a rational expression**   Make a table and a graph for
$y = (x^3 - 2x^2 - 5x + 6) \div (x + 1)$.

SOLUTION   Table 11 shows $y = (x^3 - 2x^2 - 5x + 6) \div (x + 1)$. The table lacks the symmetry found in the previous two division problems. There is an undefined output at $x = -1$ because of a zero denominator, yet instead of being a parabolic graph with a hole, the graph in Figure 28 is nearly vertical near $x = -1$.

Because the remainder is not zero, the range for this function is $(-\infty, +\infty)$.

TABLE 11

x	$\dfrac{x^3 + 2x^2 - 5x + 6}{x + 1}$
-4	23.3
-3	12
-2	0
-1	undefined
0	6
1	0
2	-1.3

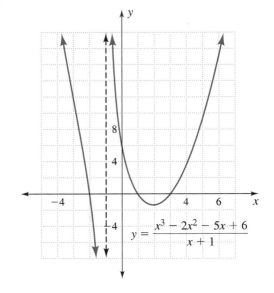

$$y = \frac{x^3 - 2x^2 - 5x + 6}{x + 1}$$

FIGURE 28

Nothing in Table 11 or Figure 28 explains the difference between the shapes of the graphs in Examples 4 and 7. The division, on the other hand, does explain why the graph in Figure 28 is nearly vertical at $x = -1$.

The remainder, 8, over the divisor, $x + 1$, forms a rational expression that is undefined at $x = -1$. The graph of $y = 8/(x + 1)$, a hyperbola, has a nearly vertical part as it approaches -1 from either the left or the right (see Figure 29). This nearly vertical behavior causes a change in the parabolic shape that we might expect from the quotient, $x^2 - 3x - 2$.

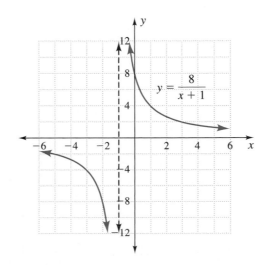

$$y = \frac{8}{x + 1}$$

FIGURE 29

■ In Example 8, we compare the graph of the quotient $x^2 - 3x - 2$ with that of the quotient and remainder to show the effect of the nonzero remainder.

EXAMPLE 8 Graphing the rational expression together with the quotient
Graph $y = (x^3 - 2x^2 - 5x + 6) \div (x + 1)$ with $y = x^2 - 3x - 2$.

SOLUTION   In Figure 30, the equations are graphed to the same scale as in Figure 28. In Figure 31, they are graphed with a zoom out. The parabolic shape of the graph seems clearer in Figure 31. The original curve follows the parabola closely except near $x = -1$. The nearly vertical behavior near $x = -1$ is due to the remainder, $8/(x + 1)$.

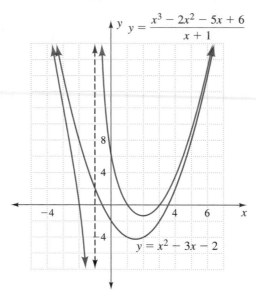

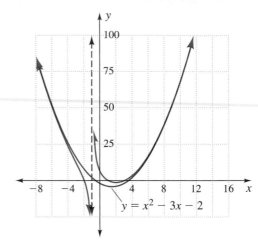

FIGURE 30   **FIGURE 31**

There is a special line associated with the graph of a hyperbola. That line is not important now, but it should be mentioned for completeness. When the number n makes the denominator zero in a rational expression that is in lowest terms (contains no common factors), we say that the line with equation $x = n$ is a *vertical asymptote*. An **asymptote** is *the name given to the line that a graph approaches.*

The vertical asymptote is *not* part of the set of points forming the graph. The vertical asymptote is a boundary. The asymptote is the dashed line, $x = -1$, drawn in Figures 28 to 31. Graphing calculators do not draw a vertical asymptote.

GRAPHING CALCULATOR TECHNIQUE:
CALCULATOR ERRORS

With certain viewing window settings, many graphing calculators draw an almost vertical line on the graph at an undefined point. *This line is an error made by the calculator. The calculator evaluates inputs to the left and right of the undefined point and connects the outputs* (see Figure 32a).

The line will disappear if the calculator is set on dot rather than connected mode or if the viewing window is adjusted to force the calculator to evaluate the input making the expression undefined (see Figure 32b).

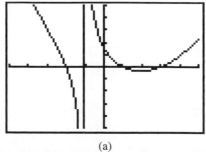

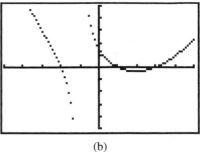

(a)   (b)

FIGURE 32 Comparing connected mode with dot mode: graph of
$Y = (X^3 - 2X^2 - 5X + 6)/(X + 1)$, -5 to 5 on X, -20 to 20 on Y

ANSWER BOX

Warm-up: 1. 121   **2.** $206\frac{3}{35}$ **3.** x **4.** $2x^2$ **5.** $4x^2$ **6.** x
Think about it 1: $x^3 + 3x^2 + 3x + 1 = (x + 1)(x^2 + 2x + 1)$
$$= (x + 1)^3$$
Think about it 2: As x gets close to 1, the output gets close to 3.

7.5   Exercises

Predict the outcome (cubic, quadratic, or linear) of each division in Exercises 1 to 12 and then divide.

1. $(x^3 - 3x^2 + 3x - 1) \div (x - 1)$ quadratic; $x^2 - 2x + 1$

2. $(x^3 + 1) \div (x + 1)$ quadratic; $x^2 - x + 1$

3. $(x^3 - 8) \div (x - 2)$ quadratic; $x^2 + 2x + 4$

4. $(x^3 + 6x^2 + 12x + 8) \div (x + 2)$ quadratic; $x^2 + 4x + 4$

5. $(2x^2 + 3x - 5) \div (x - 1)$ linear; $2x + 5$

6. $(6x^2 - x - 15) \div (2x + 3)$ linear; $3x - 5$

7. $(12x^2 - 20x - 25) \div (6x + 5)$ linear; $2x - 5$

8. $(21x^2 - 22x - 8) \div (3x - 4)$ linear; $7x + 2$

9. a. $(4x^4 - 13x^2 + 9) \div (x + 1)$ cubic; $4x^3 - 4x^2 - 9x + 9$

   b. Divide the answer by $x - 1$.   quadratic; $4x^2 - 9$

   c. Factor the answer from part b. $(2x + 3)(2x - 3)$

10. a. $(x^4 - 13x^2 + 36) \div (x - 3)$ cubic; $x^3 + 3x^2 - 4x - 12$

   b. Divide the answer by $x - 2$.   quadratic; $x^2 + 5x + 6$

   c. Factor the answer from part b. $(x + 2)(x + 3)$

11. a. $(x^4 - 37x^2 + 36) \div (x - 6)$ cubic; $x^3 + 6x^2 - x - 6$

   b. Divide the answer by $x + 6$.   quadratic; $x^2 - 1$

   c. Factor the answer from part b. $(x - 1)(x + 1)$

12. a. $(4x^4 - 37x^2 + 9) \div x - 3$ cubic; $4x^3 + 12x^2 - x - 3$

   b. Divide the answer by $x + 3$.   quadratic; $4x^2 - 1$

   c. Factor the answer from part b. $(2x - 1)(2x + 1)$

In Exercises 13 to 24, the division problems are stated in fraction form, with the denominator containing the divisor. Indicate for what values the expressions are defined, and complete the divisions.

13. $\dfrac{x^2 - 3x - 4}{x - 4}$ $\mathbb{R}$ except $x = 4$; $x + 1$

14. $\dfrac{x^2 + 3x + 2}{x - 1}$ $\mathbb{R}$ except $x = 1$; $x + 4 + \dfrac{6}{x - 1}$

15. $\dfrac{x^2 - 3x - 4}{x - 3}$ $\mathbb{R}$ except $x = 3$; $x - \dfrac{4}{x - 3}$

16. $\dfrac{x^2 + 3x + 2}{x}$ $\mathbb{R}$ except $x = 0$; $x + 3 + \dfrac{2}{x}$

17. $\dfrac{x^2 - 3x - 4}{x - 2}$ $\mathbb{R}$ except $x = 2$; $x - 1 - \dfrac{6}{x - 2}$

18. $\dfrac{x^2 + 3x + 2}{x + 1}$ $\mathbb{R}$ except $x = -1$; $x + 2$

19. $\dfrac{x^2 - 3x - 4}{x - 1}$ $\mathbb{R}$ except $x = 1$; $x - 2 - \dfrac{6}{x - 1}$

20. $\dfrac{x^2 + 3x + 2}{x + 2}$ $\mathbb{R}$ except $x = -2$; $x + 1$

21. $\dfrac{x^2 - 3x - 4}{x}$ $\mathbb{R}$ except $x = 0$; $x - 3 - \dfrac{4}{x}$

22. $\dfrac{x^2 + 3x + 2}{x + 3}$ $\mathbb{R}$ except $x = -3$; $x + \dfrac{2}{x + 3}$

23. $\dfrac{x^2 - 3x - 4}{x + 1}$ $\mathbb{R}$ except $x = -1$; $x - 4$

24. $\dfrac{x^2 + 3x + 2}{x + 4}$ $\mathbb{R}$ except $x = -4$; $x - 1 + \dfrac{6}{x + 4}$

25. a. Factor $x^2 - 3x - 4$. What was special about the quotient when the denominator was a factor of the numerator in Exercises 13 to 23 odd?
   $(x + 1)(x - 4)$; remainder of 0

   b. Make a table for $f(x) = x^2 - 3x - 4$. Use integer inputs on the interval $[-1, 4]$. How do the function values compare with the remainders in the answers to Exercises 13 to 23 odd?
   $f(x) = 0, -4, -6, -6, -4, 0$; same

   c. Use a graphing calculator to graph the expression in Exercise 15. Make a careful sketch of the graph.
   See Answer Section.

   d. Why does the graph have a nearly vertical portion? Where is it located?   approaching the restriction; at $x = 3$

   e. Describe the behavior of the graph for $x < -8$ and $x > 8$ compared with that of $y = x$.
   approaching function $y = x$

   f. Graph the expression in Exercise 13. Why is this graph different from that of the expression in Exercise 15?   See Answer Section; division has zero remainder.

   g. Graph the expression in Exercise 17. How is this graph the same as that of the expression in Exercise 15?
   See Answer Section; nearly vertical at its restriction $x = 2$

26. a. Factor $x^2 + 3x + 2$. What was special about the quotient when the denominator was not a factor in Exercises 14 to 24 even? $(x + 1)(x + 2)$; nonzero remainder

   b. Make a table for $f(x) = x^2 + 3x + 2$. Use integer inputs on the interval $[-4, 1]$. How do the function values compare with the remainders in the answers to Exercises 14 to 24 even?
   See Additional Answers; same

   c. Use a graphing calculator to graph the expression in Exercise 24. Make a careful sketch of the graph.
   See Additional Answers.

   d. Why does the graph have a nearly vertical portion? Where is it located?   approaching the restriction; at $x = -4$

   e. Describe the behavior of the graph for $x < -8$ and $x > 8$ compared with that of $y = x - 1$.
   approaching function $y = x - 1$

   f. Graph the expression in Exercise 16. How is this graph the same as that of the expression in Exercise 24?
   See Additional Answers; nearly vertical at its restriction $x = 0$

   g. Graph the expression in Exercise 20. Why is this graph different from that of the expression in Exercise 24?
   See Additional Answers; division has zero remainder.

Divide the expressions in Exercises 27 to 42. In which exercises is the denominator a factor of the numerator?

27. $\dfrac{x^3 - 1}{x - 2}$ $x^2 + 2x + 4 + \dfrac{7}{x - 2}$ **28.** $\dfrac{x^3 + 3x^2 + 3x + 1}{x + 2}$
$x^2 + x + 1 - \dfrac{1}{x + 2}$

29. $\dfrac{x^3 - 1}{x - 1}$ $x^2 + x + 1$; factor   **30.** $\dfrac{x^3 + 3x^2 + 3x + 1}{x + 1}$
$x^2 + 2x + 1$; factor

31. $\dfrac{x^3 - 1}{x}$ $x^2 - \dfrac{1}{x}$ **32.** $\dfrac{x^3 + 3x^2 + 3x + 1}{x}$ $x^2 + 3x + 3 + \dfrac{1}{x}$

33. $\dfrac{x^3 - 1}{x + 1}$ $x^2 - x + 1 + \dfrac{-2}{x + 1}$ **34.** $\dfrac{x^3 + 3x^2 + 3x + 1}{x - 1}$
$x^2 + 4x + 7 + \dfrac{8}{x - 1}$

35. $\dfrac{x^3 + 1}{x + 1}$ $x^2 - x + 1$; factor   **36.** $\dfrac{x^4 - 1}{x - 1}$
$x^3 + x^2 + x + 1$; factor

37. $\dfrac{x^4 + 1}{x + 1}$ $x^3 - x^2 + x - 1 + \dfrac{2}{x + 1}$ **38.** $\dfrac{x^3 + 8}{x + 2}$ $x^2 - 2x + 4$; factor

39. $\dfrac{x^4 - 4x^3 + 6x^2 - 4x + 1}{x - 1}$ $x^3 - 3x^2 + 3x - 1$; factor

40. $\dfrac{x^4 + 4x^3 + 6x^2 + 4x + 1}{x + 1}$ $x^3 + 3x^2 + 3x + 1$; factor

41. $\dfrac{x^2 + 3x^3 - 1}{x + 1}$
$3x^2 - 2x + 2 - \dfrac{3}{x + 1}$ **42.** $\dfrac{x + x^3 - 2x^2}{x + 1}$
$x^2 - 3x + 4 - \dfrac{4}{x + 1}$

In Exercises 43 to 46, use the calculator table feature to locate the undefined position (hole) in the graph. Start with **ΔTbl** = 1. Reset **ΔTbl** as needed to estimate the ordered pair describing the hole.

43. Exercise 35 $(-1, 3)$

44. Exercise 36 $(1, 4)$

45. Exercise 39 $(1, 0)$ **46.** Exercise 38 $(-2, 12)$

47. Use long division, $(x^3 + y^3) \div (x + y)$, to prove the factoring formula for $x^3 + y^3$.

48. Use long division, $(x^3 - y^3) \div (x - y)$, to prove the factoring formula for $x^3 - y^3$.

49. True or false: If the divisor $(x - a)$ is a factor of the dividend, there is a hole in the graph at $x = a$.   true

50. True or false: If the divisor is a factor of the dividend, the remainder is zero.   true

▪ Projects

51. Investigating Remainders

   a. Find $f(-2)$ and $f(5)$ for $f(x) = x^2 - 3x - 4$. Use long division to find the remainders for division of $x^2 - 3x - 4$ by $(x + 2)$ and $(x - 5)$. What do you observe? $f(-2) = f(5) = 6$; remainders equal 6.

   b. Find $f(-5)$ and $f(2)$ for $f(x) = x^2 + 3x + 2$. Use long division to find the remainders for division of $x^2 + 3x + 2$ by $(x + 5)$ and $(x - 2)$. What do you observe? $f(-5) = f(2) = 12$; remainders equal 12.

   c. Make a table for $f(x) = x^3 - 1$ for integer inputs from $x = -1$ to $x = 2$. Compare the outputs with the remainders in Exercises 27 to 33 odd.
   $f(x) = -2, -1, 0, 7$; same

Fill in the blanks.

   d. The output at $x = a, f(a)$, is the ___remainder___ upon division of $f(x)$ by $(x - a)$.

   e. When we divide a polynomial by one of its factors, the remainder is ___zero___.

52. Factoring Cubes List the values of these expressions.

   a. 4^3 64   **b.** 5^3 125   **c.** 6^3 216

   d. $(0.1)^3$ 0.001   **e.** $(0.4)^3$ 0.064 **f.** $(0.5)^3$ 0.125

   g. $(0.01)^3$ 0.000 001 **h.** $\left(\frac{1}{4}\right)^3$ $\frac{1}{64}$ **i.** $\left(\frac{1}{5}\right)^3$ $\frac{1}{125}$

   j. $\left(\frac{1}{6}\right)^3$ $\frac{1}{216}$

Divide these expressions.

   k. $(x^3 - 8) \div (x - 2)$ $x^2 + 2x + 4$

   l. $(x^3 + 27) \div (x + 3)$ $x^2 - 3x + 9$

   m. $27a^3 + 1 \div (3a + 1)$ $9a^2 - 3a + 1$

   n. $(y^3 + 0.001) \div (y + 0.1)$ $y^2 - 0.1y + 0.01$

   o. $\left(\frac{1}{64}x^3 + \frac{1}{27}\right) \div \left(\frac{x}{4} + \frac{1}{3}\right)$ $\frac{x^2}{16} - \frac{x}{12} + \frac{1}{9}$

   p. $(8b^3 + 125) \div (2b + 5)$ $4b^2 - 10b + 25$

   q. $(27x^3 - 64) \div (3x - 4)$ $9x^2 + 12x + 16$

   r. $(0.008a^3 + 1) \div (0.2a + 1)$ $0.04a^2 - 0.2a + 1$

   s. $(y^3 + 0.027) \div (y + 0.3)$ $y^2 - 0.3y + 0.09$

7.6 Addition and Subtraction of Rational Expressions

Objectives

▌ Find the common denominator for two or more rational expressions.

▌ Add and subtract rational expressions with like denominators.

▌ Add and subtract rational expressions with unlike denominators.

▌ Simplify complex rational expressions using least common denominators.

WARM-UP

Simplify these expressions.

1. $\frac{1}{5} + \frac{1}{3}$ $\frac{8}{15}$ **2.** $\frac{1}{6} + \frac{1}{10}$ $\frac{4}{15}$ **3.** $\frac{1}{6} - \frac{1}{10}$ $\frac{1}{15}$

4. $\frac{1}{4} + \frac{1}{4} + \frac{1}{8}$ $\frac{5}{8}$ **5.** $\frac{7}{12} - \frac{5}{18}$ $\frac{11}{36}$

Factor these expressions:

6. $x^2 + 3x + 2$ $(x+1)(x+2)$

7. $x^2 - 2x + 1$ $(x-1)^2$

8. $x^2 - 1$ $(x+1)(x-1)$

9. $x^2 + 5x + 6$ $(x+2)(x+3)$

10. $2x + 6$ $2(x+3)$

11. $xv_2 + xv_1$ $x(v_2 + v_1)$

12. Multiply and simplify:

$$\frac{6\left(\dfrac{1}{3} - x\right)}{6\left(x + \dfrac{1}{2}\right)} \qquad \frac{2 - 6x}{6x + 3} \ \text{ or } \ \frac{2(1 - 3x)}{3(2x + 1)}$$

IT IS COMMONLY ASSUMED that if a plane travels 100 miles at 500 miles per hour and another 100 miles at 300 miles per hour, the average speed for the whole trip is 400 miles per hour. This is inaccurate reasoning. This section will illustrate finding the correct answer.

Average Speed

EXAMPLE 1   **Exploring average trip speed**   An airplane travels 100 miles at 500 miles per hour and a second 100 miles at 300 miles per hour.

a. What is the total distance traveled?

b. What is the total time traveled?

c. What is the average speed for the whole trip?

SOLUTION   **a.** The total distance is 200 miles.

b. The total time is the time for the first 100 miles plus the time for the second 100 miles. For the first 100 miles,

$$t = \frac{D}{r} = \frac{100 \text{ mi}}{500 \text{ mph}} = \frac{100 \text{ mi}}{\dfrac{500 \text{ mi}}{1 \text{ hr}}} = 100 \text{ mi} \cdot \frac{1 \text{ hr}}{500 \text{ mi}} = \frac{1}{5} \text{ hr}$$

For the second 100 miles,

$$t = \frac{D}{r} = \frac{100 \text{ mi}}{300 \text{ mph}} = \frac{100 \text{ mi}}{\frac{300 \text{ mi}}{1 \text{ hr}}} = 100 \text{ mi} \cdot \frac{1 \text{ hr}}{300 \text{ mi}} = \frac{1}{3} \text{ hr}$$

The total time is

$$\frac{1}{5} \text{ hr} + \frac{1}{3} \text{ hr} = \left(\frac{3}{15} + \frac{5}{15} \right) \text{ hr} = \frac{8}{15} \text{ hr}$$

c. The average speed is the total distance divided by the total time. The average speed is

$$\text{Distance} \div \text{time} = 200 \text{ mi} \div \frac{8}{15} \text{ hr}$$

$$= \frac{200}{1} \text{ mi} \cdot \frac{15}{8} \frac{1}{\text{hr}}$$

$$= 375 \text{ miles per hour}$$

The average speed for the 200 miles is 375 miles per hour, slightly slower than the anticipated average of 400 miles per hour. One might think that the average was somehow due to the specific distances of 100 miles. In Example 2, we eliminate that possibility by letting each 100-mile distance be replaced by the variable x.

EXAMPLE 2 Exploring average trip speed Find the average speed for a trip where x miles are flown at 500 miles per hour and another x miles are flown at 300 miles per hour.

SOLUTION The total distance is $x + x = 2x$. The total time, in hours, is $\frac{x}{500} + \frac{x}{300}$. The factors used to find the least common denominator (LCD) need not be primes if common factors are readily apparent.

$$500 = 5 \cdot 100$$

$$300 = 3 \cdot 100$$

$$\text{LCD} = 3 \cdot 5 \cdot 100 = 1500$$

Student Note: See page 405 for more on the LCD.

Adding the total time expression gives

$$\frac{x}{500} + \frac{x}{300} = \frac{x \cdot 3}{5 \cdot 100 \cdot 3} + \frac{x \cdot 5}{3 \cdot 100 \cdot 5} \qquad \text{Set up the LCD.}$$

$$= \frac{3x}{1500} + \frac{5x}{1500} = \frac{3x + 5x}{1500} = \frac{8x}{1500} \text{ hr}$$

The average speed is total distance divided by total time:

$$2x \text{ mi} \div \frac{8x}{1500 \text{ hr}} = \frac{2x \text{ mi}}{1} \cdot \frac{1500}{8x} \frac{1}{\text{hr}}$$

$$= \frac{3000x}{8x} \frac{\text{mi}}{\text{hr}} = 375 \text{ miles per hour}$$

The average speed, 375 miles per hour, does not contain the variable x. Thus, this result is true for any trip where the first half is traveled at 500 miles per hour and the second half at 300 miles per hour. A round trip taken over the same route would be a typical example of such a situation.

Adding and Subtracting Rational Expressions

EQUIVALENT FRACTIONS   In order to add the fractions in Examples 1 and 2, we changed to equivalent fractions with the least common denominator. When we want to find an equivalent fraction, we multiply the numerator and denominator of the fraction by the same factor.

▪ **EQUIVALENT FRACTION PROPERTY**

> For all real numbers, $a \neq 0$, $c \neq 0$,
>
> $$\frac{b}{c} = \frac{b}{c} \cdot \frac{a}{a} = \frac{ab}{ac}$$

EXAMPLE 3   **Finding equivalent rational expressions**   Change to an equivalent expression with the indicated denominator.

a. $\dfrac{5}{6} = \dfrac{}{54}$

b. $\dfrac{3}{x} = \dfrac{}{2x^2}$

c. $\dfrac{a}{a+3} = \dfrac{}{2a(a+3)}$

d. $\dfrac{x}{x+1} = \dfrac{}{x^2+3x+2}$

SOLUTION   **a.** $\dfrac{5}{6} = \dfrac{5 \cdot 9}{6 \cdot 9} = \dfrac{45}{54}$

b. $\dfrac{3}{x} = \dfrac{3 \cdot 2x}{x \cdot 2x} = \dfrac{6x}{2x^2}$

c. $\dfrac{a}{a+3} = \dfrac{a \cdot 2a}{2a(a+3)} = \dfrac{2a^2}{2a(a+3)}$

d. $\dfrac{x}{x+1} = \dfrac{x(x+2)}{(x+1)(x+2)} - \dfrac{x(x+2)}{x^2+3x+2}$ ▪

LEAST COMMON DENOMINATORS   Although any common denominator may be used to add or subtract fractions, the advantage of using the least common denominator is that the answer is less likely to need simplifying.

▪ **LEAST COMMON DENOMINATOR (LCD)**

> To find the least common denominator:
>
> **1.** List the prime factors of each denominator.
>
> **2.** Compare the lists of prime factors.
>
> **3. a.** If the lists have no common factors, the LCD is the product of the denominators.
>
> **b.** If the lists have common factors, write each factor the highest number of times it appears in any one denominator. The LCD is the product of these factors.

EXAMPLE 4   **Adding and subtracting fractions**   Add and subtract $\frac{1}{6}$ and $\frac{1}{10}$.

SOLUTION   In addition and subtraction, we find a common denominator and change the fractions to equivalent fractions with that denominator. We factor the denominators: $6 = 2 \cdot 3$ and $10 = 2 \cdot 5$. The factors needed in the least common denominator are 2, 3, and 5, so the required denominator is $2 \cdot 3 \cdot 5 = 30$.

$$\frac{1}{6} + \frac{1}{10} = \frac{1 \cdot 5}{2 \cdot 3 \cdot 5} + \frac{1 \cdot 3}{2 \cdot 5 \cdot 3} = \frac{5}{30} + \frac{3}{30} = \frac{8}{30} = \frac{4}{15}$$

$$\frac{1}{6} - \frac{1}{10} = \frac{1 \cdot 5}{2 \cdot 3 \cdot 5} - \frac{1 \cdot 3}{2 \cdot 5 \cdot 3} = \frac{5}{30} - \frac{3}{30} = \frac{2}{30} = \frac{1}{15}$$ ▪

RATIONAL EXPRESSIONS   In the next several examples, we find the LCD and add or subtract rational expressions with variables in the denominator.

EXAMPLE 5 Adding rational expressions using a least common denominator   Find the LCD, and add:

$$\frac{2}{a^2b} + \frac{5}{abc}, \quad a \neq 0, b \neq 0, c \neq 0$$

SOLUTION   To find the LCD, we factor each denominator:

$$a^2b = a \cdot a \cdot b$$
$$abc = a \cdot b \cdot c$$

The LCD needs to be divisible by both denominators and needs two factors of a as well as one each of b and c. The product of these factors, $a \cdot a \cdot b \cdot c$, gives the LCD, a^2bc.

$$\frac{2}{a^2b} + \frac{5}{abc}$$ Change to equivalent fractions with an LCD.

$$= \frac{2 \cdot c}{a^2b \cdot c} + \frac{5 \cdot a}{abc \cdot a}$$ Multiply.

$$= \frac{2c}{a^2bc} + \frac{5a}{a^2bc}$$ Add numerators.

$$= \frac{2c + 5a}{a^2bc}$$

We also use factoring to find the LCD for rational expressions containing binomials or trinomials. Again, an LCD produces answers that are less likely to need simplification.

EXAMPLE 6 Subtracting rational expressions using an LCD   Find the LCD, and subtract:

$$\frac{x}{x^2 - 2x + 1} - \frac{2}{x^2 - 1}, \quad x \neq 1, -1$$

SOLUTION   To find the LCD, we factor the denominators:

$$x^2 - 2x + 1 = (x - 1)^2 = (x - 1)(x - 1)$$
$$x^2 - 1 = (x - 1)(x + 1)$$

The LCD will be the product of the three factors: $(x - 1)$, $(x - 1)$, and $(x + 1)$.

$$\frac{x}{x^2 - 2x + 1} - \frac{2}{x^2 - 1}$$ Factor the denominators.

$$= \frac{x}{(x - 1)(x - 1)} - \frac{2}{(x - 1)(x + 1)}$$ Set up the LCD and equivalent fractions.

$$= \frac{x(x + 1)}{(x - 1)(x - 1)(x + 1)} - \frac{2(x - 1)}{(x - 1)(x + 1)(x - 1)}$$ Combine the numerators.

$$= \frac{x(x + 1) - 2(x - 1)}{(x - 1)(x + 1)(x - 1)}$$ Apply the distributive property.

$$= \frac{x^2 + x - 2x + 2}{(x - 1)(x + 1)(x - 1)}$$ Add like terms.

$$= \frac{x^2 - x + 2}{(x - 1)^2(x + 1)}$$

Because $x^2 - x + 2$ does not factor, we know the expression cannot be simplified further. ▬

    Example 6 and the next example illustrate the fact that subtraction problems must be worked carefully because there may be a sign change when numerators are subtracted.

EXAMPLE 7   Subtracting rational expressions using an LCD   Find the LCD, and subtract:

$$\frac{3x}{x^2 + 5x + 6} - \frac{3}{2x + 6}, \quad x \neq -3, x \neq -2$$

SOLUTION   To find the LCD, we factor the denominators:

$$x^2 + 5x + 6 = (x + 2)(x + 3)$$
$$2x + 6 = 2(x + 3)$$

The LCD will be the product of the three factors: 2, $(x + 2)$, and $(x + 3)$.

$$\frac{3x}{x^2 + 5x + 6} - \frac{3}{2x + 6}$$ Factor the denominators.

$$= \frac{3x}{(x + 2)(x + 3)} - \frac{3}{2(x + 3)}$$ Set up the LCD with equivalent fractions.

$$= \frac{2 \cdot 3x}{2 \cdot (x + 2)(x + 3)} - \frac{3 \cdot (x + 2)}{2(x + 3) \cdot (x + 2)}$$ Combine the numerators.

$$\frac{6x - 3(x + 2)}{2(x + 2)(x + 3)}$$ Apply the distributive property.

$$= \frac{6x - 3x - 6}{2(x + 2)(x + 3)}$$ Add like terms.

$$= \frac{3x - 6}{2(x + 2)(x + 3)}$$ Factor the numerator.

$$= \frac{3(x - 2)}{2(x + 2)(x + 3)}$$

There are no common factors in the numerator and denominator, so the expression cannot be simplified. ▬

DENOMINATORS CONTAINING ADDITIVE INVERSES   In Section 7.4, we simplified rational expressions containing additive inverses to -1. Here, we add (or subtract) expressions containing additive inverses as denominators.

EXAMPLE 8   Adding expressions containing additive inverses   Find the common denominator, and add:

$$\frac{b}{a - 1} + \frac{a}{1 - a}$$

SOLUTION   The denominators this time are additive inverses, or opposites. We obtain a common denominator, $a - 1$, if we multiply the numerator and denominator of the second expression by -1:

$$\frac{b}{a - 1} + \frac{a}{1 - a} = \frac{b}{a - 1} + \frac{a(-1)}{(1 - a)(-1)}$$ Set up the LCD.

$$= \frac{b}{a - 1} + \frac{-a}{a - 1}$$ Combine the numerators.

$$= \frac{b - a}{a - 1}$$ ▬

If denominators of rational expressions are additive inverses, multiply the numerator and denominator of one expression by −1 to change to a common denominator.

SUMMARY: ADDING OR SUBTRACTING RATIONAL NUMBERS

1. To find the least common denominator (LCD), factor each denominator.
   a. If the denominators have no common factors, the LCD is the product of the denominators.
   b. If the denominators have common factors, the LCD is the product formed by including each factor the highest number of times it appears in any one denominator.

2. Convert to a common denominator.

3. Add or subtract the numerators, and place the result over the common denominator.

4. Factor, and simplify the answer.

Applications

THEATER EXITS   In the next two examples, we explore the rate at which a theater empties when various doors are used. This is an important consideration in theater design and in building codes.

EXAMPLE 9   Exploring theater exit rates   Suppose a theater has two doors, one of which permits emptying the theater in 4 minutes and the other in 8 minutes. The rate of exit is $\frac{1}{4}$ of the theater capacity per minute for the first door and $\frac{1}{8}$ of the theater capacity per minute for the second door. How much of the theater has been emptied after
a. 1 minute?   **b.** 2 minutes?   **c.** 3 minutes?

SOLUTION   Figure 33 shows the portion of the theater that has exited at the end of each minute.

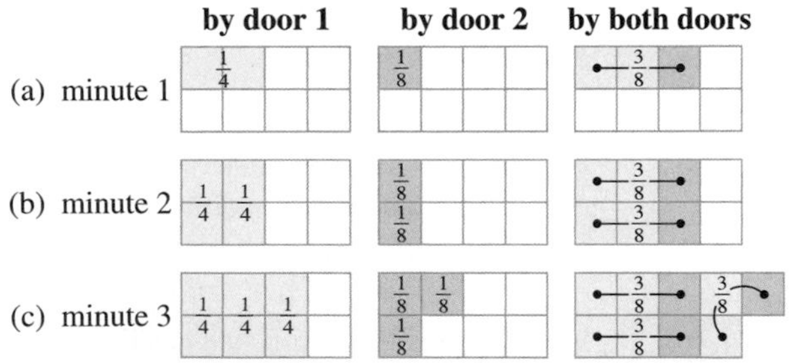

FIGURE 33

a. In the first minute, each door permits its portion of the theater to exit:

$$\left(\frac{1}{4} + \frac{1}{8}\right)\left(\frac{\text{capacity}}{\text{minute}}\right)(1 \text{ minute}) = \left(\frac{3}{8}\right)(1) = \frac{3}{8} \text{ capacity}$$

At the end of the first minute, $\frac{3}{8}$ of the theater has exited.

b. In two minutes, each door permits twice the one-minute portion to exit:

$$\left(\frac{1}{4} + \frac{1}{8}\right)\left(\frac{\text{capacity}}{\text{minute}}\right)(2 \text{ minutes}) = \left(\frac{3}{8}\right)(2) = \frac{3}{4} \text{ capacity}$$

At the end of the second minute, $\frac{3}{4}$ of the theater has exited.

c. In three minutes, each door permits three times its portion to exit:

$$\left(\frac{1}{4} + \frac{1}{8}\right)\left(\frac{\text{capacity}}{\text{minute}}\right)(3 \text{ minutes}) = \left(\frac{3}{8}\right)(3) = \frac{9}{8} = 1\frac{1}{8} \text{ capacity}$$

At the end of the third minute, $1\frac{1}{8}$ of the theater has exited. Because full capacity is $\frac{8}{8} = 1$, the theater is empty when $\frac{8}{8}$ of the people have passed through the doors. The theater will become empty during the third minute. ▬

In each step in Example 9, we multiply the rate of exit per minute by the time in minutes to obtain the fraction of the theater capacity that has exited. When the product reaches 1, the theater is empty. The process in Example 9 extends to three or more doors, as we will see in Example 10.

EXAMPLE 10 Exploring theater exit rates   Three doors (two entry doors and one emergency exit) in a theater empty the theater in 4 minutes, 4 minutes, and 8 minutes, respectively. If all doors are functioning properly, in how many minutes can the theater be emptied?

SOLUTION

$$\left(\frac{1}{4} + \frac{1}{4} + \frac{1}{8}\right)\left(\frac{\text{capacity}}{\text{minute}}\right)t = 1 \text{ capacity} \qquad \text{Simplify units.}$$

$$\left(\frac{1}{4} + \frac{1}{4} + \frac{1}{8}\right)t\left(\frac{\text{capacity}}{\text{minute}}\right)\left(\frac{\text{minute}}{\text{capacity}}\right) = (1 \text{ capacity})\left(\frac{\text{minute}}{\text{capacity}}\right)$$

$$\text{Simplify and set up LCD.}$$

$$\left(\frac{1 \cdot 2}{4 \cdot 2} + \frac{1 \cdot 2}{4 \cdot 2} + \frac{1}{8}\right)t = 1 \text{ min} \qquad \text{Simplify.}$$

$$\frac{2 + 2 + 1}{8}t = 1 \text{ min} \qquad \text{Simplify.}$$

$$\frac{5t}{8} = 1 \text{ min} \qquad \text{Multiply by } \frac{8}{5}.$$

$$t = \frac{8}{5} \text{ min}$$

$$t = 1.6 \text{ min} \qquad ▬$$

ADDING RATES FORMULA   We can summarize the theater exit process with a formula. The formula uses subscripts to distinguish the amounts of time the different doors require to empty the theater.

If the first door can empty the theater in t_1 minutes, the second in t_2 minutes, and the third in t_3 minutes, then a formula to find the number of minutes t required to empty the theater if all three doors are functioning is

$$\left(\frac{1}{t_1} + \frac{1}{t_2} + \frac{1}{t_3}\right)t = 1$$

This formula is usually divided on both sides by t:

$$\left(\frac{1}{t_1} + \frac{1}{t_2} + \frac{1}{t_3}\right) = \frac{1}{t}$$

Examples 9 and 10 and the formula above illustrate the method for adding related rates, one of the most common applications of rational expressions.

Complex Rational Expressions: Method 2, LCD

The first method was given in Section 7.4.

In Examples 11 and 12, the least common denominator of the terms in a complex rational expression is used to simplify the expression.

EXAMPLE 11 Simplifying complex rational expressions by multiplying by the LCD   Simplify $\dfrac{\frac{1}{3} - x}{x + \frac{1}{2}}$.

SOLUTION   The common denominator for terms in both the numerator and the denominator expressions is 6. If we multiply both the numerator and the denominator by 6, we clear out the fractions within them:

$$\frac{\frac{1}{3} - x}{x + \frac{1}{2}} = \frac{6(\frac{1}{3} - x)}{6(x + \frac{1}{2})} = \frac{\frac{6}{3} - 6x}{6x + \frac{6}{2}} = \frac{2 - 6x}{6x + 3}$$

After removing denominators from a complex rational expression, we factor to simplify:

$$\frac{2 - 6x}{6x + 3} = \frac{2(1 - 3x)}{3(2x + 1)}$$

There are no common factors, so the expression is simplified.   ▬

In Example 12, we return to the setting of Examples 1 and 2 and set up a general formula for the average speed.

EXAMPLE 12 Finding a formula for average trip speed   Find a formula for determining the average speed v for a trip of two equal distances, each of length x, given any two speeds v_1 and v_2.

SOLUTION   The total distance, d, for the trip is $x + x = 2x$. The total time for the trip is the sum of the times for the two parts:

$$t = t_1 + t_2, \quad \text{where } t_1 = \frac{x}{v_1} \text{ and } t_2 = \frac{x}{v_2}$$

The average speed is total distance divided by total time: $v = d/t$. Substituting for d and for total time t, which is $t_1 + t_2$, we get a complex rational expression:

$$v = \frac{2x}{\dfrac{x}{v_1} + \dfrac{x}{v_2}} \qquad \text{Multiply by the LCD, } v_1v_2.$$

$$= \frac{2x(v_1v_2)}{\left(\dfrac{x}{v_1} + \dfrac{x}{v_2}\right)(v_1v_2)} \qquad \text{Distribute the LCD.}$$

$$= \frac{2xv_1v_2}{\dfrac{x v_1 v_2}{v_1} + \dfrac{xv_1 v_2}{v_2}} \qquad \text{Simplify.}$$

$$= \frac{2xv_1v_2}{xv_2 + xv_1} \qquad \text{Factor } x \text{ in the denominator.}$$

$$= \frac{2xv_1v_2}{x(v_2 + v_1)} \qquad \text{Simplify.}$$

$$= \frac{2v_1v_2}{v_2 + v_1} \qquad \text{Note that there is no } x \text{ here.}$$

The distance traveled, x, is not part of the average speed formula for a round trip or for two segments of equal length.   ▬

ANSWER BOX

Warm-up: **1.** $\frac{8}{15}$ **2.** $\frac{4}{15}$ **3.** $\frac{1}{15}$ **4.** $\frac{5}{8}$ **5.** $\frac{11}{36}$ **6.** $(x+1)(x+2)$
7. $(x-1)^2$ **8.** $(x+1)(x-1)$ **9.** $(x+2)(x+3)$ **10.** $2(x+3)$
11. $x(v_2+v_1)$ **12.** $(2-6x)/(6x+3)$ or $2(1-3x)/[3(2x+1)]$

7.6 Exercises

In Exercises 1 to 8, change the fractions to equivalent fractions with the indicated denominator.

1. a. $\dfrac{3}{5} = \dfrac{21}{35}$ **b.** $\dfrac{5}{8} = \dfrac{30}{48}$ **c.** $\dfrac{8}{3} = \dfrac{128}{48}$

2. a. $\dfrac{5}{9} = \dfrac{30}{54}$ **b.** $\dfrac{7}{4} = \dfrac{63}{36}$ **c.** $\dfrac{5}{9} = \dfrac{20}{36}$

3. a. $\dfrac{3}{8} = \dfrac{15}{40}$ **b.** $\dfrac{5}{6} = \dfrac{45}{54}$ **c.** $\dfrac{5}{4} = \dfrac{40}{32}$

4. a. $\dfrac{6}{5} = \dfrac{54}{45}$ **b.** $\dfrac{3}{4} = \dfrac{21}{28}$ **c.** $\dfrac{5}{8} = \dfrac{35}{56}$

5. a. $\dfrac{4}{xy} = \dfrac{8x}{2\,x^2y}$ **b.** $\dfrac{3}{a^2b} = \dfrac{6b}{2\,a^2b^2}$

6. a. $\dfrac{5}{ab^2} = \dfrac{15a^2b}{3a^3b^3}$ **b.** $\dfrac{2}{x^2y} = \dfrac{10xy}{5x^3y^2}$

7. a. $\dfrac{x}{x-1} = \dfrac{x^2+x}{x^2-1}$

b. $\dfrac{a}{a+b} = \dfrac{a^2-ab}{a^2-b^2}$

8. a. $\dfrac{n}{m+2n} = \dfrac{mn+2n^2}{m^2+4mn+4n^2}$

b. $\dfrac{y}{y+1} = \dfrac{y^2+3y}{y^2+4y+3}$

Add or subtract the expressions in Exercises 9 and 10. State any restrictions on the variables.

9. a. $\dfrac{3}{4} - \dfrac{x}{4}$ $\dfrac{3-x}{4}$ **b.** $\dfrac{2}{5x} - \dfrac{7}{5x}$ $-\dfrac{1}{x}, x \neq 0$

c. $\dfrac{4}{x+2} - \dfrac{x^2}{x+2}$ $2-x, x \neq -2$

d. $\dfrac{2x}{x^2-1} - \dfrac{x-1}{x^2-1}$ $\dfrac{1}{x-1}, x \neq -1, 1$

10. a. $\dfrac{2}{5} + \dfrac{x}{5}$ $\dfrac{2+x}{5}$ **b.** $\dfrac{2}{3x} - \dfrac{5}{3x}$ $-\dfrac{1}{x}, x \neq 0$

c. $\dfrac{-9}{x+3} + \dfrac{x^2}{x+3}$ $x-3, x \neq -3$

d. $\dfrac{2}{x^2-1} - \dfrac{x+1}{x^2-1}$ $\dfrac{-1}{x+1}, x \neq -1, 1$

What is the common denominator in each of the fractions or rational expressions in Exercises 11 and 12? Add or subtract as indicated. Assume no zero denominators.

11. a. $\dfrac{1}{4} + \dfrac{7}{12}$ $12; \dfrac{5}{6}$

b. $\dfrac{2}{a} - \dfrac{5}{2a}$ $2a; -\dfrac{1}{2a}$

c. $\dfrac{b}{a^2} - \dfrac{c}{ab^2}$ $a^2b^2; \dfrac{b^3-ac}{a^2b^2}$

d. $\dfrac{5}{x^2-2x} + \dfrac{3}{x^2-4}$ $x(x-2)(x+2); \dfrac{8x+10}{x(x-2)(x+2)}$

e. $\dfrac{5x}{x^2-6x+9} - \dfrac{3x}{x^2-9}$ $(x+3)(x-3)^2; \dfrac{2x^2+24x}{(x+3)(x-3)^2}$

12. a. $\dfrac{5}{18} + \dfrac{7}{24}$ $72; \dfrac{41}{72}$

b. $\dfrac{7}{y^2} - \dfrac{2}{y}$ $y^2; \dfrac{7-2y}{y^2}$

c. $\dfrac{3}{a} - \dfrac{2}{b} + \dfrac{7}{a} - \dfrac{3}{b}$ $ab; \dfrac{10b-5a}{ab}$

d. $\dfrac{4}{x-5} - \dfrac{3x}{x^2-25}$ x^2-25 or $(x-5)(x+5); \dfrac{x+20}{(x-5)(x+5)}$

e. $\dfrac{x}{x^2+5x+6} - \dfrac{2x}{x^2+2x}$ $x(x+2)(x+3); \dfrac{-x-6}{(x+2)(x+3)}$

In Exercises 13 to 28, add or subtract the rational expressions, as indicated. Assume no zero denominators.

13. $\dfrac{5}{4a} + \dfrac{3}{6b}$ $\dfrac{2a+5b}{4ab}$

14. $\dfrac{5}{2b} + \dfrac{7}{6a}$ $\dfrac{15a+7b}{6ab}$

15. $\dfrac{x}{x-3} + \dfrac{1}{x}$ $\dfrac{x^2+x-3}{x(x-3)}$

16. $\dfrac{1}{x+2} + \dfrac{x+1}{x-2}$ $\dfrac{x^2+4x}{(x+2)(x-2)}$

17. $\dfrac{5x}{x-1} - \dfrac{8+x}{x}$ $\dfrac{4x^2-7x+8}{x(x-1)}$

18. $\dfrac{x}{x+1} - \dfrac{3}{x-1}$ $\dfrac{x^2-4x-3}{(x+1)(x-1)}$

19. $\dfrac{x}{x-3} + \dfrac{3}{x^2-6x+9}$ $\dfrac{x^2-3x+3}{(x-3)^2}$

20. $\dfrac{5x}{x+2} + \dfrac{2}{x^2+4x+4}$ $\dfrac{5x^2+10x+2}{(x+2)^2}$

21. $\dfrac{2b}{b^2-1} - \dfrac{3}{1-b}$ $\dfrac{5b+3}{(b+1)(b-1)}$

22. $\dfrac{1}{1-x} + \dfrac{x^2}{x-1}$ $x+1$

23. $\dfrac{2}{2a+ab} - \dfrac{3}{2b+b^2}$ $\dfrac{2b-3a}{ab(2+b)}$

24. $\dfrac{a}{ac-c^2} - \dfrac{c}{a^2-ac}$ $\dfrac{a+c}{ac}$

25. $\dfrac{x}{x^2-6x+9} + \dfrac{3}{x^2-3x}$ $\dfrac{x^2+3x-9}{x(x-3)^2}$

26. $\dfrac{5}{x^2+x} + \dfrac{x}{x^2-2x-3}$ $\dfrac{x^2+5x-15}{x(x+1)(x-3)}$

27. $\dfrac{x+2}{6x^2+13x+6} - \dfrac{1}{2x^2+3x}$ $\dfrac{(x-2)(x+1)}{x(2x+3)(3x+2)}$

28. $\dfrac{x+3}{6x^2-11x+5} - \dfrac{1}{6x^2-5x}$ $\dfrac{(x+1)^2}{x(6x-5)(x-1)}$

In Exercises 29 to 32, add the right side of the formula.

29. Approximating an exponential function:

$$e^x \approx 1 + x + \frac{x^2}{2} + \frac{x^3}{6} + \frac{x^4}{24}$$ $\dfrac{x^4+4x^3+12x^2+24x+24}{24}$

30. Approximating a trigonometric function:

$$\cos(x) \approx 1 - \frac{x^2}{2} + \frac{x^4}{24} - \frac{x^6}{720}$$ $\dfrac{-x^6+30x^4-360x^2+720}{720}$

31. The van der Waal equation for gases in physics:

$$P = \frac{RT}{v-b} - \frac{a}{v^2}$$ $\dfrac{RTv^2-av+ab}{v^2(v-b)}$

32. The x- and y-intercept form of a linear equation:

$$1 = \frac{x}{a} + \frac{y}{b}$$ $\dfrac{bx+ay}{ab}$

Exercises 33 to 36 give common formulas that, like the formulas in the theater examples, involve the sum of rates or other rational expressions. Add the fractions on the right side of each formula.

33. Ventilation fans in an attic:

$$\frac{1}{t} = \frac{1}{t_1} + \frac{1}{t_2}$$ $\dfrac{t_1+t_2}{t_1t_2}$

34. Days to complete a wheat harvest with two machines:

$$\frac{1}{D} = \frac{1}{D_1} + \frac{1}{D_2}$$ $\dfrac{D_1+D_2}{D_1D_2}$

35. Traffic flow through parallel doors:

$$\frac{1}{m} = \frac{1}{m_1} + \frac{1}{m_2}$$ $\dfrac{m_1+m_2}{m_1m_2}$

36. Focal distance for a lens in optics:

$$\frac{1}{F} = \frac{1}{f_1} + \frac{1}{f_2}$$ $\dfrac{f_1+f_2}{f_1f_2}$

In Exercises 37 and 38, solve the given formulas for t, the time required if the people work together. *Hint:* Add the fractions first.

37. $\dfrac{1}{t} = \dfrac{1}{t_1} + \dfrac{1}{t_2}$, where t_1 and t_2 are the two people's individual times $t = \dfrac{t_1t_2}{t_2+t_1}$

38. $\dfrac{1}{t} = \dfrac{1}{t_1} + \dfrac{1}{t_2} + \dfrac{1}{t_3}$, where t_1, t_2, and t_3 are the three people's individual times $t = \dfrac{t_1t_2t_3}{t_2t_3+t_1t_3+t_1t_2}$

39. Solve the theater-emptying formula for t. Start by adding the three fractions on the left side of $\dfrac{1}{t_1} + \dfrac{1}{t_2} + \dfrac{1}{t_3} = \dfrac{1}{t}$. Is the time required to empty the theater the sum of the individual door times? $t = \dfrac{t_1t_2t_3}{t_2t_3+t_1t_3+t_1t_2}$; no

40. Predict what the theater-emptying formula will be for four doors. $\dfrac{1}{t} = \dfrac{1}{t_1} + \dfrac{1}{t_2} + \dfrac{1}{t_3} + \dfrac{1}{t_4}$

Solve the application problems in Exercises 41 to 46. Use the formulas from the reading or from Exercises 33 to 39. Round to the nearest tenth.

41. Two ventilation fans are operating in an attic. One can change the air in 5 hours and the other in 6 hours. Show that together they will be able to change the air in 3 hours, as required by code. $\dfrac{1}{t} = \dfrac{1}{5} + \dfrac{1}{6} = \dfrac{11}{30}, t = \dfrac{30}{11}, t < 3$

42. Norman's corn harvester cuts his crop in 10 days. Karen's can cut the same crop in 8 days. If both harvesters are used together, how many days will be needed to cut Norman's crop?   4.4 days

43. The Artist-in-Residence theater has one double door and one single door. The larger door permits the audience to leave in 5 minutes. The smaller door permits the audience to leave in 9 minutes. If both doors are used at once, in how many minutes will the theater be empty?   3.2 min

44. The Hole-in-the-Wall theater has three exits. The largest door permits the audience to leave in 6 minutes. The emergency exit permits the audience to leave in 8 minutes. A third exit through the backstage permits the audience to leave in 15 minutes. If all three doors are used at once, in how many minutes will the theater be empty?   2.8 min

45. Denzel volunteers to paint a room in a Habitat for Humanity house, and it takes him 4 hours. Halle paints an identical room in 3 hours. Together, in how many hours could they paint a third identical room?
$t = 12/7$, or 1.7 hr

46. Sharon, Kelly, and Ozzy paint in another Habitat house. Sharon paints a room in 2 hours, Kelly in 4 hours, and Ozzy in 3 hours. Together, in how many hours could they paint a fourth identical room? $t = 12/13$, or 0.9 hr

In Exercises 47 to 54, simplify the complex rational expressions to eliminate the fractions from the numerator and the denominator.

47. $\dfrac{7+1}{\dfrac{1}{7}+1}$ 7

48. $\dfrac{4+1}{1+\dfrac{1}{4}}$ 4

49. $a = \dfrac{V}{\dfrac{4}{3}\pi b^2}$ $a = \dfrac{3V}{4\pi b^2}$

50. $h = \dfrac{V}{\dfrac{\pi}{8}d^2}$ $h = \dfrac{8V}{\pi d^2}$

51. $\dfrac{x - \dfrac{x}{2}}{2 + \dfrac{x}{3}}$ $\dfrac{3x}{2(x+6)}$

52. $\dfrac{\dfrac{x}{3} - x}{3 + \dfrac{x}{2}}$ $\dfrac{-4x}{3(x+6)}$

53. Parallel electric cells: $I = \dfrac{E}{R + \dfrac{r}{2}}$ $I = \dfrac{2E}{2R + r}$

54. Materials science: $\dfrac{p_2 - p_1}{\dfrac{v_1 - v_2}{v_1}}$ $\dfrac{v_1(p_2 - p_1)}{v_1 - v_2}$

55. Simplify the average speed formula from Example 12,

$$v = \dfrac{2x}{\dfrac{x}{v_1} + \dfrac{x}{v_2}} \qquad v = \dfrac{2v_1 v_2}{v_1 + v_2}$$

by first adding the fractions in the denominator.

56. Because of bad weather, Shane averages 45 miles per hour driving from Memphis to Louisville, a distance of 367 miles. On the return trip, he averages 60 miles per hour. What is his total time for the round trip? What is his average rate for the round trip?
≈14.3 hr; ≈51.4 mph

57. Heavy traffic slows Luis to an average of 55 miles per hour while driving from Amarillo to Albuquerque, a distance of 284 miles. On the return trip, he averages 65 miles per hour. What was his total time for the trip? What is his average rate for the round trip?
≈9.5 hr; ≈59.6 mph

58. What is the average speed for the round trip if Farah drives 55 miles per hour in one direction and 65 miles per hour on the return trip? ≈59.6 mph

59. What is the average speed if Azra runs 18 kilometers per hour for the first half of a race and 22 kilometers per hour for the second half of the race? 19.8 km/hr

60. If you average x miles per hour during the first half of a trip, find the speed y in miles per hour needed during the second half of the trip to reach 400 miles per hour as an overall average. $y = \dfrac{200x}{x - 200}, x > 200$

61. Describe in your own words how to find the least common denominator for two rational expressions.

62. Describe in your own words how to add two rational expressions.

▦ Error Analysis

Indicate whether the solutions to Exercises 63 to 66 are right or wrong. If wrong, explain what was done wrong. If right, state whether the method would always work and what its disadvantage is.

63. $\dfrac{3}{4} + \dfrac{2}{3} = \dfrac{3+2}{4 \cdot 3} = \dfrac{5}{12}$ wrong; find LCD before adding.

64. $\dfrac{3}{5} - \dfrac{2}{4} = \dfrac{3-2}{5-4} = \dfrac{1}{1}$ wrong; find common denominator before subtracting.

65. $\dfrac{3}{8} + \dfrac{1}{4} = \dfrac{3(4) + 8(1)}{8 \cdot 4} = \dfrac{12+8}{32} = \dfrac{20}{32} = \dfrac{5}{8}$ right; always works; not LCD; may require additional simplification.

66. $\dfrac{5}{6} + \dfrac{1}{3} = \dfrac{5+1}{6+3} = \dfrac{6}{9} = \dfrac{2}{3}$ wrong; find common denominator before adding.

Identify the missing operation sign (add, subtract, multiply, or divide) in Exercises 67 to 74.

67. a. $\dfrac{3}{5}\ \boxed{}\ \dfrac{2}{7} = \dfrac{6}{35}$ **b.** $\dfrac{3}{5}\ \boxed{}\ \dfrac{2}{7} = \dfrac{21}{10}$

68. a. $\dfrac{2}{7}\ \boxed{}\ \dfrac{3}{5} = \dfrac{10}{21}$ **b.** $\dfrac{3}{5}\ \boxed{}\ \dfrac{2}{7} = \dfrac{11}{35}$

69. a. $\dfrac{3}{5}\ \boxed{}\ \dfrac{2}{7} = \dfrac{31}{35}$ **b.** $\dfrac{5}{7}\ \boxed{}\ \dfrac{2}{3} = \dfrac{1}{21}$

70. a. $\dfrac{7}{5}\ \boxed{}\ \dfrac{2}{3} = \dfrac{11}{15}$ **b.** $\dfrac{7}{3}\ \boxed{}\ \dfrac{2}{5} = \dfrac{14}{15}$

71. a. $\dfrac{a}{b}\ \boxed{}\ \dfrac{1}{a} = \dfrac{1}{b}$ **b.** $\dfrac{a}{b}\ \boxed{}\ \dfrac{1}{a} = \dfrac{a^2}{b}$

72. a. $\dfrac{b}{a}\ \boxed{}\ \dfrac{1}{a} = \dfrac{b+1}{a}$ **b.** $\dfrac{b}{a}\ \boxed{}\ \dfrac{1}{a} = \dfrac{b}{a^2}$

73. a. $\dfrac{1}{a}\ \boxed{}\ \dfrac{1}{b} = \dfrac{a+b}{ab}$ **b.** $\dfrac{1}{a}\ \boxed{}\ \dfrac{1}{b} = \dfrac{1}{ab}$

74. a. $\dfrac{1}{a}\ \boxed{}\ \dfrac{1}{b} = \dfrac{b-a}{ab}$ **b.** $\dfrac{1}{b}\ \boxed{}\ \dfrac{1}{a} = \dfrac{a}{b}$

75. Extra Fraction Practice Add, subtract, multiply, and divide each pair of fractions in the order shown.

a. $\dfrac{1}{4}$ and $\dfrac{1}{10}$ $\dfrac{7}{20}; \dfrac{3}{20}; \dfrac{1}{40}; \dfrac{5}{2}$

b. $\dfrac{1}{8}$ and $\dfrac{1}{10}$ $\dfrac{9}{40}; \dfrac{1}{40}; \dfrac{1}{80}; \dfrac{5}{4}$

c. $\dfrac{3}{4}$ and $\dfrac{5}{6}$ $\dfrac{19}{12}; -\dfrac{1}{12}; \dfrac{5}{8}; \dfrac{9}{10}$

d. $\frac{2}{3}$ and $\frac{5}{6}$ $\frac{3}{2}$; $-\frac{1}{6}$; $\frac{5}{9}$, $\frac{4}{5}$

e. $1\frac{1}{3}$ and $2\frac{1}{2}$ $3\frac{5}{6}$; $-1\frac{1}{6}$; $3\frac{1}{3}$; $\frac{8}{15}$

f. $2\frac{1}{4}$ and $1\frac{2}{3}$ $3\frac{11}{12}$; $\frac{7}{12}$; $3\frac{3}{4}$; $1\frac{7}{20}$

g. $2\frac{1}{5}$ and $1\frac{1}{4}$ $3\frac{9}{20}$; $\frac{19}{20}$; $2\frac{3}{4}$; $1\frac{19}{25}$

h. $1\frac{3}{4}$ and $2\frac{1}{7}$ $3\frac{25}{28}$; $-\frac{11}{28}$; $3\frac{3}{4}$; $\frac{49}{60}$

■ Projects

76. Leading a Race Arturo and Brahim are the leaders in a 10-kilometer race. Arturo is ahead by a distance x at the halfway point in the race. From prior race results, Arturo knows that Brahim's average speed for the race is likely to be 21 kilometers per hour.

a. If the time required for the two runners to finish the race from their respective positions is equal, what will the result of the race be?   a tie

b. If Arturo finishes in less time, what will the result of the race be?   Arturo wins.

c. Write an expression to describe the time it will take Arturo to finish the race at 20 km/hr. $\dfrac{5 \text{ km}}{20 \text{ km/hr}}$

d. Write an expression containing x to describe the time it will take Brahim to finish the race. $\dfrac{(5 + x) \text{ km}}{21 \text{ km/hr}}$

e. Subtract the expression in part d from that in part c. Graph the result, with length of the lead on the horizontal axis and time on the vertical axis. See Additional Answers.

f. Find the point on the graph that shows the length of the lead so that the race ends in a tie. $x = 0.25$ km

g. Which points show lead lengths that give Arturo a win?   points where $x > 0.25$

h. Which points show the lead lengths that give Brahim a win?   points where $x < 0.25$

77. Lost Keys Marsha travels to work by bicycle at 15 miles per hour. Half an hour after she leaves one morning, her husband Jon finds her office keys by the door. He hopes to catch her before she reaches the bike path (at which point there is no way for him to find her). The bike path entrance is 12 miles away.

a. Suppose Jon travels an average of 35 miles per hour in the car. Show whether he catches her in time. Marsha arrives in 48 min; Jon arrives in 50.6 min.

b. Write an equation that shows Jon's travel time. $t = 0.5 \text{ hr} + \dfrac{12 \text{ mi}}{x}$; t in min, x in mph

c. At what average speed must he travel to catch her?   40 mph

---

7.7 Solving Rational Equations

Objectives

- Find the least common denominator of rational expressions in an equation.
- Solve equations containing rational expressions.
- Solve equations graphically.
- Solve application problems containing rational expressions.
- Solve application problems related to $\dfrac{1}{a} + \dfrac{1}{b} = \dfrac{1}{c}$.

WARM-UP

Factor.

1. $x^2 + x - 2$
$(x + 2)(x - 1)$

2. $2x^2 + 7x + 3$
$(2x + 1)(x + 3)$

3. $x^2 - 4x - 77$
$(x + 7)(x - 11)$

(continued)

WARM-UP (*continued*)

Complete the multiplication tables. Summarize by writing the resulting product in simplified form.

4.

Multiply	$\dfrac{1}{3}$	$+\dfrac{1}{4x}$	
$12x$	$4x$	$+3$	$4x + 3$

5.

Multiply	$\dfrac{1}{x-3}$	$+\dfrac{5}{x-2}$	
$(x-2)(x-3)$	$x-2$	$+5(x-3)$	$6x - 17$

Multiply and simplify.

6. $\dfrac{18}{1}\left(\dfrac{x}{2} + \dfrac{x}{9}\right)$ $11x$　　　　　　**7.** $10x\left(\dfrac{1}{10} + \dfrac{1}{x}\right)$ $3x + 10$

8. $x(x+1)\left(\dfrac{2}{x+1} + \dfrac{3}{x}\right)$ $5x + 3$　　　**9.** $x^2\left(\dfrac{3}{x^2} + \dfrac{2}{x} - 1\right)$ $3 + 2x - x^2$

10. $20\left(\dfrac{3}{4} + \dfrac{1}{5}\right)$ 19

11. What happened to the fractions in the expressions in Warm-up Exercises 6 to 10? Why did this happen?
No fractions remain; factor in front of parentheses is LCD.

THE PURPOSE OF THIS SECTION is to extend methods of solving equations to **rational equations**—*equations that contain rational expressions*. We solve rational equations by graphing and by multiplying both sides of the equation by the least common denominator.

Review of Equation Solving

We have already solved equations containing fractions, using both symbolic and graphical techniques. Recall that in earlier equation and formula solving, we wrote a plan in which we reversed the order of operations as applied to *x*. By doing the opposite operations, we were able to solve for *x* or other variables.

EXAMPLE 1　**Solving equations**　Solve $\frac{2}{3}x + 4 = 22$ for *x*.

SYMBOLIC SOLUTION　*Plan*: Because *x* is multiplied by $\frac{2}{3}$ and then added to 4, we subtract 4 and divide by $\frac{2}{3}$. Recall that division by $\frac{2}{3}$ is the same as multiplication by $\frac{3}{2}$.

Carry out the plan:

$$\frac{2}{3}x + 4 = 22 \qquad \text{Subtract 4 from each side.}$$

$$\frac{2}{3}x = 18 \qquad \text{Multiply by the reciprocal, } \tfrac{3}{2}.$$

$$\frac{3}{2} \cdot \frac{2}{3}x = \frac{3}{2}(18) \qquad \text{Simplify.}$$

$$x = 27$$

Check: $\frac{2}{3}(27) + 4 \stackrel{?}{=} 22$ ✓

GRAPHICAL SOLUTION   We graph the left side and the right side of $\frac{2}{3}x + 4 = 22$ separately. The graph of $y_1 = \frac{2}{3}x + 4$ is a straight line with a slope of $\frac{2}{3}$ and a y-intercept at 4. The line $y_2 = 22$ is horizontal. The graphical solution to $\frac{2}{3}x + 4 = 22$ is $x = 27$, the x-coordinate of the point of intersection of the graphs (see Figure 34).

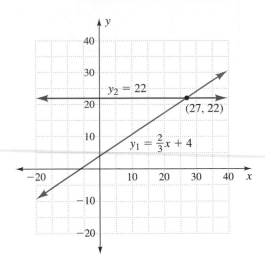

FIGURE 34

Solving Equations with the Least Common Denominator and Graphs

When we multiply both sides of an equation containing denominators by the least common denominator, we eliminate the denominators from each side.

■ SOLVING EQUATIONS CONTAINING
DENOMINATORS

> We may eliminate denominators if we multiply both sides of an equation by the least common denominator.

In Example 2, multiplying both sides of the equation by the least common denominator permits us to obtain a simpler equation.

EXAMPLE 2   Solving with an LCD   Solve $\dfrac{3}{4} + \dfrac{1}{5} = \dfrac{1}{x}$ for x by multiplying both sides of the equation by $20x$, the least common denominator. Assume $x \neq 0$.

SYMBOLIC SOLUTION

$$\frac{3}{4} + \frac{1}{5} = \frac{1}{x}$$ Multiply by the LCD.

$$20x\left(\frac{3}{4} + \frac{1}{5}\right) = 20x\left(\frac{1}{x}\right)$$ Distribute on the left side.

$$20x\left(\frac{3}{4}\right) + 20x\left(\frac{1}{5}\right) = \frac{20x}{x}$$ Factor.

$$4 \cdot 5x\left(\frac{3}{4}\right) + 4 \cdot 5x\left(\frac{1}{5}\right) = \frac{4 \cdot 5x}{x}$$ Simplify.

$$15x + 4x = 20$$ Add like terms.

$$19x = 20$$ Divide by 19.

$$x = \frac{20}{19}$$

GRAPHICAL SOLUTION We graph the left and right sides separately:

$$y_1 = \frac{3}{4} + \frac{1}{5} = \frac{19}{20} \quad \text{and} \quad y_2 = \frac{1}{x}$$

The graph of y_1 is a horizontal line. The graph of y_2 is undefined at $x = 0$ and nearly vertical on either side of the y-axis. The solution to $\frac{3}{4} + \frac{1}{5} = \frac{1}{x}$ is $x = \frac{20}{19}$, the x-coordinate of the point of intersection of the graphs in Figure 35.

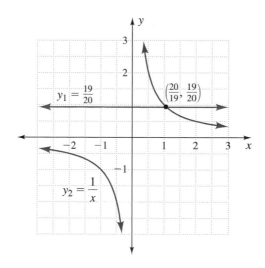

FIGURE 35

In Example 3, we obtain a quadratic equation from the multiplication.

EXAMPLE 3 **Solving with an LCD** Solve $\dfrac{7}{x + 3} = \dfrac{x - 7}{8}$ for x. Assume $x \neq -3$.

SYMBOLIC SOLUTION This least common denominator is $8(x + 3)$.

$$\frac{7}{x + 3} = \frac{x - 7}{8} \qquad \text{Multiply by the LCD.}$$

$$8(x + 3)\frac{7}{x + 3} = \frac{(x - 7)}{8} 8(x + 3) \qquad \text{Simplify.}$$

$$8(7) = (x - 7)(x + 3) \qquad \text{Multiply the sides.}$$

$$56 = x^2 - 4x - 21 \qquad \text{Change to } ax^2 + bx + c = 0.$$

$$x^2 - 4x - 77 = 0 \qquad \text{Factor (or use the quadratic formula).}$$

$$(x - 11)(x + 7) = 0 \qquad \text{Apply the zero product rule.}$$

$$\text{Either} \quad x = 11 \quad \text{or} \quad x = -7$$

Checking the solution is left as an exercise.

GRAPHICAL SOLUTION We graph the left and right sides of the equation separately:

$$y_1 = \frac{7}{x + 3} \quad \text{and} \quad y_2 = \frac{x - 7}{8}$$

At $x = -3$, y_1 is undefined. The graph turns nearly vertical near $x = -3$. The graph of y_2 is linear with a slope of $\frac{1}{8}$ and a y-intercept at $-\frac{7}{8}$.

The solutions to $\dfrac{7}{x+3} = \dfrac{x-7}{8}$ are $x = -7$ and $x = 11$, the x-coordinates of the intersections of the graphs in Figure 36.

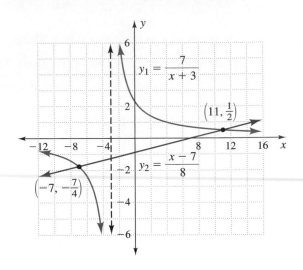

FIGURE 36

In Example 4, we obtain a quadratic equation from the multiplication.

EXAMPLE 4   Solving with an LCD   Solve $\dfrac{2}{x+1} + \dfrac{3}{x} = -2$, $x \neq -1, 0$ for x.

SYMBOLIC SOLUTION

$$\frac{2}{x+1} + \frac{3}{x} = -2 \qquad \text{Multiply by the LCD.}$$

$$x(x+1)\left(\frac{2}{x+1} + \frac{3}{x}\right) = -2 \cdot x(x+1) \qquad \text{Distribute the LCD.}$$

$$\frac{x(x+1) \cdot 2}{x+1} + \frac{x(x+1) \cdot 3}{x} = -2x(x+1) \qquad \text{Simplify fractions.}$$

$$2x + 3(x+1) = -2x^2 - 2x \qquad \text{Change to } ax^2 + bx + c = 0.$$

$$2x + 3x + 3 + 2x^2 + 2x = 0 \qquad \text{Combine like terms.}$$

$$2x^2 + 7x + 3 = 0 \qquad \text{Factor (or use the quadratic formula).}$$

$$(2x+1)(x+3) = 0 \qquad \text{Apply the zero product rule.}$$

$$\text{Either} \quad 2x+1 = 0 \quad \text{or} \quad x+3 = 0 \qquad \text{Solve the factor equations.}$$

$$x = -\tfrac{1}{2} \quad \text{or} \quad x = -3$$

Checking the solution is left as an exercise.

GRAPHICAL SOLUTION   We graph the left and right sides separately:

$$y_1 = -2 \quad \text{and} \quad y_2 = \frac{2}{x+1} + \frac{3}{x}$$

The graph of y_1 is a horizontal line. The value of y_2 is undefined at both $x = -1$ and $x = 0$, and the graph becomes nearly vertical near those values. As a result, y_2 has three pieces, with breaks at $x = -1$ and $x = 0$. The solutions are $x = -\tfrac{1}{2}$ and $x = -3$, the x-coordinates of the intersections of the graphs in Figure 37.

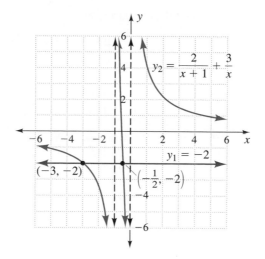

FIGURE 37

EXTRANEOUS ROOTS   Example 5 shows the importance of using the least common denominator and of checking solutions.

EXAMPLE 5   Solving proportions   Solve $\dfrac{x+1}{x-1} = \dfrac{1}{1-x}$, $x \neq 1$ for x. Instead of using the least common denominator, $-1(x-1)$, multiply both sides by $(x-1)(1-x)$.

SYMBOLIC SOLUTION   When we multiply, we find a quadratic equation with two solutions.

$$\frac{x+1}{x-1} = \frac{1}{1-x} \qquad \text{Multiply by } (x-1)(1-x).$$

$$(x-1)(1-x)\frac{(x+1)}{x-1} = \frac{1}{1-x}(x-1)(1-x) \qquad \text{Simplify.}$$

$$1 - x^2 = x - 1 \qquad \text{Change to } ax^2 + bx + c = 0.$$

$$x^2 + x - 2 = 0 \qquad \text{Factor (or use the quadratic formula).}$$

$$(x-1)(x+2) = 0 \qquad \text{Apply the zero product rule.}$$

$$\text{Either} \quad x - 1 = 0 \quad \text{or} \quad x + 2 = 0 \qquad \text{Solve the factor equations.}$$

$$x = 1 \quad \text{or} \qquad x = -2$$

Check: In checking our solutions, we find that $x = -2$ satisfies the equation:

$$\frac{-2+1}{-2-1} \stackrel{?}{=} \frac{1}{1-(-2)} \quad \checkmark$$

However, $x = 1$ gives a zero denominator. As noted in the original problem, $x = 1$ has been excluded from the set of possible inputs.

GRAPHICAL SOLUTION   We enter the left and right sides of $\dfrac{x+1}{x-1} = \dfrac{1}{1-x}$ as separate equations:

$$y_1 = \frac{x+1}{x-1} \qquad \text{and} \qquad y_2 = \frac{1}{1-x}$$

The graphs, shown in Figure 38, intersect only at $x = -2$. At $x = 1$, both sides of the equation have a zero denominator and both curves become nearly vertical. There is no point on either graph corresponding to $x = 1$.

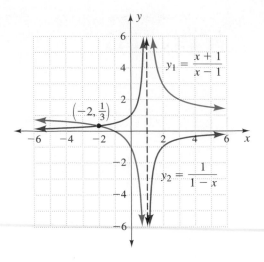

FIGURE 38

The solution $x = 1$ in Example 5 is an **extraneous root**—*a solution found algebraically that does not satisfy the original equation.* When we multiplied by $(x - 1)(1 - x)$, a quadratic expression, we introduced an extra solution, $x = 1$. Had we multiplied both sides by the least common denominator, $-(x - 1)$, a linear expression, the extraneous root would have been avoided.

> To avoid extraneous solutions, multiply both sides of a rational equation by the least common denominator.

Solving the equation in Example 5 with the least common denominator is left as an exercise.

Applications

THE GOLDEN RECTANGLE   To balance the many applications of algebra in science and engineering, this section includes an application in architecture. The Greeks believed that *rectangles with certain dimensions*, called **golden rectangles**, were more beautiful than other rectangles. The rectangle drawn around the Parthenon with its upper triangular structure intact, as shown in Figure 39, is a golden rectangle. The dimensions the Greeks had in mind were those that made it possible to remove a square from the rectangle and leave a smaller rectangle with sides in the same ratio as in the original rectangle.

FIGURE 39

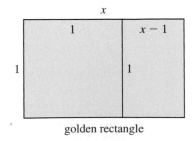

golden rectangle

FIGURE 40

The ratio of length to width in the rectangle in Figure 40 is x to 1. A square with sides of 1 unit has been marked, leaving a smaller rectangle. For the smaller rectangle, the ratio of length to width is 1 to $x - 1$. In order for the larger rectangle to be a golden rectangle, the two ratios need to be equal, creating a proportion:

$$\frac{x}{1} = \frac{1}{x - 1}$$

In Example 6, we solve the proportion for the x that creates a golden rectangle.

EXAMPLE 6 Applying the quadratic formula   Multiply by the LCD to eliminate the denominators in the proportion $\dfrac{x}{1} = \dfrac{1}{x - 1}$. Solve the resulting quadratic equation and confirm the solutions with a graphing calculator.

SOLUTION   When we multiply by the LCD in this proportion, we obtain a quadratic equation that does not factor:

$$\frac{x}{1} = \frac{1}{x - 1}$$ Multiply by $1 \cdot (x - 1)$.

$$1 \cdot (x - 1)\frac{x}{1} = \frac{1}{x - 1}1 \cdot (x - 1)$$ Simplify.

$$x^2 - x = 1$$ Change to $ax^2 + bx + c = 0$.

$$x^2 - x - 1 = 0$$ Apply the quadratic formula.

In the equation $x^2 - x - 1 = 0$, $a = 1$, $b = -1$, and $c = -1$.

$$x = \frac{-b \pm \sqrt{b^2 - 4ac}}{2a} = \frac{-(-1) \pm \sqrt{(-1)^2 - 4(1)(-1)}}{2(1)} = \frac{1 \pm \sqrt{1 + 4}}{2}$$

The solutions are

$$x = \frac{1 + \sqrt{5}}{2} \approx 1.618 \qquad \text{and} \qquad x = \frac{1 - \sqrt{5}}{2} \approx -0.618$$

When graphed, $y = x^2 - x - 1$ has x-intercepts at $x \approx 1.618$ and $x \approx -0.618$. Because x represents the length of a rectangle, the negative solution is discarded.

The *golden rectangle* has dimensions in the **golden ratio** $\dfrac{1 + \sqrt{5}}{2}$ to 1.

THEATER EXITS   Example 7 illustrates how to obtain a rational equation when we know one individual time and a total time and need to find another individual time.

EXAMPLE 7 Setting up a rational equation: movie theater exits   Suppose a movie theater has two exits. One is a single-door emergency exit that, by itself, will permit the theater to empty in 10 minutes. The other is a double-wide door. If building codes require that it be possible to empty the theater in 3 minutes, write an equation for finding the exit time through the double-wide door when it and the emergency exit are both in use.

SOLUTION   Changing exit time to a rate for emptying the theater, we say that during each minute the emergency exit and the double-wide doors permit 1/10 and 1/x of the theater to empty, respectively. The following expressions show the portion of the theater emptied after each minute:

First minute: $1\left(\dfrac{1}{10} + \dfrac{1}{x}\right)$

Second minute: $2\left(\dfrac{1}{10} + \dfrac{1}{x}\right)$

Third minute: $3\left(\dfrac{1}{10} + \dfrac{1}{x}\right)$

To satisfy building codes, the theater must be empty when the expression for the third minute equals 1 (100% of the theater capacity):

$$3\left(\frac{1}{10} + \frac{1}{x}\right) = 1$$

We divide both sides by 3 to obtain the traditional form of this equation:

$$\frac{1}{10} + \frac{1}{x} = \frac{1}{3}, \quad x \neq 0$$

■

EXAMPLE 8 Solving with an LCD Solve $\frac{1}{10} + \frac{1}{x} = \frac{1}{3}$, $x \neq 0$ for x.

SOLUTION The least common denominator for the equation is $30x$.

$$\frac{1}{10} + \frac{1}{x} = \frac{1}{3} \qquad \text{Multiply by the LCD.}$$

$$30x\left(\frac{1}{10} + \frac{1}{x}\right) = 30x\left(\frac{1}{3}\right) \qquad \text{Simplify.}$$

$$3x + 30 = 10x \qquad \text{Subtract } 3x \text{ from both sides.}$$

$$30 = 7x \qquad \text{Divide by 7.}$$

$$x = \frac{30}{7} \approx 4.3 \text{ min}$$

Check: Checking with a calculator, we have

$$\frac{1}{10} + \frac{1}{4.3} \approx 0.1 + 0.23 = 0.33 \approx \frac{1}{3} \checkmark$$

If the double-wide door by itself can empty the theater in approximately 4.3 minutes, it can do so together with the emergency exit in 3 minutes. ■

Think about it: Besides the fact that the equation would be undefined, what would $x = 0$ minutes mean in the problem setting for Example 8?

As we saw in Section 7.6, the equation in Example 8 is a common formula,

$$\frac{1}{a} + \frac{1}{b} = \frac{1}{c}$$

where a = time for one activity, b = time for second activity, c = time for both activities done together. In Example 8, $a = 10$, $c = 3$, and b is unknown. The formula may describe times for filling containers. If a and b are the times needed to fill a child's wading pool with two separate hoses, then c is the time needed to fill it with the two hoses together.

The same style formula occurs in applications that seem to have nothing to do with emptying or filling. The formula appears in optics and at least twice in basic electronics. The formula to determine the focal length in lenses is

$$\frac{1}{p} + \frac{1}{q} = \frac{1}{f}$$

The formula for calculating resistance for parallel resistors is

$$\frac{1}{R_1} + \frac{1}{R_2} = \frac{1}{R}$$

and capacitance of condensers in series is given by

$$\frac{1}{C_1} + \frac{1}{C_2} = \frac{1}{C}$$

A similar formula may apply to three doors, pieces of harvest equipment, hoses, lenses, resistors, or condensers. Each additional door, harvester, hose, lens face, resistor, or condenser adds another fraction to the left side.

Solving Systems of Equations Involving Ratios

To close this section, we solve a system of equations in which one equation contains a ratio and the solution involves a rational equation.

EXAMPLE 9   **Solving a system of equations: framing materials**   The ratio of width to length of a rectangular picture frame is 5 to 8 (see Figure 41). The amount of framing material—or the perimeter of the frame—is 130 centimeters. Write equations that permit solving for the length and width. Solve the equations.

FIGURE 41

SOLUTION   The relationships are described with two variables, w and l. The ratio information allows us to build a proportion, $w/l = 5/8$. The perimeter of the rectangle is $2w + 2l = 130$. We solve the perimeter formula for l in terms of w:

$$2w + 2l = 130 \qquad \text{Subtract } 2w.$$
$$2l = 130 - 2w \qquad \text{Divide by 2.}$$
$$l = 65 - w$$

We substitute for l in the proportion and solve for w:

$$\frac{w}{l} = \frac{5}{8} \qquad\qquad \text{Substitute for } l.$$

$$\frac{w}{65 - w} = \frac{5}{8} \qquad\qquad \text{Multiply by the LCD.}$$

$$8(65 - w)\frac{w}{65 - w} = \frac{5}{8}\, 8(65 - w) \qquad \text{Simplify.}$$

$$8w = 325 - 5w \qquad\qquad \text{Add } 5w \text{ to both sides.}$$

$$13w = 325 \qquad\qquad \text{Divide by 13 on both sides.}$$

$$w = 25$$

Then we substitute for w:

$$l = 65 - w = 65 - 25 = 40$$

Check both conditions: The ratio 25:40 simplifies to 5:8; the perimeter is $2(25) + 2(40) = 130$ centimeters. ✓

ANSWER BOX

Warm-up: 1. $(x + 2)(x - 1)$ **2.** $(2x + 1)(x + 3)$ **3.** $(x + 7)(x - 11)$ **4.** $4x + 3$ **5.** $6x - 17$ **6.** $11x$ **7.** $x + 10$ **8.** $5x + 3$ **9.** $3 + 2x - x^2$ **10.** 19 **11.** No fractions remain; the factor in front of the parentheses is the least common denominator of the expression and eliminates both denominators. **Think about it:** It would mean that people left the theater instantly, in $t = 0$ time. This is impossible without transporter technology from science fiction.

7.7 Exercises

In Exercises 1 to 8, solve for the indicated variable.

1. $\frac{3}{4} x + 5 = 23$ for x $x = 24$

2. $\frac{3}{5} x - 8 = 13$ for x $x = 35$

3. $L = \frac{\pi r \theta}{180}$ for r $r = \frac{180L}{\pi \theta}, \theta \neq 0$

4. $A = \frac{(a + b)}{2} \cdot h$ for b $b = \frac{2A}{h} - a, h \neq 0$

5. $x\left(\frac{1}{10} + \frac{1}{12}\right) = 1$ for x $x = 5\frac{5}{11}$

6. $x\left(\frac{1}{3} + \frac{1}{8}\right) = 1$ for x $x = 2\frac{2}{11}$

7. $\frac{2n + 4}{7} = \frac{3n - 7}{4}$ for n $n = 5$

8. $\frac{4 - 3k}{2} = \frac{3 - 5k}{3}$ for k $k = -6$

9. Why is it not necessary to state any restrictions on the variables in Exercises 5 to 8?   no variables in denominators

10. In the answer to Exercise 3, we state $\theta \neq 0$. Why do we not also say $\pi \neq 0$? π is a constant.

In Exercises 11 to 15, check the examples by substituting the given number into the equation for the indicated example.

11. $x = 11$ for Example 3

12. $x = -7$ for Example 3

13. $x = \frac{20}{19}$ for Example 2

14. $x = -\frac{1}{2}$ for Example 4

15. $x = -3$ for Example 4

Solve Exercises 16 to 23. Note any restrictions on the variables.

16. $\frac{x + 1}{6} = \frac{x}{9}$ $x = -3$

17. $\frac{x + 1}{2} = \frac{x - 3}{1}$ $x = 7$

18. $\frac{x - 7}{10} = \frac{-3}{x + 6}$
$\{-3, 4\}, x \neq -6$

19. $\frac{x - 8}{5} = \frac{-6}{x + 5}$
$\{-2, 5\}, x \neq -5$

20. $\frac{x + 1}{5} = \frac{4}{2x - 1}$
$\{-3.5, 3\}, x \neq \frac{1}{2}$

21. $\frac{x - 3}{6} = \frac{3}{2x - 1}$
$\{5, -1.5\}, x \neq \frac{1}{2}$

22. $\frac{x}{4} + \frac{x}{6} = 15$ $x = 36$

23. $\frac{x}{32} + \frac{x}{8} = 10$ $x = 64$

Multiply the expressions in Exercises 24 to 31, and simplify the results.

24. $12x^2\left(\frac{1}{3x^2} + \frac{1}{4x}\right)$ $4 + 3x$

25. $6x^2\left(\frac{1}{2x^2} - \frac{1}{3x}\right)$ $3 - 2x$

26. $9x^2\left(\frac{2}{3x^2} + \frac{1}{9}\right)$ $x^2 + 6$

27. $2x^2\left(\frac{1}{2x^2} + \frac{3}{x}\right)$ $6x + 1$

28. $x(x - 2)\left(\frac{1}{x - 2} - \frac{3}{x}\right)$ $-2x + 6$

29. $x(x + 1)\left(\frac{1}{x + 1} - \frac{2}{x}\right)$ $-x - 2$

30. $(x - 1)(x + 2)\left(\frac{1}{x - 1} + \frac{1}{x + 2}\right)$ $2x + 1$

31. $(x + 3)(x - 1)\left(\frac{1}{x + 3} + \frac{2}{x - 1}\right)$ $3x + 5$

32. Why are there no denominators in the answers in Exercises 24 to 31?   multiplied by a common denominator

In Exercises 33 to 60, identify any restrictions on the variables. Solve for x.

33. $\frac{3}{5} + \frac{2}{3} = \frac{1}{x}$
$x \neq 0; x = \frac{15}{19}$

34. $\frac{5}{6} + \frac{2}{5} = \frac{1}{x}$
$x \neq 0; x = \frac{30}{37}$

35. $\dfrac{1}{3} + \dfrac{1}{x} = \dfrac{1}{2}$

$x \neq 0; x = 6$

36. $\dfrac{1}{8} + \dfrac{1}{x} = \dfrac{1}{6}$

$x \neq 0; x = 24$

37. $\dfrac{5}{x} + \dfrac{2}{x} = \dfrac{4}{x}$

$x \neq 0; \{\ \}$ or $\varnothing$

38. $\dfrac{1}{x} = \dfrac{1}{2x} + \dfrac{1}{2}$

$x \neq 0; x = 1$

39. $\dfrac{1}{2} - \dfrac{1}{x} = \dfrac{1}{2x}$

$x \neq 0; x = 3$

40. $\dfrac{5}{x} - \dfrac{2}{x} = \dfrac{4}{x}$

$x \neq 0; \{\ \}$ or $\varnothing$

41. $\dfrac{4}{2x - 1} = \dfrac{1}{x - 1}$

$x \neq \frac{1}{2}, 1; x = \frac{3}{2}$

42. $\dfrac{10}{x + 1} = \dfrac{2}{x - 1}$

$x \neq \pm 1; x = \frac{3}{2}$

43. $\dfrac{x + 1}{2} = \dfrac{15}{x + 2}$

$x \neq -2; \{-7, 4\}$

44. $\dfrac{1}{x + 5} = \dfrac{x + 5}{9}$

$x \neq -5; \{-8, -2\}$

45. $\dfrac{4}{x - 1} = \dfrac{x + 2}{10}$

$x \neq 1; \{-7, 6\}$

46. $\dfrac{x - 2}{3} = \dfrac{6}{x + 5}$

$x \neq -5; \{-7, 4\}$

47. $\dfrac{18}{x^2} + \dfrac{9}{x} - 2 = 0$

$x \neq 0; \{-1.5, 6\}$

48. $2 + \dfrac{5}{x} - \dfrac{3}{x^2} = 0$ $\quad x \neq 0; \{-3, \frac{1}{2}\}$

49. $\dfrac{1}{x - 2} - 4 = \dfrac{3 - x}{x - 2}$ $\quad x \neq 2; \{\ \}$ or $\varnothing$

50. $\dfrac{4}{x - 1} = \dfrac{5 - x}{x - 1} + 2$ $\quad x \neq 1; \{\ \}$ or $\varnothing$

51. $\dfrac{3}{x - 1} + \dfrac{5}{2x + 2} = \dfrac{3}{2}$ $\quad x \neq \pm 1; \{\ \frac{1}{3}, 4\}$

52. $\dfrac{1}{x + 1} + \dfrac{1}{x - 1} = \dfrac{5}{12}$ $\quad x \neq \pm 1; \{-\frac{1}{5}, 5\}$

53. $\dfrac{x}{6 - x} = \dfrac{2}{x - 3}$ $\quad x \neq 3, 6; \{-3, 4\}$

54. $x - \dfrac{2x}{x + 3} = \dfrac{6}{x + 3}$ $\quad x \neq -3; x = 2$

55. $\dfrac{1}{x - 2} + \dfrac{x - 1}{x} = \dfrac{2x + 1}{2x}$ $\quad x \neq 0, 2; x = 6$

56. $\dfrac{1}{x} + \dfrac{x + 1}{x + 2} = \dfrac{6x - 1}{5x}$ $\quad x \neq 0, -2; \{3, -4\}$

57. $\dfrac{x - 3}{x} + \dfrac{x - 2}{x + 3} = \dfrac{6x + 1}{8x}$ $\quad x \neq 0, -3; \{5, -\frac{3}{2}\}$

58. $\dfrac{x - 2}{x - 1} + \dfrac{x}{x - 2} = \dfrac{5x + 4}{3(x - 1)}$ $\quad x \neq 1, 2; \{4, 5\}$

59. $\dfrac{2}{x - 1} + \dfrac{x - 2}{x + 1} = \dfrac{4x}{3(x + 1)}$ $\quad x \neq \pm 1; \{-3, 4\}$

60. $\dfrac{1}{x} + \dfrac{x + 1}{x + 2} = \dfrac{3x + 1}{3x}$ $\quad x \neq -2, 0; x = 4$

61. Solve the equation $\dfrac{x + 1}{x - 1} = \dfrac{1}{1 - x}$ from Example 5 by multiplying both sides by the least common denominator, $-(x - 1)$. $\quad x \neq 1; x = -2$

62. Solve $\dfrac{2}{x} + \dfrac{1}{x - 1} = 2$ from the graph in the figure.

$x \neq 0, 1; \{\frac{1}{2}, 2\}$

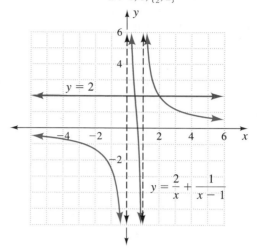

63. Solve $\dfrac{4}{x} + \dfrac{1}{x + 3} = -1$ from the graph in the figure.

$x \neq 0, -3; \{-6, -2\}$

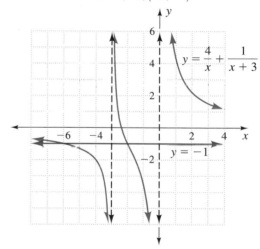

64. Solve the equations in Example 9 for l in terms of w. Graph with width on the horizontal axis and length on the vertical axis.

$l = 8w/5, l = (130 - 2w)/2;$ intersection is $(25, 40);$ see Additional Answers.

Quadratic equations appear as rational equations in equilibrium problems from first-year chemistry. In solving Exercises 65 to 68 for x, remember that $A/B = C$ is equivalent to $A = B \cdot C$. Because of physical constraints on the problems, the input variable has been restricted. Solve for x. Round to the nearest thousandth.

65. $\dfrac{x^2}{1.000 - x} = 4, 0 \leq x \leq 1$ $\quad x = -2 + 2\sqrt{2} \approx 0.828$

66. $\dfrac{x^2}{2.000 - x} = 12, 0 \leq x \leq 2$ $\quad x = -6 + 2\sqrt{15} \approx 1.746$

67. $\dfrac{(3.000 + 2x)^2}{(1.000 - x)(2.000 - x)} = 42, 0 \leq x \leq 1$ $\quad x = \dfrac{69 - 7\sqrt{39}}{38} \approx 0.665$

68. $\dfrac{(2x)^2}{(2.000 - x)(1.000 - x)} = 24, 0 \leq x \leq 1$ $\quad x = \dfrac{9 - \sqrt{21}}{5} \approx 0.883$

69. Zoning codes in a mid-size city require that a business sign in a residential neighborhood be no greater than 1.5 square feet in area. What are the dimensions for a rectangular sign with sides in the golden ratio (1.618 to 1)? Calculate the length and width of the sign in inches. Round to the nearest tenth. 18.7 in.; 11.6 in.

70. The length and width of the frame of a mirror are to be in the golden ratio (1.618 to 1). The area of the mirror is 4 square feet. Find the length and width to the nearest tenth of an inch. 30.5 in., 18.9 in.

In Exercises 71 to 78, set up equations and solve. Use the formulas from the reading material.

71. One door in a large lecture hall permits a class to exit in 6 minutes. What exit time is needed for a second door in order to drop the exit time to 2 minutes when both doors are available? $\frac{1}{6} + \frac{1}{x} = \frac{1}{2}$; $x = 3$ min

72. If $R_1 = 6000$ ohms and $R_2 = 2000$ ohms for parallel resistors in a circuit, find the total resistance R.
$\frac{1}{6000} + \frac{1}{2000} = \frac{1}{R}$; $R = 1500$ ohms

73. One hose fills a wading pool in 20 minutes. When both the original and a second hose are used, the filling time drops to 12 minutes. How long would it take the second hose to fill the pool by itself? $\frac{1}{20} + \frac{1}{x} = \frac{1}{12}$; $x = 30$ min

74. One ventilation fan changes the air in a house in 7.5 hours. How fast could a second fan vent the house by itself if, working together, the fans meet building code (a complete change of air every 3 hours)?
$\frac{1}{7.5} + \frac{1}{x} = \frac{1}{3}$; $x = 5$ hr

75. Two resistors are in parallel with $R_1 = 10,000$ ohms and $R_2 = 4000$ ohms. Find the total resistance R.
$\frac{1}{10,000} + \frac{1}{4000} = \frac{1}{R}$; $R \approx 2857$ ohms

76. Two capacitors are in series with $C_1 = 0.01$ μF and $C_2 = 0.07$ μF. Find the total capacitance C.
$\frac{1}{0.01} + \frac{1}{0.07} = \frac{1}{C}$; $C = 0.00875$ μF

77. Busloads of passengers arrive at a subway station at regular intervals. Ernest runs a toll gate where he can serve a busload in 5 minutes by himself. An exact-change automated gate could serve a busload in 3 minutes. If Ernest and two automated gates were all operating, how many minutes would be needed for each busload of passengers? $\frac{1}{5} + \frac{1}{3} + \frac{1}{3} = \frac{1}{x}$; $x \approx 1.15$ min

78. If a third automated gate were opened in the setting in Exercise 77, how many minutes would be needed for each busload of passengers? $\frac{1}{5} + \frac{1}{3} + \frac{1}{3} + \frac{1}{3} = \frac{1}{x}$; $x \approx 0.833$ min

The answers in Exercises 79 to 82 should be rounded to the nearest tenth of a centimeter. The numbers seem weird because they represent relatively accurate dimensions.

79. The ratio of length to width of a 43-gram Hershey® candy bar is 5 to 2. The perimeter is 39.9 centimeters. How long are the sides? 14.3 cm, 5.7 cm

80. The ratio of width to length of a 113-gram Hershey candy bar is 4.6 to 10. The perimeter is 50.6 centimeters. How long are the sides? 8 cm, 17.3 cm

81. The ratio of length to width of a 142-gram box of Milk Duds® is 7 to 3. The perimeter is 44.4 centimeters. What are the length and width? 15.5 cm, 6.7 cm

82. The ratio of length to width of a 45.6-gram box of Milk Duds is 8 to 3. The perimeter is 33 centimeters. What are the length and width? 12 cm, 4.5 cm

In Exercises 83 to 88, solve for the indicated letter. Assume none of the variables are zero.

83. $\frac{1}{a} + \frac{1}{b} = \frac{1}{c}$ for a $a = \frac{bc}{b-c}$

84. $\frac{1}{a} + \frac{1}{b} = \frac{1}{c}$ for b $b = \frac{ac}{a-c}$

85. $\frac{1}{a} + \frac{1}{b} = \frac{1}{c}$ for c $c = \frac{ab}{a+b}$

86. $\frac{1}{C} = \frac{1}{C_1} + \frac{1}{C_2}$ for C_2 $C_2 = \frac{C \cdot C_1}{C_1 - C}$

87. An electronic formula: $I = \frac{E}{R+r}$ for R
$R = \frac{E}{I} - r$

88. A refrigeration formula: $\frac{W}{Q} = \frac{T_2}{T_1} - 1$ for T_1
$T_1 = \frac{T_2 Q}{W+Q}$

■ Error Analysis

Exercises 89 to 92 each contain a different incorrect solution to the following problem:

$$\frac{1}{x+2} + \frac{2}{x} = \frac{11}{15} \qquad 15x + 30(x+2) = 11x(x+2)$$

Explain what was done wrong, and correct the work.

89. $15x(x+2)\left(\frac{1}{x+2} + \frac{2}{x}\right) = \left(\frac{11}{15}\right)15x(x+2)$

$15x + \frac{2}{x} = 11x(x+2)$
did not distribute $15x(x+2)$ over $2/x$

90. $15x(x+2)\left(\frac{1}{x+2} + \frac{2}{x}\right) = \left(\frac{11}{15}\right)15x(x+2)$

$15x + 2 \cdot 15 = 11x(x+2)$
did not include $(x+2)$ in distributing over $2/x$

91. $15x(x+2)\left(\frac{1}{x+2} + \frac{2}{x}\right) = \left(\frac{11}{15}\right)15x(x+2)$

$\frac{15x(x+2)}{x+2} + \frac{15x(x+2)}{x} = 11x(x+2)$
forgot to multiply by 2 in 2nd term

92. $15x(x+2)\left(\frac{1}{x+2} + \frac{2}{x}\right) = \left(\frac{11}{15}\right)15x(x+2)$

$15x + 2 \cdot 15x + 2 = 11x(x+2)$
forgot parentheses on $(x+2)$

Projects

93. Proportion Proofs If four nonzero numbers a, b, c, and d are in proportion, then several other proportions can be derived from the basic proportion. Prove any four of the five proportion statements below, using accepted equation-solving steps.

a. If $\dfrac{a}{b} = \dfrac{c}{d}$, then $\dfrac{a}{c} = \dfrac{b}{d}$.

b. If $\dfrac{a}{b} = \dfrac{c}{d}$, then $\dfrac{b}{a} = \dfrac{d}{c}$.

c. If $\dfrac{a}{b} = \dfrac{c}{d}$, then $\dfrac{a+b}{b} = \dfrac{c+d}{d}$. (*Hint:* Add 1 to both sides.)

d. If $\dfrac{a}{b} = \dfrac{c}{d}$, then $\dfrac{a-b}{b} = \dfrac{c-d}{d}$.

e. If $\dfrac{a}{b} = \dfrac{c}{d}$, then $\dfrac{a+b}{a-b} = \dfrac{c+d}{c-d}$, where $a \neq b$, $c \neq d$.
(*Hint:* Use the results in parts c and d.)

See Additional Answers.

94. Golden Ratio The *golden ratio* is $(1 + \sqrt{5})/2$ to 1. Any rectangle with this length to width ratio is a golden rectangle. The golden ratio is sometimes given its own Greek letter, ϕ (phi); ϕ is an irrational number like π (pi). The golden ratio is closely related to the Fibonacci sequence: 1, 1, 2, 3, 5, 8, 13, 21, 34, ...

a. Use differences to find the next ten terms of the Fibonacci sequence.
55, 89, 144, 233, 377, 610, 987, 1597, 2584, 4181

b. Describe in words how the sequence is formed.
Each term is the sum of the two prior terms. The sequence starts with 1, 1.

c. Divide consecutive terms of the sequence, a_n/a_{n-1}, up to 233/144. Describe what happens as n gets larger.
The quotients get closer together.

d. Extension: Write a spreadsheet to calculate the first 100 terms in the Fibonacci sequence and the quotients of consecutive terms.

7   Chapter Summary

Vocabulary

additive inverse	equivalent fraction property	least common denominator	rational expressions
asymptote	equivalent ratios	linear variation	rational functions
complex rational expressions	extraneous root	proportion	rational numbers
constant of proportionality	golden ratio	quadratic variation	similar triangles
constant of variation	golden rectangle	rate	unit analysis
corresponding angles	inverse proportion	ratio	volume of a pyramid
corresponding sides	inverse variation	rational equations	
direct variation	joint variation		

Concepts

7.1 ■ Ratios, Units of Measure, and Proportions

If a ratio contains units of measure of the same type, the units should be made the same before the ratio is simplified or is compared with another ratio.

Determine whether two ratios are equal by simplifying them to the same ratio or by dividing them to the same decimal.

To change units with a unit analysis:

1. Identify the units of measure to be changed, and identify the units needed in the answer.

2. List facts that contain the starting units and the ending units. List facts needed to relate the starting and ending units.

3. Write the starting units. Using your list of facts, set up a product of fractions so that each unit of measure appears once in the numerator and once in the denominator.

4. Use $a/a = 1$ to eliminate the unwanted units, and calculate numbers.

To change rates with a unit analysis, follow the steps above. The facts making up the product of fractions may be placed to the left or to the right of the starting rate.

Corresponding angles of similar triangles are equal. Corresponding sides of similar triangles are proportional.

7.2 ■ Proportions and Direct Variation

For linear data to be proportional, the ratio of the output to input, y/x, for each data pair must be the same. The graph of a line through proportional data passes through the origin.

7.3 ▩ Inverse Variation

The graph of a rational expression that has been simplified to lowest terms becomes nearly vertical whenever the denominator approaches zero.

For data (x_1, y_1) and (x_2, y_2), direct variation has a constant ratio, $y_1/x_1 = y_2/x_2$, and inverse variation has a constant product, $x_1y_1 = x_2y_2$.

In Figure 42, the upper graphs illustrate direct variation, $k = \dfrac{y}{x^2}, k = \dfrac{y}{x}, k = \dfrac{y}{\sqrt{x}}.$ The lower graphs illustrate inverse variation, $k = yx^2, k = yx, k = y\sqrt{x}.$

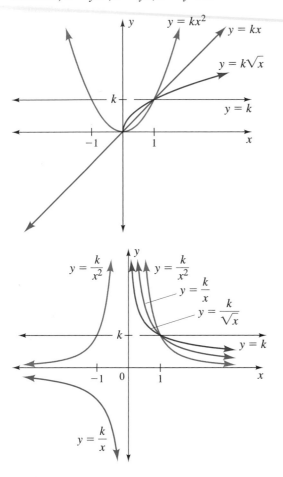

FIGURE 42

7.4 ▩ Simplification, Multiplication, and Division of Rational Expressions

When the numerator and denominator of a fraction are the same, the fraction equals 1.

When the numerator and denominator of a fraction are additive inverses, the fraction equals -1.

To multiply rational expressions, write the product with a single numerator and denominator. Factor as needed. Eliminate common factors from the numerator and denominator. Multiply remaining factors in the numerator and in the denominator.

To divide rational expressions, multiply the first expression by the reciprocal of the second expression.

Division by zero creates an undefined expression.

To simplify a rational expression, factor the numerator and denominator and eliminate common factors of the form a/a.

To simplify complex rational expressions, write the fraction bar as a division and continue as with other divisions.

7.5 ▩ Division of Polynomials and Related Graphs

To divide algebraic expressions with long division, arrange the dividend into descending order of exponents and replace missing terms with terms having a zero numerical coefficient. Then proceed as follows, repeating the five steps until the degree of the dividend is less than the degree of the divisor:

1. Estimate how many times the divisor divides into the dividend.

2. Enter the estimate in the quotient.

3. Multiply the divisor by the estimate, and place the product below the dividend.

4. Subtract.

5. Bring down the next term in the dividend.

When there is no remainder in a polynomial division, we say the divisor is a factor of the dividend.

A zero remainder indicates that the graph of the rational equation will contain a hole with an x-coordinate that makes a zero denominator in the original rational equation.

A nonzero remainder indicates that the graph of the equation will approach a vertical asymptote with an equation $x = a$. The value of a makes a zero denominator in the original rational equation.

7.6 ▩ Addition and Subtraction of Rational Expressions

To find the least common denominator, form a product by including each factor the highest number of times it appears in any denominator.

To add or subtract rational numbers, convert to a common denominator, if needed; add or subtract the numerators and place the result over the common denominator; factor and simplify the answer.

To simplify complex fractions, multiply the numerator and denominator by the least common denominator of all terms.

7.7 ▩ Solving Rational Equations

To eliminate denominators in a proportion or an equation, multiply each side by the least common denominator.

Table 12 shows the relationships among equations, sequences, and variations.

TABLE 12

Linear Equations $y = mx + b$ and $x = c$ Linear Functions $f(x) = mx + b$ Domain: Real numbers	Arithmetic Sequences $a_n = a_1 + (n - 1)d$ Domain: Positive integers Constant first differences	Direct Variation Constant ratio for x and y $\dfrac{y}{x} = k$ or $y = kx$ Linear Variation $y = mx$ Domain: Real numbers
Quadratic Equations $y = ax^2 + bx + c$ Quadratic Functions $f(x) = ax^2 + bx + c$ Domain: Real numbers	Quadratic Sequences Domain: Positive integers Constant second differences	Quadratic Variation $y = kx^2$ Domain: Real numbers
Rational Equations Example: $y = \dfrac{1}{x}$ Rational Functions Example: $f(x) = \dfrac{1}{x}$ Domain: Real numbers, $x \neq 0$ Range: Real numbers, $y \neq 0$		Inverse Variation Constant product for x and y $xy = k$ or $y = \dfrac{k}{x}$ Inverse Square Variation $y = \dfrac{k}{x^2}$

7 Review Exercises

In Exercises 1 to 4, add, subtract, multiply, and divide the fractions or rational expressions in the order shown.

1. $\frac{2}{3}$ and $\frac{4}{3}$ $2, -\frac{2}{3}, \frac{8}{9}, \frac{1}{2}$

2. $\frac{3}{4}$ and $\frac{5}{8}$ $1\frac{3}{8}, \frac{1}{8}, \frac{15}{32}, 1\frac{1}{5}$

3. $\dfrac{a}{b}$ and $\dfrac{a}{c}$ $\dfrac{ac + ab}{bc}, \dfrac{ac - ab}{bc}, \dfrac{a^2}{bc}, \dfrac{a}{b}$

4. $\dfrac{a}{c}$ and $\dfrac{c}{b}$ $\dfrac{ab + c^2}{cb}, \dfrac{ab - c^2}{cb}, \dfrac{a}{b}, \dfrac{ab}{c^2}$

Simplify the expressions in Exercises 5 to 10.

5. a. $\dfrac{27xy^2}{15x^2y}$ $\dfrac{9y}{5x}$ **b.** $\dfrac{2 - x}{x - 2}$ -1 **c.** $\dfrac{a - ac}{a}$ $1 - c$

6. a. $\dfrac{12x^3yz}{39xy^2z^2}$ $\dfrac{4x^2}{13yz}$ **b.** $\dfrac{x - 5}{5 - x}$ -1 **c.** $\dfrac{ab + b^2}{ab}$ $\dfrac{a + b}{a}$

7. a. $\dfrac{3x^2 - 12}{x + 2}$ $3(x - 2)$

b. $\dfrac{a - b}{(a + b)(a - b)}$ $\dfrac{1}{a + b}$

c. $8x\left(\dfrac{1}{4x} + \dfrac{1}{2}\right)$ $2 + 4x$

8. a. $\dfrac{21ab + 7b^2}{6a + 2b}$ $\dfrac{7b}{2}$

b. $\dfrac{(x - 2)(x + 3)}{(x + 3)(x + 2)}$ $\dfrac{x - 2}{x + 2}$

c. $x(2 - x)\left(\dfrac{1}{x - 2} + \dfrac{3}{-x}\right)$ $2x - 6$

9. a. $\dfrac{x^2 - 5x - 6}{x^2 - 4x - 5}$ $\dfrac{x - 6}{x - 5}$

b. $\dfrac{4x^2 - 1}{2x^2 + 5x + 2}$ $\dfrac{2x - 1}{x + 2}$

c. $\dfrac{x^2 + 3x - 4}{x^2 - 16}$ $\dfrac{x - 1}{x - 4}$

10. a. $\dfrac{x^2 - 7x + 12}{x^2 - 16}$ $\dfrac{x - 3}{x + 4}$

b. $\dfrac{2x^2 - 7x + 3}{2x^2 + 7x - 4}$ $\dfrac{x - 3}{x + 4}$

c. $\dfrac{3x + 6}{x^2 + 4x + 4}$ $\dfrac{3}{x + 2}$

In Exercises 11 and 12, what expression is needed in the numerator or denominator to make a true statement?

11. $\dfrac{6 - x}{x - 6} = -1$

12. $\dfrac{a - c}{a - c} = 1$

Simplify the expressions in Exercises 13 and 14.

13. 3 quarts to 4 gallons   3 to 16

14. 5 feet to 5 yards   1 to 3

In Exercises 15 and 16, for what inputs x will the expressions be undefined?

15. $\dfrac{2}{(x-1)(x+3)}$

    $x = -3, x = 1$

16. $\dfrac{-1}{(x+1)(x+1)}$

    $x = -1$

Perform the indicated operations in Exercises 17 to 30. Leave the answers in simplified form. Assume no zero denominators.

17. $\dfrac{4y^2}{9x^2} \cdot \dfrac{3x}{8y}$ $\dfrac{y}{6x}$

18. $\dfrac{x+3}{x^2-6x+9} \div \dfrac{1}{x^2-9}$ $\dfrac{(x+3)^2}{x-3}$

19. $15x\left(\dfrac{1}{3x} + \dfrac{2}{5x}\right)$ 11

20. $2x(x+5)\left(\dfrac{3}{2x} - \dfrac{1}{x+5}\right)$ $x+15$

21. $\dfrac{x^2+5x+4}{x^2-16} \cdot \dfrac{2x-8}{1-x^2}$ $\dfrac{2}{1-x}$

22. $\dfrac{2x^2+x-3}{x-3} \div \dfrac{x^2-2x+1}{3-x}$ $\dfrac{-(2x+3)}{x-1}$ or $\dfrac{2x+3}{1-x}$

23. $\dfrac{1}{x-3} + \dfrac{x}{x-3}$ $\dfrac{x+1}{x-3}$

24. $\dfrac{x}{x-3} - \dfrac{2}{x+2}$ $\dfrac{x^2+6}{(x+2)(x-3)}$

25. $\dfrac{x}{x-1} - \dfrac{1}{1-x}$ $\dfrac{x+1}{x-1}$

26. $\dfrac{x+2}{2-x} - \dfrac{x+2}{x-2}$ $\dfrac{2x+4}{2-x}$

27. $\dfrac{x-2}{x^2-1} - \dfrac{2}{x-1}$ $\dfrac{-x-4}{x^2-1}$

28. $\dfrac{1-x}{x^2-4} + \dfrac{2}{x-2}$ $\dfrac{x+5}{x^2-4}$

29. $\dfrac{1}{x^2-2x} - \dfrac{1}{3x^2-10x+8}$ $\dfrac{2}{x(3x-4)}$

30. $\dfrac{1}{3x-x^2} - \dfrac{1}{-9+9x-2x^2}$ $\dfrac{-1}{x(2x-3)}$

Simplify the expressions in Exercises 31 to 38.

31. $\dfrac{A}{\frac{1}{2}h}$ $\dfrac{2A}{h}$

32. $\dfrac{V}{\frac{1}{3}\pi r^2}$ $\dfrac{3V}{\pi r^2}$

33. $\dfrac{2500 \text{ miles}}{\dfrac{12.5 \text{ miles per gallon}}{}}$ 200 gal

34. $40 \text{ months} \cdot \dfrac{\frac{1}{150} \text{ grain}}{150 \text{ months}}$ 0.00178 grain

35. $\dfrac{x + \dfrac{x}{3}}{3 + \dfrac{2}{x}}$ $\dfrac{4x^2}{9x+6}$

36. $\dfrac{x - \dfrac{4}{x}}{1 - \dfrac{2}{x}}$ $x+2$

37. $\dfrac{\dfrac{x^2-x}{x^2-1}}{x}$ $\dfrac{x^2}{2(x+1)}$

38. $\dfrac{\dfrac{x-1}{2x^2-3x-5}}{\dfrac{x^2-1}{4x^2-20x+25}}$ $\dfrac{2x-5}{(x+1)^2}$

In Exercises 39 to 44, the long division problems are stated in fraction form with the denominator containing the divisor. Indicate restrictions on the inputs, and complete the divisions.

39. $\dfrac{x^3-3x^2+3x-1}{x-1}$ $x \neq 1; x^2-2x+1$

40. $\dfrac{x^3+1}{x+1}$ $x \neq -1; x^2-x+1$

41. $\dfrac{x^3+1}{x+3}$ $x \neq -3; x^2-3x+9 - \dfrac{26}{x+3}$

42. $\dfrac{x^3-3x^2+3x-1}{x-2}$ $x \neq 2; x^2-x+1 + \dfrac{1}{x-2}$

43. $\dfrac{x^4-1}{x+1}$ $x \neq -1; x^3-x^2+x-1$

44. $\dfrac{x^3+x-2}{x-1}$ $x \neq 1; x^2+x+2$

In Exercises 45 to 56, solve for x. Some solutions may be complex numbers. Exclude inputs that make the expressions undefined.

45. $\dfrac{6}{x} = \dfrac{15}{32}$ $x = 12.8; x \neq 0$

46. $\dfrac{x+1}{6} = \dfrac{x-2}{5}$ $x = 17$

47. $\dfrac{x}{2} = \dfrac{7}{x+5}$ $\{-7, 2\}; x \neq -5$

48. $\dfrac{x-6}{8} = \dfrac{5}{x}$ $\{-4, 10\}; x \neq 0$

49. $\dfrac{x-2}{4} = \dfrac{x+1}{3} - 3$ $x = 26$

50. $\dfrac{x+2}{4} = \dfrac{x-1}{5}$ $x = -14$

51. $\dfrac{x+3}{1} = \dfrac{-5}{2x}$ $\{-1.5 \pm 0.5i\}; x \neq 0$

52. $\dfrac{x^2+1}{x} = \dfrac{6}{5}$ $\{0.6 \pm 0.8i\}; x \neq 0$

53. $\dfrac{1}{x} + \dfrac{5}{x+2} = \dfrac{13}{3x}$ $x = 4; x \neq -2, 0$

54. $\dfrac{2}{x-4} - \dfrac{1}{x} = \dfrac{11}{3x}$ $x = 7; x \neq 0, 4$

55. $\dfrac{1}{x-1} + \dfrac{2}{x} = \dfrac{7}{6}$ $\{\frac{4}{7}, 3\}; x \neq 0, 1$

56. $\dfrac{4}{x} + \dfrac{1}{x-2} = 3$ $\{1, \frac{8}{3}\}; x \neq 0, 2$

In Exercises 57 to 60, solve for the missing sides.

57.

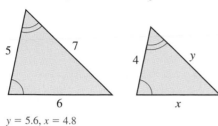

$y = 5.6, x = 4.8$

58.

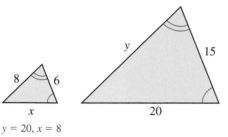

$y = 20, x = 8$

59.

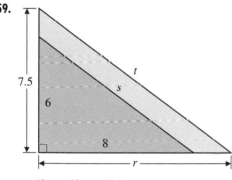

$r = 10, s = 10, t = 12.5$

60.

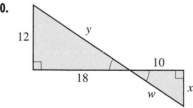

$y \approx 21.633, x \approx 6.667, w \approx 12.019$

61. Suppose a tennis serve travels 149 miles per hour, a record set by Greg Rusedski. Find how long it takes the ball to travel the 78-foot length of the court.
≈ 0.357 sec

62. Suppose a baseball pitch travels 90 miles per hour. Find how long it takes the ball to travel the 60 feet from the pitcher to home base. $\frac{5}{11}$ sec

63. Suppose a baby grows from 0 to 21 inches in the 9 months before birth. If the growth rate for the 18 years after birth were the same as the growth rate before birth, how much would the young person grow and how tall would the young person be, in inches? in feet?
504 in., ≈ 525 in.; 42 ft, ≈ 43.75 ft (the 18 years does not include the 9 months)

64. Suppose a young person grows from 21 inches at birth to 69 inches in 18 years. If the growth rate before birth were the same as that from birth to age 18, how long would it take for the unborn child to reach 21 inches?
≈ 7.9 yr

For Exercises 65 to 68, write a linear equation. State whether the relationship is a direct variation ($y = kx$). Explain the meaning of the y-intercept for those situations where the relationship is not proportional ($y = mx + b$).

65. A 64MB Memory Stick costs $29.99. A 128MB Memory Stick costs $49.99. What equation gives the cost in terms of MB?
$y = 0.3125x + 9.99$; possibly the basic cost of stick manufacture

66. An order of 6 donuts costs $3.49, and another of 12 donuts costs $4.99. What equation gives the cost in terms of the number of donuts?
$y = 0.25x + 1.99$; perhaps the box, bag, or labor

67. If 40 gallons of aviation fuel costs $114.80 and 65 gallons costs $186.55, what equation describes the cost in terms of the number of gallons? $y = 2.87x$; direct

68. A subscription to *Science News* costs $54.50 for 1 year or $98 for 2 years. What equation describes the cost in terms of the number of years? $y = 43.5x + 11$; not meaningful

Write a formula for each of the sentences in Exercises 69 to 72. Identify your variables, and use k as the constant of variation, if needed.

69. The distance you can see from a height varies directly as the square root of the height.
$D = k\sqrt{h}$; $D =$ distance, $h =$ height

70. The earnings for a week vary jointly as the hours worked and the wage per hour.
$E = kwh$; $E =$ earnings, $w =$ wage, $h =$ hours; $k = 1$

71. The volume of a box varies jointly as its length, width, and height.
$V = klwh$; $V =$ volume, $l =$ length, $w =$ width, $h =$ height; $k = 1$

72. The cost of sending a package varies directly with its weight. $C = kw$, $C =$ cost, $w =$ weight

In Exercises 73 to 76, write a formula for each of the sentences, which are based on automobile traffic accident investigation manuals.

73. The skid distance, d, required to stop a vehicle varies directly as the square of the speed, s. $d = ks^2$

74. In some auto accidents, a car hits a curb, flips into the air, and vaults some distance d through the air to a point of impact with the ground. The speed, s, of the car at the curb varies directly with the square root of the distance, d. $s = k\sqrt{d}$

75. The speed, s, of a car varies directly as the square root of the product of the skid distance, d, and coefficient of friction factor, f. $s = k\sqrt{df}$

76. The time required to skid to a stop, t, varies directly as the square root of the length, l, of the skid mark and inversely as the square root of the coefficient of friction, f. Friction is determined by the road surface and weather conditions. $t = k\sqrt{l/f}$

77. The cost of oranges varies directly with the number of pounds purchased. Compare the cost of 6 pounds with the cost of 4 pounds.   1.5 times the cost

78. The cost of a turkey varies directly with the number of pounds purchased. Compare the cost of an 18-pound turkey with the cost of a 15-pound turkey.
     1.2 times the cost

79. The energy of a wave varies directly with the square of its height. Compare the energy of a 12-foot wave with that of a 3-foot wave.   16 times the energy

80. The length of skid marks (in feet) on a concrete surface varies directly with the square of the speed, s, in miles per hour. Compare the length of skid marks at 30 miles per hour with the length of skid marks at 20 miles per hour. $\frac{9}{4}$ times the length

Describe each situation in Exercises 81 to 84 with an inverse variation equation. Identify the constant of variation.

81. An associate's degree from one community college requires a total of 90 credits. If a student takes x credits each quarter, how many quarters does the student need to complete the degree? $y = 90/x$; $k = 90$ credits

82. The output is the number of copies per press required to print 100,000 copies of a book, and the input is the number of printing presses used.
     $y = 100,000/x$; $k = 100,000$ copies

83. A school district allocates $100,000 for textbooks. If the average book price is the input, what equation describes the possible number of books purchased?
     $y = 100,000/x$; $k = \$100,000$

84. The output is the dollars per hour to earn $10,000, and the input is the number of hours worked.
     $y = 10,000/x$; $k = \$10,000$

Exercises 85 and 86 involve adding rates and use equations of the form $1/a + 1/b = 1/c$.

85. Two ventilation fans are operating in an attic. Working alone, one can change the air in 5.5 hours and the other in 7 hours. How long will they take working together? Will they be able to change the air in 3 hours, as required by code?   3.08 hr; no

86. If both the front and the back door are used, a theater can be emptied in 3 minutes. The theater can be emptied using the back door, by itself, in 7 minutes. How rapidly can the theater be emptied using only the front door?
     5.25 min

87. Mridula starts on a 300-mile trip. Almost immediately, her car is struck from behind. It takes three hours to get the accident reported and to rent a replacement car.

 a. Explain why her speed, or rate, is

$$r = \frac{300}{t - 3}$$ distance divided by total hr minus 3 hr not traveling

b. Complete the table for the rate equation.

Rate Required to Finish Trip

Total Trip Time (hours)	Rate (miles per hour)
4	300
5	150
6	100
7	75
8	60

c. The graph in the figure is a portion of a hyperbola. Explain why the graph is shifted to the right 3 units from the graph for $r = 300/t$.   Because of the 3-hr delay, we replace t by $t - 3$, causing a shift of 3 units to the right.

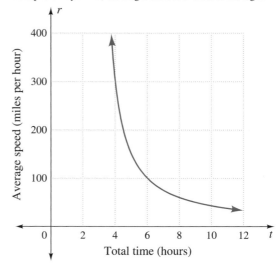

Referring to part a, calculate the rate Mridula would need to travel in order to complete her trip in a total of

d. 24 hours   14.3 mph

e. 18 hours   20 mph

f. 10 hours   42.9 mph

g. 3.5 hours   600 mph

h. 3.1 hours   3000 mph

i. 3.01 hours   30,000 mph

j. 3 hours   not possible

▮ Error Analysis

88. In solving $1 + 13/x^2 = -4/x$, a student multiplies by x^3 on both sides and finds $x = 0$ as one solution. What caused this extraneous solution? What are the correct solutions? x^3 is not the LCD; $x = -2 \pm 3i$

89. In solving $8/(x + 1) = 3/(x^2 - 1)$, a student multiplies by $(x + 1)(x^2 - 1)$ and obtains $x = 1.375$ and $x = -1$. What is the correct solution? Why is the other solution extraneous? $x = 1.375$; $x = -1$ makes the equation undefined.

90. In solving a direct variation problem, $y = kx$, a student says that the cost of 6 pounds of potatoes is 2 times more than the cost of 3 pounds. Why should "more" be omitted? (*Hint:* Selecting a price per pound might be helpful.)   "More" implies a doubling of the difference instead of the cost.

Functions and Rational Expressions

91. Given that $f(x) = 30/x^2$ and $a > 0$, explain which is larger: $f(a)$ or $f(a + 2)$.

A larger input will yield a smaller output; $f(a) > f(a + 2)$.

Graphs

92. Graph the expression in Exercise 39. What creates a hole in the graph? Where is the hole?
See Additional Answers; denominator is factor of numerator; $x = 1$

93. Graph the expression in Exercise 42. What creates a nearly vertical line in the graph? What is the equation of the asymptote?   See Answer Section; as x gets closer to 2, denominator gets closer to 0; $x = 2$

94. a. Fit a linear equation to the coordinate points $(5, 8)$ and $(10, 4)$. $y = -\frac{4}{5}x + 12$

b. Fit an inverse variation equation to the same points. $y = 40/x$

c. Graph the two equations.   See Additional Answers.

d. How are the graphs the same? How are they different?
contain $(5, 8)$ & $(10, 4)$, are decreasing; a is straight, b is curved

95. Describe the output behavior of the graph in the figure under the conditions described in parts a to d.

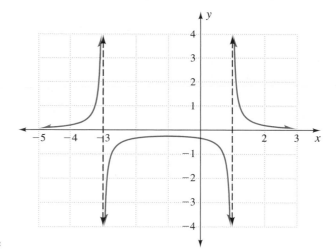

a. As x approaches 1 from the left y approaches $-\infty$.

b. As x approaches 1 from the right y approaches ∞.

c. As x approaches -3 from the right y approaches $-\infty$.

d. As x approaches -3 from the left y approaches ∞.

e. On the calculator, graph $y = 1/((x + 3)(x - 1))$, first in connected mode and then in dot mode. Adjust the window to match the figure. Compare the graphs and explain the difference.   The nearly vertical line connecting points near $x = -3$ and $x = 1$ should not be there, as shown in dot mode.

f. What are the equations of the vertical asymptotes in the figure? $x = -3, x = 1$

| **7** | **Chapter Test** | |

Simplify the expressions or perform the indicated operations in Exercises 1 to 19. Exclude any inputs that make the expression undefined.

1. $\dfrac{ab^2c}{a^2bc^2}$ $\dfrac{b}{ac}$; $a \neq 0, b \neq 0, c \neq 0$

2. $\dfrac{b + 3}{b - 3}$ simplified; $b \neq 3$

3. $\dfrac{3ac}{15ac^2}$ $\dfrac{1}{5c}$; $a \neq 0, c \neq 0$

4. $\dfrac{15b^2c^3}{10b^3c}$ $\dfrac{3c^2}{2b}$; $b \neq 0, c \neq 0$

5. $\dfrac{x^2 - 25}{20 + x - x^2}$ $-\dfrac{x + 5}{x + 4}$; $x \neq -4, 5$

6. $\dfrac{12 - 2a}{18 + 3a - a^2}$ $\dfrac{2}{a + 3}$; $a \neq -3, 6$

7. $\dfrac{3x^2 - 7x + 2}{2x^2 - 5x + 2}$ $\dfrac{3x - 1}{2x - 1}$; $x \neq \frac{1}{2}, 2$

8. $\dfrac{2x^2 + 7x + 6}{x^2 + 4x + 4}$ $\dfrac{2x + 3}{x + 2}$; $x \neq -2$

9. $\dfrac{12xy^2}{7y - y^2} \div \dfrac{6x^2}{7 - y}$ $\dfrac{2y}{x}$; $y \neq 0, 7$; $x \neq 0$

10. $\dfrac{x^2 - 9}{x^2 + 4x + 3} \cdot \dfrac{x - 3}{x + 1}$ $\dfrac{(x - 3)^2}{(x + 1)^2}$; $x \neq -3, -1$

11. $\dfrac{x^2 + 8x + 16}{x - 2} \cdot \dfrac{x^2 - 4}{x + 4}$ $(x + 4)(x + 2)$; $x \neq -4, 2$

12. $x(x + 1)\left(\dfrac{3}{x} + \dfrac{2}{x + 1}\right)$ $5x + 3$; $x \neq -1, 0$

13. $\dfrac{2 - x}{x + 2} + \dfrac{x + 2}{x - 2}$ $\dfrac{8x}{(x - 2)(x + 2)}$; $x \neq \pm 2$

14. $\dfrac{x - 2}{x + 2} - \dfrac{x}{x^2 + 2x}$ $\dfrac{x - 3}{x + 2}$; $x \neq -2, 0$

15. $\dfrac{10000 \text{ cm}^3}{10 \text{ cm}}$ 1000 cm^2

16. Change 1,000,000 hours to years. Assume 1 year = 365 days. ≈ 114.2 yr

17. $\dfrac{30 \text{ miles per gallon}}{60 \text{ miles per hour}}$ $\frac{1}{2}$ hr/gal

18. $\dfrac{\dfrac{3}{x} - x}{x + \dfrac{2}{x}}$ $\dfrac{3 - x^2}{x^2 + 2}; x \neq 0$

19. $\dfrac{\dfrac{x^2 - 3x}{3}}{\dfrac{9 - x^2}{x}}$ $\dfrac{-x^2}{3(3 + x)}; x \neq -3, 0, 3$

In Exercises 20 to 22, the division problems are stated in fraction form with the denominator containing the divisor. Indicate for what values the expressions are defined, and complete the divisions. Graph the expressions in Exercises 20 and 21; show any holes and nearly vertical portions.

20. $\dfrac{x^3 + 6x^2 + 12x + 8}{x + 2}$ $\mathbb{R}, x \neq -2; x^2 + 4x + 4;$ see Answer Section.

21. $\dfrac{x^3 + 6x^2 + 12x + 8}{x + 4}$ $\mathbb{R}, x \neq -4; x^2 + 2x + 4 - \dfrac{8}{x + 4};$ see Answer Section.

22. $\dfrac{2x^4 + x^2 - 4}{x - 1}$ $\mathbb{R}, x \neq 1; 2x^3 + 2x^2 + 3x + 3 + \dfrac{-1}{x - 1}$

In Exercises 23 to 27, solve for x using whichever method (proportion, calculator, multiplication by LCD) seems appropriate. Exclude inputs that make the expressions in the equations undefined.

23. $\dfrac{x - 2}{10} = \dfrac{2}{x - 1}$ $\{-3, 6\}; x \neq 1$

24. $\dfrac{1}{3} - \dfrac{1}{x} = \dfrac{2x}{3}$ $\{\frac{1}{4} \pm i(\sqrt{23}/4)\}; x \neq 0$

25. $\dfrac{1}{x - 4} + \dfrac{1}{x} = \dfrac{10}{3x}$ $x = 7; x \neq 0, 4$

26. $\dfrac{x - 1}{9} = \dfrac{x + 3}{10}$ $x = 37$

27. $\dfrac{3}{x} + \dfrac{2}{x + 1} = 4$ $\{-\frac{3}{4}, 1\}; x \neq -1, 0$

28. Suppose a baby grows from 0 to 7 pounds in the 9 months before birth. How many ounces per hour is this growth rate? There are 16 ounces in a pound. Assume 30 days per month.   ≈0.0173 oz/hr

29. A young person grows from 7 pounds to 140 pounds in 18 years. How many ounces per hour is this growth rate? There are 16 ounces in a pound. Assume 365 days per year.   ≈0.0135 oz/hr

In Exercises 30 and 31, the skid distance in feet for a vehicle varies directly as the square of the speed and inversely as the coefficient of friction, f (road surface and weather conditions). Suppose $d = s^2/30f$, where speed is given in miles per hour.

30. What is the skid distance at 45 miles per hour on an asphalt surface with a 0.6 coefficient of friction?   112.5 ft

31. A gravel road with a coefficient of friction of 0.4 shows a skid mark of 208 feet. How fast was the car traveling in miles per hour?   ≈50 mph

32. Finishing nails are used for cabinet work and interior wood trim. Suppose that the number of finishing nails per pound varies inversely with the square of their length, $n = k/l^2$. If a pound of 1-inch nails contains 1350 nails, how many 4-inch nails will be in a pound?   84

33. The volume of a sphere is $V = \frac{4}{3}\pi r^3$. Predict the type of variation in the volume of a sphere. What is the constant of variation in the volume of a sphere formula?   varies directly with cube of radius; $\frac{4}{3}\pi$

34. Jennifer's snowplow can clear 1000 feet of roadway of a 9-inch snowfall in 5 minutes. Michael's snowplow can clear 1000 feet of roadway of a 9-inch snowfall in 4 minutes.

 a. To the nearest tenth of a minute, in how many minutes can their two plows together clear a 1000-foot street of a 9-inch snowfall?   2.2 min

 b. A third plow, old and slow, is available. How fast must it work under the same conditions with the other two plows to clear the 1000 feet in exactly 2 minutes?   20 min per 1000 ft

35. Triangles ABC and DEF below are similar. Find the lengths of segments EF, AB, and DE.

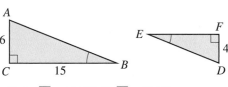

10; $3\sqrt{29} \approx 16.155$; $2\sqrt{29} \approx 10.770$

Exponents and Radicals

Before global positioning devices, navigators could plot a course around a bay known for pirates (Figure 1) by keeping at most a right angle between the sight lines to two distant lighthouses. Exactly a right angle would create a circular path around a hazard near the center of the circle. We prove this important geometrical relationship in Section 8.5.

In this chapter, we expand our definition of powers to include zero, negative-number, and rational-number exponents. We work with properties of exponents, scientific notation, and radical notation. Inverse functions provide the basis for solving power and root equations.

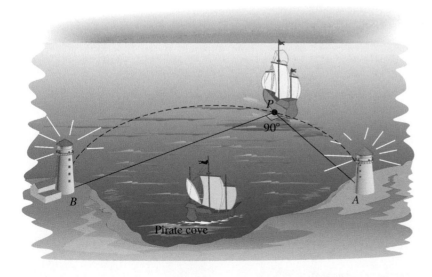

FIGURE 1

8.1   Exponents and Their Properties

Objectives

■ Find products, quotients, and powers of exponential expressions.

■ Apply properties of exponents.

■ Simplify expressions containing zero and negative numbers as exponents.

WARM-UP

Find the values of the expressions.

1. $2^6, 2^5, 2^4, 2^3, 2^2, 2^1$ 64, 32, 16, 8, 4, 2

2. $3^4, 3^3, 3^2, 3^1$ 81, 27, 9, 3

3. $4^4, 4^3, 4^2, 4^1$ 256, 64, 16, 4

4. $5^4, 5^3, 5^2, 5^1$ 625, 125, 25, 5

Power should be compared with sum, difference, product, and quotient.

THIS SECTION PRESENTS exponents and properties of exponents.

Exponents

When we apply an exponent to a base, the result is called the power. The number 8 is the power in $2^3 = 8$. We say that *8 is the 3rd power of 2*.

$$\text{base}^{\text{exponent}} = \text{power}$$

The traditional definition of exponents applies to positive integers used as exponents:

■ **DEFINITION OF POSITIVE INTEGER EXPONENT**

For b^n, n a positive integer,

$$b^n = \underbrace{b \cdot b \cdot b \cdot b \cdot \cdots \cdot b}_{n \text{ factors of } b}$$

Properties of Exponents

The sets of numbers are given color names to assist students who may be colorblind.

The properties of exponents become more apparent if we first look at sets of numbers.

x^2

Red: 1, 4, 9, 16, <u>25</u>, 36, 49, 64, 81, <u>100</u>, 121, 144, 169, 196, <u>225</u>, 256, 289, 324, 361, <u>400</u>, 441, 484, 529, 576, <u>625</u>, 676, 729, 784, 841, <u>900</u>, . . .

2^x

Blue: 2, 4, 8, 16, <u>32</u>, 64, 128, 256, 512, <u>1024</u>, 2048, 4096, 8192, 16,384, <u>32,768</u>, 65,536, 131,072, 262,144, 524,288, <u>1,048,576</u>, . . .

x^3

Orange: 1, 8, 27, 64, <u>125</u>, 216, 343, 512, 729, <u>1000</u>, 1331, 1728, 2197, 2744, <u>3375</u>, 4096, 4913, 5832, 6859, <u>8000</u>, . . .

3^x

Black: 3, 9, 27, 81, <u>243</u>, 729, 2187, 6561, 19,683, <u>59,049</u>, 177,147, 531,441, 1,594,323, 4,782,969, <u>14,348,907</u>, . . .

EXAMPLE 1   Exploring properties of exponents

a. Find two red numbers (each with at least two digits) whose product is a red number. Comment on your result. Write your result with x and y instead of numbers.

b. Which red numbers are divisible by 4? Comment on your result. Is it always true?

c. Find two orange numbers (each with at least two digits) whose product is an orange number. Comment on your result. Write your result with x and y instead of numbers.

SOLUTION See Example 4 after doing Examples 2 and 3.

EXAMPLE 2 Exploring properties of exponents
a. Are red numbers still red numbers when squared? Explain.
b. Are blue numbers still blue numbers when squared? Explain.
c. Are orange numbers still orange numbers when squared? Explain.
d. Are black numbers still black numbers when squared? Explain.

SOLUTION See Example 5 after doing Example 3.

EXAMPLE 3 Naming patterns For the red, blue, orange, and black lists, write each of the underlined numbers as an exponential expression. Then write an expression describing the list in terms of x.

SOLUTION
Red: $5^2, 10^2, 15^2, 20^2, 25^2, 30^2$; x^2
Blue: $2^5, 2^{10}, 2^{15}, 2^{20}$; 2^x
Orange: $5^3, 10^3, 15^3, 20^3$; x^3
Black: $3^5, 3^{10}, 3^{15}$; 3^x

Think about it 1: How do the expressions in Example 3 help answer the questions in Examples 1 and 2?

PROPERTIES OF EXPONENTS

The following are true when a and b are any integers (positive, negative, or zero).

To multiply numbers with like bases, add the exponents:

$$x^a \cdot x^b = x^{a+b}$$ Multiplication of like bases property

To divide numbers with like bases, subtract the exponents:

$$\frac{x^a}{x^b} = x^{a-b} \quad \text{for } x \neq 0$$ Division of like bases property

To apply an exponent to a power expression, multiply the exponents:

$$(x^a)^b = x^{a \cdot b}$$ Power of a power property

An exponent outside the parentheses applies to all parts of a product or quotient inside the parentheses:

$$(x \cdot y)^a = x^a \cdot y^a$$ Power of a product property

$$\left(\frac{x}{y}\right)^a = \frac{x^a}{y^a} \quad \text{for } y \neq 0$$ Power of a quotient property

We will work initially with positive integer exponents; however, if the bases are positive, the properties of exponents apply to all real-number exponents.

EXAMPLE 4 Identifying properties Which property explains the results in Example 1?

SOLUTION All solutions use the power of a product property: $(x \cdot y)^n = x^n \cdot y^n$.

a. All products of two red numbers are red numbers because $x^2 \cdot y^2 = (x \cdot y)^2$, which is a red number.

b. Even numbers can be represented by $2n$. $(2n)^2 = 2^2n^2 = 4n^2$, which is divisible by 4.

c. All products of two orange numbers are orange numbers because $x^3 \cdot y^3 = (xy)^3$, which is an orange number. ▬

EXAMPLE 5   Identifying properties   Which property explains the results in Example 2?

SOLUTION   To square the numbers in each set, we use the power of a power property: $(x^a)^b = x^{ab}$. The commutative property of real numbers, $a \cdot b = b \cdot a$, is also helpful.

a. The red numbers are squares; $(x^2)^2$ is a second power of x^2 and a red number.

b. The blue numbers are powers of 2; $(2^x)^2 = 2^{2x}$, which is a power of 2 and a blue number.

c. The orange numbers are cubes; $(x^3)^2 = x^{3 \cdot 2} = x^{2 \cdot 3} = (x^2)^3$, which is a third power of x^2 and an orange number.

d. The black numbers are powers of 3; $(3^x)^2 = 3^{2x}$, which is a power of 3 and a black number. ▬

Example 6 uses a different property of exponents.

EXAMPLE 6   Exploring properties of exponents

a. Find two blue numbers (with at least two digits) whose product is a blue number.

b. Find two blue numbers whose product is the 10th blue number. How many pairs are possible?

c. Find two blue numbers whose product is the 15th blue number.

d. Use powers of 2 to write the six pairs of blue numbers that multiply to the 13th blue number.

e. What do you observe about the exponents in part d?

f. What number makes $2^4 \cdot 2^x = 2^9$ true? Rewrite the statement in blue numbers.

g. Which property of exponents describes this exploration?

SOLUTION   **a.** Answers vary; for example, $32 \times 64 = 2048$.

b. Four pairs: 2 and 512, 4 and 256, 8 and 128, 16 and 64

c. Seven pairs: 128 and 256, 64 and 512, 32 and 1024, etc.

d. $2^1 \cdot 2^{12}$, $2^2 \cdot 2^{11}$, $2^3 \cdot 2^{10}$, $2^4 \cdot 2^9$, $2^5 \cdot 2^8$, $2^6 \cdot 2^7$

e. The exponents add to 13.

f. 5; $16 \cdot 32 = 512$

g. To multiply numbers with like bases, we add the exponents on the bases: $x^n \cdot x^m = x^{m+n}$. ▬

Think about it 2: Find all the pairs of black numbers that multiply to the 10th black number. Write the pairs as powers of 3. Which property describes the products?

EXAMPLE 7   Explaining the properties of quotients   To suggest how the properties of quotients are obtained, simplify with $a/a = 1$ and the power of a product property.

a. $\dfrac{a^7}{a^3}$ **b.** $\dfrac{c^8}{3c^2}$ **c.** $\left(\dfrac{a}{b}\right)^2$ **d.** $\left(\dfrac{2n}{m}\right)^3$

SOLUTION   **a.** $\dfrac{a^7}{a^3} = \dfrac{aaaaaaa}{aaa} = \dfrac{a}{a}\dfrac{a}{a}\dfrac{a}{a}\dfrac{aaaa}{1} = a^4$, or $a^{7-3} = a^4$

b. $\dfrac{c^8}{3c^2} = \dfrac{cccccccc}{3cc} = \dfrac{c}{c}\dfrac{c}{c}\dfrac{ccccc}{3} = \dfrac{c^6}{3}$, or $\dfrac{c^{8-2}}{3} = \dfrac{c^6}{3}$

c. $\left(\dfrac{a}{b}\right)^2 = \dfrac{a}{b}\dfrac{a}{b} = \dfrac{a^2}{b^2}$

d. $\left(\dfrac{2m}{n}\right)^3 = \dfrac{2m}{n}\dfrac{2m}{n}\dfrac{2m}{n} = \dfrac{8m^3}{n^3}$

Note that several of the exponents in Example 8 are variables.

EXAMPLE 8 Applying the properties of exponents Simplify these expressions.

a. $\dfrac{x^2 x^5}{x^3}$ 　　 b. $\dfrac{x^3 y^6}{x^6 y^2}$ 　　 c. $x^n (x^3)^2$

d. $\dfrac{x^n}{x^3}$ 　　 e. $\dfrac{x^n y^5}{x^2 y^k}$ 　　 f. $(x^m)^3$

SOLUTION a. $\dfrac{x^2 x^5}{x^3} = x^{2+5-3} = x^4$ 　　 Multiplication and division of like bases

b. $\dfrac{x^3 y^6}{x^6 y^2} = \dfrac{y^{6-2}}{x^{6-3}} = \dfrac{y^4}{x^3}$ 　　 Multiplication and division of like bases

c. $x^n (x^3)^2 = x^n x^6 = x^{n+6}$ 　　 Power of a power and multiplication of like bases

d. $\dfrac{x^n}{x^3} = x^{n-3}$ 　　 Division of like bases; assume $n \geq 3$

e. $\dfrac{x^n y^5}{x^2 y^k} = x^{n-2} y^{5-k}$ 　　 Division of like bases; assume $n > 2, k \leq 5$

f. $(x^m)^3 = x^{3m}$ 　　 Power of a power

Think about It 3: What is another way to state the answers to parts d and e?

EXAMPLE 9 Simplifying ratios containing exponents The radius of a basketball is about 3.5 times the radius of a baseball. Comparing their volumes with a ratio, we have

$$\frac{\text{Volume of basketball}}{\text{Volume of baseball}} = \frac{\frac{4}{3}\pi(3.5r)^3}{\frac{4}{3}\pi r^3}$$

The volume of the basketball is how many times that of the baseball?

SOLUTION $\dfrac{\text{Volume of basketball}}{\text{Volume of baseball}} = \dfrac{\frac{4}{3}\pi(3.5r)^3}{\frac{4}{3}\pi r^3}$ 　　 Power of a product property

Student Note: See Examples 10, 11, and 13 for more on zero exponents.

$= \dfrac{\frac{4}{3}}{\frac{4}{3}} \cdot \dfrac{\pi}{\pi} (3.5)^3 \cdot r^{3-3}$ 　　 Division of like bases property on the variable r

$= 1 \cdot 1 \cdot (3.5)^3 \cdot r^0$ 　　 $r^0 = 1$

≈ 43

The volume of the basketball is approximately 43 times that of the baseball.

Zero and Negative-Number Exponents

EXAMPLE 10 Finding nonpositive powers The first five numbers from the blue and black sets at the beginning of the chapter are listed below in reverse order. Use division to find the next four numbers and suggest bases and exponents that describe the numbers in each list.

Blue:　32, 16, 8, 4, 2, ___, ___, ___, ___

Black:　243, 81, 27, 9, 3, ___, ___, ___, ___

SOLUTION   For the blue numbers, we divide each term by 2 to get the next term, as shown in Table 1.

TABLE 1

Term	32	16	8	4	2	1	$\frac{1}{2}$	$\frac{1}{4}$	$\frac{1}{8}$
Power	2^5	2^4	2^3	2^2	2^1	2^0	2^{-1}	2^{-2}	2^{-3}

For the black numbers, we divide each term by 3 to get the next term, as shown in Table 2.

TABLE 2

Term	243	81	27	9	3	1	$\frac{1}{3}$	$\frac{1}{9}$	$\frac{1}{27}$
Power	3^5	3^4	3^3	3^2	3^1	3^0	3^{-1}	3^{-2}	3^{-3}

Think about it 4: Why would there not be a zero or negative-number exponent associated with the red numbers (25, 16, 9, 4, 1) or the orange numbers (125, 64, 27, 8, 1)?

Check the zero and negative power terms in Example 10 with your calculator ⌒∧ key. Also use a calculator in Example 11.

EXAMPLE 11   Exploring zero and negative one exponents   Use a calculator to investigate the meanings of the following expressions. Write decimal answers in fraction notation also.

a. 4^0 **b.** $(-3)^0$ **c.** $(2.5)^0$ **d.** 0^0 **e.** 2^{-1}

f. 3^{-1} **g.** 4^{-1} **h.** $\left(\frac{1}{2}\right)^{-1}$ **i.** $\left(\frac{2}{3}\right)^{-1}$

SOLUTION   **a.** $4^0 = 1$ **b.** $(-3)^0 = 1$ **c.** $(2.5)^0 = 1$

d. 0^0 is undefined.   **e.** $2^{-1} = 0.5 = \frac{1}{2}$ **f.** $3^{-1} = 0.3333\ldots = \frac{1}{3}$

g. $4^{-1} = 0.25 = \frac{1}{4}$ **h.** $\left(\frac{1}{2}\right)^{-1} = 2$ **i.** $\left(\frac{2}{3}\right)^{-1} = 1.5 = \frac{3}{2}$

In Example 11, the calculator gives an answer for all expressions except 0^0, which remains undefined. The example suggests meanings for exponents that are not positive integers.

DEFINITIONS

> The **zero power of a nonzero base** is defined as 1:
>
> $$b^0 = 1, \quad b \neq 0$$
>
> The expression 0^0 is not defined.
>
> The **negative one power of a nonzero base** is defined as the reciprocal of the base:
>
> $$b^{-1} = \frac{1}{b}, \quad b \neq 0$$
>
> The expression $1/0$ is not defined.
>
> The **negative one power of 1/b** is also the reciprocal:
>
> $$\left(\frac{1}{b}\right)^{-1} = \frac{b}{1}, \text{ or } b$$

In part i of Example 11, $\left(\frac{2}{3}\right)^{-1}$ was $\frac{3}{2}$. Returning to the power of a quotient property, we have $\left(\frac{2}{3}\right)^{-1} = 2^{-1}/3^{-1}$. The answer, $\frac{3}{2}$, suggests that $\left(\frac{1}{3}\right)^{-1} = 3$. In Example 12, we obtain this rule.

EXAMPLE 12   **Showing why $(1/b)^{-1} = b$**   Simplify $(1/b)^{-1}$.

$$\left(\frac{1}{b}\right)^{-1} = \frac{(b^0)^{-1}}{b^{-1}} \qquad \text{Replace 1 with } b^0\text{; power of a quotient property}$$

$$= \frac{b^0}{b^{-1}} \qquad 0(-1) = 0$$

$$= b^{0-(-1)} \qquad \text{Division of like bases property}$$

$$= b^1 = b$$

EXAMPLE 13   **Simplifying expressions**   Simplify these exponential expressions.

a. $(3 + 4)^0$ **b.** $(x + 2)^0$ **c.** $(x + 1)^{-1}$ **d.** $(0.25)^{-1}$

e. $\left(\frac{a}{b}\right)^{-1}$ **f.** $(x^{-1})^2$ **g.** $x^n x^{-1}, n \geq 1$ **h.** $\dfrac{x^n}{x^{-1}}, n \geq 0$

SOLUTION   **a.** $(3 + 4)^0 = 1$ Any nonzero base to the zero power is 1.

b. $(x + 2)^0 = 1, x \neq -2$ Exclude values making a zero base.

c. $(x + 1)^{-1} = \dfrac{1}{x + 1}$ The negative one power of a nonzero base is the reciprocal of the base.

d. $(0.25)^{-1} - \left(\frac{1}{4}\right)^{-1} = 4$ $0.25 = \frac{1}{4}$, whose reciprocal is 4.

e. $\left(\dfrac{a}{b}\right)^{-1} = \dfrac{1}{a} \cdot b = \dfrac{b}{a}$ The reciprocal of a is multiplied by the reciprocal of $1/b$ to give h/a.

f. $(x^{-1})^2 = x^{-1 \cdot 2} = x^{-2}$ Power of a power property

g. $x^n x^{-1} = x^{n+(-1)} = x^{n-1}$ Multiplication of like bases property

h. $\dfrac{x^n}{x^{-1}} = x^{n-(-1)} = x^{n+1}$ Division of like bases property

OTHER NEGATIVE EXPONENTS   We can think of the negative two exponent as the negative one exponent multiplied times two (and so on for other negative exponents).

$$a^{-2} = a^{-1 \cdot 2} \qquad \text{Let } a^{-1} = 1/a.$$

$$= \left(\frac{1}{a}\right)^2 \qquad \text{Apply the power of a quotient property.}$$

$$= \frac{1^2}{a^2} \qquad \text{Simplify the numerator.}$$

$$= \frac{1}{a^2}$$

Thus, a^{-2} gives the reciprocal of a^2.

In Example 14, look carefully at the base for each exponent.

EXAMPLE 14   **Simplfying with negative exponents**   Simplify.

a. $2a^{-2}$ **b.** $\left(-\frac{3}{4}\right)^{-2}$ **c.** -0.5^{-2}

SOLUTION   **a.** $2a^{-2} = 2 \cdot \left(\dfrac{1}{a^2}\right) = \dfrac{2}{a^2}$ The 2 is not in the base.

b. $\left(-\frac{3}{4}\right)^{-2} = \left(-\frac{4}{3}\right)^2 = \frac{16}{9}$ The negative sign is in the base.

c. $-0.5^{-2} = -\left(\frac{1}{2}\right)^{-2} = -\left(\frac{2}{1}\right)^2 = -4$ The negative sign is not in the base.

We use the term *simplify* to indicate that the definitions and properties of exponential expressions are to be used to remove zero or negative exponents, as well as to remove parentheses, to perform operations, and, where possible, to change exponential expressions into numbers or expressions without exponents.

EXAMPLE 15   **Simplifying with properties**   Simplify these expressions, replacing expressions having negative exponents with equivalent expressions having positive exponents.

a. $\left(\dfrac{-3x}{2}\right)^{-3}$ b. $\left(\dfrac{4x^3}{y^2}\right)^{-2}, y \neq 0$ c. $\dfrac{-1}{a^{-3}}, a \neq 0$

d. $\dfrac{b}{a^{-4}}, a \neq 0$ e. $\dfrac{3x^{-2}y}{6x^2y^{-3}}, x \neq 0, y \neq 0$ f. $(x^2y^{-3})^{-3}$

SOLUTION   a. $\left(\dfrac{-3x}{2}\right)^{-3} = \left(\dfrac{-2}{3x}\right)^3 = \left(\dfrac{-2}{3x}\right)\left(\dfrac{-2}{3x}\right)\left(\dfrac{-2}{3x}\right) = \dfrac{-8}{27x^3}, x \neq 0$

b. $\left(\dfrac{4x^3}{y^2}\right)^{-2} = \left(\dfrac{y^2}{4x^3}\right)^2 = \dfrac{y^{2\cdot2}}{4^2x^{3\cdot2}} = \dfrac{y^4}{16x^6}, x \neq 0$

c. $\dfrac{-1}{a^{-3}} = -a^3$

d. $\dfrac{b}{a^{-4}} = b \cdot a^4 = a^4b$

e. $\dfrac{3x^{-2}y}{6x^2y^{-3}} = \dfrac{3 \cdot y \cdot y^3}{2 \cdot 3 \cdot x^2 \cdot x^2} = \dfrac{y^4}{2x^4}, x \neq 0$

f. $(x^2y^{-3})^{-3} = x^{-6}y^9 = \dfrac{y^9}{x^6}, x \neq 0$

In Example 15, we used the reciprocal interpretation of a negative exponent. In the exercises, other approaches will be suggested.

ANSWER BOX

Warm-up: **1.** 64, 32, 16, 8, 4, 2   **2.** 81, 27, 9, 3   **3.** 256, 64, 16, 4
4. 625, 125, 25, 5   **Think about it 1:** Applying operations and writing in symbols permit us to see patterns; see also the solutions to Examples 4 and 5.
Think about it 2: All exponents add to 10: $3^1 \cdot 3^9, 3^2 \cdot 3^8, 3^3 \cdot 3^7, 3^4 \cdot 3^6$, $3^5 \cdot 3^5$; multiplication of like bases property: $x^n \cdot x^m = x^{m+n}$ **Think about it 3: d.** If $3 > n$, then we might write $1/x^{3-n}$.   **e.** If $2 > n$ or $k > 5$, then we might write $1/(x^{2-n}y^{k-5})$.   **Think about it 4:** The red numbers are the square numbers, x^2, and all have 2 as an exponent. The orange numbers are the cubic numbers, x^3, and all have 3 as an exponent.

8.1   **Exercises**

In Exercises 1 to 10, simplify the expressions.

1. a. x^4x^5 x^9 **b.** y^2y^a y^{2+a} **c.** a^na^3 a^{n+3} **4. a.** $(2d)^3$ $8d^3$ **b.** $(4a)^3$ $64a^3$ **c.** $(xy)^4$ x^4y^4

2. a. x^3x^7 x^{10} **b.** b^3b^k b^{3+k} **c.** x^nx^2 x^{n+2} **5. a.** $(x^2)^4$ x^8 **b.** $(2x^3)^5$ $32x^{15}$ **c.** $(3a^4b^2)^3$ $27a^{12}b^6$

3. a. $(3x)^4$ $81x^4$ **b.** $(ab)^3$ a^3b^3 **c.** $(4n)^2$ $16n^2$ **6. a.** $(a^3)^3$ a^9 **b.** $(3y^4)^3$ $27y^{12}$ **c.** $(3x^3y^4)^2$ $9x^6y^8$

7. a. $\dfrac{x^5}{x^2}$ x^3 **b.** $\dfrac{y^3}{y^5}$ $\dfrac{1}{y^2}$ **c.** $\dfrac{a^5 a^3}{a^6}$ a^2

8. a. $\dfrac{b^8}{b^2}$ b^6 **b.** $\dfrac{x^2}{x^6}$ $\dfrac{1}{x^4}$ **c.** $\dfrac{y^4}{y^3 y^2}$ $\dfrac{1}{y}$

9. a. $(x^n)^4$ $x^{4n}, n \geq 0$ **b.** $\dfrac{xy^3}{x^3 y}$ $\dfrac{y^2}{x^2}$ **c.** $\dfrac{x^k y^1}{xy^2}$ $\dfrac{x^{k-1}}{y}, k \geq 1$

10. a. $(y^3)^t$ $y^{3t}, t \geq 0$ **b.** $\dfrac{ab^3}{a^2 b}$ $\dfrac{b^2}{a}$ **c.** $\dfrac{a^3 b^k}{a^n b^2}$ $a^{3-n}b^{k-2}, n \leq 3, k \leq 2$

In Exercises 11 to 16, simplify the expressions. Leave no negative or zero exponents.

11. a. $2x^{-1}$ $\dfrac{2}{x}$ **b.** $\left(\dfrac{y}{x}\right)^{-1}$ $\dfrac{x}{y}$

c. $\left(\dfrac{s}{t}\right)^0$ 1 **d.** $(x^2 y^{-1})^{-1}$ $\dfrac{y}{x^2}$

e. $(x^{-1}y^3)(x^3 y^{-2})$ $x^2 y$ **f.** $(x^3 y^{-2})^{-1}$ $\dfrac{y^2}{x^3}$

12. a. $3y^{-1}$ $\dfrac{3}{y}$ **b.** $\left(\dfrac{a}{b}\right)^{-1}$ $\dfrac{b}{a}$

c. $\left(\dfrac{r}{s}\right)^0$ 1 **d.** $(x^{-1}y^3)^{-1}$ $\dfrac{x}{y^3}$

e. $(x^{-2}y^3)(xy^{-2})$ $\dfrac{y}{x}$ **f.** $(x^{-3}y^2)^{-1}$ $\dfrac{x^3}{y^2}, x \geq -0$

13. a. $(2y)^{-2}$ $\dfrac{1}{4y^2}$ **b.** $2x^{-4}$ $\dfrac{2}{x^4}$

c. $\left(\dfrac{2x}{y}\right)^{-2}$ $\dfrac{y^2}{4x^2}$ **d.** $\left(\dfrac{a}{2c}\right)^{-3}$ $\dfrac{8c^3}{a^3}$

e. $\dfrac{2x^3 y}{6x^{-1}y^2}$ $\dfrac{x^4}{3y^4}$ **f.** $b^1 b^n$ $b^{n+1}, n \geq -1$

14. a. $2a^{-2}$ $\dfrac{2}{a^2}$ **b.** $(2b)^{-3}$ $\dfrac{1}{8b^3}$

c. $\left(\dfrac{2c}{b}\right)^{-2}$ $\dfrac{b^2}{4c^2}$ **d.** $\left(\dfrac{c}{2a}\right)^{-3}$ $\dfrac{8a^3}{c^3}$

e. $\dfrac{3xy^{-3}}{12x^{-2}y^{-1}}$ $\dfrac{x^3}{4y^2}$ **f.** $a^1 a^x$ $a^{x+1}, x \geq -1$

15. a. $\left(\dfrac{3a^2}{c}\right)^{-3}$ $\dfrac{c^3}{27a^6}$ **b.** $\left(\dfrac{-a}{c^2}\right)^{-2}$ $\dfrac{c^4}{a^2}$

c. $\dfrac{-1}{c^{-3}}$ $-c^3$ **d.** $\dfrac{2}{a^{-2}}$ $2a^2$

e. $\dfrac{12a^3 b^{-2}}{4a^{-1}b^{-5}}$ $3a^4 b^3$ **f.** $\dfrac{3}{3^x}$ $3^{1-x}, x \leq 1$

16. a. $\left(\dfrac{2x^2}{y}\right)^{-3}$ $\dfrac{y^3}{8x^6}$ **b.** $\left(\dfrac{-2b}{c^2 d}\right)^{-2}$ $\dfrac{c^4 d^2}{4b^2}$

c. $\dfrac{-1}{a^{-2}}$ $-a^2$ **d.** $\dfrac{3}{x^{-4}}$ $3x^4$

e. $\dfrac{4a^{-2}b^{-3}}{20ab^{-2}}$ $\dfrac{1}{5a^3 b}$ **f.** $\dfrac{5}{5^x}$ $5^{1-x}, x \leq 1$

In Exercises 17 to 20, change to fractions as needed and evaluate without a calculator.

17. a. 5^{-1} $\frac{1}{5}$ **b.** $(-2)^{-1}$ $-\frac{1}{2}$ **c.** 0.5^{-2} 4

d. -0.25^{-2} -16 **e.** 0.2^{-1} 5 **f.** $(-3)^{-2}$ $\frac{1}{9}$

18. a. 5^{-2} $\frac{1}{25}$ **b.** $(-4)^{-1}$ $-\frac{1}{4}$ **c.** 0.05^{-1} 20

d. 0.6^{-2} $\frac{25}{9}$ **e.** -0.75^{-2} $-\frac{16}{9}$ **f.** $(-5)^{-2}$ $\frac{1}{25}$

19. a. 1.5^{-1} $\frac{2}{3}$ **b.** -2^{-1} $-\frac{1}{2}$ **c.** $(-2)^{-1}$ $-\frac{1}{2}$

d. $(-4)^{-2}$ $\frac{1}{16}$ **e.** -4^{-2} $-\frac{1}{16}$ **f.** $(-0.5)^{-2}$ 4

20. a. 2.5^{-1} $\frac{2}{5}$ **b.** -4^{-1} $-\frac{1}{4}$ **c.** $(-4)^{-1}$ $-\frac{1}{4}$

d. $(-2)^{-2}$ $\frac{1}{4}$ **e.** -2^{-2} $-\frac{1}{4}$ **f.** $(-3)^{-1}$ $-\frac{1}{3}$

In Examples 21 to 26, name the operation as adding like terms (terms), multiplying like bases (bases), or neither, and simplify, if possible.

21. a. $y^3 + 2y^3$
terms; $3y^3$
b. $x^{-1} + x^{-1}$
terms; $2x^{-1} = 2/x$
c. $y + y^2$
neither

22. a. $x^2 x^3$
bases; x^5
b. $y^2 \cdot y^{-2}$
bases; $y^0 = 1$
c. $y^2 + y^{-2}$
neither; $y^2 + 1/y^2$

23. a. $x^2 + x^{-2}$
neither; $x^2 + 1/x^2$
b. $y^{-3} + y^{-3}$
terms; $2y^{-3}$
c. $y^{-4}y^2$
bases; $y^{-2} = 1/y^2$

24. a. $x^? + x$
neither
b. $y^{-1} + y$
neither; $1/y + y$
c. $y^2 + y^2$
terms; $2y^2$

25. a. $x^2 y^2$
neither
b. $x^3 x^{-1}$
bases; x^2
c. $x^2 + 2x^2$
terms; $3x^2$

26. a. $3y^{-1} - 2y^{-1}$
terms; $y^{-1} = 1/y$
b. $x^3 y^3$
neither
c. $x^3 x^{-1}$
bases; x^2

27. Simplify the expressions from Example 15 in the suggested ways.

a. Simplify $(-3x/2)^{-3}$ by first applying the power of a quotient property. $(-3x)^{-3}/(2)^{-3} = 8/(-27x^3) = -8/27x^3$

b. Simplify $(4x^3/y^2)^{-2}$ by first applying the power of a quotient property. $(4x^3)^{-2}/(y^2)^{-2} = y^4/16x^6$

c. Simplify $3x^{-2}y/6x^2 y^{-3}$ by first applying the division of like bases property. $\frac{1}{2}x^{-2-2}y^{1-(-3)} = \frac{1}{2}x^{-4}y^4 = y^4/2x^4$

28. The diameter of a soccer ball is about 5.3 times the radius of a golf ball. Compare the ratio of their volumes, $V \approx 4\pi r^3/3$, where r = radius of a sphere. 149 times

29. If we double the radius of a sphere, how does the new volume compare with the original volume? 8 times

30. The equations $2^3 + 3^3 = 35$ and $1^3 + 4^3 = 65$ illustrate sums of cubes. The number 1729 is the smallest number expressible as the sum of two cubes in two different ways. What are the pairs of cubes adding to 1729? $9^3 + 10^3, 1^3 + 12^3$

31. Graph $y = x^0$. Describe the graph. Trace the graph. What can you conclude? Is there any x-coordinate for which the function is not defined?
See Answer Section; on $y = 1$; no output for 0^0; undefined at $x = 0$

Error Analysis

32. Is -3^{-2} positive or negative? Explain. Negative; the sign is not in the base.

33. Is $-x^2 = (-x)^2$? Explain. No; the sign is not in the left base.

34. Explain in your own words why $(-3)^4 \neq -3^4$.
$(-3)^4$ has (-3) as factor 4 times; -3^4 has 3 as factor 4 times; $81 \neq -81$.

35. Explain why $3x^{-2}$ is not $\dfrac{1}{(3x)^2}$. Exponent does not affect 3. Answer is $3/x^2$.

36. Explain why $\dfrac{2}{x^{-1}}$ is not $\dfrac{x}{2}$. Exponent does not affect 2. Answer is $2x$.

37. Explain why $5\left(\frac{5}{6}\right)^3$ does or does not equal $6\left(\frac{5}{6}\right)^4$. Show your steps carefully.
$5\left(\frac{5}{6}\right)^3 = 6\left(\frac{5}{6}\right)^4$, since $6\left(\frac{5}{6}\right)^4 = 6 \cdot \frac{5}{6} \cdot \left(\frac{5}{6}\right)^3 = 5\left(\frac{5}{6}\right)^3$

Projects

38. Rational Expressions These expressions are found in a calculus course. Rewrite the expressions by eliminating negative exponents, adding or subtracting fractions, and simplifying the resulting rational expression.

a. $\dfrac{(x+h)^{-1} - x^{-1}}{h}$ $\dfrac{-1}{x(x+h)}$ **b.** $\dfrac{(x+h)^{-2} - x^{-2}}{h}$ $\dfrac{-(2x+h)}{x^2(x+h)^2}$

c. $x^2(-1)(x+1)^{-2} + \dfrac{2x}{x+1}$ $\dfrac{x^2+2x}{(x+1)^2}$

d. $(x-1)^{-1}(2x+2) + (x^2+2x)(-1)(x-1)^{-2}$ $\dfrac{x^2-2x-2}{(x-1)^2}$

39. Negative Exponents Rewrite the formulas, equations, or units of measurement in parts a to e so that they contain no negative exponents.

a. Theater emptying: $t^{-1} = t_1^{-1} + t_2^{-1}$ $\dfrac{1}{t} = \dfrac{1}{t_1} + \dfrac{1}{t_2}$

b. Gravitational constant: $g \approx 9.81 \text{ m} \cdot \text{sec}^{-2}$ $g \approx 9.81 \text{ m/sec}^2$

c. Compound interest, continuous compounding:
$P = Ae^{-rt}$ $P = \dfrac{A}{e^{rt}}$ or $A = Pe^{rt}$

d. Density: $d = 1 \text{ g} \cdot \text{cm}^{-3}$ $d = \dfrac{1 \text{ g}}{\text{cm}^3}$

e. Tire pressure: $p = 31 \text{ lb} \cdot \text{in}^{-2}$ $p = \dfrac{31 \text{ lb}}{\text{in}^2}$

Rewrite those in parts f to j so that they contain negative exponents instead of division.

f. Jet aircraft take-off sound at 60 meters: $I = 1 \text{ W/m}^2$
$I = 1 \text{ W} \cdot \text{m}^{-2}$

g. Pressure $p = 20 \text{ kg/cm}^2$ $p = 20 \text{ kg} \cdot \text{cm}^{-2}$

h. Resistance in a parallel circuit: $\dfrac{1}{R} = \dfrac{1}{R_1} + \dfrac{1}{R_2}$
$R^{-1} = R_1^{-1} + R_2^{-1}$

i. Radioactive decay: $A = \dfrac{A_0}{e^{kt}}$ $A = A_0 e^{-kt}$

j. Speed of spine-tailed swift: $s = 106 \text{ mph}$
$s = 106 \text{ m} \cdot \text{hr}^{-1}$

40. Inverse Variation

a. The resistance of a wire of fixed length varies inversely with the square of the diameter of the wire. Use the data in the table to determine the constant of variation between diameter and resistance. $k \approx 0.0104$

American Wire Gauge Size	Diameter (inches)	Resistance at 20°C per 1000 Feet (ohms)
24	0.0201	25.67
22	0.0254	16.14
20	0.0320	10.15
18	0.0403	6.39
16	0.0508	4.02

b. Use power regression to fit a curve $y = ax^b$ to the diameter and resistance data. $R \approx 0.01034d^{-2.001}$

c. As the American wire gauge size decreases, the diameter __increases__.

41. Polynomial Evaluation A uniformly loaded cable is parabolic in shape. The length of the cable, L, is approximated by

$$L = a \left| 1 + \frac{8}{3}\left(\frac{d}{a}\right)^2 - \frac{32}{5}\left(\frac{d}{a}\right)^4 + \frac{256}{7}\left(\frac{d}{a}\right)^6 - \cdots \right|$$

where d is the sag in feet or meters and a is the span across the parabola in feet or meters. Temperature variations are ignored. Round answers to the nearest ten thousandth.

a. Suppose the span is 500 feet and the sag is 30 feet. Find the length of the cable when two, three, and four terms are used inside the absolute value.
504.8 ft; ≈ 504.7585 ft; ≈ 504.7594 ft

b. Suppose the span is 200 meters and the sag is 10 meters. Find the length of the cable when two, three, and four terms are used inside the absolute value. ≈ 201.3333 m; ≈ 201.3253 m; ≈ 201.3254 m

c. If a fifth term were used, would the result be larger or smaller than those found in parts a and b above? Why? smaller, because a negative sign follows the fourth term

42. Ramanujan and Hardy Research the work of Srinivasa Ramanujan and, if possible, describe the interaction between this mathematician from India and his colleague, Godfrey Hardy (1877–1947), related to the problem in Exercise 30.

Power regressions also appear in Project 69 in Section 8.4, Project 66 in Section 8.5, and Project 66 in Section 8.7.

8.2 Scientific Notation

Objectives

▪ Change numbers between decimal notation and scientific notation.

▪ Use a calculator to do operations with scientific notation and write answers with appropriate significant digits.

WARM-UP

Find these powers of 10 without a calculator. Write the expressions in Exercises 4 to 6 as both fractions and decimals.

1. 10^0 1   **2.** 10^1 10   **3.** 10^2 100

4. 10^{-1} $\frac{1}{10} = 0.1$ **5.** 10^{-2} $\frac{1}{100} = 0.01$ **6.** 10^{-3} $\frac{1}{1000} = 0.001$

7. Between what two integers are the negative powers of 10?   0 and 1

THIS SECTION PRESENTS scientific notation and significant digits. The exercises will include topics from prior chapters. Answers to these exercises will be in scientific notation.

Writing Scientific Notation

Scientific notation is *a short way of writing large and small numbers in which each number is written as a product of a decimal number between 1 and 10 and a power of 10.* Numbers written in scientific notation have two parts, as shown below. The first part is a decimal between 1 and 10; the second part is a power of 10.

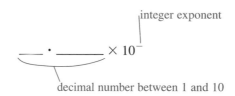

The decimal number is multiplied by the power of 10. The multiplication sign may be written as either a dot or a $\times$.

Because multiplying by 10 only changes the decimal position, writing numbers in scientific notation merely requires moving the decimal point and keeping track of how many places it is moved.

EXAMPLE 1 **Changing decimal numbers to scientific notation** Change these numbers to scientific notation. *If a number does not contain a decimal point, you can assume the decimal point is located at the right end of the number.*

a. 2000   **b.** 25,000   **c.** 0.025   **d.** 0.000 025

SOLUTION **a.** $2000 = 2 \times 10^3$
The decimal point is moved 3 places to create a number between 1 and 10. Because 2000 is larger than 10, the exponent on 10 is positive 3.

b. $25,000 = 2.5 \times 10^4$
The decimal point is moved 4 places and $25,000 > 10$, so the exponent on 10 is positive 4.

c. $0.025 = 2.5 \times 10^{-2}$
The decimal point is moved 2 places. Because 0.025 is between 0 and 1, the exponent on 10 is negative 2.

d. $0.000\ 025 = 2.5 \times 10^{-5}$

The decimal point is moved 5 places. Because $0.000\ 025$ is between 0 and 1, the exponent on 10 is negative 5.

There may be some question as to how many zeros to leave on the decimal number. Unless told otherwise, drop all zeros after the last nonzero digit to the right of the decimal point. See also Example 6 on page 448.

EXAMPLE 2 Changing to scientific notation   Change each number to scientific notation.
a. The maximum distance from the sun to Earth is 94,600,000 miles.
b. The minimum distance from the sun to Pluto is 2,756,400,000 miles.
c. The mass of an electron is
   $0.000\ 000\ 000\ 000\ 000\ 000\ 000\ 000\ 000\ 000\ 910\ 9$ kilogram.
d. The mass of a bacterium is $0.000\ 000\ 000\ 0001$ kilogram.

SOLUTION In parts a and b, the distances are numbers larger than 1, so the scientific notation will have a positive exponent on 10.

a. 9.46×10^7 miles   **b.** 2.7564×10^9 miles

In parts c and d, the masses are numbers between 0 and 1, so the scientific notation will have a negative exponent on 10.

c. 9.109×10^{-31} kilogram   **d.** 1×10^{-13} kilogram

Writing Decimal Notation

> To change scientific notation to regular decimal notation and the reverse, remember that numbers larger than 1 have a positive exponent on the 10 and small numbers between 0 and 1 have a negative exponent on the 10.

EXAMPLE 3 Changing scientific notation to decimal notation   Change these numbers to decimal notation.
a. 3.6×10^2 **b.** 3.6×10^{-3} **c.** 3.6×10^{-1}

SOLUTION   **a.** The exponent on 10 is positive 2. The decimal point is moved 2 places, making a number greater than 3.6:

$$3.6 \times 10^2 = 360$$

b. The exponent on 10 is negative 3. The decimal point is moved 3 places, making a number smaller than 3.6:

$$3.6 \times 10^{-3} = 0.0036$$

c. The exponent on 10 is negative 1. The decimal point is moved 1 place, making a number smaller than 3.6:

$$3.6 \times 10^{-1} = 0.36$$

EXAMPLE 4 Changing to decimal notation   Change each number from scientific notation to decimal notation.
a. In 1918, after World War I, the national debt of the United States was 2.6×10^{10} dollars.
b. In 1975, two years after the Vietnam War, the national debt was 5.33×10^{11} dollars.

c. Avogadro's number, 6.022×10^{23} molecules per mole, measures the number of molecules in chemistry.

d. The mass of a neutron is 1.675×10^{-27} kilogram.

SOLUTION **a.** 26,000,000,000 dollars **b.** 533,000,000,000 dollars

c. 602,200,000,000,000,000,000,000 molecules per mole

d. 0.000 000 000 000 000 000 000 000 001 675 kilogram

GRAPHING CALCULATOR TECHNIQUE:
SCIENTIFIC AND DECIMAL NOTATION

Changing to scientific notation: Look for options under MODE to change to scientific notation. In **Sci** mode, any number entered into the calculator will be changed automatically to scientific notation. The calculator displays for 25,000 and 0.000 025 are in Figure 2a. The calculator shows E to represent "times 10 raised to the exponent."

Changing to decimal notation: When in **Normal** mode, a graphing calculator will change any number in scientific notation to decimal notation if the decimal number will fit on the display. Otherwise, the number will remain in scientific notation. Use 2nd [EE]—not[e^x], [10^x], or ∧ —to enter a number in scientific notation. The EE signifies "enter exponent." Graphing calculators permit entering the negative sign before the exponent. A calculator display in **Normal** mode for 9.46×10^7 and 9.109×10^{-31} is shown in Figure 2b.

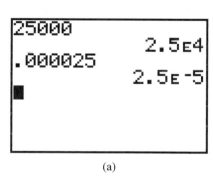   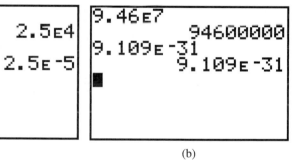

(a)   (b)

FIGURE 2

EXAMPLE 5 Estimating with scientific notation   Do these problems mentally, and then check the answers with scientific notation on a calculator.

a. $(1.2 \times 10^5)(4.0 \times 10^{-3})$ **b.** $(3.0 \times 10^2)(1.5 \times 10^{-3})$

c. $\dfrac{5.6 \times 10^{-2}}{1.4 \times 10^4}$ **d.** $\dfrac{3.6 \times 10^4}{6.0 \times 10^{-2}}$

SOLUTION **a.** $(1.2 \times 10^5)(4.0 \times 10^{-3}) = (1.2)(4) \cdot 10^{5+(-3)} = 4.8 \times 10^2$
We multiply the decimal parts and then add the exponents on the 10.

b. $(3.0 \times 10^2)(1.5 \times 10^{-3}) = (3)(1.5) \cdot 10^{2+(-3)} = 4.5 \times 10^{-1}$

c. $\dfrac{5.6 \times 10^{-2}}{1.4 \times 10^4} = \dfrac{5.6}{1.4} \cdot 10^{(-2-4)} = 4.0 \times 10^{-6}$

d. $\dfrac{3.6 \times 10^4}{6.0 \times 10^{-2}} = \dfrac{3.6}{6.0} \cdot 10^{4-(-2)} = 0.6 \times 10^6$ Change 0.6 to scientific notation.

$$= 6.0 \times 10^{-1} \times 10^6$$

$$= 6.0 \times 10^5$$

Calculators are designed to do operations with scientific notation. It is important to use [EE], because using other options may result in errors when division problems are worked.

Think about it: What is the calculator answer if you enter the expression $\dfrac{5.6 \times 10^{-2}}{1.4 \times 10^4}$ with these keystrokes: 5.6 × 10 ⌐∧⌐ ⌐(−)⌐ 2 ÷ 1.4 × 10 ⌐∧⌐ 4? Why is it incorrect?

EXPONENTIAL NOTATION   As you might expect, scientific notation is common in science. *In many chemistry and physics books, scientific notation is called exponential notation.* In mathematics, however, we use the word *exponential* more generally to describe expressions such as ab^x and the term *scientific notation* to refer specifically to those exponential expressions where a is a number between 1 and 10 and b is 10.

Significant Digits

When writing numbers in scientific notation, we look for significant digits. **Digits** are *the numbers 0 to 9*. **Significant digits** are *all nonzero digits and certain zeros:*

Zeros between nonzero digits (as in 707)

Zeros following a nonzero digit after a decimal point (as in 0.020)

Zeros that are placeholders and marked with an overbar (as in 50$\overline{0}$0)

Placeholders are *zeros that are used to properly position the decimal point*, as in 5000 or 0.0025. Placeholders are not significant, except as specified above.

EXAMPLE 6   **Finding significant digits**   Identify the significant digits in these numbers, and write the numbers in scientific notation.

a. 2500   **b.** 0.0025   **c.** 2090   **d.** 0.01090   **e.** 25$\overline{0}$0

SOLUTION   **a.** The zeros after the 5 are placeholders and are omitted; there are two significant digits in 2500.

$$2500 = 2.5 \times 10^3$$

b. The three zeros on the left are placeholders and are omitted; there are two significant digits in 0.0025.

$$0.0025 = 2.5 \times 10^{-3}$$

c. The zero on the right is a placeholder and is omitted; there are three significant digits in 2090.

$$2090 = 2.09 \times 10^3$$

d. The two zeros on the left are placeholders; there are four significant digits in 0.01090. The zero on the right is written only if it is significant, so no overbar is needed.

$$0.01090 = 1.090 \times 10^{-2}$$

e. The overbar above the zero indicates that it is significant. There are three significant digits.

$$25\overline{0}0 = 2.50 \times 10^3$$

All significant digits should be shown when numbers are written in scientific notation.

ANSWER BOX

Warm-up: **1.** 1 **2.** 10 **3.** 100 **4.** $\frac{1}{10} = 0.1$ **5.** $\frac{1}{100} = 0.01$
6. $\frac{1}{1000} = 0.001$ **7.** 0 and 1 **Think about it:** The calculator interprets
1.4×10 ⌐∧⌐ 4 as the product of two numbers, whereas it interprets
1.4 [EE] 4 as one number. Because the order of operations on a calculator is
left to right for multiplication and division, only the 1.4 will remain in the
denominator; the 10^4 will be interpreted as being in the numerator.

8.2 Exercises

Change the numbers in Exercises 1 and 2 to powers of 10.

1. a. 0.001 10^{-3}
  b. $\frac{1}{100,000}$ 10^{-5}

  c. 10,000 10^4
  d. 1 hundredth 10^{-2}

  e. 1 million 10^6
  f. 1 millionth 10^{-6}

2. a. 100,000 10^5
  b. $\frac{1}{10,000}$ 10^{-4}

  c. 0.00001 10^{-5}
  d. ten thousand 10^4

  e. 1 thousandth 10^{-3}
  f. 1 hundred thousandth 10^{-5}

In Exercises 3 and 4, write the numbers in scientific notation.

3. a. 3000 3×10^3
  b. 350 3.5×10^2

  c. 350,$\overline{0}$00 3.500×10^5
  d. 0.003 50 3.50×10^{-3}

4. a. 30$\overline{0}$ 3.00×10^2
  b. 3.50 3.50×10^0

  c. 0.000 035 3.5×10^{-5}
  d. 0.000 000 3 3×10^{-7}

5. How many significant digits are there in each part of
Exercise 3? 1; 2; 4; 3

6. How many significant digits are there in each part of
Exercise 4? 3; 3; 2; 1

Change the numbers in Exercises 7 and 8 to decimal notation.

7. a. The speed of light is 2.9979×10^{10} cm/sec.
  29,979,000,000 cm/sec
  b. The mass of a proton is 1.6726×10^{-24} g.
  0.000 000 000 000 000 000 001 672 6 g
  c. The charge on an electron is -1.6022×10^{-19} C.
  −0.000 000 000 000 000 000 160 22 C
  d. The energy of an electron volt is 1.6022×10^{-19} J.
  0.000 000 000 000 000 000 160 22 J

8. The hydrogen ion (H^+) concentration is used to find pH
in chemistry.

  a. The H^+ ion concentration of milk is 2.511×10^{-7} M.
  0.000 000 251 1 M
  b. The H^+ ion concentration of HCl (hydrochloric acid)
  is 2.0×10^{-1} M. 0.20 M

  c. The H^+ ion concentration of sodium bicarbonate
  (baking soda) is 3.981×10^{-9} M.
  0.000 000 003 981 M
  d. The H^+ ion concentration of ammonia is 2.512×10^{-12} M. 0.000 000 000 002 512 M

Estimate the scientific notation expressions in Exercises 9 to 16, and then check with
a calculator.

9. $(4.0 \times 10^{-2})(1.5 \times 10^6)$ 6.0×10^4

10. $(5.0 \times 10^2)(2.5 \times 10^{-4})$ 1.25×10^{-1}

11. $(6.0 \times 10^{-3})(2.5 \times 10^{-5})$ 1.5×10^{-7}

12. $(6.0 \times 10^{-4})(2.0 \times 10^{-1})$ 1.2×10^{-4}

13. $\frac{7.5 \times 10^{-2}}{2.5 \times 10^4}$ 3.0×10^{-6} **14.** $\frac{16 \times 10^4}{0.8 \times 10^{-2}}$ 2.0×10^7

15. $\frac{2.4 \times 10^6}{0.3 \times 10^{-2}}$ 8.0×10^8 **16.** $\frac{5.0 \times 10^{-2}}{2.5 \times 10^3}$ 2.0×10^{-5}

Write the expressions in Exercises 17 to 20 in standard form for scientific notation.
Check with a calculator.

17. a. $40\overline{0} \times 10^{-3}$
  4.00×10^{-1}
  b. $40\overline{0} \times 10^3$ 4.00×10^5

18. a. 0.6×10^2
  6×10^1
  b. 0.6×10^{-2} 6×10^{-3}

19. a. 0.050×10^{-4}
  5.0×10^{-6}
  b. $0.05\overline{0} \times 10^5$ 5.0×10^3

20. a. 5280×10^{-1}
  5.28×10^2
  b. 5280×10^4 5.28×10^7

21. Find the value of each expression with a calculator.
Remember that scientific notation uses a number
between 1 and 10 multiplied by a power of ten. Which
expressions does the calculator read as one number
value and which as the product of two numbers?

  a. 10 ⌐∧⌐ 10 1E10; one **b.** 1×10 ⌐∧⌐ 10 1E10; two

  c. 10 ⌐2nd⌐ [EE] 10 1E11; one **d.** 10×10 ⌐∧⌐ 10 1E11; two

  e. 10 ⌐∧⌐ 52 1E52; one **f.** 10 ⌐2nd⌐ [EE] 52 1E53; one

  g. Explain in your own words what ⌐2nd⌐ [EE] represents. Multiply the preceding number by 10 raised to the
  exponent that follows.

22. Many scientific calculators show the decimal part of scientific notation, a space, and the exponent (in large or
small print). What number is described by each of these
non-graphing calculator displays?

  a. 1.25 −3
  1.25×10^{-3}; 0.001 25
  b. 2.06 2
  2.06×10^2; 206
  c. 2.13 10
  2.13×10^{10}; 21,300,000,000
  d. 4.23 −15
  4.23×10^{-15};
  0.000 000 000 000 004 23

23. Refer to the national debt information in Example 4.

   a. By how many dollars did the national debt increase between 1918 and 1975? 5.07×10^{11}

   b. How many times, n, did the national debt double between 1918 and 1975? (*Hint:* $2^n(2.6 \times 10^{10}) = 5.3 \times 10^{11}$.) 4.4 by guess and check

   c. Divide n into the number of years between 1918 and 1975 to find years required for the debt to double. List the seven doubling years after 1918.
   13; 1931, 1944, 1957, 1970, 1983, 1996, 2009

   d. Predict the debt in each year in part c. Compare the growth with the facts. In 1980, the debt was $907.7 billion; in 1985, $1823.1 billion; and in 1990, $3233.3 billion.   faster than predicted

24. In science fiction space travel, warp speed is related to the speed of light. If c is the speed of light, 186,000 miles per second, warp speed is the cube of the warp number times the speed of light. Write your answers to the following in scientific notation.

   a. Use unit analysis to find the speed of light in miles per hour. 6.70×10^8 mph

   b. Calculate warp 2 in miles per hour. 5.36×10^9 mph

   c. Calculate warp 3 in miles per hour. 1.81×10^{10} mph

   d. Calculate warp 14.1 in miles per hour. (This was the highest speed ever reached by the starship Enterprise in the original Star Trek television series.)
   1.88×10^{12} mph

25. The Hoover Dam reservoir has a capacity of 2.8253×10^7 acre·feet of water. An acre·foot has an area of one acre (43,560 square feet) and a depth of 1 foot.

   a. What is the capacity of the reservoir in cubic feet?
   1.231×10^{12} ft^3

   b. To what depth, in feet, would this volume of water cover the state of Rhode Island? Rhode Island is 1045 square miles in land area. ≈ 42 ft

26. a. The formula $F = GM_1M_2/d^2$ describes the gravitational force F between two masses M_1 and M_2. The masses are a distance d apart. Describe the gravitational force as a joint and an inverse variation.
   jointly proportional to masses and inversely proportional to d^2

   b. A 60-kilogram mass on Earth's surface registers a force of 588.6 N when acted on by Earth's gravity. The radius of Earth is 6.378×10^6 meters. The gravitational constant G is 6.672×10^{-11} N·m^2/kg^2. Use the gravitational force equation to find the mass of Earth. (N is a newton, the unit of force in the metric system.) $\approx 5.98 \times 10^{24}$ kg

   c. Suppose the 60-kilogram mass is moved to a distance of 3.92×10^8 meters from Earth's center. This is approximately the distance to the center of the moon. What force, in newtons, acts on the mass at this position? (You can divide newtons by 4.4 to approximate pounds of force.) ≈ 0.155 N ≈ 0.035 lb

27. *Circular velocity*, V_{circ}, is the velocity a space vehicle needs in order to keep circling around a planet and not be pulled back to the planet by gravity. *Escape velocity*, V_{esc}, is the velocity a space vehicle needs in order to leave circular orbit around a planet and "escape" the gravitational attraction of that planet. Circular velocity is given by

$$V_{circ} = \sqrt{\frac{GM}{R}} \quad \text{for } G = 6.67 \times 10^{-11} \text{ N·m}^2\text{·kg}^{-2}$$

where G is Newton's gravitational constant, M is the mass of the planet, and R is the distance of the orbiter from the center of the planet. Escape velocity is given by

$$V_{esc} = \sqrt{2} \cdot V_{circ} \quad \text{For Exercise 27, see also Figure 1 on page 580.}$$

   Make sure the units are correct in your work. N represents the newton, a unit of force: 1 N = 1 kg · m/sec^2. Be aware of significant digits in your final answer.
   In parts a and b, find the circular velocity of a space vehicle orbiting the planet at the given distance.

   a. Mercury: $R = 3.24 \times 10^6$ m, $M = 3.30 \times 10^{23}$ kg
   $\approx 2.61 \times 10^3$ m/sec
   b. Jupiter: $R = 8.14 \times 10^7$ m, $M = 1.90 \times 10^{27}$ kg
   $\approx 3.95 \times 10^4$ m/sec
   In parts c and d, find the escape velocity of a space vehicle orbiting the planet at the given distance.

   c. Earth: $R = 7.18 \times 10^6$ m, $M = 5.98 \times 10^{24}$ kg
   $\approx 1.05 \times 10^4$ m/sec
   d. Mars: $R = 4.19 \times 10^6$ m, $M = 6.44 \times 10^{23}$ kg
   $\approx 4.53 \times 10^3$ m/sec
   e. What happens when velocity drops below circular velocity?   Vehicle is pulled back toward planet.

   f. What happens when velocity is between circular velocity and escape velocity?   Orbit becomes oval instead of circular.

28. *Terminal velocity* is the maximum falling speed attained with air resistance. The terminal velocity V of a sky diver depends on the position taken during the dive. Calculate V for each of the following three dive positions.*
   Use

$$V = \sqrt{\frac{32.2 \, \dfrac{\text{ft}}{\text{sec}^2}}{A}}$$

   a. $A = 5.18 \times 10^{-4}$ ft^{-1} in fetal position ≈ 249 ft/sec

   b. $A = 7.75 \times 10^{-4}$ ft^{-1} in nose-dive position (arms back along body) ≈ 204 ft/sec

   c. $A = 10.4 \times 10^{-4}$ ft^{-1} in horizontal position (arms and legs spread outward from body) ≈ 176 ft/sec

   d. If we increase the surface area of the diver, how does terminal velocity change?   decreases

*The data are for 135-pound female sky diver Jennifer Phillips of Orange, Massachusetts. From William Ralph Bennett, Jr., *Scientific and Engineering Problem-Solving with the Computer*, Prentice-Hall, Englewood Cliffs, NJ, 1976.

Projects

29. Growth Rates

a. Suppose a baby grows from 0 to 21 inches in the 9 months before birth. How many miles per hour is this growth rate? (1 mile = 5280 feet.) Assume 30 days per month. ≈5.115 × 10⁻⁸ mph

b. A young person grows from 21 inches at birth to 69 inches in 18 years. How many miles per hour is this growth rate? Assume 365 days per year. ≈4.805 × 10⁻⁹ mph

c. A baby grows from 15 inches at birth to 5 feet 2 inches at age 14 years. What is the rate of growth in miles per hour? 6.048 × 10⁻⁹ mph

30. Making Mountains out of Mole Hills Assume that mole hills and mountains are both shaped as cones, as shown in the figure. Suppose that Fuji-San (a mountain

in Japan) and a mole hill are of similar proportions. The formula for the volume of a cone is $V = \frac{1}{3}\pi r^2 h$.

mole hill

9 in.

mountain: Fuji-San, Japan

12,388 ft

38,150 ft

a. Write and solve a proportion to find the length of the base of the mole hill. 9/x = 12,388/38,150, x ≈ 27.7 in.

b. State the formula for the volume in terms of variation. Volume varies jointly with the height and the square of the radius.

c. Find the volume for each figure in cubic feet. State the volume of the mountain in scientific notation. mole hill, 1.046 ft³; mountain, 4.72 × 10¹² ft³

8.3 Rational Exponents

Objectives

▪ Use a calculator to explore powers with integer bases.

▪ Evaluate exponential expressions without a calculator.

▪ Apply rational exponents in settings involving compound interest.

WARM-UP

1. List the powers of 2 from 2^{-1} to 2^{10}. $\frac{1}{2}$, 1, 2, 4, 8, 16, 32, 64, 128, 256, 512, 1024

2. List the powers of 3 from 3^{-1} to 3^5. $\frac{1}{3}$, 1, 3, 9, 27, 81, 243

3. List the powers of 4 from 4^{-1} to 4^5. $\frac{1}{4}$, 1, 4, 16, 64, 256, 1024

4. List the powers of 5 from 5^{-1} to 5^4. $\frac{1}{5}$, 1, 5, 25, 125, 625

5. List the powers of 6 from 6^{-1} to 6^3. $\frac{1}{6}$, 1, 6, 36, 216

6. Factor $7.00 + 0.06(7.00)$. $7.00(1 + 0.06)$

7. Factor $7.42 + 0.06(7.42)$. $7.42(1 + 0.06)$

IN THIS SECTION, we expand our concept of exponents to include rational-number exponents and explore their meaning in terms of factors of a number. With a calculator, we apply rational exponents in compound interest.

Exponents That Are Not Integers

In Example 1, we explore solving exponential equations $y = b^x$ with a calculator. Reviewing the powers found in the Warm-up may be helpful in making guesses.

EXAMPLE 1 **Exploring powers with a calculator** Solve the exponential equations by guess and check. Use a calculator as needed. If the exponents are not integers, write them in both fraction and decimal notation.

Many solutions are fractions and motivate the discussion of rational powers.

a. $2^n = 256$ **b.** $4^n = 256$ **c.** $16^n = 256$ **d.** $8^n = 256$

e. $32^n = 256$ **f.** $4^n = 8$ **g.** $8^n = 4$ **h.** $8^n = 16$

i. $4^n = 128$ **j.** $64^n = 8$ **k.** $8^n = 128$ **l.** $8^n = 2$

SOLUTION See the Answer Box.

EXPONENTIAL EQUATION

> An exponential equation $y = b^x$ has the variable in the exponent and a constant in the base.

Example 1 suggests that exponents in exponential equations can be fractions and decimals as well as integers. We look at this idea again as we graph the exponential equation $y = 4^x$ in Example 2.

EXAMPLE 2 **Solving equations from an exponential graph** Use the numbers from Warm-up Exercise 3 to make a table and a graph of $y = 4^x$. Use the graph to find the solutions to these equations.

a. $4^x = 256$ **b.** $4^x = 128$ **c.** $4^x = 512$

SOLUTION The table for $y = 4^x$ appears in Table 3; the graph is in Figure 3. The horizontal axis is labeled *Exponents* to remind us that the inputs are exponents on a constant base, $b = 4$.

TABLE 3 $y = 4^x$

x	4^x
-1	$\frac{1}{4}$
0	1
1	4
2	16
3	64
4	256
5	1024

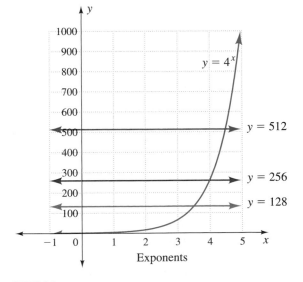

FIGURE 3

a. The solution to $4^x = 256$ is $x = 4$, the x-coordinate of the intersection of $y = 4^x$ and $y = 256$.

b. The solution to $4^x = 128$ is $x = 3.5$, the x-coordinate of the intersection of $y = 4^x$ and $y = 128$.

c. The solution to $4^x = 512$ is $x = 4.5$, the x-coordinate of the intersection of $y = 4^x$ and $y = 512$.

In the graph of $y = 4^x$ in Figure 3, the input x can take on any value along the x-axis. This suggests that there are many possible exponents—not only fractions and decimals but also irrational real numbers such as π.

Think about it 1: The number pi, π, is on the interval between what two integers? Use that interval to estimate a low and high value for 4^π. Evaluate 4^π with a calculator.

Rational Exponents

The next three examples focus on exponents that can be written as fractions or repeating decimals—that is, exponents that are rational numbers.

■ RATIONAL EXPONENT

> The expression b^x is said to have a rational exponent if the exponent can be written as the quotient of two integers $\dfrac{m}{n}$, where $n \neq 0$.

Example 3 shows how factors of numbers can be grouped to suggest rational exponents.

EXAMPLE 3 **Grouping factors** Look for patterns in these sets of facts and factors. Complete the statements in parts a and b, given that

$$4^0 = 1 \qquad 4^{1/2} = 2 \qquad 4^1 = 2 \cdot 2 = 4 \qquad 4^{3/2} = (2 \cdot 2) \cdot 2 = 8$$

a. $4^{} = (2 \cdot 2) \cdot (2 \cdot 2) = 16$
b. $4^{} = (2 \cdot 2) \cdot (2 \cdot 2) \cdot 2 = 32$

Complete the statements in parts c to g, given that

$$27^0 = 1 \qquad 27^{1/3} = 3 \qquad 27^{2/3} = 3 \cdot 3 = 9 \qquad 27^1 = (3 \cdot 3 \cdot 3) = 27$$

c. $27^{4/3} = (3 \cdot 3 \cdot 3) \cdot 3 =$
d. $27^{} = (3 \cdot 3 \cdot 3) \cdot 3 \cdot 3 - 243$
e. $27^2 = (3 \cdot 3 \cdot 3) \cdot (3 \cdot 3 \cdot 3) =$
f. $27^{} = = 2187$
g. $27^{} = = 6561$

SOLUTION
a. 4^2 or $4^{4/2} = 16$ **b.** $4^{2\frac{1}{2}}$ or $4^{5/2} = 32$
c. $27^{1\frac{1}{3}}$ or $27^{4/3} = 81$ **d.** $27^{1\frac{2}{3}}$ or $27^{5/3} = 243$

Student Note: After this example, we will tend to use improper fractions, rather than mixed numbers, for exponents.

e. $27^2 = 729$

f. $27^{2\frac{1}{3}}$ or $27^{7/3} = (3 \cdot 3 \cdot 3)(3 \cdot 3 \cdot 3) \cdot 3 = 2187$

g. $27^{2\frac{2}{3}}$ or $27^{8/3} = (3 \cdot 3 \cdot 3)(3 \cdot 3 \cdot 3) \cdot 3 \cdot 3 = 6561$ ■

Keep factors in mind as you think about Examples 4 and 5.

EXAMPLE 4 **Simplifying with rational exponents** $1/n$ Simplify these expressions mentally. List the factors of the base, as needed.
a. $8^{1/3}$ **b.** $25^{1/2}$ **c.** $64^{1/3}$ **d.** $64^{1/6}$

SOLUTION   **a.** 2   **b.** 5   **c.** 4   **d.** 2   ■

When the exponent contains a number other than 1 in the numerator, we simplify mentally by applying the numerator after the denominator:

$$8^{2/3} = (8^{1/3})^2 = 2^2 = 4$$

$$81^{3/4} = (81^{1/4})^3 = 3^3 = 27$$

$$4^{3/2} = (4^{1/2})^3 = 2^3 = 8$$

EXAMPLE 5 Simplifying with rational exponents m/n Simplify these expressions.

a. $16^{3/4}$　　**b.** $27^{2/3}$　　**c.** $8^{4/3}$　　**d.** $81^{5/4}$

SOLUTION

a. $16^{3/4} = (16^{1/4})^3 = 2^3 = 8$

b. $27^{2/3} = (27^{1/3})^2 = 3^2 = 9$

c. $8^{4/3} = (8^{1/3})^4 = 2^4 = 16$

d. $81^{5/4} = (81^{1/4})^5 = 3^5 = 243$

Rational exponent expressions may be simplified with this property:

$$b^{m/n} = (b^m)^{1/n} = (b^{1/n})^m, \quad \text{where } b^{1/n} \text{ is defined}$$

Application: Compound Interest

Radio and television advertising sometimes suggests borrowing money with a post-dated check or payday loan. Is this really a good idea? Example 10 and related exercises show just how expensive this form of borrowing can be.

COMPOUND INTEREST, COMPOUNDED ANNUALLY Compound interest is a common application of exponents on expressions. The examples show the derivation of the formulas for interest compounded annually and interest compounded twice a year. Then we will work with compound interest using fractional exponents.

EXAMPLE 6 Finding annual wages Myrna starts a job at $7.00 per hour. Suppose her hourly wage increases 6% at the end of each year she works.

a. Calculate her hourly wage at the end of each of her first three years.

b. Write a formula for the amount she will earn per hour at the end of t years.

c. Evaluate the formula for $t = 10$ years.

Student Note: In Examples 6 and 8, we assume that the employer is generous and rounds up to the nearest cent.

SOLUTION The increase is 6% of the wage for each year. Writing 6% as a decimal, we have 0.06.

a. The wages are calculated in Table 4.

TABLE 4

Year	Wage Calculation and Factored Calculation	Wage
0		7.00
1	$7.00 + 0.06(7.00) = 7.00(1 + 0.06)$	7.42
2	$7.42 + 0.06(7.42) = 7.42(1 + 0.06)$	≈ 7.87
3	$7.87 + 0.06(7.87) = 7.87(1 + 0.06)$	≈ 8.34

b. From the table, we see that each year's final wage is the prior year's wage times $(1 + 0.06)$. Over the three years, we have multiplied by $(1 + 0.06)$ three times. This suggests that we can go directly from the beginning wage to the third-year wage by multiplying by the cube of $(1 + 0.06)$:

$$7.00(1 + 0.06)^3 \approx 8.34$$

At the end of t years, Myrna's wage would be $7.00(1 + 0.06)^t$.

c. At the end of 10 years, Myrna's wage would be $7.00(1 + 0.06)^{10} \approx \$12.54$ per hour. When interest is calculated once a year, we say the interest is *compounded annually*.

■■■ ANNUAL COMPOUND INTEREST
(INTEREST COMPOUNDED ONCE A YEAR)

The compound interest formula for annual interest is

$$S = P(1 + r)^t$$

where S is the value in the future, P is the value at present, r is the annual rate of interest, and t is the time in years.

EXAMPLE 7 **Calculating annual compound interest** Find the value after 18 years of a $200 savings bond at 3.75% interest compounded annually.

SOLUTION
$$S = P(1 + r)^t \quad \text{for } P = \$200, r = 3.75\% = 0.0375, t = 18$$
$$S = 200(1 + 0.0375)^{18}$$
$$\approx \$387.99$$

■■

COMPOUND INTEREST, COMPOUNDED MORE THAN ONCE A YEAR Lenders offering postdated check cashing or payday loans charge daily interest. Savings banks may offer monthly interest. Finding interest compounded more than once a year requires a modification of our annual interest formula.

EXAMPLE 8 **Finding annual wages with two raises per year** Suppose Myrna's boss decides to calculate her wage increase so that she receives half of the 6% increase every six months.

a. Calculate her hourly wage at the end of each six months for two years.
b. Write a formula for her hourly wage after t years.
c. Evaluate the formula for $t = 2$, $t = 3$, and $t = 10$ years.
d. Compare the results for part c with those in Example 6.

SOLUTION **a.** The wages are calculated in Table 5.

TABLE 5

Year	Wage Calculation and Factored Calculation	Wage
0		7.00
$\frac{1}{2}$	$7.00 + 0.03(7.00) = 7.00(1 + 0.03)$	7.21
1	$7.21 + 0.03(7.21) = 7.21(1 + 0.03)$	≈ 7.43
$1\frac{1}{2}$	$7.43 + 0.03(7.43) = 7.43(1 + 0.03)$	≈ 7.65
2	$7.65 + 0.03(7.65) = 7.65(1 + 0.03)$	≈ 7.88

b. At the end of two years, the wage increase has been calculated four times but at half the rate, 3%. Working back up the table, we see that $(1 + 0.03)$ has been a factor two times each year, suggesting a formula for compounding twice a year:

$$P = A(1 + r/2)^{2t}$$

c. For $t = 2$ years, $\quad P = 7(1 + 0.06/2)^{2 \cdot 2} \approx \$7.88$
For $t = 3$ years, $\quad P = 7(1 + 0.06/2)^{2 \cdot 3} \approx \$8.36$
For $t = 10$ years, $\quad P = 7(1 + 0.06/2)^{2 \cdot 10} \approx \$12.64$

d. For $t = 2$ years, the gain from compounding twice a year is $0.01. For $t = 3$ years, the gain is $0.02. For $t = 10$ years, the gain is $0.10.

■■■

■ COMPOUND INTEREST

> If n is the number of times a year interest is calculated (compounded),
>
> $$S = P\left(1 + \frac{r}{n}\right)^{nt}$$
>
> where S is the future value, P is the present value (also known as the principal, or starting amount), r is the annual rate of interest (expressed as a decimal), and t is the number of years.

EXAMPLE 9   **Calculating forgotten savings**   Find the amount of money in a forgotten savings account if $1.00 is invested at 7% interest for 200 years. Assume the interest is calculated monthly, $n = 12$.

SOLUTION

$$S = P\left(1 + \frac{r}{n}\right)^{nt}$$

$$= 1.00\left(1 + \frac{0.07}{12}\right)^{(12 \cdot 200)}$$

$$\approx \$1{,}154{,}669.56$$

The answer is rounded down. The dollar has increased to over $1 million.   ■

Think about it 2: When a compound interest expression is entered into a calculator, the exponential factors must be placed in parentheses. Why would the answer to Example 9 change if the calculations were carried out without the parentheses on the factored exponent?

EXAMPLE 10   **Finding the cost of a payday loan**   If you borrow $100 from a payday lender at an interest rate of 1.5% per day and are unable to repay for a year (365 days), what will be your payment?

SOLUTION   The interest rate is $r = 1.5\%$ per day, or $0.015(365)$ annual interest.

$$S = P(1 + r/n)^{nt} \quad \text{for } P = 100, n = 365, t = 1$$

$$= 100(1 + 0.015(365)/365)^{365(1)}$$

$$= 100(1 + 0.015)^{365}$$

$$= \$22{,}914.24$$

A payday loan is a very expensive way to borrow money.   ■

COMPOUND INTEREST WITH RATIONAL EXPONENTS   We now combine rational exponents with compound interest by working problems with fractional parts of years. The compound interest formula is repeated below, for reference:

$$S = P\left(1 + \frac{r}{n}\right)^{nt}$$

EXAMPLE 11   **Finding compound interest**   Suppose we deposit $1000 at 7% interest. Find the amount of money in the account in $1\frac{2}{3}$ years with weekly compounding.

SOLUTION   Weekly compounding means $n = 52$.

$$S = P\left(1 + \frac{r}{n}\right)^{nt}$$

$$= 1000\left(1 + \frac{0.07}{52}\right)^{52\left(1\frac{2}{3}\right)}$$

$$= 1123.65$$

To evaluate the formula with a calculator, we enter the entire expression into the calculator:

$$1000 \times (1 + 0.07/52) \; \boxed{\wedge} \; (52 \times (1 + 2/3))$$

This prevents rounding errors and permits us to substitute different numbers into the same expression for subsequent calculations. *It is essential that parentheses be placed around the entire exponential expression.* ▬

Caution on rounding: If you are receiving the interest on an account balance, then you should drop digits after the hundredth place, because a bank or credit union will never round up and give you extra money. If you are paying the interest on a loan, then you should round all numbers up.

Think about it 3: Which way was the answer in Example 11 rounded? Why?

The following calculator box describes the process in the tables in Examples 6 and 8 with a **recursive formula**—*a formula that uses a prior number to calculate the next number.* $\boxed{2\text{nd}}$ [ANS] is used to obtain the answer for use within a calculation.

▬ GRAPHING CALCULATOR TECHNIQUE: FINDING
WAGES WITH A RECURSIVE FORMULA

Use $\boxed{2\text{nd}}$ [ANS] to build a recursive formula for calculating the wages in Example 6, where the starting wage is $7.00 and the interest rate is 6%.

7.00 $\boxed{\text{ENTER}}$

$\boxed{2\text{nd}}$ [ANS] $+ 0.06 \times \boxed{2\text{nd}}$ [ANS] $\boxed{\text{ENTER}}$

Repeat $\boxed{\text{ENTER}}$ to obtain successive results.

Recursion may be applied to Examples 7 and 8 and other compound interest settings with small whole-number exponents.

ANSWER BOX

Warm-up: **1.** $\frac{1}{2}$, 1, 2, 4, 8, 16, 32, 64, 128, 256, 512, 1024 **2.** $\frac{1}{3}$, 1, 3, 9, 27, 81, 243 **3.** $\frac{1}{4}$, 1, 4, 16, 64, 256, 1024 **4.** $\frac{1}{5}$, 1, 5, 25, 125, 625 **5.** $\frac{1}{6}$, 1, 6, 36, 216 **6.** 7.00(1 + 0.06) **7.** 7.42(1 + 0.06) **Example 1: a.** $n = 8$
b. $n = 4$ **c.** $n = 2$ **d.** $n = \frac{8}{3} \approx 2.67$ **e.** $n = \frac{8}{5} = 1.6$ **f.** $n = \frac{3}{2} = 1.5$
g. $n = \frac{2}{3} \approx 0.67$ **h.** $n = \frac{4}{3} \approx 1.33$ **i.** $n = \frac{7}{2} = 3.5$ **j.** $n = \frac{1}{2} = 0.5$
k. $n = \frac{7}{3} \approx 2.33$ **l.** $n = \frac{1}{3} \approx 0.33$ **Think about it 1:** π is on the interval between 3 and 4, so 4^{π} is between 4^3 and 4^4, or 64 and 256.
$4^{\pi} \approx 77.880\,233\,65$. **Think about it 2:** The calculator would calculate the exponent with only the first number and then multiply the result by the final factor (the time in years). The resulting incorrect formula is
$S = P(1 + r/n)^n \cdot t$. **Think about it 3:** The answer in Example 11 was rounded down, because it represented an account balance, not a loan payment.

8.3 Exercises

Use a calculator to guess and check for solutions to the exponential equations in Exercises 1 and 2.

1. a. $27^x = 9$
   $x = \frac{2}{3}$
   b. $9^x = 27$
   $x = \frac{3}{2}$
   c. $81^x = 27$
   $x = \frac{3}{4}$

2. a. $25^x = 125$
   $x = \frac{3}{2}$
   b. $625^x = 125$
   $x = \frac{3}{4}$
   c. $125^x = 25$
   $x = \frac{2}{3}$

In Exercises 3 to 6, consider this example. When we examine a list of factors, we find different facts, depending on how we choose to group the factors:

$$2^6 = 2 \cdot 2 \cdot 2 \cdot 2 \cdot 2 \cdot 2 = 64$$
$$4^3 = (2 \cdot 2) \cdot (2 \cdot 2) \cdot (2 \cdot 2) = 64$$
$$8^2 = (2 \cdot 2 \cdot 2) \cdot (2 \cdot 2 \cdot 2) = 64$$

Write these expressions with groups of factors to show why they are equal.

3. $81 = 3^4 = 9^2$ $3 \cdot 3 \cdot 3 \cdot 3 = (3 \cdot 3) \cdot (3 \cdot 3)$

4. $625 = 25^2 = 5^4$ $(5 \cdot 5) \cdot (5 \cdot 5) = 5 \cdot 5 \cdot 5 \cdot 5$

5. What exponents make each of the statements below true? Group the 2s in parentheses to show the relationship between the exponent and the base.

   a. $4^x = 2 \cdot 2 \cdot 2 \cdot 2 \cdot 2 \cdot 2 \cdot 2 \cdot 2 \cdot 2 \cdot 2 = 1024$
   $x = 5$

   b. $8^x = 2 \cdot 2 \cdot 2 \cdot 2 \cdot 2 \cdot 2 \cdot 2 \cdot 2 \cdot 2 \cdot 2 = 1024$
   $x = 3\frac{1}{3} = \frac{10}{3}$

   c. $16^x = 2 \cdot 2 \cdot 2 \cdot 2 \cdot 2 \cdot 2 \cdot 2 \cdot 2 \cdot 2 \cdot 2 = 1024$
   $x = 2\frac{1}{2} = \frac{5}{2}$

   d. $32^x = 2 \cdot 2 \cdot 2 \cdot 2 \cdot 2 \cdot 2 \cdot 2 \cdot 2 \cdot 2 \cdot 2 = 1024$
   $x = 2$

   e. $64^x = 2 \cdot 2 \cdot 2 \cdot 2 \cdot 2 \cdot 2 \cdot 2 \cdot 2 \cdot 2 \cdot 2 = 1024$
   $x = 1\frac{2}{3} = \frac{5}{3}$

   f. $128^x = 2 \cdot 2 \cdot 2 \cdot 2 \cdot 2 \cdot 2 \cdot 2 \cdot 2 \cdot 2 \cdot 2 = 1024$
   $x = 1\frac{3}{7} = \frac{10}{7}$

   g. $256^x = 2 \cdot 2 \cdot 2 \cdot 2 \cdot 2 \cdot 2 \cdot 2 \cdot 2 \cdot 2 \cdot 2 = 1024$
   $x = 1\frac{1}{4} = \frac{5}{4}$

6. What exponent makes each of the statements below true? Group the 3s in parentheses to show the relationship between the base and the exponent.

   a. $9^x = 3 \cdot 3 \cdot 3 \cdot 3 \cdot 3 \cdot 3 = 729$ $x = 3$

   b. $27^x = 3 \cdot 3 \cdot 3 \cdot 3 \cdot 3 \cdot 3 = 729$ $x = 2$

   c. $81^x = 3 \cdot 3 \cdot 3 \cdot 3 \cdot 3 \cdot 3 = 729$ $x = 1\frac{1}{2} = \frac{3}{2}$

   d. $243^x = 3 \cdot 3 \cdot 3 \cdot 3 \cdot 3 \cdot 3 = 729$ $x = 1\frac{1}{5} = \frac{6}{5}$

In Exercises 7 to 12, simplify the expressions without a calculator.

7. a. $64^{1/2}$
   8
   b. $16^{1/4}$
   2
   c. $32^{1/5}$
   2

8. a. $125^{1/3}$
   5
   b. $36^{1/2}$
   6
   c. $64^{1/6}$
   2

9. a. $32^{2/5}$
   4
   b. $16^{3/2}$
   64
   c. $36^{3/2}$
   216
   d. $32^{6/5}$
   64

10. a. $64^{5/6}$
   32
   b. $64^{2/3}$
   16
   c. $125^{2/3}$
   25
   d. $64^{4/3}$
   256

11. a. $16^{5/4}$
   32
   b. $64^{3/2}$
   512
   c. $25^{3/2}$
   125
   d. $81^{3/4}$
   27

12. a. $27^{4/3}$
   81
   b. $8^{5/3}$
   32
   c. $81^{3/2}$
   729
   d. $27^{2/3}$
   9

13. Use Figure 3 (page 452) to describe how to solve these equations from a graph. Estimate the solutions from the graph.

   a. $4^x = 400$
   $x \approx 4.3$
   b. $4^x = 700$
   $x \approx 4.7$
   c. $4^x = 1000$
   $x \approx 5.0$

Simplify the expressions in Exercises 14 and 15.

14. a. $16^{1.5}$
   64
   b. $36^{1.5}$
   216
   c. $9^{2.5}$
   243
   d. $0.01^{0.5}$
   0.1

15. a. $25^{1.5}$
   125
   b. $4^{2.5}$
   32
   c. $16^{2.5}$
   1024
   d. $0.09^{0.5}$
   0.3

16. Explain how a rational exponent makes it possible to simplify an expression such as $16^{1.5}$ mentally.
   $1.5 = \frac{3}{2}; 16^1 = 4 \cdot 4, 16^{1/2} = 4, 16^{3/2} = 4 \cdot 4 \cdot 4$

17. Find the final hourly wage if a $6.00 starting wage is increased 5% each year for 10 years. $9.77

18. Find the final hourly wage if a $6.00 starting wage is increased 2% each year for 10 years. $7.31

19. Find the amount of money in an account if $1000 is deposited at 7% interest compounded annually and the money is left for 5 years. $1402.55

20. Find the amount of money in an account if $1000 is deposited at 2% interest compounded annually and the money is left for 5 years. $1104.08

21. Common compounding time periods n are given special names. Match each number of compoundings with its appropriate name. Choose from
annual, daily, monthly, quarterly, semiannual, weekly

   a. once a year
   annually
   b. twice a year
   semiannual

   c. 4 times a year
   quarterly
   d. 12 times a year
   monthly

   e. 52 times a year
   weekly
   f. 365 times a year
   daily

22. The number of compounding periods, n, per year indicates how often interest is calculated. (Assume 365 days per year.)

   a. If interest were calculated each hour, what n would be used? 8760

   b. If interest were calculated each second, what n would be used?   31,536,000

In Exercises 23 and 24 enter the expressions into a calculator so that the calculator will correctly evaluate them.

23. a. $10{,}000\left(1 + \dfrac{0.04}{12}\right)^{12 \cdot 3}$ 11,272.72

   b. $10{,}000\left(1 + \dfrac{0.04}{4}\right)^{4 \cdot 3}$ 11,268.25

24. a. $10,000\left(1 + \dfrac{0.08}{52}\right)^{52 \cdot 3}$ 12,710.15

b. $10,000\left(1 + \dfrac{0.08}{2}\right)^{2 \cdot 3}$ 12,653.19

25. Describe a problem setting for each expression given in Exercise 23. $10,000 earnings, 4% compounded monthly/quarterly for 3 years

26. Describe a problem setting for each expression given in Exercise 24. $10,000 earnings, 8% compounded weekly/semiannually for 3 years

27. Find the final hourly wage if a $6.00 starting wage is increased 5% each year for 10 years. Half the raise is given every six months. $9.83

28. Find the final hourly wage if a $6.50 starting wage is increased 2% each year for 10 years. Half the raise is given every six months. $7.93

29. If you borrow $200 from a payday lender at 1% per day, what will you owe in 30 days? $269.57

30. If you borrow $100 from a payday lender at 2% per day, what will you owe in 30 days? $181.14

Interest on late property taxes is 16% per year compounded monthly. In Exercises 31 to 34, find the total to be paid for the amount and time late shown.

31. A $1350 tax, 8 months late $1500.91

32. A $670 tax, 14 months late $806.51

33. A $650 tax, 2 months late $667.45

34. A $2540 tax, 9 months late $2861.58

In Exercises 35 to 42, find the amount of money in a savings account for the given time period, annual rate of interest, and type of compounding. Start with $1000 in each account.

35. $1\frac{3}{4}$ years at 8%, with monthly compounding ≈$1,149.73

36. $10\frac{1}{2}$ years at 6%, with weekly compounding ≈$1,876.92

37. $2\frac{1}{4}$ years at 8%, with monthly compounding ≈$1,196.50

38. $8\frac{1}{3}$ years at 6%, with weekly compounding ≈$1,648.24

39. $5\frac{1}{3}$ years at 7%, with quarterly compounding ≈$1,447.88

40. 1.5 years at 5%, with monthly compounding ≈$1,077.71

41. 4.75 years at 7%, with quarterly compounding ≈$1,390.44

42. $6\frac{1}{4}$ years at 5%, with monthly compounding ≈$1,365.95

In Exercises 43 and 44, suppose that you are working with the compound interest formula and you change from compounding quarterly to compounding monthly.

43. How does the base change? smaller value

44. How does the exponent change? increased by factor of 3

45. What is the base in the compound interest formula for calculating interest n times a year? $1 + r/n$

46. What is the base in the compound interest formula for calculating interest once a year? $1 + r$

In Exercises 47 to 50, find numbers that make each statement true.

47. $a^b = b^a$ $a = b = 2; a = 2, b = 4$

48. $a^b = a^c$ (b and c are not equal) $a = 0$ or $a = 1$

49. Suppose $b \neq 0$ and $b \neq 1$. If $b^x = b^y$, then $x = y$. any $b > 0$ except $b = 1$

50. $x^a = y^a, x \neq y$ $a = 0$

51. Use calculator recursion to extend the calculation started in part a of Example 6 to the fourth and fifth years. Compare your result with that from the formula. Explain any differences in results. ≈$8.84; ≈$9.37; rounding error

52. Use calculator recursion to extend the calculation started in part a of Example 8 to the third and fourth years. Compare your result with that from the formula. Explain any differences in results. ≈$8.36; ≈$8.87; rounding error

53. Wind Chill Revisited In 2001, the National Weather Service announced a revised model for the wind-chill chart initially developed in 1941 by Paul Siple and Charley Passel. The original chart, page 1, is modeled by an equation involving square root (Project 60 in Section 1.2, page 25). The wind-chill apparent temperature, W, is now estimated with a model having rational exponents on S:

$$W = 35.74 + 0.6215T - 35.75(S^{0.16}) + 0.4275T(S^{0.16})$$

where T is the current air temperature in °F and S is the wind speed. The formula does not give the current air temperature for $S = 0$.

a. For $T = 25$°F current temperature, evaluate the formula (using [2nd] [TABLE] and $S = X$) for wind speeds of 5 to 30 mph in steps of 5. 19, 15, 13, 11, 9, 8

b. For $T = -5$°F current temperature, evaluate the formula as in part a. $-16, -22, -26, -29, -31, -33$

c. Compare the new model's temperatures with the corresponding ones on page 1. For the given current temperatures, how have the apparent temperatures changed? The model produces warmer temperatures.

54. Body Fat Percentages Steve Johnson ("Younger than Yesterday," *Bicycling Magazine*, May 1990) describes formulas for estimating the percentage of a person's body composition in fat. The formulas require weight in kilograms (divide the weight in pounds by 2.205 pounds per kilogram) and height in centimeters (multiply the height in inches by 2.54 centimeters per inch). For men, the percent body fat is

$$\frac{(\text{weight in kg})^{1.2}}{(\text{height in cm})^{3.3}} \cdot 3,000,000$$

For women, the percent body fat is

$$\frac{(\text{weight in kg})^{1.2}}{(\text{height in cm})^{3.3}} \cdot 4{,}000{,}000$$

a. Rewrite the formulas in terms of pounds and inches.
See Additional Answers.
Parts b to e give a range of normal weights for healthy adults. (Data are from the American Medical Association's *Encyclopedia of Medicine*, Random House, New York, 1989.) Find the percent body fat for both numbers in the range.

b. Woman: 5 feet 1 inch, 100 to 130 pounds
23.04 to 31.56

c. Woman: 5 feet 10 inches, 132 to 165 pounds
20.41 to 26.68

d. Man: 6 feet 2 inches, 153 to 193 pounds
15.21 to 20.10

e. Man: 5 feet 7 inches, 128 to 162 pounds
17.05 to 22.62

■ **Projects**

55. Solving for Time A commonly asked question related to compound interest is how much time is needed for borrowed or saved money to grow to a certain level. At this time, we can guess and check or solve from a table or graph.

a. To find how many years it will take for $1000 to double when saved at 7% compounded weekly, we solve the equation $2000 = 1000(1 + 0.07/52)^{52x}$. Enter the right side, substituting a guess for x (in years). Replay with (2nd) (ENTER), and keep guessing until your x, rounded to the nearest hundredth, yields an output of 2000. 9.91 yr

b. To solve part a with a table, enter $Y_1 = 1000(1 + 0.07/52)^{52x}$ in (Y=). Under (2nd) [TBLSET], begin with x = **TblStart** = 0 and Δ**Tbl** = 1. Find a pair of outputs near 2000. Adjust **TblStart** and use Δ**Tbl** = 0.1 to find x to the nearest tenth. Repeat with Δ**Tbl** = 0.01 to find x to the nearest hundredth.
9.91 yr

c. To solve with a graph, enter $Y_2 = 2000$ along with Y_1 in part b. Use (2nd) [CALC] **5 : intersect** to find x.

9.91 yr

d. If you leave $100 on a credit card at 19% annual interest compounded monthly, find the number of years needed for the amount you owe to rise to $200.
3.68 yr or 3 yr 8 mo

e. If you borrow $100 at a credit union at 10% annual interest compounded daily, find the number of years needed for the amount you owe to rise to $200.
6.93 yr or 6 yr 11 mo

f. If you borrow $100 from a payday lender at 1.5% per day, in how many days will the amount of money owed be $200? ≈46.6 days

g. If you borrow $100 from a payday lender at 1.5% per day, in how many days will the amount of money owed be $500? ≈108 days

56. Practice with Exponential Expressions These exponential expressions give practice with a variety of bases and exponents. They are intended to be simplified mentally.

a. $\left(\frac{4}{25}\right)^{1/2}$ $\frac{2}{5}$ **b.** $\left(\frac{1}{8}\right)^{1/3}$ $\frac{1}{2}$ **c.** $\left(\frac{64}{27}\right)^{2/3}$ $\frac{16}{9}$

d. $((64)^{1/2})^{1/3}$ 2 **e.** $\left(\frac{1}{4}\right)^{1/2}$ $\frac{1}{2}$ **f.** $\left(\frac{8}{27}\right)^{1/3}$ $\frac{2}{3}$

g. $\left(\frac{9}{25}\right)^{3/2}$ $\frac{27}{125}$ **h.** $0.09^{3/2}$ 0.027 **i.** $\left(\frac{1}{32}\right)^{2/5}$ $\frac{1}{4}$

j. $\left(\frac{1}{27}\right)^{2/3}$ $\frac{1}{9}$ **k.** $\left(\frac{1}{64}\right)^{-1/6}$ 2 **l.** $0.04^{3/2}$ 0.008

m. $\left(\frac{1}{81}\right)^{3/4}$ $\frac{1}{27}$ **n.** $\left(\frac{1}{8}\right)^{5/3}$ $\frac{1}{32}$ **o.** $4^{-1/2}$ $\frac{1}{2}$

p. $32^{-3/5}$ $\frac{1}{8}$ **q.** $8^{-1/3}$ $\frac{1}{2}$ **r.** $\left(\frac{3}{4}\right)^{-2}$ $\frac{16}{9}$

s. $\left(\frac{1}{3}\right)^{-2}$ 9 **t.** $0.008^{2/3}$ 0.04 **u.** $\left(\frac{9}{25}\right)^{-3/2}$ $\frac{125}{27}$

v. $\left(\frac{27}{8}\right)^{-1/3}$ $\frac{2}{3}$ **w.** $9^{-3/2}$ $\frac{1}{27}$ **x.** $0.027^{2/3}$ 0.09

57. Lottery Investment Spreadsheet You just won a million-dollar cash lottery. Suppose you deposit the money at 10% simple interest and begin spending at the end of the first year. How many years will the money last if you spend each of the following amounts annually?

a. $500,000
2 yr; with $176,000 at end of 3rd yr

b. $250,000 at end of 6th yr
5 yr; with $92,658.50

c. $200,000
7 yr; with $56,411.19 at end of 8th yr

d. $150,000 11 yr; with $80,785.81 at end of 12th yr

e. $100,000
forever (spending only the interest)

8.4 Roots and Rational Exponents

Objectives

■ Simplify *n*th root expressions.

■ Change expressions from rational exponent to radical notation and vice versa.

■ Use properties of exponents and radicals to simplify expressions.

WARM-UP

Use the product and quotient properties of square roots to evaluate these square roots. Assume $x \geq 0$, $y \geq 0$.

1. a. $\sqrt{25}$ 5   **b.** $\sqrt{2500}$ 50   **c.** $\sqrt{0.25}$ 0.5   **d.** $\sqrt{0.0025}$ 0.05

2. a. $\sqrt{144}$ 12   **b.** $\sqrt{1,440,000}$ 1200   **c.** $\sqrt{0.0144}$ 0.12   **d.** $\sqrt{14,400}$ 120

3. a. $\sqrt{x^2 y}$ $x\sqrt{y}$ **b.** $\sqrt{xy^2}$ $y\sqrt{x}$ **c.** $\sqrt{(xy)^2}$ xy **d.** $\sqrt{x/y^2}$ $\frac{1}{y}\sqrt{x}$

Use a calculator to evaluate the following, and give your conclusions about 1/2 as an exponent. $\frac{1}{2}$ is same as square root.

4. a. $36^{1/2}$ 6   **b.** $9^{1/2}$ 3   **c.** $256^{1/2}$ 16   **d.** $(-9)^{1/2}$ not a real number ($3i$ in the complex number system)

IN THIS SECTION, we study nth roots and connect the notation for rational exponents (Section 8.3) with that for square roots (Section 5.1) and nth roots.

Even and Odd nth Roots

RADICAL NOTATION AND DEFINITIONS   A **radical expression** is *an expression that can be written* $\sqrt[n]{x}$. The *sign* $\sqrt{}$ is the **radical sign**. The small n is the index (the plural is *indices*). The **index of a radical** is *the integer indicating which root is being taken.* For the square root, the index 2 is usually omitted.

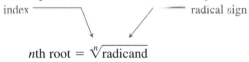

$$n\text{th root} = \sqrt[n]{\text{radicand}}$$

Cube roots, fourth roots, and higher roots have definitions similar to that of square roots. The following table summarizes those definitions.

■ nTH ROOTS

If x is a real number and n is a positive integer, then
- the **principal square root** of x is the positive number $\sqrt{x}$ such that

$$(\sqrt{x})^2 = x, \, x \geq 0$$

- the **cube root** of x is the number $\sqrt[3]{x}$ such that

$$(\sqrt[3]{x})^3 = x$$

- the principal fourth root of x is the positive number $\sqrt[4]{x}$ such that

$$(\sqrt[4]{x})^4 = x, \, x \geq 0$$

- the fifth root of x is the number $\sqrt[5]{x}$ such that

$$(\sqrt[5]{x})^5 = x$$

- the **principal nth root** of x is the positive number $\sqrt[n]{x}$, n even, such that

$$(\sqrt[n]{x})^n = x, \quad x \geq 0$$

- the **odd nth root** of x is the number $\sqrt[n]{x}$ such that

$$(\sqrt[n]{x})^n = x$$

EXAMPLE 1

Exploring *n*th roots   Given that $4096 = 2^{12} = 2 \cdot 2 \cdot 2 \cdot 2 \cdot 2 \cdot 2 \cdot 2 \cdot 2 \cdot 2 \cdot 2 \cdot$ $2 \cdot 2$, evaluate the expressions in parts a to d without a calculator. Estimate the values for parts e and f before using the calculator root option, $\sqrt[x]{}$, under $\boxed{\text{MATH}}$ (see Figure 4a).

a. $\sqrt[6]{4096}$ **b.** $\sqrt[4]{4096}$ **c.** $\sqrt[3]{4096}$

d. $\sqrt{4096}$ **e.** $\sqrt[5]{4096}$ **f.** $\sqrt[8]{4096}$

g. Explain why parts e and f do not have integer solutions.

SOLUTION   **a.** $\sqrt[6]{4096} = 2^2 = 4$ **b.** $\sqrt[4]{4096} = 2^3 = 8$

c. $\sqrt[3]{4096} = 2^4 = 16$ **d.** $\sqrt{4096} = 2^6 = 64$

e. The 5th root of 4096 is between 8 and 4 (the 4th root and the 6th root). On a calculator,

$$5 \; \boxed{\text{MATH}} \; 5 : \sqrt[x]{} \; 4096 \; \boxed{\text{ENTER}} \approx 5.278$$

as predicted (see Figure 4b).

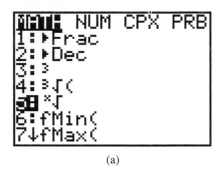 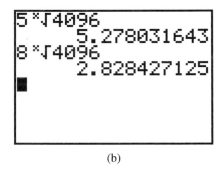

(a)   (b)

FIGURE 4

f. The 8th root of 4096 is smaller than 4, the 6th root of 4096. On a calculator,

$$8 \; \boxed{\text{MATH}} \; 5 : \sqrt[x]{} \; 4096 \; \boxed{\text{ENTER}} \approx 2.828$$

as predicted (see Figure 4b). An alternative technique is to use replay or $\boxed{\text{2nd}}$ $\boxed{\text{ENTER}}$ and replace the 5 with an 8.

g. The indices 6, 4, 3, and 2 divide evenly into the exponent 12; the indices 5 and 8 do not. ∎

Calculator note: To use the *n*th root function, enter the index before selecting *n*th root, $\sqrt[x]{}$, the number 5 choice under $\boxed{\text{MATH}}$.

ASSUMPTIONS   These ideas must be kept in mind at all times.

1. Radical expressions are functions. There is only one output for each input. In Chapter 5, $f(x) = \sqrt{x}$ was defined as the positive, or principal, square root function. Thus, although $(3)^2$ and $(-3)^2$ both equal 9, $\sqrt{9}$ equals 3. The same is true for all other even *n*th roots.

2. In Section 6.2, we defined the imaginary unit and the complex number system in order to solve square roots of negative numbers. If we continued with the ideas of Section 6.2, we would need to introduce trigonometry! To avoid trigonometry, we return to the real number system, where the even *n*th root of a negative number is undefined.

3. The expression 0^0 is undefined.

ROOTS AND NEGATIVE NUMBERS   The condition $x \geq 0$ was placed only on the definitions for the even nth roots. Odd roots of negative numbers are defined.

EXAMPLE 2   Simplifying roots of negative numbers   Simplify, if possible.

a. $\sqrt[3]{-8}$ **b.** $\sqrt[5]{-243}$ **c.** $-\sqrt{16}$ **d.** $\sqrt[4]{-16}$

SOLUTION   **a.** $\sqrt[3]{-8} = -2, (-2)^3 = -8$

b. $\sqrt[5]{-243} = -3, (-3)^5 = -243$

c. $-\sqrt{16} = -4$; this is the opposite of the square root of 16.

d. $\sqrt[4]{-16}$; this is not a real number. There is no real number that can be raised to the fourth power to obtain -16.   ▬

Connecting Rational Exponents and Roots

$1/2$ AS AN EXPONENT   The product of like bases property states that when the bases are the same, we can add the exponents:

$$x^m \cdot x^n = x^{m+n}$$

Example 3 uses the product property to suggest the meaning of $1/2$ as an exponent.

EXAMPLE 3   Exploring the exponent $1/2$ Simplify these exponential and radical expressions.

a. $16^{1/2} \cdot 16^{1/2}$ (add exponents)   **b.** $\sqrt{16} \cdot \sqrt{16}$

c. $9^{1/2} \cdot 9^{1/2}$ (add exponents)   **d.** $\sqrt{9} \cdot \sqrt{9}$

SOLUTION   **a.** $16^{1/2} \cdot 16^{1/2} = 16^{(1/2+1/2)} = 16^1 = 16$ **b.** $\sqrt{16} \cdot \sqrt{16} = 4 \cdot 4 = 16$

c. $9^{1/2} \cdot 9^{1/2} = 9^{(1/2+1/2)} = 9^1 = 9$ **d.** $\sqrt{9} \cdot \sqrt{9} = 3 \cdot 3 = 9$ ▬

As Example 3 shows, $x^{1/2}$ and $\sqrt{x}$ satisfy the same conditions, so it is natural to define the $1/2$ power as meaning the square root.

▬ THE EXPONENT $1/2$ AND THE PRINCIPAL SQUARE ROOT

> A number x raised to the $1/2$ power is said to be the principal square root of x:
>
> $$x^{1/2} = x^{0.5} = \sqrt{x}, x \geq 0$$

$1/n$ AS AN EXPONENT   It is reasonable to expect the nth root and the $1/n$ exponent to be related in the same way as the square root and the $1/2$ exponent.

▬ RATIONAL EXPONENTS $1/n$ AND RADICALS

> Let x be a nonzero real number and let n be a positive number. If $\sqrt[n]{x}$ is defined, the rational exponent $1/n$ means
>
> $$x^{1/n} = \sqrt[n]{x}$$
>
> If $1/n$ is a positive rational number, $0^{1/n} = 0$.

EXAMPLE 4   Changing notations   Write each rational exponent expression as a root. Write each root as a rational exponent expression.

a. $x^{1/3}$ **b.** $\sqrt[6]{x}$ **c.** $x^{0.25}$ **d.** $x^{0.2}$

SOLUTION   **a.** $\sqrt[3]{x}$ **b.** $x^{1/6}$ **c.** $x^{0.25} = x^{1/4} = \sqrt[4]{x}$ **d.** $x^{0.2} = x^{1/5} = \sqrt[5]{x}$ ▬

EXAMPLE 5 **Changing notation** Change the radicand into exponential notation, and simplify.

a. $\sqrt[3]{27}$ **b.** $\sqrt[4]{256}$ **c.** $\sqrt[5]{-243}$ **d.** $\sqrt[3]{-8}$ **e.** $\sqrt[4]{-81}$

SOLUTION **a.** $\sqrt[3]{27} = 27^{1/3} = (3^3)^{1/3} = 3^1 = 3$ **b.** $\sqrt[4]{256} = 256^{1/4} = (4^4)^{1/4} = 4^1$

c. $\sqrt[5]{-243} = (-243)^{1/5} = [(-3)^5]^{1/5} = (-3)^1$

d. $\sqrt[3]{-8} = (-8)^{1/3} = [(-2)^3]^{1/3} = (-2)^1$

e. The root is even and the radicand is negative; the expression is undefined. ▬

m/n AS AN EXPONENT When the exponent contains both a numerator and a denominator, $x^{m/n}$, both a power and a root are implied. The following definition shows the connection.

▬ RATIONAL EXPONENTS *m/n* AND RADICALS

> Let x be a nonzero real number. Suppose that m and n are integers with n positive and m/n in lowest terms. Then, if $\sqrt[n]{x}$ is defined in the real numbers, the rational exponent m/n means
>
> $$x^{m/n} = (x^{1/n})^m = (\sqrt[n]{x})^m \quad \text{and} \quad x^{m/n} = (x^m)^{1/n} = \sqrt[n]{x^m}$$
>
> If m/n is a positive rational number, $0^{m/n} = 0$.

EXAMPLE 6 **Changing notation** Write each rational exponent expression as a root. Write each root as a rational exponent expression.

a. $x^{2/3}$ **b.** $\sqrt[5]{x^3}$ **c.** $\sqrt[3]{x^5}$ **d.** $x^{0.75}$ **e.** $x^{0.4}$

SOLUTION **a.** $\sqrt[3]{x^2}$ **b.** $x^{3/5}$ **c.** $x^{5/3}$

d. $x^{0.75} = x^{3/4} = \sqrt[4]{x^3}, x \geq 0$ **e.** $x^{0.4} = x^{2/5} = \sqrt[5]{x^2}$ ▬

Caution: In order to use $(\sqrt[n]{x})^m = \sqrt[n]{x^m} = x^{m/n}$, the fraction m/n must be in lowest terms and $\sqrt[n]{x}$ must be defined in the real numbers.

EXAMPLE 7 **Changing notation** If possible, write each expression as a base with a rational exponent. Simplify the expression, if possible.

a. $(\sqrt[3]{-5})^3$ **b.** $(\sqrt[4]{-9})^2$ **c.** $\sqrt[3]{8^2}$

d. $\sqrt[2]{4^3}$ **e.** $\sqrt[6]{(-8)^2}$ **f.** $(\sqrt[6]{-8})^2$

SOLUTION **a.** $(\sqrt[3]{-5})^3 = (-5)^{3/3} = (-5)^1 = -5$

b. $(\sqrt[4]{-9})^2$ is not defined in the real numbers because $\sqrt[4]{-9}$ is not defined. $(\sqrt[4]{-9})^2 \neq \sqrt[4]{(-9)^2}$ because $2/4 = m/n$ is not in lowest terms.

c. $\sqrt[3]{8^2} = 8^{2/3} = (8^{1/3})^2 = 2^2 = 4$

d. $\sqrt[2]{4^3} = 4^{3/2} = (4^{1/2})^3 = 2^3 = 8$
  It is probably easier to observe that $4^3 = 64$ and $\sqrt{64} = 8$.

e. $\sqrt[6]{(-8)^2} = \sqrt[6]{64} = 2; (-8)^2 = 64$, so the expression is defined.

f. $(\sqrt[6]{-8})^2$ is not defined in the real numbers because $\sqrt[6]{-8}$ is not defined. ▬

ASSURING POSITIVE PRINCIPAL *n*TH ROOTS Recall that placing the absolute value symbol around a variable guarantees a positive output, as needed for principal *n*th roots. We extend the concept of the principal square root to square roots of variables with exponents, as follows:

$$\sqrt[n]{x^n} = \begin{cases} x & \text{if } n \text{ is odd} \\ |x| & \text{if } n \text{ is even} \end{cases}$$

EXAMPLE 8 Simplifying radicals   Simplify, where x and y are any real numbers.

a. $\sqrt{x^4}$ **b.** $\sqrt{y^2}$ **c.** $\sqrt{x^8}$ **d.** $\sqrt[4]{y^4}$

SOLUTION **a.** $\sqrt{x^4} = x^2$ **b.** $\sqrt{y^2} = |y|$ **c.** $\sqrt{x^8} = x^4$ **d.** $\sqrt[4]{y^4} = |y|$

No absolute value symbol is needed on parts a and c because the even powers are positive.

Translating Exponent Properties to Root Properties

Taking even roots of negative numbers is not permitted, but as long as we have positive bases, we can apply the exponent properties from Section 8.1 to expressions with rational exponents or to their equivalent radical expressions.

EXPONENT AND ROOT PROPERTIES, PART 1

> **Like Bases Properties**
> To multiply numbers with like bases, add the exponents:
>
> $$x^a \cdot x^b = x^{a+b} \qquad \text{Multiplication of like bases property}$$
>
> To divide numbers with like bases, subtract the exponents:
>
> $$\frac{x^a}{x^b} = x^{a-b} \quad \text{for } x \neq 0 \qquad \text{Division of like bases property}$$

In Example 9, we apply the definition $(x^m)^{1/n} = x^{m/n}$, where $x > 0$, and change from radical form into exponential form, as necessary, to use the appropriate like bases property.

EXAMPLE 9 Applying properties   Simplify. Assume the variable x is positive.

a. $\sqrt[3]{x^2} \cdot \sqrt[2]{x}$ **b.** $a^{1/2} \cdot a^{1/3}$ **c.** $\dfrac{x^{2/3}}{x^{1/2}}$ **d.** $\dfrac{\sqrt[5]{x^4}}{\sqrt[3]{x^2}}$

SOLUTION In parts a through d, common denominators are needed in order to add or subtract the exponents.

a. $\sqrt[3]{x^2} \cdot \sqrt[2]{x} = x^{2/3} \cdot x^{1/2} = x^{2/3 + 1/2} = x^{4/6 + 3/6} = x^{7/6}$

b. $a^{1/2} \cdot a^{1/3} = a^{1/2 + 1/3} = a^{3/6 + 2/6} = a^{5/6}$

c. $\dfrac{x^{2/3}}{x^{1/2}} = x^{2/3 - 1/2} = x^{4/6 - 3/6} = x^{1/6}$

d. $\dfrac{\sqrt[5]{x^4}}{\sqrt[3]{x^2}} = \dfrac{x^{4/5}}{x^{2/3}} = x^{4/5 - 2/3} = x^{12/15 - 10/15} = x^{2/15}$

It is not standard practice to write out "multiplication and division of radicals with like radicands" as a property because the resulting expressions contain fractional expressions and are awkward to memorize! See Exercises 55 and 56. The other three exponent properties may be translated to root form.

EXPONENT AND ROOT PROPERTIES, PART 2

> **Power of a Power and nth Root of an mth Root Properties**
> For positive base x and radicand x,
>
> To apply an exponent to a power expression, multiply the exponents:
>
> $$(x^a)^b = x^{a \cdot b} \qquad \text{Power of a power property}$$
>
> To apply a root to another root expression, multiply the indices:
>
> $$\sqrt[n]{\sqrt[m]{x}} = \sqrt[nm]{x}, \quad x > 0 \qquad \text{Root of a root property}$$

EXAMPLE 10   **Applying properties**   Simplify, where a, b, and y are any real numbers.

a. $(a^2)^{1/2}$ **b.** $(b^5)^{3/5}$ **c.** $\sqrt[3]{\sqrt[3]{x}}$, $x > 0$ **d.** $(y^6)^{3/2}$ **e.** $\sqrt[2]{\sqrt[3]{x}}$, $x > 0$

SOLUTION   We look at the denominator in the exponent to see whether absolute value is needed.

a. $(a^2)^{1/2} = a^{\frac{2}{1} \cdot \frac{1}{2}} = a^{2/2} = |a|$

The denominator is 2. The exponent $\frac{2}{2}$ contains an even root, so the absolute value symbol is needed.

b. $(b^5)^{3/5} = b^{\frac{5}{1} \cdot \frac{3}{5}} = b^3$

The denominator is 5. No absolute value symbol is needed on b, because the root $\left(\text{or } \frac{1}{5} \text{ exponent}\right)$ is odd.

c. $\sqrt[3]{\sqrt[3]{x}} = (x^{1/3})^{1/3} = x^{1/9}$ or $\sqrt[3 \cdot 3]{x} = \sqrt[9]{x}$

No absolute value symbol is needed, because $x > 0$.

d. $(y^6)^{3/2} = y^{\frac{6}{1} \cdot \frac{3}{2}} = |y^9|$

The denominator is 2. The exponent $\frac{3}{2}$ contains an even root, so the absolute value symbol is needed.

e. $\sqrt[2]{\sqrt[3]{x}} = (x^{1/3})^{1/2} = x^{1/6}$ or $\sqrt[2 \cdot 3]{x} = \sqrt[6]{x}$

No absolute value symbol is needed, because $x > 0$.   ▬

■ EXPONENT AND ROOT PROPERTIES, PART 3

Power of a Product, and Power of a Quotient, nth Root of a Product, and nth Root of a Quotient Properties

For positive bases, a rational exponent outside the parentheses applies to all parts of a product or quotient inside the parentheses:

$$(x \cdot y)^a = x^a \cdot y^a \qquad \text{Power of a product property}$$

$$\left(\frac{x}{y}\right)^a = \frac{x^a}{y^a} \quad \text{for } y \neq 0 \qquad \text{Power of a quotient property}$$

An index applies to all parts of a product or quotient in the radicand:

$$\sqrt[n]{x \cdot y} = \sqrt[n]{x} \cdot \sqrt[n]{y}, \quad x > 0, y > 0 \qquad \text{Root of a product}$$

$$\sqrt[n]{\frac{x}{y}} = \frac{\sqrt[n]{x}}{\sqrt[n]{y}}, \quad x > 0, y > 0 \qquad \text{Root of a quotient}$$

EXAMPLE 11   **Applying properties**   Simplify.

a. $a\sqrt{b} \cdot a\sqrt{b}$, $b > 0$

b. $\sqrt{x}(-\sqrt{x})$, $x > 0$

c. $\sqrt[4]{x} \cdot \sqrt[4]{x^3}$, $x > 0$

d. $\sqrt[3]{b}(\sqrt[3]{b})^2$

SOLUTION   In parts a to c, the variables b and x are positive, so $\sqrt[n]{b}$ and $\sqrt[n]{x}$ are defined in real numbers.

a. $a\sqrt{b} \cdot a\sqrt{b} = a \cdot a \cdot \sqrt{b^2} = a^2 b$

b. $\sqrt{x}(-\sqrt{x}) = -\sqrt{x^2} = -x$

c. Using rational exponents, we have

$$\sqrt[4]{x} \cdot \sqrt[4]{x^3} = x^{1/4} x^{3/4} = x^{1/4 + 3/4} = x$$

Using radicals yields

$$\sqrt[4]{x} \cdot \sqrt[4]{x^3} = \sqrt[4]{x \cdot x^3} = \sqrt[4]{x^4} = x$$

No absolute value symbol is needed, because $x > 0$.

d. Using rational exponents, we have

$$\sqrt[3]{b}(\sqrt[3]{b})^2 = b^{1/3} \cdot b^{2/3} = b^{1/3+2/3} = b$$

Using radicals yields

$$\sqrt[3]{b}(\sqrt[3]{b})^2 = \sqrt[3]{b} \cdot \sqrt[3]{b^2} = \sqrt[3]{b \cdot b^2} = \sqrt[3]{b^3} = b$$

No absolute value symbol is needed because of the odd root. ▬

▬ As suggested in the solution to Example 11, there are times when it is easier to work with radicals and not change to rational exponents. Here is a checklist for applying operations and simplifying radical expressions.

▬ SIMPLIFIED FORM FOR RADICAL EXPRESSIONS

A radical expression is simplified if

1. All possible roots have been taken.

2. The exponents in the radicand are smaller than the index.

3. There are no fractions under the radical sign.

4. There are no radicals in the denominator (more on this in Section 8.5).

In Example 12, we simplify radical expressions without changing to rational exponents.

EXAMPLE 12   **Simplifying radicals**   Simplify these expressions. Assume all variables represent positive numbers.

a. $\sqrt[4]{2} \cdot \sqrt[4]{8}$ **b.** $\dfrac{\sqrt[4]{32}}{\sqrt[4]{2}}$ **c.** $\sqrt[4]{x} \cdot \sqrt[4]{x^3}$ **d.** $\dfrac{\sqrt[4]{x^6}}{\sqrt[4]{x^2}}$ **e.** $\sqrt[4]{-125} \cdot \sqrt[4]{-5}$

SOLUTION   **a.** $\sqrt[4]{2} \cdot \sqrt[4]{8} = \sqrt[4]{2 \cdot 8} = \sqrt[4]{16} = 2$

b. $\dfrac{\sqrt[4]{32}}{\sqrt[4]{2}} = \sqrt[4]{\dfrac{32}{2}} = \sqrt[4]{16} = 2$

c. $\sqrt[4]{x} \cdot \sqrt[4]{x^3} = \sqrt[4]{x \cdot x^3} = \sqrt[4]{x^4} = x$

d. $\dfrac{\sqrt[4]{x^6}}{\sqrt[4]{x^2}} = \sqrt[4]{\dfrac{x^6}{x^2}} = \sqrt[4]{x^4} = x$

e. $\sqrt[4]{-125} \cdot \sqrt[4]{-5}$ cannot be simplified. The base is negative, so the properties do not apply. Multiplying the bases would give $+625$ with a fourth root of 5, a convenient but incorrect answer. ▬

When an exponent inside a radical is larger than the index, factor the expression so as to remove multiples of the factor. For example, if $x \geq 0$,

$$\sqrt{x^3} = \sqrt{x^2 \cdot x} = \sqrt{x^2} \cdot \sqrt{x} = x\sqrt{x}$$

In Example 13, we simplify without changing to rational exponents.

EXAMPLE 13 **Simplifying radicals** Simplify these expressions. Assume all variables represent positive numbers.

    a. $\sqrt{50x^7}$ **b.** $\sqrt[4]{a^7}$ **c.** $\sqrt[3]{-27x^4y^5z^6}$ **d.** $\sqrt[4]{a^4b^8c^{10}}$

SOLUTION   **a.** $\sqrt{50x^7} = \sqrt{2 \cdot 25 \cdot x^6 \cdot x} = 5x^3\sqrt{2x}$

    b. $\sqrt[4]{a^7} = \sqrt[4]{a^4 \cdot a^3} = a\sqrt[4]{a^3}$

    c. $\sqrt[3]{-27x^4y^5z^6} = \sqrt[3]{(-3)^3x^3 \cdot x \cdot y^3 \cdot y^2 \cdot z^6} = -3xyz^2\sqrt[3]{xy^2}$

The factors may be spread out within one radical sign, as in parts a, b, and c, or they may be placed in separate radical signs, as in part d.

    d. $\sqrt[4]{a^4b^8c^{10}} = \sqrt[4]{a^4}\,\sqrt[4]{b^8}\,\sqrt[4]{c^8}\,\sqrt[4]{c^2} = ab^2c^2\sqrt{c}$

The fourth root of c^2 simplifies to the square root of c.

Think about it: Although some older graphing calculators do not have an nth root key, the equivalence between $b^{1/n}$ and $\sqrt[n]{b}$, when $\sqrt[n]{b}$ is defined, makes it possible to evaluate these roots on such a calculator. Simplify each expression with a calculator. Use the exponent key $\boxed{\wedge}$ with a rational exponent. Describe the result. Indicate any expressions that are not real; explain why.

    a. $\sqrt[2]{9}$ **b.** $\sqrt[4]{16}$ **c.** $(\sqrt[3]{-5})^3$ **d.** $(\sqrt[4]{9})^4$ **e.** $(\sqrt{-4})^2$

ANSWER BOX

Warm-up: **1. a.** 5 **b.** 50 **c.** 0.5 **d.** 0.05 **2. a.** 12 **b.** 1200 **c.** 0.12 **d.** 120 **3. a.** $x\sqrt{y}$ **b.** $y\sqrt{x}$ **c.** xy **d.** $(1/y) \cdot \sqrt{x}$ **4.** $\frac{1}{2}$ is same as square root. **a.** 6 **b.** 3 **c.** 16 **d.** not a real number ($3i$ in the complex number system) **Think about it: a.** 9 $\boxed{\wedge}$ (1/2); 3; the principal square root **b.** 16 $\boxed{\wedge}$ (1/4); 2; the principal fourth root **c.** (-5) $\boxed{\wedge}$ (3/3); -5; odd roots of negative numbers are real numbers. **d.** 9 $\boxed{\wedge}$ (4/4); 9; the fourth power of the principal fourth root **e.** not a real number; $\sqrt{-4}$ is undefined in the real numbers.

8.4   Exercises

1. a. Write 256 in prime factors. $2 \cdot 2 \cdot 2 \cdot 2 \cdot 2 \cdot 2 \cdot 2 \cdot 2$

Then evaluate the expressions in parts b to g.

  b. $\sqrt[8]{256}$ 2   **c.** $\sqrt[4]{256}$ 4   **d.** $\sqrt{256}$ 16

  e. $256^{5/8}$ 32   **f.** $256^{0.75}$ 64   **g.** $256^{7/8}$ 128

2. a. Write 1024 in prime factors.
   $2 \cdot 2 \cdot 2 \cdot 2 \cdot 2 \cdot 2 \cdot 2 \cdot 2 \cdot 2 \cdot 2$
Then evaluate the expressions in parts b to g.

  b. $\sqrt[5]{1024}$ 4   **c.** $\sqrt[10]{1024}$ 2   **d.** $\sqrt{1024}$ 32

  e. $1024^{3/5}$ 64   **f.** $1024^{3/10}$ 8   **g.** $1024^{0.7}$ 128

Evaluate, if possible, the expressions in Exercises 3 to 14.

3. a. $\sqrt[3]{64}$ 4   **b.** $\sqrt[5]{32}$ 2

4. a. $\sqrt[3]{-125}$ -5 **b.** $\sqrt[7]{-128}$ -2

5. a. $\sqrt[3]{-27}$ -3 **b.** $\sqrt[4]{625}$ 5

6. a. $\sqrt[5]{-243}$ -3 **b.** $\sqrt[4]{81}$ 3

7. a. $\sqrt{-4}$
   not a real number   **b.** $\sqrt[4]{-16}$
                            not a real number

8. a. $\sqrt[3]{-8}$ -2 **b.** $-\sqrt[3]{1000}$ -10

9. a. $-\sqrt[4]{10,000}$ -10 **b.** $\sqrt[3]{-1000}$ -10

10. a. $\sqrt[6]{-64}$
   not a real number   **b.** $\sqrt{-25}$
                           not a real number

11. a. $-\sqrt[3]{64}$ -4 **b.** $-\sqrt[3]{-1000}$ 10

12. a. $\sqrt[3]{-64}$ -4 **b.** $-\sqrt[3]{27}$ -3

13. a. $-\sqrt[4]{256}$ -4 **b.** $\sqrt[4]{-625}$ not a real number

14. a. $-\sqrt[6]{64}$ -2 **b.** $\sqrt[4]{-10,000}$
                                  not a real number

In Exercises 15 to 18, write each rational expression as a root. Write each root as an expression with a rational exponent. Assume all expressions are defined.

15. a. $x^{1/6}$ $\sqrt[6]{x}$ **b.** $\sqrt[3]{x}$ $x^{1/3}$ **c.** $x^{0.5}$ $\sqrt{x}$

16. a. $x^{0.125}$ $\sqrt[8]{x}$ **b.** $x^{1/7}$ $\sqrt[7]{x}$ **c.** $\sqrt[5]{x}$ $x^{1/5}$

17. a. $x^{3/2}$ $(\sqrt{x})^3$ or $\sqrt{x^3}$ **b.** $\sqrt[4]{x^3}$ $x^{3/4}$ **c.** $x^{0.8}$ $(\sqrt[5]{x})^4$ or $\sqrt[5]{x^4}$

18. a. $\sqrt[4]{x^5}$ $x^{5/4}$ **b.** $x^{1.5}$ $(\sqrt{x})^3$ or $\sqrt{x^3}$ **c.** $x^{5/6}$ $(\sqrt[6]{x})^5$ or $\sqrt[6]{x^5}$

In Exercises 19 to 22, change to rational exponents and simplify.

19. a. $(\sqrt[3]{-8})^5$ $(-8)^{5/3} = -32$ **b.** $\sqrt[3]{27^2}$ $27^{2/3} = 9$

20. a. $\sqrt[3]{125^2}$ $125^{2/3} = 25$ **b.** $(\sqrt[3]{64})^2$ $64^{2/3} = 16$

21. a. $\sqrt[4]{16^2}$ $16^{1/2} = 4$ **b.** $(\sqrt[4]{-64})^2$ not a real number

22. a. $(\sqrt[3]{-27})^2$ $(-27)^{2/3} = 9$ **b.** $(\sqrt[4]{-16})^2$ not a real number

Simplify in Exercises 23 and 24. Assume the variables represent any real number.

23. a. $\sqrt{y^2}$ $|y|$ **b.** $\sqrt[4]{z^8}$ z^2 **c.** $\sqrt{x^6}$ $|x^3|$

24. a. $\sqrt[4]{y^4}$ $|y|$ **b.** $\sqrt{z^6}$ $|z^3|$ **c.** $\sqrt{x^2}$ $|x|$

In Exercises 25 to 30, assume $x > 0$.

25. a. $\sqrt[4]{x} \cdot \sqrt[3]{x^2}$ $x^{11/12}$ **b.** $x^{2/3} \cdot x^{1/5}$ $x^{13/15}$

26. a. $\sqrt[3]{x} \cdot \sqrt{x^3}$ $x^{11/6}$ **b.** $x^{4/3} \cdot x^3$ $x^{13/3}$

27. a. $\dfrac{x^{5/8}}{x^{1/2}}$ $x^{1/8}$ **b.** $\sqrt[8]{x^3} \cdot \sqrt{x^4}$ $x^{19/8}$

28. a. $\dfrac{x^{3/4}}{x^{3/5}}$ $x^{3/20}$ **b.** $\sqrt[5]{x^2} \cdot \sqrt[3]{x}$ $x^{11/15}$

29. a. $\dfrac{\sqrt[4]{x^5}}{\sqrt[3]{x^2}}$ $x^{7/12}$ **b.** $\dfrac{x^{4/5}}{x^{2/3}}$ $x^{2/15}$

30. a. $\dfrac{\sqrt[4]{x^3}}{\sqrt[3]{x^2}}$ $x^{1/12}$ **b.** $\dfrac{x^{7/5}}{x^{1/3}}$ $x^{16/15}$

31. Suppose $m > n$ in a^n/a^m. Explain how to write a simplified answer with a positive exponent. $\dfrac{1}{a^{m-n}}$

32. Simplify $x^{1/2}/x^{3/4}$, leaving a positive exponent. $1/x^{1/4}$

Simplify the expressions in Exercises 33 to 36. Assume the variables represent any real number. State any restrictions.

33. a. $(x^4)^{1/2}$ x^2 **b.** $(x^{2/3})^3$ x^2 **c.** $(y^6)^{1/6}$ $|y|$

34. a. $(y^6)^{1/3}$ y^2 **b.** $(y^{3/4})^4$ $y^3, y \geq 0$ **c.** $(x^4)^{1/4}$ $|x|$

35. a. $\sqrt[2]{\sqrt[4]{x}}$ $\sqrt[8]{x}, x \geq 0$ **b.** $\sqrt[4]{\sqrt[3]{x^5}}$ $\sqrt[12]{x^5}, x \geq 0$

36. a. $\sqrt[3]{\sqrt[2]{x}}$ $\sqrt[6]{x}, x \geq 0$ **b.** $\sqrt[5]{\sqrt[2]{x^3}}$ $\sqrt[10]{x^3}, x \geq 0$

Simplify the expressions in Exercises 37 and 38, and indicate any restrictions on the inputs a and b if the outputs are to be real numbers.

37. a. $\sqrt{a} \sqrt{a^3}$ $a^2, a \geq 0$ **b.** $\sqrt[3]{b} \sqrt[3]{b^2}$ b **c.** $\dfrac{\sqrt{x^5}}{\sqrt{x^3}}$ $x, x > 0$

38. a. $\sqrt[3]{b^4} \sqrt[3]{b^2}$ b^2 **b.** $\sqrt[2]{a^3} \sqrt[2]{a^3}$ $a^3, a \geq 0$ **c.** $\dfrac{\sqrt[3]{x^5}}{\sqrt[3]{x^2}}$ $x, x > 0$

Simplify the expressions in Exercises 39 and 40 using the product and quotient properties. Assume the variables are positive.

39. a. $\sqrt{100n}$ $10\sqrt{n}$ **b.** $\sqrt{0.01n}$ $0.1\sqrt{n}$ **c.** $\sqrt{\dfrac{n}{100}}$ $\dfrac{\sqrt{n}}{10}$

d. $\sqrt[3]{\dfrac{x}{1000}}$ $\dfrac{\sqrt[3]{x}}{10}$ **e.** $\sqrt[3]{8n}$ $2\sqrt[3]{n}$

40. a. $\sqrt{10,000n}$ $100\sqrt{n}$ **b.** $\sqrt{0.0001n}$ $0.01\sqrt{n}$ **c.** $\sqrt{\dfrac{n}{10,000}}$ $\dfrac{\sqrt{n}}{100}$

d. $\sqrt[3]{1000n}$ $10\sqrt[3]{n}$ **e.** $\sqrt[3]{27x}$ $3\sqrt[3]{x}$

Simplify the expressions in Exercises 41 to 44.

41. a. $\sqrt{2} \sqrt{18}$ 6   **b.** $-\sqrt{3} \sqrt{27}$ -9

c. $\sqrt{2} \sqrt{8}$ 4   **d.** $\sqrt{\sqrt[3]{64}}$ 2

42. a. $-\sqrt{12}\sqrt{3}$ -6 **b.** $\sqrt{32}\sqrt{2}$ 8

c. $\sqrt{24}\sqrt{6}$ 12   **d.** $\sqrt{\sqrt[3]{64}}$ 2

43. a. $\dfrac{\sqrt[4]{243}}{\sqrt[4]{3}}$ 3   **b.** $\sqrt[2]{\dfrac{27}{3}}$ 3

c. $\sqrt[3]{16} \cdot \sqrt[3]{4}$ 4   **d.** $\dfrac{\sqrt[4]{64}}{\sqrt[4]{4}}$ 2

44. a. $\sqrt[3]{9} \cdot \sqrt[3]{3}$ 3   **b.** $\sqrt[3]{\dfrac{81}{3}}$ 3

c. $\dfrac{\sqrt[4]{1024}}{\sqrt[4]{4}}$ 4   **d.** $\dfrac{\sqrt[5]{64}}{\sqrt[5]{2}}$ 2

In Exercises 45 to 50, change to exponential notation where appropriate. Simplify the expressions. Assume $x > 0$, $y > 0$, $a^{-1} = 1/a$.

45. a. $x^{3/4}x^{1/3}$ $x^{13/12}$ **b.** $x^{3/4}x^{2/3}$ $x^{17/12}$

c. $x^{-1/3}$ $\dfrac{1}{x^{1/3}}$ **d.** $\sqrt[3]{x^2}\sqrt[4]{x}$ $x^{2/3}x^{1/4} = x^{11/12}$

46. a. $\left(x^{\frac{3}{4}}\right)^{\frac{1}{3}}$ $x^{1/4}$ **b.** $x^{2/3}x^{1/2}$ $x^{7/6}$

c. $x^{-3/4}$ $\dfrac{1}{x^{3/4}}$ **d.** $\sqrt[4]{x^3}\sqrt[2]{x^3}$ $x^{3/4}x^{3/2} = x^{9/4}$

47. a. $\sqrt[3]{x}\sqrt[2]{x^3}$ $x^{1/3}x^{3/2} = x^{11/6}$ **b.** $\dfrac{x^{3/4}}{x^{1/3}}$ $x^{5/12}$

c. $\left(x^{\frac{3}{4}}\right)^{\frac{2}{3}}$ $x^{1/2}$ **d.** $\sqrt[3]{\sqrt[2]{x^5}}$ $x^{5/6}$

48. a. $\sqrt[4]{x}\sqrt[2]{x^5}$ $x^{1/4}x^{5/2} = x^{11/4}$ **b.** $\dfrac{x^{2/3}}{x^{3/5}}$ $x^{1/15}$

c. $\left(x^{\frac{2}{5}}\right)^{\frac{3}{2}}$ $x^{3/5}$ **d.** $\sqrt[4]{\sqrt[2]{x^3}}$ $x^{3/8}$

49. a. $\sqrt[2]{\sqrt[3]{x^2}}$ $(x^{\frac{2}{3}})^{\frac{1}{2}} = x^{1/3}$ **b.** $\left(\dfrac{16x^4}{y^8}\right)^{\frac{3}{4}}$ $\dfrac{8x^3}{y^6}$

c. $\dfrac{x^{-1/2}}{x^{3/2}}$ $\dfrac{1}{x^2}$ **d.** $\dfrac{x^{-2/3}}{x}$ $\dfrac{1}{x^{5/3}}$

50. a. $\sqrt[3]{\sqrt[2]{x}}$ $(x^{\frac{1}{2}})^{\frac{1}{3}} = x^{1/6}$ **b.** $\left(\dfrac{25x^5}{xy^4}\right)^{\frac{1}{2}}$ $\dfrac{5x^2}{y^2}$

c. $\dfrac{x^{2/3}}{x^{-1/3}}$ x **d.** $\dfrac{x^{-3/4}}{x^{1/2}}$ $\dfrac{1}{x^{5/4}}$

Simplify Exercises 51 and 52 with properties of radicals. Assume all variables represent positive numbers.

51. **a.** $\sqrt[3]{16}$ $2\sqrt[3]{2}$ **b.** $\sqrt{75a^5}$ **c.** $\sqrt[4]{32x^9}$ $2x^2\sqrt[4]{2x}$
    $5a^2\sqrt{3a}$

   d. $\sqrt[5]{-32a^4b^5c^6}$ **e.** $\sqrt[3]{x^4y^8z^9}$ $xy^2z^3\sqrt[3]{xy^2}$
    $-2bc\sqrt[5]{a^4c}$

52. **a.** $\sqrt[3]{81}$ $3\sqrt[3]{3}$ **b.** $\sqrt{18x^3}$ **c.** $\sqrt[5]{64b^{12}}$
    $3x\sqrt{2x}$ $2b^2\sqrt[5]{2b^2}$

   d. $\sqrt[4]{16x^3y^4z^5}$ **e.** $\sqrt[3]{x^4y^6z^{10}}$ $xy^2z^3\sqrt[3]{xz}$
    $2yz\sqrt[4]{x^3z}$

53. Change $\sqrt[p]{x} \cdot \sqrt[q]{x}$ into rational exponent form and find a formula for the product. $x^{(q-p)/pq}$

54. Change $\sqrt[p]{x}/\sqrt[q]{x}$ into rational exponent form and find a formula for the quotient. $x^{(p-q)/pq}$

55. Explain the difference between the radicands for the principal nth root and the odd nth root.   The former must be positive; the latter may be negative.

56. Explain what a principal nth root indicates about the index.   It is even.

57. In the definition for $x^{1/n}$, explain why we include the phrase "If $\sqrt[n]{x}$ is defined."   Even root of a negative number is undefined.

58. Explain why the phrase "principal nth root of x is the positive number" is included in definitions of even roots but not of odd roots.   Odd roots can be negative: $\sqrt[3]{-8} = -2$.

In Exercises 59 to 62, tell whether the statement is true or false.

59. A real number raised to an even power can yield a negative number.   false

60. A negative real number raised to an odd power can yield a negative number.   true

61. The principal root is always positive.   true

62. A negative rational exponent is undefined.   false

Error Analysis

63. Does the expression $\sqrt[4]{(-x)}$, $x < 0$ give a real number? Why or why not?   yes; $-x > 0$ if $x < 0$.

64. A student tries to use the product property $\sqrt{a \cdot b} = \sqrt{a} \cdot \sqrt{b}$ to take the square root of b^2 in $\sqrt{b^2 - 4ac}$. Explain why the student is wrong.   Expression contains subtraction, not multiplication.

65. Explain why $5 \cdot 5^n \neq 25^n$. $5 \cdot 5^n = 5^1 \cdot 5^n = 5^{n+1}$. Add exponents when bases are same.

66. Under what conditions is no absolute value symbol needed on x in $\sqrt{x^2} = x$? $x \geq 0$

67. Why is no absolute value symbol needed on x^2 in $\sqrt{x^4} = x^2$? x^2 is always positive.

Projects

68. The Generalized Distributive Property The power properties—such as $(a \cdot b)^n = a^n \cdot b^n$ and its corresponding radical property $\sqrt{ab} = \sqrt{a}\sqrt{b}$—look somewhat like the distributive property of multiplication over addition, $n(a + b) = na + nb$. The power properties do indeed illustrate a generalized distributive property.

Match the descriptions of the distributive property in parts a to h with one of the following expressions:

$$(b - c) \div a \qquad a(b + c) \qquad \left(\frac{b}{c}\right)^a$$

$$a(b - c) \qquad \sqrt[a]{\frac{b}{c}} \qquad (b + c) \div a \qquad \sqrt[a]{bc}$$

$$(bc)^a$$

Show the distribution from each expression. Each statement applies only if the operations are defined, so assume all the usual conditions on the variables.

a. Multiplication distributes over addition.
   $a(b + c) = ab + ac$
b. Multiplication distributes over subtraction.
   $a(b - c) = ab - ac$
c. Division on the right distributes over addition.
   $(b + c) \div a = b/a + c/a$
d. Division on the right distributes over subtraction.
   $(b - c) \div a = b/a - c/a$
e. Exponents distribute over multiplication.
   $(bc)^a = b^a c^a$
f. Exponents distribute over division. $(b/c)^a = b^a/c^a$

g. Roots distribute over multiplication. $\sqrt[a]{bc} = \sqrt[a]{b}\sqrt[a]{c}$

h. Roots distribute over division. $\sqrt[a]{b/c} = \sqrt[a]{b}/\sqrt[a]{c}$

Answer the questions in parts i to p. For parts i to l, give an example showing why or why not.

i. Can addition distribute over addition or subtraction?
   No; $3 + (4 + 5) \neq 3 + 4 + 5$.
j. Can multiplication distribute over multiplication or division?   No; $3(4 \cdot 5) \neq 3 \cdot 4 \cdot 5$.

k. Can square root distribute over addition or subtraction?   No; $\sqrt{x + 2}$ cannot be simplified.

l. Can exponents distribute over addition or subtraction?   No; $(x + 2)^2$ is not $x^2 + 4$.

m. Group the following by pairs in the sequence in which they appear in the order of operations: addition, exponents, multiplication, division, subtraction, roots   exponents and roots, multiplication and division, addition and subtraction

n. Can an operation distribute over another operation at the same level in the order of operations?   no

o. Can an operation distribute over an operation one level lower in the order of operations?   yes

p. Can an operation distribute over an operation two levels lower in the order of operations?   no

69. Water Supply The table shows the population served by a water supply pipe of the indicated inner diameter. The estimates are based on a supply of 60 gallons of water per day per person. An assumption is made that there is a drop of 50 feet per mile between the water reservoir and the city pipes. The data are from a 1909 engineer's handbook.

a. Graph the data. What shape does the graph have?
   See Additional Answers; smooth curve
b. Use the statistical features on a graphing calculator to fit a curve with the power regression option.
   $y = 19.0x^{2.49}$
c. How does the consumption of water per person per day in this 1909 table compare with your current water consumption? Find your water usage from your household utility bill, if possible.
   A utility in Oregon estimates monthly use per household at 6000 gallons.

Diameter of Pipe (inches)	Population Served
6	1647
10	5908
14	13,706
18	25,677
22	42,433
26	64,447
30	91,580
34	125,840
40	188,320
48	297,600
60	511,200
72	800,000
80	1,064,000

From John C. Trautwine, *Trautwine's Engineer's Pocket-Book*, John Wiley & Sons, 1909, p. 653.

8 Mid-Chapter Test

Write the expressions in Exercises 1 and 2 another way.

1. $x^{1/2}, x \geq 0$ $\sqrt{x}, x \geq 0$ **2.** $\dfrac{1}{x}, x \neq 0$ $x^{-1}, x \neq 0$

Simplify the expressions in Exercises 3 to 6. Assume no variables are zero.

3. a. $\dfrac{a^{-1}b^2}{c^0} \quad \dfrac{b^2}{a}$ **b.** $\dfrac{a^2b^{-1}}{c^{-2}} \quad \dfrac{a^2c^2}{b}$

4. a. $\dfrac{b^8}{b^5} \quad b^3$ **b.** $\dfrac{x^4y^5}{x^{-1}y^9} \quad \dfrac{x^5}{y^4}$

5. a. $a \cdot a^x \quad a^{x+1}$ **b.** $(x^2)^n \quad x^{2n}$

6. a. $\left(\dfrac{2x}{3}\right)^{-2} \quad \dfrac{9}{4x^2}$ **b.** $\dfrac{x}{y^{-2}} \quad xy^2$

7. a. Change 4.3×10^{-3} to decimal notation.   0.0043

   b. Change 0.000 123 to scientific notation.
   1.23×10^{-4}

   c. Simplify $\dfrac{3.6 \times 10^{-2}}{1.2 \times 10^3}$. $3 \times 10^{-5} = 0.00003$

   d. How many significant digits are in 0.060?   two

   e. Describe two examples of zero as a placeholder.

8. Solve each exponential equation for x.

   a. $4^x = 32$ $x = 2.5$ **b.** $9^x = 27$ $x = 1.5$

   c. $\left(\frac{1}{4}\right)^x = \frac{1}{8}$ $x = 1.5$ **d.** $\pi^x = 1$ $x = 0$

9. Find the amount of money in a savings account if $1000 is deposited at 7% interest compounded weekly (52 times a year) for

   a. 5 years $1418.73   **b.** $2\frac{3}{4}$ years $1212.11

10. Why is a 2% wage increase twice a year better than a 4% wage increase once a year?   2% twice a year = 4.04% annually

11. Simplify, and indicate any restrictions on the inputs n needed in order to have real-number answers.

   a. $\sqrt{4n}$ $2\sqrt{n}, n \geq 0$ **b.** $\sqrt{40n^2}$ $2|n|\sqrt{10}$

   c. $\sqrt{400n}$ $20\sqrt{n}, n \geq 0$ **d.** $\sqrt{4000n}$ $20\sqrt{10n}$

In Exercises 12 and 13, change each expression to radical notation. Simplify.

12. a. $8^{2/3}$ $(\sqrt[3]{8})^2 = 4$ **b.** $125^{1/3}$ $\sqrt[3]{125} = 5$

13. a. $16^{3/4}$ $(\sqrt[4]{16})^3 = 8$ **b.** $(a^3)^{2/3}$ $(\sqrt[3]{a^3})^2 = a^2$

In Exercises 14 to 18, simplify for $x > 0$ and $y > 0$.

14. $\sqrt[3]{x} \cdot \sqrt[4]{x}$ $x^{7/12}$

15. a. $\dfrac{x^{0.75}}{x^{2/3}}$ $x^{1/12}$ **b.** $\sqrt[3]{\sqrt[3]{x^6}}$ $x^{2/3}$

16. a. $\sqrt[4]{-10,000x^4}$ not a real number   **b.** $-\sqrt[4]{\dfrac{81x^8}{y^{-4}}}$ $-3x^2y$

17. a. $\sqrt[3]{-64}$ -4 **b.** $\sqrt[4]{16y^4}$ $2y$

18. a. $\sqrt{9x^5y^2}$ $3x^2y\sqrt{x}$ **b.** $\sqrt[3]{24x^5y^6}$ $2xy^2\sqrt[3]{3x^2}$

19. In part b of Exercise 17, is the solution defined if y is negative?   yes, because the 4th power is positive

20. What is the answer to part b of Exercise 17 if y is any real number? $2|y|$

8.5 More Operations with Radicals

Objectives

■ Add and subtract radicals.

■ Multiply two- and three-term radical expressions.

■ Identify the conjugate of a radical expression.

■ Rationalize the numerator or denominator of a radical expression.

WARM-UP

1. Solve for x using the quadratic formula: $x^2 - 2x - 2 = 0$. Leave the answer in radical notation. $x = 1 \pm \sqrt{3}$

2. Factor $x^2 - r^2$. $(x - r)(x + r)$

3. Simplify $\dfrac{r - x}{x - r}$. -1

THIS SECTION DEVELOPS skills in working with radical expressions. Although most sections in this text note applications outside mathematics, sometimes the most appropriate applications of a mathematics skill are within mathematics itself. Skills in this section are used to check solutions to equations, justify patterns, and prove geometric results.

The Quadratic Formula Revisited

One source of radical expressions is solving quadratic equations, $ax^2 + bx + c = 0$, with the quadratic formula,

$$x = \frac{-b \pm \sqrt{b^2 - 4ac}}{2a}$$

EXAMPLE 1 **Reviewing quadratic equation solutions** Solve $x^2 - 2x - 2 = 0$. Use the quadratic formula, and leave the answers in radical notation.

SOLUTION There are two real-number solutions to $x^2 - 2x - 2 = 0$:

$$x = \frac{-b \pm \sqrt{b^2 - 4ac}}{2a} = \frac{-(-2) \pm \sqrt{4 - 4(1)(-2)}}{2(1)}$$

$$= \frac{2 \pm \sqrt{12}}{2} = \frac{2 \pm 2\sqrt{3}}{2} = 1 \pm \sqrt{3}$$

To check our solutions to Example 1, we would substitute the expression $1 + \sqrt{3}$ into $x^2 - 2x - 2 = 0$:

$$(1 + \sqrt{3})^2 - 2(1 + \sqrt{3}) - 2 \overset{?}{=} 0$$

We would then square $1 + \sqrt{3}$, multiply $-2(1 + \sqrt{3})$, add like terms, and verify that the sum is zero. A similar approach is needed to check $1 - \sqrt{3}$. The next few examples summarize the operations with radical expressions that permit us to carry out these checks. You will be asked to check the solutions to Example 1 in the exercise set.

Addition and Subtraction with Radical Expressions

Just as we can add and subtract like terms in polynomials, we can add and subtract similar radicals. **Similar radicals** have *identical indices and radicands.*

EXAMPLE 2 Adding and subtracting radical expressions Add or subtract, if possible.

a. $2\sqrt{3} + 3\sqrt{3}$ **b.** $\sqrt{3} + \sqrt{2}$ **c.** $5\sqrt{3} - 2\sqrt{3}$

d. $8\sqrt[3]{5} - 6\sqrt[3]{5}$ **e.** $ab\sqrt{c} + 2ab\sqrt{c}$ **f.** $\sqrt[3]{x^2y} + \sqrt[3]{xy}$

SOLUTION **a.** The terms both contain $\sqrt{3}$ and may be added:

$$2\sqrt{3} + 3\sqrt{3} = (2 + 3)\sqrt{3} = 5\sqrt{3}$$

Using the distributive property, we factor $\sqrt{3}$ from both terms. The radical may be placed after the parentheses to prevent confusion as to whether the expression in the parentheses is under the radical.

b. $\sqrt{3} + \sqrt{2}$ cannot be added because the radicands are different.

c. The radicals contain identical radicands, so

$$5\sqrt{3} - 2\sqrt{3} = (5 - 2)\sqrt{3} = 3\sqrt{3}$$

d. $8\sqrt[3]{5} - 6\sqrt[3]{5} = (8 - 6)\sqrt[3]{5} = 2\sqrt[3]{5}$

e. The expressions have identical variable factors:

$$ab\sqrt{c} + 2ab\sqrt{c} = (1 + 2)ab\sqrt{c} = 3ab\sqrt{c}$$

f. In $\sqrt[3]{x^2y} + \sqrt[3]{xy}$, the x-variables have different exponents; the radicals cannot be added.

As noted in Section 8.4, *a square root is simplified if it contains no perfect square factor. Other radicals with index n are simplified if they contain no perfect nth power factor.*

In Example 3, we factor and simplify the radical expressions. In doing so, we obtain similar radicals.

EXAMPLE 3 Adding and subtracting radical expressions Add or subtract by changing to similar radicals.

a. $\sqrt{27} + \sqrt{3}$ **b.** $\sqrt{4x} + \sqrt{x}, x \geq 0$

c. $a\sqrt{b} + a\sqrt{9b}, b \geq 0$ **d.** $x\sqrt[3]{2} - x\sqrt[3]{16}$

SOLUTION **a.** $\sqrt{27} = \sqrt{9 \cdot 3} = 3\sqrt{3}$, so

$$\sqrt{27} + \sqrt{3} = 3\sqrt{3} + 1\sqrt{3} = 4\sqrt{3}$$

b. $\sqrt{4x} = \sqrt{4 \cdot x} = 2\sqrt{x}$, so

$$\sqrt{4x} + \sqrt{x} = 2\sqrt{x} + 1\sqrt{x} = 3\sqrt{x}, x \geq 0$$

c. $a\sqrt{9b} = a\sqrt{9}\sqrt{b} = 3a\sqrt{b}$, so

$$a\sqrt{b} + a\sqrt{9b} = a\sqrt{b} + 3a\sqrt{b} = 4a\sqrt{b}, b \geq 0$$

d. $x\sqrt[3]{16} = x\sqrt[3]{8 \cdot 2} = x\sqrt[3]{8} \cdot \sqrt[3]{2} = 2x\sqrt[3]{2}$, so

$$x\sqrt[3]{2} - x\sqrt[3]{16} = x\sqrt[3]{2} - 2x\sqrt[3]{2} = -x\sqrt[3]{2}$$

Multiplication of Expressions with Two or More Terms

Multiplying radical expressions is similar to multiplying polynomials. We use the distributive property and addition of like terms to find the product of two or more terms.

EXAMPLE 4   **Multiplying radical expressions**   Multiply these expressions.

a. $4(-2 + \sqrt{2})$ **b.** $(2 - \sqrt{3})(2 - \sqrt{3})$ **c.** $(-2 + \sqrt{2})^2$

SOLUTION   **a.** We apply the distributive property:

$$4(-2 + \sqrt{2}) = -8 + 4\sqrt{2}$$

b. Using a table, we have

Multiply	2	$-\sqrt{3}$
2	4	$-2\sqrt{3}$
$-\sqrt{3}$	$-2\sqrt{3}$	$(-\sqrt{3})^2$

The two terms $-2\sqrt{3}$ and $-2\sqrt{3}$ add to $-4\sqrt{3}$. Thus,

$$(2 - \sqrt{3})(2 - \sqrt{3}) = 4 - 4\sqrt{3} + 3 = 7 - 4\sqrt{3}$$

c. $(-2 + \sqrt{2})^2 = (-2 + \sqrt{2})(-2 + \sqrt{2}) = 4 - 2\sqrt{2} - 2\sqrt{2} + (\sqrt{2})^2$
$$= 4 - 4\sqrt{2} + 2 = 6 - 4\sqrt{2}$$

The answers in parts b and c of Example 4 both contain two terms. Products of radical expressions often simplify to fewer terms than do products of polynomials. Such simplification also may occur when variables appear in the expression, as shown in part b of Example 5.

EXAMPLE 5   **Multiplying radical expressions**   Multiply and simplify these expressions.

a. $(a + \sqrt{b})(a + \sqrt{b}), b \geq 0$ **b.** $(x - \sqrt[3]{2})(x^2 + x\sqrt[3]{2} + (\sqrt[3]{2})^2)$

SOLUTION   **a.** $(a + \sqrt{b})(a + \sqrt{b}) = a^2 + a\sqrt{b} + a\sqrt{b} + (\sqrt{b})^2$
$$= a^2 + 2a\sqrt{b} + b, b \geq 0$$

b. Using a table for $(x - \sqrt[3]{2})(x^2 + x\sqrt[3]{2} + (\sqrt[3]{2})^2)$, we have

Multiply	x^2	$+x\sqrt[3]{2}$	$(\sqrt[3]{2})^2$
x	x^3	$+x^2\sqrt[3]{2}$	$x(\sqrt[3]{2})^2$
$-\sqrt[3]{2}$	$-x^2\sqrt[3]{2}$	$-x(\sqrt[3]{2})^2$	$-(\sqrt[3]{2})^3$

Four terms in the table add to zero. Thus,

$$(x - \sqrt[3]{2})(x^2 + x\sqrt[3]{2} + (\sqrt[3]{2})^2) = x^3 - (\sqrt[3]{2})^3 = x^3 - 2$$

CONJUGATES   Section 6.2 introduced the term *complex conjugates* to describe $a + bi$ and $a - bi$. The term **real-number conjugates** describes *expressions written* $a + b$ and $a - b$. Real-number conjugates include numbers such as $a + \sqrt{b}$ and $a - \sqrt{b}$ or $\sqrt{a} + \sqrt{b}$ and $\sqrt{a} - \sqrt{b}$. When the number contains one radical sign, the change in sign is on the term containing the radical sign.

EXAMPLE 6 **Finding conjugates** Write the conjugate for each of these real numbers.

a. $2 - \sqrt{5}$ **b.** $-2 + \sqrt{3}$ **c.** $0 - \sqrt{3}$ **d.** $\sqrt{2} + \sqrt{7}$

SOLUTION **a.** $2 + \sqrt{5}$ **b.** $-2 - \sqrt{3}$ **c.** $0 + \sqrt{3}$ **d.** $\sqrt{2} - \sqrt{7}$

Think about it 1: What is another possible conjugate for part d of Example 6? Why?

In Example 7, we observe an important result from multiplying conjugates.

EXAMPLE 7 **Multiplying conjugates** Simplify these expressions, and look for patterns in the answers.

a. $(3 + \sqrt{5})(3 - \sqrt{5})$

b. $(\sqrt{5} + \sqrt{2})(\sqrt{5} - \sqrt{2})$

c. $(\sqrt{x} - \sqrt{2})(\sqrt{x} + \sqrt{2}), x \geq 0$

d. $(\sqrt{x} + \sqrt{y})(\sqrt{x} - \sqrt{y}), x \geq 0, y \geq 0$

SOLUTION **a.** $(3 + \sqrt{5})(3 - \sqrt{5})$ may be multiplied mentally or with a table.

Multiply	3	$+\sqrt{5}$
3	9	$+3\sqrt{5}$
$-\sqrt{5}$	$-3\sqrt{5}$	$-\sqrt{25}$

The terms $-3\sqrt{5}$ and $+3\sqrt{5}$ add to zero, so

$$(3 + \sqrt{5})(3 - \sqrt{5}) = 9 - \sqrt{25} = 9 - 5 = 4$$

b. $(\sqrt{5} + \sqrt{2})(\sqrt{5} - \sqrt{2}) = \sqrt{25} - \sqrt{5}\sqrt{2} + \sqrt{5}\sqrt{2} - \sqrt{4} = 5 - 2 = 3$

c. $(\sqrt{x} - \sqrt{2})(\sqrt{x} + \sqrt{2}) = \sqrt{x^2} + \sqrt{2x} - \sqrt{2x} - \sqrt{4} = x - 2$

d. $(\sqrt{x} + \sqrt{y})(\sqrt{x} - \sqrt{y}) = \sqrt{x^2} - \sqrt{xy} + \sqrt{xy} - \sqrt{y^2} = x - y$

None of the answers contain a radical sign. The number expressions simplify to a single rational number.

Think about it 2: Why are no absolute value symbols needed in parts c and d of Example 7?

The products in Example 7 suggest a property of conjugates.

CONJUGATE PRODUCTS | The product of real-number conjugates is a rational number.

Rationalizing Numerators and Denominators

The fact that conjugates multiply to a rational number (hence eliminating the square roots or radical signs) means that conjugates can be used to eliminate a radical from the numerator or denominator of a fraction. **Rationalizing** the numerator or denominator is the name given to *the process of multiplying both the numerator and the denominator by a number that eliminates radicals from one of these positions in the fraction.*

EXAMPLE 8 Exploring rationalization

a. Multiply the numerator and denominator of $\dfrac{1}{2 + \sqrt{5}}$ by $2 - \sqrt{5}$.

b. Multiply the numerator and denominator of $\dfrac{3 - \sqrt{7}}{2}$ by $3 + \sqrt{7}$.

SOLUTION **a.** $\dfrac{1}{2 + \sqrt{5}} \cdot \dfrac{2 - \sqrt{5}}{2 - \sqrt{5}} = \dfrac{2 - \sqrt{5}}{4 - 2\sqrt{5} + 2\sqrt{5} - \sqrt{25}}$

$$= \dfrac{2 - \sqrt{5}}{4 - 5} = \dfrac{2 - \sqrt{5}}{-1} = -2 + \sqrt{5}$$

b. $\dfrac{3 - \sqrt{7}}{2} \cdot \dfrac{3 + \sqrt{7}}{3 + \sqrt{7}} = \dfrac{9 + 3\sqrt{7} - 3\sqrt{7} - \sqrt{49}}{2(3 + \sqrt{7})}$

$$= \dfrac{9 - 7}{2(3 + \sqrt{7})} = \dfrac{2}{2(3 + \sqrt{7})} = \dfrac{1}{3 + \sqrt{7}}$$

In part a of Example 8, we eliminated the radical from the denominator. In part b of Example 8, we eliminated the radical from the numerator.

RATIONALIZING

To rationalize a one-term expression, multiply by a number whose product with the radical makes an exact root.

To rationalize $\sqrt{x}$, multiply by $\sqrt{x}$.
To rationalize $\sqrt[3]{x}$, multiply by $\sqrt[3]{x^2}$.
To rationalize $\sqrt[3]{x^2}$, multiply by $\sqrt[3]{x}$.

To rationalize a two-term expression, multiply by the conjugate.

To rationalize $a + \sqrt{b}$, multiply by $a - \sqrt{b}$.
To rationalize $\sqrt{a} + \sqrt{b}$, multiply by either $\sqrt{a} - \sqrt{b}$ or $-\sqrt{a} + \sqrt{b}$.

EXAMPLE 9 Rationalizing denominators Rationalize the denominators in these fractions.

a. $\dfrac{2}{\sqrt{3}}$ **b.** $\dfrac{1}{\sqrt[3]{a}}$

c. $\dfrac{3}{4 - \sqrt{2}}$ **d.** $\dfrac{1}{a - \sqrt{b}}$, $a \neq \sqrt{b}$, a and b not both zero

SOLUTION

a. $\dfrac{2}{\sqrt{3}} \cdot \dfrac{\sqrt{3}}{\sqrt{3}} = \dfrac{2\sqrt{3}}{\sqrt{9}} = \dfrac{2\sqrt{3}}{3}$

b. $\dfrac{1}{\sqrt[3]{a}} \cdot \dfrac{\sqrt[3]{a^2}}{\sqrt[3]{a^3}} = \dfrac{\sqrt[3]{a^2}}{\sqrt[3]{a^3}} = \dfrac{\sqrt[3]{a^2}}{a}$

c. $\dfrac{3}{4 - \sqrt{2}} \cdot \dfrac{4 + \sqrt{2}}{4 + \sqrt{2}} = \dfrac{3(4 + \sqrt{2})}{16 + 4\sqrt{2} - 4\sqrt{2} - \sqrt{4}} = \dfrac{3(4 + \sqrt{2})}{14}$

Because the numerator and denominator contain no common factors, the fraction cannot be simplified.

d. $\dfrac{1}{a - \sqrt{b}} \cdot \dfrac{a + \sqrt{b}}{a + \sqrt{b}} = \dfrac{a + \sqrt{b}}{a^2 + a\sqrt{b} - a\sqrt{b} - \sqrt{b^2}} = \dfrac{a + \sqrt{b}}{a^2 - b}$, $a^2 \neq b$,

$b > 0$, a and b not both zero.

Rationalizing of the denominator is needed in order to perform a division with radical expressions. The division is considered complete when the denominator no longer contains a radical.

The following checklist may be helpful in determining whether a radical expression has been simplified.

SIMPLIFIED RADICAL EXPRESSIONS

1. For a radical with index 2, all perfect square factors have been removed.
2. For a radical with index 3, all perfect cube factors have been removed.
3. The index and exponents of factors in the radicand have no common factors.
4. Exponents in the radicand are smaller than the index.
5. Radicals have been eliminated from the denominator.
6. All possible operations have been performed.

We return to the navigation setting on the chapter opener page, reproduced in Figure 5. Our geometrical proof uses rationalization.

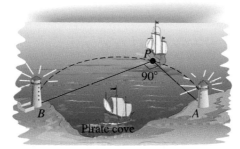

FIGURE 5

EXAMPLE 10 Applying rationalization in a proof   Prove that a right angle is formed by two line segments connecting a point (x, y) on a circle to the endpoints of the diameter of the circle.

a. Find the slopes of the two line segments.
b. Find y in terms of x and r, the radius of the circle.
c. Substitute for y, and show that the slope of one segment is the negative reciprocal of the slope of the other segment.

SOLUTION The line segments will form a right angle if they are perpendicular. We prove perpendicularity by showing that the slopes of the segments are negative reciprocals.

a. We place a circle of radius r with its center at the origin (Figure 6). The endpoints of the diameter lie at $A(r, 0)$ and $B(-r, 0)$. We place point P on the circle at (x, y).

$$\text{Slope } \overline{AP} = \frac{y_2 - y_1}{x_2 - x_1} = \frac{y - 0}{x - r} = \frac{y}{x - r}$$

$$\text{Slope } \overline{BP} = \frac{y_2 - y_1}{x_2 - x_1} = \frac{y - 0}{x - (-r)} = \frac{y}{x + r}$$

b. To write y in terms of x and r, we apply the Pythagorean theorem to the triangle formed by the radius to point P (Figure 7). The right triangle shown in the figure has base x, height y, and hypotenuse r. Thus,

$$x^2 + y^2 = r^2$$
$$y^2 = r^2 - x^2$$
$$y = \sqrt{r^2 - x^2}$$

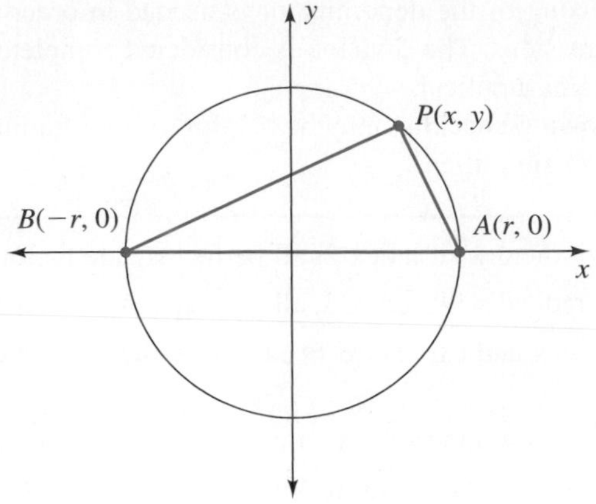

FIGURE 6

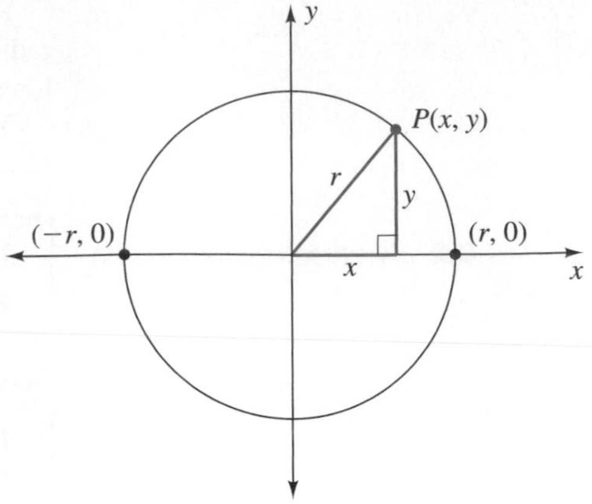

FIGURE 7

c. We start with the slope of AP and relate it to the slope of BP:

$$\text{Slope } \overline{AP} = \frac{y}{x - r} \qquad\qquad \text{Substitute } \sqrt{r^2 - x^2} \text{ for } y.$$

$$= \frac{\sqrt{r^2 - x^2}}{x - r} \qquad\qquad \text{Because } r^2 - x^2 \text{ factors, rationalize the numerator.}$$

$$= \frac{\sqrt{r^2 - x^2}}{x - r} \cdot \frac{\sqrt{r^2 - x^2}}{\sqrt{r^2 - x^2}} \qquad \text{Simplify.}$$

$$= \frac{r^2 - x^2}{(x - r)\sqrt{r^2 - x^2}} \qquad\qquad \text{Factor the numerator.}$$

$$= \frac{(r - x)(r + x)}{(x - r)\sqrt{r^2 - x^2}} \qquad\qquad \text{Simplify.}$$

$$= \frac{(-1)(r + x)}{\sqrt{r^2 - x^2}} \qquad\qquad \text{Replace } \sqrt{r^2 - x^2} \text{ by } y.$$

$$= \frac{(-1)(x + r)}{y} \qquad\qquad \text{This is the negative reciprocal of the slope of } BP.$$

Because the two segments are perpendicular, a right angle is formed by the segments connecting a point on the circle to the endpoints of the diameter. Lines PB and PA are the sight lines to the lighthouses that keep the ship on the arc of the circle. ▬

ANSWER BOX

Warm-up: **1.** $x = 1 \pm \sqrt{3}$ **2.** $(x - r)(x + r)$ **3.** -1 **Think about it 1:** If we had written $\sqrt{7} + \sqrt{2}$, then $\sqrt{7} - \sqrt{2}$ could be the conjugate. Either $\sqrt{7} - \sqrt{2}$ or $\sqrt{2} - \sqrt{7}$ will rationalize $\sqrt{2} + \sqrt{7}$. **Think about it 2:** $x \geq 0$ and $y \geq 0$ are stated. $\sqrt{x^2} = |x|$ when x is any real number. For $x \geq 0$, $\sqrt{x^2} = x$. The same applies to $\sqrt{y^2}$.

8.5 Exercises

In Exercises 1 to 24, add or subtract like terms. Assume all variables represent positive numbers.

1. $3\sqrt{2} - 4\sqrt{2} + 7\sqrt{2}$ $6\sqrt{2}$

2. $4\sqrt{5} + 5\sqrt{5} - 2\sqrt{5}$ $7\sqrt{5}$

3. a. $\sqrt{20} + \sqrt{45}$ $5\sqrt{5}$ **b.** $\sqrt{72} - \sqrt{18}$ $3\sqrt{2}$

4. a. $\sqrt{50} + \sqrt{32}$ $9\sqrt{2}$ **b.** $\sqrt{48} - \sqrt{12}$ $2\sqrt{3}$

5. a. $\sqrt{75} + \sqrt{27}$ $8\sqrt{3}$ **b.** $\sqrt{125} - \sqrt{80}$ $\sqrt{5}$

6. a. $\sqrt{98} + \sqrt{8}$ $9\sqrt{2}$ **b.** $\sqrt{320} - \sqrt{5}$ $7\sqrt{5}$

7. $\sqrt{9x} + \sqrt{16x} - \sqrt{x}$ $6\sqrt{x}$

8. $\sqrt{25x} + \sqrt{4x} + \sqrt{2.25x}$ $8.5\sqrt{x}$

9. $\sqrt{0.01x} + \sqrt{49x} - \sqrt{16x}$ $3.1\sqrt{x}$

10. $\sqrt{121x} - \sqrt{81x} + \sqrt{9x}$ $5\sqrt{x}$

11. $a\sqrt{b} + c\sqrt{d}$ not like terms

12. $a\sqrt{b} + c\sqrt{b}$ $(a + c)\sqrt{b}$

13. $\sqrt{ab} + 2\sqrt{ab}$ $3\sqrt{ab}$

14. $\sqrt{ab} + \sqrt{ac}$ not like terms

15. $3\sqrt[3]{x} - \sqrt[3]{x}$ $2\sqrt[3]{x}$

16. $5\sqrt[4]{x} - \sqrt[5]{x}$ not like terms

17. $3\sqrt[4]{x} - 2\sqrt[2]{x}$ not like terms

18. $5\sqrt[4]{x} - \sqrt[4]{x}$ $4\sqrt[4]{x}$

19. $\sqrt[3]{8x} + \sqrt[3]{27x}$ $5\sqrt[3]{x}$

20. $\sqrt[3]{64x} - \sqrt[3]{27x}$ $\sqrt[3]{x}$

21. $a\sqrt[4]{81ab} - \sqrt[4]{a^5b}$ $2a\sqrt[4]{ab}$

22. $b\sqrt[3]{125ab} + \sqrt[3]{ab^4}$ $6b\sqrt[3]{ab}$

23. $\sqrt[4]{16x^4y} + x\sqrt[4]{y}$ $3x\sqrt[4]{y}$

24. $x\sqrt[5]{32xy} + \sqrt[5]{x^6y}$ $3x\sqrt[5]{xy}$

In Exercises 25 to 28, multiply and simplify.

25. a. $(2 - \sqrt{3})(2 + \sqrt{3})$ 1

   b. $(5 - \sqrt{2})(5 - \sqrt{2})$ $27 - 10\sqrt{2}$

26. a. $(3 - \sqrt{2})(3 - \sqrt{3})$ $9 - 3\sqrt{2} - 3\sqrt{3} + \sqrt{6}$

   b. $(5 + \sqrt{2})(5 - \sqrt{2})$ 23

27. a. $(x - \sqrt{5})(x + \sqrt{5})$ $x^2 - 5$

   b. $(x - \sqrt{7})(x - \sqrt{7})$ $x^2 - 2\sqrt{7}x + 7$

28. a. $(x - \sqrt{3})(x + \sqrt{3})$ $x^2 - 3$

   b. $(x - \sqrt{3})(x - \sqrt{3})$ $x^2 - 2\sqrt{3}x + 3$

Multiply the expressions in Exercises 29 to 32. Assume $a \geq 0$, $b \geq 0$, and $x \geq 0$.

29. a. $(\sqrt{x} + 3)^2$ $x + 6\sqrt{x} + 9$

   b. $(\sqrt{a} - 3)^2$ $a - 6\sqrt{a} + 9$

30. a. $(2 - \sqrt{a})^2$ $4 - 4\sqrt{a} + a$

   b. $(\sqrt{a} - \sqrt{b})(\sqrt{a} + \sqrt{b})$ $a - b$

31. a. $(1 - \sqrt{a})^2$ $1 - 2\sqrt{a} + a$

   b. $(\sqrt{a} - \sqrt{b})(\sqrt{a} - \sqrt{b})$ $a - 2\sqrt{ab} + b$

32. a. $(\sqrt{x} - 2)^2$ $x - 4\sqrt{x} + 4$

   b. $(\sqrt{2a} - 1)^2$ $2a - 2\sqrt{2a} + 1$

33. Show that $x = 1 + \sqrt{3}$ satisfies $x^2 - 2x - 2 = 0$.

34. Show that $x = 1 - \sqrt{3}$ satisfies $x^2 - 2x - 2 = 0$.

35. Show that $x = -2 - \sqrt{2}$ satisfies $x^2 + 4x + 2 = 0$.

36. Show that $x = -2 + \sqrt{2}$ satisfies $x^2 + 4x + 2 = 0$.

37. Show that $x = 3/2 + \sqrt{3}/2$ satisfies $2x^2 - 6x + 3 = 0$.

38. Show that $x = 3/2 - \sqrt{3}/2$ satisfies $2x^2 - 6x + 3 = 0$.

In Exercises 39 and 40, multiply the expressions.

39. $(x + \sqrt[3]{2})(x^2 - x\sqrt[3]{2} + \sqrt[3]{4})$ $x^3 + 2$

40. $(x - \sqrt[3]{3})(x^2 + x\sqrt[3]{3} + \sqrt[3]{9})$ $x^3 - 3$

In Exercises 41 and 42, give the conjugate.

41. a. $3 + \sqrt{2}$ $3 - \sqrt{2}$ **b.** $3 - \sqrt{a}$ $3 + \sqrt{a}$

   c. $a - \sqrt{b}$ $a + \sqrt{b}$ **d.** $\sqrt{2} - \sqrt{3}$ $\sqrt{2} + \sqrt{3}$

42. a. $2 - \sqrt{11}$ $2 + \sqrt{11}$ **b.** $1 + \sqrt{b}$ $1 - \sqrt{b}$

   c. $\sqrt{c} - \sqrt{a}$ $\sqrt{c} + \sqrt{a}$ **d.** $5 + \sqrt{3}$ $5 - \sqrt{3}$

Rationalize the denominators in Exercises 43 to 50.

43. a. $\dfrac{4}{\sqrt{5}}$ $\dfrac{4\sqrt{5}}{5}$ **b.** $\dfrac{8}{\sqrt{6}}$ $\dfrac{4\sqrt{6}}{3}$

44. a. $\dfrac{3}{\sqrt{6}}$ $\dfrac{\sqrt{6}}{2}$ **b.** $\dfrac{5}{\sqrt{15}}$ $\dfrac{\sqrt{15}}{3}$

45. a. $\dfrac{a}{\sqrt{c}}, c > 0$ $\dfrac{a\sqrt{c}}{c}$ **b.** $\dfrac{a}{\sqrt{a}}, a > 0$ $\sqrt{a}$

46. a. $\dfrac{c}{\sqrt{c}}, c > 0$ $\sqrt{c}$ **b.** $\dfrac{x}{\sqrt{y}}, y > 0$ $\dfrac{x\sqrt{y}}{y}$

47. $\dfrac{4}{7 - \sqrt{5}}$ $\dfrac{7 + \sqrt{5}}{11}$

48. $\dfrac{6}{5 - \sqrt{2}}$ $\dfrac{30 + 6\sqrt{2}}{23}$

49. $\dfrac{x}{x - \sqrt{y}}, x \neq \sqrt{y}$, x and y not both zero $\dfrac{x^2 + x\sqrt{y}}{x^2 - y}$

50. $\dfrac{y}{y - \sqrt{x}}, y \neq \sqrt{x}$, x and y not both zero $\dfrac{y^2 + y\sqrt{x}}{y^2 - x}$

In Exercises 51 and 52, multiply the numerator and denominator by a number or variable that makes a perfect cube under the radical in the denominator. Further simplify, as needed.

51. a. $\dfrac{1}{\sqrt[3]{2}}$ $\dfrac{\sqrt[3]{4}}{2}$ **b.** $\dfrac{2}{\sqrt[3]{4}}$ $\sqrt[3]{2}$ **c.** $\dfrac{3}{\sqrt[3]{3}}$ $\sqrt[3]{9}$

52. a. $\dfrac{2}{\sqrt[3]{2}}$ $\sqrt[3]{4}$ **b.** $\dfrac{1}{\sqrt[3]{3}}$ $\dfrac{\sqrt[3]{9}}{3}$ **c.** $\dfrac{3}{\sqrt[3]{9}}$ $\sqrt[3]{3}$

53. Return to the solution of Example 10. Use steps similar to those in part c to show that the slope of $\overline{BP}$ is the negative reciprocal of the slope of $\overline{AP}$:

$$\text{Slope } \overline{AP} = \frac{y}{x - r} \quad \text{and} \quad \text{Slope } \overline{BP} = \frac{y}{x + r}$$

54. A circle with diameter d and center at $(d/2, 0)$ is shown in the figure.

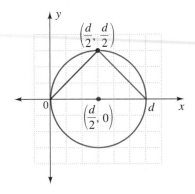

a. Find the slope of the line segment connecting $(d/2, d/2)$ with the origin $(0, 0)$.   1

b. Find the slope of the line segment connecting $(d/2, d/2)$ with the end of the diameter $(d, 0)$. -1

c. Comment on what you observe about the slopes and the angle between them.   negative reciprocals; 90°

55. Apply the Pythagorean theorem to the lengths of the sides of the triangle in the figure, to show that the **distance formula** between points A and B is given by

$$d = \sqrt{(x_2 - x_1)^2 + (y_2 - y_1)^2}$$

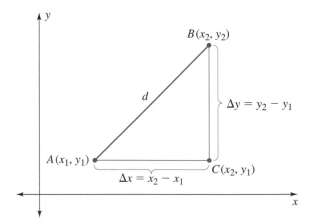

In Exercises 56 to 61, find the slope and the distance between the given points. Use the distance formula from Exercise 55.

56. $(2, 5)$ and $(6, -2)$ $-\frac{7}{4}$; $\sqrt{65} \approx 8.062$

57. $(4, -3)$ and $(-6, 3)$ $-\frac{3}{5}$; $2\sqrt{34} \approx 11.662$

58. $(-1, 5)$ and $(4, -3)$ $-\frac{8}{5}$; $\sqrt{89} \approx 9.434$

59. $(6, 7)$ and $(-3, -4)$ $\frac{11}{9}$; $\sqrt{202} \approx 14.213$

60. (a, b) and (c, d) $\dfrac{d - b}{c - a}$; $\sqrt{(c - a)^2 + (d - b)^2}$

61. $\left(\dfrac{a}{2}, \dfrac{b}{2}\right)$ and (a, b) $\dfrac{b}{a}$; $\dfrac{\sqrt{a^2 + b^2}}{2}$

62. Hero's (or Heron's) formula permits us to find the area of a triangle when we know only the lengths of its sides. The formula is named for a Greek mathematician of the 1st century A.D. or earlier. To find the area, we calculate

$$A = \sqrt{s(s - a)(s - b)(s - c)}$$

where $s = \dfrac{a + b + c}{2}$. The s is called the *semiperimeter* because it is half the perimeter of the triangle. Calculate the area of these triangles.

a. $a = 4, b = 6, c = 3$ ≈ 5.333

b. $a = 5, b = 6, c = 7$ ≈ 14.697

c. $a = 3, b = 4, c = 5$ 6

d. $a = 5, b = 12, c = 13$ 30

e. $a = 8, b = 15, c = 17$ 60

f. Which three triangles in parts a to e are right triangles? How do you know? Calculate their area in another way. Hero's formula is particularly useful in land measure, because it is easy to measure the three sides of a triangular plot of ground.

c, d, e; $a^2 + b^2 = c^2$; segments a and b are perpendicular, so $A = \frac{1}{2}ab$.

63. The golden ratio is the first of the following two expressions:

$$\frac{1 + \sqrt{5}}{2} \quad \text{and} \quad \frac{1 - \sqrt{5}}{2}$$

a. Multiply the two expressions. What do you observe?   Product is -1.

b. Evaluate each expression with a calculator. ≈ 1.6180; ≈ -0.6180

c. What is the reciprocal of each expression? What do you observe? $-\dfrac{1 - \sqrt{5}}{2}$; $-\dfrac{1 + \sqrt{5}}{2}$; negative reciprocals

d. Substitute the expressions $x = \dfrac{1 + \sqrt{5}}{2}$ and $x = \dfrac{1 - \sqrt{5}}{2}$ into the equation $x^2 - x - 1 = 0$ to check that they are solutions.

■ Projects

64. Minimum Cost This problem may be solved by table, guess and check, graph, or spreadsheet.

   A telephone company is replacing copper cable with fiber-optic cable between city A and city B (see the figure). The land above line *BC* is swamp, and the land along the line between *B* and *C* is farm land. It costs $1000 per foot to put the cable through farm land and $1500 per foot to put it through swamp.

   The distance between *B* and *C* is 12 miles. The distance between *A* and *C* is 5 miles. There are 5280 feet in a mile. The line *AC* is perpendicular to *BC*. Round answers to the nearest thousand dollars.

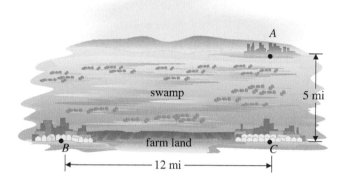

a. What is the cost of a direct route between *A* and *B* through the swamp? $102,960,000

b. What is the cost of installing the cable from *A* to *C* to *B*? $102,960,000

c. What is the cost of installation from *A* to C_1 to *B*, where C_1 is 1 mile from *C* along the line between *B* and *C*? $98,464,000

d. What is the cost of installation from *A* to C_2 to *B*, where C_2 is 2 miles from *C* along the line between *B* and *C*? $95,451,000

e. What is the cost of installation from *A* to C_n to *B*, where C_n is *x* miles from *C* along the line between *B* and *C*? $1500(5280) (\sqrt{25 + x^2}) + $1000(5280) (12 - x)$

f. Using a graph or table, find the minimum possible cost of installation. With what point along the line *BC* will the minimum cost be associated?
   At $x \approx 4.47$, cost is $92,876,000.

65. Placing a Water Pump Toni, a Peace Corps volunteer, has one water pump at *P* and two villages at *A* and *B*

that need water supplied (see the figure). One village is 10 kilometers from the river, and the other village is 20 kilometers from the river. The horizontal distance along the river bank is 20 kilometers. (A spreadsheet or graphing calculator may be useful for this project.)

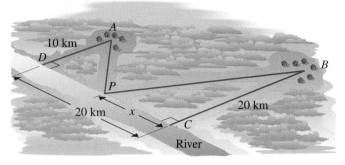

a. Write an equation for the total length of pipe $AP + PB$, where $PC = x$.
$$l = \sqrt{x^2 + 20^2} + \sqrt{(20 - x)^2 + 10^2}$$

b. Where on the river bank should she place the pump to minimize the amount of pipe needed to deliver water to the two villages? $x = 13.3\overline{3}$ km

c. If the villages were equal distances from the river, where should the pump be placed? $x = 10$ km

66. Finding the Area under the Square Root Curve This exploration will introduce you to finding the formula for the area between the *x*-axis and a square root curve.

a. On small-grid graph paper, graph $y = \sqrt{x}$ for *x* in the interval [0, 25].   See Additional Answers.

b. Make a table with *x* as the input and the area under the graph of $y = \sqrt{x}$ as the output. Your table should start with 1 as the first input. For the output, estimate the number of squares under the graph up to each input. One way to estimate the area is to count the square if half or more is under the curve and to ignore the square otherwise. Count any unmatched half squares.   See Additional Answers.

c. Enter the data from your table into calculator lists. Calculate a power regression for the lists. Write your equation as $y = ax^b$. Record the coefficient of regression *r*. The coefficient of regression indicates how well your data fit the calculated equation. An *r* of 1 or −1 indicates a perfect fit.
   $y = 0.54x^{1.58}$; $r = 0.9992$ for the sample data

8.6 Finding Inverse Functions

Objectives

▮ Find the inverse of a function from a table, from a graph, and from an equation.

▮ Identify power and root functions.

WARM-UP

Simplify these expressions:

1. $2\left(\frac{1}{2}x - 2\right) + 4$ x **2.** $\frac{1}{3}(3x - 3) + 1$ x

3. $(x - 1) + 1$ x **4.** $(x + 2) - 2$ x

5. $2\left(\frac{1}{2}x\right)$ x **6.** $\frac{1}{4}(4x)$ x

7. $(\sqrt{x})^2$ for $x \geq 0$ x **8.** $\sqrt{x^2}$ for $x \geq 0$ x

9. $\sqrt[3]{x^3}$ x **10.** $(\sqrt[3]{x})^3$ x

IN THIS SECTION, we work with a more formal way of thinking about solving equations (inverse functions) and apply it to power and root functions.

Inverse Functions

The answer to each simplification in the Warm-up is x, because the two operations in each expression are inverse (or opposite) operations. Each operation undoes the other operation in the expression.

Like operations, functions have inverses. Two functions are **inverse functions** if *they undo each other*. For example, the "add two" function $f(x) = x + 2$, in Table 6, is undone by a "subtract two" function $f^{-1}(x) = x - 2$, in Table 7.

TABLE 6 $y = f(x)$

x	$y = x + 2$
5	7
6	8
n	$n + 2$

TABLE 7 $y = $ *inverse for* $f(x)$, $y = f^{-1}(x)$

x	$y = x - 2$
7	5
8	6
$n + 2$	n

The symbol $f(x)$ represents the words *function of x*. In function notation, the symbol for the inverse function is $f^{-1}(x)$. The -1 is *not* an exponent because the f is not a base; it stands for the word *function*.

Some inverse functions are easy to find and name. Others (as we will see in Chapter 9) must be specially defined.

Finding Inverse Functions

We will now find inverse functions in three different ways: from a table, from ordered pairs on a graph, and from an equation.

INVERSE FUNCTIONS FROM A TABLE *To find the inverse from the table for a function, swap the input and output (x and y) columns and find the rule for the new table.* Check that the rule is a function.

EXAMPLE 1 Finding an inverse function from a table The function $f(x) = 4x$ is shown in Table 8. Its inverse is in Table 9. What equation describes the second table? Is it a function? What is the inverse function to $f(x) = 4x$?

TABLE 8 $y = f(x)$

Input, x	Output, $y = 4x$
1	4
2	8
3	12
4	16

TABLE 9 *Inverse*

Input, x	Output
4	1
8	2
12	3
16	4

SOLUTION The equation is $y = \frac{1}{4}x$. The rule is a function. The inverse function is $f^{-1}(x) = \frac{1}{4}x$.

INVERSE FUNCTIONS FROM A GRAPH *To find the inverse from the graph of a function, swap the numbers (x, y) in each ordered pair and plot the resulting graph.* Check that the new graph is a function.

EXAMPLE 2 Finding an inverse function from a graph The function $f(x) = \frac{1}{2}x - 2$ is shown in Figure 8. Identify four ordered pairs on the function and graph the inverse function. What equation describes the new graph? Is it a function? What is the inverse function to $f(x) = \frac{1}{2}x - 2$?

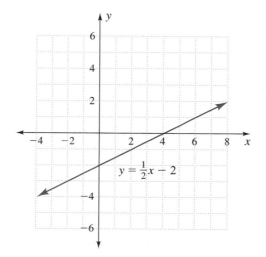

FIGURE 8

SOLUTION Ordered pairs on $y = \frac{1}{2}x - 2$ are $(-4, -4)$, $(-2, -3)$, $(0, -2)$, and $(4, 0)$.
Ordered pairs on the inverse function are $(-4, -4)$, $(-3, -2)$, $(-2, 0)$, and $(0, 4)$.

The graph is shown in Figure 9. The equation of the inverse is $y = 2x + 4$. The rule is a function by the vertical-line test. The inverse function is $f^{-1}(x) = 2x + 4$.

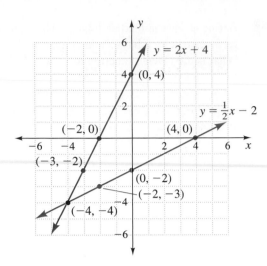

FIGURE 9

EXAMPLE 3 **Finding an inverse function from a graph** The right side of the graph of $y = x^2$ is shown in Figure 10. Identify four ordered pairs on the function and graph the inverse function. What equation describes the new graph? Is it a function? What is the inverse function to $y = x^2$, $x \geq 0$, $y \geq 0$?

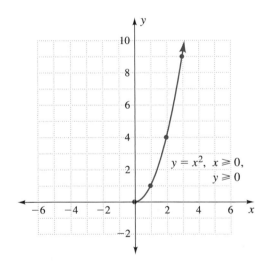

FIGURE 10

SOLUTION Ordered pairs on the function are $(0, 0)$, $(1, 1)$, $(2, 4)$, and $(3, 9)$. Ordered pairs on the inverse function are $(0, 0)$, $(1, 1)$, $(4, 2)$, and $(9, 3)$. The graph is shown in Figure 11. The equation of the inverse is $y = \sqrt{x}$, $x \geq 0$, $y \geq 0$. With the conditions, the equation is a function. The inverse function is $f^{-1}(x) = \sqrt{x}$, $x \geq 0$, $y \geq 0$.

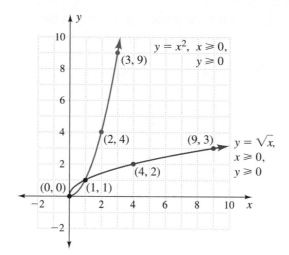

FIGURE 11

Calculator Note: To graph with conditions on the input, place the ⟨Y=⟩ expression in parentheses, as shown in Figure 12, and follow it with the condition in parentheses. The inequality signs are in the ⟨2nd⟩ [TEST] menu.

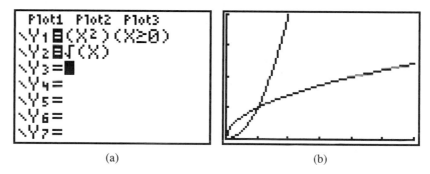

FIGURE 12   0 to 6 on X, 0 to 4 on Y

Think about it: If we draw the left side of the parabola in Figure 10 and then list the ordered pairs with x and y swapped, we obtain points below the x-axis. Why are these not part of the inverse function to $y = x^2$, $x \geq 0$, $y \geq 0$?

INVERSE FUNCTIONS FROM AN EQUATION   *To find the inverse from the equation of a function, swap x and y and solve (if possible) for y.* Check that the equation is a function. For example, to find the inverse function to $y = 4x$, we reverse x and y as follows:

$$x = 4y \qquad \text{Multiply both sides by } \tfrac{1}{4} \text{ to solve for } y.$$

$$\tfrac{1}{4}x = y \qquad \text{The equation is a function.}$$

The finding that $y = \tfrac{1}{4}x$ is the inverse function to $y = 4x$ agrees with Example 1.
To find the inverse function to $y = \tfrac{1}{2}x - 2$, we again reverse x and y:

$$x = \tfrac{1}{2}y - 2 \qquad \text{To solve for } y, \text{ first add 2.}$$

$$x + 2 = \tfrac{1}{2}y \qquad \text{Then multiply by 2.}$$

$$2x + 4 = y \qquad \text{The equation is a function.}$$

The finding that $y = 2x + 4$ is the inverse function to $y = \frac{1}{2}x - 2$ agrees with Example 2.

EXAMPLE 4 **Finding inverse functions from an equation** Find the inverse function to each of the following equations.

a. $y = 3x - 3$ **b.** $y = x - 1$

c. $y = \frac{1}{2}x$ **d.** $y = \sqrt{x}, x \geq 0, y \geq 0$

SOLUTION **a.**

$$y = 3x - 3 \qquad \text{Swap } x \text{ and } y \text{ to find the inverse.}$$

$$x = 3y - 3 \qquad \text{Add 3 to both sides.}$$

$$x + 3 = 3y \qquad \text{Multiply by } \frac{1}{3} \text{ on both sides.}$$

$$\frac{1}{3}(x + 3) = y \qquad \text{Apply the distributive property to change to } y = mx + b \text{ form.}$$

$$\frac{1}{3}x + 1 = y \qquad \text{The equation is a function.}$$

The inverse function to $f(x) = 3x - 3$ is $f^{-1}(x) = \frac{1}{3}x + 1$.

b.

$$y = x - 1 \qquad \text{Swap } x \text{ and } y \text{ to find the inverse.}$$

$$x = y - 1 \qquad \text{Add 1 to both sides.}$$

$$x + 1 = y \qquad \text{The equation is a function.}$$

The inverse function to $f(x) = x - 1$ is $f^{-1}(x) = x + 1$.

c.

$$y = \frac{1}{2}x \qquad \text{Swap } x \text{ and } y \text{ to find the inverse.}$$

$$x = \frac{1}{2}y \qquad \text{Multiply both sides by 2.}$$

$$2x = y \qquad \text{The equation is a function.}$$

The inverse function to $f(x) = \frac{1}{2}x$ is $f^{-1}(x) = 2x$.

d.

$$y = \sqrt{x}, x \geq 0, y \geq 0 \qquad \text{Swap } x \text{ and } y \text{ (in the inequality conditions, too).}$$

$$x = \sqrt{y}, y \geq 0, x \geq 0 \qquad \text{Square both sides.}$$

$$x^2 = y, y \geq 0, x \geq 0 \qquad \text{With conditions, the equation is a function.}$$

Student Note: Part d has conditions that carry through with the inverse steps.

The inverse function to $f(x) = \sqrt{x}, x \geq 0, y \geq 0$ is $f^{-1}(x) = x^2, x \geq 0, y \geq 0$. ▬

Example 3 and part d of Example 4 suggest that powers and roots are inverse functions. We now formally define power and root functions.

Power and Root Functions

For the purpose of working with inverse functions, we will define the power and root functions only in terms of positive integer powers and integer roots. However, some rational exponents and related roots will be used in selected exercises, projects, and examples of calculator regression.

The **power function** is defined by $y = x^n$, *where n is a positive integer*. The identity function ($y = x$), squaring function ($y = x^2$), cubing function ($y = x^3$), and other positive integer powers of x are all power functions.

The **root function** is defined by $y = \sqrt[n]{x}$ *with $x \geq 0$ for all functions with an even index n.* The square root function ($y = \sqrt{x}, x \geq 0$), cube root function ($y = \sqrt[3]{x}$), and fourth root function ($y = \sqrt[4]{x}, x \geq 0$) are all root functions.

GRAPHS OF ROOT FUNCTIONS   The graphs of the even root functions have limited domains and ranges. As shown in (a) and (c) of Figure 13, the square root and the fourth root have graphs only in the first quadrant. The limitations on the even roots explain why the even power functions have limitations when we are finding inverse functions. The odd roots, in (b) and (d) of Figure 13, have no limit on domain or range. Thus, the inverses to the odd roots have no limitations either.

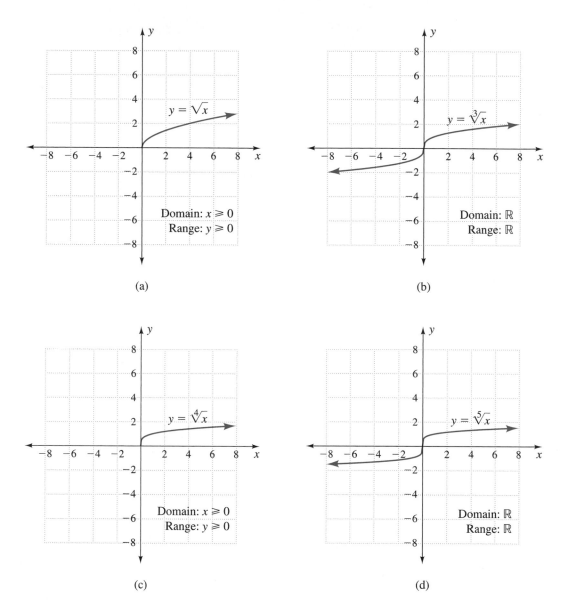

(a)

(b)

(c)

(d)

FIGURE 13

EXAMPLE 5   Finding an inverse function to an odd root

a. Use ordered pairs to find the inverse function to $y = x^3$.

b. Check by finding the inverse function from the equation.

SOLUTION   **a.** Several ordered pairs on $y = x^3$ are $(-2, -8)$, $(-1, -1)$, $(0, 0)$, $(1, 1)$, and $(2, 8)$. Ordered pairs on the inverse are $(-8, -2)$, $(-1, -1)$, $(0, 0)$, $(1, 1)$, and $(8, 2)$. The pairs $(-8, -2)$ and $(8, 2)$ suggest that the inverse function is $y = \sqrt[3]{x}$. The ordered pairs are graphed in Figure 14.

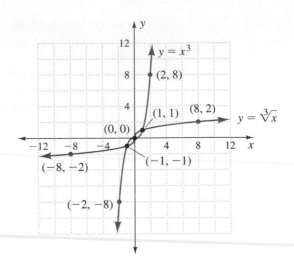

FIGURE 14

The graph suggests that the cube function and cube root are inverses, and so in part b we take the cube root to solve the equation.

b. To find the inverse function to $y = x^3$, we swap x and y.

$x = y^3$ Take the cube root of both sides.

$\sqrt[3]{x} = y$ The equation is a function.

The inverse function to $f(x) = x^3$ is $f^{-1}(x) = \sqrt[3]{x}$.   ▬

INVERSE FUNCTIONS FOR POWERS AND ROOTS

> When n is odd, the inverse function to $f(x) = x^n$ is $f^{-1}(x) = \sqrt[n]{x}$.
>
> When n is even, the inverse function to $f(x) = x^n$, $x \geq 0$, $y \geq 0$ is $f^{-1}(x) = \sqrt[n]{x}$, $x \geq 0$, $y \geq 0$.

EXAMPLE 6   Finding an inverse function to a shifted square root equation
a. Find an inverse to $y = \sqrt{x - 4}$, $x \geq 4$, $y \geq 0$.
b. Graph both the function and its inverse, along with $y = x$.

SOLUTION   **a.**

$y = \sqrt{x - 4}$, $x \geq 4$, $y \geq 0$ Swap x and y (including conditions).

$x = \sqrt{y - 4}$, $y \geq 4$, $x \geq 0$ Solve for y. Square both sides.

$x^2 = y - 4$

$x^2 + 4 = y$, where $x \geq 0$, $y \geq 4$ The inverse equation is a function.

b. The graphs are in Figure 15. The line $y = x$ is an axis of symmetry.

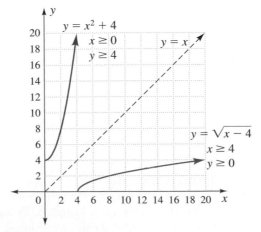

FIGURE 15   ▬

ANSWER BOX

Warm-up: Each expression simplifies to x. **Think about it:** A graph that has points above and below the x-axis, such as the one shown here, fails the vertical-line test and is not a function. The equation is $x = y^2$, $x \geq 0$.

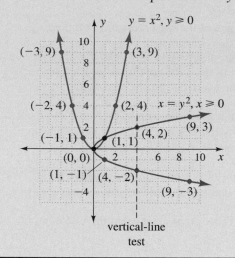

8.6 Exercises

In Exercises 1 to 4, name the ordered pairs that are in the inverse to each set.

1. $\left(1, \frac{1}{2}\right), \left(2, \frac{1}{4}\right), \left(3, \frac{1}{8}\right), \left(4, \frac{1}{16}\right)$ $\left(\frac{1}{2}, 1\right), \left(\frac{1}{4}, 2\right), \left(\frac{1}{8}, 3\right), \left(\frac{1}{16}, 4\right)$

2. $(1, 2), (2, 4), (3, 6), (4, 8)$ $(2, 1), (4, 2), (6, 3), (8, 4)$

3. $(1, 2), (2, 5), (3, 8), (4, 11)$ $(2, 1), (5, 2), (8, 3), (11, 4)$

4. $\left(1, \frac{1}{3}\right), \left(2, \frac{1}{9}\right), \left(3, \frac{1}{27}\right), \left(4, \frac{1}{81}\right)$ $\left(\frac{1}{3}, 1\right), \left(\frac{1}{9}, 2\right), \left(\frac{1}{27}, 3\right), \left(\frac{1}{81}, 4\right)$

In Exercises 5 to 16, graph the equation, show four ordered pairs on the graph and on the inverse function, and graph the inverse.

5. $y = 2^x$ See Answer Section.

6. $y = 3^x$ See Additional Answers.

7. $y = 3^x - 3$ See Answer Section.

8. $y = 2^x - 4$ See Additional Answers.

9. $y = \sqrt{x + 5}, x \geq -5, y \geq 0$ See Answer Section.

10. $y = \sqrt{x - 2}, x \geq 2, y \geq 0$ See Additional Answers.

11. $y = \sqrt{x - 3}, x \geq 3, y \geq 0$ See Answer Section.

12. $y = \sqrt{x + 3}, x \geq -3, y \geq 0$ See Additional Answers.

13. $y = \sqrt{x} - 3, x \geq 0, y \geq -3$ See Answer Section.

14. $y = \sqrt{x} + 3, x \geq 0, y \geq 3$ See Additional Answers.

15. $y = \sqrt{x} + 5, x \geq 0, y \geq 5$ See Answer Section.

16. $y = \sqrt{x} - 2, x \geq 0, y \geq -2$ See Additional Answers.

In Exercises 17 to 34, use algebra to find the inverse function.

17. $y = 3x$ $y = \frac{x}{3}$

18. $y = 2x - 2$ $y = \frac{1}{2}x + 1$

19. $y = 4x - 2$ $y = \frac{1}{4}x + \frac{1}{2}$

20. $y = x - 2$ $y = x + 2$

21. $y = x + 3$ $y = x - 3$

22. $y = \frac{1}{2}x$ $y - 2x$

23. $y = \sqrt{x}, x \geq 0, y \geq 0$ $y = x^2, x \geq 0, y \geq 0$

24. $y = \sqrt[4]{x}, x \geq 0, y \geq 0$ $y = x^4, x \geq 0, y \geq 0$

25. $y = \sqrt[5]{x}$ $y = x^5$

26. $y = x^3$ $y = \sqrt[3]{x}$

27. $y = x^4, x \geq 0, y \geq 0$ $y = \sqrt[4]{x}, x \geq 0, y \geq 0$

28. $y = x^5$ $y = \sqrt[5]{x}$

29. $y = x$ $y = x$

30. $y = \frac{1}{x}$ $y = \frac{1}{x}, x \neq 0$

31. $y = -\frac{1}{x}$ $y = -\frac{1}{x}, x \neq 0$

32. $y = -x$ $y = -x$

33. $y = 1 - x$ $y = 1 - x$

34. $y = 3 - x$ $y = 3 - x$

35. What is surprising about the inverse functions to the equations in Exercises 29 to 34 odd? same as original function

36. What do the graphs of the functions in Exercises 29 to 34 have in common? symmetric with respect to $y = x$

In Exercises 37 to 44, use algebra to find the inverse function for the equation in the given exercise.

37. Exercise 9 $f^{-1}(x) = x^2 - 5, x \geq 0, y \geq -5$

38. Exercise 10 $f^{-1}(x) = x^2 + 2, x \geq 0, y \geq 2$

39. Exercise 11 $f^{-1}(x) = x^2 + 3, x \geq 0, y \geq 3$

40. Exercise 12 $f^{-1}(x) = x^2 - 3, x \geq 0, y \geq -3$

41. Exercise 13 $f^{-1}(x) = (x + 3)^2, x \geq -3, y \geq 0$

42. Exercise 14 $f^{-1}(x) = (x - 3)^2, x \geq 3, y \geq 0$

43. Exercise 15 $f^{-1}(x) = (x - 5)^2, x \geq 5, y \geq 0$

44. Exercise 16 $f^{-1}(x) = (x + 2)^2, x \geq -2, y \geq 0$

Exercises 45 to 49 refer to the graphs and definitions of $y = \sqrt{x}, y = \sqrt[3]{x},$ $y = \sqrt[4]{x},$ and $y = \sqrt[5]{x}$ (Figure 13).

45. Why do the graphs of the even roots have no points in the second and third quadrants?
Even root of a negative number is not a real number.

46. In the definition of the even nth root of a, what phrase indicates why the even root graphs start at the origin?
if a is non-negative

47. What do we know about the outputs to even root graphs that explains why there are no points on the graphs in the fourth quadrant? Outputs are positive real numbers.

48. The even nth root of a is real for non-negative inputs. For what set of inputs is the odd nth root a real number?
all real numbers

49. How do the odd root graphs show that the odd nth root of a is real?
For any real number x, there is a corresponding real number y on the graph.

50. Why are there limitations to the set of inputs (domain) for graphs of equations containing square roots?
Square root of a negative number is not a real number.

51. Why are there limitations to the set of outputs (range) for equations containing square roots?
Principal square root is always positive.

52. Let S be the set of points $(1, 1), (1, -1), (-1, -1),$ and $(-1, 1)$.

a. Is S a function? No; it fails the vertical-line test.

b. What is unique about the inverse?
Reversing ordered pairs gives same set.

c. Is the inverse to S a function?
No; it fails the vertical-line test.

53. Let S be the set of points $(0, 0), (1, 1), (2, 2),$ and $(3, 3)$.

a. Is S a function? yes

b. What is unique about the inverse?
Reversing ordered pairs gives same set.

c. Is the inverse to S a function? yes

54. Explain why $y = x^2, x \geq 0$ has an inverse function but $y = x^2$ does not. With $x \geq 0$, no two inputs have the same output and an inverse function exists.

55. Explain why $y = (x - 2)^2, x \geq 2$ has an inverse function but $y = (x - 2)^2$ does not.
With $x \geq 2$, no two inputs have the same output and an inverse function exists.

56. Look over graphs of functions that have an inverse function. If a function must pass the vertical-line test, what line test must a function pass in order to have an inverse function? horizontal-line test

57. In order for a graph or set to be a function, there must be exactly one output for each input. In order for a function to have an inverse function, there must be exactly one input for each output . Graphs or sets having this feature are said to be **one to one**.

58. Explain why the inverse function for $f(x) = x$ is not $f^{-1}(x) = x^{-1}$. -1 in x^{-1} is an exponent; inverse $f^{-1}(x) = x$.

59. Explain why the $\boxed{x^{-1}}$ key on the calculator is not used to find inverses for a function.
It is the reciprocal key; the -1 is an exponent.

■ **Project**

60. Circles and Parts of Circles

a. The equation of a circle with radius 5 centered at the origin is $y = \pm\sqrt{25 - x^2}$. Is the circle a function?
No; it fails the vertical-line test.

b. The equation of a semicircle with radius 5 centered at the origin is $y = \sqrt{25 - x^2}$. Find ordered pairs for the integer inputs -5 to -3, 0, and 3 to 5.
$(-5, 0), (-4, 3), (-3, 4), (0, 5), (3, 4), (4, 3), (5, 0)$

c. Explain why the semicircle is a function.
For each input, there is exactly one output.

d. If we reverse the ordered pairs in part b, do we obtain a function?
No; for inputs of 3, 4, and zero, there are two outputs.

e. Does the function describing the semicircle have an inverse function? no

f. The equation of a quarter-circle with radius 5 centered at the origin is $y = \sqrt{25 - x^2}, x \geq 0$. Find ordered pairs for the integer inputs 0 to 5.
$(0, 5), (1, 2\sqrt{6}), (2, \sqrt{21}), (3, 4), (4, 3), (5, 0)$

g. Explain why the quarter-circle is a function.
For each input, there is exactly one output.

h. If we reverse the ordered pairs in part f, do we obtain a function?
Yes; there is no repeating of outputs in the original set of ordered pairs.

i. Does the function describing the quarter-circle have an inverse function? yes

8.7 Solving Root and Power Equations

Objectives

- Solve a root equation by taking the nth power of both sides.
- Solve an equation containing square roots by squaring twice.
- Solve a power equation by taking the nth root of both sides.
- Find annual rates of inflation and depreciation.

WARM-UP

1. Find the values for which $\sqrt{x - 5}$ is defined. $x \geq 5$
2. Simplify $(\sqrt{x - 5})^2$. $x - 5$
3. Find the values for which $\sqrt{2 - x}$ is defined. $x \leq 2$
4. Simplify $(\sqrt{2 - x})^2$. $2 - x$
5. Find the values for which $\sqrt{x}$ is defined. $x \geq 0$
6. Simplify $(\sqrt{x} + 3)^2$. $x + 6\sqrt{x} + 9$
7. Solve for r by guess and check: $12 = 6(1 + r)^{18}$. $r = 0.039$

IN THIS SECTION, we use powers and roots to solve equations and formulas. We return to compound interest to solve for the rate of interest.

Solving nth Root and Square Root Equations

■ SOLVING nTH ROOT EQUATIONS

To solve equations containing nth roots, take the nth power of both sides.

If $a = b$, then $a^n = b^n$ for any positive integer n.

Even powers may introduce extraneous roots, so always check answers.

Because equations containing even roots require careful checking, only our first example will involve an odd root.

EXAMPLE 1 Solving a 5th root equation   Solve $\sqrt[5]{x} = 4$ for x.

SOLUTION
$$\sqrt[5]{x} = 4 \qquad \text{Take the 5th power of both sides.}$$
$$(\sqrt[5]{x})^5 = 4^5 \qquad \text{Simplify.}$$
$$x = 1024$$

■

Think about it 1: Why are the following statements different?

If $a = b$, then $a^n = b^n$.

If $a^n = b^n$, then $a = b$.

■ In Examples 2 to 5, solutions include graphs and algebraic notation.

EXAMPLE 2 Solving equations containing square roots
a. Find inputs for which the equation $y = \sqrt{x-5}$ has real-number solutions.
b. Solve $\sqrt{x-5} = 3$ using algebraic notation.
c. Solve the equation from a graph.

SOLUTION a. The radicand, $x - 5$, is zero or positive if $x - 5 \geq 0$ or $x \geq 5$.

b.
$$\sqrt{x-5} = 3 \qquad \text{Square both sides.}$$
$$(\sqrt{x-5})^2 = 3^2 \qquad \text{Simplify.}$$
$$x - 5 = 9 \qquad \text{From part a, } x - 5 \text{ is positive.}$$
$$x = 14$$

Check: $\sqrt{14 - 5} \overset{?}{=} 3$ ✓

Also, 14 is larger than 5, satisfying $x \geq 5$.

c. The intersection of $y = \sqrt{x-5}$ with $y = 3$ in Figure 16 gives the solution to $\sqrt{x-5} = 3$. This point of intersection is $(14, 3)$, so $x = 14$. The graph also shows the condition on $y = \sqrt{x-5}$, $x \geq 5$, in that there are no points on the square root graph to the left of $x = 5$.

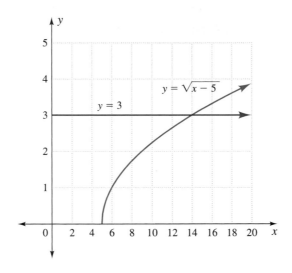

FIGURE 16

EXTRANEOUS ROOTS Example 3 illustrates the importance of checking solutions.

EXAMPLE 3 Solving equations containing square roots
a. Find inputs for which the equation $y = \sqrt{2-x}$ has real-number solutions.
b. Solve $\sqrt{2-x} = x + 1$ with algebraic notation.
c. Solve the equation by graphing.

SOLUTION a. The radicand, $2 - x$, must be zero or positive. Thus, $2 - x \geq 0$ and $2 \geq x$. The solutions are real for all $x \leq 2$.

b.
$$\sqrt{2-x} = x + 1 \qquad\qquad \text{Square both sides.}$$
$$(\sqrt{2-x})^2 = (x+1)^2 \qquad\qquad \text{Replace } (\sqrt{2-x})^2 \text{ with } (2-x).$$
$$2 - x = x^2 + 2x + 1 \qquad\qquad \text{From part a, } 2 - x \text{ is positive.}$$
$$0 = x^2 + 3x - 1 \qquad\qquad \text{Apply the quadratic formula.}$$

Either $x \approx 0.303$ or $x \approx -3.303$

The results are possible solutions. They need to be checked in the original equation.

Check:

$$\sqrt{2 - 0.303} \stackrel{?}{=} 0.303 + 1 \quad \checkmark$$

The result $x \approx 0.303$ gives a true statement. Also, $x \approx 0.303$ is less than 2, satisfying $x \leq 2$.

$$\sqrt{2 - (-3.303)} \stackrel{?}{=} -3.303 + 1$$

The result $x \approx -3.303$ gives $2.303 = -2.303$, which is false. The solution $x \approx -3.303$ must be discarded.

c. The graph in Figure 17 shows a single point of intersection, at $x \approx 0.303$, and confirms our symbolic solution.

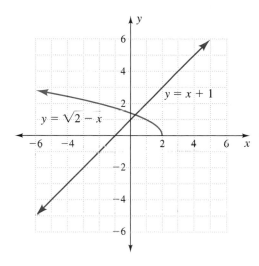

FIGURE 17

The extra result in Example 3 is an extraneous root. Extraneous roots are introduced in the squaring step, as we will see in Example 4.

EXAMPLE 4 Solving equations containing square roots   Solve $\sqrt{x} + 3 = 1$ both symbolically and graphically.

SOLUTION

$$\sqrt{x} + 3 = 1 \qquad \text{Subtract 3 from both sides.}$$

$$\sqrt{x} = -2 \qquad \text{This equation has no real-number solution.}$$

If we do not notice that $\sqrt{x} = -2$ has no real-number solution and square both sides, we discover our oversight when we check our answer.

$$(\sqrt{x})^2 = (-2)^2$$

$$x = 4$$

Check: $\sqrt{4} + 3 \stackrel{?}{=} 1$ gives a false statement, so $x = 4$ is an extraneous root. There is no real-number solution to this equation. In Figure 18, we see that the graph of $y = \sqrt{x} + 3$ has the y-intercept $(0, 3)$ as its lowest point, so it cannot pass through the line $y = 1$. Hence, there are no solutions to $\sqrt{x} + 3 = 1$.

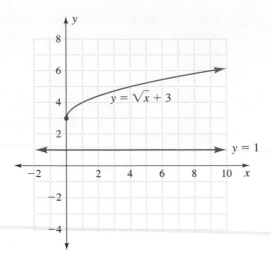

FIGURE 18

SQUARING TWICE   Example 5 illustrates that it may be necessary to square twice in order to remove all radicals from an equation and solve for x.

EXAMPLE 5 Solving equations containing square roots
a. Find inputs for which the equations $y = \sqrt{2x + 14}$ and $y = \sqrt{x} + 3$ have real-number solutions.
b. Solve $\sqrt{2x + 14} = \sqrt{x} + 3$ using symbols.
c. Solve the equation by graphing.

SOLUTION **a.** The radicand $2x + 14$ is non-negative if $2x + 14 \geq 0$. Solving for x gives $x \geq -7$. The radicand x is non-negative if $x \geq 0$. To satisfy both requirements, we limit our choices for x to $x \geq 0$.

b. In solving for x, we will need to square both sides twice.

$$\sqrt{2x + 14} = \sqrt{x} + 3 \qquad \text{Square both sides.}$$

$$(\sqrt{2x + 14})^2 = (\sqrt{x} + 3)^2$$

$$2x + 14 = x + 3\sqrt{x} + 3\sqrt{x} + 9 \qquad \text{Add like terms.}$$

$$2x + 14 = x + 6\sqrt{x} + 9 \qquad \text{Isolate the radical term.}$$

$$x + 5 = 6\sqrt{x} \qquad \text{Square both sides.}$$

$$x^2 + 10x + 25 = 36x \qquad \text{Subtract } 36x \text{ from both sides.}$$

$$x^2 - 26x + 25 = 0 \qquad \text{Apply the quadratic formula (or factor).}$$

Either $x = 1$ or $x = 25$

Check:

$$\sqrt{2(1) + 14} \overset{?}{=} \sqrt{1} + 3 \quad \checkmark$$

$$\sqrt{2(25) + 14} \overset{?}{=} \sqrt{25} + 3 \quad \checkmark$$

Both solutions check, so there should be two points of intersection on the graph.

c. The graphs of $y = \sqrt{2x + 14}$ and $y = \sqrt{x} + 3$ are shown in Figure 19. The graphs intersect at $(1, 4)$ and $(25, 8)$, so $x = 1$ and $x = 25$ are solutions.

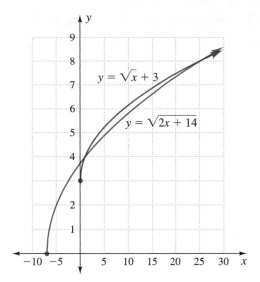

FIGURE 19

CHECKLIST FOR SOLVING RADICAL EQUATIONS

1. Look for extraneous roots.
2. Look for solutions that give negative expressions under the square root sign.
3. Look for equations that cannot be solved because of the principal square root definition—for example, $\sqrt{x + 4} = -4$.
4. Discard possible solutions that are contrary to the geometry, problem setting, or conditions.

Solving *n*th Power Equations

When the goal is to find positive real-number solutions—as it is in solving general or compound interest formulas—it is reasonable to take *n*th roots of power equations of the form $y = ab^x$. When taking *n*th roots as part of a general solution device, however, we must be careful not to miss negative or complex-number solutions; see Exercise 29.

SOLVING *N*TH POWER FORMULAS

To solve formulas containing *n*th powers, take the *n*th root of both sides.
For even *n*th roots, write any limitations on the inputs and outputs.

SOLVING FORMULAS   In Examples 6 and 7, we solve formulas by taking the *n*th root.

EXAMPLE 6   Solving a formula   The formula $V = x^3$ relates the length, x, of the side of a cube to the volume, V. Solve for x.

SOLUTION   The variable is cubed, so we take the cube root.

$$V = x^3$$
$$\sqrt[3]{V} = \sqrt[3]{x^3}$$
$$x = \sqrt[3]{V}$$

EXAMPLE 7 Solving a formula   The formula $H = M/r^3$ is related to magnets. Solve for r.

SOLUTION   We need to move r^3 out of the denominator.

$$H = \frac{M}{r^3} \qquad \text{Multiply both sides by } r^3.$$

$$H \cdot r^3 = M \qquad \text{Divide by } H.$$

$$r^3 = \frac{M}{H} \qquad \text{Take the cube root.}$$

$$r = \sqrt[3]{\frac{M}{H}}$$

COMPOUND INTEREST   In the remaining three examples, we return to settings related to compound interest.

EXAMPLE 8 Solving a power equation   Solve $12 = 6(1 + r)^{18}$ for r.

SOLUTION

$$12 = 6(1 + r)^{18} \qquad \text{Divide by 6.}$$

$$\tfrac{12}{6} = (1 + r)^{18} \qquad \text{Take the 18th root of both sides, either by raising each side to the 1/18 power or by showing an 18th root, } \sqrt[18]{}.$$

$$\left(\tfrac{12}{6}\right)^{1/18} = (1 + r)^{18 \cdot \frac{1}{18}} \qquad \text{Simplify the right side.}$$

$$\left(\tfrac{12}{6}\right)^{1/18} = (1 + r) \qquad \text{Subtract 1 from each side.}$$

$$\left(\tfrac{12}{6}\right)^{1/18} - 1 = r \qquad \text{Evaluate with a calculator, as in Figure 20.}$$

$$r \approx 0.039, \text{ or } 3.9\%$$

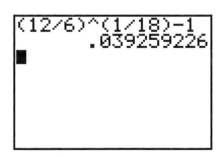

FIGURE 20 Calculation for Example 8

The variable r in the equation in Example 8 describes the annual rate of increase needed for a $6 per hour wage to double in 18 years. The solution says that an annual increase of 3.9% will cause the wage rate to double in 18 years.

ANNUAL RATE OF GROWTH AND INFLATION   The compound interest formula, with $n = 1$, describes annual growth in the value of an investment such as a house. When $n = 1$, the compound interest formula simplifies:

$$S = P\left(1 + \frac{r}{n}\right)^{nt} = P\left(1 + \frac{r}{1}\right)^{1 \cdot t} = P(1 + r)^t$$

A term often used to describe the annual growth rate of prices of houses and other goods is *inflation*. **Inflation** is *a measure of the change in prices over time.*

EXAMPLE 9 Solving a power equation: car prices and inflation   A 1968 four-door automatic Volvo sedan cost $3600. A similar style 1995 Volvo sedan cost $23,820. Suppose that the price increase was due entirely to inflation—what was the annual rate of inflation?

SOLUTION Because we want the annual rate of inflation, we let $n = 1$ in the compound interest formula. The price changes over a time period of $t = 27$ yr.

$$S = P(1 + r)^t \qquad \text{Let } S = 23{,}820, \ P = 3600, \ t = 27.$$

$$23{,}820 = 3{,}600(1 + r)^{27}$$

In solving for r, we reverse the order of operations on r, as in Example 8.

$$23{,}820 = 3{,}600(1 + r)^{27} \qquad \text{Divide by 3600.}$$

$$\frac{23{,}820}{3{,}600} = (1 + r)^{27} \qquad \text{Take the 27th root.}$$

$$\left(\frac{23{,}820}{3{,}600}\right)^{1/27} = (1 + r)^{27 \cdot \frac{1}{27}} \qquad \text{Simplify exponents.}$$

$$\left(\frac{23{,}820}{3{,}600}\right)^{1/27} = 1 + r \qquad \text{Subtract 1.}$$

$$\left(\frac{23{,}820}{3{,}600}\right)^{1/27} - 1 = r \qquad \text{Evaluate with a calculator, as in Figure 21.}$$

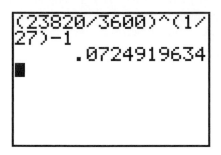

FIGURE 21 Calculation for Example 9

Evaluating, we obtain $r \approx 0.072$, or about 7% inflation per year. The Volvo increased in price at a rate of 7% per year. ■

DOWNWARD CHANGE IN VALUE AND DEPRECIATION **Depreciation** occurs when *a piece of equipment loses value with use*. The ability to calculate various forms of depreciation is a skill needed by a tax preparer or a CPA (certified public accountant). In Example 10, we use the compound interest formula with $n = 1$ to find an annual rate of depreciation for a truck.

 EXAMPLE 10 **Solving a power equation: value and depreciation** Suppose that in 5 years the value of a delivery truck drops from $40,000 to $13,107.20. What is the annual rate of depreciation?

SOLUTION We let the starting value, P, be $40,000 and the ending value, S, be $13,107.20. Replacing t with 5 years, we solve for r.

$$S = P(1 + r)^t$$

$$13{,}107.20 = 40{,}000(1 + r)^5 \qquad \text{Divide by 40,000.}$$

$$\frac{13{,}107.20}{40{,}000} = (1 + r)^5 \qquad \text{Take the 5th root.}$$

$$\left(\frac{13{,}107.20}{40{,}000}\right)^{1/5} = 1 + r \qquad \text{Subtract 1.}$$

$$\left(\frac{13{,}107.20}{40{,}000}\right)^{1/5} - 1 = r \qquad \text{Evaluate with a calculator, as in Figure 22.}$$

$$r = -0.20$$

FIGURE 22 Calculation for Example 10

The rate is negative because the value decreases over time: $r = 20\%$ depreciation. The truck loses 20% of its remaining value each year. ■

Think about it 2: What is 20% of $40,000? If the truck lost this amount each year for 5 years, what value would remain? When might this method be a good way to calculate the truck's value? When might the method in Example 10 be more appropriate?

8.7 Exercises

In Exercises 1 to 4, solve with the graph in the figure.

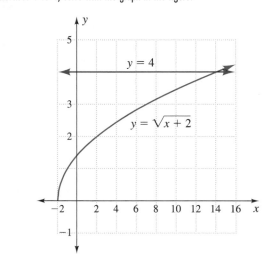

1. $\sqrt{x + 2} = 4$ $x = 14$ **2.** $\sqrt{x + 2} = 2$ $x = 2$

3. $\sqrt{x + 2} = -1$ $\{ \}$ or $\varnothing$ **4.** $\sqrt{x + 2} = -4$ $\{ \}$ or $\varnothing$

In Exercises 5 to 26, for what x are the radical equations defined? Solve using symbols, and then check your solution by substitution and by using a graph.

5. $\sqrt{2 - x} = 2$
$x \leq 2; x = -2$
6. $\sqrt{2 - x} = 3$
$x \leq 2; x = -7$
7. $\sqrt{x - 5} = 2$
$x \geq 5; x = 9$
8. $\sqrt{x - 5} = 1$
$x \geq 5; x = 6$
9. $\sqrt{x - 5} - 6 = -2$
$x \geq 5; x = 21$
10. $\sqrt{x - 5} - 4 = -1$
$x \geq 5; x = 14$
11. $\sqrt{2x + 14} = 0$
$x \geq -7; x = -7$
12. $\sqrt{2x + 14} = 2$
$x \geq -7; x = -5$
13. $\sqrt{2x + 14} = 6$
$x \geq -7; x = 11$
14. $\sqrt{2x + 14} = 4$
$x \geq -7; x = 1$
15. $\sqrt{3x - 5} = 2$
$x \geq \frac{5}{3}; x = 3$
16. $\sqrt{3x - 5} = 4$
$x \geq \frac{5}{3}; x = 7$
17. $\sqrt{3x - 5} = -1$
$x \geq \frac{5}{3}; \{ \}$ or $\varnothing$
18. $\sqrt{5x - 1} = -1$
$x \geq \frac{1}{5}; \{ \}$ or $\varnothing$
19. $\sqrt{5x - 1} = 3$
$x \geq \frac{1}{5}; x = 2$
20. $\sqrt{5x - 1} = 2$
$x \geq \frac{1}{5}; x = 1$
21. $\sqrt{2x + 7} = \sqrt{x} + 2$ $x \geq 0; \{1, 9\}$

22. $\sqrt{3x + 1} = \sqrt{2x} + 1$ $x \geq 0; \{0, 8\}$

23. $\sqrt{3x - 2} = \sqrt{x + 4}$ $x \geq 0; x = 12$

24. $\sqrt{7x + 4} = \sqrt{4x} + 4$ $x \geq 0; x = 36$

25. $\sqrt{3x - 3} = \sqrt{4x} - 1$ $x \geq 1; x = 4$

26. $\sqrt{3x - 1} = \sqrt{5x} - 3$ $x \geq \frac{1}{3}; x = \dfrac{35 + 15\sqrt{5}}{2}$

27. The graph of $y = \sqrt{3x - 5}$ intersects the graph of $y = \sqrt{2x + 1}$ once. We obtain two numbers when solving $\sqrt{3x - 5} = \sqrt{2x + 1}$ symbolically. What can you conclude about one of the two numbers?   extraneous root

28. Why is there a difference in the graphs of $y = \sqrt{x} - 7$ and $y = \sqrt{x - 7}$?   difference in order in which 7 is subtracted and square root is found

29. Find all solutions, including those containing the imaginary unit, i.

a. Solve $x^2 = 1$ by taking the square root. Note that $\sqrt{x^2} = |x|$. Solve $x^2 - 1 = 0$ by factoring.
$x = \pm 1$

b. Solve $x^4 = 1$ by taking the 4th root. Note that $\sqrt[4]{x^4} = |x|$. Solve $x^4 - 1 = 0$ by factoring.
$x = \pm 1, x = \pm i$

c. Solve $x^3 = 1$ by taking the cube root. Solve $x^3 - 1 = 0$ by factoring. If you forget the factors, divide the left side by $(x - 1)$ and use the quadratic equation on the other factor. $x = 1, x = -\frac{1}{2} \pm i\sqrt{3}/2$

Note that taking roots in parts b and c did not find all the solutions.

In Exercises 30 to 34, find the rational-number solutions. Predict the number of solutions containing the imaginary unit.

30. $8 = 27x^3$ $x = \frac{2}{3}; 2$ **31.** $4x^2 = 25$ $x = \pm 2.5;$ none

32. $1000 = 125x^3$ $x = 2; 2$ **33.** $16x^4 - 25 = 600$
$x = \pm 2.5; 2$

34. $27x^3 + 36 = 100$ $x = \frac{4}{3}; 2$

In Exercises 35 to 42, solve for *x*. Round to the nearest thousandth. Assume $1 + x > 0$.

35. $100 = 40(1 + x)^5$ $x = 0.201$

36. $50 = 25(1 + x)^8$ $x = 0.091$

37. $30 = 10(1 + x)^{15}$ $x = 0.076$

38. $56 = 8(1 + x)^{12}$ $x = 0.176$

39. $40 = 100(1 + x)^5$ $x = -0.167$

40. $25 = 50(1 + x)^8$ $x = -0.083$

41. $10 = 30(1 + x)^{15}$ $x = -0.071$

42. $8 = 56(1 + x)^{12}$ $x = -0.150$

▦ Applications

43. a. Solve $r = \sqrt{24L}$ for *L*. *L* is the length of skid marks on dry concrete at *r* mph. $L = \dfrac{r^2}{24}$

    b. Solve $r = \sqrt{12L}$ for *L*. *L* is the length of skid marks on wet concrete at *r* mph. $L = \dfrac{r^2}{12}$

    c. Evaluate the formulas in parts a and b for a 50-foot skid mark. Make a ratio of rates in mph for wet to dry pavement. What conclusion can you draw that is relevant to driving? ≈ 34.64 mph; ≈ 24.49 mph; $\dfrac{r_{wet}}{r_{dry}} = \dfrac{\sqrt{2}}{2}$; slow down when pavement is wet.

44. The distance seen in miles from a height of *h* feet on Earth is $d = \sqrt{3h/2}$.

    a. If we can see approximately 29 miles from the top of the Washington Monument in Washington, D.C., how tall is it? ≈ 560 ft

    b. If we can see approximately 35.8 miles from the top of the Transamerica Pyramid in San Francisco, estimate its height. ≈ 850 ft

    c. The distance seen in miles from a height of *h* feet on the moon is $d = \sqrt{3h/8}$. Repeat parts a and b for the moon's formula. Why are the answers so different?   2240 ft; ≈ 3420 ft; multiplying output by a factor of 8, instead of 2; moon curvature is greater

45. Solve $v = \sqrt{2gs}$ for *s*, where *v* is the final velocity of a falling object, *g* is the acceleration due to gravity, and *s* is the total distance fallen. $s = \dfrac{v^2}{2g}$

46. Solve $D = \sqrt[3]{80P/N}$ for *P*, where *D* = diameter of the driveshaft, *P* = horsepower transmitted by the shaft, and *N* = number of revolutions per second. $P = D^3 N/80$

Exercises 47 to 59 use the compound interest formula as applied to annual growth or depreciation.

47. In 1985, a house was valued at $55,000. In 1998, the same house was valued at $119,000. What annual rate of growth describes the change in value? At this rate of growth, what will the value be in 2010, to the nearest thousand? $\approx 6.1\%$; $\approx \$243,000$

48. In 2002, the house in Exercise 47 was valued at $156,000. What is the annual growth rate from 1985? from 1998? How is the rate of growth changing? 6.3%; 7%; growing faster

49. A house cost $14,000 in 1970. In 1995, the same house was worth $75,000. What is the annual rate of increase in value? What will the house be worth in 2010? $\approx 6.94\%$; $\approx \$205,315.25$

50. The house in Exercise 49 was remodeled into rental apartments, giving it a $161,500 value in 1999. In 2002,

the building was valued at $194,000. Find the annual rate of growth between 1999 and 2002. Predict the value of the house in 2010.   6.3%; $316,000

51. In 1937, the maximum yearly earnings to which Social Security taxes applied was $3000. In 1998, the maximum was $68,400. What annual rate of growth describes the change in maximum earnings? At this rate of growth, what will the maximum earnings be in 2010, to the nearest thousand? $\approx 5.3\%$; $\approx \$127,000$

52. In 2002, the maximum yearly earnings to which Social Security taxes applied (see Exercise 51) was $84,900. What is the annual rate of growth from 1937? from 1998? How is the rate of growth changing? 5.3%; 5.6%; growing faster

53. Microsoft stock valued at $6.953 per share in 1994 was sold at $52.02 in 2002. What is the annual rate of growth? 28.6%

54. Global Crossing stock purchased at $52.125 was sold two years later at $0.049. What was the annual rate of change? -96.9%

55. A DVD/video player cost $800. Three years later, a better player cost $600. What was the annual rate of change? -9.1%

56. Suppose a car was purchased in 1990 for $40,000. What annual depreciation rate was incurred if the car was worth $3000 in the year 2000? $\approx 22.8\%$

57. In 1974, a Texas Instruments SR11 calculator cost $96. The calculator could do only addition, subtraction, multiplication, division, and square root. In 2002, a similar calculator cost $5. What annual rate of interest describes the change in price? $\approx -10\%$

58. A Panasonic home fax machine cost $470 in 1994. A better Panasonic fax cost $290 in 1998. What is the annual rate of price change? $\approx -11.37\%$

59. In 2002, a Panasonic home fax machine (see Exercise 58) cost $120. What is the annual rate of price change since 1994? since 1998? -15.7%; -19.8%

60. The formula $A = \pi r^2 \theta/360$ relates the radius, *r*, and angle measure, θ, of a sector of a circle to the area of the sector, *A*. Apply this formula to answer the questions in parts a to d. Round to the nearest tenth of an inch.

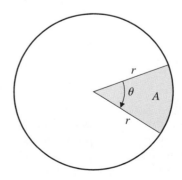

a. A pizza has a 5-inch radius. Find the area of a slice of pizza cut with a central angle of 44°. 9.6 in^2

b. A pizza slice has an area of 11 square inches. The central angle formed by the slice is 20°. What is the radius of the circle? 8.0 in.

c. A pizza slice has an area of 22 square inches. The central angle formed by the slice is 20°. What is the radius of the circle? 11.2 in.

d. A pizza has a 10-inch radius. Find the area of a slice of pizza cut with a central angle of 22°. 19.2 in^2

61. The volume of a sphere with radius r is $V = 4\pi r^3/3$. In parts a to d, find the radius of spheres with the given volumes. Round to the nearest hundredth.

a. $100\pi \text{ in}^3$ $r = 4.22 \text{ in.}$ **b.** $200\pi \text{ in}^3$ $r = 5.31 \text{ in.}$

c. $400\pi \text{ in}^3$ $r = 6.69 \text{ in.}$ **d.** $800\pi \text{ in}^3$ $r = 8.43 \text{ in.}$

e. Complete the sentence: The volume of one sphere is 8 times the volume of another sphere. The radius of the first is ___2___ times the radius of the second.

62. Solve for the radius r of a sphere given the volume, $V = \frac{4}{3}\pi r^3$. $r = \sqrt[3]{3V/4\pi}$

63. Solve for the temperature T in the Stefan-Boltzmann law for the rate of total radiant energy emission, $Q_b = \sigma A T^4$. $T = \sqrt[4]{Q_b/\sigma A}$

64. Solve for the radius r of a tube in the Poiseuille formula for the flow of liquids, $v = \pi p r^4/8Ln$. $r = \sqrt[4]{8vLn/\pi p}$

65. Solve for the velocity V in the kinetic energy equation, $K = mV^2/2$. $V = \sqrt{2K/m}$

Project

66. Planetary Motion The speed of a planet in orbit about the sun varies directly with its distance from the sun. The astronomical unit AU represents the average distance from Earth to the sun—150 million kilometers.

a. Enter the paired data from the table below into a graphing calculator. *Note:* All orbital period data must be in the same units for statistical calculation and graphing.

Planet	Average Distance from Sun	Orbital Period
Mercury	0.387 AU	88 days
Venus	0.723 AU	225 days
Earth	1.000 AU	365 days
Mars	1.52 AU	685 days
Jupiter	5.2 AU	11.9 years
Saturn	9.54 AU	29.5 years
Uranus	19.18 AU	84 years
Neptune	30.07 AU	165.2 years
Pluto	39.44 AU	248.1 years

b. Use power regression to fit a curve to the data. Record the regression equation in the form $y = ax^b$. $y \approx 365.7x^{1.5}$ days or $y \approx 1.001x^{1.5}$ yr

c. Use your equation to predict the orbital period in days or in years, as appropriate, for Mercury, Mars, Neptune, and Pluto.

---

8 Chapter Summary

Vocabulary

annual compound interest
compound interest
conjugate products
cube root
depreciation
digits
distance formula
division of like bases property
exponential expression

index of a radical
inflation
inverse function
multiplication of like bases property
negative one power
nth root
odd nth root
one to one
placeholders
power

power function
power of a power property
power of a product property
power of a quotient property
principal nth root
principal square root
radical expression
radical sign

rational exponent
rationalizing
real-number conjugates
recursive formula
root function
scientific notation
significant digits
similar radicals
zero power

Concepts

8.1 ■ Exponents and Their Properties

Positive powers can be written out as a product of factors.

To multiply numbers with like bases, add the exponents:

$$x^a \cdot x^b = x^{a+b}$$

To divide numbers with like bases, subtract the exponents:

$$\frac{x^a}{x^b} = x^{a-b} \quad \text{for } x \neq 0$$

To apply an exponent to a power expression, multiply the exponents:

$$(x^a)^b = x^{a \cdot b}$$

An exponent outside the parentheses applies to all parts of a product or quotient inside the parentheses:

$$(x \cdot y)^a = x^a \cdot y^a$$

$$\left(\frac{x}{y}\right)^a = \frac{x^a}{y^a} \quad \text{for } y \neq 0$$

The zero power of a base, b^0, is 1, $b \neq 0$.

The negative one power of a nonzero base is the reciprocal, $1/b$.

The negative one power of $1/b$, or $(1/b)^{-1}$, is b, $b \neq 0$.

"Simplify" indicates that definitions and properties of exponents are to be used to remove zero or negative exponents as well as to remove parentheses, to perform operations, and, where possible, to change exponential expressions into numbers or expressions without exponents.

8.2 ■ Scientific Notation

Scientific notation has one nonzero digit before the decimal point. This number is multiplied by a power of 10.

To change scientific notation to regular decimal notation and vice versa, remember that numbers larger than 1 have a positive exponent on the 10 and small numbers between 0 and 1 have a negative exponent on the 10.

Significant digits include zeros between nonzero digits, zeros following a nonzero digit after a decimal point, and zeros that are placeholders and are marked with an overbar.

8.3 ■ Rational Exponents

Rational exponent expressions may be simplified with this property:

$$b^{m/n} = (b^m)^{1/n} = (b^{1/n})^m, \text{ where } b^{1/n} \text{ is defined}$$

Rational exponents may appear in two types of expressions:

1. Power expressions have the variable in the base and a constant exponent: x^n.

2. Exponential expressions have a constant base and the variable in the exponent: b^x. Exponential expressions appear in Sections 8.1 to 8.3 and again in Chapter 9.

The number of compounding periods per year in the compound interest formula $A = P(1 + r/n)^{nt}$ is n. The frequency of compounding may be annual ($n = 1$), semiannual ($n = 2$), quarterly ($n = 4$), monthly ($n = 12$), weekly ($n = 52$), or daily ($n = 365$).

When rounding money accounts, round down if you are receiving the money; round up if you are paying it.

8.4 ■ Roots and Rational Exponents

For all numbers x for which $\sqrt[n]{x}$ is defined,

$$\sqrt[n]{x} = x^{1/n}$$

$x^{1/2} = \sqrt{x}$, $x \geq 0$ is the principal (positive) square root.

If n is odd, $x^{n/n} = x$. If n is even, $x^{n/n} = |x|$.

If m/n is a rational number in lowest terms and if $\sqrt[n]{x}$ is defined, then

$$x^{m/n} = (x^{1/n})^m = (\sqrt[n]{x})^m = \sqrt[n]{x^m}$$

The following power properties of radical expressions correspond to the power properties of rational exponents summarized above:

$$\sqrt[n]{\sqrt[m]{x}} = \sqrt[nm]{x}, \quad x > 0$$
$$\sqrt[n]{x \cdot y} = \sqrt[n]{x} \cdot \sqrt[n]{y}, \quad x > 0, y > 0$$
$$\sqrt[n]{\frac{x}{y}} = \frac{\sqrt[n]{x}}{\sqrt[n]{y}}, \quad x > 0, y > 0$$

8.5 ■ More Operations with Radicals

Radical expressions can be added or subtracted if they can be changed to similar radicals.

To multiply multiple-term radical expressions, apply the distributive property.

The product of real-number conjugates is a rational number.

To rationalize a one-term expression, multiply by a number whose product with the radical makes an exact root.

To rationalize a two-term expression, multiply by the conjugate.

Checklist for simplifying a radical expression:

1. For a radical with index 2, all perfect square factors have been removed.

2. For a radical with index 3, all perfect cube factors have been removed.

3. The index and exponents of factors in the radicand have no common factors.

4. Exponents in the radicand are smaller than the index.

5. Radicals have been eliminated from the denominator.

6. All possible operations have been performed.

8.6 ■ Finding Inverse Functions

To find the inverse from the table for a function, swap the input and output (x and y) columns and find the rule for the new table. Check that the inverse is a function.

To find the inverse from the graph of a function, swap the numbers (x, y) in each ordered pair and plot the resulting graph. Check that the inverse is a function.

To find the inverse from the equation of a function, swap x and y in the equation and in the restrictions and solve (if possible) for y. Check that the inverse is a function.

8.7 ■ Solving Root and Power Equations

To solve equations containing nth powers, take the nth root of both sides. For even roots, limit inputs to $x \geq 0$, as needed, and outputs to $y \geq 0$ or use absolute value.

To solve equations containing nth roots, take the nth power of both sides. It may be necessary to first isolate the radical

expression or to square the equations twice. If $a = b$, then $a^n = b^n$ for any positive integer n. Even powers may introduce extraneous roots, so always check answers.

Checklist for solving radical equations:

1. Look for extraneous roots.

2. Look for solutions that give negative expressions under the square root sign.

3. Look for equations that cannot be solved because of the principal square root definition—for example, $\sqrt{x + 4} = -4$.

4. Discard possible solutions that are contrary to the geometry, problem setting, or conditions.

To use the compound interest formula for annual rate of growth, inflation, or depreciation, let $n = 1$ in $A = P(1 + r/n)^{nt}$.

---

8 Review Exercises

In Exercises 1 to 4, simplify. Assume all variables represent positive numbers.

1. a. $\dfrac{a^0 b^{-1}}{c^2}$ $\dfrac{1}{bc^2}$ **b.** $\dfrac{a^3 b^{-2}}{c^{-1}}$ $\dfrac{a^3 c}{b^2}$ **c.** $\left(\dfrac{2x}{y^2}\right)^4$ $\dfrac{16x^4}{y^8}$

2. a. $\dfrac{2}{x^{-1}}$ $2x$ **b.** $\left(\dfrac{2x^2}{y}\right)^3$ $\dfrac{8x^6}{y^3}$ **c.** $\left(\dfrac{x^{-1}}{2y^2}\right)^{-3}$ $8x^3 y^6$

3. a. $(x^3 \cdot y^{-4})^2$ $\dfrac{x^6}{y^8}$ **b.** $(3x^3 \cdot y^2)^2(4xy^{-3})$ $36x^7 y$

4. a. $\dfrac{6x^2 y^{-3}}{3x^{-2} y^4}$ $\dfrac{2x^4}{y^7}$ **b.** $\dfrac{4a^{-3} b^2}{(6a^3 b^{-2})^2}$ $\dfrac{b^6}{9a^9}$

5. Change to decimal notation.

  a. 3.45×10^{-15}
    0.000 000 000 000 003 45

  b. 6.400×10^{-3}
    0.006 400

  c. 4.005×10^5
    400,500

  d. 4.7800×10^3
    4780.0

6. Simplify mentally and write the answer in scientific notation.

  a. $\dfrac{4.5 \times 10^{-2}}{0.9 \times 10^5}$
    5×10^{-7}

  b. $\dfrac{3.6 \times 10^4}{0.9 \times 10^{-3}}$
    4×10^7

In Exercises 7 to 10, simplify without a calculator.

7. a. $64^{1/2}$ 8 **b.** $64^{1/6}$ 2 **c.** $64^{5/6}$ 32

8. a. $32^{1/5}$ 2 **b.** $16^{3/4}$ 8 **c.** $32^{4/5}$ 16

9. a. $16^{0.75}$ 8 **b.** $25^{3/2}$ 125 **c.** $27^{4/3}$ 81

10. a. $9^{1.5}$ 27 **b.** $8^{4/3}$ 16 **c.** $4^{5/2}$ 32

11. Suppose you start at \$7 per hour and are given a 4% raise at the end of each year. Find your hourly wage after 10 years. Find your hourly wage if you get a 2% raise every six months for the 10 years.
\$10.36, \$10.40 (rounded down after 10 yr)

12. Find the amount of money in a savings account if \$1200 is deposited at 6% interest compounded daily and the money is left for 7 years. \$1826.29

13. Find the amount of money in a savings account if \$1600 is deposited at 5% interest compounded monthly and the money is left for 4 years. \$1953.43

14. Repeat Exercise 12 using $2\frac{1}{2}$ years. \$1394.18

15. Repeat Exercise 13 using $3\frac{3}{4}$ years. \$1929.21

Simplify the expressions in Exercises 16 to 19. Assume all variables represent positive numbers.

16. a. $(b^4)^{3/4}$ b^3 **b.** $a^{2/3} \cdot a^{1/3}$ a **c.** $x^{3/4} \cdot x^{1/2}$ $x^{5/4}$

  d. $\dfrac{b^{3/4}}{b^{1/2}}$ $b^{1/4}$ **e.** $b \cdot b^x$ b^{x+1} **f.** $(x^2 y^3)^{1/6}$ $x^{1/3} y^{1/2}$

17. a. $(a^2)^{3/2}$ a^3 **b.** $x^{1/3} \cdot x^{2/3}$ x **c.** $a^{1/4} a^{3/2}$ $a^{7/4}$

  d. $\dfrac{b^{2/3}}{b^{1/2}}$ $b^{1/6}$ **e.** $x \cdot x^n$ x^{n+1} **f.** $(x^3 y^4)^{1/2}$ $x^{3/2} y^2$

18. a. $(x^{1/2} \cdot y^{1/4})^2$ $xy^{1/2}$ **b.** $\dfrac{x^n}{x}$ x^{n-1} **c.** $(x^{1/4})^{2/3}$ $x^{1/6}$

  d. $\dfrac{a^{2/3}}{a^{4/3}}$ $a^{-2/3}$ or $\dfrac{1}{a^{2/3}}$ **e.** $\dfrac{a}{a^x}$ a^{1-x} or $\dfrac{1}{a^{x-1}}$ **f.** $\left(\dfrac{x^{1/2}}{y^2}\right)^{2/3}$ $\dfrac{x^{1/3}}{y^{4/3}}$

19. a. $(a^{3/4} \cdot b^{1/4})^2$ **b.** $(a^{4/3})^{3/4}$ **c.** $\dfrac{x^n}{x^{n-1}}$ x
$a^{3/2}b^{1/2}$ a

d. $x \cdot x^{n-1}$ **e.** $b^{1/2} \cdot b$ **f.** $\left(\dfrac{x^3}{y^{1/2}}\right)^{2/3}$ $\dfrac{x^2}{y^{1/3}}$
x^n $b^{3/2}$

Change each expression in Exercises 20 to 22 to exponential notation, and simplify.

20. a. $\sqrt[3]{-8}$ **b.** $\sqrt[4]{\dfrac{625}{16}}$ **c.** $\sqrt[3]{8^2}$
$(-8)^{1/3} = -2$ $\left(\frac{625}{16}\right)^{1/4} = \frac{5}{2}$ $8^{2/3} = 4$

21. a. $\sqrt[3]{\dfrac{27}{8}}$ $\left(\frac{27}{8}\right)^{1/3} = \frac{3}{2}$ **b.** $\sqrt[5]{-32}$ **c.** $\sqrt{4^3}$ $4^{3/2} = 8$
               $(-32)^{1/5} = -2$

22. a. $\dfrac{\sqrt[3]{40}}{\sqrt[3]{5}}$ $8^{1/3} = 2$ **b.** $\dfrac{-\sqrt[4]{2}}{\sqrt[4]{32}}$ $-\left(\frac{1}{16}\right)^{1/4} = -\frac{1}{2}$ **c.** $\dfrac{-\sqrt[3]{24}}{\sqrt[3]{81}}$ $-\left(\frac{8}{27}\right)^{1/3} = -\frac{2}{3}$

Change each expression in Exercises 23 to 28 to radical notation. Simplify. Radical notation may vary.

23. a. $125^{2/3}$ **b.** $64^{1/3}$ **c.** $32^{2/5}$
$(\sqrt[3]{125})^2 = 25$ $\sqrt[3]{64} = 4$ $(\sqrt[5]{32})^2 = 4$

24. a. $64^{3/2}$ **b.** $32^{3/5}$ **c.** $27^{2/3}$
$(\sqrt{64})^3 = 512$ $(\sqrt[5]{32})^3 = 8$ $(\sqrt[3]{27})^2 = 9$

25. a. $(-64)^{1/3}$ **b.** $-64^{1/2}$ **c.** $(-16)^{1/4}$
$\sqrt[3]{-64} = -4$ $-\sqrt{64} = -8$ not a real number

26. a. $-16^{0.75}$ **b.** $(-64)^{1/2}$ **c.** $-27^{1/3}$
$-(\sqrt[4]{16})^3 = -8$ not a real number $-\sqrt[3]{27} = -3$

27. a. $x^{2/3}$ **b.** $x^{1.5}$ **c.** $a^{0.75}$
$(\sqrt[3]{x})^2$ $(\sqrt{x})^3$ $(\sqrt[4]{a})^3$

28. a. $a^{3/2}$ $(\sqrt{a})^3$ **b.** $a^{1.25}$ $(\sqrt[4]{a})^5$ **c.** $x^{0.8}$ $(\sqrt[5]{x})^4$

In Exercises 29 and 30, simplify. Identify restrictions necessary to have real-number outputs.

29. a. $\sqrt{9n}$ $3\sqrt{n}, n \geq 0$ **b.** $\sqrt{90n^2}$ $3|n|\sqrt{10}$

   c. $\sqrt{0.09x}$ $0.3\sqrt{x}, x \geq 0$ **d.** $\sqrt{900x^4}$ $30x^2$

30. a. $\sqrt{0.9n^2}$ $3|n|\sqrt{0.1}$ **b.** $\sqrt{9000x^2}$ $30|x|\sqrt{10}$

   c. $\sqrt{0.009x}$ **d.** $\sqrt{90x^3}$ $3x\sqrt{10x}, x \geq 0$
    $0.3\sqrt{0.1x}, x \geq 0$

Simplify in Exercises 31 to 36. Assume all variables represent positive numbers.

31. a. $\sqrt{z}\sqrt{z^5}$ **b.** $\sqrt[3]{x^2} \cdot \sqrt[3]{x}$ **c.** $\sqrt[3]{x^5} \cdot \sqrt[3]{x^1}$
z^3 x x^2

32. a. $\sqrt[4]{32}$ **b.** $\sqrt{72b^3}$ **c.** $\sqrt[4]{32x^6y^5z^4}$
$2\sqrt[4]{2}$ $6b\sqrt{2b}$ $2xyz\sqrt[4]{2x^2y}$

33. a. $\sqrt[3]{x} \cdot \sqrt[2]{x}$ **b.** $\sqrt{x^3} \cdot \sqrt[3]{x^2}$ **c.** $\sqrt[2]{\sqrt[3]{x}}$
$x^{5/6}$ $x^{5/2}$ $x^{1/6}$

34. a. $\sqrt{x^3} \cdot \sqrt{x^4}$ **b.** $\sqrt[3]{x} \cdot \sqrt[4]{x}$ **c.** $\sqrt[3]{\sqrt[4]{x}}$
$x^{7/2}$ $x^{7/12}$ $x^{1/12}$

35. a. $\dfrac{\sqrt{x}}{\sqrt{x^3}}$ $\frac{1}{x}$ **b.** $\dfrac{\sqrt[3]{x^5}}{\sqrt[3]{x^2}}$ x **c.** $\dfrac{\sqrt[3]{x}}{\sqrt[4]{x}}$ $x^{1/12}$

36. a. $\dfrac{\sqrt[4]{x}}{\sqrt[3]{x^2}}$ $\frac{1}{x^{5/12}}$ **b.** $\dfrac{\sqrt[3]{x^4}}{\sqrt[3]{x^7}}$ $\frac{1}{x}$ **c.** $\dfrac{\sqrt{x^5}}{\sqrt{x}}$ x^2

Simplify the expressions in Exercises 37 to 40 by performing the indicated operations.

37. a. $\sqrt{3} + 2\sqrt{12}$ $5\sqrt{3}$ **b.** $\sqrt{9x} + \sqrt{x}$ $4\sqrt{x}, x \geq 0$

38. a. $\sqrt{x} + \sqrt{x^3}$ $(x+1)\sqrt{x}, x \geq 0$

   b. $\sqrt{16x} - \sqrt{x}$ $3\sqrt{x}, x \geq 0$

39. a. $(3 + \sqrt{6})(3 - \sqrt{6})$ 3

   b. $(3 - \sqrt{x})(3 - \sqrt{x})$ $9 - 6\sqrt{x} + x$

40. a. $(2 - \sqrt{2})(2 - \sqrt{2})$ $6 - 4\sqrt{2}$

   b. $(5 - \sqrt{2x})(5 + \sqrt{2x})$ $25 - 2x$

41. Rationalize the denominators.

   a. $\dfrac{1}{b - \sqrt{a}}$ $\dfrac{b + \sqrt{a}}{b^2 - a}$ **b.** $\dfrac{\sqrt{x} - 3}{\sqrt{x} + 4}$ $\dfrac{-x + 7\sqrt{x} - 12}{16 - x}$

   c. $\dfrac{1}{\sqrt{a} + \sqrt{b}}$ $\dfrac{\sqrt{a} - \sqrt{b}}{a - b}$ **d.** $\dfrac{a}{\sqrt[4]{a^2b}}$ $\dfrac{\sqrt[4]{a^2b^3}}{b}$

   e. $\dfrac{1}{\sqrt[3]{9x}}$ $\dfrac{(\sqrt[3]{9x})^2}{9x}$ or $\dfrac{\sqrt[3]{3x^2}}{3x}$ **f.** $\dfrac{3x}{\sqrt[5]{x^3}}$ $3\sqrt[5]{x^2}$

42. Use $(a + b)^2$ as a model to multiply out these problems.

   a. $(2^x + 2^{-x})^2$ **b.** $(c^x + c^{-x})^2$ $c^{2x} + 2 + c^{-2x}$
     $2^{2x} + 2 + 2^{-2x}$

   c. $(b^{1/2} + b^{1/3})^2$ **d.** $(a^{1/3} + a^{1/2})^2$
     $b + 2b^{5/6} + b^{2/3}$ $a^{2/3} + 2a^{5/6} + a$

   e. $(\sqrt{7} + \sqrt{5})^2$ **f.** $(\sqrt{a} + \sqrt{b})^2$
     $12 + 2\sqrt{35}$ $a + 2\sqrt{ab} + b$

In Exercises 43 and 44, substitute the numbers into the quadratic equation to show whether the numbers are solutions.

43. $x^2 - 2x - 1 = 0$, where $x = 1 - \sqrt{2}$. What is the other solution? $1 + \sqrt{2}$

44. $x^2 - 4x - 2 = 0$, where $x = 2 - \sqrt{6}$. What is the other solution? $2 + \sqrt{6}$

45. Show that $1/2 \pm \sqrt{7}/2$ are the solutions to $2x^2 - 2x - 3 = 0$.

46. Apply the distance formula,
$d = \sqrt{(x_2 - x_1)^2 + (y_2 - y_1)^2}$, to find the length of the line segment between these ordered pairs.

   a. $(4, 6)$ and $(-1, -6)$ 13

   b. $(-3, 2)$ and $(1, -1)$ 5

   c. $(\sqrt{2}, -\sqrt{2})$ and $(3, 3)$ $\sqrt{22}$

   d. $(\sqrt{3}, \sqrt{5})$ and $(-\sqrt{3}, -\sqrt{5})$ $\sqrt{32} = 4\sqrt{2}$

47. What is the inverse to each set of ordered pairs? Is it a function?

   a. $(3, 3)$, $(4, 5)$, $(5, 7)$ $(3, 3), (5, 4), (7, 5)$; yes

   b. $(-2, -1)$, $(0, -2)$, $(2, -3)$
            $(-1, -2), (-2, 0), (-3, 2)$; yes

48. Make a table with four entries, and graph $y = 2^x + 1$. Make a table for the inverse function, and sketch it on the graph. $y = 2, 3, 5, 9; (2, 0), (3, 1), (5, 2), (9, 3)$; see Additional Answers.

49. Find the inverse function to each equation.

a. $y = 2x - 1$ $y = \frac{1}{2}x + \frac{1}{2}$ **b.** $y = 3x$ $y = \frac{x}{3}$

c. $y = x^3$ $y = \sqrt[3]{x}$ **d.** $y = \sqrt{x}, x \geq 0, y \geq 0$
$y = x^2, x \geq 0, y \geq 0$

e. $y = -x$ $y = -x$ **f.** $y = \frac{1}{x}$ $y = \frac{1}{x}$

50. Find the inverse function algebraically. Include the conditions.

a. $y = \sqrt{x + 2}, x \geq -2, y \geq 0$ $y = x^2 - 2, x \geq 0, y \geq -2$

b. $y = \sqrt{x} - 4, x \geq 0, y \geq -4$ $y = (x + 4)^2, x \geq -4, y \geq 0$

51. State the conditions on x and y needed for these functions to have an inverse function. Find the inverse function and its conditions algebraically.
$x \geq 0, y \geq 1; y \equiv (x - 1)^2, x \geq 1, y \geq 0$

a. $f(x) = \sqrt{x} + 1$ **b.** $f(x) = \sqrt{x - 1}$
$x \geq 1, y \geq 0; y = x^2 + 1, x \geq 0, y \geq 1$

52. What does the notation $f^{-1}(x)$ mean? inverse function of $f(x)$
What does the notation x^{-1} mean? reciprocal of x

In Exercises 53 to 56, set up an equation based on $A = P(1 + r)^t$, and solve it.

53. Suppose that a starting wage of $5.50 has increased to $8.00 over a period of 10 years. What annual rate of increase does this change represent? ≈3.82%

54. Suppose that a starting wage of $5.50 has increased to $9.00 over a period of 10 years. What annual rate of increase does this change represent? ≈5.05%

55. A car cost $7000 new in 1983. A comparable car cost $17,000 in 1995. What annual rate of increase in cost (inflation) does this change represent? ≈7.67%

56. A car cost $10,000 new in 1990. A comparable car cost $15,000 in 1995. What annual rate of increase in cost (inflation) does this change represent? ≈8.45%

57. Solve for V: $r = \sqrt[3]{\dfrac{3V}{4\pi}}$. $V = \frac{4\pi r^3}{3}$

58. Solve the formula for the period of a pendulum, given by $T = 2\pi\sqrt{L/g}$, for g, the acceleration due to gravity. $g = 4L\pi^2/T^2$

59. Solve each of the following for y.

a. $x^3 + y^3 = a^3$ $y = \sqrt[3]{a^3 - x^3}$

b. $x^2 = \dfrac{8}{27k}(y - k)^3$ $y = k + \frac{3}{2}\sqrt[3]{kx^2}$

Solve the equations in Exercises 60 to 63 for the indicated variable.

60. Force of attraction between two charges separated by a distance r: $F = \dfrac{Q_1Q_2}{kr^2}$ for r $r = \sqrt{\dfrac{Q_1Q_2}{kF}}$

61. Gravitational attraction between two masses with r as the distance between their centers: $F = k\,\dfrac{M_1M_2}{r^2}$ for M_1
$M_1 = Fr^2/kM_2$

62. Lemniscate curve: $r^3 = a^2p$ for r $r = \sqrt[3]{a^2p}$

63. Gerono's lemniscate: $x^4 = a^2(x^2 - y^2)$ for y
$y = \pm x\sqrt{1 - x^2/a^2}$

64. a. Solve $\sqrt{8x} - 1 = \sqrt{5x} - 1$ from the graph below.
$x \approx 0.25, x = 2$

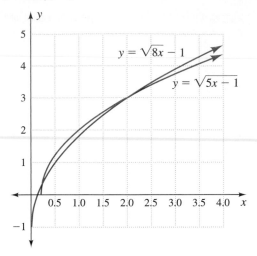

b. Solve $2 = \sqrt{5x} - 1$ with symbols and from the graph. $x = 1$

c. Solve $\sqrt{8x} - 1 = 1$ with symbols and from the graph. $x = \frac{1}{2}$

d. Solve $\sqrt{8x} - 1 = \sqrt{5x} - 1$ with symbols. $\{\frac{2}{9}, 2\}$

65. a. Solve $\sqrt{3x} + 2 = \sqrt{6x} - 8$ from the graph below.
$x = 12$

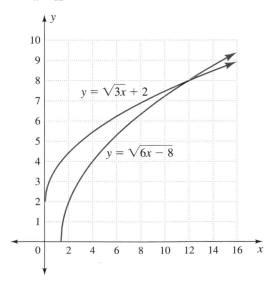

b. Solve $\sqrt{6x} - 8 = 4$ with symbols and from the graph. $x = 4$

c. Solve $\sqrt{3x} + 2 = 2$ with symbols and from the graph. $x = 0$

d. Solve $\sqrt{3x} + 2 = \sqrt{6x} - 8$ with symbols. $x = 12$

66. a. Solve $\sqrt{2x} + 1 = \sqrt{3x - 5}$ from the graph below. $x = 18$

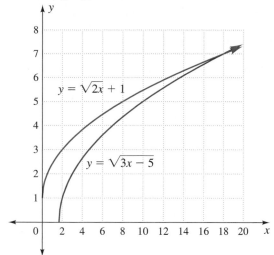

$y = \sqrt{2x} + 1$

$y = \sqrt{3x - 5}$

b. Solve $\sqrt{2x} + 1 = 5$ with symbols and from the graph. $x = 8$

c. Solve $\sqrt{3x - 5} = 0$ with symbols and from the graph. $x = \frac{5}{3}$

d. Solve $\sqrt{2x} + 1 = \sqrt{3x - 5}$ with symbols. $x = 18$

▬ Extended Problems

67. Simplify. Write your observations about patterns in the answers. Explain why the patterns hold. Show four significant digits. three sets of digits related to every third power of 10

a. $\sqrt[3]{8000}$ 20

b. $\sqrt[3]{800}$ ≈ 9.283

c. $\sqrt[3]{80}$ ≈ 4.309

d. $\sqrt[3]{8}$ 2

e. $\sqrt[3]{0.8}$ ≈ 0.9283

f. $\sqrt[3]{0.08}$ ≈ 0.4309

g. $\sqrt[3]{27,000}$ 30

h. $\sqrt[3]{2700}$ ≈ 13.92

i. $\sqrt[3]{270}$ ≈ 6.463

j. $\sqrt[3]{27}$ 3

k. $\sqrt[3]{2.7}$ ≈ 1.392

l. $\sqrt[3]{0.27}$ ≈ 0.6463

68. According to a 1995 publication of the Hanford Health Information Network, radioactivity released by the Hanford Nuclear Site into the Columbia River between World War II and 1970 included the amounts listed below. Calculate the total disintegrations per second by multiplying the amount released by the curie number. A curie is 37×10^9 disintegrations per second. Write the answers in both scientific notation and decimal notation.

a. Arsenic-76: 2,500,000 curies
$9.25 \times 10^{16} = 92,500,000,000,000,000$

b. Neptunium-239: 6,300,000 curies
$2.331 \times 10^{17} = 233,100,000,000,000,000$

c. Phosphorus-32: 230,000 curies
$8.51 \times 10^{15} = 8,510,000,000,000,000$

d. Sodium-24: 12,000,000 curies
$4.44 \times 10^{17} = 444,000,000,000,000,000$

e. Zinc-65: 490,000 curies
$1.813 \times 10^{16} = 18,130,000,000,000,000$

A disintegration is the release of an alpha, beta, or gamma particle.

---

8 Chapter Test

Simplify the expressions in Exercises 1 to 4. Assume all variables represent positive numbers.

1. a. $\dfrac{a^0 b^2}{b^{-1}}$ b^3

b. $\dfrac{a^{-2} b^3}{(bc)^2}$ $\dfrac{b}{a^2 c^2}$

c. $\left(\dfrac{x^2 y}{3x}\right)^{-2}$ $\dfrac{9}{x^2 y^2}$

2. a. $(x^{-1} y^3)^2$ $\dfrac{y^6}{x^2}$

b. $a^6 \cdot a^3$ a^9

c. $\dfrac{x}{x^n}$ $\dfrac{1}{x^{n-1}}$

3. a. $\dfrac{b^{3/4}}{b^2}$ $\dfrac{1}{b^{1.25}}$

b. $\left(\dfrac{a^{1/3}}{b^{1/2}}\right)^6$ $\dfrac{a^2}{b^3}$

c. $(x^{3/4} y^{2/3})^6$ $x^{4.5} y^4$

4. a. $(b^2)^{3/2}$ b^3

b. $b^{2/3} b^{1/3}$ b

c. $3 \cdot 3^n$ 3^{n+1}

5. Explain why $2x^2 \neq (2x)^2$. $(2x)^2 = 2^2 \cdot x^2 = 4x^2$

6. Show whether $a\left(\dfrac{b}{a}\right)^4 = b\left(\dfrac{b}{a}\right)^3$. Yes; both equal $\dfrac{b^4}{a^3}$.

7. a. Write in decimal notation: 3.450×10^{-5}.
0.000 034 50

b. Simplify $\dfrac{6.3 \times 10^{-2}}{7 \times 10^4}$ 9×10^{-7} or 0.000 000 9

8. Explain why $\sqrt{4x^2} \neq 2x$. $\sqrt{4x^2} = \sqrt{4} \cdot \sqrt{x^2} = 2|x|$

9. Simplify; note restrictions so that the outputs are real numbers.

a. $\sqrt{36x}$ $6\sqrt{x}, x \geq 0$

b. $\sqrt{0.36x^2}$ $0.6|x|$

c. $\sqrt{360x^4}$ $6x^2\sqrt{10}$

10. Change each expression to radical notation. Simplify.

a. $64^{2/3}$ $(\sqrt[3]{64})^2 = 16$

b. $32^{4/5}$ $(\sqrt[5]{32})^4 = 16$

c. $81^{3/4}$ $(\sqrt[4]{81})^3 = 27$

11. Simplify.

a. $\sqrt[4]{\dfrac{16}{81}}$ $\dfrac{2}{3}$

b. $\sqrt[3]{-27}$ -3

c. $-\sqrt[3]{-64}$ 4

12. Simplify.

a. $\sqrt[3]{\sqrt[5]{x^{10}}}$ $x^{2/3}$

b. $\sqrt[3]{x^2}\sqrt[3]{x^4}$ x^2

c. $\dfrac{\sqrt[3]{54x^2}}{\sqrt[3]{128x^{-1}}}$ $\dfrac{3x}{4}$

13. Simplify by performing the indicated operation.

a. $(2 - \sqrt{x})(2 + \sqrt{x})$ $4 - x$

b. $(8 - \sqrt{3})(3 + \sqrt{8})$ $24 - 3\sqrt{3} + 16\sqrt{2} - 2\sqrt{6}$

14. Rationalize the denominators.

a. $\dfrac{1}{x + \sqrt{y}}$ $\dfrac{x - \sqrt{y}}{x^2 - y}$

b. $\dfrac{1}{\sqrt[3]{x}}$ $\dfrac{(\sqrt[3]{x})^2}{x}$

15. Substitute $x = 4 - \sqrt{10}$ into $x^2 - 8x + 6 = 0$ to show whether it is a solution. yes; $(4 - \sqrt{10})^2 - 8(4 - \sqrt{10}) + 6 = 0$

16. Find the amount of money in an account if $2000 is deposited at 8% interest compounded quarterly and the money is left for 3 years. $2536.48

17. A car cost $12,000 new in 1990. A comparable car cost $18,000 in 1995. What annual rate of increase in cost (inflation) does this change represent? $\approx 8.45\%$

18. Solve the formula $M = \dfrac{mgl}{\pi r^2 s}$ for r. $r = \sqrt{\dfrac{mgl}{\pi s M}}$

19. Solve the formula $d = \dfrac{Wl^3}{3EI}$ for l. $l = \sqrt[3]{\dfrac{3dEI}{W}}$

20. What is the inverse to the function described by the ordered pairs? Is it a function?

a. $(-1, -1), (-2, -3)$, and $(-3, -5)$
$(-1, -1), (-3, -2), (-5, -3)$; yes

b. $(-1, -1), (-2, -1)$, and $(-3, -1)$
$(-1, -1), (-1, -2), (-1, -3)$; no

21. Make a table containing four ordered pairs, and sketch a graph of $y = \left(\frac{1}{2}\right)^x$. Make a table for the inverse function to $y = \left(\frac{1}{2}\right)^x$, and sketch its graph. See Answer Section.

22. Find the inverse function to $3x - 2y = 6$.
$3y - 2x = 6$ or $y = \frac{2}{3}x + 2$

23. a. Solve $\sqrt{x + 1} = \sqrt{2x} - 1$ from the graph below. $x = 8$

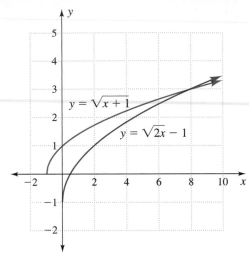

b. Solve $\sqrt{x + 1} = 2$ with symbols. Confirm with the graph. $x = 3x$

c. Solve $\sqrt{2x} - 1 = 1$ with symbols and from the graph. $x = 2$

d. Solve $\sqrt{x + 1} = \sqrt{2x} - 1$ with symbols. $x = 8$

Cumulative Review of Chapters 1 to 8

1. Solve $5 - 3(x + 2) = 32$. $x = -11$

2.. Simplify $\dfrac{6 + 9x}{15} \cdot \dfrac{2 + 3x}{5}$.

3. Solve for b: $A = \dfrac{a + b + c}{3}$. $b = 3A - a - c$

4. Solve $3 - x < 16 < 10 - x$ for $a < x < b$ form.
$-13 < x < -6$

5. In January, the County Electric bill was $68.46 for 1120 kilowatt hours of electricity. In February, the bill was $83.33 for 1410 kilowatt hours. Define variables and find the equation County Electric uses for billing. What assumptions did you make? Let y = total cost in $.
x = kilowatt hours used; assume a linear equation; $y \approx 11.03 + 0.051x$

6. If $f(x) = 3 - x$, what is $f(3)? f(-2)?$ $0; 5$

7. a. What is the slope of the line AB connecting $A(4, -2)$ and $B(-1, 1)$? $-\frac{3}{5}$

b. What is the slope of a line perpendicular to AB? $\frac{5}{3}$

8. Find a linear function through the origin that is parallel to $3y - 6x = 3$. $y = 2x$

9. What are the set of inputs (domain) and set of outputs (range) for the constant function $y = 4$? $\mathbb{R}; 4$

10. Solve the system of equations:

$$2x + 3y = z - 6$$
$$2x + z = 14 + 3y$$
$$3x + y + z = 4 \qquad (2, -3, 1)$$

11. A student has 50 credit hours and a 2.72 grade point average (GPA). How many credit hours of As at 4 points per credit hour are needed to raise the student's GPA to a 3.00? 14 credit hours

12. A credit card company charges 20% interest. A credit union loan costs 7.5%. What ratio of borrowing is needed to keep the average interest at 10%?
$4 at credit union for each $1 on credit card

13. Show whether the lengths 3.5, 12, and 12.5 are the sides of a right triangle. yes; $12.5^2 = 3.5^2 + 12^2$

14. Let $f(x) = 2x^2 - 5x + 6$.

a. Find $f(-2)$ and $f(0)$. $24; 6$

b. What is the meaning of $f(0)$ on the graph of $f(x)$?
y-intercept or $f(x)$-intercept

c. What is the vertex of $f(x)$? $(1\frac{1}{4}, 2\frac{7}{8})$

d. Show the quadratic formula and use it to find the x-intercepts. $x = \dfrac{-b \pm \sqrt{b^2 - 4ac}}{2a}$; no real x-intercepts

15. Solve each equation using the graph in the figure.

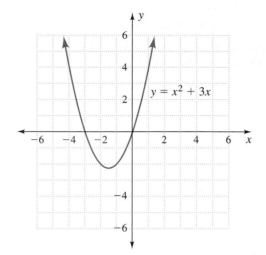

$y = x^2 + 3x$

a. $x^2 + 3x = 4$ $\{-4, 1\}$ **b.** $x^2 + 3x = 0$ $\{-3, 0\}$

c. $x^2 + 3x = -5$ no real-number solution

16. Solve the equations in parts a and b of Exercise 15 by factoring.

17. Solve $x^2 + 3x > 0$ from the figure in Exercise 15.
$x < -3$ or $x > 0$

18. Complete the square:

$$(x + \tfrac{7}{2})^2 = x^2 + 7x + \tfrac{49}{4}$$

19. Multiply $(x^3 - 3x^2 + 3x - 1)(x - 1)$. $x^4 - 4x^3 + 6x^2 - 4x + 1$

20. Give the next number in each pattern. Identify whether the sequence is from a linear function, a quadratic function, or neither. Fit an equation to the linear and quadratic sequences.

a. 2, 5, 10, 17, 26
37; quadratic; $y = x^2 + 1$

b. $-5, -2, 1, 4, 7$
10; linear; $y = 3x - 8$

c. $-3, 1, 7, 15, 25$
37; quadratic; $y = x^2 + x - 5$

d. $-2, -1, 1, 5, 13$
29; neither

21. Multiply and simplify using the properties of imaginary numbers.

a. $(8 - 2i)(8 + 2i)$ 68

b. $(5 - 4i)(3 + 2i)$ $23 - 2i$

c. $\dfrac{1}{2 + i} \cdot \dfrac{2 - i}{2 - i}$ $\dfrac{2 - i}{5}$

22. Find all the solutions, real and imaginary, to $0 = x^3 - 1$. $\left\{1, -\frac{1}{2} \pm i\sqrt{3}/2\right\}$

23. Describe how to get the graph of $y = (x - a)^2$ from that of $y = x^2$. shift $y = x^2$ a units horizontally

24. Suppose the dive path of the clavadistas (cliff divers) is approximated by the equation $y = 0.09x - 0.16x^2$, where x and y are in feet. Let the diver start at the origin and let the surface of the water be 115 feet below the diver.

   a. What is the vertex of the dive? ≈(0.28, 0.01)

   b. What is the horizontal distance the diver has traveled when he reaches the water? ≈27.1 ft

25. Tender Touch Diapers cost $8.99 per package of 56 diapers. Suppose a baby uses 10 diapers each day. Set up a unit analysis to find the cost of a 365-day supply of diapers. $\dfrac{\$8.99}{\text{package}} \cdot \dfrac{1\ \text{package}}{56\ \text{diapers}} \cdot \dfrac{10\ \text{diapers}}{1\ \text{day}} \cdot 365\ \text{days} \approx \$585.96$

Match each expression in Exercises 26 to 31 with the equivalent expression below.

(a) $\dfrac{x}{y+x}$ (b) $1 + \dfrac{b}{a-b}$ (c) $\dfrac{x}{y} + 1$

(d) $\dfrac{x}{y-x} + \dfrac{y}{x-y}$ (e) $\dfrac{1}{b} + \dfrac{1}{a}$ (f) $\dfrac{-a}{a-1}$

26. $\dfrac{x+y}{y}$ c   **27.** $\dfrac{a+b}{ab}$ e

28. $\dfrac{x}{x+y}$ a   **29.** $\dfrac{a}{1-a}$ f

30. $\dfrac{a}{a-b}$ b   **31.** $\dfrac{x-y}{y-x}$ d

32. Multiply and simplify: $\dfrac{x^2+3x}{x-1} \cdot \dfrac{x^2-1}{3x^2}$ $\frac{(x+3)(x+1)}{3x}$

33. Divide $x^3 - 1$ by $x - 1$. $x^2 + x + 1$

34. The ideal Body Mass Index (BMI) relates height and weight. The formula

$$19 \le \frac{(150)(704.5)}{x^2} \le 24$$

shows the BMIs for people weighing 150 pounds. The variable x represents height in inches. For what range of heights in inches should people weigh 150 pounds? ≈66 in. ≤ x ≤ 75 in.

35. Change to decimal notation:

   a. 3.80×10^{-16} **b.** 4.23×10^{15}
   0.000 000 000 000 000 380   4,230,000,000,000,000

36. A capsule of three types of bacteria contains at least 0.825 billion each of Lactobacillus acidophilus, Lactobacillus bifidus, and Streptococcus faecium. What is the total bacterial content of the capsule? Write the total in both decimal notation and scientific notation. $2.475 \times 10^9 = 2,475,000,000$

37. Simplify these expressions. Assume all the variables represent positive numbers.

   a. $32^{4/5}$ 16   **b.** $\sqrt[3]{125^2}$ 25   **c.** $\sqrt[4]{-81}$
                                                no real number

   d. $-\sqrt{2}\sqrt{8}$ **e.** $x^{3/5} \cdot x^{1/3}$ **f.** $\dfrac{x^{1/2}}{x^{-3/2}}$ x^2
     -4 $x^{14/15}$

   g. $\sqrt{18a^3}$ $3a\sqrt{2a}$ **h.** $\sqrt[3]{16x^3y^4z^5}$ **i.** $(x^{2/3})^{1/2}$ $x^{1/3}$
                                     $2xyz\sqrt[3]{2yz^2}$

   j. $\sqrt{8x} + \sqrt{2x} + \sqrt{50x}$ $8\sqrt{2x}$

   k. $(2 - \sqrt{2})(2 + \sqrt{2})$ 2

38. Solve for x in the following equations. Use algebraic notation, and then graph the left and right sides to check.

   a. $x + 2 = 3\sqrt{x}$ {1, 4}   **b.** $x = \sqrt{6-x}$ $x = 2$

39. Find the annual rate of interest if $1500 grows to $2500 when interest is compounded annually for 6 years. Repeat for interest compounded monthly. ≈8.89%; ≈8.54%

40. Solve for B in Steinmetz's equation, related to the magnetization of iron or steel: $W = nB^{1.6}$. (*Hint:* Change the decimal exponent to a fraction.) $B = \left(\dfrac{W}{n}\right)^{5/8}$

41. The length of the line between two ordered pairs is

$$d = \sqrt{(x_2 - x_1)^2 + (y_2 - y_1)^2}$$

Find the length of the three segments connecting $(-2, 2)$, $(4, 1)$, and $(3, -5)$. Identify the resulting figure as completely as possible.
$(-2, 2)$ & $(4, 1)$, $d = \sqrt{37}$; $(-2, 2)$ & $(3, -5)$, $d = \sqrt{74}$; $(4, 1)$ & $(3, -5)$, $d = \sqrt{37}$; isosceles right triangle

Exponential and Logarithmic Functions

Chemists rate common hair products according to their pH, a measure of acidity or basicity. Acid-balanced shampoos, bleaches, and hair straighteners all have a pH number; see Figure 1. The pH scale is based on the exponential and logarithmic functions of this chapter. We will work with pH in Section 9.4.

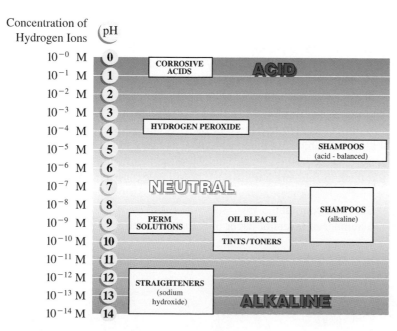

CHAPTER 9

Concentration of Hydrogen Ions — pH

10^{-0} M	**0**	CORROSIVE ACIDS — ACID
10^{-1} M	**1**	
10^{-2} M	**2**	
10^{-3} M	**3**	
10^{-4} M	**4**	HYDROGEN PEROXIDE
10^{-5} M	**5**	SHAMPOOS (acid - balanced)
10^{-6} M	**6**	
10^{-7} M	**7**	NEUTRAL
10^{-8} M	**8**	SHAMPOOS (alkaline)
10^{-9} M	**9**	PERM SOLUTIONS — OIL BLEACH
10^{-10} M	**10**	TINTS/TONERS
10^{-11} M	**11**	
10^{-12} M	**12**	STRAIGHTENERS (sodium hydroxide)
10^{-13} M	**13**	ALKALINE
10^{-14} M	**14**	

FIGURE 1

9.1 Exponential Functions

Objectives

▮ Identify the base and exponent in an exponential expression.

▮ Use differences to identify a potential exponential function.

▮ Use the first term and common ratio to find the nth term.

▮ Use calculator regression to find an exponential function.

▮ Identify nth term expressions that are not exponential functions.

▮ Use properties of exponents to show that two expressions are (or are not) equivalent.

WARM-UP

Give the next number in each sequence.

1. 3, 6, 12, 24, _48_
2. 5, 8, 11, 14, _17_
3. 1, 4, 9, 16, _25_
4. 9, 5, 1, −3, _−7_
5. 3, 10, 21, 36, _55_
6. 81, 27, 9, 3, _1_
7. 4, 12, 36, 108, _324_

IN THIS SECTION, we explore patterns created by exponential expressions. We find exponential equations and use the definition of an exponential function and properties of exponents to explain some surprising results.

Patterns

In Examples 1 and 2, we consider two applications that generate patterns.

EXAMPLE 1   **Exploring patterns: steel in Japanese swords**   The traditional Japanese sword is made from steel heated to a temperature higher than 1200°C (2200°F), hammered, folded, and hammered and folded again many times. This process removes impurities and creates a blade that does not break or bend.

How many layers of steel would there be in a sword blade created by heating and folding the steel in half ten times?

a. Model the folding process with a piece of paper (three folds are shown in Figure 2). Then make a table with number of folds as input and number of layers of paper in the stack as output. When you get to the point where the paper becomes too thick to fold, predict the number of layers of paper that would be in the stack.

b. Graph the data from your table.

c. Describe the number of layers of paper in the stack in terms of the number of folds.

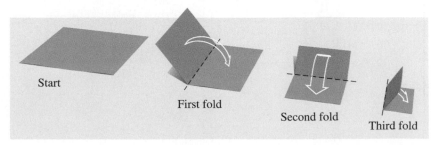

Start   First fold   Second fold   Third fold

FIGURE 2

SOLUTION **a.** Each fold doubles the number of layers of paper. Table 1 shows the numbers of layers.

b. The graph of the data from Table 1 is shown in Figure 3. The points are connected with a dashed line to show the shape of the curve.

TABLE 1

Fold, h	Layers of Paper, y
1	2
2	$2 \cdot 2 = 4$
3	$2 \cdot 2 \cdot 2 = 8$
4	$2 \cdot 2 \cdot 2 \cdot 2 = 16$
5	$2 \cdot 2 \cdot 2 \cdot 2 \cdot 2 = 32$
n	$\underbrace{2 \cdot 2 \cdot \cdots \cdot 2}_{n \text{ factors of } 2} = 2^n$

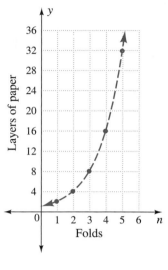

FIGURE 3

c. The number of layers of paper in the stack is always a power of 2. Thus, for n folds, the number of layers of paper in the stack is $y = 2^n$. The number of layers of steel in the sword after 10 folds is $2^{10} = 1024$.

EXAMPLE 2 **Exploring patterns: superball bounces** Suppose a superball bounces to $\frac{2}{3}$ of its previous height with each successive bounce. We start by dropping the ball from a height of 108 inches (see Figure 4).

a. Make a table for the first five heights of the ball. Use 1 to 5 as inputs and the heights of the ball as outputs.

b. Graph the data from your table.

c. Describe the nth height of the ball.

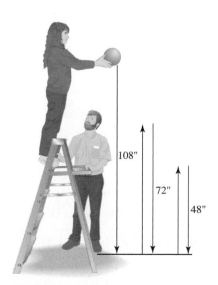

108"
72"
48"

FIGURE 4

SOLUTION   **a.** Table 2 shows the height to be $\frac{2}{3}$ of the prior height.

b. The graph of the data from Table 2 is in Figure 5. The points are connected with a dashed line to show the shape of the curve.

TABLE 2

Input, n	Output: Height, y
1	108
2	$108\left(\frac{2}{3}\right) = 72$
3	$108\left(\frac{2}{3}\right)\left(\frac{2}{3}\right) = 48$
4	$108\left(\frac{2}{3}\right)\left(\frac{2}{3}\right)\left(\frac{2}{3}\right) = 32$
5	$108\left(\frac{2}{3}\right)\left(\frac{2}{3}\right)\left(\frac{2}{3}\right)\left(\frac{2}{3}\right) = 21\frac{1}{3}$
n	$108\underbrace{\left(\frac{2}{3}\right)\left(\frac{2}{3}\right)\cdot\ \cdot\ \cdot\ \cdot\ \left(\frac{2}{3}\right)}_{n-1\ \text{factors of}\ \frac{2}{3}}$

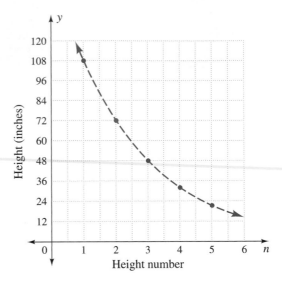

FIGURE 5

c. For the nth entry in the table, there are $n - 1$ factors of $\frac{2}{3}$. The table suggests that the nth height will be $y = 108\left(\frac{2}{3}\right)^{n-1}$.   ▬

Exponential Functions

The equations for the layers of paper and the height of bounces illustrate exponential functions.

▬ **DEFINITION OF EXPONENTIAL FUNCTION**

> A **exponential function** is written as
>
> $$f(x) = a \cdot b^x$$
>
> where the coefficient a is a constant, the base b is positive but not equal to 1, and the exponent x is a real number.

EXAMPLE 3   **Naming the parts in an exponential equation**   Identify the coefficient, the base, and the exponent in each of these exponential functions.

a. $y = 3 \cdot 2^x$

b. $y = 1000(1 + 0.065)^t$

c. $y = 19{,}000(1 - r)^t$, r constant

d. $y = 2000\left(1 + \dfrac{0.08}{12}\right)^{12t}$

SOLUTION   **a.** Coefficient is 3; base is 2; exponent is x.

b. Coefficient is 1000; base is $(1 + 0.065)$; exponent is t.

c. Coefficient is 19,000; base is $(1 - r)$; exponent is t.

d. Coefficient is 2000; base is $\left(1 + \dfrac{0.08}{12}\right)$; exponent is $12t$.   ▬

We will work with two ways to identify exponential functions: with number patterns (sequences) and with graphs (Section 9.2).

Identifying Exponential Functions

Recall that a sequence is a function with inputs from the set of positive integers. Because the inputs in Examples 1 and 2 are positive integers—the number of folds and the number of heights—we can think of the outputs as sequences. We identify linear (arithmetic) sequences with constant first differences and quadratic sequences with constant second differences.

In Example 4, we find whether any common differences occur for the sequences described by 2^n, $108\left(\frac{2}{3}\right)^{n-1}$.

EXAMPLE 4   **Finding differences**   Find the first and second differences for the sequences given by

a. 2^n
b. $108\left(\frac{2}{3}\right)^{n-1}$

SOLUTION   **a.** 2^n: 2, 4, 8, 16, 32

$\vee$ $\vee$ $\vee$ $\vee$

2, 4, 8, 16   The first differences are the original sequence.

(Check that the second differences are also the original sequence.)

Students will want to try differences first, so encourage that step. They will quickly see that it leads to a different type of sequence.

b. $108\left(\frac{2}{3}\right)^{n-1}$:   108   72   48   32 $21\frac{1}{3}$

$\vee$ $\vee$ $\vee$ $\vee$

36   24   16 $10\frac{2}{3}$ The first differences are a third of the numbers in the original sequence.

$\vee$ $\vee$ $\vee$

12   8 $5\frac{1}{3}$ The second differences are a third of the numbers in the first differences.

In Example 4, *the differences are either the same as or a multiple of the numbers in the original sequence.* This distinctive pattern suggests a new type of sequence. The repetition of the sequence in the differences occurs in various types of sequences (see Exercise 53). We need a more informative tool to identify these sequences.

FINDING RATIOS   Considering how the sequences in Examples 1 and 2 are formed—by multiplying by a constant—gives us another way to describe the sequences. Instead of subtracting terms of the sequence, we *divide* the terms.

EXAMPLE 5   **Finding ratios**   For each sequence, find the ratio of consecutive terms by dividing each term by the preceding term, a_n/a_{n-1}.

a. 2^n: 2, 4, 8, 16, 32
b. $108\left(\frac{2}{3}\right)^{n-1}$: 108, 72, 48, 32, $21\frac{1}{3}$

SOLUTION   We divide the second term by the first, the third by the second, and so forth.

a. $\dfrac{4}{2} = 2, \dfrac{8}{4} = 2, \dfrac{16}{8} = 2,$ and $\dfrac{32}{16} = 2.$ Every ratio is 2.

b. $\dfrac{72}{108} = \dfrac{2}{3}, \dfrac{48}{72} = \dfrac{2}{3}, \dfrac{32}{48} = \dfrac{2}{3},$ and $\dfrac{21\frac{1}{3}}{32} = \dfrac{2}{3}.$ Every ratio is $\dfrac{2}{3}.$

Think about it 1: Show, without a calculator, that $\dfrac{21\frac{1}{3}}{32} = \dfrac{2}{3}.$

The ratio obtained from the consecutive terms of a sequence is the **common ratio, r.** That the ratios 2 and $\frac{2}{3}$ should appear so prominently in $y = 2^n$ and $y = 108\left(\frac{2}{3}\right)^{n-1}$ is no coincidence. The second equation is in the standard nth term form.

> The ***n*th term**, a_n, of a sequence with common ratio r is
>
> $$a_n = a_1 r^{n-1}$$
>
> where a_1 is the first term in the sequence.

We show that $2^n = 2 \cdot 2^{n-1}$ in Example 6.

EXAMPLE 6 Finding the *n*th term   We return to the sequence 2^n, $n = 1, 2, 3, \ldots$.
a. Write the *n*th term for the sequence 2, 4, 8, 16, 32, ...
b. Show that the *n*th term simplifies to 2^n.

SOLUTION **a.** The first term, a_1, is 2. The common ratio, r, is 2.

$$a_n = a_1 \cdot r^{n-1} = 2 \cdot 2^{n-1}$$

b. $2 \cdot 2^{n-1} = 2^1 \cdot 2^{n-1}$ Use $a^m \cdot a^n = a^{m+n}$.

$\qquad\qquad = 2^{1+(n-1)}$ $1 + (-1) = 0$

$\qquad\qquad = 2^n$

In Example 9, we will do more work like that in Example 6.

▬ Sequences in which *the ratio of each term to the preceding term is the same for all terms* are called **geometric sequences.** From the definition of exponential functions, we observe that geometric sequences for which the ratio (the base!) is positive but not equal to 1 are also exponential functions.
  In Example 7, we find the *n*th term expressions for several sequences.

EXAMPLE 7 Finding rules with the *n*th term   Use the *n*th term expression to find the rules for these sequences. Is the rule an exponential function?
a. 4, 12, 36, 108, 324
b. 3, 6, 12, 24, 48
c. 81, 27, 9, 3, 1
d. 5, −10, 20, −40, 80

SOLUTION **a.** $\frac{12}{4} = 3, \frac{36}{12} = 3, \frac{108}{36} = 3, \frac{324}{108} = 3$. The common ratio is 3; the first term is 4. The *n*th term is $a_n = 4 \cdot 3^{n-1}$. The base, 3, is positive, so this sequence is an exponential function having a natural number domain: $f(x) = 4 \cdot 3^{x-1}$.

b. $\frac{6}{3} = 2, \frac{12}{6} = 2, \frac{24}{12} = 2, \frac{48}{24} = 2$. The common ratio is 2; the first term is 3. The *n*th term is $a_n = 3 \cdot 2^{n-1}$. The base, 2, is positive, so this sequence is an exponential function having a natural number domain: $f(x) = 3 \cdot 2^{x-1}$.

c. $\frac{27}{81} = \frac{1}{3}, \frac{9}{27} = \frac{1}{3}, \frac{3}{9} = \frac{1}{3}, \frac{1}{3} = \frac{1}{3}$. The common ratio is $\frac{1}{3}$; the first term is 81. The *n*th term is $a_n = 81 \cdot \left(\frac{1}{3}\right)^{n-1}$. The base, $\frac{1}{3}$, is positive, so this sequence is an exponential function having a natural number domain: $f(x) = 81 \cdot \left(\frac{1}{3}\right)^{x-1}$.

d. $-10/5 = -2, 20/-10 = -2, -40/20 = -2, 80/-40 = -2$. The common ratio is −2; the first term is 5. The *n*th term is $a_n = 5 \cdot (-2)^{n-1}$. The base, −2, is negative. This sequence is not an exponential function. ▬

EXPONENTIAL REGRESSION   In addition to linear, quadratic, and power regression, discussed in Chapters 2, 4, and 8, graphing calculators also have exponential regression.

■ GRAPHING CALCULATOR TECHNIQUE:
EXPONENTIAL REGRESSION

Under $\boxed{\text{STAT}}$:

Clear lists, using **4 : Clr List**.

Under EDIT:

In list 1, L1, place the sequence inputs or term numbers, such as 1, 2, 3, 4, 5.
In list 2, L2, place the terms of the sequence, such as 3, 9, 27, 81, 243.

Under $\boxed{\text{STAT}}$ [CALC]:

Calculate an exponential regression, using option **0 : ExpReg** (0 stands for the 10th item), for L1 and L2. (See Figure 6a).

The equation data are given as a and b for $y = ab^x$, followed by the correlation coefficient r. As before, if $r = 1$ or $r = -1$, the equation is a perfect fit for the data. The closer r is to 1 or -1, the better the fit. Zero indicates the worst fit.

In this example, $a = 1$ and $b = 3$, so the exponential function is $y = 1 \cdot 3^x$. (See Figure 6b.)

(a)

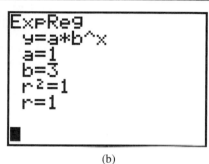

(b)

FIGURE 6

EXAMPLE 8   Using exponential regression   Explain why exponential regression can be applied to each sequence, find the function, and compare the function with the earlier result.

a. 2, 4, 8, 16, 32 (from Example 1)

b. 108, 72, 48, 32, $21\frac{1}{3}$ (from Example 2)

c. 4, 12, 36, 108, 324 (from part a, Example 7)

d. 3, 6, 12, 24, 48 (from part b, Example 7)

e. 81, 27, 9, 3, 1 (from part c, Example 7)

f. 5, -10, 20, -40, 80 (from part d, Example 7)

SOLUTION   We first check to see that the common ratio in each sequence is positive and not equal to 1. If this condition is satisfied, then we use calculator regression to find the exponential function. Assume that $x = n$.

a. The common ratio is $\frac{4}{2} = 2$. Exponential regression gives $y = 1 \cdot 2^x$. This agrees with $y = 2^n$ from Example 1.

b. The common ratio is $\frac{72}{108} = \frac{2}{3}$. Exponential regression gives $y = 162\left(\frac{2}{3}\right)^x$. This does not look like $y = 108\left(\frac{2}{3}\right)^{n-1}$ from Example 2.

c. The common ratio is $\frac{12}{4} = 3$. Exponential regression gives $y = \frac{4}{3} \cdot 3^x$. This does not look like $f(x) = 4 \cdot 3^{x-1}$ from part a of Example 7.

d. The common ratio is $\frac{6}{3} = 2$. Exponential regression gives $y = 1.5 \cdot 2^x$. This does not look like $f(x) = 3 \cdot 2^{x-1}$ from part b of Example 7.

e. The common ratio is $\frac{27}{81} = \frac{1}{3}$. Exponential regression gives $y = 243 \cdot \left(\frac{1}{3}\right)^x$. This does not look like $f(x) = 81 \cdot \left(\frac{1}{3}\right)^{x-1}$ from part c of Example 7.

f. Although we observed that this sequence is not an exponential function, we enter 1, 2, 3, 4, 5 in L1 and 5, −10, 20, −40, 80 in L2 and choose exponential regression. We obtain a "Domain Error," because the calculator detects the negative base, in violation of the definition of an exponential function (see Figure 7). We cannot fit an exponential function to the sequence.

FIGURE 7

Applying Properties of Exponents

Exponential regression gives expressions in $y = ab^x$ notation, rather than nth term form, $a_n = ar^{n-1}$. In Example 9, we show that the two expressions in parts b, c, d, and e of Example 8 are equivalent. Try the example yourself before looking up the properties or reading the solution.

EXAMPLE 9   **Showing exponential expressions to be equivalent**   Apply the definitions and properties of exponents (pages 436–440) to show that these pairs of exponential expressions are equivalent.

a. $108\left(\frac{2}{3}\right)^{x-1}$ and $162\left(\frac{2}{3}\right)^{x}$

b. $4 \cdot 3^{x-1}$ and $\frac{4}{3} \cdot 3^{x}$

c. $3 \cdot 2^{x-1}$ and $1.5 \cdot 2^{x}$

d. $81 \cdot \left(\frac{1}{3}\right)^{x-1}$ and $243 \cdot \left(\frac{1}{3}\right)^{x}$

SOLUTION   The steps are described in part a. Parts b, c, and d follow similar steps.

a. $108\left(\frac{2}{3}\right)^{x-1}$ Change subtraction to adding the opposite.

$= 108\left(\frac{2}{3}\right)^{x+(-1)}$ Apply the product property: $b^{m+n} = b^m \cdot b^n$.

$= 108\left(\frac{2}{3}\right)^{x} \cdot \left(\frac{2}{3}\right)^{-1}$ Apply the reciprocal property: $b^{-1} = \dfrac{1}{b}$.

$= 108\left(\frac{2}{3}\right)^{x} \cdot \left(\frac{3}{2}\right)$ Use the commutative property of multiplication.

$= 108 \cdot \left(\frac{3}{2}\right) \cdot \left(\frac{2}{3}\right)^{x}$ Multiply the first two factors.

$= 162\left(\frac{2}{3}\right)^{x}$

b. $4 \cdot 3^{x-1} = 4 \cdot 3^{x} \cdot 3^{-1} = 4 \cdot 3^{x} \cdot \frac{1}{3} = 4 \cdot \frac{1}{3} \cdot 3^{x} = \frac{4}{3} \cdot 3^{x}$

c. $3 \cdot 2^{x-1} = 3 \cdot 2^{x} \cdot 2^{-1} = 3 \cdot 2^{x} \cdot \frac{1}{2} = 3 \cdot \frac{1}{2} \cdot 2^{x} = 1.5 \cdot 2^{x}$

d. $81 \cdot \left(\frac{1}{3}\right)^{x-1} = 81 \cdot \left(\frac{1}{3}\right)^{x} \cdot \left(\frac{1}{3}\right)^{-1} = 81 \cdot \left(\frac{1}{3}\right)^{x} \cdot 3 = 81 \cdot 3 \cdot \left(\frac{1}{3}\right)^{x} = 243 \cdot \left(\frac{1}{3}\right)^{x}$

Application: Rate of Growth

An exponential regression equation can be the source of r, the annual rate of growth in $S = P(1 + r)^t$, as demonstrated in Example 10.

EXAMPLE 10   **Finding r, the annual rate of growth (or decline): population decline**   Use the population facts in Table 3 for Sharon, Pennsylvania, to examine a case of population decline.

TABLE 3   *Population of Sharon, Pennsylvania*

Year	Years since 1930, x	Population, y
1930	0	25,908
1940	10	25,622
1980	50	19,057
1990	60	17,533

a. Plot the data in Table 3.
b. Fit an exponential regression equation to the data. Graph the equation.
c. What is the annual population change, r?
d. Predict the population in the year 2000.

SOLUTION   **a.** We let x be the number of years after 1930. The data are plotted in Figure 8.

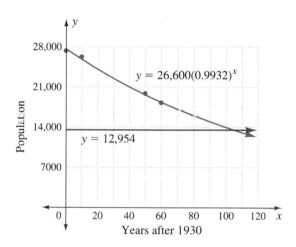

FIGURE 8

b. The exponential regression equation is $y \approx 26{,}600(0.9932)^x$. Because the exponential regression equation is fit to all four pieces of data, it does not have the 1930 population as its initial, or starting, value. Had we used only the first and last years' data, the equation would have been $y \approx 25{,}908(0.9935)^x$.

c. To estimate the annual population change, we compare the base of the exponential equation $y = 26{,}600(0.9932)^x$ with that of $S = P(1 + r)^t$. We find that $0.9932 = 1 + r$ and $-0.0068 = r$. The average annual population change is $r \approx -0.68\%$ per year.

d. For the year 2000, $x = 2000 - 1930 = 70$. Substituting $x = 70$ into the regression equation gives

$$f(70) = 26{,}600(0.9932)^{70} \approx 16{,}500$$

The actual population in the 2000 census was 16,328. Our model fit the data well. ▬

Think about it 2: Why might an exponential function represent population decline more realistically than a linear function?

■ MAKING A DISTINCTION:

There are two important differences (see Table 4) between the exponential function and the geometric sequence.

1. The domains are different. The exponential function is defined for any real-number input, whereas the geometric sequence is limited to positive integer inputs. This means that 3^{π} is a meaningful exponential expression, but the irrational number π would not be a meaningful input to n in $a_n = ar^{n-1}$ in a geometric sequence.

2. The base is more restricted for the exponential function than for the geometric sequence. A geometric sequence may have a positive or a negative base (the common ratio), whereas the base of the exponential function must be positive. The base $b = 1$ is excluded from the exponential function because it represents a *constant* function (see Section 2.6).

TABLE 4 *Differences between Geometric Sequences and Exponential Functions*

Function	Inputs (Domain)	Base
Geometric sequence $a_n = a_1 r^{n-1}$	n is a positive integer	$r < 0$ or $r > 0$
Exponential function $y = ab^x$	x is a real number	$b > 0, b \neq 1$

ANSWER BOX

Warm-up: **1.** 48 **2.** 17 **3.** 25 **4.** -7 **5.** 55 **6.** 1 **7.** 324

Think about it 1: One way to simplify this complex fraction is to multiply both numerator and denominator by the least common denominator, 3:

$$\frac{21\frac{1}{3}}{32} \cdot \frac{3}{3} = \frac{64}{96} = \frac{2 \cdot 32}{3 \cdot 32} = \frac{2}{3}$$

Think about it 2: An exponential function never goes through the x-axis. It is likely that even a ghost town will have one or two residents or a caretaker for property owners. A linear function would pass through the x-axis and into the fourth quadrant, representing a zero and then a negative population.

9.1 Exercises

In Exercises 1 to 4, identify the base and the exponent.

1. a. a^x $a; x$ **b.** $4x^3$ $x; 3$

  c. 2^{-x} $2; -x$ **d.** x^a $x; a$

2. a. $3x^5$ $x; 5$ **b.** 3^x $3; x$

  c. $(x - 2)^3$ $x - 2; 3$ **d.** 3^{x-2} $3; x - 2$

3. a. π^x $\pi; x$ **b.** $2x^4$ $x; 4$

  c. $a_1 r^{n-1}$, r constant **d.** $100(1.06)^t$ $1.06; t$
  $r; n - 1$

4. a. x^{π} $x; \pi$ **b.** $3e^x$ $e; x$

  c. Pe^{rt} $e; rt$ **d.** $4 \cdot 2^n$ $2; n$

In Exercises 5 to 16, do the following:
(a) Give the common ratio for each sequence.
(b) Find the expression for the nth term, a_n.
(c) Find the sequence's equation with exponential regression.
(d) Use the properties of exponents to show that your description of a_n is the same as that given by $y = ab^x$ from exponential regression.

5. 0.5, 1.5, 4.5, 13.5 $3; a_n = 0.5 \cdot 3^{n-1}; y = 0.167 \cdot 3^x$

6. 0.6, 1.8, 5.4, 16.2 $3; a_n = 0.6 \cdot 3^{n-1}; y = 0.2 \cdot 3^x$

7. 10, 20, 40, 80 $2; a_n = 10 \cdot 2^{n-1}; y = 5 \cdot 2^x$

8. $\frac{1}{16}, \frac{1}{8}, \frac{1}{4}, \frac{1}{2}, 1, 2$ $2; a_n = \frac{1}{16} \cdot 2^{n-1}; y = 0.03125 \cdot 2^x$

9. $\frac{1}{4}, \frac{1}{2}, 1, 2, 4$ $2; a_n = \frac{1}{4} \cdot 2^{n-1}; y = 0.125 \cdot 2^x$

10. $8, 16, 32, 64, 128$ $2; a_n = 8 \cdot 2^{n-1}; y = 4 \cdot 2^x$

11. $1, 2, 4, 8, 16$ $2; a_n = 1 \cdot 2^{n-1}; y = 0.5 \cdot 2^x$

12. $4, 2, 1, \frac{1}{2}, \frac{1}{4}$ $\frac{1}{2}; a_n = 4(\frac{1}{2})^{n-1}; y = 8 \cdot 0.5^x$

13. $32, 16, 8, 4, 2$ $\frac{1}{2}; a_n = 32(\frac{1}{2})^{n-1}; y = 64 \cdot 0.5^x$

14. $243, 81, 27, 9$ $\frac{1}{3}; a_n = 243(\frac{1}{3})^{n-1}; y = 729 \cdot 0.333^x$

15. $3, 1, \frac{1}{3}, \frac{1}{9}, \frac{1}{27}$ $\frac{1}{3}; a_n = 3 \cdot (\frac{1}{3})^{n-1}; y = 9 \cdot 0.333^x$

16. $\frac{1}{9}, \frac{1}{3}, 1, 3, 9$ $3; a_n = \frac{1}{9} \cdot 3^{n-1}; y = 0.037 \cdot 3^x$

State whether each sequence in Exercises 17 to 26 is related to a linear function (arithmetic sequence), quadratic function, or exponential function (geometric sequence). Find the equation for each.

17. $2, 8, 18, 32$ quadratic; $y = 2x^2$

18. $-3, -7, -11, -15$ linear; $y = -4x + 1$

19. $0.5, 1.5, 4.5, 13.5$ exponential; $y = 0.167 \cdot 3^x$

20. $7.5, 11, 14.5, 18, 21.5$ linear; $y = 3.5x + 4$

21. $3, 7, 12, 18$ quadratic; $y = 0.5x^2 + 2.5x$

22. $4, 9, 15, 22$ quadratic; $y = 0.5x^2 + 3.5x$

23. $30, 38, 46, 54, 62$ linear; $y = 8x + 22$

24. $1.2, 3.6, 10.8, 32.4$ exponential; $y = 0.4 \cdot 3^x$

25. $5, 10, 20, 40, 80$ exponential; $y = 2.5 \cdot 2^x$

26. $12, 24, 40, 60$ quadratic; $y = 2x^2 + 6x + 4$

In Exercises 27 to 32, change the expression into a product or a quotient (such as $a \cdot 2^x, 2^x/a, a \cdot 3^x,$ or $3^x/a$), using the properties of exponents. (Hint: In Example 9, we changed $a \cdot b^{x-1}$ expressions to $c \cdot b^x$.)

27. 2^{x+2} $4 \cdot 2^x$ **28.** 3^{x-1} $3^x/3$

29. 2^{x-1} $2^x/2$ **30.** 3^{x+2} $9 \cdot 3^x$

31. $(\frac{1}{2})^{-x}$ 2^x **32.** $(\frac{1}{3})^{-x}$ 3^x

In Exercises 33 to 40, change each expression into the form $2^{n \pm a}$ or $3^{n \pm a}$, using the properties of exponents. (Hint: Write the coefficients as powers of the given base.)

33. $2^n/2$ 2^{n-1} **34.** $2 \cdot 2^n$ 2^{n+1}

35. $4 \cdot 2^n$ 2^{n+2} **36.** $2^n/4$ 2^{n-2}

37. $3^n/3$ 3^{n-1} **38.** $3 \cdot 3^n$ 3^{n+1}

39. $9 \cdot 3^n$ 3^{n+2} **40.** $3^n/9$ 3^{n-2}

In Exercises 41 and 42, write the rule for each sequence using $f(x) = 2^{x \pm a}$.

41. a. $\frac{1}{8}, \frac{1}{4}, \frac{1}{2}, 1, 2$ $f(x) = 2^{x-4}$

   b. $4, 8, 16, 32, 64$ $f(x) = 2^{x+1}$

42. a. $\frac{1}{2}, 1, 2, 4, 8$ $f(x) = 2^{x-2}$

   b. $1, 2, 4, 8, 16$ $f(x) = 2^{x-1}$

In Exercises 43 and 44, write the rule for each sequence using $f(x) = (\frac{1}{2})^{x \pm a}$.

43. a. $4, 2, 1, \frac{1}{2}, \frac{1}{4}$ $(\frac{1}{2})^{x-3}$

   b. $8, 4, 2, 1, \frac{1}{2}$ $(\frac{1}{2})^{x-4}$

44. a. $64, 32, 16, 8, 4$ $(\frac{1}{2})^{x-7}$

   b. $\frac{1}{16}, \frac{1}{32}, \frac{1}{64}, \frac{1}{128}$ $(\frac{1}{2})^{x+3}$

Show that the numbers in Exercises 45 to 48 represent geometric sequences whose equations cannot be determined with exponential regression. Explain why.
Bases are negative, not exponential functions.

45. $2, -6, 18, -54, 162$ $r = -3, a_n = 2 \cdot (-3)^{n-1}$

46. $6, -12, 24, -48, 96$ $r = -2, a_n = 6 \cdot (-2)^{n-1}$

47. $64, -32, 16, -8, 4$ $r = -\frac{1}{2}, a_n = 64 \cdot (-\frac{1}{2})^{n-1}$

48. $81, -27, 9, -3, 1$ $r = -\frac{1}{3}, a_n = 81 \cdot (-\frac{1}{3})^{n-1}$

49. Find an exponential regression equation using the data from 1930, 1940, 1980, and 1990. Let 1930 be $x = 0$, and express the other dates in terms of the number of years after 1930. Indicate the annual rate of population growth (or decline). Use your equation to predict the population in the year 2000. Round to the nearest thousand. Compare with the year 2000 census data in the Answer Section.

   a. Albuquerque, New Mexico; 1930, 26,570; 1940, 35,449; 1980, 332,920; 1990, 384,915
   $y \approx 24{,}778(1.049)^x$; 4.9%; 705,000; actual 448,607

   b. Longview, Texas: 1930, 5036; 1940, 13,758; 1980, 65,762; 1990, 70,311
   $y \approx 6657(1.043)^x$; 4.3%; 127,000; actual 73,344

   c. Salem, Massachusetts: 1930, 43,353; 1940, 41,213; 1980, 38,276; 1990, 38,091
   $y \approx 42{,}738(0.998)^x$; -0.2%; 37,000; actual 40,407

   d. Oshkosh, Wisconsin: 1930, 40,108; 1940, 39,089; 1980, 49,620; 1990, 55,006
   $y \approx 38{,}587(1.005)^x$; 0.5%; 55,000; actual 62,916

50. The 88 keys on a piano cover a range of notes from lowest A, called A_0, to high C, called C_8. The sound frequencies in cycles per second associated with the notes form a sequence of numbers called the chromatic scale. The sequence of frequencies is most clearly seen for the A notes (see the table). What equation describes the sequence? $a_n = 27.5(2)^{n-1}$

Note	Cycles per second
A_0	27.50
A_1	55.00
A_2	110.0
A_3	220.0
A_4	440.0
A_5	880.0
A_6	1760
A_7	3520

51. a. Suppose the superball in Example 2 bounces to $\frac{1}{2}$ its previous height each time. Using both a_n and exponential regression, find the equation of the nth height. Find the tenth height. Describe how the change from $\frac{2}{3}$ to $\frac{1}{2}$ alters the equation. $a_n = 108(\frac{1}{2})^{n-1}; y = 216(\frac{1}{2})^x;$ ≈ 0.211 in.; alters the base

b. Using both a_n and exponential regression, write the equation of the nth height for the superball in Example 2 if it starts at a height of 135 inches. $a^n = 135(\frac{2}{3})^{n-1}; y = 202.5(\frac{2}{3})^x$

■ **Projects**

52. Stacking Paper Refer to Example 1, as needed.

a. If there are 500 sheets of paper in a 2-inch ream of paper, what is the thickness of one sheet? 0.004 in.

b. How thick a stack would result from folding a piece of paper ten times? (It may be easier to think in terms of the paper's being cut in half rather than folded.) 4.096 in.

c. How many folds (cuts) would be needed to make a stack of paper 1 mile high? ≈ 24

d. How many folds (cuts) would be needed to make a stack that reached from Earth to the sun, 93 million miles away? ≈ 51

53. Sequences It is tempting to say that any sequence whose differences are terms of the same sequence is geometric. Find the first and second differences for the sequences in parts a to d. Are the sequences geometric? Explain why or why not. no; no common ratios

a. 1, 1, 2, 3, 5, 8 0, 1, 1, 2, 3

b. 2, 4, 6, 10, 16, 26 2, 2, 4, 6, 10

c. 4, 5, 9, 14, 23, 37 1, 4, 5, 9, 14

d. 2, 5, 7, 12, 19, 31 3, 2, 5, 7, 12

e. Return to the project in Exercise 94 in Section 7.7 (page 427), and complete it if you have not already done so.

9.2 Exponential Graphs and Equations

Objectives

■ Identify increasing and decreasing functions.
■ Examine how the base and coefficient of an exponential expression affect the graph of an exponential function.
■ Graph exponential functions.
■ Identify the y-intercept for an exponential function.
■ Use like bases to solve an exponential equation.

WARM-UP

Complete the table for these equations.

Input, x	$y = 2^x$	$y = 2^{x-1}$	$y = 2^{x+2}$
0	1	$\frac{1}{2}$	4
1	2	1	8
2	4	2	16
3	8	4	32
4	16	8	64
5	32	16	128

IN THIS SECTION, we continue our work with exponential functions by exploring an application, examining graphs, and solving equations. If you do not do Example 1 in class, you might want to try it at home. You need 48 coins and about 10 minutes. You may be surprised by the outcome and better appreciate the nature of exponential functions.

EXAMPLE 1

Exploring patterns in coin tosses

a. *Understand the problem:* Before you carry out this activity, read the following instructions and describe, in writing, what you think will happen:

Put 48 coins in a bag, shake the bag, and then toss the coins into a low-sided box. Remove from the box the coins that are "heads up." Count the remaining coins. Repeat this procedure with the remaining coins until all the coins have been removed.

b. *Plan:* Make some predictions. In how many tosses will half the coins be removed? In how many tosses will all the coins be removed?

c. *Carry out the plan:* Make a table, with the first row showing toss 0 as input and 48 coins as output. Start the activity, and record the results after each toss. Make a graph from your table. Fit an exponential equation to your data. Do not enter the final data point (with 0 coins remaining), because the calculator function will indicate an error.

d. *Check:* Are your results reasonable? Compare your experimental results with a theoretical or expected table. State any assumptions. Find a theoretical equation describing the number of coins remaining (output) after each toss (input). Draw the theoretical graph on the same axes as the graph in part c.

Equipment needed: a set of pennies, a sandwich bag, and a box for every four students. Many students will write the theoretical table and equation in part a rather than in part d. That is perfectly acceptable. Do not hand out bags of coins until the students have written responses to parts a and b.

SOLUTION   The results from the penny toss are exponential and have a theoretical equation of the form $y = ab^x$. A possible table, the theoretical equation, and its graph are in the Answer Box. ▬

Increasing and Decreasing Functions

Recall that an increasing function rises as we trace from left to right and a decreasing function falls as we trace from left to right.

EXAMPLE 2

Identifying increasing or decreasing functions   The graphs for layers of folded paper and heights of bouncing balls are shown in Figures 9 and 10.

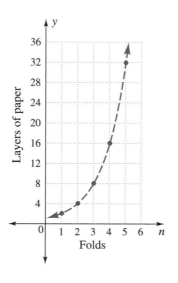

FIGURE 9

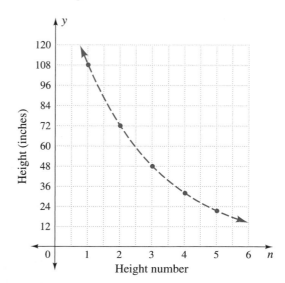

FIGURE 10

a. Which graph is an increasing function? Why?
b. Which graph is a decreasing function? Why?
c. Does the coin toss setting in Example 1 produce an increasing or decreasing function?

SOLUTION   a. Figure 9 shows an increasing function. The number of layers of paper increases as the number of folds increases.

b. Figure 10 shows a decreasing function. The height of each bounce drops as the number of bounces increases.

c. The coin toss setting produces a decreasing function. The number of coins decreases with each toss of the coins.

BASES AND GRAPHS   In Example 3, note that whether the exponential function is increasing or decreasing depends on the base.

EXAMPLE 3 **Comparing graphs and bases**   Compare the tables and graphs for $y = 2^x$ and $y = \left(\frac{1}{2}\right)^x$.

SOLUTION   The tables for $y = 2^x$ and $y = \left(\frac{1}{2}\right)^x$ contain the same output numbers, but for different inputs. In Table 5, the outputs for the two functions are the same numbers in opposite order. The change in base from 2 to $\frac{1}{2}$ changes the graph from increasing to decreasing; see Figure 11.

TABLE 5

x	$y = 2^x$	$y = \left(\frac{1}{2}\right)^x$
-2	$\frac{1}{4}$	4
-1	$\frac{1}{2}$	2
0	1	1
1	2	$\frac{1}{2}$
2	4	$\frac{1}{4}$

The patterns in the graphs focus on b^x, not b^{-x}, so as to keep the level of complexity low.

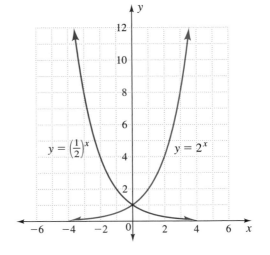

FIGURE 11

The examples suggest that when the base is a positive fraction less than 1, the graph falls from left to right. When the base is larger than 1, the graph rises from left to right.

GRAPHS OF EXPONENTIAL FUNCTIONS

> The graph of $y = b^x$ decreases from left to right for $0 < b < 1$.
>
> The graph of $y = b^x$ increases from left to right for $b > 1$.

An exponential function is defined only for positive bases, $b \neq 1$. The above summary does not mention a negative base, a base of 0, or a base of 1. You will explore these bases in Exercises 57 to 60.

Think about it 1: Write $\frac{1}{2}$ as a power of 2. Write $y = \left(\frac{1}{2}\right)^x$ in another way. Does this second way of writing $y = \left(\frac{1}{2}\right)^x$ fit our graphing rules?

SYMMETRY   Together, the graphs in Figure 11 have symmetry in that they reflect over the y-axis onto each other. Observe that the bases of the functions are reciprocals.

Features of the Exponential Graph $y = b^x$

Example 4 explores how the graph changes as the base of the exponential function changes.

EXAMPLE 4 Finding key features on an exponential graph   Making a table and a graph, compare $y = 2^x$, $y = 3^x$, and $y = 4^x$ for inputs on the interval -2 to 2. Discuss the y-intercept and the relative positions of the graphs. What is y for $x = 1$? What is the x-intercept for each graph?

SOLUTION

TABLE 6

x	$y = 2^x$	$y = 3^x$	$y = 4^x$
-2	$\frac{1}{4}$	$\frac{1}{9}$	$\frac{1}{16}$
-1	$\frac{1}{2}$	$\frac{1}{3}$	$\frac{1}{4}$
0	1	1	1
1	2	3	4
2	4	9	16

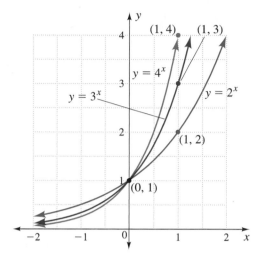

FIGURE 12

Table 6 shows a common point at $(0, 1)$. The graphs in Figure 12 all have the same y-intercept, $(0, 1)$, as we can see from the table. The y-intercept acts as a pivot point, with $y = 4^x$ highest on the right and lowest on the left, $y = 2^x$ lowest on the right and highest on the left, and $y = 3^x$ between the other two graphs on both sides.

At $x = 1$, $2^1 = 2$, $3^1 = 3$, and $4^1 = 4$. Thus, the point $(1, b)$ is on each graph.

If we trace along the graphs near the x-axis, none of the graphs touch the x-axis. The calculator rounds the output to zero if we trace far enough, but the graph never actually touches the x-axis. There are no x-intercepts.   ▬

Think about it 2: What is the meaning of the ordered pair $(2, \underline{\ \ })$ on the graph of $y = b^x$?

Example 4 suggests three conclusions about graphs of exponential functions.

▬ FEATURES OF GRAPHS OF EXPONENTIAL FUNCTIONS

> The graph of $f(x) = b^x$ passes through $(0, 1)$ and $(1, b)$.
>
> For $b > 1$, the larger the base in $y = b^x$, the steeper the graph.
>
> The graphs approach but do not touch the x-axis.

When a *graph approaches without touching the x-axis*, we say that the x-axis is a **horizontal asymptote**.

EXAMPLE 5 Graphing exponential functions   Use $(0, 1)$, $(1, b)$, $(2, b^2)$, and other facts as needed to graph $y = 5^x$ and $y = \left(\frac{1}{5}\right)^x$.

SOLUTION   The graph of $y = 5^x$ contains $(0, 1)$, $(1, 5)$, and $(2, 25)$. The graph of $y = \left(\frac{1}{5}\right)^x$ contains $(0, 1)$, $\left(1, \frac{1}{5}\right)$, and $\left(2, \frac{1}{25}\right)$.

We plot and connect the points for each equation in Figure 13. Because 5 and $\frac{1}{5}$ are reciprocals, the rest of each graph (in dashed color) can be drawn by reflecting the other graph over the y-axis.

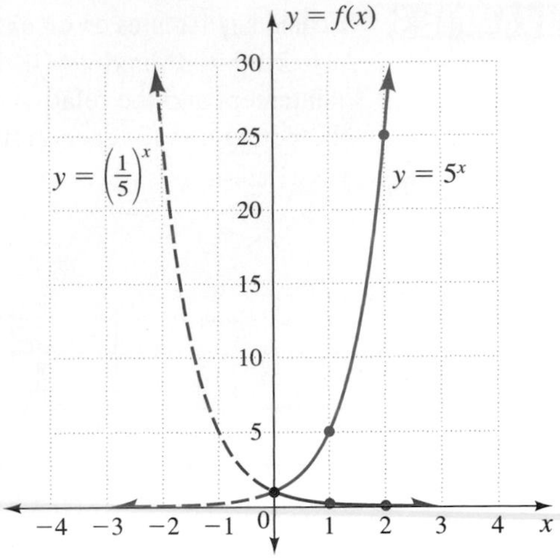

$y = f(x)$

$y = \left(\frac{1}{5}\right)^x$ $y = 5^x$

FIGURE 13

DOMAIN AND RANGE   The exponential graphs we have seen thus far help clarify the domain and range of the function $f(x) = b^x$. The graphs in Example 5 extend to the left and right without end and are entirely above the x-axis, implying the following:

■ DOMAIN AND RANGE

> The domain for $f(x) = b^x$ is all real numbers.
>
> The range for $f(x) = b^x$ is $f(x) > 0$.

Features of the Exponential Graph $y = ab^x$

y-INTERCEPTS   In Examples 6 and 7, we explore the role of the coefficient a in $f(x) = ab^x$.

EXAMPLE 6   Comparing y-intercepts   Graph $f(x) = 1 \cdot 2^x$, $f(x) = 5 \cdot 2^x$, and $f(x) = 10 \cdot 2^x$. Identify the y-intercept for each graph.

SOLUTION   The graphs, in Figure 14, show three distinct y-intercepts:

The y-intercept for $f(x) = 1 \cdot 2^x$ is 1.

The y-intercept for $f(x) = 5 \cdot 2^x$ is 5.

The y-intercept for $f(x) = 10 \cdot 2^x$ is 10.

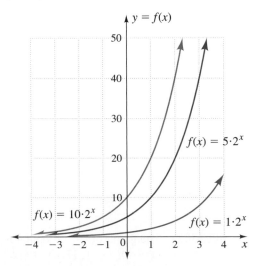

$y = f(x)$

$f(x) = 5 \cdot 2^x$

$f(x) = 10 \cdot 2^x$

$f(x) = 1 \cdot 2^x$

FIGURE 14

Think about it 3: Do the graphs in Figure 14 intersect?

Example 6 suggests that the coefficient a is the y-intercept. You will be asked to prove this in the exercises.

THE y-INTERCEPT FOR AN EXPONENTIAL FUNCTION

Because $x = 0$ at the y-intercept, the coefficient a is the y-intercept for the graph of $y = ab^x$.

EXAMPLE 7   Finding y-intercepts   Predict the y-intercept for each function and then find the intercept.

a. $y = 2b^x$ **b.** $y = 1000(1 + r)^x$ **c.** $y = 3^{x+1}$ **d.** $y = 2^{x-3}$

SOLUTION   **a.** Because $a = 2$, the y-intercept is 2. When $x = 0$,

$$y = 2b^0 = 2 \cdot 1 = 2$$

b. Because $a = 1000$, the y-intercept is 1000. When $x = 0$,

$$y = 1000(1 + r)^0 = 1000 \cdot 1 = 1000$$

c. To find the coefficient, we change to ab^x form:

$$y = 3^{x+1} = 3^x \cdot 3^1 = 3 \cdot 3^x$$

We predict that the y-intercept is 3. When $x = 0$,

$$y = 3^{0+1} = 3^1 = 3$$

d. To find the coefficient, we change to ab^x form:

$$y = 2^{x-3} = 2^x \cdot 2^{-3} = 2^x \cdot \tfrac{1}{8} = \tfrac{1}{8} \cdot 2^x$$

We predict that the y-intercept is $\tfrac{1}{8}$. When $x = 0$,

$$y = 2^{0-3} = 2^{-3} = \tfrac{1}{8}$$

Think about it 4: How does the coefficient a on $y = ab^x$ help us find an appropriate window for the graph?

HORIZONTAL SHIFTS   The functions $f(x) = 2^x$ and $f(x) = 2^{x-1}$ both represent powers of 2. We compare their graphs in Example 8.

EXAMPLE 8   Comparing graphs and exponents   Graph and compare $f(x) = 2^x$ and $f(x) = 2^{x-1}$.

SOLUTION   When the functions $f(x) = 2^x$ and $f(x) = 2^{x-1}$ are graphed on the same axes (see Figure 15), the graphs are identical in shape, but $f(x) = 2^{x-1}$ is shifted 1 unit to the right of $f(x) = 2^x$.

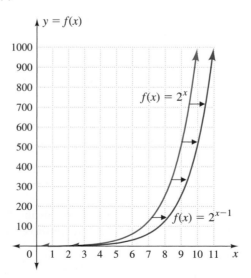

FIGURE 15

Think about it 5: How does the graph of $f(x) = (x - 1)^2$ compare with the graph of $f(x) = x^2$? How does this result compare with the one in Example 8? How does the graph of $f(x) = (x + 2)^2$ compare with the graph of $f(x) = x^2$?

EXAMPLE 9   Comparing graphs and exponents   How does the graph of $f(x) = 2^{x+2}$ compare with the graph of $f(x) = 2^x$?

SOLUTION   The graphs of $f(x) = 2^x$ and $f(x) = 2^{x+2}$ (see Figure 16) are identical in shape, but the graph of $f(x) = 2^{x+2}$ is shifted 2 units to the left of that of $f(x) = 2^x$.

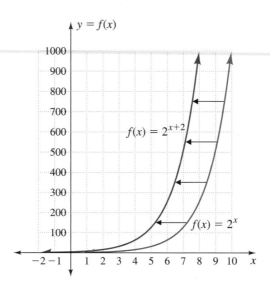

FIGURE 16

In summary, Examples 8 and 9 show the following.

HORIZONTAL SHIFTS

> For $h > 0$, the graph of $y = b^{x-h}$ is shifted h units to the right of that of $y = b^x$.
>
> For $h > 0$, the graph of $y = b^{x+h}$ is shifted h units to the left of that of $y = b^x$.

Solving with the Like Bases Property

In the introduction to rational exponents (Section 8.3), we solved exponential equations $4^x = 8$, $4^x = 128$, and $4^x = 512$ by guess and check and by graphing. Here, an algebraic technique is introduced. We will make the transition to a general method of solving exponential equations in the next section.

LIKE BASES PROPERTY

> Suppose $a > 0$ and $a \neq 1$.
>
> $$a^x = a^y \quad \text{if and only if} \quad x = y$$

The **like bases property** says that *if two exponential expressions are equal and have a common positive base, not 1, the exponents must be equal.*

■ SOLVING EQUATIONS WITH LIKE BASES

> To use the like bases property:
>
> **1.** Change the exponential expressions on both sides of an equation to the same base, if possible.
> **2.** Set the exponents equal.
> **3.** Solve for the variable.

EXAMPLE 10 Solving equations with the like bases property   Use the like bases property to solve the equations for x.

a. $4^x = 512$ **b.** $8^x = 1024$ **c.** $\left(\frac{1}{2}\right)^x = 32$

d. $\left(\frac{1}{3}\right)^x = 81$ **e.** $3^{x+2} = 9^{2x-5}$

SOLUTION **a.**

$4^x = 512$ Change to a common base.

$(2^2)^x = 2^9$ Apply the power property of exponents.

$2^{2x} = 2^9$ Set the exponents equal.

$2x = 9$ Divide by 2.

$x = \frac{9}{2}$

Check: $4^{9/2} \overset{?}{=} 512$ ✓

b.

$8^x = 1024$ Change to a common base.

$(2^3)^x = 2^{10}$ Apply the power property of exponents.

$2^{3x} = 2^{10}$ Set the exponents equal.

$3x - 10$ Divide by 3.

$x = \frac{10}{3}$

Check: $8^{10/3} \overset{?}{=} 1024$ ✓

c.

$\left(\frac{1}{2}\right)^x = 32$ Change to a common base.

$(2^{-1})^x = 2^5$ Apply the power property of exponents.

$2^{-1x} = 2^5$ Set the exponents equal.

$-1x = 5$ Divide by -1.

$x = -5$

Check: $\left(\frac{1}{2}\right)^{-5} \overset{?}{=} 32$ ✓

d.

$\left(\frac{1}{3}\right)^x = 81$ Change to a common base.

$(3^{-1})^x = 3^4$ Apply the power property of exponents.

$3^{-1x} = 3^4$ Set the exponents equal.

$-1x = 4$ Divide by -1.

$x = -4$

Check: $\left(\frac{1}{3}\right)^{-4} \overset{?}{=} 81$ ✓

e. $\qquad 3^{x+2} = 9^{2x-5}$ Change to a common base, placing the exponent on the right in parentheses.

$\qquad\qquad 3^{x+2} = 3^{2(2x-5)}$ Multiply with the distributive property on the right side.

$\qquad\qquad 3^{x+2} = 3^{4x-10}$ Set the exponents equal.

$\qquad\qquad x + 2 = 4x - 10$ Add 10 to both sides, and subtract x from both sides.

$\qquad\qquad\qquad 12 = 3x$ Divide both sides by 3.

$\qquad\qquad\qquad 4 = x$

Check: $3^{4+2} \overset{?}{=} 9^{2(4)-5}$ ✓

The like bases property is useful, even though its application is limited to equations in which bases are easily made alike.

ANSWER BOX

Warm-up: For $y = 2^x$: 1, 2, 4, 8, 16, 32; for $y = 2^{x-1}$: $\frac{1}{2}$, 1, 2, 4, 8, 16; for $y = 2^{x+2}$: 4, 8, 16, 32, 64, 128   **Example 1:** According to the table, if we start with 48 coins and the coins are equally likely to land heads up, the coins will be gone after 6 tosses. In the table, fractional coins have been rounded down to next lowest whole coin. The theoretical equation is $y = 48\left(\frac{1}{2}\right)^x$; the theoretical graph is shown in the figure.

Toss	Coins Remaining
0	48
1	24
2	12
3	6
4	3
5	1
6	0

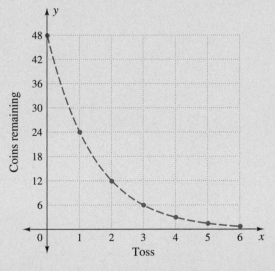

Think about it 1: $2^{-1} = \frac{1}{2}$; $y = \left(\frac{1}{2}\right)^x$ is equivalent to $y = 2^{-x}$. The rules are for $y = b^x$. Equations of the form $y = b^{-x}$ need to be changed to $y = (1/b)^x$ form before the rules are applied.   **Think about it 2:** $y = b^x$ for $x = 2$ is the square of the base. **Think about it 3:** The graphs do not intersect because $10 \cdot 2^x > 5 \cdot 2^x > 1 \cdot 2^x$ for each x. Try tracing the graphs to the left and using the up and down cursor keys to compare the functions at $x = -4$.   **Think about it 4:** The coefficient a indicates the y-intercept and suggests that the minimum y should be below the intercept and the maximum y should be above it.   **Think about it 5:** The graph of $f(x) = (x - 1)^2$ is shifted 1 unit to the right of the graph of $f(x) = x^2$. This is the same relationship found between the graphs in Example 8. The graph of $f(x) = (x + 2)^2$ is shifted 2 units to the left of the graph of $f(x) = x^2$.

9.2 **Exercises**

In Exercises 1 to 6, predict without graphing whether the graph will be increasing or decreasing from left to right.

1. a. $y = 5^x$ increasing   **b.** $y = \left(\frac{3}{4}\right)^x$ decreasing

2. a. $y = \left(\frac{2}{3}\right)^x$ decreasing   **b.** $y = 4^x$ increasing

3. a. $y = \left(\frac{3}{2}\right)^x$ increasing   **b.** $y = 0.5^x$ decreasing

4. a. $y = 1.5^x$ increasing   **b.** $y = 0.75^x$ decreasing

5. a. $y = (1 + 0.05)^x$ increasing   **b.** $y = \left(\frac{4}{3}\right)^x$ increasing

6. a. $y = 2.5^x$ increasing   **b.** $y = (1 - 0.05)^x$ decreasing

7. The figure contains graphs of $y = 2^x$, $y = 10^x$, and $y = 2.72^x$. Match each equation with its graph—a, b, or c. Do not use a calculator.   c, a, b

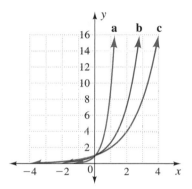

8. The figure contains graphs of $y = \pi^x$, $y = 1.5^x$, and $y = 2^x$. Match each equation with its graph—a, b, or c. Do not use a calculator.   a, c, b

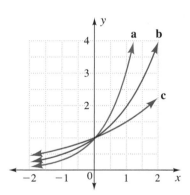

9. For what set of inputs is the graph of $y = 2^x$ below the x-axis?   { } or ∅

10. For what set of inputs x is the output to $y = 3^x$ negative?   { } or ∅

11. Evaluate $3 \cdot 2^0$, $4 \cdot 2^0$, and $\frac{1}{2} \cdot 2^0$. What do these results tell us about the y-intercept of $y = a \cdot 2^x$?   3, 4, $\frac{1}{2}$; equals a

12. Explain why the letter a in $y = ab^x$ is not part of the base.   Exponent applies only to letter b immediately to left.

13. Describe the role of the coefficient a in the graph of the exponential equation $y = ab^x$. y-intercept value

14. Show that the y-intercept for $y = ab^x$ is a.   Let $x = 0$; $y = ab^0 = a$

15. How do we find the y-intercept from the equation $y = 2^{x+h}$?   Let $x = 0$ and simplify: $y = 2^{0+h}$, or $y = 2^h$

16. Evaluate 2^1, 3^1, and 5^1. What do these results tell us about the y-coordinate of the point where $x = 1$ in $y = b^x$?   2, 3, 5; when $x = 1$, $y = b^x = b$.

17. Why does the graph of $y = b^x$ pass through $(1, b)$?   When $x = 1$, $b^x = b$.

18. Why does the graph of $y = b^x$ pass through $(0, 1)$?   When $x = 0$, $b^x = 1$.

19. Graphs of exponential functions such as $y = 2^x$ have the x-axis as a horizontal asymptote. Name a nonexponential function from a different chapter whose graph has a horizontal asymptote (the graph approaches the x-axis without touching or crossing it).   Possible answer: $y = 1/x^2$

20. Compare the effects of the exponents on the graphs of $y = 2^{-x}$ and $y = 2^x$.   starts high on left, decreases rapidly; starts low on left, increases rapidly

Name the y-intercept in Exercises 21 to 23. Change each equation to the form $y = 2^{x \pm a}$ or $y = 3^{x \pm a}$.

21. a. $y = 16 \cdot 2^x$ 16; $y = 2^{x+4}$

    b. $y = 8 \cdot 2^x$ 8; $y = 2^{x+3}$

22. a. $y = \frac{1}{2} \cdot 2^x$ $\frac{1}{2}$; $y = 2^{x-1}$

    b. $y = 3^x/9$ $\frac{1}{9}$; $y = 3^{x-2}$

23. a. $y = 27 \cdot 3^x$ 27; $y = 3^{x+3}$

    b. $y = 3^x/27$ $\frac{1}{27}$; $y = 3^{x-3}$

Change each equation in Exercises 24 to 26 to $y = ab^x$ form. Name the y-intercept.

24. a. $y = 3^{x-2}$ $y = \frac{1}{9} \cdot 3^x$; $\frac{1}{9}$

    b. $y = 3^{x+3}$ $y = 27 \cdot 3^x$; 27

25. a. $y = 2^{x-4}$ $y = \frac{1}{16} \cdot 2^x$; $\frac{1}{16}$

    b. $y = 2^{x+1}$ $y = 2 \cdot 2^x$; 2

26. a. $y = \left(\frac{1}{4}\right)^{x+1}$ $y = \frac{1}{4} \cdot \left(\frac{1}{4}\right)^x$; $\frac{1}{4}$

    b. $y = 2^{x-2}$ $y = \frac{1}{4} \cdot 2x$; $\frac{1}{4}$

In Exercises 27 and 28, complete the table for the indicated functions. Match each function with one of the graphs in the figure. Describe the effect of adding or subtracting 1 in the exponent.   shifts 1 unit to left; 1 unit to right

27.

x	$y = 2^{x-1}$	$y = 2^x$	$y = 2^{x+1}$
-2	$\frac{1}{8}$	$\frac{1}{4}$	$\frac{1}{2}$
-1	$\frac{1}{4}$	$\frac{1}{2}$	1
0	$\frac{1}{2}$	1	2
1	1	2	4
2	2	4	8
	c	b	a

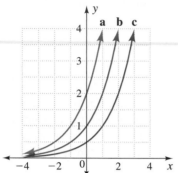

28.

x	$y = 3^{x-1}$	$y = 3^x$	$y = 3^{x+1}$
-2	$\frac{1}{27}$	$\frac{1}{9}$	$\frac{1}{3}$
-1	$\frac{1}{9}$	$\frac{1}{3}$	1
0	$\frac{1}{3}$	1	3
1	1	3	9
2	3	9	27
	c	b	a

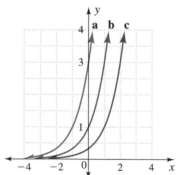

Using a graph of $y = 3^x$ as a reference, sketch a graph of each function in Exercises 29 to 32. Label the y-intercept and $f(1)$.

29. $f(x) = 3^{x+3}$
See Answer Section.

30. $f(x) = 3^{x-2}$
See Additional Answers.

31. $f(x) = 3^{x-3}$
See Answer Section.

32. $f(x) = 3^{x+2}$
See Additional Answers.

33. Where do the graphs of $y = 3^x$ and $y = 3^{x+1}$ intersect? Why?   no intersection; $3^x < 3^{x+1}$ for all x

34. Where do the graphs of $y = 2^x$ and $y = 3^x$ intersect? Why? $(0, 1)$; $2^0 = 1, 3^0 = 1$

In Exercises 35 to 50, solve the equation for n or x. Recall that $2^{-1} = \frac{1}{2}$ and $3^{-2} = \frac{1}{9}$.

35. a. $2^x = 256$ $x = 8$ **b.** $16^x = 256$ $x = 2$

36. a. $4^n = 256$ $n = 4$ **b.** $8^n = 256$ $n = 2\frac{2}{3}$

37. a. $4^n = 8$ $n = \frac{3}{2}$ **b.** $8^n = 16$ $n = \frac{4}{3}$

38. a. $8^n = 4$ $n = \frac{2}{3}$ **b.** $4^n = 128$ $n = 3\frac{1}{2}$

39. a. $16^n = 4$ $n = \frac{1}{2}$ **b.** $100^x = 1000$ $x = \frac{3}{2}$

40. a. $8^n = 2$ $n = \frac{1}{3}$ **b.** $100^x = 10$ $x = \frac{1}{2}$

41. a. $64^x = \frac{1}{8}$ $x = -\frac{1}{2}$ **b.** $125^n = \frac{1}{5}$ $n = -\frac{1}{3}$

42. a. $64^n = \frac{1}{4}$ $n = -\frac{1}{3}$ **b.** $25^x = \frac{1}{125}$ $x = -\frac{3}{2}$

43. a. $8^n = \frac{1}{2}$ $n = -\frac{1}{3}$ **b.** $81^x = \frac{1}{3}$ $x = -\frac{1}{4}$

44. a. $4^x = \frac{1}{64}$ $x = -3$ **b.** $27^x = \frac{1}{9}$ $x = -\frac{2}{3}$

45. a. $\left(\frac{1}{2}\right)^{x+5} = 32$ $x = -10$ **b.** $\left(\frac{1}{8}\right)^{x+3} = 8$ $x = -4$

46. a. $\left(\frac{1}{3}\right)^{x-5} = 81$ $x = 1$ **b.** $\left(\frac{1}{9}\right)^{x-2} = 9$ $x = 1$

47. a. $2^{x-1} = 2^{2x+5}$ $x = -6$ **b.** $3^{2x-1} = 3^{x+3}$ $x = 4$

48. a. $3^{x-2} = 3^{2x+3}$ $x = -5$ **b.** $2^{2x+3} = 2^{x-5}$ $x = -8$

49. a. $2^{x+5} = 4^{x-1}$ $x = 7$ **b.** $5^{x+1} = 125^{x-3}$ $x = 5$

50. a. $3^{x+4} = 9^{x-5}$ $x = 14$ **b.** $5^{2x-1} = 625^x$ $x = -\frac{1}{2}$

51. If the exponents in an equation are equal, are the bases equal also? Consider $x^2 = 3^2$ in your explanation.
No; $x = -3$ is another base.

52. Why is it difficult to solve $2^n = 10$ with the like bases property?   10 is not an integral power of 2.

53. Suppose two competing research groups examine bacterial growth data. The first group concludes that the population of bacteria is growing according to the rule $P = 4^{6t}$. The second group concludes that the population is growing according to the rule $P = 8^{4t}$. You have been asked to compare their results. What do you suggest?
same; $P = 2^{12t}$

54. Johanna solves $3^{x+2} = 243$ by first substituting $9 \cdot 3^x$ for 3^{x+2} and then dividing both sides by 9 to obtain $3^x = 27$. Explain whether her process is correct.
yes, since $3^{x+2} = 3^2 \cdot 3^x = 9 \cdot 3^x$

55. Refer to the figure below. The base a is not given, so answer the questions from the graph.

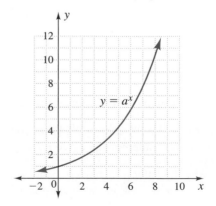

a. Estimate x for $a^x = 2$. $x \approx 2.5$

b. Estimate x for $a^x = 4$. $x \approx 4.8$

c. Estimate y corresponding to a^2. $y \approx 1.8$

d. Estimate y corresponding to a^5. $y \approx 4.5$

e. Use a graphing guess-and-check process to find a graph similar to the figure and thereby estimate the base a. $a \approx 1.34$

56. Refer to the figure below. The base d is not given, so answer the questions from the graph.

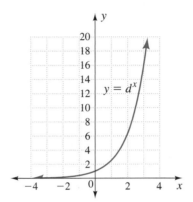

$y = d^x$

a. Estimate x for $d^x = 2$. $x \approx 0.8$

b. Estimate x for $d^x = 4$. $x \approx 1.5$

c. Estimate y corresponding to d^2. $y \approx 6$

d. Estimate y corresponding to d^3. $y \approx 16$

e. Use a graphing guess-and-check process to find a graph similar to the figure and thereby estimate the base d. $d \approx 2.5$

Use a calculator graph in Exercises 57 to 60.

57. a. Predict the graph of $y = 1^x$.

b. Graph $y = 1^x$.

c. What do you observe? horizontal line at $y = 1$

d. Is the function defined for all inputs? yes

e. What is the range of the function? $y = 1$

58. a. Predict the graph of $y = 0^x$.

b. Graph $y = 0^x$.

c. What do you observe? positive x-axis

d. Is the function defined for all inputs? no, $x > 0$

e. What is the range of the function? $y = 0$

59. a. Graph $y = (-2)^x$. Use a decimal zoom to assure decimal inputs.

b. What do you observe? points only

c. Is this equation defined for all inputs? no

d. Is this equation an exponential function?
no, negative base

60. a. Graph $y = (-1)^x$. Use a decimal zoom to assure decimal inputs.

b. What do you observe? points only

c. Is this equation defined for all inputs? no

d. Is this equation an exponential function?
no, negative base

61. The graphs in the figure are related to $y = 2^x$.

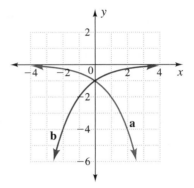

a. What is the y-intercept for each graph? 1

b. Predict the equations for graphs a and b. Check your predictions with a calculator.
$y = -2^x$; $y = -2^{-x}$, or $y = -\left(\frac{1}{2}\right)^x$

c. Complete this statement: Changing to a ___negative___ coefficient causes the exponential graph to be reflected over the x-axis.

d. What is the equation for the graph formed by reflecting the graph of $y = x^2$ over the x-axis? $y = -x^2$

Projects

62. Irrational Exponents To evaluate expressions such as 3^π, we estimate the irrational number with a rational number (a terminating or repeating decimal). Estimate 3^π by finding these values.

a. 3^3 27

b. $3^{3.1}$ 30.135 325 7

c. $3^{3.14}$ 31.489 135 65

d. $3^{3.14159}$ 31.544 188 74

e. $3^{3.141592654}$ 31.544 280 71

f. 3^π $\approx$31.544 280 7

63. Chessboard Rice A classic mathematics problem involves placing a grain of rice on the first square of a chessboard, two grains of rice on the second square, four grains on the third square, and so forth, doubling the grains for each of the 64 squares on the board. Make a table with the number of the square as the input and the number of grains of rice as the output.

a. How many grains will be on the 10th square? 512

b. What equation describes the number of grains of rice on the *n*th square? $y = 2^{n-1}$

c. How many grains will be the last (64th) square?
$2^{64-1} \approx 9.22 \times 10^{18}$

Add a column to your table. In this column, enter, for each square, the total number of grains of rice needed for it plus all the squares preceding it.

d. How does the cumulative total for each square compare with the number of grains of rice needed for that square? number of grains · 2 − 1

e. How many grains of rice will be needed on the 11th square? 1024

f. How many grains of rice will be needed in total for the first 11 squares? 2047

g. How many grains of rice will be needed in total for the whole chessboard? $2^{64} - 1 \approx 1.845 \times 10^{19}$

9.3   Solving Exponential Equations: Logarithms

Objectives

■ Find the inverse to an exponential function by table, graph, and equation.

■ Change between exponential equations and logarithmic equations.

■ Evaluate common logarithms with a calculator.

> **WARM-UP**
>
> Solve each equation for *x*. Guess and check with a calculator, as needed, to find the exponents to the nearest thousandth.
>
> **1.** $10^x = 1$ $x = 0$ **2.** $10^x = 2$ $x = 0.301$ **3.** $10^x = 10$ $x = 1$
>
> **4.** $10^x = 20$ $x = 1.301$ **5.** $10^x = 100$ $x = 2$ **6.** $10^x = 200$
> $x = 2.301$
>
> **7.** $10^x = 1000$ $x = 3$ **8.** $10^x = 2000$ $x = 3.301$

THIS SECTION INTRODUCES the inverse to the exponential function: the logarithmic function. You will learn to find base 10, or common, logarithms with a calculator.

Inverse Functions and Exponential Equations

The equations in the Warm-up containing 2, 20, 200, and 2000 could not be solved with like bases. There is no integer or rational-number power of 10 that makes these values. We now consider a solution method that is more general than guess and check or applying the like bases property.

To solve an equation such as $y = 10^x$ for *x*, we need an inverse function. From Section 8.6, we have three ways to find an inverse function: from a table, from a graph, and with symbols. We will apply each of these three methods to solve $y = 10^x$ for *x*.

INVERSE FUNCTIONS FROM A TABLE   *To find the inverse from a function listed in a table, swap the input and output (x and y) columns and find the rule for the new table.* Check that the inverse is a function.

EXAMPLE 1   Finding the inverse   Given Table 7 for $y = 10^x$, find the table for the inverse.

TABLE 7

Input, x	Output, $y = 10^x$
-2	$\frac{1}{100}$
-1	$\frac{1}{10}$
0	1
1	10
2	100
3	1000

SOLUTION   By swapping the numbers in the columns of Table 7, we can see that an inverse exists. Table 8 appears to be the table for a function, with one output for each input.

TABLE 8

Input to Inverse, x	Output to Inverse, y
$\frac{1}{100}$	-2
$\frac{1}{10}$	-1
1	0
10	1
100	2
1000	3

INVERSE FUNCTIONS FROM A GRAPH   *To find the inverse from the graph of a function, swap the numbers (x, y) in each ordered pair and plot the resulting graph.* Check that the inverse is a function.

EXAMPLE 2   Finding the inverse   Use Tables 7 and 8 to graph $y = 10^x$ for $-2 \leq x \leq 1$ and to sketch the inverse. Label both axes from -2 to 10. Is the inverse a function?

SOLUTION   The graphs of $y = 10^x$ and its inverse are in Figure 17. A vertical line placed anywhere on the graph of the inverse passes through the graph exactly once. Thus, the inverse of the exponential function $y = 10^x$ passes the vertical-line test and is also a function.

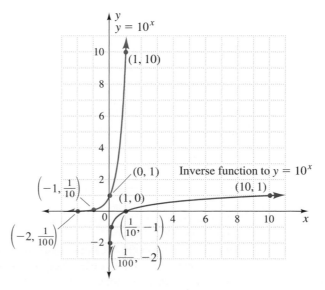

FIGURE 17

INVERSE FUNCTIONS FROM AN EQUATION   *To find the inverse from the equation of a function, swap x and y and solve (if possible) for y.* Check that the inverse is a function.

EXAMPLE 3   Finding the inverse   Given the equation $y = 10^x$, find the inverse.

SOLUTION   We swap x and y and then solve (if possible) for y.

$$y = 10^x$$

$$x = 10^y$$

There are no algebraic steps to use to solve for y.

Think about it: Show the result from doing each of the following operations to both sides of $x = 10^y$: adding y, subtracting y, multiplying by y, dividing by y, and taking the $1/y$ power (yth root).

Logarithmic Functions

Because algebra fails to solve $y = b^x$ for x or $x = b^y$ for y, mathematicians had to invent a solution. The invention of what led to the inverse to the exponential function took place shortly before 1620, when the Pilgrims landed at Plymouth, Massachusetts. In 1614, John Napier (1550–1617) called his invention *logarithms*, roughly translated from the Greek as *number words* or *number ratios*. Note the similarity between arithmetic and logarithm.

LOGARITHMIC FUNCTION

> The logarithmic function is the inverse to the exponential function, $f(x) = b^x$.
>
> The logarithmic function is $f(x) = \log_b x$, where x is a positive real number and b is a positive real number not equal to 1. Read $\log_b x$ as "log base b of x."

The graphs of exponential functions, $f(x) = b^x$, $b > 1$, and logarithmic functions, $f(x) = \log_b x$, look similar to those of $f(x) = 10^x$ and its inverse function, $f^{-1}(x) = \log_{10} x$ (Figure 17). We will do more with graphs of logarithmic functions in Section 9.4.

Changing Exponential Equations to Logarithmic Equations

In 1728, Leonhard Euler modified the definition of a logarithm. We use his version of the definition today to solve exponential equations.

SOLVING EXPONENTIAL EQUATIONS, $y = b^x$, FOR x

> The **logarithm** is the exponent to which we raise a base to obtain a given number. If $y = b^x$, where b is a fixed positive number not equal to 1, then x is the logarithm base b of y.
>
> $$y = b^x \qquad \text{is equivalent to} \qquad x = \log_b y.$$

The base b is the same for both the exponential equation and the logarithmic equation. This fact will help you keep track of where the different numbers and letters go when you change between exponential and logarithmic equations. In the next example and the related exercises, the equivalent equations are placed in tables to emphasize their relationship.

EXAMPLE 4 **Changing between exponential and logarithmic equations** Find the missing entries in Table 9, changing the logarithmic equations to exponential equations and the exponential equations to logarithmic equations.

TABLE 9

	Exponential Equation, $y = b^x$	Logarithmic Equation, $x = \log_b y$
	$10^x = 2$	$x = \log_{10} 2$
a.		$x = \log_{10} 20$
b.	$10^x = 200$	
c.		$1 = \log_{10} 10$
d.	$10^2 = 100$	
e.		$0 = \log_{10} 1$

SOLUTION The equations in parts a and b contain the variable x.

a. If $x = \log_{10} 20$, the base is 10 and

$$10^x = 20$$

b. If $10^x = 200$, the base is 10 and

$$x = \log_{10} 200$$

The equations in parts c, d, and e are common facts about powers of 10.

c. If $1 = \log_{10} 10$, the base is 10 and

$$10^1 = 10$$

This fact reminds us that $b^1 = b$; any number to the first power is itself.

d. If $10^2 = 100$, the base is 10 and

$$2 = \log_{10} 100$$

This fact states that the square of 10 is 100.

e. If $0 = \log_{10} 1$, the base is 10 and

$$10^0 = 1$$

This fact reminds us that $b^0 = 1$, $b \neq 0$.

The completed version of Table 9 appears in Table 10.

TABLE 10

	Exponential Equation, $y = b^x$	Logarithmic Equation, $x = \log_b y$
	$10^x = 2$	$x = \log_{10} 2$
a.	$10^x = 20$	$x = \log_{10} 20$
b.	$10^x = 200$	$x = \log_{10} 200$
c.	$10^1 = 10$	$1 = \log_{10} 10$
d.	$10^2 = 100$	$2 = \log_{10} 100$
e.	$10^0 = 1$	$0 = \log_{10} 1$

A common error is to place the base in an incorrect position in the logarithmic expression. *The base b is written in subscript position, small and below the main line of type containing the rest of the logarithmic expression.*

Common Logarithms

Now that we have solved $y = 10^x$ for x and obtained $x = \log_{10} y$, it would be reasonable to expect to find some numbers for x. We will work first with logarithms having base 10, because base 10 is one of the two most commonly occurring bases. The other is base e, and we will consider it in Section 9.6.

Logarithms with base 10 are called **common logarithms**. The common logarithm of a number is evaluated with the $\boxed{\text{LOG}}$ key on a calculator.

■■■ **EVALUATING LOGARITHMS BY CALCULATOR**

> To find the common logarithm, $\log_{10} x$, on a graphing calculator, press $\boxed{\text{LOG}}$ followed by the number x.

EXAMPLE 5 Finding common logarithms with a calculator   Complete Table 11.

TABLE 11

Input, x	Output, $y = \log_{10} x$
$\frac{1}{100}$	
$\frac{1}{10}$	
1	
10	
100	
1000	

SOLUTION   The solution table is same as Table 8 in Example 1. Remember: *y, the logarithm base 10 of x, is the exponent to which we raise 10 to obtain x.* ■■■

In Example 6, we look for patterns while finding the common logarithms with a calculator.

EXAMPLE 6 Finding common logarithms with a calculator   Round your answers to five decimal places. What patterns do you observe?

a. $\log_{10} 2$ b. $\log_{10} 20$ c. $\log_{10} 200$

d. $\log_{10} 3$ e. $\log_{10} 30$ f. $\log_{10} 300$

g. $\log_{10} 0.2$ h. $\log_{10} 0.02$ i. $\log_{10} 0.002$

SOLUTION   a. 0.30103   b. 1.30103   c. 2.30103

d. 0.47712   e. 1.47712   f. 2.47712

g. −0.69897   h. −1.69897   i. −2.69897

The decimal portions are the same for log base 10 of 2, 20, and 200. Similarly, the decimal portions are the same for log base 10 of 3, 30, and 300. For log base 10 of 0.2, 0.02, and 0.002, the decimal portions are the same and the numbers are negative. The decimal portion is not the same for 2 as for 0.2. Other observations are possible. ■■■

EXAMPLE 7 Evaluating common logarithms Evaluate these logarithms on the calculator. Use the
⟮ **LOG** ⟯ key for log base 10. Change each to an exponential equation to verify the
calculator result.

a. $\log_{10} 0 = x$ **b.** $\log_{10}(-2) = x$

SOLUTION Graphing calculators may give an error message for both logarithms. The message
indicates that the logarithm of zero is undefined and that the logarithm of a negative number is not a real number.

a. In $\log_{10} 0 = x$, the base is 10, so we are seeking the solution to $10^x = 0$. Since
an exponential function is always positive, there is no real-number solution to
this equation.

b. In $\log_{10}(-2) = x$, the base is 10, so $10^x = -2$ is an equivalent equation. There
is no real-number power of 10 that results in a negative number. ▬

In Example 8, we add to our table a third column in which we solve the equation.

EXAMPLE 8 Solving equations Complete Table 12, changing the exponential equations to logarithmic equations and the logarithmic equations to exponential equations. Use facts
about powers of numbers to find x before using a calculator.

TABLE 12

Exponential Equation, $y = b^x$	Logarithmic Equation, $x = \log_b y$	Solve for x
	$\log_2 128 = x$	
	$\log_{10} x = 2$	
$x^5 = 32$		
$10^x = 20$		
	$\log_{10} 40 = x$	
$10^{-3} = x$		

SOLUTION The variable x is placed in several different positions, so the definition of a logarithm must be used carefully. The completed version of Table 12 appears as
Table 13.

TABLE 13

Exponential Equation, $y = b^x$	Logarithmic Equation, $x = \log_b y$	Solve for x
$2^x = 128$	$\log_2 128 = x$	$2^x = 2^7$ $x = 7$
$10^2 = x$	$\log_{10} x = 2$	$10^2 = 100$ $x = 100$
$x^5 = 32$	$\log_x 32 = 5$	$2^5 = 32$ $x = 2$
$10^x = 20$	$\log_{10} 20 = x$	$x \approx 1.30103$
$10^x = 40$	$\log_{10} 40 = x$	$x \approx 1.60206$
$10^{-3} = x$	$\log_{10} x = -3$	$x = 0.001$

> **ANSWER BOX**
>
> **Warm-up:** **1.** $x = 0$ **2.** $x = 0.301$ **3.** $x = 1$ **4.** $x = 1.301$ **5.** $x = 2$
> **6.** $x = 2.301$ **7.** $x = 3$ **8.** $x = 3.301$ **Think about it:** $x + y = 10^y + y$;
> $x - y = 10^y - y$; $xy = y \cdot 10^y$; $x/y = 10^y/y$; $x^{1/y} = 10$. None of the expressions simplify. Thus, none of the operations both remove y from the exponent and permit us to solve for y.

9.3   Exercises

1. a. Make a table for $f(x) = 3^x$. Use integers on the interval $[-2, 4]$. See Answer Section.

   b. Show the table for the inverse.

   c. Graph $f(x) = 3^x$ and show the graph of the inverse on the same axes. Is the inverse a function? yes

   d. Start with $y = 3^x$ and find the inverse with symbols.
   $y = \log_3 x$

2. Repeat Exercise 1 for $f(x) = 2^x$.
   See Additional Answers.

Complete the tables in Exercises 3 and 4.

3.

Exponential Equation	Logarithmic Equation
$3^3 = 27$	$\log_3 27 = 3$
$2^3 = 8$	$\log_2 8 = 3$
$10^1 = 10$	$\log_{10} 10 = 1$
$5^3 = 125$	$\log_5 125 = 3$
$3^{-2} = \frac{1}{9}$	$\log_3\left(\frac{1}{9}\right) = -2$

4.

Exponential Equation	Logarithmic Equation
$6^3 = 216$	$\log_6 216 = 3$
$5^{-1} = \frac{1}{5}$	$\log_5\left(\frac{1}{5}\right) = -1$
$2^4 = 16$	$\log_2 16 = 4$
$4^{-2} = \frac{1}{16}$	$\log_4\left(\frac{1}{16}\right) = -2$
$6^1 = 6$	$\log_6 6 = 1$

Write the equations in Exercises 5 to 8 as exponential equations.

5. a. $\log_{10} 1000 = 3$
   $10^3 = 1000$
   b. $\log_{10} 1 = 0$
   $10^0 = 1$
   c. $\log_3 81 = 4$
   $3^4 = 81$
   d. $\log_{10} 100 = 2$
   $10^2 = 100$

6. a. $\log_{10} 10 = 1$
   $10^1 = 10$
   b. $\log_{10} 0.01 = -2$
   $10^{-2} = 0.01$
   c. $\log_2 32 = 5$
   $2^5 = 32$
   d. $\log_5 125 = 3$
   $5^3 = 125$

7. a. $\log_{10} 0.001 = -3$
   $10^{-3} = 0.001$
   b. $\log_4 1 = 0$
   $4^0 = 1$
   c. $\log_m n = k$
   $m^k = n$
   d. $\log_5\left(\frac{1}{25}\right) = -2$
   $5^{-2} = \frac{1}{25}$

8. a. $\log_3 3 = 1$ $3^1 = 3$
   b. $\log_{10} 10{,}000 = 4$
   $10^4 = 10{,}000$
   c. $\log_3\left(\frac{1}{27}\right) = -3$
   $3^{-3} = \frac{1}{27}$
   d. $\log_p r = m$
   $p^m = r$

Write the equations in Exercises 9 to 12 as logarithmic equations.

9. a. $2^5 = 32$ $\log_2 32 = 5$
   b. $2^1 = 2$ $\log_2 2 = 1$
   c. $2^0 = 1$ $\log_2 1 = 0$
   d. $10^2 = 100$ $\log_{10} 100 = 2$

10. a. $3^2 = 9$ $\log_3 9 = 2$
   b. $3^5 = 243$ $\log_3 243 = 5$
   c. $3^0 = 1$ $\log_3 1 = 0$
   d. $10^{-2} = 0.01$
   $\log_{10} 0.01 = -2$

11. a. $10^{-3} = 0.001$
   $\log_{10} 0.001 = -3$
   b. $f^d = g$ $\log_f g = d$
   c. $3^{-2} = \frac{1}{9}$ $\log_3\left(\frac{1}{9}\right) = -2$
   d. $4^0 = 1$ $\log_4 1 = 0$

12. a. $2^{10} = 1024$
   $\log_2 1024 = 10$
   b. $2^{-1} = \frac{1}{2}$ $\log_2\left(\frac{1}{2}\right) = -1$
   c. $10^0 = 1$
   $\log_{10} 1 = 0$
   d. $d^a = c$ $\log_d c = a$

In Exercises 13 to 16, start by finding the logs using ⎡**LOG**⎤ on a calculator. Switch to finding the logs mentally as soon as possible. Round to five decimal places.

13. a. $\log_{10} 1$ 0
   b. $\log_{10} 6$ 0.77815
   c. $\log_{10} 10$ 1
   d. $\log_{10} 60$ 1.77815
   e. $\log_{10} 100$ 2
   f. $\log_{10} 600$ 2.77815

14. a. $\log_{10} 1000$ 3
   b. $\log_{10} 7000$ 3.84510
   c. $\log_{10} 100{,}000$ 5
   d. $\log_{10} 700{,}000$ 5.84510
   e. $\log_{10} 7{,}000{,}000$ 6.84510
   f. $\log_{10} 1{,}000{,}000$ 6

15. a. $\log_{10} 0.6$ -0.22185
   b. $\log_{10} 0.1$ -1
   c. $\log_{10} 0.01$ -2
   d. $\log_{10} 0.06$ -1.22185
   e. $\log_{10} 0.006$ -2.22185
   f. $\log_{10} 0.000006$ -5.22185

16. a. $\log_{10} 0.7$ -0.15490
   b. $\log_{10} 0.07$ -1.15490
   c. $\log_{10} 0.001$ -3
   d. $\log_{10} 0.0001$ -4
   e. $\log_{10} 0.0007$ -3.15490
   f. $\log_{10} 0.00007$ -4.15490

17. a. How are the logs of 6, 60, and 6000 related?
Decimal portion same; all positive.

b. How are the logs of 0.6, 0.06, and 0.006 related?
Decimal portion same; all negative.

18. a. How are the logs of 7000, 700, 70, and 7 related?
Decimal portion same; all positive.

b. How are the logs of 0.00007, 0.0007, and 0.07 related? Decimal portion same; all negative.

Change the equations in Exercises 19 and 20 to logarithmic equations, and find x. Round to five decimal places.

19. a. $17 = 10^x$
$\log_{10} 17 = x; x \approx 1.23045$
b. $125 = 10^x$
$\log_{10} 125 = x; x \approx 2.09691$
c. $10^x = 400$
$\log_{10} 400 = x; x \approx 2.60206$
d. $10^x = 0.05$
$\log_{10} 0.05 = x; x \approx -1.30103$

20. a. $25 = 10^x$
$\log_{10} 25 = x; x \approx 1.39794$
b. $100 = 10^x$
$\log_{10} 100 = x; x = 2$
c. $10^x = 300$
$\log_{10} 300 = x; x \approx 2.47712$
d. $0.28 = 10^x$
$\log_{10} 0.28 = x; x \approx -0.55284$

Complete the tables in Exercises 21 and 22.

21.

Exponential Equation	Logarithmic Equation	Solve for x
$5^0 = x$	$\log_5 x = 0$	$x = 1$
$x^3 = 64$	$\log_x 64 = 3$	$x = 4$
$10^x = -10$	$\log_{10}(-10) = x$	$\{\ \}$ or $\varnothing$
$10^x = 0$	$\log_{10} 0 = x$	$\{\ \}$ or $\varnothing$
$10^{-1} = x$	$\log_{10} x = -1$	$x = \frac{1}{10}$

22.

Exponential Equation	Logarithmic Equation	Solve for x
$2^x = 256$	$\log_2 256 = x$	$x = 8$
$x^3 = 125$	$\log_x 125 = 3$	$x = 5$
$3^x = -4$	$\log_3(-4) = x$	$\{\ \}$ or $\varnothing$
$10^{-2} = x$	$\log_{10} x = -2$	$x = \frac{1}{100}$
$2^x = 0$	$\log_2 0 = x$	$\{\ \}$ or $\varnothing$

Write the equations in Exercises 23 to 26 as exponential equations, and find x. Do not use a calculator.

23. a. $\log_7 x = 2$ $7^2 = x, x = 49$

b. $\log_3 3 = x$ $3^x = 3, x = 1$

c. $\log_{10} 0.01 = x$ $10^x = 0.01, x = -2$

d. $\log_{10} x = 0$ $10^0 = x, x = 1$

e. $\log_x 64 = 3$ $x^3 = 64, x = 4$

f. $\log_a x = 1$ $a^1 = x, x = a$

24. a. $\log_2 512 = x$ $2^x = 512, x = 9$

b. $\log_4 x = 4$ $4^4 = x, x = 256$

c. $\log_2 x = 6$ $2^6 = x, x = 64$

d. $\log_{10} x = -1$ $10^{-1} = x, x = \frac{1}{10}$

e. $\log_x 256 = 4$ $x^4 = 256, x = 4$

f. $\log_a x = 0$ $a^0 = x, x = 1$

25. a. $\log_2 x = 8$ $2^8 = x, x = 256$

b. $\log_{10} x = -2$ $10^{-2} = x, x = \frac{1}{100}$

c. $\log_2 \left(\frac{1}{4}\right) = x$ $2^x = \frac{1}{4}, x = -2$

d. $\log_{10} 1 = x$ $10^x = 1, x = 0$

e. $\log_x 100 = 2$ $x^2 = 100, x = 10$

f. $\log_a a = x$ $a^x = a, x = 1$

26. a. $\log_3 x = 1$ $3^1 = x, x = 3$

b. $\log_x 128 = 7$ $x^7 = 128, x = 2$

c. $\log_8 x = 2$ $8^2 = x, x = 64$

d. $\log_3 1 = x$ $3^x = 1, x = 0$

e. $\log_3 27 = x$ $3^x = 27, x = 3$

f. $\log_a a^2 = x$ $a^x = a^2, x = 2$

27. What is the line of symmetry for the graph of a function and its inverse? $y = x$

28. Write how we say $\log_2 x$ in words.   log base 2 of x

Complete the statements listed in Exercises 29 to 41. In Exercises 29 to 32, choose from domain or range.

29. The set of outputs from a function is its ___range___.

30. The set of inputs of a function is its ___domain___.

31. Because we swap x and y, the domain of a function is the ___range___ of its inverse.

32. Because we swap x and y, the range of a function is the ___domain___ of its inverse.

33. If the ordered pair (a, b) belongs to a function, then the ordered pair ___(b, a)___ belongs to its inverse.

34. If the graph of an inverse to a function passes the vertical-line test, then the inverse is a ___function___.

35. The logarithm is the ___exponent___ to which we raise a base to obtain a given number.

36. The ___base___ of an exponential equation must be a positive number not equal to 1.

37. The ___base___ of a logarithmic equation must be a positive number not equal to 1.

38. The set of inputs (domain) to an exponential function, $y = b^x$, is ___$\mathbb{R}$___.

39. The set of outputs (range) to a logarithmic function, $y = \log_b x$, is ___$\mathbb{R}$___.

40. The set of inputs (domain) to a logarithmic function, $y = \log_b x$, is ___$x > 0$___.

41. The set of outputs (range) to an exponential function, $y = b^x$, is ___$y > 0$___.

▮ Error Analysis

42. What is wrong with this equation?

$$\frac{\log_{10} 2}{\log_{10}} = 2$$ Denominator is missing input value for logarithm expression.

43. Is $\dfrac{5 \text{ feet}}{\log_{10} 2} \cdot \dfrac{\log_{10} 5}{2 \text{ feet}}$ correct? No; log is not a unit.

| 9 | Mid-Chapter Test |

1. Identify the following sequences as either linear (arithmetic), quadratic, or exponential (geometric). Fit an equation to the linear and quadratic sequences. For the exponential sequences, find the nth term, a_n. Then use exponential regression on a calculator to find $f(x)$. Show, using properties of exponents, that your results are the same.

a. 3, 9, 27, 81, 243   exponential; $a_n = 3 \cdot 3^{n-1}$; $f(x) = 3^x$

b. $-1, 5, 11, 17, 23$ linear; $y = 6x - 7$

c. 81, 27, 9, 3, 1   exponential; $a_n = 81 \cdot \left(\frac{1}{3}\right)^{n-1}$; $f(x) = 243 \cdot \left(\frac{1}{3}\right)^x$

d. 1, 6, 15, 28, 45   quadratic; $y = 2x^2 - x$

e. $\frac{1}{8}, \frac{1}{4}, \frac{1}{2}, 1, 2$ exponential; $a_n = \frac{1}{8} \cdot 2^{n-1}$; $f(x) = 0.0625 \cdot 2^x$

2. Sketch an exponential curve. Mark the points on the graph illustrating these facts: See Answer Section.

a. With the exception of $b = 0$, any number b raised to the zero power is 1.

b. Any number raised to the first power is itself: $b^1 = b$.

c. Any number raised to the second power is b^2.

3. a. Explain how the two graphs $f(x) = 2^x$ and $f(x) = 2^{x-1}$ are related. 2^{x-1} is shifted right 1 unit.

b. Where do the graphs of $y = 2^x$ and $y = 2^{x-1}$ intersect? Why?   no intersection; $2^x > 2^{x-1}$

4. Refer to a graph of $y = 2^x$.

a. Estimate the exponent on 2 that gives 10 as an output. ≈ 3.3

b. Write the problem in part a as a logarithmic equation. $\log_2 10 = 3.3$

5. Compare the graphs of $y = \frac{1}{2} \cdot 3^x$ and $y = 2 \cdot \left(\frac{1}{3}\right)^x$.
Grows rapidly as x gets large; approaches zero as x gets large

a. Explain what part of the equation shows the y-intercept.   constants $\frac{1}{2}$ and 2

b. Explain how you know whether the equation, $y = b^x$, indicates an increasing or decreasing function.

$b > 1$ indicates increasing function; $0 < b < 1$ indicates decreasing function.

Solve Exercises 6 and 7 without a calculator. Show your steps.

6. a. $2^x = 32$ $x = 5$

b. $\left(\frac{1}{2}\right)^x = 0$ { } or $\varnothing$

c. $\left(\frac{1}{2}\right)^x = \frac{1}{16}$ $x = 4$

d. $5^x = \frac{1}{125}$ $x = -3$

e. $10^x = -10$ { } or $\varnothing$

f. $10^x = 0.01$ $x = -2$

7. a. $\log_4 x = -1$ $x = \frac{1}{4}$

b. $\log_{10} 0.0001 = x$ $x = -4$

c. $\log_3 x = 3$ $x = 27$

d. $\log_2 x = 0$ $x = 1$

e. $\log_4 2 = x$ $x = \frac{1}{2}$

f. $\log_4 (-2) = x$ { } or $\varnothing$

8. a. Make a table for $y = 4^x$ for integer inputs $[-2, 3]$.
See Answer Section.

b. Show the table for the inverse function.

c. Graph $y = 4^x$ and its inverse on one set of axes.

d. Use $y = 4^x$ to solve for the inverse. $y = \log_4 x$

9. Complete the table:

Exponential Equation	Logarithmic Equation	Solve for x
$10^x = 15$	$\log_{10} 15 = x$	$x \approx 1.17609$
$10^x = 13$	$\log_{10} 13 = x$	$x \approx 1.11394$
$3^x = 81$	$\log_3 81 = x$	$x = 4$
$3^9 = x$	$\log_3 x = 9$	$x = 19{,}683$
$4^{x+1} = 64$	$\log_4 64 = x + 1$	$x = 2$

9.4 Applications of Exponential and Logarithmic Functions

Objectives

- Identify the base for a common logarithm.
- Apply the change of base formula to evaluate logarithms.
- Graph logarithmic functions on a calculator.
- Solve pH and Richter scale problems.
- Solve applications with compound interest, doubling time, and half-life.

WARM-UP

Change to an exponential equation and find the value of x or y. Write the answers to Exercises 1 and 2 in scientific notation.

1. $-9.2 = \log_{10} x$ $\quad x \approx 6.310 \times 10^{-10}$ **2.** $6.9 = \log_{10} y$ $\quad y \approx 7.943 \times 10^6$

3. $x = \log_{10} 1$ $\quad x = 0$ **4.** $x = \log_{10} 10$ $\quad x = 1$

5. $y = \log_2 \left(\frac{1}{2}\right)$ $\quad y = -1$ **6.** $y = \log_2 1$ $\quad y = 0$

7. $y = \log_2 8$ $\quad y = 3$

Solve by guess and check.

8. $2 = (1.08)^t$ $\quad t \approx 9.0$ **9.** $2000 = 1000(1 + 0.07)^t$ $\quad t \approx 10.2$

IN THE PRIOR SECTION, we used common logarithms to solve exponential equations with base 10. We now continue with common logarithms but change to a more standard notation. We use the change of base formula to graph functions and solve equations containing bases other than 10. We apply our exponential and logarithmic techniques to pH, Richter numbers, compound interest, and growth applications.

Common Logarithms

As mentioned earlier, logarithms with base 10 are called *common logarithms. Base 10 logarithms are usually written without a base:*

$$y = \log x \quad \text{is} \quad y = \log_{10} x.$$

EXAMPLE 1   Identifying bases   Name the bases for these logarithms. If the base is a variable, state any restrictions.

a. $\log_2 8 = x$ **b.** $\log_4 x = 3$ **c.** $\log 100 = x$ **d.** $y = \log_5 x$

e. $\log_x 16 = 4$ **f.** $y = \log_c d$ **g.** $\log d = c$

SOLUTION   **a.** The base is 2.

b. The base is 4.

c. The base is missing, so it is 10: $\log_{10} 100 = x$.

d. The base is 5.

e. The base is x, $x > 0$, $x \neq 1$.

f. The base is c, $c > 0$, $c \neq 1$.

g. The base is missing, so it is 10: $\log_{10} d = c$.

Change of Base Formula, Base 10

The change of base formula shown below changes a logarithm with any base to logarithms with base 10. The formula can be evaluated with a calculator.

■■ CHANGE OF BASE FORMULA, BASE 10

$$\log_b a = \frac{\log a}{\log b}$$

The change of base formula can be written with any base:

$$\log_b a = \frac{\log_c a}{\log_c b}$$

The formula applies to any real numbers a, b, and c for which the logarithm is defined. The proof of the change of base formula is in Section 9.5.

In Example 2, we first change the exponential equation to a logarithmic equation and then change the logarithm to base 10 so that we can evaluate on a calculator.

EXAMPLE 2 **Applying the change of base formula**   Change these equations to logarithmic form and solve. Round to the nearest tenth.

a. $2 = (1.08)^t$ **b.** $2000 = 1000(1 + 0.07)^t$

SOLUTION   **a.** $2 = (1.08)^t$ Change to a logarithmic equation.

$t = \log_{1.08} 2$ Apply the change of base formula.

$t = \dfrac{\log 2}{\log 1.08}$ Enter the expression into a calculator. If your calculator automatically writes a parenthesis after log, close the parentheses on both the 2 and the 1.08.

$t \approx 9.0$ Remember that there will be some rounding error when you check.

Check: $(1.08)^9 \overset{?}{=} 2$ ✓

b. $2000 = 1000(1 + 0.07)^t$ Divide both sides by 1000.

$2 = (1.07)^t$ Change to a logarithmic equation.

$t = \log_{1.07} 2$ Apply the change of base formula.

$t = \dfrac{\log 2}{\log 1.07}$ Enter the expression into a calculator.

$t \approx 10.2$ Remember that there will be some rounding error when you check.

Check: $1000(1 + 0.07)^{10.2} \overset{?}{=} 2000$ ✓   ■

In order to change $y = ab^x$ to an equivalent logarithmic equation, divide both sides by a.

 ## Graphing Logarithmic Functions on a Calculator

In Example 3, we use the change of base formula to complete a table and a graph for $y = \log_2 x$.

EXAMPLE 3   Applying the change of base formula to calculator graphing   Use the change of base formula to rewrite $y = \log_2 x$ in base 10. Make a table and a graph for $y = \log_2 x$.

SOLUTION   With the change of base formula,

$$y = \log_2 x = \frac{\log x}{\log 2}$$

One possible table appears in Table 14, and the graph in Figure 18.

TABLE 14

Input, x	Output, $y = \log_2 x$
−1	No real number
0	No real number
$\frac{1}{2}$	−1
1	0
2	1
4	2
8	3

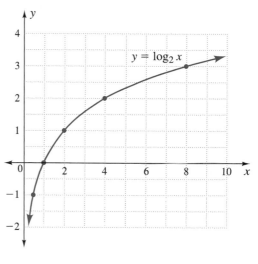

FIGURE 18

Example 4 poses questions to guide a closer examination of the graph of $y = \log_2 x$.

EXAMPLE 4   Investigating the properties of the graph of the logarithmic function
a. Describe any x- or y-intercepts of the graph of $y = \log_2 x$ in Figure 18.
b. Describe any other key points.
c. Under [TBLSET], let **TblStart** = 1 and **ΔTbl** = −0.1. Press ⬚ 2nd ⬚ [TABLE]. What happens as x approaches zero? Change **TblStart** to 0.1 and **ΔTbl** to −0.01. What happens as x approaches zero? What happens at zero? What happens below zero?

SOLUTION   **a.** The graph of $y = \log_2 x$ crosses the x-axis at (1, 0). There is no y-intercept.

b. The graph of $y = \log_2 x$ passes through the point (2, 1) because $2^1 = 2$.

c. The output becomes more negative as x passes through smaller decimal numbers between 1 and 0. We say that the graph approaches the y-axis without touching or crossing it as x approaches zero. The y-axis is a vertical asymptote to the function.
    The output is not real when x is zero or negative, so the graph does not cross the y-axis. The function is not defined in the real numbers for $x \leq 0$.

All graphs of $y = \log_b x$ have similar shapes. As with exponential functions, there are certain points and characteristics that make it possible to recognize the graphs of logarithmic functions or to sketch them with a minimum number of points.

■ SUMMARY: PROPERTIES OF THE
GRAPH OF THE LOGARITHMIC FUNCTION

- Graphs of $y = \log_b x$ have no y-intercept because $y = \log_b 0$ is undefined in the real numbers.
- Graphs of $y = \log_b x$, $b > 1$, have negative outputs for inputs between 0 and 1.
- Graphs of $y = \log_b x$ pass through the x-intercept $(1, 0)$ because $0 = \log_b 1$ implies $b^0 = 1$.
- Graphs of $y = \log_b x$ pass through $(b, 1)$ because $1 = \log_b b$ implies $b^1 = b$.
- Graphs of $y = \log_b x$ have no points in the second or third quadrants because, for negative x, $y = \log_b x$ is undefined in the real numbers. Hence, *the domain of a logarithmic function is the set of positive real numbers.*
- Graphs of $y = \log_b x$, $b > 1$, approach the y-axis as x approaches zero and rise slowly to the right as x increases. Hence, *the range of a logarithmic function is the set of all real numbers.*

Applications Using Base 10 Logarithms

pH SCALE   The pH ratings for common beauticians' supplies are shown in Figure 1 on the chapter opening (page 509). The pH scale describes the relative acidity or basicity of a substance. To the chemist, pH is a measure of the concentration of positive hydrogen ions, H^+, in a solution. The pH scale generally ranges from 1 to 14, but pH can be zero, negative, or greater than 14.

■ pH CATEGORIES

pH less than 7	acidic
pH equal to 7	neutral—pure distilled water at 25°C (about room temperature)
pH greater than 7	alkaline or basic

The formula for the pH scale depends on logarithms:

$$pH = -\log [H^+]$$

where $[H^+]$ is the symbol for hydrogen ion concentration. The pH formula is included here to provide additional practice with common logarithms and scientific notation.

EXAMPLE 5   **Evaluating a common logarithm: the pH of water**   Find the pH of water if water has an H^+ concentration of 1.0×10^{-7} M. (The M stands for *moles* and may be ignored for this application.)

SOLUTION   As mentioned above,

$$pH = -\log(1.0 \times 10^{-7}) = 7$$

No base shows in the logarithm, so it is assumed to be a base 10, or common, logarithm. Suggested keystrokes are

$$\boxed{(-)}\; \boxed{LOG}\; (1.0\; \boxed{2nd}\; [EE]\; \boxed{(-)}\; 7)\; \boxed{ENTER}$$ ■

In some chemical applications, we are given the pH of a number and need to find the hydrogen ion concentration, $[H^+]$. This involves substituting the pH number into the formula $pH = -\log[H^+]$ and solving for $[H^+]$.

> To solve $pH = -\log[H^+]$ for $[H^+]$, remember to do two things:
>
> **1.** Write the base, 10, in the equation.
>
> **2.** Multiply both sides by -1 to get the negative sign on the other side of the equation.

EXAMPLE 6

Solving a common logarithm: the pH of borax Suppose borax, a cleaning agent, has a pH of 9.2. Is borax acidic or basic? What is the hydrogen ion concentration?

SOLUTION Borax is basic because its pH is larger than the neutral value of 7 for distilled water. To find the hydrogen ion concentration in moles (M), we substitute the pH of 9.2 into the pH formula and solve for $[H^+]$.

$pH = -\log [H^+]$	Substitute 9.2 for pH and write in the base, 10.
$9.2 = -\log_{10} [H^+]$	Multiply both sides by -1.
$-9.2 = \log_{10} [H^+]$	Change to an exponential equation.
$10^{-9.2} = [H^+]$	Simplify the expression by evaluating $10^{-9.2}$.
$[H^+] \approx 6.31 \times 10^{-10} \text{ M}$	

The p in pH represents a function. The p function is *negative log of*. In function notation, pH would be written $p(H)$. The H could be replaced by the appropriate concentration of any ion, but most commonly the concentration of the hydrogen ion, H^+, is used.

RICHTER SCALE   The well-known but dated* Richter scale provides a good illustration of use of a logarithmic scale. Two earthquakes are compared by finding the ratio of their energy intensities, I. The Richter number, R, is related to the intensity of the earthquake by the formula

$$R = \log I$$

Note that this is a base 10, or common, logarithm.

EXAMPLE 7

Evaluating a common logarithm: Richter measurements Calculate the Richter number for each of these earthquakes.

a. San Francisco, 1906 (503 deaths): $I = 199,526,000$

b. Iran, 1990 (40,000 deaths): $I = 50,120,000$

c. Alaska, 1964 (131 deaths): $I = 251,190,000$

SOLUTION **a.** $R = \log I = \log 199,526,000 \approx 8.3$

b. $R = \log I = \log 50,120,000 \approx 7.7$

c. $R = \log I = \log 251,190,000 \approx 8.4$

*According to "Abandoning Richter" (R. Monastersky, *Science News*, October 15, 1994), seismologists have pretty much given up using the Richter scale to measure the relative magnitude of earthquakes but have neglected to inform the general public. Of the dozen or more available measurements of earthquakes, two have been generally adopted by scientists as replacements for the Richter number: the *preliminary magnitude* (available immediately after the earthquake is detected) and the *moment magnitude* (a measure of the total seismic wave energy released in an earthquake but not available until an hour or two after the earthquake).

To compare relative sizes, we divide the intensities, not the logarithms. In Example 8, we solve for the intensity by changing the logarithmic equation to an exponential equation, and then we compare intensities.*

EXAMPLE 8

Solving a common logarithm: San Francisco earthquake, 1989   For the San Francisco earthquake in 1989, $R = 6.9$.
a. Calculate the intensity, I, of this earthquake.
b. Divide the intensity of the 1906 quake by that of the 1989 quake to estimate the ratio of intensities.

SOLUTION **a.**

$R = \log I$	Substitute 6.9 for R and write in the base 10.
$6.9 = \log_{10} I$	Change to an exponential equation.
$10^{6.9} = I$	Calculate the intensity.

$$I \approx 7{,}943{,}000$$

b. The ratio of the intensity of the 1906 quake to that of the 1989 quake is

$$\frac{199{,}526{,}000}{7{,}943{,}000} \approx 25$$

The 1906 earthquake was rated about 25 times as intense as the 1989 earthquake.

When working with common logarithms in applications, remember:

COMMON LOGARITHM IN APPLICATIONS

1. Write the base, 10, before doing any operations or simplifications.
2. For $y = -\log_{10} x$ or $y = a \log_{10} x$, move the negative or constant a to the other side.

$$-y = \log_{10} x \qquad \text{or} \qquad \frac{y}{a} = \log_{10} x$$

3. Use the equivalence of $y = \log_b x$ and $x = b^y$ to solve equations.

Applications Requiring Change of Base

COMPOUND INTEREST   Returning to compound interest, we now look at two familiar and three new settings. The compound interest formula is

$$S = P\left(1 + \frac{r}{n}\right)^{nt}$$

where S = future value, P = present value, r = rate of annual interest, n = number of compoundings per year, and t = number of years. When interest is calculated once a year, the formula simplifies to

$$S = P(1 + r)^t$$

EXAMPLE 9

Finding the number of years to reach a target value   If Myrna starts at $5.50 per hour and receives a 6% annual wage increase, in how many years will her wage be $8.27 per hour?

*For a virtual earthquake, if you have a PC and Internet Explorer, see http://www.sciencecourseware.com/eec/Earthquake/

SOLUTION To prevent rounding error in the solution, we will enter the unsimplified expression into the calculator. The initial amount, P, is $5.50 per hour. The annual rate of growth, r, is 6%. The future amount, S, is $8.27 per hour.

$$S = P(1 + r)^t$$

$$8.27 = 5.50(1 + 0.06)^t \qquad \text{Divide each side by 5.50.}$$

$$\frac{8.27}{5.50} = (1.06)^t \qquad \text{Change to a logarithmic equation.}$$

$$\log_{1.06}\left(\frac{8.27}{5.50}\right) = t \qquad \text{Apply the change of base formula.}$$

$$t = \frac{\log\left(\dfrac{8.27}{5.50}\right)}{\log 1.06} \qquad \text{Enter the expression into a calculator.}$$

$$t \approx 7 \text{ yr}$$

It will take 7 years for her hourly wage to grow from $5.50 to $8.27 at 6% per year.

DOUBLING TIME *The time required for money to double* is called the **doubling time**. This is an important concept in studying **exponential growth**, *the behavior described by an increasing exponential function.*

EXAMPLE 10 Solving with logarithms and change of base Solve this compound interest formula for t:

$$2000 = 1000\left(1 + \frac{0.07}{52}\right)^{52t}$$

SOLUTION

We do not add within the parentheses because rounding would be required.

$$2000 = 1000\left(1 + \frac{0.07}{52}\right)^{52t} \qquad \text{Divide each side by 1000.}$$

$$2 = \left(1 + \frac{0.07}{52}\right)^{52t} \qquad \text{Change to a logarithmic equation.}$$

$$\log_{(1+0.07/52)} 2 = 52t \qquad \text{Apply the change of base formula.}$$

$$\frac{\log 2}{\log(1 + 0.07/52)} = 52t \qquad \text{Multiply each side by } \frac{1}{52}.$$

$$\frac{1}{52} \cdot \frac{\log 2}{\log(1 + 0.07/52)} = t \qquad \text{Enter into a calculator.}$$

$$t \approx 9.9$$

It will take approximately 9.9 years for the investment to double at 7% interest, compounded weekly.

Example 11 provides an example of doubling.

EXAMPLE 11 Finding doubling time: garbage production Suppose garbage production in a large city increases by 8% per year. In how many years will garbage production double? From a graph, find when it will double a second time.

SOLUTION The situation is described by compound interest calculated once a year, $n = 1$.

$$S = P(1 + r)^t$$ Doubling means $S = 2P$, so substitute $r = 0.08$ and $S = 2P$.

$$2P = P(1 + 0.08)^t$$ Divide both sides by P. Simplify the base.

$$2 = (1.08)^t$$ Change to a logarithmic equation.

$$\log_{1.08} 2 = t$$ Apply the change of base formula.

$$t = \frac{\log 2}{\log 1.08} \approx 9 \text{ yr}$$ Enter into a calculator.

It will take about 9 years for the amount of garbage produced annually to double. This assumes that the 8% change in garbage production includes changes in both population and rate of throwing things away. A sudden increase in population or in recycling could alter the doubling time.

The graph in Figure 19 shows the ratio of future garbage production (S) to the current year's production (P); that is, $y = S/P = (1.08)^t$. The graph intersects $y = 2$ at 9 years. The graph intersects $y = 4$ at 18 years—the second doubling time.

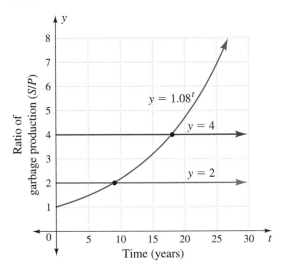

FIGURE 19

HALF-LIFE   *The time required for a value to drop to half its original size* is called the **half-life**. This is an important concept in studying **exponential decay**, *the behavior described by a decreasing exponential function.*

EXAMPLE 12   Solving with logarithms and change of base   Suppose a truck depreciates 20% each year. The initial value is $40,000. Solve the equation below to find out in how many years the truck will be worth half the original value:

$$20{,}000 = 40{,}000(1 + (-0.20))^t$$

SOLUTION $$20{,}000 = 40{,}000(1 + (-0.20))^t$$ Divide each side by 40,000.

$$\frac{20{,}000}{40{,}000} = (1 - 0.20)^t$$ Simplify.

$$0.5 = (0.80)^t$$ Change to a logarithmic equation.

$$\log_{0.80} 0.5 = t$$ Apply the change of base formula.

$$\frac{\log 0.5}{\log 0.80} = t$$ Enter into a calculator.

$$t \approx 3.1 \text{ yr}$$

In about 3.1 years, the truck will be worth half its original value.

Think about it: Why does the 0.80 in the solution to Example 12 make sense?

Note that the rate and time in Example 13 are in days rather than years. We will return to this idea in Exercise 51.

EXAMPLE 13   **Finding the half-life**   Suppose a 10-microgram mass of radioactive iodine-131 decreases by 8.3% each day. In how many days will half the mass be gone?

SOLUTION   Half the mass is 5 micrograms (5×10^{-6} gram).

$$5 = 10(1 - 0.083)^t \qquad \text{Divide by 10 on each side.}$$

$$0.5 = (1 - 0.083)^t \qquad \text{Simplify within the parentheses.}$$

$$0.5 = (0.917)^t \qquad \text{Change to a logarithmic equation.}$$

$$\log_{0.917} 0.5 = t \qquad \text{Apply the change of base formula.}$$

$$t = \frac{\log 0.5}{\log 0.917} \qquad \text{Enter into a calculator.}$$

$$t \approx 8 \text{ days}$$

Here the formula $S = P(1 + r)^t$ is used to demonstrate the concept of half-life. In physics, the formula is based on that for compounding continuously, $S = Pe^{rt}$; see Section 9.6.

The half-life of iodine-131 is 8 days.

GROWTH RATES

In exponential growth or increasing exponential functions, the base is greater than 1. The annual rate, r, is positive.

In exponential decay or decreasing exponential functions, the base is between 0 and 1. Thus, the annual rate, r, is negative.

ANSWER BOX

Warm-up: 1. $x \approx 6.310 \times 10^{-10}$ **2.** $y \approx 7.943 \times 10^6$ **3.** $x = 0$
4. $x = 1$ **5.** $y = -1$ **6.** $y = 0$ **7.** $y = 3$ **8.** $t \approx 9.0$ **9.** $t \approx 10.2$
Think about it: When 20% is lost, 80% remains. So the value remaining each year is 80% of the prior year's value.

9.4   Exercises

In Exercises 1 and 2, identify the base.

1. a. $\log_3 9 = 2$ 3

b. $\log 100 = 2$ 10

c. $4^3 = 64$ 4

d. $25 = 5^2$ 5

e. $\log x = y$ 10

f. $\log_m h = c$ m

2. a. $\log_2 32 = 5$ 2

b. $81 = 3^4$ 3

c. $\log 10 = 1$ 10

d. $2^4 = 16$ 2

e. $\log_c m = h$ c

f. $\log y = x$ 10

In Exercises 3 to 6, estimate x. Then solve, using logarithms and the change of base formula as needed. Round to two decimal places. Comment on any patterns you observe in the answers.

3. a. $2^x = 1$
$x = 0$

b. $2^x = 3$
$x \approx 1.58$

c. $2^x = 4$ $x = 2$

d. $2^x = 5$
$x \approx 2.32$

e. $2^x = 6$
$x \approx 2.58$

f. $2^x = 10$
$x \approx 3.32$

g. $2^x = 12$
$x \approx 3.58$

h. $2^x = 20$
$x \approx 4.32$

i. $2^x = 24$
$x \approx 4.58$

4. a. $3^x = 2$
$x \approx 0.63$

b. $3^x = 4$
$x \approx 1.26$

c. $3^x = 6$
$x \approx 1.63$

d. $3^x = 9$
$x = 2$

e. $3^x = 12$
$x \approx 2.26$

f. $3^x = 18$
$x \approx 2.63$

5. a. $5^x = 2$
$x \approx 0.43$

b. $5^x = 5$
$x = 1$

c. $5^x = 6$
$x \approx 1.11$

d. $5^x = 10$
$x \approx 1.43$

e. $5^x = 25$
$x = 2$

f. $5^x = 30$
$x \approx 2.11$

6. a. $4^x = 3$
$x \approx 0.79$

b. $4^x = 4$
$x = 1$

c. $4^x = 5$
$x \approx 1.16$

d. $4^x = 12$
$x \approx 1.79$

e. $4^x = 16$
$x = 2$

f. $4^x = 20$
$x \approx 2.16$

In Exercises 7 and 8, use the change of base formula, as needed, to solve the equations. Round to five decimal places.

7. a. $3^x = 10$ $x \approx 2.09590$

   b. $x = \log_4 6$ $x \approx 1.29248$

   c. $20 = 10(1.07)^x$ $x \approx 10.24477$

   d. $15 = 30(0.95)^x$ $x \approx 13.51341$

   e. $30 = 60(0.5)^x$ $x = 1$

8. a. $4^x = 20$ $x \approx 2.16096$

   b. $x = \log_5 4$ $x \approx 0.86135$

   c. $30 = 15(1.04)^x$ $x \approx 17.67299$

   d. $10 = 20(0.93)^x$ $x \approx 9.55134$

   e. $16 = 32(0.25)^x$ $x = 0.5$

9. Use a calculator to graph $y = \log_3 x$. Identify the x-intercept and $f(1)$. $1; f(1) = 0$

10. Repeat Exercise 9 for $y = \log_5 x$. $1; f(1) = 0$

11. On a calculator, graph $y = \log_3 x$ together with $y = \log_6 x$. Identify the x-intercept. Describe the relative positions of the two graphs to the right of the x-intercept and to the left of the x-intercept.
See Answer Section.

In Exercises 12 to 16, tell whether the statement is true or false.

12. For $x > 1$ and b larger than 10, the graph of $y = \log_b x$ is above the graph of $y = \log_{10} x$.   false

13. For $0 < x < 1$ and b smaller than 10, the graph of $y = \log_b x$ is above the graph of $y = \log_{10} x$.   false

14. The point $(1, 0)$ is on every graph of $y = \log_b x$.   true

15. The point $(1, b)$ is on every graph of $y = \log_b x$.   false

16. For $0 < x < 1$, $b > 1$, and $y = \log_b x$, y is negative.   true

17. On a calculator, graph $Y_1 = x$, $Y_2 = 2^x$, and $Y_3 = \log_2 x$. Select a square window. Describe the positions of Y_2 and Y_3 relative to the line $y = x$. Explain why $(0, 1)$ lies on Y_2 and $(1, 0)$ lies on Y_3. Explain why $(1, 2)$ lies on Y_2 and $(2, 1)$ lies on Y_3. Name another pair of points (a, b) and (b, a) that lie on Y_2 and Y_3, respectively.
See Answer Section.

18. On a calculator, graph together $y = x$, $y = \sqrt{x}$, and $y = x^2$ for $x \geq 0$. Select a square window. Where do the graphs cross? Why? $(0, 0), (1, 1); 1^n = 1; 0^n = 0; n \neq 0$

19. Find the pH of the following substances.

   a. NaOH (sodium hydroxide), with an H^+ concentration of 1.0×10^{-14} M   14; base

   b. HCl (hydrochloric acid), with an H^+ concentration of 0.20 M   0.7; acid

   c. Lemon juice, with $[H^+] = 5.01 \times 10^{-3}$ M   2.3; acid

   d. Milk of magnesia, with $[H^+] = 1.00 \times 10^{-10}$ M   10; base

   e. Apple cider, with $[H^+] = 1.26 \times 10^{-3}$ M   2.9; acid

20. Find the pH of the following substances.

   a. Baking soda, with $[H^+] = 3.16 \times 10^{-9}$ M   8.5; base

   b. Soft drink, with $[H^+] = 0.01$ M   2; acid

   c. Banana, with $[H^+] = 2.512 \times 10^{-5}$ M   4.6; acid

   d. Milk, with $[H^+] = 2.512 \times 10^{-7}$ M   6.6; acid

   e. Hominy, with $[H^+] = 1 \times 10^{-8}$ M   8; base

21. Write the word *base* or *acid* beside each answer in Exercise 19 to identify the nature of the substance.

22. Write the word *base* or *acid* beside each answer in Exercise 20 to identify the nature of the substance.

In Exercises 23 and 24, change the pH values back to hydrogen ion concentrations using the pH formula.

23. a. pH = 4.1 (tomatoes) 7.94×10^{-5} M

   b. pH = −0.5   3.16 M

   c. pH = 12.0 (washing soda) 1.0×10^{-12} M

   d. pH = 3.5 (acid rain) 3.16×10^{-4} M

24. a. pH = 7.4 (blood plasma) 3.98×10^{-8} M

   b. pH = −0.8   6.31 M

   c. pH = 5.8 (potatoes) 1.58×10^{-6} M

   d. pH = 3.3 (strawberries) 5.01×10^{-4} M

25. For what range of $[H^+]$ will the pH be negative? (*Hint:* Examine the graph of $y = -\log x$. For what set of inputs is the output of $y = -\log x$ negative?) For what range of $[H^+]$ will the pH be positive? $[H^+] > 1; [H^+] < 1$

26. Describe the two steps that make it difficult to solve the equation $y = -\log x$ for x.   change to $-y = \log x$; remember the base 10

In Exercises 27 and 28, write the intensity in scientific notation. Calculate the Richter number for each earthquake.

27. a. Guatemala, 1976 (23,000 deaths): $I = 31,623,000$
   $3.1623 \times 10^7; 7.5$

   b. Mexico City, 1985 (4200 deaths): $I = 125,890,000$
   $1.2589 \times 10^8; 8.1$

28. a. Armenia, 1988 (55,000 deaths): $I = 10,000,000$
   $1 \times 10^7; 7.0$

   b. Agaña, Guam, 1993: $I = 100,000,000$
   $1 \times 10^8; 8$

In Exercises 29 and 30, calculate the intensity for each earthquake. Round to the nearest thousand.

29. a. Valparaiso, Chile, 1906: $R = 8.6$ 398,100,000

   b. Gujarat, India, 2001: $R = 7.9$ 79,430,000

30. Southern California, 1992 (1 death in two earthquakes on June 28):

　　a. $R = 7.5$　31,623,000　**b.** $R = 6.6$　3,981,000

31. Compare the 1989 San Francisco earthquake, at $R = 6.9$, and the 1994 Northridge earthquake, at $R = 6.8$. First calculate the intensities, and then find how many times as strong the earlier earthquake was. $\approx 7,943,000; \approx 6,310,000; \approx 1.3$ times as strong

32. Compare the 1976 Tangshan, China, earthquake, at $R = 8.0$, and the 1993 Hokkaido, Japan, earthquake, at $R = 7.7$. First calculate the intensities, and then find how many times as strong the earlier earthquake was. $\approx 100,000,000; \approx 50,119,000; \approx 2$ times as strong

In Exercises 33 to 36, find how long (to the nearest year) it will take for $1000 to grow to $1 million at the given interest rate.

33. 2% annual interest　349 yr

34. 8% annual interest　90 yr

35. 10% annual interest　73 yr

36. 14% annual interest　53 yr

In Exercises 37 to 40, identify the annual growth rate, r.

37. a. $y = 1000(1 + 0.08)^t$　8%

　　b. $y = 50(1.06)^x$　6%

38. a. $y = 400(1 + 0.05)^t$　5%

　　b. $y = 30(1.04)^x$　4%

39. a. $y = 150(1 - 0.03)^t$　-3%

　　b. $y = 200(0.95)^x$　-5%

40. a. $y = 40(1 - 0.10)^t$　-10%

　　b. $y = 100(0.93)^x$　-7%

41. Suppose water consumption in a large city increases by 6% per year. In how many years will water consumption double?　≈ 11.9 yr

42. Suppose electricity usage in a large city increases by 4% per year. In how many years will electricity usage double?　≈ 17.7 yr

43. Suppose a $30,000 car loses 10% of its value each year. Write an exponential equation and use it to find in how many years (to the nearest hundredth) the value will be half of the original.　$(0.9)^t = 0.5, t \approx 6.58$ yr

44. Suppose $5000 software loses 40% of its value each year. Write an exponential equation and use it to find in how many years (to the nearest hundredth) the value will be less than $100.　$100 = 5000(0.6)^t, t \approx 7.66$ yr

45. A bouncing ball's height is given by $h = 108\left(\frac{2}{3}\right)^{n-1}$, where $n = 1$ is the starting height of 108 inches.

　　a. What is n when the height is half the original height? $n \approx 2.7$

b. What is n when the height is 32 inches?　$n = 4$

c. What is n when the height is 1 inch?　$n \approx 12.5$

46. In Exercise 45, suppose we start the ball at 144 inches and it bounces to $\frac{3}{4}$ of its previous height each time. The starting height is where $n = 1$.

　　a. Write a formula for the ball's height in terms of n.　$h = 144\left(\frac{3}{4}\right)^{n-1}$

　　b. Find what n must be for the height to be 60.75 inches.　$n = 4$

　　c. Find what n must be for the height to be 20 inches.　$n \approx 7.86$

47. When an initial amount P is doubled, the ending amount is $2P$. Using this fact and $S = P(1 + r)^t$, solve for t to find a formula for doubling time in terms of the rate, r. Does P appear in your formula? What does this say about doubling time and the initial amount of money?　$t = \frac{\log 2}{\log(1 + r)}$; no; doubling time is not dependent on P.

48. When an initial amount P is halved, the ending amount is $\frac{1}{2}P$. Using this fact and $S = P(1 + r)^t$, solve for t to find a formula for half-life in terms of the rate, r. Does P appear in your formula? What does this say about half-life and the initial amount?　$t = \frac{\log 0.5}{\log(1 + r)}$; no; half-life is not dependent on P.

49. Five radioactive substances released in greatest quantity by the Hanford Nuclear Site into the Columbia River between World War II and 1970 are listed below. The release data are in curies* and represent the initial values.

　　In parts a to c, estimate the number of half-lives in 1 year.

　　a. Arsenic-76: 2,500,000 curies, half-life of 26.3 hours ≈ 333

　　b. Neptunium-239: 6,300,000 curies, half-life of 2.4 days　≈ 152

　　c. Phosphorus-32: 230,000 curies, half-life of 14.3 days ≈ 26

In parts d to f, fit an equation using exponential regression. Use $(0, S)$ and $(\text{half-life}, \frac{1}{2}S)$ for data points.

　　d. Neptunium decays into plutonium-239, with a half-life of 24,000 years. The amount of neptunium-239 released decayed into 1.7 curies of plutonium-239. $y = 1.7(0.99997)^x$, x in years

　　e. Sodium-24: 12,000,000 curies, half-life of 15 hours $y = 12,000,000(0.955)^x$, x in hr

　　f. Zinc-65: 490,000 curies, half-life of 245 days $y = 490,000(0.997)^x$, x in days

50. **Rule of 72** Dividing 72 by the annual interest rate (expressed as a whole number) provides an estimate of

---

*A curie is 37×10^9 disintegrations per second. A disintegration is the release of an alpha, beta, or gamma particle. The data are from *Radionuclides in the Columbia River: Possible Health Problems in Humans and Effects on Fish*, a publication of the Hanford Health Information Network, 1995.

the number of years required for the initial amount P to double. For 4% interest, $72 \div 4 \approx 18$ yr.

a. Use the rule of 72 to give the doubling times for 1%, 2%, 5%, 8%, 10%, and 20% interest.
72 yr, 36 yr, 14.4 yr, 9 yr, 7.2 yr, 3.6 yr

b. Does the rule of 72 describe a direct variation or an inverse variation between interest rate and doubling time? Explain why. inverse variation; 72 is the constant product for interest rate and doubling time

The article "A Population Exploding," in the December 1988 *National Geographic Magazine*, gives the following doubling times for the populations of the given countries if the populations continue to rise at their current rates. Use the rule of 72 to estimate to the nearest tenth the percent annual growth rate being experienced in each country.

c. Kenya, 17 years 4.2%   **d.** Brazil, 34 years 2.1%

e. India, 35 years 2.1%   **f.** China, 49 years 1.5%

51. In Example 13, the compound interest formula, $S = P(1 + r)^t$, was used when the rate of interest and time were in the same units. Consider why this worked:

a. If r is rate per day and t is number of days, then what expressions describe rate per year and time in years?
365r; $t/365$

b. Substitute your answers from part a into the formula $S = P(1 + r/n)^{nt}$ for $n = 365$ and show that it simplifies to the annual formula.

c. What is repaid if $100 is borrowed at 1.5% per day for 20 days? $134.69

d. What is repaid if $100 is borrowed at 1.5% per day for 30 days? $156.31

▉ Exponential Modeling

52. The following data sets have the same outputs and different inputs. Find the regression equation for each set. Compare the results.

a. (45, 50), (50, 25), (55, 12.5), (60, 6.25)
$y \approx 25,600(0.87)^x$

b. (0, 50), (5, 25), (10, 12.5), (15, 6.25) $y \approx 50(0.87)^x$

c. (0, 50), (1, 25), (2, 12.5), (3, 6.25) $y \approx 50(0.5)^x$

Each output is half the output from the prior data point, yet the regression equations for parts a and b contain base 0.87. Why is the base not 0.5?
x values are in 5-unit intervals and slow the rate of growth.

53. In 1986, it was possible to find Glue Stics® on sale at 2 for $1.00. In 1995, Glue Stics cost $0.99 each; in 2001, they were $2 each.

a. Estimate the cost of Glue Stics in 2005.

b. Fit a linear regression equation to the cost data. Interpret the slope of the line. $y = 0.0964x + 0.392$; 9.64¢ increase per year

c. Fit an exponential regression equation to the cost data. What is the y-intercept?
$y = 0.4808(1.095)^x$; $\approx$cost in 1986

d. From the exponential equation, find the rate of inflation (growth rate, r). $\approx 9.5\%$

e. What factors are assumed with the linear model and with the exponential model?
constant price change each year; percent increase each year

f. Use each equation to predict the price of Glue Stics in the year 2005. $\approx$2.22, $\approx$2.72

54. Suppose a house that cost $28,000 in 1973 doubles in value every 10 years.

a. Estimate the value of the house in the year 2003.
$\approx$224,000

b. Use exponential regression to fit an equation to the house values calculated in order to complete part a.
$y \approx 28,000(1.072)^x$

c. From the exponential regression equation, find the annual growth rate, r. $\approx 7.2\%$

▉ Projects

55. **Wage Options (spreadsheet optional)** Suppose that there are three wage options for food servers.

Option 1: The base salary is $5.00 per hour. An annual raise of 25 cents per hour is given at the end of the year.

Option 2: The base salary is $4.00 per hour. A raise is given annually. The raise is 25 cents per hour at the end of the first year, 30 cents per hour at the end of the second year, 35 cents per hour at the end of the third year, 40 cents per hour at the end of the fourth year, and so on.

Option 3: The base salary is $4.00 per hour. At the end of each year, a 5% increase is made in the hourly wage.

a. Evaluate each wage option for the first five years.

b. Identify the type of function for each.

c. Fit an equation to each.

d. Discuss the circumstances under which each option would be the best.

e. In how many years will the wage double under each option?   See Additional Answers.

56. **Birthday Gift Options (spreadsheet optional)** When each of her grandchildren is born, a grandmother offers three options for birthday gifts.

Option 1: A dollar for each year of the child's age.

Option 2: A dollar at age 1, with a 10% increase on each subsequent birthday.

Option 3: A dollar at age 1, with a $0.10 increase at age 2, a $0.20 increase at age 3, a $0.30 increase at age 4, a $0.40 increase at age 5, and so forth.

a. Evaluate each option for the first five years.

b. Identify the type of function for each.

c. Describe each option with an equation.

d. In how many years will the birthday gift double under each option?

e. Discuss the circumstances under which each option would be the best.

f. Spreadsheet extension: Show the cumulative totals for each year in order to compare total gifts received under each option.   See Additional Answers.

57. Decibel Ratings Research the subject of sound intensity and decibels. Find a chart showing the relative decibel ratings for various sounds, such as those of a jet aircraft and a gasoline-powered lawn mower. Create five exercises based on your research, and solve them.

9.5   Properties of Logarithms and the Logarithmic Scale

Objectives

▪ Prove the properties of logarithms.

▪ Apply properties of logarithms in writing expressions.

▪ Solve equations by taking the logarithm of both sides.

▪ Prove the change of base formula.

▪ Graph data on semilog graph paper.

WARM-UP

Use a calculator to find the logarithms in Exercises 1 to 3. Round to three decimal places. Explain how the expressions in parts a and b might be related.

1. a. $\log_{10} 5 + \log_{10} 8$ 1.602   **b.** $\log_{10} (5 \cdot 8)$ 1.602; equal

2. a. $3 \log_{10} 8$ 2.709   **b.** $\log_{10} 8^3$ 2.709; equal

3. a. $\log_{10} 60 - \log_{10} 6$ 1   **b.** $\log_{10} \left(\frac{60}{6}\right)$ 1; equal

4. Solve $3^x = \frac{1}{27}$ with the like bases property. $x = -3$

5. Solve $3^x = 15$ with the definition of logarithms and the change of base formula. $x \approx 2.465$

IN THIS SECTION, we prove properties of logarithms, solve equations containing logarithms, and solve equations by taking the logarithm of both sides.

We examine exponential functions plotted on semilogarithmic graphs.

Properties of Logarithms

The Warm-up suggests three properties of logarithms involving multiplication, exponents, and division.

PROPERTIES OF LOGARITHMS

If m, n, and b are positive numbers and $b \neq 1$, then

1. $\log_b (m \cdot n) = \log_b m + \log_b n$ Logarithm of a product property

2. $\log_b m^n = n \cdot \log_b m$ Logarithm of a power property

3. $\log_b \left(\frac{m}{n}\right) = \log_b m - \log_b n$ Logarithm of a quotient property

Logarithms are exponents, so it should not be surprising that the properties of exponents are important in proving the properties of logarithms.

In Examples 1 and 2, we prove the first two of the three properties of logarithms. To prove the third property, we return to the definitions and follow a line of thinking similar to that used in the proof of property 1. This proof is left as an exercise.

EXAMPLE 1 Proving property 1   Prove $\log_b (m \cdot n) = \log_b m + \log_b n$.

SOLUTION We start by writing equations containing logarithmic expressions like those on the right side of property 1.

$x = \log_b m$ and $y = \log_b n$	Change to exponential equations.
$m = b^x$ and $n = b^y$	Multiply the left sides and the right sides of the exponential equations.
$mn = b^x \cdot b^y$	Apply the property of like bases: $a^m \cdot a^n = a^{m+n}$.
$mn = b^{x+y}$	Change to a logarithmic equation.
$\log_b mn = x + y$	Substitute for x and y.
$\log_b mn = \log_b m + \log_b n$	

Think about it 1: What would change in Example 1 if we changed the multiplication of m and n to a division?

To prove the logarithm of a power property, we use the fact that a positive integer exponent means repeated multiplication of the base and then apply property 1.

EXAMPLE 2 Proving property 2   Prove $\log_b m^n = n \cdot \log_b m$.

SOLUTION We start by rewriting m^n as n factors of m:

$\log_b m^n = \log_b (m \cdot m \cdot m \cdot \cdots \cdot m)$	Apply property 1, to obtain n terms.
$\quad = \log_b m + \log_b m + \log_b m + \cdots + \log_b m$	Change n terms to n times the term.
$\log_b m^n = n \cdot \log_b m$	

John Napier invented logarithms to aid in arithmetic calculation. Because calculators now do complex arithmetic operations, the properties of logarithms are used primarily for proofs and for working with expressions such as those in Examples 3 and 4.

EXAMPLE 3 Expanding logarithmic expressions   Use properties of logarithms to write each expression in terms of the logarithms of x, y, and z, as needed. Assume that x, y, z, and b are positive numbers ($b \neq 1$).

a. $\log_b \dfrac{xz}{y}$ b. $\log_b (y^3 z^2)$ c. $\log_b x\sqrt{y}$

SOLUTION

a. $\log_b \dfrac{xz}{y} = \log_b x + \log_b z - \log_b y$	Logarithm of product and quotient properties
b. $\log_b (y^3 z^2) = \log_b y^3 + \log_b z^2$	Logarithm of product property
$\quad = 3 \log_b y + 2 \log_b z$	Logarithm of power property
c. $\log_b x\sqrt{y} = \log_b x + \log_b \sqrt{y}$	Logarithm of product property
$\quad = \log_b x + \tfrac{1}{2} \log_b y$	$\sqrt{y} = y^{1/2}$ and power property

EXAMPLE 4 Simplifying logarithmic expressions   Use the properties of logarithms to write each expression as a logarithm of a single expression. Assume that x, y, z, and b are positive numbers ($b \neq 1$).

a. $\frac{1}{3} \log_b x + 2 \log_b y - 3 \log_b z$

b. $2 \log_b (x + y) - \log_b (x + y)$

SOLUTION   **a.** $\frac{1}{3} \log_b x + 2 \log_b y - 3 \log_b z = \log_b \left(\dfrac{x^{1/3} \cdot y^2}{z^3} \right)$ Logarithm of power, product, and quotient properties

$$= \log_b \left(\sqrt[3]{x} \cdot \dfrac{y^2}{z^3} \right) \quad \tfrac{1}{3} \text{ exponent} = \text{cube root}$$

b. $2 \log_b (x + y) - \log_b (x + y) = \log_b \dfrac{(x + y)^2}{x + y}$ Logarithm of power and quotient properties

$$= \log_b (x + y)$$

Solving Equations by Taking the Logarithm of Both Sides

We have solved exponential equations such as $3^x = \frac{1}{27}$ and $3^x = 15$ with a variety of methods: guess and check, graphing, the like bases property (if $a^x = a^y$, then $x = y$), changing to a logarithm (if $y = b^x$, then $x = \log_b y$), and the change of base formula.

In addition to using these equation-solving methods, we may also take the logarithm, to any base, of both sides of an equation.

SOLVING EXPONENTIAL EQUATIONS

> We may take the logarithm base b of both sides of an equation.

In Example 5, we solve the equations from the Warm-up by taking the log base 10 of both sides.

EXAMPLE 5 Taking the logarithm base b of both sides   Solve these equations by taking the log base 10 of both sides.

a. $3^x = \frac{1}{27}$ **b.** $3^x = 15$

SOLUTION   **a.** $3^x = \frac{1}{27}$ Take the log base 10 of both sides.

$\log_{10} 3^x = \log_{10} \left(\frac{1}{27}\right)$ Use the logarithm of a power property to move the exponent x.

$x \log_{10} 3 = \log_{10} \left(\frac{1}{27}\right)$ Divide both sides by $\log_{10} 3$.

$x = \dfrac{\log_{10} \left(\frac{1}{27}\right)}{\log_{10} 3}$ This is the change of base formula, applied to $x = \log_3 \left(\frac{1}{27}\right)$.

$x = -3$

Check: $3^{-3} = \frac{1}{27}$ ✓

b. $3^x = 15$ Take the log base 10 of both sides.

$\log_{10} 3^x = \log_{10} 15$ Use the logarithm of a power property to move the exponent x.

$x \log_{10} 3 = \log_{10} 15$ Divide both sides by $\log_{10} 3$.

$x = \dfrac{\log_{10} 15}{\log_{10} 3}$ This is the change of base formula, applied to $x = \log_3 15$.

$x \approx 2.46497$

Check: $3^{2.46497} \approx 15$ ✓

Think about it 2: Could we solve the equations in Example 5 by taking the logarithm base 3 of both sides?

EXAMPLE 6 Taking the logarithm base 10 of both sides   Solve the equation $1000(1 + 0.08)^t = 2000$ for t, the time in years needed for $1000 to double to $2000 at 8% interest.

SOLUTION

$$1000(1 + 0.08)^t = 2000$$ Divide both sides by 1000.

$$(1 + 0.08)^t = \frac{2000}{1000}$$ Take log base 10 of both sides.

$$\log_{10}(1 + 0.08)^t = \log_{10} 2$$ Apply the logarithm of a power property to move the exponent t.

$$t \log_{10}(1 + 0.08) = \log_{10} 2$$ Divide both sides by the coefficient of t.

$$t = \frac{\log_{10} 2}{\log_{10}(1 + 0.08)}$$ Enter into a calculator.

$$t \approx 9 \text{ yr}$$ ■

In parts a and b of Example 5 and in Example 6, our last step was evaluating an expression that was in fact the change of base formula. This suggests that we can prove the change of base formula by taking the log of both sides of an equation.

Proof of the Change of Base Formula

By taking the log of both sides of an equation and applying the properties of logarithms, we prove the change of base formula in Example 7.

EXAMPLE 7 Proving change of base   Prove the change of base formula,

$$\log_b a = \frac{\log_c a}{\log_c b}$$

SOLUTION Our plan is to work with one side of the equation and show that it equals the other side of the equation. We start by setting the left side of the formula equal to the variable, x.

$$x = \log_b a$$ Change to an exponential equation.

$$b^x = a$$ Take the log base c of both sides.

$$\log_c b^x = \log_c a$$ Apply the logarithm of a power property to move the exponent x.

$$x \log_c b = \log_c a$$ Divide both sides by $\log_c b$.

$$x = \frac{\log_c a}{\log_c b}$$ This equation contains the right side of the formula. Substitute $\log_b a$ for x.

$$\log_b a = \frac{\log_c a}{\log_c b}$$ The change of base formula is true.   ■

As you will find in the exercises, by replacing the log base c with the log base 10, we can prove the change of base formula that we have used on the calculator.

$$\log_b a = \frac{\log a}{\log b}$$

Application: Logarithmic Scale

We now turn to a visual representation of logarithms—the logarithmic scale. Have you ever wondered how illustrators fit small numbers (such as the growth rate of a child in miles per hour) and large numbers (such as the speed of light in miles per hour) on the same graph? Example 8 illustrates the problems encountered in graphing small and large numbers on the same scale.

This material is optional. Important in science, forestry, and technical fields, it is related to Exercises 53 to 59.

EXAMPLE 8 **Exploring the scale on the axes** In a table, list the outputs for $y = 2^x$ if the inputs are the integers from 1 to 10. Graph the table values, and discuss the limitations of the graph.

SOLUTION The input-output table appears in Table 15, and the graph in Figure 20.

TABLE 15

x	$y = 2^x$
1	2
2	4
3	8
4	16
5	32
6	64
7	128
8	256
9	512
10	1024

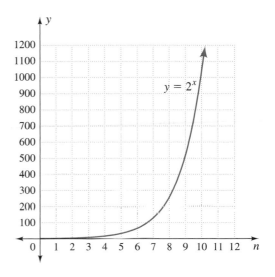

FIGURE 20

Finding a scale for the vertical axis in Figure 20 is a challenge. When we count by 100 on the vertical axis, the outputs between 0 and 100 are hard to read because they are so near the x-axis. If we enlarged the distance between 0 and 100 to, say, an inch, the vertical axis would need to be 12 inches tall and the graph would not fit on the page. ◼

The solution to graphing widely spaced numbers is to use a logarithmic scale. *When the spacing on an axis of a line graph is proportional to the logarithms of numbers,* the scale is called a **logarithmic scale**. The pH scale for hair products in Figure 1 on the chapter opener shows the negative powers of 10 forming a logarithmic scale on the vertical axis.

◼ When we place a logarithmic scale on one axis of a coordinate graph, we create a semilog graph. On a **semilog graph**, *one scale is uniform and the other scale is based on logarithms.* The logarithmic scale can be on either the horizontal or the vertical axis.

In Example 9, we create a semilog graph of the equation $y = 2^x$.

EXAMPLE 9 Graphing on semilog axes
a. Construct a semilog graph of $y = 2^x$ from the values shown in Table 15. Use special semilog paper, with the logarithmic scale on the vertical axis.
b. What is distinctive about the vertical axis?
c. Compare the graph of $y = 2^x$ with that in Figure 20.
d. Which outputs are most easily read?

SOLUTION **a.** Figure 21 shows a semilog graph of $y = 2^x$.

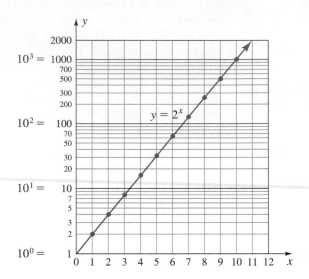

FIGURE 21

b. The spacing between the horizontal lines varies, but in a regular pattern. On the vertical axis, the distance between 2 and 20 is the same as that between 20 and 200. The distance between 10 and 100 is the same as that between 100 and 1000. The value of the logarithm increases by one when we multiply by 10.

c. The exponential curve in Figure 20 has been replaced by a straight line in Figure 21.

d. The smaller outputs, 2, 4, 8, and 16, are clearly shown in Figure 21, the semilog graph, whereas in Figure 20 they are all near the x-axis and hard to read. The larger outputs, 512 and 1024, are harder to read in the semilog graph. ▄▄▄

Think about it 3: What are the values of log 1, log 10, log 100, log 1000, and log 10,000?

On the logarithmic scale in Figure 21 (vertical axis), we multiply by 10 between numbers on the number line.

Figure 22 is a line with 0.1 spacing between the numbers 0 to 1.0. In Figure 22, dots are placed at the logarithms of the numbers 1 to 10. Observe that the spacing between the logarithms of the numbers 1 to 10 is similar to the spacing between the lines on the vertical axis in Figure 21. If we changed the numbers below the number line to 1.0 to 2.0, the spacing between dots for the logarithms of 10 to 100 would be identical to that for the logarithms of 1 to 10 (see Exercise 59).

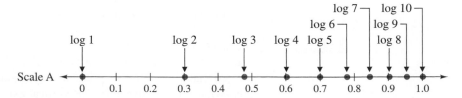

FIGURE 22

> **ANSWER BOX**
>
> **Warm-up:** **1. a.** 1.602 **b.** 1.602; equal **2. a.** 2.709 **b.** 2.709; equal **3. a.** 1 **b.** 1; equal **4.** $x = -3$ **5.** $x \approx 2.465$ **Think about it 1:** The exponents in b^x and b^y would be subtracted instead of added. **Think about it 2:** Yes, any base could be used. **Think about it 3:** $\log 1 = 0$; $\log 10 = 1$; $\log 100 = 2$; $\log 1000 = 3$; $\log 10{,}000 = 4$.

9.5 Exercises

In Exercises 1 and 2, match an expression from the list of choices to each logarithmic expression. Write the original expression as well as the chosen expression.

1. Choose from

$\log (x - 2)$, $\log (x/2)$, $\log (2 - x)$, $\log (2 + x)$, $\log 2 - \log x$, $2 \log x$, $x \log 2$, $\log 2 + x$, $\log 2x$

  a. $\log 2 + \log x$ $\log 2x$

  b. $\log x^2$ $2 \log x$

  c. $\log (2/x)$ $\log 2 - \log x$

  d. $\log 2^x$ $x \log 2$

  e. $\log x - \log 2$ $\log \left(\frac{x}{2}\right)$

2. Choose from

$\log 3 + \log x$, $\log x^3$, $\log (3 - x)$, $\log (3/x)$, $\log 3^x$, $\log x - \log 3$, $\log 3 + x$, $\log (x - 3)$, $\log (3 + x)$

  a. $\log (x/3)$ $\log x - \log 3$ **b.** $\log 3x$ $\log 3 + \log x$

  c. $\log 3 - \log x$ $\log \left(\frac{3}{x}\right)$ **d.** $x \log 3$ $\log 3^x$

  e. $3 \log x$ $\log x^3$

Identify the property of logarithms that explains each fact in Exercises 3 to 6.

3. If $\log 2 \approx 0.30103$ and $\log 3 \approx 0.47712$, then $\log 6 \approx 0.77815$.   log of product property

4. If $\log 2 \approx 0.30103$, then $\log (10 \cdot 2) \approx 1 + 0.30103$.   log of product property

5. If $\log 3 \approx 0.47712$, then $\log 9 \approx 0.95424$.   log of power property

6. If $\log 2 \approx 0.30103$, then $\log 8 \approx 0.90309$.   log of power property

In Exercises 7 to 14, use properties of logarithms to write each expression in terms of the logarithms of x, y, and z. Assume that x, y, z, and b are positive numbers ($b \neq 1$).

7. a. $\log_b xz^3$
   $\log_b x + 3 \log_b z$
     b. $\log_b \dfrac{x^2}{y}$ $2 \log_b x - \log_b y$

8. a. $\log_b \dfrac{x}{3y}$
   $\log_b x - (\log_b 3 + \log_b y)$
     b. $\log_b \left(\dfrac{5}{x}\right)^2$ $2(\log_b 5 - \log_b x)$

9. a. $\log_b (xy)^2$
   $2(\log_b x + \log_b y)$
     b. $\log_b \dfrac{5}{x^2}$ $\log_b 5 - 2 \log_b x$

10. a. $\log_b (25x)^{1/2}$
   $\frac{1}{2}(\log_b 25 + \log_b x)$
     b. $\log_b 25x^{1/2}$ $\log_b 25 + \frac{1}{2} \log_b x$

11. a. $\log_b \dfrac{\sqrt{x}}{y}$ $\frac{1}{2} \log_b x - \log_b y$
     b. $\log_b (x - z)^2$ $2 \log_b (x - z)$

12. a. $\log_b \sqrt{\dfrac{x}{y}}$ $\frac{1}{2}(\log_b x - \log_b y)$
     b. $\log_b (x + z)^2$ $2 \log_b (x + z)$

13. a. $\log_b \dfrac{xz^3}{y^2}$
   $\log_b x + 3 \log_b z - 2 \log_b y$
     b. $\log_b \dfrac{y^3 z^2}{x}$ $3 \log_b y + 2 \log_b z - \log_b x$

14. a. $\log_b \dfrac{x\sqrt{y}}{z^2}$
   $\log_b x + \frac{1}{2} \log_b y - 2 \log_b z$
     b. $\log_b \dfrac{y^{1/3} x^2}{z}$ $\frac{1}{3} \log_b y + 2 \log_b x - \log_b z$

In Exercises 15 to 18, use the properties of logarithms to write each expression as a logarithm of a single expression. Assume that x, y, z, and b are positive numbers ($b \neq 1$).

15. a. $\log_b x + 2 \log_b y$ $\log_b (xy^2)$
     b. $\log_b y - 2 \log_b x$ $\log_b (y/x^2)$

16. a. $2(\log_b 2 + \log_b x)$ $\log_b (2x)^2$
     b. $3(\log_b x - \log_b y)$ $\log_b (x/y)^3$

17. a. $\frac{1}{2}(\log_b y + 3 \log_b z)$ $\log_b \sqrt{y/z^3}$
     b. $\frac{1}{2} \log_b x - \log_b 3$ $\log_b (\sqrt{x}/3)$

18. a. $\frac{1}{3} \log_b y - 2 \log_b x$ $\log_b (\sqrt[3]{y}/x^2)$
     b. $\frac{1}{3}(\log_b x + 2 \log_b z)$ $\log_b \sqrt[3]{xz^2}$

In Exercises 19 to 22, write the expression as a single logarithm and simplify where possible. Assume that all logarithms are defined. (*Hint:* Factoring may be helpful.)

19. a. $\log (x + 1) + \log (x - 1)$ $\log (x^2 - 1)$

     b. $\log (x^2 - 1) - \log (x - 1)$ $\log (x + 1)$

20. a. $\log (x + 1) - \log (x^2 - 1)$ $\log \left(\dfrac{1}{x - 1}\right) = -\log (x - 1)$

     b. $\log (x - 2) + \log (x + 4)$ $\log (x^2 + 2x - 8)$

21. a. $\log (x^2 + 3x + 2) - \log (x + 1)$ $\log (x + 2)$

     b. $\log (x + 3) + \log (x - 2)$ $\log (x^2 + x - 6)$

22. a. $\log (x - 3) + \log (x + 3)$ $\log (x^2 - 9)$

     b. $\log (x^2 - 9) - \log (x^2 + 6x + 9)$ $\log \left(\dfrac{x - 3}{x + 3}\right)$

23. Show that $\log \sqrt{x} = \frac{1}{2} \log x$.
   $\log \sqrt{x} = \log(x^{1/2}) = \frac{1}{2} \log x$

24. Why is $\log 3 = \frac{1}{2} \cdot \log 9$? $\log 3 = \log\sqrt{9} = \log 9^{1/2} = \frac{1}{2}\log 9$

25. Show that $\log (1/x) = -\log x$. $\log (1/x) = \log (x^{-1}) = -\log x$

26. Prove that $\log_b (m/n) = \log_b m - \log_b n$. Model your proof after Example 1.

27. In Example 7, replace the log base c with the log base 10, and prove the change of base formula $\log_b a = (\log a)/(\log b)$. (*Hint:* Let $y = \log_b a$.)

28. One reference book gives the pH formula as

$$pH = \log \frac{1}{[H^+]}$$

Another reference has

$$pH = -\log [H^+]$$

Are these formulas the same or different? Why?
same; $\log \left(1/[H^+]\right) = \log [H^+]^{-1} = -\log [H^+]$

Simplify the left side of each equation in Exercises 29 to 36, and then solve.

29. $\log_2 x^2 + \log_2 x = 6$ $x = 4$

30. $\log (x^2 - x) - \log (x - 1) = 2$ $x = 100$

31. $\log x + \log x = 2$ $x = 10$

32. $\log_5 x^3 - \log_5 x^2 = 1$ $x = 5$

33. $\log (x^2 + x) - \log (x + 1) = 1$ $x = 10$

34. $\log_4 x + \log_4 1 = 2$ $x = 16$

35. $\log_3 x^3 - \log_3 x = 4$ $x = 9$

36. $\log_6 x + \log_6 x = 4$ $x = 36$

Solve the equations in Exercises 37 to 52 in two ways:
(a) Take the log of both sides.
(b) Change to logarithmic equations and apply the change of base formula. Round to four decimal places.

37. $5^x = 20$ $x \approx 1.8614$

38. $6^x = 24$ $x \approx 1.7737$

39. $3^{x+2} = 48$ $x \approx 1.5237$

40. $8^{x+1} = 36$ $x \approx 0.7233$

41. $4^{x-1} = 28$ $x \approx 3.4037$

42. $7^{x-1} = 35$ $x \approx 2.8271$

43. $9^{x+1} = 42$ $x \approx 0.7011$

44. $10^{x-1} = 25$ $x \approx 2.3979$

45. $1000(1.08)^t = 2000$ $t \approx 9.0065$

46. $1000(1.09)^t = 3000$ $t \approx 12.7482$

47. $15(1.07)^t = 45$ $t \approx 16.2376$

48. $10(1.06)^t = 20$ $t \approx 11.8957$

49. $1000(1 - 0.08)^t = 500$ $t \approx 8.3130$

50. $900(1 - 0.09)^t = 300$ $t \approx 11.6489$

51. $45(1 - 0.07)^t = 15$ $t \approx 15.1385$

52. $20(1 - 0.06)^t = 10$ $t \approx 11.2023$

For Exercises 53 and 54, make a table for integer inputs -1 to 6. Graph the data from the tables on both regular graph paper and semilog graph paper. Photocopy the semilog grid below if needed.

53. $y = 3^x$

54. $y = 4^x$

See Answer Section.   See Additional Answers.

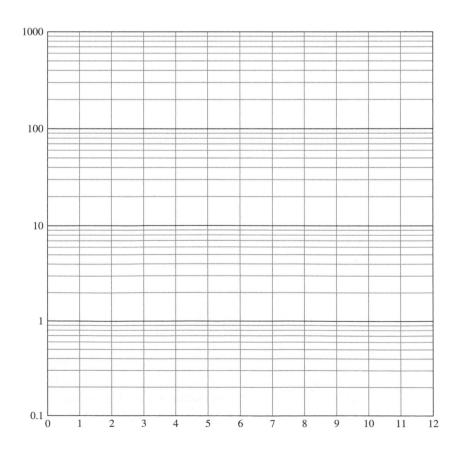

55. What shape is formed by exponential data on a semilog graph? straight line

56. The spacing between the powers of 10 on the vertical axis of the semilog graphs in Exercises 53 and 54 is equal. The spacing represents the integers formed by the __logarithms__ of the numbers shown on the axis.

57. For parts a to f below, refer to the graph of the snag "recruitment" process. The graph shows the decay of a dead fir tree that has been left standing to serve as a wildlife habitat.

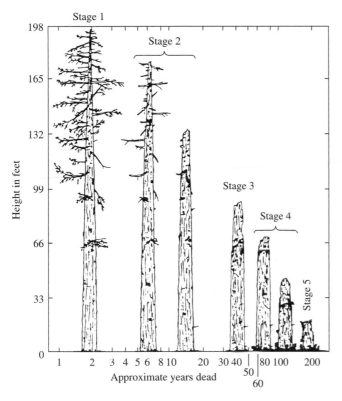

Snag "recruitment" process

Source: *Rising from the Ashes, Shady Beach—A New Perspective on Recovery*, U.S. Forest Service (Willamette National Forest), Department of Agriculture, 1990.

a. Approximately how many feet of height are lost between the second year and the sixth year? ≈22 ft

b. Estimate how many years it takes for the tree to lose its top 66 feet. ≈12 yr

c. Estimate how many years until the tree loses its second 66 feet of height. 60 to 70 yr

d. Estimate how many years until the last 66 feet of height are lost. over 200 yr

e. Is the height loss more rapid between year 2 and year 12 or between year 100 and year 200? What natural events might explain this? 2 and 12; wind breakage

f. Estimate the height of the stump in year 200. ≈16 ft

58. The graph below is a plot of $y = \log_{10} x$. Trace or photocopy the graph, including axes. Draw a horizontal line through each point on the graph of $y = \log_{10} x$ associated with one of the inputs x from 1 to 10. (You can trace the horizontal lines for the first three outputs, $f(1) = \log 1$, $f(2) = \log 2$, and $f(3) = \log 3$. You will need to draw the remaining horizontal lines, for $f(4)$, $f(5), f(6), \ldots$, and $f(10)$.) Compare the spacing of these horizontal lines with that of the lines in the second graph for Exercises 53 and 54. See Additional Answers; spacing is same (on larger scale).

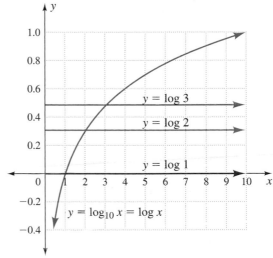

59. a. Trace or photocopy scales A, B, and C. Mark the results from scale A in Figure 22 onto the copy of scale A below. Let $Y_1 = \log x$. See Answer Section.

b. Use the table feature of your calculator to find $\log x$ for $x = 10, 20, 30, \ldots, 100$. Place a dot on scale B below for each log value.

c. Use the table feature of your calculator to find $\log x$ for $x = 100, 200, 300, \ldots, 1000$. Place a dot on scale C below for each log value.

d. Describe the pattern formed by the dots on scales A, B, and C. Why does this occur?
Dots are in same relative positions because log 20 = 1 + log 2, etc.

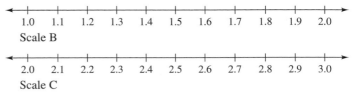

60. Place the depth data from the table on page 562 into calculator list L1 and the percents of light penetration into calculator list L2. Let a third calculator list be the log of the second list.

Penetration of Light through the Clearest Ocean Water

Depth (in meters)	Light Penetration (in percents)
0	100
1	45
2	39
10	22
50	5
100	0.5

Source: D. Ingmanson and W. J. Wallace, *Oceanography: An Introduction,* 4th ed. (Belmont, CA: Wadsworth Publishing, 1989), p. 114.

a. Fit an exponential regression equation to the data in L1 and L2. $y = 52.2(0.95)^x$

b. Write a sentence that describes the relationship between depth of water and light penetration.
Deeper the water, less the amount of light penetration.

c. Fit a linear regression equation to L1 and L3. The resulting equation will be in the form $\log_{10} y = ax + b$. $\log_{10} y = -0.02x + 1.72$

d. Change the equation in part c into exponential form, and show that it is the same as the equation in part a.
$y = 10^{-0.02x+1.72}$; error due to rounding

▮ Projects

61. Scientific Notation and Logarithms Properties of exponents and $3 \approx 10^{0.47712}$ permit us to write these numbers in scientific notation and then as powers of 10.

$$30{,}000 = 3 \times 10^4 \approx 10^{0.47712} \times 10^4$$
$$= 10^{0.47712+4} = 10^{4.47712}$$
$$0.3 = 3 \times 10^{-1} \approx 10^{0.47712} \times 10^{-1}$$
$$= 10^{0.47712-1} = 10^{-0.52288}$$

In parts a to l, write the given number in scientific notation. Then use the following facts to find the exponent n, the logarithm of the given number:

$$4 \approx 10^{0.60206}$$
$$5 \approx 10^{0.69897}$$

a. 400, $10^n = 400$ 4×10^2; 2.60206

b. 5000, $10^n = 5000$ 5×10^3; 3.69897

c. 5,000,000, $10^n = 5{,}000{,}000$ 5×10^6; 6.69897

d. 500,000, $10^n = 500{,}000$ 5×10^5; 5.69897

e. 4000, $10^n = 4000$ 4×10^3; 3.60206

f. 500, $10^n = 500$ 5×10^2; 2.69897

g. 40,000, $10^n = 40{,}000$ 4×10^4; 4.60206

h. 4,000,000, $10^n = 4{,}000{,}000$ 4×10^6; 6.60206

i. 0.4, $10^n = 0.4$ 4×10^{-1}; $-1 + 0.60206$, or -0.39794

j. 0.005, $10^n = 0.005$ 5×10^{-3}; $-3 + 0.69897$, or -2.30103

k. 0.0004, $10^n = 0.0004$ 4×10^{-4}; $-4 + 0.60206$, or -3.39794

l. 0.5, $10^n = 0.5$ 5×10^{-1}; $-1 + 0.69897$, or -0.30103

m. What property of exponents allows us to add or subtract the exponents in parts a to l? $x^a \cdot x^b = x^{a+b}$

n. Explain why each n in parts a to l is a logarithm.
Logarithm is exponent to which we raise base to obtain given number.

o. How can scientific notation tell us the first number of a logarithm of a number greater than 1?
Exponent on 10 is first number.

p. When a number is written in scientific notation, $a \times 10^n$, the number a is always between what numbers? Write those numbers as powers of 10.
1 and 10, $1 = 10^0$ and $10 = 10^1$

q. What numbers have negative logarithms? Why?
numbers between 0 and 1; their logarithms are sum of number between 0 and 1 and negative exponent on 10.

62. Antilogs Changing back to the original number—finding $x = 900$ when $\log x = 2.95424$—is called *finding the antilog*. Antilog 2.95424 means $10^{2.95424}$ and is approximately 900. Evaluate the expressions or solve the equations in parts a to p.

a. antilog 3.29907 ≈ 1991

b. antilog 0.9132 ≈ 8.188

c. $10^{2.5}$ ≈ 316.2278

d. $\log x = 1.69897$ $x \approx 50$

e. antilog 3.28892 ≈ 1945

f. $\log x = 0.49715$ $x \approx 3.1416$

g. $10^{1.5}$ ≈ 31.62278

h. antilog 2.000 100

i. 1.30103 is log x. $x \approx 20$

j. 2.30103 is log x. $x \approx 200$

k. antilog $(-3) = x$ $x \approx 0.001$

l. antilog 1 10

m. antilog (-2.000) 0.01

n. antilog (-2.5) $\approx 0.003\ 162\ 28$

o. antilog (-1) 0.1

p. antilog 0 1

9.6   The Natural Number *e* in Exponential and Logarithmic Functions

Objectives

▪ Calculate powers and roots of *e*.

▪ Apply the formula for continuously compounded interest.

▪ Find logarithms for base *e*.

▪ Change natural logarithmic equations to exponential equations and solve.

▪ Change exponential equations to natural logarithmic equations and solve.

WARM-UP

The exclamation point, !, is used to denote "factorial." The **factorial** of a number is *the number itself multiplied by every lower positive integer.* Thus,

$$3! = 3 \cdot 2 \cdot 1 = 6$$

and

$$7! = 7 \cdot 6 \cdot 5 \cdot 4 \cdot 3 \cdot 2 \cdot 1 = 5040$$

Student Note: On the TI83+, factorial is (MATH) PRB **4 : !**

Look in your calculator manual to see how to find the factorial, !.
    Evaluate these factorial expressions by hand.

1. 2!   2

2. 5!   120

3. 8!   40,320

Use a calculator to do the next two problems.

8.066×10^{67}

4. Evaluate 52!, the number of ways a deck of 52 cards can be arranged.

5. Suppose 0! and 1! are defined as 1. Evaluate the following expression by adding one term at a time to the prior answer. Record your sum term by term.

1, 2, 2.5, 2.666 . . . ,
2.708 333 . . . ,
2.716 66 . . . ,
2.718 055 5 . . . ,
≈2.718 253 968,
≈2.718 278 77

$$\frac{1}{0!} + \frac{1}{1!} + \frac{1}{2!} + \frac{1}{3!} + \frac{1}{4!} + \frac{1}{5!} + \frac{1}{6!} + \frac{1}{7!} + \frac{1}{8!}$$

THIS SECTION INTRODUCES the constant *e*—the natural number—and its role in exponential and logarithmic functions. Of particular importance are the roles of *e* in calculating compound interest and as a base for many exponential and logarithmic functions.

Compound Interest

EXAMPLE 1   Exploring compound interest calculations   An investor has $1000 to deposit for a year. Shopping around, she discovers that the current interest rate is 5%, but it is being calculated in four different ways: annually, monthly, daily, and hourly. Calculate the amount of money she would have with each method after 1 year.

SOLUTION   Table 16 summarizes the calculations.

TABLE 16

n	$S = P(1 + r/n)^{nt}$
Annually: 1	$\$1000(1 + 0.05/1)^{1\cdot1} = \$1050$
Monthly: 2	$\$1000(1 + 0.05/12)^{12\cdot1} \approx \$1051.16$
Daily: 365	$\$1000(1 + 0.05/365)^{365\cdot1} \approx \$1051.26$
Hourly: 8760	$\$1000(1 + 0.05/8760)^{8760\cdot1} \approx \$1051.27$

Student Note: Continue to round down if interest is received, and up if interest is paid.

Rounded to the nearest cent, there is a one-cent difference between daily interest and hourly interest. This may surprise you, because there were larger differences between annual and monthly interest and between monthly and daily interest.

The results in Example 1 are even more apparent in a graph. Figure 23 shows the graph of $y = 1000(1 + 0.05/x)^{x\cdot1}$ for x between 0 and 400 and y between 1050 and 1052. As x gets large, the curve flattens out, so there is little change in y as we trace.

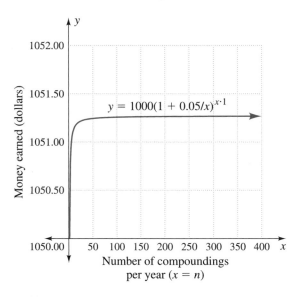

FIGURE 23

Think about it 1: Is there a y-intercept in Figure 23?

The Natural Number e

Mathematicians looked for an upper limit to the compound interest formula as the number of compoundings increased toward infinity. They held the interest rate constant ($r = 1$) and studied $(1 + 1/n)^n$.

EXAMPLE 2   Looking for an upper limit to $(1 + 1/n)^n$ Evaluate $(1 + 1/n)^n$ for $n = 1, 2, 12, 52,$ 365, and 8760.

SOLUTION   Table 17 contains values of $(1 + 1/n)^n$ for several inputs. As we found in Example 1, there is considerable change in the output for lower inputs and little change in the output for higher inputs. This suggests that the expression is approaching a limit near 2.718.

TABLE 17

n	$(1 + 1/n)^n$
1	2.0000
2	2.2500
12	2.613035
52	2.692597
365	2.714567
8760	2.718127

Although the process of finding the upper limit requires calculus, Example 2 suggests that we can observe the limit by looking at the behavior of $(1 + 1/n)^n$ for large n. For $n = 8760$, we are within two ten-thousandths of the limit.

The limiting number is approximately

$$2.718\ 281\ 828\ 459\ 045$$

The limit is irrational (it cannot be written in the form a/b, where a and b are integers and $b \neq 0$). This limit is called the **natural number** and is named with the letter e.

The letter e was chosen by Leonhard Euler, a noted Swiss mathematician (1707–1783) whose image appears on the Swiss 10-franc note (Figure 24). It is believed unlikely that Euler named the number for himself; however, considering his work with the number, it would have been appropriate.

FIGURE 24

The factorial expression in Warm-up Exercise 5 adds up to approximately 2.718 278 8. The sum is approaching e. We can get as close as we like to e by adding more terms: $1/9!$, $1/10!$, and so forth.

Think about it 2: How are pi and the golden ratio (phi) like e?

Calculating with *e*

The number e is so important that it appears on every scientific and business calculator. It usually is obtained in a shifted position with a key labeled ⎡ **LN** ⎤ (more about ⎡ **LN** ⎤ later).

EXAMPLE 3 Evaluating expressions containing e Evaluate these expressions with a calculator.

a. e^2 **b.** $e^{0.5 \cdot 3}$ **c.** e **d.** 2^e **e.** $e^{-\pi/2}$ **f.** $\sqrt{e}$

SOLUTION   **a.** To evaluate e^2, we enter e to the exponent 2. Using ⎡ **2nd** ⎤ $[e^x]$ 2 ⎡ **ENTER** ⎤, we obtain 7.389 056 099. Because e is almost always used with an exponent, the $[e^x]$ key has the exponent sign, ⎡ **^** ⎤, built in to save keystrokes.

b. To evaluate $e^{0.5 \cdot 3}$, we enter e to the exponent $(0.5 \cdot 3)$. The parentheses must enclose the exponent. Using ⟨2nd⟩ $[e^x]$ (0.5×3) ⟨ENTER⟩, we obtain 4.481 689 07.

c. To evaluate e, we enter e to the exponent 1. Using ⟨2nd⟩ $[e^x]$ 1 ⟨ENTER⟩, we obtain 2.718 281 828.

d. To evaluate 2^e, we enter 2 to the exponent e to the exponent 1. Using 2 ⟨^⟩ ⟨2nd⟩ $[e^x]$ 1 ⟨ENTER⟩, we obtain the calculator display 2^e^(1), followed by the answer 6.580 885 991.

e. To evaluate $e^{-\pi/2}$, we use parentheses to enclose the exponent. Using ⟨2nd⟩ $[e^x]$ $(-\pi/2)$ ⟨ENTER⟩, we obtain 0.207 879 576 4.

f. To evaluate $\sqrt{e}$, we enter the square root followed by e to the exponent 1. The square root of e is approximately 1.648 721 271. ■

Practice doing the problems in Example 3 to make sure you are using the calculator correctly.

Continuously Compounded Interest

Because $(1 + 1/n)^n$ has a limiting value, e, we are able to calculate interest continuously. With **continuously compounded interest**, *the number of compounding periods approaches infinity* (and the length of the compounding period approaches zero). The limiting value, e, is the base in the formula for continuously compounded interest.

■ CONTINUOUSLY COMPOUNDED INTEREST

> The formula for continuously compounded interest is
>
> $$S = Pe^{rt}$$
>
> where S is the future value, P is the present value (principal), r is the annual rate of interest, and t is the time in years.

EXAMPLE 4 **Finding the future value for continuously compounded interest**   Calculate the value of $1000 after 1 year, with interest compounded continuously at 5%, and compare the result with that for compounding every hour.

SOLUTION $S = Pe^{rt} = 1000 \cdot e^{0.05 \cdot 1} \approx \$1051.271\ 096$

$$S = 1000\left(1 + \frac{0.05}{8760}\right)^{8760} \approx \$1051.270\ 947$$

Continuously compounding interest yields a slightly higher value than compounding each hour. ■

As you work through Example 5, practice enclosing the exponent in parentheses when evaluating on a calculator.

EXAMPLE 5 **Finding the future value for continuously compounded interest**   Find the value of $1000 if interest is compounded continuously at the given rate for the given time.

a. 5% for 3 years   **b.** 3% for 5 years

c. 6.93% for 10 years   **d.** 6% for 11.55 years

SOLUTION   **a.** $S = Pe^{rt} = 1000 \cdot e^{0.05 \cdot 3} = \$1161.83$

b. $S = Pe^{rt} = 1000 \cdot e^{0.03 \cdot 5} = \$1161.83$

Parts a and b give the same answer.

c. $S = Pe^{rt} = 1000 \cdot e^{0.0693 \cdot 10} = \$1999.70$

d. $S = Pe^{rt} = 1000 \cdot e^{0.06 \cdot 11.55} = \$1999.70$

Parts c and d give the same answer.

Example 6 illustrates a typical use of $S = Pe^{rt}$.

EXAMPLE 6

Finding the initial or present value with continuously compounded interest What amount of money would you need to invest at 8% interest compounded continuously to have $1,000,000 in 20 years?

SOLUTION

We are missing the initial amount of money P (the present value or principal). *Because interest is compounded continuously, we use the $S = Pe^{rt}$ formula.*

$$S = Pe^{rt} \qquad \text{Substitute the facts.}$$

$$1,000,000 = Pe^{0.08 \cdot 20} \qquad \text{Divide by } e^{0.08 \cdot 20} \text{ to solve for } P.$$

$$\frac{1,000,000}{e^{0.08 \cdot 20}} = P$$

$$P \approx \$201,896.52$$

In 20 years, at 8% interest compounded continuously, $202,000 will grow to approximately $1,000,000.

Logarithms with Base *e*

Because both time and rate of interest are in the exponent in the $S = Pe^{rt}$ formula, we need logarithms to solve for t or r. In Example 7, we solve for the length of time needed for money to double at 8% interest compounded continuously.

EXAMPLE 7

Finding doubling time with continuous compounding What length of time is needed for money to double at 8% interest compounded continuously?

SOLUTION

When money doubles, the future value is double the starting value, and $S = 2P$. Thus, we can proceed as follows:

$$S = Pe^{rt} \qquad \text{Substitute the facts.}$$

$$2P = Pe^{0.08t} \qquad \text{Divide by } P.$$

$$2 = e^{0.08t} \qquad \text{Change to a logarithmic equation.}$$

$$\log_e 2 = 0.08t \qquad \text{Divide by 0.08.}$$

$$\frac{\log_e 2}{0.08} = t \qquad \text{Apply the change of base formula.}$$

$$t = \frac{\dfrac{\log 2}{\log e^1}}{0.08}$$

$$t \approx 8.7 \text{ yr}$$

Caution: On a calculator, correct use of parentheses in the change of base formula and with other logarithmic expressions is essential.

It is not necessary to use the change of base formula on $\log_e 2$, as we did in Example 7, because log base *e* has its own calculator key. The key for log base *e* is $\boxed{\text{LN}}$ (the same key as is used in shifted position for e^x). LN stands for **natural logarithm**, *the logarithm with the natural number as its base.*

▮ SPECIAL NOTATION FOR LOGARITHMS WITH BASE e

The natural logarithm of x has a special notation:

$$\log_e x = \ln x$$

Unfortunately, ln x looks like In x or 1n x. The yellow highlighting is to help you read ln x in this section.

EXAMPLE 8 Using the natural logarithm   Write $t = (\log_e 2)/0.08$ with a natural logarithm, and evaluate it with the natural logarithm key, $\boxed{\text{LN}}$.

SOLUTION

$$t = \frac{\log_e 2}{0.08}$$

$$= \frac{\ln 2}{0.08}$$

Student Note: On the calculator, use extra parentheses in Examples 8 and 9.

$$\approx 8.7 \text{ yr}$$ ▬

Caution: Until you have considerable experience with logarithms, write *every* logarithm with its base. For example, in solving equations, use $\log_{10} x$ and $\log_e x$ when doing steps with algebraic notation and change to log x and ln x just before evaluating with a calculator.

EXAMPLE 9 Solving for the rate, r A house cost \$55,000 in 1985. Its value 14 years later is \$119,000. Assuming the value of the house has grown continuously, what is the rate?

SOLUTION

The word *continuously* points to the continuous compounding formula.

$$S = Pe^{rt} \qquad \text{Let } S = 119{,}000, P = 55{,}000, \text{ and } t = 14.$$

$$119{,}000 = 55{,}000e^{r\cdot 14} \qquad \text{Divide by 55,000.}$$

$$\frac{119{,}000}{55{,}000} = e^{r\cdot 14} \qquad \text{Change to a logarithmic equation and simplify the fraction.}$$

$$\log_e \frac{119}{55} = r \cdot 14 \qquad \text{Change to natural log notation and divide by 14.}$$

$$\frac{\ln \frac{119}{55}}{14} = r \qquad \text{Evaluate.}$$

$$r \approx 0.055$$

The rate is approximately 5.5% growth. ▬

▮ SUMMARY: CALCULATORS AND LOGARITHMS

- The common logarithm is abbreviated log x and means $\log_{10} x$. Use $\boxed{\text{LOG}}$ on a calculator.
- The natural logarithm is abbreviated ln x and means $\log_e x$. Use $\boxed{\text{LN}}$ on a calculator.
- All other logarithms are written $\log_b x$ and are entered on a calculator with the change of base formula, $(\log x)/(\log b)$.

Changing between Exponential and Logarithmic Form

Example 10 illustrates the importance of including the bases when we change a logarithmic equation to its corresponding exponential equation.

EXAMPLE 10 Changing forms of equations   Complete Table 18, changing the exponential equations to logarithmic equations and the logarithmic equations to exponential equations. Where appropriate, solve for x.

TABLE 18

Exponential Equation	Logarithmic Equation	Solve for x (as needed)
	$\ln e = 1$	
	$\log 100 = 2$	
$e^0 = 1$		
$10^x = 1000$		
	$\ln 2 = x$	

SOLUTION   We write the base on the logarithm before changing to an exponential equation. The completed table appears in Table 19.

TABLE 19

Exponential Equation	Logarithmic Equation	Solve for x (as needed)
$e^1 = e$	$\ln e = 1$ $\log_e e = 1$	
$10^2 = 100$	$\log 100 = 2$ $\log_{10} 100 = 2$	
$e^0 = 1$	$\log_e 1 = 0$ $\ln 1 = 0$	
$10^x = 1000$	$\log_{10} 1000 = x$ $\log 1000 = x$	$x = 3$
$e^x = 2$	$\ln 2 = x$ $\log_e 2 = x$	$x \approx 0.693$

Application: Natural Logarithms in Chemistry

In the next examples, we apply properties of exponents and logarithms to solving chemical equations commonly found in first-year chemistry courses. As before, remember to write the bases in all logarithmic expressions.

EXAMPLE 11 Solving logarithmic equations: Boltzmann equation   Solve for Ω in the Boltzmann equation, $S = k \ln \Omega$.

SOLUTION

$S = k \ln \Omega$ Divide both sides by k.

$\dfrac{S}{k} = \ln \Omega$ Write ln as log base e.

$\dfrac{S}{k} = \log_e \Omega$ Change to an exponential equation.

$e^{S/k} = \Omega$

Some logarithmic equations in applications may seem intimidating because they contain many variables or unfamiliar symbols (such as the Greek letter omega, Ω, in Example 11). Often, difficulties stem from forgetting that the base is e.

EXAMPLE 12   Solving logarithmic equations   Solve $E = E^0 - \dfrac{RT}{nF} \ln Q$ for the reactant quotient Q.

(*Note:* Many chemistry texts use a zero superscript to represent standard conditions such as 1 atm pressure and 25°C. In this formula, the E^0 should not be interpreted as the zeroth power of E.)

SOLUTION

$$E = E^0 - \frac{RT}{nF} \ln Q \qquad \text{Subtract } E^0.$$

$$E - E^0 = -\frac{RT}{nF} \ln Q \qquad \text{Multiply by } -\frac{nF}{RT}.$$

$$-\frac{nF}{RT}(E - E^0) = \ln Q \qquad \text{Write ln as log base } e.$$

$$-\frac{nF}{RT}(E - E^0) = \log_e Q \qquad \text{Change to exponential form.}$$

$$e^{-\frac{nF}{RT}(E - E^0)} = Q$$

ANSWER BOX

Warm-up: 1. 2   **2.** 120   **3.** 40,320   **4.** 8.066×10^{67} **5.** 1, 2, 2.5, 2.666 . . . , 2.708 333. . . , 2.716 66 . . . , 2.718 055 5. . . , $\approx 2.718\ 253\ 968$, $\approx 2.718\ 278\ 77$ **Think about it 1:** There is no y-intercept because the expression $0.05/x$ is undefined if $x = 0$.   **Think about it 2:** All three numbers are irrational.

9.6   Exercises

In Exercises 1 and 2, assume a deposit of $1000 is made to a savings account at the rate of interest shown. Find the amount of money in the account after 3 years if interest is compounded annually, monthly, and daily (assume 365 days). How much is gained by compounding monthly and daily rather than annually?

1. 4% $\approx\$1,124.86$, $\approx\$1,127.27$, $\approx\$1,127.48$; $\approx\$2.41$, $\approx\$2.62$

2. 7% $\approx\$1,225.04$, $\approx\$1,232.92$, $\approx\$1,233.65$; $\approx\$7.88$, $\approx\$8.61$

Evaluate the expressions in Exercises 3 to 6 with a calculator. Round to the nearest hundredth.

3. a. e^2 7.39   **b.** e^π 23.14   **c.** $2\sqrt{e}$ 3.30

4. a. $e^{1.5}$ 4.48   **b.** π^e 22.46   **c.** $\frac{1}{2}\sqrt{e}$ 0.82

5. a. 3^e 19.81   **b.** $e^{0.4 \cdot 2}$ 2.23   **c.** $e^{\pi/2}$ 4.81

6. a. 4^e 43.31   **b.** $e^{2/\pi}$ 1.89   **c.** $e^{0.6 \cdot 3}$ 6.05

7. Describe how to get e^e on your calculator. $e\wedge(e\wedge(1))$

8. Between which two of the numbers 1^2, 2^2, 3^2, and 4^2 does the expression e^2 fall? $2 < e < 3$, so $2^2 < e^2 < 3^2$

In Exercises 9 to 14, find the value of the principal after one year, with interest compounded continuously. Round to the nearest cent.

9. $P = \$1000$, 8% annual interest $1083.28

10. $P = \$1000$, 6% annual interest $1061.83

11. $P = \$1000$, 10% annual interest $1105.17

12. $P = \$1000$, 12% annual interest $1127.49

13. $P = \$1000$, $4\frac{1}{2}$% annual interest $1046.02

14. $P = \$1000$, $21\frac{1}{2}$% annual interest $1239.86

15. What amount of money would need to be invested at 10% compounded continuously to have $500,000 in 12 years? $\approx\$150,597.11$

16. What amount of money would need to be invested at 10% compounded continuously to have $500,000 in 10 years? $\approx\$183,939.73$

17. What amount of money would need to be invested at 10% compounded continuously to have $500,000 in 20 years? $\approx\$67,667.65$

18. Would the answer to Exercise 17 change if the money were invested at 20% compounded continuously for 10 years? Why?   No; in $S = Pe^{rt}$, r and t are multiplied: 10%(20 years) = 20%(10 years)

Complete the tables in Exercises 19 and 20.

19.

Logarithmic Equation	Show base *e* in logarithm	Exponential Equation	$y = ?$
$\ln y = 1$	$\log_e y = 1$	$y = e^1$	$y = e$
$\ln y = 0$	$\log_e y = 0$	$y = e^0$	$y = 1$
$y = \ln(-1)$	$y = \log_e(-1)$	$e^y = -1$	{ } or ∅
$y = \ln e^2$	$y = \log_e e^2$	$e^y = e^2$	$y = 2$
$y = \ln e^e$	$y = \log_e e^e$	$e^y = e^e$	$y = e$

20.

Logarithmic Equation	Show base *e* in logarithm	Exponential Equation	$y = ?$
$y = \ln e$	$y = \log_e e$	$e^y = e$	$y = 1$
$y = \ln 1$	$y = \log_e 1$	$e^y = 1$	$y = 0$
$y = \ln 0$	$y = \log_e 0$	$e^y = 0$	{ } or ∅
$\ln y = -1$	$\log_e y = -1$	$y = e^{-1}$	$y = \frac{1}{e} \approx 0.3679$
$\ln y = e$	$\log_e y = e$	$y = e^e$	$y \approx 15.154$

In Exercises 21 to 24, solve for *x*. Round to four decimal places.

21. a. $e^x = 4$ $x = 1.3863$ **b.** $\ln x = -2$ $x = 0.1353$

22. a. $\ln x = -1$ $x = 0.3679$ **b.** $e^x = 1.5$ $x = 0.4055$

23. a. $\ln x = 1.5$ $x = 4.4817$ **b.** $e^x = 2$ $x = 0.6931$

24. a. $e^x = 0.5$ $x = -0.6931$ **b.** $\ln x = 0.5$ $x = 1.6487$

25. Using $S = Pe^{rt}$, find the rate of interest required for money to double in 8 years. ≈8.66%

26. Using $S = Pe^{rt}$, find the rate of interest required for money to double in 11 years. ≈6.30%

27. Using $S = Pe^{rt}$, find an expression for the rate of interest required for money to double in *t* years. $r = \frac{\ln 2}{t}$

28. Using $S = Pe^{rt}$, find the amount of time required for money to triple at 8% interest. ≈13.7 yr

29. Using $S = Pe^{rt}$, find the amount of time required for money to triple at 6% interest. ≈18.3 yr

30. Using $S = Pe^{rt}$, find an expression for the amount of time required for money to triple at a rate *r* of interest. $t = \frac{\ln 3}{r}$

31. Find the amount of time required for a city's population to halve when its rate of decrease is 8% per year. Use $S = Pe^{rt}$. $t \approx 8.7$ yr

32. Find the amount of time required for a city's population to halve when its rate of decrease is 4% per year. Use $S = Pe^{rt}$. $t \approx 17.3$ yr

The doubling time for continuously compounded interest can be estimated by dividing 69 by the interest rate (as a whole number). For 10% interest, $69 \div 10 = 6.9$ years. Use this **rule of 69** in Exercises 33 to 35.

33. Estimate each rate of interest, using the rule of 69.

   a. The rate of interest required for money to double in 8 years. ≈8.625%

   b. The rate of interest required for money to double in 11 years. ≈6.27%

   c. An expression for the rate of interest required for money to double in *t* years. ≈69/*t*

   d. Compare your answers from parts a and b with those from Exercises 25 and 26. Rule of 69 is accurate to within 0.1%.

34. Is the relation of interest rate and years to doubling time, as expressed in the rule of 69, inversely proportional? If so, what is the constant of variation? yes; $\ln 2 \approx 0.693$

35. Prove the rule of 69 by solving $2P = Pe^{rt}$ for *t*. One of your last steps will be $t \approx 0.693/r$. Multiplying by 100/100 will change the fraction to approximately $69/100r$, which gives the rate of interest as a whole number.

36. When we solve $S = Pe^{rt}$ for *P*, we find two answers that appear to be different: $P = Se^{-rt}$ and $P = S/e^{rt}$. What definition of exponents explains why they are the same? $b^{-1} = 1/b$

37. a. To investigate the change of base formula using $\ln x$ instead of $\log x$, evaluate $\log_6 8$ using the ratios $(\log a)/(\log b)$ and $(\ln a)/(\ln b)$. ≈1.1606; ≈1.1606

   b. Evaluate log 8, log 6, ln 8, and ln 6 separately. ≈0.903; ≈0.778; ≈2.079; ≈1.792

   c. What do you observe? Ratios are equal; logarithms are not.

38. The change of base formula may contain the natural logarithm instead of the common logarithm:

$$\log_b a = \frac{\ln a}{\ln b}$$

Using as a model the proof shown in Example 7 of Section 9.5, write a proof for the natural logarithm change of base formula.

Solve each chemistry formula in Exercises 39 to 42 for the variable indicated. The zero superscript represents standard conditions, not a zero exponent.

39. Solve $\ln K = -\dfrac{Ea}{RT} + C$ for *K*, the rate constant. $K = e^{\left(-\frac{Ea}{RT} + C\right)}$

40. Solve $\Delta G^0 = -RT \ln K$ for *K*, the equilibrium constant. $K = e^{\left(-\frac{\Delta G^0}{RT}\right)}$

41. Solve $E = -\dfrac{RT}{nF} \ln [H^+]$ for $[H^+]$, the hydrogen ion concentration. $[H^+] = e^{\left(-\frac{EnF}{RT}\right)}$

42. Solve $\ln P = \dfrac{-\Delta H_{vap}}{RT} + B$ for *P*, the pressure. $P = e^{\left(-\frac{\Delta H_{vap}}{RT} + B\right)}$

43. The standard normal probability curve in statistics is described by

$$y = \frac{1}{\sqrt{2\pi}} e^{-x^2/2}$$

The curve describes the distribution of a set of numbers with a mean of 0 and a standard deviation of 1.

a. Graph the normal probability equation for $x = -3$ to $x = 3$ and $y = -1$ to $y = 1$. See Answer Section.

b. Estimate the highest point on the normal curve.
$y \approx 0.4$ at $x = 0$

c. What part of the equation shows the y-intercept? $1/\sqrt{2\pi}$

d. If the total area under the curve equals 1 square unit, estimate the area under the curve between $x = -1$ and $x = 1$. ≈ 0.65

44. Velocity is a function of time, t. The velocity in feet per second of a sky diver in free fall is given by

$$v = \frac{g}{a}(1 - e^{-at})$$

where g is the acceleration due to gravity, 32.2 ft/sec². The factor a contains information on the mass of the sky diver and the resisting force of the air. The a values in parts a to c have been calculated for a 135-pound female sky diver. Substitute a into the equation. Use a calculator graph to find the terminal (maximum) velocity and the time t required to be within 1 foot per second of terminal velocity. Use [0, 60] for X and [0, 300] for Y.

a. Horizontal position (arms and legs spread from body): $a = 0.183$ ≈ 176 ft/sec; ≈ 28 sec

b. Fetal position: $a = 0.129$ ≈ 250 ft/sec; ≈ 47 sec

c. Nose dive (arms along body): $a = 0.1586$ ≈ 203 ft/sec; ≈ 34 sec

d. Change the terminal velocity in part c to miles per hour. ≈ 138 mph

45. Suppose the population of Detroit, Michigan, in t years after 1950 can be estimated by the function $f(t) = 1{,}849{,}568e^{-0.0147t}$. Round answers to three significant digits.

a. What was the population in 1950? 1,850,000

b. What was the population 20 years later, in 1970? 1,380,000

c. What was the population 40 years later, in 1990? 1,030,000

d. Which part of the function indicates whether the population is increasing or decreasing? sign of exponent

e. When would the population be half the population in 1950? ≈ 1997

f. Predict the population in the year 2000. Compare with the actual 2000 population of 951,270.
887,000; decline in population slowing

46. Suppose the population of El Paso, Texas, in t years after 1950 can be estimated by the function $f(t) = 130{,}485e^{0.0343t}$. Round answers to three significant digits.

a. What was the population in 1950? 130,000

b. What was the population 20 years later, in 1970? 259,000

c. What was the population 40 years later, in 1990? 515,000

d. Which part of the function indicates whether the population is increasing or decreasing?
sign on exponent

e. When would the population be twice the population in 1950? ≈ 1970

f. Predict the population in the year 2000. Compare with the actual 2000 population of 563,662.
725,000; overestimated rate of growth

47. Evaluate the factorials, as defined in the Warm-up.

a. 4! 24

b. 10! 3,628,800

c. 6! 720

d. 9! 362,880

e. Evaluate the expression for the nine terms shown, and compare the result with e^2. ≈ 7.3873; $e^2 \approx 7.3891$

$$2^0 + \frac{2^1}{1!} + \frac{2^2}{2!} + \frac{2^3}{3!} + \frac{2^4}{4!} + \frac{2^5}{5!} + \frac{2^6}{6!} + \frac{2^7}{7!} + \frac{2^8}{8!} + \cdots$$

■ Projects

48. Graphing with e^x and $\ln x$ Graph the equations with a calculator, and record sketches on paper. Label the curves with their equations, and label intercepts and horizontal or vertical lines that the graphs approach. Mention any graphing shortcuts that you recall from earlier graphing exercises.

a. Graph $y = e^{-x}$ and $y = -e^{-x}$.

b. Graph $y = -e^x$ and $y = e^x$.

c. Graph $y = e^x + 1$ and $y = e^{x+1}$.

d. Graph $y = e^{x-1}$ and $y = e^x - 1$.

e. Graph $y = \ln x$ and $y = \ln x - 1$.

f. Graph $y = \ln x + 1$ and $y = \ln(x + 1)$.

g. Graph $y = e^x$ and $y = \ln x$, together with $y = x$. Change to a square window. Describe the relative positions of the graphs. Symmetric over line $y = x$; see Additional Answers.

49. Exceeding the Accuracy of the Calculator In evaluating compound interest for n larger than 8760 (hourly), it is possible to exceed the capacity of a graphing calculator and obtain results that are wrong.

a. Calculate the amount in an account at 5% interest compounded every hour for 1 year. Start with $1000. $1051.270 947

b. Calculate the amount in an account at 5% interest compounded every minute for 1 year, $n = 60 \times 8760 = 525,600$. Start with $1000. $1051.271 107

c. Calculate the maximum amount that should be in the account. Use $S = Pe^{rt}$. $1051.271 096

d. Verify that your graphing calculator gives a number above the limit in part b. How far above the limit is the number? 0.000 011, or 1.1×10^{-5}

e. Set $Y_1 = 1000(1 + 0.05/x)^{x \cdot 1}$ and $Y_2 = 1000e^{0.05}$. Set **TblStart** = 525,600 and **ΔTbl** = 1. Move the cursor down the Y_1 column, and comment on what happens to the outputs. go above and below $1051.271 096

f. Set **Xmin** = 525600, **Xmax** = 525605, **Ymin** = 1051.271, and **Ymax** = 1051.2712, and graph Y_1 and Y_2. Compare the results. sawtooth pattern for Y_1

---

9 Chapter Summary

Vocabulary

change of base formula	exponential decay	horizontal asymptote	nth term of a geometric
common logarithm	exponential function	like bases property	sequence
common ratio	exponential growth	logarithm	quadratic sequence
continuously compounded	factorial	logarithmic function	rule of 72
interest	geometric sequence	logarithmic scale	rule of 69
doubling time	half-life	natural logarithm	semilog graph
e, the natural number			

Concepts

This chart summarizes compound interest, where S = future value, P = present value (or principal), r = annual rate of interest, t = time in years (or whatever other unit matches that of r), n is the number of times a year the interest is calculated (compounded), and e, the natural number, is an irrational constant.

Compound Interest	$S = P\left(1 + \dfrac{r}{n}\right)^{nt}$ $n = 1$ (annual) $n = 2$ (semiannual) $n = 4$ (quarterly) $n = 12$ (monthly) $n = 52$ (weekly) $n = 365$ (daily)
Annual Compounding $n = 1$	$S = P(1 + r)^t$ Inflation Depreciation
Continuously Compounded Interest	$S = Pe^{rt}$

This chart summarizes important ideas about linear, quadratic, and exponential functions, sequences, and variation. See the related chart in Chapter 7 for inverse variation.

Linear Equations $y = mx + b$ and $x = c$ Linear Functions $f(x) = mx + b$ Domain: Real numbers	Arithmetic Sequences Domain: Positive integers Constant first differences	Direct Variation Direct Proportion $y = kx$ Linear Variation $y = mx + b$ Domain: Real numbers
Quadratic Equations $y = ax^2 + bx + c$ Quadratic Functions $f(x) = ax^2 + bx + c$ Domain: Real numbers	Quadratic Sequences Domain: Positive integers Constant second differences	Quadratic Variation $y = kx^2$ Domain: Real numbers
Exponential Equations $y = ab^x, b > 0, b \neq 1$ Exponential Functions $f(x) = ab^x$ Domain: Real numbers Range: Positive real numbers	Geometric Sequences $a_n = a_1 r^{n-1}, r < 0$ or $r > 0$ Domain: Positive integers Differences are the same as or a multiple of the original sequence. Constant ratio of terms	

9.1 ▨ Exponential Functions

Geometric sequences have differences that are either the same as or a multiple of the original sequence. However, they are not the only sequence with this pattern. A common ratio uniquely defines a geometric sequence.

Any exponential function with the domain limited to the positive integers can be written as a geometric sequence.

Use properties of exponents to change exponential expressions b^{x+c} to ab^x, where $a = b^c$.

9.2 ▨ Exponential Graphs and Equations

If $0 < b < 1$, the graph of $y = b^x$ decreases from left to right.

If $b > 1$, the graph of $y = b^x$ increases from left to right.

The graph of $y = b^x$ passes through $(0, 1)$ and $(1, b)$.

For $b > 1$, the larger the base in $y = b^x$, the steeper the graph.

The x-axis is a horizontal asymptote for the graph of $y = b^x$.

The domain for $f(x) = b^x$ is all real numbers.

The range for $f(x) = b^x$ is $f(x) > 0$.

The coefficient a is the y-intercept of the graph of $y = ab^x$.

For $h > 0$, the graph of $y = b^{x-h}$ is shifted h units to the right of the graph of $y = b^x$.

For $h > 0$, the graph of $y = b^{x+h}$ is shifted h units to the left of the graph of $y = b^x$.

To use the like bases property, change the exponential expressions on both sides of an equation to the same base, set the exponents equal, and solve for the variable.

9.3 ▨ Solving Exponential Equations: Logarithms

The logarithmic function, $y = \log_b x$, is the inverse function to the exponential function, $y = b^x, b > 0, b \neq 1, x \geq 0$.

The equations $y = b^x$ and $x = \log_b y$ are equivalent.

The base b for an equivalent pair of exponential and logarithmic equations is the same.

Use the [LOG] key on a calculator to obtain the common logarithm of a number.

9.4 ▨ Applications of Exponential and Logarithmic Functions

Common, or base 10, logarithms are usually written without the base: $\log_{10} x = \log x$.

The change of base formula may be written with any base. Base 10 is usually convenient:

$$\log_b a = \frac{\log a}{\log b}$$

In order to change $y = ab^x$ to an equivalent logarithmic equation, divide both sides by a.

Graphs of logarithmic functions have no y-intercept because $y = \log_b 0$ is undefined.

Graphs of logarithmic functions $y = \log_b x$ or $b^y = x$ pass through the x-intercept $(1, 0)$ because $b^0 = 1$ and pass through $(b, 1)$ because $b^1 = b$.

When data involve years, let the first year be zero and the other dates be in terms of the number of years after the first year.

The rule of 72 gives a way to estimate doubling time for interest compounded annually. The doubling time formula for rate r is $t = 72/(100r)$.

9.5 ■ Properties of Logarithms and the Logarithmic Scale

Properties of logarithms: If m, n, and b are positive numbers and $b \neq 1$, then

1. $\log_b (m \cdot n) = \log_b m + \log_b n$

2. $\log_b m^n = n \cdot \log_b m$

3. $\log_b \dfrac{m}{n} = \log_b m - \log_b n$

The graph of an exponential function becomes a straight line on semilog paper.

9.6 ■ The Natural Number e in Exponential and Logarithmic Functions

When the number of compoundings per year, n, is increased without bound, the compound interest formula $S = P(1 + r/n)^{nt}$ becomes $S = Pe^{rt}$.

The formula $S = Pe^{rt}$ is used whenever interest is compounded continuously.

The number e is the natural number, $e \approx 2.718\ 281\ 828\ 459\ 045$. To get e on a calculator, find e^1 with $[e^x]$ 1.

Natural logarithms with base e, $\log_e x$, are written $\ln x$. The calculator key for natural logarithms is $\boxed{\text{LN}}$.

When working with logarithmic expressions involving either base 10 or base e, rewrite the expressions to show the base.

The rule of 69 applies to continuously compounded growth. The formula $t = 69/(100r)$ permits us to estimate the time required for a quantity to double at rate r or the rate needed for a quantity to double in a specified time.

Various techniques can be used to solve exponential and logarithmic equations:

1. Guess and check (Sections 9.1 and 9.2)

2. Graphing (Section 9.2)

3. Changing to like bases and applying the like bases property (Section 9.2)

4. Changing exponential equations, $y = b^x$, to logarithmic equations (Sections 9.3, 9.4, and 9.6)

5. Dividing $y = ab^x$ on both sides by a and changing to a logarithmic equation (Sections 9.3, 9.4, and 9.6)

6. Changing logarithmic equations to exponential equations (Sections 9.3, 9.4, and 9.6)

7. Taking the logarithm of both sides (Section 9.5)

---

9   Review Exercises

In Exercises 1 to 6, find the next number in each sequence. Find whether the sequence is arithmetic, quadratic, or geometric, and then fit a linear, quadratic, or exponential equation.

1. $9, 27, 81, 243$ 729; geometric; $y = 3 \cdot 3^x$ or $y = 9 \cdot 3^{x-1}$

2. $10, 13, 16, 19, 22$ 25; arithmetic; $y = 3x + 7$

3. $1, 3, 7, 13, 21$ 31; quadratic; $y = x^2 - x + 1$

4. $-7, -4, 1, 8, 17$ 28; quadratic; $y = x^2 - 8$

5. $5, 8, 11, 14, 17$ 20; arithmetic; $y = 3x + 2$

6. $\frac{1}{8}, \frac{1}{4}, \frac{1}{2}, 1, 2$ 4; geometric; $y = \frac{1}{8} \cdot 2^{x-1}$ or $y = \frac{1}{16} \cdot 2^x$

Find the nth term a_n for the sequences in Exercises 7 to 10. Then use exponential regression to find $f(x)$. Using properties of exponents, show that your results are the same.

7. $\frac{1}{4}, \frac{1}{2}, 1, 2, 4$ $a_n = \frac{1}{4}(2)^{n-1}$; $f(x) = 0.125 \cdot 2^x$

8. $8, 16, 32, 64, 128$ $a_n = 8(2)^{n-1}$; $f(x) = 4 \cdot 2^x$

9. $\frac{1}{16}, \frac{1}{8}, \frac{1}{4}, \frac{1}{2}, 1$ $a_n = \frac{1}{16}(2)^{n-1}$; $f(x) = 0.03125 \cdot 2^x$

10. $6, 2, \frac{2}{3}, \frac{2}{9}, \frac{2}{27}$ $a_n = 6(\frac{1}{3})^{n-1}$; $f(x) = 18 \cdot (\frac{1}{3})^x$

In Exercises 11 and 12, change the equations to $y = ab^x$ form and identify the y-intercept of the graph of each equation.

11. a. $y = 2^{x-3}$ **b.** $y = 3^{x+1}$ **c.** $y = 3^{x-3}$
     $y = \frac{1}{8} \cdot 2^x; \frac{1}{8}$ $y = 3 \cdot 3^x; 3$ $y = \frac{1}{27} \cdot 3^x; \frac{1}{27}$

12. a. $y = 2^{x+4}$ **b.** $y = 2^{x-2}$ **c.** $y = 3^{x+2}$
     $y = 16 \cdot 2^x; 16$ $y = \frac{1}{4} \cdot 2^x; \frac{1}{4}$ $y = 9 \cdot 3^x; 9$

In Exercises 13 and 14, name the y-intercept of the graph of each equation. Change the equation to $y = b^{x \pm n}$ form.

13. a. $y = 2 \cdot 2^x$ 2; $y = 2^{x+1}$ **b.** $y = \frac{1}{9} \cdot 3^x$ $\frac{1}{9}; y = 3^{x-2}$

14. a. $y = 81 \cdot 3^x$ **b.** $y = \frac{1}{16} \cdot 2^x$ $\frac{1}{16}; y = 2^{x-4}$
     81; $y = 3^{x+4}$

Solve each equation in Exercises 15 to 20 for n or for x.

15. a. $4^x = 64$ $x = 3$ **b.** $2^x = 2$ $x = 1$

     c. $a^x = \dfrac{1}{a}, a \neq 0$ **d.** $b^n = 1, b \neq 0$ $n = 0$
         $x = -1$

16. a. $\left(\frac{1}{2}\right)^x = \frac{1}{8}$ $x = 3$ **b.** $16^x = 8$ $x = \frac{3}{4}$

     c. $a^x = 1, a \neq 0$ **d.** $\pi^x = \dfrac{1}{\pi^2}$ $x = -2$
       $x = 0$

17. a. $25^n = 125$ $n = \frac{3}{2}$ **b.** $27^n = \frac{1}{3}$ $n = -\frac{1}{3}$

   c. $\left(\frac{1}{25}\right)^n = 125$ $n = -\frac{3}{2}$ **d.** $\left(\frac{1}{16}\right)^n = 16$ $n = -1$

18. a. $49^n = 343$ $n = \frac{3}{2}$ **b.** $64^n = \frac{1}{4}$ $n = -\frac{1}{3}$

   c. $\left(\frac{1}{4}\right)^n = 64$ $n = -3$ **d.** $(0.1)^n = 100$ $n = -2$

19. a. $\left(\frac{1}{100}\right)^n = 10$ $n = -\frac{1}{2}$ **b.** $100^n = 10$ $n = \frac{1}{2}$

   c. $4^{x-4} = 64$ $x = 7$ **d.** $27^{x+1} = 81$ $x = \frac{1}{3}$

20. a. $\left(\frac{1}{10}\right)^n = 100$ $n = -2$ **b.** $\left(\frac{1}{100}\right)^n = 100$ $n = -1$

   c. $4^{0.5x+2} = 1$ $x = -4$ **d.** $25^{x-2} = \frac{1}{5}$ $x = \frac{3}{2}$

For Exercises 21 and 22, complete the table.

21.

Exponential Equation	Logarithmic Equation	Solve for x
$2^x = 16$	$\log_2 16 = x$	$x = 4$
$x^2 = 25$	$\log_x 25 = 2$	$x = 5$
$3^x = 81$	$\log_3 81 = x$	$x = 4$
$10^{1/2} = x$	$\log_{10} x = \frac{1}{2}$	$x = \sqrt{10} \approx 3.162$
$10^x = 19$	$\log_{10} 19 = x$	$x \approx 1.2788$
$4^0 = x$	$\log_4 x = 0$	$x = 1$

22.

Exponential Equation	Logarithmic Equation	Solve for x
$10^3 = x$	$\log x = 3$	$x = 1000$
$x^4 = 16$	$\log_x 16 = 4$	$x = 2$
$5^x = 1$	$\log_5 1 = x$	$x = 0$
$10^x = 10$	$\log_{10} 10 = x$	$x = 1$
$4^x = 9$	$\log_4 9 = x$	$x \approx 1.585$
$4^x = 4$	$\log_4 4 = x$	$x = 1$

Solve the equations in Exercises 23 to 46. Round to three decimal places.

23. $10^x = 0.1$ $x = -1$ **24.** $10^x = 1$ $x = 0$

25. $36 = 10^x$ $x = 1.556$ **26.** $15 = 10^x$ $x = 1.176$

27. $10^x = 0.75$ $x = -0.125$ **28.** $10^{1.5} = x$ $x = 31.623$

29. $4^{x+1} = 32$ $x = 1.5$ **30.** $4^{x+2} = 1$ $x = -2$

31. $3^{2x} = 6$ $x = 0.815$ **32.** $4^{2x} = 0.2$ $x = -0.580$

33. $\log_{10} x = -1$ $x = 0.1$ **34.** $\log_{10} 0.0001 = x$ $x = -4$

35. $\log_2 x = 1$ $x = 2$ **36.** $\log_3 x = 0$ $x = 1$

37. $\log_2 2 = x$ $x = 1$ **38.** $\log_3 x = -2$ $x = \frac{1}{9}$

39. $\log 10{,}000 = x$ $x = 4$ **40.** $\log 0.01 = x$ $x = -2$

41. $\log x = 2$ $x = 100$ **42.** $\log_2 x = 3$ $x = 8$

43. $\log_3 9 = x$ $x = 2$ **44.** $\log_7 7 = x$ $x = 1$

45. $\log_{27} 9 = x$ $x = \frac{2}{3}$ **46.** $\log_2 x = 0$ $x = 1$

In Exercises 47 to 52, assume x, y, z, and b are positive, $b \neq 1$. Write each as logarithms of x, y, and z.

47. $\log_b xyz$ $\log_b x + \log_b y + \log_b z$

48. $\log_b \dfrac{xy^2}{z}$ $\log_b x + 2\log_b y - \log_b z$

49. $\log_b \dfrac{x}{y^2}$ $\log_b x - 2\log_b y$

50. $\log_b \dfrac{\sqrt{x}}{z}$ $\frac{1}{2}\log_b x - \log_b z$

51. $\log_b \sqrt{\dfrac{z}{x}}$ $\frac{1}{2}\log_b z - \frac{1}{2}\log_b x$

52. $\log_b x^{1/3}y^3$ $\frac{1}{3}\log_b x + 3\log_b y$

In Exercises 53 and 54, assume x, y, z, and b are positive, $b \neq 1$. Write each expression as a single logarithm.

53. $\log_b x + \frac{1}{2}\log_b y$ $\log_b (x\sqrt{y})$

54. $\frac{1}{2}\log_b y - 2\log_b z$ $\log_b (\sqrt{y}/z^2)$

Write each expression in Exercises 55 to 58 as a single logarithm. Simplify, where possible.

55. $\log (x - 1) + \log (x - 2)$ $\log (x^2 - 3x + 2)$

56. $\log (x^2 + 4x + 4) - \log (x + 2)$ $\log (x + 2)$

57. $\log (x^2 + x) - \log x$ $\log (x + 1)$

58. $\log (x^2 + x + 2) + \log x$ $\log (x^3 + x^2 + 2x)$

In Exercises 59 and 60, simplify the left side and then solve the equation.

59. $\log_3 x^2 - \log_3 x = 2$ $x = 9$

60. $\log (x^2 - 4) - \log (x + 2) = 2$ $x = 102$

61. Evaluate, rounding to three significant digits.

   a. e^e 15.2   **b.** 3^e 19.8   **c.** e^3 20.1

62. Evaluate, rounding to three significant digits.

   a. $\ln 3$ 1.10   **b.** $\ln \pi$ 1.14   **c.** $\ln e$ 1

In Exercises 63 to 70, solve for x.

63. $e^x = 1/e^2$ $x = -2$ **64.** $e^x = 1$ $x = 0$

65. $(1/e)^x = e$ $x = -1$ **66.** $e^x = e$ $x = 1$

67. $\ln x = 3$ $x = e^3$ **68.** $\ln x = -2$ $x = e^{-2}$

69. $\ln x = 1/e$ $x = e^{1/e}$ **70.** $\ln x = e$ $x = e^e$

71. Explain why $f(x) = \left(\frac{1}{2}\right)^x$ and $f(x) = 2^{-x}$ have the same graph. $\left(\frac{1}{2}\right)^x = (2^{-1})^x = 2^{-x}$

72. Explain why $y = -3^x$ and $y = (-3)^x$ are not the same.
$-3^x = -1 \cdot 3^x; (-3)^x = (-1)^x \cdot 3^x$

73. Graph the equations, and label the graphs with their rules: See Answer Section.

$$y = 2^x, \quad y = 2^{-x}, \quad y = -2^x, \quad y = -2^{-x}$$

74. Graph the equations, and label the graphs with their rules: See Additional Answers.

$$y = 3^x, \quad y = 3^{-x}, \quad y = -3^x, \quad y = -3^{-x}$$

75. Graph $y = 2^x$, and explain how to obtain the graph of these equations. Show the graph. See Answer Section.

a. $y = 2^{x+1}$
Shift 1 unit to left.

b. $y = 2^{x-2}$
Shift 2 units to right.

76. Graph $y = 3^x$, and explain how to obtain the graph of these equations. Show the graph. See Additional Answers.

a. $y = 3^{x-1}$
Shift 1 unit to right.

b. $y = 3^{x+3}$
Shift 3 units to left.

77. Graph $y = 3^x$ and $y = \log_3 x$ on a calculator with a square window.

a. Explain what facts make the points $(0, 1)$ and $(1, 0)$ appear on the two graphs. $3^0 = 1, \log_3 1 = 0$

b. Explain what facts make the points $(1, 3)$ and $(3, 1)$ appear on the two graphs. $3^1 = 3, \log_3 3 = 1$

c. Graph $y = x$. What do you observe about the graphs now? mirror each other across $y = x$

78. Graph $y = 2^x$ and $y = \log_2 x$ on a calculator with a square window.

a. Explain what facts make the points $(0, 1)$ and $(1, 0)$ appear on the two graphs. $2^0 = 1, \log_2 1 = 0$

b. Explain what facts make the points $(1, 2)$ and $(2, 1)$ appear on the two graphs. $2^1 = 2, \log_2 2 = 1$

c. Graph $y = x$. What do you observe about the graphs now? mirror each other across $y = x$

79. What is the shape of the graph obtained when we plot exponential data on a semilog graph? straight line

80. Sketch a logarithmic scale with labels 10^0 to 10^4. Locate these points on the scale: 3, 8, 30, 300, 800, and 3000.
See Additional Answers.

81. If we know $\log 5 = 0.69897$, how do we get $\log 500$ without a calculator? $\log (5 \cdot 100) = \log 5 + \log 100 = \log 5 + 2 \approx 2.69897$

82. If we know $\log 3000 = 3.47712$, how do we get $\log 30$ without a calculator? $\log 3000 - 2 \approx 1.47712$

83. What do $\log 300$, $\log 400$, $\log 500$, and $\log 600$ have in common? $2 + $ a decimal

84. Explain how $\log x$ and $\log (x/10)$ are related?
$\log x$ is 1 greater.

85. Explain how $\log x$ and $\log (1000x)$ are related.
$\log (1000x)$ is 3 greater.

86. Show, using symbols, that $y = \log_3 x$ is the inverse function to $y = 3^x$. Inverse is $x = 3^y$ or $y = \log_3 x$.

87. Show, using symbols, that $y = \log_2 x$ is the inverse function to $y = 2^x$. Inverse is $x = 2^y$ or $y = \log_2 x$.

88. Show that $\log (x^2 + 2x + 1) = 2 \log (x + 1)$.
$\log (x^2 + 2x + 1) = \log (x + 1)^2 = 2 \log (x + 1)$

89. Show that $\log (8x^3) = 3 \log (2x)$.
$\log (8x^3) = \log (2^3x^3) = \log (2x)^3 = 3 \log (2x)$

90. Show that $\log \sqrt{x} = \frac{1}{2} \log x$. $\log \sqrt{x} = \log x^{1/2} = \frac{1}{2} \log x$

91. Show that $\frac{1}{3} \log x = \log \sqrt[3]{x}$. $\frac{1}{3} \log x = \log x^{1/3} = \log \sqrt[3]{x}$

In Exercises 92 to 94, calculate the amount in a savings account if $P = \$1000$, $r = 6\%$ interest, and $t = 2$ years. Then calculate how long it will take for the money to double at 6% interest.

92. $n = 4$ ≈$1126.49, ≈11.64 yr

93. $n = 12$ ≈$1127.15, ≈11.58 yr

94. $n = 365$ ≈$1127.48, ≈11.55 yr

95. Use the rule of 72 to estimate how long it will take money to double at 6% interest, 8% interest, and 12% interest, compounded annually. ≈12 yr, 9 yr, 6 yr

96. If you start with $1000, find the amount of money in a savings account after 5 years with continuous compounding at these interest rates.

a. 6%
≈$1,349.85

b. 8%
≈$1,491.82

c. 12%
≈$1,822.11

97. A savings account holds $1000 after 5 years. Find the starting amount of money under continuous compounding at these interest rates.

a. 6%
≈$740.82

b. 8%
≈$670.32

c. 12%
≈$548.81

98. Find the number of years required for $1000 to double at these interest rates.

a. 6%
≈11.55 yr

b. 8%
≈8.66 yr

c. 12%
≈5.78 yr

99. Using the rule of 69, estimate the interest rate required for money to double every 5 years with continuous compounding. Confirm your answer with an exponential equation. ≈13.8%; $2 = e^{5r}$, $r \approx 0.1386$

100. A 200-sheet, 5-subject notebook cost $1.99 in 1986. In 1995, the same notebook cost $2.99. Let the input be x, the number of years since 1986.

a. Fit an exponential regression equation to the data.
$y \approx 1.99(1.0463)^x$ with $x = 0$ for 1986

b. Use your equation to predict the cost of the notebook in the year 2005. ≈$4.70

101. The yearly economic impact of salmon fishing on one West Coast state has been estimated as follows: 1989, $46.5 million; 1990, $29.6 million; 1991, $18.8 million; 1992, $15.4 million; 1993, $8.0 million; 1994, $4.0 million. Fit an exponential equation to the data for the years 1989 to 1994. Predict the economic impact in the year 2005.
$y \approx 49.3(0.626)^x$, with $x = 0$ for 1989; ≈$0.027 million

102. The cost of a movie ticket in 1960 was $1.00. If the cost doubles every 12 years, estimate the cost in the year 2000. Write an exponential equation. Assume annual interest. $\approx\$10; C \approx (1.0595)^x$

In Exercises 103 and 104, find a population equation with exponential regression. Let the input be the number of years after 1980. State the rate of growth or decay. Predict the population in the year 2010 with your equation. State whether your estimate seems reasonable.

103. Euclid, Ohio: 1980, 59,999; 1990, 54,875; 2000, 52,717 $P = 59,509(0.9936)^x$; -0.64%; 49,011; reasonable

104. Pleasantville, New Jersey: 1980, 13,435; 1990, 16,027; 2000, 19,012 $P = 13,447(1.0175)^x$; $+1.75\%$; 22,638; reasonable

105. The interest on the national debt of the United States consumes a sizable portion of the annual federal budget. Historical national debt data follow: 1910, $1.1 billion; 1920, $24 billion; 1930, $16 billion; 1940, $43 billion; 1945, $258 billion; 1950, $256 billion; 1960, $284 billion; 1970, $370 billion; 1980, $908 billion; 1985, $1,823 billion; 1990, $3,233 billion; 1995, $4,974 billion; 2000, $5,674 billion.

a. Graph the data. Start with 1910 as year zero, and mark the graph in decades. Use $200 billion as the scale on the vertical axes. See Answer Section.

b. Fit an appropriate regression equation to the data. $y \approx (4.340 \times 10^9)(1.086)^x$

c. History question: What events caused the large jumps in the debt? wars, spending in 1980s

d. Use the graph and the equation to predict the current debt. Compare with the actual current debt.

e. Create a semilog-style graph by graphing the log of the debt on the vertical axis. How does this change the graph? See Answer Section; graph becomes somewhat linear.

---

9 Chapter Test

In Exercises 1 and 2, do the following:
(a) Identify the type of sequence (arithmetic, quadratic, or geometric).
(b) For geometric sequences, write the equation using a_n and calculator regression. Show that the results are the same with both methods.
(c) For quadratic sequences, use quadratic regression to find the equation.

1. $\frac{1}{8}, \frac{1}{4}, \frac{1}{2}, 1, 2$ geometric; $a_n = \frac{1}{8} \cdot 2^{(n-1)}$; $f(x) = 0.0625 \cdot 2^x$

2. 3, 12, 27, 48, 75 quadratic; $f(x) = 3x^2$

3. Solve the equations for x or for n.

a. $3^x = 27$ $x = 3$ **b.** $\left(\frac{1}{16}\right)^n = 64$ $n = -\frac{3}{2}$

c. $\frac{1}{1000} = 0.1^x$ $x = 3$ **d.** $4^{0.5x} = 16$ $x = 4$

e. $4^x = \frac{1}{64}$ $x = -3$ **f.** $100^n = 10$ $n = \frac{1}{2}$

g. $27^x = 9$ $x = \frac{2}{3}$ **h.** $4^{x+1} = 8$ $x = \frac{1}{2}$

4. Solve the equations for x.

a. $10^x = 3$ $x \approx 0.477$ **b.** $5^{2x} = 0.25$ $x \approx -0.431$

c. $10^x = -1$ $\{\}$ or $\varnothing$ **d.** $e^x = 1/e$ $x = -1$

e. $5^{x-2} = \frac{1}{125}$ $x = -1$ **f.** $\log_2 x = -2$ $x = \frac{1}{4}$

g. $\log_5 x = 3$ $x = 125$ **h.** $\log_x 32 = 5$ $x = 2$

i. $\log x = 1$ $x = 10$ **j.** $\ln x = 1.5$ $x \approx 4.482$

5. Evaluate or simplify these expressions.

a. $\ln e^2$ 2   **b.** $\log 1$ 0

c. $\log_2 x + \log_2 (x + 1)$ $\log_2 (x^2 + x)$

d. $\log_3 (x^2 - 9) - \log_3 (x + 3)$ $\log_3 (x - 3)$

6. Assuming positive x, y, z, and b, $b \neq 1$, write with logarithms of x, y, and z:

a. $\log_b \dfrac{z^2}{y}$ $2 \log_b z - \log_b y$ **b.** $\log \dfrac{\sqrt{x}}{y}$ $\frac{1}{2} \log_b x - \log_b y$

7. Suppose $y = b^{x+2}$. Change the equation to the form $y = ab^x$. Find the y-intercept. $y = b^2 b^x$; b^2

8. How long will it take $1000 to increase to $1360 at 7.5% interest compounded annually? ≈ 4.25 yr

9. How long will it take $1000 to increase to $1906 at 7.5% interest compounded continuously? ≈ 8.6 yr

10. Plot (1, 2), (5, 32), (8, 256), and (9, 512) on a semilog graph with positive integers 1 to 10 on the horizontal axis and powers of 10 (10^0 to 10^3) in a logarithmic scale on the vertical axis. See Answer Section.

11. What is the equation of the exponential graph in the figure? (*Hint:* What is y when $x = 0$? What is it when $x = 1$?) Sketch the inverse function. $f(x) = 3^x$; see Answer Section.

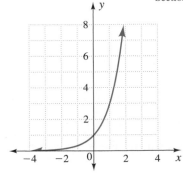

12. What is the equation of the logarithmic graph in the figure? (*Hint:* What is x when $y = 0$? What is x when $y = 1$?) Sketch the inverse function. $f(x) = \log_2 x$; see Answer Section.

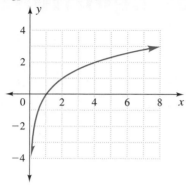

13. Sketch a graph of $y = b^{-x}$, $b > 1$.   See Answer Section.

14. Give two pairs of points such that (a, b) is on the graph of $y = 10^x$ and (b, a) is on the graph of $y = \log x$. Explain why this is true.   Possible answer: (0, 1), (1, 0); (1, 10), (10, 1); if $y = 10^x$, $x = \log y$, $y = \log x$ have reversed coordinates.

15. Explain how to find r from $S = 100(0.93)^x$.
$1 + r = 0.93$; $r = -0.07$

16. Explain how $\log x$ and $\log (100x)$ are related.
$\log(100x) = 2 + \log x$

17. The formula for pH is $pH = -\log [H^+]$.

a. Find the pH for sea water with $[H^+] = 5.01 \times 10^{-9}$ M.   8.3

b. Find $[H^+]$, given that the pH of household ammonia is 11.9. $\approx 1.26 \times 10^{-12}$ M

18. Explain the reasoning behind the proof of the logarithm of a power property:

$$\log_b x^n = n \log_b x$$

Use n factors of x: $\log_b x \cdot x \cdot x \cdot \cdots \cdot x = \log_b x + \log_b x + \log_b x + \cdots + \log_b x = n \log_b x$

19. Historical gasoline prices for a West Coast city are as follows: 1920, \$0.27; 1930, \$0.21; 1940, \$0.21; 1950, \$0.28; 1960, \$0.34; 1970, \$0.35; 1980, \$1.19; 1990, \$1.38.

a. Graph the data.   See Answer Section.

b. Would you use a linear, quadratic, or exponential equation to model these data? Explain your choice.
Possible answer: exponential; data start off slowly, increase rapidly

c. Fit the chosen type of equation to the data with calculator regression, and use the graph and the equation to predict the current price. $f(x) \approx 0.161(1.0263)^x$

20. Population data for Klamath Falls, Oregon, are as follows: 1930, 16,093; 1940, 16,497; 1950, 15,875; 1960, 16,949; 1970, 15,775; 1980, 16,661; 1990, 17,737; 2000, 19,462. This is a small college town with older timber and agriculture resources.

a. Set inputs as years since 1930 and outputs as population. Fit an exponential equation to the data. What is the annual growth rate? Project the population for 2010. $y = 15,649(1.0021)^x$; 0.2%; 18,516

b. An outside "consultant" suggests dropping data before 1970 and making projections on more recent data, with inputs set as years since 1970. Fit an exponential equation to the data, find the annual growth rate, and project the population for 2010.
$15,642(1.00695)^x$; 0.7%; 20,637

c. (Extra) Suggest which projection you might believe and why. Suggest other questions you might want to ask.
how much the college grew; how much it is expected to grow

CHAPTER 10

Systems of Nonlinear Equations and Matrices

The exploration of space has made more familiar the conic sections (that is, the circle, ellipse, hyperbola, and parabola). Figure 1 illustrates the role of conic sections in spacecraft orbit. The spaceflight technique known as *gravity assist* makes use of the gravity of a planet to change the velocity of a spacecraft, thus permitting it to move from one elliptical orbit to another. The fundamental drama of the Apollo 13 space mission was whether the support team could calculate the speed and angle of return to Earth needed to achieve orbit and not a parabolic or hyperbolic path returning to space.

We explore the conic sections in Section 10.1 and then include them in solving systems of nonlinear equations in Section 10.2 and systems of nonlinear inequalities in Section 10.3. The chapter closes with matrices (Sections 10.4 and 10.5), which provide a method of solving linear equations.

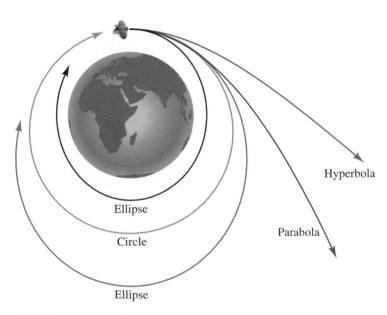

FIGURE 1

10.1 Conic Sections

Objectives

∎ Use specific parameters and formulas to find equations for conic sections.

∎ Identify the standard forms of conic sections centered at the origin.

∎ Graph conic sections on a calculator, using two functions when required.

∎ Explore, with a graphing calculator, how changing parameters changes the shape or orientation of the curve.

WARM-UP

Using the Pythagorean theorem, find which sets of numbers describe the lengths of sides of a right triangle.

1. $\{8, 15, 17\}$ right triangle; $8^2 + 15^2 = 17^2$

2. $\{1.5, 2, 2.5\}$ right triangle, $1.5^2 + 2^2 = 2.5^2$

3. $\{10, \sqrt{21}, 11\}$ right triangle; $10^2 + (\sqrt{21})^2 = 11^2$

4. $\{9, 16, 25\}$ $9^2 + 16^2 \neq 25^2$; does not even make a triangle

Find the distance between these coordinates.

5. $(3, 4), (7, 2)$ $\sqrt{20} \approx 4.472$

6. $(5, -2), (-3, 4)$ 10

7. $(x, y), (0, 0)$ $\sqrt{x^2 + y^2}$

8. $(x, y), (h, k)$ $\sqrt{(x - h)^2 + (y - k)^2}$

The conic sections in this text are centered at the origin. Although conics shifted away from the origin are not part of the instructional material, the project in Exercise 53 derives the formula for circles centered at (h, k).

IN THIS SECTION, we explore a set of curves called conic sections. The discussion starts with an overview of conic sections and then introduces the equations of circles and ellipses. Two of the conic section curves were presented earlier: parabolas in Chapters 4 to 6 and hyperbolas in Chapter 7. We revisit the parabola and hyperbola and examine standard forms of their equations.

Conical Surfaces and Conic Sections

A **conical surface** is *a double cone*. We generate a conical surface with a line. Model the formation of a conical surface with a pencil as a representation of a line. Hold the pencil gently in the middle (lengthwise). With the other hand, move the lower end of the pencil in a circular motion. The upper half of the pencil will also move in a circular motion. The surface modeled by the motion of the two halves of the pencil is the conical surface shown in each part of Figure 2.

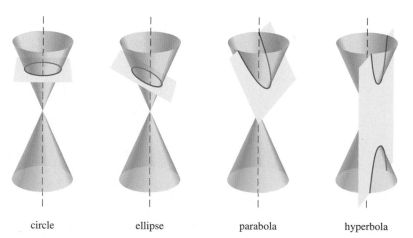

circle   ellipse   parabola   hyperbola

FIGURE 2

Conic sections are *the set of curves obtained by slicing the conical surface with a plane. The vertical line through the center of the conical surface* is the **axis**. The orientation of the slice relative to the axis or to the surface itself determines the type of curve obtained; see Figure 2.

When we slice the conical surface perpendicular to the axis, we obtain a *circle* as the intersection. A slice tilted slightly relative to the axis gives an *ellipse* as the intersection. A slice parallel to the conical surface gives a *parabola*. A slice parallel to the axis intersects the conical surface in two places and forms the two branches of a *hyperbola*. The hyperbola is the only curved conic section in two parts.

A slice along the axis itself forms two straight lines (a "degenerate" conic section); see Figure 3. A slice that just touches the conical surface forms a straight line. A slice through the center of the surface—where the points of the cones meet—forms a point (another degenerate conic section).

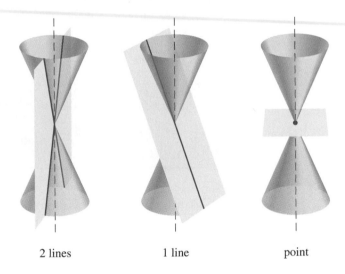

2 lines   1 line   point

FIGURE 3

The four conic sections are shown together in the sketch of paths of a spacecraft in Figure 4. If the orbital velocity is less than the escape velocity, the spacecraft will follow an elliptical or circular orbit. If the orbital velocity is greater than or equal to the escape velocity, the spacecraft will follow a parabolic or hyperbolic path and leave Earth's orbit.

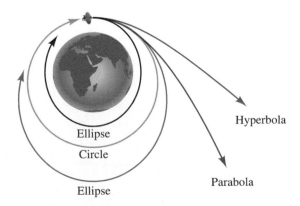

FIGURE 4

We begin a more detailed look at conic sections by examining the circle. We will then consider the ellipse, hyperbola, and parabola.

Circle

A **circle** is the set of all points $P(x, y)$ equidistant from a point called the center C. The constant distance between the point P and the center C is the **radius** r.

We obtain the equation of a circle directly from its definition by applying the distance formula to two points: P with coordinates (x, y) on the circle and C with coordinates $(0, 0)$ at the origin (Figure 5).

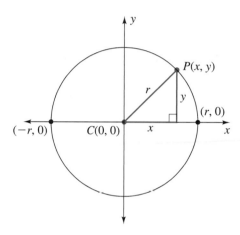

FIGURE 5

EXAMPLE 1 **Finding the equation of a circle** Suppose a circle has radius r and center at the origin; find its equation.

SOLUTION By definition, the point $P(x, y)$ located anywhere on the circle must be a constant distance r from the center $C(0, 0)$. We substitute the coordinates of P and C and the distance r into the distance formula:

$$d = \sqrt{(x_2 - x_1)^2 + (y_2 - y_1)^2}$$

$$r = \sqrt{(x - 0)^2 + (y - 0)^2} \qquad \text{Square both sides to eliminate the radical.}$$

$$r^2 = x^2 + y^2$$

The equation of a circle with center at the origin is $x^2 + y^2 = r^2$. ■

Think about it 1: If we had used the Pythagorean theorem directly to find the equation of the circle from Figure 5, what are the sides of the right triangle that we would have used?

A circle with center at the origin and radius r has the equation

$$x^2 + y^2 = r^2, \quad r > 0$$

In Example 2, we graph first with a sketch and then with equations. Sketches are important because they prevent many errors.

EXAMPLE 2   Sketching and graphing circles   Make a sketch of a circle with center at the origin and radius 4. Write the equation of the circle, and then show the graph.

SOLUTION   The circle with center at the origin and radius 4 will pass through points exactly 4 units from the origin in each horizontal and vertical direction. Four such points on the axes are $(4, 0)$, $(-4, 0)$, $(0, 4)$, and $(0, -4)$. These four points provide the basis of a good sketch of the circle, shown in Figure 6.

The equation will have $r = 4$ in $x^2 + y^2 = r^2$:

$$x^2 + y^2 = 4^2$$

$$x^2 + y^2 = 16$$

We solve for y:

$$y^2 = 16 - x^2$$

$$y = \pm\sqrt{16 - x^2}$$

It takes two calculator equations to graph the circle, as shown in Figure 7.

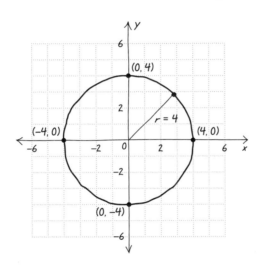

FIGURE 6   Hand-sketched graph   **FIGURE 7**

GRAPHING CALCULATOR TECHNIQUE:
GRAPHING A CIRCLE

Two problems may arise when you view the graphs of the two equations needed for a circle. First, the two pieces of the circle may not meet at or near the x-axis if the points for which the calculator evaluates the equations are not "close enough" to the x-intercepts to complete the circle. Second, the circle may appear oval on the calculator because the horizontal scale spreads the numbers over a longer distance than the vertical scale does. To correct the latter problem, choose the square window option **5 : ZSquare** under (ZOOM). Unfortunately, fixing the appearance of the circle may reopen the gap between the upper and lower halves.

Think about it 2: Why does it take two equations to graph a circle? *Hint:* Is the equation of a circle a function? (See Section 2.2.)

Ellipse

In the following example, we observe what happens when we place a numerical coefficient on one of the variable terms in the equation of a circle.

EXAMPLE 3 Comparing graphs

a. Graph and compare these equations:

$$x^2 + y^2 = 1$$

$$\frac{x^2}{4} + y^2 = 1$$

$$\frac{x^2}{8} + y^2 = 1$$

Hint: Write the equation as $x^2/L1 + y^2 = 1$, solve for y, and enter 1, 4, and 8 into calculator list L1.

b. Compare the results in part a with those for $x^2 + y^2/L1 = 1$.

Use the approach in Example 3 to explore the properties of ellipses with students. A similar exploration of hyperbolas is outlined in Exercises 45 and 46. Using L1 produces multiple graphs with one equation.

SOLUTION

a. $\dfrac{x^2}{L1} + y^2 = 1$

$$y^2 = 1 - \frac{x^2}{L1}$$ We need two equations to solve for y.

$$y_1 = \sqrt{1 - \frac{x^2}{L1}}$$

$$y_2 = -\sqrt{1 - \frac{x^2}{L1}}$$

The first equation, $x^2 + y^2 = 1$, forms a circle of radius 1. The second equation, $x^2/4 + y^2 = 1$, forms an ellipse; and the third, $x^2/8 + y^2 = 1$, forms a still wider ellipse. Placing a number under the x^2 term seems to stretch the circle horizontally (see Figure 8).

b. To graph $x^2 + y^2/L1 = 1$, we solve for y, obtaining

$$y_1 = \sqrt{L1}\sqrt{1 - x^2} \quad \text{and} \quad y_2 = -\sqrt{L1}\sqrt{1 - x^2}$$

This time, we have a vertical stretching as a result of dividing the y^2 term by 4 and 8 (see Figure 9).

Student Note: Try these on your calculator, using [-3, 3] for X and [-3, 3] for Y. View the graph and then regraph with **ZOOM** **5 : ZSquare**.

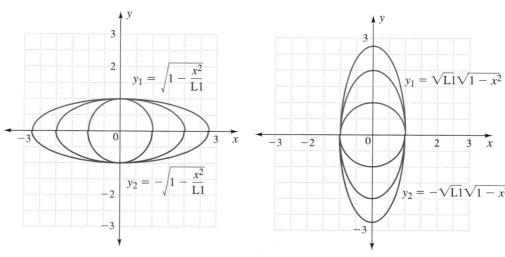

FIGURE 8   **FIGURE 9**

Example 3 suggests that slight variations in the equation of a circle will produce an oval shape called an ellipse.

■ EQUATION OF AN ELLIPSE

> An equation written in the form $\dfrac{x^2}{a^2} + \dfrac{y^2}{b^2} = 1$, where a and b are nonzero real numbers, has a graph that is an ellipse.

The similarity in structure between the equation for a circle and that for an ellipse is more apparent when we divide the circle equation by r^2 on both sides:

$$x^2 + y^2 = r^2$$

$$\frac{x^2}{r^2} + \frac{y^2}{r^2} = 1$$

If a and b are equal, then the equation of an ellipse is that of a circle with radius $a = b = r$. Thus, *the key distinction between the circle and the ellipse is that the coefficients on x^2 and y^2 are equal for a circle and different for an ellipse.*

■ Although Example 3 suggests how we might predict whether the ellipse stretches horizontally or vertically, it is usually easier to calculate the x- and y-intercepts of the equation and sketch the graph.

EXAMPLE 4   Sketching ellipses   Sketch these ellipses by calculating the x- and y-intercepts.

a. $\dfrac{x^2}{4} + \dfrac{y^2}{9} = 1$ **b.** $\dfrac{x^2}{25} + \dfrac{y^2}{1} = 1$

SOLUTION   **a.** Substituting $y = 0$, we find the x-intercepts to be $(\pm2, 0)$. Similarly, for $x = 0$, the y-intercepts are $(0, \pm3)$. The graph is in Figure 10.

b. Substituting $y = 0$, we find the x-intercepts to be $(\pm5, 0)$. Similarly, for $x = 0$, the y-intercepts are $(0, \pm1)$. The graph is in Figure 11.

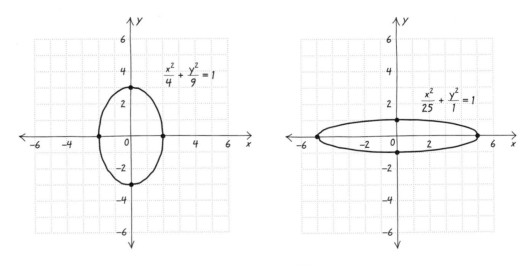

FIGURE 10   Hand-sketched graph   **FIGURE 11**   Hand-sketched graph   ■

For the circle, the x- and y-intercepts are $\pm r$. For the ellipse centered at the origin, the x-intercepts are $\pm a$ and the y-intercepts are $\pm b$. We can graph the ellipse by marking the x- and y-intercepts and drawing a smooth curve that passes through each of the intercepts.

EXAMPLE 5   Finding equations   What is the equation of an ellipse with intercepts $(-5, 0)$, $(5, 0)$, $(0, -4)$, and $(0, 4)$?

SOLUTION The x-intercepts are ± 5, so $a = \pm 5$ and $a^2 = 25$. The y-intercepts are ± 4, so $b = \pm 4$ and $b^2 = 16$. We substitute into

$$\frac{x^2}{a^2} + \frac{y^2}{b^2} = 1$$

obtaining

$$\frac{x^2}{25} + \frac{y^2}{16} = 1$$

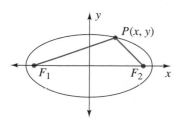

FIGURE 12

Before reading the formal definition of an ellipse, imagine splitting the center of a circle into two points and attaching a radius from a point $P(x, y)$ on the circle to each of the "centers," F_1 and F_2, in Figure 12.

DEFINITION OF AN ELLIPSE

> An **ellipse** is the set of all points $P(x, y)$ such that the sum of the distances from P to two fixed points F_1 and F_2 is constant.

The points F_1 and F_2 are foci of the ellipse (see Figure 12). In a circle, the radius is a constant distance. In an ellipse, the two distances P to F_1 and P to F_2 add to a constant.

Hyperbola

The hyperbola has an equation similar to that of an ellipse, but its graph is an entirely different shape. Hyperbolas are used in the design of cooling towers for many nuclear power plants. See Figure 13 in Example 6 for a graph of the "cooling tower" style hyperbola.

In Example 6, we explore the effect of changing the addition operation in the ellipse formula to a subtraction.

EXAMPLE 6 Graphing hyperbolas   Solve for y, and graph the resulting equations.

a. $\dfrac{x^2}{4} - \dfrac{y^2}{9} = 1$ **b.** $\dfrac{y^2}{16} - \dfrac{x^2}{9} = 1$

SOLUTION **a.**
$$\frac{x^2}{4} - \frac{y^2}{9} = 1$$

$$\frac{x^2}{4} - 1 = \frac{y^2}{9}$$

$$9\left(\frac{x^2}{4} - 1\right) = y^2$$

$$\pm 3\sqrt{\frac{x^2}{4} - 1} = y$$

The graph, in Figure 13, is a hyperbola centered at the origin.

b.
$$\frac{y^2}{16} - \frac{x^2}{9} = 1$$

$$\frac{y^2}{16} = 1 + \frac{x^2}{9}$$

$$y^2 = 16\left(1 + \frac{x^2}{9}\right)$$

$$y = \pm 4\sqrt{1 + \frac{x^2}{9}}$$

The graph, in Figure 14, is a hyperbola centered at the origin.

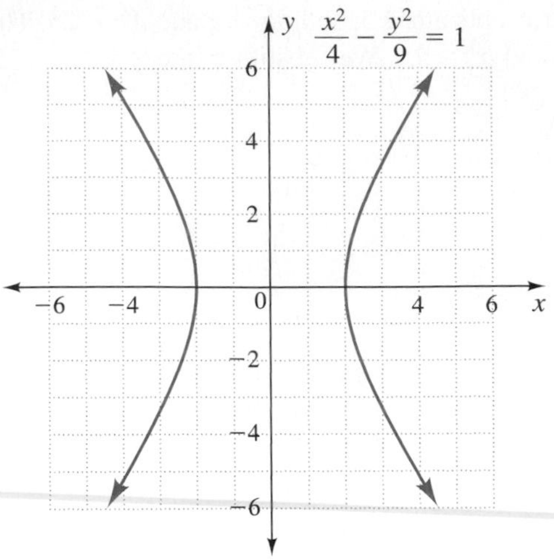

FIGURE 13

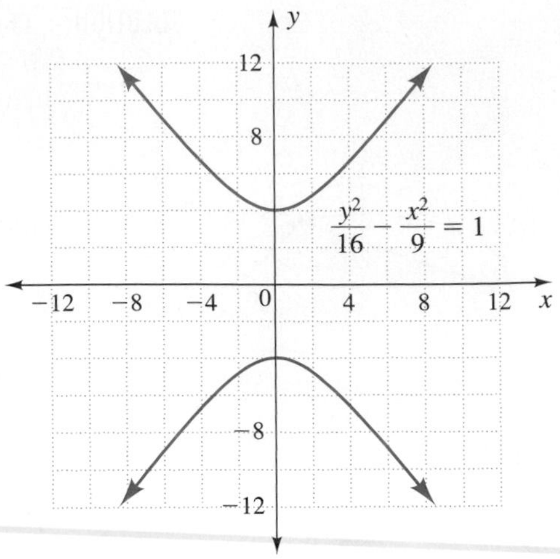

FIGURE 14

EQUATIONS OF HYPERBOLAS

An equation written in the form $\dfrac{x^2}{a^2} - \dfrac{y^2}{b^2} = 1$ or $\dfrac{y^2}{b^2} - \dfrac{x^2}{a^2} = 1$, where a and b are nonzero real numbers, has a graph that is a hyperbola.

The hyperbola may cross either the x- or the y-axis, depending on its equation.

EXAMPLE 7 **Finding intercepts** Determine the x- and y-intercepts of these hyperbolas.

$$\textbf{a.}\ \frac{x^2}{4} - \frac{y^2}{9} = 1 \qquad\qquad \textbf{b.}\ \frac{y^2}{16} - \frac{x^2}{9} = 1$$

SOLUTION **a.**

$$\frac{x^2}{4} - \frac{y^2}{9} = 1 \qquad \text{Let } y = 0 \text{ for } x\text{-intercepts.}$$

$$\frac{x^2}{4} - \frac{0}{9} = 1 \qquad \text{Solve for } x.$$

$$x = \pm 2$$

$$\frac{x^2}{4} - \frac{y^2}{9} = 1 \qquad \text{Let } x = 0 \text{ for } y\text{-intercepts.}$$

$$\frac{0}{4} - \frac{y^2}{9} = 1 \qquad \text{Solve for } y^2.$$

$$y^2 = -9$$

There is no real-number solution to $y^2 = -9$ and no y-intercept (see Figure 13).

b.

$$\frac{y^2}{16} - \frac{x^2}{9} = 1 \qquad \text{Let } y = 0.$$

$$\frac{0}{16} - \frac{x^2}{9} = 1 \qquad \text{Solve for } x^2.$$

$$x^2 = -9$$

There is no real-number solution to $x^2 = -9$ and no x-intercept (see Figure 14).

$$\frac{y^2}{16} - \frac{x^2}{9} = 1 \qquad \text{Let } x = 0.$$

$$\frac{y^2}{16} - \frac{0}{9} = 1 \qquad \text{Solve for } y.$$

$$y = \pm 4$$

For completeness, we include a formal definition of the hyperbola. Like the ellipse, it has two foci, F_1 and F_2 (see Figure 15), but its definition refers to a difference in distances rather than the sum of distances found in the definition of the ellipse.

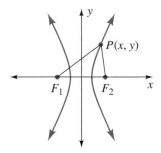

FIGURE 15

■ **DEFINITION OF A HYPERBOLA**

> A **hyperbola** is the set of all points $P(x, y)$ such that the absolute value of the difference of the distances from P to two fixed points F_1 and F_2 is a constant positive number.

■ Section 7.3 introduced a different type of hyperbola. Such a hyperbola, which *approaches but does not intersect either of the rectangular coordinate axes*, is called a **rectangular hyperbola**. The orientation of a rectangular hyperbola is turned, or rotated, from the position of a conic section hyperbola. The graph of the reciprocal function $f(x) = 1/x$, shown in Figure 16, is a rectangular hyperbola. The equation of a rectangular hyperbola, $y = k/x$, can be written in conic section form, but doing the mathematics requires trigonometry.

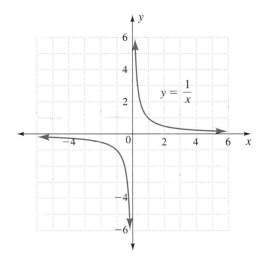

$$y = \frac{1}{x}$$

FIGURE 16

Think about it 3: Is the rectangular hyperbola a function? Is the conic section hyperbola, as shown in this section, a function?

Parabola

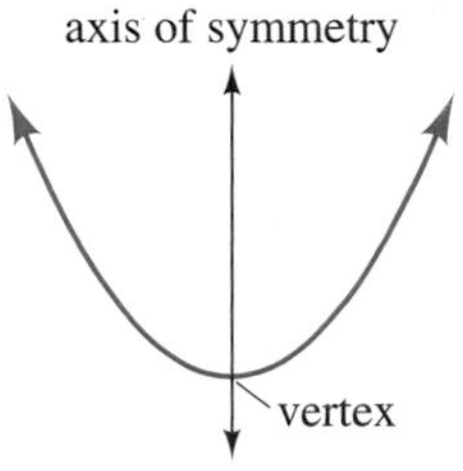

axis of symmetry

vertex

FIGURE 17

In Chapters 4 to 6, we worked extensively with quadratic functions $f(x) = ax^2 + bx + c$. The graphs of these quadratic equations were parabolas with a vertical axis of symmetry passing through the vertex (see Figure 17). We saw parabolas in the paths of objects in motion, roadbed transition curves, museum arches, and support cables for suspension bridges.

Although the graphs of quadratic functions are parabolas, not all parabolas are quadratic functions. The conic sections expand our set of parabolas to include those with a horizontal axis of symmetry. The simplest of these curves is given by $x = y^2$.

EXAMPLE 8   **Graphing parabolas**   Make a table and a graph of $y = x^2$ and $x = y^2$.

SOLUTION   The tables for the two equations appear as Table 1 and Table 2. The graphs of the parabolas $y = x^2$ and $x = y^2$ are shown in Figure 18.

TABLE 1

x	$y = x^2$
-2	4
-1	1
0	0
1	1
2	4

TABLE 2

$x = y^2$	y
4	-2
1	-1
0	0
1	1
4	2

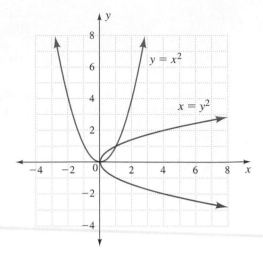

FIGURE 18 ▬

Think about it 4: If we solve $x = y^2$ for y, what two functions together describe this parabola with a horizontal axis of symmetry?

Section 6.3 introduced the vertex form of a parabola, $y = a(x - h)^2 + k$, with $a \neq 0$ and vertex (h, k).

EXAMPLE 9 Comparing conic section and vertex forms of a parabola

a. What effect does locating the vertex at the origin have on the vertex form of a parabola, $y = a(x - h)^2 + k$?

b. Describe the parabola if $a > 0$.

c. Describe the parabola if $a < 0$.

SOLUTION **a.** $y = a(x - h)^2 + k$

$= a(x - 0)^2 + 0$

$= ax^2$

b. If $a > 0$, the parabola opens up.

c. If $a < 0$, the parabola opens down. ▬

In our work with conic sections, we have focused on vertices at the origin. Examples 8 and 9 suggest the conclusions in Table 3.

TABLE 3

Way Parabola Opens	Equation with Vertex at the Origin	Function?
Upward	$y = ax^2$, a is positive	Yes
Downward	$y = ax^2$, a is negative	Yes
Right	$x = ay^2$, a is positive	No
Left	$x = ay^2$, a is negative	No

EXAMPLE 10   Finding equations   What is the equation for each of these parabolas?

a. Vertex at origin, opens left, passes through $(-4, 2)$

b. Vertex at origin, opens downward, passes through $(2, -3)$

SOLUTION   **a.** Because the parabola has a vertex at the origin and opens to the left, the equation is $x = ay^2$. To find a, we substitute $(-4, 2)$ and solve for a.

$$x = ay^2 \qquad \text{Substitute } (-4, 2).$$

$$-4 = a(2)^2 \qquad \text{Simplify.}$$

$$-4 = a \cdot 4 \qquad \text{Divide by 4.}$$

$$a = -1$$

The equation is $x = -1y^2$.

b. Because the parabola has a vertex at the origin and opens downward, the equation is $y = ax^2$. To find a, we substitute $(2, -3)$ and solve for a.

$$y = ax^2 \qquad \text{Substitute } (2, -3).$$

$$-3 = a(2)^2 \qquad \text{Simplify.}$$

$$-3 = a \cdot 4 \qquad \text{Divide by 4.}$$

$$a = -\tfrac{3}{4}$$

The equation is $y = -\tfrac{3}{4}x^2$.   ■

For completeness, we include the formal definition of a parabola.

DEFINITION OF A PARABOLA

> A **parabola** is the set of all points $P(x, y)$ such that the distance from P to a fixed point F is equal to the distance from P to a fixed line l.

The point F is the focus, and the line l is the directrix (see Figure 19).

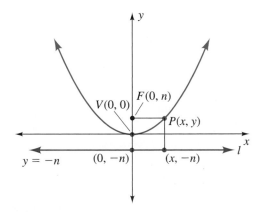

FIGURE 19

The distance from a point to a line is always the perpendicular distance. Consequently, in Figure 19, the coordinates of the point directly below $P(x, y)$ on the directrix are $(x, -n)$. We return to this definition and illustration in Exercise 48.

■ IDENTIFYING A CONIC SECTION
FROM ITS EQUATION

First, arrange the equation in standard form, and then look carefully at the squared terms.

The equation of a *circle* centered at the origin with radius r is $x^2 + y^2 = r^2$, or

$$\frac{x^2}{r^2} + \frac{y^2}{r^2} = 1$$

The x^2 and y^2 terms have the same coefficients.

The equation of an *ellipse* centered at the origin is

$$\frac{x^2}{a^2} + \frac{y^2}{b^2} = 1$$

The coefficients on the x^2 and y^2 terms are different but have the same sign.

The equation of a *hyperbola* centered at the origin is

$$\frac{x^2}{a^2} - \frac{y^2}{b^2} = 1 \qquad \text{or} \qquad \frac{y^2}{b^2} - \frac{x^2}{a^2} = 1$$

The x^2 and y^2 terms have opposite signs.

The equation of a *parabola* with vertex at the origin has one of the following forms:

$$y = ax^2 \qquad \text{or} \qquad x = ay^2$$

The parabola has either x or y squared but not both.

Work done in earlier chapters allows us to recognize other parabolic and hyperbolic forms:

• The vertex form of a *parabola* is $y = a(x - h)^2 + k$, and the quadratic form is $y = ax^2 + bx + c$. In either case, the parabola still has a square on only one of the two variables, x or y.

• A *rectangular hyperbola* has equation $y = k/x$ or $xy = k$. The x and y are multiplied, and no squared term is present.

ANSWER BOX

Warm-up: **1.** right triangle; $8^2 + 15^2 = 17^2$ **2.** right triangle; $1.5^2 + 2^2 = 2.5^2$ **3.** right triangle; $10^2 + (\sqrt{21})^2 = 11^2$ **4.** $9^2 + 16^2 \neq 25^2$; does not even make a triangle. **5.** $\sqrt{20} \approx 4.472$ **6.** 10 **7.** $\sqrt{x^2 + y^2}$
8. $\sqrt{(x - h)^2 + (y - k)^2}$ **Think about it 1:** x, y, and r **Think about it 2:** When we solve for y, we get two equations. The graphs of the equations create the upper and lower halves of the circle. The equation of a circle, $x^2 + y^2 = r^2$, is not a function because the circle fails the vertical-line test. However, each of the two equations whose graphs form the circle is a function.
Think about it 3: The rectangular hyperbola is a function; the conic section hyperbolas shown here are not functions. **Think about it 4:** $y = \sqrt{x}$ and $y = -\sqrt{x}$.

10.1 Exercises

In Exercises 1 and 2, making a sketch on graph paper may be helpful.

1. List four points at a distance of 3 units from $(0, 0)$.
$(\pm 3, 0), (0, \pm 3)$

2. List four points at a distance of 4 units from $(0, 0)$.
$(\pm 4, 0), (0, \pm 4)$

Write an equation fitting each of the descriptions in Exercises 3 to 20.

3. Circle, center at origin, radius $= 5$ $x^2 + y^2 = 25$

4. Circle, center at origin, radius $= 1.5$ $x^2 + y^2 = 2.25$

5. Ellipse, center at origin, x-intercepts $(\pm 5, 0)$, y-intercepts $(0, \pm 2)$ $\dfrac{x^2}{25} + \dfrac{y^2}{4} = 1$

6. Ellipse, center at origin, x-intercepts $(\pm 4, 0)$, y-intercepts $(0, \pm 1)$ $\dfrac{x^2}{16} + y^2 = 1$

7. Hyperbola, center at origin, passes through $(0, \pm 4)$, no x-intercept, $a = 3$ $\dfrac{y^2}{16} - \dfrac{x^2}{9} = 1$

8. Hyperbola, center at origin, passes through $(\pm 5, 0)$, no y-intercept, $b = 4$ $\dfrac{x^2}{25} - \dfrac{y^2}{16} = 1$

9. Parabola, vertex at origin, opens to the right, passes through $(3, -1)$ $x = 3y^2$

10. Parabola, vertex at origin, opens to the left, passes through $(-2, -2)$ $x = -\frac{1}{2}y^2$

11. Parabola, vertex at origin, opens downward, passes through $(3, -1)$ $y = -\frac{1}{9}x^2$

12. Parabola, vertex at origin, opens downward, passes through $(-2, -2)$ $y = -\frac{1}{2}x^2$

13. Circle, center at origin, radius $= 4$ $x^2 + y^2 = 16$

14. Ellipse, center at origin, x-intercepts $(\pm 3, 0)$, y-intercepts $(0, \pm 4)$ $\dfrac{x^2}{9} + \dfrac{y^2}{16} = 1$

15. Ellipse, center at origin, x-intercepts $(\pm 1, 0)$, y-intercepts $(0, \pm 5)$ $x^2 + \dfrac{y^2}{25} = 1$

16. Parabola, vertex at origin, opens to the left, passes through $(-2, 1)$ $x = -2y^2$

17. Hyperbola, center at origin, passes through $(0, \pm 3)$, no x-intercept, $a = 5$ $\dfrac{y^2}{9} - \dfrac{x^2}{25} = 1$

18. Hyperbola, center at origin, passes through $(\pm 4, 0)$, no y-intercepts, $b = 2$ $\dfrac{x^2}{16} - \dfrac{y^2}{4} = 1$

19. Parabola, vertex at origin, opens upward, passes through $(-2, 1)$ $y = \frac{1}{4}x^2$

20. Circle, center at origin, radius $= 25$ $x^2 + y^2 = 625$

21. What happens to the equation of a circle if $r < 0$? if $r = 0$? $r < 0$ has no meaning; circle becomes point.

22. What happens to the equation of an ellipse if a and b are equal? It describes a circle.

In Exercises 23 to 44, identify the conic section (circle, ellipse, hyperbola, parabola, or straight line). Find any x- or y-intercepts. Write the equations needed to graph the conic section on a calculator. Sketch the graph.

23. $y = x^2 + 2$ parabola; $(0, 2)$; see Answer Section.

24. $y = -\dfrac{1}{x}$ rectangular hyperbola; see Additional Answers.

25. $x^2 + y^2 = 9$ circle; $(\pm 3, 0), (0, \pm 3)$; $y = \pm\sqrt{9 - x^2}$; see Answer Section.

26. $x = y^2 + 2$ parabola; $(2, 0)$; $y = \pm\sqrt{x - 2}$; see Additional Answers.

27. $x^2 - y^2 = 4$ hyperbola; $(\pm 2, 0)$; $y = \pm\sqrt{x^2 - 4}$; see Answer Section.

28. $y = x^2 - 4x + 4$ parabola; $(2, 0), (0, 4)$; see Additional Answers.

29. $\dfrac{x^2}{4} + y^2 = 1$ ellipse; $(\pm 2, 0), (0, \pm 1)$; $y = \pm\sqrt{1 - \dfrac{x^2}{4}}$ or $y = \pm\frac{1}{2}\sqrt{4 - x^2}$; see Answer Section.

30. $x^2 + \dfrac{y^2}{4} = 1$ ellipse; $(\pm 1, 0), (0, \pm 2)$; $y = \pm 2\sqrt{1 - x^2}$; see Additional Answers.

31. $4x^2 + y^2 = 100$ ellipse; $(\pm 5, 0), (0, \pm 10)$; $y = \pm\sqrt{100 - 4x^2}$ or $y = \pm 2\sqrt{25 - x^2}$; see Answer Section.

32. $-2y - 10 + 5x = 0$ straight line; $(2, 0), (0, -5)$; $y = 2.5x - 5$; see Additional Answers.

33. $2x^2 + 2y^2 - 8 = 0$ circle, $(\pm 2, 0), (0, \pm 2)$; $y = \pm\sqrt{4 - x^2}$; see Answer Section.

34. $xy = 5$ rectangular hyperbola; $y = \dfrac{5}{x}$; see Additional Answers.

35. $y = x$ straight line; $(0, 0)$; see Answer Section.

36. $y = 2x^2$ parabola; $(0, 0)$; see Additional Answers.

37. $x^2 - 4y^2 = 1$ hyperbola; $(\pm 1, 0)$; $y = \pm\frac{1}{2}\sqrt{x^2 - 1}$; see Answer Section.

38. $y = 2x - 1$ straight line; $(\frac{1}{2}, 0), (0, -1)$; see Additional Answers.

39. $x = 4y^2$ parabola; $(0, 0)$; $y = \pm\frac{1}{2}\sqrt{x}$; see Answer Section.

40. $25y^2 - 4x^2 = 100$ hyperbola; $(0, \pm 2)$; $y = \pm 2\sqrt{1 + \dfrac{x^2}{25}}$ or $y = \pm\frac{2}{5}\sqrt{x^2 + 25}$; see Additional Answers.

41. $\dfrac{y^2}{1} - \dfrac{x^2}{4} = 1$ hyperbola; $(0, \pm 1)$; $y = \pm\sqrt{1 + \dfrac{x^2}{4}}$ or $y = \pm\frac{1}{2}\sqrt{x^2 + 4}$; see Answer Section.

42. $x^2 + y^2 = 2$ circle; $(\pm\sqrt{2}, 0), (0, \pm\sqrt{2})$; $y = \pm\sqrt{2 - x^2}$; see Additional Answers.

43. $x = -y^2$ parabola; $(0, 0)$; $y = \pm\sqrt{-x}$; see Answer Section.

44. $\dfrac{x^2}{4} + \dfrac{y^2}{25} = 1$ ellipse; $(\pm 2, 0), (0, \pm 5)$; $y = \pm 5\sqrt{1 - \dfrac{x^2}{4}}$ or $y = \pm\frac{5}{2}\sqrt{4 - x^2}$; see Additional Answers.

45. How does the shape of the hyperbola $\dfrac{x^2}{1} - \dfrac{y^2}{b^2} = 1$ change as b^2 gets large? (*Hint:* Use $b^2 = 1$, $b^2 = 5$, and $b^2 = 10$, and graph.) Branches become steeper; see Answer Section.

46. How does the shape of the hyperbola $\dfrac{x^2}{a^2} - \dfrac{y^2}{1} = 1$ change as a^2 gets large? (*Hint:* Use $a^2 = 1$, $a^2 = 5$, and $a^2 = 10$, and graph.) Branches become flatter, x-intercepts move to $(\pm a, 0)$; see Additional Answers.

47. **a.** Make a table and graph for $x = -y^2$. Let x be $\{-9, -4, -1, 0, 1, 4, 9\}$. See Answer Section.

b. Why might a student think that the equation $y = \sqrt{-x}$ has imaginary outputs and yet produces a graph? Opposite of x makes radicand appear negative.

c. Under what circumstances is $y = -\sqrt{-x}$ meaningful? when x is a negative number

48. **a.** Referring to Figure 19, write an expression for the distance between the focus (point F) and the arbitrary point P on the parabola. $\sqrt{x^2 + (y-n)^2}$

b. Referring to Figure 19, write an expression for the distance between the arbitrary point P on the parabola and the point $(x, -n)$ on the directrix. $\sqrt{(x-x)^2 + [y-(-n)]^2}$ or $|y + n|$

c. Explain how the origin, lying between points $F(0, n)$ and $(0, -n)$, satisfies the definition of a parabola. Why is the origin labeled V? same distance from $F(0, n)$ as from directrix; vertex

d. Simplify the expressions in the equation

$$\sqrt{(x-0)^2 + (y-n)^2} = \sqrt{(x-x)^2 + (y-(-n))^2}$$

and derive the formula

$$y = \frac{1}{4n} x^2$$

for a parabola with vertex at the origin. The parameter n is the distance between the vertex and the focus of a parabola.

What are the coordinates of the focus for each of the parabolas in parts e to h?

e. $y = x^2$ $\left(0, \frac{1}{4}\right)$

f. $y = -2x^2$ $\left(0, -\frac{1}{8}\right)$

g. $y = 0.5x^2$ $\left(0, \frac{1}{2}\right)$

h. $y = -0.25x^2$ $(0, -1)$

▇ Projects

49. Happy Face Calculating

a. Given the shifted equation of a circle $(x - h)^2 + (y - k)^2 = r^2$, with center at (h, k), create a set of equations that can be used to produce the happy face shown in the figure on a calculator.

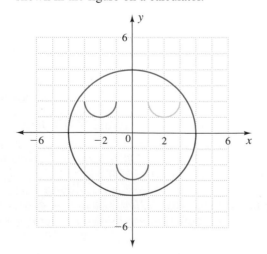

b. By applying the distance formula to the coordinates for C and P on the circle in the figure, derive the equation of a circle given in part a.

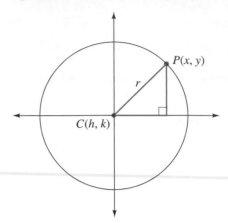

50. Waxed Paper Parabola Construction

a. On a sheet of waxed paper about the size of a notebook page, draw a horizontal line segment 20 centimeters long, positioned about 5 centimeters from the bottom edge of the paper.

b. Mark the line segment into centimeters.

c. Place a point A about 15 centimeters above the midpoint of the segment.

d. Fold the paper so that the first centimeter mark on the line segment lies on point A. Crease the paper along the fold. Reopen the paper.

e. Fold the paper so that the second centimeter mark lies on point A. Crease the paper again. Reopen the paper.

f. Repeat the folding and creasing for each of the centimeter marks on the line segment.

The resulting shape is called an *envelope curve*. Explain how the envelope curve you have created satisfies the definition of a conic section. What is the conic section? Fold represents distance halfway between directrix line and focal point A; parabola

51. More Waxed Paper Constructions

a. Start by drawing a large circle (about 6 inches in diameter), with center C, on a piece of waxed paper. Mark points evenly around the circle (a protractor can be used to get 36 equally spaced points).

b. Place a point P somewhere outside the circle. Fold the paper so that P touches one point on the circumference of the circle. Crease the paper and unfold. Fold again so that P touches the next point on the circumference of the circle. Crease and unfold. Repeat for each point on the circumference. What is the conic section? hyperbola

c. Repeat part a for a point P somewhere inside the circle. What is the conic section now? ellipse

52. String Conic Section Tie a piece of string in a loop so that the circumference of the loop is about 10 inches. Place a piece of notebook paper on a sheet of cardboard, and insert two thumbtacks about 2 inches apart near the center of the paper. Lay the string so that it loops around the tacks. Place a pencil point inside the loop, and pull the string tight. Keeping the string tight, move the pencil along the paper, tracing out a curve. What conic section is formed? Explain how the string satisfies the definition of that conic section. *ellipse; string keeps pencil point at a constant sum of distances from the tacks.*

10.2 Solving Systems of Nonlinear Equations

Objectives

- Predict the number of solutions to a system including nonlinear equations by identifying the conic sections within the system.
- Solve systems including nonlinear equations with a graph.
- Solve systems algebraically by substitution.

WARM-UP

Solve the equations in Exercises 1 to 3 with the quadratic formula or factoring.

1. $2x^2 - 5x - 3 = 0$ $x = 3, -0.5$ **2.** $x^2 + 3x - 10 = 0$ $x = 2, -5$

3. $n^2 - 4n + 1 = 0$ $n \approx 3.732, 0.268$

Solve the exponential and logarithmic equations in Exercises 4 and 5.

4. $2^x = 2.5$ $x \approx 1.322$ **5.** $\log_2 x = \frac{1}{2}$ $x \approx 1.414$

6. Divide $x^3 - 2x - 1$ by $(x + 1)$. $x^2 - x - 1$

The examples and exercises are arranged so that you have the option of covering exponential and logarithmic equations without covering the conic sections or vice versa. Rational equations such as $y = 1/x$ are included with the conic sections.

IN THIS SECTION, we identify conic sections within systems of equations, predict the number of solutions to systems, and solve systems graphically and algebraically. The systems of equations include exponential and logarithmic equations as well as conic sections.

Nonlinear Systems Including Conic Sections

This section extends the solution of systems of equations to several combinations of conic sections, called **nonlinear systems** because *at least one equation is not linear*.

Applications of systems of conic sections are common. In trapshooting, for a shot to be a "hit," the parabolic path of the shot must intersect the parabolic path of the clay target. In tennis, the circular motion of the tennis racket must intersect the parabolic path of the tennis ball. When a dog catches a Frisbee or a baseball player makes a running catch, the intersection is accomplished intuitively. In contrast, ensuring that the flight of a space vehicle on a path around the sun intersects the elliptical orbit of a planet requires calculation by computer.

In the examples in this section, we will solve systems by starting with substitution and then employing a variety of algebraic techniques to complete the solution. Suggestions for graphical solutions to the examples are included in the exercises.

EXAMPLE 1 **Solving a nonlinear system** Identify the conic equations, predict the number of solutions, and solve the system by substitution:

$$y = 2x^2 - 3x - 5$$
$$2x - y = 2$$

SOLUTION The system contains a parabola and a line. There may be up to two points of intersection (see Figure 20).

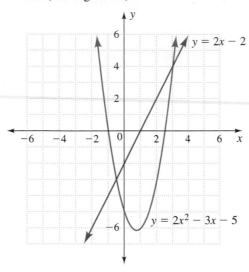

FIGURE 20

For the substitution, we solve the linear equation for y:

$$2x - y = 2 \qquad \text{Add } y \text{ to both sides.}$$
$$2x = 2 + y \qquad \text{Subtract 2.}$$
$$2x - 2 = y$$

We then substitute $y = 2x - 2$ into the quadratic equation:

$$y = 2x^2 - 3x - 5$$
$$2x - 2 = 2x^2 - 3x - 5 \qquad \text{Add 2 and subtract } 2x \text{ on both sides.}$$
$$0 = 2x^2 - 5x - 3 \qquad \text{Solve with the quadratic formula.}$$

With $a = 2$, $b = -5$, and $c = -3$, we have

$$x = 3 \qquad \text{and} \qquad x = -0.5$$

Now we substitute each value of x into the linear equation:

$$y = 2(3) - 2$$
$$= 4 \qquad \text{At } x = 3, y = 4.$$
$$y = 2(-0.5) - 2$$
$$= -3 \qquad \text{At } x = -0.5, y = -3.$$

Check: We must check both coordinates in both equations:

$$4 \overset{?}{=} 2(3)^2 - 3(3) - 5 \checkmark \qquad\qquad 2(3) - 4 \overset{?}{=} 2 \checkmark$$
$$-3 \overset{?}{=} 2(-0.5)^2 - 3(-0.5) - 5 \checkmark \qquad 2(-0.5) - (-3) \overset{?}{=} 2 \checkmark$$

The ordered pairs $(3, 4)$ and $(-0.5, -3)$ are the points of intersection. ■

In Example 1, we used the quadratic formula to solve for x. Example 2 will also require use of the quadratic formula within its solution.

EXAMPLE 2 **Solving a nonlinear system**   Identify the conic sections, predict the number of solutions, and solve the system:

$$y = 2x^2 + 3x - 6$$

$$x^2 - y + 4 = 0$$

SOLUTION   Because only one variable is squared in each equation, the system contains two parabolas. The x^2 indicates vertical axes of symmetry. There are two intersections (see Figure 21).

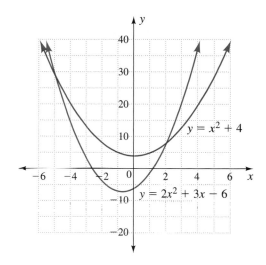

FIGURE 21

We solve $x^2 - y + 4 = 0$ for y:

$$x^2 + 4 = y$$

Then we substitute $y = x^2 + 4$ into the other quadratic equation for y:

$$y = 2x^2 + 3x - 6$$

$$x^2 + 4 = 2x^2 + 3x - 6 \qquad \text{Subtract } x^2 + 4 \text{ from both sides.}$$

$$0 = x^2 + 3x - 10 \qquad \text{Solve with the quadratic formula.}$$

With $a = 1$, $b = 3$, and $c = -10$, we have

$$x = 2 \qquad \text{and} \qquad x = -5$$

Now we find the corresponding y for each x. At $x = 2$,

$$y = (2)^2 + 4$$

$$y = 8$$

At $x = -5$,

$$y = (-5)^2 + 4$$

$$y = 29$$

Check: We must check both coordinates in both equations:

$$8 \stackrel{?}{=} 2(2)^2 + 3(2) - 6 \checkmark \qquad (2)^2 - 8 + 4 \stackrel{?}{=} 0 \checkmark$$

$$29 \stackrel{?}{=} 2(-5)^2 + 3(-5) - 6 \checkmark \quad (-5)^2 - 29 + 4 \stackrel{?}{=} 0 \checkmark$$

The coordinates of the points of intersection are $(2, 8)$ and $(-5, 29)$. ▬

Think about it 1: What is the greatest possible number of intersections for a system of two quadratic equations?

In Example 3, we apply reasoning and long division to find the solutions to a fourth-degree equation. Repeat the problem with a graphing calculator.

EXAMPLE 3 Solving a nonlinear system   Identify the conic sections, predict the number of solutions, and solve the system:

$$y = x^2 - 1$$
$$x = y^2 - 1$$

SOLUTION These are both parabolas, but one has a vertical axis and the other has a horizontal axis. There may be up to four points of intersection (see Figure 22).

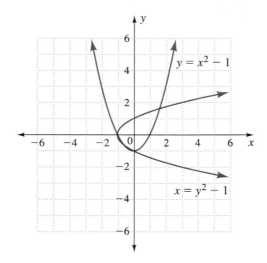

FIGURE 22

If we substitute $y = x^2 - 1$ into $x = y^2 - 1$, we obtain

$$x = (x^2 - 1)^2 - 1 \qquad \text{Square the expression } x^2 - 1.$$

$$x = x^4 - 2x^2 + 1 - 1 \qquad \text{Change to "= 0" form.}$$

$$x^4 - 2x^2 - x = 0 \qquad x \text{ is a common factor.}$$

$$x(x^3 - 2x - 1) = 0$$

We must guess and check to factor the expression on the left. Because the last term in $x^3 - 2x - 1$ is -1, we try $(x - 1)$ and $(x + 1)$. By polynomial long division (Section 7.5), we find that $(x + 1)$ is a factor, and

$$\frac{x^3 - 2x - 1}{x + 1} = x^2 - x - 1$$

Thus, our fourth-degree equation becomes

$$x(x + 1)(x^2 - x - 1) = 0$$

Two solutions are $x = 0$ and $x = -1$. The quadratic formula applied to $x^2 - x - 1 = 0$ gives $x \approx 1.618$ and $x \approx -0.618$ as the other two solutions. We substitute each value of x into $y = x^2 - 1$ to find the corresponding y.

$$y = (-1)^2 - 1 \qquad \text{For } x = -1$$
$$= 0$$
$$y = (-0.618)^2 - 1 \qquad \text{For } x \approx -0.618$$
$$\approx -0.618$$
$$y = 0^2 - 1 \qquad \text{For } x = 0$$
$$= -1$$
$$y = (1.618)^2 - 1 \qquad \text{For } x \approx 1.618$$
$$\approx 1.618$$

There are four points of intersection: $(-1, 0)$, $(-0.618, -0.618)$, $(0, -1)$, and $(1.618, 1.618)$.

Check: The coordinates agree with the graph in Figure 22, so we will not check them in the equations. ✓

▬ In solving systems where one expression must be squared, we substitute so as to square the simplest possible expression.

EXAMPLE 4 Solving a nonlinear system   Identify the conic sections, estimate the number of solutions, and solve the system:

$$y = 1/x$$
$$x^2 + y^2 = 4$$

SOLUTION   The equations describe a rectangular hyperbola and a circle. There could be up to four points of intersection and, hence, four solutions (see Figure 23).

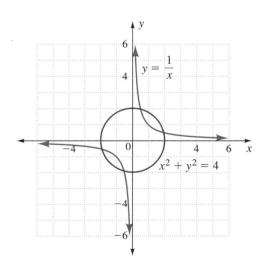

FIGURE 23

We substitute $y = 1/x$ into the equation of a circle:

$$x^2 + y^2 = 4 \qquad \text{Substitute } y = 1/x.$$

$$x^2 + \left(\frac{1}{x}\right)^2 = 4 \qquad \text{Simplify.}$$

$$x^2 + \frac{1}{x^2} = 4 \qquad \text{Multiply both sides by } x^2.$$

$$x^4 + 1 = 4x^2 \qquad \text{Subtract } 4x^2 \text{ from both sides.}$$

$$x^4 - 4x^2 + 1 = 0$$

We can solve $x^4 - 4x^2 + 1 = 0$ with the quadratic formula. We let $n = x^2$ and $n^2 = x^4$:

$$n^2 - 4n + 1 = 0$$

Evaluating the quadratic equation for $a = 1$, $b = -4$, and $c = 1$ gives

$$n \approx 3.73205 \qquad \text{and} \qquad n \approx 0.26795$$

We replace n with x^2:

$$x^2 \approx 3.73205 \qquad \text{and} \qquad x^2 \approx 0.26795$$
$$x \approx \pm 1.932 \qquad \text{and} \qquad x \approx \pm 0.518$$

We substitute each x into $y = 1/x$ and obtain four ordered pairs representing the four points of intersection of the graphs:

$$(1.932, 0.518), (-1.932, -0.518), (0.518, 1.932), \text{ and } (-0.518, -1.932)$$

Check: The coordinates agree with the graph in Figure 23, so we will not check them in the equations. ✓

Think about it 2: What is the reciprocal of 1.932? Why are the x and y in Example 4 thus related?

EXAMPLE 5   **Solving a nonlinear system**   Identify the conic sections, predict the number of solutions, and solve the system:

$$\frac{x^2}{9} + \frac{y^2}{4} = 1$$

$$\frac{x^2}{1} - \frac{y^2}{4} = 1$$

SOLUTION   By comparing the x^2 and y^2 terms, we observe that the first equation is an ellipse and the second is a hyperbola. They could intersect in at most four places (see Figure 24).

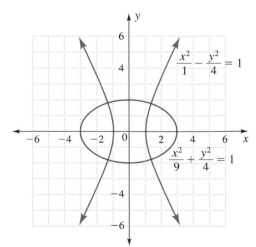

FIGURE 24

Because $y^2/4$ is in both equations, we solve the hyperbola equation for $y^2/4$:

$$\frac{x^2}{1} - \frac{y^2}{4} = 1$$

$$x^2 - 1 = \frac{y^2}{4}$$

We next substitute $x^2 - 1$ for $y^2/4$ in the ellipse equation:

$$\frac{x^2}{9} + \frac{y^2}{4} = 1$$

$$\frac{x^2}{9} + x^2 - 1 = 1 \qquad \text{Using } x^2 = \frac{9x^2}{9}, \text{ add the } x^2 \text{ terms.}$$

$$\frac{10x^2}{9} = 2$$

$$x^2 = 1.8$$

$$x = \pm\sqrt{1.8}$$

$$x \approx \pm 1.342$$

Then we substitute $x^2 = 1.8$ in the hyperbola equation:

$$\frac{y^2}{4} = 1.8 - 1$$

$$y^2 = 3.2$$

$$y = \pm\sqrt{3.2}$$

$$y \approx \pm 1.789$$

The solutions, described as ordered pairs, are $(-1.342, -1.789)$, $(-1.342, 1.789)$, $(1.342, -1.789)$, and $(1.342, 1.789)$.

Check: The coordinates agree with the intersections in Figure 24. ✓   ▬

Nonlinear Systems Including Exponential and Logarithmic Equations

In Examples 6 and 7, we return to exponential and logarithmic equations.

EXAMPLE 6   **Solving a nonlinear system**   Solve this system of equations by substitution.

$$y = 2^x$$

$$y = 5 - 2^x$$

SOLUTION   The equations are graphed in Figure 25. There is one point of intersection.

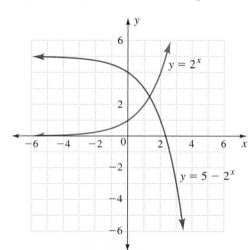

FIGURE 25

The equations are both solved for y. Substitution for y gives

$$2^x = 5 - 2^x \qquad \text{Add } 2^x \text{ to both sides.}$$

$$2^x + 2^x = 5 \qquad \text{Add like terms.}$$

$$2 \cdot 2^x = 5 \qquad \text{Divide by 2.}$$

$$2^x = 2.5 \qquad \text{Change to a logarithmic equation.}$$

$$\log_2 2.5 = x \qquad \text{Apply the change of base formula.}$$

$$x = \frac{\log 2.5}{\log 2}$$

$$\approx 1.322$$

Then we solve for y:

$$y \approx 2^{1.322}$$

$$\approx 2.5$$

The solution to the system of equations is $x \approx 1.322$, $y \approx 2.5$.

Check: $2.5 \overset{?}{=} 2^{1.322}$ ✓

$2.5 \overset{?}{=} 5 - 2^{1.322}$ ✓

Think about it 3: Which step in the solution to Example 6 indicates that y is exactly 2.5 and not approximately 2.5?

The system in Example 7 contains logarithms. Look for similarities between the solution of the exponential system in Example 6 and the solution of the logarithmic system in Example 7.

EXAMPLE 7   **Solving a nonlinear system**   Graph this system of equations and then solve by substitution.

$$y = \log_2 x$$

$$y = 1 - \log_2 x$$

SOLUTION   The system is graphed in Figure 26. There is one point of intersection.

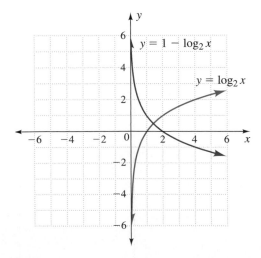

FIGURE 26

We use substitution to replace y in one of the equations:

$$\log_2 x = 1 - \log_2 x \qquad \text{Add } \log_2 x \text{ to both sides.}$$

$$\log_2 x + \log_2 x = 1 \qquad \text{Add like terms.}$$

$$2 \log_2 x = 1 \qquad \text{Divide both sides by 2.}$$

$$\log_2 x = \tfrac{1}{2} \qquad \text{Change to an exponential equation.}$$

$$2^{1/2} = x$$

$$x \approx 1.414$$

Because $y = \log_2 x$ and the solution above indicates that $\log_2 x = \frac{1}{2}$, we have $y = \frac{1}{2}$. Thus, the solution to the system is $x \approx 1.414$ and $y = \frac{1}{2}$.

Check: $\frac{1}{2} \stackrel{?}{=} \log_2 1.414$ ✓

$\qquad\quad \frac{1}{2} \stackrel{?}{=} 1 - \log_2 1.414$ ✓

Think about it 4: What is another way of writing $2^{1/2}$?

The closing example is based on the observation that the values of y in Examples 6 and 7 were rational. Sometimes we can evaluate expressions without knowing x.

EXAMPLE 8   Evaluating expressions without knowing x

a. If $2^x = \frac{4}{3}$, what is 2^{x+2}?

b. If $2^x = \frac{1}{3}$, what is 2^{x+1}?

c. If $3^x = \frac{1}{10}$, what is 3^{x+2}?

SOLUTION **a.** $2^{x+2} = 2^x \cdot 2^2 = \frac{4}{3} \cdot 4 = \frac{16}{3}$

b. $2^{x+1} = 2^x \cdot 2^1 = \frac{1}{3} \cdot 2 = \frac{2}{3}$

c. $3^{x+2} = 3^x \cdot 3^2 = \frac{1}{10} \cdot 9 = \frac{9}{10}$

ANSWER BOX

Warm-up: **1.** $x = 3, -0.5$ **2.** $x = 2, -5$ **3.** $n \approx 3.732, 0.268$
4. $x \approx 1.322$ **5.** $x \approx 1.414$ **6.** $x^2 - x - 1$ **Think about it 1:** Two parabolas (one with a horizontal axis and one with a vertical axis) could have up to four points of intersection, as shown in Example 3. **Think about it 2:** The reciprocal is 0.518. The hyperbola $y = 1/x$ is the reciprocal function, so each point (x, y) must satisfy that condition. **Think about it 3:** When we divided by 2 in $2 \cdot 2^x = 5$, we obtained $2^x = 2.5$. Because one of the original equations was $y = 2^x$, we know at this point that $y = 2.5$.
Think about it 4: $2^{1/2} = \sqrt{2}$

10.2 Exercises

In Exercises 1 to 20, identify the conic sections (straight line, parabola, circle, ellipse, or hyperbola), and solve the system. Write the solution as ordered pairs. Leave answers in radical form or round to three decimal places. Sketches are helpful!

1. $y = x^2 - 3x + 2$ parabola, line; $(-1, 6), (3, 2)$

$\quad y = -x + 5$

2. $y = 2x^2 - 3x + 2$ parabola, line; $(-1, 7), (3, 11)$

$\quad y = x + 8$

3. $\dfrac{x^2}{4} - \dfrac{y^2}{9} = 1$ hyperbola, parabola; $(\pm 2.5, 2.25), (\pm 2, 0)$

$\quad y = x^2 - 4$

4. $y = x^2 - 2$

$x^2 + y^2 = 4$ parabola, circle; $(\pm\sqrt{3}, 1)$, $(0, -2)$

5. $y = 5x^2 + 2x$

$y = -x^2 + x + 2$ parabolas; $\left(-\frac{2}{3}, \frac{8}{9}\right)$ $\left(\frac{1}{2}, \frac{9}{4}\right)$

6. $y = 2x^2 + 4x$

$y = x^2 + 2x + 3$ parabolas; $(-3, 6)$, $(1, 6)$

7. $x^2 + \dfrac{y^2}{4} = 1$

$y = x^2 - 1$ ellipse, parabola; $(\pm 1, 0)$

8. $y = x^2 - 2$

$y = \dfrac{1}{x}$ parabola, hyperbola; $(-1, -1)$, $\approx(-0.618, -1.618)$, $\approx(1.618, 0.618)$

9. $x^2 + y^2 = 16$

$y = x^2 - 4$ circle, parabola; $(\pm\sqrt{7}, 3)$, $(0, -4)$

10. $y = 15x^2 + 5x - 1$

$y = -5x^2 + 4x$ parabolas; $(-0.25, -1.3125)$, $(0.2, 0.6)$

11. $y = -x^2 + 2$

$y = x^2 - 6$ parabolas; $(\pm 2, -2)$

12. $x^2 + y^2 = 16$

$y = x^2 - 2$ circle, parabola; $\approx(\pm 2.297, 3.275)$

13. $\dfrac{x^2}{4} + y^2 = 1$

$y = x^2 - 1$ ellipse, parabola; $(0, -1)$, $\approx(\pm 1.323, 0.75)$

14. $y = \dfrac{2}{x}$

$x^2 + y^2 = 4$ hyperbola, circle; $(\sqrt{2}, \sqrt{2})$, $(-\sqrt{2}, -\sqrt{2})$

15. $x^2 + y^2 = 4$

$x^2 - y^2 = 4$ circle, hyperbola; $(\pm 2, 0)$

16. $x^2 + y^2 = 4$

$x^2 - y^2 = 2$ circle, hyperbola; $(\pm\sqrt{3}, 1)$, $(\pm\sqrt{3}, -1)$

17. $y = \dfrac{1}{x}$

$x + y = 4$ hyperbola, line; $\approx(3.732, 0.268)$, $\approx(0.268, 3.732)$

18. $x^2 + y^2 = 4$

$y = 2x + 1$ circle, line; $\approx(0.472, 1.943)$, $\approx(-1.272, -1.543)$

19. $x^2 + y^2 = 9$

$\dfrac{x^2}{16} + \dfrac{y^2}{4} = 1$ circle, ellipse; $\left(\pm\dfrac{2\sqrt{15}}{3}, \dfrac{\sqrt{21}}{3}\right)$, $\left(\pm\dfrac{2\sqrt{15}}{3}, -\dfrac{\sqrt{21}}{3}\right)$

20. $\dfrac{x^2}{4} - \dfrac{y^2}{1} = 1$

$\dfrac{x^2}{9} + \dfrac{y^2}{4} = 1$ hyperbola, ellipse; $\left(\pm\dfrac{6\sqrt{5}}{5}, \dfrac{2\sqrt{5}}{5}\right)$, $\left(\pm\dfrac{6\sqrt{5}}{5}, -\dfrac{2\sqrt{5}}{5}\right)$

21. A graph of the system in Example 3 requires three equations:

$$y = x^2 - 1$$
$$y = \sqrt{x + 1}$$
$$y = -\sqrt{x + 1}$$

Find the dimensions of a viewing window that clearly shows these three (of the four) points of intersection: $(-1, 0)$, $(-0.618, -0.618)$, and $(0, -1)$.
Possible answer: **Xmin** -1.1, **Xmax** 0.1, **Ymin** -1.1, **Ymax** 0.1

22. Solve $x^4 - 4x^2 + 1 = 0$ by graphing, and show on a sketch that the solutions also are the x-coordinates of the points of intersection for the system in Example 4:

$$y = \dfrac{1}{x}$$
$$x^2 + y^2 = 4$$ See Additional Answers.

23. Are $(-1, 0)$ and $(1, 0)$ solutions to the following system? Why or why not?

$$y = -2x^2 + 2$$
$$x^2 + (y - 2)^2 = 4$$ No; ordered pairs do not satisfy second equation.

24. Are $(\pm 2, 2)$ and $(\pm\sqrt{3}, 1)$ solutions to the following system? Why or why not?

$$y = x^2 - 2$$
$$x^2 + (y - 2)^2 = 4$$ Yes; all make both equations true.

In Exercises 25 to 32, solve the system of equations, and write the solution as an ordered pair.

25. $y = 3^x$

$y = 2 - 3^x$ $(0, 1)$

26. $y = 3^{x+2}$

$y = 1 - 3^x$ $(-2.096, 0.9)$

27. $y = 2^{x+2}$

$y = 4 + 2^x$ $(0.415, 5.333)$

28. $y = 2^{x+1}$

$y = 1 - 2^x$ $(-1.585, 0.667)$

29. $y = \log x$

$y = 3 - \log x$ $(31.6, 1.5)$

30. $y = \log_3 x$

$y = 1 - 2\log_3 x$ $(1.442, 0.333)$

31. $y = \log_4 x$

$y = 2 - \log_4 x$ $(4, 1)$

32. $y = \frac{1}{2}\log_2 x$

$y = 6 - \log_2 x$ $(16, 2)$

In Exercises 33 to 40, answer the question without solving for *x*.

33. If $2^x = \frac{1}{3}$, what is 2^{x+2}? $\frac{4}{3}$

34. If $3^x = 4$, what is 3^{x+2}? 36

35. If $3^x = 2$, what is 3^{x+1}? 6

36. If $2^x = \frac{3}{4}$, what is 2^{x+1}? $\frac{3}{2}$

37. If $2^x = 5$, what is 2^{x-1}? $\frac{5}{2}$

38. If $3^x = 6$, what is 3^{x-2}? $\frac{2}{3}$

39. If $4^x = 3$, what is 4^{x-2}? $\frac{3}{16}$

40. If $4^x = 2$, what is 4^{x-1}? $\frac{1}{2}$

41. In the solution to Example 7, we divided both sides of the equation $2 \log_2 x = 1$ by 2 and obtained $\log_2 x = \frac{1}{2}$, with the single solution $x = 2^{1/2}$. If we had applied properties of logarithms, we would have obtained

$\log_2 x^2 = 1$. Change this new equation to exponential notation, and solve for *x*. How is the new solution the same? How is it different? What must we conclude?
$x^2 = 2, x = \pm\sqrt{2}$; root is same, $-\sqrt{2}$ must be discarded

42. Think back to the meaning of the degree of an equation. Explain why the conic sections (including the rectangular hyperbola, $xy = k$) are second-degree equations.
Highest power is 2 or the product, x^1y^1, has a total power of 2.

▌ **Projects**

43. Intersections of Circle and Parabola What are the possible numbers of intersections of a parabola and the circle $x^2 + y^2 = 4$? Make sketches and give equations of the parabolas. 0 with $y = x^2 + 3$; 1 with $y = x^2 + 2$; 2 with $y = x^2 + 1$; 3 with $y = x^2 - 2$; 4 with $y = x^2 - 3$

44. Intersections of Hyperbola and Straight Line What are the possible numbers of intersections of the hyperbola $x^2/4 - y^2/4 = 1$ with a straight line? Make sketches and give equations of the lines.
0 with $y = x$; 1 with $y = x - 2$; 2 with $y = 0$

▌10.3 Solving Systems of Inequalities

Objectives

▌ Identify quadrants described by inequalities.

▌ Describe quadrants with inequalities.

▌ Solve linear inequalities in two variables.

▌ Solve systems of linear inequalities.

▌ Find inequalities for conic sections.

▌ Solve systems of nonlinear inequalities.

WARM-UP

Sketch the graphs of these equations.

1. $y = x + 4$
$y = 4 - x$ See Figure 34.

2. $x^2 + y^2 = 4$ See Figure 35.

3. $\frac{x^2}{4} - \frac{y^2}{9} = 1$ See Figure 36.

4. $x = 2y^2$
$x + y = 3$ See Figure 40.

5. $y = \frac{1}{3}x^2$
$x^2 + y^2 = 9$ See Figure 41.

IN THIS SECTION, we extend our work with linear and nonlinear equations to inequalities.

Inequalities and Quadrants

In many applications, we make assumptions about the inputs or outputs that limit the graphs to the first quadrant or to the first and second quadrants. We describe these limitations with inequalities.

In Examples 1 and 2, we examine the inequalities needed to describe quadrants, as we review the logical statements "*a* or *b*" and "*a* and *b*." Recall that "*a* or *b*" means either *a is true or b is true or both are true.*

EXAMPLE 1 Identifying quadrants   Which quadrants are described by the statement $x > 0$ or $y > 0$?

SOLUTION   The first and fourth quadrants contain $x > 0$; sec Figure 27a. The first and second quadrants contain $y > 0$; see Figure 27b. Because either statement may be true, the quadrants described by $x > 0$ or $y > 0$ are the first, second, and fourth quadrants; see Figure 27c.

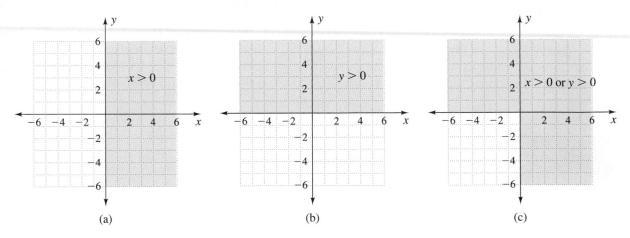

(a)   (b)   (c)

FIGURE 27

Recall that "*a* and *b*" means *both a and b must be true.* Example 2 gives two applications of the logical *and.*

EXAMPLE 2 Describing quadrants   Describe these quadrants with inequalities and a graph.
a. First quadrant
b. Third quadrant

SOLUTION   **a.** Both *x* and *y* are positive in the first quadrant, so we write $x > 0$ and $y > 0$. The first quadrant is shaded in Figure 28 to indicate all points where $x > 0$ and $y > 0$.

b. Both *x* and *y* are negative in the third quadrant, so we write $x < 0$ and $y < 0$. The third quadrant is shaded in Figure 29.

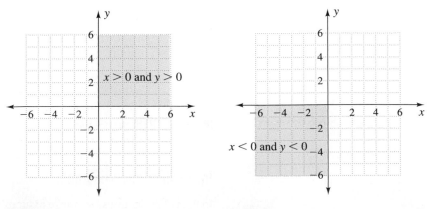

FIGURE 28   **FIGURE 29**

Solving Linear Inequalities in Two Variables

■ LINEAR INEQUALITY IN TWO VARIABLES

> A linear inequality in two variables can be written $ax + by < c$, where a, b, and c are real numbers and a and b are not both zero.

The definition holds for all other inequalities: $\leq$, $>$, and $\geq$.

The solution set to a linear inequality in two variables is a half-plane. A **half-plane** is *the region on one side of a line*. The graph of every straight line creates two half-planes. *The line between the half-planes* is called the **boundary line**. The name *linear inequality* comes from the fact that the boundary is a line.

In Figure 30, the region to the right of the boundary line $3x + 2y = 6$ is the half-plane showing the solution to $3x + 2y > 6$. Together the boundary line and the shaded region are the solution set to $3x + 2y \geq 6$.

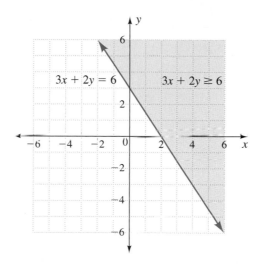

FIGURE 30

■ SOLVING A LINEAR INEQUALITY IN TWO VARIABLES

> 1. Graph the boundary line formed by replacing the inequality sign with an equal sign. Use a dashed line if the inequality is $<$ or $>$. Use a solid line if the inequality is $\leq$ or $\geq$.
>
> 2. Select a test point, not on the boundary line. Substitute the ordered pair for the test point into the inequality.
>
> 3. **a.** If the test point makes a true statement, shade the half-plane that contains the test point.
>
>   **b.** If the test point makes a false statement, shade the half-plane that does not contain the test point.

The inequality $ax + by < c$ is generally for noncalculator use. When we use a graphing calculator, we must solve the inequality for y.

▬ **GRAPHING CALCULATOR TECHNIQUE:**
GRAPHING AN INEQUALITY IN TWO VARIABLES

Graphing calculators have a graphing option for inequalities, which starts with the equation of the boundary line and adds shading. Solve the boundary equation for y. Enter the equation in $\boxed{Y=}$. Move the cursor to the far left of Y_1 and press $\boxed{ENTER}$ until you see the correct shading option. Set an appropriate viewing window. Graph. Figure 31 shows the $\boxed{Y=}$ screen for parts a and b of Example 3. The equation and inequality in part c of Example 3 cannot be solved for y, and so you cannot use the above calculator options. For more practice, obtain Figure 30 on your calculator.

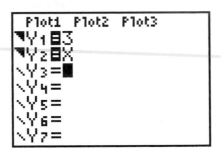

FIGURE 31

EXAMPLE 3   **Shading the half-plane**   Draw the indicated boundary line and shade the region described.

a. $y = 3, y \geq 3$ **b.** $y = x, y > x$ **c.** $x = -6, x \leq -6$

SOLUTION

a.
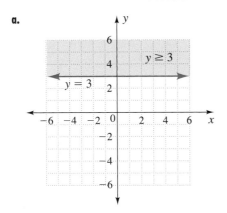

b.
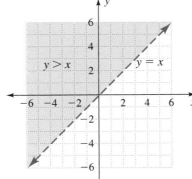

c.
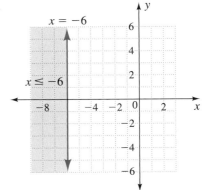

The line $y = x$ must be dashed to fit the inequality $y > x$.

▬

APPLICATION: MEETING A GOAL   When we have two different ways to meet a goal (such as a goal for dollars, calories, or enrollment), we can describe the possible outcomes with an inequality.

EXAMPLE 4   **Writing and solving a two-variable inequality**   The university has more than 2400 first-year students who need to enroll in history courses. The school can offer large lectures of 300 students each or small research-based study courses of 40 students each.

a. Write an inequality that shows the numbers of each course the university can offer.

b. Solve the inequality with a graph.

c. Explain the meaning of the boundary line and half-plane in the problem setting.

SOLUTION **a.** Let x = number of large lectures. Let y = number of study courses. The total students served needs to be larger than 2400. The inequality is

$$300x + 40y > 2400$$

b. The solution is shown in Figure 32.

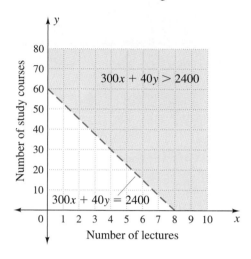

FIGURE 32

c. The boundary line shows that the university can serve 2400 students by offering 8 lectures or 60 study courses. Any point to the right shows a combination of lectures and study courses serving more than 2400 students. Only whole numbers of lectures and study courses would be meaningful. ▬

Systems of Linear Inequalities

For a system of inequalities, we graph equations and shade the appropriate regions. The region where the half-planes (shadings) overlap is the solution to the system.

The shaded region in Figure 33 highlights the upper and lower limits on the antibiotic level required in the bloodstream to effectively combat a bacterial infection. The vertical axis is labeled with names for the limits rather than with numbers. The horizontal axis is time in hours. In this model of the antibiotic level, the patient should have taken a pill every 8 hours after an initial injection but did not. The antibiotic level is dropping close to the point where the antibiotic becomes ineffective.

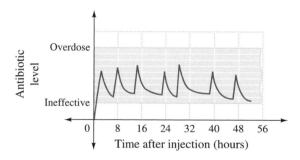

FIGURE 33

EXAMPLE 5 **Solving a system of linear inequalities**   Graph each inequality and find the region that satisfies both inequalities.

$$y \geq 4 + x$$

$$y \leq 4 - x$$

SOLUTION   The graphs are shown in Figure 34. The area where the two half-planes overlap is the solution. The test point $(-4, 4)$ makes each inequality true and confirms the solution.

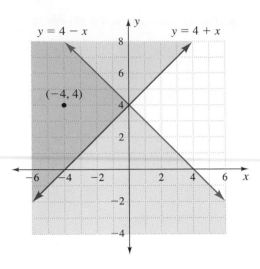

FIGURE 34

Second-Degree Inequalities

Linear inequalities are characterized by a boundary line and a half-plane. When inequalities have boundary lines that are curves (such as parabolas or circles), the shape of the solution set varies with the shape of the curve.

GRAPHING SECOND-DEGREE INEQUALITIES

> To graph a second-degree inequality, graph the equation as a boundary line. Then use a test point to find which side of the boundary to shade.

As before, a dashed line excludes the boundary and a solid line includes the boundary in the solution set.

EXAMPLE 6   Graphing second-degree inequalities   Graph each inequality.

a. $x^2 + y^2 > 4$

b. $\dfrac{x^2}{4} - \dfrac{y^2}{9} \leq 1$

SOLUTION   **a.** The equation $x^2 + y^2 = 4$ is the equation of a circle with radius 2. The inequality symbol $>$ indicates that the graph of the boundary should be dashed. The origin makes the inequality false. The graph is shown in Figure 35.

b. The equation $\dfrac{x^2}{4} - \dfrac{y^2}{9} = 1$ is the equation of a hyperbola with x-intercepts $(\pm 2, 0)$. The inequality symbol $\leq$ indicates that the graph of the boundary should be solid. The origin makes the inequality true. The graph is shown in Figure 36.

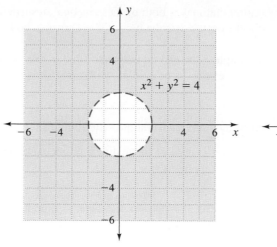

FIGURE 35   **FIGURE 36**

At times, we want to describe regions in the coordinate plane with quadratic inequalities. After graphing the equation corresponding to the inequality, we can use test points to locate regions to be shaded.

EXAMPLE 7   Finding regions described by inequalities   Match the description with the shaded region in Figure 37, 38, or 39. Use test points as needed.

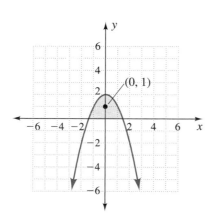

FIGURE 37

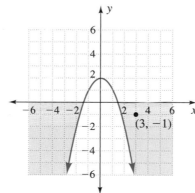

FIGURE 38

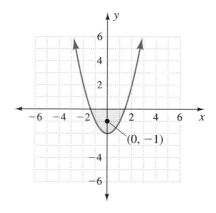

FIGURE 39

a. $y \leq 0$ and $y \geq x^2 - 2$ **b.** $y \geq 0$ and $y \leq -x^2 + 2$

c. $y \leq 0$ and $y \geq -x^2 + 2$

SOLUTION   **a.** The inequality $y \leq 0$ represents the region on or below the x-axis. The inequality $y \geq x^2 - 2$ represents the region above the curve $y = x^2 - 2$. The region that makes both of these inequalities true is shaded in Figure 39. As a check, we see that the point $(0, -1)$ makes both inequalities true.

b. The inequality $y \geq 0$ represents the region on or above the x-axis. The inequality $y \leq -x^2 + 2$ represents the region below the curve $y = -x^2 + 2$. The region that makes both of these inequalities true is shaded in Figure 37. As a check, we see that the point $(0, 1)$ makes both inequalities true.

c. The inequality $y \leq 0$ represents the region on or below the x-axis. The inequality $y \geq -x^2 + 2$ represents the region above the curve $y = -x^2 + 2$. The region that makes both of these inequalities true is shaded in Figure 38. As a check, we see that the point $(3, -1)$ makes both inequalities true. ■

Example 7 showed on informal solution to a system of inequalities. We now consider two examples of systems with more shading.

Solving Systems of Nonlinear Inequalities

In solving a system of inequalities, graph each inequality separately and then look for the overlapping region.

EXAMPLE 8 **Solving a system of nonlinear inequalities** Find the solution set for each inequality, and then identify the solution to the system.

$$x \geq 2y^2$$

$$x + y \leq 3$$

SOLUTION This system has a parabola (vertex at the origin) opening to the right and a line with intercepts at $(3, 0)$ and $(0, 3)$. Using $(1, 0)$ as a test point, we shade to the left of the line and inside the parabola. The overlap of regions, shown in Figure 40, is the solution to the system.

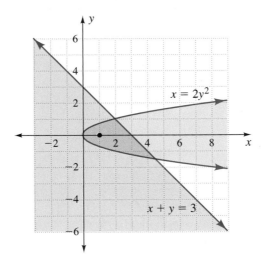

FIGURE 40 ■

EXAMPLE 9 **Solving a system of nonlinear inequalities** Find the solution set for each inequality, and then identify the solution to the system.

$$x^2 + y^2 \leq 9$$

$$y \geq \tfrac{1}{3}x^2$$

SOLUTION This system has a circle of radius 3 and a parabola (vertex at the origin) opening upward. Using $(0, 1)$ as a test point, we shade inside the circle and inside the parabola. The overlap of regions, shown in Figure 41, is the solution to the system.

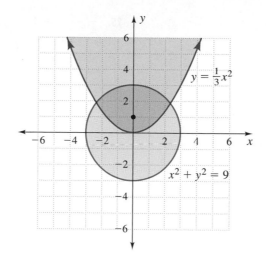

FIGURE 41

ANSWER BOX

Warm-up: **1.** See the lines graphed in Figure 34. **2.** See the circle graphed in Figure 35. **3.** See the hyperbola graphed in Figure 36.
4. See the parabola and line graphed in Figure 40. **5.** See the parabola and circle graphed in Figure 41.

10.3 Exercises

Name the quadrants that satisfy the inequalities in Exercises 1 to 8.

1. $x < 0$ and $y > 0$ 2

2. $x > 0$ and $y < 0$ 4

3. $x < 0$ and $y < 0$ 3

4. $y > 0$ 1 and 2

5. $y < 0$ 3 and 4

6. $x > 0$ 1 and 4

7. $x < 0$ 2 and 3

8. $x < 0$ and $y > 0$ or $x > 0$ and $y < 0$ 2 or 4

Write inequalities that describe the quadrants listed in Exercises 9 to 12.

9. Quadrant 2 $x < 0$ and $y > 0$

10. Quadrant 4 $x > 0$ and $y < 0$

11. Quadrants 2 and 4 $(x < 0$ and $y > 0)$ or $(x > 0$ and $y < 0)$

12. Quadrants 1 and 4 $x > 0$

In Exercises 13 to 16, name the inequalities shown in the graph.

13.

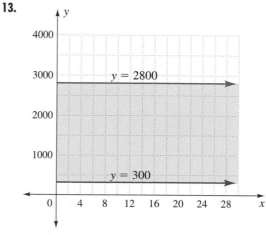

$y \le 2800$, $y \ge 300$, and $x \ge 0$

14.

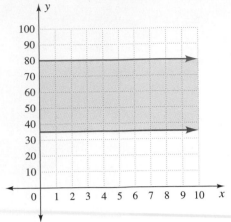

$y \leq 80$, $y \geq 35$, and $x \geq 0$

15.

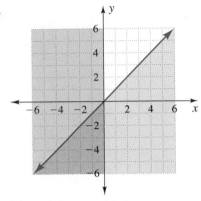

$y \leq x$ and $x \leq 0$

16.

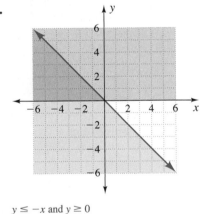

$y \leq -x$ and $y \geq 0$

In Exercises 17 to 20, graph each inequality, and find the region that satisfies both inequalities.

17. $y \geq x - 3$

$2x - y \leq 5$
See Answer Section.

19. $2x + 2y < 5$

$3x - 2y > 6$
See Answer Section.

18. $x + y \leq 5$

$\dfrac{x}{2} + 2 \geq y$
See Additional Answers.

20. $y - 6 < 3x$

$y > -3x$
See Additional Answers.

In Exercises 21 to 28, graph each inequality.

21. $y \geq \dfrac{1}{x}$
See Answer Section.

22. $y \leq x^2$
See Additional Answers.

23. $x^2 + \dfrac{y^2}{4} \leq 1$
See Answer Section.

24. $x^2 + y^2 \leq 4$
See Additional Answers.

25. $\dfrac{x^2}{4} + y^2 \geq 1$
See Answer Section.

26. $\dfrac{x^2}{9} - \dfrac{y^2}{4} \geq 1$
See Additional Answers.

27. $x > y^2$
See Answer Section.

28. $x < -y^2$
See Additional Answers.

Match each of the descriptions in Exercises 29 to 34 with one of the graphs a–f.

(a)

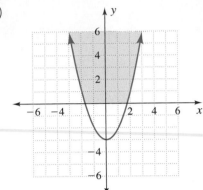

(b)

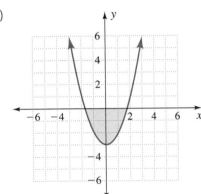

(c)

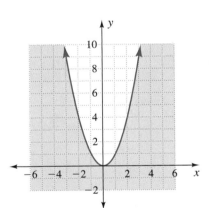

(d)

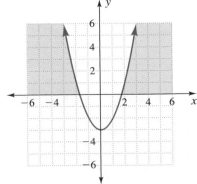

(e)

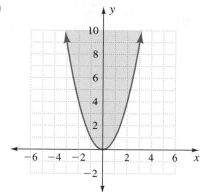

(f)

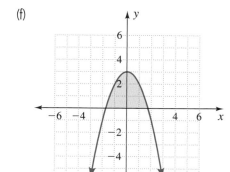

29. $y \leq x^2$ c

30. $y \geq x^2$ e

31. $y \geq 0$ and $y \leq -x^2 + 3$
  f

32. $y \leq 0$ and $y \geq x^2 - 3$
  b

33. $y \geq 0$ and $y \geq x^2 - 3$
  a

34. $y \geq 0$ and $y \leq x^2 - 3$
  d

35. Why do no points make $y < 0$ and $y \geq x^2$ true?
  x^2 is always ≥ 0, so if $y \geq x^2$, $y \geq 0$

36. Why do no points make $y \leq -x^2$ and $y > 0$ true?
  $-x^2$ is always ≤ 0, so if $y \leq -x^2$, $y \leq 0$

In Exercises 37 to 42, graph each inequality and indicate the solution to the system.

37. $y \leq \dfrac{x^2}{2}$

  $x^2 + 4y^2 \geq 4$
  See Answer Section.

38. $x^2 + y^2 \geq 9$

  $4x^2 + y^2 \leq 4$
  See Additional Answers.

39. $y \geq \dfrac{1}{x}$

  $y \leq x$
  See Answer Section.

40. $y \geq -\dfrac{1}{x}$

  $x^2 + y^2 \leq 4$
  See Additional Answers.

41. $y \leq -\dfrac{1}{x}$

  $x^2 + 4y^2 \leq 4$
  See Answer Section.

42. $y \geq -\dfrac{x^2}{2}$

  $x^2 + y^2 \leq 9$
  See Additional Answers.

43. To meet its bid on an after-season bowl game, a school must earn $500,000 from ticket sales. Student tickets are $15, and regular tickets are $50. Write and graph an inequality showing the set of possible sales that would meet the goal. Explain the solution in terms of the problem setting. x = no. student tickets, y = no. regular tickets; $15x + 50y \geq 500,000$; see Answer Section.

44. A daily diet calls for 3500 mg of potassium. A serving of dried apricots contains 480 mg of potassium. A serving of dates contains 240 mg of potassium. Suppose these are the only sources of potassium in the diet. Write and graph an inequality to show the set of possible servings. Explain the solution in terms of the problem setting. x = no. servings apricots, y = no. servings dates; $480x + 240y \geq 3500$; see Additional Answers.

45. A maximum 2500-calorie diet in a hospital is from two sources: intravenous (I.V.) and a "total nutrition" liquid by mouth. If each I.V. solution contains 500 calories and the nutrition liquid contains 250 calories, write and graph an inequality that shows the possible sources of the calories. x = no. units I.V., y = no. units liquid; $500x + 250y \leq 2500$; see Answer Section.

46. A monthly budget permits spending a maximum of $75 on movies and books. Books cost an average of $15. Movies cost $5. Write and graph an inequality that shows the possible ways to spend up to $75. Explain the solution in terms of the problem setting. x = no. books, y = no. movies; $15x + 5y \leq 75$; see Additional Answers.

47. Explain why some boundary lines on graphs are dashed and others are solid. dashed lines for $>$, $<$; solid lines for $\geq$, $\leq$

48. Explain how to find which side of a boundary line to shade when graphing the solution set to an inequality. Select a test point. If test point makes inequality true, shade half-plane that contains point. Otherwise shade other half-plane.

---

10   **Mid-Chapter Test**

Write an equation fitting each of the descriptions in Exercises 1 to 3.

1. Ellipse, center at origin, x-intercepts (± 2, 0), y-intercepts (0, ± 4) $\dfrac{x^2}{4} + \dfrac{y^2}{16} = 1$

2. Hyperbola, center at origin, passes through (± 3, 0), $b = 5$ $\dfrac{x^2}{9} - \dfrac{y^2}{25} = 1$

3. Parabola, vertex at origin, opens to the left, passes through (-4, -2) $x = -y^2$

In Exercises 4 and 5, name the conic section. Find any x- or y-intercepts. Write the equations needed to graph the conic section on a calculator.

4. $x = -2y^2$ parabola; (0, 0), (0, 0); $Y = \pm\sqrt{(-X/2)}$

5. $x^2 + y^2 = 625$ circle; (± 25, 0), (0, ± 25); $Y = \pm\sqrt{(625 - X^2)}$

In Exercises 6 and 7, solve the system of equations. A sketch may be helpful.

6. $x^2 + \dfrac{y^2}{4} = 1$

  $y = 4 - 2^x$
  (-1, 0), (0, 2), (1, 0)

7. $y = 2^x$

  $y = 4 - 2^x$ (1, 2)

8. Which quadrant is described by $x < 0$ and $y > 0$?
  second

9. Explain how to solve $y > x - 2$ and $y < -x - 3$ by graphing. See below.

10. Explain how to solve $\dfrac{x^2}{4} + y^2 \leq 1$ and $y \geq x$ by graphing. Graph ellipse and shade inside, graph line and shade above; solution is overlap.

9. Graph each equation with dashed lines, using test point to find half-plane to shade; solution is overlap.

10.4 Solving Systems of Two Linear Equations with Matrices

Objectives

- Multiply matrices by hand and with a calculator.
- Find the determinant of a matrix by hand and with a calculator.
- Translate a system of linear equations into a matrix equation.
- Use the determinant to identify systems of linear equations with a unique solution.
- Use a calculator to solve a matrix equation, if a solution exists.
- Use algebra and/or graphing to find whether a system without a unique solution represents parallel or coincident lines.

Students will be highly motivated if they are first shown a calculator matrix solution of Exercise 46 on page 625.

WARM-UP

Change these equations into standard form, $ax + by = c$. Eliminate any fractions in the equations.

1. $\frac{2}{3}x + \frac{1}{3} = y$ $2x - 3y = -1$ **2.** $y = \frac{3}{4}x - \frac{1}{2}$ $3x - 4y = 2$

3. $4 + y = \frac{7}{2}x$ $7x - 2y = 8$ **4.** $y = \frac{12}{5}x + \frac{4}{5}$ $12x - 5y = -4$

IN THIS SECTION, we explore matrices and the application of calculator matrix operations to solving a system of two linear equations. These examples use systems of two linear equations in two variables so that the properties of the systems can be verified graphically. In Section 10.5, matrix operations are extended to systems of three or more linear equations.

Matrices

In this text, we will use matrices to provide a general procedure for solving a system of linear equations, just as the quadratic formula

$$x = \frac{-b \pm \sqrt{b^2 - 4ac}}{2a}$$

provides a general procedure for solving equations of the form $ax^2 + bx + c = 0$. After a brief introduction to matrix operations, we will use the operations on a calculator to solve linear equations.

A **matrix** is *an array of numbers in rows and columns placed between brackets. The number of rows and the number of columns* determine the **shape of a matrix**, $r \times c$. The matrix below has two rows and three columns.

$$\begin{array}{cc} & \text{column 1} \quad \text{column 2} \quad \text{column 3} \\ \begin{array}{c} \text{row 1} \\ \text{row 2} \end{array} & \begin{bmatrix} a & b & c \\ d & e & f \end{bmatrix} \end{array}$$

The matrix is said to be a 2 by 3, or 2×3, matrix. When we refer to a matrix without the word *matrix*, we place its name in brackets. Thus, the array of numbers for matrix A will have the symbol $[A]$.

ADDITION AND SUBTRACTION OF MATRICES   As with other sets of numbers, operations such as addition, subtraction, and multiplication are defined for matrices.

In order *to be added (or subtracted), matrices must have the same shape*—that is, the same number of rows and columns. We add by adding numbers in the same position inside the matrix. We subtract by subtracting numbers in the same position inside the matrix.

Here is how we add 2×1 matrices:

$$\begin{bmatrix} a \\ b \end{bmatrix} + \begin{bmatrix} c \\ d \end{bmatrix} = \begin{bmatrix} a + c \\ b + d \end{bmatrix}$$

Here is how we subtract 1×3 matrices:

$$[a \quad b \quad c] - [d \quad e \quad f] = [a - d \quad b - e \quad c - f]$$

We will not need addition and subtraction in order to solve a system of equations. The operations are included here only for completeness of presentation.

MULTIPLICATION OF MATRICES   The expression $[A][X]$ implies multiplication of $[A]$ and $[X]$. The product $[A][X]$, where

$$[A] = \begin{bmatrix} a & b \\ c & d \end{bmatrix} \quad \text{and} \quad [X] = \begin{bmatrix} x \\ y \end{bmatrix}$$

is precisely the left side of the system of equations

$$ax + by = e$$
$$cx + dy = f$$

Thus,

$$[A][X] = \begin{bmatrix} a & b \\ c & d \end{bmatrix} \begin{bmatrix} x \\ y \end{bmatrix} = \begin{bmatrix} ax + by \\ cx + dy \end{bmatrix}$$

EXAMPLE 1   Multiplying matrices   Multiply these matrices.

a. $\begin{bmatrix} 1 & 1 \\ 1 & -2 \end{bmatrix} \begin{bmatrix} x \\ y \end{bmatrix}$ **b.** $\begin{bmatrix} -2 & 1 \\ -0.5 & 1 \end{bmatrix} \begin{bmatrix} x \\ y \end{bmatrix}$

SOLUTION   **a.** $\begin{bmatrix} 1 & 1 \\ 1 & -2 \end{bmatrix} \begin{bmatrix} x \\ y \end{bmatrix} = \begin{bmatrix} 1x + 1y \\ 1x - 2y \end{bmatrix}$

b. $\begin{bmatrix} -2 & 1 \\ -0.5 & 1 \end{bmatrix} \begin{bmatrix} x \\ y \end{bmatrix} = \begin{bmatrix} -2x + 1y \\ -0.5x + 1y \end{bmatrix}$

To multiply a 2×2 matrix by another 2×2 matrix, we use the following rule:

$$\begin{bmatrix} a & b \\ c & d \end{bmatrix} \begin{bmatrix} e & f \\ g & h \end{bmatrix} = \begin{bmatrix} ae + bg & af + bh \\ ce + dg & cf + dh \end{bmatrix}$$

EXAMPLE 2   Multiplying matrices   Multiply:

$$\begin{bmatrix} 1 & 2 \\ 3 & 4 \end{bmatrix} \begin{bmatrix} p & q \\ r & s \end{bmatrix}$$

SOLUTION

$$\begin{bmatrix} 1 & 2 \\ 3 & 4 \end{bmatrix} \begin{bmatrix} p & q \\ r & s \end{bmatrix} = \begin{bmatrix} 1p + 2r & 1q + 2s \\ 3p + 4r & 3q + 4s \end{bmatrix}$$

Note that we *multiply a row in the first matrix times a column in the second matrix*. The product of the first numbers in the row and column is added to the product of the second numbers.

IDENTITIES AND INVERSES   In arithmetic, when we multiply any number by 1, we get the same number back. When we multiply a number by its multiplicative inverse, or reciprocal, we get 1. That is,

$$a \cdot 1 = a \qquad 4 \cdot 1 = 4 \qquad \text{1 is the identity.}$$

$$a \cdot \frac{1}{a} = 1 \qquad 4 \cdot \frac{1}{4} = 1 \qquad \frac{1}{a} \text{ is the inverse.}$$

In matrices, when we multiply a matrix by an **identity matrix**, I, we *get the same matrix back*. When we multiply a matrix by its **matrix inverse**, we *get an identity matrix, I*. The 2×2 identity matrix is written

$$I = \begin{bmatrix} 1 & 0 \\ 0 & 1 \end{bmatrix}$$

The symbol for the inverse of a matrix will have a superscript -1, just as the symbol for the inverse of the function $f(x)$ is $f^{-1}(x)$. *The matrix inverse of $[A]$ is $[A]^{-1}$*, if the inverse exists. In most cases, we will use a calculator to find the matrix inverse. The instructions to Exercises 29 to 32 show how to find the inverse of a 2×2 matrix by hand.

In Example 3, we check that multiplying a matrix by an identity matrix I yields the original matrix and that multiplying a matrix by its inverse gives an identity matrix I.

EXAMPLE 3 Multiplying matrices For the matrices $[A] = \begin{bmatrix} 1 & -1 \\ -3 & 2 \end{bmatrix}$, $[A]^{-1} = \begin{bmatrix} -2 & -1 \\ -3 & -1 \end{bmatrix}$, and $[I] = \begin{bmatrix} 1 & 0 \\ 0 & 1 \end{bmatrix}$, calculate

a. $[A][I]$ **b.** $[A][A]^{-1}$ **c.** $[A]^{-1}[A]$

SOLUTION **a.** $[A][I] = \begin{bmatrix} 1 & -1 \\ -3 & 2 \end{bmatrix}\begin{bmatrix} 1 & 0 \\ 0 & 1 \end{bmatrix}$

$$= \begin{bmatrix} 1 \cdot 1 + -1 \cdot 0 & 1 \cdot 0 + -1 \cdot 1 \\ -3 \cdot 1 + 2 \cdot 0 & -3 \cdot 0 + 2 \cdot 1 \end{bmatrix}$$

$$= \begin{bmatrix} 1 & -1 \\ -3 & 2 \end{bmatrix} = [A]$$

b. $[A][A]^{-1} = \begin{bmatrix} 1 & -1 \\ -3 & 2 \end{bmatrix}\begin{bmatrix} -2 & -1 \\ -3 & -1 \end{bmatrix}$

$$= \begin{bmatrix} 1 \cdot (-2) + (-1) \cdot (-3) & 1 \cdot (-1) + (-1) \cdot (-1) \\ -3 \cdot (-2) + 2 \cdot (-3) & -3 \cdot (-1) + 2 \cdot (-1) \end{bmatrix}$$

$$= \begin{bmatrix} 1 & 0 \\ 0 & 1 \end{bmatrix} = [I]$$

c. $[A]^{-1}[A] = \begin{bmatrix} -2 & -1 \\ -3 & -1 \end{bmatrix}\begin{bmatrix} 1 & -1 \\ -3 & 2 \end{bmatrix}$

$$= \begin{bmatrix} -2 \cdot 1 + (-1) \cdot (-3) & -2 \cdot (-1) + (-1) \cdot 2 \\ -3 \cdot 1 + (-1) \cdot (-3) & -3 \cdot (-1) + (-1) \cdot 2 \end{bmatrix}$$

$$= \begin{bmatrix} 1 & 0 \\ 0 & 1 \end{bmatrix} = [I]$$

Repeat Example 3 with a calculator, using [A], [B], and [C] for $[A]$, $[A]^{-1}$, and $[I]$. First enter the 2×2 matrix [A] into the calculator. Using $\boxed{\text{2nd}}$ [MATRX] EDIT to get to the entry screen, choose A and enter the size and matrix numbers. Press $\boxed{\text{2nd}}$ [QUIT]. Find $[A]^{-1}$ with $\boxed{\text{2nd}}$ [MATRX] [A] $\boxed{x^{-1}}$ $\boxed{\text{ENTER}}$. To store this inverse in [B], use $\boxed{\text{STO}}$ $\boxed{\text{2nd}}$ [MATRX] **2 : [B]** $\boxed{\text{ENTER}}$. Next, enter the 2×2 identity matrix in [C]. To bring any matrix onto the working screen, use $\boxed{\text{2nd}}$ [MATRX], choose the matrix, and then press $\boxed{\text{ENTER}}$.

The results in Example 3 are consistent with our definitions:

• The product of a matrix and an identity matrix is the original matrix:

$$[A] \, [I] = [A]$$

- The product of a matrix and its inverse is an identity matrix:

$$[A]\,[A]^{-1} = [A]^{-1}[A] = I$$

CALCULATING THE DETERMINANT There is no reciprocal (or multiplicative inverse) for zero because $\frac{1}{0}$ is undefined. We have a similar situation with matrices. In the introduction to the matrix inverse, note the phrase *if the inverse exists*. There are times when the matrix inverse has the matrix equivalent of a zero in the denominator. Thus, we say that

If [A] *has a determinant of zero, then* $[A]^{-1}$ *is undefined.*

The **determinant** is a *value*, defined only for square matrices, *that determines whether the corresponding system of equations has a unique solution.* How we calculate the determinant, det [A], depends on the size of the matrix. For a 2×2 matrix,

$$\text{If} \quad [A] = \begin{bmatrix} a & b \\ c & d \end{bmatrix}, \quad \text{then det } [A] = ad - bc.$$

A nonzero value for the determinant of matrix A tells us that the inverse, $[A]^{-1}$, exists.

EXAMPLE 4 Finding the determinant Calculate the determinant for these matrices:

a. $\begin{bmatrix} 1 & 1 \\ 1 & -2 \end{bmatrix}$ **b.** $\begin{bmatrix} -2 & 1 \\ -0.5 & 1 \end{bmatrix}$ **c.** $\begin{bmatrix} 2 & 1 \\ 2 & 1 \end{bmatrix}$

SOLUTION **a.** The value of the determinant is $1(-2) - 1(1) = -3$.

b. The value of the determinant is $-2(1) - (-0.5)(1) = -1.5$.

c. The value of the determinant is $2(1) - 2(1) = 0$.

Repeat Example 4 with a calculator. After entering the matrices, evaluate the determinant with ⎡2nd⎤ [MATRX] MATH **1 : det(**, followed by the matrix name, such as [A], and then ⎡ENTER⎤.

Solving a System of Equations

As in solving quadratic equations, the first step in the matrix solution of linear equations is to place the equations in standard form.

■ STANDARD FORM OF A LINEAR EQUATION

All linear equations in two variables may be written as

$$ax + by = e$$

where a, b, and e are real numbers.

(The letter c has been changed here to the letter e so that we can more clearly distinguish the left and right sides of equations in subsequent steps.)

We arrange the equations in standard form, with the variable terms on the left and the constant terms on the right:

$$ax + by = e$$
$$cx + dy = f$$

We are now ready to change the system of equations in standard form to matrix form.

SETTING UP THE MATRIX EQUATION   A **matrix equation** *contains matrices instead of coefficients, variables, and constants.* We use the symbols $[A] [X] = [B]$ to describe the matrix equation in which $[A]$ is the matrix of coefficients on the variables, $[X]$ is the matrix of variables, and $[B]$ is the matrix of constants.

Matrix A is the **matrix of coefficients** and *contains the coefficients of the variable terms from the left side of each equation.* If one variable is missing, we write zero as its coefficient. The shape of matrix A is 2 by 2, meaning two rows and two columns. We write matrix A by placing the coefficients of the variables in the form

$$[A] = \begin{bmatrix} a & b \\ c & d \end{bmatrix}$$

Matrix X is the **matrix of variables** and *contains the two variables x and y:*

$$[X] = \begin{bmatrix} x \\ y \end{bmatrix}$$

The shape of matrix X is 2 by 1, meaning two rows and one column.

Matrix B is the **matrix of constants** and *contains the constant term from the right side of each equation.* The shape of matrix B is 2 by 1. We write matrix B by placing the constant terms in the form

$$[B] = \begin{bmatrix} e \\ f \end{bmatrix}$$

We can combine the matrices above into the matrix equation

$$\begin{bmatrix} a & b \\ c & d \end{bmatrix} \begin{bmatrix} x \\ y \end{bmatrix} = \begin{bmatrix} e \\ f \end{bmatrix}$$

A shorter way to write the matrix equation is $[A] [X] = [B]$.

EXAMPLE 5   **Writing matrix equations**   Write this system of equations as a matrix equation.

$$y = -2x + 3$$
$$y + 2x = 4$$

SOLUTION   The equations may be written in standard form as

$$2x + y = 3$$
$$2x + y = 4$$

The matrix equation is

$$\begin{bmatrix} 2 & 1 \\ 2 & 1 \end{bmatrix} \begin{bmatrix} x \\ y \end{bmatrix} = \begin{bmatrix} 3 \\ 4 \end{bmatrix}$$

SOLVING THE MATRIX EQUATION   To solve a matrix equation, we must do the matrix equivalent of dividing both sides of $[A] [X] = [B]$ by the matrix A. It is nearly that simple on a calculator, but we do need to look at the steps leading to the calculator solution. The calculator solution relies on *inverses* and *identities.*

As an introduction to the matrix process, look carefully at the role played by the reciprocal, or multiplicative inverse, in Example 6.

EXAMPLE 6   **Solving with inverses**   Solve $\frac{2}{3} x = 50$.

SOLUTION

$$\frac{2}{3} x = 50 \qquad \text{Multiply both sides by the reciprocal, } \frac{3}{2}.$$
$$\frac{3}{2} \cdot \frac{2}{3} x = \frac{3}{2} \cdot 50 \qquad \text{Simplify.}$$
$$x = 75$$

In Example 6, we multiplied by the multiplicative inverse of $\frac{2}{3}$ to solve the equation. To solve a matrix system $[A][X] = [B]$, we multiply both sides by the inverse of matrix A, $[A]^{-1}$, if the inverse exists.

$$[A][X] = [B]$$

$$[A]^{-1}[A][X] = [A]^{-1}[B]$$

$$[I][X] = [A]^{-1}[B]$$

$$[X] = [A]^{-1}[B]$$

Student Note: The multiplication by $[A]^{-1}$ is on the left on both sides—more about this in the exercises.

Thus, the solution to $[A][X] = [B]$ is $[X] = [A]^{-1}[B]$.

Using a Calculator to Solve $[A][X] = [B]$

To solve a system of linear equations with matrices, we take the numerical coefficients and constants from the equations, place them in matrices, and then solve on the calculator. Here is the process:

■ **MATRIX SOLUTION OF A SYSTEM OF LINEAR EQUATIONS (SUMMARY)**

1. Set up the matrix equation $[A][X] = [B]$.
2. Enter $[A]$ into the calculator.
3. Enter $[B]$ into the calculator.
4. Calculate det $[A]$. If det $[A] = 0$, stop; if det $[A] \neq 0$, proceed to step 5.
5. Calculate $[A]^{-1}[B]$, and record the answer.

EXAMPLE 7 **Solving with matrices** Solve the system of equations with matrices.

$$x - y = 3$$

$$2y - 3x = -6$$

SOLUTION To set up the matrix equation, we first arrange the equations in standard form. In this case, only the second equation must be rearranged.

$$x - y = 3$$

$$-3x + 2y = -6$$

We select the coefficients from the variables and place them in matrix A:

$$[A] = \begin{bmatrix} 1 & -1 \\ -3 & 2 \end{bmatrix} \qquad [A] \text{ is a 2-row by 2-column matrix.}$$

We write the variables in matrix X:

$$[X] = \begin{bmatrix} x \\ y \end{bmatrix} \qquad [X] \text{ is a 2-row by 1-column matrix.}$$

We take the constants from the right side and place them in matrix B:

$$[B] = \begin{bmatrix} 3 \\ -6 \end{bmatrix} \qquad [B] \text{ is a 2-row by 1-column matrix.}$$

We then place $[A]$, $[X]$, and $[B]$ into the matrix equation $[A][X] = [B]$:

$$\begin{bmatrix} 1 & -1 \\ -3 & 2 \end{bmatrix} \begin{bmatrix} x \\ y \end{bmatrix} = \begin{bmatrix} 3 \\ -6 \end{bmatrix}$$

After entering [A] and [B] into the calculator, we calculate det [A]. The determinant of A, det [A], is -1. Because it is not equal to zero, we can continue with the solution process.

The next step is to calculate $[A]^{-1}[B]$ and set the answer equal to the variable matrix [X]:

$$[X] = \begin{bmatrix} x \\ y \end{bmatrix} = \begin{bmatrix} 0 \\ -3 \end{bmatrix}$$

The solution to the system is $x = 0$ and $y = -3$. This means that the graphs of the equations intersect at $(0, -3)$.

The following box gives more details on the graphing calculator steps for the matrix solution.

GRAPHING CALCULATOR TECHNIQUE: MATRIX SOLUTION OF A SYSTEM OF LINEAR EQUATIONS (DETAILED FORM)

To solve the system $x - y = 3$ and $2y - 3x = -6$ with matrices, follow this general procedure, modifying it as necessary to fit your calculator.

1. Set up the matrix equation $[A][X] = [B]$ by hand. (See Example 7.)

2. Enter matrix A into the calculator.

   a. Under the matrix menu, choose EDIT; press 1.

   b. Enter the shape of matrix A as 2×2 with 2 (**ENTER**) 2 (**ENTER**).

   c. Enter the matrix A numbers in the order $1, -1, -3, 2$. Usually (**ENTER**) is needed after each number.

   d. Leave with (**2nd**) [QUIT].

3. Enter matrix B into the calculator.

   a. Under the matrix menu, choose EDIT; press 2.

   b. Enter the shape of matrix B as 2×1 with 2 (**ENTER**) 1 (**ENTER**).

   c. Enter the matrix B numbers in the order $3, -6$. Usually (**ENTER**) is needed after each number.

   d. Leave with (**2nd**) [QUIT].

4. Calculate det [A].

   a. Under the matrix MATH menu, find the determinant, **1 : det(**.

   b. Under the matrix menu, select [A], and press (**ENTER**) to evaluate the determinant.

   c. If det [A] = 0, stop; if det [A] $\neq$ 0, proceed to step 5.

5. Calculate $[A]^{-1}[B]$.

   a. Under the matrix menu, select [A].

   b. Obtain $^{-1}$ with the (x^{-1}) key. **Do not use (^) ((−)) 1; it will not work.**

   c. Under the matrix menu, select [B].

   d. Press (**ENTER**) to evaluate.

   e. Record the answer.

EXAMPLE 8   **Solving with matrices**   Solve these systems of equations with calculator matrix operations.

a. $x + y = 3$
   $x - 2y = 6$

b. $y = 2x + 4$
   $y = 0.5x - 2$

SOLUTION   **a.** For $x + y = 3$ and $x - 2y = 6$, the matrix equation $[A]\,[X] = [B]$ is

$$\begin{bmatrix} 1 & 1 \\ 1 & -2 \end{bmatrix} \begin{bmatrix} x \\ y \end{bmatrix} = \begin{bmatrix} 3 \\ 6 \end{bmatrix}$$

We enter matrix A, the 2×2 matrix

$$[A] = \begin{bmatrix} 1 & 1 \\ 1 & -2 \end{bmatrix}$$

Then we enter matrix B, the 2×1 matrix

$$[B] = \begin{bmatrix} 3 \\ 6 \end{bmatrix}$$

Calculation of det $[A]$ yields

$$\det [A] = -3$$

Because det $[A] \neq 0$, we may continue with the solution process.
Calculation of $[A]^{-1}[B]$ gives a 2×1 matrix equal to the matrix of variables:

$$[X] = \begin{bmatrix} x \\ y \end{bmatrix} = \begin{bmatrix} 4 \\ -1 \end{bmatrix}$$

Matrix X shows that $x = 4$ and $y = -1$. The graphs of the equations intersect at $(4, -1)$.

b. For $y = 2x + 4$ and $y = 0.5x - 2$, the matrix equation $[A]\,[X] = [B]$ is

$$\begin{bmatrix} -2 & 1 \\ -0.5 & 1 \end{bmatrix} \begin{bmatrix} x \\ y \end{bmatrix} = \begin{bmatrix} 4 \\ -2 \end{bmatrix}$$

We enter matrix A, the 2×2 matrix

$$[A] = \begin{bmatrix} -2 & 1 \\ -0.5 & 1 \end{bmatrix}$$

Then we enter matrix B, the 2×1 matrix

$$[B] = \begin{bmatrix} 4 \\ -2 \end{bmatrix}$$

Calculation of det $[A]$ yields

$$\det [A] = -1.5$$

Because det $[A] \neq 0$, we may continue with the solution process.
Calculation of $[A]^{-1}[B]$ gives a 2×1 matrix equal to the matrix of variables:

$$[X] = \begin{bmatrix} x \\ y \end{bmatrix} = \begin{bmatrix} -4 \\ -4 \end{bmatrix}$$

Matrix X shows that $x = -4$ and $y = -4$. The graphs of the equations intersect at $(-4, -4)$. ■

■ What can go wrong? To solve the matrix equation on a calculator, we need the expression $[A]^{-1}[B]$. For a matrix of reasonable size, the calculator will find $[A]^{-1}$, if it exists. Unfortunately, not all matrices have inverses.

In order to have an inverse, a matrix A must meet two conditions:

1. $[A]$ must be a **square matrix**, *having the same number of rows and columns.*

2. The determinant of $[A]$, det $[A]$, cannot be zero.

Thus, as indicated above, it is important to always find whether or not a system has a unique solution by checking for a zero determinant value.

In Example 9, we return to the system in Example 5.

EXAMPLE 9   Solving with matrices

a. Set up a matrix solution of

$$y = -2x + 3$$

$$y + 2x = 4$$

b. Enter the matrices A and B into a calculator.

c. Evaluate the determinant, $ad - bc$, by hand and with det $[A]$.

d. Find the solution, if possible, with the calculator. If not, solve by hand in order to find the type of system.

SOLUTION   **a.** The matrix system is

$$\begin{bmatrix} 2 & 1 \\ 2 & 1 \end{bmatrix} \begin{bmatrix} x \\ y \end{bmatrix} = \begin{bmatrix} 3 \\ 4 \end{bmatrix}$$

b. Matrix A is 2×2, and matrix B is 2×1. These sizes are entered, along with the matrix numbers.

c. Calculated by hand, the determinant is

$$ad - bc = 2(1) - 2(1) = 0$$

On the calculator, det $[A] = 0$.

d. Since the determinant is zero, we have to solve by hand.

$$y = -2x + 3$$

$$y + 2x = 4 \qquad \text{Substitute } -2x + 3 \text{ from the first equation for } y.$$

$$-2x + 3 + 2x = 4 \qquad \text{Simplify the left side.}$$

$$3 = 4 \qquad \text{Contradiction}$$

There is no solution. The graphs of the lines are parallel, which we could observe by changing the second equation into $y = mx + b$ form.   ▬

Recall that if the solution is an identity, $a = a$, the graphs of the equations are coincident lines. The possibilities will be summarized more extensively in the next section.

ANSWER BOX

Warm-up: **1.** $2x - 3y = -1$ **2.** $3x - 4y = 2$ **3.** $7x - 2y = 8$
4. $12x - 5y = -4$

10.4   Exercises

For Exercises 1 to 10, use the following matrices.

$$[A] = [2 \quad 3] \qquad [B] = [2 \quad 4] \qquad [C] = \begin{bmatrix} 1 & 2 & 3 \\ 4 & 5 & 6 \\ 7 & 8 & 9 \end{bmatrix}$$

$$[D] = \begin{bmatrix} 6 & 7 & 8 \\ 9 & 10 & 11 \\ 12 & 13 & 14 \end{bmatrix} \qquad [E] = \begin{bmatrix} 1 \\ 2 \\ 3 \end{bmatrix} \qquad [F] = \begin{bmatrix} 4 \\ 5 \\ 6 \end{bmatrix}$$

1. Describe the shape of matrix A. 1×2

2. Describe the shape of matrix C. 3 × 3

3. Describe the shape of matrix F. 3 × 1

4. How do we describe the shape of a matrix—row by column or column by row? row by column

5. Add $[A]$ and $[B]$. [4 7]

6. Add $[C]$ and $[D]$.
$$\begin{bmatrix} 7 & 9 & 11 \\ 13 & 15 & 17 \\ 19 & 21 & 23 \end{bmatrix}$$

7. Add $[E]$ and $[F]$.
$$\begin{bmatrix} 5 \\ 7 \\ 9 \end{bmatrix}$$

8. Find $[B] - [A]$. [0 1]

9. Find $[D] - [C]$.
$$\begin{bmatrix} 5 & 5 & 5 \\ 5 & 5 & 5 \\ 5 & 5 & 5 \end{bmatrix}$$

10. Find $[F] - [E]$.
$$\begin{bmatrix} 3 \\ 3 \\ 3 \end{bmatrix}$$

For Exercises 11 to 34, use the following matrices.

$$[A] = \begin{bmatrix} 1 & 3 \\ -1 & -2 \end{bmatrix} \quad [B] = \begin{bmatrix} 2 & 3 \\ 4 & 5 \end{bmatrix} \quad [C] = \begin{bmatrix} 4 & -2 \\ 1 & 2 \end{bmatrix}$$

$$[D] = \begin{bmatrix} -2 & -1 \\ 1 & 2 \end{bmatrix} \quad [E] = \begin{bmatrix} 2 \\ -1 \end{bmatrix} \quad [X] = \begin{bmatrix} x \\ y \end{bmatrix}$$

Find the product of the matrices in Exercises 11 and 12 without using a calculator.

11. $[A][X]$ $\begin{bmatrix} x + 3y \\ -x - 2y \end{bmatrix}$

12. $[B][X]$ $\begin{bmatrix} 2x + 3y \\ 4x + 5y \end{bmatrix}$

Find the product of the matrices in Exercises 13 to 16 by hand or using a calculator.

13. $[A][E]$ $\begin{bmatrix} -1 \\ 0 \end{bmatrix}$

14. $[B][E]$ $\begin{bmatrix} 1 \\ 3 \end{bmatrix}$

15. $[C][E]$ $\begin{bmatrix} 10 \\ 0 \end{bmatrix}$

16. $[D][E]$ $\begin{bmatrix} -3 \\ 0 \end{bmatrix}$

Find the products in Exercises 17 to 24. Do at least one without using a calculator.

17. $[A][B]$ $\begin{bmatrix} 14 & 18 \\ -10 & -13 \end{bmatrix}$

18. $[A][C]$ $\begin{bmatrix} 7 & 4 \\ -6 & -2 \end{bmatrix}$

19. $[B][A]$ $\begin{bmatrix} -1 & 0 \\ -1 & 2 \end{bmatrix}$

20. $[C][A]$ $\begin{bmatrix} 6 & 16 \\ -1 & -1 \end{bmatrix}$

21. $[C][D]$ $\begin{bmatrix} -10 & -8 \\ 0 & 3 \end{bmatrix}$

22. $[B][C]$ $\begin{bmatrix} 11 & 2 \\ 21 & 2 \end{bmatrix}$

23. $[D][C]$ $\begin{bmatrix} -9 & 2 \\ 6 & 2 \end{bmatrix}$

24. $[C][B]$ $\begin{bmatrix} 0 & 2 \\ 10 & 13 \end{bmatrix}$

In Exercises 25 to 28, find the determinant of each matrix by hand and then by calculator.

25. det $[A]$ 1

26. det $[B]$ −2

27. det $[C]$ 10

28. det $[D]$ −3

In Exercises 29 to 32, find the inverse by hand, and then check with a calculator. For a 2 × 2 matrix, the inverse of $[K]$, $[K]^{-1}$, may be calculated as follows:

If $[K] = \begin{bmatrix} a & b \\ c & d \end{bmatrix}$ and det $[K] = ad - bc$,

then $[K]^{-1} = \begin{bmatrix} \dfrac{d}{\det [K]} & \dfrac{-b}{\det [K]} \\ \dfrac{-c}{\det [K]} & \dfrac{a}{\det [K]} \end{bmatrix}$

29. $[A]^{-1}$ $\begin{bmatrix} -2 & -3 \\ 1 & 1 \end{bmatrix}$

30. $[B]^{-1}$ $\begin{bmatrix} -2.5 & 1.5 \\ 2 & -1 \end{bmatrix}$

31. $[C]^{-1}$ $\begin{bmatrix} 0.2 & 0.2 \\ -0.1 & 0.4 \end{bmatrix}$

32. $[D]^{-1}$ $\begin{bmatrix} -\frac{2}{3} & -\frac{1}{3} \\ \frac{1}{3} & \frac{2}{3} \end{bmatrix}$

33. When we multiply a matrix by its inverse, we obtain the identity matrix, I. Calculate the matrix obtained from

a. $[A][A]^{-1}$ $\begin{bmatrix} 1 & 0 \\ 0 & 1 \end{bmatrix}$

b. $[A]^{-1}[A]$ $\begin{bmatrix} 1 & 0 \\ 0 & 1 \end{bmatrix}$

34. Why is there no inverse for matrix E? not a square matrix

Solve Exercises 35 to 42 with matrices. If no unique solution is possible, explain why.

35. $1x + 2y = 5$
$4x + 4y = 6$
$x = -2, y = 3.5$

36. $1x + 2y = 5$
$3x + 4y = 6$
$x = -4, y = 4.5$

37. $1x + 2y = 5$
$2x + 4y = 6$
det $[A] = 0$; parallel lines

38. $1x + 2y = 5$
$1x + 4y = 6$
$x = 4, y = 0.5$

39. $1x + 2y = 5$
$0x + 4y = 6$
$x = 2, y = 1.5$

40. $1x + 2y = 5$
$-1x + 4y = 6$
$x \approx 1.33, y \approx 1.83$

41. $1x + 2y = 5$
$-2x + 4y = 6$
$x = 1, y = 2$

42. $1x + 2y = 5$
$-3x + 4y = 6$
$x = 0.8, y = 2.1$

43. Describe what Exercises 35 to 42 have in common. For the line $ax + 4y = 6$, how will changing the coefficient a on x change the graph? What part of the graph of $ax + 4y = 6$ stays the same? $1x + 2y = 5$, $ax + 4y = 6$; slope changes; y-intercept

44. What will all the solutions (if they exist) to the systems in Exercises 35 to 42 have in common? on $1x + 2y = 5$

Solve Exercises 45 to 50 with matrices. If no unique solution is possible, explain why.

45. $5x = 28.2 + 2y$
$3x + 5y = -5.4$
$x = 4.2, y = -3.6$

46. $4x + 3y = 25$
$3x + 2.5 = 2y$
$x = 2.5, y = 5$

47. $3y = 2x + 12$
$y + 2 = 2x/3$
det $[A] = 0$; parallel lines

48. $y = 0.25x + 2$
$4y = x + 8$
det $[A] = 0$; coincident lines

49. $x - 1.5y = 3.5$
$2x - 7 = 3y$
det $[A] = 0$; coincident lines

50. $4x + 2 = y$
$x - 0.25y = 0.5$
det $[A] = 0$; parallel lines

51. a. Evaluate the determinant for $[E] = \begin{bmatrix} 2 & -1 \\ -4 & 2 \end{bmatrix}$. 0

b. Evaluate $[E]^{-1}$ without a calculator. Describe the result. zero denominators

c. Evaluate $[E]^{-1}$ with a calculator. Explain the probable meaning of the phrase *singular matrix*. matrix with no inverse

52. What do we call the property of multiplication indicating that $4 \cdot 5 = 5 \cdot 4$? Examine the answers to Exercises 17 to 24. Is there a similar property for the multiplication of matrices? Explain. commutative property of multiplication; no

▇▇ **Project**

53. Cramer's Rule Research Cramer's rule, and explain how determinants are used to find the solutions to a system of equations.

10.5 ▮ Solving Systems of Three or More Linear Equations with Matrices

Objectives

▮ Solve a system of three or more linear equations with matrices on a calculator.

▮ Describe possible implications of a zero determinant.

WARM-UP

For each pair of values, evaluate $ax^4 + bx^3 + cx^2 + dx^1 + e = y$.

1. $(-2, 1)$ **2.** $(0, 2)$ **3.** $(1, -1)$

4. $(2, 4)$ **5.** $(3, 3)$ See Example 3.

THIS SECTION EXTENDS the calculator matrix solution of linear equations to systems of three or more equations.

Solving Systems of Three Equations

Example 1 provides a reminder that any letters can be variables.

EXAMPLE 1 Solving a system of equations Solve the system of equations

$$a + b + c = 3$$

$$2a + 3b + c = 13$$

$$2a - b = 0$$

a. Set up the matrix equation $[A][X] = [B]$.

b. Calculate the determinant of the matrix of coefficients, det $[A]$, and solve the system if possible.

SOLUTION **a.** The equations are already in standard form. We write the matrix equation directly from the equations:

$$\begin{bmatrix} 1 & 1 & 1 \\ 2 & 3 & 1 \\ 2 & -1 & 0 \end{bmatrix} \begin{bmatrix} a \\ b \\ c \end{bmatrix} = \begin{bmatrix} 3 \\ 13 \\ 0 \end{bmatrix}$$

$[A]$ is 3×3, while $[X]$ and $[B]$ are 3×1. We enter $[A]$ and $[B]$ into the calculator.

b. Because det $[A] = -5$, a nonzero number, it is possible to solve the system. We calculate $[A]^{-1}[B]$ to find the variable matrix, $[X]$.

$$[X] = \begin{bmatrix} a \\ b \\ c \end{bmatrix} = \begin{bmatrix} 2 \\ 4 \\ -3 \end{bmatrix}$$

Thus, $a = 2$, $b = 4$, and $c = -3$. There is a unique solution, so the equations in the system are consistent. ▬

When we solve a system of equations on a calculator, *we obtain a zero determinant, det $[A] = 0$, for both inconsistent and dependent equations. To find which type—inconsistent or dependent—the equations are, we will solve the system by* *hand.* Table 4 summarizes the geometry and algebra associated with solving a system of three linear equations (see also Section 3.4).

TABLE 4

	Consistent Equations, det $[A] \neq 0$	Inconsistent Equations, det $[A] = 0$	Dependent Equations, det $[A] = 0$
Geometry	Planes intersect in one common point.	There is no common intersection.	There is linear or planar intersection.
Algebra	Equations can be solved to yield unique x, y, and z values.	The variables drop out, and the resulting statement is false.	The variables drop out, and the resulting statement is true.
Numerical Result	The unique solution is given by the ordered triple (x, y, z).	A contradiction results.	An identity, $a = a$, results.
Solution	There is one solution, an ordered triple.	There is no solution.	An infinite number of ordered triples satisfy the system.

Applications

CANDY PRODUCTION   In Example 2, we return to the chocolate rabbit setting from Section 3.4. Fenton and Lee are Janelle's sons.

EXAMPLE 2   Solving applications with matrices: more chocolate rabbits   As he begins production, Fenton finds that the small rabbits with their hand-painted eggs take longer to make than he had planned. He consults with his brother Lee, who suggests still making 620 rabbits, but dropping to three times as many small rabbits as medium rabbits and limiting the large rabbits to 20. Set up a system of equations, prepare a matrix equation, and solve.

SOLUTION   Again, we let s = number of small rabbits, m = number of medium rabbits, and l = number of large rabbits. The equations are $s + m + l = 620$, $s = 3m$, and $l = 20$. The system of equations is

$$s + \quad m + l = 620$$
$$s - 3m + 0 = \quad 0$$
$$0 + \quad 0 + l = \quad 20$$

The matrix equation $[A][X] = [B]$ is

$$\begin{bmatrix} 1 & 1 & 1 \\ 1 & -3 & 0 \\ 0 & 0 & 1 \end{bmatrix} \begin{bmatrix} s \\ m \\ l \end{bmatrix} = \begin{bmatrix} 620 \\ 0 \\ 20 \end{bmatrix}$$

$[A]$ is 3×3, while $[X]$ and $[B]$ are 3×1. The value of the determinant, det $[A]$, is -4. The solution is

$$[X] = [A]^{-1}[B] = \begin{bmatrix} 450 \\ 150 \\ 20 \end{bmatrix} = \begin{bmatrix} s \\ m \\ l \end{bmatrix}$$

This plan calls for making 450 small rabbits, 150 medium rabbits, and 20 large rabbits.

We will return to the candy-making scenario in the exercises, as we explore the meaning of multiplication of matrices.

FITTING POLYNOMIAL EQUATIONS TO DATA Matrices have traditionally been used to fit polynomial equations to sets of data. Most graphing calculators can fit up to a fourth-degree, or quartic, equation with their regression options. The method in Example 3 can be employed to fit equations of higher degree by using $n + 1$ points for an nth-degree equation. In Example 3, we find a fourth-degree equation, so we can check our results with the calculator regression option. The first step in Example 3 will be used again in Example 4 when we substituting an ordered pair into the roadbed transition curve equation.

EXAMPLE 3 Solving applications by matrices Fit a quartic (fourth-degree polynomial) equation to the ordered pairs $(-2, 1)$, $(0, 2)$, $(1, -1)$, $(2, 4)$, and $(3, 3)$. Use $y = ax^4 + bx^3 + cx^2 + dx + e$ and each ordered pair (x, y) to generate a system of five linear equations. Compare your results with those from quartic regression.

SOLUTION We substitute each x-coordinate and y-coordinate for x and y:

$$a(16) + b(-8) + c(4) + d(-2) + e = 1 \qquad \text{For } (-2, 1)$$
$$a(0) + b(0) + c(0) + d(0) + e = 2 \qquad \text{For } (0, 2)$$
$$a(1) + b(1) + c(1) + d(1) + e = -1 \qquad \text{For } (1, -1)$$
$$a(16) + b(8) + c(4) + d(2) + e = 4 \qquad \text{For } (2, 4)$$
$$a(81) + b(27) + c(9) + d(3) + e = 3 \qquad \text{For } (3, 3)$$

Then we set up a matrix equation $[A][X] = [B]$:

$$\begin{bmatrix} 16 & -8 & 4 & -2 & 1 \\ 0 & 0 & 0 & 0 & 1 \\ 1 & 1 & 1 & 1 & 1 \\ 16 & 8 & 4 & 2 & 1 \\ 81 & 27 & 9 & 3 & 1 \end{bmatrix} \begin{bmatrix} a \\ b \\ c \\ d \\ e \end{bmatrix} = \begin{bmatrix} 1 \\ 2 \\ -1 \\ 4 \\ 3 \end{bmatrix}$$

There is a subtle difference between [A] in Example 2 and [A] in Example 3. In Example 2, the variables are s, m, and l and the coefficients are entered into [A]. In Example 3, we are solving for the coefficients a, b, c, d, and e. We evaluate the variables and enter the evaluations into [A]. Students may sense that there is something slightly different but may not be able to formulate an appropriate question.

$[A]$ is 5×5, while $[X]$ and $[B]$ are 5×1. The value of the determinant of the matrix of coefficients, det $[A]$, is 1440. Evaluating $[X] = [A]^{-1}[B]$ gives

$$[X] = \begin{bmatrix} -0.725 \\ 2.0167 \\ 3.025 \\ -7.3167 \\ 2 \end{bmatrix} = \begin{bmatrix} a \\ b \\ c \\ d \\ e \end{bmatrix}$$

The equation is

$$y = -0.725x^4 + 2.0167x^3 + 3.025x^2 - 7.3167x + 2$$

With a calculator, we place the x-coordinates in calculator list L1 and the y-coordinates in calculator list L2. Applying quartic regression to L1 and L2 yields the same equation.

Think about it: What are the parameters a, b, c, d, and e in fraction notation? (*Hint:* Obtain $[X] = [A]^{-1}[B]$ on the display, press ENTER to obtain the matrix containing the parameters, and then convert to fraction notation with MATH **1 : ▶ Frac.**)

This application reminds students about quadratic equations, in addition to providing practice in solving a system of linear equations.

ROADBED TRANSITION CURVES The number of unknowns in a system determines the number of equations needed to solve the system. When we have three unknowns in a system, we need three equations. A matrix solution may or may not be necessary.

In Example 4, we return to the *transition curve* parabolic model for a highway roadbed (Section 6.4). To find the equation of the parabolic transition curve, we need to find the three unknowns a, b, and c in $y = ax^2 + bx + c$. We therefore need three equations. One of the equations is $y = ax^2 + bx + c$. The other two equations come from relationships within the problem setting. Those relationships are based on the formula for the slope m of the parabolic roadbed at any point (x, y):

$$m = 2ax + b \qquad \text{The slope of a parabola at } (x, y)$$

The parameters a and b are the same as in $y = ax^2 + bx + c$, because this slope formula is from a calculus operation known as the derivative of $y = ax^2 + bx + c$.

EXAMPLE 4 Finding an equation: roadbed transition curve   A roadbed is to pass through point G on one hill and point H on the next hill. A side view of the required roadbed is shown in Figure 42. There is to be a 3% downgrade slope at G and a 2% upgrade slope at H. Suppose point G has coordinates $(0, 1000)$ with units in feet. Point H has an x-coordinate of 1500 feet because it is 1500 feet away from the y-axis.

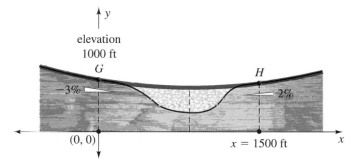

FIGURE 42

a. Use the slope formula and the x-coordinate at point G to write one equation.
b. Use the slope formula and the x-coordinate at point H to write a second equation.
c. Use $y = ax^2 + bx + c$ and the ordered pair at point G to write a third equation.
d. Use the three equations from parts a, b, and c to write the quadratic equation that meets the roadbed requirements. Show a matrix equation for the three equations.
e. Find the elevation (y-coordinate) at point H.

SOLUTION   In $y = ax^2 + bx + c$, we have three unknowns: a, b, and c. The facts give us information for three equations:

$$m = 2ax + b \qquad \text{The slope at } G$$

$$m = 2ax + b \qquad \text{The slope at } H$$

$$y = ax^2 + bx + c \qquad \text{The parabola through } (0, 1000)$$

a. The slope formula permits us to enter the x-coordinate and the slope at that point and solve for a or b. Point G at $(0, 1000)$ has $x = 0$ and a 3% downgrade slope; thus, $m = -\frac{3}{100}$.

$$m = 2ax + b \qquad \text{Substitute } m = -\frac{3}{100} \text{ and } x = 0.$$

$$-\frac{3}{100} = 2a(0) + b$$

$$b = -\frac{3}{100}$$

b. We use the slope formula at point H with $x = 1500$ and a 2% upgrade slope, $m = \frac{2}{100}$.

$$m = 2ax + b \qquad \text{Substitute } m = \frac{2}{100} \text{ and } x = 1500.$$

$$\frac{2}{100} = 2a(1500) + b$$

$$\frac{2}{100} = 3000a + b$$

c. Because the road passes through point G, we may obtain another equation by substituting the ordered pair at G, (0, 1000), into the parabola's equation:

$$y = ax^2 + bx + c \qquad \text{Substitute } x = 0 \text{ and } y = 1000.$$

$$1000 = a(0)^2 + b(0) + c$$

$$c = 1000$$

d. From part a, $b = -\frac{3}{100}$. Substituting this value into the equation from part b, we have

$$\frac{2}{100} = 3000a + b$$

$$\frac{2}{100} = 3000a - \frac{3}{100} \qquad \text{Add } \frac{3}{100} \text{ to both sides.}$$

$$\frac{5}{100} = 3000a \qquad \text{Divide by 3000.}$$

$$\frac{5}{300,000} = a \qquad \text{Simplify.}$$

$$a = \frac{1}{60,000}$$

We substitute a, b, and c into $y = ax^2 + bx + c$ to obtain the quadratic equation describing the roadbed:

$$y = \frac{1}{60,000} x^2 - \frac{3}{100} x + 1000$$

A matrix equation for the three equations is possible:

$$\begin{bmatrix} 0 & 1 & 0 \\ 3000 & 1 & 0 \\ 0 & 0 & 1 \end{bmatrix} \begin{bmatrix} a \\ b \\ c \end{bmatrix} = \begin{bmatrix} -\frac{3}{100} \\ \frac{2}{100} \\ 1000 \end{bmatrix}$$

e. To find the elevation at H, we place $x = 1500$ into the transition curve equation:

$$y = \frac{1}{60,000} x^2 - \frac{3}{100} x + 1000$$

$$= \frac{1}{60,000} (1500)^2 - \frac{3}{100} (1500) + 1000$$

$$= 992.5 \text{ ft} \qquad \blacksquare$$

The placement of the origin in Example 4 helped us find the coefficients a, b, and c. It also made the computation of a matrix solution unnecessary.

■ ORDERED PAIRS AND EQUATIONS

1. The location of the origin, (0, 0), in an application is arbitrary. Place the origin at any convenient, reasonable position. Because $f(0)$ is the y-intercept, a convenient origin makes it easier to find equations or solve for the constant term in an equation.

2. If a point lies on a graph, its ordered pair makes the equation true for that graph. Thus, the ordered pair for the point may be substituted into the equation to find an equation or to solve for an unknown.

ANSWER BOX

Warm-up: See Example 3. **Think about it:** $a = -\frac{29}{40}, b = \frac{121}{60},$ $c = \frac{121}{40}, d = -\frac{439}{60}, e = 2.$

10.5   Exercises

For Exercises 1 to 10, set up a system of matrices and solve with a calculator. If the system cannot be solved with a calculator, use substitution or elimination to find the reason.

1. $3x + 2y + 2z = -2$
$4x - 3y - z = -16$
$2x + 4y + z = 7$
$(-2, 3, -1)$

2. $3x + 5y - 2z = 10$
$x + 3y + 3z = -2$
$2x - 4y - 5z = 3$
$\left(-\frac{1}{2}, 1\frac{1}{2}, -2\right)$

3. $2x + 3y - z = 8$
$4x - 2y + 2z = -6$
$x + 4y - \frac{1}{2}z = 9$
$\left(\frac{1}{4}, 2, -1\frac{1}{2}\right)$

4. $3x + y + 2z = -1$
$2x - y - 2z = 11$
$x - 3y + z = 9$
$(2, -3, -2)$

5. $-x + 2y + 3z = 6$
$x - 2y - 3z = 4$
$2x + y - z = -2$
no solution, inconsistent equations

6. $x - 3y + z = 0$
$-3x + 2y + z = 2$
$2x + y - 2z = 2$
infinite number of solutions, dependent equations

7. $4x + 5y - 2z = 3$
$-3x - 4y + z = -3$
$x + y - z = 0$ infinite number of solutions, dependent equations

8. $2x - 3y + z = -5$
$-3x + 4y - 2z = -2$
$2x + 3y + z = 4$
no solution, inconsistent equations

9. $2x + 3y + 4z = 13$
$-x + 2y + 5z = 4$
$-x + y - 3z = -11$
$(4, -1, 2)$

10. $2x - y + 2z = 1$
$3x + 4y - z = -5$
$-x + 3y + z = 8$
$(-2, 1, 3)$

In Exercises 11 and 12, use matrices to find the second-degree polynomial whose graph passes through the points located by these ordered pairs. Check with quadratic regression.

11. $(0, 3), (2, 1), (1, 0)$ $y = 2x^2 - 5x + 3$

12. $(0, -2), (-1, 2), (2, 8)$ $y = 3x^2 - x - 2$

In Exercises 13 and 14, use matrices to find the third-degree polynomial whose graph passes through the points located by these ordered pairs. Check with cubic regression.

13. $(-2, -27), (-1, -7), (0, -1), (1, 3)$
$y = 2x^3 - x^2 + 3x - 1$

14. $(-1, 0), (0, 4), (1, 4), (2, 18)$
$y = 3x^3 - 2x^2 - x + 4$

In Exercises 15 and 16, use matrices to find the fourth-degree polynomial whose graph passes through the points located by these ordered pairs. Check with quartic regression.

15. $(-2, -7), (-1, -9), (0, -5), (1, -1), (2, 45)$
$y = 2x^4 + 3x^3 - 2x^2 + x - 5$

16. $(-2, -32), (-1, -3), (1, 1), (3, 13), (4, -38)$
$y = -x^4 + 3x^3 + 2x^2 - x - 2$

17. Lee is surprised that his plan in Example 2 reduced the number of small rabbits by only 50. He changes to an equal number of small and medium rabbits and keeps total rabbits at 620, with 20 large rabbits. Write the new matrix equation. See Answer Section.

18. To solve his problem in Example 2, Fenton consults with Janelle, who suggests making twice as many small rabbits as medium rabbits, 20 large rabbits, and a total of 620 rabbits. Write the new matrix equation. See Additional Answers.

Exercises 19 to 21 examine applications of matrix multiplication.

19. Janelle's accountant, Amelie, wants to find the total income from various combinations of products. Small rabbits sell for $6, medium for $12, and large for $30. The prices are shown in matrix A. Four production options are listed in matrix B; each column gives the numbers of small, medium, and large rabbits for one option. The product $[A][B]$ gives a matrix of total revenue for the various options.

$$[A][B] = [6 \quad 12 \quad 30] \begin{bmatrix} 500 & 450 & 400 & 300 \\ 100 & 150 & 200 & 300 \\ 20 & 20 & 20 & 20 \end{bmatrix}$$

a. Identify the shapes of matrices A and B, and predict the shape of the product $[A][B]$. $1 \times 3, 3 \times 4, 1 \times 4$

b. Enter the matrices into a calculator in $[A]$ and $[B]$, use the multiplication key to multiply, and store the result in matrix C. $4800, $5100, $5400, $6000

20. Oliko, Janelle's production assistant, wants to calculate the manufacturing time for each production option. She estimates 6 minutes to make a small rabbit and 4 minutes to make a medium or large rabbit.

a. Express the time required to make one of each rabbit as a fraction of an hour. $\frac{1}{10}$ hr, $\frac{1}{15}$ hr, $\frac{1}{15}$ hr

b. Change the elements in matrix A in Exercise 19 so that they represent the time in hours required to make each rabbit instead of the sale price of each rabbit. Enter the times as fractions (the calculator will change the entries to decimals). Find the total production time $[A][B]$. 58 hr, 56.3 hr, 54.7 hr, 51.3 hr

c. Find the cost of each production option, assuming that workers cost $9 per hour (minimum wage plus benefits). *Hint:* Replay the prior operation, multiply by 9, and store the result in matrix D.
$522, $507, $492, $462

21. Janelle's purchasing agent, Juan, reports that chocolate costs $3.50 per pound. The small rabbit contains 4.5 ounces of chocolate; the medium rabbit, 12 ounces; and the large rabbit, 3 pounds.

a. Write an expression for the weight, in pounds, of each rabbit. $\frac{4.5}{16}$ lb $= \frac{9}{32}$ lb, $\frac{3}{4}$ lb, 3 lb

b. Calculate the weight of the chocolate used for each production option. Round to the nearest tenth of a pound. 275.6 lb, 299.1 lb, 322.5 lb, 369.4 lb

c. Calculate the cost of chocolate used for each production option. Round to the nearest dollar. *Hint:* Replay the prior operation, multiply by $3.50 (the cost per pound), and store the result in matrix E.
$965, $1047, $1129, $1293

In Exercises 22 and 23, use addition or subtraction of matrices found as answers to prior exercises. *Hint:* $D = [522\ 507\ 492\ 462]$.

22. To obtain the total cost for each production option, add the costs for labor (from Exercise 20) and for chocolate (from Exercise 21): $[D] + [E]$. $1487, $1554, $1621, $1755

23. To obtain the profit for each production option, subtract the costs from the revenue, using the matrices from prior exercises: $[C] - [D] - [E]$. From this profit, Janelle must pay sales clerks, rent, utilities, taxes, and so forth.
$3313, $3546, $3779, $4245

In Exercises 24 to 27, the road grades connecting two hills are to be designed with a parabolic transition curve. The slope of the parabolic transition curve at any point is given by $m = 2ax + b$. The curve itself is $y = ax^2 + bx + c$.

24. In the figure, point K has a 4% downgrade slope and point L has a 3% upgrade slope. The horizontal distance between the y-axis and L is 2000 feet. The elevation of point K is 700 feet, and its coordinates are $(0, 700)$. Find the equation for the road grade. $y = \frac{7}{400,000}x^2 - \frac{1}{25}x + 700$

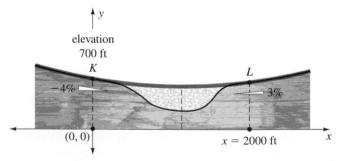

25. In the figure for Exercise 24, suppose that point K has a 3% downgrade slope and point L has a 5% upgrade slope. As before, the horizontal distance between the y-axis and L is 2000 feet. The elevation of point K is 700 feet, and its coordinates are $(0, 700)$. Find the equation for the road grade. $y = \frac{1}{50,000}x^2 - \frac{3}{100}x + 700$

26. Point M in the figure has a 5% downgrade slope, and point N has a 6% upgrade slope. The horizontal distance between the y-axis and N is 2400 feet. The elevation of point N is 1200 feet, and its coordinates are $(2400, 1200)$. Find the equation for the road grade.

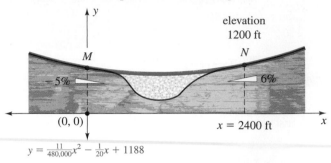

$y = \frac{11}{480,000}x^2 - \frac{1}{20}x + 1188$

27. In the figure for Exercise 26, suppose that point M has a 4% downgrade slope and point N has a 2% upgrade slope. As before, the horizontal distance between the y-axis and N is 2400 feet. The elevation of point N is 1200 feet, and its coordinates are $(2400, 1200)$. Find the equation for the road grade. $y = \frac{1}{80,000}x^2 - \frac{1}{25}x + 1224$

Projects

28. Computer (and Calculator) Testing During the 1970s, the calculation of the inverse, $[A]^{-1}$, of a so-called Hilbert matrix was used as a test of the inversion program. Test your calculator inversion package by calculating $[A]^{-1}$ for successively larger Hilbert matrices and multiplying $[A]\,[A]^{-1}$. The product $[A]\,[A]^{-1}$ must be an identity matrix with ones on the main diagonal (upper left to lower right) and zeros elsewhere. Summarize your results by listing the Hilbert matrix and its inverse for a 2×2, 3×3, 4×4, and 5×5 matrix. The $n \times n$ Hilbert matrix is

$$\begin{bmatrix} 1 & \frac{1}{2} & \frac{1}{3} & \frac{1}{4} & \cdots & \frac{1}{n} \\ \frac{1}{2} & \frac{1}{3} & \frac{1}{4} & \frac{1}{5} & \cdots & \frac{1}{n+1} \\ \frac{1}{3} & \frac{1}{4} & \frac{1}{5} & \frac{1}{6} & \cdots & \frac{1}{n+2} \\ \frac{1}{4} & \frac{1}{5} & \frac{1}{6} & \frac{1}{7} & \cdots & \frac{1}{n+3} \\ \vdots & \vdots & \vdots & \vdots & & \vdots \\ \frac{1}{n} & \frac{1}{n+1} & \frac{1}{n+2} & \frac{1}{n+3} & \cdots & \frac{1}{2n-1} \end{bmatrix}$$

It may come as a surprise to you that the computation of pi to millions of digits is also useful in checking the computational accuracy of a computer.

29. Roadbed Spreadsheet Design a spreadsheet to do Example 4 and the related exercises. When given slopes m_1 and m_2 at points $(0, y_1)$ and (x_2, y_2), respectively, the spreadsheet should calculate the coefficients a, b, and c for the transition curve $y = ax^2 + bx + c$ between the two hills. Design the spreadsheet to also calculate the height of the roadbed every 50 feet along the horizontal between the two reference points (between $x = 0$ and $x = x_2$).

30. **Circle Equations** We may use matrices to fit any three noncollinear points to a circle. A general equation of a circle, not necessarily centered at the origin, is

$$x^2 + y^2 + Dx + Ey + F = 0$$

In parts a to d, use the coordinates given to write three equations containing D, E, and F; solve the system; and state the resulting equation of the circle.

a. $(-12, 5), (12, 5), (5, -12)$
   $D = 0, E = 0, F = -169; x^2 + y^2 = 169$

b. $(6, -8), (6, 8), (-8, 6)$
   $D = 0, E = 0, F = -100; x^2 + y^2 = 100$

c. $(-4, -1), (-5, 6), (3, 6)$
   $D = 2, E = -6, F = -15; x^2 + y^2 + 2x - 6y - 15 = 0$

d. $(-1, -5), (6, 2), (-1, 3)$
   $D = -4, E = 2, F = -20; x^2 + y^2 - 4x + 2y - 20 = 0$

---

10 Chapter Summary

Vocabulary

axis	ellipse	matrix equation	radius
boundary line	half-plane	matrix inverse	rectangular hyperbola
circle	hyperbola	matrix of coefficients	shape of a matrix
conic sections	identity matrix	matrix of constants	square matrix
conical surface	inconsistent equations	matrix of variables	standard form of a linear
consistent equations	linear inequality in two	nonlinear system	equation
dependent equations	variables	ordered triples	
determinant	matrix	parabola	

Concepts

10.1 ▉ Conic Sections

To identify a conic section from its equation, arrange the equation in standard form.

For a circle, the x^2 and y^2 terms have the same sign and the same coefficients.

For an ellipse, the coefficients on the x^2 and y^2 terms are different but have the same sign.

For a hyperbola, the x^2 and y^2 terms have opposite signs.

A rectangular hyperbola has equation $y = k/x$ or $xy = k$. The x and y are multiplied, and no squared term is present.

The parabola has either x or y squared, but not both.

10.2 ▉ Solving Systems of Nonlinear Equations

Systems of nonlinear equations may be solved by substitution and by graphing.

10.3 ▉ Solving Systems of Inequalities

Systems of linear and nonlinear inequalities may be solved by graphing the inequalities separately and identifying the overlapping region.

The solution set to linear inequalities in two variables is a half-plane.

A test point is necessary to decide which half-plane is in the solution set.

10.4 ▉ Solving Systems of Two Linear Equations with Matrices

Matrices must have the same shape in order to be added or subtracted.

The product of a matrix and an identity matrix is the original matrix:

$$[A][I] = [A]$$

The product of a matrix and its inverse is an identity matrix:

$$[A][A]^{-1} = [I]$$

Matrices must be square and have a nonzero determinant in order to have an inverse.

Use the value of the determinant to verify that a unique matrix solution exists before actually solving the system. For a 2×2 matrix:

$$\text{If} \quad [A] = \begin{bmatrix} a & b \\ c & d \end{bmatrix}, \quad \text{then} \quad \det [A] = ad - bc.$$

To solve a system of linear equations on a calculator:

1. Change the system of equations into a matrix equation.

2. Specify the shape of matrix A, and enter matrix A into the calculator.

3. Specify the shape of matrix B, and enter matrix B into the calculator.

4. Calculate the value of the determinant of matrix A, det $[A]$. If det $[A] = 0$, stop; if det $[A] \neq 0$, go on to step 5.

5. Calculate $[A]^{-1}[B]$. The numbers are the values of the variables for the system of equations and appear in the same order as in the matrix of variables $[X]$.

10.5 ▪ Solving Systems of Three or More Linear Equations with Matrices

If a system of equations has exactly one solution, the system is consistent. The determinant of a matrix for a consistent system is nonzero.

If the determinant is zero, the system is either inconsistent or dependent. A system of equations is inconsistent if it has no solution. A system of equations is dependent if it has an infinite number of solutions.

To find what is causing a zero determinant value in a system of three or more equations, solve the system algebraically.

10 Review Exercises

Write equations for the conic sections described in Exercises 1 to 4.

1. A parabola with vertex at the origin, opening to the left, and passing through $(-3, 1)$ $x = -3y^2$

2. A circle with center at the origin and radius $= 16$
   $x^2 + y^2 = 16^2$ or $x^2 + y^2 = 256$

3. An ellipse centered at the origin with x-intercepts $(\pm 3, 0)$ and y-intercepts $(0, \pm 2)$ $\frac{x^2}{9} + \frac{y^2}{4} = 1$

4. A hyperbola centered at the origin with y-intercepts $(0, \pm 4)$, no x-intercepts, and $a = 3$ $\frac{y^2}{16} - \frac{x^2}{9} = 1$

In Exercises 5 to 14, name the conic sections (circles, ellipses, hyperbolas, and parabolas). Solve the equation for the indicated value. Graph.

5. $y = x^2 + 5$ for x^2 parabola; $x^2 = y - 5$; see Answer Section.

6. $x = 3y^2 - 4$ for y^2 parabola; $y^2 = \frac{1}{3}x + \frac{4}{3}$; see Additional Answers.

7. $-3 = -2y^2 + x^2$ for y^2 hyperbola; $y^2 = \frac{1}{2}x^2 + \frac{3}{2}$; see Answer Section.

8. $5 = -x^2 - y^2$ for x^2 cannot be circle, negative r^2; $x^2 = -y^2 - 5$; no graph

9. $x^2 + y^2 = 8$ for y circle; $y = \pm\sqrt{8 - x^2}$; see Answer Section.

10. $x^2 - y^2 = 4$ for y hyperbola; $y = \pm\sqrt{x^2 - 4}$; see Additional Answers.

11. $x^2 - y^2 = 10$ for x hyperbola; $x = \pm\sqrt{10 + y^2}$; see Answer Section.

12. $x^2 + y^2 = 5$ for x circle; $x = \pm\sqrt{5 - y^2}$; see Additional Answers.

13. $x^2 = 4 - 2y^2$ for y ellipse; $y = \pm\sqrt{2 - x^2/2}$; see Answer Section.

14. $2x^2 + y^2 = 9$ for y ellipse; $y = \pm\sqrt{9 - 2x^2}$; see Additional Answers.

In Exercises 15 to 24, name the straight lines, parabolas, hyperbolas, circles, and ellipses. Solve the system of equations.

15. $y = x^2 + x - 2$
    $y = -x - 3$
    parabola, line; $(-1, -2)$

16. $y = -x^2 - x - 2$
    $y = x + 2$
    parabola, line; { } or $\varnothing$

17. $y = x^2 - 4x - 1$
    $y = -x + 3$
    parabola, line; $(4, -1), (-1, 4)$

18. $y = x^2 + 2x - 5$
    $y = -x - 1$
    parabola, line; $(-4, 3), (1, -2)$

19. $x^2 + y^2 = 16$
    $y = x^2 + 1$
    circle, parabola; $(\pm 1.629, 3.653)$

20. $y = \frac{1}{x}$
    $x^2 + y^2 = 2$
    hyperbola, circle; $(1, 1), (-1, -1)$

21. $y = x^2 - x$
    $y = -x^2 + 3$
    parabolas; $(\frac{3}{2}, \frac{3}{4}), (-1, 2)$

22. $y = -x^2 + 1$
    $y = x^2 - 5$
    parabolas; $(\pm 1.732, -2)$ or $(\pm\sqrt{3}, -2)$

23. $y = x^2 + 1$
    $y = x^2 - 5$
    parabolas; { } or $\varnothing$

24. $y = x^2 - 1$
    $\frac{x^2}{4} + y^2 = 1$
    parabola, ellipse; $(\pm 1.323, 0.75)$, $(0, -1)$

Solve the systems of equations in Exercises 25 to 28.

25. $y = 2^x$
    $y = 4 - 2^x$
    $x = 1, y = 2$

26. $y = 3^x$
    $y = 7 - 3^{x+1}$
    $x \approx 0.509, y = 1.75$

27. $y = \log x$
    $y = 3 - 2 \log x$
    $x = 10, y = 1$

28. $y = 5 - \log x$
    $y = \log x$
    $x = 10^{5/2}, y = 2.5$

In Exercises 29 to 34, draw the indicated line, and shade the half-plane described.

29. $x = 3, x \geq 3$
    See Answer Section.

30. $y = -2, y \geq -2$
    See Additional Answers.

31. $y = 1, y \geq 1$
    See Answer Section.

32. $x = -2, x \leq -2$
    See Additional Answers.

33. $y = x, y \leq x$
    See Answer Section.

34. $y = -x, y \geq -x$
    See Additional Answers.

In Exercises 35 to 42, graph the inequalities.

35. $x + y > 3$
    See Answer Section.

36. $x - y < 2$
    See Additional Answers.

37. $x - y \leq 4$
    See Answer Section.

38. $x + y \geq -2$
    See Additional Answers.

39. $2x + 3y < 6$
    See Answer Section.

40. $2x - 3y > 6$
    See Additional Answers.

41. $3x - 2y \geq 12$
    See Answer Section.

42. $3x + 2y \leq 12$
    See Additional Answers.

Use a graph to solve the inequalities or systems of inequalities given in Exercises 43 to 48.

43. $x^2 + y^2 \geq 4$
    See Answer Section.

44. $\dfrac{x^2}{4} \geq y$
    See Additional Answers.

45. $y \geq x$
    $\dfrac{x^2}{4} + \dfrac{y^2}{9} \leq 1$ See Answer Section.

46. $y \leq -\dfrac{1}{x}$
    $x \geq y$ See Additional Answers.

47. $x^2 + y^2 \leq 9$
    $y \geq 2x^2$
    See Answer Section.

48. $y \leq x^2 + 2x + 1$
    $y \geq x + 2$ See Additional Answers.

49. Match each of the descriptions in parts a to d with one of Graphs 1 to 4.

Graph 1:

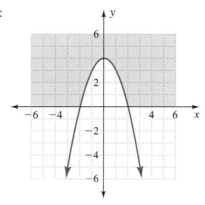

Graph 2:

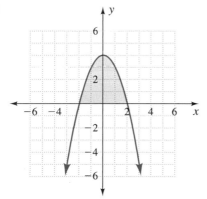

Graph 3:

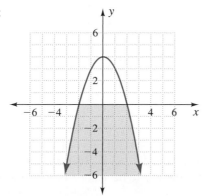

Graph 4:

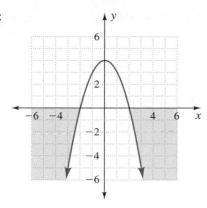

a. $y < 0, y < -x^2 + 4$
    Graph 3

b. $y > 0, y > -x^2 + 4$
    Graph 1

c. $y < 0, y > -x^2 + 4$
    Graph 4

d. $y > 0, y < -x^2 + 4$
    Graph 2

50. Write a set of equations whose graphs are two parallel lines, set up matrix A, and evaluate det $[A]$.
    Answers will vary; det $[A] = 0$

Use these matrices to evaluate the expressions in Exercises 51 to 56:

$$[A] = \begin{bmatrix} 2 & 4 \\ 1 & 3 \end{bmatrix}, \quad [B] = \begin{bmatrix} 1 & 2 \\ 4 & 3 \end{bmatrix}, \quad [C] = \begin{bmatrix} 1.5 & -2 \\ -0.5 & 1 \end{bmatrix}, \quad [I] = \begin{bmatrix} 1 & 0 \\ 0 & 1 \end{bmatrix}$$

51. $[A] + [B]$ $\begin{bmatrix} 3 & 6 \\ 5 & 6 \end{bmatrix}$

52. det $[A]$ $\quad 2$

53. $[A][B]$ $\begin{bmatrix} 18 & 16 \\ 13 & 11 \end{bmatrix}$

54. $[A][C]$ $\begin{bmatrix} 1 & 0 \\ 0 & 1 \end{bmatrix}$

55. $[C][A]$ $\begin{bmatrix} 1 & 0 \\ 0 & 1 \end{bmatrix}$

56. $[A][I]$ $\begin{bmatrix} 2 & 4 \\ 1 & 3 \end{bmatrix}$

Solve the systems in Exercises 57 to 66 with matrices. If no solution is possible, use algebra to show the reason.

57. $3y = 2x + 12$ $\quad (6, 8)$
    $2x - y = 4$

58. $y = 3x/2 - 3$ $\quad (2, 0)$
    $3x - y = 6$

59. $2x = y + 13$ $\quad (4, -5)$
    $2y + x + 6 = 0$

60. $3x + 4y + 4 = 0$ $\quad (-2, 0.5)$
    $5x + 11 = 2y$

61. $2x + y = 5$ $\quad$ infinite number of solutions, coincident lines
    $4x - 10 = -2y$

62. $3y/2 = x + 3$ $\quad$ no solution, parallel lines
    $y - 4 = 2x/3$

63. $2x - 3y - 4z = -4$
    $3x - 2y + z = -17$
    $4x - y - 3z = -3$
    $(-2, 4, -3)$

64. $3x - y - 4z = 9$

$2x - 3y + z = 19$

$-2x + y + 3z = -7$ (3, −4, 1)

65. $4x + 2y - 3z = 5$

$-3x - 3y + 2z = 7$

$x - y - z = -2$ no solution, inconsistent equations

66. $5x - 3y + z = 12$

$2x - y + 2z = 7$

$3x - 2y - z = 5$ infinite number of solutions, dependent equations

In Exercises 67 and 68, $y = ax^2 + bx + c$ is the equation of a transition curve between two hills. The slope of the curve at any point (x, y) is $m = 2ax + b$.

67. At (2800, 300), there is to be a slope of 4%. The slope is to be −3% at a horizontal distance of 2800 feet to the left. What are the coefficients a, b, and c of the transition curve? $a = \frac{1}{80,000}, b = -\frac{3}{100}, c = 286$

68. At (3000, 500), there is to be a slope of 2%. The slope is to be −6% at a horizontal distance of 3000 feet to the left. What are the coefficients a, b, and c of the transition curve? $a = \frac{1}{75,000}, b = -\frac{3}{50}, c = 560$

69. Calculator Quirks As the solution to the system

$3x - y = 4$ and $-6x + 2y = -8$

one older calculator gives (10, 10), a false solution, while another model gives (−2, −10), one of an infinite

number of solutions. Set up the matrix equation, solve, and describe how your calculator reacts to the problem. Give three ordered pairs that do satisfy the system.

det [A] = 0; graphs coincident; possible points: (0, −4), (1, −1), (2, 2)

In Exercises 70 and 71, use matrices to find the second-degree polynomial whose graph passes through the points located by these ordered pairs. Check with quadratic regression.

70. $(-1, -10), (0, -5), (1, -4)$ $y = -2x^2 + 3x - 5$

71. $(0, -2), (1, -1), (2, -8)$ $y = -4x^2 + 5x - 2$

In Exercises 72 and 73, use matrices to find the third-degree polynomial whose graph passes through the points located by these ordered pairs. Check with cubic regression.

72. $(-2, 0), (-1, -5), (1, -3), (2, -8)$ $y = -x^3 + 2x - 4$

73. $(-1, 0), (1, 2), (2, 0), (3, 4)$ $y = x^3 - 3x^2 + 4$

In Exercises 74 and 75, use matrices to find the fourth-degree polynomial whose graph passes through the points located by these ordered pairs. Check with quartic regression.

74. $(-3, 50), (-1, -6), (0, -1), (1, 2), (2, 15)$ $y = x^4 - 2x^2 + 4x - 1$

75. $(-2, 8), (-1, 1), (0, 2), (2, 4), (3, 53)$ $y = x^4 - 3x^2 - x + 2$

76. A unique linear equation fits through __2__ points. A unique quadratic equation fits through __3__ points. A unique nth-degree polynomial fits through a set of _$n+1$_ points.

---

10   Chapter Test

1. What is the equation of a circle with center at the origin and radius 16? $x^2 + y^2 = 256$

2. What is the equation of an ellipse centered at the origin with x-intercepts (±1, 0) and y-intercepts (0, ±4)? $x^2/1 + y^2/16 = 1$

3. What is the equation of a parabola with vertex at the origin, opening to the left, and passing through (−4, 1)? $x = -4y^2$

4. What is the equation of a hyperbola centered at the origin with x-intercepts (±3, 0), no y-intercepts, and $b = 2$? $x^2/9 - y^2/4 = 1$

Solve the systems of equations in Exercises 5 to 10. In Exercises 5 to 8, also identify the straight lines, parabolas, hyperbolas, circles, and ellipses.

5. $y = 3x^2 - 4x - 5$

$y = x - 3$

parabola, line; $(-\frac{1}{3}, -\frac{10}{3})$, (2, −1)

6. $y = x^2 - 3x + 2$

$y = \frac{1}{2}x - 2$

parabola, line; { }

7. $x^2 + y^2 = 4$

$-x^2 + y^2 = 4$

circle, hyperbola; (0, ±2)

8. $y = x^2 + 1$

$x^2 + y^2 = 4$

parabola, circle; (±0.890, 1.791)

9. $y = 3^x$

$y = 5 - 3^x$

(0.834, 2.5)

10. $y = \log_3 x$

$y = 4 - \log_3 x$

(9, 2)

In Exercises 11 to 14, graph the inequalities.

11. $x - y < 3$

See Answer Section.

12. $2x + y \geq 4$

See Answer Section.

13. $2x^2 + y^2 > 9$

See Answer Section.

14. $y^2 < x$

See Answer Section.

In Exercises 15 and 16, solve the systems of inequalities with a graph.

15. $x + 2y < 1$

$y - 2x \geq 2$ See Answer Section.

16. $x^2 + y^2 \geq 16$

$6x \leq -y^2$ See Answer Section.

17. The equation $y = ax^2 + bx + c$ is the equation of a transition curve between two hills. The slope of the curve at any point (x, y) is given by $m = 2ax + b$. At (0, 1500), there is to be a slope of −3%. The slope is to be 6% at $x = 3200$ ft. What are the coefficients a, b, and c of the transition curve? $a = \frac{9}{640,000}, b = -\frac{3}{100}, c = 1500$

18. If $[A] = \begin{bmatrix} 3 & -5 \\ -1 & 2 \end{bmatrix}$ and $[B] = \begin{bmatrix} 2 & 5 \\ 1 & 3 \end{bmatrix}$, find

   a. $[A][B]$ $\begin{bmatrix} 1 & 0 \\ 0 & 1 \end{bmatrix}$

   b. $[A]^{-1}$ $\begin{bmatrix} 2 & 5 \\ 1 & 3 \end{bmatrix}$

   c. det $[B]$ 1

19. Set up and solve a matrix equation:

   $2x = 6 + 3y$
   $4x + 5y + 32 = 0$

   $\begin{bmatrix} 2 & -3 \\ 4 & 5 \end{bmatrix} \begin{bmatrix} x \\ y \end{bmatrix} = \begin{bmatrix} 6 \\ -32 \end{bmatrix}; \begin{bmatrix} x \\ y \end{bmatrix} = \begin{bmatrix} -3 \\ -4 \end{bmatrix}$

Solve the systems in Exercises 20 and 21 with matrices. If no solution is possible, use algebra to show the reason.

20. $3x - 2y + z = 4$
   $2x - 3y + 4z = 0$
   $x - z = 2$

   det $[A] = 0$, inconsistent equations

21. $x - 2y + z = 8$
   $2x - 3y = 11$
   $x - z = 2$ $(4, -1, 2)$

Final Exam Review

PART I: SHORT ANSWERS

In exercises where words are enclosed in brackets, [ ], choose the correct word to complete the sentence.

Chapter 1

1. Evaluate $-x^2$ for $x = -2$. -4

2. Simplify $5 - 3(2 - x)$. $3x - 1$

3. Simplify $\dfrac{xy^2}{x^2 y}$. $\dfrac{y}{x}$

4. Evaluate $\sqrt{(x + 2)^2 - 9}$ for $x = -2$. no real-number solution

5. Solve for x: $x - 6 + 3(2 - x) = 8$. $x = -4$

6. Solve for x: $2x + 3 = -4x$. $x = -\frac{1}{2}$

7. The -4 in $-4xyz$ is the [coordinate, coefficient] of xyz.

Chapter 2

8. Sketch a line graph for $x \le -4$ or $x > -1$. See Additional Answers.

9. If $f(x) = 3x + 4$, what is $f(-2)$? -2

10. A line parallel to the x-axis may be described by [a constant, an identity] function.

11. The graphs of $y = 2x - 3$ and $y = -2x + 3$ are [parallel, perpendicular, intersecting] lines.

12. Which equation in Exercise 11 is a decreasing function? $y = -2x + 3$

13. What is the equation of a line perpendicular to $y = 1.5x$ passing through the origin? $y = -\frac{2}{3}x$

14. The range of the absolute value function is _____. $f(x) \ge 0$

15. What is the slope of the line passing through (x_1, y_1) and (x_2, y_2)? $m = (y_2 - y_1)/(x_2 - x_1)$

16. What is the equation of a line passing through the points $(-40, -40)$, $(0, 32)$, and $(100, 212)$? Show steps, and check with linear regression. $y = \frac{9}{5}x + 32$

Chapter 3

Solve the systems of equations in Exercises 17 and 18.

17. $y = 79 - x$
$2x - 3y = 28$
$x = 53, y = 26$

18. $5x + 3y = 17.8$
$4x - 2y = 5$
$x = 2.3, y = 2.1$

19. In solving a system of two equations whose graphs are parallel, we may expect a result that is [false, such as $0 = 4$; true, such as $0 = 0$].

20. A system of two equations that has an infinite number of solutions is represented by graphs that are [parallel lines, perpendicular lines, coincident lines].

21. In the first Little League World Series, in 1947, Williamsport, PA beat Lock Haven, PA by 9 points. The sum of their scores was 23 points. Write and solve a system of equations that gives the scores.
$w = $ Wmpt's score, $l = $ LH's score; $w - l = 9$, $w + l = 23$; $w = 16, l = 7$

22. Solve and sketch a line graph of the solution:
$3 - x \le 8$. $x \ge -5$; see Additional Answers.

23. Solve with algebraic notation and with a graph:
$x + 7 \le 4x - 8$. $x \ge 5$; see Answer Section.

24. Solve $|5 - 3x| < 2$. $1 < x < \frac{7}{3}$

Chapter 4

25. What name is given to the graph of a quadratic function $y = ax^2 + bx + c$? parabola

26. If $f(x) = 2x^2 + x$, what is $f(-2)$? 6

27. If $f(x) = 2x^2$, what is $f(a + b)$? $2a^2 + 4ab + 2b^2$

28. Factor $4a^2 - b^2$. $(2a - b)(2a + b)$

29. Factor $6x^2 - 5x - 4$. $(2x + 1)(3x - 4)$

30. Factor $9x^2 - 12x + 4$. $(3x - 2)^2$

31. Factor $8x^3 - 27$. $(2x - 3)(4x^2 + 6x + 9)$

32. Solve by factoring: $3x^2 - x - 10 = 0$. $x = -\frac{5}{3}$ or $x = 2$

33. Solve by factoring: $4x^2/3 = 15x + 12$. $x = -\frac{3}{4}$ or $x = 12$

Chapter 5

34. Complete the statement: The solutions of a quadratic equation $ax^2 + bx + c = 0$ are _____. $x = (-b \pm \sqrt{b^2 - 4ac})/2a$

35. State the Pythagorean theorem formula. $a^2 + b^2 = c^2$

36. Solve $ax^2 + b = 0$ for x. $x = \pm\sqrt{-b/a}$, $a > 0$ and $b \le 0$ or $a < 0$ and $b \ge 0$

37. $(x + \underline{\frac{5}{2}})^2 = x^2 + 5x + \underline{\frac{25}{4}}$.

38. Simplify $\sqrt{0.36x^2}$ for any real number x. $0.6|x|$

39. Give an example of an irrational number. Possible answers: π, e, $\sqrt{2}$, $\sqrt{3}$

40. What is the vertex of $y = 2x^2 - 8x - 5$? $(2, -13)$

In Exercises 41 and 42, solve the equation by at least three different methods.

41. $0 = 5x^2 - x - 4$ $\{-0.8, 1\}$

42. $0 = 5x^2 - 8x - 4$ $\{-0.4, 2\}$

43. For what x is $5x^2 + 8x - 4 \le 0$? $-2 \le x \le \frac{2}{5}$

Chapter 6

In Exercises 44 to 46, simplify the expressions.

44. $(3 + 2i)^2$ $5 + 12i$

45. $(3 + 2i)(3 - 2i)$ 13

46. $(2 - i)(3 + i)$ $7 - i$

47. What is the conjugate of $-3 - 9i$? $-3 + 9i$

48. Simplify $\sqrt{-16}$. $4i$

49. The vertex tells us the __minimum__ or __maximum__ value of a quadratic function.

50. What are the solutions to $x^2 + 2x + 5 = 0$? $-1 \pm 2i$

51. The expression $b^2 - 4ac$ is called the [determinant, __discriminant__, discriminate].

52. What is the maximum number of possible solutions to the equation $ax^4 + bx^3 + cx^2 + dx + e = 0$? 4

Chapter 7

53. What is the additive inverse, or opposite, of $x - 2$?
$2 - x$ or $-x + 2$

54. $x^3 - x$ is the least common denominator for fractions with denominators of $x^2 + x$, x, and what other expression? $x - 1$; other answers are possible

55. $y = k/x$ is an equation for [direct, __inverse__] variation.

56. Solve for the missing sides of the similar triangles in the figure. $x \approx 11.9$, $y \approx 20.5$

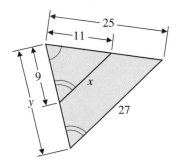

57. Without graphing, finding its equation, or using linear regression, how can we use two ordered pairs to determine whether a linear function passes through the origin?   Ratios of the coordinates, y to x, will be equal.

58. The intensity of illumination from a light source is given by $I = P/r^2$, where P is in watts and r is distance from the light source in meters. Solve for r. $r = \sqrt{P/I}$

In Exercises 59 to 67, simplify or perform the indicated operations.

59. $\dfrac{4x^2}{28x}$ $\dfrac{x}{7}$

60. $12x\left(\dfrac{1}{12} + \dfrac{2}{3x}\right)$ $x + 8$

61. $\dfrac{5x^2 + 15x}{6x + 18}$ $\dfrac{5x}{6}$

62. $\dfrac{a^2 + 2a}{2a - 6} \cdot \dfrac{a^2 + a - 12}{a + 2}$ $\dfrac{a(a + 4)}{2}$

63. $\dfrac{x + 1}{x - 3} + \dfrac{x}{x^2 - 9}$ $\dfrac{x^2 + 5x + 3}{(x - 3)(x + 3)}$

64. $\dfrac{x - 2}{x^2 - 3x + 2} - \dfrac{2x}{x - 2}$ $\dfrac{-2x^2 - 3x + 2}{(x - 2)(x - 1)}$

65. $\dfrac{x^2 - 8x + 12}{x^2 - 1} \div \dfrac{2x - x^2}{x^2 - 5x - 6}$ $\dfrac{-(x - 6)^2}{x(x - 1)}$

66. Divide $x^3 + 2x^2 - 3x + 4$ by $x + 1$. $x^2 + x - 4 + \dfrac{8}{x + 1}$

67. Divide $x^3 - 2x^2 + 3x + 4$ by $x - 1$. $x^2 - x + 2 + \dfrac{6}{x - 1}$

In Exercises 68 to 73, indicate for what values the expressions in each equation are defined, and solve the equation.

68. $\dfrac{x - 5}{10} = \dfrac{x + 3}{12}$ $\mathbb{R}$; $x = 45$

69. $\dfrac{x + 1}{8} = \dfrac{2}{x + 1}$ $\mathbb{R}$; $x \neq -1$; $\{-5, 3\}$

70. $\dfrac{x + 3}{x - 2} = \dfrac{2}{1}$ $\mathbb{R}$; $x \neq 2$; $x = 7$

71. $\dfrac{x}{2} = \dfrac{3}{5 - x}$ $\mathbb{R}$; $x \neq 5$; $\{2, 3\}$

72. $\dfrac{2}{x - 4} - \dfrac{1}{x} = \dfrac{x + 4}{3x}$ $\mathbb{R}$; $x \neq 0$, $x \neq 4$; $\{-4, 7\}$

73. $\dfrac{x}{3} = \dfrac{3 - x}{x - 3}$ $\mathbb{R}$; $x \neq 3$; $x = -3$ (discard $x = 3$)

Chapter 8

74. The principal square root of a number is always __positive or greater than zero__.

75. Simplify $\sqrt[3]{8x^6y^{12}}$. $2x^2y^4$

76. Change 34×10^{-2} to correct scientific notation. 3.4×10^{-1}

77. Write 0.000 000 567 in correct scientific notation. 5.67×10^{-7}

78. Simplify $64^{1/3}$.   4

79. What is the distance between (x_1, y_1) and (x_2, y_2)?
$d = \sqrt{(x_2 - x_1)^2 + (y_2 - y_1)^2}$

80. We use a __-1__ exponent to indicate the reciprocal of a number.

81. If $x = 3^y$ and $x = 1$, what is y? 0

Simplify the expressions in Exercises 82 to 84.

82. $\sqrt{(3x + 2)^2}$ $|3x + 2|$

83. $(3x + 2)^2$ $9x^2 + 12x + 4$

84. $(\sqrt{3} + 2)^2$ $7 + 4\sqrt{3}$

85. For $x > 0$ and $y > 0$, change these expressions to radical form:

a. $x^{1.5}$ $(\sqrt{x})^3$ or $\sqrt{x^3}$ **b.** $x^{n/m}$ $(\sqrt[m]{x})^n$ or $\sqrt[m]{x^n}$ **c.** $x^{1/2}(y^2)^{1/4}$ $\sqrt{x}\sqrt{y} = \sqrt{xy}$

86. Change these expressions to exponent form:

a. $\sqrt[3]{x^2}$ $x^{2/3}$ **b.** $\sqrt[4]{a^3b^4}$, $a > 0$, $b > 0$ $a^{3/4}b$

87. Simplify and leave in radical form:

a. $\sqrt[3]{a^3b^4c^5}$ $abc\sqrt[3]{bc^2}$ **b.** $\sqrt[4]{w^4x^5y^8z^{12}}$, $x > 0$ $|wz^3|xy^2\sqrt[4]{x}$

Chapter 9

88. Describe how to distinguish among arithmetic sequences (linear functions), quadratic sequences, and geometric sequences (exponential functions).   same first differences, same second differences, constant ratio of consecutive terms

89. Complete this statement of the like bases property: If $a^x = a^y$ and $a > 0$, $a \neq 1$, then __$x = y$__.

Solve the equations in Exercises 90 to 96 for *n* or for *x*.

90. $2^x = 2$ $x = 1$

91. $16^x = 8$ $x = \frac{3}{4}$

92. $\left(\frac{1}{2}\right)^x = 8$ $x = -3$

93. $0.01^x = 100$ $x = -1$

94. $25^n = 0.04$ $n = -1$

95. $9^{x+1} = \frac{1}{27}$ $x = -\frac{5}{2}$

96. $4^{n-2} = 2$ $n = \frac{5}{2}$

97. Write $y = \log x$ in exponential form. $10^y = x$

98. What is the growth rate in $y = 4(1.082)^x$? 8.2%

99. Write $\log_b a = c$ in exponential form. $b^c = a$

100. In geometric sequences, consecutive terms have a common _____ratio_____.

101. Apply the change of base formula to $y = \log_b a$ using log base 10. $y = (\log_{10} a)/(\log_{10} b)$

102. Write $y = 5^{x-2}$ as a logarithmic equation. $\log_5 y = x - 2$

103. Name the base in $2^6 = 64$. 2

104. Name the base in $\log_4 1024 = 5$. 4

105. When we write a logarithm without a base, we mean log base ____10____.

106. The natural logarithm has base ____e____.

107. Calculating $y = $ antilog 2.3 means raising ____10____ to the power 2.3.

In Exercises 108 to 123, solve for *x*. Round decimals to the nearest thousandth.

108. $3^x = 19$ $x = 2.680$

109. $2^{x+1} = 19$ $x = 3.248$

110. $17^x = 10$ $x = 0.813$

111. $10^x = 17$ $x = 1.230$

112. $\log_8 1 = x$ $x = 0$

113. $\log_2 x = 1$ $x = 2$

114. $\log_4 x = 0$ $x = 1$

115. $\log_{10} 2 = x$ $x = 0.301$

116. $\log_{10} x = -2$ $x = 0.01$

117. $\log_{10} 1000 = x$ $x = 3$

118. $\log_{10} 0.001 = x$ $x = -3$

119. $\log_9 27 = x$ $x = 1.5$

120. $\log_8 \sqrt[3]{2} = x$ $x = \frac{1}{9}$

121. $\ln x = -1$ $x = 0.368$

122. $\ln x = 2$ $x = 7.389$

123. $\ln x = e$ $x = 15.154$

In Exercises 124 and 125, find the time needed for $1000 to grow to $2500 under the given conditions.

124. Interest is compounded semiannually at 9.4%. $\approx$10 yr

125. Interest is continuously compounded at 8%. $\approx$11.5 yr

▄ Chapter 10

In Exercises 126 and 127, identify the conic section represented by each equation and then solve the system of nonlinear equations. Write the solutions as coordinate points.

126. $x^2 + y^2 = 11$
$x^2 - y^2 = 7$
circle, hyperbola; $(\pm 3, \pm\sqrt{2})$

127. $2x^2 + y^2 = 31$
$y = x^2 + 2$
ellipse, parabola; $(\pm\sqrt{3}, 5)$
(Note: discard $y = -7$)

128. Sketch the solution set for this system of inequalities:
$x^2 + y^2 \le 4$ See Additional Answers.
$x \ge \frac{1}{4} y^2$

129. Set up a matrix equation for the following system and solve with a calculator:
$2x + 3y - z = 10.7$ $x = 3.5, y = 2.6, z = 4.1$
$3x - 4y + z = 4.2$
$4x - y + 2z = 19.6$

▄ Chapter 11

130. How many different outcomes are possible in choosing

a. a consonant or a vowel from the letters in the word EXPONENTIAL? 9

b. a consonant and a vowel from the letters in the word EXPONENTIAL? 20

131. If a letter is chosen randomly from the word EXPONENTIAL, what is the probability of choosing

a. an E? $\frac{2}{11}$

b. a vowel or an N? $\frac{7}{11}$

c. an N followed by another N without replacement? $\frac{2}{11} \cdot \frac{1}{10} = \frac{1}{55}$

d. an N followed by another N with replacement? $\frac{2}{11} \cdot \frac{2}{11} = \frac{4}{121}$

132. Evaluate the following:

a. $6!$ 720 **b.** $_6P_4$ 360 **c.** $_8C_5$ 56

▄ PART 2: EXTENDED PROBLEMS

▄ Chapter 1

1. a. A league bowling handicap *h* is calculated in terms of the bowler's average *a*, where $h = 0.8(200 - a)$. Bowling scores (and hence averages) vary from 0 to 300. People with averages over 200 are considered excellent bowlers. A handicap is never negative. What is the handicap for an average of 100? 150? 200? 250? 80; 40; 0; −40 (not possible)

b. Write and graph an equation for the scores if the average and the handicap are added. Then write and graph an equation for the average scores without the handicap. $y = a + 0.8(200 - a)$ for $a < 200$; $y = a$; see Answer Section.

▄ Chapter 2

2. Any person in Sweden finding an ancient gold "treasure" must sell the object to the Swedish government at the current price of gold by weight plus 10%.

a. Give the equation for the finder's sale price.
$r = $ payment received, $p = $ price/oz, $w = $ weight in oz; $r = pw + 0.10pw$.

b. In building the equation, does it matter whether we add the 10% to the weight or to the price per ounce?
no; $r = p(w + 0.10w)$ and $r = w(p + 0.10p)$ give same payment.

c. In July 1995, a Swede announced he had found a gold necklace weighing about 17 ounces. The necklace was estimated to be 2000 years old. At the time, gold was selling for about $384 per ounce. What should he receive from the sale? $7180.80

d. The newspaper article estimated that the man would get about $7450. Does this match the calculations in part *c*? no

▪ Chapter 3

3. A student is attempting to solve the system of equations

$$2x - 3y = 18$$
$$5x + 2y = 26$$

The student correctly solves the first equation for *x*. The student then substitutes for *x* in the first equation and from the results concludes that the two original equations have graphs that are coincident lines. Work through each step of the student's solution. Explain why the student thought the lines were coincident, and then solve the system correctly. Substitution created system in which first line was coincident with itself; $x = 6$; $y = -2$.

4. Jesse Owens won an Olympic gold medal by completing a 100-meter run in 10.3 seconds. Florence Griffith Joyner set a world record of 10.49 seconds in the 100-meter run 52 years later. Joyner's time would have won the 1956 men's 100-meter run. The sum of the years for Owens's and Joyner's races is 3924. Write and solve a system of equations that gives the year for each race. f = year of Florence's race, j = year of Jesse's race, $f - j = 52, f + j = 3924; f = 1988, j = 1936$

5. Ward bought 11 grocery items, with a total weight of 13 pounds 15 ounces and a total cost of $13.58. A can of tomatoes weighs 17 ounces, a package of pizza sauce weighs 16 ounces, and a jar of salsa weighs 2 pounds 4.5 ounces. A can of tomatoes costs $0.50, a package of pizza sauce costs $1.62, and a jar of salsa costs $2.86. Set up and solve the system of equations needed to find how many of each product Ward bought. (*Note:* 16 ounces = 1 pound.) 6 tomatoes, 3 pizza sauce, 2 salsa

▪ Chapters 4 to 6

$\mathbb{R}$; $y \leq -4.75$; $x = 0.25$; $(0.25, -4.75)$; highest point determines limit to range; x-coordinate of vertex determines equation of symmetry.

6. For the equation $y = -4x^2 + 2x - 5$, give the domain, range, axis of symmetry, and vertex. Explain why some of the answers are related.

7. Give two different quadratic functions whose graphs pass through the point $(-2, 0)$ and $(3, 0)$. Replace *a* with any real number: $a(x - 3)(x + 2) = y$

8. Find a quadratic function describing a graph that passes through $(-1, -6)$, $(1, -2)$, and $(5, 30)$. $y = x^2 + 2x - 5$

In Exercises 9 to 11, find the original quadratic equation and give the solutions, real or complex. State what the solutions indicate about the graph of the quadratic function related to the original equation.

9. $t = \dfrac{-6 \pm \sqrt{36 + 64}}{32}$ $16t^2 + 6t - 1 = 0$; $\{-0.5, 0.125\}$; two horizontal axis intercepts

10. $x = \dfrac{-2 \pm \sqrt{4 + 60}}{2}$ $x^2 + 2x - 15 = 0$; $\{-5, 3\}$; two horizontal axis intercepts

11. $x = \dfrac{8 \pm \sqrt{64 - 80}}{10}$ $5x^2 - 8x + 4 = 0$; $\frac{4}{5} \pm \frac{2}{5}i$; no horizontal axis intercepts

12. Use the graph of $f(x) = x^2 - 2x - 3$ in the figure to answer the following questions.

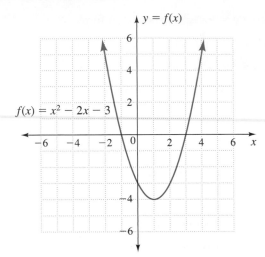

a. For what inputs *x*, if any, is $f(x)$ less than zero? Use both an inequality and an interval to answer. $-1 < x < 3$, $(-1, 3)$

b. As we trace the graph from left to right, for what inputs *x* is the graph increasing (rising)? Use both an inequality and an interval to answer. $x > 1$, $(1, +\infty)$

13. Explain what is the same and what is different about each of these algebraic statements. Be as specific as possible. Tables or graphs may help.

$$x^2 - 3x - 10$$
$$y = x^2 - 3x - 10$$
$$0 = x^2 - 3x - 10$$

expression; function with parabolic graph—infinite number of ordered pairs make it true; equation whose two solutions, $x = -2$ and $x = 5$, are x-intercepts for function

14. Explain differences between the meanings of $(1, 3)$ and $\{1, 3\}$. Give examples. See Additional Answers.

15. What is the significance of *c* in the graph of $y = ax^2 + bx + c$? vertical axis intercept

16. The graph of a quadratic function $f(x) = ax^2 + bx + c$ does not cross the x-axis. What does this tell us about the solutions to $f(x) = 0$? two complex-number solutions

17. Show by substitution whether $-3 - 2i$ is a solution to $x^2 + 6x + 13 = 0$. If so, what is the other solution? yes, $-3 + 2i$, the conjugate

18. Solve the equations and tell what kinds of numbers the solutions are.

a. $3x = 1$ $x = \frac{1}{3}$; rational

b. $x^2 = 3$ $x = \pm\sqrt{3}$; real or irrational

c. $x^3 = 8$ $x = 2, x = -1 \pm i\sqrt{3}$; 1 real, 2 complex

d. $x^2 + 4 = 0$ $x = \pm 2i$; complex or imaginary

19. The cost of fiberglass roof panels depends on the length of the panel as follows: 8 ft, $9.95; 10 ft, $11.99; 12 ft, $14.49; 14 ft, $17.50; 16 ft, $19.95. Do both a linear regression and quadratic regression on the data. Graph the data with both regression equations. Give one reason why each regression might be appropriate.
See Answer Section.

▉ Chapter 7

In Exercises 20 to 22, simplify.

20. $\dfrac{\dfrac{55 \text{ mi}}{1\text{hr}} \cdot \dfrac{10 \text{ hr}}{1 \text{ day}}}{\dfrac{7 \text{ mi}}{1 \text{ gal}}}$
78.6 gal/day

21. $\dfrac{y + \dfrac{1}{y}}{y - \dfrac{1}{y}}$ $\dfrac{y^2 + 1}{y^2 - 1}$

22. $\dfrac{3 - \dfrac{x}{2}}{x + \dfrac{6}{x}}$ $\dfrac{x(6 - x)}{2(x^2 + 6)}$

23. What is the reciprocal of $\dfrac{x}{2} + \dfrac{3}{x}$? $\dfrac{2x}{x^2 + 6}$

24. In 2003, the record for the fastest women's tennis serve was 127 miles per hour. This record was held by Venus Williams. How long does it take a 127-mile-per-hour serve to travel the 78-foot length of the court?
0.42 sec

25. Play the teacher. Look at the solutions below, presented by four different students attempting to solve the formula $A = \frac{1}{2} h(a + b)$ for b. Show whether each student has or has not used correct algebra to get to the given answer. Which might be the most correct answer? Why?

a. $b = \dfrac{2A}{h} - a$

b. $b = \dfrac{A}{\frac{1}{2} h} - a$

c. $b = \dfrac{A - \frac{1}{2} ah}{\frac{1}{2} h}$

d. $b = \dfrac{2A - ah}{h}$

All are correct; part d contains no fractions in the numerator or denominator and is a single term.

▉ Chapter 8

26. Solve $\sqrt{3x - 11} = \sqrt{5x} - 3$ from the graph in the figure below. $x = 5$ or $x = 20$

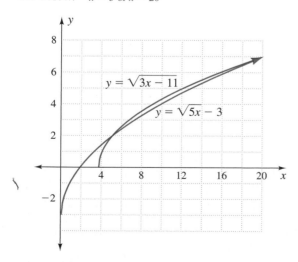

27. Solve $\sqrt{3x - 11} = 1$ with symbols and from the graph in the figure in Exercise 26. $x = 4$

28. Solve $\sqrt{5x} - 3 = 5$ with symbols. $x = 12.8$

29. Solve $\sqrt{x - 9} = 9 - \sqrt{x}$ for x. $x = 25$

30. How much will be in a savings account if it starts with $1000 and receives 5% interest compounded annually for 10 years? compounded monthly?   ≈$1628.89; ≈$1647.00

31. At what rate of interest, compounded annually, will $1000 double in 10 years? in 8 years? in $5\frac{3}{8}$ years?   ≈7.2%; ≈9.1%; ≈13.8%

▉ Chapter 9

In Exercises 32 to 36,
(a) Identify the type of sequence: arithmetic, quadratic, or geometric.
(b) For arithmetic and geometric sequences, write the equation using a_n and using linear or exponential regression on a calculator. Show that the results are the same with both methods.
(c) For quadratic sequences, use quadratic regression to find the equation.
(d) Find the sum to 10 terms for each geometric sequence. (See Appendix 3.)
(e) Find the infinite sum if it is defined for the sequence. (See Appendix 3.)

32. $\frac{1}{16}, \frac{1}{32}, \frac{1}{64}, \frac{1}{128}$ geometric; $a_n = \frac{1}{16}\left(\frac{1}{2}\right)^{n-1}$, $y = 0.125(0.5)^x$; 0.1249; 0.125

33. 27, 9, 3, 1   geometric; $a_n = 27\left(\frac{1}{3}\right)^{n-1}$, $y = 81\left(\frac{1}{3}\right)^x$; 40.4993; 40.5

34. 4, 8, 16, 32, 64   geometric; $a_n = 4(2)^{n-1}$, $y = 2(2)^x$; 4092

35. 1, 6, 15, 28, 45   quadratic; $y = 2x^2 - x$

36. 2, 6, 10, 14, 18   arithmetic; $a_n = 2 + (n - 1)4$, $y = 4x - 2$

37. The numbers of students being home schooled in Lane County, Oregon, in September of the years 1985–1994 are as follows: 1985, 220; 1986, 190; 1987, 300; 1988, 432; 1989, 425; 1990, 486; 1991, 577; 1992, 700; 1993, 906; 1994, 1167. Let 1985 = 1, 1986 = 2, etc. Enter the data into lists and graph.

a. Fit an equation with an appropriate regression.
exponential, $y = 164.6(1.208)^x$, $r \approx 0.9793$
b. What is the rate of growth per year?   20.8%

c. Predict the number of home-schooled students in 2001.   4088 (actual number was 1736)

38. The yearly sales figures for Rich Products Corporation (Associated Press, June 25, 1995) are as follows: 1945, $28,000; 1946, $474,000; 1952, $1.4 million; 1960, $3.86 million; 1976, $100 million; 1994, $1 billion. Graph the data. Fit an exponential equation to the data. Discuss your results, and include annual growth rate. Does $1.7 billion in sales for 2001 fit the pattern?
$y \approx 175{,}200(1.206)^x$, where x is years since 1945; annual growth rate ≈ 20.6%; no, predicted value is $6.2 billion

▉ Chapter 10

39. Solve this system of equations:

$$2a - 4b + c = 13$$
$$3a + 2b - c = 6$$
$$a + 2b - c = 0$$

$(3, -2, -1)$

40. The equation of a transition curve between two hills is $y = ax^2 + bx + c$. The slope of the curve at any point (x, y) is $m = 2ax + b$. At $(0, 800)$, there is to be a slope of -2%. At a horizontal distance of 2500 feet to the right, the slope is to be 4%. What are the coefficients a, b, and c of the transition curve? $a = \frac{3}{250,000}$, $b = -\frac{1}{50}$, $c = 800$

41. The equation of a transition curve between two hills is $y = ax^2 + bx + c$. The slope of the curve at any point (x, y) is $m = 2ax + b$. At $(3200, 1000)$, there is to be a slope of 4%. At a horizontal distance of 3200 feet to the left, the slope is to be -5%. What are the coefficients a, b, and c of the transition curve? $a = \frac{9}{640,000}$, $b = -\frac{1}{20}$, $c = 1016$

Chapter 11

42. Draw a tree diagram that shows the outcomes from choosing first one letter from the word OHIO and then a second letter, without replacement. See Additional Answers.

a. What is the probability of choosing an O? $\frac{5}{6}$

b. What is the probability of choosing only the two letters in the word HI? $\frac{1}{6}$

c. What do you observe about the answers to parts a and b? They add to 1 and describe all possible outcomes.

43. Sweet Treats has six kinds of chocolate cake and ten flavors of ice cream. How many ways could you select

a. cake and ice cream? 60

b. cake or ice cream? 16

c. cake and two flavors of ice cream? $6 \cdot {}_{10}C_2 = 270$

d. three flavors of ice cream? ${}_{10}C_3 = 120$

44. In *Star Wars, Episode I: The Phantom Menace*, the character Yoda exemplifies wisdom. His sentence structure in "More to say have you?" is similar to that used by Shakespeare.

a. How many ways can these five words be written? $5! = 120$

b. List four other sensible ways this five-word question could be written. Example: Have you more to say?

45. For what value of n is ${}_xC_n = {}_xC_5$? $n = x - 5$

46. When 3 coins are tossed, what is the probability of tossing

a. exactly 2 heads or at least 1 tail? $\frac{7}{8}$

b. at least 2 heads or at least 1 tail? $\frac{8}{8} = 1$

Appendix I: Selected Formulas

Triangle $\qquad\qquad$ Area $= \frac{1}{2}$ base $\cdot$ height $= \frac{1}{2} bh$*

Square $\qquad\qquad$ Area $=$ side $\cdot$ side $= s^2$

Perimeter $= 4 \cdot$ side $= 4s$

Rectangle $\qquad\qquad$ Area $=$ length $\cdot$ width $= lw$

Perimeter $= 2l + 2w$

Parallelogram $\qquad\qquad$ Area $=$ base $\cdot$ height $= bh$

Trapezoid $\qquad\qquad$ Area $= \frac{1}{2}$ height $\cdot$ (sum of parallel sides)

$\qquad\qquad\qquad\qquad = \frac{1}{2} h(a + b)$

Circle $\qquad\qquad$ Area $= \pi r^2$, $r =$ radius

Circumference $= 2\pi r = \pi d$

Diameter $= d = 2r$

Rectangular prism (box) $\qquad\qquad$ Surface area $= 2lw + 2hl + 2hw$

Volume $= lwh$

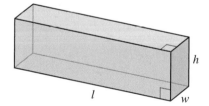

Cylinder $\qquad\qquad$ Surface area $= 2\pi r^2 + 2\pi rh$

Volume $= \pi r^2 h$, $r =$ radius, $h =$ height

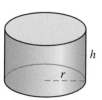

Sphere $\qquad\qquad$ Surface area $= 4\pi r^2$

Volume $= \frac{4}{3} \pi r^3$

*In all formulas, the base and height (or length and width) refer to dimensions that are perpendicular.

Cone

Surface area $= \pi r l + \pi r^2$
Volume $= \frac{1}{3}\pi r^2 h$

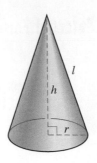

Simple interest

$I = prt$, where I is the interest earned on p dollars at an annual interest rate r for t years

Annual compound interest

$S = P(1 + r)^t$, where S is the future value, P is the dollars put in (principal or present value), r is the annual interest rate, and t is the time in years

Compound interest
(n times a year)

$S = P(1 + r/n)^{nt}$, where S is the future value, P is the dollars put in (principal or present value), r is the annual interest rate, n is the number of compoundings per year, and t is the time in years

Distance, rate, and time

$d = rt$

Distance formula (length of a line segment on a graph)

$d = \sqrt{(x_2 - x_1)^2 + (y_2 - y_1)^2}$

Pythagorean theorem
(a and b are legs and c the hypotenuse of a right triangle)

$a^2 + b^2 = c^2$

Quadratic formula

If $ax^2 + bx + c = 0$, then

$$x = \frac{-b \pm \sqrt{b^2 - 4ac}}{2a}$$

Slope formula

$$m = \frac{y_2 - y_1}{x_2 - x_1}$$

Arithmetic sequence, nth term

$a_n = a_1 + (n - 1)d$

Arithmetic sequence, sum

$$S_n = \frac{n}{2}(a_1 + a_n)$$

Geometric sequence, nth term

$a_n = a_1 r^{n-1}$

Geometric sequence, sum

$$S_n = \frac{a_1(1 - r^n)}{1 - r} \text{ for } n \text{ terms}$$

$$S_n = \frac{a_1}{1 - r} \text{ for an infinite number of terms}$$

Appendix 2: Percent

Changing Numbers to Percents

EXAMPLE 1 Exploring 100   Find the values of these expressions.
  a. $100(25)$ **b.** $100(2.5)$ **c.** $100(0.025)$ **d.** $100\left(\frac{1}{100}\right)$

SOLUTION   **a.** 2500 **b.** 250 **c.** 2.5 **d.** 1

The key to understanding percent is understanding multiplication or division by 100. The expression $n\%$ means the ratio of n to 100—that is, n divided by 100. Thus, 18% means 18 per hundred or $\frac{18}{100}$. Most importantly, $100\% = 100$ per 100 or $\frac{100}{100} = 1$.

Part d of Example 1 showed that $100\left(\frac{1}{100}\right) = 1$. Because multiplication by 1 does not change the value of a number, multiplication by $100 \cdot \frac{1}{100}$, or 100%, does not change the value of a number.

To change numbers to percent form, we multiply by 100%; that is, we multiply by a special form of 1:

$$0.5 = 0.5(100\%) = (0.5 \cdot 100)\% = 50\%$$

EXAMPLE 2 Changing decimals to percents   Write the following numbers as percents.
  a. 25 **b.** 2.5 **c.** 0.025

SOLUTION   **a.** $25(1) = 25(100\%) = 2500\%$

  b. $2.5(1) = 2.5(100\%) = 250\%$

  c. $0.025(1) = 0.025(100\%) = 2.5\%$

EXAMPLE 3 Changing fractions to percents   Write the following numbers as percents.
  a. $\frac{1}{2}$ **b.** $\frac{3}{4}$ **c.** $\frac{1}{3}$

SOLUTION   **a.** $\frac{1}{2}(1) = \frac{1}{2}(100\%) = 50\%$

  b. $\frac{3}{4}(1) = \frac{3}{4}(100\%) = \left(\frac{300}{4}\right)\% = 75\%$

  c. $\frac{1}{3}(1) = \frac{1}{3}(100\%) = \left(\frac{100}{3}\right)\% = 33\frac{1}{3}\%$

Changing Percents to Decimal or Fractional Numbers

Because $\%$ means $\frac{1}{100}$, to change a percent back to decimal or fraction notation, replace the percent sign with $\frac{1}{100}$ and simplify.

EXAMPLE 4 Changing percents to decimal or fractional notation   Write the following numbers as both decimals and fractions.
  a. 45% **b.** 150% **c.** 0.5%

SOLUTION   **a.** $45\% = 45 \cdot \frac{1}{100} = \frac{45}{100}$
  $\frac{45}{100} = 0.45$ and $\frac{45}{100} = \frac{9}{20}$

  b. $150\% = 150 \cdot \frac{1}{100} = \frac{150}{100}$
  $\frac{150}{100} = 1.5$ and $\frac{150}{100} = \frac{15}{10} = \frac{3}{2}$

  c. $0.5\% = 0.5\left(\frac{1}{100}\right) = 0.5/100$
  $0.5/100 = 0.005$ and $0.5/100 = \frac{5}{1000} = \frac{1}{200}$

Finding Percent Change

Percent change can be either a percent increase (a positive change) or a percent decrease (a negative change). We always find the change by subtracting the beginning amount from the ending amount.

$$\text{Percent change} = \frac{(\text{ending amount} - \text{beginning amount})(100\%)}{\text{beginning amount}}$$

EXAMPLE 5 **Finding percent change** Describe the relationship between the two values in each set with a percent change.
a. Postage: 1994, $0.29; 2003, $0.37
b. Minimum wage: 1991, $4.25; 1997, $5.15
c. Savings by elderly: 2000, $100,000; 2010, $10,000

SOLUTION **a.** $\dfrac{(\$0.37 - \$0.29)(100\%)}{\$0.29} = \dfrac{0.08(100)}{0.29}\% \approx 27.6\%$

b. $\dfrac{(\$5.15 - \$4.25)(100\%)}{\$4.25} = \dfrac{0.90(100)}{4.25}\% \approx 21.2\%$

c. $\dfrac{(\$10,000 - \$100,000)(100\%)}{\$100,000} = \dfrac{-90,000(100)}{100,000}\% \approx -90\%$

Appendix 2 Exercises

1. Write as percents.

  a. 0.30 30%   **b.** 0.20 20%   **c.** 0.075 7.5%

  d. 0.0005 0.05%   **e.** 2.0 200%   **f.** 5.0 500%

  g. 20 2000%   **h.** 100 10,000%   **i.** 2000 200,000%

2. Write as percents.

  a. $\frac{3}{5}$ 60%   **b.** $\frac{2}{3}$ $66\frac{2}{3}\%$ **c.** $\frac{1}{20}$ 5%

  d. $\frac{7}{10}$ 70%   **e.** $\frac{7}{8}$ 87.5%   **f.** $\frac{12}{5}$ 240%

  g. $\frac{17}{8}$ 212.5%   **h.** $\frac{1}{200}$ 0.5%   **i.** $\frac{57}{40}$ 142.5%

3. Write as both fractions and decimals.

  a. 15% $\frac{3}{20}$, 0.15   **b.** 80% $\frac{4}{5}$, 0.8   **c.** 0.25% $\frac{1}{400}$, 0.0025

  d. 10% $\frac{1}{10}$, 0.10   **e.** 1000% 10   **f.** 17.5% $\frac{7}{40}$, 0.175

  g. 0.15% $\frac{3}{2000}$, 0.0015   **h.** 35% $\frac{7}{20}$, 0.35   **i.** 200% 2

4. Describe the relationship between the two values in each set with a percent change.

  a. Price before discount, $79.99; price after discount, $48 −40%

  b. Value before store mark-up, $125; value after mark-up, $200 60%

  c. Height: 1995, 21 inches; 2010, 5 feet 8 inches 223.8%

  d. Weight: 1990, 7 pounds; 2010, 150 pounds 2042.9%

  e. Weight: 1990, 185 pounds; 2010, 140 pounds −24.3%

  f. Population of Scottsdale, AZ: 1990, 130,099; 2000, 202,705 55.8%

5. If, in the percent change examples and exercises, the dates were inputs and the other data were outputs to a linear function, how would the slope description differ from percent change? change per year versus total change

6. Why does the following rule work? To change a decimal number to a percent, move the decimal point two places to the right and place a percent sign at the end. Moving decimal is multiplying by 100; placing percent sign is multiplying by $\frac{1}{100}$.

7. Why does the following rule work? To change a fraction to a percent, divide the numerator by the denominator and move the decimal point two places to the right. Dividing changes the fraction to a decimal; then see answer to Exercise 6.

Appendix 3   Sequences and Their Sums

Arithmetic Sequences

Section 2.4 introduced arithmetic sequences. Arithmetic sequences are characterized by a constant first difference, d. We use the first term, a_1, of the sequence and the constant difference to build the nth term, a_n.

■ *n*TH TERM OF AN ARITHMETIC SEQUENCE

$$a_n = a_1 + (n - 1)d$$

ARITHMETIC SEQUENCES AND LINEAR FUNCTIONS   An arithmetic sequence is associated with a linear function, where the constant difference of the arithmetic sequence is the slope of the linear function. Subtracting the common difference from the first term gives the y-intercept for the linear function. The domain for arithmetic sequences is the positive integers, whereas the domain for linear functions is all real numbers.

SUMS OF ARITHMETIC SEQUENCES   The story is told of an eighteenth-century teacher in Germany who wanted to keep his young students busy for a considerable period. He asked them to add the numbers from 1 to 100. Within a few minutes, before the teacher could get comfortable doing something else, one student raised his hand and said that he had the answer. The teacher was amazed at the 10-year-old boy's technique. The teacher supposedly took the boy, Johann Friederich Carl Gauss (1777–1855), to a private tutor who was able to help Gauss get into college at age 15. Gauss became one of history's most highly regarded mathematicians.

Gauss was a superb problem solver. While his classmates assumed that they should start adding numbers one at a time, Gauss reconsidered the assumptions and came up with a pairing of numbers, as shown in Example 1.

EXAMPLE 1 Solving Gauss's problem   Add the sequence of numbers from 1 to 100,

$$1 + 2 + 3 + 4 + 5 + \cdots + 97 + 98 + 99 + 100$$

SOLUTION   Gauss's technique was to add pairs of numbers:

$$1 + 2 + 3 + 4 + 5 + \cdots + 97 + 98 + 99 + 100$$

Because $1 + 100 = 101$, $2 + 99 = 101$, $3 + 98 = 101$, and so forth, all Gauss needed to finish the problem was the number of such pairs, which is 50. The sum is $50 \cdot 101 = 5050$. ■

Gauss's technique for addition was based on adding the first and last terms and multiplying by half the number of terms. The process works for adding any arithmetic sequence.

EXAMPLE 2 Adding arithmetic sequences
a. To identify the arithmetic sequence shown below, give the first term, a_1, the common difference, d, and the number of terms, n.

$$8, 12, 16, 20, 24, 28, 32, 36, 40, 44, 48, 52$$

b. Use Gauss's pairing method to add the numbers.

SOLUTION   **a.** *Sequence:*   8, 12, 16, 20, 24, . . .   The common difference is 4.

Differences: 4, 4, 4, 4, . . .   Arithmetic

$a_1 = 8$, $d = 4$, and $n = 12$ terms.

b. 8, 12, 16, 20, 24, 28, 32, 36, 40, 44, 48, 52

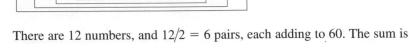

There are 12 numbers, and $12/2 = 6$ pairs, each adding to 60. The sum is $6(60) = 360$.

Example 3 shows that Gauss's method also works with an odd number of numbers.

EXAMPLE 3   **Adding arithmetic sequences**   Suppose you earn $20,000 during the first year at a new job and receive a $1000 pay increase at the end of each subsequent year.
a. Write the arithmetic sequence for your earnings over the first 5 years. Give the first term, a_1, the common difference, d, and the number of terms, n.
b. Use Gauss's pairing method to add the numbers.

SOLUTION   **a.** $a_i = \$20,000$, $d = \$1000$, and $n = 5$. The sequence is

$20,000, $21,000, $22,000, $23,000, $24,000

b. $20,000 + $21,000 + $22,000 + $23,000 + $24,000

There are $2\frac{1}{2}$ pairs, each adding to $44,000. The sum is

$2\frac{1}{2}$ ($44,000) = $110,000

Gauss's pairing method leads us to a general formula for the sum of n terms in an arithmetic sequence.

SUM OF n TERMS OF AN ARITHMETIC SEQUENCE
To add an arithmetic sequence, add the first and last terms and multiply by half the number of terms. The sum, S_n, of n terms is

$$S_n = \frac{n}{2}(a_1 + a_n)$$

EXAMPLE 4   **Finding the number of terms**   Add 5, 8, 11, 14, 17, . . . , 77.

SOLUTION   In the sequence 5, 8, 11, 14, 17, . . . , 77, $a_1 = 5$, $d = 3$, and $a_n = 77$. To find the number of terms n, we substitute into

$$a_n = a_1 + (n - 1)d$$

and solve for n.

$77 = 5 + (n - 1) \cdot 3$ Subtract 5 from both sides.

$72 = (n - 1) \cdot 3$ Divide by 3.

$24 = n - 1$ Add 1 to both sides.

$25 = n$

Thus, the sequence contains 25 terms. Substituting a_1, d, a_n, and n into the sum formula

$$S_n = \frac{n}{2}(a_1 + a_n)$$

we obtain

$$\frac{25}{2}(5 + 77) = 1025$$

       ■

The *sum of a sequence* is called a **series**. This is an awkward word choice, because in ordinary English *series* means one event right after another or an order of a number of things. This ordinary English meaning is very close to the sequence concept, so it is easy to confuse the two words. We will emphasize the word *sum* and the adding concept rather than the word *series*.

Geometric Sequences

Section 9.1 introduced geometric sequences. Geometric sequences are characterized by a constant ratio, r, of consecutive terms. We use the first term, a_1, of the sequence and the constant ratio to build the nth term, a_n.

■ *n*TH TERM OF A GEOMETRIC SEQUENCE

$$a_n = a_1 r^{n-1}$$

GEOMETRIC SEQUENCES AND EXPONENTIAL FUNCTIONS   A geometric sequence is associated with an exponential function, where the constant ratio of the geometric sequence is the base of the exponential function. Dividing the common ratio into the first term of the sequence gives the y-intercept for the exponential function. The domain for geometric sequences is the positive integers, whereas the domain for exponential functions is all real numbers. (See also page 518.)

SUMS OF GEOMETRIC SEQUENCES

EXAMPLE 5   Finding sums: doubling sequence   One at a time, add the terms of the sequence $a_n = 2^{n-1}$, giving the grains of rice on the chessboard in Exercise 63 of Section 9.2.

SOLUTION   The sequence $a_n = 2^{n-1}$ is 1, 2, 4, 8, 16, 32, 64, ... The sums of the terms, one at a time, are

$$S_1 = 1$$
$$S_2 = 1 + 2 = 3$$
$$S_3 = 1 + 2 + 4 = 7$$
$$S_4 = 1 + 2 + 4 + 8 = 15$$
$$S_5 = 1 + 2 + 4 + 8 + 16 = 31$$

The sums are 1 less than 2^n, or $S_n = 2^n - 1$.   ■

In Example 5, both the sequence and the sum were based on powers of 2. This suggests that there is a formula for finding the sum of the terms.

■ SUM OF *n* TERMS OF A GEOMETRIC SEQUENCE

The **sum of** n terms of **a geometric sequence** is given by

$$S_n = \frac{a_1(1 - r^n)}{1 - r}$$

In Example 6, we apply the formula to the doubling sequence in Example 5.

EXAMPLE 6   Finding sums   Use the sum formula $S_n = \dfrac{a_1(1 - r^n)}{1 - r}$ on the sequence in Example 5.

SOLUTION   The first term, a_1, is 1. The common ratio, r, is 2.

$$S_n = \frac{a_1(1 - r^n)}{1 - r}$$

$$= \frac{1(1 - 2^n)}{1 - 2} = \frac{1 - 2^n}{-1} = 2^n - 1$$

The equation is the same as the one in Example 5. For 64 chessboard squares, the total number of grains of rice is $2^{64} - 1$.   ▬

In Example 7, we return to the superball bounce (Example 2 of Section 9.1).

EXAMPLE 7   Finding sums: superball bounce, continued   The first 10 heights reached by the superball form the sequence 108, 72, 48, 32, 21.3, . . . , 2.8 (see Figure 1). What is the total of these 10 heights?

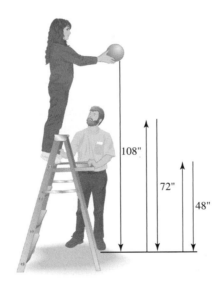

108"

72"

48"

FIGURE 1

SOLUTION   In order to find the sum of the first 10 heights, we add the numbers in the sequence 108, 72, 48, 32, 21.3, . . . , 2.8 using the sum formula, with $a_1 = 108$, $r = \frac{2}{3}$, and $n = 10$:

$$S_n = \frac{a_1(1 - r^n)}{1 - r}$$

$$S_{10} = \frac{108\left[1 - \left(\frac{2}{3}\right)^{10}\right]}{1 - \frac{2}{3}}$$

$$\approx 318.38 \text{ in.}$$

(*Note:* When entering the sum expression in the calculator, remember to put parentheses around the denominator.)   ▬

There is an important difference between the geometric sums in Examples 6 and 7. The common ratio for the rice is 2. The sum grows larger faster and faster with each term added. The common ratio for the superball heights is $\frac{2}{3}$. The sum grows, but more slowly with each bounce.

When a geometric sequence has a common ratio between 0 and 1, the exponential expression describes exponential decay. The diminishing value of terms in the geometric sequence makes it possible to *add all the terms in an infinite sequence*, for an **infinite sum**.

Figure 2 shows an area model of the sum of the first five terms of $a_n = \left(\frac{1}{2}\right)^n$. Observe that the sum of the first five terms is less than 1. Because we are adding half the remaining area each time, the sum will stay less than 1.

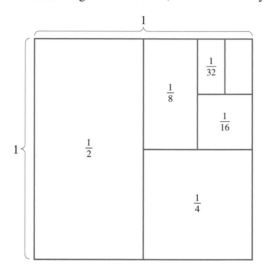

FIGURE 2 Adding $\frac{1}{2} + \frac{1}{4} + \frac{1}{8} + \frac{1}{16} + \frac{1}{32} + \cdots$

In Example 8, we examine a graphical model of the sum of the sequence described by $a_n = \left(\frac{1}{2}\right)^n$. The sum to n terms is $S_n = 1 - \left(\frac{1}{2}\right)^n$.

EXAMPLE 8 Exploring sums with a graph   Graph the sum formula $S_n = 1 - \left(\frac{1}{2}\right)^n$ with x replacing n (see Figure 2). Trace or use a table to find the sum at 10 and 15 terms. Does there seem to be an upper limit?

SOLUTION   We change the sum formula to a formula in x and y: $y = 1 - \left(\frac{1}{2}\right)^x$. Figure 3 is the graph. The sum for 10 terms is at (10, 0.99902); for 15 terms, it is at (15, 0.99997). The graph rises toward but does not cross $y = 1$. The number 1 appears to be an upper limit for the graph and for the sum of the terms of the sequence. We say that the line $y = 1$ is a horizontal asymptote for the graph of $y = 1 - \left(\frac{1}{2}\right)^x$.

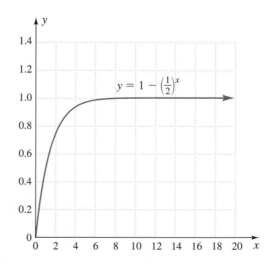

FIGURE 3

To add all the terms of a geometric sequence, we use the sum to infinity formula.

■ SUM TO INFINITY FORMULA

> If the common ratio, r, is between 0 and 1, the sum to an infinite number of terms for a geometric sequence with first term a_1 is
>
> $$S_\infty = \frac{a_1}{1 - r}$$

EXAMPLE 9 Finding sums   What is the sum to an infinite number of terms for the sequence $a_n = \left(\frac{1}{2}\right)^n$ from Example 8?

SOLUTION   With $a_1 = \frac{1}{2}$ and $r = \frac{1}{2}$,

$$S_\infty = \frac{\frac{1}{2}}{1 - \frac{1}{2}} = 1$$

The formula is consistent with the graphical and tabular results.

Appendix 3   Exercises

■ **ARITHMETIC SEQUENCES**

In Exercises 1 to 6, what is the sum of the first 20 terms of each arithmetic sequence?

1. 18, 16, 14, 12, 10, . . . −20

2. 31, 27, 23, 19, 15, . . . −140

3. 10, 18, 26, 34, 42, . . . 1720

4. 4, 10, 18, 28, 40, . . . not arithmetic

5. 3, 9, 27, 81, . . . not arithmetic

6. −16, −9, −2, 5, 12, . . . 1010

7. Explain why Gauss's system of adding pairs of numbers would not be helpful in adding a sequence such as 1, 3, 6, 10, 15, 21, 28, 36, 45, 55, 66.
   not arithmetic; pairs have different sums.

In Exercises 8 to 13, assume the setting is a sequence, write a_n, and then answer the question. The questions do not ask for sums.

8. A population census of the United States is required by the Constitution. The census is taken every 10 years. How many times was the census taken between 1790 and 2000? $a_n = 1790 + (n − 1)10$; 22 times

9. Dominique bought a block of basketball tickets numbered consecutively from 47 to 113. How many seats does she have? $a_n = 47 + (n − 1)$; 67 seats

10. Ravi Tej bought a block of tickets to a play. The tickets are numbered consecutively from 53 to 99. How many tickets did he receive? $a_n = 53 + (n − 1)$; 47 tickets

11. The Summer Olympic Games are held every 4 years. If the first modern Olympics was held in 1896, what should be the total number of games held between 1896 and 2000? World events bonus: Why were the games canceled in 1916, 1940, and 1944?
    $a_n = 1896 + (n − 1)4$; 27 games; WWI and WWII

12. Shareen changes her car's oil every 3000 miles. She bought the car at 22,000 miles and immediately changed the oil. At 109,000 miles, she has just changed the oil again. How many oil changes has she made altogether?
    $a_n = 22{,}000 + (n − 1)3000$; 30 oil changes

13. At the end of the first hour of a treatment crisis, Lari tests a sample of the water leaving the city waste water system and directs that the water be tested every 3 hours thereafter. At the 145th hour, a test is made. How many tests have been made altogether? $a_n = 1 + (n − 1)3$; 49 tests

14. Solve the formula $a_n = a_1 + (n − 1)d$ for n. If you were to solve Exercises 12 and 13 by reasoning, how would that reasoning explain the form of the answer in this exercise? $n = \dfrac{a_n − a_1}{d} + 1$

■ **Projects**

15. **Sum of Odd Numbers**

    a. Find a_n for the sequence of odd numbers, 1, 3, 5, 7, 9, . . . using the formula for the nth term of an arithmetic sequence. $a_n = 2n − 1$

    b. Use linear regression on the ordered pairs (1, 1), (2, 3), (3, 5), (4, 7), . . . to find an expression for the nth term. $y = 2x − 1$ or $a_n = 2n − 1$

c. What pattern is formed by adding odd numbers as follows?

1

1 + 3

1 + 3 + 5

1 + 3 + 5 + 7

and so on. 1, 4, 9, 16, . . . , the square numbers

d. How can we relate the pattern observed in part a to the odd numbers pictured in the figure?

$$1 \quad 3 \quad 5 \quad 7 \quad 9$$

The corner shapes fit together to form a square.

e. Find the sum of n odd numbers using the formula for the sum of an arithmetic sequence. $S_n = \frac{n}{2}[1 + (2n - 1)] = n^2$

16. Trapezoid Area Build a model or draw a figure to suggest why the formula for the sum of an arithmetic sequence,

$$\text{sum} = \tfrac{1}{2}\, n(a_i + a_n)$$

and the formula for the area of a trapezoid,

$$A = \tfrac{1}{2}\, h(a + b)$$

have the same form. Blocks or grid paper may be helpful.

GEOMETRIC SEQUENCES

In Exercises 17 to 22, find the sum to 10 terms, and find the infinite sum, if appropriate.

17. 9, 27, 81, 243, . . . 265,716

18. 3.5, 7, 14, 28, . . . 3580.5

19. 64, 32, 16, 8, 4, . . . 127.875; 128

20. 8, 4, 2, 1, $\frac{1}{2}$, . . . 15.984; 16

21. 3, 9, 27, 81, . . . 88,572

22. $\frac{1}{2}, \frac{1}{4}, \frac{1}{8}, \frac{1}{16}$, . . . ≈0.9990; 1

23. Find the sum of the first 20 heights of the superball in Example 7. ≈323.9 in.

24. Find the sum in Exercise 23 to an infinite number of bounces. 324 in.

25. Find the sum of the first 20 heights of the superball if the initial height is 135 inches and each rebound is $\frac{2}{3}$ of the previous height. ≈404.88 in.

26. Find the sum in Exercise 25 to an infinite number of bounces. 405 in.

27. Find the infinite sum of the sequence 1, $\frac{1}{2}, \frac{1}{4}, \frac{1}{8}, \frac{1}{16}$, ... 2

28. Find the infinite sum of the sequence 1, $\frac{1}{3}, \frac{1}{9}, \frac{1}{27}$, ... $\frac{3}{2}$

29. What is the sum to infinity of the distances traveled by a child on a swing if the swing starts with an initial distance of 18 feet and goes $\frac{5}{6}$ of the prior distance on each subsequent swing? 108 ft

30. Explain why

$$S_n = \frac{a_1(1 - r^n)}{1 - r}$$

yields the same value as

$$S_n = \frac{a_1(r^n - 1)}{r - 1} \qquad \frac{r^n - 1}{r - 1} = \frac{-(1 - r^n)}{-(1 - r)} = \frac{1 - r^n}{1 - r}$$

Projects

31. Repeating Decimals The repeating decimal 0.3333333 . . . may be written as the sum of a geometric sequence:

$$0.3 + 0.03 + 0.003 + 0.0003 + \cdots$$

a. Write this sequence in fractions. Determine the first term and the common ratio, and find the sum to an infinite number of terms. Write the sum as a fraction. $\frac{3}{10}, \frac{3}{100}, \frac{3}{1000}, \ldots; \frac{3}{10}, \frac{1}{10}, \frac{1}{3}$

b. Convert 0.4444444 . . . to a fraction, as you did the repeating decimal in part a. $\frac{4}{10}, \frac{4}{100}, \frac{4}{1000}, \ldots; \frac{4}{10}, \frac{1}{10}, \frac{4}{9}$

c. Convert 0.12121212 . . . to a fraction, as you did the repeating decimal in part a. $\frac{12}{100}, \frac{12}{10,000}, \frac{12}{1,000,000}, \ldots; \frac{12}{100}, \frac{1}{100}, \frac{4}{33}$

d. Convert 0.42424242 . . . to a fraction, as you did the repeating decimal in part a. $\frac{42}{100}, \frac{42}{10,000}, \frac{42}{1,000,000}, \ldots; \frac{42}{100}, \frac{1}{100}, \frac{14}{33}$

e. Convert 0.99999999 . . . to a fraction, as you did the repeating decimal in part a. $\frac{9}{10}, \frac{9}{100}, \frac{9}{1000}, \ldots; \frac{9}{10}, \frac{1}{10}, 1$

32. Allowance You normally give your teenage daughter $10 per week. On January 1, she asks that her weekly allowance for the coming year be 1¢ the first week, 2¢ the second week, 4¢ the third week, 8¢ the fourth week, and so on, doubling each week as the year progresses.

a. How much will she be paid in the 10th week? $5.12

b. How much will she be paid in the 20th week? $5242.88

c. How much will she be paid altogether in the first 20 weeks? $10,485.75

d. Make a graph of her allowance for the first 12 weeks. See Additional Answers.

e. Is this plan a good deal for her? Yes!

f. What is the total allowance paid for the entire year? ≈$4.5 × 10^{13}

Answers to Selected Odd-Numbered Exercises and Tests

As you compare your answers to those listed here, keep these hints in mind:

- Don't give up too quickly.
- Have confidence that you worked the exercise correctly.
- Check that you looked up the right answer.
- Check that you copied the exercise correctly.
- See if you can use algebraic notation or simplification to change your answer to match the text's answer.

If the answers still don't match, try working the exercise again:

- Make sure you thoroughly understand the exercise. Read the exercise aloud. Shut the book and say it in your own words.
- On a separate piece of paper, copy the exercise from the text.
- Work the exercise without looking at your first attempt.
- Let the problem rest for an hour or two or overnight. Sometimes the solutions to problems become clear when you step away from them.
- Compare your work with that of another student. (Do this only after you have tried the problem twice.)
- Review the text material and related examples.
- Go on to another problem. You can continue doing homework without having completed each and every exercise.
- At the next opportunity, ask your instructor to review your work. Note that this does not mean that you should ask him or her to show you how to do the exercise.

Only after trying the above steps should you assume either that your work is wrong or that the four to six human beings who worked every exercise for this book made an error. The latter is possible, and the author and publisher would appreciate corrections.

EXERCISES 1.1

3. **a.** real numbers **b.** integers **c.** irrational numbers **d.** multiplicative inverses **e.** additive inverses
5. **a.** distributive property of multiplication over addition **b.** factors **c.** commutative property for multiplication **d.** factoring **e.** associative property for addition
7. **a.** negative **b.** one **c.** positive **d.** one **e.** zero
9. integer **11.** irrational **13.** natural **15.** natural numbers
17. **a.** $-8, 16, -48, -\frac{1}{3}$ **b.** $\frac{5}{6}, \frac{1}{6}, \frac{1}{6}, \frac{3}{2}$ **c.** $-\frac{1}{2}, -\frac{5}{6}, -\frac{1}{9}, -4$ **d.** $0.55, 0.95, -0.15, 3.75$ **e.** $-1.95, 1.05, 0.675, 0.3$ **f.** $-1\frac{53}{72}, \frac{35}{72}, \frac{25}{36}, \frac{9}{16}$
19. **a.** 12 **b.** 14 **c.** 5.5 **d.** -10 **e.** 1 **f.** $7\frac{1}{4}$

21. **a.** undefined **b.** 3 **c.** -3
23. **a.** opposite of negative three **b.** negative two **c.** five subtract negative three
25. **a.** 0 **b.** 210 **c.** 30 **d.** 190 **27. a.** 1900 **b.** 14 **c.** 1495 **d.** 2900 **e.** 34
29. **a.** $A \approx 11,689.9$ cm^2 **b.** $S \approx 81.7$ in^2 **c.** $d = 193$ **d.** $x = -27$ **31. a.** $-\frac{1}{2}$ **b.** 5 **c.** 212 **d.** 100 **e.** 12.8
33. **a.** $x = -3$ **b.** $d = 10$ **c.** $m = \frac{3}{4}$ **d.** $x = 1\frac{2}{3}$ **e.** $d = 2.0\overline{46}$
35. Possible answer: Subtraction can be restated as addition of the opposite.
37. slope, Pythagorean theorem, Celsius to Fahrenheit temperature, Fahrenheit to Celsius temperature, distance formula between two ordered pairs, slope, quadratic formula, discriminant for quadratic formula

EXERCISES 1.2

1. **a.** $x, 2, 3$ **b.** r, π, none **c.** $x, -1, 4$ **d.** $x, 1$ and $1, -1$
3. Visualize x columns of two dots with one dot missing.
5. **a.**

Input x	Output y
1	3
2	6
3	9
4	12

b. 3 **c.** $y = 3x$

7. **a.**

Input x	Output y
1	2
2	5
3	8
4	11
5	14

b. 3 **c.** $y = 3x - 1$

9. **a.** four subtract the opposite of c **b.** the opposite of the cube of z (or z-cubed) **c.** negative z-squared or the opposite of z-squared **d.** The outputs for $-z^3$ are $+8, +1, 0, -1, -8$. The output may be either positive or negative. **e.** For any real number z, z^2 is positive.
11. **a.** 9 **b.** -8 **c.** -9 **d.** $9x^2$ **e.** $3x^2$ **f.** $-3x^2$
13. **a.** -3 **b.** -2 **c.** 3 **d.** $3x$ **e.** x **f.** x
15. **a.** $-8x^3$ **b.** $-8x^3$ **c.** $81y^4$ **d.** $-x^4y^4$ **e.** x^4y^4 **f.** $8y^3$

17. a.

Multiply	x	-2
$3x$	$3x^2$	$-6x$

b.

Factor	a	$-b$
$2a$	$2a^2$	$-2ab$

19. a. $6x + 12$ **b.** $5, 3y + 2$ **21. a.** $x^2 + 5x$ **b.** $b, a + c$
23. a. $-6 + 8x$ **b.** $-ab + ac$ **c.** $x^3 - 2x^2 - 3x$
25. a. $6, 6(x - 9)$ **b.** $15, 15(x - 15)$ **c.** $2a, 2a(3a - 4)$
27. a. $b(a - b)$ **b.** $xy(1 - x)$ **c.** $xy^2(x - 1)$
29. a. $3(2ab - 4a + 1)$ **b.** $4(ac + 3a - 4)$ **31.** $3x^2 + 5x + 6$
33. $4a^2 + 3ab + 2b^2$ **35. a.** $a + 2b$ **b.** $2x^2 + 2x - 3$
37. a. $8ac - 3ab + 7bc$ **b.** $x^2 + 8x$
39. a. $15, \frac{1}{3}$ **b.** $16, \frac{3}{10}$ **c.** $240, \frac{22}{15}$

41. a. $be, \dfrac{a}{n}$ **b.** $b, a + c$ **c.** cannot be simplified

43. a. $x, \dfrac{y + z}{y}$ **b.** $x, \dfrac{x + 2}{2}$ **c.** $xy, \dfrac{y - 2x}{2}$

45. a. 64 **b.** 64 **47. a.** 32 **b.** 4 **49. a.** -9 **b.** 36
51. a. $C = 0$ **b.** $C = -40$ **c.** $C = 37$
53. a. $V = 14.1 \text{ cm}^3$ **b.** $V = 113.1 \text{ cm}^3$ **c.** $V = 904.8 \text{ cm}^3$
55. a. $S = 10,000$ **b.** $S = 10,100$ **c.** $S = 25,250$
57. a. 10 **b.** 13 **c.** 17
59. If a is positive, $a > -a$; if a is negative, $-a > a$.

EXERCISES 1.3

1. $y = 3x - 1$ **3.** $y = 2x + 1$ **5.** $y = 2x$ **7.** $y = x^2 + 1$
9. $2, \frac{1}{2}n$ **11.** $8, n^3$ **13.** $x + y = 7$ **15.** $y - x = -8$
17. the difference between x and seven
19. the quotient of x and seven
21. the difference between seven and x
23. the quotient of seven and x
25. the quotient of seven and the sum of x and two
27. the quotient of the sum of x and two and the difference between x and two

29. $10 - 2x = y$ **31.** $y = \frac{1}{2}x + 10$ **33.** $\dfrac{x}{15} + 2 = y$

35. $y = 7x - 14$ **37.** The output is five less than twice the input.
39. The input added to the output is eleven.
41. The difference between twice the input and the output is five.
43. The quotient of the input and eight is the output.
45. $s; A$ **47.** $s; A$ **49.** $A; E$
51. weight of package; cost of shipping
53. A cooler day might bring more kittens. Let temperature be the independent variable and number of kittens be the dependent variable.
55. $y = 1.29x - 0.45$ **57.** $y = 1.25 - 0.05x$
59. $y = 125x + 85$ **61. a.** 16.7% **b.** 60% **c.** 2000% **d.** 5%
63. a. 0.2 **b.** 0.125 **c.** 0.00005 **d.** 5
65. $y = 0.05x$, $x = $ season ticket cost, $y = $ daily cost
67. $y = 0.062x$, $x = $ amount of wages, $y = $ amount of tax
69. $y = 0.01x$, $x = $ amount of loan, $y = $ one borrowing cost
71. $y = x - \left(0.0765x + 0.20x + \dfrac{n}{100}x\right)$, $x = $ amount of wages, $y = $ net pay
73. a woman with her mother and her daughter

MID-CHAPTER 1 TEST

1. a. -2 **b.** $-\frac{3}{4}$ **c.** 7 **d.** -0.6
2. a. 1700 **b.** 21 **c.** -160 **d.** $\frac{3}{22}$
3. a. -16 **b.** 16 **c.** -16 **d.** $20\frac{1}{2}$ **e.** $-\frac{1}{3}$ **f.** $-\frac{3}{5}$
  g. $3\sqrt{10} \approx 9.5$ **h.** $\frac{37}{30} \approx 1.2$
4. a. 97.43 **b.** 590 **c.** 67 **d.** 390 **e.** $\frac{1}{3}$ **f.** 127
5. a. $-4x^2$ will be negative because $4x^2$ is always positive
  b. because x could be either positive or negative
  c. because x could be either positive or negative
  d. $-|x|$ will be negative because $|x|$ is always positive
6. a. $3x + 6$ **b.** $-3x + 6$ **c.** $-4x - 13$ **d.** $2x$
7. a. $2, 2(2x - 9)$ **b.** $x, x(x + 5)$ **c.** $n, n(mn - p^2)$
  d. $7xy, 7xy(9x - 7y)$
8. a. $x; -4; -1$ **b.** $x;$ none; 2 and -1
9. a. $-x^3 + 13x - 12$ **b.** $a^3 - 3a^2b - ab^2 + b^3$

10. a. $\dfrac{x}{y}$ **b.** $\dfrac{1}{y}$ **c.** $\dfrac{x - z}{x}$

11. a. $s; h; h = 2\sqrt{3}$ yd
  b. $d; A; A = \frac{25}{4}\pi \text{ cm}^2$ or $A = 6.25\pi \text{ cm}^2$
12. a. number of quarters (independent variable), parking time (dependent variable).
  b. speed (independent variable), distance traveled (dependent variable)
  c. size of the luggage (independent variable), cost (dependent variable)
13. a. $C = x + 0.0875x$ or $C = 1.0875x$
  b. $A = x + 0.03x$ or $A = 1.03x$
  c. $C = 1.49x - 0.30$
14. a.

Number of Pairs of Pens	Total Number of Panels
1	7
2	12
3	17
4	22
10	52
50	252
100	502

  b. $y = 5x + 2$ **15.** j **16.** a **17.** g **18.** h **19.** i
20. b **21.** e **22.** c

EXERCISES 1.4

1.

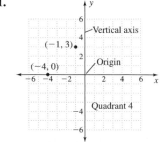

3. a. solution set **b.** evaluate **c.** scale **d.** dependent variable
  e. independent variable **5. a.** J **b.** E **c.** G **d.** K
7. a. $(-18, 0)$ **b.** $A(-40, -40)$

9.

x	$y = 3x - 2$
−3	−11
−2	−8
−1	−5
0	−2
1	1
2	4
3	7

11.

x	$y = 4 - 2x$
−3	10
−2	8
−1	6
0	4
1	2
2	0
3	−2

13.

x	$y = -2 - x$
−3	1
−2	0
−1	−1
0	−2
1	−3
2	−4
3	−5

15.

x	$y = 2x - x^2$
−3	−15
−2	−8
−1	−3
0	0
1	1
2	0
3	−3

17.

x	$y = x^2 + 2$
−3	11
−2	6
−1	3
0	2
1	3
2	6
3	11

19.

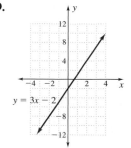

21.

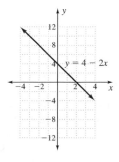

23.

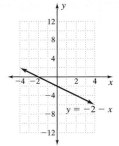

25.

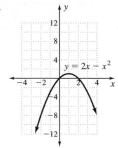

27.

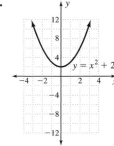

29. a. 19, 21, 23   **b.** 25, 27   **c.** $y = ax + b$

31.

x	$y = 3^x$
−3	0.037
−2	0.111
−1	0.333
0	1
1	3
2	9
3	27

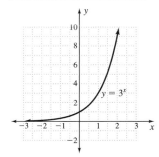

33.

x	y	$x^2 + y^2 = 25$
±5	0	$25 + 0 = 25$
0	±5	$0 + 25 = 25$
±4	±3	$16 + 9 = 25$
±3	±4	$9 + 16 = 25$

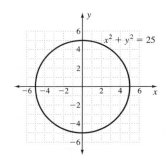

35.

x	y	$x = y^2$
4	± 2	$4 = 4$
9	± 3	$9 = 9$
16	± 4	$16 = 16$
25	± 5	$25 = 25$

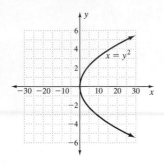

37.

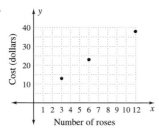

39.

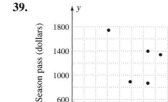

EXERCISES 1.5

1. $x = 1, x = 3$ **3.** $x = 6, x = -4$ **5. a.** $x = -4$ **b.** $x = 1$
7. a. $x = 1.5$ **b.** $x = 3.5$ **9. a.** $x = -2$ **b.** $x = -3$
11. a. $x = 0$ or $x = -1$ **b.** $x = -3$ or $x = 2$
  c. no real-number solution   **d.** $x = -0.5$
13. a. $x = 2$ or $x = -2$ **b.** no real-number solution
  c. $x = -3$ or $x = 3$ **d.** $x = 0$
15. a. $x = -1$ or $x = 0$ **b.** $x = -2$ or $x = 1$
  c. no real-number solution   **d.** $x = -3$ or $x = 2$
17. a. $\left\{0, \frac{1}{2}\right\}$ **b.** approximately $\{1.3, -0.8\}$ **c.** $\{-1.5, 2\}$
  d. $\{\ \}$
19. a. $\left\{-1, \frac{1}{3}\right\}$ **b.** $\left\{-1\frac{2}{3}, 1\right\}$ **c.** $\left\{-\frac{1}{3}\right\}$ **d.** $\left\{-\frac{2}{3}, 0\right\}$
21.

Weight (lb)	Index
100	18.3
110	20.2
120	22.0
130	23.8
140	25.7

23.

Weight (lb)	Index
130	18.7
140	20.1
150	21.6
160	23.0
170	24.4

25. a. $W = 130$ lb   **b.** $W = 110$ lb
27. a. $W = 160$ lb   **b.** $W = 170$ lb
29. Find output for input 2.
31. Find x-coordinate for point of intersection of $y = 3x + 4$
  and $y = 7$.
33. finding the independent variable
35. finding the dependent variable
37. finding the independent variable
39. $303.79   **41.** 464 miles   **43.** $360,000; $310,000; $330,000
45. a. $\{-3, -1, 1\}$ **b.** $\{-2.3, -2, 1.3\}$ **c.** $\{-3.3, 0, 0.3\}$
  d. $\{1.4\}$
47. a. $\{-3, 0\}$ **b.** $\{-2, 1\}$ **c.** $\{-2.7, -1, 0.7\}$
  d. $\{-2.9, -0.7, 0.5\}$
49.

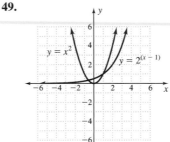

Approximate answers are $\{-0.58, 1, 6.32\}$.

51.

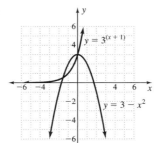

Approximate answers are $\{-1.57, 0\}$.

EXERCISES 1.6

1. a. -10 **b.** -2 **c.** 1   **d.** 25   **e.** 1   **f.** -18
3. a. 3   **b.** $\frac{1}{5}$ **c.** 3   **d.** 0   **e.** $\frac{8}{7}$ **f.** $-\frac{3}{8}$
5. a. $x = 4$ **b.** $x = -9$ **c.** $x = -2$ **d.** $x = -12$ **e.** $x = 28$
  f. $x = 32$ **g.** $x = -36$ **h.** $x = -32$ **i.** $x = 40$
7. Take off jacket, take off vest, take off shirt; dressing and
  undressing
9. Inverse not meaningful; taking pictures
11. Order not important (unless the hole and firewood are in the
  path of the sprinkler)
13. a. $x = 6$ **b.** $x = 5$ **c.** $x = -1$ **d.** $x = 2$
15. a. $x = -13$ **b.** $x = 50$ **c.** $x = 41$ **d.** $x = -22$
17. $x = -1$ **19.** $x = 15$ **21.** $x = 24$ **23.** $x = 32$
25. a. I   **b.** C   **27. a.** C   **b.** I   **29. a.** I   **b.** I   **31.** $x = -3$
33. $x = -9$ **35.** $x = 12$ **37.** $x = 1.5$ **39.** $x = -5.5$
41. $x = 35$ **43.** $x = 38$ **45.** $x = 2.5$ **47.** $x = -2$
49. $x = \frac{1}{2}$ **51.** $t = \dfrac{D}{r}$ **53.** $b = y - mx$ **55.** $r^3 = \dfrac{3V}{4\pi}$
57. $p = \dfrac{1}{N}$ **59.** $b = \dfrac{2A}{h} - a$ **61.** $a_1 = a_n - (n - 1)d$
63. $a = 3A - b - c$ **65.** $a_1 = \dfrac{2S}{n} - a_n$ **67.** $T_c = -ET_h + T_h$
69. $T = \dfrac{V - 344}{0.6} + 20$ **71.** $R = \dfrac{E}{I}$ **73.** $S_e = \dfrac{0.16V}{T}$

75. $r = \dfrac{a_1 + S}{S}$ or $r = \dfrac{a_1}{S} + 1$

CHAPTER 1 REVIEW EXERCISES

1. associative properties for addition and multiplication, commutative properties for addition and multiplication, distributive property for multiplication over addition
3. absolute value, braces, brackets, fraction bar, parentheses, square root
5. independent variable, dependent variable
7. sum, difference, product, quotient
9. input-output relationships, independent variable, dependent variable
11. multiplicative inverses, reciprocals
13. a. $-4, -10$ **b.** $-4, 10$ **c.** $3, 10$
15. a. 7 **b.** $-6, 1$ **c.** $-3, -10$ **d.** $-15, 2$
17. $2\frac{2}{3}$; associative property for addition
19. 1300; commutative property for multiplication
21. 4 **23.** 16.5π in^2 **25.** \$7.75
27. a. The difference between three and the product of two and x
  b. three times the difference between x and five

29. $\dfrac{x}{15} + 8$ **31.** $-26x + 27y + 9$ **33.** -21

35. $A = 12.25\pi$ in^2

37.

x	$y = -x$
-3	3
2	2
-1	1
0	0
1	-1
2	-2
3	-3

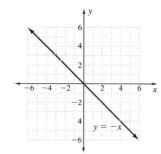

39.

x	$y = 30 - 3.5x$
-3	40.5
-2	37
-1	33.5
0	30
1	26.5
2	23
3	19.5

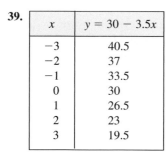

41. $x = 4$ **43.** $F = 104$ **45.** $n = 12$ **47.** $r = \dfrac{1}{Pt}$

49. $b = \dfrac{C - a}{Y}$ **51.** $3x, 3x(2x + 5)$ **53.** $x = 0$
55. $x = -1$ **57. a.** $y = -3x + 10$ **b.** $y = 7x - 27$
59. $x = -3$ and $x = 1$ **61.** $E = 0.167x$; $x = \$2200$
63. ≈ 114 lb, ≈ 144 lb

CHAPTER 1 TEST

1. opposite **2.** ordered pair **3.** independent

4.

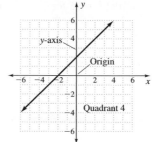

5. a. $-30, -24, 81, 9$ **b.** $-1, -2, -0.75, -3$
  c. $1\frac{1}{4}, 1\frac{3}{4}, -\frac{3}{8}, -6$
6. a. 748; associative property for addition
  b. 989; commutative property for addition
7. a. 160 **b.** -9 **c.** 9 **d.** 500 **e.** $6x - 8$
8. four times the difference between x and three

9. $b = 3A - a - c$ **10.** $b = \dfrac{2A}{h} - a$ **11.** $3xy$

12. a. $x = -5$ **b.** $x = 2$ **c.** $x = -2$ **14.** zero
15. 2.2 m^2 **16.** $x = 7$ **17.** $y = -4x + 3$
18. 1 year = 12 months, so multiply the decimal part by 12 to obtain 8 months
19. $C = 3675x$; 68 months, or 5 years 8 months
20. a. 1 **b.** 7
  c. In part b, the denominator set of parentheses is incomplete, so the calculator assumes a parenthesis at the end of the expression, dividing $\sqrt{4}$ by $2 - 3 = -1$ instead of by 2.
21. ~ 59 miles

EXERCISES 2.1

1. a. $1 > \frac{1}{4}$ **b.** $1 = 1$ **c.** $\frac{1}{2} > \frac{1}{4}$ **d.** $\frac{1}{4} < \frac{1}{2}$ **e.** $0.15 < 0.16$
  f. $-2 > -3$
3. $x \le 2$ **5.** $-1 < x < 5$; x is between -1 and 5
7. 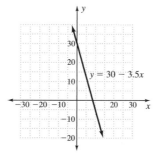 x is greater than or equal to -1

9. $-2 < x < 4$; x is between -2 and 4
11. $x < 2$ or $x > 4$; x is less than 2 or x is greater than 4
13. $x < -3$ or $x > 2$; x is less than -3 or x is greater than 2
15. $x \le -4$ or $x \ge 1$; x is less than or equal to -4 or x is greater than or equal to 1
17. $x > -3$ and $x < 4$ **19.** not appropriate **21.** $-1 < x \le 5$
23. $x \ge -1$ and $x < 1$ **25.** $3 < x < 4$
27. a. $-25 \le x \le 15$, x on the interval $[-25, 15]$
  b. $-10 \le y \le 20$, y on the interval $[-10, 20]$
29. $-3 < x < 5$; $(-3, 5)$
31. $-4 < x \le 2$; $(-4, 2]$
33. $x > 5$; $(5, \infty)$; set of numbers greater than 5
35. $(-\infty, -2)$; set of numbers less than -2
37. $x \le -3$; set of numbers less than or equal to -3
39. $x \ge 4$; set of numbers greater than or equal to 4
41. $y = 4 + 0.5(x - 2)$ for $x > 2$; $y = 4$ for $0 < x \le 2$
43. $y = 20 + 5(x - 3)$ for $x > 3$, x rounded up to next integer; $y = 20$ for $0 < x \le 3$

45. $y = 65 + 0.15(x - 100)$ for $x > 100$; $y = 65$ for $0 < x \leq 100$
47. $y = 85 + 4.75(x - 10)$ for $x > 10$; $y = 85$ for $0 < x \leq 10$
49. $y = 19.95 + 0.25(x - 100)$ for $x > 100$, x rounded up to next
  integer; $y = 19.95$ for $0 < x \leq 100$
51. **a.** $<$ **b.** inequality **c.** $\geq$ **d.** logic phrase "a or b"

EXERCISES 2.2

1. a.

x	$y = 15x - 4$
1	11
2	26
3	41
4	56

b. function; $f(x) = 15x - 4$

3. a.

$x = y^2 + 2$	y
18	-4
6	-2
2	0
6	2
18	4

b. not a function

5. a.

x	$y = 25 - x^2$
-5	0
-2	21
0	25
2	21
5	0

b. function; $f(x) = 25 - x^2$

7. function **9.** not a function **11.** not a function
13. not a function **15.** function **17.** not a function
19. $y = x + 2$ **21.** $y = \frac{1}{3}x$
23. a. $\sqrt{(100 - 36)}$ or $\sqrt{(100 - 36)}$ **b.** $\sqrt{(100)} - 36$;
  b is correct
25. a. $(2 + 3)^2$ **b.** $2 + (3^2)$ or $2 + 3^2$; b is correct
27. a. $abs(3) - 4$ **b.** $abs(3 - 4)$ or $abs(3 - 4$; b is correct
29.

x	$g(x) = 5 + 2(x - 3)$
-2	-5
-1	-3
0	-1
1	1
2	3
3	5
4	7

31.

x	$f(x) = x^2 - 2x - 3$
-2	5
-1	0
0	-3
1	-4
2	-3
3	0
4	5

33.

x	$g(x) = 8 - x^2$
-2	4
-1	7
0	8
1	7
2	4
3	-1
4	-8

35. a. 3 **b.** 11 **c.** $2n + 1$ **d.** $2n + 2m + 1$
37. a. 10 **b.** 0 **c.** $\square^2 + \square - 2$ **d.** $n^2 + n - 2$
  e. $n^2 - 2nm + m^2 + n - m - 2$
39. Exercise 14 **41.** Exercise 15
43. x is any real number; $f(x) \geq 0$
45. x is any real number; $h(x) \geq -2$
47. x is any real number; $g(x) \leq 6$
49. a. domain **b.** negative numbers plus zero **c.** $(-\infty, 0]$
51. a. range **b.** positive numbers **c.** $(0, \infty)$
53. a. domain **b.** positive numbers **c.** $(0, \infty)$
55. a. range **b.** negative numbers plus zero **c.** $(-\infty, 0]$
57.

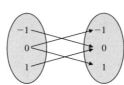

function; for each input, x, there is exactly one output, y.

59.

not a function; the input 0 has two outputs.

61. $r > 0$ **63.** $x > 0$
65. Answers will vary, should be within $0 \leq x \leq 24$
67. 10 **69.** 9
71. In Exercises 1 and 18, inputs are CDs and dress sizes.
  Fractional and decimal values are meaningless, so the domain
  is the positive integers (or natural numbers). In Exercises 14
  and 15, inputs may be fractional or decimal values so the more
  general domain of real numbers is appropriate.
73. $g(a + b)$ **75. a.** 0 **b.** -2
77. a. $x = 3$ or $x = -1$ **b.** $x = 6$ or $x = -4$
79. a. $x = -2$ or $x = 2$ **b.** $x = -1$ or $x = 1$

EXERCISES 2.3

1. $3x - y = 8$ **3.** $C - 1.21x = 0$ **5.** not linear
7. $2\pi r - 11 = 0$ **9.** $10n - t = 2$ **11.** linear **13.** linear
15. not linear **17.** linear **19.** not linear
21. a. $-1, 2$ **b.** $3, -2$ **23.** -1 and $1, -1$ **25.** Exercise 22b
27. Exercise 21a **29.** Exercise 24 **31. a.** 2 **b.** $\frac{2}{3}$ **33.** $-\frac{1}{3}$
35. $\frac{1}{2}$ **37.** $\frac{3}{5}$ **39.** $-\frac{3}{2}$ **41.** $-\frac{95}{11}$ **43.** $-\frac{23}{17}$ **45.** $-\frac{4}{9}$
47. a. \$0.06 tax/\$ sales **b.** $(0, 0)$; for \$0 of sales, tax is \$0
  c. $(0, 0)$; there is 0 sales tax if there are 0 sales
49. a. $-\$0.75$/trip **b.** $\left(26\frac{2}{3}, 0\right)$, maximum number of trips is 26
  c. $(0, 20)$, original value of ticket is \$20.00
51. not linear

53. a. and b.

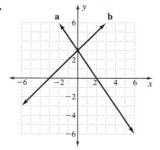

55. a. and b.

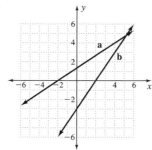

57. slope is $-\dfrac{b}{a}$ **59.** (7, 3)   **61.** (1, 4)

63. a. slope $= -0.05$ **b.** slope $= -0.15$

MID-CHAPTER 2 TEST

1. a. $(-4, +\infty)$

b. $(-\infty, 6)$

c. $[-3, 2]$

d. $(3, 6]$

e. $(-\infty, -2)$ or $(3, +\infty)$

f. $\mathbb{R}$ or $(-\infty, +\infty)$

2. a. $-1 \le x \le 1$; $[-1, 1]$ **b.** $x \ge -3$; $[-3, +\infty)$
   c. $y \ge -2$; $[-2, +\infty)$ **d.** all real numbers, $\mathbb{R}$; $(-\infty, +\infty)$
   e. $-2 < y \le 4$; $(-2, 4]$
3. a. the set of numbers between -1 and 3, including -1
   b. the set of inputs between -4 and -1
   c. the set of numbers less than or equal to -2
   d. the set of outputs less than or equal to -1
4. $y = 16.45 + 0.29(x - 30)$ for $x > 30$; $y = 16.45$
   for $0 < x \le 30$
5. non-negative   **6.** domain   **7.** $(0, y)$
8. a. -2 **b.** 4   **c.** -20 **d.** $3a - 5$
   e. $3(a + b) - 5$ or $3a + 3b - 5$
9. a. 0   **b.** 6   **c.** 30   **d.** $a^2 - a$
   e. $(a + b)^2 - (a + b)$ or $a^2 + 2ab + b^2 - a - b$
10. a. $\mathbb{R}$; $-\infty < x < +\infty$; $(-\infty, +\infty)$
   b. non-negative; $y \ge 0$; $[0, +\infty)$ **c.** yes
11. a. $\mathbb{R}$; $-\infty < x < +\infty$; $(-\infty, +\infty)$
   b. $\mathbb{R}$; $-\infty < y < +\infty$; $(-\infty, +\infty)$ **c.** yes
12. a. $-5 \le x \le 1$; $[-5, 1]$ **b.** $-3 \le y \le 3$; $[-3, 3]$ **c.** no
13. a. $0x + 1y = 5$ **b.** $1x + 0y = 4$ **c.** $2\pi r + 0y = 7$
   d. not linear   **14. a.** $(4, 0)$; $(0, 3)$ **b.** $(-17.78, 0)$; $(0, 32)$

15. a. Slope is $-3/1$; intercept is $(0, 5)$.
   b. Slope is \$3.50 per foot; intercept is $(0, 5)$.
16. a. and b.

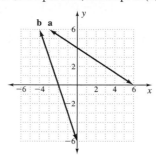

EXERCISES 2.4

1. \$0.055 per dollar; \$0   **3.** \$3.00 per person; \$10   **5.** 2π; 0
7. μ; 0   **9.** b; a **11.** $y = 8x - 4$ **13.** $y = \frac{1}{2}x - 8$
15. $y = -2x$ **17.** $y = \frac{8}{3}x - 2$ **19.** $y = -x + 2$
21. a. Pulse rate is a function of age.   **b.** Answers vary.
   c. $P = 0.5(220 - x)$ **d.** $P = 0.7(220 - x)$ **e.** 85 to 119
   f. 30
23. \$300; \$0.025; $C = 0.025x + 300$, C in \$
25. $y = -\frac{1}{3}x + \frac{17}{3}$ **27.** $y = 3x - 11$ **29.** $y = -2x + 9$
31. $C = 43.70x + 75$, C in \$; \$75; \$43.70   **33.** $F = \frac{9}{5}C + 32$
35. $y = 0.5x - 1.01$ **37.** $y = \frac{1}{3}x + 2.66$
39. average slope 0.60, $y = 0.60x + 2.19$; $y = 0.578x + 2.247$,
   $r = 0.9962$
41. average slope 0.35, intercept $3.99 - 5(0.35) = 2.24$,
   $y = 0.35x + 2.24$; $y = 0.359x + 2.818$, $r = 0.9889$
43. average slope 0.38, intercept $4.99 - 0.38(4) = 3.47$,
   $y = 0.38x + 3.47$; $y = 0.314x + 4.051$, $r = 0.9882$
45. $y = 5x - 13$ **47.** $y = 2x + 7$ **49.** $y = 6x - 4$
51. $y = 0.4x - 0.21$, $r \approx 1$
53. \$0.10 is added to the price for each quarter-inch increase in
   diameter. Each ordered pair exactly fits the price increase rule.
   Note that 0.10 to $\frac{1}{4}$ equals 0.40 to 1, which is the slope.
55. linear; $y = -x + 17$
57. $y \approx 0.135x + 1.29$, x in oz, y in \$

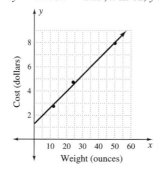

59. $y \approx 1051.3x - 32.9$, x in carats, y in \$
61. The slope is $m = \Delta y/\Delta x$, and the output for $x = 0$ is b
   in $y = mx + b$.

EXERCISES 2.5

1. undefined, $x = 4$ **3.** 0, $y = -3$ **5.** $y = -3$ **7.** $x = -1$
9. $y = 0$
11. *Hint:* Write numbers as fractions or improper fractions.

13. perpendicular **15.** perpendicular **17.** parallel **19.** neither
21. $C = 78x$, $C = 98x$, $C = 108x$; not parallel
23. $V = 5.00 - 0.15x$, $V = 10.00 - 0.15x$, $V = 20.00 - 0.15x$; parallel
25. Although the postage cost appears to be the slope, buying in groups of stamps changes the slope. If the postage cost is $0.37, $C = 100(0.37)x$, $C = 50(0.37)x$, $C = 20(0.37)x$; not parallel
27. $y = -\frac{2}{3}x$ **29.** $y = 2x$ **31.** $y = -\frac{8}{5}x + \frac{31}{5}$
33. $y = \frac{4}{3}x + \frac{11}{3}$ **35.** $y = -\frac{2}{5}x - \frac{1}{5}$ **37.** $\frac{1}{2}$, -2, $\frac{1}{2}$, -2; opposite lines are parallel; adjacent lines are perpendicular

39.

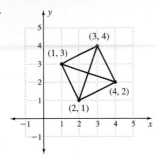

a. 3 **b.** $-\frac{1}{3}$
Diagonals are perpendicular

41. Opposite sides should have same slopes; adjacent sides, negative reciprocal slopes; is not rectangle; is rectangle
43. Nonvertical lines are perpendicular if their slopes multiply to -1. If nonvertical lines are perpendicular, then their slopes multiply to -1.

EXERCISES 2.6

1. domain $\mathbb{R}$, range 2 **3.** domain $\mathbb{R}$, range -2

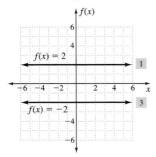

5. constant function **7.** constant function
9. constant function **11.** identity function
13. identity **15.** neither
17. identity function **19.** identity
21. identity function
23. a. 5, 3, 2, 1, 1, 3

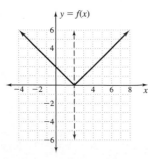

b. $f(0) = f(4) = 2$ **c.** $x = 2$

25.

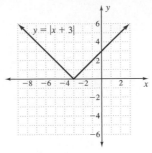

domain $\mathbb{R}$, range $y \geq 0$

27.

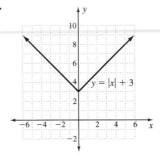

domain $\mathbb{R}$, range $y \geq 3$

29.

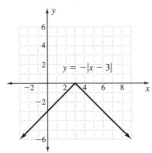

domain $\mathbb{R}$, range $y \leq 0$

31.

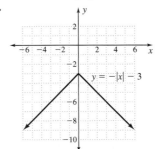

domain $\mathbb{R}$, range $y \leq -3$
33. a. $x = -1$ or $x = 4$ **b.** $x = 0$ or $x = 3$ **c.** $x = \frac{3}{2}$
  d. no solution
35. a. -2 **b.** 2 **c.** 3, 0 **d.** $y = -2x + 3$, $x \leq 1.5$ **e.** 0
  f. $y = 2x - 3$, $x \geq 1.5$ **g.** 0
37. a. $x = -10$ or $x = 2$ **b.** no solution **c.** $x = -6$ or $x = -2$
  d. $x = 4$ or $x = -12$
39. $\{\pm 4\}$ **41.** $\{-5, 1\}$ **43.** $\{3, 7\}$ **45.** $\{2, 6\}$
47. a. 2 **b.** 3 **c.** 4 **d.** 1
49. a. 270 mi **b.** 252 mi
  $D = |x_1 - x_2|$

51. a. dot graph
b.

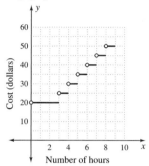

53. a. step graph
b.

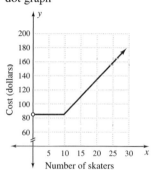

55. a. dot graph
b.

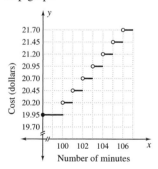

Note: Dots in graph appear as a solid line.

57. a. step graph
b.

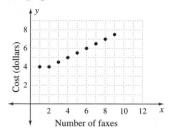

59. part of an hour, portion of a minute

CHAPTER 2 REVIEW EXERCISES

1. vertical-line test   **3.** dot graph   **5.** relevant domain
7. linear function   **9.** identity function
11. absolute value function (squaring function is not in the list)
13. subscripts
15. regression, point-slope, slope-intercept, sequence
17. a. $-8 < x \le -4$; $(-8, -4]$

b. $-\infty < x < +\infty$; $(-\infty, +\infty)$

c. $-2 < x < 7$; $(-2, 7)$

d. $x > -3$; $(-3, +\infty)$

e. $x > 0$; $(0, +\infty)$

f. $x \ge 0$; $[0, +\infty)$

19. a. $-6.2 \le x \le 2.2$ **b.** $-3 \le y \le 3$ **c.** no
21. a. all real numbers, $\mathbb{R}$ **b.** $y \ge 0$ **c.** yes
23. a. 1   **b.** 4   **c.** $\frac{1}{2}$ **d.** 8   **e.** 2   **f.** $x \ge 1$ **g.** none
   h. $x \le 2$ **i.** 1   **j.** all real numbers   **k.** $y > 0$
25. 0   **27.** 0   **29.** 15   **31.** $2\square^2 - 3\square + 1$
33. a. $y = -\frac{1}{3}x$ **b.** $y = -\frac{1}{3}x + 5$
35. -1; $y = -x + 4$; neither; 4; 4   **37.** -1; $y = -x$; neither; 0; 0
39. 2; $y = 2x + 1$; neither; $-\frac{1}{2}$; 1
41. 0; $y = 3$; horizontal; none; 3
43. Exercises 35 and 37 are parallel; none are perpendicular.
45. a. $y = 0.065x$
   b. $0.065 tax/$1 purchased; $0, no tax if nothing is purchased
47. a. $y = 500 + 45x$
   b. $45/hour of repair; $500, basic inspection cost
49. $y \approx 11,528 - 42x$ **51.** $y = 1.50x + 8$ **53.** $y = -4x + 35$
55. $C = \$350$; constant   **57.** $C = 5.95x$, C in $; increasing
59. $V = 350 - 5x$, V in $; decreasing
61. $x = $ no. of people, $y = $ total cost; $y = 85 + 4.75(x - 10)$
   for $x > 10$, $y = 85$ for $0 < x \le 10$; dot graph
63. $x = $ no. of min, $y = $ cost; $y = 0.26 + 0.19(x - 1)$ for $x > 1$,
   $y = 0.26$ for $0 < x \le 1$; step graph
65. a. $\{-3, 5\}$ **b.** $\{-1, 3\}$ **c.** $\{1\}$ **d.** $\{\ \}$

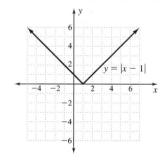

67. all real numbers; 365   **69.** all real numbers; $y \ge 0$
71. all real numbers; all real numbers
73. all real numbers; $y \ge 1$ **75.** $x = -2, x = 14$

CHAPTER 2 TEST

1. a. $x \le 5$; $(-\infty, 5]$

b. $-2 < x < 5$; $(-2, 5)$

c. all real numbers; $(-\infty, +\infty)$

2. **a.** not a function   **b.** function   **c.** function
   d. not a function
3. **a.** 12   **b.** −4   **c.** 4
4. **a.** $-\frac{2}{7}$ **b.** $y = -\frac{2}{7}x + \frac{24}{7}$ **c.** $-\frac{2}{7}$ **d.** $y = \frac{7}{2}x - 8$
5. **a.** zero   **b.** negative; decreasing   **c.** linear   **d.** constant
   e. constant   **f.** positive integers or natural numbers
6. **a.** $y = 7x + 2.5$, y in $ **b.** $7 per mile
7. **a.** time in minutes $= x$; cost in dollars $= y$
   b. $(1, 0.13)$, $(19, 2.11)$ **c.** $y = 0.11x + 0.02$
8. $y \approx 10.1x - 13.8$
9. **a.** 50; $y = 8x + 2$ **b.** 19; $y = 7x - 23$

10.
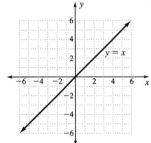

11.
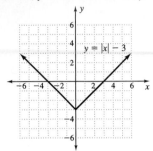

12.
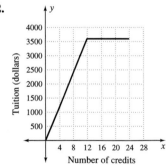

13.
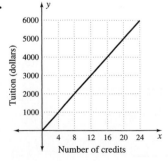

14.
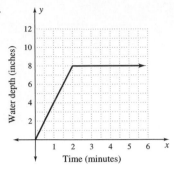

Water overflows bucket when $y = 8$.

15.
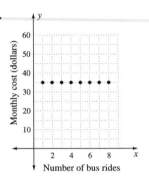

16.

Transcripts	Total Cost ($)
1	5
2	5
3	7
4	9
5	11
6	13

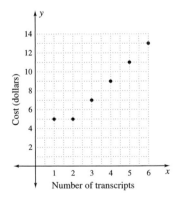

We are not purchasing part of a transcript, so points should not be connected.

17. $x = -3.75$, $x = 1.25$

CUMULATIVE REVIEW OF CHAPTERS 1 AND 2

1.

Input x	Input y	Output xy	Output $x + y$	Output $x - y$
-2	4	-8	2	-6
-3	7	-21	4	-10
2	-3	-6	-1	5
-3	-2	6	-5	-1
-1	-6	6	-7	5
-5	-2	10	-7	-3
3	-2	-6	1	5
2	-9	-18	-7	11

3. a. opposites **b.** factors **c.** reciprocals **d.** factoring
e. sets
5. a sum to a product
7. division to multiplication by the reciprocal
9. $2ac - 2bc$ **11.** 56.25π ft^2 **13.** does not simplify
15. $x = 15$ **17.** $3x = x + 15; x = 7.5$ **19.** 1, 2, 3, 4
21. 1, 0, 1, 4 **23.** $-\frac{2}{5}; y = -\frac{2}{5}x - \frac{7}{5}$ **25.** $y = -\frac{1}{3}x$
27. $(4, 4); f(x) = x$ **29.** $(4, 4); f(x) = 4$
31. $y = 65x + 500$, x in no. of workers, y in \$
33. a. $\{-1\}$ **b.** $\{-4, 2\}$ **c.** $\{-5, 3\}$ **d.** $\{-3, 1\}$ **e.** $\{\ \}$

EXERCISES 3.1

1. $y = -3x + 4$ **3.** $x = 3y + 7$ **5.** $x - 2y - \frac{5}{7}$
7. $y = 2x + \frac{3}{2}$ **9.** $x = 3, y = 5$ **11.** $x = -7.5, y = 7.5$
13. $x = 5, y = 2.2$ **15.** $x = 3, y = -1.2$ **17.** $x = -4, y = 2$
19. $x = 1.5, y = -1.2$ **21.** $x = 6, y = -3$ **23.** $x = -1, y = 2$
25. $A = -20, B = 38$ **27.** $C = -13, D = 15$
29. $E = 1.8, F = 7.2$ **31.** $G = -1.6, H = 2.4$
33. What is total value of money? Dimes are \$0.10; quarters, \$0.25.
35. What is measure of each angle? Sum of the measures of three interior angles in triangle is 180°.
37. What is the perimeter? What are the measures of the other two sides? Lengths of opposite sides of a rectangle are equal.
39. \$5000, \$750 **41.** \$10,000, \$1150 **43.** \$1.05, \$0.05
45. 12 in., 8 in. **47.** 110 yd, 80 yd **49.** 57.5°, 32.5°
51. 135°, 45° **53.** 90°, 0°; impossible with nonzero angles
55. $y = \dfrac{ce - af}{bc - ad}$ or $y = \dfrac{af - ce}{ad - bc}$
57. One variable has coefficient 1 and is easy to solve for.

EXERCISES 3.2

1.

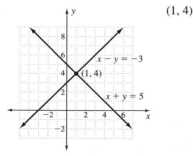

$(1, 4)$

3.

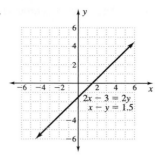

coincident; variables drop out, identity

5.

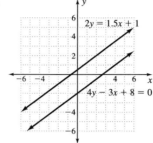

parallel; variables drop out, contradiction

7.

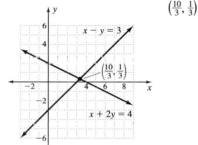

$\left(\frac{10}{3}, \frac{1}{3}\right)$

9.

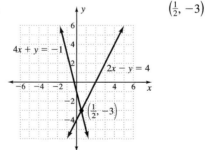

$\left(\frac{1}{2}, -3\right)$

11.

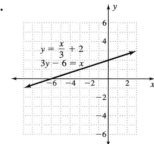

coincident; variables drop out, identity

13.

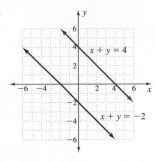

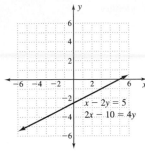

parallel; variables drop out, contradiction

coincident; variables drop out, identity

17. $5000; $750   **19.** none; 4 is always better
21. becomes steeper   **23.** packages 3 and 5
25. Write equations for pair of parallel lines.   **27.** yes
29. $0x + y = 3; a = 0, b = 1, c = 3$
31. a. $n = 0$ **b.** $n = 1$ **c.** the identity, 1 **d.** $y = x$ or $f(x) = x$

EXERCISES 3.3

1. $1600, $12,000; 5% and 8% annual interest
3. 20 hr, 15 hr; $6.50/hr and $7.25/hr
5. 100 mL, 1000 mL; 8% and 0% solution
7. 3.09 points/credit hour   **9.** 17 credit hours
11. ≈156 kg, ≈44 kg   **13.** ≈0.1 oz, ≈0.2 oz
15. $25,000, $50,000   **17.** $2000, $8000   **19.** $5600, $18,000
21. 0.45 L, 0.05 L   **23.** 800 L, 200 L   **25.** 850 gal, 150 gal
27. 400 lb, 600 lb

MID-CHAPTER 3 TEST

1. $y = \frac{5}{2}x - 2$ **2.** $y = 4x - 10$ **3.** contradiction; parallel lines
4. $\left(2, -\frac{2}{3}\right)$ **5.** $(12, -11.4)$ **6.** identity; coincident lines
7. $x = \frac{5}{8}, y = -\frac{1}{8}$ **8.** $x = -30, y = 24$
9. $y = -x + 2, y = \frac{2}{3}x + 2; x = 0, y = 2$ **10.** 135°, 45°
11. 31,419; 23,149 **12.** 10 hr **13.** $\frac{1}{3}$ pt, $\frac{2}{3}$ pt

EXERCISES 3.4

1. $w = 2c, s = 3c, c + s + w = 60; c = 10$ hr, $s = 30$ hr, $w = 20$ hr
3. $x = y, z - 21 = x, x + y + z = 180; 53°, 53°, 74°$
5. $c = 2p, c + f + p = 28, 4c + 9f + 4p = 222; c = 4$ g, $f = 22$ g, $p = 2$ g **7.** $a = 3, b = -2, c = \frac{1}{2}$
9. $a = 5, b = -1, c = -2$ **11.** $x = -9, y = 8, z = 3$
13. $x = -1, y = 6, z = 5$ **15.** $x = 2.5, y = 3, z = 3.5$
17. $a = -1, b = -3, c = 7$ **19.** $a = 2, b = 6, c = 2$
21. $x = 5, y = -2, z = 1$ **23.** contradiction, inconsistent
25. identity, dependent **27.** identity, dependent

29. contradiction, inconsistent
31. a. plane; dependent   **b.** plane; dependent
  c. no common point of intersection; inconsistent
  d. no common point of intersection; inconsistent
33. 29 pennies, 24 dimes, 8 quarters
35. $O + S = P + 412,545, S = O + 459,907,$
$O + P = 1,445,464$; Okamoto $466,034, Strange $925,941,
Pavin $979,430
37. $f = p, c + f + p = 29, 4c + 9f + 4p = 121; c = 27$ g,
$f = 1$ g, $p = 1$ g
39. $p + f + c = 32, c = f + 12, 4p + 9f + 4c = 148; p = 12$ g,
$f = 4$ g, $c = 16$ g
41. $n + d + q = 44, q = d + 3, 0.05n + 0.10d + 0.25q = 5.80$;
17 nickels, 12 dimes, 15 quarters
43. cannot have both no solution and an infinite number of
solutions

EXERCISES 3.5

1. $y = -2$ is below $y = -2x + 3$ for $x < 2.5$
3. $y = -2x + 3$ is above $y = 0$ for $x < 1.5$
5. $y = x + 4$ is above $y = -\frac{1}{2}x + 1$ for $x > -2$
7. $y = 0$ is below $y = x + 4$ for $x > -4$
9. $y = -\frac{1}{2}x + 1$ is below or the same as $y = 0$ for $x \geq 2$
11. $y = -2x + 3$ is above or the same as $y = 3$ for $x \leq 0$
13. $y = x + 4$ is below or the same as $y = 4$ for $x \leq 0$
15. $x > -1$
17. $x < \frac{3}{2}$
19. $x < \frac{1}{3}$
21. $x \geq \frac{1}{3}$
23. $x \leq -\frac{1}{3}$
25.

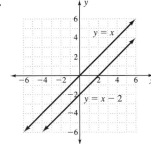

$y = x - 2$ is below $y = x$ for all real numbers.
27. for example, $x - 2 > x$ **29.** $x > \frac{1}{3}$ **31.** $x < 6$
33. $x \leq 25$ **35.** $x < -4$ **37.** $x > 2$ **39.** $x \leq -1.25$
41. $2 < x < 4$ **43.** $-2 > x > -4$ **45.** $5 > x > -2$
47. $-5 \leq x \leq 4$
49. a. $x = -2$ **b.** all real numbers **c.** $-6 < x < 2$
  d. all real numbers **e.** no solution **f.** $x = 1$ or $x = -5$
  g. $x < -5$ or $x > 1$ **51.** $x < -2$ or $x > 5$
53. $x > -1.5$ and $x < 4; -1.5 < x < 4$ **55.** $x < -3$ or $x > 15$
57. $x \geq -6$ and $x \leq 10; -6 \leq x \leq 10$ **59.** $0.025x \leq 5, x \leq 200$
61. $5 \leq 0.02x \leq 25, 250 \leq x \leq 1250$
63. $3.75 + 0.50x < 9 + 0.25x; x < 21$ photos
65. $30 + 6x < 50 + 5x; x < 20$ guests

67. There is a \$20 difference in basic costs. The per-guest charge differs by \$1. It will take $20 \div 1 = 20$ guests for the party costs to be equal.

69. Choose a number satisfying the answer and see if it works.

71. Negative changes relative value of sides; for example, $20 > 10 \Rightarrow (-1)(20) \boxed{?} (-1)(10) \Rightarrow -20 < -10$

73. Dot indicates that value is a solution; small circle indicates that value is excluded.

CHAPTER 3 REVIEW EXERCISES

1. $x = -0.8, y = -1.6$ **3.** identity (infinite number of solutions), coincident lines

5. contradiction ($\{\ \}$ or $\varnothing$), parallel lines **7.** $x = 6, y = -8$

9. $x = 5, y = -2$ **11.** $x = 6, y = 1$ **13.** $x = \frac{1}{2}, y = 2\frac{1}{3}$

15. $a = 2, b = -3, c = 1$

17. identity, infinite number of solutions (dependent)

19. $x = 6.2, y = -7, z = 2.4$

21. $y = 2x - 3, y = -x - 6; x = -1, y = -5$

23. $y = \frac{1}{2}x + 3, y = -\frac{1}{4}x; x = -4, y = 1$ **25.** 16 credit hours

27. 200 pounds, 800 pounds **29.** 5 in., 4 in., 3 in.

31. 10 in., 6 in., 4 in. **33.** Jeanne 21, Tish 19 **35.** 150°, 30°

37. There are an infinite number of solutions (any pair of angle measures adding to 90°).

39. $p = 19$ g, $f = 1$ g, $c = 29$ g

41. 23 nickels, 10 dimes, 41 quarters

43. 5 textbooks, 3 paperbacks, 8 notebooks

45. a. $x > 2$
-4 -3 -2 -1 0 1 2 3 4

b. $x < -2$
-4 -3 -2 -1 0 1 2 3 4

c. $x < 1$
-4 -3 -2 -1 0 1 2 3 4

47. $x \geq -1$ **49.** $x < 1$ **51.** $x \leq 1$

53. a. $147.7 < x < 186.5$ **b.** $103.7 < x < 131.0$

55. $8 < 0.05x < 9; 160 < x < 180$ **57.** $-\frac{3}{4} < x < \frac{9}{4}$

59. $x < \frac{16}{3}$ or $x > 8$

CHAPTER 3 TEST

1. $x = 108, y = 78$ **2.** contradiction ($\{\ \}$ or $\varnothing$), parallel lines

3. $x = -3, y = -4$

4. identity (infinite number of solutions), coincident lines

5. contradiction ($\{\ \}$), inconsistent

6. $x = 4.5, y = -3, z = -2.5$ **7.** 12 in., 10 in., 2 in.

8. 1956, 1966 **9.** 365 gal, 35 gal

10. 25 pennies, 12 nickels, 22 quarters

11. 6 juice, 3 apples, 2 Oreos®

12. a. true **b.** true **c.** false **d.** false

13. $y = x - 1, y = -\frac{1}{2}x - 4; (-2, -3)$

14. $-8 < x < 4$ **15.** $-4 < x < 12$

16. 124.7 lb $\leq x \leq 157.5$ lb

EXERCISES 4.1

1. a. 729; neither **b.** 46; linear **c.** 11; linear **d.** 49; quadratic

3. a. neither **b.** neither **c.** quadratic **d.** linear **5. a.** r; π, $2\pi(3)$, 0 **b.** T; $g/4\pi^2$, 0, 0 **c.** x; $\frac{1}{2}$, $\frac{1}{2}$, 0 **7. a.** not quadratic **b.** quadratic; 1, 3, 4 **c.** quadratic; 1, −3, −8 **d.** not quadratic

9.

x	$y = 2x^2 + x + 1$
-3	16
-2	7
-1	2
0	1
1	4
2	11
3	22

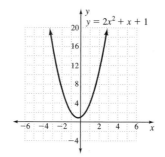

11.

x	$y = -x^2 - 2$
-3	-11
-2	-6
-1	-3
0	-2
1	-3
2	-6
3	-11

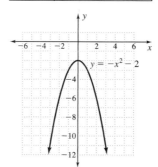

13. x-intercepts $(1, 0)$ and $(3, 0)$; y-intercept $(0, 3)$; vertex $(2, -1)$

15.

17. 15 **19. a.** $\{-1, 5\}$ **b.** $\{1, 3\}$ **c.** $\{-2, 6\}$ **d.** $\{\ \}$ or $\varnothing$

21. a. $(-2, 0), (4, 0)$ **b.** $(0, -8)$ **c.** $x = 1$ **d.** $(1, -9)$ **e.** $\{1\}$ **f.** $\{-1, 3\}$ **g.** $\{-2, 4\}$ **h.** $y \geq -9$

23.

x	$y = -x^2 - 2x + 8$
-4	0
-3	5
-2	8
-1	9
0	8
1	5
2	0

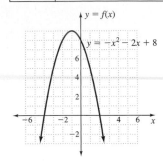

a. $(-4, 0), (2, 0)$ **b.** $(0, 8)$ **c.** $x = -1$ **d.** $(-1, 9)$
e. $\{-4, 2\}$ **f.** $\{-2, 0\}$ **g.** $\{\ \}$ or $\varnothing$ **h.** $\{-3, 1\}$ **i.** $y \leq 9$

25.

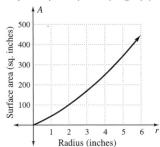

r	$A = 2\pi r^2 + 12\pi r$
0	0
1	44
2	101
3	170
4	251
5	346
6	452

a. ≈ 3.4 in.   **b.** ≈ 5.5 in.
c. Top and bottom surface area varies with square of radius.
27. a linear equation
29. Range is bounded by vertex's y-coordinate.   **31.** $x = -4$

EXERCISES 4.2

1. quadratic; $y = x^2 + 6x$ **3.** linear; $y = 4x$
5. quadratic; $y = 2x^2 + 13x - 7$ **7.** quadratic; $y = x^2 + 4x$
9. linear; $y = 12x + 1$ **11.** neither
13. quadratic; $y = -x^2 + 10x$ **15.** $y = x^2 - 2x + 3$; 38, 51, 66
17. $y = 4x^2 - 1$; 143, 195, 255
19. a. Unless there is a reason (such as that larger containers cost
relatively more to make) to price larger amounts relatively
higher, the linear equation is reasonable.
b. $y = 0.245x + 0.891$ **c.** $y = 0.0140x^2 + 0.0233x + 1.424$
d. \$4.81 and \$5.38; respectively. The parabola is quite flat; so
the costs for 16 ounces are within \$0.60.

21. a. Quadratic; a linear equation would suggest that pots of 2 or
4 inches were free.
b. $y = 0.5x - 2.16$
c. $y = 0.0375x^2 - 0.175x + 0.69$
d. \$5.84 and \$7.49, respectively. The cost of plastic may make
the quadratic equation more reasonable for larger pots.
Larger pots take up relatively more shelf space and require
relatively more material.
23. a. $y = -80x + 656$, $y = -16x^2 + 576$
b. drop of 80 ft per sec from starting height of 656 ft
c. 0 ft per sec. It cannot travel as far in the first second as in
later seconds; graph is not linear.
d. starting height, 576 ft   **e.** 6 sec
25. a. $B(-50, 0), C(50, 0), D(0, 30)$; $y = -0.012x^2 + 30$
b. For $B(0, 0)$: $C(100, 0), D(50, 30)$; $y = -0.012x^2 + 1.2x$.
For $C(0, 0)$: $B(-100, 0), D(-50, 30)$; $y = -0.012x^2 - 1.2x$. For $D(0, 0)$: $B(-50, -30), C(50, -30)$;
$y = -0.012x^2$.
27. For $A(0, 0)$: $B(200, 150), C(400, 0)$, $y = -0.00375x^2 + 1.5x$.
For $B(0, 0)$: $A(-200, -150), C(200, -150)$; $y = -0.00375x^2$.
For $C(0, 0)$: $A(-400, 0), B(-200, 150)$;
$y = -0.00375x^2 - 1.5x$. For $D(0, 0)$: $A(-200, 0)$,
$B(0, 150), C(200, 0)$; $y = -0.00375x^2 + 150$.
29. a.

x	$f(x)$	1st diff.	2nd diff.
1	$a + b + c$		
2	$4a + 2b + c$	$3a + b$	
3	$9a + 3b + c$	$5a + b$	$2a$
4	$16a + 4b + c$	$7a + b$	$2a$
5	$25a + 5b + c$	$9a + b$	$2a$

b. Yes; the second difference, $2a$, is constant.
c. Yes; could use this to find b.
d. and e.

x	$f(x) = ax + b$	1st diff.
1	$a + b$	
2	$2a + b$	a
3	$3a + b$	a
4	$4a + b$	a
5	$5a + b$	a

f. slope $= a =$ first difference; work backward in the table to
find $f(x)$ for $x = 0$ to get b.

EXERCISES 4.3

1. a. $1 \cdot 48, 2 \cdot 24, 3 \cdot 16, 4 \cdot 12, 6 \cdot 8$
b. $1 \cdot 36, 2 \cdot 18, 3 \cdot 12, 4 \cdot 9, 6 \cdot 6$
c. $1 \cdot 72, 2 \cdot 36, 3 \cdot 24, 4 \cdot 18, 6 \cdot 12, 8 \cdot 9$
3. a. not polynomial   **b.** binomial
5. a. monomial   **b.** trinomial
7.

m	n	$m + n$	$m \cdot n$
-3	-5	-8	15
3	4	7	12
2	6	8	12
3	5	8	15
-4	-6	-10	24
-2	-12	-14	24
-3	-8	-11	24
2	-6	-4	-12

9. a. $x^3 - 8$ **b.** $x^3 - 9x^2 + 27x - 27$
11. a. $3x - 1$ **b.** $7x - 6$ **13. a.** $x^2 - 4$ **b.** $12 - x - x^2$
15. a. $x^3 + 3x^2 + 3x + 1$ **b.** $a^3 + b^3$
17. $x^3 - 3x^2y + 3xy^2 - y^3$ **19.** $6x^2 + 7x - 3$
21. $9x^2 + 6x + 1$ **23.** $x^3 + 8$ **25.** $x^2 + 3x - 18$
27. $x^2 - 3x - 18$ **29.** $x^2 - 11x + 18$ **31.** $x^2 - 16$
33. $x^2 - 10x + 25$ **35.** $2x^2 + 5x - 12$ **37.** $2x^2 - 5x - 12$
39. $9x^2 + 3x - 2$ **41.** $x^2 + 4x + 4$ **43.** $4x^2 - 1$ **45.** $x^3 - 8$
47. $x^4 - 13x^2 + 4$ **49.** $(2x - 3)(6x + 1)$ **51.** $(3x - 4)(x - 3)$
53. $(x + 4)(x + 3)$ **55.** $(x - 3)(x + 4)$ **57.** $(x - 3)(x - 4)$
59. $(x - 12)(x + 1)$ **61.** $(x + 7)(x - 4)$ **63.** $(2x + 5)(3x + 2)$
65. $(2x + 5)(3x - 2)$ **67.** $(6x - 5)(x - 2)$
69. $(3x - 2)(5x + 3)$ **71.** $(2x + 1)(3x + 5)$
73. $(10x + 1)(x + 6)$ **75.** $2(3x - 1)(x - 5)$
77. $x(x^2 + 2x + 4)$ **79.** $x(x - 3)^2$ **81.** $a(a^2 - ab + b^2)$
83. $a(a^2 + ab + b^2)$
85. All are in the form $(a \pm b)^2$; Exercises 32 and 33
87. $3(x - 4)(x + 4)$ **89.** $3(2x - 3)(2x + 3)$
91. $5(2x - 3)(2x + 3)$
93. $7(2 - 3x)(2 + 3x)$; all are in the form $(a + b)(a - b)$
95. a. factors **b.** monomial **c.** binomial **d.** trinomial
  e. square **f.** perfect cube
97. a. $x^4 - x^2 - 2x - 1$ **b.** $x^4 - x^2 + 2x - 1$
  c. $x^4 + 2x^3 + x^2 - 1$ **d.** $x^4 - 2x^3 + x^2 - 1$
  e. $x^4 + 2x^3 + 3x^2 + 2x + 1$

MID-CHAPTER 4 TEST

1. a. vertex **b.** y-axis or $x = 0$ **c.** binomial
  d. $y = ax^2 + bx + c$ **e.** y
2. a. $x = -1$ or $x = -5$ **b.** $x = 1$ or $x = 3$ **c.** $x = 2$
3. $a = \frac{1}{2}, b = \frac{1}{2}, c = 1$ **4.** $a = 1000, b = 2000, c = 1000$
5. a.

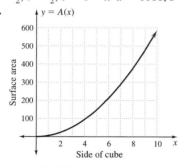

x	$f(x) = 6x^2$
0	0
1	6
2	24
3	54
4	96
5	150
6	216
7	294
8	384

  b. It is 4 times the original.
6. a. quadratic; $y = x^2 - 10x + 24$ **b.** linear; $y = 5x - 1$
  c. linear; $y = 7x - 10$
  d. neither; $(-1)^n$ or $y = 1$ for odd x, $y = -1$ for even x
7. $f(x) \approx 4.9x^2$; quadratic regression because second differences
  are a constant 9.81

8. a. $x^3 - 7x^2 + 27x - 27$ **b.** $x^3 - 27$ **c.** $7a + b$
  d. $5a + b$
9. a. $x^2 - 25$ **b.** $x^2 - 2x + 1$ **c.** $-x^2 - 2x + 3$
  d. $-4x^2 + 9$
10. a. $(3x - 4)(x + 3)$ **b.** $x(x - 1)$ **c.** $(2x - 1)(3x + 4)$
  d. $3x(x + 1)^2$
11. a. $h = -16t^2 + 60t$
  b.

t	$h = -16t^2 + 60t$
0	0
1	44
2	56
3	36
4	-16

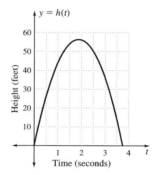

  c. $(0, 0), (3.75, 0)$; first is at the instant the ball is thrown
  d. $\{1.75, 2\}$; on the way up and down **e.** 56.25 ft

EXERCISES 4.4

1. $x^2 + x - 12$ **3.** $2x^2 - 5x - 12$ **5.** $x^2 + 6x + 9$; pst
7. $x^2 - 36$; ds **9.** $4x^2 - 20x + 25$; pst **11.** 8 **13.** 14, 7
15. 6.25 **17.** 12.25, 3.5 **19.** 144, 12 **21.** $(x + 2)^2$
23. $(n + 1)^2$ **25.** $(7x + 1)^2$ **27.** $(3x - y)^2$ **29.** $(x + 2.5)^2$
31. $(x + 2)(x - 2)$ **33.** cannot be factored
35. $(7x + 1)(7x - 1)$ **37.** $(y - 6z)(y + 6z)$
39. $(3x - y)(3x + y)$ **41.** $(x + 12)(x - 12)$
43. $(0.5x + 0.1)(0.5x - 0.1)$
45. The diagonal must both add to zero and multiply to $+16n^2$;
  both conditions cannot be met.
47. shift left 5 units; $(-5, 0)$ **49.** shift $y = (2x)^2$ right 2.5 units
51. a. $(0,0), (-2,0)$; let $sx - t = 0$, solve for x; on x-axis at t/s
  b. $\left(-\frac{1}{2}, 0\right)$ **c.** $\left(\frac{4}{3}, 0\right)$
53. a. right $|t/s|$ units when $t > 0$, left $|t/s|$ units when $t < 0$
  b. shift right $\frac{1}{2}$ unit **c.** shift left $\frac{3}{2}$ units
55. shift down 64 units **57.** $x^3 + 1$ **59.** $x^3 - 9x^2 + 27x - 27$
61. $x^3 - x^2 - x + 1$ **63.** $x^2 - 4x + 16$
65. a. $(x + 3)(x^2 - 3x + 9)$ **b.** $(x + 2)(x^2 - 2x + 4)$
  c. $(x - 4)(x^2 + 4x + 16)$
67. a. $(x - 0.1)(x^2 + 0.1x + 0.01)$ **b.** $\left(x - \frac{1}{3}\right)\left(x^2 + \frac{1}{3}x + \frac{1}{9}\right)$
69. no; shifted left 2 units; moved up 8 units
71. a. $\pi(R - r)(R + r)$
  b. $R - r$, the distance between the outer and inner radii
73. Answers will vary. **75. a.** 399 **b.** 1599
77. $x^2 - 4 = (x - 2)(x + 2)$

EXERCISES 4.5

1. $\{-2, 3\}$ **3.** $\left\{-1, \frac{1}{2}\right\}$ **5.** $\{\pm 11\}$; ds **7.** $\{\pm 4\}$; ds **9.** $\left\{0, \frac{1}{2}\right\}$
11. $\{0, 3\}$ **13.** $\{1, 2\}$ **15.** $\{3\}$; pst, double root **17.** $\{-2, 5\}$

19. $\{-6, 4\}$ **21.** $\{-1, \frac{1}{2}\}$ **23.** $\{-\frac{1}{3}, 2\}$ **25.** $4, -3$ **27.** $\frac{5}{2}, -3$
29. $\frac{4}{3}, -\frac{1}{2}$ **31.** $a = -1$ **33.** $a = 1.5$ **35.** $y = 2x^2 - 2x - 12$
37. $y = -2x^2 + 2x + 12$ **39.** $y = -0.5x^2 + x + 7.5$
41. $y = 2x^2 - 6x - 20$ **43.** $y = -0.12x^2 + 4.8x - 36$
45. $y = -\frac{3}{40}(x - 40)(x - 10)$; average x-intercept values, substitute to find y
47. Use zero product rule and solve factor equations.
49. $0 = a(x + 2)(x + 1)$, a any real number
51. $0 = a(2x + 3)(2x - 1)$, a any real number
53. **a.** $t = 5$ sec; $v = -144$ ft/sec; baseball catcher's mitt or bucket of foam, protective face mask, and helmet are suggested.
  b. $t = 4$ sec; $v = -144$ ft/sec; same precautions; speed is same in both settings.
55. **a.** $(8.5 - 2x)(11 - 2x)$ **b.** $x = 4.25$, $x = 5.5$
  c. $x = 4.25$ uses the whole width of the paper; it is impossible to make squares with $x = 5.5$.
  d. $x = 1.25$ in. (reasonable) and $x = 8.5$ in. (impossible)
57. $x = 7.5$ and $x = 12.5$

CHAPTER 4 REVIEW EXERCISES

1. $y = 3x^2 - 18x + 23$; $a = 3, b = -18, c = 23$
3. $x^2 - 4x - 7 = 0$; $a = 1, b = -4, c = -7$
5. true **7.** true
9. **a.**

x	$f(x) = x^2 - x - 6$
-2	0
-1	-4
0	-6
1	-6
2	-4

  b. $(-2, 0), (3, 0)$ **c.** $x = 0.5$ **d.** $(0.5, -6.25)$
  e.

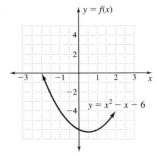

  f. $\{-3, 4\}$ **g.** $\{-2, 3\}$ **h.** $\{\ \}$ or $\varnothing$ **i.** $y \geq -6.25$
11.

r	$A = 4\pi r^2$
0	0
1	13
2	50
3	113
4	201
5	314
6	452
7	616
8	804
9	1018
10	1257

13. **a.** $\approx \{0.5, 3.5\}$ **b.** $\{0, 4\}$
15. **a.** $2x^2 - 8x + 4 = -4$ **b.** any $2x^2 - 8x + 4 < -4$
  c. not possible
17. $y = 2x^2 + 3x + 1$ **19.** $y = 3x - 4$ **21.** $y = 16 - x^2$
23. neither
25. **a.** $y \approx 99.93 - 0.000989x$
  b. $y \approx 0.00000000276x^2 - 0.00107x + 100.2$
  c.

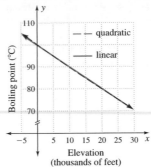

For the interval, both are good.
  d. $y \approx 101.2°C$; $y \approx 101.6°C$
  e. The straight line does not fit this new fact. The parabola fits the new fact but then turns upward after about 190,000 ft, which is also not realistic.
27. $y = -0.111x^2 + 6.667x$
29. **a.** binomial **b.** not a polynomial **c.** not a polynomial
  d. binomial **e.** trinomial **f.** not a polynomial
31. **a.** $x^2 - 9$ **b.** $4x^2 - 20x + 25$ **c.** $x^3 - 1$
  d. $n^2 + 8n + 16$ **e.** $6x^2 - x - 12$ **f.** $x^4 - 3x^2 + 1$
33. **a.** $(x + 4)(x - 1)$ **b.** $x(2x - 3)$ **c.** $(2x + 3)(x - 1)$
  d. $(3x + 2)^2$ **e.** $x(x - 1)$ **f.** $3(x + 1)^2$
35. **a.** $4x^2 - 9$; difference of squares
  b. $4x^2 - 12x + 9$; perfect square trinomial
  c. $4x^2 + 12x + 9$; perfect square trinomial
  d. $6x^2 + 5x - 6$ **37.** $25, 5$ **39.** $26, 13$ **41.** $(x + 7)(x - 7)$
43. $(2x - 1)^2$ **45.** cannot be factored **47.** $(2x + 3)^2$
49. $a^3 - 3a^2b + 3ab^2 - b^3$ **51.** $a^3 + b^3$
53. $(a - 2)(a^2 + 2a + 4)$ **55.** $(x + 3)(x^2 - 3x + 9)$
57. $(x - 10)(x^2 + 10x + 100)$
59. $A = \dfrac{\sqrt{3}}{4}X^2 - \dfrac{\sqrt{3}}{4}x^2 = \dfrac{\sqrt{3}}{4}(X - x)(X + x)$
61. $\{-4, 1\}$ **63.** $\{4\}$ **65.** $\{-\frac{2}{3}\}$ **67.** $\{-1\frac{1}{2}, 1\}$ **69.** $\{-\frac{3}{2}, \frac{2}{3}\}$
71. $\{-12\}$ **73.** $\{\pm\frac{3}{2}\}$ **75.** Answers will vary.
77. $y = -\frac{1}{3}x^2 + \frac{2}{3}x + \frac{8}{3}$
79. $0 = a(x + 2)(x - 4)$, a any real number
81. $0 = a(5x + 2)(5x - 1)$, a any real number
83. $y = x^2 - 9$ crosses x-axis twice; $y = x^2 + 9$ does not cross x-axis.
85. Factors are the same; double root
87. 2 units to left; $(-2, 0)$ vs. $(0, 0)$; $(0, 4)$ vs. $(0, 0)$

CHAPTER 4 TEST

1. **a.** $\{-4, -1\}$ **b.** $\{-7, 2\}$ **c.** $\{-6, 1\}$ **d.** $\{\ \}$ **e.** $x = -2.5$
  f. $(-2.5, -12.25)$ **g.** $\mathbb{R}$; $y \geq -12.25$
2. **a.** 54; quadratic; $y = x^2 + 5x - 12$
  b. -17; linear; $y = -8x + 31$
3. $x^2 - 8x + 12$; other **4.** $x^2 - 11x - 12$; other
5. $x^2 + 13x + 12$; other **6.** $2x^2 + 5x - 12$; other

7. $x^2 - 4x + 4$; perfect square trinomial
8. $4x^2 - 1$; difference of squares   **9.** $x^3 - 8$; other
10. $(2x - 3)(x - 2)$ **11.** $(x - 3)^2$ **12.** $(x - 7)(x + 7)$
13. $6(2x - 1)(x + 3)$ **14.** $(x - 3)(x^2 + 3x + 9)$
15. $(x + 5)(x^2 - 5x + 25)$ **16.** $\{\frac{3}{2}, 2\}$ **17.** $\{-2, 14\}$
18. $\{-\frac{5}{2}, 3\}$ **19.** $y = -\frac{1}{6}x^2 + \frac{1}{6}x + 2$
20. a. $y = -0.298x + 59.6$; voter turnout will decline to 46.5%.
   b. $y = 0.0111x^2 - 0.697x + 61.7$; voter turnout will have
   bottomed out and will rise to 52.5%.
21. Answers will vary.

CUMULATIVE REVIEW OF CHAPTERS 1 TO 4

1. $-4x + 17$ **3.** $y = \frac{3}{2}x$ **5. a.** $n = \dfrac{c}{t^2}$ **b.** $n = \dfrac{s - 2a + d}{d}$
7. a. 10   **b.** 2   **c.** 4   **d.** $3(x + 1)^2 - 5(x + 1) + 2$ or $3x^2 + x$
9. $y = \frac{2}{3}x^2 + \frac{2}{3}x - 8$ **11.** Rule is $f(x) = x^2 + 3$.
13. $x < -2.5$ or $x > 3.5$ **15.** 10°, 40°, 130°
17. a. $\{-1, 3\}$ **b.** $\{0, 2\}$ **19. a.** $15x^2 - 34x + 15$ **b.** $9 - x^2$

EXERCISES 5.1

1. $x = \pm 15$ **3.** $x = \pm 11$ **5.** $x = \pm 100$ **7.** no real number
9. 225   **11.** 7   **13.** 121   **15.** 0.1   **17.** 1.4   **19.** $5\sqrt{3}$
21. $2\sqrt{10}$ **23.** $10\sqrt{2}$ **25.** $5\sqrt{6}$ **27.** $4\sqrt{5}$ **29.** $4\sqrt{2}$
31. 2.414   **33.** -0.366 **35.** 2.414   **37.** -0.366 **39.** $\dfrac{2 + \sqrt{2}}{2}$
41. simplified   **43.** $2\sqrt{2}$ **45. a.** 0.9   **b.** 0.05   **c.** $\frac{1}{6}$ **d.** $\frac{11}{7}$
   e. 1.5   **f.** 800   **g.** 4   **h.** 9
47. a. identical for $y \geq 0$; $y = \sqrt{x}$ is a function, $x = y^2$ is not.
   b. $4 = 2^2, 2 = \sqrt{4}; 9 = 3^2, 3 = \sqrt{9}; 16 = 4^2, 4 = \sqrt{16}$
   c. $y = \sqrt{x}$ has restriction $y \geq 0$; $x = y^2$ does not.
   d. x cannot be a negative number if y is a real number.
49. 8
51. Factor 2 from numerator before changing to lowest terms.
53. a. $x = \sqrt{14}$ **b.** $x = 2\sqrt{14}$
55. a. $x = 3\sqrt{3}$ **b.** $x = 4\sqrt{3}$
57. a. $x = 8\sqrt{2}$ **b.** $x = 12\sqrt{2}$ **c.** $x = 4\sqrt{2}$ **59.** $\sqrt{2}$
61. right triangle   **63.** right triangle   **65.** not a right triangle
67. right triangle   **69.** right triangle   **71. a.** 11.6 in.   **b.** 0.6 in.
73. $\approx$16 ft 11.6 in.   **75.** $\approx$23 ft 0.3 in.   **77.** 29.7 ft   **79.** 16.5 ft
81. A triangle is a right triangle if and only if the sum of the
   squares of the lengths of the two shorter sides is equal to the
   square of the length of the longest side.

EXERCISES 5.2

1. a. $x = -5, x = -1$ **b.** $x = -5, x = -1$
   c.

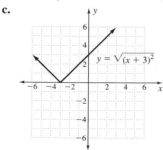

The graph is the same as that of $y = |x + 3|$; $\sqrt{w^2} = |w|$

3.

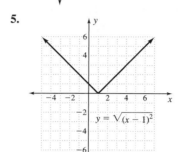

5.

7. $\{-5, 1\}$ **9.** $\{-1, 11\}$ **11.** $\{-7, -1\}$ **13.** $\{-2, 6\}$
15. $\{-\frac{1}{2}, \frac{1}{2}\}$ **17.** $\{-\frac{1}{2}\}$ **19.** $\{-2, 2\}$ **21.** $\{\pm 15\}$ **23.** $\{\pm 11\}$
25. $\{\pm 4\}$ **27.** $\{\pm 3\}$ **29.** $\{\pm 1.7\}$ **31.** $\{\pm 2.2\}$ **33.** $20\sqrt{2}$ m
35. 2 ft   **37.** 6 in.   **39.** $4\sqrt{3}$ in.   **41.** $\dfrac{5\sqrt{3}}{2}$ in.
43. $A = \dfrac{x^2\sqrt{3}}{4}$ **45.** $x^2 + x^2 = h^2, 2x^2 = h^2, h = x\sqrt{2}$
47. a. $d = 2\sqrt{\dfrac{A}{\pi}}$ **b.** 0.7854; keys form a square.
49. 80 in.; 6.7 ft   **51.** $r \approx 2.8$ in.
53. a. Let $h = 0$ and solve for t.   **b.** 5.9 sec   **c.** 3.2 sec
   d. 11.5 sec
55. a. $h = \dfrac{2d^2}{3}$ **b.** 27.4 mi   **c.** 13.7 mi   **d.** 54 ft
   e. 216 ft   **f.** half, 4
57. It is placed at the end of the expression.

MID-CHAPTER 5 TEST

1. a. $5\sqrt{3}$ **b.** $3\sqrt{2}$ **c.** $4\sqrt{2}$ **d.** $2\sqrt{5}$
2. a. $\sqrt{8}$ **b.** $\sqrt{50}$ **c.** $\sqrt{48}$ **d.** $\sqrt{72}$ **e.** $\sqrt{27}$ **f.** $\sqrt{12}$
3. a. 50   **b.** 0.05   **c.** 1.5   **d.** 150
4. a. $\frac{7}{8}$ **b.** 2   **c.** $\frac{4}{3}$ **d.** $\frac{1}{9}$
5. $x^2 = 16$ has two solutions, $\{-4, 4\}$; $x = \sqrt{16}$ has one, $\{4\}$.
6. a. 1.816   **b.** 0.048
7. A closing parenthesis is needed after the $\sqrt{(6)}$ in the
   numerator, not after the 3 in the denominator.
8. a. right triangle   **b.** right triangle   **c.** not a right triangle
9. a. $x = 12$ **b.** $x = 8\sqrt{6}$ **c.** $x = 6.5$
10. a. $\{-1, 5\}$ **b.** $\{-7, -3\}$
11. a. $\{\pm 11\}$ **b.** $\{\pm 0.2\}$ **c.** $\{-3, 1\}$ **d.** $\{-9, 1\}$ **e.** $\{-5, 1\}$
   f. $\{-4, -2\}$
12. 4   **13. a.** true   **b.** true   **c.** false   **d.** $3\sqrt{3}$
14. $t = \sqrt{\dfrac{-2h}{g}}$ **15.** $A = \dfrac{\pi d^2}{4}$

EXERCISES 5.3

1. $x^2 + 6x + 9$ **3.** $x^2 - 12x + 36$ **5.** $x^2 - x + \frac{1}{4}$
7. $4x^2 - 12x + 9$ **9.** $(x - 2)^2$ **11.** $(x + 4)^2$ **13.** $(x + 9)^2$
15. $(x - 8)^2$
17. $(a + 7)^2 = a^2 + 14a + 49$; missing entries: $+7, +7, 49$
19. $\left(x - \frac{7}{2}\right)^2 = x^2 - 7x + \frac{49}{4}$; missing entries: $-\frac{7}{2}, -\frac{7}{2}, \frac{49}{4}$
21. $x^2 + 22x + 121 = 144$; $(x + 11)^2 = 144$
23. $x^2 + 12x + 36 = 49$; $(x + 6)^2 = 49$
25. $x^2 - 11x + \frac{121}{4} = \frac{49}{4}$; $\left(x - \frac{11}{2}\right)^2 = \frac{49}{4}$
27. $x^2 + 15x + \frac{225}{4} = \frac{169}{4}$; $\left(x + \frac{15}{2}\right)^2 = \frac{169}{4}$
29. $\{-3 \pm \sqrt{17}\}$; irrational **31.** $\{4, 2\}$; rational
33. $\{\frac{1}{2}\}$; rational **35.** $\{-1, -\frac{1}{2}\}$; rational
37. $\left\{\dfrac{-3 \pm \sqrt{-7}}{4}\right\}$; not real **39.** $\{1, \frac{3}{2}\}$; rational
41. $\{5\}$; rational **43.** $x = -1$; $1, -4, -5$; $x^2 - 4x - 5 = 0$
45. $x = \frac{3}{2}$; $2, 5, -12$; $2x^2 + 5x - 12 = 0$
47. $x = 1$; $3, 1, -4$; $3x^2 + x - 4 = 0$
49. $x = -b/2a$, leading term of quadratic formula
51. $(-0.75, 3.875)$ **53.** $(-0.3, -3.45)$
55. $\{4.226, 15.774\}$ **57.** $\{5.918, 14.082\}$
59. a. initial height of 32.8 ft **b.** no **c.** 0 and ≈0.4 sec
61. ≈1.55 sec **63.** $0 = -\frac{1}{2}(9.81)t^2 + 35$, $t = 2.7$ sec
65. $0 = -\frac{1}{2}(9.81)t^2 + 1.5t + 35$, $t = 2.8$ sec
67. a. false; $\sqrt{w^2} = |w|$, not w

    b. false; there is no square root of sums property.

    c. false; there is no square root of differences property.

    d. true

69. Step 1: Standard form. Step 2: Multiply both sides by $4a$. Step 3: Subtract $4ac$ from both sides. Step 4: Add b^2 to both sides. Step 5: Factor left side. Step 6: Take square root of both sides. Step 7: Simplify left side, giving both solutions. Step 8: Subtract b from both sides. Step 9: Divide both sides by $2a$.

EXERCISES 5.4

1. $x < -3$ or $x > 2$; $(-\infty, -3)$ or $(2, +\infty)$

3. $x \geq 4$; $[4, +\infty)$; set of numbers greater than or equal to 4
5. $(-3, +\infty)$; set of numbers greater than -3

7. $x \leq 4$; set of numbers less than or equal to 4

9. $x < 2$ or $x > 3$ **11.** $-2 \leq x \leq 7$ **13.** $x \leq -1$ or $x \geq 6$
15. $\mathbb{R}, x \neq -3$ **17.** $\{\ \}$ **19.** $-5 < x < 3$
21. $x \leq -12$ or $x \geq 2$ **23.** $-3 \leq x \leq 6$ **25.** $x < -18$ or $x > 1$
27. $\mathbb{R}$ **29.** $\{\ \}$ or $\varnothing$ **31.** $(-\infty, -1)$ or $(3, +\infty)$ **33.** $[-5, 2]$
35. $\{\ \}$ or $\varnothing$ **37.** $(-\infty, +\infty)$ **39.** $(-2, 1)$
41. $(-\infty, -\frac{1}{2}]$ or $[\frac{3}{2}, +\infty)$
43. a. $\approx12 < x < 100$ **b.** $\approx0 < x < 12$ or $100 < x < 112$
    c. ≈28 ft
    d. No; it shows the ball's path; both axes are in feet.
    e. 50 ft **f.** $\approx25 < x < 85$

45. a. seasonal products such as ski equipment or heating oil
    b. $\approx$Nov. 1 to March 31 **c.** Dec. 29 **d.** Oct. 28
    e. \$6,600,000 **f.** \$4,250,000 **g.** yes

CHAPTER 5 REVIEW EXERCISES

1. a. $5\sqrt{3}$ **b.** $2\sqrt{2}$ **c.** $4\sqrt{2}$ **3. a.** $1 - \sqrt{6}$ **b.** $1 + \sqrt{6}$
5. a. 120 **b.** 1.2 **c.** 0.11 **d.** 0.0001
7. a. $12.5^2 = 7.5^2 + 10^2$
    b. $30^2 = 18^2 + 24^2$
    c. $(\sqrt{13})^2 = (\sqrt{5})^2 + (\sqrt{8})^2$
    d. $6^2 = 4^2 + (\sqrt{20})^2$
9. $z = x\sqrt{2}$; $\dfrac{z}{x} = \sqrt{2}$
11. false; there is no graph in the third or fourth quadrant.
13. $a = \sqrt{\dfrac{C^2 - 2\pi^2 b^2}{2\pi^2}}$ **15. a.** $x^2 + \dfrac{bx}{a} + \dfrac{b^2}{4a^2}$

    b. $\dfrac{b^2 - 4ac}{4a^2}$ **c.** $\left|x + \dfrac{b}{2a}\right|$ **17.** $\{\pm5\}$ **19.** $\approx\{0.199, 1.024\}$
21. $\{4, 6\}$ **23.** $\{-2, 4\}$ **25.** $\{-3, 2\}$ **27.** $\{1\}$
29. no real-number solution; no x-intercepts; negative number under the radical sign
31. a. $h = -16t^2 + 72t$
    b. $(0, 0)$, $(4.5, 0)$; first is when ball is thrown, second is when it comes back down.
    c. $\{2, 2.5\}$; on the way up and down **d.** 81 ft
    e. $-16t^2 + 72t > 80$; $2 < t < 2.5$; 2 ft (1 up and 1 down)
    f. $-16t^2 + 72t < 32$; $0 < t < 0.5$ and $4 < t < 4.5$; 64 ft
    g. 2, 32; speed slows at the top.
33. a. $\left(-\frac{3}{2}, \frac{7}{4}\right)$ **b.** $(2, -1)$ **c.** $\left(\frac{2}{3}, -\frac{4}{3}\right)$
35. $-7.123 \leq x \leq 1.123$ **37.** $x \leq -\frac{1}{2}$ or $x \geq \frac{1}{2}$
39. $x < -3.667$ or $x > 2$ **41.** for all x, $\mathbb{R}$
43. no real-number solutions **45.** $-6 \leq x \leq 0$
47. $\approx3.7 < x < 36.3$

CHAPTER 5 TEST

1. $x = \dfrac{-b \pm \sqrt{b^2 - 4ac}}{2a}$ **2. a.** $3\sqrt{5}$ **b.** $7\sqrt{2}$ **3.** $\dfrac{2 + \sqrt{2}}{4}$
4. $\{\pm14\}$ **5.** $\{-1, 9\}$ **6.** 25, 25; $\{-12, 2\}$ **7.** $x = \pm65$
8. $\{-5, 3\}$ **9.** $\{1, 3\}$ **10.** $\{1.5, 2\}$
11. a. $x = 3\sqrt{3}$ **b.** $x = \dfrac{7\sqrt{2}}{2}$ **12.** $\{-1, 2\}$; $\left(\frac{1}{2}, -2\frac{1}{4}\right)$
13. $-6 \leq x \leq 1$ **14.** $x < -6$ or $x > 1$ **15.** $-1 < x < 1$
16. for all x, $\mathbb{R}$ **17.** $x \leq -4$ or $x \geq \frac{5}{6}$
18. a.

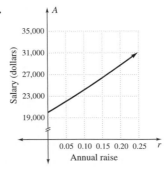

r	$A = \$20,\!000(1 + r)^2$
0	$20,000
0.02	$20,808
0.04	$21,632
0.06	$22,472
0.08	$23,328
0.10	$24,200

  b. 7% **c.** $\approx 22.5\%$

19. $s = \dfrac{v^2}{2g}$ **20.** 300 steps

21.

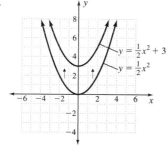

$\sqrt{x^2} = 2, x = \pm 2; x = \sqrt{4}, x = 2$

EXERCISES 6.1

  1. a is negative
  3. $|a| = 0.00375, |a| < 1$
  5. 0; 0 **7.** 0; 0
  9. $y = \frac{4}{125}x^2$; positive coefficient of x^2
11.

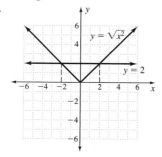

13.

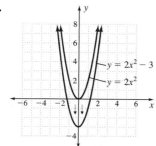

15. $g(x)$ **17.** $j(x)$ **19.** $h(x)$
21. $p(x)$ **23.** steeper

25. and 27.

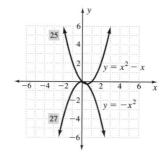

29. and 31.

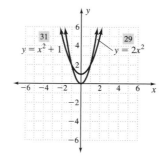

33. and 35.

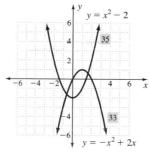

37. and 39.

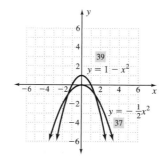

41. and 43.

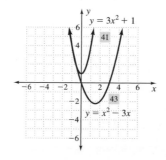

45.

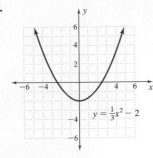

$y = \frac{1}{3}x^2 - 2$

47. $h(x)$, $m(x)$, $q(x)$ **49.** $f(x)$, $g(x)$, $h(x)$, $j(x)$, $m(x)$, $q(x)$ **51.** $j(x)$

53. $\left(\dfrac{v_0}{g}, \dfrac{v_0{}^2}{2g}\right)$ **55.** $y = x^2 - 3$ or $y = -x^2 - 3$

57. a. $(0, 20)$, $(200, 170)$, $(400, 20)$
b. $y = -0.00375x^2 + 1.5x + 20$

EXERCISES 6.2

1. a. $2i\sqrt{5}$ **b.** $2i\sqrt{10}$ **c.** $6i\sqrt{2}$ **3. a.** $i\sqrt{3}$ **b.** $3i\sqrt{6}$
   c. $3i\sqrt{3}$ **5. a.** $16 + 0i$ **b.** $0 + 4i$ **c.** $6 + i\sqrt{6}$
7. a. $3 + i\sqrt{3}$ **b.** $\frac{1}{2} - \frac{1}{2}i\sqrt{6}$ **9.** $2 - 3i$ **11.** $1 - i$
13. $-3 - 2i$ **15.** $4 + 2i$ **17.** $-8i$ **19.** $11 - 15i$ **21.** $-2a$
23. $-17 + 4i$ **25.** $3 + 6i$ **27.** $13 - 6i$ **29.** $-15 + 7i$
31. a. 25 **b.** 25 **33. a.** 3 **b.** 3 **35.** 2 real solutions
37. 2 complex solutions   **39.** 1 real solution
41. 2 complex solutions   **43.** 2 real solutions
45. $\{\pm 2\sqrt{2}\} \approx \{\pm 2.8\}$ **47.** $\{-1 \pm i\sqrt{3}\} \approx \{-1 \pm 1.7i\}$
49. $\{2 \pm 2i\}$ **51.** $\{-4, 1\}$ **53.** $\{2\}$ **55.** $\{-1 \pm i\}$
57. $\{-2.5 \pm 0.5i\}$ **59.** $(x - 2)(x^2 + 2x + 4)$
61. $\{-2, 1 \pm i\sqrt{3}\} \approx \{-2, 1 \pm 1.7i\}$ **63.** $\{\pm 3, \pm 3i\}$
65. $\{-2, -1, 0\}$ **67.** $\left\{3, -\frac{3}{2} \pm \dfrac{3i\sqrt{3}}{2}\right\}$ or $\{3, -1.5 \pm 2.6i\}$
69. $\{0 \text{ (double root)}, 4\}$ **71.** $\left\{0 \text{ (double root)}, 0.5, -\frac{1}{4} \pm \dfrac{i\sqrt{3}}{4}\right\}$
   or $\{0 \text{ (double root)}, 0.5, -0.25 \pm 0.43i\}$

MID-CHAPTER 6 TEST

1. opens down and is steeper
2. 4 units below; $(0, -4)$ vs. $(0, 0)$; $(-2, 0)$, $(2, 0)$ vs. $(0, 0)$
3. and 4.   **5. and 6.**

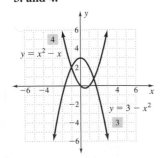

$y = x^2 - x$
$y = 3 - x^2$

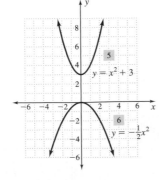

$y = x^2 + 3$
$y = -\frac{1}{2}x^2$

7. a. $0 + 4i$ **b.** $4 + 0i$ **c.** $3 + 2i$ **d.** $0 + 2i\sqrt{14}$
8. a. $2 - 2i\sqrt{3}$ **b.** $\dfrac{3}{2} + \dfrac{i\sqrt{2}}{2}$ **9. a.** -3 **b.** $-2 + 6i$

10. a. 13 **b.** 5 **c.** 13
11. a. both 13; reversing a and b in $(a + bi)(a - bi)$ gives same
   real-number product.   **b.** both 5
12. 16; 2; $\{-1, 3\}$ **13.** $\{2, -1 \pm i\sqrt{3}\}$ or $\{2, -1 \pm 1.7i\}$
14. $\left\{-\frac{1}{2} \pm \dfrac{i\sqrt{3}}{2}\right\}$ or $\{-0.5 \pm 0.87i\}$ **15.** $\{0 \text{ (double root)}, \pm\frac{1}{3}\}$

EXERCISES 6.3

1. a. $y = (x + 2)^2$ **b.** $y = (x - 3)^2$
3. a. $y = x^2 + 2$ **b.** $y = x^2 - 1$
5. a. $y = x^2 - 1$ **b.** $y = -x^2 - 3$
7. a. $y = (x + 1)^2 - 2$ **b.** $y = (x - 2)^2 + 1$
9. Shift left 2 units.   **11.** Shift right 4 units.
13. Shift right 3 units and up 4 units.
15. Shift left 4 units and up 3 units.   **17.** $(1, 0)$ **19.** $(-3, 4)$
21. $(2, -3)$ **23.** $y = \frac{1}{8}(x - 3)^2 + 1$ **25.** $y = 2(x + 1)^2 - 1$
27. $y = -2(x - 3)^2 - 2$
29. Possible answer: $(10, 0)$; $y = -0.24x^2 + 2.4x$
31. a. $(0, -25)$, $(20, -15)$
   b. $y = 0.0375(x - 20)^2 - 15$ or $y = 0.0375x^2 - 1.5x$
   c. vertical shift of 25
33. $y = -\frac{1}{9}(x - 30)^2 + 100$ or $y = -\frac{1}{9}x^2 + \frac{20}{3}x$
35. $y = (x + 5)^2$ **37.** $y = (x - 3)^2$ **39.** $y = (x + 5)^2 + 5$
41. $y = (x - 3)^2 - 1$
43. a. $y = -0.025(x - 200)^2 + 1025$
   b. $y = -0.125(x + 20)^2 + 60$
   c. $y = 0.15(x - 20)^2 - 45$
   d. $y = 0.75(x + 3)^2 - 11.75$
   e. $y = 0.9(x - 1.5)^2 - 17.025$

EXERCISES 6.4

1. $(0, 0)$, $(2, 0)$; $(1, -1)$ **3.** $(-7, 0)$, $(3, 0)$; $(-2, 25)$
5. $(-4, -1)$ **7.** $(2, 1)$ **9.** $\left(\frac{1}{4}, -\frac{25}{8}\right)$ or $(0.25, -3.125)$
11. $(-1, -5)$ **13.** $(1.5, 0.25)$ **15.** $(75, 60)$ **a.** 60 ft   **b.** 50 ft
17. $(200, 10)$ **a.** 10 ft   **b.** 15 ft
19. factor to find x-intercepts, use quadratic formula, double
   horizontal distance to vertex; 200 ft
21. 62.9 ft; 251.6 ft   **23.** no; 310.6 ft
25.

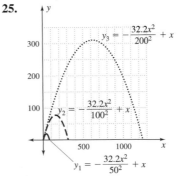

$y_3 = -\dfrac{32.2x^2}{200^2} + x$

$y_2 = -\dfrac{32.2x^2}{100^2} + x$

$y_1 = -\dfrac{32.2x^2}{50^2} + x$

   a. quadrupled   **b.** quadrupled
27. 30.3 ft   **29.** At the vertex: $(3.6 \text{ sec}, 205.4 \text{ ft})$
31. At the vertex, $t = -b/2a$ and does not contain $h_0 = c$.
33. 690 ft; $(1143, 677)$ **35.** 1188 ft; $(1091, 1161)$ **37.** ≈ 155 ft
39. $A = 450 \text{ ft}^2$; 30 ft, 15 ft   **41.** $A = 600 \text{ ft}^2$, 20 ft, 30 ft
43. distance $|h - x|$ from vertex on opposite side

CHAPTER 6 REVIEW EXERCISES

1. steeper **3.** shifted down 9 units
5, 7, and 9.

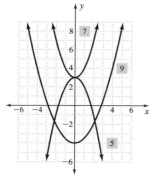

11. a. Olympic diver; $a \approx -4.9$ implies a very steep curve, and $c \neq 0$ is diver's initial height in vertical motion equation.
  b. fireworks; $a \approx -4.9$ implies a very steep curve, $c = 0$ starts at ground level, and $b = 115$ is a large initial velocity.
  c. bridge cable; parabola turns up for $a > 0$, and $c = 60$ ft is reasonable location for cable anchor point.
  d. golf ball; curve turns down for $a < 0$, a implies a very flat curve; $c = 10$ is reasonable location for the tee.
13. a. $4i$ **b.** $5i\sqrt{2} \approx 7.07i$ **15. a.** $4 + 3i$ **b.** $2 + i$
17. a. -3 **b.** $-2 + 6i$ **19. a.** $7 + 24i$ **b.** 25 **c.** 25
21. c **23.** b **25.** b **27.** 25; 2 real; 2 **29.** -23; 2 complex; 0
31. 0; 1 double root; 1 **33.** $\{1\}$ (double root)

35. $\left\{-\dfrac{1}{3} + \dfrac{i\sqrt{2}}{3}\right\} \approx \{-0.33 \pm 0.47i\}$

37. a. not possible **b. to d.** Answers vary.
  e. not possible

39. a. $\{-3, -1, 2\}$ **b.** $\{-3.5, 1.5, 0\}$ **41.** $\left\{3, -\dfrac{3}{2} \pm \dfrac{3i\sqrt{3}}{2}\right\}$

43. $\left\{0, -\dfrac{1}{2} \pm \dfrac{i\sqrt{3}}{2}\right\}$ **45.** $\{0$ (double root), $1\}$ **47.** $\{\pm 2, \pm 2i\}$

49.

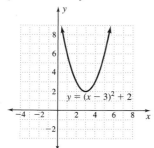

51.

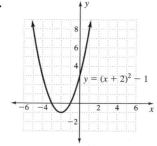

53. $y = -\dfrac{1}{32}(x - 8)^2 + 2$ **55.** $y = x^2/32$
57. $y = -\dfrac{3}{8}(x + 2)^2 + 3$ **59.** $y = -0.025(x - 100)^2 + 250$

61. $y = (x + 2)^2 + 5$ **63.** $\{-1, 2\}$; $(0.5, -2.25)$
65. no x-intercepts; $(1, 3)$
67. a. $\approx(271, 135)$ **b.** $\approx(752, 376)$ **c.** $\approx(2484, 1242)$
  d. $\approx(3882, 1941)$
69. a. 100 ft **b.** $\approx x < 8.8$ ft or $x > 51.2$ ft
71. a. $(480, 695.2)$
  b. between $x = 360$ ft and $x = 600$ ft, or 240 ft
73. $A = 40w - w^2$; vertex $(20, 400)$; maximum area is square with $w = l = 20$; $20^2 = 400$

CHAPTER 6 TEST

1. shifted up 2 units; no x-intercepts; vertex at $(0, 2)$
2. upside down **3.** the first, because its graph opens down
4. a. $6i$ **b.** $5i\sqrt{3} \approx 8.7i$ **5. a.** $1 + 2i$ **b.** $4 + 2i$
6. a. 26 **b.** 26 **c.** $-24 - 10i$ **7.** $-8 + 6i$

8. a. 16 **b.** 2 real; 2 **c.** $\{-1, 3\}$ **9.** $\left\{1, -\dfrac{1}{2} \pm \dfrac{i\sqrt{3}}{2}\right\}$

10. shift right 3 units and down 1 unit **11.** $y = 3x^2 + 6x - 5$
12. $y = (x + 2)^2 - 5$ **13. a.** $(0.61$ sec, 4.83 m$)$ **b.** 1.6 sec
14. a. $y = lw$ **b.** $y = 15w - 2w^2$
  c. $(3.75, 28.125)$; area, y, is largest at peak of graph.

CUMULATIVE REVIEW OF CHAPTERS 1 TO 6

1. $24 + 2 = 26$ **3.** $x = 2\frac{1}{3}$ **5. a.** 2 **b.** $2i$ **c.** $i\sqrt{2}$ **d.** -2
7. 10 **9. a.** $x^2 + 2x + 1$ **b.** $a^2 + 2ab + b^2$
11. a. 28; quadratic; $y = 2x^2 - 9x + 10$
  b. 23; linear; $y = 3x + 5$ **c.** 29; other
13. $x \leq 0$ or $x \geq 2$ **15.** $x \leq -1$ or $x \geq 4$
17. $-11 < x < -1$ **19.** $a = 0.5, b = 3$ **21.** 5.5% **23.** $\frac{9}{4}, \frac{3}{2}$

EXERCISES 7.1

1. 31 cm **3.** $\dfrac{1}{48 \text{ in}^2}$ **5.** 60 lb **7.** 50 kw

9. $300 \text{ mL} \cdot \dfrac{1 \text{ L}}{1000 \text{ mL}} = 0.3 \text{ L}$

11. $25 \text{ ft} \cdot \dfrac{12 \text{ in.}}{1 \text{ ft}} \cdot \dfrac{1 \text{ m}}{39.37 \text{ in.}} \approx 7.62 \text{ m}$

13. $150 \text{ ft}^3 \cdot \left(\dfrac{1 \text{ yd}}{3 \text{ ft}}\right)^3 \approx 5.6 \text{ yd}^3$

15. $100 \text{ in.}^3 \cdot \left(\dfrac{1 \text{ ft}}{12 \text{ in.}}\right)^3 \approx 0.058 \text{ ft}^3$

17. $200 \text{ mL} \cdot \dfrac{1 \text{ kg}}{1000 \text{ mL}} \cdot \dfrac{1000 \text{ g}}{1 \text{ kg}} = 200 \text{ g}$

19. a. $\frac{3}{1}$ **b.** $\frac{1}{2}$ **c.** $\frac{1}{3}$ **21. a.** $\frac{1}{100}$ **b.** $\frac{2}{15}$ **c.** $\frac{8}{3}$

23. $\dfrac{55 \text{ ft}}{1 \text{ sec}} \cdot \dfrac{1 \text{ mi}}{5280 \text{ ft}} \cdot \dfrac{60 \text{ sec}}{1 \text{ min}} \cdot \dfrac{60 \text{ min}}{1 \text{ hr}} = 37.5 \text{ mph}$

25. $\dfrac{1 \text{ gal}}{5 \text{ mi}} \cdot \dfrac{55 \text{ mi}}{1 \text{ hr}} \cdot \dfrac{\$1.35}{1 \text{ gal}} \cdot \dfrac{10 \text{ hr}}{1 \text{ day}} = \$148.50/\text{day}$

27. $1 \text{ yr} \cdot \dfrac{12 \text{ mo}}{1 \text{ yr}} \cdot \dfrac{\text{adult}}{150 \text{ mo}} \cdot \dfrac{500 \text{ mg}}{\text{adult}} = 40 \text{ mg}$

29. a. $x = 8.4$ **b.** $x = 2.4$ **c.** $x = 54.9$ **31.** 15 ft
33. $a = 6\frac{2}{3}, b = 6$ **35.** $d = 6\sqrt{3}, f = 12$
37. $e = 10\sqrt{3}/3, f = 20\sqrt{3}/3$ **39.** $d = \sqrt{15}, f = 2\sqrt{5}$
41. $n = 2\sqrt{2}, x = 4\sqrt{2}, y = 4\sqrt{2}$ **43.** 4.5 ft **45.** 5.9 ft
47. Unit conversion is needed.
49. did not multiply each side by 15

EXERCISES 7.2

1. a. x = weight in lb, y = distance stretched in inches; $y = \frac{2}{5}x$
  b. direct variation   **c.** Intercept is zero.
3. a. x = no. of credit hr, y = cost in \$; $y = 80x + 25$
  b. not proportional   **c.** student fees
5. a. x = purchase price, y = tax; $y = 0.075x$
  b. direct variation   **c.** Intercept is zero.
7. k = 55 mph; 440 mi   **9.** k = \$14.49 per CD; \$144.90
11. $k \approx 0.367$ oz per in^2; ≈ 0.827 oz   **13.** are   **15.** $y = mx$
17. a. $k = \frac{1}{2}$ **b.** $k = 1$ **c.** $k = \frac{1}{3}\pi$ **19.** $C = kd$; $k = \pi$
21. $A = kr^2$; $k = \pi$ **23.** $A = klw$; $k = 1$ **25.** $V = khr^2$; $k = \pi$
27. $\frac{8}{3}$ the length   **29.** 25 times the energy
31. 2.56 times the length   **33.** $\approx 2{,}594{,}576$ m^3

EXERCISES 7.3

1. $y = 100/x$; y = points per problem, x = no. of problems;
  k = 100 points
3. $y = 1{,}000{,}000/x$; y = yr, x = yd^3 per yr; k = 1,000,000 yd^3
5. $y = 650{,}000{,}000/x$; y = yr, x = metric tons per yr;
  k = 650,000,000 metric tons
7. inverse variation; $k = 5600 \frac{\text{tons}}{\text{day}} \cdot \text{years}$
9. direct variation; k = \$0.21875 per item
11. inverse variation; k = 90 mi
13. inverse variation; k = 12 worker$\cdot$days
15. direct variation; k = \$1100 per semester
17. x = 52.5 lb; k = 210 lb$\cdot$ft   **19.** x = 60 people; k = 60 cups
21. x = 99.4 yr; k = 129,220,000 metric tons
23. x = 1200 ft per gal; k = 300 ft^2 per gal
25.

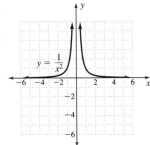

  a. undefined   **b.** $x = 0$ **c.** approaches x-axis
27. reflected over x-axis

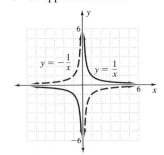

29. $y_1 = \dfrac{k}{x_1}$, $y_2 = \dfrac{k}{x^2}$; $k = x_2 y_2$; multiply $y_1 = \dfrac{x_2 y_2}{x_1}$ by x_1.
31. $x_2 y_1 = x_1 y_2$; subscripts are not alike on each side.
33. No; first is inverse variation, and second is direct.
35. b. $H = M/r^3$ **37.** 9
39. a. (10 lb, 3 in.), $(x, 1$ in.)
  b. (10 lb)(3 in.) = $(x)(1$ in.), 30 lb   **c.** 30 lb$\cdot$in.

EXERCISES 7.4

1. $x = -3$ **3.** $a = 4$ **5.** defined for all $\mathbb{R}$
7. a. $-x - y$ **b.** $x - y$ **c.** $x - y$
9. a. $x - 3, x \neq 3$ **b.** $-2 - x, x \neq -2$ **c.** $b - a, b \neq a$
  d. $x - 3, x \neq 3$
11. a. $\dfrac{8a}{5}, a \neq 0$ **b.** $\dfrac{x}{5y}, x \neq 0, y \neq 0$ **c.** $-\dfrac{7xy^3}{9}, x \neq 0, y \neq 0$
  d. $-\dfrac{4c}{3a^2b}, a \neq 0, b \neq 0$
13. a. simplified; $y \neq -2x$ **b.** $\dfrac{y}{2 + y}, x \neq 0, y \neq -2$
15. a. $-\dfrac{1}{x + 3}, x \neq \pm 3$ **b.** $\dfrac{x - 3}{x + 2}, x \neq -3, -2$
17. a. $\dfrac{c}{2d}, a \neq -2b, d \neq 0$ **b.** $\dfrac{2x + 1}{2(2x - 1)}, x \neq 0, \frac{1}{2}$
19. a. $-(x + 3), x \neq 2$ **b.** $-\frac{1}{2}, x \neq 3$
21. a. x **b.** $\dfrac{1}{a^3b^2}$ **c.** $\dfrac{1}{a}$ **d.** $\dfrac{b^3}{a^3}$ **e.** $\dfrac{y}{x^2}$ **f.** $\dfrac{81}{2d^3}$
23. a. $\dfrac{a + 3}{a(a + 2)}$ **b.** $x + 1$
25. a. $x + 2$ **b.** $\dfrac{-2a - 1}{2a(2 + 3a)}$
27. a. $\dfrac{x}{x - 2}$ **b.** $\dfrac{a}{(a + 5)(a - 5)}$
29. a. $\dfrac{1}{a^2 + ab + b^2}$ **b.** $\dfrac{-(x - y)}{x + y}$ or $\dfrac{y - x}{x + y}$
31. a. $a^2 + ab + b^2$
  b. $\dfrac{(x + y)(x - y)}{x^2 - xy + y^2}$ or $\dfrac{x^2 - y^2}{x^2 - xy + y^2}$
33. a. $\dfrac{-(x + 5)}{x}$ **b.** $\dfrac{-2(x - 1)}{2x - 1}$
35. a. $\dfrac{\frac{1}{a}}{\frac{1}{b}} = \dfrac{b}{a}$ **b.** $\dfrac{\frac{1}{b}}{a} = \dfrac{1}{ab}$ **c.** $\dfrac{\frac{1}{b}}{b} = \dfrac{1}{b^2}$
  d. $\dfrac{\frac{a}{b}}{\frac{1}{b}} = a$
37. $\dfrac{x(x - 2)}{x + 1}$ or $\dfrac{x^2 - 2x}{x + 1}$ **39.** $\dfrac{-1}{x + 4}$ **41.** 60 sec
43. $4\frac{1}{3}$ gal/hr   **45.** $\approx$6 cans/\$ **47.** 96 stitches/ft
49. $\approx$0.3 page/min   **51.** division, \$0.20/mi or 5 mi/\$
53. multiplication; spend 8 hr/wk having coffee
55. Terms, not factors, were simplified.
57. Numerator factors to $(x - 1)(x - 1)$; 1 is also divided by $x - 2$.
59. The result has an extra factor of a; multiplication is not
  distributive over multiplication.
61. true; always works; may give fraction in numerator or
  denominator

MID-CHAPTER 7 TEST

1. $\frac{4}{3}$ **2.** $\frac{1}{12}$ **3.** $\dfrac{b}{es}$ **4.** $1\frac{2}{25}$ **5.** $\dfrac{3d}{2c}$ **6.** $x - 2$ **7.** $\dfrac{1}{a + 3}$
8. $1 + c$ **9.** $\dfrac{12x + 5}{x + 4}$ **10.** $\dfrac{b^3}{c^3}$ **11.** $\dfrac{x - 4}{x - 2}$
12. $x(x - 1)$ or $x^2 - x$ **13.** $\dfrac{-(x^2 + 4x + 16)}{x + 4}$
14. $x = -2, x = \frac{1}{2}, x = 3$ **15.** $\frac{8}{5}$ **16.** $\frac{5}{6}$ **17.** 3906.25 gal

18. 0.3 mi/$ and $3.33/mi   **19.** $a = 5.14$ **20.** $x = 11$
21. $n = 8, p = 13.2$ **22.** $y = \frac{3}{2}x$; proportional
23. $y = 0.45x + 0.15$; not proportional
24. a. 400 lb·ft   **b.** 5 ft from pivot point
25. $y = 100/x$; y = no. of cookies per child, x = no. of children;
$k = 100$ cookies   **26.** $18.18

EXERCISES 7.5

1. quadratic; $x^2 - 2x + 1$ **3.** quadratic; $x^2 + 2x + 4$
5. linear; $2x + 5$ **7.** linear; $2x - 5$
9. a. cubic; $4x^3 - 4x^2 - 9x + 9$ **b.** quadratic; $4x^2 - 9$
c. $(2x + 3)(2x - 3)$
11. a. cubic; $x^3 + 6x^2 - x - 6$ **b.** quadratic; $x^2 - 1$
c. $(x - 1)(x + 1)$
13. $\mathbb{R}$ except $x = 4$; $x + 1$
15. $\mathbb{R}$ except $x = 3$; $x - \dfrac{4}{x - 3}$
17. $\mathbb{R}$ except $x = 2$; $x - 1 - \dfrac{6}{x - 2}$
19. $\mathbb{R}$ except $x = 1$; $x - 2 - \dfrac{6}{x - 1}$
21. $\mathbb{R}$ except $x = 0$; $x - 3 - \dfrac{4}{x}$ **23.** $\mathbb{R}$ except $x = -1$; $x - 4$
25. a. $(x + 1)(x - 4)$; remainder of 0
b.

x	$x^2 - 3x - 4$
-1	0
0	-4
1	-6
2	-6
3	-4
4	0

same
c.

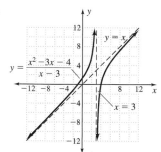

d. approaching the restriction; at $x = 3$
e. approaching function $y = x$
f.

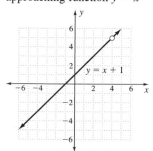

linear with hole at $x = 4$; division has zero remainder

g.

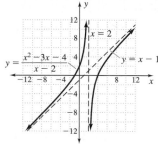

nearly vertical at its restriction $x = 2$

27. $x^2 + 2x + 4 + \dfrac{7}{x - 2}$ **29.** $x^2 + x + 1$; factor
31. $x^2 - \dfrac{1}{x}$ **33.** $x^2 - x + 1 + \dfrac{-2}{x + 1}$ **35.** $x^2 - x + 1$; factor
37. $x^3 - x^2 + x - 1 + \dfrac{2}{x + 1}$ **39.** $x^3 - 3x^2 + 3x - 1$; factor
41. $3x^2 - 2x + 2 - \dfrac{3}{x + 1}$ **43.** $(-1, 3)$ **45.** $(1, 0)$
47. $(x^3 + y^3) \div (x + y) = x^2 - xy + y^2$ **49.** true

EXERCISES 7.6

1. a. 21   **b.** 30   **c.** 128   **3. a.** 15   **b.** 45   **c.** 40
5. a. $8x$ **b.** $6b$ **7. a.** $x^2 + x$ **b.** $a^2 - ab$
9. a. $\dfrac{3 - x}{4}$ **b.** $-\dfrac{1}{x}, x \neq 0$ **c.** $2 - x, x \neq -2$
d. $\dfrac{1}{x - 1}, x \neq -1, 1$ **11. a.** $12; \frac{5}{6}$ **b.** $2a; -\dfrac{1}{2a}$
c. $a^2b^2; \dfrac{b^3 - ac}{a^2b^2}$ **d.** $x(x - 2)(x + 2); \dfrac{8x + 10}{x(x - 2)(x + 2)}$
e. $(x + 3)(x - 3)^2; \dfrac{2x^2 + 24x}{(x + 3)(x - 3)^2}$
13. $\dfrac{2a + 5b}{4ab}$ **15.** $\dfrac{x^2 + x - 3}{x(x - 3)}$ **17.** $\dfrac{4x^2 - 7x + 8}{x(x - 1)}$
19. $\dfrac{x^2 - 3x + 3}{(x - 3)^2}$ **21.** $\dfrac{5b + 3}{(b + 1)(b - 1)}$ **23.** $\dfrac{2b - 3a}{ab(2 + b)}$
25. $\dfrac{x^2 + 3x - 9}{x(x - 3)^2}$ **27.** $\dfrac{(x - 2)(x + 1)}{x(2x + 3)(3x + 2)}$
29. $\dfrac{x^4 + 4x^3 + 12x^2 + 24x + 24}{24}$
31. $\dfrac{RTv^2 - av + ab}{v^2(v - b)}$ **33.** $\dfrac{t_1 + t_2}{t_1 t_2}$ **35.** $\dfrac{m_1 + m_2}{m_1 m_2}$
37. $t = \dfrac{t_1 t_2}{t_1 + t_2}$ **39.** $t = \dfrac{t_1 t_2 t_3}{t_2 t_3 + t_1 t_3 + t_1 t_2}$; no
41. $\dfrac{1}{t} = \dfrac{1}{5} + \dfrac{1}{6} = \dfrac{11}{30}, t = \dfrac{30}{11}, t < 3$ **43.** 3.2 min
45. $t = 12/7$, or 1.7 hr   **47.** 7   **49.** $a = \dfrac{3V}{4\pi b^2}$ **51.** $\dfrac{3x}{2(x + 6)}$
53. $I = \dfrac{2E}{2R + r}$ **55.** $v = \dfrac{2v_1 v_2}{v_2 + v_1}$ **57.** ≈ 9.5 hr; ≈ 59.6 mph
59. 19.8 km/hr   **61.** Answers will vary.
63. wrong; find LCD before adding
65. right; always works; not LCD; may require additional
simplification
67. a. ·   **b.** ÷   **69. a.** +   **b.** −   **71. a.** ·   **b.** ÷
73. a. +   **b.** ·

75. a. $\frac{7}{20}; \frac{3}{20}; \frac{1}{40}; \frac{5}{2}$ **b.** $\frac{9}{40}; \frac{1}{40}; \frac{1}{80}; \frac{5}{4}$ **c.** $\frac{19}{12}; -\frac{1}{12}; \frac{5}{8}; \frac{9}{10}$

d. $\frac{3}{2}; -\frac{1}{6}; \frac{5}{9}; \frac{4}{5}$ **e.** $3\frac{5}{6}; -1\frac{1}{6}; 3\frac{1}{3}; \frac{8}{15}$ **f.** $3\frac{11}{12}; \frac{7}{12}; 3\frac{3}{4}; 1\frac{7}{20}$

g. $3\frac{9}{20}; \frac{19}{20}; 2\frac{3}{4}; 1\frac{19}{25}$ **h.** $3\frac{25}{28}; -\frac{11}{28}; 3\frac{3}{4}; \frac{49}{60}$

EXERCISES 7.7

1. $x = 24$ **3.** $r = \dfrac{180L}{\pi\theta}, \theta \neq 0$ **5.** $x = 5\frac{5}{11}$ **7.** $n = 5$

9. no variables in denominators **17.** $x = 7$

19. $\{-2, 5\}, x \neq -5$ **21.** $\{-1.5, 5\}, x \neq \frac{1}{2}$ **23.** $x = 64$

25. $3 - 2x$ **27.** $6x + 1$ **29.** $-x - 2$ **31.** $3x + 5$

33. $x \neq 0; x = \frac{15}{19}$ **35.** $x \neq 0; x = 6$ **37.** $x \neq 0; \{ \ \}$ or $\varnothing$

39. $x \neq 0; x = 3$ **41.** $x \neq \frac{1}{2}, 1; x = \frac{3}{2}$ **43.** $x \neq -2; \{-7, 4\}$

45. $x \neq 1; \{-7, 6\}$ **47.** $x \neq 0; \{-1.5, 6\}$ **49.** $x \neq 2; \{ \ \}$ or $\varnothing$

51. $x \neq \pm 1; \{-\frac{1}{3}, 4\}$ **53.** $x \neq 3, 6; \{-3, 4\}$ **55.** $x \neq 0, 2; x = 6$

57. $x \neq 0, -3; \{-\frac{3}{2}, 5\}$ **59.** $x \neq \pm 1; \{-3, 4\}$

61. $x \neq 1; x = -2$ **63.** $x \neq 0, -3; \{-6, -2\}$

65. $x = -2 + 2\sqrt{2} \approx 0.828$ **67.** $x = \dfrac{69 - 7\sqrt{39}}{38} \approx 0.665$

69. 18.7 in.; 11.6 in. **71.** $\dfrac{1}{6} + \dfrac{1}{x} = \dfrac{1}{2}; x = 3$ min

73. $\dfrac{1}{20} + \dfrac{1}{x} = \dfrac{1}{12}; x = 30$ min

75. $\dfrac{1}{10,000} + \dfrac{1}{4000} = \dfrac{1}{R}; R = 2857$ ohms

77. $\dfrac{1}{5} + \dfrac{1}{3} + \dfrac{1}{3} = \dfrac{1}{x}; x \approx 1.15$ min **79.** 14.3 cm, 5.7 cm

81. 15.5 cm, 6.7 cm **83.** $a = \dfrac{bc}{b - c}$ **85.** $c = \dfrac{ab}{a + b}$

87. $R = \dfrac{E}{I} - r$ **89.** did not distribute $15x(x + 2)$ over $2/x$

91. forgot to multiply by 2 in 2nd term

CHAPTER 7 REVIEW EXERCISES

1. $2, -\frac{2}{3}, \frac{8}{9}, \frac{1}{2}$ **3.** $\dfrac{ac + ab}{bc}, \dfrac{ac - ab}{bc}, \dfrac{a^2}{bc}, \dfrac{c}{b}$

5. a. $\dfrac{9y}{5x}$ **b.** -1 **c.** $1 - c$

7. a. $3(x - 2)$ **b.** $\dfrac{1}{a + b}$ **c.** $2 + 4x$

9. a. $\dfrac{x - 6}{x - 5}$ **b.** $\dfrac{2x - 1}{x + 2}$ **c.** $\dfrac{x - 1}{x - 4}$ **11.** $x - 6$

13. 3 to 16 **15.** $x = -3, x = 1$ **17.** $\dfrac{y}{6x}$ **19.** 11 **21.** $\dfrac{2}{1 - x}$

23. $\dfrac{x + 1}{x - 3}$ **25.** $\dfrac{x + 1}{x - 1}$ **27.** $\dfrac{-x - 4}{x^2 - 1}$ **29.** $\dfrac{2}{x(3x - 4)}$

31. $\dfrac{2A}{h}$ **33.** 200 gal **35.** $\dfrac{4x^2}{9x + 6}$ **37.** $\dfrac{x^2}{2(x + 1)}$

39. $x \neq 1; x^2 - 2x + 1$ **41.** $x \neq -3; x^2 - 3x + 9 - \dfrac{26}{x + 3}$

43. $x \neq -1; x^3 - x^2 + x - 1$ **45.** $x = 12.8; x \neq 0$

47. $\{-7, 2\}; x \neq -5$ **49.** $x = 26$ **51.** $\{-1.5 \pm 0.5i\}; x \neq 0$

53. $x = 4; x \neq -2, 0$ **55.** $\{\frac{4}{7}, 3\}; x \neq 0, 1$ **57.** $y = 5.6, x = 4.8$

59. $r = 10, s = 10, t = 12.5$ **61.** ≈ 0.357 sec

63. grows 504 in.; ≈ 525 in.; 42 ft, ≈ 43.75 ft (the 18 years does not include the 9 months)

65. $y = 0.3125x + 9.99$; possibly the basic cost of stick manufacture

67. $y = 2.87x$; direct **69.** $D = k\sqrt{h}$; $D = $ distance, $h = $ height

71. $V = klwh$; $V = $ volume, $l = $ length, $w = $ width, $h = $ height; $k = 1$

73. $d = ks^2$ **75.** $s = k\sqrt{df}$ **77.** 1.5 times the cost

79. 16 times the energy **81.** $y = 90/x; k = 90$ credits

83. $y = 100,000/x; k = \$100,000$ **85.** 3.08 hr; no

87. a. distance divided by total hours minus 3 hours not traveling

b.

Total Trip Time (hr)	Rate (mph)
4	300
5	150
6	100
7	75
8	60

c. Because of the 3-hr delay, we replace t with $t - 3$, causing a shift of 3 units to the right.

d. 14.3 mph **e.** 20 mph **f.** 42.9 mph **g.** 600 mph

h. 3000 mph **i.** 30,000 mph **j.** not possible

89. $x = 1.375; x = -1$ makes equation undefined.

91. A larger input will yield a smaller output; $f(a) > f(a + 2)$.

93.

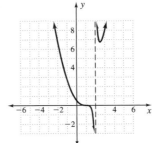

As x gets closer to 2, denominator gets closer to 0; $x = 2$

95. a. y approaches $-\infty$ **b.** y approaches ∞

c. y approaches $-\infty$ **d.** y approaches ∞

e. The nearly vertical line connecting points near $x = -3$ and $x = 1$ should not be there, as shown in dot mode.

f. $x = -3, x = 1$

CHAPTER 7 TEST

1. $\dfrac{b}{ac}; a \neq 0, b \neq 0, c \neq 0$ **2.** simplified; $b \neq 3$

3. $\dfrac{1}{5c}; a \neq 0, c \neq 0$ **4.** $\dfrac{3c^2}{2b}; b \neq 0, c \neq 0$

5. $-\dfrac{x + 5}{x + 4}; x \neq -4, 5$ **6.** $\dfrac{2}{a + 3}; a \neq -3, 6$

7. $\dfrac{3x - 1}{2x - 1}; x \neq \frac{1}{2}, 2$ **8.** $\dfrac{2x + 3}{x + 2}; x \neq -2$

9. $\dfrac{2y}{x}; y \neq 0, 7; x \neq 0$ **10.** $\dfrac{(x - 3)^2}{(x + 1)^2}; x \neq -3, -1$

11. $(x + 4)(x + 2); x \neq -4, 2$ **12.** $5x + 3; x \neq -1, 0$

13. $\dfrac{8x}{(x - 2)(x + 2)}; x \neq \pm 2$ **14.** $\dfrac{x - 3}{x + 2}; x \neq -2, 0$

15. 1000 cm^2 **16.** ≈ 114.2 yr **17.** 0.5 hr/gal

18. $\dfrac{3 - x^2}{x^2 + 2}; x \neq 0$ **19.** $\dfrac{-x^2}{3(3 + x)}; x \neq -3, 0, 3$

20. $\mathbb{R}$, $x \neq -2$; $x^2 + 4x + 4$

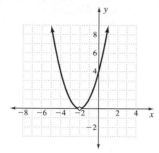

21. $\mathbb{R}$, $x \neq -4$; $x^2 + 2x + 4 - \dfrac{8}{x+4}$

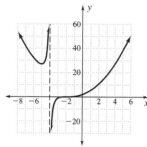

22. $\mathbb{R}$, $x \neq 1$; $2x^3 + 2x^2 + 3x + 3 + \dfrac{-1}{x-1}$ **23.** $\{-3, 6\}$; $x \neq 1$

24. $\left\{\dfrac{1}{4}, \pm i\left(\dfrac{\sqrt{23}}{4}\right)\right\}$; $x \neq 0$ **25.** $x = 7$; $x \neq 0, 4$ **26.** $x = 37$

27. $\left\{-\dfrac{3}{4}, 1\right\}$; $x \neq -1, 0$ **28.** ≈ 0.0173 oz/hr **29.** ≈ 0.0135 oz/hr

30. 112.5 ft **31.** ≈ 50 mph **32.** 84

33. varies directly with cube of radius; $\frac{4}{3}\pi$

34. a. 2.2 min **b.** 20 min per 1000 ft

35. 10; ≈ 16.155; ≈ 10.770

EXERCISES 8.1

1. a. x^9 **b.** y^{2+a} **c.** a^{n+3} **3. a.** $81x^4$ **b.** a^3b^3 **c.** $16n^2$

5. a. x^8 **b.** $32x^{15}$ **c.** $27a^{12}b^6$ **7. a.** x^3 **b.** $\dfrac{1}{y^2}$ **c.** a^2

9. a. x^{4n}, $n \geq 0$ **b.** $\dfrac{y^2}{x^2}$ **c.** $\dfrac{x^{k-1}}{y}$, $k \geq 1$

11. a. $\dfrac{2}{x}$ **b.** $\dfrac{x}{y}$ **c.** 1 **d.** $\dfrac{y}{x^2}$ **e.** x^2y **f.** $\dfrac{y^2}{x^3}$

13. a. $\dfrac{1}{4y^2}$ **b.** $\dfrac{2}{x^4}$ **c.** $\dfrac{y^2}{4x^2}$ **d.** $\dfrac{8c^3}{a^3}$ **e.** $\dfrac{x^4}{3y^4}$ **f.** b^{n+1}, $n \geq -1$

15. a. $\dfrac{c^3}{27a^6}$ **b.** $\dfrac{c^4}{a^2}$ **c.** $-c^3$ **d.** $2a^2$ **e.** $3a^4b^3$ **f.** 3^{1-x}, $x \leq 1$

17. a. $\frac{1}{5}$ **b.** $-\frac{1}{2}$ **c.** 4 **d.** -16 **e.** 5 **f.** $\frac{1}{9}$

19. a. $\frac{2}{3}$ **b.** $-\frac{1}{2}$ **c.** $-\frac{1}{16}$ **d.** $\frac{1}{16}$ **e.** $-\frac{1}{16}$ **f.** 4

21. a. terms; $3y^3$ **b.** terms; $2x^{-1} = \dfrac{2}{x}$ **c.** neither

23. a. neither; $x^2 + \dfrac{1}{x^2}$ **b.** terms; $2y^{-3} = \dfrac{2}{y^3}$

   c. bases; $y^{-2} = \dfrac{1}{y^2}$

25. a. neither **b.** bases; x^2 **c.** terms; $3x^2$

27. a. $\dfrac{(-3x)^{-3}}{(2)^{-3}} = \dfrac{8}{(-27x^3)} = -\dfrac{8}{27x^3}$ **b.** $\dfrac{(4x^3)^{-2}}{(y^2)^{-2}} = \dfrac{y^4}{16x^6}$

   c. $\frac{1}{2}x^{-2-2}y^{1-(-3)} = \frac{1}{2}x^{-4}y^4 = \dfrac{y^4}{2x^4}$ **29.** 8 times

31.

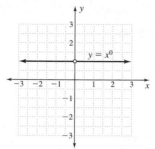

on $y = 1$; no output for 0^0; undefined at $x = 0$

33. no, the sign is not in the left base

35. Exponent does not affect 3. Answer is $3/x^2$.

37. $5\left(\frac{5}{6}\right)^3 = 6\left(\frac{5}{6}\right)^4$, since $6\left(\frac{5}{6}\right)^4 = 6 \cdot \frac{5}{6} \cdot \left(\frac{5}{6}\right)^3 = 5\left(\frac{5}{6}\right)^3$

EXERCISES 8.2

1. a. 10^{-3} **b.** 10^{-5} **c.** 10^4 **d.** 10^{-2} **e.** 10^6 **f.** 10^{-6}

3. a. 3×10^3 **b.** 3.5×10^2 **c.** 3.500×10^5 **d.** 3.50×10^{-3}

5. 1; 2; 4; 3 **7. a.** 29,979,000,000 cm/sec

   b. 0.000 000 000 000 000 000 000 001 672 6 g

   c. $-0.000\ 000\ 000\ 000\ 000\ 000\ 160\ 22$ C

   d. 0.000 000 000 000 000 000 160 22 J

9. 6.0×10^4 **11.** 1.5×10^{-7} **13.** 3.0×10^{-6} **15.** 8.0×10^8

17. a. 4.00×10^{-1} **b.** 4.00×10^5

19. a. 5.0×10^{-6} **b.** 5.0×10^3

21. a. 1E10; one **b.** 1E10; two **c.** 1E11; one **d.** 1E11; two

   e. 1E52; one **f.** 1E53; one

   g. Multiply the preceding number by 10 raised to the exponent that follows.

23. a. 5.07×10^{11} **b.** 4.4

   c. 13; 1931, 1944, 1957, 1970, 1983, 1996, 2009

   d. faster than predicted

25. a. 1.231×10^{12} ft^3 **b.** ≈ 42 ft

27. a. $\approx 2.61 \times 10^3$ m/sec **b.** $\approx 3.95 \times 10^4$ m/sec

   c. $\approx 1.05 \times 10^4$ m/sec **d.** $\approx 4.53 \times 10^3$ m/sec

   e. The vehicle is pulled back toward the planet.

   f. The orbit becomes oval instead of circular.

EXERCISES 8.3

1. a. $x = \frac{2}{3}$ **b.** $x = \frac{3}{2}$ **c.** $x = \frac{3}{4}$

3. $3 \cdot 3 \cdot 3 \cdot 3 = (3 \cdot 3) \cdot (3 \cdot 3)$

5. a. $x = 5$ **b.** $x = 3\frac{1}{3} = \frac{10}{3}$ **c.** $x = 2\frac{1}{2} = \frac{5}{2}$ **d.** $x = 2$

   e. $x = 1\frac{2}{3} = \frac{5}{3}$ **f.** $x = 1\frac{3}{7} = \frac{10}{7}$ **g.** $x = 1\frac{1}{4} = \frac{5}{4}$

7. a. 8 **b.** 2 **c.** 2 **9. a.** 4 **b.** 64 **c.** 216 **d.** 64

11. a. 32 **b.** 512 **c.** 125 **d.** 27

13. a. $x \approx 4.3$ **b.** $x \approx 4.7$ **c.** $x \approx 5.0$

15. a. 125 **b.** 32 **c.** 1024 **d.** 0.3 **17.** $9.77 **19.** $1402.55

21. a. annual **b.** semiannual **c.** quarterly **d.** monthly

   e. weekly **f.** daily **23. a.** 11,272.72 **b.** 11,268.25

25. $10,000 earnings, 4% compounded monthly/quarterly for 3 years

27. $9.83 **29.** $269.57 **31.** $1500.91 **33.** $667.45

35. $\approx$$1,149.73 **37.** $\approx$$1,196.50 **39.** $\approx$$1,447.88

41. $\approx$$1,390.44 **43.** smaller value **45.** $1 + r/n$

47. Two possible answers are $a = b = 2$ and $a = 2$, $b = 4$.

49. any $b > 0$ except $b = 1$ **51.** $\approx$$8.84; $\approx$$9.37; rounding error

53. a. 19, 15, 13, 11, 9, 8 **b.** $-16, -22, -26, -29, -31, -33$

   c. The model produces warmer temperatures.

EXERCISES 8.4

1. a. $2 \cdot 2 \cdot 2 \cdot 2 \cdot 2 \cdot 2 \cdot 2 \cdot 2$ **b.** 2 **c.** 4 **d.** 16
e. 32 **f.** 64 **g.** 128 **3. a.** 4 **b.** 2
5. a. -3 **b.** 5 **7. a.** not a real number **b.** not a real number
9. a. -10 **b.** -10 **11. a.** -4 **b.** 10
13. a. -4 **b.** not a real number **15. a.** $\sqrt[6]{x}$ **b.** $x^{1/3}$ **c.** $\sqrt{x}$
17. a. $(\sqrt{x})^3$ or $\sqrt{x^3}$ **b.** $x^{3/4}$ **c.** $(\sqrt[5]{x})^4$ or $\sqrt[5]{x^4}$
19. a. $(-8)^{5/3} = -32$ **b.** $27^{2/3} = 9$
21. a. $16^{1/2} = 4$ **b.** not a real number
23. a. $|y|$ **b.** z^2 **c.** $|x^3|$ **25. a.** $x^{11/12}$ **b.** $x^{13/15}$
27. a. $x^{1/8}$ **b.** $x^{19/8}$ **29. a.** $x^{7/12}$ **b.** $x^{2/15}$ **31.** $\dfrac{1}{a^{m-n}}$
33. a. x^2 **b.** x^2 **c.** $|y|$ **35. a.** $\sqrt[8]{x}, x \geq 0$ **b.** $\sqrt[12]{x^5}, x \geq 0$
37. a. $a^2, a \geq 0$ **b.** b **c.** $x, x > 0$
39. a. $10\sqrt{n}$ **b.** $0.1\sqrt{n}$ **c.** $\dfrac{\sqrt{n}}{10}$ **d.** $\dfrac{\sqrt[3]{x}}{10}$ **e.** $2\sqrt[3]{n}$
41. a. 6 **b.** -9 **c.** 4 **d.** 2 **43. a.** 3 **b.** 3 **c.** 4 **d.** 2
45. a. $x^{13/12}$ **b.** $x^{17/12}$ **c.** $\dfrac{1}{x^{1/3}}$ **d.** $x^{11/12}$
47. a. $x^{11/6}$ **b.** $x^{5/12}$ **c.** $x^{1/2}$ **d.** $x^{5/6}$
49. a. $x^{1/3}$ **b.** $\dfrac{8x^3}{y^6}$ **c.** $\dfrac{1}{x^2}$ **d.** $\dfrac{1}{x^{5/3}}$
51. a. $2\sqrt[3]{2}$ **b.** $5a^2\sqrt{3a}$ **c.** $2x^2\sqrt[4]{2x}$ **d.** $-2bc\sqrt[5]{a^4c}$
e. $xy^2z^3\sqrt[3]{xy^2}$
53. $x^{(q-p)/pq}$
55. The former must be positive; the latter may be negative.
57. The even root of a negative number is undefined.
59. false **61.** true **63.** yes; $-x > 0$ if $x < 0$
65. $5 \cdot 5^n = 5^1 \cdot 5^n = 5^{n+1}$. Add exponents when bases are the same.
67. x^2 is always positive.

MID-CHAPTER 8 TEST

1. $\sqrt{x}, x \geq 0$ **2.** $x^{-1}, x \neq 0$ **3. a.** $\dfrac{b^2}{a}$ **b.** $\dfrac{a^2c^2}{b}$
4. a. b^3 **b.** $\dfrac{x^5}{y^4}$ **5. a.** a^{x+1} **b.** x^{2n} **6. a.** $\dfrac{9}{4x^2}$ **b.** xy^2
7. a. 0.0043 **b.** 1.23×10^{-4} **c.** $3 \times 10^{-5} = 0.00003$
d. two
8. a. $x = 2.5$ **b.** $x = 1.5$ **c.** $x = 1.5$ **d.** $x = 0$
9. a. \$1418.73 **b.** \$1212.11
10. 2% twice a year = 4.04% annually
11. a. $2\sqrt{n}, n \geq 0$ **b.** $2|n|\sqrt{10}$ **c.** $20\sqrt{n}, n \geq 0$
d. $20\sqrt{10n}$
12. a. $(\sqrt[3]{8})^2 = 4$ **b.** $\sqrt[3]{125} = 5$
13. a. $(\sqrt[4]{16})^3 = 8$ **b.** $(\sqrt[3]{a^3})^2 = a^2$
14. $x^{7/12}$ **15. a.** $x^{1/12}$ **b.** $x^{2/3}$
16. a. not a real number **b.** $-3x^2y$ **17. a.** -4 **b.** $2y$
18. a. $3x^2y\sqrt{x}$ **b.** $2xy^2\sqrt[3]{3x^2}$
19. yes, because the 4th power is positive **20.** $2|y|$

EXERCISES 8.5

1. $6\sqrt{2}$ **3. a.** $5\sqrt{5}$ **b.** $3\sqrt{2}$ **5. a.** $8\sqrt{3}$ **b.** $\sqrt{5}$ **7.** $6\sqrt{x}$

9. $3.1\sqrt{x}$ **11.** not like terms **13.** $3\sqrt{ab}$ **15.** $2\sqrt[3]{x}$
17. not like terms **19.** $5\sqrt[3]{x}$ **21.** $2a\sqrt[4]{ab}$ **23.** $3x\sqrt[4]{y}$
25. a. 1 **b.** $27 - 10\sqrt{2}$ **27. a.** $x^2 - 5$ **b.** $x^2 - 2x\sqrt{7} + 7$
29. a. $x + 6\sqrt{x} + 9$ **b.** $a - 6\sqrt{a} + 9$
31. a. $1 - 2\sqrt{a} + a$ **b.** $a - 2\sqrt{ab} + b$
33. $(1 + \sqrt{3})^2 - 2(1 + \sqrt{3}) - 2 = 0$; then simplify
35. $(-2 - \sqrt{2})^2 + 4(-2 - \sqrt{2}) + 2 = 0$; then simplify
37. $2\left(\dfrac{3}{2} + \dfrac{\sqrt{3}}{2}\right)^2 - 6\left(\dfrac{3}{2} + \dfrac{\sqrt{3}}{2}\right) + 3 = 0$ **39.** $x^3 + 2$
41. a. $3 - \sqrt{2}$ **b.** $3 + \sqrt{a}$ **c.** $a + \sqrt{b}$ **d.** $\sqrt{2} + \sqrt{3}$
43. a. $\dfrac{4\sqrt{5}}{5}$ **b.** $\dfrac{4\sqrt{6}}{3}$ **45. a.** $\dfrac{a\sqrt{c}}{c}$ **b.** $\sqrt{a}$ **47.** $\dfrac{7 + \sqrt{5}}{11}$
49. $\dfrac{x^2 + x\sqrt{y}}{x^2 - y}$ **51. a.** $\dfrac{\sqrt[3]{4}}{2}$ **b.** $\sqrt[3]{2}$ **c.** $\sqrt[3]{9}$
57. $-\dfrac{3}{5}; 2\sqrt{34} \approx 11.662$ **59.** $\dfrac{11}{9}; \sqrt{202} \approx 14.213$
61. $\dfrac{b}{a}; \dfrac{\sqrt{a^2 + b^2}}{2}$
63. a. Product is -1. **b.** $\approx 1.6180; \approx -0.6180$
c. $\dfrac{-1 - \sqrt{5}}{2}; \dfrac{1 + \sqrt{5}}{2}$; negative reciprocals

EXERCISES 8.6

1. $\left(\frac{1}{2}, 1\right), \left(\frac{1}{4}, 2\right), \left(\frac{1}{8}, 3\right), \left(\frac{1}{16}, 4\right)$
3. $(2, 1), (5, 2), (8, 3), (11, 4)$
5.

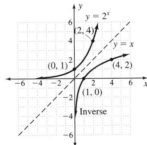

7.

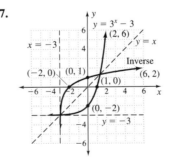

9.

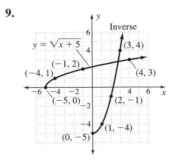

11.

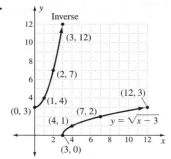

13.

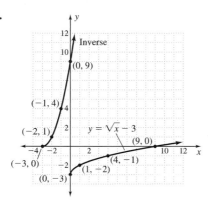

15.

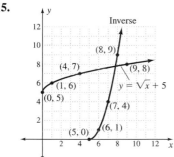

17. $y = x/3$ **19.** $y = \frac{1}{4}x + \frac{1}{2}$ **21.** $y = x - 3$
23. $y = x^2, x \geq 0, y \geq 0$ **25.** $y = x^5$ **27.** $y = \sqrt[4]{x}, x \geq 0, y \geq 0$
29. $y = x$ **31.** $y = -1/x, x \neq 0$ **33.** $y = 1 - x$
35. same as original function
37. $f^{-1}(x) = x^2 - 5, x \geq 0, y \geq -5$
39. $f^{-1}(x) = x^2 + 3, x \geq 0, y \geq 3$
41. $f^{-1}(x) = (x + 3)^2, x \geq -3, y \geq 0$
43. $f^{-1}(x) = (x - 5)^2, x \geq 5, y \geq 0$
45. Even root of a negative number is not a real number.
47. Outputs are positive real numbers.
49. For any real number x, there is a corresponding real number y on the graph.
51. Principal square root is always positive.
53. a. yes **b.** Reversing ordered pairs gives same set. **c.** yes
55. With $x \geq 2$, no two inputs have the same output; an inverse function exists.
57. input, output
59. It is the reciprocal key; the -1 is an exponent.

EXERCISES 8.7

1. $x = 14$ **3.** $\{\ \}$ or $\varnothing$ **5.** $x \leq 2; x = -2$ **7.** $x \geq 5; x = 9$
9. $x \geq 5; x = 21$ **11.** $x \geq -7; x = -7$ **13.** $x \geq -7; x = 11$

15. $x \geq \frac{5}{3}; x = 3$ **17.** $x \geq \frac{5}{3}; \{\ \}$ or $\varnothing$ **19.** $x \geq \frac{1}{5}; x = 2$
21. $x \geq 0; \{1, 9\}$ **23.** $x \geq 0; x = 12$ **25.** $x \geq 1; x = 4$
27. extraneous root

29. a. $x = \pm 1$ **b.** $x = \pm 1, x = \pm i$ **c.** $x = 1; x = -\frac{1}{2} \pm \frac{i\sqrt{3}}{2}$

31. $x = \pm 2.5$; none **33.** $x = \pm 2.5; 2$ **35.** $x = 0.201$
37. $x = 0.076$ **39.** $x = -0.167$ **41.** $x = -0.071$

43. a. $L = \dfrac{r^2}{24}$ **b.** $L = \dfrac{r^2}{12}$

c. ≈ 34.64 mph; ≈ 24.49 mph; $\dfrac{r_{\text{wet}}}{r_{\text{dry}}} = \dfrac{\sqrt{2}}{2}$;
slow down when pavement is wet

45. $s = \dfrac{v^2}{2g}$ **47.** $\approx 6.1\%; \approx \$243,000$

49. $\approx 6.94\%; \approx \$205,315.25$ **51.** $\approx 5.3\%; \approx \$127,000$
53. 28.6% **55.** -9.1% **57.** $\approx -10\%$ **59.** $-15.7\%; -19.8\%$
61. a. $r = 4.22$ in. **b.** $r = 5.31$ in. **c.** $r = 6.69$ in.
  d. $r = 8.43$ in. **e.** two

63. $T = \sqrt[4]{\dfrac{Q_b}{\sigma A}}$ **65.** $V = \sqrt{\dfrac{2K}{m}}$

CHAPTER 8 REVIEW EXERCISES

1. a. $\dfrac{1}{bc^2}$ **b.** $\dfrac{a^3 c}{b^2}$ **c.** $\dfrac{16x^4}{y^8}$ **3. a.** $\dfrac{x^6}{y^8}$ **b.** $36x^7 y$

5. a. 0.000 000 000 000 003 45 **b.** 0.006 400
  c. 400,500 **d.** 4780.0
7. a. 8 **b.** 2 **c.** 32 **9. a.** 8 **b.** 125 **c.** 81
11. \$10.36; \$10.40 (rounded down after 10 yr)
13. \$1953.43 **15.** \$1929.21
17. a. a^3 **b.** x **c.** $d^{7/4}$ **d.** $b^{1/6}$ **e.** x^{n+1} **f.** $x^{3/2} y^2$

19. a. $a^{3/2} b^{1/2}$ **b.** a **c.** x **d.** x^n **e.** $b^{3/2}$ **f.** $\dfrac{x^2}{y^{1/3}}$

21. a. $\left(\dfrac{27}{8}\right)^{1/3} = \dfrac{3}{2}$ **b.** $(-32)^{1/5} = -2$ **c.** $4^{3/2} = 8$

23. a. $\left(\sqrt[3]{125}\right)^2 = 25$ **b.** $\sqrt[3]{64} = 4$ **c.** $\left(\sqrt[5]{32}\right)^2 = 4$

25. a. $\sqrt[3]{-64} = -4$ **b.** $-\sqrt[4]{64} = -8$ **c.** not a real number

27. a. $\left(\sqrt[3]{x}\right)^2$ **b.** $\left(\sqrt{x}\right)^3$ **c.** $\left(\sqrt[4]{a}\right)^3$

29. a. $3\sqrt{n}, n \geq 0$ **b.** $3|n|\sqrt{10}$ **c.** $0.3\sqrt{x}, x \geq 0$ **d.** $30x^2$

31. a. z^3 **b.** x **c.** x^2 **33. a.** $x^{5/6}$ **b.** $x^{5/2}$ **c.** $x^{1/6}$

35. a. $1/x$ **b.** x **c.** $x^{1/12}$ **37. a.** $5\sqrt{3}$ **b.** $4\sqrt{x}, x \geq 0$
39. a. 3 **b.** $9 - 6\sqrt{x} + x$

41. a. $\dfrac{b + \sqrt{a}}{b^2 - a}$ **b.** $\dfrac{-x + 7\sqrt{x} - 12}{16 - x}$ **c.** $\dfrac{\sqrt{a} - \sqrt{b}}{a - b}$

  d. $\dfrac{\sqrt[4]{a^2 b^3}}{b}$ **e.** $\dfrac{\left(\sqrt[3]{9x}\right)^2}{9x}$ or $\dfrac{\sqrt[3]{3x^2}}{3x}$ **f.** $3\sqrt[5]{x^2}$

43. $1 + \sqrt{2}$
47. a. (3, 3), (5, 4), (7, 5); yes
  b. $(-1, -2), (-2, 0), (-3, 2)$; yes
49. a. $y = \frac{1}{2}x + \frac{1}{2}$ **b.** $y = x/3$ **c.** $y = \sqrt[3]{x}$
  d. $y = x^2, x \geq 0, y \geq 0$ **e.** $y = -x$ **f.** $y = 1/x$
51. a. $x \geq 0, y \geq 1; y = (x - 1)^2, x \geq 1, y \geq 0$
  b. $x \geq 1, y \geq 0; y = x^2 + 1, x \geq 0, y \geq 1$

53. $\approx 3.82\%$ **55.** $\approx 7.67\%$ **57.** $V = \dfrac{4\pi r^3}{3}$

59. a. $y = \sqrt[3]{a^3 - x^3}$ **b.** $y = k + \frac{3}{2}\sqrt[3]{kx^2}$

61. $M_1 = \dfrac{Fr^2}{kM_2}$ **63.** $y = \pm x\sqrt{1 - \dfrac{x^2}{a^2}}$

65. a. $x = 12$ **b.** $x = 4$ **c.** $x = 0$ **d.** $x = 12$
67. a. 20 **b.** ≈ 9.283 **c.** ≈ 4.309 **d.** 2 **e.** ≈ 0.9283
 f. ≈ 0.4309 **g.** 30 **h.** ≈ 13.92 **i.** ≈ 6.463 **j.** 3
 k. ≈ 1.392 **l.** ≈ 0.6463

CHAPTER 8 TEST

1. a. b^3 **b.** $\dfrac{b}{a^2c^2}$ **c.** $\dfrac{9}{x^2y^2}$ **2. a.** $\dfrac{y^6}{x^2}$ **b.** a^9 **c.** $\dfrac{1}{x^{n-1}}$

3. a. $\dfrac{1}{b^{1.25}}$ **b.** $\dfrac{a^2}{b^3}$ **c.** $x^{4.5}y^4$ **4. a.** b^3 **b.** b **c.** 3^{n+1}

5. $(2x)^2 = 2^2 \cdot x^2 = 4x^2$ **6.** Yes, both equal b^4/a^3.
7. a. $0.000\ 034\ 50$ **b.** 9×10^{-7} or $0.000\ 000\ 9$
8. $\sqrt{4x^2} = \sqrt{4} \cdot \sqrt{x^2} = 2|x|$
9. a. $6\sqrt{x}, x \geq 0$ **b.** $0.6|x|$ **c.** $6x^2\sqrt{10}$
10. a. $\left(\sqrt[3]{64}\right)^2 = 16$ **b.** $\left(\sqrt[5]{32}\right)^4 = 16$ **c.** $\left(\sqrt[4]{81}\right)^3 = 27$

11. a. $\frac{2}{3}$ **b.** -3 **c.** 4 **12. a.** $x^{2/3}$ **b.** x^2 **c.** $\dfrac{3x}{4}$

13. a. $4 - x$ **b.** $24 - 3\sqrt{3} + 16\sqrt{2} - 2\sqrt{6}$

14. a. $\dfrac{x - \sqrt{y}}{x^2 - y}$ **b.** $\dfrac{\left(\sqrt[3]{x}\right)^2}{x}$

15. yes; $(4 - \sqrt{10})^2 - 8(4 - \sqrt{10}) + 6 = 0$ **16.** 2536.48

17. $\approx 8.45\%$ **18.** $r = \sqrt{\dfrac{mgl}{\pi s M}}$ **19.** $l = \sqrt[3]{\dfrac{3dEI}{W}}$

20. a. $(-1, -1), (-3, -2), (-5, -3)$; yes
 b. $(-1, -1), (-1, -2), (-1, -3)$; no

21.

x	0	1	2	3
$f(x)$	1	0.5	0.25	0.125

x	1	0.5	0.25	0.125
$f^{-1}(x)$	0	1	2	3

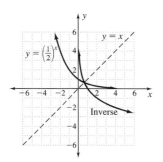

22. $3y - 2x = 6$ or $y = \frac{2}{3}x + 2$
23. a. $x = 8$ **b.** $x = 3$ **c.** $x = 2$ **d.** $x = 8$

CUMULATIVE REVIEW OF CHAPTERS 1 TO 8

1. $x = -11$ **3.** $b = 3A - a - c$
5. Let $y =$ total cost in \$, $x =$ kilowatt hours used; assume a linear equation; $y \approx 11.03 + 0.051x$
7. a. $-\frac{3}{5}$ **b.** $\frac{5}{3}$ **9.** $\mathbb{R}$; 4 **11.** 14 credit hours
13. yes; $12.5^2 = 3.5^2 + 12^2$
15. a. $\{-4, 1\}$ **b.** $\{-3, 0\}$ **c.** no real-number solution
17. $x < -3$ or $x > 0$ **19.** $x^4 - 4x^3 + 6x^2 - 4x + 1$

21. a. 68 **b.** $23 - 2i$ **c.** $\dfrac{2 - i}{5}$

23. Shift $y = x^2$ a total of a units horizontally.

25. $\dfrac{\$8.99}{\text{package}} \cdot \dfrac{1\ \text{package}}{56\ \text{diapers}} \cdot \dfrac{10\ \text{diapers}}{1\ \text{day}} \cdot 365\ \text{days} \approx \$585.96$

27. e **29.** f **31.** d **33.** $x^2 + x + 1$
35. a. $0.000\ 000\ 000\ 000\ 000\ 380$ **b.** $4{,}230{,}000{,}000{,}000{,}000$
37. a. 16 **b.** 25 **c.** no real number **d.** -4 **e.** $x^{14/15}$
 f. x^2 **g.** $3a\sqrt{2a}$ **h.** $2xyz\sqrt[3]{2yz^2}$ **i.** $x^{1/3}$ **j.** $8\sqrt{2x}$ **k.** 2
39. $\approx 8.89\%$; $\approx 8.54\%$
41. $(-2, 2)$ and $(4, 1)$, $d = \sqrt{37}$; $(-2, 2)$ and $(3, -5)$, $d = \sqrt{74}$;
 $(4, 1)$ and $(3, -5)$, $d = \sqrt{37}$; isosceles right triangle

EXERCISES 9.1

1. a. $a; x$ **b.** $x; 3$ **c.** $2; -x$ **d.** $x; a$
3. a. $\pi; x$ **b.** $x; 4$ **c.** $r; n - 1$ **d.** $1.06; t$
5. a. 3 **b.** $a_n = 0.5 \cdot 3^{n-1}$ **c.** $y = 0.167 \cdot 3^x$
7. a. 2 **b.** $a_n = 10 \cdot 2^{n-1}$ **c.** $y = 5 \cdot 2^x$
9. a. 2 **b.** $a_n = \frac{1}{4} \cdot 2^{n-1}$ **c.** $y = 0.125 \cdot 2^x$
11. a. 2 **b.** $a_n = 1 \cdot 2^{n-1}$ **c.** $y = 0.5 \cdot 2^x$
13. a. $\frac{1}{2}$ **b.** $a_n = 32\left(\frac{1}{2}\right)^{n-1}$ **c.** $y = 64 \cdot 0.5^x$
15. a. $\frac{1}{3}$ **b.** $a_n = 3 \cdot \left(\frac{1}{3}\right)^{n-1}$ **c.** $y = 9 \cdot 0.333^x$
17. quadratic; $y = 2x^2$ **19.** exponential; $y = 0.167 \cdot 3^x$
21. quadratic; $y = 0.5x^2 + 2.5x$ **23.** linear; $y = 8x + 22$
25. exponential; $y = 2.5 \cdot 2^x$ **27.** $4 \cdot 2^x$ **29.** $2^x/2$ **31.** 2^x
33. 2^{n-1} **35.** 2^{n+2} **37.** 3^{n-1} **39.** 3^{n+2}
41. a. $f(x) = 2^{x-4}$ **b.** $f(x) = 2^{x+1}$
43. a. $\left(\frac{1}{2}\right)^{x-3}$ **b.** $\left(\frac{1}{2}\right)^{x-4}$ **45.** $r = -3, a_n = 2 \cdot (-3)^{n-1}$
47. $r = -\frac{1}{2}, a_n = 64 \cdot \left(-\frac{1}{2}\right)^{n-1}$
 45 and 47 are not exponential functions, the bases are negative.
49. a. $y \approx 24{,}778(1.049)^x$; 4.9%; $\approx 705{,}000$; actual 448,607
 b. $y \approx 6657(1.043)^x$; 4.3%; $\approx 127{,}000$; actual 73,344
 c. $y \approx 42{,}738(0.998)^x$; -0.2%; $\approx 37{,}000$; actual 40,407
 d. $y \approx 38{,}587(1.005)^x$; 0.5%; $\approx 55{,}000$; actual 62,916
51. a. $a_n = 108\left(\frac{1}{2}\right)^{n-1}$; $y = 216\left(\frac{1}{2}\right)^x$; ≈ 0.211 in.; alters the base
 b. $a_n = 135\left(\frac{2}{3}\right)^{n-1}$; $y = 202.5\left(\frac{2}{3}\right)^x$

EXERCISES 9.2

1. a. increasing **b.** decreasing
3. a. increasing **b.** decreasing
5. a. increasing **b.** increasing **7.** c, a, b, respectively
9. $\{\ \}$ or $\varnothing$ **11.** $3, 4, \frac{1}{2}$; equals a **13.** y-intercept value
15. Let $x = 0$ and simplify: $y = 2^{0+h}$, or $y = 2^h$
17. When $x = 1$, $b^x = b$. **19.** Possible answer: $y = 1/x^2$
21. a. 16; $y = 2^{x+4}$ **b.** 8; $y = 2^{x+3}$
23. a. 27; $y = 3^{x+3}$ **b.** $\frac{1}{27}$; $y = 3^{x-3}$
25. a. $y = \frac{1}{16} \cdot 2^x$; $\frac{1}{16}$ **b.** $y = 2 \cdot 2^x$; 2
27.

x	$y = 2^{x-1}$	$y = 2^x$	$y = 2^{x+1}$
-2	$\frac{1}{8}$	$\frac{1}{4}$	$\frac{1}{2}$
-1	$\frac{1}{4}$	$\frac{1}{2}$	1
0	$\frac{1}{2}$	1	2
1	1	2	4
2	2	4	8
graph	c	b	a

Adding or subtracting 1 in the exponent shifts the graph 1 unit to the left or right.

29.

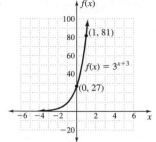

31.

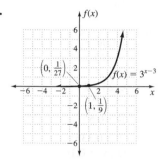

33. no intersection; $3^x < 3^{x+1}$ for all x **35. a.** $x = 8$ **b.** $x = 2$
37. a. $n = \frac{3}{2}$ **b.** $n = \frac{4}{3}$ **39. a.** $n = \frac{1}{2}$ **b.** $x = \frac{3}{2}$
41. a. $x = -\frac{1}{2}$ **b.** $n = -\frac{1}{3}$ **43. a.** $n = -\frac{1}{3}$ **b.** $x = -\frac{1}{4}$
45. a. $x = -10$ **b.** $x = -4$ **47. a.** $x = -6$ **b.** $x = 4$
49. a. $x = 7$ **b.** $x = 5$ **51.** No; $x = -3$ is another base.
53. same; $P = 2^{12t}$
55. a. $x \approx 2.5$ **b.** $x \approx 4.8$ **c.** $y \approx 1.8$ **d.** $y \approx 4.5$
   e. $a \approx 1.34$
57. c. horizontal line at $y = 1$ **d.** yes   **e.** $y = 1$
59. b. points only   **c.** no   **d.** no; negative base
61. a. -1 **b.** $y = -2^x$; $y = -2^{-x}$, or $y = -\left(\frac{1}{2}\right)^x$
   c. negative   **d.** $y = -x^2$

EXERCISES 9.3

1. a.

x	$f(x) = 3^x$
-2	$\frac{1}{9}$
-1	$\frac{1}{3}$
0	1
1	3
2	9
3	27
4	81

b.

x	$y = f^{-1}(x)$
$\frac{1}{9}$	-2
$\frac{1}{3}$	-1
1	0
3	1
9	2
27	3
81	4

c.   yes

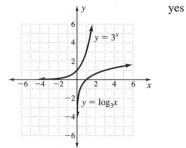

d. $y = \log_3 x$

3.

Exponential Equation	Logarithmic Equation
$3^3 = 27$	$\log_3 27 = 3$
$2^3 = 8$	$\log_2 8 = 3$
$10^1 = 10$	$\log_{10} 10 = 1$
$5^3 = 125$	$\log_5 125 = 3$
$3^{-2} = \frac{1}{9}$	$\log_3\left(\frac{1}{9}\right) = -2$

5. a. $10^3 = 1000$ **b.** $10^0 = 1$ **c.** $3^4 = 81$ **d.** $10^2 = 100$
7. a. $10^{-3} = 0.001$ **b.** $4^0 = 1$ **c.** $m^k = n$ **d.** $5^{-2} = \frac{1}{25}$
9. a. $\log_2 32 = 5$ **b.** $\log_2 2 = 1$ **c.** $\log_2 1 = 0$
   d. $\log_{10} 100 = 2$
11. a. $\log_{10} 0.001 = -3$ **b.** $\log_f g = d$ **c.** $\log_3\left(\frac{1}{9}\right) = -2$
   d. $\log_4 1 = 0$
13. a. 0 **b.** 0.77815 **c.** 1 **d.** 1.77815 **e.** 2 **f.** 2.77815
15. a. -0.22185 **b.** -1 **c.** -2 **d.** -1.22185 **e.** -2.22185
   f. -5.22185
17. a. Decimal portion same; all positive.
   b. Decimal portion same; all negative.
19. a. $\log_{10} 17 = x$; $x \approx 1.23045$ **b.** $\log_{10} 125 = x$; $x \approx 2.09691$
   c. $\log_{10} 400 = x$; $x \approx 2.60206$
   d. $\log_{10} 0.05 = x$; $x \approx -1.30103$

21.

Exponential Equation	Logarithmic Equation	Solve for x
$5^0 = x$	$\log_5 x = 0$	$x = 1$
$x^3 = 64$	$\log_x 64 = 3$	$x = 4$
$10^x = -10$	$\log_{10}(-10) = x$	$\{\ \}$ or $\varnothing$
$10^x = 0$	$\log_{10} 0 = x$	$\{\ \}$ or $\varnothing$
$10^{-1} = x$	$\log_{10} x = -1$	$x = 0.1$

23. a. $7^2 = x$, $x = 49$ **b.** $3^x = 3$, $x = 1$
   c. $10^x = 0.01$, $x = -2$ **d.** $10^0 = x$, $x = 1$
   e. $x^3 = 64$, $x = 4$ **f.** $a^1 = x$, $x = a$
25. a. $2^8 = x$, $x = 256$ **b.** $10^{-2} = x$, $x = \frac{1}{100}$
   c. $2^x = \frac{1}{4}$, $x = -2$ **d.** $10^x = 1$, $x = 0$ **e.** $x^2 = 100$, $x = 10$
   f. $a^x = a$, $x = 1$
27. $y = x$ **29.** range   **31.** range   **33.** (b, a)
35. exponent   **37.** base   **39.** $\mathbb{R}$ **41.** $y > 0$
43. No; log is not a unit.

MID-CHAPTER 9 TEST

1. a. exponential; $a_n = 3 \cdot 3^{n-1}$; $f(x) = 3x$
   b. linear; $y = 6x - 7$
   c. exponential; $a_n = 81 \cdot \left(\frac{1}{3}\right)^{n-1}$; $f(x) = 243 \cdot \left(\frac{1}{3}\right)^x$

d. quadratic; $y = 2x^2 - x$

e. exponential; $a_n = \frac{1}{8} \cdot 2^{n-1}$; $f(x) = 0.0625 \cdot 2^x$

2.

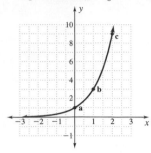

3. a. 2^{x-1} is shifted right 1 unit.   **b.** no intersection; $2^x > 2^{x-1}$

4. a. ≈ 3.3 **b.** $\log_2 10 \approx 3.3$

5. Grows rapidly as x gets large; approaches zero as x gets large
   a. constants $\frac{1}{2}$ and 2
   b. $b > 1$ indicates increasing function; $0 < b < 1$ indicates decreasing function

6. a. $x = 5$ **b.** $\{\ \}$ or $\varnothing$ **c.** $x = 4$ **d.** $x = -3$
   e. $\{\ \}$ or $\varnothing$ **f.** $x = -2$

7. a. $x = \frac{1}{4}$ **b.** $x = -4$ **c.** $x = 27$ **d.** $x = 1$
   e. $x = \frac{1}{2}$ **f.** $\{\ \}$ or $\varnothing$

8. a.

x	$y = 4^x$
-2	$\frac{1}{16}$
-1	$\frac{1}{4}$
0	1
1	4
2	16
3	64

b.

x	$y = f^{-1}(x)$
$\frac{1}{16}$	-2
$\frac{1}{4}$	-1
1	0
4	1
16	2
64	3

c.

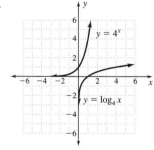

d. $y = \log_4 x$

9.

Exponential Equation	Logarithmic Equation	Solve for x
$10^x = 15$	$\log_{10} 15 = x$	$x \approx 1.17609$
$10^x = 13$	$\log_{10} 13 = x$	$x \approx 1.11394$
$3^x = 81$	$\log_3 81 = x$	$x = 4$
$3^9 = x$	$\log_3 x = 9$	$x = 19,683$
$4^{x+1} = 64$	$\log_4 64 = x + 1$	$x = 2$

EXERCISES 9.4

1. a. 3   **b.** 10   **c.** 4   **d.** 5   **e.** 10   **f.** m

3. a. $x = 0$ **b.** $x \approx 1.58$ **c.** $x = 2$ **d.** $x \approx 2.32$
   e. $x \approx 2.58$ **f.** $x \approx 3.32$ **g.** $x \approx 3.58$ **h.** $x \approx 4.32$
   i. $x \approx 4.58$

5. a. $x \approx 0.43$ **b.** $x = 1$ **c.** $x \approx 1.11$ **d.** $x \approx 1.43$
   e. $x = 2$ **f.** $x \approx 2.11$

7. a. $x \approx 2.09590$ **b.** $x \approx 1.29248$ **c.** $x \approx 10.24477$
   d. $x \approx 13.51341$ **e.** $x = 1$ **9.** 1; $f(1) = 0$

11. x-intercept $= 1$ for both graphs; the graph of $\log_6 x$ is above that of $\log_3 x$ to the left of the x-intercept and is below it to the right of the x-intercept.

13. false   **15.** false

17. Graphs mirror each other across $y = x$. $2^0 = 1$ is the same fact as $\log_2 1 = 0$; $2^1 = 2$ is the same fact as $\log_2 2 = 1$. Answers may vary: $(2, 4)$ is on Y2, $(4, 2)$ is on Y3.

19. a. 14   **b.** 0.7   **c.** 2.3   **d.** 10   **e.** 2.9

21. a. base   **b.** acid   **c.** acid   **d.** base   **e.** acid

23. a. 7.94×10^{-5} M   **b.** 3.16 M   **c.** 1.0×10^{-12} M
   d. 3.16×10^{-4} M

25. $[H^+] > 1$; $[H^+] < 1$

27. a. 3.1623×10^7; 7.5   **b.** 1.2589×10^8; 8.1

29. a. 398,100,000   **b.** 79,430,000

31. $\approx 7,943,000$; $\approx 6,310,000$; ≈ 1.3 times as strong

33. 349 yr   **35.** 73 yr   **37. a.** 8%   **b.** 6%

39. a. -3% **b.** -5% **41.** ≈ 11.9 yr

43. $(0.9)^t = 0.5$, $t \approx 6.58$ yr

45. a. $n \approx 2.7$ **b.** $n = 4$ **c.** $n \approx 12.5$

47. $t = \dfrac{\log 2}{\log (1 + r)}$; no; doubling time is not dependent on P.

49. a. ≈ 333 **b.** ≈ 152 **c.** ≈ 26
   d. $y = 1.7(0.99997)^x$, x in years
   e. $y = 12,000,000(0.955)^x$, x in hr
   f. $y = 490,000(0.997)^x$, x in days

51. a. $365r$; $t/365$ **c.** $\$134.69$ **d.** $\$156.31$

53. a. Answers will vary.
   b. $y = 0.0964x + 0.392$; $9.64\cent$ increase per year
   c. $y = 0.4808(1.095)^x$; $\approx$cost in 1986
   d. $\approx 9.5\%$
   e. constant price change each year; percent increase each year
   f. $\approx \$2.22$; $\approx \$2.72$

EXERCISES 9.5

1. a. $\log 2x$ **b.** $2 \log x$ **c.** $\log 2 - \log x$ **d.** $x \log 2$
   e. $\log (x/2)$

3. log of product property

5. log of power property

7. a. $\log_b x + 3 \log_b z$ **b.** $2 \log_b x - \log_b y$

9. a. $2(\log_b x + \log_b y)$ **b.** $\log_b 5 - 2 \log_b x$

11. a. $\frac{1}{2} \log_b x - \log_b y$ **b.** $2 \log_b (x - z)$

13. a. $\log_b x + 3 \log_b z - 2 \log_b y$
   b. $3 \log_b y + 2 \log_b z - \log_b x$

15. a. $\log_b (xy^2)$ **b.** $\log_b (y/x^2)$

17. a. $\log_b \sqrt{y/z^3}$ **b.** $\log_b \sqrt{x}/3$

19. a. $\log (x^2 - 1)$ **b.** $\log (x + 1)$

21. a. $\log (x + 2)$ **b.** $\log (x^2 + x - 6)$

23. $\log \sqrt{x} = \log (x^{1/2}) = \frac{1}{2} \log x$

25. $\log (1/x) = \log (x^{-1}) = -\log x$

29. $x = 4$ **31.** $x = 10$ **33.** $x = 10$ **35.** $x = 9$

37. $x \approx 1.8614$ **39.** $x \approx 1.5237$ **41.** $x \approx 3.4037$

43. $x \approx 0.7011$ **45.** $t \approx 9.0065$ **47.** $t \approx 16.2376$

49. $t \approx 8.3130$ **51.** $t \approx 15.1385$

53.

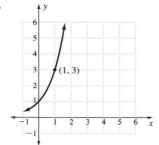

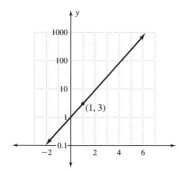

55. straight line
57. a. $\approx$22 ft **b.** $\approx$12 yr **c.** 60 to 70 yr **d.** over 200 yr
  e. 2 and 12; wind breakage **f.** $\approx$16 ft
59. Let $Y_1 = \log x$.

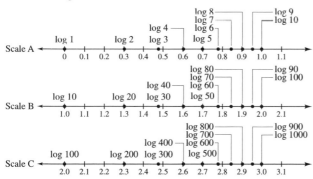

  d. Dots are in same relative positions because
    $\log 20 = 1 + \log 2$, etc.

EXERCISES 9.6

1. $\approx$\$1,124.86, $\approx$\$1,127.27, $\approx$\$1,127.48; $\approx$\$2.41, $\approx$\$2.62
3. a. 7.39 **b.** 23.14 **c.** 3.30 **5. a.** 19.81 **b.** 2.23 **c.** 4.81
7. e^(e^(1)) **9.** \$1083.28 **11.** \$1105.17 **13.** \$1046.02
15. $\approx$\$150,597.11 **17.** $\approx$\$67,667.65
19.

Log. Eqn.	Show Base e	Exp. Eqn.	$y = ?$
$\ln y = 1$	$\log_e y = 1$	$y = e^1$	$y = e$
$\ln y = 0$	$\log_e y = 0$	$y = e^0$	$y = 1$
$y = \ln(-1)$	$y = \log_e(-1)$	$e^y = -1$	$\{\ \}$ or $\varnothing$
$y = \ln e^2$	$y = \log_e e^2$	$e^y = e^2$	$y = 2$
$y = \ln e^e$	$y = \log_e e^e$	$e^y = e^e$	$y = e$

21. a. $x = 1.3863$ **b.** $x = 0.1353$
23. a. $x = 4.4817$ **b.** $x = 0.6931$ **25.** $\approx$8.66%
27. $r = (\ln 2)/t$ **29.** $\approx$18.3 yr **31.** $t \approx 8.7$ yr
33. a. $\approx$8.625% **b.** $\approx$6.27% **c.** $\approx$69/t
  d. Rule of 69 is accurate to within 0.1%.
37. a. $\approx$1.1606; $\approx$1.1606
  b. $\approx$0.903; $\approx$0.778; $\approx$2.079; $\approx$1.792
  c. Ratios are equal; logarithms are not.
39. $K = e^{(-Ea/RT+C)}$ **41.** $[H^+] = e^{(-EnF/RT)}$
43. a.

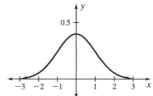

  b. $y \approx 0.4$ at $x = 0$ **c.** $1/\sqrt{2\pi}$ **d.** $\approx$0.65
45. a. 1,850,000 **b.** 1,380,000 **c.** 1,030,000
  d. sign of exponent **e.** $\approx$1997
47. a. 24 **b.** 3,628,800 **c.** 720 **d.** 362,880
  e. $\approx$7.3873; $e^2 \approx 7.3891$

CHAPTER 9 REVIEW EXERCISES

1. 729; geometric; $y = 3 \cdot 3^x$ or $y = 9 \cdot 3^{x-1}$
3. 31; quadratic; $y = x^2 - x + 1$
5. 20; arithmetic; $y = 3x + 2$
7. $u_n - \frac{1}{4}(2)^{n-1}$; $f(x) = 0.125 \cdot 2^x$
9. $u_n = \frac{1}{16}(2)^{n-1}$; $f(x) = 0.03125 \cdot 2^x$
11. a. $y = \frac{1}{8} \cdot 2^x$; $\frac{1}{8}$ **b.** $y = 3 \cdot 3^x$; 3 **c.** $y = \frac{1}{27} \cdot 3^x$; $\frac{1}{27}$
13. a. 2; $y = 2^{x+1}$ **b.** $\frac{1}{9}$; $y = 3^{x-2}$
15. a. $x = 3$ **b.** $x = 1$ **c.** $x = -1$ **d.** $n = 0$
17. a. $n = \frac{3}{2}$ **b.** $n = -\frac{1}{3}$ **c.** $n = -\frac{3}{2}$ **d.** $n = -1$
19. a. $n = -\frac{1}{2}$ **b.** $n = \frac{1}{2}$ **c.** $x = 7$ **d.** $x = \frac{1}{3}$
21.

Exponential Equation	Logarithmic Equation	Solve for x
$2^x = 16$	$\log_2 16 = x$	$x = 4$
$x^2 = 25$	$\log_x 25 = 2$	$x = 5$
$3^x = 81$	$\log_3 81 = x$	$x = 4$
$10^{1/2} = x$	$\log_{10} x = \frac{1}{2}$	$x \approx 3.162$
$10^x = 19$	$\log_{10} 19 = x$	$x \approx 1.2788$
$4^0 = x$	$\log_4 x = 0$	$x = 1$

23. $x = -1$ **25.** $x = 1.556$ **27.** $x = -0.125$ **29.** $x = 1.5$
31. $x = 0.815$ **33.** $x = 0.1$ **35.** $x = 2$ **37.** $x = 1$
39. $x = 4$ **41.** $x = 100$ **43.** $x = 2$ **45.** $x = \frac{2}{3}$
47. $\log_b x + \log_b y + \log_b z$ **49.** $\log_b x - 2\log_b y$
51. $\frac{1}{2}\log_b z - \frac{1}{2}\log_b x$ **53.** $\log_b (x\sqrt{y})$ **55.** $\log(x^2 - 3x + 2)$
57. $\log(x + 1)$ **59.** $x = 9$ **61. a.** 15.2 **b.** 19.8 **c.** 20.1
63. $x = -2$ **65.** $x = -1$ **67.** $x = e^3$ **69.** $x = e^{1/e}$
71. $\left(\frac{1}{2}\right)^x = (2^{-1})^x = 2^{-x}$

73.

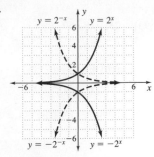

75. a. Shift 1 unit to left.   **b.** Shift 2 units to right.

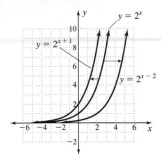

77. a. $3^0 = 1$, $\log_3 1 = 0$ **b.** $3^1 = 3$, $\log_3 3 = 1$
   c. mirror each other across $y = x$
79. straight line
81. $\log (5 \cdot 100) = \log 5 + \log 100 = \log 5 + 2 \approx 2.69897$
83. $2 + $ a decimal   **85.** $\log (1000x)$ is 3 greater
87. Inverse is $x = 2^y$ which is $y = \log_2 x$.
89. $\log (8x^3) = \log (2^3x^3) = \log (2x)^3 = 3 \log (2x)$
91. $\frac{1}{3} \log x = \log x^{1/3} = \log \sqrt[3]{x}$ **93.** $\approx\$1127.15$, ≈ 11.58 yr
95. ≈ 12 yr, 9 yr, 6 yr
97. a. $\approx\$740.82$ **b.** $\approx\$670.32$ **c.** $\approx\$548.81$
99. $\approx 13.8\%$; $2 = e^{5r}$, $r \approx 0.1386$
101. $y \approx 49.3(0.626)^x$, with $x = 0$ for 1989; $\approx\$0.027$ million
103. $P = 59,509(0.9936)^x$; -0.64%; 49,011; reasonable
105. a.

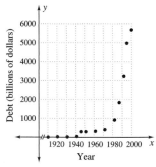

b. $y \approx (4.340 \times 10^9)(1.086)^x$ **c.** wars, spending in 1980s
e.

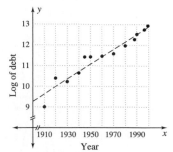

Graph becomes somewhat linear.

CHAPTER 9 TEST

1. geometric; $a_n = \frac{1}{8} \cdot 2^{n-1}$; $f(x) = 0.0625 \cdot 2^x$
2. quadratic; $f(x) = 3x^2$
3. a. $x = 3$ **b.** $n = -\frac{3}{2}$ **c.** $x = 3$ **d.** $x = 4$ **e.** $x = -3$
   f. $n = \frac{1}{2}$ **g.** $x = \frac{2}{3}$ **h.** $x = \frac{1}{2}$
4. a. $x \approx 0.477$ **b.** $x \approx -0.431$ **c.** $\{\ \}$ or $\varnothing$ **d.** $x = -1$
   e. $x = -1$ **f.** $x = \frac{1}{4}$ **g.** $x = 125$ **h.** $x = 2$
   i. $x = 10$ **j.** $x \approx 4.482$
5. a. 2   **b.** 0   **c.** $\log_2 (x^2 + x)$ **d.** $\log_3 (x - 3)$
6. a. $2 \log_b z - \log_b y$ **b.** $\frac{1}{2} \log_b x - \log_b y$ **7.** $y = b^2 b^x$; b^2
8. ≈ 4.25 yr   **9.** ≈ 8.6 yr
10.

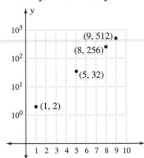

11. $f(x) = 3^x$;

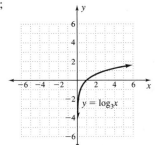

12. $f(x) = \log_2 x$;

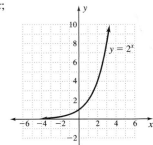

13.

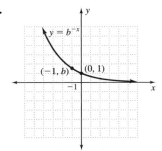

14. Possible answer: (0, 1), (1, 0); (1, 10), (10, 1);
   if $y = 10^x$, $x = \log y$, $y = \log x$ have reversed coordinates.
15. $1 + r = 0.93$; $r = -0.07$ **16.** $\log (100x) = 2 + \log x$
17. a. 8.3   **b.** $\approx 1.26 \times 10^{-12}$ M

18. Use n factors of x: $\log_b x \cdot x \cdot x \cdot \ldots \cdot x =$
$\log_b x + \log_b x + \log_b x + \cdots + \log_b x = n \log_b x$

19. a.

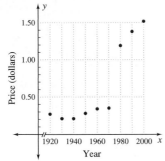

b. Possible answer: exponential: data start off slowly, increase rapidly.

c. $f(x) \approx 0.161(1.0263)^x$

20. a. $y = 15,649(1.0021)^x$; 0.2%; 18,516

b. $y = 15,642(1.00695)^x$; 0.7%; 20,637

c. how much the college grew; how much it is expected to grow

EXERCISES 10.1

1. $(\pm3, 0), (0, \pm3)$ **3.** $x^2 + y^2 = 25$ **5.** $\dfrac{x^2}{25} + \dfrac{y^2}{4} = 1$

7. $\dfrac{y^2}{16} - \dfrac{x^2}{9} = 1$ **9.** $x = 3y^2$ **11.** $y = -\dfrac{1}{9}x^2$

13. $x^2 + y^2 = 16$ **15.** $x^2 + \dfrac{y^2}{25} = 1$ **17.** $\dfrac{y^2}{9} - \dfrac{x^2}{25} = 1$

19. $y = \dfrac{1}{4}x^2$

21. $r < 0$ has no meaning; circle becomes a point.

23. parabola; $(0, 2)$

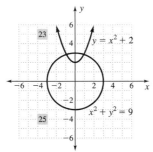

25. circle; $(\pm3, 0), (0, \pm3)$; $y = \pm\sqrt{9 - x^2}$

27. hyperbola; $(\pm2, 0)$; $y = \pm\sqrt{x^2 - 4}$

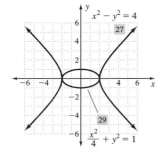

29. ellipse; $(\pm2, 0), (0, \pm1)$; $y = \pm\sqrt{1 - \dfrac{x^2}{4}}$ or
$y = \pm\dfrac{1}{2}\sqrt{4 - x^2}$

31. ellipse; $(\pm5, 0), (0, \pm10)$;
$y = \pm\sqrt{100 - 4x^2}$ or
$y = \pm2\sqrt{25 - x^2}$

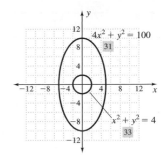

33. circle; $(\pm2, 0), (0, \pm2)$; $y = \pm\sqrt{4 - x^2}$

35. straight line; $(0, 0)$

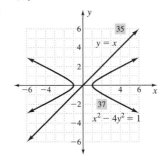

37. hyperbola; $(\pm1, 0)$; $y = \pm\dfrac{1}{2}\sqrt{x^2 - 1}$

39. parabola; $(0, 0)$; $y = \pm\dfrac{1}{2}\sqrt{x}$

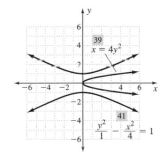

41. hyperbola; $(0, \pm1)$; $y = \pm\sqrt{1 + \dfrac{x^2}{4}}$ or $y = \pm\dfrac{1}{2}\sqrt{x^2 + 4}$

43. parabola; $(0, 0)$;
$y = \pm\sqrt{-x}$

45. Branches become steeper.

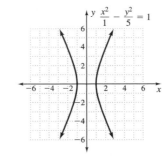

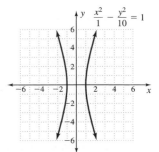

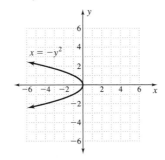

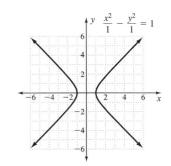

47. a.

x	$x = -y^2$
-9	± 3
-4	± 2
-1	± 1
0	0
1	not a real number
4	not a real number
9	not a real number

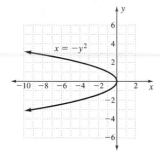

b. Opposite of x makes radicand appear negative.
c. when x is a negative number

EXERCISES 10.2

1. parabola, line; $(-1, 6)$, $(3, 2)$
3. hyperbola, parabola; $(\pm 2.5, 2.25)$, $(\pm 2, 0)$
5. parabolas; $\left(-\frac{2}{3}, \frac{8}{9}\right)$, $\left(\frac{1}{2}, \frac{9}{4}\right)$ **7.** ellipse, parabola; $(\pm 1, 0)$
9. circle, parabola; $\left(\pm \sqrt{7}, 3\right)$, $(0, -4)$
11. parabolas; $(\pm 2, -2)$
13. ellipse, parabola; $(0, -1)$, $\approx (\pm 1.323, 0.75)$
15. circle, hyperbola; $(\pm 2, 0)$
17. hyperbola, line; $\approx (3.732, 0.268)$, $\approx (0.268, 3.732)$
19. circle, ellipse; $\left(\pm \dfrac{2\sqrt{15}}{3}, \dfrac{\sqrt{21}}{3}\right)$, $\left(\pm \dfrac{2\sqrt{15}}{3}, -\dfrac{\sqrt{21}}{3}\right)$
23. No; ordered pairs do not satisfy second equation.
25. $(0, 1)$ **27.** $(0.415, 5.333)$ **29.** $(31.6, 1.5)$ **31.** $(4, 1)$
33. $\frac{4}{3}$ **35.** 6 **37.** $\frac{5}{2}$ **39.** $\frac{3}{16}$
41. $x^2 = 2$, $x = \pm \sqrt{2}$; root is same, $-\sqrt{2}$ must be discarded.

EXERCISES 10.3

1. 2 **3.** 3 **5.** 3 and 4 **7.** 2 and 3 **9.** $x < 0$ and $y > 0$
11. $(x < 0$ and $y > 0)$ or $(x > 0$ and $y < 0)$
13. $y \le 2800$, $y \ge 300$, and $x \ge 0$ **15.** $y \le x$ and $x \le 0$
17.

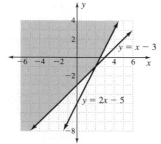

19.

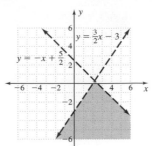

21.

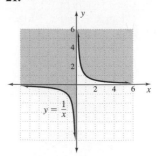

23.

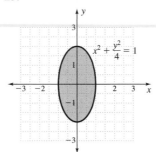

25.

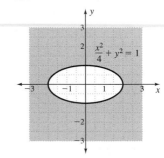

27.

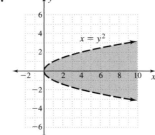

29. c **31.** f **33.** a **35.** x^2 is always ≥ 0, so if $y \ge x^2$, $y \ge 0$
37.

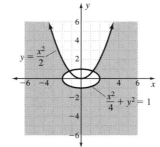

39.

41.

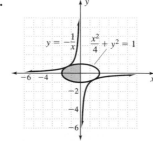

43. $x = $ no. of student tickets, $y = $ no. of regular tickets;
$15x + 50y \ge 500{,}000$.
If only student tickets are sold, sales of 33,334 tickets are

needed. If only regular tickets are sold, sales of 10,000 tickets are needed. Other ticket sales on the boundary line represent possible combinations exactly meeting the goal.

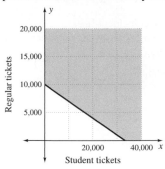

45. x = units of I.V., y = units of liquid; $500x + 250y \leq 2500$

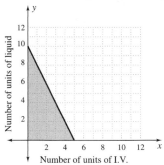

MID-CHAPTER 10 TEST

1. $\dfrac{x^2}{4} + \dfrac{y^2}{16} = 1$ **2.** $\dfrac{x^2}{9} + \dfrac{y^2}{25} = 1$ **3.** $x = -y^2$

4. parabola; $(0, 0, (0, 0); Y = \pm\sqrt{(-X/2)}$
5. circle; $(\pm 25, 0), (0, \pm 25), Y = \pm\sqrt{(625 - X^2)}$
6. $(-1, 0), (0, 2), (1, 0)$ **7.** $(1, 2)$ **8.** second
9. Graph $y = x - 2$ and $y = -x - 3$ with dashed lines. Select a test point for each inequality to find a half-plane to shade. Solution is overlap of shading.
10. Graph ellipse and shade inside. Graph line $y = x$ and shade above. Solution is overlap of shading.

EXERCISES 10.4

1. 1×2 **3.** 3×1 **5.** $[4 \quad 7]$
7. $\begin{bmatrix} 5 \\ 7 \\ 9 \end{bmatrix}$ **9.** $\begin{bmatrix} 5 & 5 & 5 \\ 5 & 5 & 5 \\ 5 & 5 & 5 \end{bmatrix}$
11. $\begin{bmatrix} x + 3y \\ -x - 2y \end{bmatrix}$ **13.** $\begin{bmatrix} -1 \\ 0 \end{bmatrix}$ **15.** $\begin{bmatrix} 10 \\ 0 \end{bmatrix}$
17. $\begin{bmatrix} 14 & 18 \\ -10 & -13 \end{bmatrix}$ **19.** $\begin{bmatrix} -1 & 0 \\ -1 & 2 \end{bmatrix}$ **21.** $\begin{bmatrix} -10 & -8 \\ 0 & 3 \end{bmatrix}$
23. $\begin{bmatrix} -9 & 2 \\ 6 & 2 \end{bmatrix}$ **25.** 1 **27.** 10 **29.** $\begin{bmatrix} -2 & -3 \\ 1 & 1 \end{bmatrix}$
31. $\begin{bmatrix} 0.2 & 0.2 \\ -0.1 & 0.4 \end{bmatrix}$ **33. a.** $\begin{bmatrix} 1 & 0 \\ 0 & 1 \end{bmatrix}$ **b.** $\begin{bmatrix} 1 & 0 \\ 0 & 1 \end{bmatrix}$
35. $x = -2, y = 3.5$ **37.** det $[A] = 0$; parallel lines
39. $x = 2, y = 1.5$ **41.** $x = 1, y = 2$
43. $1x + 2y = 5, ax + 4y = 6$; slope changes; y-intercept
45. $x = 4.2, y = -3.6$ **47.** det $[A] = 0$; parallel lines
49. det $[A] = 0$; coincident lines
51. a. 0 **b.** zero denominators **c.** matrix with no inverse

EXERCISES 10.5

1. $(-2, 3, -1)$ **3.** $(0.25, 2, -1.5)$
5. no solution, inconsistent equations
7. infinite number of solutions, dependent equations
9. $(4, -1, 2)$ **11.** $y = 2x^2 - 5x + 3$
13. $y = 2x^3 - x^2 + 3x - 1$ **15.** $y = 2x^4 + 3x^3 - 2x^2 + x - 5$
17. $[A][X] = [B]$, where $[A] = \begin{bmatrix} 1 & 1 & 1 \\ 1 & -1 & 0 \\ 0 & 0 & 1 \end{bmatrix}, [X] = \begin{bmatrix} s \\ m \\ l \end{bmatrix}, [B] = \begin{bmatrix} 620 \\ 0 \\ 20 \end{bmatrix}$
19. a. $1 \times 3, 3 \times 4, 1 \times 4$ **b.** $4800, 5100, 5400, 6000$
21. a. $\frac{9}{32}$ lb, $\frac{3}{4}$ lb, and 3 lb
  b. 275.6 lb, 299.1 lb, 322.5 lb, and 369.4 lb
  c. $965, $1047, $1129, $1293
23. $3313, $3546, $3779, $4245 **25.** $y = \frac{1}{50,000}x^2 - \frac{3}{100}x + 700$
27. $y = \frac{1}{80,000}x^2 - \frac{1}{25}x + 1224$

CHAPTER 10 REVIEW EXERCISES

1. $x = -3y^2$ **3.** $\dfrac{x^2}{9} + \dfrac{y^2}{4} = 1$
5. parabola; $x^2 = y - 5$

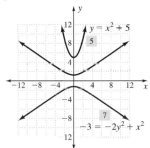

7. hyperbola; $y^2 = \frac{1}{2}x^2 + \frac{3}{2}$
9. circle; $y = \pm\sqrt{8 - x^2}$

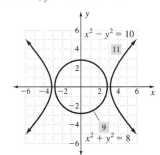

11. hyperbola; $x = \pm\sqrt{10 + y^2}$
13. ellipse; $y = \pm\sqrt{2 - \dfrac{x^2}{2}}$

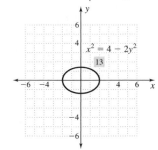

15. parabola, line; $(-1, -2)$
17. parabola, line; $(4, -1), (-1, 4)$
19. circle, parabola; $(\pm 1.629, 3.653)$
21. parabolas; $(1.5, 0.75), (-1, 2)$
23. parabolas; $\{\ \}$
25. $x = 1, y = 2$
27. $x = 10, y = 1$

29.

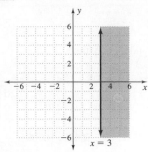

$x = 3$

31.

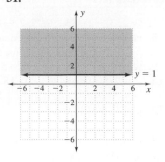

$y = 1$

33.

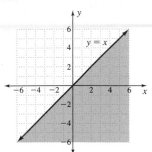

$y = x$

35.

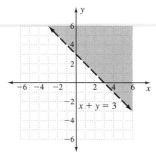

$x + y = 3$

37.

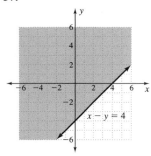

$x - y = 4$

39.

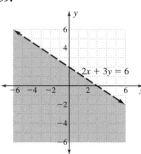

$2x + 3y = 6$

41.

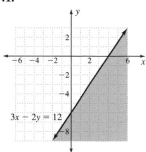

$3x - 2y = 12$

43.

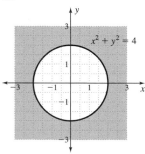

$x^2 + y^2 = 4$

45.

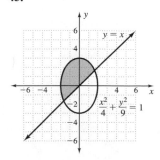

$y = x$

$\dfrac{x^2}{4} + \dfrac{y^2}{9} = 1$

47.

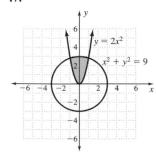

$y = 2x^2$

$x^2 + y^2 = 9$

49. a. Graph 3 **b.** Graph 1 **c.** Graph 4 **d.** Graph 2

51. $\begin{bmatrix} 3 & 6 \\ 5 & 6 \end{bmatrix}$ **53.** $\begin{bmatrix} 18 & 16 \\ 13 & 11 \end{bmatrix}$ **55.** $\begin{bmatrix} 1 & 0 \\ 0 & 1 \end{bmatrix}$

57. $(6, 8)$ **59.** $(4, -5)$

61. infinite number of solutions, coincident lines

63. $(-2, 4, -3)$ **65.** no solution, inconsistent equations

67. $a = \frac{1}{80,000}, b = -\frac{3}{100}, c = 286$

69. $\det [A] = 0$; graphs coincident

71. $y = -4x^2 + 5x - 2$ **73.** $y = x^3 - 3x^2 + 4$

75. $y = x^4 - 3x^2 - x + 2$

CHAPTER 10 TEST

1. $x^2 + y^2 = 256$ **2.** $\dfrac{x^2}{1} + \dfrac{y^2}{16} = 1$ **3.** $x = -4y^2$

4. $\dfrac{x^2}{9} - \dfrac{y^2}{4} = 1$ **5.** parabola, line; $\left(-\frac{1}{3}, -\frac{10}{3}\right), (2, -1)$

6. parabola, line; $\{\ \}$ **7.** circle, hyperbola; $(0, \pm 2)$

8. parabola, circle; $(\pm 0.890, 1.791)$ **9.** $(0.834, 2.5)$ **10.** $(9, 2)$

11.

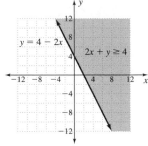

$x - y < 3$

$y = x - 3$

12.

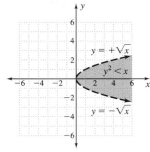

$y = 4 - 2x$

$2x + y \geq 4$

13.

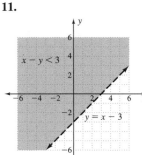

$2x^2 + y^2 = 9$

$2x^2 + y^2 > 9$

14.

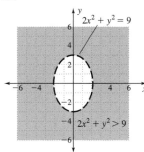

$y = +\sqrt{x}$

$y^2 < x$

$y = -\sqrt{x}$

15.

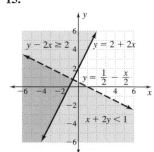

$y - 2x \geq 2$

$y = 2 + 2x$

$y = \dfrac{1}{2} - \dfrac{x}{2}$

$x + 2y < 1$

16.

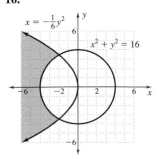

$x = -\dfrac{1}{6}y^2$

$x^2 + y^2 = 16$

17. $a = \dfrac{9}{640,000}, b = -\dfrac{3}{100}, c = 1500$

18. a. $\begin{bmatrix} 1 & 0 \\ 0 & 1 \end{bmatrix}$ **b.** $\begin{bmatrix} 2 & 5 \\ 1 & 3 \end{bmatrix}$ **c.** 1

19. $\begin{bmatrix} 2 & -3 \\ 4 & 5 \end{bmatrix} \begin{bmatrix} x \\ y \end{bmatrix} = \begin{bmatrix} 6 \\ -32 \end{bmatrix}$; $\begin{bmatrix} x \\ y \end{bmatrix} = \begin{bmatrix} -3 \\ -4 \end{bmatrix}$
20. det $[A] = 0$, inconsistent equations   **21.** $(4, -1, 2)$

EXERCISES 11.1

1. a. GG, GB, BG, BB
   b. GGG, GGB, GBG, GBB, BGG, BGB, BBG, BBB
   c. GGGG, GGGB, GGBG, GGBB, GBGG, GBGB, GBBG,
     GBBB, BGGG, BGGB, BGBG, BGBB, BBGG, BBGB,
     BBBG, BBBB
3. GWS, GSW, SWG, SGW, WSG, WGS
5.

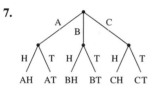

7.

A   C
  B
H T H T H T
AH AT BH BT CH CT

9. a. 32 **b.** 64 **c.** 1024 **d.** 2^n
11. Two choices each time; number of outcomes and sample
   spaces are the same.
13. 10^9 **15.** 4^{20} **17. a.** 40^3 **b.** 59,280 **19.** 8
21. a. 10 **b.** not possible **c.** 8 **d.** 4
23. a. 1352 **b.** 35,152 **c.** 36,504
25. a. 405 **b.** 42 **c.** 702 **d.** 210
27. a. 17 **b.** 45 **c.** 19,800 **d.** 27 **e.** 2730 **f.** 75
29. a. 6 **b.** 11 **c.** 2 **d.** 20 (by counting) **e.** 11
31. Place a 1 behind each of the 36 given outcomes, then a 2
   behind each of the 36 outcomes, and repeat for a 3, 4, 5, and 6.
   The total will be $6 \cdot 36 = 216$.
33. a. 160 **b.** 800 **c.** $8^2 \cdot 10^8$ **d.** $800 \cdot 799 \cdot 10^4$

EXERCISES 11.2

1. a. 40,320 **b.** 3,628,800 **c.** 1 **d.** 3,628,800
   e. 40,322 **f.** 720
3. a. false **b.** true **c.** true **d.** false **e.** false **f.** true
5. a. 240 **b.** 6 **c.** 24 **d.** 15 **e.** 28 **f.** 3
7. $6! = 720$; also MP **9.** $7^7 = 823,543$; MP
11. $7! = 5040$; also MP **13.** $_4P_3$
15. $_4P_2$; ab, ac, ad, ba, bc, bd, ca, cb, cd, da, db, dc
17. a. 120 **b.** 120 **c.** 12
19. a. 60 **b.** 132,600 **c.** 19,958,400
21. a. 60 **b.** 125 **23. a.** 64 **b.** 56 **25.** 2730
27. $_4C_3$; wxy, wxz, wyz, xyz
29. $_3C_2$; bc, bd, cd **31. a.** 15 **b.** 6 **c.** 495
33. a. 56 **b.** 1326 **c.** 12 **35.** 12,271,512 **37.** 120
39. a. 1; sentence 3 **b.** 1; sentence 2 **c.** 12; sentence 1

41. Permutations are also MP. **a.** MP; 1,679,616
   b. permutations; 1,413,720 **c.** combinations; 58,905
   d. permutations; 24
43. Permutations are also MP. **a.** permutations; 336
   b. combinations; 56 **c.** permutations; $8! = 40,320$
   d. MP; $8^5 = 32,768$
45. a. permutations; 165,765,600
   b. MP; $26^6 = 308,915,776$
47. a. combinations and AP; 9
   b. combinations and MP; 36
49. a. combinations; 1,623,160
   b. combinations and MP; 125,400
51. a. No; order makes for twice as many permutations.
   b. Yes; order has no meaning when only 1 item is selected.
53. a. $_nP_1 = \dfrac{n!}{(n-1)!}$; $_nC_1 = \dfrac{n!}{(n-1)!1!} = \dfrac{n!}{(n-1)!}$
   b. $_nP_1 = \dfrac{n!}{(n-1)!} = \dfrac{n(n-1)!}{(n-1)!} = n$
   c. $_nC_1 = \dfrac{n!}{(n-1)!1!} = \dfrac{n(n-1)!}{(n-1)!} = n$
   d. no
   e. $_nP_n = \dfrac{n!}{(n-n)!} = \dfrac{n!}{0!} = \dfrac{n!}{1} = n!$
   f. $_nC_n = \dfrac{n!}{(n-n)!n!} = \dfrac{n!}{0!n!} = \dfrac{n!}{1n!} = 1$
   g. $_nP_0 = \dfrac{n!}{(n-0)!}$;
     $_nC_0 = \dfrac{n!}{(n-0)!0!} = \dfrac{n!}{(n-0)!1} = \dfrac{n!}{(n-0)!}$
   h. $_nP_0 = \dfrac{n!}{(n-0)!} = \dfrac{n!}{n!} = 1$
   i. $_nC_0 = \dfrac{n!}{(n-0)!0!} = \dfrac{n!}{(n-0)!1} = \dfrac{n!}{n!} = 1$

EXERCISES 11.3

1. $\frac{2}{5}$ **3. a.** $\frac{8}{17}$ **b.** $\frac{7}{17}$ **c.** $\frac{2}{17}$ **d.** $\frac{4}{17}$ **e.** $\frac{7}{17}$
5. a. $\frac{1}{11}$ **b.** $\frac{2}{11}$ **c.** $\frac{5}{11}$; y is a vowel in this case **d.** $\frac{4}{11}$ **e.** $\frac{5}{11}$
7.

A   C
  B
$\frac{1}{3}$ $\frac{1}{3}$ $\frac{1}{3}$
H T H T H T
$\frac{1}{2}$ $\frac{1}{2}$ $\frac{1}{2}$ $\frac{1}{2}$ $\frac{1}{2}$ $\frac{1}{2}$
AH AT BH BT CH CT
$\frac{1}{6}$ $\frac{1}{6}$ $\frac{1}{6}$ $\frac{1}{6}$ $\frac{1}{6}$ $\frac{1}{6}$

   a. $\frac{1}{6}$ **b.** $\frac{1}{2}$ **c.** $\frac{1}{3}$

9. a. AP; 1 **b.** MP; $\frac{1}{4}$ **c.** AP; $\frac{1}{3}$ **d.** MP; $\frac{1}{12}$ **e.** neither; $\frac{7}{12}$
11. Answers in order from left to right, top to bottom: Boy, $\frac{1}{2}$, $\frac{1}{2}$;
   GG, $\frac{1}{4}$, $\frac{3}{4}$; BB, $\frac{3}{4}$, $\frac{1}{4}$; $\{2, 4\}$, $\frac{3}{5}$; z, $\frac{2}{3}$; odd sum, $\frac{1}{2}$, $\frac{1}{2}$
13. a. $\frac{5}{6}$ **b.** 1 **c.** $\frac{5}{6}$ **d.** 0 **e.** $\frac{5}{36}$ **15. a.** $\frac{1}{4}$ **b.** $\frac{1}{4}$ **c.** $\frac{1}{7}$
   d. $\frac{4}{7}$, 0 is even **e.** $\frac{3}{4}$ **f.** $\frac{3}{14}$ **g.** 0
17. a. 128 **b.** $\frac{21}{128}$ **c.** $\frac{1}{128}$ **d.** $\frac{15}{16}$ **e.** $\frac{7}{16}$ **f.** 0 **g.** $\frac{29}{128}$
19. $\frac{1}{4096}$ **21. a.** 24 **b.** $\frac{1}{24}$ **23. a.** 6 **b.** $\frac{1}{6}$
25. domain: any event; range: any rational number p, $0 \le p \le 1$
27. $\frac{1}{10,000}$

MID-CHAPTER 11 TEST

1. a.

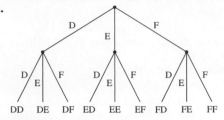

b. DD, DE, DF, ED, EE, EF, FD, FE, FF
2. a. DD **b.** DF or FD **3. a.** 135 **b.** 24 **c.** 72
4. $26 \cdot 10 \cdot 26 \cdot 10 \cdot 26 \cdot 10 = 17,576,000$
5. a. 40,320 **b.** 6 **c.** 15 **d.** 360 **e.** 26 **f.** 8
6. 56; combination **7.** 56; combination **8.** 6720; permutation
9. a. 20,358,520; combination **b.** $\approx 6.35 \times 10^{11}$; combination
10. a. careful was I, careful I was, I was careful, I careful was,
was I careful, was careful I **b.** likely "I was careful"
11. a. $\frac{1}{2}$ **b.** $\frac{3}{8}$ **c.** $\frac{1}{2}$ **d.** $\frac{1}{2}$ **e.** $\frac{1}{4}$ **f.** $\frac{1}{4}$
12. $\dfrac{1}{{}_{52}C_6} = \dfrac{1}{20,358,520}$
13. a. $\frac{1}{10,000}$
b. No, the numbers are permutations because order counts.
14. very close to 1: $\frac{9999}{10,000}$

EXERCISES 11.4

1. a. 4:5, 4/5 **b.** 3/7, 3 to 7 **c.** 3 to 2, 3:2
3. a. $2:4 = 1:2$ **b.** $\dfrac{\frac{2}{6}}{\frac{4}{6}} = \dfrac{2}{4} = \dfrac{1}{2}$
5. a. 1:3 **b.** $\dfrac{\frac{1}{4}}{\frac{3}{4}} = \dfrac{1}{3}$ **7. a.** 4 **b.** 1:3 **c.** 1:3 **d.** 1:1
9. a. $H_1H_{1/2}, H_1T_{1/2}, H_1H_{1/2}, H_1T_{1/2}$ **b.** 1:1 **c.** 0
d. 0:4 **e.** 4:0 **f.** 1
11. must be a ratio of 2 numbers, 1:1
13. must be a ratio of 2 numbers, 3:4
15. a. No; 1:1 means $P(A) = \frac{1}{2}$. **b.** 2:0 **c.** 0 **d.** 0:2 **e.** true
17. $\frac{1}{2}$ **19.** $\frac{2}{5}$ **21.** $\frac{2}{3}$ **23. a.** $\frac{11}{36}$ **b.** $\frac{11}{36}$ **c.** $\frac{1}{18}$ **d.** $\frac{5}{9}$ **e.** $\frac{4}{9}$
25. a. $\frac{1}{12}$ **b.** $\frac{1}{3}$ **c.** $\frac{5}{18}$ **d.** $\frac{7}{18}$ **27. a.** $\frac{11}{26}$ **b.** $\frac{4}{13}$ **c.** $\frac{1}{2}$ **d.** $\frac{25}{52}$
29. a. 1176 **b.** $\frac{1176}{10,001}$ **31. a.** 490 **b.** $\frac{25}{232}$ **c.** $\frac{86}{155}$ **d.** $\frac{116}{245}$
33. 0
35. a.

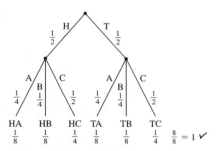

b. HA: $\frac{1}{8}$, HB: $\frac{1}{8}$, HC: $\frac{1}{4}$, TA: $\frac{1}{8}$, TB: $\frac{1}{8}$, TC: $\frac{1}{4}$ **c.** $\frac{1}{2}$
d. $\frac{5}{8}$, second AP **e.** $\frac{3}{4}$ **f.** $\frac{1}{8}$

37. a.

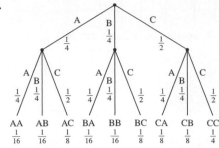

b. $\frac{1}{16}$ **c.** $\frac{7}{16}$ **d.** $\frac{1}{4}$ **e.** $P(\text{not AA}) = \frac{15}{16}$ **f.** $P(\text{not CC}) = \frac{3}{4}$

39. a.

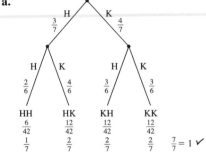

b. $\frac{2}{7}$ **c.** $1 - P(2 \text{ Kisses}) = \frac{5}{7}$ **d.** $\frac{4}{7}$ **e.** $\frac{6}{7}$

41. a.

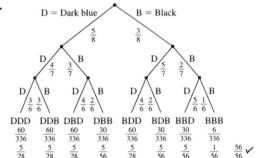

b. $\frac{5}{7}$ **c.** $\frac{2}{7}$ **d.** 1
e. With 3 socks in 2 colors, one pair must match.

43.

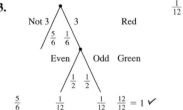

45.

47. a.

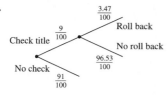

b. $\frac{3.47}{100} \cdot \frac{9}{100} \approx \frac{31}{10,000}$

49. a.

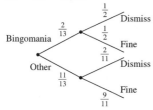

b. $\frac{2}{13} \cdot \frac{1}{2} = \frac{1}{13}$ **c.** $\frac{11}{13} \cdot \frac{9}{11} = \frac{9}{13}$

51. a. Yes, if the survey first asked about being tired and then asked those people if they were exhausted.

 b. $\frac{1}{10}$

53. There is overlap with (3, 3); probability is $\frac{11}{36}$.

55. $a + b = 1$ and $c + d = 1$

$$ac + ad + b = 1$$
$$a(c + d) + b = 1$$
$$a(1) + b = 1$$
$$a + b = 1$$
$$1 = 1$$

EXERCISES 11.5

1. 1, 10, 45, 120, 210, 252, 210, 120, 45, 10, 1

3. $128 = 2^7$; $256 = 2^8$ **5.** 1, 5, 10, 10, 5, 1

7. 1, 12, 66, 220, 495, 792, 924, 792, . . .

9. a. 35 **b.** 15 **c.** 6 **11. a.** 20 **b.** 1 **c.** 1 **d.** 16

13. The first two are 1 and n, and the last two are n and 1.

15. a. $_3C_3$ **b.** $_8C_6$ **c.** $_7C_4$ **d.** $_nC_{n-b}$ **17.** odd n **19.** all n

21. The row beginning 1, 7, . . . is the 8th row. **23.** $\frac{5}{32}$ **25.** $\frac{1}{32}$

27. $\frac{3}{32}$ **29.** $2^{12} = 4096$ **31.** $\frac{33}{2048}$ **33.** $\frac{99}{512}$ **35.** $\frac{33}{2048}$

37. $_{12}C_{10} = {_{12}C_2}$ **39. a.** 10 ways **b.** 20 ways **c.** 21 ways

 d. 66 ways

CHAPTER 11 REVIEW EXERCISES

1. No; not all months have 31 days. **3. a.** 2 **b.** 4 **c.** 8 **d.** 16

5. a. 5040 **b.** 42 **c.** 360 **d.** 3024 **e.** 1 **f.** 56

7. a. Hard to see the dark side is. To see the dark side is hard. The dark side is hard to see. Hard the dark side to see is. Hard the dark side is to see. Hard to see is the dark side.

 b. 5040

9. I love you. I you love. You I love. You love I. Love you I. Love I you.

11. a. P; 32,760 **b.** C; 455

13. Permutation is also MP. **a.** C; 120 **b.** P; 3,628,800

 c. P; 720 **d.** C; 210 **e.** MP; 10^5

15. a. 20 **b.** 96 **c.** 28 **d.** 132

17. Answers from left to right, top to bottom: {HT, TH}, $\frac{1}{2}$, $\frac{1}{2}$; {HH, HT, TH}, $\frac{1}{4}$, $\frac{3}{4}$; odd numbers, $\frac{2}{5}$, $\frac{3}{5}$; {white, blue}, $\frac{2}{3}$, $\frac{1}{3}$; {b, d}, $\frac{1}{2}$, $\frac{1}{2}$; {a, b, c}, $\frac{3}{4}$, $\frac{1}{4}$

19. a. AP; 1 **b.** MP; $\frac{1}{4}$

21. $\frac{7}{12}$

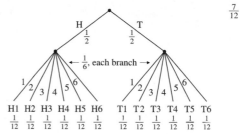

23. a. $1/10^3$ **b.** $\frac{1}{120}$ **c.** no

25. a. 36 **b.** 1:5 **c.** 11:25 **d.** 25:11

27. a. overlap; 15 **b.** overlap; 14 **c.** overlap; 20

 d. no overlap; $\frac{4}{9}$ **e.** overlap; $\frac{2}{9}$ **f.** no overlap; $\frac{2}{3}$

29. a. **b.** $\frac{4}{27}$ **c.** $\frac{65}{81}$ **d.** $\frac{65}{81}$ **e.** $\frac{25}{81}$

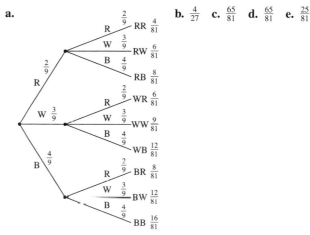

31. a. **b.** $\frac{1}{6}$ **c.** $\frac{5}{6}$

 d. $\frac{5}{6}$ **e.** $\frac{5}{18}$

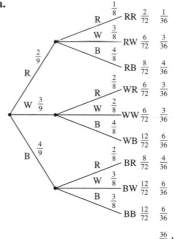

33. a.

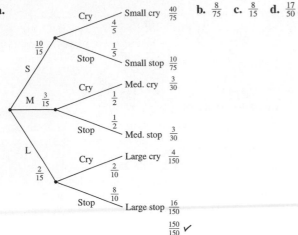

b. $\frac{8}{75}$ **c.** $\frac{8}{15}$ **d.** $\frac{17}{50}$

8. a.

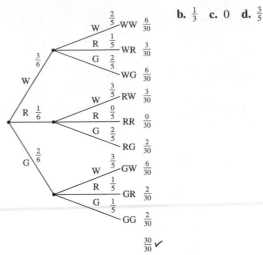

b. $\frac{1}{3}$ **c.** 0 **d.** $\frac{3}{5}$

35.

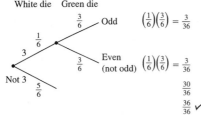

9. $\dfrac{1}{{}_{30}C_5} = \dfrac{1}{142{,}506}$

10. a. 1, 8, 27, 56, 72, 56, 27, 8, 1 **b.** yes; $256 = 2^8$ **c.** 20
d. $\frac{36}{128} = \frac{9}{32}$
e. Row represents 0, 1, 2, 3, . . . boys, so fourth number is 36.
f. Total of the row is $2^7 = 128$.

37.

| | $\frac{16}{52^2}$ $\frac{16}{2704}$ | $\frac{1}{169}$ |

Five — Five $\frac{4}{52}$ — $\frac{16}{52^2}$ $\frac{16}{2704}$

Not five $\frac{48}{52}$ — Not five $\frac{48}{52}$ — $\frac{192}{2704}$

$\frac{2496}{2704}$

$\frac{2704}{2704} = 1$ ✓

41. a. 1, 7, 20, 36, 36, 20, 7, 1 **b.** yes; $128 = 2^7$ **c.** 14 **d.** $\frac{11}{32}$
e. number of ways of obtaining 0 items from a set of n items

43. ${}_aC_b = \dfrac{a!}{(a-b)!\,b!}$; 1, a, $\dfrac{a!}{(a-2)!\,2!}$, . . . , $\dfrac{a!}{(a-2)!\,2!}$, a, 1

CHAPTER 11 TEST

1. a.

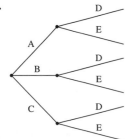

b. AD, AE, BD, BE, CD, CE
c. AD or BD **d.** BD or BE
e. $\frac{2}{3}$ **f.** $\frac{1}{6}$

2. a. 25 **b.** 150 **c.** 210 **d.** $\frac{3}{5}$ **e.** 1 **f.** $\frac{1}{210}$
3. a. 6250 **b.** 10^5
4. ${}_nP_0$, ${}_nC_0$, and ${}_nC_n$ equal 1; ${}_nC_1$ and ${}_nC_{n-1}$ equal n. **a.** 52
b. $52!/0!$ **c.** 1 **d.** 1 **e.** 1 **f.** 52
5. a. ${}_{10}C_7$ **b.** ${}_{10}P_5$ **c.** 10^5 **d.** ${}_{10}P_7$ **e.** ${}_{10}C_3$
6. a. $\frac{3}{8}$ **b.** $\frac{5}{8}$ **c.** $\frac{3}{8}$ **7. a.** $\frac{1}{6}$ **b.** $\frac{1}{3}$ **c.** $\frac{1}{2}$ **d.** $\frac{1}{2}$ **e.** 0

FINAL EXAM REVIEW—PART 1

1. -4 **3.** y/x **5.** $x = -4$ **7.** coefficient **9.** -2
11. intersecting **13.** $y = -\frac{2}{3}x$ **15.** $m = \dfrac{y_2 - y_1}{x_2 - x_1}$
17. $x = 53$, $y = 26$ **19.** false, such as $0 = 4$
21. $w = $ Wmpt's score, $l = $ LH's score; $w - l = 9$, $w + l = 23$;
$w = 16$, $l = 7$
23. $x \geq 5$

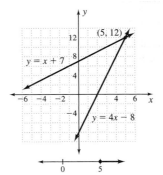

25. parabola **27.** $2a^2 + 4ab + 2b^2$ **29.** $(2x + 1)(3x - 4)$
31. $(2x - 3)(4x^2 + 6x + 9)$ **33.** $x = -0.75$ or $x = 12$
35. A triangle is a right triangle if and only if the sum of the squares of the lengths of the two shorter sides equals the square of the length of the longest side. $a^2 + b^2 = c^2$
37. $\frac{5}{2}$; $\frac{25}{4}$
39. Possible answers: π, e, $\sqrt{2}$, $\sqrt{3}$
41. $\{-0.8, 1\}$
43. $-2 \leq x \leq \frac{2}{5}$ **45.** 13 **47.** $-3 + 9i$
49. minimum, maximum **51.** discriminant
53. $2 - x$ or $-x + 2$ **55.** inverse
57. Ratios of the coordinates, y to x, will be equal.
59. $\dfrac{x}{7}$ **61.** $\dfrac{5x}{6}$ **63.** $\dfrac{x^2 + 5x + 3}{(x - 3)(x + 3)}$

65. $\dfrac{-(x-6)^2}{x(x-1)}$ **67.** $x^2 - x + 2 + \dfrac{6}{x-1}$

69. $\mathbb{R}, x \neq -1; \{-5, 3\}$ **71.** $\mathbb{R}, x \neq 5; \{2, 3\}$

73. $\mathbb{R}, x \neq 3; x = -3$ (discard $x = 3$) **75.** $2x^2y^4$

77. 5.67×10^{-7} **79.** $d = \sqrt{(x_2 - x_1)^2 + (y_2 - y_1)^2}$

81. 0 **83.** $9x^2 + 12x + 4$

85. a. $(\sqrt{x})^3$ or $\sqrt{x^3}$ **b.** $(\sqrt[m]{x})^n$ or $\sqrt[m]{x^n}$ **c.** $\sqrt{x}\sqrt{y} = \sqrt{xy}$

87. a. $abc\sqrt[3]{bc^2}$ **b.** $|wz^3|xy^2\sqrt[4]{x}$ **89.** $x = y$ **91.** $x = \frac{3}{4}$

93. $x = -1$ **95.** $x = -\frac{5}{2}$ **97.** $10^y = x$ **99.** $b^c = a$

101. $y = \dfrac{\log_{10} a}{\log_{10} b}$ **103.** 2 **105.** 10 **107.** 10 **109.** $x = 3.248$

111. $x = 1.230$ **113.** $x = 2$ **115.** $x = 0.301$ **117.** $x = 3$

119. $x = 1.5$ **121.** $x = 0.368$ **123.** $x = 15.154$

125. ≈ 11.5 yr

127. ellipse, parabola; $(\pm\sqrt{3}, 5)$ (Note: discard $y = -7$)

129. $x = 3.5, y = 2.6, z = 4.1$ **131. a.** $\frac{2}{11}$ **b.** $\frac{7}{11}$

   c. $\frac{2}{11} \cdot \frac{1}{10} = \frac{2}{110} = \frac{1}{55}$ **d.** $\frac{2}{11} \cdot \frac{2}{11} = \frac{4}{121}$

FINAL EXAM REVIEW—PART 2

1. a. $80; 40; 0; -40$ (not possible)
   b. $y = a + 0.8(200 - a)$ for $a < 200$; $y = a$

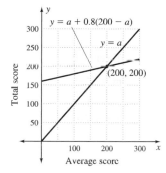

3. Substitution created a system in which the first line was coincident with itself; $x = 6, y = -2$

5. 6 tomatoes, 3 pizza sauce, 2 salsa

7. Replace a with any real number; $y = a(x - 3)(x + 2)$

9. $16t^2 + 6t - 1 = 0; \{-0.5, 0.125\}$; two horizontal axis intercepts

11. $5x^2 - 8x + 4 = 0; \frac{4}{5} \pm \frac{2}{5}i$; no horizontal axis intercepts

13. expression; function with parabolic graph—infinite number of ordered pairs make it true; equation whose two solutions, $x = -2$ and $x = 5$, are x-intercepts for the function

15. vertical axis intercept **17.** yes; $-3 + 2i$, the conjugate

19. $y = 1.2755x - 0.53$; $y = 0.02375x^2 + 0.7055x + 2.7$. The data look linear, so linear regression is logical. Shorter panels might cost relatively more, so the positive y-intercept of the quadratic equation makes sense.

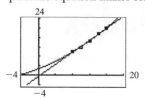

21. $\dfrac{y^2 + 1}{y^2 - 1}$ **23.** $\dfrac{2x}{x^2 + 6}$

25. All are correct; part d contains no fractions in the numerator or denominator and is a single term.

27. $x = 4$ **29.** $x = 25$ **31.** $\approx 7.2\%; \approx 9.1\%; \approx 13.8\%$

33. geometric; $a_n = 27\left(\frac{1}{3}\right)^{n-1}, y = 81\left(\frac{1}{3}\right)^x$; 40.4993; 40.5

35. quadratic; $y = 2x^2 - x$

37. a. exponential; $y = 164.6(1.208)^x$ **b.** 20.8% **c.** 4088

39. $(3, -2, -1)$ **41.** $a = \frac{9}{640,000}, b = -\frac{1}{20}, c = 1016$

43. a. 60 **b.** 16 **c.** $6 \cdot {}_{10}C_2 = 270$ **d.** ${}_{10}C_3 = 120$

45. $n = x - 5$

APPENDIX 2

1. a. 30% **b.** 20% **c.** 7.5% **d.** 0.05% **e.** 200% **f.** 500%
   g. 2000% **h.** 10,000% **i.** 200,000%

3. a. $\frac{3}{20}$, 0.15 **b.** $\frac{4}{5}$, 0.8 **c.** $\frac{1}{400}$, 0.0025 **d.** $\frac{1}{10}$, 0.10 **e.** 10
   f. $\frac{7}{40}$, 0.175 **g.** $\frac{3}{2000}$, 0.0015 **h.** $\frac{7}{20}$, 0.35 **i.** 2

5. Slope gives change per year; percent change describes the total change. Without time stated, percent change can be misleading.

7. Dividing changes the fraction to a decimal. Then use the rule in Exercise 6.

APPENDIX 3

1. -20 **3.** 1720 **5.** not arithmetic

7. not arithmetic; pairs have different sums.

9. $a_n = 47 + (n - 1)$; 67 seats

11. $a_n = 1896 + (n - 1)4$; 27 games; WWI and WWII

13. $a_n = 1 + (n - 1)3$; 49 tests **17.** 265,716 **19.** 127.875; 128

21. 88,572 **23.** ≈ 323.9 in. **25.** ≈ 404.88 in. **27.** 2

29. 108 ft

Glossary/Index

Absolute value The distance a number is from zero. The absolute value of x is x whenever the input is zero or positive and the opposite of x whenever the input is negative. 134
  inequalities and, 190
  and square roots, 275

Absolute value function A function that gives the distance of a number from zero on a number line. 133

Absolute value symbol The symbol $|\ |$, which signifies the absolute value of the quantity within. 9

Adding like terms, 19

Adding rates formula, 409

Addition, 5
  of complex numbers, 320
  of matrices, 616–617
  of polynomials, 225–226
  words that describe, 31

Addition principle of counting, 643–644
  for probability, 666, 677

Addition property of equations If $a = b$, then $a + c = b + c$. 60

Addition property of inequalities When a number is added to (or subtracted from) each side of an inequality, the result is an equivalent inequality, and the direction of the inequality sign $(<, >, \le, \ge)$ is not changed. 187

Additive inverses Numbers that are the same distance from zero on a number line and that add to zero; also known as opposites. 5, 59, 386
  denominators containing, 407

Algebraic notation, vocabulary for, 22–24

Altitude of a triangle The perpendicular distance from a vertex (corner) to the opposite side; also known as the height. 277

Annual compound interest, 455

Antilog, 562

Arithmetic sequence A function with the natural numbers as its set of inputs and numbers with a constant difference between them as its outputs. 116, 120

Associative property for addition
$a + (b + c) = (a + b) + c$. 6

Associative property for multiplication
$a \cdot (b \cdot c) = (a \cdot b) \cdot c$. 6

Asymptote The line that a graph approaches. 400
  horizontal, 523

At least 1 One or more. 664

Axes Two number lines placed at right angles so that they cross at zero. 38

Axis The vertical line through the center of a conical surface. 582

Axis of symmetry A line across which a graph can be folded so that points on one side of the graph match with points on the other side of the graph; also known as a line of symmetry. 205
  of hyperbola, 348
  of parabola, 206

Babbage, Charles, 676

Base The number to which an exponent is applied. 16
  for logarithm, 534, 536

Binomial A polynomial with two terms. 224
  multiplying, 226–227
  squaring, 235

Binomial square A two-term expression in the form $(a + b)(b + a)$ or $(a + b)^2$, obtained by squaring a binomial; also known as the square of a binomial. 235, 237

Boltzman equation, 569

Boundary line The line between two half-planes in a coordinate plane. 607

Braces Grouping symbols that resemble little wires, $\{\ \}$. 9

Brackets Square shaped grouping symbols, $[\ ]$. 9, 77

Breaking mind set Thinking outside the expected. 28–29

Byron, Lord (George Gordon), 676

Calculator
  absolute value on, 2nd [CATALOG], 9, 89
  changing decimal to fraction on, MATH 1, 89
  combinations on, MATH PRB3, 654–655
  editing on, DEL, 2nd [INS], 25–26
  factorials on, MATH PRB4, 563, 649
  grouping symbols and, 9
  logarithms and, LOG, 536, 542–543, 568
  natural logarithm and, LN, 567, 568
  natural number e function and, 2nd $[e^x]$, 565–566
  negative sign on, $[(-)]$, 6
  nth root on, n MATH 5, 462
  order of operations on, 8
  permutations on, MATH PRB2, 652
  reciprocal function on, $[x^{-1}]$, 89
  scientific notation on,
    entering, 2nd [EE], 447
    selecting as output, MODE Sci, 447
  square root and parentheses on, 2nd $\sqrt{\ }$, 9, 89
  squaring function on, $[x^2]$, 89
  storing a number in, STO▶, 89
  10^x function on, 2nd $[10^x]$, 89

Additional Answers

EXERCISES 1.2

6. a.

Input x	Output y
1	3
2	5
3	7
4	9

b. 2 **c.** $y = 2x + 1$

8. a.

Input x	Output y
1	3
2	7
3	11
4	15

b. 4 **c.** $y = 4x - 1$

12.

x	$y = 3 - x$
-3	6
-2	5
-1	4
0	3
1	2
2	1
3	0

14.

x	$y = 3 - 2x$
-3	9
-2	7
-1	5
0	3
1	1
2	-1
3	-3

16.

x	$y = x^2 - x$
-3	12
-2	6
-1	2
0	0
1	0
2	2
3	6

18.

x	$y = 2 - x^2$
-3	-7
-2	-2
-1	1
0	2
1	1
2	-2
3	-7

EXERCISES 1.3

74. Start

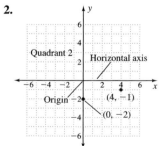

20.

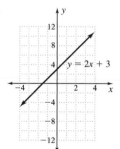

22.

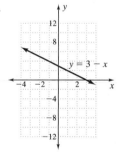

EXERCISES 1.4

2.

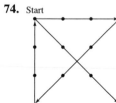

10.

x	$y = 2x + 3$
-3	-3
-2	-1
-1	1
0	3
1	5
2	7
3	9

24.

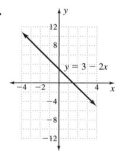

26.

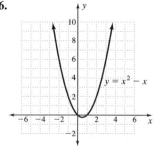

28.

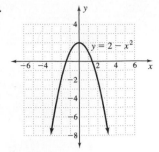

40.

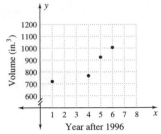

The volume of bags gets larger over time.

32.

x	$y = x^3$
-3	-27
-2	-8
-1	-1
0	0
1	1
2	8
3	27

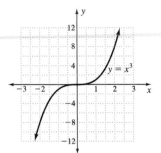

42. Wind-chill apparent temperature (°F)

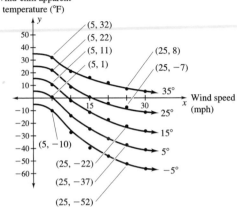

34.

x	y
0	13
0	-13
13	0
-13	0
5	12
5	-12
-5	12
-5	-12
12	5
12	-5
-12	5
-12	-5

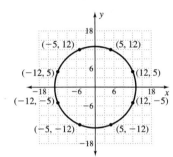

EXERCISES 1.5

50.

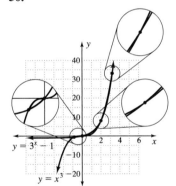

52.

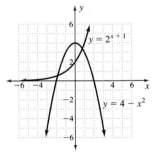

36.

x	$y = |x|$
-3	3
-2	2
-1	1
0	0
1	1
2	2
3	3

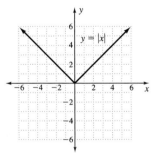

CHAPTER 1 REVIEW EXERCISES

38.

x	$y = 4 - 2x$
-3	10
-2	8
-1	6
0	4
1	2
2	0
3	-2

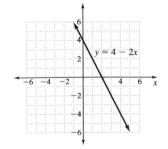

38.

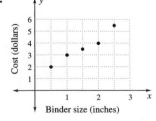

The $2\frac{1}{2}$-inch binder is relatively expensive, as costs increase less rapidly for the $1\frac{1}{2}$-inch and 2-inch binders than for smaller ones.

40.

x	$y = 150x + 300$
-3	-150
-2	0
-1	150
0	300
1	450
2	600
3	750

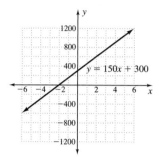

EXERCISES 2.2

30.

x	$f(x)$
-2	-23
-1	-19
0	-15
1	-11
2	-7
3	-3
4	1

32.

x	$g(x)$
-2	-9
-1	-8
0	-5
1	0
2	7
3	16
4	27

34.

x	$f(x)$
-2	4
-1	6
0	6
1	4
2	0
3	-6
4	-14

58.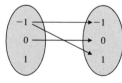

60.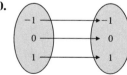

80.

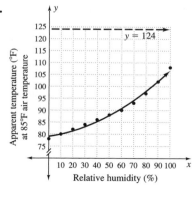

Student graph will contain two additional air temperature plots from Table 1.

EXERCISES 2.3

54.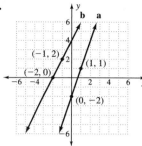

56.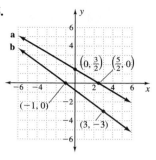

EXERCISES 2.5

40.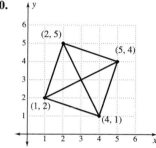

44. a. 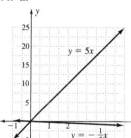 **b.**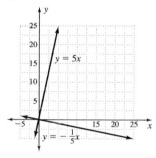

EXERCISES 2.6

2. and 4.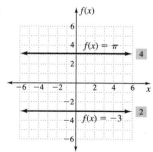

24.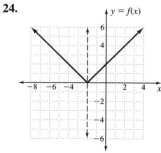

26.

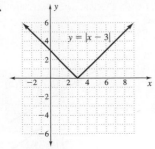

$\mathbb{R}; y \geq 0$

28.

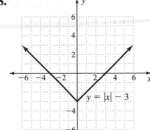

$\mathbb{R}; y \geq -3$

30.

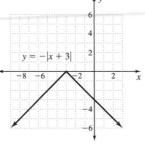

$\mathbb{R}; y \leq 0$

32.

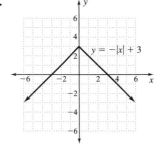

$\mathbb{R}; y \leq 3$

52. b.

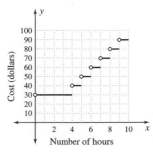

54. b.

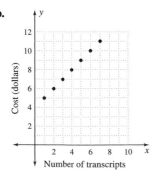

56. b.

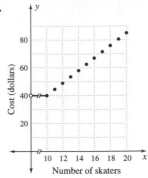

58. b.

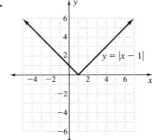

CHAPTER 2 REVIEW EXERCISES

66.

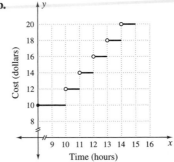

CUMULATIVE REVIEW OF CHAPTERS 1 AND 2

18. a.

No. of Cans	Total Cost
1	$ 5.98
2	$11.96
3	$21.94
4	$31.92
5	$41.90

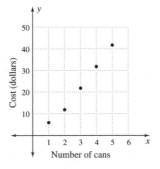

EXERCISES 3.2

2.

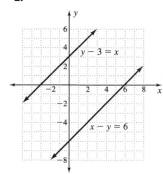

4.

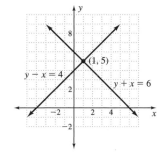

6.

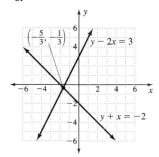

8.

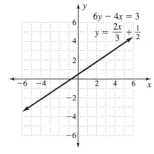

10.

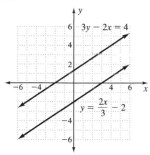

12.

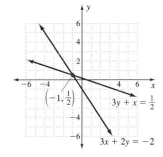

14.

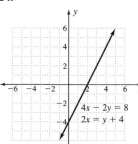

16.

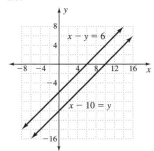

32. a. Draw axes, with time of day on the x-axis and distance from the ranch on the y-axis. Uniform rate permits drawing straight lines. The line starting at the rim height with negative slope shows the trip down. That line ends when it intersects the x-axis, the range level. The second line starts on the x-axis at departure time and rises. This line ends at the point describing rim level. If departure from ranch is at a time of day before arrival at ranch on prior day, the two lines will cross. Where they cross is the time and position required.

b. Departure time on second day is after arrival time the prior day, or the round trip is done in 1 day.

EXERCISES 3.5

16.

18.

20.

22.

24.

26.

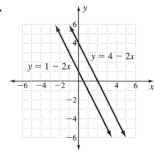

EXERCISES 4.1

10.

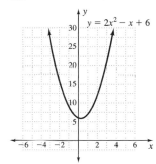

12.

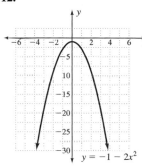

16.

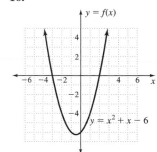

24.

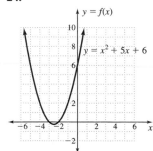

26. For r in [0, 0.20], A is 1000, 1020, 1040, 1061, 1082, 1103, 1124, 1145, 1166, 1188, 1210, 1232, 1254, 1277, 1300, 1323, 1346, 1369, 1392, 1416, and 1440.

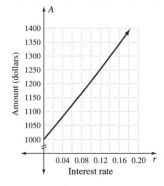

EXERCISES 4.2

20. a. Linear; the price does not increase rapidly with a change in volume.
   b. $y = 0.3715x + 3.678$, $r = 0.9797$
   c. $y = -0.0304x^2 + 0.803x + 2.737$
   d. \$6.65 and \$7.21, respectively. Both prices are reasonable. Note that the parabola peaks near 12 cups and using it would imply that larger sizes would cost less than a 12-cup bowl.

22. a. Linear; the points lie along a line.
   b. $y = 1.88x + 1.29$, $r = 0.9977$
   c. $y = 0.1909x^2 + 1.1027x + 1.972$, $R^2 = 0.999$
   d. For $\frac{1}{2}$-inch width: \$2.23 and \$2.57, respectively. For 4-inch width: \$8.81 and \$9.44, respectively. (Actual cost of the $\frac{1}{2}$-inch width was \$3.29.) There is very little difference in materials between a $\frac{1}{2}$-inch-wide and a 1-inch-wide notebook. Larger durable notebooks should use extra material and should cost relatively more. The quadratic may be most reasonable in the $\frac{1}{2}$-inch to 4-inch range.

26. b. For $B(0, 0)$: $C(25, -20)$, $D(50, 0)$; $y = 0.032x^2 - 1.6x$
   For $C(0, 0)$: $B(-25, 20)$, $D(25, 20)$; $y = 0.032x^2$
   For $D(0, 0)$: $B(-50, 0)$, $C(-25, -20)$; $y = 0.032x^2 + 1.6x$
   For $E(0, 0)$: $B(-50, 20)$, $C(-25, 0)$; $D(0, 20)$;
   $y = 0.032x^2 + 1.6x + 20$

28. For $A(0, 0)$: $B(10, 12)$, $C(20, 0)$; $y = -0.12x^2 + 2.4x$
   For $B(0, 0)$: $A(-10, -12)$, $C(10, -12)$; $y = -0.12x^2$
   For $C(0, 0)$: $A(-20, 0)$, $B(-10, 12)$; $y = -0.12x^2 - 2.4x$
   For $D(0, 0)$: $A(-10, 0)$, $B(0, 12)$, $C(10, 0)$; $y = -0.12x^2 + 12$

30. d. First row: x^2, $(x + 1)x$, $(x + 2)x$, $(x + 3)x$, etc.; second row: $x(x + 1)$, $(x + 1)^2$, $(x + 2)(x + 1)$, $(x + 3)(x + 1)$, etc.; third row: $x(x + 2)$, $(x + 1)(x + 2)$, $(x + 2)^2$, $(x + 3)(x + 2)$, etc.

EXERCISES 4.3

96. From the table, the product of one diagonal is $(acx^2)(bd)$, or $abcdx^2$. The product of the other diagonal is $(bcx)(adx)$, or $abcdx^2$. The diagonal products are the same for any real numbers a, b, c, and d.

98. Possible answers and strategies for three whole numbers: Divisible by 3, always even, always a whole number, always divisible by 6. Consider cases. If x is even, then is $x^3 + 3x^2 + 2x$ even? If x is odd, then is it even? For more ideas, try other ways of writing the consecutive numbers—that is, $3k$, $3k + 1$, $3k + 2$ or $(x - 1)x(x + 1)$ or $(x - 2)(x - 1)(x)$.

100. $C_1 + 1$ foot $= C_2$, $2\pi r_1 + 1$ foot $= 2\pi r_2$, 1 foot $= 2\pi r_2 - 2\pi r_1$, 1 foot $= 2\pi(r_2 - r_1)$, 1 foot$/2\pi = r_2 - r_1$. On the right is the difference in radii, or the clearance between the Earth and the string. On the left is 1 foot (or 12 inches) divided by 2π, which is 1.9 inches—plenty of room for a mouse to run under the string, whether the sphere is the Earth or a basketball or a medicine ball!

EXERCISES 4.4

79. a. $(x + 1)^2 = x^2 + 2x + 1$
   $(x + 1)^3 = x^3 + 3x^2 + 3x + 1$
   $(x + 1)^4 = x^4 + 4x^3 + 6x^2 + 4x + 1$
   b. $(x - 1)^2 = x^2 - 2x + 1$
   $(x - 1)^3 = x^3 - 3x^2 + 3x - 1$
   $(x - 1)^4 = x^4 - 4x^3 + 6x^2 - 4x + 1$

c. $(x + y)^2 = x^2 + 2x^1y^1 + y^2$
   $(x + y)^3 = x^3 + 3x^2y^1 + 3x^1y^2 + y^3$
   $(x + y)^4 = x^4 + 4x^3y^1 + 6x^2y^2 + 4x^1y^3 + y^4$
   d. Possible answers: Among sets, the same powers have the same coefficients. The first letter appears with descending exponents. The powers on $(x + y)^n$ terms add to n.
   e. The coefficients of the powers are found in the second through fourth rows of Pascal's triangle.

80. a. Sequence: 1, 8, 27, 64, 125. First differences: 7, 19, 37, 61. Second differences: 12, 18, 24. Third differences: 6, 6.
   Sequence: 3, 18, 57, 132, 255. First differences: 15, 39, 75, 123. Second differences: 24, 36, 48. Third differences: 12, 12.
   Sequence: 1.5, 8, 22.5, 48, 87.5. First differences: 6.5, 14.5, 25.5, 39.5. Second differences: 8, 11, 14. Third differences: 3, 3. A constant difference appears in the third difference row. The constant difference seems to be $6a$ in $y = ax^3 + bx^2 + cx + d$.
   b. A constant difference will appear in the nth row. The difference is n factorial times a.
   c. The first and third sequences are from cubic polynomials.
   d.

x	$y = ax^3 + bx^2 + cx + d$	
1	$1a + b + c + d$	$7a + 3b + c$
2	$8a + 4b + 2c + d$	$19a + 5b + c$
3	$27a + 9b + 3c + d$	$37a + 7b + c$
4	$64a + 16b + 4c + d$	$61a + 9b + c$
5	$125a + 25b + 5c + d$	

   Second differences: $12a + 2b$, $18a + 2b$, $24a + 2b$.
   Third differences: $6a$, $6a$.
   Set the third difference equal to $6a$ and solve for a. Set the second difference first term equal to $12a + 2b$, substitute for a, and solve for b. Set the first difference first term equal to $7a + 3b + c$, substitute for a and b, and solve for c. Set the first term equal to $a + b + c + d$, substitute for a, b, and c, and solve for d. Set up equation $y = ax^3 + bx^2 + cx + d$.

EXERCISES 4.5

54. a. $4(2 - x)(4 - x)$ **b.** $x = 2$, $x = 4$
   c. $x = 2$ uses the whole width of the plywood; it is impossible to make squares with $x = 4$.
   d. For an area of 45 square feet, $x = 6.5$ and $x = -0.5$; both are impossible in the problem setting.

CHAPTER 4 REVIEW EXERCISES

10. a.

x	$f(x)$
-2	-10
-1	-12
0	-12
1	-10
2	-6

e.

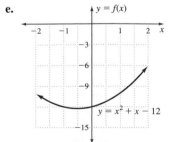

12.

r	A
0.00	20,000
0.01	19,602
0.02	19,208
0.03	18,818
0.04	18,432
0.05	18,050
0.06	17,672
0.07	17,298
0.08	16,928

EXERCISES 5.2

2. c.

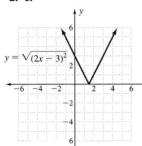

$y = \sqrt{(2x - 3)^2}$

4.

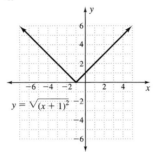

$y = \sqrt{(x + 1)^2}$

6.

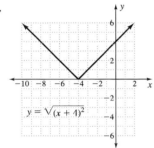

$y = \sqrt{(x + 4)^2}$

EXERCISES 5.4

47. With 6 feet of water above the hole, the water travels horizontally 19.7 ft/sec as it exits the hole. For $h - h_0 = -9$, $v_y = 0$, vertical motion suggests $t = 0.75$ second for water to travel between hole and rustler. Time does not change because it depends on the constant height to the rustler's head, 9 feet. Rustler could be 14.8 feet from tank and get wet. As water level drops, v_x drops. At 4 feet of water, $v_x \approx 16$ ft/sec and distance traveled is 12 feet. 12 ft ≤ wet region ≤ 14.8 ft

EXERCISES 6.1

12.

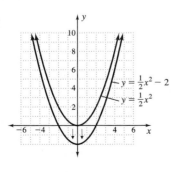

$y = \frac{1}{2}x^2 - 2$
$y = \frac{1}{2}x^2$

14.

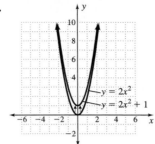

$y = 2x^2$
$y = 2x^2 + 1$

26. and 28.

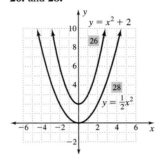

$y = x^2 + 2$
$y = \frac{1}{2}x^2$

30. and 32.

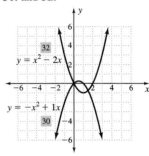

$y = x^2 - 2x$
$y = -x^2 + 1x$

34. and 36.

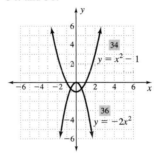

$y = x^2 - 1$
$y = -2x^2$

38. and 40.

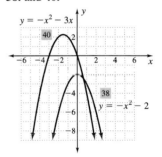

$y = -x^2 - 3x$
$y = -x^2 - 2$

42. and 44.

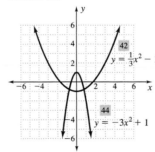

$y = \frac{1}{3}x^2 - 1$
$y = -3x^2 + 1$

46.

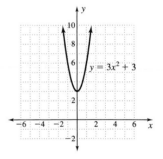

$y = 3x^2 + 3$

EXERCISES 6.4

26. 40 ft/sec

x	0	10	20	30	40	50
y	0	8.0	11.95	11.88	7.8	−0.3

vertex (24.8, 12.4); x-intercepts (0, 0), (49.7, 0)

80 ft/sec

x	0	40	80	120	160	200
y	0	31.95	47.8	47.55	31.2	−1.25

vertex (99.4, 49.7); x-intercepts (0, 0), (198.8, 0)

160 ft/sec

x	0	160	320	480	640	800
y	0	127.8	191.2	190.2	124.8	−5

vertex (397.5, 198.8); x-intercepts (0, 0), (795.0, 0)

CHAPTER 6 REVIEW EXERCISES

6., 8., and **10.**

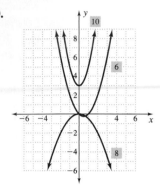

50.

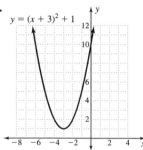

$y = (x + 3)^2 + 1$

52.

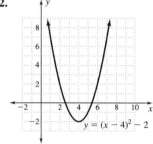

$y = (x - 4)^2 - 2$

EXERCISES 7.3

26.

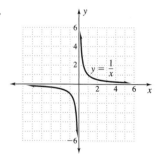

$y = \dfrac{1}{x}$

28. a.

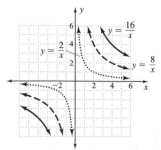

$y = \dfrac{2}{x}$ $y = \dfrac{16}{x}$ $y = \dfrac{8}{x}$

EXERCISES 7.4

62. For the expressions in Exercise 15a:

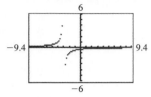

Graph is in dot mode.

a., d. hole at $x = 3$ before simplifying   **b.** not linear
c. nearly vertical around $x = -3$

For the expressions in Exercise 15b:

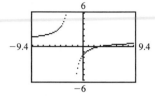

Graph is in dot mode.

a., d. hole at $x = -3$ before simplifying   **b.** not linear
c. nearly vertical around $x = -2$

For the expressions in Exercise 19a:

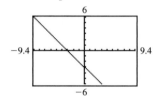

a., d. hole at $x = 2$ before simplifying
b. linear; $m = -1$ **c.** no

EXERCISES 7.5

26. b.

x	f(x)
−4	6
−3	2
−2	0
−1	0
0	2
1	6

c.

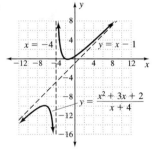

$x = -4$ $y = x - 1$

$y = \dfrac{x^2 + 3x + 2}{x + 4}$

f.

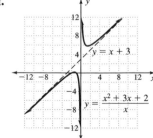

$y = x + 3$

$y = \dfrac{x^2 + 3x + 2}{x}$

g.

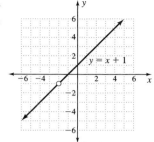

EXERCISES 7.6

76. e.

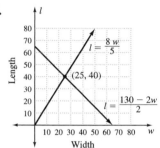

EXERCISES 7.7

64.

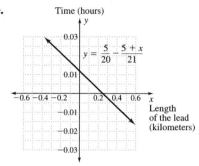

93. a. $\dfrac{a}{b} = \dfrac{c}{d}$

$ad = bc$

$\dfrac{ad}{cd} = \dfrac{bc}{cd}$

$\dfrac{a}{c} = \dfrac{b}{d}$

b. $\dfrac{a}{b} = \dfrac{c}{d}$

$ad = bc$

$\dfrac{ad}{ac} = \dfrac{bc}{ac}$

$\dfrac{d}{c} = \dfrac{b}{a}$

$\dfrac{b}{a} = \dfrac{d}{c}$

c. $\dfrac{a}{b} = \dfrac{c}{d}$

$\dfrac{a}{b} + 1 = \dfrac{c}{d} + 1$

$\dfrac{a}{b} + \dfrac{b}{b} = \dfrac{c}{d} + \dfrac{d}{d}$

$\dfrac{a+b}{b} = \dfrac{c+d}{d}$

d. $\dfrac{a}{b} = \dfrac{c}{d}$

$\dfrac{a}{b} - 1 = \dfrac{c}{d} - 1$

$\dfrac{a}{b} - \dfrac{b}{b} = \dfrac{c}{d} - \dfrac{d}{d}$

$\dfrac{a-b}{b} = \dfrac{c-d}{d}$

e. $\dfrac{a}{b} = \dfrac{c}{d}$

From part c,

$$\dfrac{a+b}{b} = \dfrac{c+d}{d} \qquad \text{(I)}$$

From part d,

$$\dfrac{a-b}{b} = \dfrac{c-d}{d} \qquad \text{(II)}$$

Divide I by II:

$$\dfrac{\dfrac{a+b}{b}}{\dfrac{a-b}{b}} = \dfrac{\dfrac{c+d}{d}}{\dfrac{c-d}{d}}$$

$$\dfrac{a+b}{b} \cdot \dfrac{b}{a-b} = \dfrac{c+d}{d} \cdot \dfrac{d}{c-d}$$

$$\dfrac{a+b}{a-b} = \dfrac{c+d}{c-d}$$

Please: Let the author know if you find a more direct proof.

CHAPTER 7 REVIEW EXERCISES

92.

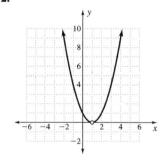

94. c.

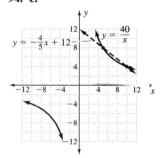

EXERCISES 8.3

54. a. Men: $\dfrac{\left(\dfrac{\text{pounds}}{2.205}\right)^{1.2} \cdot 3{,}000{,}000}{(\text{height in inches} \times 2.54)^{3.3}}$

Women: Replace 3,000,000 with 4,000,000 in men's formula.

EXERCISES 8.4

69. a.

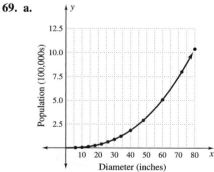

EXERCISES 8.5

66. a.

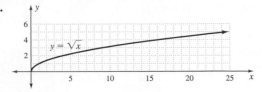

b. One possible result

x	Area under $y = \sqrt{x}$
1	0.5 square
4	5.5 squares
9	18 squares
16	42.5 squares
25	83 squares

EXERCISES 8.6

6.

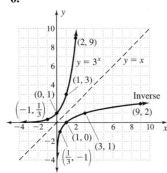

8.

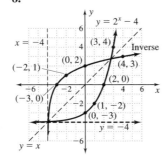

10.

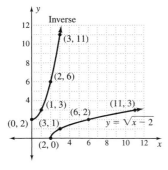

12.

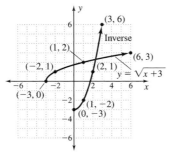

14.

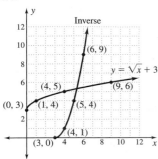

16.

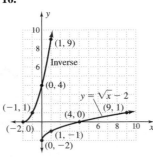

CHAPTER 8 REVIEW EXERCISES

48.

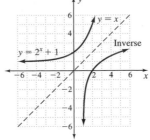

EXERCISES 9.2

30.

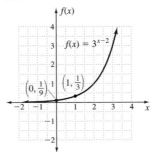

32.

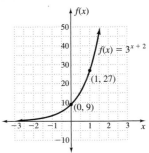

EXERCISES 9.3

2. a.

x	$f(x) = 2^x$
-2	$\frac{1}{4}$
-1	$\frac{1}{2}$
0	1
1	2
2	4
3	8
4	16

b.

x	$y = f^{-1}(x)$
$\frac{1}{4}$	-2
$\frac{1}{2}$	-1
1	0
2	1
4	2
8	3
16	4

c.

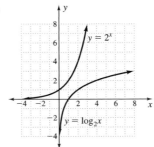

d. $y = \log_2 x$

EXERCISES 9.4

55. a.

Yr.	Option 1	Option 2	Option 3
1	5.25	4.25	4.20
2	5.50	4.55	4.41
3	5.75	4.90	4.63
4	6.00	5.30	4.86
5	6.25	5.75	5.11

b. Option 1: linear; Option 2: quadratic; Option 3: exponential
c. Option 1: $y = 0.25x + 5.00$ **e.** 20 yr
    Option 2: $y = 0.025x^2 + 0.225x + 4$ 9 yr
    Option 3: $y = 4(1.05)^x$ 15 yr

56. a.

Yr.	Option 1	Option 2	Option 3
1	1	1	1
2	2	1.10	1.10
3	3	1.21	1.30
4	4	1.33	1.60
5	5	1.46	2.00

b. Option 1 is linear; it has constant slope. Option 2 is exponential; the ratio of consecutive terms is 1.1 to 1. Option 3 is quadratic; the second differences are constant.
c. Let x = year after birth and y = birthday gift amount. $y = x$; $y = 1(1 + 0.10)^{(x-1)}$; $y = 0.05x^2 - 0.05x + 1$
d. 1 year, 9 years, 5 years (time to double again varies)
e. Option 1 for ages 1 to 20; Options 1 and 3 are tied at $20 at age 20; Option 3 for ages 21 to 52; Opton 2 for ages 53 and above.

b.

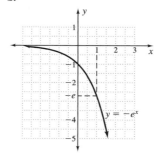

c.

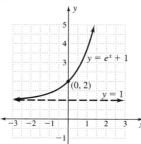

d.

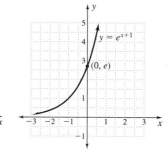

EXERCISES 9.5

54.

x	$y = 4^x$
-1	$\frac{1}{4}$
0	1
1	4
2	16
3	64
4	256
5	1024
6	4096

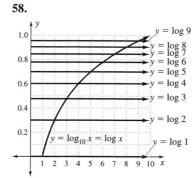

e.

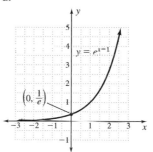

58.

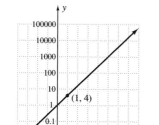

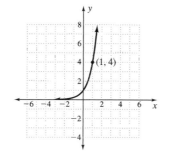

f.

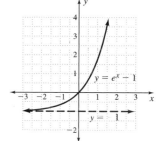

EXERCISES 9.6

48. a.

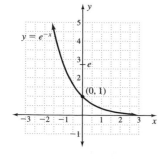

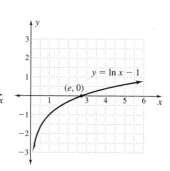

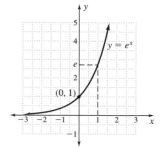

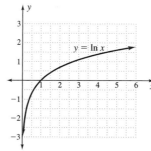

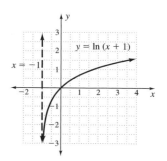

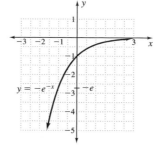

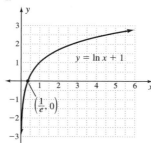

CHAPTER 9 REVIEW EXERCISES

74.

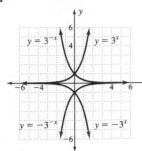

76.

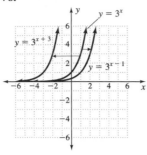

80.

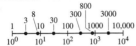

EXERCISES 10.1

24. and **26.**

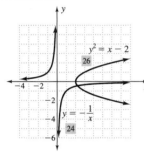

28. and **30.**

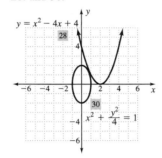

32. and **34.**

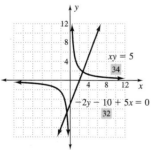

36. and **38.**

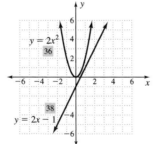

40. and **42.**

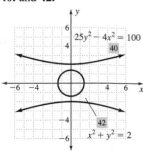

44.

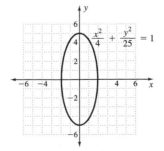

46. $a^2 = 1$ $a^2 = 5$

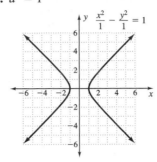

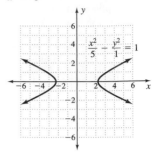

$a^2 = 10$

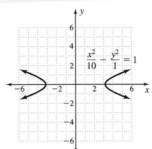

EXERCISES 10.2

22.

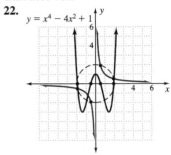

EXERCISES 10.3

18.

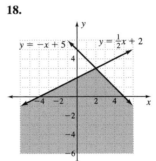

20.

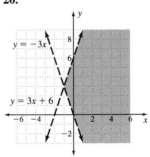

22.

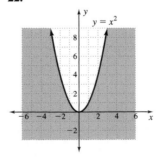

24.

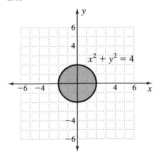

26.

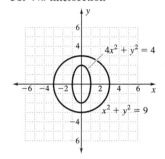

28.

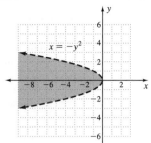

CHAPTER 10 REVIEW EXERCISES

6.

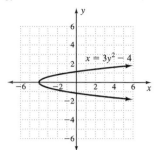

10. and **12.**

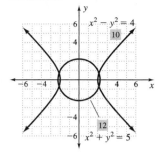

38. No intersection

40.

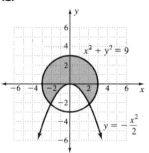

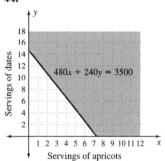

14.

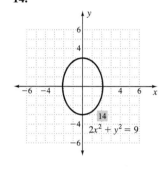

30.

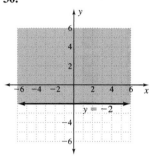

42.

44.

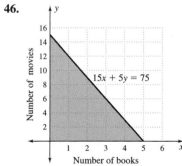

32.

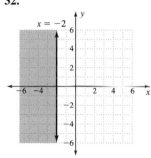

34.

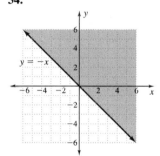

46.

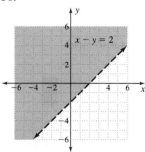

36.

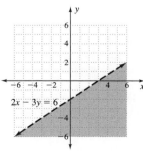

38.

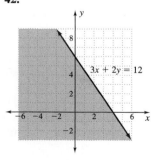

EXERCISES 10.5

18. $[A][X] = [B]$, where

$$[A] = \begin{bmatrix} 1 & 1 & 1 \\ 1 & -2 & 0 \\ 0 & 0 & 1 \end{bmatrix},$$

$$[X] = \begin{bmatrix} s \\ m \\ l \end{bmatrix}, \text{ and } [B] = \begin{bmatrix} 620 \\ 0 \\ 20 \end{bmatrix}$$

40.

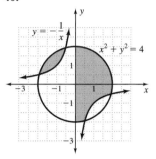

42.

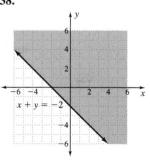

44.

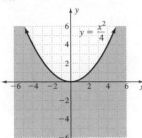

46.

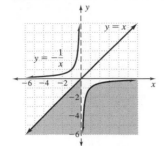

48.

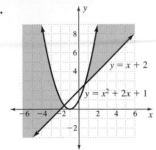

EXERCISES 11.1

2.

HHHH	HHHT	HHTT	HTTT	TTTT
	HHTH	HTHT	THTT	
	HTHH	HTTH	TTHT	
	THHH	TTHH	TTTH	
		THTH		
		THHT		

6.

8.

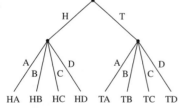

28. Because $a(b + c) = ab + ac$, for Exercise 26e we can write eggs and meat with toast or eggs and meat with potato, for 26f we can write eggs with meat or eggs with toast or eggs with potato, and for 27f we can write psychology with history or psychology with economics.

34. g. There are more different sums. There are up to ten ways to get a difference of 1. The most for any given sum is six ways. The number of ways to get sums is symmetric; the differences are not symmetric.

   h. If (1, 6) is an outcome, then (6, 1) has the same outcome. The sample space for looking at both sums and differences contains 36 sets of numbers.

EXERCISES 11.2

54. h. There would be one way to get 6 heads and 0 tails in 6 trials. There would be 6 ways to get 5 heads and 1 tail in 6 trials. There would be $6!/[(6 - r)!(r!)]$ ways to get $6 - r$ heads and r tails in 6 trials. There would be $n!/[(n - r)!(r!)]$ ways to get $n - r$ heads and r tails in n trials.

EXERCISES 11.3

8.

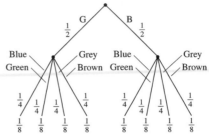

EXERCISES 11.4

36. a.

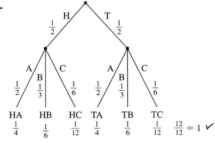

38. a.

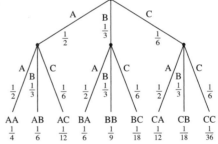

40. a.

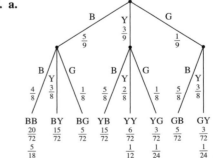

42. a.

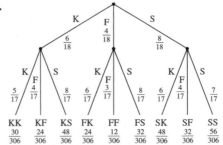

44.

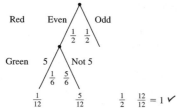

$\frac{1}{12}$ $\frac{5}{12}$ $\frac{1}{2}$ $\frac{12}{12} = 1$ ✓

34. a.

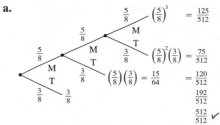

$\left(\frac{5}{8}\right)^3 = \frac{125}{512}$

$\left(\frac{5}{8}\right)^2\left(\frac{3}{8}\right) = \frac{75}{512}$

$\left(\frac{5}{8}\right)\left(\frac{3}{8}\right) = \frac{15}{64} = \frac{120}{512}$

$\frac{192}{512}$

$\frac{512}{512}$ ✓

46.

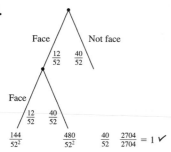

$\frac{144}{52^2}$ $\frac{480}{52^2}$ $\frac{40}{52}$ $\frac{2704}{2704} = 1$ ✓

36.

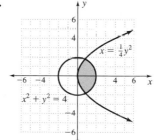

$\frac{4}{4} = 1$ ✓

38.

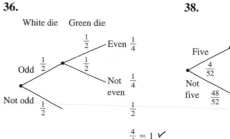

$\frac{2652}{2652} = 1$ ✓

48. a.

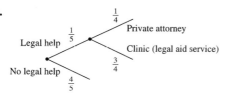

50. a.

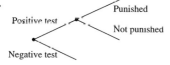

FINAL EXAM REVIEW, PART 1

8.

22. ← → line with points at -5 and 0

128.

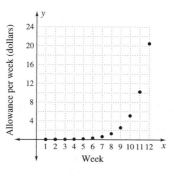

$x = \frac{1}{4}y^2$

$x^2 + y^2 = 4$

CHAPTER 11 REVIEW EXERCISES

2.

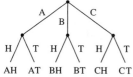

AH   AT   BH   BT   CH   CT

8. Revealed your opinion is (original). Revealed is your opinion. Your opinion is revealed. Your opinion revealed is. Is revealed your opinion. Is your opinion revealed. (The latter sounds like a question.)

FINAL EXAM REVIEW, PART 2   APPENDIX 3

42.

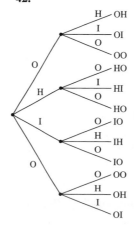

32. d.

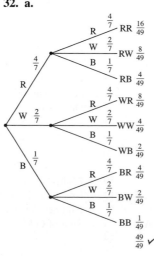

30. a.

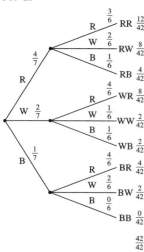

32. a.

RR $\frac{16}{49}$
RW $\frac{8}{49}$
RB $\frac{4}{49}$
WR $\frac{8}{49}$
WW $\frac{4}{49}$
WB $\frac{2}{49}$
BR $\frac{4}{49}$
BW $\frac{2}{49}$
BB $\frac{1}{49}$

$\frac{49}{49}$ ✓

Index of Projects: It is recommended that individual students be required to do three projects each term or semester. If you have three midterm tests, require that one project be completed before each midterm. You might want to require that a project be completed within two weeks of the date given for the section on the course syllabus or—if you do not provide a section by section plan for the term—two weeks after the section is covered in class. All projects can be carried out by students working in pairs or groups. Those projects ideal for in-class group work are marked with an asterisk (*).